Spectra of Ionized Atoms: From Laboratory to Space

Special Issue Editor

Joseph Reader

MDPI • Basel • Beijing • Wuhan • Barcelona • Belgrade

Special Issue Editor
Joseph Reader
National Institute of Standards and Technology
USA

Editorial Office
MDPI
St. Alban-Anlage 66
Basel, Switzerland

This edition is a reprint of the Special Issue published online in the open access journal *Atoms* (ISSN 2218-2004) from 2016–2017 (available at: http://www.mdpi.com/journal/atoms/special_issues/laboratory_space).

For citation purposes, cite each article independently as indicated on the article page online and as indicated below:

Lastname, F.M.; Lastname, F.M. Article title. *Journal Name* **Year**, *Article number*, page range.

First Edition 2018

ISBN 978-3-03842-869-5 (Pbk)
ISBN 978-3-03842-870-1 (PDF)

Central portion of the image of the Crab Nebula taken by the Hubble Space Telescope (HST). The Crab Nebula is among the most interesting and well-studied objects in astronomy. The image on the cover was assembled from 24 individual exposures taken with the Wide Field and Planetary Camera 2 (WFPC2). It is the largest image ever taken with WFPC2. WFPC2 was built by the Jet Propulsion Laboratory in Pasadena, California. It was installed on Hubble in 1993 and removed during the servicing mission in 2009. WFPC2 is currently on display at the National Air and Space Museum in Washington, DC.

Cover photo courtesy of National Aeronautics and Space Administration (NASA) and the Space Telescope Science Institute (STScI) for permission to use this photograph. A more complete image can be found at https://cdn.spacetelescope.org/archives/images/screen/heic0515a.jpg.

Table of Contents

About the Special Issue Editor v

Joseph Reader
Spectrum and Energy Levels of Four-Times Ionized Yttrium (Y V)
doi:10.3390/atoms4040031 1

Jorge Reyna Almandos and Mónica Raineri
Spectral Analysis of Moderately Charged Rare-Gas Atoms
doi:10.3390/atoms5010012 37

Alexander Ryabtsev and Edward Kononov
Resonance Transitions in the Spectra of the Ag^{6+}–Ag^{8+} Ions
doi:10.3390/atoms5010011 51

Alexander Kramida, Gillian Nave and Joseph Reader
The Cu II Spectrum
doi:10.3390/atoms5010009 91

Swapnil and Tauheed Ahmad
The Third Spectrum of Indium: In III
doi:10.3390/atoms5020023 193

Roshani Silwal, Endre Takacs, Joan M. Dreiling, John D. Gillaspy and Yuri Ralchenko
Identification and Plasma Diagnostics Study of Extreme Ultraviolet Transitions in Highly Charged Yttrium
doi:10.3390/atoms5030030 216

Ali Meftah, Mourad Sabri, Jean-François Wyart and Wan-Ü Lydia Tchang-Brillet
Spectrum of Singly Charged Uranium (U II) : Theoretical Interpretation of Energy Levels, Partition Function and Classified Ultraviolet Lines
doi:10.3390/atoms5030024 237

Laurentius Windholz
The Role of the Hyperfine Structure for the Determination of Improved Level Energies of Ta II, Pr II and La II
doi:10.3390/atoms5010010 273

Dylan F. Del Papa, Richard A. Holt and S. David Rosner
Hyperfine Structure and Isotope Shifts in Dy II
doi:10.3390/atoms5010005 287

Stefan Gustafsson, Per Jönsson, Charlotte Froese Fischer and Ian Grant
Combining Multiconfiguration and Perturbation Methods: Perturbative Estimates of Core–Core Electron Correlation Contributions to Excitation Energies in Mg-Like Iron
doi:10.3390/atoms5010003 297

Gediminas Gaigalas, Charlotte Froese Fischer, Pavel Rynkun and Per Jönsson
`JJ2LSJ` Transformation and Unique Labeling for Energy Levels
doi:10.3390/atoms5010006 309

Charlotte Froese Fischer, Gediminas Gaigalas, and Per Jösson
Core Effects on Transition Energies for $3d^k$ Configurations in Tungsten Ions
doi:10.3390/atoms5010007 . . . . **320**

Alan Hibbert
Calculation of Rates of 4p–4d Transitions in Ar II
doi:10.3390/atoms5010008 . . . . **354**

Per Jönsson, Gediminas Gaigalas, Pavel Rynkun, Laima Radžiūtė,Jörgen Ekman, Stefan Gustafsson, Henrik Hartman, Kai Wang,Michel Godefroid, Charlotte Froese Fischer, Ian Grant, Tomas Brage and Giulio Del Zanna
Multiconfiguration Dirac-Hartree-Fock Calculations with Spectroscopic Accuracy: Applications to Astrophysics
doi:10.3390/atoms5020016 . . . . **361**

Elena Ivanova
Wavelengths of the Self-Photopumped Nickel-Like $4f\ ^1P_1 \rightarrow 4d\ ^1P_1$ X-ray Laser Transitions
doi:10.3390/atoms5030025 . . . . **385**

Luning Liu, Deirdre Kilbane, Padraig Dunne, Xinbing Wang and Gerry O'Sullivan
Configuration Interaction Effects in Unresolved $5p^6 5d^{N+1}-5p^5 5d^{N+2}+5p^6 5d^N 5f^1$ Transition Arrays in Ions Z = 79–92
doi:10.3390/atoms5020020 . . . . **395**

About the Special Issue Editor

Joseph Reader is Scientist Emeritus at the National Institute of Standards and Technology (NIST) in Gaithersburg, Maryland, U.S.A. After receiving his Ph.D. from the University of California, Berkeley in 1962 and serving as Postdoctoral Research Associate at the Argonne National Laboratory, in 1963 he joined the staff at NIST. During his years at NIST he has specialized in the observation and interpretation of spectra of highly ionized atoms and the design of optical spectrometers. His measurements of the spectra of neutral and singly ionized platinum have served to calibrate wavelengths of spectra observed with the Hubble Space Telescope. He is a Fellow of the American Physical Society and the Optical Society, and has received the William F. Meggers Award of the Optical Society of America. Before retiring from NIST in 2014, he directed the NIST Atomic Spectroscopy Data Center.

atoms

MDPI

Article

Spectrum and Energy Levels of Four-Times Ionized Yttrium (Y V)

Joseph Reader

National Institute of Standards and Technology, Gaithersburg, MD 20899-8422, USA; joseph.reader@nist.gov; Tel.: +1-301-975-3222

Academic Editor: James Babb
Received: 18 November 2016; Accepted: 6 December 2016; Published: 21 December 2016

Abstract: The analysis of the spectrum of four-times-ionized yttrium, Y V, was extended to provide a large number of new spectrum lines and energy levels. The new analysis is based on spectrograms made with sliding-spark discharges on 10.7 m normal- and grazing-incidence spectrographs. The measurements cover the region 184–2549 Å. The results revise levels for this spectrum by Zahid-Ali et al. (1975) and by Ateqad et al. (1984). Five hundred and seventy lines were classified as transitions between 23 odd-parity and 90 even-parity levels. The $4s^24p^5$, $4s4p^6$, $4s^24p^44d$, 5s, 5p, 5d, 6s configurations are now complete. Results for the $4s^24p^46d$ and 7s configurations are tentative. Ritz-type wavelengths were determined from the optimized energy levels, with uncertainties as low as ± 0.0004 Å. The observed configurations were interpreted with Hartree-Fock calculations and least-squares fits of the energy parameters to the observed levels. Oscillator strengths for all classified lines were calculated with the fitted parameters. The results are compared with values for the level energies, percentage compositions, and transition probabilities from recent ab initio theoretical calculations. The ionization energy was revised to 607,760 $\pm$ 300 cm^{-1} (75.353 $\pm$ 0.037 eV).

Keywords: yttrium; ionic spectrum; vacuum ultraviolet; wavelengths; energy levels; transition probabilities; parametric calculations; ionization energy

---

1. Introduction

The four-times ionized yttrium atom, Y V, has a Br-like electronic structure with ground configuration $4s^24p^5$ and excited states $4s4p^6$ and $4s^24p^4nl$. The spectrum has a somewhat checkered past. It was first analyzed in 1939 by Paul and Rense [1], who, from a set of transitions to the $4s^24p^5$ ^2P ground term, determined levels of the $4s4p^6$ $^2S_{1/2}$, $4s^24p^44d$, and $4s^24p^45s$ configurations. Unfortunately, an isoelectronic plot published by Edlén [2] in 1964 showed that the $4s^24p^5$ $^2P_{3/2}$-$^2P_{1/2}$ interval of Paul and Rense [1] (12,068 cm^{-1}) was inconsistent with the known intervals for the rest of the isoelectronic sequence. From his plot, Edlén predicted an interval of 12,470 $\pm$ 20 cm^{-1}. Since essentially all of their levels were based on transitions to the $4s^24p^5$ ^2P term, Edlén concluded that the analysis would have to be completely revised. A start on this revision came in 1970, when Reader and Epstein [3] observed the true $4s^24p^5$ $^2P_{1/2,3/2}$-$4s4p^6$ $^2S_{1/2}$ transitions, thus obtaining the position of $4s4p^6$ $^2S_{1/2}$ and a revised value for the ^2P term splitting. Their splitting of 12,459.9 $\pm$ 3.0 cm^{-1} was indeed close to the value predicted by Edlén. In 1972 Reader and Epstein [4] observed further transitions to the $4s^24p^5$ ^2P ground term and established nearly all levels of the $4s^24p^44d$ and 5s configurations. Only the levels of $4p^44d$ with $J = 7/2$ and $9/2$, which do not combine with $4p^5$ $^2P_{1/2,3/2}$, and the $4p^44d$ $(^3P)^4D_{1/2}$ level could not be located.

In 1975, Zahid-Ali et al. [5] observed the spectrum at lower wavelengths and reported levels of the $4s^24p^45d$, 6s, 6d, and 7s configurations. Since all lines terminated on the $4s^24p^5$ ^2P term, again only levels with $J = 1/2$, $3/2$, and $5/2$ could be found. Finally, in 1984 Ateqad and Chaghtai [6] reported levels of the $4s^24p^44f$ and 5p configurations. From transitions to these new configurations they were able to report levels of $4s^24p^44d$ and 5d having $J = 7/2$ and 9/2.

In the present work we observed the spectrum of Y V in the ultraviolet and determined a new set of energy levels. About half the $4s^24p^45d$ levels of [5] were found to be spurious. Several of the $4s^24p^46s$ levels in this paper had incorrect J-values and in fact belong to $4s^24p^45d$. Nearly all of the $4s^24p^45p$ levels of [6] were spurious, as were all of the reported $J = 7/2$, 9/2 levels of $4s^24p^44d$ and 5d.

2. Experiment

The observations were the same as used for earlier work in our laboratory on yttrium [4,7,8]. Briefly, the light source was a low-voltage sliding-spark with metallic yttrium electrodes. The source was operated as described by Reader et al. [9]. From 500 to 2549 Å we used the NIST 10.7-m normal-incidence vacuum spectrograph; from 184 to 500 Å we used the NIST 10.7-m grazing-incidence spectrograph. Both instruments had gratings with 1200 lines/mm. The plate factor for the normal-incidence spectrograph was about 0.78 Å/mm. The plate factor for the grazing-incidence spectrograph at 350 Å was 0.25 Å/mm. From 600 to 2549 Å the spectra were calibrated by spectra of Cu II excited in a hollow cathode discharge. Below 600 Å calibration was obtained from lines of Y in various stages of ionization. Shifts between the reference spectra and the yttrium spectra were removed by use of impurity lines of oxygen, nitrogen, carbon, and silicon. Complete references for the calibration spectra are given in Reference [8].

Ionization stages were distinguished by comparing the intensities of the lines at various peak currents in the spark. The spectra of Y V were relatively enhanced at a peak current of about 2000 A.

The wavelengths, intensities, and classifications of the observed lines of Y V are given in Table 1. All wavelengths are in vacuum. The intensities are estimates of photographic plate blackening. The intensities range from 1 to 5,000,000. The system used to obtain this extensive scale of intensities is described in a recent paper on Mo VI [10]. No attempt was made to account for spectrograph or plate emulsion response. The strongest lines in the spectrum appear as a group of $4p^45p$-5d transitions around 1350 Å.

The general uncertainty of the wavelengths is ±0.007 Å. Hazy lines (h) were given an uncertainty of ±0.010 Å; perturbed (p), complex (c), or asymmetric lines (s, *l*) an uncertainty of ±0.020 Å; unresolved (u) or doubly classified (dc) lines an uncertainty of ±0.030 Å. All uncertainties are reported at the level of one standard deviation.

Table 1. Observed spectral lines of Y V. Wavelengths and wave numbers are in vacuum. Wavelength values in parentheses are Ritz values. General uncertainty of the observed wavelengths is ±0.007 Å. Uncertainties for less certain wavelengths are given in Section 2 of the text. |CF| is the cancellation factor (see text). Unc (Å) is the uncertainty of the Ritz wavelength.

λ_{obs} (Å)	Int [a]		σ_{obs} (cm^{-1})	Even Level [b]	Odd Level [b]	λ_{Ritz} (Å)	Unc (Å)	g_uA (s^{-1})	log(g_lf)	|CF|
184.144	1		543053	6d83	p5 3	184.1443	0.0049	1.13E+08	−3.25	0.05
187.849	25		532342	7s31	p5 1	187.8490	0.0070	1.99E+09	−1.99	0.69
188.469	3		530591	6d83	p5 1	188.4687	0.0051	3.71E+09	−1.71	0.75
191.571	50		522000	7s25	p5 3	191.5710	0.0070	7.75E+09	−1.38	0.69
193.843	3		515881	6d51	p5 3	193.8426	0.0049	2.45E+08	−2.87	0.03
193.888	3		515762	6d65	p5 3	193.8880	0.0070	2.19E+09	−1.91	0.40
194.165	10		515026	6d73	p5 3	194.1705	0.0061	2.59E+09	−1.84	0.61
194.457	10		514253	6d41	p5 3	194.4567	0.0048	4.97E+09	−1.56	0.54
196.148	30		509819	7s21	p5 3	196.1490	0.0049	4.20E+09	−1.62	0.75
196.206	25		509668	7s33	p5 1	196.2060	0.0070	5.65E+09	−1.49	0.66
196.444	40	u	509051	7s23	p5 3	196.4440	0.0070	3.06E+09	−1.76	0.75
198.495	3		503791	6d53	p5 3	198.4961	0.0049	1.30E+09	−2.12	0.07
198.640	20		503423	6d51	p5 1	198.6404	0.0051	1.42E+10	−1.08	0.65
198.753	60		503137	6d55	p5 3	198.7530	0.0070	4.59E+09	−1.57	0.50
198.990	3		502538	6d73	p5 1	198.9847	0.0064	9.56E+09	−1.25	0.65
199.285	3		501794	6d41	p5 1	199.2853	0.0051	1.89E+09	−1.96	0.48
199.461	2	u	501351	6d43	p5 3	199.4610	0.0300	1.47E+09	−2.07	0.26
200.392	80		499022	7s13	p5 3	200.3926	0.0049	1.25E+10	−1.13	0.74
200.694	1		498271	7s15	p5 3	200.6940	0.0300	4.57E+08	−2.57	0.60
201.064	30		497354	7s21	p5 1	201.0631	0.0051	5.12E+09	−1.51	0.57
202.784	30	u	493136	6d23	p5 3	202.7798	0.0065	1.76E+10	−0.97	0.66
202.792	40	p	493116	6d25	p5 3	202.7920	0.0200	9.35E+09	−1.25	0.64
203.531	10		491326	6d53	p5 1	203.5300	0.0052	9.88E+09	−1.22	0.53
205.525	1		486559	7s13	p5 1	205.5244	0.0051	9.84E+08	−2.21	0.18
205.731	10		486072	6s31	p5 3	205.7327	0.0004	1.48E+09	−2.03	0.14
208.036	2		480686	6d23	p5 1	208.0362	0.0069	1.25E+09	−2.10	0.08
211.144	20		473610	6s31	p5 1	211.1453	0.0005	3.52E+09	−1.63	0.50
212.318	20		470992	5d85	p5 3	212.3188	0.0004	3.67E+09	−1.61	0.30
217.564	100		459635	6s25	p5 3	217.5632	0.0005	1.36E+10	−1.02	0.55
217.853	80		459025	5d83	p5 1	217.8535	0.0005	1.23E+10	−1.06	0.54
222.825	30		448783	6s21	p5 3	222.8289	0.0005	5.90E+08	−2.36	0.02
223.032	70		448366	5d73	p5 3	223.0363	0.0004	5.03E+09	−1.43	0.23

Table 1. *Cont.*

λ_{obs} (Å)	Int [a]		σ_{obs} (cm^{-1})	Even Level [b]	Odd Level [b]	λ_{Ritz} (Å)	Unc (Å)	g_UA (s^{-1})	log(g_Lf)	|CF|
223.561	90	p	447305	6s33	p5 1	223.5630	0.0005	1.04E+10	−1.11	0.54
223.569	80	u	447289	5d51	p5 3	223.5749	0.0004	8.34E+09	−1.20	0.19
223.857	200	dc	446714	6s23	p5 3	223.8547	0.0005	2.27E+09	−1.77	0.12
223.857	200	dc	446714	5d75	p5 3	223.8594	0.0004	2.66E+10	−0.70	0.50
224.107	2		446215	6s11	p5 3	224.1101	0.0005	4.26E+06	−4.49	0.00
224.566	100		445303	5d65	p5 3	224.5678	0.0004	8.77E+09	−1.18	0.34
224.726	150		444986	5d63	p5 3	224.7272	0.0004	2.93E+10	−0.65	0.48
225.159	100		444131	5d41	p5 3	225.1599	0.0004	1.22E+10	−1.03	0.27
228.743	200		437172	6s13	p5 3	228.7434	0.0005	2.18E+10	−0.77	0.56
229.191	3		436317	6s21	p5 1	229.1924	0.0006	8.65E+07	−3.17	0.00
229.411	100		435899	5d73	p5 1	229.4118	0.0006	3.44E+10	−0.57	0.47
229.530	20		435673	6s15	p5 3	229.5312	0.0005	1.52E+08	−2.92	0.03
229.845	70		435076	5d53	p5 3	229.8474	0.0005	4.58E+09	−1.44	0.07
229.981	100	p, x	434819	5d51	p5 1	229.9816	0.0006	2.83E+10	−0.65	0.49
230.071	150		434648	5d55	p5 3	230.0718	0.0005	4.05E+10	−0.49	0.44
230.277	2		434260	6s23	p5 1	230.2778	0.0006	2.09E+08	−2.78	0.04
231.120	20	p, x	432676	5d45	p5 3	231.1214	0.0005	9.45E+08	−2.12	0.27
231.201	50		432524	5d63	p5 1	231.2011	0.0006	7.06E+09	−1.25	0.37
231.659	20		431669	5d41	p5 1	231.6592	0.0006	4.99E+08	−2.40	0.02
231.784	70		431436	5d43	p5 3	231.7843	0.0005	5.06E+09	−1.39	0.27
232.269	20		430535	5d35	p5 3	232.2697	0.0005	1.04E+09	−2.08	0.14
232.370	20		430348	5d33	p5 3	232.3723	0.0005	7.48E+08	−2.22	0.33
232.800	20		429553	5d31	p5 3	232.8009	0.0005	7.21E+08	−2.23	0.15
235.250	200		425080	5d25	p5 3	235.2511	0.0005	3.08E+10	−0.59	0.49
235.386	150	u	424834	5d23	p5 3	235.3880	0.0005	1.72E+10	−0.84	0.46
235.452	25		424715	6s13	p5 1	235.4543	0.0006	8.34E+08	−2.16	0.07
236.208	70	u	423356	5d21	p5 3	236.2106	0.0005	3.24E+09	−1.57	0.25
236.623	80		422613	5d53	p5 1	236.6241	0.0006	1.21E+10	−1.00	0.24
238.675	50		418980	5d43	p5 1	238.6775	0.0006	2.96E+09	−1.60	0.10
238.711	10		418917	5d13	p5 3	238.7123	0.0005	7.10E+07	−3.22	0.02
239.298	2		417889	5d33	p5 1	239.3010	0.0006	2.49E+08	−2.67	0.13
239.754	10		417094	5d31	p5 1	239.7555	0.0006	7.80E+08	−2.17	0.10
242.501	60		412369	5d23	p5 1	242.5005	0.0006	2.55E+09	−1.65	0.05
243.375	3		410889	5d21	p5 1	243.3736	0.0006	4.26E+08	−2.42	0.03
245.389	5		407516	5d11	p5 1	245.3885	0.0006	1.81E+08	−2.79	0.04

Table 1. *Cont.*

λ_{obs} (Å)	Int [a]		σ_{obs} (cm^{-1})	Even Level [b]	Odd Level [b]	λ_{Ritz} (Å)	Unc (Å)	$g_U A$ (s^{-1})	log($g_L f$)	|CF|
289.182	90	H, x	345803	5s31	p5 3	289.1826	0.0008	1.16E+09	−1.84	0.03
(299.992)		A, x	333334	5s31	p5 1	299.9920	0.0010	7.43E+09	−1.00	0.28
312.888	50		319603.2	5s33	p5 3	312.8874	0.0008	2.02E+09	−1.53	0.04
313.349	1000		319133.0	5s25	p5 3	313.3494	0.0008	3.24E+10	−0.32	0.42
320.467	500		312044.6	4d83	p5 3	320.4676	0.0009	9.44E+09	−0.84	0.03
321.691	700		310857.3	5s21	p5 3	321.6905	0.0009	1.57E+10	−0.61	0.42
325.580	2000		307144.2	5s33	p5 1	325.5805	0.0011	1.23E+10	−0.71	0.15
326.567	3000		306215.9	5s23	p5 3	326.5675	0.0009	4.03E+10	−0.19	0.83
(328.337)		A, x	304566.1	5s11	p5 3	328.3372	0.0009	1.06E+08	−2.77	0.00
330.398	300		302665.3	4d51	p5 3	330.3989	0.0009	1.06E+10	−0.76	0.05
333.084	10000		300224.6	4d85	p5 3	333.0844	0.0010	7.88E+11	1.12	0.82
333.796	5000		299584.2	4d83	p5 1	333.7963	0.0012	5.17E+11	0.94	0.83
335.125	200		298396.1	5s21	p5 1	335.1232	0.0012	3.32E+10	−0.25	0.68
335.143	800		298380.1	5s13	p5 3	335.1445	0.0010	1.33E+10	−0.65	0.13
336.621	5000		297070.0	4d73	p5 3	336.6197	0.0010	4.71E+11	0.91	0.88
339.023	3000		294965.2	4d41	p5 3	339.0225	0.0010	2.34E+11	0.61	0.71
340.016	1000		294103.8	5s15	p5 3	340.0176	0.0010	7.67E+09	−0.88	0.92
340.419	75		293755.6	5s23	p5 1	340.4194	0.0012	1.61E+09	−1.55	0.11
342.342	50		292105.6	5s11	p5 1	342.3429	0.0012	5.06E+09	−1.05	0.32
344.583	2000		290205.8	4d51	p5 1	344.5848	0.0013	1.97E+11	0.55	0.82
349.648	800		286001.9	4d75	p5 3	349.6483	0.0011	9.34E+08	−1.77	0.00
349.752	300		285916.9	5s13	p5 1	349.7498	0.0013	4.20E+08	−2.11	0.01
351.355	800		284612.4	4d73	p5 1	351.3567	0.0013	3.58E+09	−1.18	0.02
353.976	1500		282505.0	4d41	p5 1	353.9753	0.0013	1.12E+10	−0.68	0.05
355.564	1500		281243.3	4d63	p5 3	355.5625	0.0011	9.16E+09	−0.76	0.09
372.047	4000		268783.2	4d63	p5 1	372.0454	0.0015	1.22E+10	−0.60	0.06
379.963	1000		263183.5	4d65	p5 3	379.9623	0.0013	2.40E+09	−1.29	0.14
397.767	1000		251403.5	4d55	p5 3	397.7663	0.0015	6.86E+08	−1.79	0.01
403.452	1500		247861.0	4d45	p5 3	403.4517	0.0015	2.13E+09	−1.28	0.01
408.806	10		244614.8	4d53	p5 3	408.8086	0.0015	1.31E+08	−2.49	0.00
409.312	1500		244312.4	4d35	p5 3	409.3134	0.0015	1.28E+09	−1.49	0.02
415.027	1500		240948.2	4d43	p5 3	415.0250	0.0016	1.22E+09	−1.50	0.01
418.179	600		239132.0	4d33	p5 3	418.1776	0.0016	1.66E+09	−1.36	0.02
418.591	1800		238896.7	4d31	p5 3	418.5882	0.0017	1.12E+09	−1.53	0.14
419.792	400		238213.2	4d23	p5 3	419.7887	0.0016	3.12E+08	−2.08	0.01
420.737	1500		237678.2	4d25	p5 3	420.7379	0.0017	8.42E+08	−1.65	0.74

Table 1. *Cont.*

λ_{obs} (Å)	Int [a]		σ_{obs} (cm^{-1})	Even Level [b]	Odd Level [b]	λ_{Ritz} (Å)	Unc (Å)	$g_U A$ (s^{-1})	log($g_L f$)	|CF|
427.875	40		233713.1	4d21	p5 3	427.8764	0.0017	1.41E+07	−3.42	0.00
430.753	500		232151.6	4d53	p5 1	430.7502	0.0020	7.11E+07	−2.71	0.00
437.661	500	p	228487.3	4d43	p5 1	437.6574	0.0021	1.30E+09	−1.43	0.01
441.161	2		226674.6	4d33	p5 1	441.1647	0.0021	1.65E+07	−3.32	0.00
441.622	25		226438.0	4d31	p5 1	441.6217	0.0022	7.60E+07	−2.65	0.01
442.947	300	d, x	225760.6	4d23	p5 1	442.9581	0.0022	3.74E+07	−2.96	0.00
451.974	25		221251.7	4d21	p5 1	451.9728	0.0023	9.54E+07	−2.54	0.00
452.911	5		220793.9	4d11	p5 3	452.9095	0.0024	1.26E+07	−3.41	0.00
455.846	35		219372.3	4d13	p5 3	455.8429	0.0020	2.74E+07	−3.07	0.00
(457.838)		A, x	218417.9	4d15	p5 3	457.8395	0.0021	4.23E+07	−2.88	0.00
479.994	1	x	208335.9	4d11	p5 1	479.9972	0.0030	9.83E+06	−3.47	0.00
481.827	20		207543.4	4p61	5p51	481.8281	0.0022	6.46E+08	−1.65	0.21
491.807	5		203331.8	4p61	5p73	491.8074	0.0023	1.69E+08	−2.21	0.21
498.642	90	*l*	200544.7	4p61	5p63	498.6395	0.0024	1.15E+09	−1.37	0.26
550.483	1	x	181658.7	4p61	5p21	550.4803	0.0029	6.84E+07	−2.51	0.10
573.075	2		174497.2	4p61	5p11	573.0754	0.0031	1.02E+08	−2.30	0.16
584.982	50000		170945.4	4p61	p5 3	584.9815	0.0044	1.66E+09	−1.07	0.03
585.101	5		170910.7	4p61	5p13	585.0990	0.0033	1.09E+08	−2.25	0.19
630.973	30000		158485.4	4p61	p5 1	630.9727	0.0056	7.94E+08	−1.33	0.04
690.718	25		144776.9	4d21	5p51	690.7217	0.0018	7.67E+07	−2.26	0.02
693.434	20		144209.8	6s31	5p13	693.4292	0.0028	2.73E+06	−3.71	0.00
702.063	30		142437.4	4d15	5p53	702.0699	0.0026	6.24E+07	−2.34	0.02
706.816	25		141479.5	4d13	5p53	706.8173	0.0023	5.71E+07	−2.37	0.02
709.527	75		140939.0	4d27	5p55	709.5328	0.0024	1.28E+08	−2.02	0.07
709.676	75		140909.4	4d15	5p43	709.6772	0.0026	3.91E+08	−1.53	0.06
711.410	70		140565.9	4d21	5p73	711.4154	0.0019	4.61E+08	−1.45	0.18
713.977	4	H, x	140060.5	4d11	5p53	713.9878	0.0040	3.35E+07	−2.59	0.04
714.521	5	H, x	139953.9	4d13	5p43	714.5284	0.0023	2.89E+07	−2.66	0.01
715.114	5	H, x	139837.8	4d11	5p41	715.1005	0.0040	3.00E+06	−3.64	0.00
717.580	8		139357.3	4d33	5p51	717.5885	0.0016	2.71E+07	−2.68	0.02
721.365	150		138626.1	4d15	5p35	721.3673	0.0027	3.19E+08	−1.60	0.04
722.091	500		138486.7	4d17	5p35	722.0949	0.0034	1.06E+09	−1.08	0.08
726.376	50		137669.7	4d13	5p35	726.3802	0.0024	1.72E+08	−1.87	0.07
731.760	30		136656.8	4d15	5p33	731.7646	0.0028	9.51E+07	−2.12	0.02
732.991	75		136427.3	4d23	5p55	732.9955	0.0021	1.97E+07	−2.80	0.02
734.949	40		136063.9	4d23	5p73	734.9584	0.0021	2.81E+08	−1.65	0.12

Table 1. *Cont.*

λ_{obs} (Å)	Int [a]		σ_{obs} (cm^{-1})	Even Level [b]	Odd Level [b]	λ_{Ritz} (Å)	Unc (Å)	$g_U A$ (s^{-1})	log($g_L f$)	|CF|
736.921	125		135699.8	4d13	5p33	736.9236	0.0025	3.78E+08	−1.51	0.10
737.953	100		135510.0	4d33	5p55	737.9597	0.0018	3.79E+08	−1.51	0.24
738.669	250	dc	135378.6	4d31	5p73	738.6674	0.0022	8.80E+07	−2.14	0.17
738.669	250	dc	135378.6	4d13	5p31	738.6715	0.0025	6.41E+08	−1.28	0.14
739.566	15		135214.4	4d27	5p27	739.5717	0.0027	2.08E+07	−2.77	0.02
(739.949)		B, x	135144.4	4d33	5p73	739.9494	0.0017	7.27E+07	−2.22	0.04
743.145	400		134563.2	4d15	5p23	743.1465	0.0029	8.96E+08	−1.13	0.12
744.719	50		134278.8	4d11	5p33	744.7212	0.0044	1.76E+08	−1.83	0.13
746.501	25		133958.3	4d11	5p31	746.5064	0.0044	9.33E+07	−2.11	0.03
(747.308)		A, x	133815.4	4d25	5p63	747.3082	0.0022	3.28E+07	−2.56	0.06
747.983	20		133692.9	4d43	5p55	747.9865	0.0020	5.43E+08	−1.34	0.23
(750.032)		C, x	133330.0	4d43	5p73	750.0306	0.0020	6.58E+08	−1.26	0.08
(750.322)		A, x	133276.8	4d23	5p63	750.3217	0.0022	6.03E+07	−2.30	0.02
750.571	150		133231.9	4d13	5p21	750.5742	0.0026	1.12E+08	−2.03	0.02
753.083	20		132787.5	4d27	5p45	753.0857	0.0027	1.13E+08	−2.02	0.02
754.179	30		132594.5	4d31	5p63	754.1877	0.0023	1.14E+08	−2.01	0.09
755.513	10		132360.4	4d33	5p63	755.5242	0.0018	4.84E+07	−2.38	0.01
755.822	150		132306.3	4d37	5p55	755.8279	0.0032	6.53E+08	−1.25	0.11
756.525	30	p	132183.3	4d11	5p23	756.5130	0.0045	4.63E+07	−2.40	0.03
758.670	2000		131809.6	4d11	5p21	758.6650	0.0045	7.25E+08	−1.20	0.17
767.276	800		130331.2	4d35	5p55	767.2829	0.0020	5.58E+08	−1.31	0.28
769.054	600		130029.9	4d53	5p55	769.0631	0.0020	5.91E+08	−1.28	0.23
771.212	1200	p	129666.0	4d53	5p73	771.2243	0.0019	3.31E+08	−1.53	0.07
778.674	2000		128423.4	4d15	5p17	778.6707	0.0033	7.86E+08	−1.15	0.52
779.517	10000		128284.6	4d17	5p17	779.5184	0.0040	5.04E+09	−0.34	0.68
780.633	5000		128101.2	4d17	5p25	780.6328	0.0039	1.92E+09	−0.76	0.29
785.186	50		127358.4	4d33	5p45	785.1885	0.0020	1.60E+07	−2.83	0.06
785.647	40		127283.6	4d13	5p25	785.6435	0.0028	3.18E+07	−2.53	0.03
786.287	1000		127180.0	4d35	5p63	786.2890	0.0020	5.38E+08	−1.30	0.14
787.875	50		126923.7	4d21	5p41	787.8806	0.0023	1.02E+08	−2.02	0.03
788.155	1200		126878.6	4d53	5p63	788.1586	0.0020	1.08E+09	−1.00	0.15
788.759	3000		126781.4	4d45	5p55	788.7654	0.0023	2.47E+09	−0.64	0.31
791.036	50		126416.5	4d45	5p73	791.0389	0.0022	1.35E+08	−1.90	0.03
793.220	3000		126068.4	4d13	5p11	793.2170	0.0029	1.14E+09	−0.97	0.26
796.085	75		125614.7	4d21	5p43	796.0902	0.0024	2.37E+08	−1.64	0.13

Table 1. *Cont.*

λ_{obs} (Å)	Int [a]		σ_{obs} (cm^{-1})	Even Level [b]	Odd Level [b]	λ_{Ritz} (Å)	Unc (Å)	$g_U A$ (s^{-1})	$\log(g_L f)$	|CF|
796.549	150	p	125541.6	4d43	5p45	796.5496	0.0023	6.33E+08	−1.22	0.34
802.263	4500	s	124647.4	4d11	5p11	802.2588	0.0051	1.84E+09	−0.75	0.56
802.529	35		124606.1	4d35	5p27	802.5321	0.0023	1.11E+08	−1.97	0.14
805.446	3000		124154.8	4d37	5p45	805.4484	0.0036	1.09E+09	−0.97	0.09
807.914	30000		123775.6	4d15	5p15	807.9081	0.0035	5.66E+09	−0.26	0.66
808.825	50000		123636.1	4d17	5p15	808.8207	0.0043	8.06E+09	−0.10	0.43
810.116	10000		123439.1	4d15	5p13	810.1120	0.0034	3.75E+09	−0.43	0.35
810.775	8000		123338.8	4d27	5p35	810.7736	0.0031	3.34E+09	−0.48	0.20
811.434	1000		123238.6	4d55	5p55	811.4400	0.0027	7.59E+08	−1.12	0.23
811.845	10		123176.2	4d25	5p53	811.8503	0.0026	2.02E+07	−2.70	0.02
813.842	700		122874.0	4d55	5p73	813.8463	0.0026	6.21E+08	−1.21	0.16
814.203	8000		122819.5	4d13	5p15	814.2011	0.0032	1.23E+09	−0.91	0.64
815.407	700	p	122638.1	4d23	5p53	815.4080	0.0025	1.05E+07	−2.98	0.01
816.442	30000		122482.7	4d13	5p13	816.4396	0.0031	4.07E+09	−0.39	0.65
816.852	50		122421.2	4d23	5p41	816.8597	0.0025	3.10E+07	−2.51	0.01
819.971	500		121955.5	4d31	5p53	819.9759	0.0028	5.13E+08	−1.29	0.18
820.496	15		121877.5	4d53	5p45	820.4957	0.0022	2.46E+07	−2.60	0.03
821.555	450		121720.4	4d33	5p53	821.5560	0.0021	7.68E+08	−1.11	0.22
822.039	450		121648.7	4d25	5p43	822.0399	0.0026	4.72E+08	−1.32	0.26
823.031	900		121502.1	4d33	5p41	823.0296	0.0021	8.60E+08	−1.06	0.38
823.990	450		121360.7	4d21	5p33	823.9898	0.0026	4.57E+08	−1.33	0.34
825.685	1200		121111.6	4d23	5p43	825.6877	0.0026	7.20E+08	−1.14	0.26
826.035	30000	dc	121060.2	4d11	5p13	826.0217	0.0054	1.08E+09	−0.96	0.67
826.035	30000	dc	121060.2	4d45	5p27	826.0640	0.0026	8.57E+08	−1.06	0.45
826.178	90		121039.3	4d21	5p31	826.1758	0.0026	2.36E+08	−1.61	0.18
830.375	10		120427.5	4d31	5p43	830.3718	0.0028	2.04E+07	−2.68	0.01
831.998	50		120192.6	4d33	5p43	831.9922	0.0022	8.61E+08	−1.05	0.19
837.767	1500		119364.9	4d25	5p35	837.7658	0.0028	1.19E+09	−0.90	0.37
838.455	40		119267.0	4d21	5p23	838.4497	0.0026	2.09E+07	−2.65	0.01
841.095	1800		118892.6	4d21	5p21	841.0940	0.0027	1.82E+09	−0.71	0.65
847.209	1600		118034.6	4d47	5p27	847.2103	0.0035	1.54E+09	−0.78	0.69
848.107	90		117909.7	4d33	5p35	848.1050	0.0023	1.95E+08	−1.67	0.11
848.430	10000		117864.8	4d29	5p27	848.4300	0.0072	1.73E+10	0.27	0.74
850.972	100		117512.7	4d55	5p27	850.9677	0.0030	3.09E+08	−1.47	0.51
851.819	30000		117395.8	4d25	5p33	851.8219	0.0028	3.61E+09	−0.41	0.60

Table 1. *Cont.*

λ_{obs} (Å)	Int [a]		σ_{obs} (cm^{-1})	Even Level [b]	Odd Level [b]	λ_{Ritz} (Å)	Unc (Å)	$g_U A$ (s^{-1})	log($g_L f$)	|CF|
854.303	1000000		117054.5	4d19	5p17	854.3030	0.0072	1.71E+10	0.27	0.79
855.746	30000		116857.1	4d23	5p33	855.7394	0.0028	7.09E+08	−1.11	0.19
856.665	10000		116731.7	5d83	5p31	856.6622	0.0027	2.05E+07	−2.65	0.00
858.081	30000	dc, p	116539.1	4d35	5p53	858.0633	0.0024	2.03E+09	−0.65	0.25
858.081	30000	dc, p	116539.1	4d23	5p31	858.0974	0.0028	3.19E+09	−0.46	0.54
859.152	10000		116393.8	4d63	5p83	859.1583	0.0024	5.20E+08	−1.24	0.30
860.292	1800		116239.6	4d53	5p53	860.2902	0.0024	1.00E+09	−0.95	0.28
860.771	75		116174.9	4d31	5p33	860.7718	0.0030	1.43E+08	−1.80	0.16
861.905	2000		116022.1	4d53	5p41	861.9062	0.0024	2.03E+09	−0.65	0.68
862.511	3000		115940.6	4d33	5p33	862.5132	0.0023	8.94E+08	−1.00	0.22
864.926	2000	p	115616.8	4d33	5p31	864.9086	0.0024	8.39E+07	−2.02	0.02
864.993	5000		115607.9	4d47	5p45	864.9915	0.0035	1.10E+10	0.09	0.66
867.282	10000		115302.8	4d25	5p23	867.2843	0.0029	3.95E+09	−0.35	0.76
868.911	1600		115086.6	4d55	5p45	868.9086	0.0030	8.38E+08	−1.02	0.35
869.451	10000		115015.1	4d35	5p43	869.4541	0.0025	2.86E+09	−0.49	0.33
871.358	25		114763.4	4d23	5p23	871.3457	0.0029	3.53E+07	−2.40	0.01
871.460	2000		114750.0	4d63	5p61	871.4691	0.0030	3.02E+09	−0.46	0.55
871.787	60000		114706.9	4d37	5p35	871.7907	0.0042	8.68E+09	0.00	0.80
872.792	30	c	114574.8	6s31	5p63	872.7799	0.0044	1.07E+08	−1.91	0.15
874.202	4000		114390.0	4d23	5p21	874.2019	0.0029	8.67E+08	−1.01	0.22
874.741	100		114319.6	6s33	5p11	874.7481	0.0022	1.50E+08	−1.76	0.07
876.241	2000		114123.9	4d43	5p33	876.2416	0.0027	9.74E+08	−0.95	0.25
876.565	2500		114081.7	4d31	5p23	876.5639	0.0031	7.29E+08	−1.08	0.39
878.373	600		113846.9	4d33	5p23	878.3698	0.0024	4.76E+08	−1.26	0.13
878.713	3000		113802.8	4d43	5p31	878.7141	0.0027	1.05E+09	−0.92	0.37
879.466	75		113705.4	4d31	5p21	879.4544	0.0032	1.28E+07	−2.83	0.01
881.272	5000		113472.3	4d33	5p21	881.2723	0.0025	6.82E+08	−1.10	0.20
883.887	5000		113136.6	4d27	5p17	883.8812	0.0039	7.24E+08	−1.07	0.26
885.016	5000		112992.3	4d45	5p53	885.0191	0.0027	1.37E+09	−0.80	0.25
885.312	60000		112954.5	4d27	5p25	885.3143	0.0037	7.71E+09	−0.04	0.80
887.067	75		112731.1	4d35	5p35	887.0659	0.0026	1.23E+08	−1.84	0.05
889.448	60		112429.3	4d53	5p35	889.4461	0.0026	1.14E+08	−1.86	0.08
892.613	8000		112030.6	4d43	5p23	892.6118	0.0028	1.53E+09	−0.74	0.42
895.015	3000		111730.0	4d21	5p11	895.0120	0.0030	1.95E+08	−1.63	0.08
895.609	3000		111655.9	4d43	5p21	895.6094	0.0028	3.68E+08	−1.36	0.08
895.769	8000		111635.9	4d75	5p83	895.7698	0.0029	7.97E+09	−0.02	0.77

Table 1. *Cont.*

λ_{obs} (Å)	Int [a]		σ_{obs} (cm^{-1})	Even Level [b]	Odd Level [b]	λ_{Ritz} (Å)	Unc (Å)	$g_U A$ (s^{-1})	log($g_L f$)	|CF|
897.155	5000		111463.5	4d45	5p43	897.1419	0.0028	1.10E+09	−0.88	0.13
900.135	8000		111094.4	4d65	5p73	900.1453	0.0022	3.88E+09	−0.33	0.63
902.141	15	H, x	110847.4	5d83	5p41	902.1284	0.0030	3.83E+07	−2.33	0.01
902.843	4000		110761.2	4d35	5p33	902.8405	0.0027	7.25E+08	−1.05	0.33
905.307	1000		110459.8	4d53	5p33	905.3063	0.0026	4.17E+08	−1.29	0.12
907.946	1000		110138.7	4d53	5p31	907.9457	0.0027	1.65E+08	−1.69	0.10
913.665	8000		109449.3	4d55	5p53	913.6659	0.0033	3.42E+09	−0.36	0.70
915.904	5000		109181.7	4d45	5p35	915.9053	0.0030	6.60E+08	−1.08	0.26
917.601	3000		108979.8	4d25	5p25	917.5967	0.0033	4.78E+08	−1.22	0.51
920.232	3000		108668.2	4d35	5p23	920.2295	0.0027	4.74E+08	−1.22	0.11
921.750	7000		108489.3	4d27	5p15	921.7454	0.0041	1.20E+09	−0.82	0.24
922.118	20000	dc	108446.0	4d57	5p55	922.1279	0.0034	8.24E+09	0.02	0.89
922.118	20000	dc	108446.0	4d23	5p25	922.1442	0.0033	5.10E+06	−3.19	0.02
922.785	700		108367.6	4d53	5p23	922.7913	0.0027	2.08E+08	−1.57	0.06
923.292	3000		108308.1	4d65	5p63	923.2994	0.0023	1.75E+09	−0.65	0.64
924.692	300		108144.1	4d21	5p13	924.6888	0.0033	1.95E+08	−1.59	0.18
925.999	8000		107991.5	4d53	5p21	925.9954	0.0027	7.02E+08	−1.04	0.22
926.587	10000		107922.9	4d55	5p43	926.5919	0.0033	2.82E+09	−0.44	0.59
930.021	25		107524.5	4d33	5p25	930.0148	0.0027	1.44E+08	−1.73	0.16
932.598	3000		107227.3	4d23	5p11	932.5955	0.0033	4.43E+08	−1.24	0.28
932.728	5000		107212.4	4d45	5p33	932.7321	0.0030	1.41E+09	−0.74	0.32
938.575	2000		106544.5	4d31	5p11	938.5756	0.0036	4.74E+08	−1.21	0.75
940.648	1500		106309.7	4d33	5p11	940.6463	0.0028	7.87E+08	−0.98	0.40
941.972	500		106160.3	4d47	5p35	941.9739	0.0042	3.28E+08	−1.36	0.14
946.000	1500		105708.2	4d43	5p25	945.9961	0.0032	1.44E+08	−1.72	0.18
946.619	60		105639.1	4d55	5p35	946.6212	0.0036	7.32E+07	−2.00	0.09
951.300	7000		105119.3	4d45	5p23	951.3034	0.0031	9.14E+08	−0.91	0.19
956.356	15		104563.6	6s25	5p33	956.3568	0.0058	1.12E+09	−0.81	0.78
956.797	75	dc	104515.4	4d25	5p15	956.7919	0.0038	1.26E+08	−1.77	0.11
956.797	75	dc	104515.4	5d75	5p15	956.7967	0.0025	2.17E+07	−2.53	0.00
956.904	75		104503.7	4d37	5p17	956.8938	0.0052	1.06E+08	−1.83	0.08
956.999	5000		104493.3	4d43	5p11	956.9983	0.0032	5.29E+08	−1.14	0.17
958.295	10		104352.0	6s11	5p13	958.2918	0.0030	3.98E+08	−1.26	0.07
958.572	20000		104321.8	4d37	5p25	958.5736	0.0050	2.35E+09	−0.49	0.22
959.888	300		104178.8	4d25	5p13	959.8846	0.0036	1.51E+08	−1.68	0.12
961.737	1500		103978.5	4d23	5p15	961.7373	0.0038	1.31E+08	−1.74	0.23

Table 1. *Cont.*

λ_{obs} (Å)	Int [a]		σ_{obs} (cm^{-1})	Even Level [b]	Odd Level [b]	λ_{Ritz} (Å)	Unc (Å)	g_UA (s^{-1})	$\log(g_Lf)$	|CF|
964.612	75		103668.6	4d55	5p33	964.6066	0.0036	1.07E+07	−2.82	0.01
964.862	1800		103641.8	4d23	5p13	964.8620	0.0036	1.60E+08	−1.66	0.11
970.302	1000		103060.7	4d33	5p15	970.3014	0.0032	3.12E+08	−1.35	0.22
971.262	8000		102958.8	4d31	5p13	971.2645	0.0039	1.10E+09	−0.81	0.47
973.506	8000		102721.5	4d57	5p27	973.5164	0.0039	2.37E+09	−0.47	0.63
987.710	60	s	101244.3	4d43	5p15	987.7102	0.0037	3.38E+06	−3.31	0.01
997.069	200		100294.0	4d57	5p45	997.0683	0.0039	1.91E+08	−1.54	0.08
1001.401	8000	D, x	99860.1	4d37	5p15	1001.4294	0.0056	2.23E+06	−3.47	0.00
1007.459	8	x	99259.62	5s13	5p83	1007.4683	0.0024	1.68E+07	−2.59	0.01
1012.182	300		98796.46	4d45	5p25	1012.1782	0.0036	1.13E+08	−1.76	0.05
1013.316	2	x	98685.90	5d41	5p11	1013.3143	0.0021	8.97E+07	−1.86	0.03
1013.765	2		98642.19	5d65	5p25	1013.7694	0.0023	1.21E+08	−1.73	0.03
1021.642	8000		97881.65	4d35	5p15	1021.6380	0.0037	1.02E+09	−0.80	0.34
1023.866	300		97669.03	4d65	5p53	1023.8658	0.0028	2.65E+08	−1.38	0.25
1024.417	5	x	97616.50	5s13	5p61	1024.4382	0.0033	1.60E+06	−3.60	0.00
1024.800	2000		97580.02	4d53	5p15	1024.7965	0.0036	2.12E+08	−1.47	0.19
1025.165	8000		97545.27	4d35	5p13	1025.1648	0.0035	4.94E+08	−1.11	0.13
1026.554	20		97413.29	4d85	5p83	1026.5538	0.0025	4.97E+08	−1.10	0.18
1028.336	7000	dc	97244.48	4d63	5p51	1028.3415	0.0028	9.37E+08	−0.83	0.64
1028.336	7000	dc	97244.48	4d53	5p13	1028.3452	0.0034	5.34E+07	−2.07	0.02
1040.119	200	p	96142.85	4d65	5p43	1040.1257	0.0029	2.31E+08	−1.43	0.21
1043.889	3		95795.63	6s21	5p23	1043.8995	0.0040	2.43E+08	−1.40	0.07
1044.113	500		95775.07	4d47	5p25	1044.1106	0.0051	3.02E+08	−1.31	0.06
1044.349	3	x	95753.43	5d73	5p21	1044.3602	0.0028	6.83E+07	−1.95	0.01
1049.162	25		95314.17	6s13	5p13	1049.1621	0.0026	7.41E+08	−0.91	0.23
1062.555	500		94112.78	6s23	5p21	1062.5512	0.0035	2.15E+09	−0.44	0.67
1063.881	75		93995.48	4d45	5p13	1063.8787	0.0040	5.32E+07	−2.05	0.02
1065.429	35		93858.91	4d65	5p35	1065.4310	0.0032	8.84E+07	−1.82	0.30
1065.938	200	p, x	93814.09	6s15	5p13	1065.9406	0.0028	3.71E+09	−0.20	0.63
1069.777	8000		93477.43	6s15	5p15	1069.7805	0.0032	7.12E+09	0.09	0.77
1072.143	500	*l*	93271.14	6s33	5p45	1072.1497	0.0034	5.69E+09	−0.01	0.74
1072.627	90		93229.05	6s11	5p23	1072.6265	0.0037	1.19E+09	−0.69	0.41
(1073.586)		A, x	93145.95	6s25	5p45	1073.5858	0.0073	1.50E+09	−0.59	0.95
1074.887	500		93033.04	4d63	5p73	1074.8909	0.0030	5.18E+08	−1.05	0.33
1077.696	3		92790.55	5d55	5p13	1077.7006	0.0022	5.91E+07	−1.99	0.00
1086.914	8		92003.60	5d63	5p23	1086.9113	0.0022	5.29E+08	−1.03	0.13

Table 1. *Cont.*

λ_{obs} (Å)	Int [a]		σ_{obs} (cm^{-1})	Even Level [b]	Odd Level [b]	λ_{Ritz} (Å)	Unc (Å)	$g_U A$ (s^{-1})	$\log(g_L f)$	|CF|
(1087.355)		A, x	91967.64	6s23	5p31	1087.3552	0.0037	5.74E+07	−1.99	0.03
1088.274	30		91888.62	4d65	5p33	1088.2688	0.0032	8.46E+07	−1.82	0.14
(1090.176)		E, x	91728.48	6s13	5p11	1090.1761	0.0028	1.55E+09	−0.56	0.55
1091.166	500		91645.08	6s23	5p33	1091.1651	0.0036	2.58E+09	−0.34	0.52
1093.406	500		91457.34	6s11	5p31	1093.4074	0.0040	2.09E+09	−0.43	0.89
1097.106	2	H, x	91148.90	5d41	5p23	1097.1087	0.0024	3.23E+08	−1.24	0.31
(1097.260)		A, x	91136.19	6s11	5p33	1097.2599	0.0039	8.52E+08	−0.81	0.43
1100.762	500		90846.16	4d57	5p35	1100.7638	0.0048	2.33E+08	−1.37	0.17
1104.810	8000		90513.30	6s13	5p25	1104.8136	0.0029	6.37E+09	0.07	0.83
1105.225	8		90479.31	5d45	5p15	1105.2225	0.0030	5.86E+08	−0.97	0.05
1112.211	5		89911.00	5d63	5p33	1112.2129	0.0024	4.42E+08	−1.09	0.07
1113.632	500		89796.27	4d65	5p23	1113.6345	0.0033	2.86E+08	−1.27	0.26
1115.130	8000		89675.64	6s23	5p35	1115.1319	0.0039	6.41E+09	0.08	0.88
1115.714	35		89628.70	5d53	5p11	1115.7157	0.0021	5.57E+08	−0.98	0.06
1117.949	5000	H, x	89449.52	6s21	5p43	1117.9617	0.0046	1.22E+09	−0.64	0.30
1122.891	40		89055.84	5d41	5p33	1122.8930	0.0025	1.48E+09	−0.55	0.67
1123.198	15		89031.50	5d73	5p43	1123.2007	0.0032	3.51E+08	−1.18	0.10
1123.432	5	x	89012.95	6s15	5p25	1123.4351	0.0031	4.85E+07	−2.04	0.02
1125.741	8000		88830.38	6s15	5p17	1125.7511	0.0037	8.69E+09	0.22	0.93
(1127.686)		F, x	88676.22	5d35	5p13	1127.6865	0.0027	5.36E+08	−0.99	0.03
(1128.160)		C, x	88640.07	4d75	5p55	1128.1600	0.0041	6.70E+08	−0.89	0.61
1130.833	200		88430.39	6s31	5p83	1130.8356	0.0079	4.02E+09	−0.11	0.98
1131.984	40		88340.47	5d35	5p15	1131.9850	0.0032	2.81E+08	−1.27	0.02
(1133.055)		A, x	88257.20	5d65	5p35	1133.0546	0.0030	1.67E+08	−1.50	0.10
1134.495	75	p	88144.95	6s25	5p63	1134.4903	0.0082	1.51E+09	−0.54	0.38
1134.562	200	p	88139.74	6s21	5p41	1134.5635	0.0048	2.27E+09	−0.35	0.50
1136.509	5		87988.74	5d55	5p25	1136.5057	0.0024	1.55E+08	−1.52	0.01
1136.993	20		87951.29	5d51	5p43	1136.9939	0.0030	4.84E+08	−1.03	0.13
1137.123	3		87941.23	5d63	5p35	1137.1238	0.0027	6.51E+07	−1.90	0.04
1137.375	200		87921.75	6s21	5p53	1137.3759	0.0048	1.31E+09	−0.60	0.49
(1142.799)		E, x	87504.68	5d73	5p53	1142.7989	0.0033	3.50E+08	−1.17	0.17
1150.971	150	p	86883.16	6s11	5p43	1150.9739	0.0043	1.73E+09	−0.46	0.80
1157.079	40		86424.52	5d51	5p53	1157.0806	0.0032	4.32E+08	−1.06	0.24
1161.657	20	p	86083.93	6s23	5p41	1161.6681	0.0041	2.25E+08	−1.34	0.17
1162.889	3		85992.73	5d43	5p11	1162.8868	0.0020	2.63E+08	−1.27	0.02
1163.149	8		85973.51	5d65	5p43	1163.1491	0.0030	1.42E+08	−1.54	0.03

Table 1. *Cont.*

λ_{obs} (Å)	Int [a]		σ_{obs} (cm^{-1})	Even Level [b]	Odd Level [b]	λ_{Ritz} (Å)	Unc (Å)	$g_U A$ (s^{-1})	log($g_L f$)	|CF|
1164.616	200		85865.21	6s23	5p53	1164.6166	0.0042	1.10E+09	−0.65	0.40
1167.437	5		85657.73	5d63	5p43	1167.4378	0.0026	2.34E+08	−1.32	0.04
1168.328	200	p	85592.40	4d83	5p83	1168.3133	0.0031	2.55E+08	−1.28	0.40
1168.565	4	x	85575.04	6s11	5p41	1168.5784	0.0045	7.92E+07	−1.79	0.24
1169.735	150		85489.45	4d75	5p63	1169.7331	0.0042	8.55E+07	−1.76	0.05
1169.818	200		85483.38	6s33	5p73	1169.8078	0.0040	2.78E+09	−0.25	0.65
1176.540	300	p	84994.99	6s25	5p55	1176.5399	0.0089	4.82E+09	0.00	0.81
1177.817	200	u	84902.83	5d33	5p11	1177.8398	0.0027	4.71E+08	−1.01	0.16
1179.214	8		84802.25	5d41	5p43	1179.2104	0.0028	3.64E+08	−1.12	0.11
1185.554	20		84348.75	5d63	5p41	1185.5535	0.0027	3.41E+08	−1.14	0.13
1187.777	200		84190.89	6s13	5p23	1187.7767	0.0033	1.70E+09	−0.45	0.44
1188.940	5		84108.53	5d31	5p11	1188.9343	0.0027	1.23E+08	−1.59	0.02
1191.195	500		83949.31	4d83	5p61	1191.1959	0.0044	1.20E+09	−0.59	0.61
1191.640	20		83917.96	5d37	5p17	1191.6238	0.0100	1.31E+08	−1.55	0.02
1197.981	300		83473.78	4d65	5p25	1197.9782	0.0039	1.40E+08	−1.52	0.25
1201.630	800		83220.29	5d25	5p13	1201.6217	0.0041	1.28E+09	−0.56	0.12
1209.328	100		82690.55	6s15	5p23	1209.3272	0.0035	2.40E+08	−1.28	0.10
1210.115	500		82636.77	5d23	5p15	1210.1135	0.0033	7.42E+08	−0.79	0.28
1213.322	40		82418.35	6s13	5p31	1213.3120	0.0036	1.98E+08	−1.36	0.19
1218.066	3		82097.36	6s13	5p33	1218.0577	0.0035	5.86E+07	−1.89	0.02
1221.502	150		81866.42	5d73	5p45	1221.4970	0.0039	1.45E+09	−0.49	0.64
1227.090	200		81493.61	5d21	5p13	1227.0824	0.0024	1.32E+09	−0.53	0.24
1230.424	150		81272.80	6s33	5p51	1230.4229	0.0045	1.03E+09	−0.63	0.64
1240.739	3	x	80597.13	6s15	5p33	1240.7314	0.0037	3.48E+07	−2.10	0.03
1241.643	50		80538.45	5s15	5p55	1241.6320	0.0027	3.43E+07	−2.10	0.01
1242.844	10000	D, x	80460.62	4d57	5p25	1242.8343	0.0060	1.83E+08	−1.37	0.12
1246.600	15000		80218.19	5d75	5p45	1246.5996	0.0038	5.57E+09	0.11	0.63
1248.301	75	dc	80108.88	5d57	5p45	1248.2886	0.0022	4.67E+07	−1.96	0.01
1248.301	75	dc	80108.88	5s13	5p51	1248.2891	0.0048	2.70E+07	−2.20	0.01
1250.044	200	p	79997.18	5d53	5p33	1250.0283	0.0025	8.30E+08	−0.71	0.12
1254.821	100		79692.64	5d45	5p23	1254.8152	0.0032	5.76E+08	−0.87	0.03
1256.150	60		79608.33	4d63	5p53	1256.1462	0.0041	3.84E+07	−2.04	0.27
1256.701	12000		79573.42	5d55	5p33	1256.6933	0.0028	2.23E+09	−0.28	0.20
1259.640	100000		79387.76	5d23	5p11	1259.6400	0.0029	4.28E+09	0.01	0.59
1260.859	12000		79311.01	4d41	5p73	1260.8507	0.0024	2.93E+08	−1.15	0.67
1262.106	15000		79232.65	5d27	5p15	1262.1007	0.0051	1.69E+09	−0.39	0.28

Table 1. *Cont.*

λ_{obs} (Å)	Int [a]		σ_{obs} (cm^{-1})	Even Level [b]	Odd Level [b]	λ_{Ritz} (Å)	Unc (Å)	g_UA (s^{-1})	log(g_Lf)	|CF|
1265.676	75		79009.16	4d65	5p15	1265.6696	0.0048	1.78E+07	−2.37	0.09
1268.550	300000		78830.16	5d43	5p21	1268.5452	0.0024	5.09E+09	0.09	0.48
1271.817	60		78627.66	6s15	5p35	1271.8124	0.0041	1.91E+08	−1.34	0.06
1273.809	200	p	78504.71	5s25	5p83	1273.8054	0.0038	2.51E+08	−1.21	0.35
1274.001	10	H, x	78492.87	5d63	5p45	1273.9965	0.0034	2.21E+08	−1.27	0.36
(1275.190)		G, x	78419.26	5d25	5p25	1275.1895	0.0046	5.59E+09	0.14	0.70
1278.197	5	x	78235.20	5d25	5p17	1278.1743	0.0053	4.48E+07	−1.96	0.07
1279.227	200		78172.21	5d23	5p25	1279.2227	0.0030	4.74E+08	−0.94	0.25
1280.076	300000		78120.36	5d11	5p13	1280.0759	0.0033	4.17E+09	0.01	0.91
1280.713	50		78081.51	4d63	5p43	1280.7090	0.0043	2.47E+07	−2.22	0.06
1281.503	40		78033.37	5s33	5p83	1281.4978	0.0039	6.24E+07	−1.81	0.29
1283.561	200000		77908.26	5d21	5p11	1283.5609	0.0026	2.83E+09	−0.15	0.88
1284.588	90		77845.97	6s13	5p43	1284.6081	0.0039	1.04E+08	−1.59	0.04
1285.486	400		77791.59	5d75	5p27	1285.4819	0.0045	1.42E+09	−0.46	0.74
1287.278	300000		77683.30	5d57	5p27	1287.2785	0.0055	7.52E+09	0.27	0.77
1288.606	400	p	77603.24	5d55	5p35	1288.5894	0.0033	8.01E+08	−0.70	0.15
1288.660	700	p	77599.99	5d45	5p33	1288.6595	0.0034	1.63E+09	−0.39	0.28
1289.155	200000		77570.19	4d73	5p55	1289.1509	0.0028	3.51E+08	−1.06	0.33
1289.429	250000		77553.71	5d35	5p23	1289.4261	0.0034	3.15E+09	−0.11	0.21
1292.599	700		77363.51	5d33	5p23	1292.5942	0.0031	7.63E+08	−0.72	0.17
1295.238	300000		77205.89	4d73	5p73	1295.2349	0.0022	5.46E+08	−0.87	0.23
1297.729	2000000		77057.69	5d13	5p13	1297.7315	0.0030	8.63E+09	0.34	0.88
1299.606	300000		76946.4	5d31	5p21	1299.6041	0.0032	4.84E+09	0.09	0.92
1303.441	300000	p	76720.00	5d13	5p15	1303.4275	0.0038	3.11E+09	−0.10	0.88
1306.580	25	p	76535.69	6s13	5p41	1306.5769	0.0040	3.93E+08	−0.99	0.21
1306.759	300000		76525.20	4d41	5p63	1306.7526	0.0026	7.33E+08	−0.73	0.53
(1307.988)		G, x	76453.71	5d15	5p13	1307.9875	0.0076	6.73E+09	0.24	0.42
1309.081	10		76389.47	5s33	5p61	1309.0811	0.0055	4.03E+07	−1.98	0.14
1309.538	300000		76362.81	5d43	5p33	1309.5431	0.0024	3.93E+09	0.01	0.58
1309.854	150		76344.39	6s15	5p43	1309.8529	0.0042	5.56E+08	−0.84	0.21
1310.312	15		76317.70	6s13	5p53	1310.3081	0.0041	9.53E+07	−1.61	0.06
1313.774	250		76116.59	5d15	5p15	1313.7740	0.0080	1.12E+10	0.46	0.90
(1314.550)		J, x	76071.89	5d17	5p15	1314.5505	0.0055	1.57E+10	0.61	0.64
1315.851	500000	D, x	75996.45	4d57	5p15	1315.8442	0.0071	2.90E+07	−2.12	0.06

Table 1. *Cont.*

λ_{obs} (Å)	Int [a]		σ_{obs} (cm^{-1})	Even Level [b]	Odd Level [b]	λ_{Ritz} (Å)	Unc (Å)	$g_U A$ (s^{-1})	log($g_L f$)	|CF|
1317.557	90		75898.04	5s13	5p73	1317.5505	0.0022	5.68E+06	−2.83	0.00
1317.946	2000000		75875.64	5d47	5p45	1317.9551	0.0055	2.43E+10	0.80	0.96
1318.842	500000		75824.09	4d51	5p51	1318.8428	0.0028	8.90E+08	−0.64	0.67
1320.219	250		75745.01	5d53	5p43	1320.2188	0.0028	6.57E+08	−0.77	0.14
1322.222	500000		75630.26	5d45	5p35	1322.2205	0.0039	3.42E+09	−0.05	0.58
1322.889	3000000		75592.13	5d33	5p31	1322.8927	0.0034	9.64E+09	0.40	0.94
1324.659	500000		75491.13	5d83	5p61	1324.6640	0.0082	1.01E+10	0.42	0.95
1325.186	3000000		75461.11	5d35	5p33	1325.1898	0.0036	8.78E+09	0.36	0.74
1327.651	3000000		75321.00	5d55	5p43	1327.6556	0.0032	9.45E+09	0.40	0.70
1328.540	50		75270.60	5d33	5p33	1328.5363	0.0033	3.35E+08	−1.05	0.08
1335.958	50	H, x	74852.65	4d75	5p53	1335.9802	0.0054	2.64E+07	−2.15	0.05
1336.587	40		74817.43	6s15	5p53	1336.5834	0.0044	1.68E+08	−1.35	0.08
1336.645	25		74814.18	5s15	5p27	1336.6350	0.0035	2.00E+07	−2.27	0.00
1336.909	25		74799.41	5d31	5p31	1336.9043	0.0035	1.82E+08	−1.32	0.06
1337.456	2000000		74768.81	5d27	5p25	1337.4604	0.0053	2.23E+10	0.78	0.98
1340.744	500		74585.45	5d27	5p17	1340.7443	0.0060	3.68E+08	−1.00	0.22
1341.665	500000		74534.25	5d11	5p11	1341.6604	0.0036	1.85E+09	−0.30	0.46
1342.673	500		74478.3	5d31	5p33	1342.6683	0.0033	6.55E+08	−0.76	0.40
1343.434	1000000		74436.11	5d53	5p41	1343.4334	0.0030	7.36E+09	0.30	0.80
1343.738	1000000	p	74419.27	4d73	5p63	1343.7226	0.0024	1.22E+09	−0.49	0.43
1344.216	50		74392.81	5d43	5p35	1344.2154	0.0029	4.96E+08	−0.87	0.21
1347.381	200		74218.06	5d53	5p53	1347.3784	0.0030	8.35E+08	−0.65	0.31
1349.888	500000		74080.22	5d73	5p73	1349.8860	0.0046	5.01E+09	0.13	0.64
1350.389	900000		74052.74	4d85	5p73	1350.3838	0.0025	6.66E+08	−0.73	0.38
1352.760	2	x	73922.94	5s11	5p51	1352.7493	0.0027	1.50E+07	−2.38	0.11
1354.162	200		73846.41	5d83	5p83	1354.1583	0.0076	1.89E+09	−0.28	0.83
1354.861	500000		73808.31	5d65	5p63	1354.8575	0.0041	7.07E+09	0.29	0.76
1355.128	100000		73793.77	5d55	5p53	1355.1252	0.0034	2.73E+09	−0.12	0.25
1355.245	1800000		73787.40	5d29	5p27	1355.2451	0.0081	3.01E+10	0.92	1.00
1356.518	5000000		73718.15	5d37	5p35	1356.5305	0.0127	2.36E+10	0.81	0.99
1359.388	4500000		73562.51	5d19	5p17	1359.3879	0.0084	2.97E+10	0.91	0.99
1360.408	25		73507.36	4d63	5p31	1360.4060	0.0049	1.84E+07	−2.29	0.04
1360.689	500000	dc	73492.18	5d63	5p63	1360.6799	0.0036	5.35E+09	0.17	0.60
1360.689	500000	dc	73492.18	5d35	5p35	1360.7068	0.0041	8.34E+08	−0.64	0.11
1361.069	150000		73471.66	5d13	5p11	1361.0686	0.0033	1.01E+09	−0.55	0.17

Table 1. *Cont.*

λ_{obs} (Å)	Int [a]		σ_{obs} (cm^{-1})	Even Level [b]	Odd Level [b]	λ_{Ritz} (Å)	Unc (Å)	$g_U A$ (s^{-1})	log($g_L f$)	|CF|
1361.502	300		73448.29	5d47	5p27	1361.4937	0.0063	1.37E+09	−0.42	0.95
1363.284	500000		73352.29	5d85	5p83	1363.2841	0.0090	1.68E+10	0.67	0.98
1367.758	500000		73112.35	5s13	5p63	1367.7557	0.0024	3.08E+08	−1.06	0.09
1369.857	75		73000.32	5d51	5p73	1369.8579	0.0045	1.04E+09	−0.54	0.51
1376.698	100000		72637.57	5d41	5p63	1376.6992	0.0039	2.64E+09	−0.12	0.58
1380.608	500000		72431.86	5d75	5p73	1380.6093	0.0045	6.15E+09	0.25	0.51
1383.652	75		72272.51	5s23	5p51	1383.6407	0.0026	5.21E+07	−1.83	0.06
1383.966	10		72256.11	5d13	5p25	1383.9606	0.0035	3.45E+07	−2.01	0.02
1384.557	8		72225.27	5d23	5p21	1384.5558	0.0035	6.43E+07	−1.73	0.01
1386.785	100		72109.23	5d43	5p43	1386.7829	0.0027	4.73E+08	−0.87	0.07
1387.003	500000		72097.90	5d25	5p23	1387.0083	0.0053	8.83E+09	0.41	0.58
1387.587	100000		72067.55	5d75	5p55	1387.5896	0.0050	3.69E+09	0.03	0.43
1389.683	600000		71958.86	5d57	5p55	1389.6831	0.0062	1.52E+10	0.65	0.97
1391.789	600000		71849.97	5d23	5p23	1391.7813	0.0033	3.99E+09	0.06	0.68
1392.363	600000		71820.35	5d45	5p53	1392.3692	0.0041	1.03E+10	0.48	0.76
1396.402	300		71612.62	4d51	5p73	1396.3986	0.0030	2.01E+08	−1.24	0.33
1396.507	10		71607.23	5d17	5p25	1396.5072	0.0057	1.64E+09	−0.32	0.19
1400.088	500		71424.08	5d17	5p17	1400.0878	0.0064	7.65E+09	0.35	0.99
1403.173	500		71267.05	4d85	5p63	1403.1726	0.0026	1.42E+09	−0.37	0.66
1404.344	400		71207.62	5d35	5p43	1404.3421	0.0041	4.46E+09	0.12	0.60
1408.005	75		71022.48	5d65	5p73	1408.0035	0.0044	3.23E+09	−0.02	0.37
1408.101	60		71017.63	5d33	5p43	1408.1009	0.0037	9.19E+08	−0.56	0.48
1412.421	25		70800.42	5d43	5p41	1412.4202	0.0029	4.78E+07	−1.84	0.01
1414.290	15		70706.86	5d63	5p73	1414.2927	0.0039	4.44E+08	−0.88	0.14
1415.262	800		70658.30	5d65	5p55	1415.2643	0.0049	3.92E+09	0.07	0.38
1416.781	500		70582.54	5d43	5p53	1416.7814	0.0029	1.63E+09	−0.31	0.34
1421.043	400		70370.85	5d21	5p23	1421.0425	0.0030	8.43E+08	−0.60	0.38
1421.641	50	u	70341.25	5d63	5p55	1421.6187	0.0045	2.32E+09	−0.15	0.75
1422.759	3		70285.97	6s21	5p51	1422.7526	0.0077	4.07E+08	−0.91	0.17
1423.983	5		70225.56	5d31	5p43	1423.9864	0.0038	1.40E+08	−1.37	0.09
1426.967	5		70078.71	5d23	5p31	1426.9714	0.0038	9.71E+07	−1.53	0.04
1428.475	2		70004.73	5d25	5p33	1428.4769	0.0057	2.47E+08	−1.12	0.03
1431.245	600		69869.24	5d73	5p51	1431.2485	0.0053	3.38E+09	0.01	0.72
1431.605	60		69851.67	5d41	5p73	1431.6073	0.0042	8.36E+08	−0.59	0.84

Table 1. *Cont.*

λ_{obs} (Å)	Int [a]		σ_{obs} (cm^{-1})	Even Level [b]	Odd Level [b]	λ_{Ritz} (Å)	Unc (Å)	$g_U A$ (s^{-1})	$\log(g_L f)$	|CF|
1433.535	35		69757.63	5d23	5p33	1433.5401	0.0036	5.20E+08	−0.80	0.13
1434.541	3		69708.71	5d33	5p41	1434.5399	0.0039	6.91E+07	−1.67	0.04
1435.114	75		69680.88	5d35	5p53	1435.1135	0.0043	4.03E+08	−0.91	0.09
1439.035	15		69491.01	5d33	5p53	1439.0390	0.0039	9.91E+07	−1.51	0.10
1440.524	60		69419.18	4d73	5p45	1440.5145	0.0030	9.41E+07	−1.54	0.22
1451.028	20		68916.66	5d31	5p41	1451.0311	0.0039	2.33E+08	−1.13	0.10
1453.723	400		68788.90	5d51	5p51	1453.7206	0.0052	2.52E+09	−0.10	0.77
1455.629	15		68698.82	5d31	5p53	1455.6344	0.0040	1.70E+08	−1.27	0.18
1455.759	400		68692.69	4d85	5p27	1455.7530	0.0040	2.37E+08	−1.11	0.70
1457.750	15		68598.87	5d21	5p31	1457.7474	0.0035	1.87E+08	−1.23	0.14
1464.603	8		68277.89	5d21	5p33	1464.6032	0.0032	1.39E+08	−1.35	0.10
1469.257	60		68061.61	5s23	5p73	1469.2521	0.0026	7.48E+07	−1.62	0.05
1476.575	12		67724.29	5d47	5p55	1476.5744	0.0072	2.05E+08	−1.17	0.13
(1478.619)		G, x	67630.67	5s21	5p51	1478.6191	0.0031	6.50E+08	−0.67	0.55
1498.117	800		66750.46	5s15	5p53	1498.1125	0.0032	6.00E+08	−0.69	0.19
1503.863	60		66495.42	5d63	5p51	1503.8621	0.0046	6.82E+08	−0.64	0.38
1505.017	1200		66444.43	4d83	5p51	1505.0166	0.0030	1.10E+09	−0.42	0.61
1508.089	150		66309.08	5d13	5p21	1508.0852	0.0041	3.24E+06	−2.96	0.00
1509.064	40		66266.24	4d85	5p45	1509.0561	0.0034	7.31E+07	−1.59	0.21
1522.539	60		65679.76	6s13	5p63	1522.5393	0.0055	3.29E+08	−0.94	0.29
1523.456	12		65640.23	5d41	5p51	1523.4543	0.0050	1.92E+08	−1.17	0.11
(1526.619)		J, x	65503.71	5d23	5p43	1526.6194	0.0041	3.63E+08	−0.90	0.07
1531.962	1200		65275.77	5s23	5p63	1531.9592	0.0028	1.08E+09	−0.42	0.66
1533.188	700		65223.57	5s15	5p43	1533.1817	0.0032	2.77E+08	−1.01	0.07
1540.731	3		64904.26	5d11	5p33	1540.7340	0.0046	7.53E+07	−1.57	0.07
1557.043	10	*l*	64224.30	5d25	5p53	1557.0348	0.0068	1.45E+08	−1.28	0.02
(1561.446)		L, x	64043.37	5d35	5p45	1561.4460	0.0054	1.29E+05	−4.33	0.00
1561.896	12		64024.75	5d21	5p43	1561.8970	0.0037	1.02E+08	−1.43	0.04
1563.049	150		63977.52	5d23	5p53	1563.0522	0.0043	6.08E+08	−0.65	0.15
1566.381	25		63841.43	5d13	5p33	1566.3841	0.0042	1.34E+08	−1.31	0.05
1573.219	800		63563.94	4d73	5p41	1573.2175	0.0032	3.31E+08	−0.92	0.21
1576.805	400		63419.38	5s21	5p73	1576.8043	0.0032	1.47E+08	−1.26	0.66
1583.386	300		63155.79	5d55	5p63	1583.3874	0.0047	6.74E+08	−0.60	0.24
1588.810	400		62940.19	5s15	5p35	1588.8062	0.0040	1.39E+08	−1.28	0.03
1589.983	75	p	62893.75	6s13	5p73	1589.9818	0.0060	1.46E+07	−2.26	0.02

Table 1. *Cont.*

λ_{obs} (Å)	Int [a]		σ_{obs} (cm^{-1})	Even Level [b]	Odd Level [b]	λ_{Ritz} (Å)	Unc (Å)	$g_U A$ (s^{-1})	log($g_L f$)	|CF|
1594.488	3		62716.06	5d21	5p41	1594.4938	0.0039	8.65E+07	−1.48	0.06
1600.054	300		62497.89	5d21	5p53	1600.0540	0.0040	3.85E+08	−0.83	0.23
1600.662	1200		62474.15	5s13	5p53	1600.6584	0.0032	6.22E+08	−0.62	0.20
1606.264	600	dc	62256.27	5s13	5p41	1606.2619	0.0032	2.53E+08	−1.02	0.12
1606.264	600	dc	62256.27	4d73	5p43	1606.2936	0.0032	2.94E+07	−1.95	0.03
1606.856	200		62233.33	4d83	5p73	1606.8596	0.0030	4.72E+07	−1.73	0.04
1609.689	800		62123.80	6s33	5p83	1609.6880	0.0090	2.84E+06	−2.96	0.08
1616.245	30		61871.81	5d13	5p35	1616.2496	0.0049	7.19E+07	−1.55	0.05
1634.471	60		61181.87	5d45	5p63	1634.4717	0.0056	1.16E+08	−1.33	0.12
(1640.133)		M, x	60970.52	5s15	5p33	1640.1329	0.0037	2.31E+08	−1.03	0.28
1640.759	600		60947.40	5s13	5p43	1640.7572	0.0032	1.09E+08	−1.36	0.03
1643.670	5		60839.46	4d75	5p17	1643.6701	0.0093	9.07E+06	−2.43	0.13
1644.896	150		60794.12	5d53	5p73	1644.8973	0.0046	3.11E+08	−0.90	0.16
1648.776	20		60651.05	5d11	5p43	1648.7785	0.0053	8.11E+07	−1.48	0.05
1649.261	600	p	60633.22	5s21	5p63	1649.2544	0.0035	1.87E+08	−1.12	0.63
1649.378	400	p	60628.92	4d85	5p53	1649.3787	0.0035	1.48E+08	−1.21	0.39
(1656.458)		A, x	60369.56	5d55	5p73	1656.4576	0.0051	1.71E+08	−1.15	0.08
1659.064	5	x	60274.95	5s23	5p45	1659.0513	0.0038	8.14E+06	−2.47	0.01
1663.695	600		60107.17	4d41	5p33	1663.6884	0.0040	9.46E+07	−1.41	0.31
1666.513	50		60005.53	5d55	5p55	1666.5161	0.0059	3.03E+08	−0.90	0.31
1667.463	2		59971.35	4d73	5p35	1667.4554	0.0041	3.71E+07	−1.81	0.10
1668.214	60		59944.35	5d43	5p63	1668.2142	0.0042	2.16E+08	−1.04	0.13
1682.168	8		59447.09	4d83	5p63	1682.1639	0.0033	2.99E+05	−3.90	0.00
1685.140	12	H, x	59342.25	5d11	5p41	1685.1448	0.0056	4.37E+07	−1.72	0.05
1691.351	5	H, x	59124.33	5d11	5p53	1691.3565	0.0057	4.24E+07	−1.74	0.04
1691.986	500		59102.14	4d85	5p43	1691.9881	0.0036	1.68E+08	−1.13	0.20
1698.233	1500		58884.73	5s33	5p51	1698.2352	0.0042	1.51E+09	−0.19	0.69
1698.443	700		58877.45	5s15	5p23	1698.4365	0.0039	3.19E+08	−0.86	0.19
1704.624	400		58663.96	5s13	5p35	1704.6238	0.0041	1.67E+08	−1.13	0.03
1722.320	20	dc	58061.22	5d13	5p53	1722.3173	0.0052	4.20E+07	−1.73	0.02
1722.320	20	dc	58061.22	5d31	5p63	1722.3446	0.0057	3.80E+07	−1.77	0.08
1724.080	60		58001.95	4d73	5p33	1724.0800	0.0037	4.41E+07	−1.71	0.04
1725.009	400		57970.71	4d51	5p41	1725.0088	0.0044	1.44E+08	−1.21	0.22
1734.923	200		57639.45	4d41	5p21	1734.9226	0.0046	7.25E+07	−1.49	0.29
1759.984	500		56818.70	4d85	5p35	1759.9880	0.0046	1.57E+08	−1.13	0.49
1763.854	3	H, x	56694.03	5s13	5p33	1763.8457	0.0037	3.31E+07	−1.81	0.01

Table 1. *Cont.*

λ_{obs} (Å)	Int [a]		σ_{obs} (cm^{-1})	Even Level [b]	Odd Level [b]	λ_{Ritz} (Å)	Unc (Å)	$g_U A$ (s^{-1})	log($g_L f$)	|CF|
1764.854	25		56661.91	4d51	5p43	1764.8562	0.0046	7.38E+07	−1.47	0.18
1767.320	12		56582.85	5d53	5p51	1767.3212	0.0056	9.91E+07	−1.34	0.05
1773.897	400		56373.06	5s13	5p31	1773.8929	0.0042	1.80E+08	−1.07	0.17
1776.577	600		56288.02	5s11	5p53	1776.5732	0.0042	4.41E+08	−0.68	0.37
1783.481	200		56070.12	5s11	5p41	1783.4786	0.0042	1.54E+08	−1.14	0.50
1788.622	50		55908.96	4d73	5p23	1788.6221	0.0039	2.19E+08	−0.98	0.34
1800.697	35		55534.05	4d73	5p21	1800.6987	0.0043	5.29E+07	−1.60	0.08
1801.507	2000		55509.08	5s25	5p55	1801.5093	0.0054	4.44E+09	0.34	0.76
1813.410	1000		55144.73	5s25	5p73	1813.4129	0.0044	9.62E+08	−0.32	0.81
(1816.934)		K, x	55038.23	5s33	5p55	1816.9341	0.0056	1.00E+09	−0.30	0.85
1823.186	200		54849.04	4d85	5p33	1823.1906	0.0042	1.32E+08	−1.18	0.12
1826.103	1800		54761.42	5s11	5p43	1826.1065	0.0043	1.31E+09	−0.18	0.75
1829.037	2000		54673.58	5s33	5p73	1829.0430	0.0045	2.44E+09	0.09	0.64
1830.236	1500		54637.76	5s23	5p53	1830.2378	0.0039	7.41E+08	−0.43	0.35
1831.453	3000		54601.46	5s13	5p23	1831.4579	0.0039	1.94E+09	−0.01	0.57
1836.654	20		54446.84	4d83	5p45	1836.6566	0.0045	1.70E+08	−1.06	0.57
1837.566	300		54419.81	5s23	5p41	1837.5675	0.0039	1.64E+08	−1.09	0.15
1866.128	5	H, x	53586.89	5d25	5p63	1866.1431	0.0098	7.76E+07	−1.39	0.05
1874.773	3	H, x	53339.79	5d23	5p63	1874.7934	0.0063	6.94E+07	−1.44	0.04
1882.840	250		53111.26	5s23	5p43	1882.8530	0.0040	1.61E+08	−1.07	0.07
1888.675	2		52947.17	5d43	5p51	1888.6765	0.0057	4.22E+07	−1.65	0.04
1895.509	300	H, x	52756.28	4d85	5p23	1895.5222	0.0044	7.02E+07	−1.41	0.13
1896.139	30000		52738.75	5s15	5p17	1896.1404	0.0076	6.28E+09	0.53	0.96
1902.748	1		52555.57	5s15	5p25	1902.7475	0.0054	1.79E+07	−2.01	0.01
1909.897	1500		52358.84	5s25	5p63	1909.9026	0.0048	1.24E+09	−0.17	0.42
1919.844	25		52087.57	4d51	5p31	1919.8442	0.0058	9.34E+07	−1.30	0.45
1927.247	1500		51887.49	5s33	5p63	1927.2483	0.0050	8.58E+08	−0.32	0.47
1929.191	1500		51835.20	5s31	5p83	1929.1910	0.0133	2.92E+09	0.21	0.96
1967.435	5000		50827.60	5s23	5p35	1967.4431	0.0053	4.06E+09	0.37	0.96
1979.871	500		50508.34	5s11	5p33	1979.8785	0.0050	5.20E+08	−0.51	0.47
1981.092	20		50477.21	4d41	5p11	1981.0991	0.0060	4.85E+07	−1.55	0.30
1992.386	500	p	50191.08	5d23	5p55	1992.3903	0.0165	2.37E+06	−2.85	0.01
1992.546	1200	p	50187.05	5s11	5p31	1992.5463	0.0056	1.11E+09	−0.18	0.86
2008.655	5000		49784.56	5s25	5p27	2008.6537	0.0075	5.31E+09	0.51	0.97
2008.939	1200		49777.52	5s21	5p41	2008.9462	0.0049	9.32E+08	−0.26	0.92
2016.657	4		49587.01	4d73	5p25	2016.6633	0.0056	3.60E+07	−1.67	0.12

Table 1. *Cont.*

λ_{obs} (Å)	Int [a]		σ_{obs} (cm^{-1})	Even Level [b]	Odd Level [b]	λ_{Ritz} (Å)	Unc (Å)	$g_U A$ (s^{-1})	log($g_L f$)	|CF|
2046.757	1500		48857.78	5s23	5p33	2046.7594	0.0047	1.40E+09	−0.06	0.68
2048.791	8		48809.27	4d83	5p53	2048.8000	0.0048	4.94E+07	−1.50	0.20
2057.983	20		48591.27	4d83	5p41	2057.9892	0.0047	4.55E+07	−1.54	0.11
2060.302	12		48536.57	5s23	5p31	2060.3005	0.0054	5.08E+07	−1.49	0.08
2063.186	1500		48468.73	5s21	5p43	2063.1974	0.0051	1.02E+09	−0.19	0.88
2067.325	20		48371.69	4d73	5p11	2067.3300	0.0057	4.43E+05	−3.56	0.00
2071.281	5000		48279.30	5s13	5p25	2071.2849	0.0057	3.56E+09	0.36	0.96
2079.377	8000		48091.33	5s15	5p15	2079.3838	0.0084	3.53E+09	0.36	0.93
2094.038	4000		47754.63	5s15	5p13	2094.0466	0.0064	1.75E+09	0.06	0.89
2111.562	1200		47358.31	5s25	5p45	2111.5663	0.0065	6.65E+08	−0.35	0.95
2112.086	1	x	47346.56	5d63	5p83	2112.0917	0.0124	6.57E+07	−1.36	0.22
2114.962	8		47282.17	4d83	5p43	2114.9590	0.0049	2.42E+07	−1.79	0.07
2124.761	3000		47064.12	5s13	5p11	2124.7699	0.0057	1.02E+09	−0.17	0.88
2132.594	20		46891.25	4d41	5p13	2132.5977	0.0071	2.17E+07	−1.83	0.18
2132.785	1200		46887.05	5s33	5p45	2132.7887	0.0068	2.62E+09	0.26	0.77
2138.358	15		46764.85	5s23	5p23	2138.3637	0.0050	4.77E+06	−2.49	0.00
2153.595	1200		46433.99	4d85	5p25	2153.6032	0.0064	1.63E+08	−0.93	0.55
2155.636	2400		46390.02	5s23	5p21	2155.6477	0.0057	9.32E+08	−0.19	0.92
2222.293	1	x	44998.57	4d83	5p35	2222.2844	0.0066	6.15E+06	−2.33	0.15
2261.667	100		44215.17	5s21	5p33	2261.6611	0.0061	1.91E+07	−1.84	0.06
2278.201	100	p	43894.28	5s21	5p31	2278.2064	0.0069	1.15E+07	−2.05	0.10
2300.017	3000		43477.94	5s13	5p13	2300.0106	0.0069	3.20E+08	−0.60	0.32
2374.041	400		42122.27	5s21	5p23	2374.0394	0.0066	1.24E+08	−0.98	0.18
2395.369	150		41747.22	5s21	5p21	2395.3622	0.0074	5.30E+07	−1.34	0.18
2401.966	5		41632.56	4d85	5p13	2401.9604	0.0079	1.70E+07	−1.82	0.08
2424.306	5	x	41248.92	5s33	5p53	2424.2853	0.0077	9.67E+06	−2.07	0.06
2437.169	400		41031.21	5s33	5p41	2437.1620	0.0077	1.06E+08	−1.04	0.43
2442.837	12		40936.01	4d83	5p23	2442.8328	0.0063	1.75E+07	−1.80	0.08
2446.321	300		40877.71	5s11	5p11	2446.3174	0.0081	3.70E+07	−1.48	0.21
2465.420	3		40561.04	4d83	5p21	2465.4152	0.0072	8.56E+06	−2.11	0.03
2472.650	400		40442.44	5s23	5p25	2472.6384	0.0077	6.44E+07	−1.23	0.03
2487.961	400		40193.56	5s25	5p43	2487.9527	0.0077	1.38E+08	−0.89	0.32
2549.253	1		39227.18	5s23	5p11	2549.2424	0.0079	9.27E+06	−2.05	0.02

[a] Symbols: dc, doubly classified; p, perturbed; u, unresolved from close line; s, shaded to shorter wavelength; *l*, shaded to longer wavelength; x, not included in level optimization; d, double line; c, complex. A, blended or obscured by Y VI; B, perturbed by O II; C, perturbed by Si IV; D, intensity much higher than expected; E, perturbed by second order line; F, perturbed ghost of Si IV line; G, perturbed by Si IV; H, uncertain stage of ionization; J, perturbed by Y III; K, perturbed by Si II; L, perturbed by C I; M, perturbed by unknown impurity; [b] Level codes are explained in Table 2.

3. Spectrum Analysis and Level Values

The analysis was carried out in a manner similar to that used for the recent analysis of Mo V [11]. As described there "Interpretation of the spectrum was guided by calculations of the level structures and transition probabilities with the Hartree-Fock code of Cowan [12]. Further guidance was provided by construction of two-dimensional transition arrays with the computer spreadsheet method described by Reader [13]."

The odd parity energy levels are given in Table 2, the even levels in Table 3. In addition to the usual spectroscopic designations in either LS or J_1l (pair) coupling, the levels are given shorthand designations that are used in the classification of the spectral lines. The shorthand designations are explained in the footnotes to Tables 2 and 3. As described in [11] "The values of the energy levels were optimized with the computer program ELCALC, an iterative procedure in which the observed wave numbers are weighted according to the inverse square of their uncertainties. The uncertainties of the level values given by this procedure are also listed." (The program ELCALC was written by L. J. Radziemski of the Research Corporation, Tucson, Arizona 85712. The procedure and definition of level value uncertainties have been described by Radziemski and Kaufman [14].) For the level optimization only the most reliably classified lines were used. That is, lines that were very weak or that appeared with suspiciously high intensities were excluded.

Figure 1 shows a schematic overview of the positions of the $4s^24p^5$, $4s4p^6$, $4s^24p^44d$, 5s, 5p, 5d, and 6s, configurations. It also shows the calculated positions of the $4s^24p^44f$ and $4s4p^54d$ configurations.

Table 2. Odd parity levels (cm^{-1}) of Y V.

Configuration	Term	J	Desig. [a]	Energy	Uncert.	No. Trans.
$4s^24p^5$	^2P	3/2	p5 3	0.00	0.86	70
		1/2	p5 1	12460.12	1.05	46
$4s^24p^45p$	$(^3P_2)[1]$	3/2	5p13	341856.85	0.09	26
	$(^3P_2)[2]$	5/2	5p15	342193.59	0.16	23
	$(^3P_2)[1]$	1/2	5p11	345442.71	0.09	24
	$(^3P_2)[3]$	5/2	5p25	346658.00	0.10	26
	$(^3P_2)[3]$	7/2	5p17	346841.13	0.18	13
	$(^3P_1)[0]$	1/2	5p21	352605.14	0.09	20
	$(^3P_2)[2]$	3/2	5p23	352980.10	0.07	31
	$(^3P_0)[1]$	1/2	5p31	354751.98	0.10	21
	$(^3P_1)[2]$	3/2	5p33	355073.09	0.08	41
	$(^3P_1)[2]$	5/2	5p35	357042.76	0.11	29
	$(^3P_0)[1]$	3/2	5p43	359326.26	0.08	40
	$(^3P_1)[1]$	1/2	5p41	360635.14	0.08	25
	$(^3P_1)[1]$	3/2	5p53	360853.08	0.09	39
	$(^1D_2)[3]$	5/2	5p45	366490.78	0.11	22
	$(^1D_2)[3]$	7/2	5p27	368917.16	0.16	14
	$(^1D_2)[1]$	3/2	5p63	371491.26	0.09	30
	$(^1D_2)[2]$	3/2	5p73	374277.21	0.09	31
	$(^1D_2)[2]$	5/2	5p55	374641.58	0.14	23
	$(^1D_2)[1]$	1/2	5p51	378488.47	0.11	19
	$(^1S_0)[1]$	1/2	5p61	395993.26	0.30	5
	$(^1S_0)[1]$	3/2	5p83	397637.50	0.22	13

[a] Designations are given with a short form of the configuration (two places) followed by the ordinal number of the calculated J-value for the configuration (one place) and the J value (one place). For example, 5p73 indicates the seventh level with $J = 3$ for the $4p^45p$ configuration. p5 3 and p5 1 indicate the $J = 3/2$ and $1/2$ levels of the $4p^5$ configuration, respectively.

Table 3. Even parity energy levels (cm^{-1}) of Y V.

Configuration	Term	J	Desig. [a]	Energy	Uncert.	No. Trans.
$4s4p^6$	^2S	1/2	4p61	170945.58	0.95	8
$4s^24p^44d$	$(^3P)^4D$	5/2	4d15	218417.13	0.51	9
	$(^3P)^4D$	7/2	4d17	218556.80	0.64	4
	$(^3P)^4D$	3/2	4d13	219373.81	0.45	11
	$(^3P)^4D$	1/2	4d11	220794.66	0.78	10
	$(^3P)^4F$	9/2	4d19	229786.64	0.98	1
	$(^3P)^4F$	7/2	4d27	233703.76	0.46	7
	$(^1D)^2P$	1/2	4d21	233712.36	0.37	12
	$(^3P)^4F$	5/2	4d25	237677.66	0.37	10
	$(^3P)^4F$	3/2	4d23	238215.09	0.37	16
	$(^3P)^4P$	1/2	4d31	238898.28	0.40	11
	$(^3P)^4P$	3/2	4d33	239132.83	0.30	18
	$(^1D)^2D$	3/2	4d43	240949.32	0.34	12
	$(^3P)^2F$	7/2	4d37	242336.33	0.54	6
	$(^3P)^4P$	5/2	4d35	244311.56	0.32	11
	$(^1D)^2P$	3/2	4d53	244613.24	0.31	15
	$(^1D)^2D$	5/2	4d45	247861.17	0.34	11
	$(^1D)^2G$	7/2	4d47	250882.71	0.46	4
	$(^1D)^2G$	9/2	4d29	251052.40	0.99	1
	$(^3P)^2F$	5/2	4d55	251403.88	0.38	9
	$(^1D)^2F$	5/2	4d65	263184.02	0.26	10
	$(^1D)^2F$	7/2	4d57	266196.75	0.38	6
	$(^1S)^2D$	3/2	4d63	281244.51	0.25	9
	$(^1S)^2D$	5/2	4d75	286001.66	0.29	6
	$(^1D)^2S$	1/2	4d41	294965.68	0.12	8
	$(^3P)^2P$	3/2	4d73	297071.14	0.10	14
	$(^3P)^2D$	5/2	4d85	300224.19	0.10	13
	$(^3P)^2P$	1/2	4d51	302664.42	0.12	7
	$(^3P)^2D$	3/2	4d83	312044.02	0.08	13
$4s^24p^45s$	$(^3P_2)[2]$	5/2	5s15	294102.42	0.11	12
	$(^3P_2)[2]$	3/2	5s13	298378.79	0.09	17
	$(^3P_0)[0]$	1/2	5s11	304564.94	0.10	9
	$(^3P_1)[1]$	3/2	5s23	306215.37	0.08	16
	$(^3P_1)[1]$	1/2	5s21	310857.80	0.09	11
	$(^1D_2)[2]$	5/2	5s25	319132.57	0.10	8
	$(^1D_2)[2]$	3/2	5s33	319603.81	0.10	11
	$(^1S_0)[0]$	1/2	5s31	345802.30	0.29	3
$4s^24p^45d$	$(^3P_2)[3]$	7/2	5d17	418265.33	0.27	3
	$(^3P_2)[2]$	5/2	5d15	418310.18	0.44	2
	$(^3P_2)[2]$	3/2	5d13	418914.39	0.15	9
	$(^3P_2)[1]$	1/2	5d11	419977.22	0.18	7
	$(^3P_2)[4]$	9/2	5d19	420403.65	0.42	1
	$(^3P_2)[4]$	7/2	5d27	421426.57	0.28	3
	$(^3P_2)[0]$	1/2	5d21	423350.97	0.13	10
	$(^3P_2)[1]$	3/2	5d23	424830.47	0.16	13
	$(^3P_2)[3]$	5/2	5d25	425077.72	0.27	8
	$(^3P_1)[1]$	1/2	5d31	429551.65	0.17	10
	$(^3P_0)[2]$	3/2	5d33	430343.90	0.17	9
	$(^3P_0)[2]$	5/2	5d35	430533.98	0.19	9
	$(^3P_1)[3]$	7/2	5d37	430760.23	0.68	2
	$(^3P_1)[1]$	3/2	5d43	431435.60	0.12	11
	$(^3P_1)[2]$	5/2	5d45	432673.11	0.19	7
	$(^3P_1)[3]$	5/2	5d55	434647.00	0.16	10
	$(^3P_1)[2]$	3/2	5d53	435071.28	0.14	9
	$(^1D_2)[4]$	7/2	5d47	442365.90	0.30	3
	$(^1D_2)[4]$	9/2	5d29	442704.55	0.41	1
	$(^1D_2)[0]$	1/2	5d41	444128.77	0.18	9
	$(^1D_2)[1]$	3/2	5d63	444983.93	0.18	13
	$(^1D_2)[2]$	5/2	5d65	445299.76	0.20	7

Table 3. *Cont.*

Configuration	Term	J	Desig. [a]	Energy	Uncert.	No. Trans.
	$(^1D_2)[3]$	7/2	5d57	446600.43	0.29	3
	$(^1D_2)[3]$	5/2	5d75	446709.00	0.22	6
	$(^1D_2)[1]$	1/2	5d51	447277.48	0.22	6
	$(^1D_2)[2]$	3/2	5d73	448357.54	0.24	8
	$(^1S_0)[2]$	5/2	5d85	470989.78	0.43	2
	$(^1S_0)[2]$	3/2	5d83	471484.11	0.35	5
$4s^24p^46s$	$(^3P_2)[2]$	5/2	6s15	435670.71	0.23	10
	$(^3P_2)[2]$	3/2	6s13	437171.00	0.22	13
	$(^3P_0)[0]$	1/2	6s11	446209.20	0.32	7
	$(^3P_1)[1]$	3/2	6s23	446718.25	0.29	8
	$(^3P_1)[1]$	1/2	6s21	448774.76	0.36	7
	$(^1D_2)[2]$	5/2	6s25	459636.57	0.63	5
	$(^1D_2)[2]$	3/2	6s33	459761.34	0.27	6
	$(^1S_0)[0]$	1/2	6s31	486067.68	0.58	5
$4s^24p^46d$	$(^3P_2)[3]$	5/2	6d25 [c]	493116	49	1
	$(^3P_2)[1]$	3/2	6d23 [b]	493146	16	2
	$(^3P_1)[1]$	3/2	6d43 [c]	501351	75	1
	$(^3P_1)[3]$	5/2	6d55 [c]	503137	18	1
	$(^3P_1)[2]$	3/2	6d53 [b]	503788	12	2
	$(^1D_2)[0]$	1/2	6d41 [b]	514253	13	2
	$(^1D_2)[2]$	3/2	6d73 [b]	515011	16	2
	$(^1D_2)[2]$	5/2	6d65 [c]	515762	19	1
	$(^1D_2)[1]$	1/2	6d51 [b]	515882	13	2
	$(^1S_0)[2]$	3/2	6d83 [b]	543052	14	2
$4s^24p^47s$	$(^3P_2)[2]$	5/2	7s15 [c]	498271	74	1
	$(^3P_2)[2]$	3/2	7s13 [b]	499020	12	2
	$(^3P_1)[1]$	3/2	7s23 [c]	509051	18	1
	$(^3P_1)[1]$	1/2	7s21 [b]	509817	13	2
	$(^1D_2)[2]$	5/2	7s25 [c]	522000	19	1
	$(^1D_2)[2]$	3/2	7s33 [c]	522129	18	1
	$(^1S_0)[0]$	1/2	7s31 [c]	544803	20	1

[a] Designations are explained in Table 2; 4p61 indicates the $J = 1/2$ level of $4s4p^6$; [b] Tentative designation; not included in LSF; [c] Tentative level with tentative designation; not included in LSF.

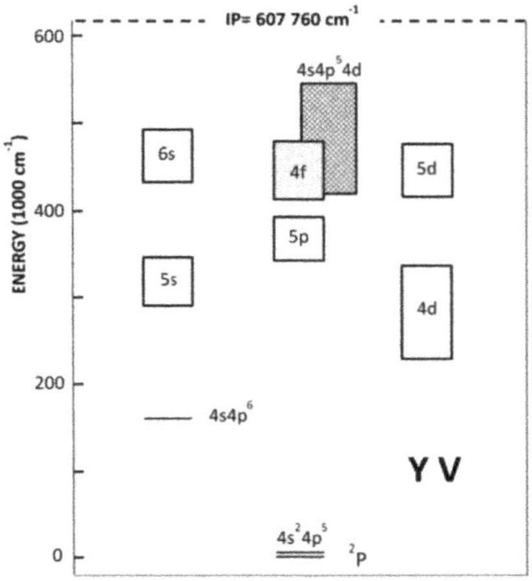

Figure 1. Schematic overview of the observed configurations of Y V. The calculated positions of the $4s^24p^44f$ and $4s4p^54d$ configurations are also shown.

3.1. $4s^24p^44d$ Levels

Nearly all levels of this configuration were given in [4]. Remaining as unknown were $(^3P)^4D_{1/2,7/2}$, $(^3P)^4F_{7/2,9/2}$, $(^3P)^2F_{7/2}$, $(^1D)^2G_{7/2,9/2}$, and $(^1D)^2F_{7/2}$. These levels have now been established based on their transitions to $4p^45p$. All values for these levels reported in [6] are spurious.

The $4p^44d$ $(^3P)^4F_{9/2}$ (4d19) and $(^1D)^2G_{9/2}$ (4d29) levels are necessarily based on only a single transition. However, the lines assigned to these transitions are both very strong and place the $J = 9/2$ levels close to their predicted positions. There is no doubt as to their identifications.

The structure of the $4p^44d$ configuration is shown in Figure 2. This is similar to Figure 1 of [4], except that we show here the observed positions of levels that were previously unknown.

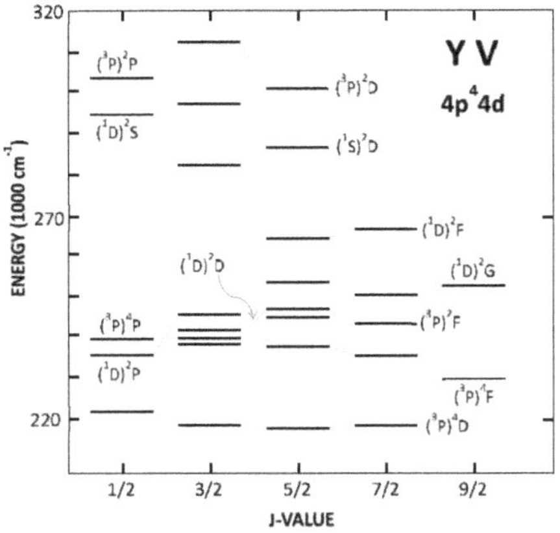

Figure 2. Structure of the $4s^24p^44d$ configuration of Y V.

3.2. $4s^24p^45s$ Levels

The levels of the $4s^24p^45s$ configuration, which were complete in [4], have improved values as a result of their combinations with $4p^45p$. In Figure 3 we give the structure of the $4p^45s$ configuration. This is the same as Figure 2 of [4], except that here we designate the levels in J_1l-coupling, rather than J_1j-coupling.

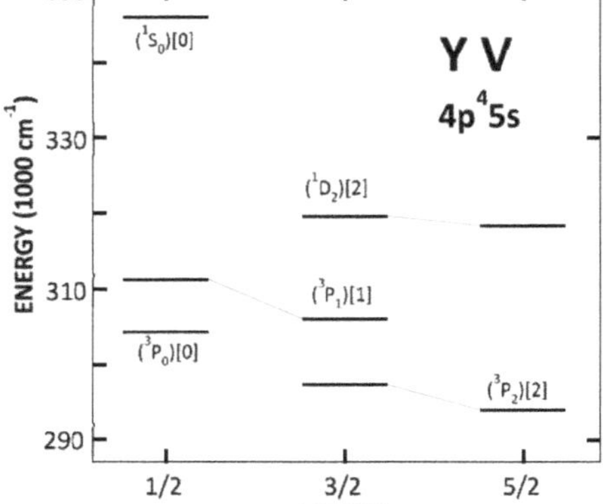

Figure 3. Structure of the $4s^24p^45s$ configuration of Y V.

3.3. $4s^2 4p^4 5p$ Levels

All levels of this configuration have been located. Of the 21 levels of this configuration given in [6], only three could be confirmed (345444, 360851, and 374278 cm^{-1}). The levels at 342193 and 355076 cm^{-1} were confirmed, but were found to have incorrect J-values. The structure of the $4p^4 5p$ levels is shown in Figure 4. The levels are designated in $J_1 l$-coupling.

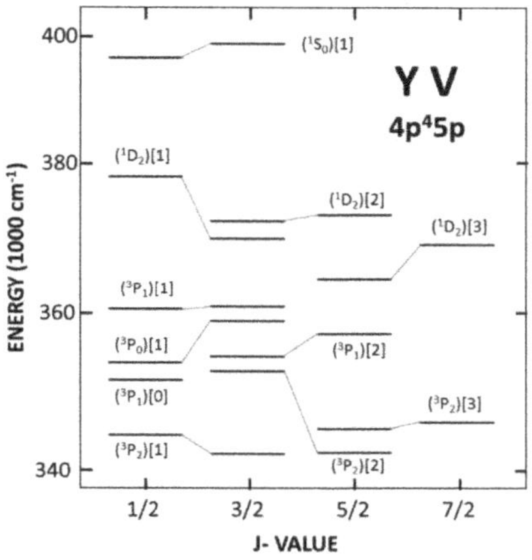

Figure 4. Structure of the $4s^2 4p^4 5p$ configuration of Y V.

3.4. $4s^2 4p^4 5d$ and $4s^2 4p^4 6s$ Levels

The $4p^4 5d$ and 6s configurations lie very close in energy and are treated together. The levels are shown in Figure 5; they are designated in $J_1 l$-coupling. As with $4p^4 4d$, the $J = 9/2$ levels could be established by only a single line. However, there is little doubt as to the identifications. A few of the levels of these configurations given in [5] could be confirmed, although some of the J-values and configuration assignments had to be revised. All of the $4p^4 5d$ levels of [6] were found to be spurious.

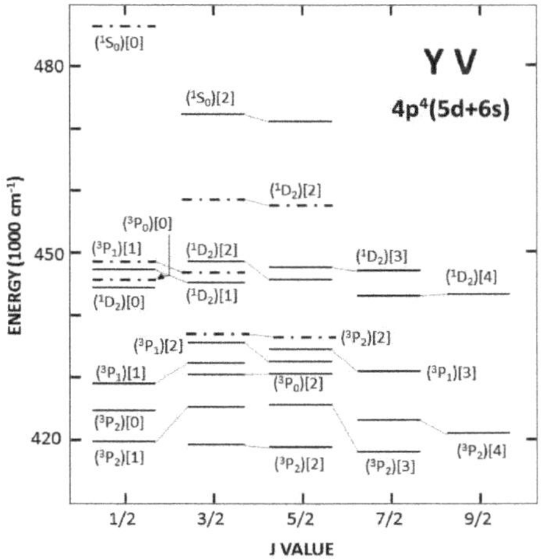

Figure 5. Structures of the $4s^2 4p^4 5d$ and $4s^2 4p^4 6s$ configurations of Y V. The $4s^2 4p^4 6s$ levels are shown as dashed.

3.5. $4s^2 4p^4 6d$ and $4s^2 4p^4 7s$ Levels

Based on our calculations, we were able to assign a number of low wavelength lines with clear Y V character as transitions to the ground term from levels of $4p^4 6d$ and 7s. For pairs of lines with wave number differences that closely match the $4p^5$ ^2P interval, the implied levels are relatively certain. However, the designations are considered to be tentative. Where the levels are based on single transitions, the line and level identifications are even less certain. None of these levels were included in the least-squares-fits, described below.

None of the information for the $4p^5$-$4p^4 6d$, 7s transitions and $4p^4 6d$,7s levels of Zahid-Ali et al. [5] could be confirmed.

3.6. $4s^2 4p^4 4f$ and $4s4p^5 4d$ Configurations

Extensive efforts to find levels of these configurations were not successful. Levels of $4p^4 4f$ were given in [6], but it is almost certain that all of them are spurious.

4. Theoretical Interpretation

4.1. Odd Parity Configurations

As in [11] "The observed configurations were interpreted theoretically by making least-squares fits of the energy parameters to the observed levels with the Cowan suite of codes, RCN (Hartree-Fock), RCG (energy matrix diagonalization), and RCE (least-squares parameter fitting) [12]. The Hartree-Fock code was run in a relativistic mode (HFR) with a correlation term in the potential. Breit energies were not included. For the initial calculations the HFR values were scaled by factors of 0.85 for the direct electrostatic parameters F^k, the exchange electrostatic parameters G^k, and the configuration interaction parameters R^k." The odd configurations $4s^2 4p^5$, $4s^2 4p^4 5p$, $4s^2 4p^4 4f$, and $4s4p^5 4d$ were treated as a single group.

The Hartree-Fock and least-squares fitted parameters for the odd configurations are given in Table 4. For these calculations, the $4p^4 5p$ exchange electrostatic parameters, $G^0(4p5p)$ and $G^2(4p5p)$, were linked at their HFR ratio. The LSF/HFR ratio of 0.836 is satisfactory. The configuration interaction (CI) parameters for the $4s^2 4p^5$-$4s^2 4p^4 5p$ interaction were held fixed at their scaled HFR values. All other CI parameters and parameters for $4s^2 4p^4 4f$ and $4s4p^5 4d$ were fixed at their scaled HFR values. The value of the effective interaction parameter $\alpha(4p4p)$ for the $4p^4 5p$ configuration was fixed at the value observed for the $4p^4$ core of Y VI [7]. In Table 4, only values for the observed configurations $4s^2 4p^5$ and $4s^2 4p^4 5p$ are given.

Table 4. Hartree-Fock and least–squares fitted parameters for the odd configurations of Y V. Mean error of fit 179 cm^{-1}.

Configuration	Parameter	HFR	LSF	Unc.	LSF/HFR
$4s^2 4p^5$	$E_{av}(4s^2 4p^5)$	8182	8400	134	
	ζ_{4p}	7941	8369	170	1.054
$4s^2 4p^4 5p$	$E_{av}(4s^2 4p^4 5p)$	364894	360966	40	0.989
	$F^2(4p4p)$	78434	65400	358	0.834
	$\alpha(4p4p)$		-56 [a]		
	ζ_{4p}	8458	8679	108	1.026
	ζ_{5p}	1688	2016	85	1.194
	$F^2(4p5p)$	22406	20623	364	0.920
	$G^0(4p5p)$	4773	3988 [b]	49	0.836
	$G^2(4p5p)$	6387	5337 [b]	66	0.836
Config. Interaction					
$4s^2 4p^5$-$4s^2 4p^4 5p$	$R^0(4p4p,4p5p)$	2217	1885 [c]		0.850
	$R^2(4p4p,4p5p)$	10661	9062 [c]		0.850

[a] Fixed at value from $4p^4$ of Y VI [7]; [b] Linked in LSF fit; [c] Fixed at scaled HFR value.

The calculated level values and eigenvector compositions for the odd configurations are given in Table 5. This table gives the percentage compositions for the three leading eigenvector states in LS-coupling and the percentage for the leading eigenvector state in J_1l-coupling. As can be seen there is not much mixing between the $4s^24p^5$ and the $4s^24p^45p$ configurations, and $4s^24p^45p$ has essentially no mixture of either $4s^24p^44f$ or $4s4p^54d$.

Table 5. Calculated energy levels (cm^{-1}) and percentage compositions for the odd levels of Y V.

J	Observed	Calculated	O−C	$\%J_1l$	Percentage Composition (LS-Coupling)								
3/2	0	0	0		99%	$4p^5$	$(^1S)^2P$						
1/2	12460	12460	0		99%	$4p^5$	$(^1S)^2P$	1%	$4s4p^54d$	$(^1P)^2P$			
3/2	341857	341900	−43	41% $(^3P_2)[1]$	66%	$4p^45p$	$(^3P)^4P$	9%	$4p^45p$	$(^3P)^4S$	8%	$4p^45p$	$(^1D)^2P$
5/2	342194	342140	54	83% $(^3P_2)[2]$	73%	$4p^45p$	$(^3P)^4P$	21%	$4p^45p$	$(^3P)^4D$	3%	$4p^45p$	$(^1D)^2D$
1/2	345443	345735	−292	55% $(^3P_2)[1]$	50%	$4p^45p$	$(^3P)^4P$	21%	$4p^45p$	$(^3P)^2P$	17%	$4p^45p$	$(^1D)^2P$
5/2	346658	346626	32	78% $(^3P_2)[3]$	62%	$4p^45p$	$(^3P)^2D$	16%	$4p^45p$	$(^3P)^4D$	12%	$4p^45p$	$(^3P)^4P$
7/2	346841	346682	159	92% $(^3P_2)[3]$	92%	$4p^45p$	$(^3P)^4D$	8%	$4p^45p$	$(^1D)^2F$			
1/2	352605	352708	−103	59% $(^3P_1)[0]$	37%	$4p^45p$	$(^3P)^4P$	22%	$4p^45p$	$(^3P)^2P$	17%	$4p^45p$	$(^3P)^4D$
3/2	352980	352884	96	36% $(^3P_2)[2]$	35%	$4p^45p$	$(^3P)^4D$	24%	$4p^45p$	$(^3P)^2D$	18%	$4p^45p$	$(^3P)^2P$
1/2	354752	354622	130	62% $(^3P_0)[1]$	70%	$4p^45p$	$(^3P)^4D$	11%	$4p^45p$	$(^3P)^4P$	10%	$4p^45p$	$(^3P)^2S$
3/2	355073	355061	12	35% $(^3P_1)[2]$	47%	$4p^45p$	$(^3P)^4D$	35%	$4p^45p$	$(^3P)^2P$	9%	$4p^45p$	$(^1D)^2P$
5/2	357043	356868	175	95% $(^3P_1)[2]$	59%	$4p^45p$	$(^3P)^4D$	27%	$4p^45p$	$(^3P)^2D$	13%	$4p^45p$	$(^3P)^4P$
3/2	359326	359369	−43	70% $(^3P_0)[1]$	29%	$4p^45p$	$(^3P)^2D$	25%	$4p^45p$	$(^3P)^4S$	15%	$4p^45p$	$(^3P)^4P$
3/2	360853	360766	87	65% $(^3P_1)[1]$	47%	$4p^45p$	$(^3P)^4S$	41%	$4p^45p$	$(^3P)^2D$	5%	$4p^45p$	$(^3P)^4P$
1/2	360635	361119	−484	61% $(^3P_1)[1]$	72%	$4p^45p$	$(^3P)^2S$	14%	$4p^45p$	$(^3P)^2P$	6%	$4p^45p$	$(^3P)^4D$
5/2	366491	366367	124	87% $(^1D_2)[3]$	87%	$4p^45p$	$(^1D)^2F$	7%	$4p^45p$	$(^3P)^2D$	4%	$4p^45p$	$(^1D)^2D$
7/2	368917	368795	122	91% $(^1D_2)[3]$	91%	$4p^45p$	$(^1D)^2F$	8%	$4p^45p$	$(^3P)^4D$			
3/2	371491	371556	−65	62% $(^1D_2)[1]$	62%	$4p^45p$	$(^1D)^2P$	17%	$4p^45p$	$(^1D)^2D$	11%	$4p^45p$	$(^3P)^2P$
3/2	374277	374172	105	76% $(^1D_2)[2]$	76%	$4p^45p$	$(^1D)^2D$	16%	$4p^45p$	$(^3P)^2P$	7%	$4p^45p$	$(^1D)^2P$
5/2	374642	374641	1	92% $(^1D_2)[2]$	92%	$4p^45p$	$(^1D)^2D$	3%	$4p^45p$	$(^1D)^2F$	2%	$4p^45p$	$(^3P)^4P$
1/2	378488	378486	2	64% $(^1D_2)[1]$	64%	$4p^45p$	$(^1D)^2P$	34%	$4p^45p$	$(^3P)^2P$	1%	$4p^45p$	$(^3P)^2S$
1/2	395993	395986	7	84% $(^1S_0)[1]$	84%	$4p^45p$	$(^1S)^2P$	6%	$4p^45p$	$(^3P)^2P$	5%	$4p^45p$	$(^3P)^4D$
3/2	397637	397712	−75	86% $(^1S_0)[1]$	86%	$4p^45p$	$(^1S)^2P$	3%	$4p^45p$	$(^3P)^2D$	3%	$4p^45p$	$(^3P)^4D$

4.2. *Even Parity Configurations*

The parameters for the even configurations are given in Table 6. Here, the $4s4p^6$, $4p^44d$, 5s, 5d, 6s, 6d, and 7s configurations were treated as single group. For the initial calculations the HFR values were scaled by factors of 0.85 for the direct electrostatic parameters F^k, the exchange electrostatic parameters G^k, and the configuration interaction parameters R^k. All the parameters that were allowed to vary were well defined in the fit and have reasonable ratios to the HFR values. The exchange parameters G^1(4p5d) and G^3(4p5d) were linked at their HFR ratio. The CI parameters for the $4s4p^6$-$4s^24p^44d$ and $4s4p^6$-$4s^24p^45d$ interactions were also linked at their HFR ratio. The fitted values are reasonable. The other CI parameters and all of the parameters for $4p^46d$ and $4p^47s$ were held fixed at their scaled HFR values. As described in [4] the interaction of $4s4p^6\ ^2S_{1/2}$ with the $4s^24p^44d\ (^1D)^2S$ level is great, with a mutual repulsion of ~31,000 cm^{-1}. On the other hand, interaction between $4s4p^6$ and $4s^24p^45d$ is negligible. The value of the effective interaction parameter α(4p4p) for the $4p^44d$, 5s, 5d, and 6s configurations was again fixed at the value observed for the $4p^4$ core of Y VI [7]. The calculated level values and eigenvector compositions for the even levels are given in Table 7. This table gives the percentage compositions for the three leading eigenvector states in LS-coupling and the percentage for the leading eigenvector state in J_1l-coupling, where appropriate. As can be seen, the purity of the states of the $4p^44d$ configuration in LS-coupling is low, leading to low leading percentages for many of the levels. Even though the $4p^45d$ and $4p^46s$ configurations are practically coincident, there is not much mixing of states.

Table 6. Hartree-Fock and least-squares fitted parameters for the even configurations of Y V. Mean error of fit 273 cm^{-1}.

Configuration	Parameter	HF	LSF	Unc.	LSF/HFR
$4s4p^6$	$E_{av}(4s4p^6)$	215344	203602	511	0.942
$4s^24p^44d$	$E_{av}(4s^24p^44d)$	255431	251213	55	0.982
	$F^2(4p4p)$	77057	63230	629	0.821
	$\alpha(4p4p)$		−56 [a]		
	ζ_{4p}	8132	8494	156	1.045
	ζ_{4d}	507	612	73	1.209
	$F^2(4p4d)$	62105	53714	481	0.865
	$G^1(4p4d)$	76519	61136	169	0.799
	$G^3(4p4d)$	47207	39325	921	0.833
$4s^24p^45s$	$E_{av}(4s^24p^45s)$	314448	309938	101	0.985
	$F^2(4p4p)$	78065	64882	811	0.831
	$\alpha(4p4p)$		−56 [a]		
	ζ_{4p}	8391	8647	254	1.031
	$G^1(4p5s)$	7780	6747	374	0.867
$4s^24p^45d$	$E_{av}(4s^24p^45d)$	438648	434854	54	0.991
	$F^2(4p4p)$	78487	65109	485	0.830
	$\alpha(4p4p)$		−56 [a]		
	ζ_{4p}	8452	8827	119	1.044
	ζ_{5d}	146	214	61	1.463
	$F^2(4p5d)$	16322	13742	557	0.842
	$G^1(4p5d)$	10162	6965 [b]	260	0.685
	$G^3(4p5d)$	7247	4967 [b]	185	0.685
$4s^24p^46s$	$E_{av}(4s^24p^46s)$	453803	450227	103	0.991
	$F^2(4p4p)$	78549	64962	784	0.827
	$\alpha(4p4p)$		−56 [a]		
	ζ_{4p}	8477	8833	223	1.042
	$G^1(4p6s)$	2422	2041	372	0.843
Config. Interaction					
$4s4p^6$-$4s^24p^44d$	$R^1(4p4p,4s4d)$	86708	66719 [c]	419	0.769
$4s4p^6$-$4s^24p^45d$	$R^1(4p4p,4s5d)$	30749	23660 [c]	149	0.769
$4s4p^6$-$4s^24p^45s$	$R^1(4p4p,4s5s)$	2884	2452 [d]		0.850
$4s4p^6$-$4s^24p^46s$	$R^1(4p4p,4s6s)$	668	567 [d]		0.850
$4s^24p^44d$-$4s^24p^45s$	$R^2(4p4d,4p5s)$	−9422	−8009 [d]		0.850
	$R^1(4p4d,5s4p)$	−1919	−1631 [d]		0.850
$4s^24p^44d$-$4s^24p^46s$	$R^2(4p4d,4p6s)$	−5249	−4462 [d]		0.850
	$R^1(4p4d,6s4p)$	−1888	−1605 [d]		0.850

[a] Fixed at value from $4p^4$ of Y VI [7]; [b,c] Linked in groups in LSF fit; [d] Fixed at scaled HFR value.

Table 7. Calculated energy levels (cm^{-1}) and percentage compositions for the even levels of Y V.

J	Obs.	Calc.	O−C	% J_1l	Percentage Composition (LS-Coupling)								
1/2	170946	170944	2		75%	$4s4p^6$	$(^2S)^2S$	24%	$4p^44d$	$(^1D)^2S$			
5/2	218417	218286	131		90%	$4p^44d$	$(^3P)^4D$	3%	$4p^44d$	$(^3P)^4F$	2%	$4p^44d$	$(^3P)^4P$
7/2	218557	218521	36		92%	$4p^44d$	$(^3P)^4D$	5%	$4p^44d$	$(^3P)^4F$	2%	$4p^44d$	$(^1D)^2F$
3/2	219374	219237	137		88%	$4p^44d$	$(^3P)^4D$	4%	$4p^44d$	$(^3P)^4P$	3%	$4p^44d$	$(^1D)^2D$
1/2	220795	220814	−19		88%	$4p^44d$	$(^3P)^4D$	5%	$4p^44d$	$(^1D)^2P$	4%	$4p^44d$	$(^3P)^2P$
9/2	229787	229647	140		91%	$4p^44d$	$(^3P)^4F$	9%	$4p^44d$	$(^1D)^2G$			
7/2	233704	233513	191		72%	$4p^44d$	$(^3P)^4F$	14%	$4p^44d$	$(^3P)^2F$	11%	$4p^44d$	$(^1D)^2G$
1/2	233712	234734	−1022		45%	$4p^44d$	$(^1D)^2P$	39%	$4p^44d$	$(^3P)^2P$	11%	$4p^44d$	$(^3P)^4D$
5/2	237678	237425	253		94%	$4p^44d$	$(^3P)^4F$	3%	$4p^44d$	$(^3P)^4D$	2%	$4p^44d$	$(^1S)^2D$
3/2	238215	237945	270		63%	$4p^44d$	$(^3P)^4F$	11%	$4p^44d$	$(^1S)^2D$	10%	$4p^44d$	$(^3P)^4P$
1/2	238898	238811	87		91%	$4p^44d$	$(^3P)^4P$	4%	$4p^44d$	$(^3P)^2P$	3%	$4p^44d$	$(^1D)^2P$
3/2	239133	239284	−151		44%	$4p^44d$	$(^3P)^4P$	22%	$4p^44d$	$(^3P)^4F$	20%	$4p^44d$	$(^1D)^2P$
3/2	240949	240804	145		38%	$4p^44d$	$(^1D)^2D$	24%	$4p^44d$	$(^3P)^2D$	10%	$4p^44d$	$(^3P)^4F$
7/2	242336	242673	−337		48%	$4p^44d$	$(^3P)^2F$	20%	$4p^44d$	$(^3P)^4F$	20%	$4p^44d$	$(^1D)^2G$
5/2	244312	244201	111		77%	$4p^44d$	$(^3P)^4P$	7%	$4p^44d$	$(^1S)^2D$	6%	$4p^44d$	$(^3P)^2D$
3/2	244613	245018	−405		39%	$4p^44d$	$(^3P)^4P$	24%	$4p^44d$	$(^1D)^2P$	22%	$4p^44d$	$(^3P)^2P$
5/2	247861	247600	262		40%	$4p^44d$	$(^1D)^2D$	23%	$4p^44d$	$(^3P)^2D$	16%	$4p^44d$	$(^3P)^4P$
7/2	250883	250572	311		68%	$4p^44d$	$(^1D)^2G$	22%	$4p^44d$	$(^3P)^2F$	8%	$4p^44d$	$(^1D)^2F$
9/2	251052	250641	411		91%	$4p^44d$	$(^1D)^2G$	9%	$4p^44d$	$(^3P)^4F$			
5/2	251404	252009	−605		67%	$4p^44d$	$(^3P)^2F$	17%	$4p^44d$	$(^1D)^2F$	10%	$4p^44d$	$(^1D)^2D$
5/2	263184	263228	−44		80%	$4p^44d$	$(^1D)^2F$	11%	$4p^44d$	$(^3P)^2F$	7%	$4p^44d$	$(^1D)^2D$
7/2	266197	266306	−109		82%	$4p^44d$	$(^1D)^2F$	15%	$4p^44d$	$(^3P)^2F$	1%	$4p^44d$	$(^1D)^2G$
3/2	281245	281192	53		62%	$4p^44d$	$(^1S)^2D$	26%	$4p^44d$	$(^1D)^2D$	4%	$4p^44d$	$(^1D)^2P$
5/2	286002	286024	−22		72%	$4p^44d$	$(^1S)^2D$	16%	$4p^44d$	$(^1D)^2D$	4%	$4p^44d$	$(^3P)^2F$
5/2	294102	294060	42	93%$(^3P_2)$[2]	93%	$4p^45s$	$(^3P)^4P$	6%	$4p^45s$	$(^1D)^2D$			
1/2	294966	295069	−103		62%	$4p^44d$	$(^1D)^2S$	20%	4s4p6	$(^2S)^2S$	8%	$4p^44d$	$(^1D)^2P$
3/2	297071	296670	401		47%	$4p^44d$	$(^3P)^2P$	36%	$4p^44d$	$(^1D)^2P$	7%	$4p^44d$	$(^1D)^2D$
3/2	298379	298417	−38	79%$(^3P_2)$[2]	47%	$4p^45s$	$(^3P)^2P$	43%	$4p^45s$	$(^3P)^4P$	8%	$4p^45s$	$(^1D)^2D$
5/2	300224	300747	−523		62%	$4p^44d$	$(^3P)^2D$	20%	$4p^44d$	$(^1D)^2D$	14%	$4p^44d$	$(^1S)^2D$
1/2	302664	301985	679		44%	$4p^44d$	$(^3P)^2P$	38%	$4p^44d$	$(^1D)^2P$	10%	$4p^44d$	$(^1D)^2S$
1/2	304565	304602	−37	57%$(^3P_0)$[0]	90%	$4p^45s$	$(^3P)^4P$	5%	$4p^45s$	$(^1S)^2S$	2%	$4p^44d$	$(^1D)^2S$
3/2	306215	306149	66	89%$(^3P_1)$[1]	56%	$4p^45s$	$(^3P)^4P$	42%	$4p^45s$	$(^3P)^2P$	2%	$4p^45s$	$(^1D)^2D$
1/2	310858	310839	19	66%$(^3P_1)$[1]	95%	$4p^45s$	$(^3P)^2P$	4%	$4p^45s$	$(^1S)^2S$			
3/2	312044	312297	−253		54%	$4p^44d$	$(^3P)^2D$	20%	$4p^44d$	$(^1S)^2D$	13%	$4p^44d$	$(^1D)^2D$
5/2	319133	319168	−35	93%$(^1D_2)$[2]	93%	$4p^45s$	$(^1D)^2D$	6%	$4p^45s$	$(^3P)^4P$			
3/2	319604	319699	−95	88%$(^1D_2)$[2]	88%	$4p^45s$	$(^1D)^2D$	10%	$4p^45s$	$(^3P)^2P$	1%	$4p^45s$	$(^3P)^4P$
1/2	345802	345752	50	88%$(^1S_0)$[0]	88%	$4p^45s$	$(^1S)^2S$	6%	$4p^45s$	$(^3P)^4P$	5%	$4p^45s$	$(^3P)^2P$

Table 7. *Cont.*

J	Obs.	Calc.	O−C	$\% J_1 l$	Percentage Composition (LS-Coupling)								
7/2	418265	418341	−76	$91\%(^3P_2)[3]$	73%	$4p^45d$	$(^3P)^4D$	20%	$4p^45d$	$(^3P)^4F$	5%	$4p^45d$	$(^1D)^2F$
5/2	418310	418347	−37	$57\%(^3P_2)[2]$	71%	$4p^45d$	$(^3P)^4D$	10%	$4p^45d$	$(^3P)^4P$	10%	$4p^45d$	$(^3P)^4F$
3/2	418914	418953	−39	$60\%(^3P_2)[2]$	60%	$4p^45d$	$(^3P)^4D$	22%	$4p^45d$	$(^3P)^4P$	5%	$4p^45d$	$(^1D)^2D$
1/2	419977	420043	−66	$78\%(^3P_2)[1]$	44%	$4p^45d$	$(^3P)^4D$	29%	$4p^45d$	$(^3P)^4P$	15%	$4p^45d$	$(^3P)^2P$
9/2	420404	420469	−65	$91\%(^3P_2)[4]$	91%	$4p^45d$	$(^3P)^4F$	9%	$4p^45d$	$(^1D)^2G$			
7/2	421427	421205	222	$89\%(^3P_2)[4]$	66%	$4p^45d$	$(^3P)^2F$	23%	$4p^45d$	$(^3P)^4F$	9%	$4p^45d$	$(^1D)^2G$
1/2	423351	423374	−23	$83\%(^3P_2)[0]$	55%	$4p^45d$	$(^3P)^4P$	28%	$4p^45d$	$(^3P)^2P$	9%	$4p^45d$	$(^1D)^2S$
3/2	424830	424828	2	$64\%(^3P_2)[1]$	41%	$4p^45d$	$(^3P)^4P$	30%	$4p^45d$	$(^3P)^2D$	13%	$4p^45d$	$(^3P)^2P$
5/2	425078	425015	63	$53\%(^3P_2)[3]$	33%	$4p^45d$	$(^3P)^2D$	27%	$4p^45d$	$(^3P)^2F$	17%	$4p^45d$	$(^3P)^4P$
1/2	429552	429752	−200	$86\%(^3P_1)[1]$	49%	$4p^45d$	$(^3P)^4D$	34%	$4p^45d$	$(^3P)^2P$	9%	$4p^45d$	$(^1D)^2P$
3/2	430344	430325	19	$67\%(^3P_0)[2]$	71%	$4p^45d$	$(^3P)^4F$	13%	$4p^45d$	$(^3P)^4D$	8%	$4p^45d$	$(^1S)^2D$
5/2	430534	430524	10	$51\%(^3P_0)[2]$	58%	$4p^45d$	$(^3P)^4F$	15%	$4p^45d$	$(^3P)^4P$	14%	$4p^45d$	$(^3P)^4D$
7/2	430760	430623	137	$97\%(^3P_1)[3]$	53%	$4p^45d$	$(^3P)^4F$	24%	$4p^45d$	$(^3P)^2F$	22%	$4p^45d$	$(^3P)^4D$
3/2	431436	431423	13	$53\%(^3P_1)[1]$	23%	$4p^45d$	$(^3P)^4P$	23%	$4p^45d$	$(^3P)^4D$	21%	$4p^45d$	$(^3P)^2D$
5/2	432673	432560	113	$97\%(^3P_1)[2]$	48%	$4p^45d$	$(^3P)^4P$	33%	$4p^45d$	$(^3P)^2F$	11%	$4p^45d$	$(^3P)^4F$
5/2	434647	434607	40	$51\%(^3P_1)[3]$	47%	$4p^45d$	$(^3P)^2D$	34%	$4p^45d$	$(^3P)^2F$	4%	$4p^45d$	$(^3P)^4P$
3/2	435071	435299	−228	$39\%(^3P_1)[2]$	61%	$4p^45d$	$(^3P)^2P$	19%	$4p^45d$	$(^3P)^2D$	8%	$4p^45d$	$(^1D)^2P$
5/2	435671	435664	7	$92\%(^3P_2)[2]$	92%	$4p^46s$	$(^3P)^4P$	7%	$4p^46s$	$(^1D)^2D$	1%	$4p^45d$	$(^3P)^2D$
3/2	437171	437167	4	$89\%(^3P_2)[2]$	69%	$4p^46s$	$(^3P)^2P$	21%	$4p^46s$	$(^3P)^4P$	8%	$4p^46s$	$(^1D)^2D$
7/2	442366	442275	91	$90\%(^1D_2)[4]$	90%	$4p^45d$	$(^1D)^2G$	7%	$4p^45d$	$(^3P)^2F$	2%	$4p^45d$	$(^3P)^4F$
9/2	442705	442669	36	$91\%(^1D_2)[4]$	91%	$4p^45d$	$(^1D)^2G$	9%	$4p^45d$	$(^3P)^4F$			
1/2	444129	444050	79	$79\%(^1D_2)[0]$	79%	$4p^45d$	$(^1D)^2S$	11%	$4p^45d$	$(^1D)^2P$	8%	$4p^45d$	$(^3P)^4P$
3/2	444984	444917	67	$80\%(^1D_2)[1]$	80%	$4p^45d$	$(^1D)^2P$	6%	$4p^45d$	$(^3P)^4P$	5%	$4p^45d$	$(^3P)^2P$
5/2	445300	445422	−122	$59\%(^1D_2)[2]$	59%	$4p^45d$	$(^1D)^2D$	36%	$4p^45d$	$(^1D)^2F$	1%	$4p^45d$	$(^3P)^4P$
1/2	446210	446244	−34	$59\%(^3P_0)[0]$	92%	$4p^46s$	$(^3P)^4P$	7%	$4p^46s$	$(^1S)^2S$			
7/2	446600	446452	148	$93\%(^1D_2)[3]$	93%	$4p^45d$	$(^1D)^2F$	3%	$4p^45d$	$(^3P)^4D$	2%	$4p^45d$	$(^3P)^2F$
5/2	446709	446619	90	$56\%(^1D_2)[3]$	56%	$4p^45d$	$(^1D)^2F$	32%	$4p^45d$	$(^1D)^2D$	7%	$4p^45d$	$(^3P)^2D$
3/2	446718	446692	26	$96\%(^3P_1)[1]$	74%	$4p^46s$	$(^3P)^4P$	22%	$4p^46s$	$(^3P)^2P$	2%	$4p^45d$	$(^1D)^2P$
1/2	447277	447284	−7	$41\%(^1D_2)[1]$	41%	$4p^45d$	$(^1D)^2P$	34%	$4p^46s$	$(^3P)^2P$	13%	$4p^45d$	$(^3P)^2P$
3/2	448358	448574	−216	$78\%(^1D_2)[2]$	78%	$4p^45d$	$(^1D)^2D$	18%	$4p^45d$	$(^3P)^2D$	1%	$4p^45d$	$(^1D)^2P$
1/2	448775	448783	−8	$45\%(^3P_1)[1]$	61%	$4p^46s$	$(^3P)^2P$	25%	$4p^45d$	$(^1D)^2P$	8%	$4p^45d$	$(^3P)^2P$
5/2	459637	459629	8	$92\%(^1D_2)[2]$	93%	$4p^46s$	$(^1D)^2D$	7%	$4p^46s$	$(^3P)^4P$			
3/2	459761	459765	−4	$91\%(^1D_2)[2]$	91%	$4p^46s$	$(^1D)^2D$	7%	$4p^46s$	$(^3P)^2P$	1%	$4p^46s$	$(^3P)^4P$
5/2	470990	471048	−58	$88\%(^1S_0)[2]$	88%	$4p^45d$	$(^1S)^2D$	3%	$4p^45d$	$(^3P)^2F$	3%	$4p^45d$	$(^3P)^4P$
3/2	471484	471481	3	$86\%(^1S_0)[2]$	86%	$4p^45d$	$(^1S)^2D$	5%	$4p^45d$	$(^3P)^4F$	4%	$4p^45d$	$(^3P)^2D$
1/2	486068	486067	1	$88\%(^1S_0)[0]$	88%	$4p^46s$	$(^1S)^2S$	7%	$4p^46s$	$(^3P)^4P$	4%	$4p^46s$	$(^3P)^2P$

5. 4s4p^6-4s^{2}4p^{4}5p Transitions

Transitions between the 4s4p^6 and 4s^{2}4p^{4}5p configurations are normally forbidden as two electron jumps. However, because of configuration interaction between 4s4p^6 and 4s^{2}4p^{4}4d, they can in fact take place. We observe six of them in Y V. The wavelengths for these transitions are long relative to the resonance lines and serve to improve the accuracy of the excited levels.

6. Ritz Wavelengths

We determined Ritz wavelengths for all of the lines by differencing the energy level values in Tables 2 and 3. The Ritz wavelengths are given in Table 1. The uncertainties of the calculated wavelengths correspond to the square root of the sum of the squares of the uncertainties of the combining levels. The Ritz values have uncertainties that are as low as ±0.0004 Å. Those lines with uncertainties in the Ritz wavelengths of ±0.0020 Å or less should serve well as wavelength standards in the deep VUV.

7. Oscillator Strengths

Table 1 lists the transition probabilities $g_U A$ and log $g_L f$ for each observed line as calculated with wavefunctions obtained from the fitted energy parameters. Here, f is the oscillator strength, g_U is the statistical weight of the upper level $2J_U + 1$ and g_L is the statistical weight of the lower level $2J_L + 1$. The A-values are compared with recently published ab initio values in Section 9 below.

Since there are no experimental values for the transition probabilities of Y V, it is difficult to estimate the uncertainty of the calculated values. One guide is the cancellation factor. This is the ratio of the calculated transition probability to a value calculated with all parts of the wave function taken as positive [12]. Low cancellation factors generally indicate a larger uncertainty in the calculated values. Indeed, many of the values in Table 1 have low cancellation factors. The present calculated transition probabilities can be considered as qualitative estimates of the relative intensities of the lines. Based on general experience, we estimate the uncertainties to be about ±50%.

8. Ionization Energy

An ionization energy of 605,000 ± 4000 cm^{-1} was obtained in [4] by estimating a value for n*(4p^{4}5s) of 2.98 ± 0.02. On the basis of their observed 4s^{2}4p^4ns(n = 5–7) and nd(n = 4–6) series, Zahid-Ali, Chaghtai, and Singh [5] revised this downward slightly to 604,700 ± 2500 cm^{-1}. Since many of the levels used in their determination are now known to be spurious, this value must be re-determined.

For our new determination, we use the centers-of-gravity of the 4p^{4}5s and 4p^{4}6s configurations together with an estimated value for the change in effective quantum number Δn*(4p^{4}6s-4p^{4}5s) = n*(4p^{4}6s)-n*(4p^{4}5s). This allows us to find the limit of the 4p^4ns series, which is the center-of-gravity of the 4p^4 configuration of Y VI.

From the observed levels in Table 3, we find the centers-of-gravity of the 4p^{4}5s and 4p^{4}6s configurations as 309,955.06 and 450,284.98 cm^{-1}, respectively. Our value for Δn*(4p^{4}6s-4p^{4}5s) is taken from Δn*(4p^{6}6s-4p^{6}5s) for the one-electron atom Nb V [15], 1.03577. We use Cowan's Hartree-Fock code to estimate the change in going from Nb V to Y V. For Nb V we calculate Δn*(4p^{6}6s-4p^{6}5s) as 1.0394 and for Y V we calculate Δn*(4p^{4}6s-4p^{4}5s) as 1.0369, a difference of 0.0025. We thus estimate Δn*(4p^{4}6s-4p^{4}5s) for Y V as 1.03577 − 0.00251 = 1.0333, with an estimated uncertainty of ±0.0015. This produces a limit of 621,810 ± 300 cm^{-1}. The effective quantum numbers for Y V are n*(5s) = 2.966(1) and n*(6s) = 3.999(3). Correcting for the energy of the center-of-gravity of 4p^4 in Y VI, 14 051 cm^{-1} [7], we obtain for the ionization energy of Y V 607,760 ± 300 cm^{-1} (75.353 ± 0.037 eV). (Conversion from cm^{-1} to eV was done with the factor 8065.54429(18) cm^{-1}/eV [16].)

9. Comparison with ab Initio Calculations

Recently, two sets of ab initio calculations for the levels and oscillator strengths of Y V have appeared. Singh et al. [17] used a multiconfiguration Dirac-Fock (MCDF) approach to make calculations for transitions within the n = 4 complex; $4s^24p^5$, $4s4p^6$, $4s^24p^44d$. Aggarwal and Keenan [18] used the General-purpose Relativistic Atomic Structure Package (GRASP) for calculations within the same complex of n = 4 configurations. Both calculations are based on new versions of the Grant atomic structure code. Froese Fischer [19] has discussed the accuracy that might be expected from calculations for complex atoms with GRASP, in particular as applied to the Br-like ion W^{39+}.

Comparisons of our present results with those of the ab initio calculations of [17,18] are given in Tables 8–10. The index numbers for the levels in these tables are those used in [17,18]. The wavelengths for Aggarwal and Keenan [18] in Table 8 are differences of the GRASP3 energies in their Table 3. It should be noted that the level with index 25 in [17] is misprinted $4s^24p^4(^1D)4d\ ^2P_{3/2}$; it should be $4s^24p^4(^1S)4d\ ^2D_{3/2}$, as given in [18].

The main difference between the results of [17,18] and our present results is that the energies of the levels designated $4s^24p^4(^3P)4d\ ^2P_{1/2}$ (index 28) and $4s^24p^4(^1D)4d\ ^2S_{1/2}$ (index 30) are reversed in order of energy. That is, the level with index 28 corresponds to our level 4d51, and the level with index 30 corresponds to our level 4d41.

That our present order is correct can be seen from the fact that $(^3P)^2P$ has little interaction with $4s4p^6\ ^2S$, and its position is largely fixed by the internal parameters of $4p^44d$. If omitted from the LSF calculation, the calculated energy is very close to the observed value. So, there is no doubt about this assignment. This leaves the level at 294,965 cm^{-1} as the only possibility for $(^1D)^2S$. The position of $(^1D)^2S$ is harder to pin down, because it is affected not only by the internal parameters of $4p^44d$, but also by the amount of its upward displacement due to interaction with $4s4p^6\ ^2S$. In our present calculations this uncertainty is removed, because when the level is included in the LSF, the CI parameter R^1(4p4p,4s4d) takes a fitted value that has a reasonable ratio to HFR. This conclusion is supported by the observed line intensities, which follow the predicted pattern for these two levels. See for example the lines at 339.023, 353.976, 330.398, and 344.583 Å in Table 8. It is clear that in the MCDF calculations the upward displacement of $4s^24p^4(^1D)4d\ ^2S_{1/2}$ due to interaction with $4s4p^6\ ^2S_{1/2}$ is a little too large. The LSF/HFR scale factor of 0.769 for this interaction in Table 6 also reflects this circumstance.

In Table 8 we compare the wavelengths and transition probabilities A (s^{-1}) found from GRASP with our present results. The values of A(present) in this table are those given in Table 1 divided by the statistical weight of the upper level $2J_u$+1. A notable disagreement for the transition probabilities for the $4s^24p^5\ ^2P_{3/2}$-$4s^24p^44d\ (^3P)^4F_{3/2}$ transition (indices 1–12), observed at 419.792 Å. Both Singh et al. [17] and Aggarwal and Keenan [18] find an extremely low transition probability for this transition. However, we obtain a somewhat higher A-value, and it is indeed observed as a reasonably strong line. This transition is nominally forbidden as an inter-combination line in LS-coupling because of the change of spin. However, although the $4p^44d$ level (238,215 cm^{-1} observed value) has a leading percentage composition in LS coupling of 63% $4p^44d\ (^3P)^4F_{3/2}$, the full percentage compositions show that it actually has a total doublet character of about 31%. This accounts for our calculated transition probability and observed line strength. Singh et al. [17] report a composition of 88% $4p^44d\ (^3P)^4F_{3/2}$ for this level, with no secondary percentage mentioned. Percentage compositions were not reported by Aggarwal and Keenan [18]. The present percentage compositions for Y V are practically the same as were given in [4]. This paper was not cited in either [17] or [18].

Other striking differences can be seen in Table 8. The values found by all three calculations for the $4s^24p^5\ ^2P_{3/2}$-$4s^24p^44d(^1S)\ ^4D_{5/2}$ transition (indices 1–26) are extremely discrepant. The value of Aggarwal [18] is a little closer to our present value. The values for the $4s^24p^5\ ^2P_{1/2}$-$4s^24p^44d\ (^3P)^4D_{3/2}$ transition (indices 2–6) also disagree by a large amount. Still, they all predict that this will be a very weak line, and in fact it has not been observed.

Table 8. Comparison of wavelengths λ (Å) and transition probabilities A (s^{-1}) for Y V calculated with the MCDF2 method of Singh et al. [17] and the GRASP3 method of Aggarwal and Keenan [18] with present values. Index numbers are those used in [17,18]. Blank spaces indicate that line was not observed. Designations are for the upper levels in the transition.

Lower Level	Upper Level	Desig.	Index	λ [17]	λ [18]	λ (obs)	A [17]	A [18]	A (Pres.)	CF	Int (obs)
$4s^24p^5\ ^2P_{3/2}$	$4s4p^6\ ^2S_{1/2}$	4p61	3	594	555.817	584.982	3.41E+08	6.4885E+08	8.30E+08	0.03	50000
(index = 1)	$4s^24p^44d(^3P)^4D_{5/2}$	4d15	4	466	447.189	457.838	8.49E+06	1.4695E+07	7.05E+06	0.00	
	$(^3P)^4D_{3/2}$	4d13	6	464	445.053	455.846	5.17E+06	5.7124E+06	6.85E+06	0.00	35
	$(^3P)^4D_{1/2}$	4d11	7	461	442.088	452.911	2.69E+06	1.9727E+06	6.30E+06	0.00	5
	$(^1D)^2P_{1/2}$	4d21	10	427	410.339	451.974	1.10E+07	2.4808E+06	4.77E+07	0.00	25
	$(^3P)^4F_{5/2}$	4d25	11	425	410.674	420.737	9.04E+07	1.3428E+08	1.40E+08	0.74	1500
	$(^3P)^4F_{3/2}$	4d23	12	423	409.298	419.792	2.44E+06	2.4683E+06	7.80E+07	0.01	400
	$(^3P)^4P_{3/2}$	4d33	14	418	403.024	418.179	3.73E+08	4.1741E+08	4.15E+08	0.02	600
	$(^1D)^2D_{3/2}$	4d43	15	414	399.946	415.027	2.77E+08	2.3712E+08	3.05E+08	0.01	1500
	$(^3P)^4P_{5/2}$	4d35	17	410	395.482	409.312	1.10E+08	1.2373E+08	2.13E+08	0.02	1500
	$(^1D)^2P_{3/2}$	4d53	18	408	393.779	408.806	8.38E+07	6.7013E+06	3.28E+07	0.00	10
	$(^1D)^2D_{5/2}$	4d45	19	403	389.248	403.452	3.66E+08	2.4821E+08	3.55E+08	0.01	1500
	$(^3P)^2F_{5/2}$	4d55	22	396	382.946	397.767	1.49E+08	2.2968E+08	1.14E+08	0.01	1000
	$(^1D)^2F_{5/2}$	4d65	23	374	362.156	379.963	2.34E+08	3.3501E+08	4.00E+08	0.14	1000
	$(^1S)^2D_{3/2}$	4d63	25	346	345.203	355.564	1.06E+09	3.1288E+09	2.29E+09	0.09	1500
	$(^1S)^2D_{5/2}$	4d75	26	341	341.242	349.648	5.28E+07	1.4890E+09	1.56E+08	0.00	800
	$(^3P)^2P_{3/2}$	4d73	27	319	324.443	336.621	1.27E+11	1.0669E+11	1.18E+11	0.88	5000
	$(^3P)^2P_{1/2}$	4d51	28	315	320.652	330.398	1.07E+11	6.8148E+10	5.30E+09	0.05	300
	$(^3P)^2D_{5/2}$	4d85	29	311	318.592	333.084	1.55E+11	1.2220E+11	1.31E+11	0.82	10000
	$(^1D)^2S_{1/2}$	4d41	30	309	310.266	339.023	3.70E+10	5.8584E+10	1.17E+11	0.71	3000
	$(^3P)^2D_{3/2}$	4d83	31	301	308.303	320.467	6.68E+09	4.2526E+09	2.36E+09	0.03	500
$4p^5\ ^2P_{1/2}$	$4s4p^6\ ^2S_{1/2}$	4p61	3	640	595.852	630.973	1.59E+08	3.0194E+08	5.45E+07	0.04	30000
(index = 2)	$4s^24p^44d(^3P)^4D_{3/2}$	4d13	6	491	470.358		1.47E+05	7.9784E+04	1.18E+06	0.00	
	$(^3P)^4D_{1/2}$	4d11	7	488	467.049	479.994	3.11E+06	1.7391E+06	4.92E+06	0.00	1
	$(^1D)^2P_{1/2}$	4d21	10	450	431.756	451.974	5.58E+07	2.7075E+07	4.77E+07	0.00	25
	$(^3P)^4F_{3/2}$	4d23	12	446	430.603	442.947	2.89E+07	4.0470E+07	9.35E+06	0.00	300
	$(^3P)^4P_{1/2}$	4d31	13	441	423.647	441.622	1.30E+07	1.1362E+07	3.80E+07	0.01	25
	$(^3P)^4P_{3/2}$	4d33	14	441	423.664	441.161	2.24E+07	2.7072E+06	4.13E+06	0.00	2
	$(^1D)^2D_{3/2}$	4d43	15	436	420.265	437.661	3.20E+08	3.2991E+08	3.25E+08	0.01	500
	$(^1D)^2P_{3/2}$	4d53	18	430	413.461	430.753	2.86E+07	1.7387E+07	1.78E+07	0.00	500
	$(^1S)^2D_{3/2}$	4d63	25	361	360.235	372.047	1.92E+09	5.6428E+08	3.05E+09	0.06	4000
	$(^3P)^2P_{3/2}$	4d73	27	332	337.687	351.355	3.32E+09	2.3159E+09	8.95E+08	0.02	800
	$(^3P)^2P_{1/2}$	4d51	28	328	333.582	344.583	1.03E+11	3.9796E+10	9.85E+10	0.82	2000
	$(^1D)^2S_{1/2}$	4d41	30	321	322.356	353.976	2.85E+10	4.1111E+10	5.60E+09	0.05	1500
	$(^3P)^2D_{3/2}$	4d83	31	312	320.238	333.796	1.46E+11	1.2065E+11	1.29E+11	0.83	5000

Both Singh et al. [17] and Aggarwal and Keenan [18] compare their calculated level values with the observed values given in the NIST Atomic Spectra Database [20]. Since we have made a number of revisions to the $4p^4 4d$ levels, a new comparison is called for. This is given in Table 9.

Table 9. Comparison of level energies E (cm^{-1}) for Y V calculated with the MCDF2 method of Singh et al. [17] and the GRASP3 method of Aggarwal and Keenan [18] with present experimental energies. Index numbers are those used in [17,18].

Configuration	Term	Desig.	Index	J	E [17]	E [18]	E (Present)
$4s^2 4p^5$	^2P	p5 3	1	3/2	0	0.00	0.00
	^2P	p5 1	2	1/2	12147.85	12088.59	12460.12
$4s4p^6$	^2S	4p61	3	1/2	168478.74	179915.44	170945.58
$4s^2 4p^4 4d$	$(^3P)^4D$	4d15	4	5/2	214469.38	223619.16	218417.13
	$(^3P)^4D$	4d17	5	7/2	214524.25	223608.19	218556.80
	$(^3P)^4D$	4d13	6	3/2	215610.64	224692.39	219373.81
	$(^3P)^4D$	4d11	7	1/2	217092.09	226199.07	220794.66
	$(^3P)^4F$	4d19	8	9/2	227122.02	235877.84	229786.64
	$(^3P)^4F$	4d27	9	7/2	231522.46	240231.10	233703.76
	$(^1D)^2P$	4d21	10	1/2	234342.70	243700.97	233712.36
	$(^3P)^4F$	4d25	11	5/2	235024.04	243502.35	237677.66
	$(^3P)^4F$	4d23	12	3/2	236230.17	244320.98	238215.09
	$(^3P)^4P$	4d31	13	1/2	238754.11	248134.33	238898.28
	$(^3P)^4P$	4d33	14	3/2	239017.48	248124.45	239132.83
	$(^1D)^2D$	4d43	15	3/2	241420.71	250033.87	240949.32
	$(^3P)^2F$	4d37	16	7/2	242276.66	251012.72	242336.33
	$(^3P)^4P$	4d35	17	5/2	244197.05	252856.30	244311.56
	$(^1D)^2P$	4d53	18	3/2	244932.29	253949.27	244613.24
	$(^1D)^2D$	4d45	19	5/2	248290.23	256905.58	247861.17
	$(^1D)^2G$	4d29	20	9/2	251033.65	259335.15	251052.40
	$(^1D)^2G$	4d47	21	7/2	251351.88	259728.01	250882.71
	$(^3P)^2F$	4d55	22	5/2	252383.41	261133.73	251403.88
	$(^1D)^2F$	4d65	23	5/2	267274.68	276123.76	263184.02
	$(^1D)^2F$	4d57	24	7/2	270072.96	278936.31	266196.75
	$(^1S)^2D$	4d63	25	3/2	289079.36	289685.02	281244.51
	$(^1S)^2D$	4d75	26	5/2	293095.72	293047.35	286001.66
	$(^3P)^2P$	4d73	27	3/2	313792.06	308220.63	297071.14
	$(^3P)^2P$	4d51	28	1/2	317095.13	311864.99	302664.42
	$(^3P)^2D$	4d85	29	5/2	321122.47	313880.85	300224.19
	$(^1D)^2S$	4d41	30	1/2	323306.23	322304.24	294965.68
	$(^3P)^2D$	4d83	31	3/2	332502.17	324356.31	312044.02

The percentage compositions for $4s4p^6$ and $4s^2 4p^4 4d$ obtained in the present work are compared with those obtained in the MCDF calculations of Singh et al. [17] in Table 10. The general agreement is qualitatively reasonable.

Table 10. Comparison of the present percentage compositions for the $4s4p^6$ and $4s^24p^44d$ configurations of Y V (in bold type) with those of Singh et al. [17] (in parentheses). Level values are in cm^{-1}.

Index	Desig.	*J*	*E* (obs) [a]	Percentage Composition							
3	4p61	1/2	170946	**75**(69)%	4s ^2S	**24**(30)%	$(^1D)^2S$				
4	4d15	5/2	218417	**90**(92)%	$(^3P)^4D$	**3**%	$(^3P)^4F$	**2**%	$(^3P)^4P$		
5	4d17	7/2	218557	**92**(94)%	$(^3P)^4D$	**5**%	$(^3P)^4F$	**2**%	$(^1D)^2F$		
6	4d13	3/2	219374	**88**(91)%	$(^3P)^4D$	**4**%	$(^3P)^4P$	**3**%	$(^1D)^2D$		
7	4d11	1/2	220795	**88**(91)%	$(^3P)^4D$	**5**%	$(^1D)^2P$	**4**%	$(^3P)^2P$		
8	4d19	9/2	229787	**91**(93)%	$(^3P)^4F$	**9**%	$(^1D)^2G$				
9	4d27	7/2	233704	**72**(81)%	$(^3P)^4F$	**14**%	$(^3P)^2F$	**11**%	$(^1D)^2G$		
10	4d21	1/2	233712	**45**(45)%	$(^1D)^2P$	**39**(40)%	$(^3P)^2P$	**11**%	$(^3P)^4D$		
11	4d25	5/2	237678	**94**(94)%	$(^3P)^4F$	**3**%	$(^3P)^4D$	**2**%	$(^1S)^2D$		
12	4d23	3/2	238215	**63**(88)%	$(^3P)^4F$	**11**%	$(^1S)^2D$	**10**%	$(^3P)^4P$		
13	4d31	1/2	238898	**91**(92)%	$(^3P)^4P$	**4**%	$(^3P)^2P$	**3**%	$(^1D)^2P$		
14	4d33	3/2	239133	**44**(56)%	$(^3P)^4P$	**22**%	$(^3P)^4F$	**20**(22)%	$(^1D)^2P$		
15	4d43	3/2	240949	**38**(40)%	$(^1D)^2D$	**24**(25)%	$(^3P)^2D$	**10**%	$(^3P)^4F$		
16	4d37	7/2	242336	**48**(57)%	$(^3P)^2F$	**20**%	$(^3P)^4F$	**20**(19)%	$(^1D)^2G$		
17	4d35	5/2	244312	**77**(89)%	$(^3P)^4P$	**7**%	$(^1S)^2D$	**6**%	$(^3P)^2D$		
18	4d53	3/2	244613	**39**(24)%	$(^3P)^4P$	**24**(33)%	$(^1D)^2P$	**21**(27)%	$(^3P)^2P$		
19	4d45	5/2	247861	**40**(42)%	$(^1D)^2D$	**23**(25)%	$(^3P)^2D$	**16**%	$(^3P)^4P$	**12**(17%)	$(^3P)^2F$
20	4d29	9/2	251052	**91**(93)%	$(^1D)^2G$	**9**%	$(^3P)^4F$				
21	4d47	7/2	250883	**68**(73)%	$(^1D)^2G$	**22**(19)%	$(^3P)^2F$	**8**%	$(^1D)^2F$		
22	4d55	5/2	251404	**67**(65)%	$(^3P)^2F$	**17**(16)%	$(^1D)^2F$	**10**%	$(^1D)^2D$		
23	4d65	5/2	263184	**80**(83)%	$(^1D)^2F$	**11**%	$(^3P)^2F$	**7**%	$(^1D)^2D$		
24	4d57	7/2	266197	**82**(83)%	$(^1D)^2F$	**15**%	$(^3P)^2F$	**1**%	$(^1D)^2G$		
25	4d63	3/2	281245	**62**(67)%	$(^1S)^2D$	**26**(26)%	$(^1D)^2D$	**4**%	$(^1D)^2P$		
26	4d75	5/2	286002	**72**(74)%	$(^1S)^2D$	**16**(16)%	$(^1D)^2D$	**4**%	$(^3P)^2F$		
27	4d73	3/2	297071	**47**(51)%	$(^3P)^2P$	**36**(41)%	$(^1D)^2P$	**7**%	$(^1D)^2D$		
28	4d51	1/2	302664	**44**(38)%	$(^3P)^2P$	**38**(37)%	$(^1D)^2P$	**10**(17)%	$(^1D)^2S$	**3**%	4s ^2S
29	4d85	5/2	300224	**62**(65)%	$(^3P)^2D$	**20**(21)%	$(^1D)^2D$	**14**%	$(^1S)^2D$		
30	4d41	1/2	294966	**62**(53)%	$(^1D)^2S$	**20**(22)%	4s ^2S	**8**%	$(^1D)^2P$	**6**%	$(^3P)^2P$
31	4d83	3/2	312044	**54**(60)%	$(^3P)^2D$	**20**(18)%	$(^1S)^2D$	**13**(17)%	$(^1D)^2D$		

[a] Present value from Table 3.

Acknowledgments: The spectrograms for this analysis were made in collaboration with Romuald Zalubas and Charles Corliss. They were used for our early work on Y VI [7]. I thank Craig Sansonetti and Haris Kunari for their careful reading of the manuscript.

Conflicts of Interest: The author declares no conflicts of interest.

References

1. Paul, F.W.; Rense, W.A. The spectra of Y V and Zr VI. *Phys. Rev.* **1939**, *56*, 1110–1113. [CrossRef]
2. Edlén, B. Atomic Spectra. In *Handbuch der Phyisk*; Flügge, S., Ed.; Springer: Berlin, Germany, 1964; Volume XXVII, p. 111.
3. Reader, J.; Epstein, G.L. Revised $4p^5\ ^2P_{1/2,3/2}$ splitting of Y V. *J. Opt. Soc. Am.* **1970**, *60*, 140. [CrossRef]
4. Reader, J.; Epstein, G.L. Analysis of the spectrum of quadruply ionized yttrium Y V. *J. Opt. Soc. Am.* **1972**, *62*, 619–622. [CrossRef]
5. Zahid-Ali; Chaghtai, M.S.Z.; Singh, S.P. Term analysis of Y V. *J. Phys. B* **1975**, *8*, 185–193. [CrossRef] [PubMed]
6. Ateqad, N.; Chaghtai, M.S.Z. The 4f and 5p configurations of Y V. *J. Phys. B* **1984**, *17*, 1727–1734. [CrossRef]
7. Zalubas, R.; Reader, J.; Corliss, C.H. $4s^24p^4$–$4s4p^5$ transitions in five-times-ionized yttrium (Y VI). *J. Opt. Soc. Am.* **1976**, *66*, 35–36. [CrossRef]
8. Epstein, G.L.; Reader, J. Spectrum and energy levels of triply ionized yttrium (Y IV). *J. Opt. Soc. Am.* **1982**, *72*, 476–492. [CrossRef]
9. Reader, J.; Epstein, G.L.; Ekberg, J.O. Spectra of Rb II, Sr III, Y IV, Zr V, Nb VI, and Mo VII in the vacuum ultraviolet. *J. Opt. Soc. Am.* **1972**, *62*, 273–284. [CrossRef]
10. Reader, J. Spectrum and energy levels of five-times ionized molybdenum, Mo VI. *J. Phys. B* **2010**, *43*, 1–16. [CrossRef]
11. Reader, J.; Tauheed, A. Spectrum and energy levels of quadruply-ionized molybdenum, Mo V. *J. Phys. B* **2015**, *48*, 144001. [CrossRef]
12. Cowan, R.D. Cowan programs RCN, RCN2, RCG, and RCE. In *The Theory of Atomic Structure and Spectra*; University of California Press: Berkeley, CA, USA, 1981.
13. Reader, J. Transition arrays in atomic spectroscopy with a commercial spreadsheet program. *Comp. Phys.* **1997**, *11*, 190–193. [CrossRef]
14. Radziemski, L.J., Jr.; Kaufman, V. Wavelengths, energy levels, and analysis of neutral atomic chlorine (Cl I). *J. Opt. Soc. Am.* **1969**, *59*, 424–443. [CrossRef]
15. Kagan, D.T.; Conway, J.G.; Meinders, E. Spectrum and energy levels of four-times ionized niobium. *J. Opt. Soc. Am.* **1981**, *71*, 1193–1196. [CrossRef]
16. Mohr, P.J.; Taylor, B.N.; Newell, D.B. CODATA recommended values of the fundamental physical constants: 2010a). *Rev. Mod. Phys.* **2012**, *84*, 1527–1605. [CrossRef]
17. Singh, A.K.; Aggarwal, S.; Mohan, M. Level energies, lifetimes and radiative rates in the $4p^44d$ configurations of bromine-like ions. *Phys. Scr.* **2013**, *88*, 035301. [CrossRef]
18. Aggarwal, K.M.; Keenan, F.P. Energy levels, radiative rates and lifetimes for transitions in Br-like ions with $38 \leq Z \leq 42$. *Phys. Scr.* **2014**, *89*, 125404. [CrossRef]
19. Froese Fischer, C. Evaluation and comparison of configuration interaction calculations for complex atoms. *Atoms* **2014**, *2*, 1–14. [CrossRef]
20. Kramida, A.; Ralchenko, Yu.; Reader, J.; NIST ASD Team. *NIST Atomic Spectra Database, Version 5.2*; National Institute of Standards and Technology: Gaithersburg, MD, USA, 2014. Available online: http://physics.nist.gov/asd (accessed on 17 July 2015).

atoms

MDPI

Review

Spectral Analysis of Moderately Charged Rare-Gas Atoms

Jorge Reyna Almandos * and Mónica Raineri

Centro de Investigaciones Ópticas, M. B. Gonnet, P.O. Box 3 (1897), La Plata 20646, Argentina; monicar@ciop.unlp.edu.ar

* Correspondence: jreyna@ciop.unlp.edu.ar; Tel.: +54-221-484-2957

Academic Editor: Joseph Reader

Received: 27 December 2016; Accepted: 21 February 2017; Published: 7 March 2017

Abstract: This article presents a review concerning the spectral analysis of several ions of neon, argon, krypton and xenon, with impact on laser studies and astrophysics that were mainly carried out in our collaborative groups between Argentina and Brazil during many years. The spectra were recorded from the vacuum ultraviolet to infrared regions using pulsed discharges. Semi-empirical approaches with relativistic Hartree–Fock and Dirac-Fock calculations were also included in these investigations. The spectral analysis produced new classified lines and energy levels. Lifetimes and oscillator strengths were also calculated.

Keywords: atomic spectra; energy levels; relativistic Hartree-Fock calculations

1. Introduction

There is great interest in spectroscopy data from rare gases due to applications in collision physics, fusion diagnostics, photo-electron spectroscopy, astrophysics, and to help understanding laser emission mechanisms. Information on wavelengths and intensities of the spectral lines and energy levels of moderately charged rare-gas atoms is important for these studies. Many processes must be considered in few times ionized atoms related with the widths and shapes of the spectral lines due to the presence of electric and magnetic fields. New spectral analysis of the p^2, p^3, p^4 configurations of moderately ionized noble gases provide us with relevant information to understand the behavior of the atomic parameters in the intermediate type of coupling, when neither a pure electrostatic nor spin-orbit scheme exists.

In plasma physics, the radiative properties of atoms and ions are important for temperature determination and to calculate the concentrations of different plasma components. In the development of future tokamaks, damage by excessive heat load on the plasma-facing components is a major problem. In the ITER (International Thermonuclear Experimental Reactor) Tokamak, injection of krypton has been proposed to produce a peripheral radiating mantle that spreads the heat load, cooling the outermost plasma region and reducing erosion. In the plasma edge and in the divertor region, electron energy ranges from a few to about 100 eV corresponding to Kr I to Kr VIII spectra.

In astrophysics, the spectra of lower stages of ionization are found in space in nebulae, interstellar clouds, chemically peculiar stars, and in the sun. Several atomic parameters, such as energy levels, oscillator strengths, transition probabilities, and radiative lifetimes are important to determine many features of cosmic objects such as element abundance and electronic temperature. Transition probabilities are also needed to calculate the energy transport through the star in model atmospheres.

In this article, a comprehensive review is presented concerning the spectral analysis of several ions of neon, argon, krypton, and xenon, with deep and extended implications for astrophysical and laser studies that were mainly carried out in our collaborative groups between Argentina and Brazil over many years. Using pulsed discharges, the spectra were recorded from the vacuum ultraviolet (VUV) to infrared regions. Semi-empirical approaches with relativistic Hartree–Fock (HFR) and multi-configurational Dirac-Fock (MCDF) calculations were used in the studies. A great number of new energy levels and classified lines were established, along with lifetimes and weighted oscillator strengths being reported in local and international meetings, distributed through internal reports, and published elsewhere. Regularities and systematic trends from the atomic structure were also used for the interpretation of the spectra.

2. Brief History

At the beginning of the 1960s, the activities related with Atomic Spectroscopy carried on by some of the present members of the Centro de Investigaciones Opticas (CIOp), were centered on the measuring by interferometric methods of secondary standards in the thorium 232 wavelength. These works were done at the Physics Department of the Universidad Nacional de La Plata, under the direction of Athos Giacchetti [1].

From 1964, with the laser already known, research orientation was directed to subjects that can be defined as spectroscopy of the laser. The first works on this new field were made abroad, thanks to fellowships granted to the group members (currently M. Garavaglia, M. Gallardo and C.A. Massone) by the Swedish government. They dealt with noble gases and molecular nitrogen lasers. These works included the spectroscopy study of the emitted radiation, lines assignment, and the analysis of the excitation mechanisms of laser transitions [2–7].

When CIOp was constituted in 1977, investigations of noble gases were centered on studies about the spontaneous and laser spectroscopy of xenon. A pulsed electrical discharge tube was employed in two ways: as generator of stimulated emission in the blue-green and secondly in spontaneous emission developing very rich-in-lines spectra of the first ions of the gas [8,9].

In 1985, A. G. Trigueiros, from Brazil, and J. Reyna Almandos made their postdoctoral studies at the Lund Institute of Technology, Sweden, under the direction of Willy Persson. They were working on the atomic emission spectroscopy of moderately rare gas ions, in particular in obtaining krypton spectra in the VUV region using a θ-pinch as a spectral light source. Using these data an extended spectral analysis of the $4s^2$, 4s4d, $4p^2$, and 4s4p configurations in Zn-like six times ionized krypton was completed [10]. The interest in data belonging to this isoelectronic sequence is due to observations of impurity-lines from highly ionized heavy ions with few valence electrons in high temperature plasmas in which such lines have been used for diagnostic purposes. In particular, the resonant transition $4s^{2}\,{}^1S_0$–$4s4p\ {}^1P_1$ has been observed for a large range of Z values in the Zn I isoelectronic sequence.

Since 1986, when A.G.T. returned to the University of Campinas (UNICAMP), Brazil and J.R.A. to CIOp, Argentina, a strong collaboration began on the spectral studies of noble gases that involved both groups. At this point, it is important to mention the support provided to these first activities was given by W. Persson and I. Martinson from the University of Lund, Sweden, which included the donation of one spectrograph for work in the VUV region to the CIOp and two to the UNICAMP. With this equipment, it was possible to extend the spectral range for our studies and new and extended analysis of different ions of Kr, Xe, Ar, and Ne were carried out; some examples of these results were reported in [11–16] (and references therein).

3. Experimental Methods

We have used two different light sources in our experiments, a pulsed electrical discharge tube and a θ-pinch discharge.

The pulsed discharge tube was built at CIOp [16]. This light source consists of a Pyrex tube 100 cm in length and with an inner diameter 0.5 cm in which the gas excitation is produced by discharging a bank of low-inductance capacitors ranging from 20 to 280 nF, charged with voltages up to 20 kV. The schematic of the electric circuit is shown in Figure 1.

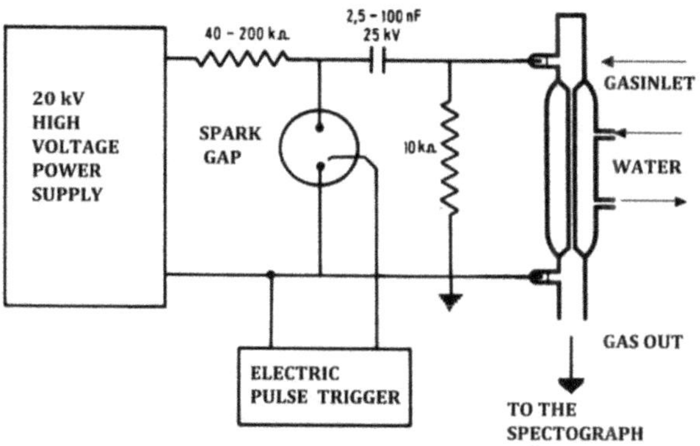

Figure 1. Schematic of the electric circuit. Reproduced and modified by permission of IOP Publishing from [16]. (© The Royal Swedish Academy of Sciences. All rights reserved.)

The second experiment was performed using a θ-pinch discharge built at the Department of Physics, UNICAMP, where the energy was also fed into the plasma using a capacitor bank of 7.7 μF, charged from a high-voltage power supply. A discharge tube with a length of 100 cm and an outside diameter of 8 cm was used. Figure 2 shows a schematic of the electric circuit and more details of the experiment can be found in Ref. [17].

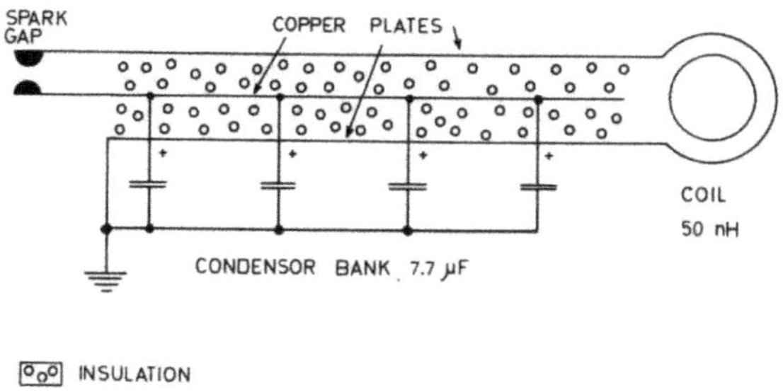

Figure 2. Schematic of the θ-pinch electric circuit. Reproduced and modified by permission of IOP Publishing from [18]. (© The Royal Swedish Academy of Sciences. All rights reserved.)

The wavelength range above 2000 Å was observed in La Plata using a diode array detector coupled to the 3.4 m Ebert plane-grating spectrograph with 600 lines/mm and a plate factor of 5 Å/mm in the first diffraction order. Photographic plates were used to record the spectra in the first, second, and third diffraction orders. The spectral lines observed were compared with interferometrically measured ^{232}Th wavelengths [19] and with known lines of noble gas spectra. The positions of spectral lines were determined by means of a rotating prism photoelectric semiautomatic Grant comparator. For sharp lines, the settings are reproducible to within ±1 μm. A third-order interpolation formula was used to reduce comparator settings to wavelength values. Most of the spectral lines from this region used in the analysis were recorded in the third diffraction order. The accuracy of the thorium standard wavelength values used was on the order of 0.005 Å, and the determination of the overall wavelength values of lines presented in this work was estimated to be correct to ±0.01 Å.

In the wavelength range below 2100 Å light radiation emitted axially was analyzed using a 3 m normal incidence vacuum spectrograph with a concave diffraction grating of 1200 lines mm^{-1} and a plate factor in the first order of 2.77 Å mm^{-1}. Kodak SWR and Ilford Q plates were used to record the spectra. C, N, O, and known lines of noble gas spectra served as internal wavelength standards. The uncertainty of the wavelength below 2100 Å was estimated to be ±0.02 Å. By observing the behavior of the spectral line intensity as a function of pressure and discharge voltage, we were able to distinguish the different ionic states of spectra.

Energy level values derived from the observed lines were determined by means of an iterative procedure that takes into account the wave numbers of the lines, weighted by their estimated uncertainties. The uncertainty of the adjusted experimental energy level values was assumed to be lower than 2 cm^{-1}.

With the above mentioned experimental devices, it was possible to record the spectra of Ne II-VII, Ar II-VIII, Kr II-VIII, and Xe II-IX in the region between 250 and 7000 Å.

4. Atomic Calculations

Calculation of energy levels and transition probability in different spectral analysis has been carried out using the HFR approach [20]. Weighted oscillator strengths (gf), weighted transition probabilities (gA), and lifetimes were done for the experimentally known dipole transition and levels. Values were determined by using the Hartree–Fock method, including relativistic corrections and core-polarization effects with electrostatic parameters optimized by a least squares procedure in order to obtain energy levels adjusted to the corresponding experimental values in which core polarization refers to the deformation of the internal atomic orbitals due to the orbit of the active electron, which repels the remaining electrons [21]. In our work, we modified the electric dipole matrix elements to take core polarization effects into account. Other extensive calculation and studies on noble gas ions including this effect were also carried out by Biémont and co-workers [22–25]. In this last case, all the radial wave functions were also modified by a model potential, including one- and two-body core-polarization contributions, together with a core-penetration correction. In some of our studies, the fully relativistic MCDF approach was also used, more precisely the general-purpose relativistic atomic structure package (GRASP) [26]. These computations were done with an extended average level (EAL) option.

5. Previous Works and Laser Studies

The spectral analysis of moderately charged rare-gas atoms has been carried out in our groups for many years. Studies on line shifts of Xe II, Xe III, Kr II, and Kr III in high current pinched discharges were also conducted. These shifts were observed when comparing the experimental values obtained through the pulsed discharge tube with those coming from a different kind of discharge [27]. By using a simple collision model, it was possible to show that these shifts may be attributed to the microscopic Stark effect [28,29].

Low-pressure xenon plasma excited by pulsed high-current-high-voltage electrical discharges produces high-gain laser transitions in the near UV and visible. Thus, knowledge of the spectral analysis corresponding to different involved ions, is necessary to understand the population mechanisms responsible for most of the classified laser xenon transitions. Pulsed discharges have been used in La Plata to produce laser action at UV, visible, and infrared (IR) wavelengths corresponding to Xe III, Xe V, Xe VII, Xe VIII [30–33], Xe VI, and Xe IX [16,25]. In these investigations, time-resolved experiments were also done and relativistic Hartree-Fock calculations were also performed to obtain weighted lifetimes and radiative transition rates.

6. Some Recent Results on Xe, Ar and Kr Ions

The spectral analysis of several ions of xenon has a relevant impact on astronomy and laser studies. Various atomic parameters such as energy levels, oscillator strengths, transition probabilities and radiative lifetimes have many important astrophysical applications. Transition probabilities are needed for calculating the energy transport through the star in model atmospheres [34] and for direct analysis of stellar chemical compositions [35]. Xenon is a very rare element in the cosmos, observed in chemically peculiar stars [36] and in planetary nebulae [37]. Emission lines of Kr III–V, Xe III–V in a sample of planetary nebulae were identified [38] and Kr VI, Kr VII, Xe VI and Xe VII lines were recently observed in the ultraviolet spectrum of the hot DO-type white dwarf RE 0503-289 [39].

6.1. Xe V

In a recent spectral analysis of four times ionized xenon, Xe V, 12 new energy levels belonging to the $5s^25p6d$ and $5s^25p7s$ configurations and 81 new classified lines were reported [40]. Using relativistic Hartree–Fock and multi-configurational Dirac–Fock calculations, the lifetimes and weighted oscillator strengths were calculated. Table 1 shows the new energy levels belonging to the configurations $5s^25p6d$ and $5s^25p7s$ and the percentage composition of these levels obtained in the least-squares fit. The calculated radiative lifetimes are also presented. The configurations involved in these transitions were $5s^25p6p$, $5s^25p4f$ and $5s5p^3$, $5s^25p5d$, $5s^25p6d$, $5s^25p6s$, and $5s^25p7s$ for the even and odd parities. The weighted oscillator strengths for the observed lines were calculated considering fitted values for the energy parameters. The lifetime values calculated with the GRASP program are presented in the Babushkin gauge since this one, in the non-relativistic limits and in many situations, has been found to be the most stable value, in the sense that it converges smoothly as more correlation is included and is less sensitive to the details of the computational method [41].

Table 1. Experimental and calculated energy levels established for the $5s^25p6d$ and $5s^25p7s$ configurations of Xe V. Calculated radiative lifetimes are also given. Reproduced with permission from [40]. (© IOP Publishing. All rights reserved.)

Level		Energy Obs (cm^{-1})	Percentage Composition [a]	Lifetime (ns)	
				HFR	GRASP
$5s^2$ 5p6d	$^3F^o_2$	284871 [b]	72 $5s^2$ 5p6d ^3F + 17 $5s^2$ 5p6d ^1D	0.48	0.41
	$^3D^o_2$	287391 (5)	20 $5s^2$ 5p6d ^3P + 20 $5s^2$ 5p6d ^3D + 20 $5s^2$ 5p6d ^1D	0.29	0.23
	$^3F^o_3$	287696 (6)	36 $5s^2$ 5p6d ^3F + 17 $5s^2$ 5p6d ^3D + 14 $5s^2$ 5p6d ^1F	0.53	0.27
	$^3D^o_1$	288830 (6)	51 $5s^2$ 5p6d ^3D + 16 $5s^2$ 5p6d ^1P + 13 $5s^2$ 5p6d ^3P	0.28	0.16
	$^3F^o_4$	298739 (3)	88 $5s^2$ 5p6d ^3F	0.65	0.55
	$^1D^o_2$	299596 (3)	46 $5s^2$ 5p6d ^1D + 17 $5s^2$ 5p6d ^3D + 16 $5s^2$ 5p6d ^3F	0.34	0.21
	$^3D^o_3$	300327 (6)	54 $5s^2$ 5p6d ^3D + 23 $5s^2$ 5p6d ^3F	0.28	0.21
	$^3P^o_2$	301483 [b]	21 $5s^2$ 5p6d ^3P + 25 $5s^2$ 5p6d ^3D + 14 $5s5p^26p$ ^5P	0.37	0.23
	$^3P^o_1$	301555 (5)	51 $5s^2$ 5p6d ^3P + 20 $5s^2$ 5p6d ^3D + 15 $5s5p^26p$ ^3D	0.31	0.21
	$^3P^o_0$	301998 (2)	90 $5s^2$ 5p6d ^3P	0.38	0.25
	$^1F^o_3$	304985 [b]	68 $5s^2$ 5p6d ^1F + 14 $5s^2$ 5p6d ^3D	0.24	0.14
	$^1P^o_1$	306065 (7)	41 $5s^2$ 5p6d ^1P + 19 $5s5p^26p$ ^5P + 16 $5s^2$ 5p6d ^3P	0.35	0.21
$5s^2$ 5p7s	$^3P^o_0$	297673 [b]	96 $5s^2$ 5p7s ^3P	0.16	0.19
	$^3P^o_1$	298053 (4)	69 $5s^2$ 5p7s ^3P + 28 $5s^2$ 5p7s ^1P	0.15	0.18
	$^3P^o_2$	312956 (3)	97 $5s^2$ 5p7s ^3P	0.18	0.23
	$^1P^o_1$	313883 (4)	67 $5s^2$ 5p7s ^1P + 27 $5s^2$ 5p7s ^3P	0.14	0.15

Notes: [a] Percentages below 5% have been omitted. [b] Calculated value. ( ) Number of transitions used for establishing the levels. HFR, relativistic Hartree–Fock calculations. GRASP, general-purpose relativistic atomic structure package.

6.2. Xe VI

A new laser line at 33224.0 Å, corresponding to the five-times ionized xenon spectrum and classified as $5s^24f\ ^2F_{7/2}$–$5s^25d\ ^2D_{5/2}$ was recently observed. Semi-empirical calculations using energy parameters adjusted from least-squares were done and the obtained lifetimes were 10.55 ns and 0.055 ns for the upper and lower levels, respectively. The calculated transition probability value was $2.4 \times 10^5\ s^{-1}$. In Figure 3 the gross structure of the low configurations and the laser transition observed in Xe VI is shown.

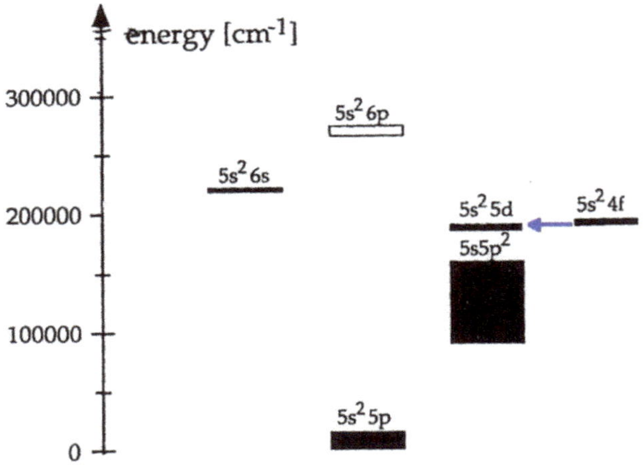

Figure 3. New $5s^2\ 4f\ ^2F_{7/2}$–$5s^2\ 5d\ ^2D_{5/2}$ laser line at 33224.0 Å. Reproduced by permission of IOP Publishing from [42]. (© The Royal Swedish Academy of Sciences. All rights reserved.)

The first detection of krypton and xenon in a white dwarf [39], encouraged us to extend the spectral analysis of five times ionized xenon, Xe VI. In our work the xenon spectra were recorded in the 400–5700 Å region and 243 lines were observed in the experiments, 146 of which were determined for the first time. 32 new and 33 revised energy levels were reported [43]. The gf, gA and lifetimes were calculated using the HFR approach and HFR plus core polarization (HFR+CP) effects. All Xe VI lines observed in the spectrum of the hot DO-type white dwarf RE 0503-289 by Werner et al. [39] were confirmed. The Xe VI analysis is difficult because of the strong mixing of level composition and the non-smooth behavior of the structure which together result in changes in level positions and composition along the isoelectronic sequence. By using all this research, a new paper on the Xe VI ultraviolet spectrum and the xenon abundance in the hot DO-type white dwarf RE 0503-289 was also reported [44].

6.3. Xe VII and Xe VIII

Considering that the red laser line at 6699.40 Å remains unclassified and taking into account the detection of Xe VII in a white dwarf [39], we have decided to make a new spectral analysis of the six- and seven-times ionized xenon spectra. Thus, by using our experimental data covering the wavelength range 430–4640 Å, 40 new transitions of Xe VII and 25 of Xe VIII were classified. We have also revised the values for the previously known energy levels and extended the analysis for Xe VII to 10 new energy levels belonging to 5s6d, 5s7s and 5s7p, $4d^95s^25p$ even and odd configurations, respectively. Seven new energy levels of the $4d^95s5d$ core excited configuration of Xe VIII were determined [45]. Relativistic Hartree–Fock and least-squares-fitted parametric calculations were used in the interpretation of the observed spectra.

6.4. Ar VI

The analysis of the atomic spectra of five times ionized argon (Ar VI), that belongs to the Al I isoelectronic sequence, can be used for studies related with electron-correlation calculations of excited-state structures. As another example of noble gas studies that also have astrophysical interest, an extended analysis of the $3s^24p$, $3s^25d$, $3s^25s$, $3s3d^2$, 3s3p3d, 3s3p4p, and $3p^23d$ configurations in Ar VI, including the determination of new classified lines and energy levels belonging to these configurations, using Al I isoelectronic data and HFR calculations, was completed. Atomic transitions are expected to be prominent in the absorption spectra of interstellar gas clouds [46], and Ar ions have been identified in the Markarian 3 galaxy [47] Chandra HETGS spectrum. In our analysis 14 new energy levels for the configurations 3s3p3d, $3s^24p$, $3s^25s$, $3p^23d$, $3s3d^2$ were established, and two adjusted energy levels of the configuration $3s^25d$ and 68 new spectral lines were classified [48]. Rydberg series interactions for the 3s3p3d configuration were included as their configuration interaction integrals have shown significant values.

Figure 4 shows the gross structure of the Ar VI configurations and in Figure 5 the behavior of the $\lambda_{obs}-\lambda_{cal}$ versus the net charge for the $3s3p(^1P)3d\ ^2F_{5/2}$–$3s3d^2\ ^2G_{7/2}$ transition in the Al I sequence is shown.

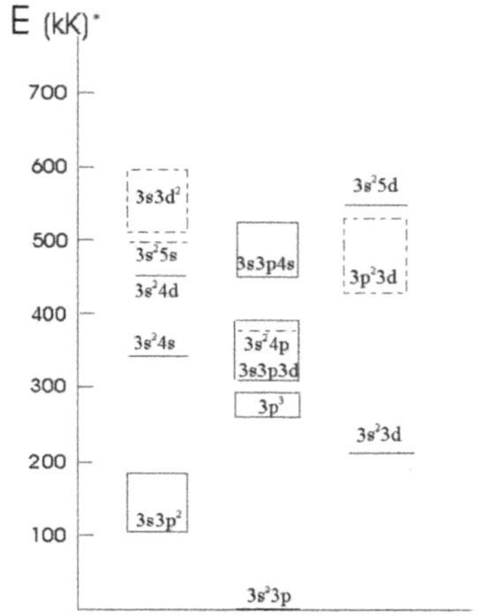

Figure 4. Gross structure of the experimental and theoretical (dotted line) Ar VI configurations. 'kK' means 10^3 cm^{-1}. Reproduced by permission of IOP Publishing from [48]. (© The Royal Swedish Academy of Sciences. All rights reserved.)

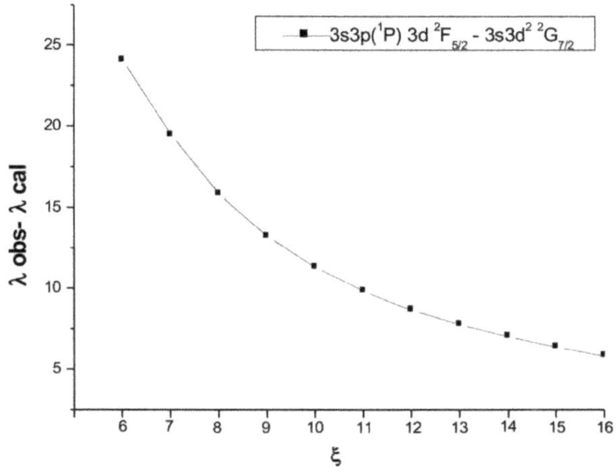

Figure 5. $\lambda_{obs}-\lambda_{cal}$ versus the net charge for the $3s3p(^1P)3d\ ^2F_{5/2}$–$3s3d^2\ ^2G_{7/2}$ transition in the Al I sequence. Reproduced by permission of IOP Publishing from [48]. (© The Royal Swedish Academy of Sciences. All rights reserved.)

6.5. *Kr V*

Forbidden lines belonging to Kr V transitions were found in many nebulae, such as NGC 2440 [49], IC2501 [49], IC4191 [49], and Hen2-436 [50]. The forbidden Kr V lines have not been directly observed at the laboratory, and their wavelengths are presumed from energy differences between the ground configuration levels. Therefore, a precise determination of such levels is crucial for the establishment of the wavelengths of these lines.

Using experimental data from a θ-pinch and a pulsed discharge, an extended analysis of the Kr V spectrum was conducted. The spectrum was recorded in the 230–4900 Å wavelength range, resulting in 91 new classified lines. We were able to identify 21 new energy levels belonging to the $4s^24p5d$, $4s^24p5s$, $4s^24p6s$, $4s^24p5p$, and $4s4p^24d$ configurations [51]. Relativistic Hartree–Fock calculations were used to predict energy levels and transitions and, at this stage, it is important to mention the strong interaction that exists between the ground configuration with the $4p^4$ and the $4s4p^24d$ configurations in this ion. This behavior was also observed with the s^2p^2, p^4, and sp^2d configurations in the spectral analysis of Xe V and Ar V [31,52].

After this research, a new study on lifetimes and transition probabilities in four times ionized krypton was completed. The gf, gA and lifetimes were calculated for all experimentally known dipole transitions and levels of Kr V. The values were determined by four methods. Three of them were based on the Hartree–Fock method including relativistic corrections (the first including a small set of configurations, the second a large set, and the third including core-polarization effects) with electrostatic parameters optimized by a least-squares procedure, in order to obtain energy levels adjusted to the corresponding experimental values. The fourth method was based on a relativistic multiconfigurational Dirac–Fock approach. The 313 dipole electric lines reported in the 294–3615 Å region, included 47 new classified lines [53]. In this investigation, we have also related the observed line intensity with $\log_{10}$ (gA) by the statistical correlation coefficient. The analysis was disaggregated by transition arrays between configurations, i.e., between levels with the same set of dominant configurations in their eigenvector composition. Thirteen sets of transitions were identified, and results are shown in Table 2. Comparing the correlation factors, the best results are for HFR large set of configurations, with or without CP, with an average value of 0.40. These calculations display the best values for more than half of the arrays analyzed. The inclusion of CP effects does not cause significant differences in results.

Table 2. Statistical correlation between $\log_{10}$ gA and experimental line intensity. Maximum correlation for each transition array in boldface. Reproduced with permission from [53]. (© IOP Publishing. All rights reserved.)

Transitions Arrays	HFR Small Set	HFR Large Set	HFR Large Set + CP	GRASP Babushkin Gauge	Number of Lines	Interval (Å)
$4s^24p^2 : 4p5s$	**0.77**	**0.77**	**0.77**	0.62	12	404.45–495.72
$4p5s : 4p5p$	0.59	**0.62**	**0.62**	0.52	21	1498.28–3058.63
$4p4d : 4p^4$	0.27	**0.58**	**0.58**	0.23	6	780.27–3058.63
$4s4p^3 : 4s4p^24d$	0.19	0.48	**0.55**	0.03	11	650.27–1131.28
$4s4p^3 : 4p^4$	0.36	0.45	0.45	**0.53**	10	526.57–1131.28
$4s^24p^2 : 4s4p^3$	**0.35**	**0.39**	**0.39**	0.35	29	515.35–909.63
$4s4p^24d : 4p5d$	**0.33**	**0.33**	**0.33**	−0.55	12	909.63–1735.48
$4s4p^3 : 4p5p$	0.41	0.39	0.39	**0.50**	39	561.80–1735.48
$4p4d : 4p5p$	0.44	**0.46**	**0.46**	0.31	52	940.55–2426.06
$4p5p : 4p6s$	0.17	0.21	021	**0.44**	16	1184.84–2426.06
$4p5p : 4p5d$	0.30	0.27	0.27	**0.38**	40	1240.02–2069.37
$4p4d : 4s4p^24d$	−0.13	**0.33**	−0.27	0.10	11	1131.57–2442.34
$4s^24p^2 : 4p4d$	−0.10	−0.09	−0.10	**0.15**	31	420.67–2442.34
Average	**0.30**	**0.40**	**0.40**	**0.28**		

Notes: CP, core polarization.

6.6. Kr VI

In the first work carried out on this ion the spectrum of five-time ionized krypton was recorded in the 240–2600 Å using both light sources previously described. The study involved the 4s4p4d and the 4s4p5s configurations resulting in 109 new line classifications and 22 new energy levels [54]. This analysis also showed the strong interaction between the 4s4p4d configurations and the other $n = 4$ complex configurations. Oscillator strengths calculated from fitted values of the energy parameters that give gf values that are in better agreement with line intensity observations, were also reported [55].

In a recent analysis [56], 61 lines as transitions between levels of configurations $4p^3$, $4s^25p$, 4s4p4d, 4s4p5s, and 4s4p5p were classified for the first time and all the 18 energy levels belonging to 4s4p5p configuration except one were determined. Eight new energy level values corresponding to configurations 4s4p4f and $4p^24d$, supported by 26 new classified lines, were also determined and used in the interpretation of the observed 4s4p5p configuration. In Figure 6, the gross structure of the observed Kr VI configurations is shown. The dashed arrows indicate the 4s4p5p–$4p^3$, 4s4p5p–$4s^25p$, 4s4p5p–4s4p4d, and 4s4p5p–4s4p5s transitions identified in this work.

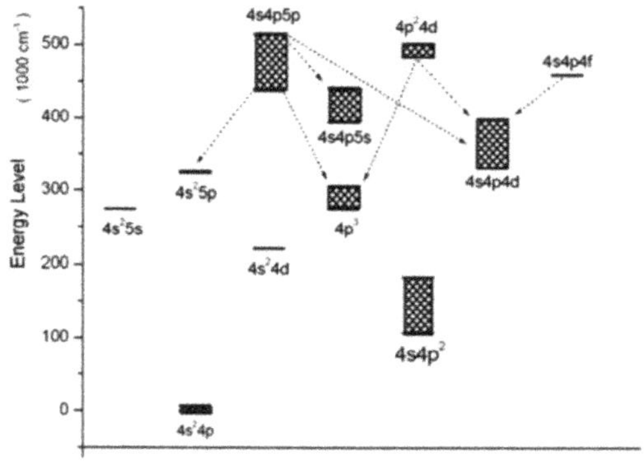

Figure 6. Gross structure of the observed Kr VI configurations. Reproduced by permission of IOP Publishing from [56]. (© The Royal Swedish Academy of Sciences. All rights reserved.)

6.7. Kr VII

Six times ionized krypton belongs to the Zn I isoelectronic sequence and has the ground configuration $3d^{10}4s^2$. Due to the considerable interest in the diagnosis of high-temperature laboratory and astrophysical plasmas, the spectra of this sequence has been extensively studied and the first detection of Kr VII in a white dwarf was recently reported [39]. The spectra of the Zn-like Kr ion has been studied by the use of different light sources and by using a θ-pinch and a pulsed electrical discharge tube [16] in the region from 300 to 4800 Å. The values for the previously known energy levels were revised and a new extended analysis resulting in 115 new classified lines and 38 new energy levels belonging to 4s5s, 4s6s, 4p4f, 4s6d and 4p4d, 4s5p, 4s4f, 4p5s, 4s5f, 4s6p, 4s6f even and odd configurations were completed [57]. For the prediction of the energy level values we studied the behavior of the difference between the observed and calculated energy values along the isoelectronic sequence. As an example, Figure 7 shows the interpolated value for the 4p5s $^3P_1$ E_{obs}–E_{cal} of Kr VII in the Zn-like sequence. The adjusted energy difference value represented in A using our new energy level value in 587029 cm^{-1} fits better than the value in 587068 cm^{-1} reported in Ref. [58], as shown in B.

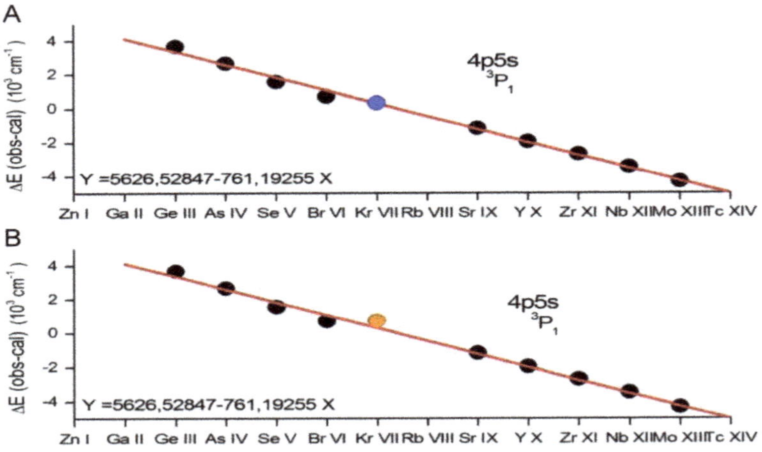

Figure 7. Interpolated value for the 4p5s $^3P_1 E_{obs}–E_{cal}$ of Kr VII in the Zn-like sequence. Our new adjusted energy difference value represented in (**A**) fits better than the value reported in Ref. [58]; as shown in (**B**). Reproduced by permission of IOP Publishing from [57]. (© The Royal Swedish Academy of Sciences. All rights reserved.)

We have recently reported the study of the $4p^2$, $5p^2$, and 5s5f excited configurations of the Zn I and Cd I isoelectronic sequences, using relativistic and non-relativistic semi-empirical approaches [59]. The configuration 5s5f was also analyzed taking into account the Landé's interval rule. In this research, different semi-empirical approaches considering the linearity of the Slater integrals for large Z, the smoothness of the sF screening parameters, the energy values in terms of Z, and the differences of the $E_{obs} - E_{cal}$ values were used. E_{cal} values means the energies calculated with HFR, and E_{obs} are the experimental values.

It is important to mention that most of our results on the studied xenon, krypton, and argon gases were critically compiled and reported by E.B. Saloman [60–62] and an extension and new level optimization of the Ne IV spectrum, that includes our line identification on this gas, was recently made by A. Kramida [63]. Also our previous results a new Kr IV–VII oscillator strengths and an improved spectral analysis of the hot, hydrogen-deficient DO-type white dwarf RE 0503-289, and a new Zr IV–VII, Xe IV-V, and Xe VII oscillator strengths and the Al, Zr, and Xe abundances in the hot white dwarfs G191-B2B and RE 0503-289 were very recently reported [64,65].

7. Conclusions

A comprehensive review concerning the spectral analysis of several moderately charged rare-gas atoms carried out by our international collaborative groups was presented. New and earlier analyses for these ions were revised and extended. Semi-empirical approaches with relativistic Hartree–Fock and Dirac-Fock calculations were used in the studies. The spectra were recorded from the VUV to the near-IR regions using two different light sources, a pulsed electrical discharge tube, and a θ-pinch discharge.

It is important to point out that both spectral sources, together with the detection and measuring systems previously described, allowed us to generate a large amount of new spectral data with very good accuracy in a wide spectral range. Some experimental data used for the study of moderately charged rare-gas atoms, obtained with other kinds of spectral sources such as beam-foil or collision-based spectroscopy, were improved.

Most of our experimental data were obtained using the pulsed electrical discharge tube source. An important characteristic of this light source is that it generates very rich ion spectra, requiring much less capacity and current in the electric discharge, and producing ion spectra like those generated in the θ-pinch.

The pulsed electrical discharge tube also showed that it is a very suitable spectral source for the studies related with the population mechanisms involved in the laser emission of different xenon ions. It is important to continue these works using both time-resolved and frequency spectroscopy under different experimental conditions, together with lifetimes and transition probability calculations for the involved energy levels and lines responsible for the laser emission. Specifically, it is important to extend the theoretical and experimental studies in eight times ionized xenon, where stimulated emission in the VUV was observed on the $4d^95p\ ^1P_1$–$4d^9\ 5d\ ^1S_0$ transition in the Pd-like Xe ion. It is necessary to analyze how the $4d^9$ 4f configuration affects the plasma dynamics in the Pd-like ions, where the $4d^94f\ ^1P_1$–$4d^9\ 5d\ ^1S_0$ transition has conditions for laser action in high ions on this sequence. Similar studies on this subject in the Ne-like and Ni-like ions are also important to conduct.

The experimental device that we used to obtain noble gas spectra can also be applied for the study of non noble gases. In fact, the pulsed electrical discharge tube was used in the works on the N_2 laser, showing very high accuracy in the obtained results [3–5]. The θ-pinch discharge was also used for oxygen ion studies [66] (and references therein).

Interpreting the atomic observations theoretically and by testing computational predictions by experimental data are interactive processes. In the comparison, the high resolution spectral data show better computational results. This was used in all our works.

Discrepancies between experimental measurements and the theoretical calculations led us to carry out more accurate predictions of the allowed and forbidden transitions. MCDF calculations including more active orbitals enabled dealing with a larger amount of excited configurations. The differences of the gauge values in length and velocity of the E1 oscillator strengths obtained showed the accuracy of our calculations GRASP in the performed studies on Kr V, Xe V, and Xe IX.

In HFR calculations for Kr V, Xe V, Xe VI, and Xe IX, CP effects were included, which, combined with a semi-empirical optimization of radial parameters, minimized the discrepancies between the calculated levels of energy and those experimentally obtained. In this type of calculation, the cancellation factors [20] were also assessed, indicating when the oscillator intensities and transition probabilities were affected by great uncertainty.

The spectral analysis of ionized noble gases related with astrophysics and laser studies will be continued in our future work by using the obtained experimental material, different computational calculations, and semi-empirical approximations.

During all these years of fruitful work, fluid collaborations involved groups working on similar subjects in Argentina (Centro de Investigaciones Opticas, Universidad Nacional del Centro, and Universidad Nacional de Mar del Plata); in Brazil (University of Campinas, Universidad Federal Fluminense, and Universidad Federal de Roraima); in Colombia (Universidad del Atlántico); and in Sweden (University of Lund). These collaborations generated mutual visits performed by scientists and students for courses and/or joint work, the exchange of scientific material, the production of more than a dozen of PhD theses, more than seventy international papers, three book chapters, and various reports to international conferences.

Acknowledgments: The careful reading and comments of the manuscript by M. Garavaglia, and Dra. M. Tebaldi are gratefully acknowledged. The Comisión de Investigaciones Científicas de la Provincia de Buenos Aires (CICPBA), Argentina—where J.G. Reyna Almandos and M. Raineri are researchers—is also gratefully acknowledged.

Author Contributions: J.R.A. conceived and wrote the paper. M.R. had important participation in the discussions on the final version.

Conflicts of Interest: The authors declare no conflict of interest.

References

1. Giacchetti, A.; Gallardo, M.; Garavaglia, M.; González, Z.; Valero, F.P.J.; Zakowicz, E. Interferometrically measured thorium wavelengths. *J. Opt. Soc. Am.* **1964**, *54*, 957–959. [CrossRef]
2. Andrade, O.; Gallardo, M.; Bockasten, K. High-gain laser lines in noble gases. *Appl. Phys. Lett.* **1967**, *11*, 99–100. [CrossRef]
3. Andrade, O.; Gallardo, M.; Bockasten, K. New lines in a pulsed N_2 laser. *Appl. Opt.* **1967**, *6*, 2006. [CrossRef] [PubMed]
4. Gallardo, M.; Massone, C.A.; Garavaglia, M. Superradiant and laser spectroscopy in the second positive system of N_2. *Appl. Opt.* **1968**, *7*, 2418. [CrossRef] [PubMed]
5. Garavaglia, M.; Gallardo, M.; Massone, C.A. On the interaction between the first and the second positive laser system in N_2. *Phys. Lett.* **1969**, *28A*, 787–788. [CrossRef]
6. Gallardo, M.; Garavaglia, M.; Tagliaferri, A.A.; Gallego Lluesma, E. About unidentified ionized Xe laser lines. *IEEE J. Quantum Electron.* **1970**, *6*, 745–747. [CrossRef]
7. Gallego Lluesma, E.; Tagliaferri, A.A.; Massone, C.A.; Garavaglia, M.; Gallardo, M. Ionic assignment of unidentified xenon laser lines. *J. Opt. Soc. Am.* **1973**, *63*, 362–364. [CrossRef]
8. Gallardo, M.; Massone, C.A.; Tagliaferri, A.A.; Garavaglia, M.; Persson, W. $5s^2\,5p^3\,(^4S)$ nl levels of Xe III. *Phys. Scr.* **1979**, *19*, 538–544. [CrossRef]
9. Hansen, J.E.; Meijer, F.G.; Outred, M.; Persson, W.; Di Rocco, H.O. Identification of the $4d^{10}\,5p^{6}\,{}^1S_0$ level in Xe III using optical spectroscopy. *Phys. Scr.* **1983**, *27*, 254–255. [CrossRef]
10. Trigueiros, A.; Pettersson, S.-G.; Reyna Almandos, J.G. Transitions within the n = 4 complex of Kr VII obtained from a θ-pinch light source. *Phys. Scr.* **1986**, *34*, 164–166. [CrossRef]
11. Bredice, F.; Reyna Almandos, J.G.; Gallardo, M.; DiRocco, H.O.; Trigueiros, A.G. Revised and extended analysis of the low configurations in Kr III. *J. Opt. Soc. Am. B* **1988**, *5*, 222–235. [CrossRef]
12. Trigueiros, A.G.; Pagan, C.J.B.; Reyna Almandos, J.G. Energy levels of the configurations $4s^24p$, $4s4p^2$, $4s^24d$, and $4s^25p$ in Kr VI, obtained from a θ-pinch light source. *Phys. Rev. A* **1988**, *38*, 166–169. [CrossRef]
13. Persson, W.; Wahlström, C.G.; Bertuccelli, G.; Di Rocco, H.O.; Reyna Almandos, J.G.; Gallardo, M. Spectrum of doubly ionized xenon (Xe III). *Phys. Scr.* **1988**, *38*, 347–369. [CrossRef]
14. Reyna Almandos, J.G.; Bredice, F.; Gallardo, M.; Pagan, C.J.B.; Di Rocco, H.O.; Trigueiros, A.G. $5s^25p^2$ (5d+6s) configurations in triply ionized xenon (Xe IV). *Phys. Rev. A* **1991**, *43*, 6098–6103. [CrossRef] [PubMed]
15. Raineri, M.; Bredice, F.; Gallardo, M.; Reyna Almandos, J.G.; Pagan, C.J.B.; Trigueiros, A.G. Revised and extended analysis of five times ionized argon (Ar VI). *Phys. Scr.* **1992**, *45*, 584–589. [CrossRef]
16. Reyna Almandos, J.; Bredice, F.; Raineri, M.; Gallardo, M. Spectral analysis of ionized noble gases and implications for astronomy and laser studies. *Phys. Scr. T* **2009**, *134*, 014018. [CrossRef]
17. Trigueiros, A.G.; Machida, M.; Pagan, C.J.B.; Reyna Almandos, J.G. The spectroscopic study of radiation produced in a θ-pinch. *Nucl. Instrum. Methods Phys. Res. A* **1989**, *280*, 589–592. [CrossRef]
18. Pettersson, S.-G. The spectrum of O III. *Phys. Scr.* **1982**, *26*, 296–318. [CrossRef]
19. Valero, F.P.J. Improved Values for energy Levels, Ritz Standards, Interferometrically Measured Wavelengths in Th I. *J. Opt. Soc. Am.* **1968**, *58*, 1048–1053. [CrossRef]
20. Cowan, R.D. *The Theory of Atomic Structure and Spectra*; Cambridge University Press: Cambridge, MA, USA, 1981; pp. 178–202.
21. Curtis, L.J. *Atomic Structure and Lifetimes: A Conceptual Approach*; Cambridge University Press: Cambridge, MA, USA, 2003; pp. 54–58.
22. Biémont, E.; Quinet, P.; Zeippen, C.J. Transition Probabilities in Xe V. *Phys. Scr.* **2005**, *71*, 163–169. [CrossRef]
23. Biémont, E.; Buchard, V.; Garnir, H.-P.; Lefebvre, P.-H.; Quinet, P. Radiative lifetime and oscillator strength determinations in Xe VI. *Eur. Phys. J. D* **2005**, *33*, 181–191.
24. Biemont, E.; Clar, M.; Fivet, V.; Garnir, H.-P.; Palmeri, P.; Quinet, P.; Rostohar, D. Lifetime and transition probability determination in xenon ions. The cases of Xe VII and Xe VIII. *Eur. Phys. J. D* **2007**, *44*, 23–33. [CrossRef]
25. Gallardo, M.; Raineri, M.; Reyna Almandos, J.; Biémont, É. New energy levels, calculated lifetimes and transition probabilities in Xe IX. *J. Phys. B* **2011**, *44*, 045001. [CrossRef]
26. Dyall, K.G.; Grant, I.P.; Johnson, C.T.; Parpia, F.A.; Plummer, E.P. GRASP: A general-purpose relativistic atomic structure program. *Comput. Phys. Commun.* **1989**, *55*, 425–456. [CrossRef]

27. Humphreys, C.J. The Third Spectrum of Krypton. *Phys. Rev.* **1935**, *47*, 712–717. [CrossRef]
28. Di Rocco, H.O.; Bertuccelli, G.; Reyna Almandos, J.G.; Gallardo, M. Line shift of singly ionized xenon in high current pinched discharges. *J. Quant. Spectrosc. Radiat. Transf.* **1986**, *35*, 443–446. [CrossRef]
29. Di Rocco, H.O.; Bertuccelli, G.; Reyna Almandos, J.; Bredice, F.; Gallardo, M. Line shift of Kr II, Kr III, and Xe III in high-current, pinched discharges. *J. Quant. Spectrosc. Radiat. Transf.* **1989**, *41*, 161–165. [CrossRef]
30. Duchowicz, R.; Schinca, D.; Gallardo, M. New analysis for the assignment of UV-visible ionic Xe laser lines. *IEEE J. Quant. Elect.* **1994**, *30*, 155–159. [CrossRef]
31. Gallardo, M.; Raineri, M.; Reyna Almandos, J.G.; Sobral, H.; Callegari, F. Revised and extended analysis in four times ionized xenon, Xe V. *J. Quant. Spectrosc. Radiat. Transf.* **1999**, *61*, 319–327. [CrossRef]
32. Sobral, H.; Schinca, D.; Gallardo, M.; Duchowicz, R. Time dependent study of a Multi-Ionic Xenon plasma. *J. Appl. Phys.* **1999**, *85*, 69–73. [CrossRef]
33. Sobral, H.; Raineri, M.; Schinca, D.; Gallardo, M.; Duchowicz, R. Excitation Mechanisms and Characterization of a Multi-Ionic Xenon Laser. *IEEE J. Quantum Electron.* **1999**, *35*, 1308–1313. [CrossRef]
34. Gustafsson, B. The Future of Stellar Spectroscopy and its Dependence on YOU. *Phys. Scr.* **1991**, *34*, 14–19. [CrossRef]
35. Biémont, E.; Blagoev, K.; Campos, J.; Mayo, R.; Malcheva, G.; Ortíz, M.; Quinet, P. Radiative parameters for some transitions in Cu(II) and Ag(II) spectrum. *J. Electron. Spectrosc. Relat. Phenom.* **2005**, *27*, 144–147. [CrossRef]
36. Cowley, C.R.; Hubrig, S.; Palmeri, P.; Quinet, P.; Biémont, É.; Wahlgren, G.M.; Schütz, O.; González, H.D. 65949: Rosetta stone or red herring. *Mon. Not. R. Astron. Soc.* **2010**, *405*, 1271–1284. [CrossRef]
37. Otsuka, M.; Tajitsu, A. Chemical abundances in the extremely carbon-rich and xenon-rich halo planetary nebula H4-1. *Astrophys. J.* **2013**, *778*, 146–158. [CrossRef]
38. Zhang, Y.; Liu, X.-W. Fe/Ni ratio in the Ant Nebula Mz3. *Proc. IAU Symp.* **2006**, *2*, 547–548. [CrossRef]
39. Werner, K.; Rauch, T.; Ringat, E.; Kruk, J.W. First detection of krypton and xenon in a White dwarf. *Astrophys. J.* **2012**, *753*, L7. [CrossRef]
40. Raineri, M.; Gallardo, M.; Padilla, S.; Reyna Almandos, J. New energy levels, lifetimes and transition probabilities in four times ionized xenon (Xe V). *J. Phys. B* **2009**, *42*, 205004. [CrossRef]
41. Froese Fischer, C.; Rubin, R.H. Transition rates for some forbidden lines in Fe IV. *J. Phys. B* **1998**, *31*, 1657–1669. [CrossRef]
42. Larsson, M.O.; Gonzalez, A.M.; Hallin, R.; Heijkenskjöld, F.; Nyström, B.; O´Sullivan, G.; Weber, C.; Wännström, A. Wavelengths and Energy Levels of Xe V and Xe VI Obtained by Collision-based Spectroscopy. *Phys. Scr.* **1996**, *53*, 317–324. [CrossRef]
43. Gallardo, M.; Raineri, M.; Reyna Almandos, J.; Pagan, C.J.B.; Abrahao, R.A. Revised and extended analysis of five times ionized xenon, Xe VI. *Astophys. J. Suppl.* **2015**, *216*, 11. [CrossRef]
44. Rauch, T.; Hoyer, D.; Quinet, P.; Gallardo, M.; Raineri, M. The Xe VI ultraviolet spectrum and the xenon abundance in the hot DO-type white dwarf RE 0503-289. *Astron. Astrophys.* **2015**, *577*, A88. [CrossRef]
45. Raineri, M.; Gallardo, M.; Reyna Almandos, J.; Pagan, C.J.B.; Sarmiento, R. Extended analysis of Xe VII and Xe VIII. *Can. J. Phys.* **2017**, in press. [CrossRef]
46. Morton, D.C.; Smith, W.H. A Summary of Transition Probabilities for Atomic Absorption Lines Formed in Low-Density Clouds. *Astrophys. Suppl. Scr.* **1973**, *26*, 333–363. [CrossRef]
47. Sako, M.; Kahn, S.M.; Paerels, F.; Liedahl, D.A. The Chandra High-Energy Transmission Grating Observation of an X-Ray Ionization Cone in Markarian 3. *Astrophys. J.* **2000**, *543*, L115–L118. [CrossRef]
48. Raineri, M.; Gallardo, M.; Borges, F.O.; Trigueiros, A.G.; Reyna Almandos, J. Extended analysis of the Al-like argon spectrum. *Phys. Scr.* **2009**, *79*, 025302. [CrossRef]
49. Sharpee, B.; Zhang, Y.; Williams, R.; Pellegrini, E.; Cavagnolo, K.; Baldwin, J.A.; Phillips, M.; Liu, X.W. Photoionization cross sections for the trans-iron element Se^{+} from 18 to 31 eV. *Astrophys. J.* **2007**, *659*, 1265–1290. [CrossRef]
50. Otsuka, M.; Meixner, M.; Riebel, D.; Hyung, S.; Tajitsu, A.; Izumiura, H. Dust and chemical abundances of the Sagittarius dwarf galaxy planetary nebula Hen 2–436. *Astrophys. J.* **2011**, *729*, 39. [CrossRef]
51. Rezende, D.C.J.; Borges, F.O.; Cavalcanti, G.H.; Raineri, M.; Gallardo, M.; Reyna Almandos, J.; Trigueiros, A.G. Extended analysis of the Kr V spectrum. *J. Quant. Spectrosc. Radiat. Transf.* **2010**, *111*, 2000–2006. [CrossRef]
52. Cavalcanti, G.; Trigueiros, A.; Gallardo, M.; Reyna Almandos, J. Study of the $3p^4$ configuration in four times ionized argon, Ar V. *J. Phys. B* **1996**, *29*, 6049–6053. [CrossRef]

53. Raineri, M.; Gallardo, M.; Pagan, C.J.B.; Trigueiros, A.G.; Reyna Almandos, J. Lifetimes and transition probabilities in Kr V. *J. Quant. Spectrosc. Radiat. Transf.* **2012**, *113*, 1612–1627. [CrossRef]
54. Pagan, C.J.B.; Reyna Almandos, J.; Gallardo, M.; Pettersson, S.-G.; Cavalcanti, G.; Trigueiros, A.G. Study of the 4s4p4d and 4s4p5s configurations of Kr VI. *J. Opt. Soc. Am. B* **1995**, *12*, 203–211. [CrossRef]
55. Pagan, C.J.B.; Raineri, M.; Bredice, F.; Reyna Almandos, J.; Gallardo, M.; Pettersson, S.-G.; Cavalcanti, G.; Trigueiros, A.G. Weighted oscillator strengths for Kr VI spectrum. *J. Quant. Spectrosc. Radiat. Transf.* **1996**, *55*, 163–168. [CrossRef]
56. Farias, E.E.; Raineri, M.; Gallardo, M.; Reyna Almandos, J.; Cavalcanti, G.H.; Borges, F.O.; Trigueiros, A.G. New energy levels and transitions for the 4s4p5p configuration in Kr VI. *J. Quant. Spectrosc. Radiat. Transf.* **2011**, *112*, 2463–2468. [CrossRef]
57. Raineri, M.; Farías, E.E.; Souza, J.O.; Amorim, E.; Gallardo, M.; Reyna Almandos, J. Revised and extended analysis of the Zn-like Kr íon. *J. Quant. Spectrosc. Radiat. Transf.* **2014**, *148*, 90–98. [CrossRef]
58. Churilov, S.S. Analysis of the spectrum of the Zn-like Kr VII Ion: Highly excited 4p4d and 4p5s configurations. *Opt. Spectrosc.* **2002**, *93*, 826–832. [CrossRef]
59. Di Rocco, H.O.; Raineri, M.; Reyna Almandos, J. Study of the $4p^2$, $5p^2$ and 5s5f excited configurations of the Zn and Cd isoelectronic sequences, using relativistic and non-relativistic semi empirical approaches. *Eur. Phys. J. D* **2016**, *70*, 239. [CrossRef]
60. Saloman, E.B. Energy levels and Observed spectral lines of xenon, Xe I through Xe LIV. *J. Phys. Chem. Ref. Data* **2004**, *33*, 765–921. [CrossRef]
61. Saloman, E.B. Energy levels and Observed spectral lines of krypton, Kr I through Kr XXXVI. *J. Phys. Chem. Ref. Data* **2007**, *36*, 215–386. [CrossRef]
62. Saloman, E.B. Energy levels and Observed spectral lines of argon, Ar II through Ar XVIII. *J. Phys. Chem. Ref. Data* **2010**. [CrossRef]
63. Kramida, A.; Brown, C.M.; Feldman, U.; Reader, J. Extension and new level optimization of The Ne IV spectrum. *Phys. Scr.* **2012**, *85*, 025303. [CrossRef]
64. Rauch, T.; Quinet, P.; Hoyer, D.; Werner, K.; Richter, P.; Kruk, J.W.; Demleitner, M. New Kr IV–VII oscillator strengths and an improved spectral analysis of the hot, hydrogen-deficient DO-type white dwarf RE 0503-289. *Astron. Astrophys.* **2016**. [CrossRef]
65. Rauch, T.; Gamrath, S.; Quinet, P.; Löbling, L.; Hoyer, D.; Werner, K.; Kruk, J.W.; Demleitner, M. New Zr IV–VII, Xe IV–V, and Xe VII oscillator strengths and the Al, Zr, and Xe abundances in the hot white dwarfs G191-B2B and RE 0503-289. *Astron. Astrophys.* **2016**. [CrossRef]
66. Reyna Almandos, J.; Hutton, R. *Light Sources for Atomic Spectroscopy. Handbook for Highly Charged Ion Spectroscopic Research*, 1st ed.; Zou, Y., Hutton, R., Eds.; CRC: New York, NY, USA, 2012; pp. 3–20.

atoms

MDPI

Article

Resonance Transitions in the Spectra of the Ag^{6+}–Ag^{8+} Ions

Alexander Ryabtsev * and Edward Kononov

Institute of Spectroscopy, Russian Academy of Sciences, Troitsk, Moscow 108840, Russia; kononov@isan.troitsk.ru

* Correspondence: ryabtsev@isan.troitsk.ru; Tel.: +7-495-851-0225

Academic Editor: Joseph Reader

Received: 16 January 2017; Accepted: 21 February 2017; Published: 4 March 2017

Abstract: The spectrum of silver, excited in a vacuum spark, was recorded in the region 150–350 Å on a 3-m grazing incidence spectrograph. The resonance $4d^k$–($4d^{k-1}5p$ + $4d^{k-1}4f$ + $4p^54d^{k+1}$) was studied in the Ag^{6+}–Ag^{8+} spectra (Ag VII–Ag IX) with k = 5–3, respectively. Several hundred lines were identified with the aid of the Cowan code and orthogonal operator technique calculations. The energy levels were found and the transition probabilities were calculated.

Keywords: vacuum ultraviolet; ion spectra; wavelengths; energy levels; transition probabilities; parametric calculations

1. Introduction

Six- through eight-times ionized silver atoms are the members of the isonuclear sequence of the silver ions with the unfilled $4d^k$ (k = 5–3) ground-state configuration. The spectra of these ions have not been investigated previously. The excitation of the 4d electron leads to the lowest odd configurations $4d^{k-1}5p$ and $4d^{k-1}4f$. The third odd configuration $4p^54d^{k+1}$ is formed by the excitation of the inner shell 4p electron. The resonance transitions are represented by the transitions from these odd configurations to the ground-state configuration. Out of all resonance transitions only the $4d^k$–$4d^{k-1}4p$ (k = 9–6) ones were previously studied in the silver spectra of the lower ionization stages: Ag III [1], Ag IV [2], Ag V [3] and Ag VI [4]. On the other hand, all three resonance transition arrays were investigated in rather simple spectra of ions having 4d and $4d^2$ ground-state configurations: Ag XI [5,6] and Ag X [7]. In this article we report the results of the study of the Ag VII, Ag VIII and Ag IX to fill the gap between the Ag VI and Ag X isonuclear spectra.

This study is part of a project to get atomic data for the ions of lighter than tin chemical elements isoelectronic with Sn IX–Sn XIV which are relevant to a development of bright source for projection lithography at the 135 Å wavelength. The results for the palladium isonuclear spectra were recently published (see [8] and references therein). Such isoelectronic data are necessary for validation of previously reported analyses of the corresponding tin ion spectra [9,10]. Research on these spectra is also of general interest to atomic physics for improving of theoretical methods of calculations of multi-electron heavy atom spectra.

2. Experiment

The experimental technique and the theoretical approaches for spectrum calculations were the same as in our previous publications [7,8]. Briefly, the light source was a low-inductance vacuum spark operated with an additional inductance up to 2.5 μH. A 150 or 12 μF capacitor was charged up to 4.5 kV resulting in the spark peak current in a range of ~10–20 kA. Ionization stages were distinguished by comparing the intensities of the lines at various peak currents. A 3 m grazing incidence spectrograph (85° angle of incidence) equipped with a gold coated holographic grating having 3600 lines/mm was used for taking the spectra. A plate factor of the spectrograph in the region 160–350 Å was 0.32–0.46 Å mm^{-1} respectively. The spectra were recorded on Kodak SWR photographic plates (Eastman Kodak Company, Rochester, NY, USA) and measured on an EPSON EXPRESSION 10000XL scanner (Seiko Epson Corporation, Suwa, Japan). Wavelengths were calibrated using titanium ion lines [11] as the standards. The titanium spectrum was superimposed on some silver exposures. The measured wavelength uncertainty is estimated as ±0.005 Å for the unperturbed lines of moderate intensity. General view of the silver spectrum in the region 150–350 Å is shown in Figure 1, where the lines identified in this work, previously identified, and remaining unidentified are marked by different colours depending on particular spectrum.

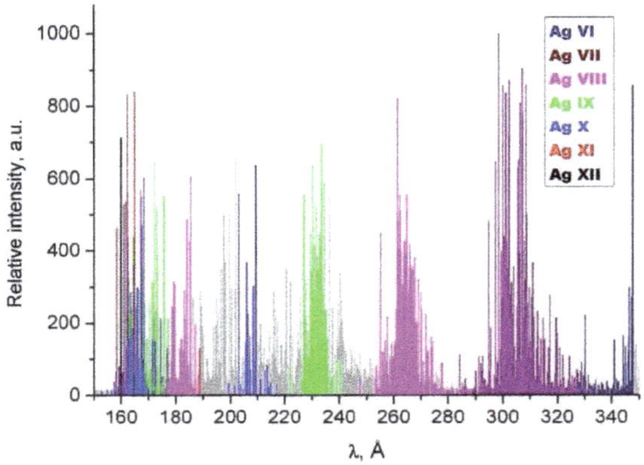

Figure 1. Spectrum of silver in the region 150–350 Å excited in a vacuum spark. Lines of different ion spectra are marked by different colours: Ag VI—royal blue, Ag VII—wine, Ag VIII—magenta, Ag IX—green, Ag X—blue, Ag XI—red, Ag XII—black and unidentified lines—gray.

The relative line intensities were obtained as described in our previous article [8] "from the measured optical densities using an approximate photoplate response curve estimated from different experiments. They should be considered mostly as qualitative ones because of some uncertainty of used photoplate response curve and neglect of the wavelength dependence of the spectrograph efficiency and photoplate sensitivity. Also the saturation effects resulting from the photoplate response nonlinearity can significantly influence the intensity ratios of the weak to strong lines." The intensity I = 1000 was attributed to the strongest line of the $4d^k$–$4d^{k-1}5p$ transition array in each ion spectrum.

The program IDEN [12] was used for the spectrum identification. As in [8], ab initio calculations were performed with the use of the Dirac–Fock (DF) code of Parpia et al. [13], or by the Hartree–Fock method with relativistic corrections (HF) with the use of the Cowan code (Cowan programs RCN, RCN2, RCG, and RCE) [14]. Semiempirical correction of ab initio values of Slater parameters was made with the RCE Cowan code or by using a technique of orthogonal operators [15–18].

The energies derived after the identification of spectral lines were optimized using the program LOPT [19].

3. Results

In the following, the results of the analyses of silver ions in the charge states Ag^{6+}, Ag^{7+} and Ag^{8+} are presented. Line identifications are summarized in Tables A1–A4 (see Appendix A at the end of the document) and energy levels are collected in Tables A5–A11. The data were interpreted using semi-empirical orthogonal parameters and Cowan code calculations resulting in calculated values for the energy levels, wave-function composition, transition probabilities and energy parameters. The semi-empirical energy parameters and their comparison with the corresponding ab initio values are shown in Tables A12–A14.

3.1. *Ag VII*

A diagram of the low lying configurations of Ag VII with the ground-state configuration $4d^5$ is shown in Figure 2. As in the case of silver ions in lower stages of ionization (Ag III–Ag VI) we were able to make the analysis of only the $4d^5$–$4d^45p$ transition array (Table A1). The lines of these transitions are represented by a compact group in the region 271–343 Å mostly isolated from the other transitions in Ag VII as well as in the neighboring ions (see Figure 1). Three hundred and seventy-eight lines were identified in this transition array, 47 of them were doubly and one trebly identified. Eight lines are probably blended with previously identified Ag VI transitions. Thirty-four levels out of 37 possible $4d^5$ ones were found (Table A5) and 142 levels of the $4d^45p$ configurations were located out of 180 possible levels (Table A6). The relative uncertainty of the level energies given by least-squares optimization [19] ranges from 1 to 4 cm^{-1} for $4d^5$ levels and from 3 to 8 cm^{-1} for $4d^45p$ depending on the number of lines used for the level optimization and on their wavelength uncertainties. Identification was performed with the help of the semi-empirical calculations based on the orthogonal operators. The initial orthogonal energy parameters were extrapolated along the sequence Ag IV–Ag V. Final energy parameters of Ag VII after a fitting of the calculated levels to the found levels are listed in Tables A12 and A13. They are compared with the values from the Parpia et al. code [13]. Only the parameters of the $4d^5$ and $4d^45p$ configurations are listed in the tables although the matrixes of the interacting $4d^5 + 4d^45s + 4d^35s^2$ (even) and $4d^45p + 4d^35s5p + 4d^25s^25p$ (odd) configurations were used in the fittings. The parameters of the unknown configurations were fixed on extrapolated values; the interaction parameters were fixed on values obtained with scaling by 0.85 of the ab initio integrals.

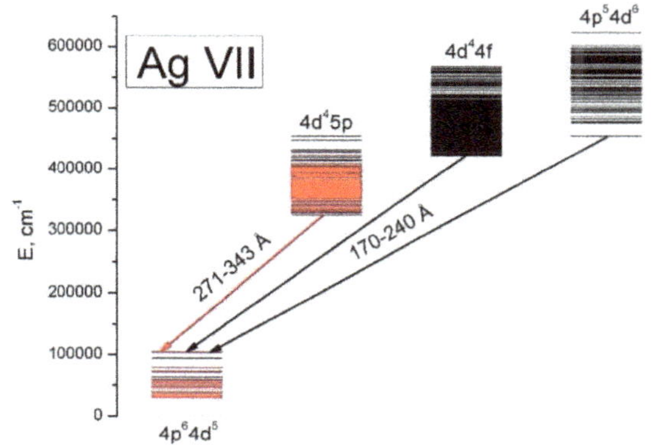

Figure 2. Energy levels of Ag VII. The arrows show electric dipole transitions. The levels found in this work and studied transitions are marked by red color. Black color indicates unknown levels and transitions.

In a treatment of the $4d^5$ shell (as well as of the other $4d^k$ shells) by the orthogonal parameter technique O2, O2′, Ea′ and Eb′ are the orthogonal counterparts of the traditional parameters F^2(4d,4d,), F^4(4d,4d), α and β. The one-electron magnetic (spin-orbit) operator ζ(4d) and the effective 3-particle electrostatic operators T1 and T2 are the same as in Cowan code and (Ac...A0) are additional 2-body magnetic parameters. The $4d^45p$ configuration and the other $4d^{k-1}5p$ configurations contain additional parameters: C1dp–C3pd are the orthogonal counterparts of the Slater exchange integrals

$G^1(4d,5p)$–$G^3(4d,5p)$; S1dp, S2dp are the effective electrostatic 2-body dp-parameters; Sd.Lp ... SS(dp)20 are magnetic 2-body dp-parameters [15], and T16 to T35 are the electrostatic 3-body ddp-parameters. In case of Ag VII 2-body magnetic parameters were varied at the fitting on one bunch keeping the ratios of the corresponding ab initio values. Root mean square deviations of the fitting σ were 14 and 19 cm^{-1} in even and odd configurations, respectively.

Almost all levels of the $4d^5$ configuration can be well designated with the leading member of their eigenvector composition. Only 48,086 cm^{-1} ($J = 3/2$) and 47119 cm^{-1} ($J = 5/2$) were designated with the second term. For $4d^4 5p$ configuration, in many cases two wave functions have the same first component leading to non-unique labels for the energy levels. Therefore, the level energies are listed in Table A6 along with the level designations to avoid the ambiguities.

According to our predictions the most intense lines of the $4d^5$–($4d^4 4f$ + $4p^5 4d^6$) transitions are expected in the 170–240 Å wavelength range. As it is seen in Figure 1 there are many unknown lines in this region but we were not able to make reliable identification of these lines.

3.2. Ag VIII

The low lying configurations of Ag VIII are shown in Figure 3. The transitions from all low odd configurations decaying to the ground-state $4d^4$ configuration are identified in this spectrum. The $4d^4$–$4d^3 5p$ transitions are overlapped with some unknown lines of moderate intensity. But nevertheless, 118 lines were identified in this transition array (Table A2). Twenty-one lines were doubly and two lines trebly identified. Twenty-nine (out of 34 possible) $4d^4$ levels were found with the relative uncertainty from 3 to 7 cm^{-1} and collected in Table A7. The levels of the $4d^3 5p$ configurations are contained in Table A8. It was possible to locate 83 out of 110 possible levels of this configuration. Their uncertainties are from 4 to 14 cm^{-1}. As in Ag VII the identification of the $4d^4$ $4d^3 5p$ transitions was performed with by means of the semi-empirical calculations based on the orthogonal operators.

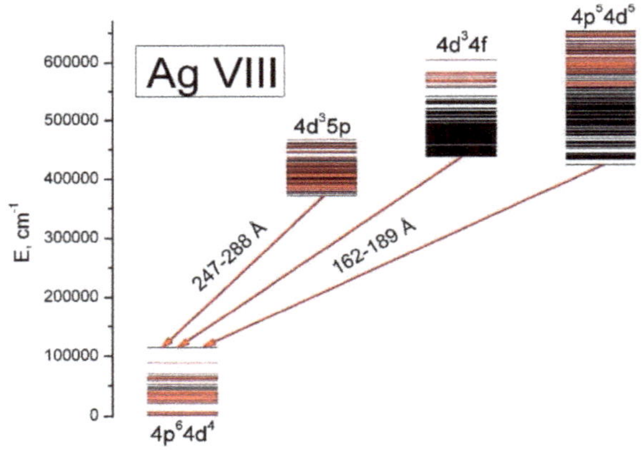

Figure 3. Energy levels of Ag VIII. The arrows show electric dipole transitions studied in this article. The levels found in this work are marked by red color. Black color indicates unknown levels.

The energy parameters obtained in the final fitting are collected in Table A12 for $4d^4$ and Table A13 for $4d^3 5p$. For the meaning of the energy parameters and the procedure of the calculations see Section 3.1. Root mean square deviations of the fitting were 26 and 47 cm^{-1} in the $4d^4$ and $4d^3 5p$ configurations, respectively. In case of the $4d^3 5p$ configuration the fitting is affected by the interaction with the levels of the $4p^5 4d^5$ configuration. The $4d^3 5p$ levels above ~424,000 cm^{-1} overlap with the low lying $4p^5 4d^5$ levels. Their interaction cannot be taken into account in the orthogonal operator code. The LS-coupling scheme is good approximation for the $4d^4$ levels. The value of the first component of the eigenvector composition for all levels is not less than 50%, thus a unique label by the name of the first component can be assigned to all energy levels. To differentiate $4d^4$ terms with the same LS values (recurring terms) the seniority numbers are used in the orthogonal operator code, whereas the Nielson and Koster sequential indices [20] are employed in the Cowan code [14]. Both labels are retained in Table A7 for the $4d^4$ levels because the $4d^4$–($4d^35p$ + $4d^34f$ + $4p^54d^5$) transitions were analyzed with the aid of the Cowan code as described below. Contrary to $4d^4$, the percentage of the first component of the eigenvector composition is less than 50% for many of the $4d^35p$ levels. It goes down to 16%. It makes LS-labeling of many levels meaningless in many cases. Therefore, the energy level values are listed in Table A2 along with the LSJ labels for the wavelength identification.

The identification of the $4d^4$–($4d^35p$ + $4d^34f$ + $4p^54d^5$) transitions using the Cowan code resulted in 118 classified lines in the region 162–189 Å (Table A3). Seventeen lines were doubly classified. The wavelengths and intensities of 10 lines are affected by blending with the Ag IX lines. Table A9 contains the ($4d^34f$ + $4p^54d^5$) levels above 556,000 cm^{-1}. It was possible to find 58 levels of these configurations with the uncertainties from 7 to 19 cm^{-1}. Cowan's calculations of the odd level system were performed for a matrix of interacting configurations $4d^35p$ + $4d^36p$ + $4d^25s5p$ + $4d5s^25p$ + $4d^3$(4f-6f) + $4p^54d^5$ + $4p^54d^45s$. Starting energy parameters for the $4d^35p$, $4d^34f$ and $4p^54d^5$ configurations in Ag VIII were estimated by extrapolation of the scaling factors (the ratios of the fitted to the corresponding Hartree - Fock energy parameters) from Pd VII [21] and Pd VIII [8]. The ab initio electrostatic parameters in the unknown configurations were multiplied by 0.85 scaling factor. The configuration interaction parameters were scaled by 0.8 and the average energies along with the spin-orbit parameters were fixed at the corresponding HF values. Final energy parameters for the $4d^35p$, $4d^34f$ and $4p^54d^5$ configurations obtained in the fitting of the calculated energy levels to the experimental ones using the Cowan code are presented in Table A14. Standard deviation of the fitting σ was 213 cm^{-1}. It should be noted, that for the $4d^3$ levels alone, the fitting by the Cowan code results in σ = 129 cm^{-1} what is 2.7 times larger than at the fitting using the orthogonal parameter code (see Table A13).

All found levels belong to the upper part of configurations ("emissive zone" [22]) from 557,000 to 669,000 cm^{-1}. Only the levels for this energy range are listed in Table A9. According to our calculations full spread of the $4d^34f$ + $4p^54d^5$ levels cover the range up to 424,000 cm^{-1} overlapping with the $4d^35p$ levels. Because of significant uncertainty in prediction of the low lying $4d^34f$ + $4p^54d^5$ levels they are omitted from Table A9.

Examination of Table A9 shows that the percentage contribution of the leading eigenvector component never exceeds 41% and can be as low as 9%. Moreover, the $4d^34f$ wave function can be found as the leading component only at 13 levels with the largest contribution 31%, second component being mostly $4p^54d^5$. Therefore, not only LS-assignment of many levels in Table A3, but also configuration attributions are arbitrary in many cases. Therefore, in Table A9, the upper levels of the transitions are designated by their energies and *J* values, whereas for convenience, a configuration name and LS-label are given according to the output files from the Cowan code in spite of possible ambiguity in many cases.

3.3. *Ag IX*

The scheme of the $4d^3$, $4d^25p$, $4d^24f$ and $4p^54d^4$ levels for Ag IX is shown in Figure 4. It shows that in comparison with Ag VII and Ag VIII the $4d^25p$ levels are almost fully imbedded within the widely spread $4d^24f$ + $4p^54d^4$ levels. The levels of all odd configurations strongly interact. Their initial prediction in the framework of the Cowan code was performed by cross-extrapolation of the scaling factors and effective parameters from isonuclear Ag VIII (this work) and isoelectronic Pd VIII [8]. The $4d^3$ energies were calculated in the framework of the orthogonal parameters by extrapolation from Ag VII and Ag VIII (Table A12) and used as an input to Cowan's calculations of the $4d^3$–($4d^25p$ + $4d^24f$ + $4p^54d^4$) transition probabilities. Thus predicted energy levels and transition probabilities were then used for the spectrum analysis by the IDEN code [12].

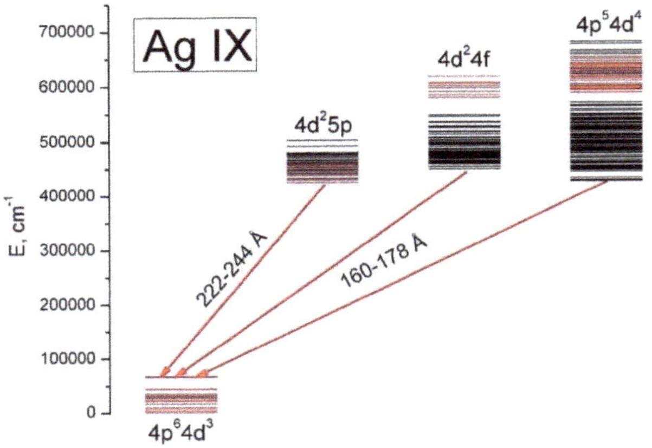

Figure 4. Energy levels of Ag IX. The arrows show electric dipole transitions studied in this article. The levels found in this work are marked by red color. Black color indicates calculated positions of unknown levels.

As a result, 132 lines were identified in the $4d^3$–($4d^25p$ + $4d^24f$ + $4p^54d^4$) transition array (Table A4). Nine lines were doubly classified and one line was trebly classified. The $4d^3$–$4d^25p$ part of this transition array lying in the 221–244 Å region is overlapped by unidentified lines (see Figure 1) discussed in Section 3.1. Nevertheless, it was possible to select the majority of the Ag IX lines by observation of their intensities with the change of the vacuum spark excitation conditions. The other $4d^3$–($4d^24f$ + $4p^54d^4$) part falls in the middle of the region where the spectrum consists of many overlapping lines in Ag VIII–Ag XII. Therefore 10 lines of Ag IX are found to be blended with Ag VIII and 8 with Ag X. In total, 17 levels of the $4d^3$ configuration and 78 levels of the $4d^25p$ + $4d^24f$ + $4p^54d^4$ configurations were established and collected in Tables A10 and A11, respectively. The uncertainty of relative positions of the levels after optimization by LOPT [19] ranges from 4 to 17 cm^{-1} for the ground-state configuration and from 6 to 19 cm^{-1} for the excited configurations.

As was mentioned above the energy levels of the $4d^3$ configuration were treated by orthogonal operator technique. As in Ag VII and Ag VIII calculated matrix consisted of three interacting configurations: $4d^3$ + $4d^25s$ + $4d5s^2$ with similar scaling of the energy parameters for unknown configurations. The levels of the $4d^3$ configuration are presented in Table A10 along with the eigenvector compositions and deviations from the orthogonal parameter calculations. Standard deviation of the fitting was 27 cm^{-1}. The resulting energy parameters of this configuration are collected in Table A12 in comparison with those of $4d^4$ (Ag VIII) and $4d^5$ (Ag VII). Table A12 shows regular behavior of the parameters and scaling factors along this part of the isonuclear sequence of silver ions. The labeling of the $4d^3$ energy levels by the fist component of their eigenvectors is unambiguous.

Table A11 contains all 306 levels of the $4d^25p$ + $4d^24f$ + $4p^54d^4$ configurations. Because of the numerous blends only 78 levels were found. Similar to Ag VIII, a set of the interacting configurations ($4d^25p$ + $4d^26p$ + $4d5s5p$ + $5s^25p$ + $4d^2(4f-6f)$ + $4p^54d^4$ + $4p^55d^35s$) with the same treatment of the unknown configurations was used in the Cowan code calculations. The energy parameters for these configurations in Ag IX are listed in Table A14. The standard deviation of the fitting σ was 327 cm^{-1}, to be compared with σ = 213 cm^{-1} in Ag VIII. It should be noted that in Ag IX more energy parameters than in Ag VIII were fixed on the estimated values for stability of the fitting. Similar considerations are applied to the eigenvector composition of the Ag IX odd levels. There are ambiguities in the LS-labeling and configuration assignment of the levels. Only the level energy and J value can serve as unique label, what is used in the list of the identified lines in Table A4.

4. Discussion

The spectra reported in this article are relevant to the verification of the identifications of the EUV spectra of Sn ions [9,10] which are used as a "fuel" in the radiation sources for the projection lithography at the 135 Å wavelength. The previous analyses in [9,10] were performed without any isoelectronic or isoionic support. The isoelectronic sequence Rh VIII–Cd XI was recently studied in [7]. It was found by extrapolation to Sn XIII that the identification of this spectrum should be revised. Similar conclusion was made after the identification of the M1 transitions between the levels of ground-state configurations in Sn XIII and other ions with open 4d- shell [23]. More data on the VUV spectra of the neighboring to Sn elements are needed. The analyses of Ag VII, Ag VIII and Ag IX were performed in this work for the first time and all Ag ion spectra with the $4d^k$ (k = 1–10) ground-state configuration now became known. After the studies of spectra of the 4d- palladium ions ([8] and references therein) the present work on Ag ion spectra is the next step in the study of the ion spectra isoelectronic with Sn IX–Sn XIII. The work on Cd- and In- ion spectra is in progress at this laboratory.

Author Contributions: A.R. recorded the spectra, performed their analyses and wrote the paper; E.K. made spectrum measurements and wrote the paper.

Conflicts of Interest: The authors declare no conflict of interest.

Appendix A

Table A1. Identified lines of the $4d^5-4d^45p$ transitions in the spectrum of Ag^{6+}.

λ (Å) [a]	o-c, (Å) [b]	ν (cm^{-1})	I [c]	gA, (10^8 s^{-1})	5d^5		5d^{4}5p	
					Term [e]	E (cm^{-1})	Term [e]	E (cm^{-1})
271.910	0.000	367,768.8	36	31	$5^4G_{11/2}$	30,662	$(2^3F)^4F_{9/2}$	398,431
277.706	0.003	360,092.8	11	42	$5^4G_{11/2}$	30,662	$(4^1F)^2G_{9/2}$	390,759
277.852	−0.001	359,904.2	19	37	$5^4G_{5/2}$	29,390	$(2^3F)^4F_{3/2}$	389,293
277.969	0.003	359,752.2	13	51	$5^4D_{7/2}$	36,485	$(2^3P)^4D_{7/2}$	396,241
280.155	−0.002	356,944.6	12	11	$5^4D_{5/2}$	39,299	$(2^3P)^4D_{7/2}$	396,241
280.826	0.000	356,092.1	5	48	$3^2H_{9/2}$	53,797	$(2^1G)^2F_{7/2}$	409,889
281.791	−0.003	354,872.4	13	47	$5^6S_{5/2}$	0	$(4^5D)^4D_{7/2}$	354,869
283.387	0.002	352,874.3	9	44	$3^4F_{3/2}$	53,796	$(2^3F)^4D_{1/2}$	406,673
284.236	−0.003	351,820.2	62	518	$3^2H_{11/2}$	57,962	$(2^1G)^2G_{9/2}$	409,779
284.511	−0.001	351,479.9	12	93	$3^2H_{11/2}$	57,962	$(2^1G)^2H_{11/2}$	409,441
285.168	−0.003	350,670.3	30	276	$5^2G_{9/2}$	59,223	$(2^1G)^2F_{7/2}$	409,889
285.727	0.002	349,984.1	6	153	$5^2F_{7/2}$	59,792	$(2^1G)^2G_{9/2}$	409,779
285.785	0.004	349,912.9	5	55	$3^4F_{9/2}$	49,104	$(2^3F)^4G_{11/2}$	399,022
286.212		349,391.3	32	377	$3^2G_{9/2}$	79,131	$(2^1D)^2F_{7/2}$	428,522
286.254	−0.010	349,339.5	30	159	$3^2H_{9/2}$	53,797	$(2^3F)^4D_{7/2}$	403,133
286.276	0.011	349,313.3	23	76	$3^4F_{9/2}$	49,104	$(2^3F)^4F_{9/2}$	398,431
286.411	−0.002	349,148.6	8	28	$3^4F_{7/2}$	48,712	$(2^3F)^4G_{7/2}$	397,858
286.744	0.009	348,743.4	15	132	$3^4F_{9/2}$	49,104	$(2^3F)^4G_{7/2}$	397,858
287.950	0.001	347,282.9	15	113	$3^2F_{7/2}$	53,353	$(4^1F)^2D_{5/2}$	400,637
288.074	0.003	347,133.0	10	88	$3^4F_{9/2}$	49,104	$(2^3P)^4D_{7/2}$	396,241
288.806	0.002	346,253.7	16	36	$5^4G_{9/2}$	30,907	$(4^3D)^4F_{7/2}$	377,163
288.970	0.003	346,057.0	12	33	$5^6S_{5/2}$	0	$(4^5D)^4F_{7/2}$	346,061
289.226	0.005	345,750.6	16	25	$5^4G_{11/2}$	30,662	$(4^1I)^2I_{13/2}$	376,419
289.431	0.001	345,506.0	5	36	$5^4G_{9/2}$	30,907	$(4^3G)^4G_{11/2}$	376,414
289.575	0.002	345,334.0	18	95	$3^4F_{7/2}$	48,712	$(2^3F)^4F_{5/2}$	394,049
289.665	−0.001	345,225.9	7	28	$3^2H_{9/2}$	53,797	$(2^3F)^4G_{11/2}$	399,022
289.944	0.000	344,894.1	18	114	$5^2G_{9/2}$	59,223	$(2^3F)^2F_{7/2}$	404,117
290.169	0.006	344,626.6	47	326	$3^2H_{9/2}$	53,797	$(2^3F)^4F_{9/2}$	398,431
290.203	−0.003	344,586.3	9	24	$3^4F_{7/2}$	48,712	$(2^3F)^4G_{7/2}$	393,295
290.265	−0.006	344,512.5	12	72	$3^2F_{7/2}$	53,353	$(2^3F)^4G_{7/2}$	397,858
290.415	0.002	344,335.1	18	53	$5^2D_{3/2}$	48,086	$(2^3P)^4D_{1/2}$	392,424
291.263	0.003	343,332.2	52	39	$5^6S_{5/2}$	0	$(4^5D)^4F_{5/2}$	343,336

Table A1. *Cont.*

λ (Å) [a]	o-c, (Å) [b]	ν (cm^{-1})	I [c]	gA, (10^8 s^{-1})	5d^5		5d^{4}5p	
					Term [e]	E (cm^{-1})	Term [e]	E (cm^{-1})
291.310	0.002	343,277.0	15	113	$3^2G_{7/2}$	79,705	$(2^1D)^2F_{5/2}$	422,985
291.817	0.003	342,680.2	14	79	$3^4F_{7/2}$	48,712	$(2^3F)^4G_{5/2}$	391,396
291.862	−0.002	342,627.8	19	132	$3^4F_{7/2}$	48,712	$(2^3F)^4F_{7/2}$	391,338
291.915	−0.004	342,565.6	32	138	$3^4F_{5/2}$	51,049	$(4^1D)^2P_{3/2}$	393,610
292.019	0.001	342,443.3	46	390	$3^2H_{9/2}$	53,797	$(2^3P)^4D_{7/2}$	396,241
292.260	0.000	342,160.6	44	235	$5^4G_{11/2}$	30,662	$(4^3G)^4H_{13/2}$	372,822
292.700	0.007	341,647.2	53	450	$3^4F_{9/2}$	49,104	$(4^1F)^2G_{9/2}$	390,759
292.860	−0.002	341,460.2	37	80	$5^6S_{5/2}$	0	$(4^5D)^6D_{7/2}$	341,458
293.002	0.001	341,294.3	17	43	$5^4G_{7/2}$	30,378	$(4^3G)^4G_{7/2}$	371,673
293.077	0.000	341,206.8	31	89	$5^2D_{3/2}$	48,086	$(2^3F)^4F_{3/2}$	389,293
293.201	−0.002	341,062.7	2	30	$3^2H_{11/2}$	57,962	$(2^3F)^4G_{11/2}$	399,022
293.388	0.000	340,845.3	17	225	$5^2F_{7/2}$	59,792	$(4^1F)^2D_{5/2}$	400,637
293.457	0.001	340,764.8	8	27	$5^4G_{9/2}$	30,907	$(4^3G)^4G_{7/2}$	371,673
293.516	−0.001	340,696.8	32	201	$3^2F_{7/2}$	53,353	$(2^3F)^4F_{5/2}$	394,049
293.681	−0.002	340,506.1	33	126	$3^4F_{5/2}$	51,049	$(2^3P)^4D_{3/2}$	391,553
293.711	−0.001	340,470.6	14	178	$3^2H_{11/2}$	57,962	$(2^3F)^4F_{9/2}$	398,431
293.822	0.004	340,342.2	5	31	$3^4F_{5/2}$	51,049	$(2^3F)^4G_{5/2}$	391,396
293.865	−0.003	340,292.7	4	101	$3^4F_{5/2}$	51,049	$(2^3F)^4F_{7/2}$	391,338
293.954	−0.001	340,189.4	9	35	$5^4G_{5/2}$	29,390	$(4^3F)^2F_{5/2}$	369,578
294.467	−0.002	339,596.6	24	43	$5^4G_{9/2}$	30,907	$(4^3G)^2H_{9/2}$	370,501
294.526	−0.001	339,529.1	18	44	$5^2I_{11/2}$	44,011	$(4^1I)^2H_{11/2}$	383,539
294.551	−0.001	339,499.6	13	109	$3^2H_{9/2}$	53,797	$(2^3F)^4G_{7/2}$	393,295
294.798	−0.006	339,214.9	26	119	$5^2G_{9/2}$	59,223	$(2^3F)^4F_{9/2}$	398,431
294.834	−0.002	339,174.3	197	342	$5^6S_{5/2}$	0	$(4^5D)^6D_{5/2}$	339,172
294.872	0.003	339,130.2	37	202	$3^2F_{7/2}$	53,353	$(2^3F)^4G_{9/2}$	392,486
294.932	−0.002	339,061.6	23	117	$3^4F_{9/2}$	49,104	$(4^1F)^2G_{7/2}$	388,163
294.980 [d]	0.004	339,006.0	43	199	$3^4F_{5/2}$	51,049	$(2^3F)^4F_{5/2}$	390,060
294.980 [d]	−0.006	339,006.0	43	134	$5^4G_{11/2}$	30,662	$(4^1I)^2I_{11/2}$	369,661
295.091	0.002	338,878.2	19	34	$5^4D_{7/2}$	36,485	$(4^3G)^2G_{7/2}$	375,365
295.253	−0.003	338,692.6	22	61	$3^2H_{9/2}$	53,797	$(2^3F)^4G_{9/2}$	392,486
295.304	0.001	338,633.8	47	203	$5^2G_{9/2}$	59,223	$(2^3F)^4G_{7/2}$	397,858
295.304	−0.005	338,633.8	47	159	$3^4F_{3/2}$	53,796	$(2^3P)^4D_{1/2}$	392,424
295.537	−0.001	338,367.6	90	657	$5^2I_{11/2}$	44,011	$(4^1I)^2H_{9/2}$	382,377
295.648	0.004	338,240.1	12	32	$3^4F_{5/2}$	51,049	$(2^3F)^4F_{3/2}$	389,293
295.801	0.001	338,064.9	20	35	$5^2F_{7/2}$	59,792	$(2^3F)^4G_{7/2}$	397,858
295.944	0.002	337,901.3	22	85	$3^4F_{7/2}$	48,712	$(4^1F)^2G_{7/2}$	386,615
295.980	0.003	337,861.1	10	79	$5^4D_{5/2}$	39,299	$(4^3D)^4F_{7/2}$	377,163
296.056	−0.006	337,773.4	22	55	$5^4G_{11/2}$	30,662	$(4^3G)^4H_{9/2}$	368,429
296.579	0.005	337,177.8	14	87	$5^4G_{5/2}$	29,390	$(4^3F)^4F_{3/2}$	366,573
296.716 [d]	0.004	337,023.2	18	52	$5^4D_{1/2}$	38,685	$(4^3D)^4D_{3/2}$	375,706
296.716 [d]	−0.004	337,023.2	18	117	$5^2G_{9/2}$	59,223	$(2^3P)^4D_{7/2}$	396,241
296.779	0.003	336,950.9	12	116	$3^2D_{5/2}$	72,934	$(2^1G)^2F_{7/2}$	409,889
296.892	−0.001	336,822.6	37	160	$5^4G_{11/2}$	30,662	$(4^3F)^4G_{11/2}$	367,483
296.910	0.001	336,802.8	17	103	$5^4D_{1/2}$	38,685	$(4^3D)^4P_{1/2}$	375,489
296.944	−0.008	336,763.9	17	103	$5^4D_{3/2}$	39,788	$(4^3D)^4F_{5/2}$	376,543
297.106	−0.004	336,580.1	22	55	$5^4G_{9/2}$	30,907	$(4^3F)^4G_{11/2}$	367,483
297.124	0.004	336,559.4	22	35	$3^4P_{5/2}$	32,005	$(4^3F)^4D_{3/2}$	368,569
297.216	−0.002	336,456.2	29	174	$3^4P_{3/2}$	32,994	$(4^3D)^4P_{3/2}$	369,448
297.251	−0.008	336,415.8	12	54	$5^4D_{5/2}$	39,299	$(4^3D)^4D_{3/2}$	375,706
297.324		336,333.8	307	565	$5^6S_{5/2}$	0	$(4^5D)^6P_{3/2}$	336,333
297.504	−0.004	336,130.1	8	44	$5^4D_{1/2}$	38,685	$(4^3P)^2P_{1/2}$	374,810
297.702	−0.007	335,906.6	36	79	$5^4G_{9/2}$	30,907	$(4^3F)^4F_{7/2}$	366,806
297.882	−0.002	335,702.9	16	188	$5^4D_{3/2}$	39,788	$(4^3D)^4P_{1/2}$	375,489
297.911	−0.002	335,670.5	55	299	$3^4F_{9/2}$	49,104	$(4^1D)^2F_{7/2}$	384,772
297.951	−0.003	335,625.9	18	143	$5^2G_{7/2}$	55,773	$(2^3F)^4G_{5/2}$	391,396
298.066 [d]	0.001	335,495.9	40	272	$3^4F_{3/2}$	53,796	$(2^3F)^4F_{3/2}$	389,293
298.066 [d]	−0.008	335,495.9	41	68	$5^4G_{5/2}$	29,390	$(4^3P)^4P_{5/2}$	364,877
298.311	−0.005	335,220.5	52	378	$5^2I_{11/2}$	44,011	$(4^3G)^2G_{9/2}$	379,226
298.367	−0.001	335,157.4	35	58	$3^2D_{3/2}$	71,517	$(2^3F)^4D_{1/2}$	406,673

Table A1. *Cont.*

λ (Å) [a]	o-c, (Å) [b]	ν (cm^{-1})	I [c]	gA, (10^8 s^{-1})	5d^5		5d^{4}5p	
					Term [e]	E (cm^{-1})	Term [e]	E (cm^{-1})
298.564 [d]		334,936.4	660	266	$5^4D_{5/2}$	39,299	$(4^3D)^4P_{3/2}$	374,236
298.564 [d]	−0.001	334,936.4	660	239	$5^4G_{9/2}$	30,907	$(4^3F)^2G_{7/2}$	365,842
298.591 [d]		334,905.8	1000	1405	$5^6S_{5/2}$	0	$(4^5D)^6P_{7/2}$	334,906
298.591 [d]	0.001	334,905.8	1000	621	$5^4G_{11/2}$	30,662	$(4^3F)^4F_{9/2}$	365,569
298.651	0.003	334,839.3	23	100	$3^4P_{1/2}$	34,605	$(4^3D)^4P_{3/2}$	369,448
299.048	0.002	334,394.9	12	251	$3^4F_{7/2}$	48,712	$(4^3D)^2F_{5/2}$	383,109
299.073	−0.011	334,367.0	79	244	$5^4G_{11/2}$	30,662	$(4^3H)^4H_{11/2}$	365,017
299.301 [d]	−0.002	334,111.7	82	115	$5^4G_{11/2}$	30,662	$(4^3F)^4G_{9/2}$	364,772
299.301 [d]	−0.001	334,111.7	82	221	$5^4G_{9/2}$	30,907	$(4^3H)^4H_{11/2}$	365,017
299.346	0.002	334,061.9	26	140	$5^4G_{5/2}$	29,390	$(4^3D)^4F_{3/2}$	363,454
299.396	0.002	334,006.1	12	97	$3^4F_{7/2}$	48,712	$(4^1G)^2G_{9/2}$	382,720
299.432	−0.002	333,966.1	15	119	$3^4P_{1/2}$	34,605	$(4^3F)^4D_{3/2}$	368,569
299.526	0.004	333,861.1	127	498	$5^4G_{9/2}$	30,907	$(4^3F)^4G_{9/2}$	364,772
299.719	0.001	333,645.6	63	419	$5^4G_{7/2}$	30,378	$(4^3D)^2D_{5/2}$	364,025
299.740	0.001	333,622.6	32	180	$3^4F_{9/2}$	49,104	$(4^1G)^2G_{9/2}$	382,728
299.772	−0.002	333,586.8	35	71	$5^4G_{5/2}$	29,390	$(4^3P)^4D_{5/2}$	362,975
299.803	0.000	333,551.8	15	72	$5^4G_{5/2}$	29,390	$(4^3G)^4H_{7/2}$	362,942
299.941	−0.004	333,399.3	32	185	$5^4D_{7/2}$	36,485	$(4^3P)^4D_{7/2}$	369,879
299.954	−0.003	333,384.9	32	46	$5^2I_{11/2}$	44,011	$(4^3F)^2G_{9/2}$	377,393
300.006		333,327.1	315	783	$5^6S_{5/2}$	0	$(4^5D)^6P_{5/2}$	333,327
300.065 [d]	0.001	333,261.5	41	193	$5^2G_{9/2}$	59,223	$(2^3F)^4G_{9/2}$	392,486
300.065 [d]	0.001	333,261.5	41	114	$3^2F_{7/2}$	53,353	$(4^1F)^2G_{7/2}$	386,615
300.216	−0.001	333,093.9	36	178	$5^4D_{7/2}$	36,485	$(4^3F)^2F_{5/2}$	369,578
300.224	0.000	333,084.2	36	113	$3^4P_{3/2}$	32,993	$(4^3F)^4F_{3/2}$	366,077
300.305	−0.003	332,994.5	56	319	$5^4G_{5/2}$	29,390	$(4^3H)^4G_{5/2}$	362,381
300.342	−0.001	332,953.5	74	267	$5^4G_{11/2}$	30,662	$(4^3H)^4I_{13/2}$	363,614
300.419	0.004	332,868.1	12	66	$3^4P_{5/2}$	32,005	$(4^3P)^4P_{5/2}$	364,877
300.473 [d]	0.000	332,808.5	144	274	$5^2I_{13/2}$	45,546	$(4^1I)^2K_{15/2}$	378,355
300.473 [d]	−0.010	332,808.5	144	641	$3^2H_{11/2}$	57,962	$(4^1F)^2G_{9/2}$	390,759
300.693	−0.001	332,564.7	39	203	$5^4G_{7/2}$	30,378	$(4^3G)^4H_{7/2}$	362,942
300.760	−0.007	332,490.7	39	208	$5^4D_{7/2}$	36,485	$(4^3P)^4D_{7/2}$	368,968
300.833	−0.006	332,410.1	26	57	$5^2I_{11/2}$	44,011	$(4^3G)^4G_{11/2}$	376,414
300.855	0.004	332,385.9	26	96	$5^2G_{7/2}$	55,773	$(4^1F)^2G_{7/2}$	388,163
301.047	−0.003	332,174.2	342	1899	$5^2I_{13/2}$	45,546	$(4^1G)^2H_{11/2}$	377,717
301.098	−0.003	332,118.1	16	95	$5^2G_{9/2}$	59,223	$(2^3F)^4F_{7/2}$	391,338
301.141	−0.001	332,070.5	15	141	$5^4G_{7/2}$	30,378	$(4^3G)^4F_{9/2}$	362,447
301.178 [d]	0.006	332,029.0	34	149	$5^4G_{9/2}$	30,907	$(4^3G)^4H_{7/2}$	362,942
301.178 [d]	−0.008	332,029.0	34	192	$3^4P_{5/2}$	32,005	$(4^3D)^2D_{5/2}$	364,025
301.262	0.007	331,936.7	16	143	$5^4D_{7/2}$	36,485	$(4^3G)^4H_{9/2}$	368,429
301.314	−0.006	331,879.4	140	547	$5^4G_{11/2}$	30,662	$(4^3H)^2I_{11/2}$	362,535
301.404 [d]	0.002	331,781.0	147	92	$5^4G_{5/2}$	29,390	$(4^3F)^4G_{7/2}$	361,173
301.404 [d]	0.004	331,781.0	147	712	$5^4G_{11/2}$	30,662	$(4^3G)^4F_{9/2}$	362,447
301.585	0.001	331,581.0	9	152	$5^2I_{11/2}$	44,011	$(4^3D)^4F_{9/2}$	375,593
301.618 [d]	0.001	331,545.3	73	415	$5^2F_{7/2}$	59,792	$(2^3F)^4F_{7/2}$	391,338
301.618 [d]	−0.008	331,545.3	73	58	$5^2G_{9/2}$	59,223	$(4^1F)^2G_{9/2}$	390,759
301.701	−0.004	331,453.7	42	243	$3^4P_{5/2}$	32,005	$(4^3D)^4F_{3/2}$	363,454
301.720	0.001	331,433.2	89	388	$5^4D_{7/2}$	36,485	$(4^3D)^4P_{5/2}$	367,919
301.866	−0.001	331,272.8	28	65	$5^4G_{7/2}$	30,378	$(4^3H)^4G_{9/2}$	361,650
301.920	0.001	331,213.5	60	333	$5^4G_{7/2}$	30,378	$(4^3G)^4F_{7/2}$	361,593
301.932		331,200.4	147	211	$5^6S_{5/2}$	0	$(4^5D)^4P_{3/2}$	331,200
302.140	−0.005	330,972.2	13	41	$5^2F_{7/2}$	59,792	$(4^1F)^2G_{9/2}$	390,759
302.191	0.002	330,916.4	155	680	$5^4G_{7/2}$	30,378	$(4^3F)^4F_{5/2}$	361,296
302.226 [d]	−0.009	330,878.0	351	441	$5^2I_{13/2}$	45,546	$(4^3G)^4G_{11/2}$	376,414
302.226 [d]	−0.005	330,878.0	351	768	$5^2I_{13/2}$	45,546	$(4^1I)^2I_{13/2}$	376,419
302.307 [d]	0.005	330,789.4	146	453	$5^4G_{7/2}$	30,378	$(4^3F)^4G_{7/2}$	361,173
302.307 [d]	0.001	330,789.4	146	248	$5^2D_{5/2}$	47,119	$(4^1S)^2P_{3/2}$	377,910
302.350 [d]	0.001	330,742.4	565	1173	$5^4G_{9/2}$	30,907	$(4^3H)^4G_{9/2}$	361,650
302.350 [d]	0.014	330,742.4	565	619	$3^2G_{9/2}$	79,131	$(2^1G)^2F_{7/2}$	409,889
302.401	−0.001	330,687.0	192	881	$5^4G_{9/2}$	30,907	$(4^3G)^4F_{7/2}$	361,593

λ (Å) [a]	o-c, (Å) [b]	ν (cm⁻¹)	I [c]	gA, (10⁸ s⁻¹)	5d⁵ Term [e]	5d⁵ E (cm⁻¹)	5d⁴5p Term [e]	5d⁴5p E (cm⁻¹)

Table A1. *Cont.*

λ (Å) [a]	o-c, (Å) [b]	ν (cm^{-1})	I [c]	gA, (10^8 s^{-1})	5d^5		5d^{4}5p	
					Term [e]	E (cm^{-1})	Term [e]	E (cm^{-1})
302.687 [d]	0.002	330,374.0	21	108	$3^4P_{5/2}$	32,005	$(4^3H)^4G_{5/2}$	362,381
302.687 [d]	−0.008	330,374.0	21	186	$3^4F_{7/2}$	48,712	$(4^1F)^2F_{5/2}$	379,077
302.748 [d]	0.002	330,307.5	57	208	$3^2G_{9/2}$	79,131	$(2^1G)^2H_{11/2}$	409,441
302.748 [d]	0.012	330,307.5	57	154	$5^4D_{7/2}$	36,485	$(4^3F)^4F_{7/2}$	366,806
302.784 [d]	−0.001	330,268.8	49	87	$5^2F_{7/2}$	59,792	$(2^3F)^4F_{5/2}$	390,060
302.784 [d]	−0.003	330,268.8	49	84	$5^4G_{9/2}$	30,907	$(4^3F)^4G_{7/2}$	361,173
302.836	−0.004	330,212.1	110	309	$5^4D_{7/2}$	36,485	$(4^3D)^4D_{5/2}$	366,693
302.865	0.008	330,180.0	26	413	$5^2G_{7/2}$	55,773	$(4^3D)^2F_{5/2}$	385,962
302.936 [d]	0.001	330,103.2	34	150	$5^4G_{7/2}$	30,378	$(4^3F)^4G_{9/2}$	360,482
302.936 [d]	0.003	330,103.2	34	225	$3^2F_{5/2}$	59,954	$(2^3F)^4F_{5/2}$	390,060
303.045	−0.003	329,983.8	72	299	$3^4P_{3/2}$	32,994	$(4^3P)^4D_{5/2}$	362,975
303.063	−0.007	329,964.7	29	442	$3^4F_{9/2}$	49,104	$(4^1G)^2G_{7/2}$	379,061
303.133	−0.004	329,888.5	9	47	$5^4D_{1/2}$	38,685	$(4^3F)^4D_{3/2}$	368,569
303.191 [d]	−0.001	329,824.7	95	131	$5^2D_{3/2}$	48,086	$(4^1S)^2P_{3/2}$	377,910
303.191 [d]	−0.005	329,824.7	95	332	$5^4G_{11/2}$	30,662	$(4^3H)^4I_{11/2}$	360,481
303.252	−0.003	329,758.8	20	60	$3^2F_{7/2}$	53,353	$(4^3D)^2F_{5/2}$	383,109
303.340	−0.003	329,663.2	11	147	$5^4D_{3/2}$	39,788	$(4^3D)^4P_{3/2}$	369,448
303.421	0.000	329,575.0	46	170	$5^4G_{9/2}$	30,907	$(4^3F)^4G_{9/2}$	360,482
303.505	−0.002	329,483.4	55	893	$3^2G_{7/2}$	79,705	$(2^1G)^2F_{5/2}$	409,186
303.557	−0.003	329,427.2	12	64	$5^2D_{5/2}$	47,119	$(4^3D)^4F_{5/2}$	376,543
303.670	0.004	329,304.9	103	391	$5^4G_{7/2}$	30,378	$(4^3G)^4G_{7/2}$	359,687
303.791	−0.004	329,173.5	46	331	$5^2F_{5/2}$	57,413	$(4^1S)^2P_{3/2}$	386,582
303.852	0.001	329,107.7	69	335	$5^2I_{11/2}$	44,011	$(4^3G)^4G_{9/2}$	373,120
303.950 [d]	−0.002	329,001.2	111	371	$5^2G_{7/2}$	55,773	$(4^1D)^2F_{7/2}$	384,772
303.950 [d]		329,001.2	111	353	$5^4G_{5/2}$	29,390	$(4^3F)^4G_{5/2}$	358,392
304.087	0.002	328,852.8	35	264	$5^2F_{5/2}$	57,413	$(4^1D)^2D_{5/2}$	386,268
304.087	−0.004	328,852.8	35	61	$3^4P_{1/2}$	34,605	$(4^3D)^4F_{3/2}$	363,454
304.154	−0.001	328,781.1	12	103	$5^4G_{9/2}$	30,907	$(4^3G)^4G_{7/2}$	359,687
304.261	0.000	328,665.6	41	348	$5^2I_{11/2}$	44,011	$(4^3H)^2H_{11/2}$	372,676
304.302	−0.001	328,621.4	28	210	$5^4D_{5/2}$	39,299	$(4^3D)^4P_{5/2}$	367,919
304.338	−0.002	328,582.2	53	421	$3^2H_{9/2}$	53,797	$(4^1I)^2H_{9/2}$	382,377
304.515	0.001	328,391.4	17	97	$5^4D_{7/2}$	36,485	$(4^3P)^4P_{5/2}$	364,877
304.568		328,334.2	12	215	$3^2D_{3/2}$	71,517	$(4^1F)^2D_{3/2}$	399,850
304.608 [d]	−0.004	328,290.9	36	163	$5^4D_{7/2}$	36,485	$(4^3F)^4G_{9/2}$	364,772
304.608 [d]	−0.002	328,290.9	36	55	$3^4F_{9/2}$	49,104	$(4^3F)^2G_{9/2}$	377,393
304.684	0.000	328,208.5	12	112	$3^2F_{5/2}$	59,954	$(4^1F)^2G_{7/2}$	388,163
304.798	0.000	328,086.6	44	281	$3^4P_{5/2}$	32,005	$(4^3P)^4P_{3/2}$	360,092
304.831	0.008	328,050.5	8	50	$3^4F_{9/2}$	49,104	$(4^3D)^4F_{7/2}$	377,162
305.046 [d]	0.000	327,819.4	18	121	$5^2D_{5/2}$	47,119	$(4^3D)^2P_{3/2}$	374,938
305.046 [d]	0.011	327,819.4	18	101	$3^4F_{7/2}$	48,712	$(4^3D)^4F_{5/2}$	376,543
305.152	−0.002	327,705.2	91	266	$3^2D_{5/2}$	72,934	$(4^1F)^2D_{5/2}$	400,637
305.172	−0.002	327,684.1	91	223	$3^4P_{5/2}$	32,005	$(4^3G)^4G_{7/2}$	359,687
305.249	0.001	327,601.6	87	395	$5^4G_{5/2}$	29,390	$(4^3F)^2D_{3/2}$	356,993
305.305	−0.003	327,541.3	32	49	$3^2F_{7/2}$	53,353	$(4^1G)^2F_{5/2}$	380,891
305.341	0.004	327,503.1	49	406	$5^4D_{5/2}$	39,299	$(4^3F)^4F_{7/2}$	366,806
305.441 [t]	−0.003	327,395.2	47	490	$5^2G_{9/2}$	59,223	$(4^1F)^2G_{7/2}$	386,615
305.441 [t]	−0.001	327,395.2	47	234	$5^4D_{5/2}$	39,299	$(4^3D)^4D_{5/2}$	366,693
305.441 [t]	−0.002	327,395.2	47	108	$5^4D_{1/2}$	38,685	$(4^3F)^4F_{3/2}$	366,078
305.469	−0.002	327,365.7	58	111	$3^2G_{9/2}$	79,131	$(2^3F)^2G_{7/2}$	406,495
305.524 [d]	0.000	327,306.8	54	145	$5^2D_{5/2}$	47,119	$(4^3G)^4F_{5/2}$	374,426
305.524 [d]	0.003	327,306.8	54	154	$3^4F_{9/2}$	49,104	$(4^3G)^4G_{11/2}$	376,414
305.554 [d]	0.001	327,274.8	262	868	$5^2I_{13/2}$	45,546	$(4^3G)^4H_{13/2}$	372,822
305.554 [d]	−0.001	327,274.8	262	92	$5^4D_{5/2}$	39,299	$(4^3F)^4F_{3/2}$	366,573
305.689	0.000	327,129.8	295	1616	$5^2I_{13/2}$	45,546	$(4^3H)^2H_{11/2}$	372,676
305.787	0.002	327,025.0	265	1038	$5^4G_{11/2}$	30,662	$(4^3H)^4G_{11/2}$	357,689
305.902	0.003	326,901.5	19	143	$5^4D_{3/2}$	39,788	$(4^3D)^4D_{5/2}$	366,693
305.961	0.004	326,838.8	107	133	$5^4G_{5/2}$	29,390	$(4^3F)^4D_{5/2}$	356,233
306.009 [d]	0.002	326,787.6	173	936	$3^2G_{7/2}$	79,705	$(2^3F)^2G_{7/2}$	406,495
306.009 [d]	−0.002	326,787.6	173	266	$5^4D_{3/2}$	39,788	$(4^3F)^4F_{3/2}$	366,573

Table A1. *Cont.*

λ (Å) [a]	o-c, (Å) [b]	ν (cm^{-1})	I [c]	gA, (10^8 s^{-1})	$5d^5$		$5d^45p$	
					Term [e]	E (cm^{-1})	Term [e]	E (cm^{-1})
306.020	0.006	326,775.9	31	397	$5^4G_{9/2}$	30,907	$(4^3H)^4G_{11/2}$	357,689
306.072	0.004	326,720.1	21	146	$5^2D_{3/2}$	48,086	$(4^3P)^2P_{1/2}$	374,810
306.159	0.000	326,627.8	46	361	$3^2F_{5/2}$	59,954	$(4^1S)^2P_{3/2}$	386,582
306.288 [d]	−0.001	326,490.6	372	167	$3^4F_{9/2}$	49,104	$(4^3D)^4F_{9/2}$	375,593
306.288 [d]	−0.001	326,490.6	372	1477	$5^2I_{11/2}$	44,011	$(4^3G)^2H_{9/2}$	370,501
306.303	0.002	326,474.1	77	436	$5^2F_{7/2}$	59,792	$(4^1D)^2D_{5/2}$	386,268
306.425	−0.004	326,344.3	16	55	$5^2D_{3/2}$	48,086	$(4^3G)^4F_{5/2}$	374,426
306.479	0.003	326,286.8	12	110	$5^4D_{3/2}$	39,788	$(4^3F)^4F_{3/2}$	366,078
306.712	−0.001	326,039.2	12	46	$5^4G_{7/2}$	30,378	$(4^3H)^4I_{9/2}$	356,416
306.737	−0.004	326,011.7	23	294	$3^2F_{5/2}$	59,954	$(4^3D)^2F_{5/2}$	385,962
306.781	−0.003	325,965.8	62	360	$5^4D_{7/2}$	36,485	$(4^3G)^4F_{9/2}$	362,447
306.857	0.002	325,884.4	18	51	$5^4D_{7/2}$	36,485	$(4^3H)^4G_{5/2}$	362,381
306.876	0.002	325,864.4	34	417	$5^2F_{5/2}$	57,413	$(4^3D)^2D_{3/2}$	383,280
306.981	0.001	325,753.5	133	394	$5^4G_{11/2}$	30,662	$(4^3H)^4I_{9/2}$	356,416
307.013	0.004	325,719.5	131	656	$3^2F_{7/2}$	53,353	$(4^1F)^2F_{5/2}$	379,077
307.086 [d]	0.008	325,641.9	505	1764	$5^2I_{11/2}$	44,011	$(4^1I)^2I_{11/2}$	369,661
307.086 [d]	0.001	325,641.9	505	213	$5^2I_{11/2}$	44,011	$(4^1I)^2K_{13/2}$	369,654
307.147 [d]	0.001	325,577.1	393	172	$5^4D_{5/2}$	39,299	$(4^3P)^4P_{5/2}$	364,877
307.147 [d]	0.000	325,577.1	393	1711	$3^2H_{11/2}$	57,962	$(4^1I)^2H_{11/2}$	383,539
307.178	0.005	325,543.6	100	607	$5^2G_{9/2}$	59,223	$(4^1D)^2F_{7/2}$	384,772
307.230	−0.002	325,488.8	31	185	$3^4P_{1/2}$	34,605	$(4^3P)^4P_{3/2}$	360,092
307.290	0.003	325,425.5	3	139	$3^2H_{9/2}$	53,797	$(4^3G)^2G_{9/2}$	379,226
307.448	0.006	325,257.7	70	609	$3^2H_{9/2}$	53,797	$(4^1G)^2G_{7/2}$	379,061
307.579	−0.002	325,120.2	83	598	$5^2G_{7/2}$	55,773	$(4^1G)^2F_{5/2}$	380,891
307.608	0.000	325,088.7	14	74	$5^4D_{3/2}$	39,788	$(4^3P)^4P_{5/2}$	364,877
307.707	0.001	324,984.9	49	518	$3^2G_{9/2}$	79,131	$(2^3F)^2F_{7/2}$	404,117
307.871		324,811.2	111	987	$3^2G_{9/2}$	79,131	$(2^3F)^2G_{9/2}$	403,942
307.918 [d]	0.007	324,761.4	147	198	$5^4D_{1/2}$	38,685	$(4^3D)^4F_{3/2}$	363,454
307.918 [d]	−0.003	324,761.4	147	786	$3^2H_{11/2}$	57,962	$(4^1G)^2G_{9/2}$	382,720
307.981	−0.001	324,695.8	53	394	$5^2G_{7/2}$	55,773	$(4^1F)^2F_{7/2}$	380,468
308.024	0.007	324,649.9	11	89	$3^4F_{5/2}$	51,049	$(4^3D)^4D_{3/2}$	375,706
308.114	−0.001	324,555.0	16	110	$5^2D_{5/2}$	47,119	$(4^3G)^4G_{7/2}$	371,673
308.180	0.002	324,486.0	96	453	$3^4P_{5/2}$	32,005	$(4^3F)^4D_{7/2}$	356,493
308.247 [d]	0.003	324,415.1	66	179	$5^2I_{11/2}$	44,011	$(4^3G)^4H_{9/2}$	368,429
308.247 [d]	−0.007	324,415.1	66	167	$3^4F_{7/2}$	48,712	$(4^3G)^4G_{9/2}$	373,120
308.538 [d]	0.005	324,109.7	515	155	$5^2I_{13/2}$	45,546	$(4^1I)^2I_{11/2}$	369,661
308.538 [d]	−0.002	324,109.7	515	1122	$5^2I_{13/2}$	45,546	$(4^1I)^2K_{13/2}$	369,654
308.588 [d]		324,056.3	220	408	$5^4G_{11/2}$	30,662	$(4^3H)^4H_{13/2}$	354,719
308.588 [d]	0.007	324,056.3	220	196	$5^4G_{5/2}$	29,390	$(4^3G)^4H_{7/2}$	353,454
308.640	−0.004	324,001.7	62	284	$5^4D_{7/2}$	36,485	$(4^3F)^4G_{9/2}$	360,482
308.685	0.007	323,954.5	16	31	$5^4G_{9/2}$	30,907	$(4^5D)^4D_{7/2}$	354,869
308.820	−0.003	323,813.2	58	157	$3^2F_{7/2}$	53,353	$(4^3D)^4F_{7/2}$	377,163
308.885	−0.001	323,745.4	59	255	$5^4G_{7/2}$	30,378	$(4^3H)^4H_{9/2}$	354,122
308.922	−0.002	323,706.6	20	195	$5^2G_{7/2}$	55,773	$(4^3G)^2F_{5/2}$	379,478
309.038	0.011	323,584.5	15	156	$3^2H_{9/2}$	53,797	$(4^3F)^2G_{9/2}$	377,393
309.123	0.001	323,496.0	9	163	$5^2G_{9/2}$	59,223	$(4^1G)^2G_{9/2}$	382,720
309.159	0.002	323,458.5	78	311	$5^4G_{11/2}$	30,662	$(4^3H)^4H_{9/2}$	354,122
309.190	−0.005	323,426.0	12	86	$3^2G_{7/2}$	79,705	$(2^3F)^4D_{7/2}$	403,133
309.243 [d]	0.007	323,370.1	45	219	$3^4F_{5/2}$	51,049	$(4^3G)^4F_{5/2}$	374,426
309.243 [d]	−0.004	323,370.1	45	165	$3^2H_{9/2}$	53,797	$(4^3D)^4F_{7/2}$	377,163
309.286	0.001	323,325.0	13	216	$3^2F_{5/2}$	59,954	$(4^3D)^2D_{3/2}$	383,280
309.318	0.002	323,292.4	17	67	$5^4G_{5/2}$	29,390	$(4^3G)^4G_{5/2}$	352,685
309.366	−0.002	323,241.3	55	212	$3^4P_{3/2}$	32,994	$(4^3F)^4D_{5/2}$	356,233
309.426	0.008	323,178.8	90	168	$5^4D_{3/2}$	39,788	$(4^3P)^4D_{5/2}$	362,975
309.450	0.000	323,153.9	90	593	$5^2G_{9/2}$	59,223	$(4^1I)^2H_{9/2}$	382,377
309.633	−0.002	322,963.2	107	582	$3^4F_{7/2}$	48,712	$(4^3G)^4G_{7/2}$	371,673
309.732	0.004	322,859.6	19	119	$3^4P_{5/2}$	32,005	$(4^5D)^4D_{7/2}$	354,869
309.793		322,796.5	32	192	$3^4P_{3/2}$	32,994	$(4^3P)^4P_{1/2}$	355,790

Table A1. *Cont.*

λ (Å) [a]	o-c, (Å) [b]	ν (cm^{-1})	I [c]	gA, (10^8 s^{-1})	$5d^5$		$5d^4 5p$	
					Term [e]	E (cm^{-1})	Term [e]	E (cm^{-1})
309.976	0.011	322,605.2	23	148	$3^2H_{9/2}$	53,797	$(4^3G)^4G_{11/2}$	376,414
310.029	−0.003	322,550.5	73	151	$5^4G_{9/2}$	30,907	$(4^3G)^4H_{7/2}$	353,454
310.140	−0.005	322,434.9	52	310	$3^2G_{9/2}$	79,131	$(2^1G)^2H_{9/2}$	401,561
310.211	−0.011	322,360.8	15	111	$3^4F_{7/2}$	48,712	$(4^3H)^2H_{9/2}$	371,061
310.259	−0.005	322,311.8	37	90	$5^4G_{7/2}$	30,378	$(4^3G)^4G_{5/2}$	352,685
310.340	0.003	322,227.0	48	75	$3^4P_{5/2}$	32,005	$(4^5D)^4D_{5/2}$	354,235
310.604	0.004	321,953.2	51	303	$3^4F_{9/2}$	49,104	$(4^3H)^2H_{9/2}$	371,061
310.663	−0.003	321,892.2	89	407	$3^4F_{9/2}$	49,104	$(4^3F)^2F_{7/2}$	370,993
310.707 [d]	0.002	321,847.2	21	124	$5^2D_{5/2}$	47,119	$(4^3P)^4D_{7/2}$	368,968
310.707 [d]	0.009	321,847.2	21	194	$3^2G_{7/2}$	79,705	$(2^1G)^2H_{9/2}$	401,561
310.885	0.002	321,662.3	5	165	$5^2F_{5/2}$	57,412	$(4^1F)^2F_{5/2}$	379,076
310.923	−0.003	321,622.5	2	86	$5^2G_{7/2}$	55,773	$(4^3F)^2G_{9/2}$	377,393
310.985 [d]	0.009	321,559.0	65	194	$3^2H_{9/2}$	53,797	$(4^3G)^2G_{7/2}$	375,365
310.985 [d]	−0.001	321,559.0	65	225	$5^2I_{11/2}$	44,011	$(4^3F)^4F_{9/2}$	365,569
311.058	0.003	321,483.3	122	687	$3^2F_{7/2}$	53,353	$(4^3H)^4G_{7/2}$	374,839
311.121	0.002	321,418.0	17	38	$5^4G_{11/2}$	30,662	$(4^3H)^4H_{11/2}$	352,082
311.161	0.007	321,376.6	43	223	$3^4F_{9/2}$	49,104	$(4^3G)^4H_{11/2}$	370,488
311.272	0.002	321,262.0	49	513	$3^2H_{11/2}$	57,962	$(4^3G)^2G_{9/2}$	379,226
311.355	−0.002	321,177.0	66	185	$5^4G_{9/2}$	30,907	$(4^3H)^4H_{11/2}$	352,082
311.414	−0.001	321,115.8	8	169	$3^2D_{5/2}$	72,934	$(2^3F)^4F_{5/2}$	394,049
311.453	−0.003	321,076.2	18	81	$3^2F_{7/2}$	53,353	$(4^3G)^4F_{5/2}$	374,426
311.518 [d]	−0.006	321,008.4	60	381	$3^4F_{5/2}$	51,049	$(4^3H)^4G_{5/2}$	372,051
311.518 [d]	−0.002	321,008.4	60	59	$5^2I_{11/2}$	44,011	$(4^3H)^4H_{11/2}$	365,017
311.594 [d]	0.006	320,930.4	36	412	$3^2F_{5/2}$	59,954	$(4^1G)^2F_{5/2}$	380,891
311.594 [d]	0.002	320,930.4	36	154	$3^2G_{7/2}$	79,705	$(4^1F)^2D_{5/2}$	400,637
311.845	0.003	320,672.7	8	146	$3^2D_{5/2}$	72,934	$(4^1D)^2P_{3/2}$	393,610
311.891	0.005	320,624.5	20	135	$3^4F_{3/2}$	53,796	$(4^3G)^4F_{5/2}$	374,426
311.963	0.004	320,551.2	24	164	$3^4P_{3/2}$	32,994	$(4^5D)^4D_{3/2}$	353,549
312.007	0.001	320,505.8	9	78	$5^2I_{11/2}$	44,011	$(4^3H)^2I_{13/2}$	364,518
312.030	0.001	320,481.5	26	118	$5^2D_{3/2}$	48,086	$(4^3F)^4D_{3/2}$	368,569
312.465	0.000	320,035.6	29	140	$3^2D_{3/2}$	71,517	$(2^3P)^4D_{3/2}$	391,553
312.497	0.000	320,002.7	59	466	$5^2G_{9/2}$	59,223	$(4^3G)^2G_{9/2}$	379,226
312.716	0.001	319,779.1	36	152	$3^2H_{9/2}$	53,797	$(4^3F)^4D_{7/2}$	373,577
312.742	0.003	319,752.1	36	149	$3^2H_{11/2}$	57,962	$(4^1G)^2H_{11/2}$	377,717
312.807	0.000	319,685.6	39	345	$5^2F_{7/2}$	59,792	$(4^3G)^2F_{5/2}$	379,478
312.836		319,656.8	99	248	$5^4G_{5/2}$	29,390	$(4^3H)^4H_{7/2}$	349,047
312.895	−0.004	319,596.1	46	332	$5^2G_{7/2}$	55,773	$(4^3G)^2G_{7/2}$	375,365
312.966	0.000	319,523.7	12	123	$3^2F_{5/2}$	59,954	$(4^3G)^2F_{5/2}$	379,478
313.025	0.008	319,463.4	31	228	$5^2I_{13/2}$	45,546	$(4^3H)^4H_{11/2}$	365,017
313.058	0.002	319,429.4	59	513	$3^2H_{11/2}$	57,962	$(4^3F)^2G_{9/2}$	377,393
313.143	0.000	319,343.1	48	115	$5^4G_{7/2}$	30,378	$(4^3H)^4I_{9/2}$	349,721
313.164	0.001	319,321.8	48	199	$3^2H_{9/2}$	53,797	$(4^3G)^4G_{9/2}$	373,120
313.208	0.000	319,276.3	23	81	$3^4P_{3/2}$	32,994	$(4^5D)^4D_{1/2}$	352,270
313.375	0.000	319,106.8	21	191	$3^2F_{5/2}$	59,954	$(4^1G)^2G_{7/2}$	379,061
313.423	0.001	319,057.6	48	155	$5^4G_{11/2}$	30,662	$(4^3H)^4I_{9/2}$	349,721
313.522		318,957.4	32	150	$3^2G_{7/2}$	79,705	$(2^3F)^2F_{5/2}$	398,662
313.599	0.000	318,879.0	9	82	$3^2H_{9/2}$	53,797	$(4^3H)^2H_{11/2}$	372,676
313.641	−0.006	318,835.7	9	72	$3^4F_{5/2}$	51,049	$(4^3P)^4D_{7/2}$	369,878
313.957	0.009	318,514.8	17	92	$5^2I_{11/2}$	44,011	$(4^3H)^2I_{11/2}$	362,535
313.981 [d]	0.003	318,490.5	17	147	$5^2G_{9/2}$	59,223	$(4^1G)^2H_{11/2}$	377,717
313.981 [d]	−0.003	318,490.5	17	165	$5^2D_{3/2}$	48,086	$(4^3F)^4F_{3/2}$	366,573
314.013	−0.001	318,458.5	66	339	$3^2H_{11/2}$	57,962	$(4^1I)^2I_{13/2}$	376,419
314.152	0.003	318,316.8	7	60	$3^2F_{7/2}$	53,353	$(4^3G)^4G_{7/2}$	371,673
314.212	−0.001	318,256.6	22	197	$3^4F_{3/2}$	53,796	$(4^3H)^4G_{5/2}$	372,051
314.298 [d]	−0.006	318,169.6	18	206	$1^2D_{3/2}$	104,821	$(2^1D)^2F_{5/2}$	422,985
314.298 [d]	0.000	318,169.6	18	220	$5^2G_{9/2}$	59,223	$(4^3F)^2G_{9/2}$	377,393
314.370	−0.003	318,096.9	19	75	$3^4F_{7/2}$	48,712	$(4^3F)^4F_{7/2}$	366,806
314.400	0.002	318,065.8	8	69	$5^2I_{13/2}$	45,546	$(4^3H)^4I_{13/2}$	363,614
314.545	0.000	317,919.3	19	184	$3^4F_{5/2}$	51,049	$(4^3P)^4D_{7/2}$	368,968

Table A1. *Cont.*

λ (Å) [a]	o-c, (Å) [b]	ν (cm^{-1})	I [c]	gA, (10^8 s^{-1})	$5d^5$		$5d^45p$	
					Term [e]	E (cm^{-1})	Term [e]	E (cm^{-1})
314.650	−0.010	317,813.5	9	113	$5^2G_{7/2}$	55,773	$(4^3F)^4D_{7/2}$	373,577
314.712	−0.001	317,750.9	23	81	$5^4D_{7/2}$	36,485	$(4^5D)^4D_{5/2}$	354,235
314.749	0.002	317,713.4	89	236	$5^4G_{11/2}$	30,662	$(4^5D)^6D_{9/2}$	348,377
314.797	0.000	317,665.5	23	109	$3^4P_{1/2}$	34,605	$(4^5D)^4D_{1/2}$	352,270
314.856	−0.005	317,605.8	22 VI	156	$5^2F_{7/2}$	59,792	$(4^3F)^2G_{9/2}$	377,393
315.100	−0.012	317,359.3	14	145	$5^2G_{7/2}$	55,773	$(4^3G)^4G_{9/2}$	373,120
315.192	−0.003	317,266.6	11	79	$3^2H_{9/2}$	53,797	$(4^3H)^2H_{9/2}$	371,061
315.264 [d]	0.001	317,194.8	83	181	$3^2H_{9/2}$	53,797	$(4^3F)^2F_{7/2}$	370,993
315.264 [d]	−0.004	317,194.8	83	310	$5^2G_{9/2}$	59,223	$(4^3G)^4G_{11/2}$	376,414
315.327 [d]	−0.001	317,130.8	37	284	$3^4F_{7/2}$	48,712	$(4^3F)^2G_{7/2}$	365,842
315.327 [d]	−0.005	317,130.8	37	125	$3^2D_{5/2}$	72,934	$(2^3F)^4F_{5/2}$	390,060
315.552	0.001	316,905.2	16	119	$5^2D_{5/2}$	47,119	$(4^3D)^2D_{5/2}$	364,025
315.751	−0.002	316,705.6	11	124	$3^2H_{9/2}$	53,797	$(4^3G)^2H_{9/2}$	370,501
315.758	−0.007	316,698.2	18	97	$3^2H_{9/2}$	53,797	$(4^3G)^4H_{11/2}$	370,488
315.988	0.003	316,467.3	4	41	$5^2I_{11/2}$	44,011	$(4^3H)^4I_{11/2}$	360,481
316.291	0.000	316,164.3	18	148	$5^2F_{5/2}$	57,413	$(4^3F)^4D_{7/2}$	373,577
316.384	0.010	316,071.5	9	106	$3^2H_{9/2}$	53,797	$(4^3P)^4D_{7/2}$	369,878
316.548	0.006	315,907.6	44	248	$3^4F_{9/2}$	49,104	$(4^3H)^4H_{11/2}$	365,017
316.588	−0.004	315,867.8	12	103	$3^2H_{9/2}$	53,797	$(4^1I)^2I_{11/2}$	369,661
316.780	−0.009	315,676.9	21	149	$3^4F_{9/2}$	49,104	$(4^3F)^4G_{9/2}$	364,772
316.804	−0.001	315,652.9	29	26	$3^4F_{3/2}$	53,796	$(4^3D)^4P_{3/2}$	369,448
316.843	0.002	315,613.8	83	249	$5^2G_{9/2}$	59,223	$(4^3H)^4G_{7/2}$	374,839
317.044	−0.002	315,413.3	8	46	$3^2F_{5/2}$	59,954	$(4^3G)^2G_{7/2}$	375,365
317.146	−0.008	315,312.5	4	18	$5^4G_{5/2}$	29,389	$(4^5D)^4F_{5/2}$	344,697
317.305	−0.001	315,154.6	126	362	$5^4G_{9/2}$	30,907	$(4^5D)^4F_{7/2}$	346,061
317.408	−0.005	315,051.9	14	200	$5^2F_{7/2}$	59,792	$(4^3H)^4G_{7/2}$	374,839
317.572	0.000	314,889.0	15	31	$5^2D_{3/2}$	48,086	$(4^3P)^4D_{5/2}$	362,975
317.594	−0.003	314,867.0	26	65	$5^4D_{1/2}$	38,685	$(4^5D)^4D_{3/2}$	353,549
317.710	−0.001	314,752.3	20	62	$3^2D_{3/2}$	71,517	$(4^1D)^2D_{5/2}$	386,268
317.735	0.000	314,728.1	24	91	$5^2G_{7/2}$	55,773	$(4^3G)^2H_{9/2}$	370,501
317.823	−0.008	314,640.2	31	135	$3^2H_{9/2}$	53,797	$(4^3G)^4H_{9/2}$	368,429
318.013	−0.005	314,452.3	30 VI	156	$5^4D_{3/2}$	39,788	$(4^5D)^4D_{5/2}$	354,235
318.174	0.001	314,293.4	30	125	$5^2D_{3/2}$	48,086	$(4^3H)^4G_{5/2}$	362,381
318.202	−0.003	314,265.9	19	92	$3^4F_{7/2}$	48,712	$(4^3P)^4D_{5/2}$	362,975
318.307	0.001	314,162.5	17	224	$3^2G_{9/2}$	79,131	$(2^3F)^4G_{7/2}$	393,295
318.411	−0.004	314,059.9	11	37	$3^4P_{5/2}$	32,005	$(4^5D)^4F_{7/2}$	346,061
318.791 [d]	−0.004	313,685.3	9	200	$3^2D_{5/2}$	72,934	$(4^1F)^2G_{7/2}$	386,615
318.791 [d]	0.001	313,685.3	9	53	$3^2H_{9/2}$	53,797	$(4^3F)^4G_{11/2}$	367,483
318.832	0.003	313,645.3	26	135	$3^2D_{5/2}$	72,934	$(4^1S)^2P_{3/2}$	386,582
319.143 [d]	−0.005	313,338.9	14	126	$3^2D_{5/2}$	72,934	$(4^1D)^2D_{5/2}$	386,268
319.143 [d]	0.004	313,338.9	14	184	$3^4F_{9/2}$	49,104	$(4^3G)^4F_{9/2}$	362,447
319.273	−0.002	313,211.9	14	46	$5^2D_{3/2}$	48,086	$(4^3F)^4F_{5/2}$	361,296
319.534	0.002	312,955.9	102	256	$5^4G_{7/2}$	30,378	$(4^5D)^4F_{5/2}$	343,336
319.902	−0.003	312,595.6	65	194	$5^4G_{5/2}$	29,390	$(4^5D)^4F_{3/2}$	341,983
320.035	−0.001	312,465.5	12	57	$5^2F_{5/2}$	57,413	$(4^3P)^4D_{7/2}$	369,878
320.251	0.004	312,255.5	14	30	$5^2F_{7/2}$	59,792	$(4^3H)^4G_{5/2}$	372,051
320.366	0.000	312,142.8	18	43	$5^2I_{13/2}$	45,546	$(4^3H)^4G_{11/2}$	357,689
320.625	0.001	311,891.2	28	89	$5^4D_{7/2}$	36,485	$(4^5D)^6D_{9/2}$	348,377
320.752 [d]	0.003	311,767.4	14	62	$5^2G_{9/2}$	59,223	$(4^3F)^2F_{7/2}$	370,993
320.752 [d]	−0.005	311,767.4	9	248	$3^2D_{3/2}$	71,517	$(4^3D)^2D_{3/2}$	383,280
320.823 [d]	0.000	311,698.5	25	82	$3^2H_{11/2}$	57,962	$(4^1I)^2I_{11/2}$	369,661
320.823 [d]	−0.008	311,698.5	25	216	$3^2G_{7/2}$	79,705	$(2^3F)^4G_{5/2}$	391,396
320.900	0.003	311,623.9	60	163	$5^4G_{11/2}$	30,662	$(4^5D)^6D_{9/2}$	342,289
321.154	0.001	311,377.5	15 VI	66	$3^4F_{9/2}$	49,104	$(4^3F)^4G_{9/2}$	360,482
321.198	−0.003	311,334.3	26	136	$3^4P_{5/2}$	32,005	$(4^5D)^4F_{5/2}$	343,336
321.257	0.001	311,277.1	17	19	$5^2G_{9/2}$	59,223	$(4^3G)^2H_{9/2}$	370,501
321.312	−0.004	311,224.0	18	42	$3^2H_{9/2}$	53,797	$(4^3H)^4H_{11/2}$	365,017

Table A1. *Cont.*

λ (Å) [a]	o-c, (Å) [b]	ν (cm^{-1})	I [c]	gA, (10^8 s^{-1})	5d^5		5d^{4}5p	
					Term [e]	E (cm^{-1})	Term [e]	E (cm^{-1})
322.007	−0.001	310,552.4	41	91	$5^4G_{9/2}$	30,907	$(4^5D)^6D_{7/2}$	341,458
322.338		310,233.1	29	579	$1^2D_{5/2}$	103,996	$(2^3F)^2D_{5/2}$	414,230
322.612	0.002	309,969.4	8	57	$3^4P_{5/2}$	32,005	$(4^5D)^4F_{3/2}$	341,976
322.852	0.006	309,739.4	13	74	$5^2G_{9/2}$	59,223	$(4^3P)^4D_{7/2}$	368,968
323.598	0.007	309,025.3	12 VI	118	$3^2G_{9/2}$	79,131	$(4^1F)^2G_{7/2}$	388,163
323.626	0.000	308,998.8	11	5	$5^2G_{7/2}$	55,773	$(4^3F)^4G_{9/2}$	364,772
323.721	−0.001	308,907.5	17	87	$5^2D_{3/2}$	48,086	$(4^3F)^2D_{3/2}$	356,993
323.897	−0.002	308,740.0	22	204	$3^2H_{9/2}$	53,797	$(4^3H)^2I_{11/2}$	362,535
324.055	−0.005	308,589.8	25	199	$3^4F_{9/2}$	49,104	$(4^3H)^4G_{11/2}$	357,689
324.404	0.003	308,257.6	26 VI	79	$5^2G_{9/2}$	59,223	$(4^3F)^4G_{11/2}$	367,483
324.427	−0.003	308,235.7	82	59	$3^4P_{3/2}$	32,994	$(4^5D)^6D_{1/2}$	341,227
324.458	0.003	308,206.7	38	87	$5^4D_{7/2}$	36,485	$(4^5D)^4F_{5/2}$	344,695
324.718	0.002	307,959.3	19	194	$3^2D_{3/2}$	71,517	$(4^3G)^2F_{5/2}$	379,478
325.135	−0.005	307,564.4	5	92	$3^2D_{3/2}$	71,516	$(4^1F)^2F_{5/2}$	379,077
325.170	0.003	307,531.5	15	74	$3^2D_{5/2}$	72,934	$(4^1F)^2F_{7/2}$	380,468
325.319	−0.001	307,390.2	32	161	$3^4F_{9/2}$	49,104	$(4^3F)^4D_{7/2}$	356,493
325.829	0.001	306,909.1	5	51	$3^2G_{7/2}$	79,705	$(4^1F)^2G_{7/2}$	386,615
325.943	0.008	306,802.0	13	4	$3^2H_{11/2}$	57,962	$(4^3F)^4G_{9/2}$	364,772
326.139	0.004	306,618.0	17	19	$3^4P_{1/2}$	34,605	$(4^5D)^6D_{1/2}$	341,227
326.204	−0.001	306,556.6	26 VI	122	$3^2H_{11/2}$	57,962	$(4^3H)^2I_{13/2}$	364,518
326.519	−0.004	306,260.7	9	129	$3^2G_{7/2}$	79,705	$(4^3D)^2F_{5/2}$	385,962
326.609	0.002	306,176.4	13	12	$3^4P_{3/2}$	32,994	$(4^5D)^6D_{5/2}$	339,172
326.915 [d]	0.000	305,890.1	24	41	$3^2H_{9/2}$	53,797	$(4^3G)^4G_{7/2}$	359,687
326.915 [d]	0.003	305,890.1	24	318	$1^2D_{5/2}$	103,996	$(2^1G)^2F_{7/2}$	409,889
327.004	−0.003	305,806.4	11	20	$5^4D_{7/2}$	36,485	$(4^5D)^6D_{9/2}$	342,289
327.048	0.000	305,765.2	39	219	$3^4F_{9/2}$	49,104	$(4^5D)^4D_{7/2}$	354,869
327.307	−0.001	305,523.8	38 VI	119	$3^4F_{7/2}$	48,712	$(4^5D)^4D_{5/2}$	354,235
327.438	−0.002	305,401.0	31	133	$5^4D_{5/2}$	39,299	$(4^5D)^4F_{5/2}$	344,697
327.850	0.001	305,017.1	8	20	$3^4F_{9/2}$	49,104	$(4^3H)^4H_{9/2}$	354,122
328.555	0.002	304,362.8	31 VI	228	$1^2D_{3/2}$	104,821	$(2^1G)^2F_{5/2}$	409,186
329.712	0.003	303,295.2	12	36	$5^4D_{1/2}$	38,685	$(4^5D)^4F_{3/2}$	341,983
329.830	0.000	303,186.1	21	94	$3^4F_{5/2}$	51,049	$(4^5D)^4D_{5/2}$	354,235
330.391	0.000	302,672.1	16	202	$3^2G_{7/2}$	79,705	$(4^1I)^2H_{9/2}$	382,377
330.560	0.004	302,516.8	15	50	$3^2H_{11/2}$	57,962	$(4^3F)^4G_{9/2}$	360,482
332.140	−0.006	301,077.9	12	65	$3^2H_{9/2}$	53,797	$(4^5D)^4D_{7/2}$	354,869
332.353	−0.003	300,885.1	11	69	$3^2F_{7/2}$	53,353	$(4^5D)^4D_{5/2}$	354,235
332.485 [d]	−0.003	300,765.3	12	133	$3^2G_{7/2}$	79,705	$(4^1F)^2F_{7/2}$	380,468
332.485 [d]	0.004	300,765.3	12	76	$3^2F_{7/2}$	53,353	$(4^3H)^4H_{9/2}$	354,122
333.198	−0.001	300,121.8	4	266	$1^2D_{5/2}$	103,996	$(2^3F)^2F_{7/2}$	404,117
334.140	−0.003	299,275.8	12	52	$3^4F_{9/2}$	49,104	$(4^5D)^6D_{9/2}$	348,377
334.911	−0.001	298,587.1	1	53	$3^2G_{9/2}$	79,131	$(4^1G)^2H_{11/2}$	377,717
338.225	−0.001	295,661.2	15	92	$3^2G_{7/2}$	79,705	$(4^3G)^2G_{7/2}$	375,365
341.089	0.008	293,178.3	20	30	$3^4F_{9/2}$	49,104	$(4^5D)^6D_{9/2}$	342,289
342.628	0.000	291,861.7	30	9	$3^2G_{9/2}$	79,131	$(4^3F)^2F_{7/2}$	370,993

[a] Observed wavelengths, d—doubly identified, t—trebly identified; [b] Difference between the observed wavelength and the wavelength derived from the final level energies (Ritz wavelength). A blank value indicates that the upper level is derived only from that line; [c] Relative intensity; VI—line is also identified as Ag VI; [e] The number preceding the terms is seniority number.

Table A2. Identified lines of the $4d^4$–$4d^35p$ transitions in the spectrum of Ag^{7+}.

λ (Å) [a]	o-c, (Å) [b]	ν (cm^{-1})	I [c]	gA, (10^8 s^{-1})	5d^4		5d^{3}5p	
					Term [e]	E (cm^{-1})	Term [f]	E (cm^{-1})
247.539	−0.005	403,976.8	56	150	4^3H_6	28,185	$(3^2F)^3G_5$	432,154
253.534	−0.002	394,424.4	54	114	4^3H_4	23,302	$(3^2H)^3G_3$	417,724
253.627	0.001	394,279.8	98	383	2^3F_4	60,980	$(1^2D)^3F_4$	455,261
253.737	−0.001	394,108.9	60	284	4^5D_3	5292	$(3^4P)^5P_3$	399,399

Table A2. *Cont.*

λ (Å) [a]	o-c, (Å) [b]	ν (cm^{-1})	I [c]	gA, (10^8 s^{-1})	5d^4		5d^{3}5p	
					Term [e]	E (cm^{-1})	Term [f]	E (cm^{-1})
254.226	−0.003	393,350.8	20	174	4^1G_4	43,104	$(3^2F)^1F_3$	436,451
254.451	0.001	393,003.0	23	75	4^3H_6	28,185	$(3^2H)^1I_6$	421,189
254.641	0.000	392,709.8	17	41	2^3F_3	62,552	$(1^2D)^3F_4$	455,261
255.131	0.000	391,955.5	550	1307	4^5D_4	7443	$(3^4P)^5P_3$	399,399
255.214		391,828.0	140	596	2^3F_4	60,980	$(1^2D)^3D_3$	452,808
255.570	0.001	391,282.2	28	99	4^5D_0	0	$(3^4P)^5P_1$	391,283
255.891	0.000	390,791.4	38	78	2^3F_3	62,552	$(1^2D)^1D_2$	453,343
255.891		390,791.4	38	210	4^3G_3	26,967	$(3^2D)^3D_2$	417,758
255.919		390,748.6	115	455	4^5D_3	5292	$(3^4P)^5P_2$	396,041
256.069		390,519.8	61	435	2^1G_4	69,584	$(1^2D)^1F_3$	460,104
256.138	0.002	390,414.5	49	231	4^5D_2	3212	$(3^4P)^5P_2$	393,630
256.446	−0.002	389,945.7	70	320	4^5D_1	1,340	$(3^4P)^5P_1$	391,283
256.757		389,473.3	180	347	4^3H_6	28,185	$(3^2H)^3I_7$	417,658
257.391	−0.001	388,514.0	110	354	4^5D_4	7443	$(3^4F)^5F_5$	395,955
257.505	−0.003	388,342.0	57	314	4^5D_3	5292	$(3^4P)^5P_2$	393,630
257.603	0.003	388,194.3	35	194	4^3G_3	26,967	$(3^4P)^3D_3$	415,165
257.627	0.006	388,158.1	160d	114	4^3H_4	23,302	$(3^2D)^3D_3$	411,469
257.627	0.007	388,158.1	160d	567	4^3F_4	29,555	$(3^2H)^3G_3$	417,724
257.645	−0.004	388,131.0	83	482	4^3G_4	32,698	$(3^2H)^3G_4$	420,822
257.685	0.000	388,070.7	260	437	4^5D_2	3212	$(3^4P)^5P_1$	391,283
257.740	0.001	387,987.9	33	230	4^3P_2	32,134	$(3^2F)^3F_3$	420,123
257.796	−0.002	387,903.6	43	172	4^5D_3	5292	$(3^4F)^5F_4$	393,193
257.947	−0.001	387,676.5	68	289	4^5D_2	3212	$(3^4F)^5F_3$	390,887
258.332	0.004	387,098.8	26	165	4^3F_3	32926	$(3^2P)^3P_2$	420,031
258.347	0.002	387,076.3	29	107	4^3H_5	26,250	$(3^2G)^1H_5$	413,329
258.583	0.003	386,723.0	120	492	4^3H_5	26,250	$(3^2F)^1G_4$	412,977
258.855		386,316.7	22	82	4^5D_0	0	$(3^4F)^5F_1$	386,317
259.104	−0.001	385,945.4	170	944	4^3G_5	35,077	$(3^2H)^3G_5$	421,021
259.192	0.000	385,814.4	130	320	4^3H_6	28,185	$(3^2H)^3I_6$	413,999
259.239	0.000	385,744.4	60d	412	4^3G_5	35,077	$(3^2H)^3G_4$	420,822
259.239	0.004	385,744.4	60d	259	4^5D_4	7443	$(3^4F)^5F_4$	393,193
259.283	−0.001	385,679.0	56	106	2^1G_4	69,584	$(1^2D)^3F_4$	455,261
259.342	0.001	385,591.2	71d	163	4^3P_1	26,526	$(3^2D)^3D_1$	412,118
259.342	0.011	385,591.2	71d	125	4^3H_4	23,302	$(3^2G)^3H_5$	408,910
259.432	0.002	385,457.5	39	120	4^5D_1	1340	$(3^4F)^5F_2$	386,800
259.643	0.000	385,144.2	120	672	4^3H_6	28,185	$(3^2G)^1H_5$	413,329
259.878	−0.001	384,795.9	110d	155	4^3G_4	32,698	$(3^2D)^3F_4$	417,492
259.878	0.001	384,795.9	110d	680	4^3F_3	32,926	$(3^2H)^3G_3$	417,724
259.978	0.003	384,647.9	130	554	4^5D_3	5292	$(3^4F)^3G_3$	389,944
260.115	0.001	384,445.3	37	337	4^1F_3	52,004	$(3^2F)^1F_3$	436,451
260.280	0.002	384,201.6	180	746	4^3H_4	23,302	$(3^2G)^3F_3$	407,506
260.376	0.005	384,060.0	32	172	4^3F_2	28,051	$(3^2D)^3D_1$	412,118
260.443	0.001	383,961.2	81	398	4^5D_1	1340	$(3^4F)^5D_2$	385,302
260.575	0.000	383,766.7	41	302	2^3F_3	62,552	$(1^2D)^3D_3$	446,318
260.695	−0.001	383,590.0	170	198	4^5D_2	3212	$(3^4F)^5F_2$	386,800
260.709	−0.001	383,569.4	130	319	4^5D_0	0	$(3^4F)^5D_1$	383,568
260.989	−0.004	383,157.9	59	220	4^3D_3	36,879	$(3^2P)^3P_2$	420,031
261.086	0.000	383,015.6	290	561	4^5D_4	7443	$(3^4F)^5G_5$	390,458
261.173	0.000	382,888.0	32	207	2^3F_2	61,644	$(1^2D)^3D_2$	444,532
261.247	0.002	382,779.5	26	75	4^3G_3	26,967	$(3^2G)^1F_3$	409,750
261.323	−0.001	382,668.2	1000	1735	4^3H_6	28,185	$(3^2G)^3H_6$	410,851
261.328	0.000	382,660.9	958	1568	4^3H_5	26,250	$(3^2G)^3H_5$	408,910
261.400		382,555.5	94	486	4^3G_3	26,967	$(3^2D)^3F_2$	409,522
261.458	−0.002	382,470.6	99	547	4^3G_4	32,698	$(3^4P)^3D_3$	415,165
261.496	0.000	382,415.0	590	1655	4^3G_5	35,077	$(3^2D)^3F_4$	417,492
261.625	0.001	382,226.5	37	267	4^5D_1	1340	$(3^4F)^5D_1$	383,568

Table A2. *Cont.*

λ (Å) [a]	o-c, (Å) [b]	ν (cm^{-1})	I [c]	gA, (10^8 s^{-1})	$5d^4$		$5d^35p$	
					Term [e]	E (cm^{-1})	Term [f]	E (cm^{-1})
261.718	0.000	382,090.7	250	720	4^5D_2	3212	$(3^4F)^5D_2$	385,302
261.752	−0.003	382,041.0	490	1121	4^5D_3	5292	$(3^4F)^5F_3$	387,329
261.793	0.000	381,981.2	65d	264	2^3F_3	62,552	$(1^2D)^3D_2$	444,532
261.793	0.003	381,981.2	65d	399	4^3H_5	26,250	$(3^4P)^5D_4$	408,236
261.836	−0.003	381,918.4	160	695	4^3F_4	29,555	$(3^2D)^3D_3$	411,469
262.106	−0.003	381,525.0	420	950	4^5D_4	7443	$(3^4F)^5F_4$	388,964
262.160	−0.001	381,446.4	670	1886	4^1I_6	39,744	$(3^2H)^1I_6$	421,189
262.276	−0.001	381,277.8	620	2055	4^1I_6	39,744	$(3^2H)^3G_5$	421,021
262.323	0.000	381,209.4	110	397	4^3P_2	32,134	$(3^2P)^3P_1$	413,344
262.340		381,184.7	100	468	4^5D_1	1340	$(3^4F)^5D_0$	382,525
262.520	−0.004	380,923.3	280d	359	4^3G_5	35,077	$(3^2H)^1H_5$	415,995
262.520		380,923.3	280d	492	4^3H_5	26,250	$(3^4P)^5D_4$	407,173
262.535		380,901.6	490	1502	4^1G_4	43,104	$(3^2P)^3D_3$	424,006
262.653	−0.004	380,730.5	27	162	4^3H_6	28,185	$(3^2G)^3H_5$	408,910
262.787	0.002	380,536.3	51	240	4^3G_3	26,967	$(3^2G)^3F_3$	407,506
262.910		380,358.3	250d	553	4^3F_2	28,051	$(3^2G)^3F_2$	408,409
262.910	−0.002	380,358.3	250d	688	4^5D_2	3212	$(3^4F)^5D_1$	383,568
262.966	0.001	380,277.3	50	392	4^3G_4	32,698	$(3^2F)^1G_4$	412,977
263.020	−0.003	380,199.2	87	269	4^3F_4	29,555	$(3^2G)^1F_3$	409,750
263.150	−0.001	380,011.4	130	687	4^5D_3	5292	$(3^4F)^5D_2$	385,302
263.212	0.003	379,921.9	370	1258	4^5D_3	5292	$(3^4F)^5D_4$	385,218
263.240	0.003	379,881.5	73	306	4^5D_4	7443	$(3^4F)^5F_3$	387,329
263.303		379,790.6	46	267	4^1G_4	43,104	$(3^2F)^3F_3$	422,895
263.336	0.001	379,743.0	35	303	4^3P_1	26,526	$(3^2D)^3D_2$	406,270
263.510	−0.004	379,492.3	250	1114	4^3F_3	32,926	$(3^2P)^3D_2$	412,412
263.811	0.001	379,059.3	520	1894	4^3H_6	28,185	$(3^2H)^3I_5$	407,246
263.914	−0.012	378,911.3	500t	372	4^3H_5	26,250	$(3^2H)^3H_5$	405,144
263.914	0.007	378,911.3	500 t	279	4^3G_5	35,077	$(3^2H)^3I_6$	413,999
263.914	0.001	378,911.3	500 t	1145	4^3H_5	26,250	$(3^2G)^3G_4$	405,162
264.072	−0.003	378,684.6	84	627	4^3F_4	29,555	$(3^4P)^5D_4$	408,236
264.159		378,559.9	140	338	4^3P_0	21,309	$(3^2P)^3P_1$	399,869
264.293	0.000	378,367.9	470	1252	4^5D_2	3212	$(3^4F)^5D_3$	381,580
264.374	0.000	378,252.0	50	384	4^3G_5	35,077	$(3^2G)^1H_5$	413,329
264.409	−0.005	378,201.9	33	130	4^3G_3	26,967	$(3^2G)^3G_4$	405,162
264.546	−0.001	378,006.1	23	208	4^3P_1	26,526	$(3^2P)^3D_1$	404,530
264.581	−0.004	377,956.1	96	406	4^3F_4	29,555	$(3^2G)^3F_3$	407,506
264.705	−0.003	377,779.0	680	2033	4^5D_4	7443	$(3^4F)^5D_4$	385,218
264.752	0.004	377,712.0	65	753	4^1G_4	43,104	$(3^2H)^3G_4$	420,822
264.769	0.002	377,687.7	180	415	4^3F_4	29,555	$(3^2H)^3I_5$	407,246
264.802		377,640.7	73	288	2^1D_2	88,586	$(1^2D)^1P_1$	466,227
265.169	−0.006	377,118.0	83	467	4^3H_5	26,250	$(3^2H)^3H_6$	403,359
265.283	0.002	376,955.9	340	936	4^3H_6	28,185	$(3^2H)^3H_5$	405,144
265.439		376,734.4	320d	1405	4^3G_4	32,698	$(3^2D)^3F_3$	409,432
265.439	0.000	376,734.4	320d	317	2^1G_4	69,584	$(1^2D)^3D_3$	446,318
265.620	0.001	376,477.7	78	366	4^3F_2	28,051	$(3^2P)^3D_1$	404,530
265.728	−0.002	376,324.7	390	725	4^5D_1	1340	$(3^4F)^3D_2$	377,661
265.782	0.002	376,248.2	500	1790	4^1I_6	39,744	$(3^2H)^1H_5$	415,995
265.948	0.001	376,013.3	260	394	4^3D_3	36,879	$(3^4P)^3D_2$	412,894
265.977	−0.007	375,972.3	440d	1413	4^3H_4	23,302	$(3^2G)^1G_4$	399,265
265.977	0.008	375,972.3	440d	291	4^5D_4	7443	$(3^4F)^5G_5$	383,426
266.238	0.002	375,603.8	170	816	4^3F_4	29,555	$(3^2G)^3G_4$	405,162
266.285	0.001	375,537.5	140	292	4^3G_4	32,698	$(3^4P)^5D_4$	408,236
266.384	0.003	375,397.9	420	1410	4^3H_4	23,302	$(3^2G)^3G_3$	398,704
266.545	0.002	375,171.2	320	831	4^3H_6	28,185	$(3^2H)^3H_6$	403,359
266.744		374,891.3	77	345	4^5D_0	0	$(3^4F)^3D_1$	374,891

Table A2. *Cont.*

λ (Å) [a]	o-c, (Å) [b]	ν (cm^{-1})	I [c]	gA, (10^8 s^{-1})	5d^4		5d^{3}5p	
					Term [e]	E (cm^{-1})	Term [f]	E (cm^{-1})
266.964	0.006	374,582.3	430t	921	4^3D_3	36,879	$(3^2D)^3D_3$	411,469
266.964		374,582.3	430t	529	4^3D_3	36,879	$(3^4P)^5S_2$	411,461
266.964	−0.002	374,582.3	430t	762	4^3F_3	32,926	$(3^2G)^3F_3$	407,506
267.197	0.000	374,255.7	120	277	4^1I_6	39,744	$(3^2H)^3I_6$	413,999
267.281	−0.001	374,138.1	250d	602	4^3P_2	32,134	$(3^2D)^3D_2$	406,270
267.281	−0.001	374,138.1	250d	326	4^5D_4	7443	$(3^4F)^5D_3$	381,580
267.426	0.001	373,935.2	130	462	4^3G_3	26,967	$(3^4P)^5D_2$	400,904
267.676	−0.001	373,586.0	460d	1579	4^1I_6	39,744	$(3^2G)^1H_5$	413,329
267.676	−0.003	373,586.0	460d	310	4^3H_5	26,250	$(3^4F)^3G_5$	399,831
267.942	0.004	373,215.1	28	222	4^3D_1	39,191	$(3^2P)^3D_2$	412,412
268.145	−0.004	372,932.6	48	324	4^3D_1	39,191	$(3^2D)^3D_1$	412,118
268.201	−0.001	372,854.7	81	412	4^3F_2	28,051	$(3^4P)^5D_2$	400,904
268.347	0.001	372,651.8	120	332	4^3H_4	23,302	$(3^4F)^5F_5$	395,955
268.362		372,631.0	250	1228	4^1F_3	52,004	$(3^2D)^1D_2$	424,635
268.487	−0.008	372,457.5	48d	274	4^3G_4	32,698	$(3^2H)^3H_5$	405,144
268.487	0.005	372,457.5	48d	205	4^3G_4	32,698	$(3^2G)^3G_4$	405,162
268.553	0.002	372,366.0	24	110	4^5D_3	5292	$(3^4F)^3D_2$	377,661
268.597	−0.005	372,305.0	28	109	4^3G_3	26,967	$(3^2G)^1G_4$	399,265
268.648	0.001	372,234.3	29	308	4^3F_3	32,926	$(3^2G)^3G_4$	405,162
268.691	−0.004	372,174.7	31	270	4^3G_5	35,077	$(3^2H)^3I_5$	407,246
269.009	0.002	371,734.8	71	618	4^3G_3	26,967	$(3^2G)^3G_3$	398,704
269.074	0.001	371,645.0	390	1249	4^3H_6	28,185	$(3^4F)^3G_5$	399,831
269.420	0.004	371,167.7	24	201	2^3F_4	60,980	$(3^2F)^3G_5$	432,154
269.465	0.001	371,105.7	41	164	4^1I_6	39,744	$(3^2G)^3H_6$	410,851
269.790	−0.004	370,658.7	33	364	4^3F_2	28,051	$(3^2G)^3G_3$	398,704
270.102	0.001	370,230.5	350d	995	4^3H_5	26,250	$(3^4F)^3G_4$	396,481
270.102	−0.004	370,230.5	350d	281	4^1G_4	43,104	$(3^2G)^1H_5$	413,329
270.227	0.006	370,059.3	50	255	4^3G_5	35,077	$(3^2H)^3H_5$	405,144
270.265		370,007.2	37	356	4^3D_2	38,402	$(3^2G)^3F_2$	408,409
270.358		369,879.9	290d	395	2^3F_3	62,552	$(3^2F)^3D_2$	432,431
270.358	−0.005	369,879.9	290d	858	4^1G_4	43,104	$(3^2F)^1G_4$	412,977
270.518		369,661.2	32	330	2^3F_4	60,980	$(3^2F)^3D_3$	430,642
270.958		369,060.9	110	382	2^3P_1	63,371	$(3^2F)^3D_2$	432,431
271.536	0.005	368,275.3	37	205	4^3G_5	35,077	$(3^2H)^3H_6$	403,359
271.933		367,737.7	270	1089	2^1G_4	69,584	$(3^2F)^1G_4$	437,322
272.381	0.000	367,132.8	50	336	4^3G_4	32,698	$(3^4F)^3G_5$	399,831
272.579	0.001	366,866.1	150	1131	2^1G_4	69,584	$(3^2F)^1F_3$	436,451
272.743	−0.003	366,645.5	99	473	4^3H_4	23,302	$(3^4F)^3G_3$	389,944
272.808	0.007	366,558.2	81	500	4^3G_4	32,698	$(3^2G)^1G_4$	399,265
273.055	0.000	366,226.6	48	307	4^3G_3	26,967	$(3^4F)^5F_4$	393,193
273.866	0.000	365,142.1	200	595	4^3G_5	35,077	$(3^4F)^3F_4$	400,219
273.912		365,080.8	43	325	4^3G_4	32,698	$(3^4F)^3F_3$	397,779
273.967		365,007.5	15	109	4^3P_0	21,309	$(3^4F)^5F_1$	386,317
274.155	−0.002	364,757.2	160 d	415	4^3G_5	35,077	$(3^4F)^3G_5$	399,831
274.155	0.000	364,757.2	160 d	663	2^1D_2	88,586	$(1^2D)^1D_2$	453,343
274.705	0.000	364,026.9	31	200	4^3H_4	23,302	$(3^4F)^5F_3$	387,329
275.061	−0.001	363,555.7	27	292	4^3F_3	32,926	$(3^4F)^3G_4$	396,481
275.224	0.000	363,340.4	41	454	4^3D_3	36,879	$(3^4F)^3F_4$	400,219
275.408		363,097.7	24	301	2^3P_1	63,371	$(3^4P)^3S_1$	426,468
275.702	0.003	362,710.5	35	243	4^3H_5	26,250	$(3^4F)^5F_4$	388,964
275.902		362,447.5	22	229	2^3F_3	62,552	$(3^2F)^3F_4$	424,999
276.325	0.000	361,892.7	18	175	4^3F_2	28,051	$(3^4F)^3G_3$	389,944
276.754	0.000	361,331.7	22	251	4^3F_4	29,555	$(3^4F)^5F_3$	390,887
277.091	−0.001	360,892.3	24	140	4^1F_3	52,004	$(3^4P)^3D_2$	412,894
277.749	0.003	360,037.3	110 d	376	2^3F_4	60,980	$(3^2H)^3G_5$	421,021
277.749		360,037.3	110 d	594	4^1F_3	52,004	$(3^2H)^1G_4$	412,041

Table A2. *Cont.*

λ (Å) [a]	o-c, (Å) [b]	ν (cm^{-1})	I [c]	gA, (10^8 s^{-1})	5d^4		5d^{3}5p	
					Term [e]	E (cm^{-1})	Term [f]	E (cm^{-1})
278.440		359,143.8	24 d	390	2^1D_2	88,586	$(1^2D)^3F_2$	447,730
278.440	−0.001	359,143.8	20 d	237	2^3F_4	60,980	$(3^2F)^3F_3$	420,123
279.517	0.000	357,760.0	17	34	4^3G_4	32,698	$(3^4F)^5G_5$	390,458
280.308	−0.005	356,750.4	22	285	2^3F_4	60,980	$(3^2H)^3G_3$	417,724
280.775	0.003	356,157.1	17	188	4^1G_4	43,104	$(3^2G)^1G_4$	399,265
281.501	0.002	355,238.5	27	130	4^3H_6	28,185	$(3^4F)^5G_5$	383,426
284.333	0.000	351,700.3	12	73	2^3F_2	61,644	$(3^2P)^3P_1$	413,344
284.791	0.000	351,134.7	17	147	4^3P_1	26,526	$(3^4F)^3D_2$	377,661
287.065	−0.003	348,353.2	18	73	4^3G_5	35,077	$(3^4F)^5G_5$	383,426
287.468	0.000	347,864.8	17	227	2^1D_2	88,586	$(3^2F)^1F_3$	436,451

[a] Observed wavelengths, d—doubly identified, t—trebly identified; [b] Difference between the observed wavelength and the wavelength derived from the final level energies (Ritz wavelength). A blank value indicates that the upper level is derived only from that line; [c] Relative intensity; [e] The number preceding the terms is seniority number; [f] Term attribution is arbitrary in a few cases (see text) for the level composition, see Table A8. The number preceding the terms of the 5d^3 configuration is seniority number.

Table A3. Identified lines of the 4d^4–(4d^{3}4f + 4p^{5}4d^5) transitions in the spectrum of Ag^{7+}.

λ, Å [a]	o-c, (Å) [b]	ν (cm^{-1})	I [c]	gA, (10^9 s^{-1})	4d^4		(4d^{3}4f + 4p^{5}4d^5)		
					Term [e]	E (cm^{-1})	Config. [f]	Term [f]	E (cm^{-1})
162.528	0.001	615,277	32	394	$^3G_4$	32,698	4p^{5}4d^5	$(^4F)^3F_3$	647,982
162.554	−0.001	615,182	26	564	$^3G_5$	35,078	4p^{5}4d^5	$(^4F)^3F_4$	650,256
164.321	−0.001	608,564	54	1260	$^3H_5$	26,249	4p^{5}4d^5	$(^4G)^3G_4$	634,810
164.542	0.001	607,749	534 IX	4247	$^3H_6$	28,185	4p^{5}4d^5	$(^4G)^3G_5$	635,935
165.158	0.004	605,482	122	567	$^3F_42$	29,555	4p^{5}4d^5	$(^4G)^3G_3$	635,050
165.481		604,300	294	5122	$^1I_6$	39,744	4p^{5}4d^5	$(^2H)^1H_5$	644,043
166.744	0.003	599,721	355 IX	2821	$^3G_5$	35,078	4p^{5}4d^5	$(^4G)^3G_4$	634,810
166.917	0.006	599,099	72	1966	$^1G_41$	69,585	4p^{5}4d^5	$(^2G1)^1F_3$	668,708
167.156	−0.005	598,244	693	1429	$^3F_42$	29,555	4p^{5}4d^5	$(^4P)^3D_3$	627,779
167.460		597,158	190 IX	2888	$^5D_4$	7,442	4p^{5}4d^5	$(^6S)^5P_3$	604,600
167.731	0.000	596,193	111	275	$^1I_6$	39,744	4p^{5}4d^5	$(^4G)^3G_5$	635,935
167.835	−0.003	595,824	83	1261	$^3H_5$	26,249	4p^{5}4d^5	$(^4D)^3F_4$	622,060
168.436	0.000	593,698	254 IX	827	$^3G_3$	26,968	4p^{5}4d^5	$(^4G)^3F_2$	620,664
168.741	−0.003	592,625	20	509	$^3F_22$	28,051	4p^{5}4d^5	$(^4G)^3F_2$	620,664
168.932	−0.002	591,953	109 IX	2031	$^1G_42$	43,105	4p^{5}4d^5	$(^4G)^3G_3$	635,050
169.010	−0.003	591,683	24	591	$^5D_2$	3,212	4p^{5}4d^5	$(^6S)^5P_2$	594,881
169.235	0.002	590,894	37	432	$^3D_3$	36,878	4p^{5}4d^5	$(^4P)^3D_3$	627,779
169.613	0.003	589,578	72 IX	831	$^5D_3$	5292	4p^{5}4d^5	$(^6S)^5P_2$	594,881
169.676	0.001	589,357	49	597	$^3G_4$	32,698	4p^{5}4d^5	$(^4D)^3F_4$	622,060
170.070		587,994	48	1080	$^3F_31$	62,552	4p^{5}4d^5	$(^4F)^3D_2$	650,545
170.271		587,298	39	513	$^1G_42$	43,105	4p^{5}4d^5	$(^2D3)^1F_3$	630,389
170.359d	0.002	586,994	312 IX	1490	$^3F_41$	60,981	4p^{5}4d^5	$(^4F)^3F_3$	647,982
170.359d	−0.003	586,994	312 IX	664	$^3G_5$	35,078	4p^{5}4d^5	$(^4D)^3F_4$	622,060
170.595	0.002	586,184	57	628	$^3G_5$	35,078	4p^{5}4d^5	$(^2I)^1I_6$	621,268
170.812	−0.003	585,439	23	881	$^3F_31$	62,552	4p^{5}4d^5	$(^4F)^3F_3$	647,982
171.075	0.001	584,539	35	643	$^3F_21$	61,645	4p^{5}4d^5	$(^4D)^3P_2$	646,185
171.339	−0.001	583,638	46	492	$^3F_31$	62,552	4p^{5}4d^5	$(^4D)^3P_2$	646,185
171.512	−0.001	583,050	34	337	$^1F_3$	52,004	4p^{5}4d^5	$(^4G)^3G_3$	635,050
171.539		582,957	383 IX	2207	$^3F_41$	60,981	4p^{5}4d^5	$(^2H)^1G_4$	643,939
171.587	0.003	582,795	27	211	$^1F_3$	52,004	4p^{5}4d^5	$(^4G)^3G_4$	634,810
171.748	0.004	582,247	20	574	$^3D_2$	38,403	4p^{5}4d^5	$(^4G)^3F_2$	620,664
171.960	−0.002	581,530	267	6420	$^1I_6$	39,744	4p^{5}4d^5	$(^2I)^1I_6$	621,268
172.171	−0.001	580,818	66	606	$^5D_1$	1340	4p^{5}4d^5	$(^6S)^5P_1$	582,154
172.215	0.000	580,671	787	2558	$^1G_41$	69,585	4p^{5}4d^5	$(^4F)^3F_4$	650,256
172.372	−0.006	580,140	61	1512	$^1D_21$	88,587	4p^{5}4d^5	$(^2G1)^1F_3$	668,708
172.460		579,846	71	1498	$^3F_21$	61,645	4p^{5}4d^5	$(^4F)^3F_2$	641,516
172.730	0.001	578,937	92	549	$^5D_2$	3212	4p^{5}4d^5	$(^6S)^5P_1$	582,154
172.968	−0.002	578,143	20	1138	$^3H_5$	26,249	4p^{5}4d^5	$(^2G1)^3G_4$	604,385
173.327	0.000	576,943	295	5775	$^3H_6$	28,185	4p^{5}4d^5	$(^2I)^3H_6$	605,128
173.525	0.000	576,287	28	323	$^3D_3$	36,878	4p^{5}4d^5	$(^4G)^3F_4$	613,166
173.682	0.003	575,766	45	1187	$^1F_3$	52,004	4p^{5}4d^5	$(^4P)^3D_3$	627,779

Table A3. *Cont.*

λ, Å [a]	o-c, (Å) [b]	ν (cm^{-1})	I [c]	gA, (10^9 s^{-1})	4d^4		(4d^{3}4f + 4p^{5}4d^5)		
					Term [e]	E (cm^{-1})	Config. [f]	Term [f]	E (cm^{-1})
174.487	−0.004	573,110	104 IX	2384	$^3H_5$	26,249	$4p^54d^5$	$(^4F)^3G_5$	599,345
174.558	−0.002	572,875	16	605	$^3H_4$	23,304	$4p^54d^5$	$(^2I)^3H_4$	596,171
174.618	−0.003	572,678	70 IX	343	$^3F_42$	29,555	$4p^54d^5$	$(^4F)^3F_4$	602,225
174.742	−0.004	572,271	40	486	$^3F_31$	62,552	$4p^54d^5$	$(^4G)^3G_4$	634,810
174.993		571,450	63	2290	$^3G_5$	35,078	$4p^54d^5$	$(^4G)^3G_5$	606,529
175.054d	−0.004	571,253	48	464	$^3G_3$	26,968	$4p^54d^5$	$(^4F)^3G_4$	598,206
175.054d	0.004	571,253	39	1651	$^1G_42$	43,105	$4p^54d^5$	$(^2G1)^1H_5$	614,370
175.086	0.004	571,149	31	1104	$^3H_6$	28,185	$4p^54d^5$	$(^4F)^3G_5$	599,345
175.419d	−0.001	570,063	30	2899	$^1G_42$	43,105	$4p^54d^5$	$(^4G)^3F_4$	613,166
175.419d	−0.004	570,063	37	946	$^3G_5$	35,078	$4p^54d^5$	$(^2I)^3H_6$	605,128
175.580	−0.004	569,541	41	722	$^3G_4$	32,698	$4p^54d^5$	$(^4F)^3F_4$	602,225
175.652	−0.001	569,306	29	470	$^5D_4$	7442	$4d^34f$	$(^4F)^3H_5$	576,746
175.855	0.000	568,651	46	1521	$^3F_42$	29,555	$4p^54d^5$	$(^4F)^3G_4$	598,206
176.130	−0.006	567,761	20	350	$^5D_3$	5292	$4d^34f$	$(^4F)^3H_4$	573,036
176.489	0.002	566,609	40	1205	$^3F_42$	29,555	$4p^54d^5$	$(^2I)^3H_4$	596,171
176.526	−0.002	566,490	18	162	$^3H_4$	23,304	$4p^54d^5$	$(^2H)^3G_4$	589,786
176.699d	−0.021	565,933	160	1440	$^3H_5$	26,249	$4p^54d^5$	$(^2H)^1H_5$	592,118
176.699d	−0.002	565,933	160	2788	$^3H_4$	23,304	$4p^54d^5$	$(^2I)^3H_4$	589,229
176.833	0.001	565,504	52	1171	$^3G_4$	32,698	$4p^54d^5$	$(^4F)^3G_4$	598,206
176.872	0.001	565,381	33	652	$^1I_6$	39,744	$4p^54d^5$	$(^2I)^3H_6$	605,128
177.156	0.000	564,474	33	906	$^3G_3$	26,968	$4p^54d^5$	$(^4F)^3G_3$	591,442
177.484	0.002	563,431	47	688	$^3P_12$	26,525	$4p^54d^5$	$(^2D1)^3P_2$	589,963
177.625	−0.001	562,983	24	435	$^3H_5$	26,249	$4p^54d^5$	$(^2I)^3H_4$	589,229
177.674	−0.003	562,827	15	151	$^3G_3$	26,968	$4p^54d^5$	$(^2H)^3G_4$	589,786
178.240w	0.013	561,041	61	685	$^3F_41$	60,981	$4p^54d^5$	$(^4D)^3F_4$	622,060
178.280	0.004	560,916	60	1099	$^3P_22$	32,137	$4p^54d^5$	$(^2F1)^3D_3$	593,064
178.292	−0.001	560,878	113	1859	$^5D_4$	7442	$4d^34f$	$(^4F)^3I_5$	568,319
178.695d		559,614	299	2756	$^5D_2$	3212	$4p^54d^5$	$(^4G)^5F_3$	562,825
178.695d	−0.001	559,614	299	3408	$^5D_3$	5292	$4p^54d^5$	$(^4F)^5D_4$	564,902
178.728	0.002	559,508	294	1956	$^5D_1$	1340	$4p^54d^5$	$(^4G)^5F_2$	560,857
178.765	0.008	559,394	50	870	$^3G_4$	32,698	$4p^54d^5$	$(^2H)^1H_5$	592,118
178.864d	0.003	559,083	50	587	$^5D_3$	5292	$4p^54d^5$	$(^4F)^5D_4$	564,383
178.864d	0.00	559,083	50	14	$^5D_2$	3212	$4p^54d^5$	$(^4F)^5D_3$	562,295
178.916	−0.003	558,922	50	362	$^5D_1$	1340	$4d^34f$	$(^4F)^5D_2$	560,251
179.217	0.000	557,983	383	3390	$^5D_4$	7442	$4p^54d^5$	$(^4G)^5F_5$	565,427
179.296	−0.002	557,738	68	1058	$^5D_0$	0	$4d^34f$	$(^4F)^5D_1$	557,732
179.323	−0.003	557,653	22	752	$^5D_2$	3212	$4p^54d^5$	$(^4G)^5F_2$	560,857
179.383	−0.003	557,468	17	268	$^5D_4$	7442	$4p^54d^5$	$(^4F)^5D_4$	564,902
179.431	−0.001	557,319	135	903	$^5D_1$	1340	$4p^54d^5$	$(^4G)^5F_1$	558,654
179.531d	0.010	557,007	260	958	$^5D_2$	3212	$4d^34f$	$(^4F)^5D_2$	560,251
179.531d	−0.001	557,007	260	2739	$^5D_3$	5292	$4p^54d^5$	$(^4F)^5D_3$	562,295
179.550	−0.002	556,949	375	3857	$^5D_4$	7442	$4p^54d^5$	$(^4F)^5D_4$	564,383
179.574	−0.005	556,872	111	2462	$^3F_32$	32,925	$4p^54d^5$	$(^2H)^3G_4$	589,786
179.642	−0.009	556,663	35	525	$^3G_3$	26,968	$4p^54d^5$	$(^2G1)^3F_3$	583,603
179.795	−0.002	556,190	18	479	$^3D_3$	36,878	$4p^54d^5$	$(^2F1)^3D_3$	593,064
180.038	0.001	555,437	60	527	$^5D_2$	3212	$4p^54d^5$	$(^4G)^5F_1$	558,654
180.194	0.001	554,959	60	1111	$^5D_3$	5292	$4d^34f$	$(^4F)^5D_2$	560,251
180.229	0.001	554,849	45	770	$^5D_4$	7442	$4p^54d^5$	$(^4F)^5D_3$	562,295
180.288	−0.002	554,667	20	515	$^3D_2$	38,403	$4p^54d^5$	$(^2F1)^3D_3$	593,064
180.338	0.002	554,513	22	408	$^5D_2$	3212	$4d^34f$	$(^4F)^5D_1$	557,732
180.455		554,156	28	1276	$^3P_12$	26,525	$4p^54d^5$	$(^2F2)^3D_2$	580,680
180.685	−0.002	553,449	26	134	$^3H_4$	23,304	$4d^34f$	$(^4F)^3H_5$	576,746
180.819	0.000	553,038	31	748	$^3D_2$	38,403	$4p^54d^5$	$(^4F)^3G_3$	591,442
180.869	0.006	552,888	28	525	$^3D_3$	36,878	$4p^54d^5$	$(^2H)^3G_4$	589,786
180.919	0.001	552,735	55	936	$^3H_6$	28,185	$4d^34f$	$(^4F)^3H_6$	580,923
181.004	−0.002	552,475	50	872	$^3F_42$	29,555	$4d^34f$	$(^4P)^3G_5$	582,024
181.047	0.003	552,343	33	826	$^3D_3$	36,878	$4p^54d^5$	$(^2I)^3H_4$	589,229
181.236	−0.005	551,768	33	611	$^3F_42$	29,555	$4p^54d^5$	$(^2G2)^3G_4$	581,306
181.302	−0.002	551,565	14	364	$^3D_2$	38,403	$4p^54d^5$	$(^2D1)^3P_2$	589,963
181.341	0.007	551,448	46	1007	$^3P_22$	32,137	$4p^54d^5$	$(^2G1)^3F_3$	583,603
181.563	0.002	550,772	60	867	$^3H_6$	28,185	$4d^34f$	$(^2H)^1K_7$	578,963
181.654	0.000	550,496	37	216	$^3H_5$	26,249	$4d^34f$	$(^4F)^3H_5$	576,746
181.752	0.007	550,200	159	2125	$^1F_3$	52,004	$4p^54d^5$	$(^4F)^3F_4$	602,225
181.781	0.001	550,114	17	256	$^3G_3$	26,968	$4p^54d^5$	$(^2H)^3G_3$	577,085
182.043	0.002	549,321	192	2799	$^3G_4$	32,698	$4d^34f$	$(^4P)^3G_5$	582,024
182.137d	−0.008	549,038	157	1393	$^1G_42$	43,105	$4p^54d^5$	$(^2H)^1H_5$	592,118
182.137d	−0.002	549,038	157	2609	$^3F_22$	28,051	$4p^54d^5$	$(^2H)^3G_3$	577,085

Table A3. *Cont.*

λ, Å [a]	o-c, (Å) [b]	ν (cm^{-1})	I [c]	gA, (10^9 s^{-1})	4d^4		(4d^{3}4f + 4p^{5}4d^5)		
					Term [e]	E (cm^{-1})	Config. [f]	Term [f]	E (cm^{-1})
182.354	−0.001	548,384	65	1175	$^3F_32$	32,925	4p^{5}4d^5	(^{2}G2)$^3G_4$	581,306
182.749	−0.003	547,199	145	2520	$^3F_42$	29,555	4d^{3}4f	(^{4}F)$^3H_5$	576,746
183.131	0.004	546,058	234	3198	$^3G_3$	26,968	4d^{3}4f	(^{4}F)$^3H_4$	573,036
183.202	0.000	545,845	355	4991	$^3G_5$	35,078	4d^{3}4f	(^{4}F)$^3H_6$	580,923
183.421	0.002	545,194	27	454	$^3D_2$	38,403	4p^{5}4d^5	(^{2}G1)$^3F_3$	583,603
183.482	0.001	545,014	260	2889	$^3H_4$	23,304	4d^{3}4f	(^{4}F)$^3I_5$	568,319
183.503	0.000	544,949	62	115	$^3P_22$	32,137	4p^{5}4d^5	(^{2}H)$^3G_3$	577,085
183.556	−0.003	544,792	30	721	$^1G_41$	69,585	4p^{5}4d^5	(^{2}G1)$^1H_5$	614,370
183.685	0.006	544,411	41	801	$^3D_3$	36,878	4p^{5}4d^5	(^{2}G2)$^3G_4$	581,306
183.808	0.001	544,047	76	1418	$^3G_4$	32,698	4d^{3}4f	(^{4}F)$^3H_5$	576,746
184.001	0.002	543,474	24	288	$^3F_42$	29,555	4d^{3}4f	(^{4}F)$^3H_4$	573,036
184.113	−0.003	543,144	594	6224	$^3H_5$	26,249	4d^{3}4f	(^{4}F)$^3I_6$	569,385
184.456	−0.003	542,133	102	2095	$^3H_4$	23,304	4p^{5}4d^5	(^{4}G)$^5F_5$	565,427
184.481	0.003	542,061	29	403	$^3H_5$	26,249	4d^{3}4f	(^{4}F)$^3I_5$	568,319
184.560	0.002	541,830	44	1623	$^3F_31$	62,552	4p^{5}4d^5	(^{2}G1)3G4	604,385
184.620	0.005	541,655	32	545	$^3G_5$	35,078	4d^{3}4f	(^{4}F)3H5	576,746
184.641	0.003	541,592	29	74	$^3H_4$	23,304	4p^{5}4d^5	(^{4}F)5D4	564,902
184.780d	0.006	541,184	34	264	$^3H_6$	28,185	4d^{3}4f	(^{4}F)3I6	569,385
184.780d	−0.002	541,184	34	255	$^1I_6$	39,744	4d^{3}4f	(^{4}F)3H6	580,923
185.452	−0.001	539,223	520	6429	$^1I_6$	39,744	4d^{3}4f	(^{2}H)1K7	578,963
185.553		538,930	741	6779	$^3H_6$	28,185	4d^{3}4f	(^{2}H)1K7	567,115
185.607	−0.003	538,773	62	536	$^3F_42$	29,555	4d^{3}4f	(^{4}F)3I5	568,319
185.897	0.001	537,931	31	306	$^3G_3$	26,968	4p^{5}4d^5	(^{4}F)5D4	564,902
185.959	0.010	537,753	27	613	$^1F_3$	52,004	4p^{5}4d^5	(^{2}H)3G4	589,786
186.615	0.003	535,862	24	121	$^3F_42$	29,555	4p^{5}4d^5	(^{4}G)5F5	565,427
187.168		534,280	243	2386	$^1G_41$	69,585	4d^{3}4f	(^{2}D1)1H5	603,864
188.279	0.004	531,127	68	1100	$^3F_41$	60,981	4p^{5}4d^5	(^{2}H)1H5	592,118

[a] Observed wavelengths: d—doubly identified, w—wide; [b] Difference between the observed wavelength and the wavelength derived from the final level energies (Ritz wavelength). A blank value indicates that the upper level is derived only from that line; [c] Relative intensity; IX—line is also identified as Ag IX; [e] Numbers following the term values display Nielson and Koster sequential indices [20]; [f] Designation and configuration attribution is arbitrary in a few cases (see text), for the level composition, see Table A9. Numbers following the term values of the 4d^5 configuration display Nielson and Koster sequential indices [20].

Table A4. Identified lines of the 4d^3–(4d^{2}5p + 4d^{2}4f + 4p^{5}4d^4) transitions in the spectrum of Ag^{8+}.

λ, Å [a]	o-c, (Å) [b]	ν (cm^{-1})	I [c]	gA, (10^9 s^{-1})	4d^3	4d^{2}5p + 4d^{2}4f + 4p^{5}4d^4		
					Term [e]	Config. [f]	Term [f]	E (cm^{-1})
160.837	0.005	621,749	100 X	1092	$^4F_{7/2}$	4p^{5}4d^4	(^{5}D)$^4D_{5/2}$	628,302
161.466	−0.005	619,327	50	3141	$^4F_{9/2}$	4p^{5}4d^4	(^{5}D)$^4D_{7/2}$	629,173
162.302		616,135	250 X	1091	$^2D_{5/2}2$	4p^{5}4d^4	(^{3}P1)$^2P_{3/2}$	649,186
162.411	−0.002	615,723	190	1327	$^2F_{7/2}$	4p^{5}4d^4	(^{1}G1)$^2F_{7/2}$	659,798
162.500		615,383	130	645	$^4F_{3/2}$	4p^{5}4d^4	(^{5}D)$^4D_{1/2}$	615,385
162.644	0.002	614,839	540	2516	$^2H_{9/2}$	4p^{5}4d^4	(^{3}H)$^2G_{7/2}$	647,197
163.010	−0.003	613,458	370	4684	$^2H_{11/2}$	4p^{5}4d^4	(^{3}H)$^2G_{9/2}$	644,980
163.132	0.003	613,001	90	1863	$^2F_{5/2}$	4p^{5}4d^4	(^{3}G)$^2F_{5/2}$	657,606
163.228	−0.003	612,641	80	464	$^2H_{9/2}$	4p^{5}4d^4	(^{3}H)$^2G_{9/2}$	644,980
163.267	0.001	612,494	70	577	$^2G_{7/2}$	4p^{5}4d^4	(^{3}F2)$^2F_{7/2}$	634,657
163.416	0.003	611,937	180	356	$^2D_{5/2}1$	4p^{5}4d^4	(^{3}F1)$^2D_{3/2}$	680,196
163.532	−0.004	611,503	220	2272	$^2D_{3/2}1$	4p^{5}4d^4	(^{3}F1)$^2D_{3/2}$	680,196
163.589	0.004	611,287	110 X	1237	$^2G_{9/2}$	4p^{5}4d^4	(^{3}F2)$^2F_{7/2}$	634,657
163.839		610,356	140	1254	$^2D_{3/2}2$	4p^{5}4d^4	(^{5}D)$^4P_{1/2}$	637,178
163.949	−0.003	609,947	400	1702	$^4P_{5/2}$	4p^{5}4d^4	(^{5}D)$^4P_{5/2}$	632,322
164.099		609,387	110	668	$^4F_{5/2}$	4p^{5}4d^4	(^{5}D)$^4D_{3/2}$	612,491
164.397		608,284	310	2866	$^2D_{5/2}1$	4p^{5}4d^4	(^{3}F1)$^2D_{5/2}$	676,533
164.493		607,930	80	578	$^4P_{5/2}$	4p^{5}4d^4	(^{5}D)$^4P_{3/2}$	630,316
164.542d	0.013	607,749	740 VIII	1677	$^2F_{7/2}$	4p^{5}4d^4	(^{3}P2)$^2D_{5/2}$	651,880
164.542d	−0.005	607,749	740	7777	$^2H_{11/2}$	4p^{5}4d^4	(^{3}H)$^2H_{11/2}$	639,262
164.772	0.004	606,900	170	576	$^2H_{9/2}$	4p^{5}4d^4	(^{3}H)$^2H_{11/2}$	639,262

Table A4. *Cont.*

λ, Å [a]	o-c, (Å) [b]	ν (cm^{-1})	I [c]	gA, (10^9 s^{-1})	$4d^3$	$4d^25p + 4d^24f + 4p^54d^4$		
					Term [e]	Config. [f]	Term [f]	E (cm^{-1})
164.808	0.005	606,766	370 X	1781	${}^4P_{5/2}$	$4p^54d^4$	$({}^5D){}^4D_{7/2}$	629,173
164.952	−0.000	606,239	70	725	${}^4F_{7/2}$	$4p^54d^4$	$({}^5D){}^4D_{5/2}$	612,769
165.036	−0.003	605,927	170	831	${}^4P_{5/2}$	$4p^54d^4$	$({}^5D){}^4D_{5/2}$	628,302
165.418	0.001	604,531	100	1245	${}^2G_{7/2}$	$4p^54d^4$	$({}^1D1){}^2F_{5/2}$	626,693
165.945	−0.002	602,610	90	1679	${}^2F_{5/2}$	$4p^54d^4$	$({}^3H){}^2G_{7/2}$	647,197
165.999	0.003	602,412	510 X	1403	${}^2P_{3/2}$	$4p^54d^4$	$({}^3P2){}^2P_{3/2}$	638,784
166.023	−0.005	602,325	200	1348	${}^2H_{9/2}$	$4p^54d^4$	$({}^3F2){}^2F_{7/2}$	634,657
166.423	0.005	600,880	90	1650	${}^2F_{7/2}$	$4p^54d^4$	$({}^3H){}^2G_{9/2}$	644,980
166.518	−0.000	600,537	60	509	${}^2H_{11/2}$	$4p^54d^4$	$({}^3H){}^2H_{9/2}$	632,069
166.744	0.000	599,721	490 VIII	5298	${}^2H_{9/2}$	$4p^54d^4$	$({}^3H){}^2H_{9/2}$	632,069
166.872	0.002	599,262	50	862	${}^2D_{5/2}2$	$4p^54d^4$	$({}^5D){}^4P_{5/2}$	632,322
167.307		597,704	620	5028	${}^2G_{9/2}$	$4p^54d^4$	$({}^1G2){}^2G_{9/2}$	621,061
167.460	−0.002	597,158	270 VIII	2912	${}^2F_{7/2}$	$4p^54d^4$	$({}^1G2){}^2F_{7/2}$	641,234
167.492	−0.001	597,042	100	883	${}^4F_{7/2}$	$4p^54d^4$	$({}^5D){}^4F_{9/2}$	603,570
167.814		595,896	120	1278	${}^2G_{7/2}$	$4p^54d^4$	$({}^3D){}^2F_{5/2}$	618,058
167.898	−0.000	595,601	170 X	1248	${}^4P_{3/2}$	$4p^54d^4$	$({}^5D){}^4D_{5/2}$	612,769
168.022	−0.004	595,162	160	1520	${}^2D_{5/2}1$	$4p^54d^4$	$({}^3P1){}^2P_{3/2}$	663,396
168.130		594,778	40	964	${}^4P_{1/2}$	$4p^54d^4$	$({}^5D){}^4D_{3/2}$	612,491
168.159	0.004	594,677	30	322	${}^2D_{3/2}1$	$4p^54d^4$	$({}^3P1){}^2P_{3/2}$	663,396
168.294	−0.002	594,197	90	469	${}^2F_{5/2}$	$4p^54d^4$	$({}^3P2){}^2P_{3/2}$	638,784
168.414	−0.001	593,775	90	707	${}^4F_{5/2}$	$4p^54d^4$	$({}^5D){}^4F_{7/2}$	596,876
168.436d	0.007	593,698	430 VIII	1955	${}^2G_{7/2}$	$4p^54d^4$	$({}^1G2){}^2G_{7/2}$	615,884
168.436d	0.001	593,698	430 VIII	3827	${}^4F_{9/2}$	$4p^54d^4$	$({}^5D){}^4F_{9/2}$	603,570
168.479	−0.002	593,545	110	909	${}^2F_{5/2}$	$4p^54d^4$	$({}^3F2){}^2F_{5/2}$	638,134
168.605	0.005	593,101	50	1153	${}^2G_{9/2}$	$4p^54d^4$	$({}^3F2){}^2G_{9/2}$	616,475
168.766	−0.003	592,537	80	1585	${}^2G_{9/2}$	$4p^54d^4$	$({}^1G2){}^2G_{7/2}$	615,884
168.808	0.002	592,389	110 X	759	${}^4F_{7/2}$	$4p^54d^4$	$({}^3F1){}^4D_{7/2}$	598,929
168.932d	−0.003	591,953	150 VIII	967	${}^2P_{3/2}$	$4p^54d^4$	$({}^5D){}^4D_{5/2}$	628,302
168.932d	−0.000	591,953	150 VIII	1267	${}^2G_{9/2}$	$4p^54d^4$	$({}^3H){}^2H_{11/2}$	615,308
169.049	0.002	591,543	140	3139	${}^2D_{5/2}1$	$4p^54d^4$	$({}^1G1){}^2F_{7/2}$	659,798
169.392d	−0.004	590,346	270	1157	${}^2P_{3/2}$	$4p^54d^4$	$({}^1D1){}^2F_{5/2}$	626,693
169.392d	−0.001	590,346	270	2701	${}^4F_{7/2}$	$4p^54d^4$	$({}^5D){}^4F_{7/2}$	596,876
169.475	0.001	590,058	50 X	548	${}^2F_{5/2}$	$4p^54d^4$	$({}^3F2){}^2F_{7/2}$	634,657
169.595	0.001	589,642	50	428	${}^4F_{3/2}$	$4p^54d^4$	$({}^5D){}^4F_{5/2}$	589,644
169.613	0.000	589,578	100 VIII	765	${}^4F_{9/2}$	$4p^54d^4$	$({}^1I){}^2H_{9/2}$	599,446
169.762	0.000	589,062	50	787	${}^4F_{9/2}$	$4p^54d^4$	$({}^3F1){}^4D_{7/2}$	598,929
169.806	−0.002	588,907	20	1462	${}^2D_{3/2}1$	$4p^54d^4$	$({}^3G){}^2F_{5/2}$	657,606
170.192	0.002	587,572	30	80	${}^4F_{7/2}$	$4p^54d^4$	$({}^1D1){}^2F_{5/2}$	594,111
170.306	−0.001	587,180	60	620	${}^4F_{5/2}$	$4d^24f$	$({}^3P){}^4D_{3/2}$	590,280
170.359t	0.007	586,994	440 VIII	1882	${}^4F_{9/2}$	$4d^24f$	$({}^3F){}^2I_{11/2}$	596,885
170.359t	0.004	586,994	440 VIII	769	${}^4F_{9/2}$	$4p^54d^4$	$({}^5D){}^4F_{7/2}$	596,876
170.359t	−0.001	586,994	440 VIII	1501	${}^2G_{7/2}$	$4p^54d^4$	$({}^3G){}^2G_{7/2}$	609,152
170.490	−0.001	586,545	280	2414	${}^4F_{5/2}$	$4p^54d^4$	$({}^5D){}^4F_{5/2}$	589,644
170.933		585,026	120	2598	${}^2D_{5/2}2$	$4d^24f$	$({}^3P){}^2F_{7/2}$	618,075
171.286		583,818	260	2018	${}^4F_{3/2}$	$4d^24f$	$({}^3F){}^4F_{3/2}$	583,819
171.339	−0.002	583,638	60	848	${}^2D_{5/2}1$	$4p^54d^4$	$({}^3P2){}^2D_{5/2}$	651,880
171.493	−0.000	583,114	40	409	${}^4F_{7/2}$	$4p^54d^4$	$({}^5D){}^4F_{5/2}$	589,644
171.539	0.001	582,957	530 VIII	5000	${}^2H_{9/2}$	$4p^54d^4$	$({}^3H){}^2H_{11/2}$	615,308
171.996	0.001	581,408	40	482	${}^2G_{7/2}$	$4p^54d^4$	$({}^5D){}^4F_{9/2}$	603,570
172.215d	−0.001	580,671	1100	4948	${}^4F_{7/2}$	$4p^54d^4$	$({}^3H){}^4G_{9/21}$	587,200
172.215d	−0.001	580,671	1100	3951	${}^4F_{5/2}$	$4d^24f$	$({}^3F){}^4G_{7/21}$	583,771
172.372		580,140	90	1120	${}^2D_{3/2}2$	$4d^24f$	$({}^1D){}^2D_{3/2}$	606,963
172.556	−0.005	579,521	390	3149	${}^4F_{3/2}$	$4d^24f$	$({}^3F){}^4G_{5/2}$	579,505
173.077	−0.001	577,778	890	5063	${}^4F_{9/2}$	$4p^54d^4$	$({}^3H){}^4G_{11/2}$	587,642
173.224d	0.014	577,288	670	765	${}^4F_{9/2}$	$4p^54d^4$	$({}^3H){}^4G_{9/2}$	587,200
173.224d	−0.000	577,288	670	4944	${}^2G_{7/2}$	$4p^54d^4$	$({}^1I){}^2H_{9/2}$	599,446
173.236	−0.002	577,246	380	932	${}^4F_{7/2}$	$4d^24f$	$({}^3F){}^4G_{7/2}$	583,771
173.445	−0.003	576,551	150	1869	${}^4P_{5/2}$	$4p^54d^4$	$({}^3F1){}^4D_{7/2}$	598,929
173.495	0.005	576,386	60	528	${}^4F_{5/2}$	$4d^24f$	$({}^3F){}^4G_{5/2}$	579,505
173.578	−0.003	576,110	110	1309	${}^2D_{5/2}2$	$4p^54d^4$	$({}^3G){}^2G_{7/2}$	609,152

Table A4. *Cont.*

λ, Å [a]	o-c, (Å) [b]	ν (cm^{-1})	I [c]	gA, (10^9 s^{-1})	4d^3	4d^{2}5p + 4d^{2}4f + 4p^{5}4d^4		
					Term [e]	Config. [f]	Term [f]	E (cm^{-1})
174.248	0.003	573,895	30	20	$^4F_{9/2}$	4d^{2}4f	(^{3}F)$^4G_{7/2}$	583,771
174.357	−0.002	573,535	360 VIII	3568	$^2G_{9/2}$	4d^{2}4f	(^{3}F)$^2I_{11/2}$	596,885
174.487	0.000	573,110	140	1302	$^4P_{3/2}$	4d^{2}4f	(^{3}P)$^4D_{3/2}$	590,280
174.526	0.002	572,979	40	690	$^2D_{5/2}$1	4p^{5}4d^4	(^{1}G2)$^2F_{7/2}$	641,234
174.618		572,678	100 VIII	1088	$^4P_{5/2}$	4p^{5}4d^4	(^{3}P1)$^4S_{3/2}$	595,066
174.701	−0.004	572,407	340	2551	$^2F_{7/2}$	4p^{5}4d^4	(^{3}F2)$^2G_{9/2}$	616,475
174.908	−0.002	571,730	60	94	$^4P_{5/2}$	4p^{5}4d^4	(^{1}D1)$^2F_{5/2}$	594,111
175.617	0.001	569,422	30	1317	$^2D_{3/2}$1	4p^{5}4d^4	(^{3}F2)$^2F_{5/2}$	638,134
175.738		569,029	940	7991	$^2H_{11/}$	4d^{2}4f	(^{3}F)$^2I_{13/2}$	600,561
176.977	−0.001	565,044	40	328	$^2G_{7/2}$	4p^{5}4d^4	(^{3}H)$^4G_{9/2}$	587,200
177.136d	0.006	564,539	100	1114	$^2F_{5/2}$	4p^{5}4d^4	(^{3}G)$^2G_{7/2}$	609,152
177.216d	0.000	564,283	360	2077	$^2G_{9/2}$	4p^{5}4d^4	(^{3}H)$^4G_{11/2}$	587,642
221.880	−0.001	450,694	80	22	$^4F_{9/2}$	4d^{2}5p	(^{3}P)$^4D_{7/2}$	460,559
223.061	0.001	448,307	40	12	$^2G_{9/2}$	4d^{2}5p	(^{1}G)$^2H_{11/2}$	471,667
226.030	0.003	442,420	120	17	$^4F_{9/2}$	4d^{2}5p	(^{1}G)$^2G_{9/2}$	452,292
226.226	0.000	442,036	60	15	$^2G_{9/2}$	4d^{2}5p	(^{3}P)$^4D_{7/2}$	465,394
227.202	−0.001	440,137	700	144	$^2H_{11/2}$	4d^{2}5p	(^{1}G)$^2H_{11/2}$	471,667
227.723	−0.004	439,130	400	81	$^4F_{9/2}$	4d^{2}5p	(^{3}F)$^2F_{7/2}$	448,989
228.487		437,662	320	67	$^4F_{3/2}$	4d^{2}5p	(^{3}F)$^4D_{1/2}$	437,662
228.963	0.004	436,751	140	45	$^4P_{5/2}$	4d^{2}5p	(^{3}P)$^4P_{3/2}$	459,146
229.415		435,891	520	172	$^4F_{7/2}$	4d^{2}5p	(^{3}F)$^4D_{5/2}$	442,423
229.557		435,621	400	91	$^4F_{5/2}$	4d^{2}5p	(^{3}F)$^4D_{3/2}$	438,725
229.783	0.001	435,193	340	36	$^4F_{5/2}$	4d^{2}5p	(^{3}F)$^4F_{5/2}$	438,297
229.820		435,124	200	32	$^2F_{5/2}$	4d^{2}5p	(^{1}G)$^2F_{5/2}$	479,718
229.877		435,015	100	20	$^2F_{5/2}$	4d^{2}5p	(^{1}G)$^2F_{5/2}$	479,610
230.176	0.004	434,450	230	61	$^2H_{11/2}$	4d^{2}5p	(^{1}G)$^2H_{9/2}$	465,990
230.241		434,328	630	157	$^4F_{9/2}$	4d^{2}5p	(^{3}F)$^4F_{9/2}$	444,195
230.601	−0.005	433,650	350	113	$^2H_{9/2}$	4d^{2}5p	(^{1}G)$^2H_{9/2}$	465,990
230.922	−0.001	433,047	500	89	$^2H_{9/2}$	4d^{2}5p	(^{3}P)$^4D_{7/2}$	465,394
231.179	0.002	432,565	150	16	$^4F_{3/2}$	4d^{2}5p	(^{3}F)$^4F_{5/2}$	432,569
231.545		431,882	620	106	$^2H_{9/2}$	4d^{2}5p	(^{1}D)$^2F_{7/2}$	464,230
231.602d	−0.005	431,774	580	19	$^4F_{7/2}$	4d^{2}5p	(^{3}F)$^4F_{5/2}$	438,297
231.602d	−0.002	431,774	580	149	$^4F_{9/2}$	4d^{2}5p	(^{3}F)$^4D_{7/2}$	441,637
231.819		431,370	300	32	$^2F_{7/2}$	4p^{5}4d^4	(^{3}G)$^4G_{7/2}$	475,453
231.946	−0.003	431,135	450	154	$^4F_{7/2}$	4d^{2}5p	(^{3}F)$^4F_{7/2}$	437,662
232.127		430,799	130	13	$^2D_{3/2}$2	4d^{2}5p	(^{3}P)$^4D_{5/2}$	457,621
232.259	−0.005	430,554	410	251	$^2G_{9/2}$	4d^{2}5p	(^{1}G)$^2G_{7/2}$	453,902
232.337		430,410	330	159	$^2G_{7/2}$	4d^{2}5p	(^{1}D)$^2F_{5/2}$	452,569
232.845	−0.002	429,470	550	102	$^4F_{5/2}$	4d^{2}5p	(^{3}F)$^4F_{5/2}$	432,569
233.133	−0.003	428,940	450	89	$^2G_{9/2}$	4d^{2}5p	(^{1}G)$^2G_{9/2}$	452,292
233.206	−0.005	428,806	50	16	$^4P_{5/2}$	4d^{2}5p	(^{3}P)$^4S_{3/2}$	451,183
233.531	0.000	428,209	750 m	103	$^2H_{9/2}$	4d^{2}5p	(^{3}P)$^4D_{7/2}$	460,559
233.691		427,915	310	91	$^4F_{3/2}$	4d^{2}5p	(^{3}F)$^4F_{3/2}$	427,916
233.759	0.003	427,790	300	17	$^4F_{9/2}$	4d^{2}5p	(^{3}F)$^4F_{7/2}$	437,662
233.877		427,576	170	64	$^4F_{9/2}$	4d^{2}5p	(^{3}F)$^4G_{9/2}$	437,442
234.539		426,368	1000	424	$^2H_{11/2}$	4d^{2}5p	(^{3}F)$^2G_{9/2}$	457,900
234.686	−0.003	426,101	150	22	$^2D_{5/2}$2	4d^{2}5p	(^{3}P)$^4P_{3/2}$	459,146
234.945	0.001	425,631	350	36	$^2G_{9/2}$	4d^{2}5p	(^{3}F)$^2F_{7/2}$	448,989
235.240	−0.002	425,098	360	88	$^2G_{7/2}$	4d^{2}5p	(^{3}F)$^2G_{7/2}$	447,255
235.532d		424,570	200	19	$^4F_{3/2}$	4d^{2}5p	(^{3}F)$^4G_{5/2}$	424,571
235.532d		424,570	200	17	$^2D_{5/2}$2	4d^{2}5p	(^{3}P)$^4D_{5/2}$	457,621
235.581		424,482	190	51	$^2G_{7/2}$	4d^{2}5p	(^{3}F)$^2F_{5/2}$	446,642
235.649		424,360	200	71	$^2D_{3/2}$2	4d^{2}5p	(^{3}P)$^4S_{3/2}$	451,183
235.726		424,222	230	101	$^2F_{5/2}$	4d^{2}5p	(^{3}P)$^2D_{3/2}$	468,817
235.907	0.002	423,895	100	21	$^2G_{9/2}$	4d^{2}5p	(^{3}F)$^2G_{7/2}$	447,255
236.756		422,376	110	32	$^2F_{7/2}$	4d^{2}5p	(^{1}D)$^2D_{5/2}$	466,458
237.222	0.004	421,546	70	33	$^2H_{9/2}$	4d^{2}5p	(^{1}G)$^2G_{7/2}$	453,902

Table A4. *Cont.*

λ, Å [a]	o-c, (Å) [b]	ν (cm^{-1})	I [c]	gA, (10^9 s^{-1})	$4d^3$	$4d^25p + 4d^24f + 4p^54d^4$		
					Term [e]	Config. [f]	Term [f]	E (cm^{-1})
237.458	0.001	421,127	80	15	$^4P_{3/2}$	$4d^25p$	$(^3F)^4F_{5/2}$	438,297
239.077	0.003	418,275	160	34	$^2G_{9/2}$	$4d^25p$	$(^3F)^4D_{7/2}$	441,637
239.161	0.002	418,129	60	19	$^2D_{5/2}2$	$4d^25p$	$(^3P)^4S_{3/2}$	451,183
240.423	0.002	415,934	160	32	$^2D_{5/2}2$	$4d^25p$	$(^3F)^2F_{7/2}$	448,989
243.734	−0.001	410,284	160	30	$^2P_{3/2}$	$4d^25p$	$(^3F)^2F_{5/2}$	446,642

[a] Observed wavelengths: d—doubly identified, t—trebly identified; [b] Difference between the observed wavelength and the wavelength derived from the final level energies (Ritz wavelength). A blank value indicates that the upper level is derived only from that line; [c] Relative intensity: X, VIII—line is also identified as respectively Ag X or Ag VIII; m masked by O IV; [e] Numbers following the term values display Nielson and Koster sequential indices [20]; [f] Designation and configuration attribution is arbitrary in a few cases (see text), for the level composition, see Table A11. Numbers following the term values of the $4d^4$ configuration display Nielson and Koster sequential indices [20].

Table A5. Energies (in cm^{-1}) of the $4d^5$ configuration of Ag VII.

E [a]	o-c [b]	Eigenvector Composition [c]		
		$J = 1/2$		
94,730 *		98% 3^2P	2% 5^2S	
63,467 *		96% 5^2S	2% 3^4P	2% 3^2P
38,685	6	67% 5^4D	32% 3^4P	1% 5^2S
34,605	−20	66% 3^4P	33% 5^4D	1% 5^2S
		$J = 3/2$		
104,821	−27	74% 1^2D	18% 5^2D	6% 3^2P
93,529 *		94% 3^2P	4% 1^2D	1% 5^2D
71,517	−2	97% 3^2D	2% 1^2D	1% 5^4D
53,796	0	55% 3^4F	37% 5^2D	6% 1^2D
48,086	2	42% 5^2D	43% 3^4F	13% 1^2D
39,788	−1	52% 5^4D	44% 3^4P	1% 3^4F
32,994	−4	53% 3^4P	44% 5^4D	1% 3^4F
		$J = 5/2$		
103,996	34	80% 1^2D	19% 5^2D	1% 3^2D
72,934	−26	91% 3^2D	6% 5^2F	2% 5^4D
59,954	−5	30% 3^2F	29% 5^2F	19% 5^2D
57,413	2	53% 5^2F	18% 3^2F	17% 5^2D
51,049	11	74% 3^4F	12% 3^2F	8% 5^2F
47,119	2	31% 5^2D	32% 3^2F	13% 3^4P
39,299	10	54% 5^4D	32% 3^4P	7% 5^2D
32,005	−19	44% 3^4P	35% 5^4D	13% 5^4G
29,390	27	79% 5^4G	8% 3^2F	4% 3^4P
0	4	98% 5^6S	2% 3^4P	
		$J = 7/2$		
79,705	4	96% 3^2G	2% 5^2F	1% 5^2G
59,792	11	75% 5^2F	15% 3^4F	6% 3^2F
55,773	8	74% 5^2G	17% 3^2F	8% 5^2F
53,353	2	47% 3^2F	34% 3^4F	8% 5^2G
48,712	10	45% 3^4F	26% 3^2F	15% 5^2G
36,485	9	92% 5^4D	4% 3^4F	3% 5^4G
30,378	0	93% 5^4G	3% 3^2F	2% 3^4F

Table A5. *Cont.*

E [a]	o-c [b]	Eigenvector Composition [c]		
		$J = 9/2$		
79,131	−13	99% 3^2G	1% 5^2G	
59,223	2	47% 5^2G	42% 3^2H	10% 3^4F
53,797	−11	52% 3^2H	30% 3^4F	18% 5^2G
49,104	−11	59% 3^4F	34% 5^2G	6% 3^2H
30,907	−2	97% 5^4G	1% 3^4F	1% 3^2H
		$J = 11/2$		
57,962	2	86% 3^2H	11% 5^2I	2% 5^4G
44,011	1	88% 5^2I	10% 3^2H	1% 5^4G
30,662	−14	96% 5^4G	3% 3^2H	
		$J = 13/2$		
45,546	3	100% 5^2I		

[a] The star * indicates a calculated value for the level; [b] The difference between the observed and the calculated energies; [c] For the eigenvector composition, up to three components with the largest percentages in the LS-coupling scheme are listed. The number preceding the terms is the seniority number.

Table A6. Energies (in cm^{-1}) of the $4d^45p$ configuration of Ag VII.

E [a]	o-c [b]	Eigenvector Composition [c]		
		$J = 1/2$		
448,199 *		68% $(0^1S)^2P$	15% $(^1S)^2P$	14% $(2^1D)^2P$
419,718 *		64% $(2^1D)^2P$	12% $(0^1S)^2P$	11% $(^1D)^2P$
409,158 *		47% $(^3P)^2S$	27% $(^3P)^2S$	12% $(2^3P)^2P$
406,673	37	39% $(2^3F)^4D$	20% $(^3F)^4D$	13% $(2^3P)^2P$
405,606 *		28% $(2^3P)^2P$	13% $(2^3F)^4D$	10% $(^3P)^2P$
394,516 *		41% $(2^3P)^4P$	11% $(^3P)^4P$	9% $(2^3P)^4D$
392,424	−57	36% $(2^3P)^4D$	21% $(2^3P)^4D$	11% $(2^3P)^4P$
389,336 *		59% $(^1D)^2P$	9% $(^3P)^4P$	7% $(2^1D)^2P$
380,449 *		48% $(^3D)^2P$	23% $(^1S)^2P$	9% $(^3D)^4D$
375,489	17	64% $(^3D)^4P$	10% $(2^3P)^4P$	6% $(^3D)^4D$
374,810	−24	35% $(^3P)^2P$	16% $(^1S)^2P$	9% $(^3P)^4P$
370,586 *		22% $(^1S)^2P$	19% $(^3P)^2S$	15% $(^3P)^2S$
368,200 *		32% $(^3F)^4D$	30% $(^3D)^4D$	11% $(^3D)^2P$
366,416 *		32% $(^3D)^4D$	31% $(2^3F)^4D$	13% $(2^3F)^4D$
364,278 *		19% $(^3P)^4P$	19% $(^3P)^2P$	11% $(2^3P)^2S$
355,790	−19	32% $(^3P)^4P$	16% $(^3P)^4P$	13% $(2^3P)^2S$
352,270	26	77% $(^5D)^4D$	9% $(^3P)^4D$	4% $(^3D)^4D$
347,791 *		32% $(^3P)^4D$	23% $(2^3P)^4D$	10% $(^3P)^2P$
341,227	−1	65% $(^5D)^6D$	30% $(^5D)^4P$	1% $(2^3P)^2S$
328,555 *		54% $(^5D)^4P$	33% $(2^5D)^6D$	5% $(^3P)^4P$
325,860 *		89% $(^5D)^6F$	3% $(^3P)^4D$	3% $(^5D)^4D$
		$J = 3/2$		
455,917 *		74% $(0^1S)^2P$	16% $(^1S)^2P$	7% $(2^1D)^2P$
429,981 *		65% $(2^1D)^2D$	24% $(^1D)^2D$	6% $(2^1D)^2P$
416,277 *		29% $(2^3F)^2D$	18% $(^3P)^2D$	14% $(^3P)^2D$
415,087 *		56% $(2^1D)^2P$	6% $(^3P)^2D$	6% $(^1D)^2P$
406,960 *		38% $(2^3F)^4D$	17% $(2^3F)^4D$	11% $(2^3P)^4D$
404,310 *		23% $(2^3P)^2P$	14% $(2^3P)^4S$	13% $(2^3P)^4S$

Table A6. *Cont.*

E [a]	o-c [b]	Eigenvector Composition [c]		
403,643 *		31% $(2^3P)^2P$	22% $(^3P)^4S$	19% $(^3P)^4S$
399,850	2	41% $(^1F)^2D$	13% $(^3F)^2D$	8% $(2^3P)^2D$
395,638 *		39% $(^1F)^2D$	18% $(2^3F)^4F$	12% $(2^3P)^2D$
393,610	−2	22% $(^1D)^2P$	15% $(2^3P)^4P$	8% $(2^3F)^4D$
391,553	7	19% $(2^3P)^4D$	14% $(2^3P)^4D$	14% $(2^3F)^2D$
389,293	26	45% $(2^3F)^4F$	7% $(2^3P)^4D$	7% $(^3D)^2D$
387,806 *		21% $(2^3P)^4P$	10% $(^3P)^4P$	10% $(^1D)^2P$
386,582	−38	31% $(^1S)^2P$	10% $(2^1D)^2D$	8% $(^3D)^2P$
383,280	4	46% $(^3D)^2D$	19% $(^3F)^2D$	9% $(^1D)^2D$
377,910	34	19% $(^1S)^2P$	18% $(^1D)^2D$	17% $(^1D)^2P$
375,706	21	30% $(^3D)^4D$	13% $(^3D)^2P$	11% $(^3D)^4F$
374,938	0	22% $(^3D)^2P$	11% $(^3P)^2P$	11% $(^3D)^4F$
374,236	31	24% $(^3D)^4P$	18% $(^3G)^4F$	9% $(^3D)^4F$
371,218 *		25% $(^3P)^2D$	14% $(2^3P)^2P$	13% $(2^3P)^2D$
369,448	−4	27% $(^3D)^4P$	18% $(^3D)^4D$	15% $(^3P)^2P$
368,569	0	22% $(^3F)^4D$	10% $(^3D)^4D$	8% $(^3D)^4F$
366,573	3	25% $(^3F)^4F$	14% $(^3F)^4D$	9% $(^3F)^2D$
366,078	−4	17% $(^3F)^4F$	16% $(^3P)^4S$	14% $(2^3P)^4S$
363,454	−23	31% $(^3D)^4F$	29% $(^3G)^4F$	12% $(^3P)^4P$
361,846 *		20% $(^3P)^2P$	9% $(^3P)^4S$	9% $(^5D)^4D$
360,092	−51	34% $(^3P)^4P$	14% $(^3P)^4P$	11% $(^3D)^4D$
356,993	−2	20% $(^3F)^2D$	18% $(^3F)^4F$	10% $(^3P)^4D$
353,549	22	44% $(^5D)^4D$	22% $(2^3P)^4D$	8% $(2^3P)^4D$
351,904 *		34% $(^5D)^4D$	16% $(^3P)^4D$	14% $(2^3P)^4D$
342,965 *		48% $(^5D)^4F$	25% $(^5D)^6D$	16% $(^5D)^4P$
341,983	4	31% $(^5D)^4F$	27% $(^5D)^4P$	26% $(^5D)^6D$
336,333	−24	72% $(^5D)^6P$	16% $(^5D)^6D$	6% $(^5D)^4P$
331,200	−4	41% $(^5D)^4P$	31% $(^5D)^6D$	22% $(^5D)^6P$
327,322 *		91% $(^5D)^6F$	3% $(^5D)^4D$	2% $(2^3P)^4D$
		$J = 5/2$		
432,566 *		64% $(2^1D)^2D$	23% $(^1D)^2D$	8% $(2^1D)^2F$
422,985	−10	43% $(2^1D)^2F$	17% $(^1G)^2F$	13% $(^1G)^2F$
414,230	8	46% $(2^3F)^2D$	17% $(^3F)^2D$	14% $(2^3P)^2D$
409,186	50	51% $(2^1G)^2F$	21% $(2^1D)^2F$	8% $(2^3F)^2F$
405,964 *		45% $(2^3F)^4D$	16% $(^3F)^4D$	12% $(2^3P)^4D$
400,637	16	28% $(^1F)^2D$	10% $(2^3P)^4D$	10% $(2^3F)^2F$
398,662	−16	32% $(2^3F)^2F$	22% $(^3F)^4G$	8% $(^3F)^4G$
395,990 *		44% $(2^3P)^4P$	15% $(2^3P)^4P$	10% $(2^3P)^2D$
394,049	7	28% $(2^3F)^4F$	20% $(2^1F)^2D$	12% $(2^3P)^4D$
391,396	28	29% $(2^3F)^4G$	11% $(^3F)^2F$	11% $(^3F)^4G$
390,060	−66	38% $(2^3F)^4F$	11% $(^3P)^4D$	6% $(2^3F)^4D$
387,358 *		20% $(2^3P)^2D$	12% $(2^3F)^2D$	11% $(^1F)^2D$
386,268	−6	25% $(^1D)^2D$	17% $(2^1D)^2F$	9% $(2^3F)^2F$
385,962	10	21% $(^3D)^2F$	13% $(2^3D)^2D$	12% $(^1G)^2F$
383,109	22	25% $(^3D)^2F$	8% $(^1F)^2F$	7% $(^3P)^2D$
380,891	−18	28% $(^1G)^2F$	18% $(^3D)^2D$	8% $(^3F)^2F$
379,478	−25	25% $(^3G)^2F$	21% $(^1D)^2F$	12% $(^3P)^2D$
379,077	−7	29% $(^1F)^2F$	13% $(^3P)^2D$	12% $(2^3P)^2D$
376,543	17	23% $(^3D)^4F$	17% $(^3D)^4D$	9% $(^3G)^4F$
374,426	−0	13% $(^3G)^4F$	11% $(^1G)^2F$	11% $(^3P)^2D$
372,051	−6	24% $(^3H)^4G$	20% $(^3G)^4G$	16% $(^3F)^2F$
370,940 *		16% $(^3D)^4D$	15% $(^3F)^4D$	12% $(^3F)^4F$

Table A6. *Cont.*

E [a]	o-c [b]	Eigenvector Composition [c]		
369,578	12	21% $(^3F)^2F$	15% $(^3D)^4P$	11% $(^3H)^4G$
367,919	13	24% $(^3D)^4P$	23% $(^3P)^4P$	9% $(^3G)^4G$
366,693	−15	14% $(^3D)^4D$	13% $(^3F)^4F$	13% $(^3F)^4D$
364,877	−3	17% $(^3P)^4P$	13% $(^3P)^4P$	8% $(^3P)^2D$
364,025	−28	15% $(^3D)^2D$	14% $(^3D)^4P$	10% $(^3H)^4G$
362,975	3	25% $(^3P)^4D$	17% $(2^3F)^4F$	13% $(^3G)^4F$
362,381	56	20% $(^3H)^4G$	11% $(^3D)^2F$	10% $(^3D)^4F$
361,296	−14	22% $(^3F)^4F$	13% $(^3G)^4F$	11% $(^3F)^2D$
358,392	29	21% $(^3F)^4G$	12% $(^3G)^2F$	8% $(^3P)^4D$
356,233	1	14% $(^3F)^4D$	13% $(^3F)^4G$	12% $(^3P)^4D$
354,235	16	78% $(^5D)^4D$	4% $(^3D)^4D$	2% $(^3F)^4F$
352,685	45	26% $(^3G)^4G$	22% $(^3F)^4G$	14% $(2^3F)^4G$
344,695	6	32% $(^5D)^4F$	30% $(^5D)^4P$	27% $(^5D)^6D$
343,336	−17	42% $(^5D)^4F$	33% $(^5D)^4P$	5% $(^5D)^6P$
339,172	1	50% $(^5D)^6D$	30% $(^5D)^6P$	8% $(^5D)^4P$
333,327	18	61% $(^5D)^6P$	18% $(^5D)^4P$	17% $(^5D)^6D$
329,356 *		89% $(^5D)^6F$	4% $(^5D)^4F$	2% $(^5D)^4D$
		$J = 7/2$		
428,522	−6	70% $(2^1D)^2F$	15% $(^1D)^2F$	7% $(^1G)^2F$
409,889	5	39% $(2^1G)^2F$	14% $(^1G)^2G$	10% $(2^3F)^2G$
406,495	−4	46% $(2^3F)^2G$	17% $(^3F)^2G$	12% $(2^1G)^2G$
404,117	−8	27% $(2^3F)^2F$	19% $(2^3F)^4D$	9% $(2^1G)^2G$
403,133	−5	25% $(2^3F)^4D$	17% $(^1G)^2G$	10% $(2^3P)^4D$
397,858	12	26% $(2^3F)^4G$	22% $(2^3F)^4F$	9% $(^3F)^4G$
396,241	−22	32% $(2^3P)^4D$	21% $(2^3F)^2F$	15% $(^3P)^4D$
393,295	−10	20% $(2^3F)^4G$	15% $(2^3F)^4D$	13% $(2^3F)^2F$
391,338	32	43% $(2^3F)^4F$	12% $(2^3F)^4G$	12% $(^1D)^2F$
388,163	−3	23% $(^1F)^2G$	14% $(2^3D)^2F$	13% $(^1D)^2F$
386,615	15	26% $(^1F)^2G$	21% $(2^1F)^2F$	7% $(2^3F)^4F$
384,772	25	24% $(^1D)^2F$	16% $(^1G)^2G$	11% $(^3F)^2F$
380,468	−6	20% $(^1F)^2F$	19% $(^3G)^2G$	10% $(^3D)^2F$
379,061	−3	21% $(^1G)^2G$	18% $(^1F)^2F$	15% $(^3D)^2F$
377,163	−3	27% $(^3D)^4F$	17% $(^3D)^4D$	17% $(^3G)^4F$
375,365	3	37% $(^3G)^2G$	9% $(^3D)^2F$	8% $(^3D)^4D$
374,839	−5	14% $(^3H)^4G$	13% $(^1G)^2F$	11% $(^3G)^4F$
373,577	−30	13% $(^3F)^4D$	10% $(^3D)^4D$	10% $(^3G)^2F$
371,673	0	25% $(^3G)^4G$	24% $(^1G)^2F$	7% $(2^1G)^2F$
370,993	−50	33% $(^3F)^2F$	25% $(^3G)^2F$	10% $(^3D)^4D$
369,879	−23	28% $(^3P)^4D$	13% $(^3D)^4F$	11% $(2^3P)^4D$
368,968	3	15% $(^3P)^4D$	10% $(^3H)^4G$	10% $(2^3P)^4D$
366,806	21	29% $(^3F)^4F$	11% $(^3G)^4F$	8% $(^3H)^4G$
365,842	3	16% $(^3F)^2G$	15% $(^3F)^4F$	14% $(^3D)^2F$
362,942	−4	25% $(^3G)^4H$	25% $(^3H)^4H$	10% $(^3H)^2G$
361,593	−18	25% $(^3G)^4F$	18% $(^3D)^4F$	9% $(^5D)^4F$
361,173	−0	39% $(^3F)^4G$	12% $(^3H)^4G$	11% $(^3G)^4H$
359,687	−10	16% $(^3G)^4G$	12% $(^3F)^4D$	12% $(^3H)^4G$
356,493	−0	17% $(^3F)^4D$	11% $(^3G)^4G$	10% $(^3H)^4H$
354,869	−9	68% $(^5D)^4D$	6% $(^3D)^4D$	5% $(^3F)^4F$
353,454	−22	19% $(^3G)^4H$	18% $(^3H)^2G$	9% $(^3F)^4G$
349,047	10	47% $(^3H)^4H$	23% $(^3G)^4H$	9% $(^3H)^2G$
346,061	−1	55% $(^5D)^4F$	28% $(^5D)^6D$	3% $(^5D)^6F$
341,458	1	59% $(^5D)^6D$	18% $(^5D)^4F$	12% $(^5D)^6F$
334,906	25	85% $(^5D)^6P$	6% $(^5D)^6D$	4% $(^5D)^4D$
331,831 *		82% $(^5D)^6F$	8% $(^5D)^4F$	4% $(^5D)^6D$

Table A6. *Cont.*

E [a]	o-c [b]	Eigenvector Composition [c]		
		$J = 9/2$		
409,779	3	34% $(2^1G)^2G$	17% $(^1G)^2G$	16% $(2^1G)^2H$
403,942	−32	49% $(2^3F)^2G$	12% $(^3F)^2G$	11% $(2^3F)^4G$
401,561	−26	33% $(2^1G)^2H$	14% $(^1G)^2H$	14% $(2^1G)^2G$
398,431	16	37% $(2^3F)^4F$	25% $(^3F)^4G$	7% $(2^1G)^2H$
392,486	12	29% $(2^3F)^4G$	21% $(^1F)^2G$	17% $(2^3F)^4F$
390,759	19	45% $(^1F)^2G$	31% $(2^3F)^4F$	5% $(2^3F)^2G$
382,720	4	31% $(^1G)^2G$	22% $(^1G)^2G$	18% $(^3G)^2G$
382,377	−27	57% $(^1I)^2H$	13% $(2^3G)^2H$	10% $(2^1G)^2H$
379,226	12	20% $(^3G)^2G$	12% $(2^3G)^4F$	11% $(^1G)^2H$
377,393	24	22% $(^3F)^2G$	13% $(^3H)^2G$	11% $(^3D)^4F$
375,593	8	53% $(^3D)^4F$	14% $(^3G)^2G$	9% $(^3H)^2G$
373,120	4	22% $(^3G)^4G$	14% $(^3H)^2H$	11% $(^3H)^4G$
371,061	−6	22% $(^3H)^2H$	21% $(^3G)^2H$	13% $(^3H)^4G$
370,501	1	13% $(^3G)^2H$	11% $(^3G)^2G$	10% $(^3H)^2H$
368,429	−8	25% $(^3G)^4H$	17% $(^3F)^4G$	14% $(^3F)^4F$
365,569	−33	43% $(^3F)^4F$	10% $(^3G)^2H$	8% $(^3G)^4H$
364,772	−14	17% $(^3F)^4G$	17% $(^3G)^4F$	13% $(^3H)^2G$
362,447	−2	30% $(^3G)^4F$	22% $(^3H)^4H$	6% $(^3G)^4H$
361,650	0	34% $(^3H)^4G$	22% $(^3G)^4G$	13% $(^3F)^2G$
360,482	0	15% $(^3F)^4G$	13% $(^3G)^4F$	12% $(^1G)^2H$
356,416	−9	33% $(^3H)^4I$	11% $(^3G)^4H$	11% $(^3G)^4G$
354,122	−6	29% $(^3H)^4H$	14% $(^3H)^4I$	13% $(^3H)^2G$
349,721	−13	27% $(^3H)^4I$	27% $(^3H)^4H$	18% $(^3G)^4H$
348,377	−14	43% $(^5D)^6D$	29% $(^5D)^4F$	5% $(^3H)^4I$
342,289	−12	41% $(^5D)^6D$	27% $(^5D)^6F$	24% $(^5D)^4F$
334,849 *		69% $(^5D)^6F$	15% $(^5D)^4F$	9% $(^5D)^6D$
		$J = 11/2$		
409,441	8	64% $(2^1G)^2H$	26% $(^1G)^2H$	5% $(^1I)^2H$
399,022	−1	78% $(2^3F)^4G$	14% $(^3F)^4G$	4% $(2^1G)^2H$
383,539	−14	24% $(^1I)^2H$	20% $(^1G)^2H$	17% $(2^1G)^2H$
377,717	14	38% $(^1G)^2H$	24% $(^1I)^2H$	14% $(^3G)^2H$
376,414	−1	32% $(^3G)^4G$	23% $(^3G)^2H$	18% $(^3H)^4G$
372,676	6	45% $(^3H)^2H$	14% $(^1I)^2H$	14% $(^3G)^4G$
370,488	−1	20% $(^3G)^4H$	18% $(^3G)^2H$	15% $(^3G)^4G$
369,661	6	33% $(^1I)^2I$	21% $(^3H)^2I$	15% $(^3H)^2H$
367,483	−2	41% $(^3F)^4G$	21% $(^3H)^2I$	9% $(^1I)^2I$
365,017	−2	28% $(^3H)^4H$	23% $(^3G)^4H$	17% $(^3G)^2H$
362,535	−1	35% $(^3H)^2I$	14% $(^3F)^4G$	13% $(^3G)^2H$
360,481	1	52% $(^3H)^4I$	14% $(^3G)^4G$	14% $(^3H)^4H$
357,689	2	45% $(^3H)^4G$	12% $(^3H)^2H$	12% $(^3G)^4G$
352,082	−7	39% $(^3H)^4H$	30% $(^3H)^4I$	17% $(^3G)^4H$
340,947 *		96% $(^5D)^6F$	3% $(^3F)^4G$	1% $(2^3F)^4G$
		$J = 13/2$		
376,419	14	55% $(^1I)^2I$	38% $(^1I)^2K$	4% $(^3H)^4H$
372,822	−2	52% $(^3G)^4H$	29% $(^3H)^2I$	8% $(^3H)^4H$
369,654	7	39% $(^1I)^2K$	23% $(^1I)^2I$	19% $(^3H)^2I$
364,518	12	34% $(^3H)^2I$	22% $(^3G)^4H$	20% $(^3H)^4I$
363,614	−26	49% $(^3H)^4I$	37% $(^3H)^4H$	6% $(^1I)^2K$
354,719	3	48% $(^3H)^4H$	24% $(^3H)^4I$	13% $(^3H)^2I$
		$J = 15/2$		
378,355	9	94% $(^1I)^2K$	6% $(^3H)^4I$	
365,186 *		94% $(^3H)^4I$	6% $(^1I)^2K$	

[a] The star * indicates a calculated value for the level; [b] The difference between the observed and the calculated energies; [c] For the eigenvector composition, up to three components with the largest percentages in the LS-coupling scheme are listed. The number preceding the terms is the seniority number.

Table A7. Energies (in cm^{-1}) of the $4d^4$ configuration of Ag VIII.

E [a]	o-c [b]	Eigenvector Composition [c]		
		$J = 0$		
114,851 *		82% 0^1S1	17% 4^1S2	1% 2^3P1
65,762 *		62% 2^3P1	35% 4^3P2	3% 4^1S2
45,338 *		72% 4^1S2	16% 0^1S1	11% 4^3P2
21,309	−37	50% 4^3P2	32% 2^3P1	8% 4^5D
0	−26	92% 4^5D	4% 2^3P1	4% 4^3P2
		$J = 1$		
63,371	−50	63% 2^3P1	37% 4^3P2	0% 4^3D
39,191	−7	95% 4^3D	3% 2^3P1	1% 4^3P2
26,526	19	60% 4^3P2	32% 2^3P1	4% 4^3D
1,340	−10	96% 4^5D	2% 2^3P1	2% 4^3P2
		$J = 2$		
88,586	34	80% 2^1D1	19% 4^1D2	0% 4^3D
61,644	6	68% 2^3F1	24% 4^3F2	5% 4^1D2
58,958 *		64% 2^3P1	28% 4^3P2	3% 4^1D2
48,505 *		61% 4^1D2	13% 2^1D1	10% 4^3D
38,402	17	63% 4^3D	12% 2^3P1	8% 4^3P2
32,134	−19	55% 4^3P2	22% 4^3D	21% 2^3P1
28,051	4	70% 4^3F2	21% 2^3F1	5% 4^1D2
3,212	−7	98% 4^5D	1% 2^3P1	1% 4^3D
		$J = 3$		
62,552	−10	72% 2^3F1	17% 4^3F2	10% 4^1F
52,004	31	85% 4^1F	7% 2^3F1	5% 4^3D
36,879	−39	92% 4^3D	3% 4^1F	1% 2^3F1
32,926	23	50% 4^3F2	42% 4^3G	7% 2^3F1
26,967	11	55% 4^3G	31% 4^3F2	12% 2^3F1
5,292	1	98% 4^5D	1% 4^3F2	1% 4^3D
		$J = 4$		
69,584	−60	66% 2^1G1	28% 4^1G2	4% 2^3F1
60,980	42	83% 2^3F1	10% 4^3F2	6% 2^1G1
43,104	−4	52% 4^1G2	18% 2^1G1	16% 4^3F2
32,698	−11	56% 4^3G	20% 4^3F2	16% 4^1G2
29,555	17	42% 4^3F2	25% 4^3H	21% 4^3G
23,302	−4	68% 4^3H	15% 4^3G	6% 4^3F2
7,443	−5	95% 4^5D	3% 4^3F2	1% 2^3F1
		$J = 5$		
35,077	11	84% 4^3G	16% 4^3H	
26,250	12	84% 4^3H	16% 4^3G	
		$J = 6$		
39,744	−4	91% 4^1I	9% 4^3H	
28,185	−5	91% 4^3H	9% 4^1I	

[a] The star * indicates a calculated value for the level; [b] The difference between the observed and the calculated energies; [c] For the eigenvector composition, up to three components with the largest percentages in the LS-coupling scheme are listed. The number preceding the terms is the seniority number. The number following the terms displays Nielson and Koster sequential indices [20].

Table A8. Energies (in cm^{-1}) of the $4d^35p$ configuration of Ag VIII.

E [a]	o-c [b]	Eigenvector Composition [c]		
		$J = 0$		
460,564 *		72% $(1^2D)^3P$	26% $(3^2D)^3P$	1% $(3^2P)^3P$
422,372 *		44% $(3^2D)^3P$	30% $(3^2P)^3P$	12% $(3^2P)^1S$
410,445 *		48% $(3^4P)^3P$	27% $(3^2P)^1S$	8% $(3^2D)^3P$
400,473 *		53% $(3^2P)^3P$	15% $(3^4P)^3P$	14% $(3^2D)^3P$
398,971 *		39% $(3^2P)^1S$	39% $(3^4P)^5D$	10% $(3^4F)^5D$
391,021 *		34% $(3^4P)^5D$	27% $(3^4P)^3P$	17% $(3^2P)^1S$
382,525	−78	73% $(3^4F)^5D$	20% $(3^4P)^5D$	2% $(3^2D)^3P$
		$J = 1$		
466,227	−16	73% $(1^2D)^1P$	18% $(3^2D)^1P$	4% $(1^2D)^3P$
459,403 *		64% $(1^2D)^3P$	22% $(3^2D)^3P$	6% $(1^2D)^3D$
444,840 *		53% $(1^2D)^3D$	19% $(3^2F)^3D$	15% $(3^2D)^3D$
433,769 *		60% $(3^2F)^3D$	26% $(1^2D)^3D$	8% $(3^2D)^1P$
430,256 *		45% $(3^2P)^1P$	13% $(3^4P)^3S$	10% $(3^2D)^1P$
426,468	33	69% $(3^4P)^3S$	10% $(3^2D)^3P$	5% $(3^4P)^3P$
419,422 *		33% $(3^2D)^3P$	22% $(3^2D)^1P$	11% $(3^4P)^3S$
415,902 *		47% $(3^4P)^3D$	27% $(3^2P)^3S$	9% $(3^2P)^3P$
413,344	47	33% $(3^2P)^3P$	25% $(3^4P)^3D$	22% $(3^2P)^3S$
412,118	−9	55% $(3^2D)^3D$	11% $(3^2P)^3S$	7% $(1^2D)^3D$
411,728 *		35% $(3^2P)^1P$	13% $(3^2D)^1P$	10% $(3^2P)^3S$
404,530	9	54% $(3^2P)^3D$	13% $(3^4P)^3P$	9% $(3^2P)^3S$
403,127 *		44% $(3^4P)^3P$	30% $(3^4P)^5D$	8% $(3^2P)^3D$
399,869	32	16% $(3^2P)^3P$	15% $(3^2D)^3P$	14% $(3^4P)^5P$
393,703 *		37% $(3^4P)^5D$	23% $(3^4P)^3P$	19% $(3^4F)^5D$
391,283	81	72% $(3^4P)^5P$	7% $(3^2P)^3S$	6% $(3^2P)^3P$
386,317	−2	53% $(3^4F)^5F$	26% $(3^4F)^3D$	7% $(3^4P)^3D$
383,568	−48	64% $(3^4F)^5D$	21% $(3^4P)^5D$	6% $(3^4F)^5F$
374,891	13	47% $(3^4F)^3D$	36% $(3^4F)^5F$	7% $(3^4F)^5D$
		$J = 2$		
457,523 *		59% $(1^2D)^3P$	14% $(3^2D)^3P$	12% $(1^2D)^3D$
453,343	−75	36% $(1^2D)^1D$	23% $(1^2D)^3F$	12% $(3^2F)^1D$
447,730	−20	31% $(1^2D)^3F$	21% $(3^2F)^1D$	10% $(3^2D)^3F$
444,532	−36	44% $(1^2D)^3D$	12% $(1^2D)^3F$	12% $(1^2D)^3P$
432,431	−90	54% $(3^2F)^3D$	16% $(1^2D)^3D$	11% $(1^2D)^1D$
431,074 *		30% $(1^2D)^1D$	26% $(3^2F)^1D$	12% $(3^2F)^3D$
424,635	−19	33% $(3^2D)^1D$	26% $(3^2P)^1D$	17% $(3^2F)^3F$
421,941 *		26% $(3^2F)^3F$	22% $(3^2P)^1D$	12% $(3^2P)^3P$
420,031	58	25% $(3^2P)^3P$	22% $(3^4P)^3D$	14% $(3^2D)^3P$
417,758	31	26% $(3^2D)^3D$	16% $(3^2F)^3F$	16% $(3^2P)^3P$
412,894	30	28% $(3^4P)^3D$	21% $(3^4P)^5S$	16% $(3^2D)^3P$
412,412	−44	50% $(3^2P)^3D$	18% $(3^2G)^3F$	11% $(3^2D)^3F$
411,461	−29	48% $(3^4P)^5S$	20% $(3^4P)^3P$	11% $(3^2D)^3P$
409,522	−14	19% $(3^2D)^3F$	15% $(3^2F)^1D$	14% $(3^2D)^1D$
408,409	−16	52% $(3^2G)^3F$	12% $(3^2P)^3D$	9% $(3^2P)^1D$
406,270	−12	25% $(3^2D)^3D$	23% $(3^2P)^3P$	22% $(3^2D)^3P$
402,086 *		25% $(3^4P)^3P$	23% $(3^4P)^5D$	16% $(3^4P)^5P$
400,904	31	23% $(3^4P)^5D$	21% $(3^4F)^3F$	16% $(3^2D)^3F$
396,041	59	36% $(3^4P)^5P$	18% $(3^4P)^5D$	15% $(3^4F)^5D$
394,552 *		52% $(3^4F)^3F$	8% $(3^2D)^3F$	7% $(3^4F)^5G$
393,630	−4	34% $(3^4P)^5P$	27% $(3^4P)^3P$	13% $(3^4F)^3D$
386,800	−9	55% $(3^4F)^5F$	12% $(3^4F)^3D$	5% $(3^4P)^3D$
385,302	−69	55% $(3^4F)^5D$	22% $(3^4P)^5D$	5% $(3^4F)^5F$
377,661	4	38% $(3^4F)^3D$	33% $(3^4F)^5F$	17% $(3^4F)^5D$
371,899 *		86% $(3^4F)^5G$	7% $(3^4F)^3F$	3% $(3^2D)^3F$

Table A8. *Cont.*

E [a]	o-c [b]	Eigenvector Composition [c]		
		J = 3		
460,104	100	62% $(1^2D)^1F$	15% $(3^2D)^1F$	11% $(1^2D)^3F$
452,808	172	41% $(1^2D)^3D$	35% $(1^2D)^3F$	9% $(3^2D)^3F$
446,318	−13	33% $(1^2D)^3D$	26% $(1^2D)^3F$	13% $(1^2D)^1F$
436,451	−25	74% $(3^2F)^1F$	11% $(3^2F)^3D$	5% $(3^2F)^3G$
430,642	2	26% $(3^2F)^3D$	19% $(3^2F)^3F$	19% $(3^2F)^3G$
427,542 *		21% $(3^2H)^3G$	19% $(3^2F)^3D$	18% $(3^2F)^3G$
424,006	−25	24% $(3^2P)^3D$	23% $(3^2D)^1F$	11% $(3^2G)^1F$
422,895	77	25% $(3^2F)^3F$	22% $(3^2D)^3D$	17% $(3^2H)^3G$
420,123	36	24% $(3^2F)^3F$	17% $(3^4P)^3D$	11% $(3^2P)^3D$
417,724	43	24% $(3^2H)^3G$	23% $(3^2F)^3G$	16% $(3^4P)^3D$
415,165	114	24% $(3^4P)^3D$	14% $(3^2G)^1F$	10% $(3^2H)^3G$
411,469	−67	24% $(3^2D)^3D$	18% $(3^2P)^3D$	15% $(3^2F)^3D$
409,750	−9	25% $(3^2G)^1F$	13% $(3^2D)^1F$	11% $(3^2D)^3F$
409,432	−14	31% $(3^2D)^3F$	20% $(3^2G)^3F$	15% $(3^2D)^3D$
407,506	39	23% $(3^2G)^3F$	22% $(3^2G)^3G$	11% $(3^4F)^3F$
402,520 *		62% $(3^4P)^5D$	11% $(3^4F)^5D$	8% $(3^2P)^3D$
399,399	46	72% $(3^4P)^5P$	9% $(3^4F)^3D$	5% $(3^2P)^3D$
398,704	20	35% $(3^2G)^3G$	21% $(3^2G)^1F$	16% $(3^4F)^3G$
397,779	1	59% $(3^4F)^3F$	7% $(3^2D)^3F$	6% $(3^4P)^5P$
390,887	−10	29% $(3^4F)^5F$	21% $(3^4F)^3D$	18% $(3^4F)^3G$
389,944	−8	31% $(3^4F)^3G$	15% $(3^4F)^3D$	14% $(3^4F)^5D$
387,329	−2	34% $(3^4F)^5F$	25% $(3^4F)^5D$	19% $(3^4F)^3G$
381,580	−5	40% $(3^4F)^5D$	29% $(3^4F)^5F$	21% $(3^4F)^3D$
375,456 *		87% $(3^4F)^5G$	5% $(3^4F)^3F$	2% $(3^4F)^3G$
		J = 4		
455,261	29	77% $(1^2D)^3F$	15% $(3^2D)^3F$	3% $(3^2F)^1G$
437,322	112	61% $(3^2F)^1G$	28% $(3^2H)^1G$	6% $(3^2F)^3G$
431,659 *		48% $(3^2F)^3G$	25% $(3^2F)^3F$	18% $(3^2H)^3G$
424,999	1	47% $(3^2F)^3F$	13% $(3^2H)^3G$	11% $(3^2F)^3G$
420,822	37	35% $(3^2H)^3G$	16% $(3^2F)^3G$	14% $(3^2G)^3F$
417,492	−62	67% $(3^2D)^3F$	10% $(1^2D)^3F$	7% $(3^4F)^3F$
412,977	6	22% $(3^2F)^1G$	17% $(3^2G)^1G$	13% $(3^2F)^3G$
412,041	−47	33% $(3^2H)^1G$	28% $(3^2G)^1G$	13% $(3^2H)^3G$
408,236	37	30% $(3^4P)^5D$	14% $(3^2H)^3H$	13% $(3^2G)^3H$
407,173	−50	57% $(3^4P)^5D$	14% $(3^2G)^3F$	9% $(3^2H)^3H$
405,162	−21	52% $(3^2G)^3G$	16% $(3^2G)^3F$	13% $(3^2H)^3H$
400,219	−10	58% $(3^4F)^3F$	18% $(3^2G)^3F$	3% $(3^2H)^3G$
399,265	14	34% $(3^2G)^1G$	18% $(3^2G)^3H$	13% $(3^2H)^3H$
396,481	18	43% $(3^4F)^3G$	19% $(3^2H)^3H$	18% $(3^2G)^3H$
393,193	−2	34% $(3^4F)^5F$	32% $(3^2G)^3H$	11% $(3^2H)^3H$
388,964	4	33% $(3^4F)^5F$	23% $(3^4F)^3G$	16% $(3^4F)^5G$
385,218	−11	67% $(3^4F)^5D$	15% $(3^4F)^5F$	6% $(3^4F)^3F$
379,386 *		80% $(3^4F)^5G$	7% $(3^4F)^3G$	6% $(3^4F)^5F$
		J = 5		
432,154	32	85% $(3^2F)^3G$	13% $(3^2H)^3G$	1% $(3^2G)^1H$
421,021	4	45% $(3^2H)^3G$	16% $(3^2G)^1H$	13% $(3^2H)^1H$
415,995	−15	52% $(3^2H)^1H$	24% $(3^2G)^3G$	17% $(3^2H)^3I$
413,329	57	61% $(3^2G)^1H$	20% $(3^2H)^3G$	10% $(3^2H)^3I$
408,910	1	54% $(3^2G)^3H$	31% $(3^2H)^3H$	9% $(3^2G)^3G$
407,246	−19	38% $(3^2H)^3I$	36% $(3^2G)^3G$	8% $(3^2H)^3G$
405,144	26	26% $(3^2H)^3H$	23% $(3^2H)^1H$	11% $(3^2H)^3I$
399,831	−1	47% $(3^4F)^3G$	23% $(3^4F)^5F$	16% $(3^2H)^3H$
395,955	−2	32% $(3^4F)^5F$	22% $(3^2G)^3H$	16% $(3^2H)^3I$
390,458	−6	40% $(3^4F)^5G$	28% $(3^4F)^5F$	18% $(3^4F)^3G$
383,426	−48	58% $(3^4F)^5G$	15% $(3^4F)^3G$	13% $(3^4F)^5F$

Table A8. *Cont.*

E [a]	o-c [b]	Eigenvector Composition [c]		
		$J = 6$		
421,189	52	67% $(3^2H)^1I$	19% $(3^2G)^3H$	7% $(3^2H)^3I$
413,999	−7	66% $(3^2H)^3I$	24% $(3^2H)^3H$	5% $(3^2H)^1I$
410,851	20	57% $(3^2G)^3H$	25% $(3^2H)^1I$	14% $(3^2H)^3H$
403,359	−13	55% $(3^2H)^3H$	25% $(3^2H)^3I$	14% $(3^2G)^3H$
391,512 *		94% $(3^4F)^5G$	6% $(3^2G)^3H$	
		$J = 7$		
417,658	−35	100% $(3^2H)^3I$		

[a] The star * indicates a calculated value for the level; [b] The difference between the observed and the calculated energies; [c] For the eigenvector composition, up to three components with the largest percentages in the LS-coupling scheme are listed. The number preceding the terms is the seniority number.

Table A9. Energies (in cm^{-1}) of the $4d^34f + 4p^54d^5$ configurations of Ag VIII.

E [a]	o-c [b]	J	Eigenvector Composition [c]		
557,732	258	1	24% $4d^34f$ $(^4F)^5D$	20% $4p^54d^5$ $(^4F)^5D$	10% $4p^54d^5$ $(^4D)^5D$
557,962 *		4	40% $4p^54d^5$ $(^2G1)^3H$	10% $4p^54d^5$ $(^2D1)^3F$	6% $4p^54d^5$ $(^2F2)^3G$
558,654	182	1	22% $4p^54d^5$ $(^4G)^5F$	18% $4d^34f$ $(^4F)^5F$	17% $4d^34f$ $(^4P)^5F$
560,251	257	2	31% $4d^34f$ $(^4F)^5D$	27% $4p^54d^5$ $(^4F)^5D$	13% $4p^54d^5$ $(^4D)^5D$
560,857	186	2	27% $4p^54d^5$ $(^4G)^5F$	21% $4d^34f$ $(^4F)^5F$	20% $4d^34f$ $(^4P)^5F$
560,877 *		4	24% $4p^54d^5$ $(^2D1)^3F$	13% $4p^54d^5$ $(^2D3)^3F$	11% $4p^54d^5$ $(^2G1)^3F$
562,295	122	3	21% $4d^34f$ $(^4F)^5D$	21% $4p^54d^5$ $(^4F)^5D$	13% $4p^54d^5$ $(^4G)^5F$
562,825	92	3	14% $4p^54d^5$ $(^4G)^5F$	10% $4d^34f$ $(^4P)^5F$	10% $4d^34f$ $(^4F)^5F$
564,383	−53	4	20% $4p^54d^5$ $(^4F)^5D$	17% $4d^34f$ $(^4F)^5D$	16% $4p^54d^5$ $(^4G)^5F$
565,233 *		3	14% $4p^54d^5$ $(^2D1)^3F$	11% $4p^54d^5$ $(^2D1)^1F$	9% $4d^34f$ $(^2D1)^1F$
564,902	−365	4	14% $4p^54d^5$ $(^4G)^5F$	13% $4p^54d^5$ $(^4F)^5D$	10% $4d^34f$ $(^4F)^5F$
565,427	−49	5	17% $4p^54d^5$ $(^4G)^5F$	10% $4d^34f$ $(^4F)^5F$	9% $4d^34f$ $(^4P)^5F$
567,115	119	7	21% $4d^34f$ $(^2H)^1K$	18% $4d^34f$ $(^4F)^3I$	15% $4d^34f$ $(^2H)^3I$
568,319	12	5	15% $4d^34f$ $(^4F)^3I$	13% $4p^54d^5$ $(^2I)^3I$	13% $4p^54d^5$ $(^4G)^5F$
568,352 *		2	26% $4p^54d^5$ $(^2P)^3D$	14% $4p^54d^5$ $(^2D1)^3D$	10% $4p^54d^5$ $(^2D1)^3F$
569,385	138	6	22% $4d^34f$ $(^4F)^3I$	18% $4p^54d^5$ $(^2I)^3I$	18% $4d^34f$ $(^2H)^3I$
571,903 *		1	38% $4p^54d^5$ $(^2P)^3P$	20% $4p^54d^5$ $(^2D1)^3P$	18% $4p^54d^5$ $(^2D3)^3P$
573,036	−36	4	17% $4d^34f$ $(^4F)^3H$	11% $4p^54d^5$ $(^2H)^3H$	8% $4d^34f$ $(^2G)^3H$
576,375 *		1	20% $4p^54d^5$ $(^2F2)^3D$	10% $4d^34f$ $(^2P)^3D$	10% $4d^34f$ $(^2D2)^3D$
576,746	−170	5	17% $4d^34f$ $(^4F)^3H$	10% $4p^54d^5$ $(^2H)^3H$	8% $4d^34f$ $(^2G)^3H$
577,085	−795	3	14% $4p^54d^5$ $(^2H)^3G$	12% $4d^34f$ $(^2G)^3G$	11% $4p^54d^5$ $(^2G2)^3G$
578,963	209	7	29% $4d^34f$ $(^2H)^1K$	18% $4d^34f$ $(^2G)^1K$	14% $4p^54d^5$ $(^2I)^1K$
578,767 *		2	21% $4p^54d^5$ $(^2P)^3P$	15% $4p^54d^5$ $(^2D1)^3P$	12% $4p^54d^5$ $(^2D2)^3P$
579,661 *		1	36% $4p^54d^5$ $(^2P)^3D$	11% $4p^54d^5$ $(^2D1)^3D$	7% $4d^34f$ $(^2D1)^3D$
580,373 *		3	15% $4p^54d^5$ $(^2D2)^1F$	15% $4p^54d^5$ $(^2D1)^3F$	8% $4d^34f$ $(^2D2)^1F$
580,680 *		2	13% $4p^54d^5$ $(^2F2)^3D$	9% $4d^34f$ $(^2P)^3D$	6% $4p^54d^5$ $(^2F1)^3D$
580,568 *		0	39% $4p^54d^5$ $(^2P)^3P$	20% $4p^54d^5$ $(^2D3)^3P$	12% $4p^54d^5$ $(^2P)^1S$
581,094 *		1	32% $4d^34f$ $(^2D1)^1P$	19% $4d^34f$ $(^2D2)^1P$	12% $4p^54d^5$ $(^2P)^1P$
580,923	−221	6	22% $4d^34f$ $(^4F)^3H$	14% $4p^54d^5$ $(^4G)^3H$	14% $4p^54d^5$ $(^2H)^3H$
581,306	22	4	11% $4p^54d^5$ $(^2G2)^3G$	11% $4d^34f$ $(^4P)^3G$	7% $4d^34f$ $(^2G)^3G$
582,024	−42	5	13% $4d^34f$ $(^4P)^3G$	11% $4p^54d^5$ $(^2G2)^3G$	7% $4d^34f$ $(^2G)^3G$
582,154	−19	1	30% $4p^54d^5$ $(^6S)^5P$	10% $4d^34f$ $(^4F)^5P$	10% $4p^54d^5$ $(^4P)^3S$
583,603	46	3	15% $4p^54d^5$ $(^2G1)^3F$	10% $4d^34f$ $(^2D2)^3F$	10% $4p^54d^5$ $(^4F)^3F$
585,291 *		3	10% $4p^54d^5$ $(^2D2)^3D$	9% $4p^54d^5$ $(^2F2)^3D$	6% $4d^34f$ $(^2G)^3D$
587,262 *		2	10% $4p^54d^5$ $(^2P)^3D$	9% $4p^54d^5$ $(^2D1)^3F$	7% $4p^54d^5$ $(^2G1)^3F$
589,229	−70	4	12% $4p^54d^5$ $(^2I)^3H$	8% $4d^34f$ $(^2H)^3H$	5% $4p^54d^5$ $(^4F)^3G$
589,963	238	2	12% $4p^54d^5$ $(^2D1)^3F$	10% $4p^54d^5$ $(^2D1)^3P$	8% $4p^54d^5$ $(^2D2)^3D$

Table A9. *Cont.*

E [a]	o-c [b]	*J*	Eigenvector Composition [c]		
589,786	−107	4	9% $4p^54d^5\,(^2H)^3G$	4% $4p^54d^5\,(^2I)^3H$	4% $4d^34f\,(^2D2)^1G$
590,761 *		1	9% $4p^54d^5\,(^2D2)^3D$	8% $4p^54d^5\,(^2F1)^3D$	8% $4p^54d^5\,(^2D2)^1P$
591,056 *		2	13% $4p^54d^5\,(^2D1)^3F$	11% $4p^54d^5\,(^2G2)^3F$	8% $4p^54d^5\,(^2G1)^3F$
591,442	124	3	17% $4p^54d^5\,(^4F)^3G$	6% $4d^34f\,(^4F)^3G$	6% $4p^54d^5\,(^4G)^3G$
591,781 *		1	11% $4p^54d^5\,(^6S)^5P$	11% $4p^54d^5\,(^2D1)^3P$	7% $4p^54d^5\,(^4D)^3P$
592,118	−49	5	9% $4p^54d^5\,(^2H)^1H$	9% $4d^34f\,(^2G)^1H$	8% $4d^34f\,(^2D1)^3G$
593,449 *		2	19% $4p^54d^5\,(^6S)^5P$	6% $4p^54d^5\,(^2D2)^3F$	6% $4p^54d^5\,(^2D1)^3P$
593,578 *		0	17% $4p^54d^5\,(^4P)^3P$	16% $4p^54d^5\,(^2D1)^3P$	15% $4p^54d^5\,(^4D)^3P$
593,064	−698	3	15% $4p^54d^5\,(^2F1)^3D$	11% $4p^54d^5\,(^4D)^3D$	11% $4d^34f\,(^2P)^3D$
594,881	−367	2	27% $4p^54d^5\,(^6S)^5P$	8% $4p^54d^5\,(^4D)^5P$	8% $4p^54d^5\,(^4P)^5P$
596,171	−15	4	9% $4p^54d^5\,(^2I)^3H$	9% $4p^54d^5\,(^2G2)^1G$	8% $4d^34f\,(^2H)^3H$
598,210	207	4	16% $4p^54d^5\,(^4F)^3G$	7% $4p^54d^5\,(^2G2)^3G$	6% $4p^54d^5\,(^2F1)^1G$
598,784 *		3	17% $4p^54d^5\,(^2G1)^3G$	15% $4d^34f\,(^2D1)^3G$	5% $4p^54d^5\,(^2F2)^3D$
599,018	15	1	11% $4p^54d^5\,(^2P)^3S$	10% $4d^34f\,(^2F)^3P$	9% $4p^54d^5\,(^2S)^3P$
599,345	109	5	15% $4p^54d^5\,(^4F)^3G$	10% $4p^54d^5\,(^2I)^3H$	8% $4p^54d^5\,(^2G2)^3G$
602,225	723	4	8% $4p^54d^5\,(^4F)^3F$	7% $4p^54d^5\,(^2G2)^1G$	7% $4d^34f\,(^2F)^1G$
601,789 *		2	11% $4p^54d^5\,(^2D3)^3P$	8% $4d^34f\,(^2F)^3P$	7% $4p^54d^5\,(^4P)^3P$
602,116 *		3	9% $4p^54d^5\,(^4F)^3G$	8% $4d^34f\,(^2G)^3F$	8% $4p^54d^5\,(^2G2)^3F$
603,864	132	5	20% $4d^34f\,(^2D1)^1H$	11% $4p^54d^5\,(^2I)^3H$	9% $4p^54d^5\,(^2G1)^1H$
604,385	−19	4	24% $4p^54d^5\,(^2G1)^3G$	13% $4d^34f\,(^2D1)^3G$	7% $4d^34f\,(^2H)^3G$
604,600	3	3	43% $4p^54d^5\,(^6S)^5P$	12% $4p^54d^5\,(^4D)^5P$	9% $4p^54d^5\,(^4P)^5P$
605,132	−102	6	25% $4p^54d^5\,(^2I)^3H$	14% $4p^54d^5\,(^2I)^1I$	13% $4p^54d^5\,(^4G)^3H$
606,529	91	5	17% $4p^54d^5\,(^4G)^3G$	16% $4p^54d^5\,(^4F)^3G$	8% $4d^34f\,(^4F)^3G$
607,167 *		1	16% $4p^54d^5\,(^2P)^3S$	14% $4p^54d^5\,(^4P)^3S$	6% $4p^54d^5\,(^2D3)^1P$
607,348 *		2	16% $4p^54d^5\,(^4D)^3P$	9% $4p^54d^5\,(^2D1)^3P$	7% $4p^54d^5\,(^4P)^3P$
607,709 *		0	27% $4p^54d^5\,(^2P)^3P$	20% $4p^54d^5\,(^2S)^3P$	19% $4d^34f\,(^2F)^3P$
608,132 *		1	12% $4p^54d^5\,(^2D2)^1P$	10% $4p^54d^5\,(^2D2)^3D$	6% $4d^34f\,(^2G)^3D$
607,628	−663	3	10% $4p^54d^5\,(^2F2)^3D$	10% $4p^54d^5\,(^2F1)^1F$	8% $4d^34f\,(^2F)^1F$
613,166	−109	4	12% $4p^54d^5\,(^4G\ )^3F$	9% $4p^54d^5\,(^2F2)^1G$	7% $4d^34f\,(^2G)^1G$
614,370	76	5	19% $4p^54d^5\,(^2G1)^1H$	13% $4p^54d^5\,(^2G1)^3G$	6% $4p^54d^5\,(^2I)^1H$
615,653 *		3	11% $4p^54d^5\,(^2D1)^3D$	8% $4p^54d^5\,(^4D)^3D$	7% $4p^54d^5\,(^2F1)^3D$
615,971 *		1	19% $4p^54d^5\,(^4D)^3D$	16% $4p^54d^5\,(^2D3)^3D$	11% $4p^54d^5\,(^4P)^3D$
616,249 *		2	11% $4p^54d^5\,(^4D)^3D$	11% $4p^54d^5\,(^2D2)^1D$	11% $4p^54d^5\,(^2D3)^3D$
617,860 *		1	34% $4p^54d^5\,(^2D1)^3D$	7% $4d^34f\,(^2D1)^3D$	6% $4d^34f\,(^2D2)^1P$
619,061 *		2	26% $4p^54d^5\,(^2D1)^3D$	11% $4p^54d^5\,(^4P)^3D$	9% $4d^34f\,(^2D1)^3D$
620,379 *		3	15% $4p^54d^5\,(^4G)^3F$	13% $4p^54d^5\,(^4D)^3F$	5% $4p^54d^5\,(^4G)^3G$
620,664	128	2	13% $4p^54d^5\,(^4G)^3F$	10% $4p^54d^5\,(^2F1)^1D$	9% $4p^54d^5\,(^2D3)^1D$
621,268	−115	6	41% $4p^54d^5\,(^2I)^1I$	16% $4p^54d^5\,(^2I)^3H$	11% $4d^34f\,(^2H)^1I$
622,060	−146	4	20% $4p^54d^5\,(^4D)^3F$	12% $4p^54d^5\,(^4G)^3G$	8% $4p^54d^5\,(^2G1)^3F$
622,489 *		3	15% $4p^54d^5\,(^4D)^3D$	9% $4p^54d^5\,(^4P)^3D$	7% $4p^54d^5\,(^4G)^3G$
624,693 *		2	15% $4p^54d^5\,(^2D2)^1D$	10% $4p^54d^5\,(^2P)^1D$	8% $4p^54d^5\,(^4P)^3D$
627,779	−430	3	12% $4p^54d^5\,(^4P)^3D$	11% $4p^54d^5\,(^2G2)^1F$	8% $4p^54d^5\,(^2F2)^1F$
630,389	405	3	14% $4p^54d^5\,(^2D3)^1F$	11% $4p^54d^5\,(^4G)^3G$	11% $4p^54d^5\,(^2F1)^1F$
630,848 *		2	14% $4p^54d^5\,(^2P)^1D$	10% $4p^54d^5\,(^2F2)^1D$	7% $4p^54d^5\,(^4G)^3F$
634,810	230	4	33% $4p^54d^5\,(^4G)^3G$	8% $4p^54d^5\,(^4F)^3G$	8% $4p^54d^5\,(^4G)^3F$
635,050	−166	3	22% $4p^54d^5\,(^4G)^3G$	11% $4p^54d^5\,(^2F2)^1F$	7% $4p^54d^5\,(^2F1)^1F$
635,935	510	5	41% $4p^54d^5\,(^4G)^3G$	18% $4p^54d^5\,(^2H)^3G$	13% $4p^54d^5\,(^4F)^3G$
635,583 *		1	29% $4p^54d^5\,(^2D3)^1P$	15% $4p^54d^5\,(^2S)^1P$	6% $4p^54d^5\,(^2D3)^3D$
636,245 *		1	39% $4p^54d^5\,(^4P)^3S$	21% $4p^54d^5\,(^2P)^3S$	8% $4d^34f\,(^2F)^3S$
641,516	56	2	16% $4p^54d^5\,(^4F)^3F$	7% $4p^54d^5\,(^4F)^3D$	5% $4p^54d^5\,(^4D)^3P$
643,491 *		3	18% $4p^54d^5\,(^4F)^3D$	11% $4p^54d^5\,(^2D1)^3D$	9% $4p^54d^5\,(^2D1)^1F$
644,043	065	5	28% $4p^54d^5\,(^2H)^1H$	26% $4p^54d^5\,(^2I)^1H$	21% $4p^54d^5\,(^2G2)^1H$
643,939	−245	4	18% $4p^54d^5\,(^2H)^1G$	15% $4p^54d^5\,(^4F)^3F$	13% $4p^54d^5\,(^2G1)^1G$
644,760 *		1	25% $4p^54d^5\,(^4D)^3P$	10% $4p^54d^5\,(^2D2)^3P$	9% $4p^54d^5\,(^4F)^3D$

Table A9. *Cont.*

E [a]	o-c [b]	*J*	Eigenvector Composition [c]		
645,906	221	2	15% $4p^5 4d^5$ (^{2}D2)^{1}D	11% $4p^5 4d^5$ (^{2}D1)^{1}D	7% $4d^3 4f$ (^{2}D1)^{1}D
646,185	348	2	11% $4p^5 4d^5$ (^{4}D)^{3}P	9% $4p^5 4d^5$ (^{4}F)^{3}F	7% $4p^5 4d^5$ (^{2}D2)^{3}P
647,898 *		0	40% $4p^5 4d^5$ (^{4}D)^{3}P	14% $4p^5 4d^5$ (^{2}D2)^{3}P	10% $4p^5 4d^5$ (^{4}P)^{3}P
647,982	30	3	22% $4p^5 4d^5$ (^{4}F)^{3}F	12% $4p^5 4d^5$ (^{4}F)^{3}D	5% $4p^5 4d^5$ (^{2}G1)^{3}F
650,256	171	4	19% $4p^5 4d^5$ (^{4}F)^{3}F	13% $4p^5 4d^5$ (^{2}G1)^{1}G	11% $4p^5 4d^5$ (^{2}G1)^{3}F
650,545	−375	2	23% $4p^5 4d^5$ (^{4}F)^{3}D	7% $4p^5 4d^5$ (^{2}P)^{3}D	5% $4p^5 4d^5$ (^{4}F)^{3}F
651,358 *		3	18% $4p^5 4d^5$ (^{2}F1)^{1}F	15% $4p^5 4d^5$ (^{2}G2)^{1}F	14% $4p^5 4d^5$ (^{2}D1)^{1}F
664,590 *		1	40% $4p^5 4d^5$ (^{2}P)^{1}P	15% $4p^5 4d^5$ (^{2}D1)^{1}P	9% $4d^3 4f$ (^{2}D1)^{1}P
668,708	−344	3	32% $4p^5 4d^5$ (^{2}G1)^{1}F	15% $4p^5 4d^5$ (^{2}D2)^{1}F	13% $4p^5 4d^5$ (^{2}F2)^{1}F

[a] The star * indicates a calculated value for the level; [b] The difference between the observed and the calculated energies; [c] For the eigenvector composition, up to three components with the largest percentages in the LS-coupling scheme are listed. The number following the terms displays Nielson and Koster sequential indices [20].

Table A10. Energies (in cm^{-1}) of the $4d^3$ configuration of Ag IX.

E [a]	o-c [b]	Eigenvector Composition [c]		
		J = 1/2		
28,502 *		87% ^{2}P	13% ^{4}P	
17,851 *		87% ^{4}P	13% ^{2}P	
		J = 3/2		
68,707	48	76% ^{2}D1	24% ^{2}D2	
36,360	−6	57% ^{2}P	27% ^{2}D2	10% ^{2}D1
26,823	24	41% ^{2}D2	25% ^{4}P	20% ^{2}P
17,169	56	69% ^{4}P	23% ^{2}P	5% ^{2}D2
0	12	95% ^{4}F	3% ^{2}D2	1% ^{2}D1
		J = 5/2		
68,249	−22	83% ^{2}D1	12% ^{2}D2	3% ^{2}F
44,595	−41	96% ^{2}F	2% ^{2}D1	2% ^{2}D2
33,051	10	84% ^{2}D2	12% ^{2}D1	2% ^{4}P
22,387	−10	97% ^{4}P	2% ^{2}D1	1% ^{2}D2
3103	0	98% ^{4}F	1% ^{2}D2	0% ^{2}D1
		J = 7/2		
44,082	−3	98% ^{2}F	1% ^{2}G	0% ^{4}F
22,160	−24	96% ^{2}G	2% ^{4}F	1% ^{2}F
6532	4	97% ^{4}F	2% ^{2}G	0% ^{2}F
		J = 9/2		
32,349	19	57% ^{2}H	41% ^{2}G	2% ^{4}F
23,357	−22	50% ^{2}G	43% ^{2}H	7% ^{4}F
9867	−13	91% ^{4}F	8% ^{2}G	1% ^{2}H
		J = 11/2		
31,532	15	100% ^{2}H		

[a] The star * indicates a calculated value for the level; [b] The difference between the observed and the calculated by orthogonal parameter technique energies; [c] For the eigenvector composition, up to three components with the largest percentages in the LS-coupling scheme are listed. The number following the terms displays Nielson and Koster sequential indices [20].

Table A11. Energies (in cm^{-1}) of the 4d^{2}5p + 4d^{2}4f + 4p^{5}4d^4 configurations of Ag IX.

E [a]	o-c [b]	*J*	Config. [c]	Eigenvector Composition [d]		
424,571	90	5/2	4d^{2}5p	71% 4d^{2}5p (^{3}F)^{4}G	17% 4d^{2}5p (^{3}F)^{2}F	7% 4d^{2}5p (^{1}D)^{2}F
427,386 *		7/2	4p^{5}4d^4	81% 4p^{5}4d^4 (^{5}D)^{6}D	11% 4p^{5}4d^4 (^{5}D)^{6}F	6% 4p^{5}4d^4 (^{5}D)^{6}P
427,916	−112	3/2	4d^{2}5p	61% 4d^{2}5p (^{3}F)^{4}F	25% 4d^{2}5p (^{3}F)^{2}D	7% 4d^{2}5p (^{3}F)^{4}D
428,156 *		5/2	4p^{5}4d^4	78% 4p^{5}4d^4 (^{5}D)^{6}D	9% 4p^{5}4d^4 (^{5}D)^{6}P	8% 4p^{5}4d^4 (^{5}D)^{6}F
428,306 *		9/2	4p^{5}4d^4	85% 4p^{5}4d^4 (^{5}D)^{6}D	12% 4p^{5}4d^4 (^{5}D)^{6}F	1% 4p^{5}4d^4 (^{5}D)^{4}F
430,308 *		3/2	4p^{5}4d^4	76% 4p^{5}4d^4 (^{5}D)^{6}D	9% 4p^{5}4d^4 (^{5}D)^{6}P	5% 4p^{5}4d^4 (^{5}D)^{6}F
431,109 *		7/2	4d^{2}5p	77% 4d^{2}5p (^{3}F)^{4}G	9% 4d^{2}5p (^{3}F)^{4}F	9% 4d^{2}5p (^{3}F)^{2}F
432,569	157	5/2	4d^{2}5p	56% 4d^{2}5p (^{3}F)^{4}F	16% 4d^{2}5p (^{3}F)^{2}D	16% 4d^{2}5p (^{3}F)^{4}D
433,690 *		1/2	4p^{5}4d^4	84% 4p^{5}4d^4 (^{5}D)^{6}D	6% 4p^{5}4d^4 (^{5}D)^{4}P	4% 4p^{5}4d^4 (^{3}D)^{4}P
437,442	34	9/2	4d^{2}5p	58% 4d^{2}5p (^{3}F)^{4}G	25% 4d^{2}5p (^{3}F)^{4}F	11% 4d^{2}5p (^{3}F)^{2}G
437,662	68	1/2	4d^{2}5p	49% 4d^{2}5p (^{3}F)^{4}D	35% 4d^{2}5p (^{3}P)^{4}D	6% 4d^{2}5p (^{3}P)^{2}S
437,662	−478	7/2	4d^{2}5p	53% 4d^{2}5p (^{3}F)^{4}F	27% 4d^{2}5p (^{3}F)^{4}D	10% 4d^{2}5p (^{3}F)^{2}F
438,297	−107	5/2	4d^{2}5p	29% 4d^{2}5p (^{3}F)^{4}F	22% 4d^{2}5p (^{3}F)^{4}G	22% 4d^{2}5p (^{3}F)^{2}F
438,725	−103	3/2	4d^{2}5p	36% 4d^{2}5p (^{3}F)^{4}D	28% 4d^{2}5p (^{3}F)^{4}F	14% 4d^{2}5p (^{3}F)^{2}D
440,957 *		3/2	4d^{2}5p	25% 4d^{2}5p (^{3}P)^{4}D	25% 4d^{2}5p (^{3}F)^{2}D	18% 4d^{2}5p (^{3}P)^{2}D
441,473 *		1/2	4d^{2}5p	73% 4d^{2}5p (^{3}P)^{2}S	15% 4d^{2}5p (^{3}P)^{4}P	6% 4d^{2}5p (^{3}F)^{4}D
441,637	−122	7/2	4d^{2}5p	26% 4d^{2}5p (^{3}F)^{4}D	23% 4d^{2}5p (^{3}F)^{2}F	20% 4d^{2}5p (^{3}F)^{4}F
442,423	120	5/2	4d^{2}5p	46% 4d^{2}5p (^{3}F)^{4}D	25% 4d^{2}5p (^{3}P)^{4}D	13% 4d^{2}5p (^{1}D)^{2}F
444,195	220	9/2	4d^{2}5p	43% 4d^{2}5p (^{3}F)^{4}F	41% 4d^{2}5p (^{3}F)^{4}G	10% 4d^{2}5p (^{3}F)^{2}G
444,319 *		11/2	4p^{5}4d^4	92% 4p^{5}4d^4 (^{5}D)^{6}F	5% 4p^{5}4d^4 (^{3}F2)^{4}G	1% 4p^{5}4d^4 (^{3}F1)^{4}G
444,355 *		3/2	4d^{2}5p	50% 4d^{2}5p (^{3}P)^{4}S	15% 4d^{2}5p (^{1}D)^{2}P	6% 4d^{2}5p (^{3}F)^{4}D
445,529 *		5/2	4p^{5}4d^4	41% 4p^{5}4d^4 (^{3}H)^{4}G	32% 4d^{2}4f (^{3}F)^{4}G	10% 4p^{5}4d^4 (^{3}G)^{4}G
446,642	−4	5/2	4d^{2}5p	41% 4d^{2}5p (^{3}F)^{2}F	20% 4d^{2}5p (^{3}F)^{2}D	18% 4d^{2}5p (^{3}P)^{2}D
447,243 *		7/2	4d^{2}5p	43% 4d^{2}5p (^{3}F)^{2}G	18% 4d^{2}5p (^{1}G)^{2}G	10% 4d^{2}5p (^{3}F)^{4}F
447,255	−279	7/2	4p^{5}4d^4	33% 4p^{5}4d^4 (^{3}H)^{4}G	30% 4d^{2}4f (^{3}F)^{4}G	11% 4p^{5}4d^4 (^{3}G)^{4}G
447,694 *		11/2	4d^{2}5p	97% 4d^{2}5p (^{3}F)^{4}G	1% 4d^{2}5p (^{1}G)^{2}H	1% 4d^{2}4f (^{3}F)^{4}G
448,369 *		7/2	4d^{2}4f	32% 4d^{2}4f (^{3}F)^{2}G	10% 4p^{5}4d^4 (^{3}H)^{2}G	8% 4d^{2}5p (^{3}F)^{2}G
448,623 *		3/2	4p^{5}4d^4	28% 4p^{5}4d^4 (^{3}P2)^{4}S	11% 4d^{2}5p (^{1}D)^{2}P	8% 4p^{5}4d^4 (^{3}P2)^{4}P
448,989	232	7/2	4d^{2}5p	31% 4d^{2}5p (^{3}F)^{2}F	22% 4d^{2}5p (^{3}F)^{4}D	12% 4d^{2}5p (^{3}P)^{4}D
448,995 *		1/2	4d^{2}5p	54% 4d^{2}5p (^{3}P)^{4}D	36% 4d^{2}5p (^{3}F)^{4}D	3% 4d^{2}5p (^{3}P)^{2}P
449,006 *		9/2	4p^{5}4d^4	16% 4p^{5}4d^4 (^{5}D)^{6}F	16% 4d^{2}4f (^{3}F)^{4}G	14% 4p^{5}4d^4 (^{3}H)^{4}G
450,334 *		3/2	4p^{5}4d^4	16% 4p^{5}4d^4 (^{3}G)^{4}F	12% 4d^{2}4f (^{3}F)^{4}F	10% 4p^{5}4d^4 (^{3}F2)^{4}D
451,183	−53	3/2	4d^{2}5p	31% 4d^{2}5p (^{3}P)^{4}S	23% 4d^{2}5p (^{1}D)^{2}P	13% 4d^{2}5p (^{1}D)^{2}D
451,424 *		9/2	4p^{5}4d^4	28% 4p^{5}4d^4 (^{5}D)^{6}F	14% 4p^{5}4d^4 (^{3}H)^{4}G	13% 4d^{2}4f (^{3}F)^{4}G
451,430 *		5/2	4p^{5}4d^4	17% 4p^{5}4d^4 (^{5}D)^{6}F	16% 4p^{5}4d^4 (^{3}F2)^{4}D	12% 4p^{5}4d^4 (^{5}D)^{6}P
451,946 *		1/2	4p^{5}4d^4	41% 4p^{5}4d^4 (^{3}F2)^{4}D	12% 4p^{5}4d^4 (^{5}D)^{4}D	12% 4p^{5}4d^4 (^{3}D)^{4}D
452,041 *		7/2	4p^{5}4d^4	28% 4p^{5}4d^4 (^{5}D)^{6}F	11% 4p^{5}4d^4 (^{3}F2)^{4}D	9% 4p^{5}4d^4 (^{5}D)^{6}P
452,292	198	9/2	4d^{2}5p	22% 4d^{2}5p (^{1}G)^{2}G	21% 4d^{2}5p (^{3}F)^{4}F	18% 4d^{2}5p (^{3}F)^{2}G
452,569	249	5/2	4d^{2}5p	59% 4d^{2}5p (^{1}D)^{2}F	12% 4d^{2}5p (^{3}F)^{2}F	11% 4d^{2}5p (^{3}F)^{2}D
452,857 *		3/2	4d^{2}5p	28% 4d^{2}5p (^{3}P)^{4}D	15% 4d^{2}5p (^{3}F)^{4}D	11% 4p^{5}4d^4 (^{5}D)^{6}P
452,951 *		9/2	4p^{5}4d^4	21% 4p^{5}4d^4 (^{5}D)^{6}F	20% 4d^{2}4f (^{3}F)^{2}G	8% 4p^{5}4d^4 (^{3}H)^{2}G
453,404 *		3/2	4d^{2}5p	24% 4d^{2}5p (^{3}P)^{4}D	13% 4p^{5}4d^4 (^{3}G)^{4}F	11% 4d^{2}5p (^{3}F)^{4}D
453,632 *		11/2	4p^{5}4d^4	26% 4d^{2}4f (^{3}F)^{4}G	22% 4p^{5}4d^4 (^{3}H)^{4}G	17% 4p^{5}4d^4 (^{3}F2)^{4}G
453,902	−209	7/2	4d^{2}5p	52% 4d^{2}5p (^{1}G)^{2}G	14% 4d^{2}5p (^{1}D)^{2}F	11% 4d^{2}5p (^{3}F)^{2}G
454,133 *		5/2	4p^{5}4d^4	27% 4p^{5}4d^4 (^{3}G)^{4}F	12% 4d^{2}4f (^{3}F)^{4}F	10% 4p^{5}4d^4 (^{3}F2)^{4}D
455,093 *		7/2	4p^{5}4d^4	13% 4p^{5}4d^4 (^{3}F2)^{4}D	13% 4p^{5}4d^4 (^{5}D)^{6}F	13% 4p^{5}4d^4 (^{3}G)^{4}F
455,362 *		3/2	4p^{5}4d^4	31% 4p^{5}4d^4 (^{5}D)^{6}P	18% 4p^{5}4d^4 (^{3}D)^{4}P	15% 4p^{5}4d^4 (^{3}F2)^{4}D
455,506 *		5/2	4p^{5}4d^4	20% 4p^{5}4d^4 (^{3}P1)^{4}P	15% 4d^{2}4f (^{3}F)^{4}P	10% 4p^{5}4d^4 (^{3}P2)^{4}D
457,009 *		1/2	4d^{2}5p	50% 4d^{2}5p (^{1}D)^{2}P	39% 4d^{2}5p (^{3}P)^{4}P	5% 4d^{2}5p (^{3}F)^{4}D
457,621	41	5/2	4d^{2}5p	49% 4d^{2}5p (^{3}P)^{4}D	19% 4d^{2}5p (^{3}F)^{4}D	12% 4d^{2}5p (^{3}F)^{2}D
457,734 *		7/2	4d^{2}4f	80% 4d^{2}4f (^{3}F)^{4}H	3% 4p^{5}4d^4 (^{3}G)^{4}H	3% 4p^{5}4d^4 (^{3}H)^{4}G
457,900	111	9/2	4d^{2}5p	52% 4d^{2}5p (^{3}F)^{2}G	33% 4d^{2}5p (^{1}G)^{2}G	10% 4d^{2}5p (^{1}G)^{2}H
457,882 *		9/2	4p^{5}4d^4	22% 4d^{2}4f (^{3}F)^{4}H	9% 4d^{2}4f (^{3}F)^{4}I	9% 4p^{5}4d^4 (^{3}H)^{4}H
458,434 *		1/2	4d^{2}4f	26% 4d^{2}4f (^{3}F)^{2}S	18% 4p^{5}4d^4 (^{3}D)^{4}P	16% 4d^{2}4f (^{3}F)^{4}P
458,503 *		9/2	4d^{2}4f	34% 4d^{2}4f (^{3}F)^{4}H	33% 4d^{2}4f (^{3}F)^{4}I	7% 4d^{2}4f (^{1}D)^{2}H
458,762 *		11/2	4p^{5}4d^4	39% 4d^{2}4f (^{3}F)^{4}H	23% 4p^{5}4d^4 (^{3}H)^{4}H	15% 4p^{5}4d^4 (^{3}G)^{4}H
459,044 *		7/2	4p^{5}4d^4	19% 4d^{2}4f (^{3}F)^{4}F	16% 4p^{5}4d^4 (^{5}D)^{6}F	15% 4p^{5}4d^4 (^{3}P2)^{4}D
459,146	−45	3/2	4d^{2}5p	59% 4d^{2}5p (^{3}P)^{4}P	11% 4d^{2}5p (^{1}D)^{2}P	9% 4d^{2}5p (^{1}D)^{2}D
460,347 *		13/2	4p^{5}4d^4	31% 4d^{2}4f (^{3}F)^{4}H	23% 4p^{5}4d^4 (^{3}H)^{4}H	22% 4p^{5}4d^4 (^{3}G)^{4}H
460,559	−65	7/2	4d^{2}5p	37% 4d^{2}5p (^{3}P)^{4}D	15% 4d^{2}5p (^{1}D)^{2}F	12% 4d^{2}5p (^{3}F)^{4}D
460,753 *		5/2	4p^{5}4d^4	11% 4p^{5}4d^4 (^{3}F2)^{4}F	11% 4d^{2}4f (^{3}F)^{2}F	11% 4d^{2}4f (^{3}F)^{4}F
460,784 *		9/2	4d^{2}4f	25% 4p^{5}4d^4 (^{3}D)^{4}F	20% 4d^{2}4f (^{3}F)^{4}I	13% 4d^{2}4f (^{3}F)^{4}H
460,974 *		11/2	4d^{2}4f	57% 4d^{2}4f (^{3}F)^{4}I	13% 4d^{2}4f (^{3}F)^{4}H	7% 4p^{5}4d^4 (^{3}H)^{4}I

Table A11. *Cont.*

E [a]	o-c [b]	*J*	Config. [c]	Eigenvector Composition [d]		
461,009 *		1/2	4d^{2}5p	29% 4d^{2}5p (^{3}P)^{4}P	22% 4d^{2}5p (^{1}D)^{2}P	10% 4d^{2}4f (^{3}F)^{2}S
461,370 *		5/2	4p^{5}4d^4	15% 4p^{5}4d^4 (^{5}D)^{6}P	12% 4d^{2}5p (^{3}P)^{4}P	12% 4d^{2}5p (^{1}D)^{2}D
461,715 *		3/2	4p^{5}4d^4	15% 4d^{2}5p (^{1}D)^{2}D	11% 4p^{5}4d^4 (^{5}D)^{6}F	9% 4d^{2}5p (^{3}P)^{4}P
461,846 *		5/2	4d^{2}5p	30% 4d^{2}5p (^{1}D)^{2}D	22% 4d^{2}5p (^{3}P)^{4}P	10% 4d^{2}5p (^{3}P)^{4}D
461,926 *		3/2	4d^{2}5p	21% 4d^{2}5p (^{1}D)^{2}D	18% 4d^{2}5p (^{3}P)^{4}P	8% 4d^{2}5p (^{1}D)^{2}P
462,202 *		9/2	4d^{2}4f	29% 4p^{5}4d^4 (^{1}I)^{2}H	18% 4d^{2}4f (^{3}F)^{4}I	16% 4d^{2}4f (^{1}G)^{2}H
462,353 *		1/2	4d^{2}5p	17% 4d^{2}4f (^{3}F)^{2}S	14% 4d^{2}5p (^{3}P)^{4}P	11% 4d^{2}4f (^{3}F)^{4}P
462,396 *		5/2	4p^{5}4d^4	16% 4p^{5}4d^4 (^{5}D)^{6}P	15% 4p^{5}4d^4 (^{3}D)^{4}P	10% 4p^{5}4d^4 (^{5}D)^{4}P
462,523 *		7/2	4p^{5}4d^4	20% 4p^{5}4d^4 (^{3}F2)^{4}F	12% 4p^{5}4d^4 (^{3}F1)^{4}F	7% 4p^{5}4d^4 (^{3}D)^{4}F
462,635 *		3/2	4p^{5}4d^4	18% 4p^{5}4d^4 (^{5}D)^{6}F	7% 4p^{5}4d^4 (^{3}P2)^{4}S	7% 4d^{2}4f (^{3}F)^{4}F
463,146 *		9/2	4p^{5}4d^4	31% 4d^{2}4f (^{3}F)^{4}F	20% 4p^{5}4d^4 (^{3}F2)^{4}F	13% 4p^{5}4d^4 (^{3}F1)^{4}F
463,748 *		13/2	4d^{2}4f	53% 4d^{2}4f (^{3}F)^{4}I	26% 4d^{2}4f (^{3}F)^{4}H	16% 4p^{5}4d^4 (^{3}H)^{4}I
464,491 *		15/2	4p^{5}4d^4	58% 4p^{5}4d^4 (^{3}H)^{4}I	40% 4d^{2}4f (^{3}F)^{4}I	1% 4p^{5}4d^4 (^{1}I)^{2}K
464,230	−294	7/2	4d^{2}5p	21% 4d^{2}5p (^{1}D)^{2}F	16% 4d^{2}4f (^{3}F)^{4}D	8% 4p^{5}4d^4 (^{3}P2)^{4}D
464,852 *		11/2	4p^{5}4d^4	21% 4p^{5}4d^4 (^{1}I)^{2}H	18% 4d^{2}4f (^{1}D)^{2}H	16% 4d^{2}4f (^{1}G)^{2}H
465,027 *		1/2	4d^{2}4f	41% 4d^{2}4f (^{3}F)^{4}D	13% 4p^{5}4d^4 (^{5}D)^{6}F	10% 4d^{2}4f (^{3}F)^{2}S
465,143 *		3/2	4p^{5}4d^4	19% 4d^{2}4f (^{3}F)^{4}D	9% 4p^{5}4d^4 (^{1}F)^{2}D	7% 4d^{2}4f (^{1}D)^{2}D
465,394	−335	7/2	4d^{2}5p	28% 4d^{2}5p (^{3}P)^{4}D	17% 4d^{2}5p (^{1}D)^{2}F	11% 4p^{5}4d^4 (^{3}P2)^{4}D
465,872 *		5/2	4d^{2}5p	24% 4d^{2}5p (^{3}P)^{4}P	17% 4d^{2}5p (^{1}D)^{2}D	8% 4d^{2}5p (^{3}P)^{2}D
465,990	62	9/2	4d^{2}5p	68% 4d^{2}5p (^{1}G)^{2}H	28% 4d^{2}5p (^{1}G)^{2}G	1% 4d^{2}5p (^{3}F)^{4}F
466,458	26	5/2	4d^{2}5p	19% 4d^{2}5p (^{1}D)^{2}D	16% 4d^{2}5p (^{3}P)^{4}P	7% 4p^{5}4d^4 (^{3}F2)^{4}F
467,537 *		7/2	4p^{5}4d^4	11% 4d^{2}4f (^{3}F)^{2}G	10% 4p^{5}4d^4 (^{3}H)^{4}H	9% 4p^{5}4d^4 (^{3}G)^{4}G
467,914 *		3/2	4p^{5}4d^4	15% 4d^{2}4f (^{3}F)^{4}D	10% 4d^{2}5p (^{3}P)^{2}D	6% 4d^{2}4f (^{3}F)^{4}P
467,929 *		1/2	4p^{5}4d^4	26% 4p^{5}4d^4 (^{5}D)^{6}F	17% 4p^{5}4d^4 (^{3}F2)^{4}D	14% 4d^{2}4f (^{3}F)^{4}D
468,817	138	3/2	4d^{2}5p	46% 4d^{2}5p (^{3}P)^{2}D	15% 4d^{2}5p (^{3}P)^{2}P	11% 4d^{2}5p (^{3}F)^{2}D
468,895 *		11/2	4p^{5}4d^4	34% 4p^{5}4d^4 (^{3}G)^{4}G	24% 4p^{5}4d^4 (^{3}F2)^{4}G	13% 4d^{2}4f (^{3}F)^{4}H
469,199 *		5/2	4d^{2}4f	45% 4d^{2}4f (^{3}F)^{4}D	6% 4p^{5}4d^4 (^{3}P1)^{4}P	6% 4p^{5}4d^4 (^{5}D)^{6}F
469,249 *		11/2	4d^{2}4f	23% 4d^{2}4f (^{3}F)^{4}H	21% 4d^{2}4f (^{3}F)^{4}I	10% 4p^{5}4d^4 (^{3}H)^{4}H
469,387 *		3/2	4p^{5}4d^4	13% 4d^{2}4f (^{3}P)^{4}D	10% 4p^{5}4d^4 (^{3}F2)^{4}F	6% 4d^{2}4f (^{3}F)^{4}D
469,401 *		7/2	4p^{5}4d^4	19% 4d^{2}4f (^{3}P)^{4}D	14% 4p^{5}4d^4 (^{5}D)^{6}P	13% 4p^{5}4d^4 (^{3}P2)^{4}D
470,136 *		9/2	4p^{5}4d^4	21% 4p^{5}4d^4 (^{3}H)^{4}H	18% 4p^{5}4d^4 (^{3}G)^{4}G	15% 4d^{2}4f (^{3}F)^{4}H
470,867 *		7/2	4d^{2}4f	14% 4d^{2}4f (^{3}P)^{4}D	14% 4d^{2}4f (^{3}F)^{2}F	13% 4p^{5}4d^4 (^{3}H)^{4}H
471,042 *		5/2	4p^{5}4d^4	20% 4d^{2}4f (^{3}P)^{4}D	11% 4d^{2}4f (^{1}D)^{2}D	9% 4p^{5}4d^4 (^{3}D)^{4}D
471,491 *		1/2	4d^{2}5p	46% 4d^{2}5p (^{3}P)^{2}P	17% 4d^{2}4f (^{3}F)^{2}P	8% 4d^{2}4f (^{1}D)^{2}P
471,667	31	11/2	4d^{2}5p	92% 4d^{2}5p (^{1}G)^{2}H	2% 4p^{5}4d^4 (^{3}F2)^{4}G	1% 4d^{2}5p (^{3}F)^{4}G
471,960 *		5/2	4d^{2}5p	25% 4d^{2}5p (^{3}P)^{2}D	13% 4d^{2}5p (^{3}F)^{2}D	11% 4d^{2}5p (^{3}P)^{4}P
472,224 *		3/2	4d^{2}4f	23% 4d^{2}4f (^{3}F)^{2}P	21% 4d^{2}4f (^{1}D)^{2}P	6% 4d^{2}4f (^{3}F)^{4}D
472,431 *		5/2	4p^{5}4d^4	19% 4d^{2}5p (^{3}P)^{2}D	12% 4d^{2}4f (^{1}D)^{2}D	8% 4p^{5}4d^4 (^{1}F)^{2}D
472,982 *		9/2	4p^{5}4d^4	15% 4p^{5}4d^4 (^{3}F2)^{4}G	12% 4p^{5}4d^4 (^{3}G)^{4}G	12% 4d^{2}4f (^{3}F)^{2}G
473,124 *		3/2	4p^{5}4d^4	16% 4p^{5}4d^4 (^{1}S2)^{2}P	14% 4d^{2}4f (^{1}D)^{2}P	11% 4p^{5}4d^4 (^{1}F)^{2}D
473,235 *		13/2	4d^{2}4f	33% 4d^{2}4f (^{3}F)^{4}H	28% 4d^{2}4f (^{3}F)^{4}I	26% 4p^{5}4d^4 (^{3}H)^{4}H
474,478 *		11/2	4d^{2}4f	36% 4d^{2}4f (^{1}G)^{2}I	28% 4d^{2}4f (^{3}F)^{2}I	6% 4p^{5}4d^4 (^{3}F2)^{4}G
474,797 *		7/2	4d^{2}5p	61% 4d^{2}5p (^{1}G)^{2}F	7% 4d^{2}5p (^{1}G)^{2}G	6% 4d^{2}5p (^{1}D)^{2}F
474,983 *		1/2	4d^{2}4f	31% 4d^{2}4f (^{1}D)^{2}P	27% 4d^{2}5p (^{3}P)^{2}P	10% 4p^{5}4d^4 (^{1}D2)^{2}P
475,453	150	7/2	4p^{5}4d^4	19% 4p^{5}4d^4 (^{3}G)^{4}G	13% 4d^{2}5p (^{1}G)^{2}F	7% 4p^{5}4d^4 (^{3}D)^{4}F
475,428 *		9/2	4p^{5}4d^4	32% 4d^{2}4f (^{3}P)^{4}G	18% 4p^{5}4d^4 (^{3}F2)^{4}G	12% 4p^{5}4d^4 (^{1}G2)^{2}G
475,874 *		3/2	4d^{2}5p	23% 4d^{2}5p (^{3}P)^{2}P	10% 4d^{2}4f (^{3}P)^{4}D	8% 4d^{2}5p (^{3}P)^{2}D
476,126 *		5/2	4d^{2}4f	17% 4d^{2}4f (^{3}P)^{4}G	12% 4d^{2}4f (^{3}P)^{2}D	9% 4d^{2}4f (^{3}F)^{4}G
476,268 *		1/2	4p^{5}4d^4	14% 4d^{2}5p (^{3}P)^{2}P	12% 4p^{5}4d^4 (^{1}D2)^{2}P	11% 4d^{2}4f (^{3}F)^{2}P
476,658 *		13/2	4d^{2}4f	39% 4d^{2}4f (^{1}G)^{2}I	19% 4d^{2}4f (^{3}F)^{2}I	12% 4p^{5}4d^4 (^{1}I)^{2}I
476,896 *		15/2	4d^{2}4f	55% 4d^{2}4f (^{3}F)^{4}I	33% 4p^{5}4d^4 (^{3}H)^{4}I	11% 4p^{5}4d^4 (^{1}I)^{2}K
477,099 *		7/2	4p^{5}4d^4	13% 4d^{2}4f (^{3}P)^{4}G	11% 4p^{5}4d^4 (^{1}D2)^{2}F	11% 4p^{5}4d^4 (^{3}F2)^{4}G
477,494 *		1/2	4p^{5}4d^4	28% 4p^{5}4d^4 (^{5}D)^{6}F	24% 4d^{2}4f (^{3}P)^{4}D	14% 4p^{5}4d^4 (^{3}F1)^{4}D
477,525 *		5/2	4p^{5}4d^4	19% 4p^{5}4d^4 (^{3}G)^{4}G	17% 4d^{2}4f (^{1}D)^{2}F	8% 4p^{5}4d^4 (^{3}H)^{4}G
477,730 *		3/2	4d^{2}5p	35% 4d^{2}5p (^{3}P)^{2}P	8% 4d^{2}4f (^{3}P)^{4}D	6% 4d^{2}4f (^{1}D)^{2}D
478,297 *		7/2	4d^{2}4f	31% 4d^{2}4f (^{3}P)^{4}G	10% 4p^{5}4d^4 (^{1}D2)^{2}F	9% 4d^{2}4f (^{1}D)^{2}F
478,563 *		11/2	4p^{5}4d^4	23% 4p^{5}4d^4 (^{3}H)^{4}I	16% 4d^{2}4f (^{3}P)^{4}G	10% 4p^{5}4d^4 (^{3}F1)^{4}G
478,577 *		5/2	4d^{2}4f	28% 4d^{2}5p (^{1}G)^{2}F	14% 4d^{2}4f (^{3}P)^{2}D	6% 4d^{2}4f (^{3}F)^{2}D
479,341 *		9/2	4d^{2}4f	21% 4d^{2}4f (^{3}F)^{2}H	21% 4d^{2}4f (^{1}D)^{2}H	10% 4p^{5}4d^4 (^{3}G)^{2}H
479,610	45	5/2	4p^{5}4d^4	22% 4d^{2}5p (^{1}G)^{2}F	19% 4d^{2}4f (^{3}P)^{4}G	6% 4p^{5}4d^4 (^{1}G1)^{2}F
479,719 *		3/2	4p^{5}4d^4	23% 4p^{5}4d^4 (^{3}P2)^{2}D	9% 4p^{5}4d^4 (^{3}P1)^{2}D	7% 4d^{2}4f (^{3}F)^{2}D
479,718	−233	5/2	4d^{2}5p	35% 4d^{2}5p (^{1}G)^{2}F	14% 4d^{2}4f (^{3}P)^{4}G	5% 4p^{5}4d^4 (^{3}F2)^{4}G
480,710 *		13/2	4p^{5}4d^4	39% 4p^{5}4d^4 (^{3}H)^{4}I	18% 4p^{5}4d^4 (^{3}H)^{2}I	13% 4p^{5}4d^4 (^{3}G)^{4}H
570,288 *		9/2	4p^{5}4d^4	24% 4p^{5}4d^4 (^{1}G2)^{2}H	22% 4p^{5}4d^4 (^{1}G1)^{2}H	16% 4p^{5}4d^4 (^{3}F1)^{2}G
570,992 *		7/2	4p^{5}4d^4	47% 4p^{5}4d^4 (^{1}D1)^{2}F	17% 4p^{5}4d^4 (^{1}G2)^{2}F	12% 4p^{5}4d^4 (^{1}D2)^{2}F

Table A11. *Cont.*

E [a]	o-c [b]	J	Config. [c]	Eigenvector Composition [d]		
579,505	472	5/2	$4d^24f$	32% $4d^24f$ (^{3}F)^{4}G	28% $4p^54d^4$ (^{3}H)^{4}G	15% $4d^24f$ (^{3}P)^{4}G
583,771	278	7/2	$4p^54d^4$	32% $4d^24f$ (^{3}F)^{4}G	31% $4p^54d^4$ (^{3}H)^{4}G	14% $4d^24f$ (^{3}P)^{4}G
583,819	21	3/2	$4p^54d^4$	22% $4d^24f$ (^{3}F)^{4}F	20% $4p^54d^4$ (^{5}D)^{4}F	18% $4p^54d^4$ (^{3}F1)^{4}F
587,200	100	9/2	$4p^54d^4$	29% $4p^54d^4$ (^{3}H)^{4}G	27% $4d^24f$ (^{3}F)^{4}G	12% $4d^24f$ (^{3}P)^{4}G
587,642	141	11/2	$4d^24f$	17% $4p^54d^4$ (^{3}H)^{4}G	15% $4d^24f$ (^{3}F)^{4}G	12% $4d^24f$ (^{3}F)^{2}H
589,644	137	5/2	$4p^54d^4$	24% $4p^54d^4$ (^{5}D)^{4}F	23% $4d^24f$ (^{3}F)^{4}F	18% $4p^54d^4$ (^{3}G)^{4}F
590,280	472	3/2	$4p^54d^4$	14% $4d^24f$ (^{3}P)^{4}D	10% $4p^54d^4$ (^{3}F1)^{4}D	10% $4p^54d^4$ (^{3}F2)^{4}D
590,209 *		1/2	$4p^54d^4$	21% $4d^24f$ (^{3}P)^{4}D	16% $4p^54d^4$ (^{3}F1)^{4}D	15% $4d^24f$ (^{3}F)^{4}D
594,111	557	5/2	$4p^54d^4$	38% $4p^54d^4$ (^{1}D1)^{2}F	16% $4p^54d^4$ (^{1}G2)^{2}F	14% $4p^54d^4$ (^{1}D2)^{2}F
594533 *		5/2	$4p^54d^4$	17% $4p^54d^4$ (^{3}F1)^{4}D	14% $4d^24f$ (^{3}P)^{4}D	11% $4d^24f$ (^{3}F)^{4}D
595,066	−292	3/2	$4p^54d^4$	27% $4p^54d^4$ (^{3}P1)^{4}S	21% $4d^24f$ (^{3}F)^{4}S	15% $4p^54d^4$ (^{3}P2)^{4}S
596,876	171	7/2	$4p^54d^4$	25% $4p^54d^4$ (^{5}D)^{4}F	21% $4d^24f$ (^{3}F)^{4}F	19% $4p^54d^4$ (^{3}G)^{4}F
596,885	−16	11/2	$4d^24f$	22% $4d^24f$ (^{3}F)^{2}I	20% $4d^24f$ (^{1}G)^{2}I	15% $4p^54d^4$ (^{3}H)^{4}G
598,929	−85	7/2	$4p^54d^4$	20% $4p^54d^4$ (^{3}F1)^{4}D	11% $4d^24f$ (^{3}P)^{4}D	10% $4d^24f$ (^{3}F)^{4}D
599,446	190	9/2	$4p^54d^4$	18% $4p^54d^4$ (^{1}I)^{2}H	15% $4d^24f$ (^{3}F)^{2}H	13% $4d^24f$ (^{1}D)^{2}H
600,561	−15	13/2	$4d^24f$	36% $4d^24f$ (^{3}F)^{2}I	29% $4d^24f$ (^{1}G)^{2}I	21% $4p^54d^4$ (^{1}I)^{2}I
601,905 *		3/2	$4p^54d^4$	30% $4p^54d^4$ (^{1}S1)^{2}P	26% $4p^54d^4$ (^{1}D1)^{2}P	15% $4p^54d^4$ (^{1}S2)^{2}P
603,570	205	9/2	$4p^54d^4$	22% $4p^54d^4$ (^{5}D)^{4}F	17% $4p^54d^4$ (^{3}G)^{4}F	15% $4d^24f$ (^{3}F)^{4}F
605,674 *		5/2	$4p^54d^4$	11% $4d^24f$ (^{1}D)^{2}F	10% $4p^54d^4$ (^{1}G1)^{2}F	8% $4p^54d^4$ (^{3}F2)^{2}D
606,963	−39	3/2	$4p^54d^4$	22% $4d^24f$ (^{1}D)^{2}D	22% $4p^54d^4$ (^{1}F)^{2}D	6% $4p^54d^4$ (^{3}F2)^{2}D
609,152	153	7/2	$4p^54d^4$	14% $4p^54d^4$ (^{3}G)^{2}G	13% $4p^54d^4$ (^{3}F2)^{2}G	11% $4d^24f$ (^{3}F)^{2}G
612,491	−463	3/2	$4p^54d^4$	21% $4p^54d^4$ (^{5}D)^{4}D	8% $4p^54d^4$ (^{3}F2)^{4}D	7% $4p^54d^4$ (^{5}D)^{4}P
612,769	−650	5/2	$4p^54d^4$	17% $4p^54d^4$ (^{5}D)^{4}D	14% $4p^54d^4$ (^{5}D)^{4}P	7% $4p^54d^4$ (^{3}F2)^{4}D
613,700 *		1/2	$4p^54d^4$	21% $4p^54d^4$ (^{3}P2)^{2}P	19% $4p^54d^4$ (^{1}D2)^{2}P	8% $4d^24f$ (^{3}F)^{2}P
615,308	−172	11/2	$4p^54d^4$	21% $4p^54d^4$ (^{3}H)^{2}H	14% $4d^24f$ (^{3}F)^{2}H	12% $4p^54d^4$ (^{1}I)^{2}H
615,385	−297	1/2	$4p^54d^4$	40% $4p^54d^4$ (^{5}D)^{4}D	10% $4p^54d^4$ (^{3}F2)^{4}D	9% $4p^54d^4$ (^{3}P2)^{2}S
615,884	−639	7/2	$4p^54d^4$	19% $4p^54d^4$ (^{1}G2)^{2}G	18% $4p^54d^4$ (^{3}F1)^{2}G	13% $4d^24f$ (^{1}G)^{2}G
616,475	−614	9/2	$4p^54d^4$	15% $4p^54d^4$ (^{3}F2)^{2}G	14% $4p^54d^4$ (^{3}G)^{2}G	13% $4p^54d^4$ (^{1}G1)^{2}G
618,075	152	7/2	$4p^54d^4$	14% $4d^24f$ (^{3}P)^{2}F	12% $4d^24f$ (^{1}D)^{2}F	11% $4p^54d^4$ (^{1}G1)^{2}F
618,058	−125	5/2	$4p^54d^4$	13% $4p^54d^4$ (^{3}D)^{2}F	12% $4p^54d^4$ (^{1}D1)^{2}F	11% $4p^54d^4$ (^{1}F)^{2}D
618,894 *		1/2	$4p^54d^4$	20% $4p^54d^4$ (^{1}S1)^{2}P	15% $4p^54d^4$ (^{1}D1)^{2}P	12% $4p^54d^4$ (^{3}P2)^{2}S
620,387 *		3/2	$4p^54d^4$	20% $4p^54d^4$ (^{5}D)^{4}D	11% $4p^54d^4$ (^{3}D)^{2}P	6% $4d^24f$ (^{3}F)^{2}P
621,061	107	9/2	$4p^54d^4$	17% $4p^54d^4$ (^{1}G2)^{2}G	17% $4p^54d^4$ (^{3}H)^{2}H	11% $4p^54d^4$ (^{3}H)^{2}G
622,698 *		1/2	$4p^54d^4$	15% $4p^54d^4$ (^{1}S1)^{2}P	14% $4p^54d^4$ (^{1}D1)^{2}P	10% $4p^54d^4$ (^{3}D)^{2}P
625,187 *		3/2	$4p^54d^4$	12% $4p^54d^4$ (^{3}D)^{2}P	10% $4p^54d^4$ (^{5}D)^{4}P	8% $4p^54d^4$ (^{5}D)^{4}D
626,693	543	5/2	$4p^54d^4$	10% $4p^54d^4$ (^{1}D1)^{2}F	10% $4p^54d^4$ (^{3}D)^{2}F	9% $4p^54d^4$ (^{3}F2)^{2}F
628,302	215	5/2	$4p^54d^4$	27% $4p^54d^4$ (^{5}D)^{4}D	10% $4p^54d^4$ (^{3}F2)^{4}D	8% $4p^54d^4$ (^{3}F2)^{2}D
629,173	363	7/2	$4p^54d^4$	56% $4p^54d^4$ (^{5}D)^{4}D	15% $4p^54d^4$ (^{3}F2)^{4}D	6% $4d^24f$ (^{3}P)^{4}D
630,316	−277	3/2	$4p^54d^4$	26% $4p^54d^4$ (^{5}D)^{4}P	11% $4p^54d^4$ (^{3}D)^{4}P	7% $4d^24f$ (^{3}F)^{4}P
632,069	−311	9/2	$4p^54d^4$	29% $4p^54d^4$ (^{3}H)^{2}H	12% $4p^54d^4$ (^{3}F1)^{2}G	10% $4p^54d^4$ (^{3}G)^{2}G
632,322	−117	5/2	$4p^54d^4$	27% $4p^54d^4$ (^{5}D)^{4}P	14% $4p^54d^4$ (^{3}D)^{4}P	8% $4d^24f$ (^{3}F)^{4}P
634,657	4	7/2	$4p^54d^4$	24% $4p^54d^4$ (^{3}F2)^{2}F	19% $4p^54d^4$ (^{3}D)^{2}F	12% $4p^54d^4$ (^{1}G2)^{2}F
635,225 *		1/2	$4p^54d^4$	24% $4p^54d^4$ (^{3}P1)^{2}P	17% $4p^54d^4$ (^{3}P2)^{2}S	14% $4p^54d^4$ (^{1}S2)^{2}P
637,178	286	1/2	$4p^54d^4$	20% $4p^54d^4$ (^{5}D)^{4}P	17% $4p^54d^4$ (^{3}P2)^{2}S	9% $4p^54d^4$ (^{3}P1)^{2}P
638,134	− 70	5/2	$4p^54d^4$	20% $4p^54d^4$ (^{3}F2)^{2}F	14% $4p^54d^4$ (^{3}F1)^{2}F	13% $4d^24f$ (^{3}P)^{2}F
638,784	81	3/2	$4p^54d^4$	18% $4p^54d^4$ (^{3}P2)^{2}P	12% $4p^54d^4$ (^{1}D2)^{2}P	11% $4p^54d^4$ (^{3}F2)^{2}D
639,262	308	11/2	$4p^54d^4$	36% $4p^54d^4$ (^{3}H)^{2}H	32% $4p^54d^4$ (^{1}I)^{2}H	12% $4d^24f$ (^{1}G)^{2}H
641,234	− 78	7/2	$4p^54d^4$	16% $4p^54d^4$ (^{3}F1)^{2}F	16% $4p^54d^4$ (^{1}G2)^{2}F	13% $4p^54d^4$ (^{3}G)^{2}F
644,980	−248	9/2	$4p^54d^4$	33% $4p^54d^4$ (^{3}H)^{2}G	17% $4p^54d^4$ (^{3}G)^{2}G	14% $4p^54d^4$ (^{3}F2)^{2}G
647,197	−45	7/2	$4p^54d^4$	35% $4p^54d^4$ (^{3}H)^{2}G	15% $4p^54d^4$ (^{3}G)^{2}G	11% $4p^54d^4$ (^{3}F2)^{2}G
649,186	144	3/2	$4p^54d^4$	20% $4p^54d^4$ (^{3}P1)^{2}P	14% $4p^54d^4$ (^{1}D2)^{2}P	12% $4p^54d^4$ (^{3}P2)^{2}P
651,880	259	5/2	$4p^54d^4$	19% $4p^54d^4$ (^{3}P2)^{2}D	12% $4p^54d^4$ (^{3}F1)^{2}D	11% $4p^54d^4$ (^{1}D1)^{2}D
657,606	122	5/2	$4p^54d^4$	38% $4p^54d^4$ (^{3}G)^{2}F	11% $4p^54d^4$ (^{3}F1)^{2}F	10% $4p^54d^4$ (^{1}G1)^{2}F
659,798	−619	7/2	$4p^54d^4$	24% $4p^54d^4$ (^{1}G1)^{2}F	20% $4p^54d^4$ (^{3}G)^{2}F	17% $4p^54d^4$ (^{3}F1)^{2}F
664,136 *		1/2	$4p^54d^4$	37% $4p^54d^4$ (^{1}S1)^{2}P	16% $4p^54d^4$ (^{3}P1)^{2}P	13% $4p^54d^4$ (^{1}S2)^{2}P
663,396	−939	3/2	$4p^54d^4$	30% $4p^54d^4$ (^{3}P1)^{2}P	22% $4p^54d^4$ (^{1}D1)^{2}P	8% $4d^24f$ (^{1}G)^{2}P
676,533	735	5/2	$4p^54d^4$	36% $4p^54d^4$ (^{3}F1)^{2}D	13% $4p^54d^4$ (^{3}D)^{2}D	8% $4p^54d^4$ (^{1}F)^{2}D
680,196	360	3/2	$4p^54d^4$	42% $4p^54d^4$ (^{3}F1)^{2}D	13% $4p^54d^4$ (^{3}D)^{2}D	11% $4p^54d^4$ (^{1}D1)^{2}D

[a] The star * indicates a calculated value for the level; [b] The difference between the observed and the calculated energies; [c] Configuration attribution is arbitrary in a few cases (see text); [d] For the eigenvector composition, up to three components with the largest percentages in the LS-coupling scheme are listed. The number following the terms of the $4d^4$ configuration displays Nielson and Koster sequential indices [20].

Table A12. Energy parameters (in cm^{-1}) of the ground configuration in Ag VII, Ag VIII and Ag IX calculated by orthogonal parameter technique in comparison with the Dirac-Fock (DF) parameters.

Name	AgVII ($4d^5$)				AgVIII ($4d^4$)				AgIX ($4d^3$)			
	FIT	Error [a]	DF	FIT/DF	FIT	Error [a]	DF	FIT/DF	FIT	Error [a]	DF	FIT/DF
E_{av}	51914	3	60,863	0.853	37,710	6	42,776	0.882	27,795	8	31,440	0.884
O2	8652	2	10,175	0.850	8978	6	10,515	0.854	9295	8	10,834	0.858
O2′	5512	3	6923	0.796	5701	7	7128	0.800	5892	8	7323	0.805
Ea′	213	2			223	3			251	6		
Eb′	38	2			45	6			50	f		
ζ	2493	2	2428	1.027	2655	5	2603	1.020	2830	7	2782	1.017
T1	−4.62	0.08			−4.62	0.19			−4.85	0.36		
T2	0.40	f			0.50	f						
Ac	7.80	1.5	13.21	0.6	7.46	f	12.43	0.6	7.08	f	11.81	0.6
A3	1.93	r	3.27	0.6	1.90	f	3.18	0.6	1.87	f	3.13	0.6
A4	3.28	r	5.56	0.6	3.31	f	5.52	0.6	3.32	f	5.53	0.6
A5	3.16	r	5.36	0.6	3.31	f	5.51	0.6	3.42	f	5.71	0.6
A6	0.96	r	1.63	0.6	0.47	f	0.78	0.6	0.00	f	−0.00	1.0
A1	−0.10	r	−0.16	0.6	−0.05	f	−0.08	0.6	0	f	0	0.6
A2	−0.32	r	−0.55	0.6	−0.43	f	−0.72	0.6	−0.53	f	−0.88	0.6
A0	−0.49	r	0.29	0.6	−0.28	f	−0.46	0.6	−0.25	f	−0.42	0.6
σ	14				26				27			

[a] r—parameters are fixed at DF ratio to Ac, f- fixed parameter.

Table A13. Energy parameters (in cm^{-1}) of the $4d^45p$ configuration in Ag VII and $4d^35p$ configuration in Ag VIII calculated by orthogonal parameter technique in comparison with the DF parameters.

Name	Ag VII				Ag VIII			
	FIT	Error [a]	DF	FIT/DF	FIT	Error [b]	DF	FIT/DF
E_{av}	37,3862	2	384,168	0.973	41,1052	5	417,702	0.984
O2dd	8849	2	10,418	0.849	9147	8	10,738	0.852
O2′dd	5581	4	7070	0.789	5747	12	7265	0.791
Ea′	216	1			221	6		
Eb′	36	3			31	6		
T1	−4.86	0.09			−5.08	0		
T2	0.48	0.09			0.50	f		
ζ(4d)	2631	4	2572	1.023	2809	8	2747	1.022
Ac	8.01	1.8	12.68	0.63	11.76	f	11.76	1.0
A3	1.97	r1	3.12	0.63	3.44	f	3.44	1.0
A4	3.25	r1	5.15	0.63	5.40	f	5.40	1.0
A5	3.46	r1	5.48	0.63	5.68	f	5.68	1.0
A6	0.64	r1	1.02	0.62	0.18	f	0.18	1.0
A1	−0.17	r1	−0.28	0.63	−0.23	f	−0.23	1.0
A2	−0.63	r1	−1.00	0.63	−0.35	f	−0.35	1.0
A0	−0.69	r1	−1.10	0.63	−0.18	f	−0.19	1.0
C1dp	3702	4	4270	0.867	4012	11	4621	0.868
C2dp	2605	3	2998	0.869	2755	10	3176	0.867
C3dp	1258	4	1411	0.891	1321	10	1471	0.898
S1dp	67	3			63	8		
S2dp	−118	3			−129	8		
ζ(5p)	6317	5	5975	1.057	7268	14	6933	1.048
Sd.Lp	−27.49	1.5	−34.81	0.79	−27.2	f	−34.04	0.8
Sp.Ld	−2.77	r2	−3.52	0.79	−2.7	f	−3.46	0.8
Zp2ppa	−19.98	r2	−25.29	0.79	−20.0	f	−25.01	0.8
Zp2dda	13.74	r2	17.42	0.79	13.5	f	16.93	0.8
Zp1ppa	41.56	r2	52.62	0.79	41.2	f	51.56	0.8
Zp1dda	−2.94	r2	−3.71	0.79	−2.3	f	−2.99	0.8
Zp3ppa	10.85	r2	13.74	0.79	11.0	f	13.81	0.8
Zp3dda	−2.22	r2	−2.82	0.79	−2.3	f	−2.95	0.8
SS(dp)02	−1.53	r2	−1.95	0.79	−1.8	f	−2.32	0.8
SS(dp)20	−0.52	r2	−0.66	0.79	−0.3	f	−0.40	0.8

Table A13. *Cont.*

Name	Ag VII				Ag VIII			
	FIT	Error [a]	DF	FIT/DF	FIT	Error [b]	DF	FIT/DF
t16′	−23.8	2.8			−23.8	f		
t17′	8.0	2.8			8.0	f		
t18′	−10.4	2.9			−10.4	f		
t19′	−8.9	2.1			−8.9	f		
t20′	−42.2	3.6			−42.2	f		
t21′	−3.4	2.3			−3.4	f		
t22′	−14.4	4.6			−14.4	f		
t23′	−4.2	3.9			−4.2	f		
t24′	−7.5	2.9			−7.5	f		
t25′	3.6	2.5			3.6	f		
t26′	−33.5	3.0			−33.5	f		
t27′	18.5	2.4			18.5	f		
t28′	35.6	3.5			35.6	f		
t29′	−12.1	2.5			−12.1	f		
t30′	−45.5	2.7			−45.5	f		
t31′	−4.4	3.0			−4.4	f		
t32′	−0.3	2.2			−0.3	f		
t33′	11.9	2.8			11.9	f		
t34′	−30.2	3.2			−30.2	f		
t35′	−32.3	3.4			−32.3	f		
σ	19				47			

[a] r1—parameters are fixed at DF ratio to Ac, r1 parameters are fixed at DF ratio to Sd.Lp; [b] f—parameter is fixed on predetermined value.

Table A14. Fitted (FIT) with their uncertainties (Unc.) and Hartree - Fock (HF) energy parameters in cm^{-1} of the odd $4d^35p$, $4d^34f$, and $4p^54d^4$ configurations in Ag VIII and $4d^25p$, $4d^24f$ and $4p^54d^3$ configurations in Ag IX calculated with the Cowan code.

Name [a]	Ag VIII				Ag IX			
	HF	FIT	Unc. [b]	FIT/HF [c]	HF	FIT	Unc. [b]	FIT/HF [c]
$E_{av}(5p)$	417,702	412,728	26	−4974	459,300	455,716	88	−3584
$F^2(4d,4d)$	95,978	80,972	237	0.844	98,603	83,046	1159	0.842
$F^4(4d,4d)$	64,366	56,750	484	0.882	66,314	54,364	3227	0.820
α		49	5			71	26	
β		−627	−99			−600	f	
T1		−4	−1					
ζ(4d)	2702	2812	36	1.041	2870	2919	65	1.017
ζ (5p)	6510	7299	66	1.121	7426	8338	165	1.123
$F^1(4d,5p)$		−2072	−265			−2000	f	
$F^2(4d,5p)$	39,205	32,005	275	0.816	41,777	35,950	898	0.861
$G^1(4d,5p)$	12,359	10,505	137	0.850 [d]	12,926	11,649	342	0.901 [d]
$G^3(4d,5p)$	12,110	10,293	134	0.850 [d]	12,836	11,568	340	0.901 [d]
$E_{av}(4f)$	508,665	496,302	569	−12,363	522,389	507,990	380	−14,399
$F^2(4d,4d)$	95,126	80,381	f	0.845	97,640	80,359	f	0.823
$F^4(4d,4d)$	63,728	55,443	f	0.87	65,594	54,443	f	0.83
α		48	f			62	f	
β		−600	f					
T1		−4	f					
ζ(4d)	2652	2732	f	1.03	2808	2910	f	1.036
ζ(4f)	95	95	f	1.0	124	124	f	1.0

Table A14. *Cont.*

Name [a]	Ag VIII				Ag IX			
	HF	FIT	Unc. [b]	FIT/HF [c]	HF	FIT	Unc. [b]	FIT/HF [c]
F^2(4d,4f)	70,569	64,433	1452	0.913 [d]	78,433	71,374	f	0.91
F^4(4d,4f)	44,636	40,755	919	0.913 [d]	50,344	45,814	f	0.91
G^1(4d,4f)	83,516	72,648	572	0.87	93,840	85,394	f	0.91
G^3(4d,4f)	51,477	47,876	377	0.930 [d]	58,481	53,218	f	0.91
G^5(4d,4f)	36,169	33,640	265	0.930 [d]	41276	37,562	f	0.91
E_{av}(pd)	538,566	526,473	194	−12,093	529,361	52,3487	344	−5874
F^2(4d,4d)	94,305	79,014	393	0.838	97,018	85,702	496	0.883
F^4(4d,4d)	63,119	51,318	598	0.813	65,133	50,607	1080	0.777
α		48	f			60	f	
β		−600	f			−600	f	
T1		−4	f			−4	f	
ζ(4p)	29,355	29,355	f	1	30,239	30,576	415	1.011
ζ(4d)	2602	2849	f	1.095	2767	2782	116	1.005
F^2(4p,4d)	100,314	83,916	1065	0.837	102,723	79,518	973	0.774
G^1(4p,4d)	127,225	101,162	192	0.795 [d]	130,315	101,056	342	0.775 [d]
G^3(4p,4d)	79,353	63097	120	0.795 [d]	81,523	63,219	214	0.775 [d]
σ		213				327		

[a] E_{av}(5p), E_{av}(4f) and E_{av}(pd) stand for $E_{av}(4d^{k-1}5p)$, $E_{av}(4d^{k-1}4f)$ and $E_{av}(4p^5 4d^{k+1})$ for Ag VIII and Ag IX where k = 4 and 3, respectively; [b] f- parameter is fixed on predetermined value; [c] For E_{av} the FIT-HF difference is listed; [d] Adjacent pairs of parameters are linked at their HF ratios.

References

1. Benschop, H.; Joshi, Y.N.; Van Kleef, Th.A.M. The spectrum of doubly ionized silver: Ag III. *Can. J. Phys.* **1975**, *53*, 498–503. [CrossRef]
2. Van Kleef, Th.A.M.; Joshi, Y.N. Analysis of $4d^8$–$d^7 5p$ transitions in trebly ionized silver: Ag IV. *Can. J. Phys.* **1981**, *59*, 1930–1939. [CrossRef]
3. Van Kleef, Th.A.M.; Raassen, A.J.J.; Joshi, Y.N. Analysis of the $4d^7$–$4d^6 5p$ transitions in the fifth spectrum of silver (Ag V). *Phys. Scr.* **1987**, *36*, 140–148. [CrossRef]
4. Joshi, Y.N.; Raassen, A.J.J.; Van Kleef, Th.A.M.; Van der Valk, A.A. The sixth spectrum of silver: Ag VI, and a study of the parameter values in 4d-spectra. *Phys. Scr.* **1988**, *38*, 677–698. [CrossRef]
5. Sugar, J.; Kaufman, V.; Rowan, W.L. Rb-like spectra: Pd X to Nd XXIV. *J. Opt. Soc. Am. B* **1992**, *9*, 1959–1961. [CrossRef]
6. Ryabtsev, A.N.; Kononov, E.Y.; Churilov, S.S. Spectra of rubidium-like Pd X-Sn XIV ions. *Opt. Spectr.* **2008**, *105*, 844–850. [CrossRef]
7. Ryabtsev, A.N.; Kononov, E.Y. Resonance transitions in Rh VIII, Pd IX, Ag X and Cd XI spectra. *Phys. Scr.* **2011**, *84*, 015301. [CrossRef]
8. Ryabtsev, A.N.; Kononov, E.Ya. Eighth spectrum of palladium: Pd VIII. *Phys. Scr.* **2016**, *91*, 025402. [CrossRef]
9. Churilov, S.S.; Ryabtsev, A.N. Analysis of the spectra of In XII-XIV and Sn XIII-XV in the far-VUV region. *Opt. Spectr.* **2006**, *101*, 169–178. [CrossRef]
10. Churilov, S.S.; Ryabtsev, A.N. Analyses of the Sn IX-Sn XII spectra in the EUV region. *Phys. Scr.* **2006**, *73*, 614–619. [CrossRef]
11. Svensson, L.A.; Ekberg, J.O. The titanium vacuum-spark spectrum from 50 to 425 Å. *Ark. Fys.* **1969**, *40*, 145–164.
12. Azarov, V.I. Formal approach to the solution of the complex-spectra identification problem. 2. Implementaton. *Phys. Scr.* **1993**, *48*, 656–667. [CrossRef]
13. Parpia, F.A.; Froese Fischer, C.; Grant, I.P. GRASP92: A package for large-scale relativistic atomic structure calculations. *Comput. Phys. Commun.* **1996**, *94*, 249–271. [CrossRef]
14. Cowan, R.D. *The Theory of Atomic Structure and Spectra*; University of California Press: Berkeley, CA, USA, 1981.
15. Hansen, J.E.; Uylings, P.H.M.; Raassen, A.J.J. Parametric fitting with orthogonal operators. *Phys. Scr.* **1988**, *37*, 664–672. [CrossRef]

16. Hansen, J.E.; Raassen, A.J.J.; Uylings, P.H.M.; Lister, G.M.S. Parametric fitting to dn configurations using ortogonal operators. *Nucl. Instrum. Methods Phys. Res. B* **1988**, *31*, 134–138. [CrossRef]
17. Uylings, P.H.M.; Raassen, A.J.J.; Wyart, J.-F. Calculations of $5d^{N-1}6s$ systems using orthogonal operators: do orthogonal operators survive configuration interaction? *J. Phys. B* **1993**, *26*, 4683–4693. [CrossRef]
18. Uylings, P.H.M.; Raassen, A.J.J. High precision calculation of odd iron-group systems with orthogonal operators. *Phys. Scr.* **1996**, *54*, 505–513. [CrossRef]
19. Kramida, A.E. The program LOPT for least-squares optimization of energy levels. *Comput. Phys. Commun.* **2010**, *182*, 419–434. [CrossRef]
20. Nielson, C.W.; Koster, G.F. *Spectroscopic Coefficients for the* p^n, d^n, *and* f^n *Configurations*; The M.I.T. Press: Cambridge, MA, USA, 1963.
21. Ryabtsev, A.N.; Kononov, E.Y. Resonance transitions in the Pd VII spectrum. *Phys. Scr.* **2012**, *85*, 025301. [CrossRef]
22. Bauche, J.; Bauche-Arnoult, C.; Luc-Koenig, E.; Wyart, J.-F.; Klapisch, M. Emissive zones of complex atomic configurations in highly ionized atoms. *Phys. Rev. A* **1983**, *28*, 829–835. [CrossRef]
23. Windberger, A.; Torretti, F.; Borschevsky, A.; Ryabtsev, A.; Dobrodey, S.; Bekker, H.; Eliav, E.; Kaldor, U.; Ubachs, W.; Hoekstra, R.; et al. Analysis of the fine structure of $Sn^{11+...14+}$ ions by optical spectroscopy in an electron beam ion trap. *Phys. Rev. A* **2016**, *94*, 012506. [CrossRef]

atoms

MDPI

Article

The Cu II Spectrum

Alexander Kramida *, Gillian Nave and Joseph Reader

National Institute of Standards and Technology, Gaithersburg, MD 20899, USA; gillian.nave@nist.gov (G.N.); joseph.reader@nist.gov (J.R.)

* Correspondence: alexander.kramida@nist.gov; Tel.: +1-301-975-8074; Fax: +1-301-975-5560

Academic Editor: James Babb

Received: 16 December 2016; Accepted: 8 February 2017; Published: 24 February 2017

Abstract: New wavelength measurements in the vacuum ultraviolet (VUV), ultraviolet and visible spectral regions have been combined with available literature data to refine and extend the description of the spectrum of singly ionized copper (Cu II). In the VUV region, we measured 401 lines using a concave grating spectrograph and photographic plates. In the UV and visible regions, we measured 276 lines using a Fourier-transform spectrometer. These new measurements were combined with previously unpublished data from the thesis of Ross, with accurate VUV grating measurements of Kaufman and Ward, and with less accurate older measurements of Shenstone to construct a comprehensive list of ≈2440 observed lines, from which we derived a revised set of 379 optimized energy levels, complemented with 89 additional levels obtained using series formulas. Among the 379 experimental levels, 29 are new. Intensities of all lines observed in different experiments have been reduced to the same uniform scale by using newly calculated transition probabilities (A-values). We combined our calculations with published measured and calculated A-values to provide a set of 555 critically evaluated transition probabilities with estimated uncertainties, 162 of which are less than 20%.

Keywords: atomic spectra; singly ionized copper; energy levels; wavelengths; Ritz standards; transition probabilities; critical compilation

1. Introduction

The spectrum of singly ionized copper, belonging to the Ni isoelectronic sequence, has a long history of research. The most significant contributions to the analysis were made by Shenstone [1] in 1936 and by Ross [2] in 1969. Interest in this spectrum was mainly due to the fact that it has a large number of sharp distinct lines in the vacuum ultraviolet (VUV) region, as well as an equally large number of lines in the ultraviolet (UV), visible, and infrared (IR) regions, from which accurate wavelengths of the VUV lines can be established using the Ritz combination principle, thus providing a large set of lines usable as secondary VUV wavelength standards. This spectrum is also of considerable interest for astrophysics. Lines of Cu II were observed in the spectra of nebulae by Thackeray [3], Aller et al. [4], McKenna et al. [5], and by Wallerstein et al. [6], and also in interstellar H I clouds, as well as in Ap, Be, and Bp stars (Jaschek, and Jaschek [7], Danezis and Theodossiou [8]) and in the Sun (Samain [9]). Cu^+ was successfully used as an active lasing medium in various hollow cathodes. Continuous-wave or pulsed lasing has been reported for 28 lines of Cu II in a wide range from UV to IR (McNeil et al. [10,11], Jain [12], Zinchenko and Ivanov [13]). Since copper is an important impurity in tokamaks, the Cu II spectrum has potential applications in fusion research, where its lines can be used for diagnostic purposes.

For several decades after the original analysis by Shenstone [1] the VUV wavelengths of Cu II calculated from the energy levels (i.e., Ritz wavelengths) were widely used as auxiliary wavelength standards. After the refinements of the measured UV wavelengths by Reader et al. [14] and by Kaufman and Ward [15], and after improved and greatly extended measurements by Ross [2] in the UV and visible ranges, the quality of the VUV Ritz wavelengths of Cu II seemed to be so good that no further investigations of this spectrum were needed. The only inconvenience stemmed from the fact that Ross's thesis [2] was never published, although a large part of it was released with small modifications as a report of the Los Alamos National Laboratory, see Ross [16]. Energy levels and a few strongest lines from Ross [2] were included in the compilations by Sugar and Musgrove [17] and by Sansonetti and Martin [18], but the bulk of the wavelength and line intensity data remains nearly inaccessible. Thus, the initial goal of the present work was simply to digitize Ross's line lists and include them in the Atomic Spectra Database (ASD; see Kramida et al. [19]) of the National Institute of Standards and Technology (NIST) with consistent energy-level identifications. However, a close examination revealed several problems with Ross's data.

The first problem is with internal consistency of Ross's data. An important aspect of ASD is that it requires the observed wavelengths for an atom to be consistent with the energy levels tabulated for that atom. That is, the wavelengths derived from the energy levels (Ritz wavelengths) must agree with the observed wavelengths to within the stated uncertainties. However, in attempting to incorporate the data of Ross into ASD, we found that for a large number of lines the differences between observed and Ritz wavelengths were much greater than the uncertainties. We decided that, because of the importance of Ritz wavelengths for Cu II, this had to be investigated.

The second problem with Ross's data is the possible presence of systematic shifts in his wavelengths. Nave and Sansonetti [20] showed that Ritz wavelengths below 2400 Å (41667 cm^{-1}) derived from Ross's Cu II energy levels are systematically too low, the mean relative deviation being about 4×10^{-7}. This error is significantly greater than the average relative uncertainty of Ritz wavenumbers given by Ross, 1.2×10^{-7}. Near 2000 Å (50000 cm^{-1}), it corresponds to an error of 0.02 cm^{-1}, compared to the mean stated uncertainty of 0.006 cm^{-1}. This puts the usability of Ross's Ritz wavenumbers in the VUV into question.

The third problem is the absence of a consistent description of line intensities throughout the entire range of observed wavelengths. While Ross's own measurements cover the large range above 1980 Å, most of the observed intensities in the VUV region are furnished by old measurements of Shenstone [1], which were made with seven different instruments and with two different light sources. Additional measurements for some selections of strongest lines were made by Reader et al. [14] and by Kaufman and Ward [15], each with their own intensity scale.

Finally, Ross's theoretical interpretation of the energy levels, adopted by Sugar and Musgrove [17] and by Sansonetti and Martin [18], is inadequate. Ross used the *LS* coupling scheme in his analysis, as well as in his final level list. However, the observed fine-structure intervals indicate that the level structure of most configurations is best described by the $J_1 l$ (a.k.a. *JK*) coupling scheme, similar to the isoelectronic Ni I spectrum interpreted by Litzén et al. [21].

Another problem that needs to be addressed is the scarcity of available critically evaluated data on radiative transition probabilities (*A*-values). At present, the NIST ASD [19] includes only seven *A*-values of Cu II from the compilation of Wiese and Martin [22]. All of them have large estimated uncertainties ($\leq$40% for six transitions and $\leq$50% for one transition).

The main purpose of the present work is to solve the above-mentioned problems and construct a self-consistent set of recommended energy levels and wavelengths of Cu II supplemented with a uniform description of line intensities. Re-interpretation of the energy levels, as well as the search for possible classifications of previously unidentified lines, required new calculations, which necessarily produce radiative transition probabilities (*A*-values). Thus, we also critically evaluate all available data on *A*-values and extend them with our calculated ones found to be of sufficiently good accuracy.

The usability of the Cu II wavelengths as standards is limited by substantial hyperfine structure (HFS) splitting and presence of two stable isotopes, ^{63}Cu (69.15%) and ^{65}Cu (30.85%) in natural copper (Coursey et al. [23]). HFS and isotope shifts (IS) of a few tens of Cu II lines between 2218 Å and 8096 Å were first observed and roughly estimated by Shenstone [1] and then more accurately investigated by Elbel et al. [24,25]. The measured IS was in the range (0.001 to 0.101) cm^{-1}. For singly-excited $3d^9nl$ configurations the IS is smaller than 0.05 cm^{-1}, while for the doubly-excited $4d^84s^2$ configuration it is about 0.1 cm^{-1}. The HFS constants A_{hfs} for several [$3d^9$]4s, 4p, 4d, 5s, 5p, 5d, 6s and [$3d^8$]$4s^2$ and 4s5p levels were found to range from 0.022 cm^{-1} to 0.075 cm^{-1} except for one level, $3d^94s\ ^3D_1$, having a very small $A_{hfs} = -0.004$ cm^{-1}. The nuclear spin I of both ^{63}Cu and ^{65}Cu is $I = 3/2$ [18]. Thus, depending on the total angular momentum J of the levels, the width of the HFS varies between 0.016 cm^{-1} and 0.5 cm^{-1}. The ionization energy of Cu^+ is 163669.2 cm^{-1} (20.292 39 eV) (Ross [2]). Thus, excitation of this spectrum requires temperatures in excess of 6000 K (0.5 eV). Such temperatures lead to Doppler broadening $\Delta\sigma_{Dop}/\sigma > 4 \times 10^{-6}$, resulting in line widths >0.4 cm^{-1} at wavelengths near 1000 Å, 0.2 cm^{-1} near 2000 Å, and 0.1 cm^{-1} near 4000 Å. As a result, most observed lines possess unresolved or partially resolved HFS and IS. The smallness of HFS and IS explains the scarcity of studies of these effects in Cu II. However, these effects should be relatively easy to observe in the infrared region. Such observations would be very valuable.

All wavelength measurements described in the present paper were made with natural copper samples. As we know, there is no such physical entity as an atom of natural copper. Thus, the energy levels and Ritz wavelengths derived from these measurements do not pertain to a real atom, but rather are empirical values that best describe the spectrum as normally observed.

2. Wavelength Measurements

The unfortunate consequence of the thorough analysis made by Ross [2,16] is that, in the 46 years after his thesis, there were no studies devoted specifically to Cu II. Instead, his Cu II Ritz wavelengths were widely used as secondary standards for investigation of other species. In particular, the archive of the NIST Atomic Spectroscopy Group has several tens of photographic plates with VUV spectra obtained with the 10.7 m normal incidence vacuum spectrograph with a 1200 lines/mm concave grating. These plates were recorded for studies of Y, Zr, Ge, La, and other elements. For calibration purposes, separate tracks were exposed on these plates with spectra of copper hollow cathodes. The latter, in addition to copper lines, contain lines of carrier gases (helium, neon, and argon), as well as hydrogen, carbon, oxygen, nitrogen, silicon, and germanium that were present as intrinsic or deliberately introduced impurities. These spectra were obtained in different years from 1969 to 1974. In addition to that, we have several high-resolution spectra recorded with two NIST vacuum Fourier transform spectrometers (FTS) in years 2002 to 2008 using Cu/Ge/Pt/Fe/Ne and Cu/Re/Ar/He hollow cathodes. The "new" measurements described in the Abstract were all made with these old recordings. However, we carried out new reductions of the wavelengths and re-analyzed the uncertainties, so it is in this sense that the measurements can be considered as new. To obtain a consistent and comprehensive description of the Cu II spectrum, we combined these new measurements with the old published data from Shenstone [1], Kaufman and Ward [15], and Ross [2,16], which we re-analyzed to evaluate the wavelength calibration, measurement uncertainties, and observed line intensities. Since this re-analysis depends on our new measurements, we describe them first.

Table 1 lists the exposures used in the present analysis. Since exposures used in grating measurements were taken over a period of several years, several different lamp designs were employed. The design of these lamps was similar to the one described in detail by Reader and Davis [26], except that a special fitting was made to attach them to the NIST vacuum spectrograph. The cathode was solid copper with a cylindrical hole of about 7 mm in diameter. Following Kaufman and Ward [15], small pieces of Ge and Si were placed in the cathode to obtain good reference lines in regions that were not covered well by Cu II. The carrier gas was a combination of flowing helium at a pressure of about 0.5 kPa (4 Torr) and neon or argon at about 70 Pa (0.5 Torr). The entrance slit of the spectrometer separated the lamp from the spectrometer chamber, which was maintained at a residual pressure of about 7×10^{-3} Pa (5×10^{-5} Torr). The demountable high-current lamp made at the University of Hannover, used several tens of years later in the FTS measurements, had a similar design described by Danzmann et al. [27]. Ar at a pressure of 200 Pa (1.5 Torr) with an addition of He at 70 Pa (0.5 Torr) was used as a carrier gas in spectrum 14, while Ne at a pressure of 270 Pa (2 Torr) was used in spectra 12 and 13. Pieces of Ge, Pt, Fe, or Re were placed inside the cathode. The grating spectrograms covered the region (636 to 2682) Å, while the FTS spectrograms covered the region (1792 to 11733) Å.

Table 1. Spectrograms used in the present measurements.

Exposure	Track	Region (Å)	Hollow Cathode Used	Impurities	Equip. [a]	Wavelength Standards [b]	Track Label	Date	Fit Poly [c]	St. Dev. [d]
1	1	636–827	Cu/He/Ar/Ne		NIVS	Ar II (11), Ne I (1), Cu II (TW-1) [e]	418	4/3/1974	2	0.0009
2	2	822–1000	Cu/Ge/Si/He/Ar	H, C, O	NIVS	H I (2), O I (9), O II (1), Ar II (2), Ge II (3)	419_tr42	4/4/1974	2	0.0013
	6	977–1166		H, C, N, O	NIVS	C I (2), C II (1), N I (2), O I (7), Ar I (2), Si II (2), Ge II (3)	419_tr42	4/4/1974	2	0.0013
3	5	824–1020	Cu/He	H, O	NIVS	H I (1), O I (5), Cu II (KW66-4, TW-10) [f]		12/17/1969	2	0.0012
4	7	1181–1360	Cu/Ge/Si/He/Ar	H, C, N, O	NIVS, LiF	C I (6), N I (3), Si II (6), Ge II (1), Cu II (KW66-6, TW-1) [g]	X431_tr27	10/24/1974	2	0.0024
	8	1339–1527			NIVS, LiF	C I (4), O I (1), Cu II (KW66-48)	X431_tr27	10/24/1974	2, 3 [k]	0.0021
5	9	1399–1590	Cu/He/Ar	C, N	NIVS	C I (1), Cu II (KW66-62, TW-4) [h]		12/16/1969	2, 2 [l]	0.0010
6	10	1398–1584	Cu/Si/He/Ne	C, N	NIVS	C I (3), Si II (1), Cu II (KW66-44, TW-1) [i]		1/15/1970	2	0.0012
	17	1920–2106			NIVS	C I (1), Cu II (R69-20)		1/15/1970	2	0.0021
7	11	1399–1587	Cu/Ge/Si/He	C, N	NIVS	C I (3), N I (1), Si II (2), Ge II (2), Cu II (KW66-66)	X434_tr52	11/11/1974	3	0.0014
	14	1560–1743			NIVS	C I (5), N I (1), Ge II (3), Cu II (KW66-21)	X434_tr52	11/11/1974	3	0.0016
8	12	1394–1584	Cu/Ge/Si/He	C, N	NIVS	C I (3), Si II (2), Ge II (3), Cu II (KW66-52)	X434_tr27	11/11/1974	2	0.0016
	13	1559–1743			NIVS	C I (1), Si I (3), Ge II (1), Cu I (1), Cu II (KW66-14)	X434_tr27	11/11/1974	2	0.0019
	15	1751–1939			NIVS	Si I (7), Si II (1), Cu I (1), Cu II (KW66-5)	X434_tr27	11/11/1974	2	0.0017
9	16	1790–1980	Cu/Si/He	C	NIVS	C I (1), Cu I (1), Cu II (KW66-4, R69-1, TW-2) [j]		11/10/1969	4	0.0012
10	18	1977–2162	Cu/Ge/Si/He	H, C, O	NIVS	Si I (2), Ge I (6), Ge II (2), Cu I (I), Cu II (R69-11)	X433_tr27	11/7/1974	2	0.0016
	20	2139–2315			NIVS	Si I (4), Ge I (1), Cu I (4), Cu II (R69-26)	X433_tr27	11/7/1974	5	0.0021
	21	2336–2517			NIVS	Si I (3), Cu I (1), Cu II (R69-13)	X433_tr27	11/7/1974	3	0.0017
	22	2490–2682			NIVS	O I (3), Si I (7), Ge I (3), Ge II (1), Cu II (KW66-2, R69-5)	X433_tr27	11/7/1974	2	0.0011
11	19	2117–2310	Cu/Ge/Si/He		NIVS	Si I (2), Ge I (1), Cu I (4), Cu II (R69-20)	E16-3	6/3/1969	2	0.0019
12	–	1792–3324	Cu/Ge/Pt/Fe/Ne		FTS	Ge I-II (37)		2002	1	1.7×10^{-8}
13	–	1825–3324	Cu/Ge/Pt/Fe/Ne		FTS	Ge I-II (37)		2002	1	2.0×10^{-8}
14	–	2769–11733	Cu/Re/Ar/He		FTS	Ar II (68)		2009	1	2.0×10^{-9}

[a] Equipment used: NIVS = Normal Incidence Vacuum Spectrograph (NIST, 10.7 m grating, reciprocal linear dispersion 0.78 Å); FTS = Fourier Transform Spectrometer (NIST, FT700 0.2 m vacuum FTS for spectra 12 and 13, and a 2 m vacuum FTS for spectrum 14); LiF = a LiF window was used to remove higher diffraction orders;
[b] The number of spectral lines of each spectrum used as standards is given in parentheses. Unless otherwise indicated, the standard wavelengths used in grating measurements were the Ritz wavelengths taken from the ASD database [19]. For Cu II lines used as standards, the references are as follows: KW66—Kaufman and Ward [15]; R69—Ross [2]; TW—this work. Standards used in the FTS measurements are described in the text;
[c] Power of the polynomial used to fit standard lines;
[d] For grating spectra (exposures 1–11), standard deviation of the measured wavenumbers of the standard lines from the fitted polynomial (Å); For FTS spectra (exposures 12–14), standard deviation of the correction factor from the linear fit (dimensionless);
[e] Cu II line at 826.9946 Å measured in 1st order on track 2 was used as standard on track 1;
[f] Ten Cu II lines measured in 1st order on track 2 were used as standards on track 5;
[g] Cu II line at 1275.5713 Å measured in 2nd order on track 22 was used as standard on track 7;
[h] Four Cu II lines measured in 1st order on tracks 10–12 were used as standards on track 9;
[i] Cu II line at 1407.1688 Å measured in 1st order on tracks 8, 11, and 12 was used as standard on track 10;
[j] Two Cu II lines measured in 1st order on track 15 were used as standards on track 16;
[k] Long-wavelength end of track 8 was fitted separately with a cubic polynomial;
[l] Short-wavelength end of track 9 was fitted separately with a 2nd degree polynomial.

The grating spectra were photographed on Kodak SWR plates and measured with a Grant semiautomatic comparator. (Commercial products are identified in this paper for adequate specification of the experimental procedure. This identification does not imply recommendation or endorsement by NIST.) Repeated measurements of the same plates were made, from which the measurement uncertainty of line positions was estimated to be 1.6 μm for isolated well-resolved lines, corresponding to statistical wavelengths uncertainties of 0.0012 Å. Most of the lines were measured on two to six different tracks. Thirty-eight Cu II lines were measured in the second order of diffraction, and one (861.9932 Å) in the third order, while the rest of them were measured in the first order. Below 1980 Å, most of the tracks had a sufficient number of impurity lines (H, He, C, N, O, Ne, Si, Ar, Ge), which were used as standards. Cu II lines accurately measured by Kaufman and Ward [15] were also used as standards in the region (860 to 1663) Å. In the region above 1980 Å, we used Ross's [2] interferometric and grating measurements as standards. Although later we found that Ross's wavelengths are systematically too long (see below), the average error in the wavelengths we used as standards in the grating spectra was only (-0.0004 ± 0.0008) Å, well below our total measurement uncertainties. Therefore, we did not remove the systematic errors from Ross's wavelengths used at this stage. The systematic uncertainty of each measured wavelength on our plates was estimated as a combination in quadrature of the standard deviation of the fitted polynomial (given in Table 1) and a mean uncertainty of standard wavelengths in the vicinity of the measured line. These systematic uncertainties were combined in quadrature with the statistical uncertainty (0.0012 Å for sharp isolated lines and up to 0.005 Å for blended or overexposed lines). Multiple measurements of the same line on different tracks were averaged with weights inversely proportional to squares of total uncertainties. The uncertainty of the mean was calculated as a combination in quadrature of the reduced statistical uncertainty and the straight average systematic uncertainty. In total, we have measured 1217 unique spectral lines in the grating spectra between 636 Å and 2682 Å, of which 938 were either identified as Cu II lines or did not have any identification, and the rest were identified as belonging to impurities noted above. The wavelengths determined in the present work have uncertainties ranging from 0.0008 Å to 0.010 Å. A number of lines reported by Kaufman and Ward [15] were overexposed on our plates, which led to significant systematic shifts in determining their positions. Such overexposed lines were not used in the current wavelength measurements.

Our FTS measurements were calibrated assuming a common calibration factor for all wavenumbers in each spectrogram. The value of the calibration factor was determined as a weighted mean of the ratio $\sigma/\sigma_{\text{std}}$, where σ is the measured wavenumber, and σ_{std} is the tabulated wavenumber of the reference line. For the Ar II and Ge I-II lines used as standards in the FTS spectra, we used the Ritz values from ASD as σ_{std}. We used reciprocal squared uncertainties of σ_{std}, combined in quadrature with our measurement uncertainties (see below), as weights in this averaging. Uncertainties of thus obtained calibration factors (given in Table 1) represent the systematic uncertainties in our FTS measurements. Positions of the line centers σ, line widths W, signal-to-noise ratios S/N, and total integrated intensities were evaluated using the XGREMLIN code by Nave et al. [28] either automatically, assuming a Lorentzian profile for isolated symmetrical lines, Gaussian profile for nearly symmetrical but visibly perturbed lines, or manually with a Gaussian profile with a line width fixed at a visually estimated value.

For well-resolved symmetrical lines, statistical uncertainties of FTS measurements can be approximated by Equation (1) of Brault [29]:

$$\delta\sigma_{\text{stat}} = \frac{k}{\sqrt{N_w}} \cdot \frac{W}{S/N}, \tag{1}$$

where σ is the wavenumber, W is the full width at half maximum of the line, N_w is the number of statistically-independent points in the line width, and k is a constant depending on the line shape and the algorithm used for fitting the line. For Gaussian profiles, $k = 0.693$. For an optimally sampled spectrum, the interferogram is recorded to a path difference such that N_w is between 3 and 4 for the majority of the spectral lines. This gives the commonly-used approximation of

$$\delta\sigma_{\text{stat}} = \frac{W}{2 \cdot S/N} \tag{2}$$

Using $N_w = W/r$, where r is the resolution of the spectrum in cm^{-1}, Equation (1) can be re-written as:

$$\delta\sigma_{\text{stat}} = \frac{k}{S/N}\sqrt{rW} \tag{3}$$

Our spectra were taken with $r = 0.015\ \text{cm}^{-1}$ for the Cu/Re/Ar/He spectrum and $r = 0.13\ \text{cm}^{-1}$ for the Cu/Ge/Pt/Fe/Ne spectra. Typical line widths of symmetrical lines in our spectra range from $0.04\ \text{cm}^{-1}$ around $10000\ \text{cm}^{-1}$ to $0.25\ \text{cm}^{-1}$ around $30000\ \text{cm}^{-1}$, giving values of N_w between 2.5 and 3. The instrumental line shape of the FTS comes from the finite length of the interferogram and is a sinc function of width r. When this is convolved with the Gaussian lines in our spectra, the increase in the width is negligible for a line with $N_w = 3$ and only 0.2% for a line with $N_w = 2$ and can thus be ignored in the calculation of the statistical uncertainty.

Many of the Cu lines are affected by HFS and cannot be adequately fitted using Gaussian profiles. For example, Figure 1 shows the profile of the Cu II line at $12852\ \text{cm}^{-1}$.

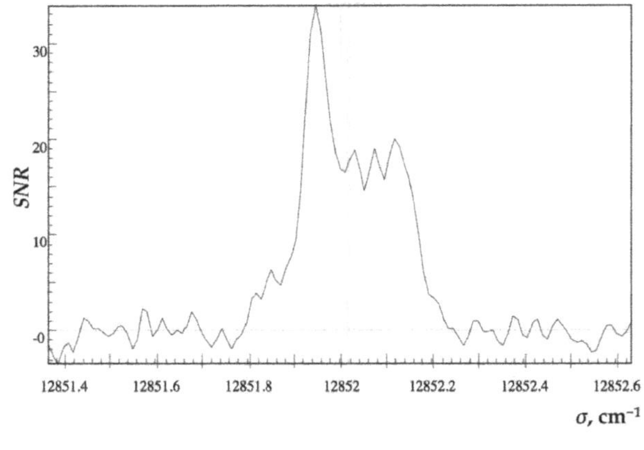

Figure 1. Profile of the Cu II line at $12852\ \text{cm}^{-1}$ (7778.7 Å) as measured in our Fourier transform spectrometers (FTS) spectrum. This line has a partially resolved hyperfine structure (HFS). The vertical line indicates the center of gravity.

The position of such lines was estimated using the center of gravity. For strong isolated lines ($S/N > 150$), where it is easy to estimate the range for calculation of the center of gravity, Equation (3) can be used to estimate the uncertainty using $k = 1$ [29]. However, this uncertainty increases for weaker lines where it is not as easy to estimate the range of calculation for the center of gravity. The measurements are further complicated by the asymmetric HFS. Its precise shape and width is not studied so far for the Cu II spectrum. In our high-resolution Cu/Re/Ar/He spectrogram, HFS patterns are partially resolved in 230 out of total 352 Cu II lines. The average width of the HFS of those lines is $W_{\text{ave}} = 0.14\ \text{cm}^{-1}$. Since the shapes of the HFS are unknown, an additional uncertainty in the measured centers of gravity arises from the possible omission (or erroneous inclusion) of a weak HFS component at the far wing of the line profile. The possible error caused by this effect of noise is easy to estimate by calculating the shift of the center of gravity of the structure:

$$\delta\sigma_{hfs} = 0.5W_{hfs}I_{noise}/(I_{tot} + I_{noise}), \quad (4)$$

where I_{tot} is the measured integrated intensity, W_{hfs} is the maximum of W_{ave} (see above) and the fitted width of the structure, and I_{noise} is the estimated possible integrated intensity of a hypothetical missed or erroneously included HFS component, equal to the root of mean square of the noise amplitude in the vicinity of the structure multiplied by the average width of sharp Cu II lines (equal to 5×10^{-6} times wavenumber for this spectrum).

The calibration of exposures 12 and 13, measured with a Cu/Ge/Pt/Ne hollow cathode, was made using 37 Ge I and II lines interferometrically measured by Kaufman and Andrew [30]. The reference wavenumbers from the latter paper have been decreased by 1.4 parts in 10^8 to put them on the scale of a recent measurement of ^{198}Hg by Sansonetti and Veza [31], as described in Nave and Sansonetti [32]. Exposure 14, made with a Cu/Re/Ar/He hollow cathode, was calibrated using 68 Ar II standards from Whaling et al. [33]. Uncertainties of the calibration factors were determined as the sum in quadrature of the statistical uncertainty (coming from the different values of the calibration factor derived from different standard lines) and the mean uncertainty of the standard lines.

The total uncertainty of Cu II measurements was determined as the sum in quadrature of statistical and systematic uncertainties. The former was taken as the sum in quadrature of Equations (3) and (4), and the latter is the uncertainty in the calibration factor (given in Table 1) times the wavenumber. Total uncertainties of our Cu II wavelengths measured by FTS range from 0.000025 Å to 0.023 Å.

Kaufman and Ward [15] photographed the VUV spectrum of a water-cooled hollow cathode discharge containing germanium and silicon in the first, second, and third orders of diffraction of the same 10.7 m Eagle-mounting vacuum grating spectrograph as used in the present work (designated as NIVS in Table 1). They reported 141 measured wavelengths of Cu II in the range (861 to 1663) Å. As wavelength standards, they used lines of Cu II, Ge I, and Si I that were either calculated (Ritz) or interferometrically measured. The reciprocal linear dispersion was 0.78 Å/mm in the first order of diffraction. Many of the reported lines were measured on several (up to 11) spectrograms. We estimated their measurement uncertainties by comparing the measured wavelengths with the Ritz values (using Ross's [2] energy levels) separately for lines measured only in the first order (0.003 Å for a single measurement), measured several times in the 1st order, but only once in the 2nd and/or 3rd order (0.0010 Å), and for lines measured on several spectrograms in several orders of diffraction (varying from 0.0004 Å to 0.0018 Å). These estimates agree well with those given explicitly by Kaufman and Ward for a few lines. Careful work was required to reconstruct the list of observed wavelengths, since some of the values given in Table III of Kaufman and Ward [15] are Ritz wavelengths, but many of them are also given in their Table I including the value of the residual $\lambda_{Ritz} - \lambda_{obs}$. Since there is no information regarding the wavelength standards actually used in various wavelength regions, all wavelengths reported by Kaufman and Ward [15] were adopted without corrections.

Ross [2,16] investigated the Cu II spectrum photographically from 1979 Å to 11217 Å using plane and concave grating spectrographs and Fabry-Perot interferometers. The light source was a water-cooled hollow cathode discharge with helium or neon as a buffer gas. Ross [2] noted that intensity of most Cu II lines was greatly enhanced when He at a pressure of about 0.9 kPa (7 Torr) was used as the carrier gas, while if Ne is used, only lines below 3000 Å arising from relatively low levels are excited. The standards for the Fabry-Perot interferograms were the lines of ^{198}Hg emitted by a water-cooled sealed electrodeless-discharge lamp containing argon carrier gas at a pressure of 33 Pa (0.25 Torr), with vacuum wavelengths 5462.27055 Å and 2537.2687 Å referred to Kaufman [34]. For grating measurements, a 9.2 m concave grating spectrograph with Paschen-Runge mounting was used. The grating had 600 lines/mm, was blazed at 10000 Å, and provided a reciprocal dispersion of ≈1.7 Å/mm. For calibration, interferometrically measured Cu II lines were used as standards.

As noted in the Introduction, subsequent studies reported some systematic errors in Ross's short-wavelength measurements. This can be explained by the fact that below 2200 Å, Ross encountered severe technical difficulties in his interferometric measurements, caused by strong absorption in the crystalline quartz of the plates and windows. He partially solved the problem by removing the windows, which exposed the interferometer to spectroscopically non-standard air and did not permit thorough temperature control. In addition, Kaufman's values for the ^{198}Hg lines Ross used as standards, 5462.27046 Å and 2537.26877 Å [34], have since been re-evaluated. Their currently recommended vacuum wavelengths are 5462.27062(3) Å and 2537.268755(17) Å [35]. Additional errors could have been due to a possibly imperfect match between filling the interferometer's aperture with light from the hollow cathode and from the mercury lamp. To verify the calibration of Ross's wavelengths, we compared his reported wavelengths with those measured in our FTS spectra. Similar to FTS measurements, interferometric measurements of Ross can be corrected by a multiplicative factor [20]:

$$\sigma_c = (1 + k_{eff})\sigma_u, \qquad (5)$$

where σ_c is the corrected wave number, σ_u is the uncorrected wave number, and the correction factor k_{eff} is determined from one or more internal standard lines in the spectrum, in this case from our wavenumbers measured by FTS. Dependence of thus derived k_{eff} on wavenumber is plotted in Figure 2.

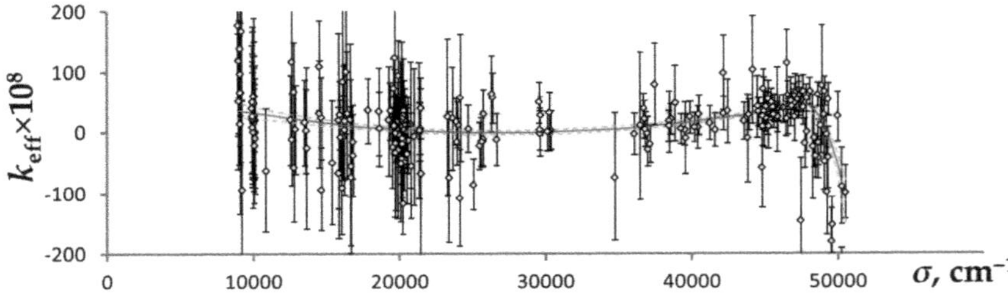

Figure 2. Dependence of calibration factor for interferometric measurements of Ross [2] on wavenumber. The error bars are measurement uncertainties of Ross [2] and our FTS data, combined in quadrature. The solid line is a weighted fit with two cubic polynomials stitched together at about 45,000 cm^{-1}. The dashed lines are 68% confidence intervals (±1 standard deviation) of the fit.

It should be noted that Figure 2 includes all Ross's lines that were also measured in our FTS spectra. Many of those lines were weak in our FTS spectra, resulting in large error bars in Figure 2. However, there was a sufficiently large number of strong lines with small error bars, which explain the very small uncertainties of the fitted curve in Figure 2.

Unlike FTS measurements, where k_{eff} is a constant over the entire range of measured lines, the calibration correction of Ross's measurements depends on wavenumber. This dependence is rather weak, and the value of k_{eff} is small (<10^{-7}) in the interval (17000 to 38000) cm^{-1} (6000 Å to 2700 Å), but k_{eff} notably increases (up to 3.5×10^{-7}) for infrared lines, and varies rapidly for the far-UV lines below 2700 Å. The accuracy of the fitted k_{eff} values varies from 3×10^{-8} to 3.5×10^{-7}, depending on the measurement uncertainties in Ross [2] and in our FTS data. It determines the systematic uncertainties in the Ross's wavenumbers as now re-calibrated.

This re-calibration and increased uncertainty associated with it allowed us to explain only a small part of numerous lines strongly deviating from Ritz wavelengths in Ross's list, noted in the Introduction. To explain the remaining problematic lines, additional factors contributing to uncertainties, not accounted for by Ross, must be considered. One such factor can be pressure shifts caused by relatively high pressure of He in his discharge, 0.9 kPa (7 Torr). Such shifts could cause quasi-random deviations of measured wavelengths in both directions from unperturbed values. Another contributing factor could be partially resolved HFS. The Doppler width reported by Ross for the sharp line at 2473 Å (40419 cm^{-1}) was about 0.1 cm^{-1}, which is of the same order of magnitude as most of the known HFS widths. He noted that partially-resolved HFS was indeed a problem in his wavelength measurements, and that his tabulated results are centers of gravity of observed HFS structures. However, he did not give any details about the method with which these centers of gravity were determined. Such determination depends on the measurement accuracy of the relative intensities of the HFS components, which might have been distorted by non-linearity of response of photographic plates used by Ross. The shift in the measured center-of-gravity wavenumber caused by errors in intensity measurements would be some fraction of the HFS width. We accounted for both types of such possible shifts (caused by pressure and HFS) by adding in quadrature a constant quantity to the wavenumber-measurement uncertainties of Ross [2]. The value of this constant was found empirically to be 0.007 cm^{-1}. These increased uncertainties are statistically consistent with residuals $\sigma_{\text{obs}} - \sigma_{\text{Ritz}}$ of the level-optimization procedure (see Section 3).

Shenstone [1] made the most comprehensive study of the Cu II spectrum prior to Ross [2]. He photographed spectra of a copper hollow cathode (Schuler tube) filled with helium or neon using several spectrographs: 6.4 m and 3 m normal incidence grating spectrographs and two Hilger prism spectrographs for the UV and visible regions, a 2 m normal incidence vacuum grating spectrograph for the VUV region, and another concave grating spectrograph (in NBS) for the infrared region. Since Ross could not observe lines shorter than 1979 Å, Shenstone's line list remained the only source of information about observed Cu II lines and their intensities in the VUV. We have re-measured many of VUV lines listed by Shenstone with much greater accuracy. Most of his lines in the UV, visible, and infrared regions were re-measured by Ross [2]. However, 94 lines from Shenstone [1] ranging from 836 Å to 6870 Å were not observed in our and Ross's spectra. For 15 of them, Shenstone did not give a measured wavelength (apparently, because low resolution did not permit accurate measurement), but gave the observed intensity. Wavelength calibration and uncertainties of Shenstone's measurements were investigated by comparing them with Ritz wavelengths calculated from Ross's [2] energy levels. Because of the relatively low resolution of Shenstone's measurements, the small differences of Ross's Ritz wavelengths from more accurate ones obtained from our final level optimization (see Section 3) did not have any effect on this analysis. Since most of Shenstone's lines used in the final line list were measured with the vacuum grating spectrograph, we compare his measurements made on this instrument with Ritz wavelength in Figure 3.

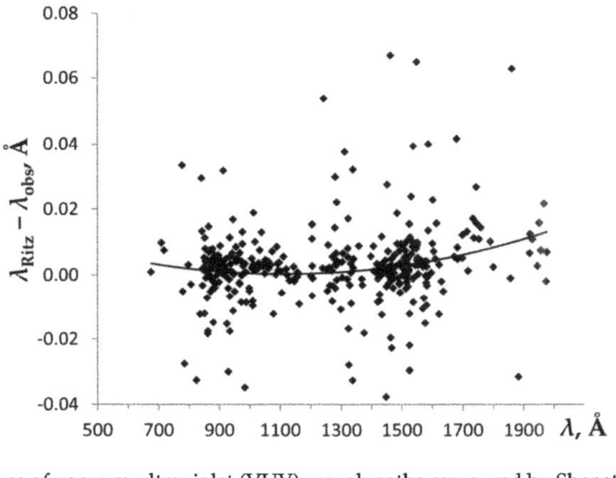

Figure 3. Deviations of vacuum ultraviolet (VUV) wavelengths measured by Shenstone [1] from Ritz values from Ross [2]. The solid line is a 2nd degree polynomial fit.

Although Figure 3 includes all wavelengths reported by Shenstone (excluding a few lines with extremely large deviations), assessment of uncertainties was made separately for "good" wavelengths given with three figures after the decimal point (in Å) and having no indication of blending or other perturbations, for wavelengths given with two decimal figures (apparently, deemed less accurate by Shenstone), and for lines indicated as blended or perturbed, or having multiple classifications. The trend indicated in Figure 3 by a solid line was derived from accurately measured lines only. It resulted in a correction to Shenstone's VUV wavelength varying from +0.003 Å for the shortest wavelengths near 700 Å to +0.010 Å near 1980 Å. After this correction, the uncertainty for each category of lines was estimated as a root-mean-square (rms) of the residuals. These uncertainties were found to vary from 0.006 Å for the best measurements to 0.07 Å for the worst. A similar analysis was made separately for each spectrograph used by Shenstone.

For completeness, we included in our line list one line observed by Wagatsuma and Hirokawa [36] at 4485.3 Å, which was not reported by other observers. Since wavelengths reported by Wagatsuma and Hirokawa are the Ritz ones (owing to low precision of their measurements), we also give only the Ritz wavelength for this line.

In all the laboratory studies described above, only electric-dipole allowed (E1) lines could be observed, since densities in the discharge light sources are high enough to collisionally depopulate metastable levels, from which forbidden lines could emerge. However, four forbidden lines of Cu II were observed in other studies. The first observation known to us is that of Thackeray [3] who identified the line of Cu II at 3806.3 Å as the electric-quadrupole (E2) $3d^{10}\ ^1S_0$–$3d^94s\ ^1D_2$ transition in emission spectra of two nebulae, η Carinae and RR Telescopii. Another E2 transition, $3d^{10}\ ^1S_0$–$3d^94s\ ^3D_2$ at 4375.8 Å, as well as the hyperfine-induced $3d^{10}\ ^1S_0$–$3d^94s\ ^3D_3$ transition at 4558.7 Å, were observed but not identified in emission of RR Telescopii by Aller et al. [4]. The above-mentioned E2 transition at 3806.3 Å, as well as the magnetic-dipole (M1) $3d^{10}\ ^1S_0$–$3d^94s\ ^3D_1$ transition at 4165.7 Å, were observed and identified by McKenna et al. [5] in emission of the same nebula, RR Telescopii. In the laboratory, the two E2 transitions at 3806.3 Å and 4375.8 Å were observed (with an accuracy inferior to astrophysical observations) and their radiative decay rates were measured by Prior [37] in an electrostatic ion trap.

All observed and identified spectral lines of Cu II are collected in Appendix A, Table A1. In addition to observed lines, this table includes several predicted transitions, in particular, six M1 and E2 transitions between the levels of the first excited configuration, $3d^94s$. This table lists 2557 transitions corresponding to 2494 unique measured spectral lines, 677 of which were measured in this work, and includes 50 additional lines that are either predicted (not observed), observed but not measured, or were masked by stronger neighboring lines in observed spectra. In addition to observed wavelengths, Table A1 gives for each transition the Ritz wavelength with standard uncertainty obtained in the level optimization procedure (see Section 3), energy-level classification, intensity on a unified scale (see Section 5), and a reference to the source of the observed wavelength. For 555 transitions, we also give a critically evaluated transition probability (*A*-value) with its uncertainty estimate and a reference. Lines on which laser action was reported in the literature are marked with "L" in the Notes column.

In Table A1, wavelengths between 2000 Å and 20000 Å are given in standard air; outside of this region, they are in vacuum. Conversion from air to vacuum was made using the five-parameter formula from Peck and Reeder [38]. Ritz wavelengths and their uncertainties were obtained using the LOPT (level optimization) code [39] as described below in Section 3. Transition probabilities are either calculated in the present work or critically compiled from references [40–51] as described in Section 6.

3. Energy Levels

The precise positions of energy levels, as well as the Ritz wavelengths and their uncertainties, were derived from the identified lines listed in Table A1 using the least-squares level optimization code LOPT [39]. The resulting energy levels are listed in Table A2.

Of all 2557 transitions in Table A1, 113 were excluded from the level optimization procedure. Among those, there are 61 transitions that were either predicted (e.g., far-IR forbidden transitions), had poorly measured wavelengths (e.g., astrophysically observed forbidden lines), were severely blended, or were masked by much stronger nearby lines. The remaining 52 lines were excluded because their observed wavelengths deviated too much from the Ritz values. In total, we used 2443 observed transitions in our level-optimization procedure. For comparison, Ross [2,16] used 1691 observed transitions in his level optimization.

If one compares the presently found energy levels with those given by Ross [2,16], the average agreement is good. The mean difference ($E_{TW} - E_{Ross}$) is only 0.007 cm^{-1} with a standard deviation of 0.016 cm^{-1} (the subscript "TW" means "this work"). However, a more detailed comparison shows that, of the 347 levels correctly identified by Ross, 156 deviate from those given in Table A2 by more than two (up to 11) combined uncertainties. Uncertainties of our level values are on average greater than those of Ross by a factor of 1.4, owing to our increased estimate of Ross's wavelength uncertainties. For only 32 levels our uncertainties are smaller than those of Ross (by up to a factor of eight). This improvement is due to our FTS measurements. Similar conclusions can be made for the Ritz wavelengths. Among about 500 Ritz VUV wavelengths listed by Ross, only a few are incorrect because of erroneous identifications described below. If these are excluded, the mean difference ($\sigma_{TW} - \sigma_{Ross}$) is 0.012 cm^{-1} with a standard deviation of 0.019 cm^{-1}. However, for 273 VUV lines (more than half of all given by Ross), Ross's Ritz wavelengths deviate from ours by more than twice the combined uncertainty. Our Table A1 includes 632 Ritz wavelengths for VUV lines below 2000 Å. Most of the new lines were observed and measured in this work.

The analysis that led to Tables A1 and A2 was made in an iterative manner. In the first step, after an initial list of all observed lines was constructed, it contained more than 600 unclassified lines, 143 of which were listed by Ross [2,16], 54 by Shenstone [1], one by Kaufman and Ward [15], and the rest were observed in our VUV grating spectra. Twenty-four of these lines (18 in Ross [2,16] and 6 in Shenstone [1]) were found to be due to previously identified transitions in Cu I. To find possible identifications for the remaining unknown lines and to find proper designations for the energy levels, we made a parametric analysis of the Cu II spectrum using Cowan's suite of atomic codes RCN/RCN2/RCG/RCE [52] (A version of the codes adapted by A. Kramida for Windows-based personal computers is available online: http://das101.isan.troitsk.ru/COWAN). In this analysis, we included the following sets of configurations: 3d^{10}, [3d^9](ns, nd) (n = 4–10), ng (n = 5–10), [3d^8](4s^2, 4p^2, 4s4d, 4s5s, 4d^2, 4s5d) in the even parity, and [3d^9](np, nf) (n = 4–8), nh (n = 6–8), [3d^8](4s4p, 4s4f, 4p4d, 4s5p, 4s5f) in the odd parity, 25 configurations in total. All known energy levels from Sugar and Musgrove [17] were included in the least-squares parametric fitting (LSF) with the RCE code. The calculations were made in the relativistic mode (HFR), including Breit corrections and correlation term (explained in the RCN manual). The fitted parameters were substituted as input for the RCG code to obtain a list of predicted lines with calculated A-values. The calculated A-values and energy levels, as well as known experimental levels, were used in the input files for the visual line-identification code IDEN1 originally designed by Azarov [53] and later programmed for Windows-based computers by one of the present authors (AK). Using this tool, we identified 117 previously unclassified lines with transitions between known energy levels, revised four levels and found 28 new energy levels. The revised and newly found energy levels explained 79 previously unclassified lines and involved revised classifications of 24 lines previously identified by Ross [2] and Shenstone [1].

At this stage of the analysis, it was also found that two levels listed by Ross [2] and included in the compilation of Sugar and Musgrove [17] had to be rejected. One of them, an undesignated odd-parity level with J = 1 at 144240.6 cm^{-1}, was retained by Ross from Shenstone's level list. It was identified by one line measured by Shenstone at 823.802 Å, which we re-measured to be at 823.8361(18) Å and identified with a transition between other previously known levels, 3d^{9}4s ^{1}D$_2$–3d^9(^{2}D$_{5/2}$)7p ^{2}[5/2]°$_2$. It was also found that there are no possible unknown odd-parity levels with J = 1 sufficiently close to the value given by Ross to explain this level. Another rejected level is also of odd parity, designated by Ross as 3d^{9}6p ^{3}P°$_0$ (at 141154.164 cm^{-1}). It was based by Ross on two lines, one observed by Shenstone at 853.56 Å and identified by Ross as a transition from this level to 3d^{9}4s ^{3}D$_1$, and another observed by Ross at 3217.641 Å, which he interpreted as a transition from this level to 3d^9(^{2}D$_{3/2}$)5s

$^2[3/2]_1$ (to be consistent with our tables, we use our new *JK* designations). Both these transitions were predicted to be extremely weak, certainly much weaker than other possible transitions from this level. We re-measured the first line to be at 853.544(2) Å and re-classified both lines as transitions from other levels. In addition, the energy given by Ross deviates too much from that predicted by our parametric fitting.

Two levels listed by Ross as $3d^96p$ $^3F°_2$ and $^3D°_1$ at 141244.576 cm^{-1} and 141734.167 cm^{-1}, respectively, were each based by Ross on several transitions to lower-lying even levels with $J = 1$ or 2. Most of the corresponding lines were previously observed and similarly identified by Shenstone. However, we found that observed line intensities agree much better with those predicted if assignments of these two levels are interchanged. Further proof for this revision was provided by our identification of two transitions from the lower of these two levels down to levels with $J = 3$. One of the corresponding lines was newly observed in our VUV spectrum, and another one was present among Ross's unclassified lines.

Following Shenstone [1], Ross interpreted the level at 137212.765 cm^{-1} as $3d^84s4p$ $^1P°_1$. Sugar and Musgrove [17] gave a more specific designation, $3d^8(^3P)4s4p(^3P°)$ $^1P°_1$ based on a parametric calculation made by Roth [54]. We found three transitions connecting this level with even $J = 3$ levels. Since other combining levels have $J = 1$ and 2, the only possible J value for this level is 2. We labeled this strongly mixed level as $3d^8(^1G)4s4p(^3P°)$ $^3F°_2$, although it represents only 36% of the percentage composition; an equal contribution is from $3d^8(^3F)4s4p(^1P°)$ $^3F°_2$, followed by 14% of $3d^8(^3P)4s4p(^3P°)$ $^1D°_2$.

Ross found the level at 147491.888 cm^{-1}, which he designated as $3d^97p$ $^3F°_3$ based on three weak lines at 3360.9941 Å, 9310.353 Å, and 9569.11 Å, identified as transitions from this level down to $3d^94d$ $^3G_3$, $3d^95d$ $^3P_2$, and $3d^95d$ $^3F_4$. Although these three lines indeed perfectly satisfy the Ritz combination principle, our calculations indicate that their intensities should be negligibly small, while much stronger transitions to other even-parity levels should occur, but were not observed by Ross. The strongest predicted transitions from this level are in the VUV region inaccessible to Ross. We have found three lines in our VUV exposures at 824.663 Å, 1609.342 Å, and 1924.548 Å, which have wavelengths and intensities consistent with this level, placing it at our new revised position, 147525.93 cm^{-1}. All other odd-parity levels with $J = 3$ predicted to occur within ± 2000 cm^{-1} from this energy are experimentally known, which leads us to conclude that the combination of three lines observed by Ross is spurious. We note that the density of observed lines in the UV, visible, and IR regions is sufficiently high to produce several tens of spurious combinations of two, and a few of three lines, which we observed in the combined line list. We could not find alternate identifications for the three lines assigned previously by Ross to this revised level.

Three other levels in Ross's list were found to be questionable, either because they were based on only one or two weak lines, or because the energy strongly disagrees with our parametric calculation. These levels are at 154838.963 cm^{-1}, 155244.833 cm^{-1}, and 156958.096 cm^{-1}, designated by Ross as $3d^98d$ $^3P_1$, $^3P_0$, and $^3F_2$, respectively.

After the new identifications were made, the new experimental energies were incorporated in the LSF procedure, and the questionable levels removed. We also inserted in the LSF several tens of levels belonging to the $[3d^9]$9d, 10d, 8g, 9g, 10g, 7f, 8f, 6h, 7h, and 8h levels whose positions were accurately extrapolated by Ritz-type quantum defect or polarization formulas (see Section 4). These levels were found to be sufficiently pure, unperturbed by interactions with other configurations, and thus the series formulas provided dependable results.

In this final LSF, 218 known even-parity levels were fitted with 38 free parameters with a standard deviation of 40 cm^{-1}, and 241 known odd-parity levels were fitted with 35 free parameters with a standard deviation of 72 cm^{-1}. The fitted (LSF) and ab initio Hartree-Fock (HF) values of parameters are given in Table 3. In both parities, the ζ_{3d} parameters of the $3d^n$ core were linked together in one group for all configurations, so that their LSF/HF ratio was the same for all members of the group, but was allowed to vary in the fitting. Other parameters, similar in different configurations, such as the effective parameter α_{3d}, electrostatic parameter F^2(3d,3d) of the $3d^8$ core, and configuration-interaction parameters, were also linked in similar groups, which decreased the number of free parameters and made the fitting more stable.

Using the transformation procedures implemented in Cowan's RCE code, eigenvector compositions were calculated in several coupling schemes. Similar to the spectrum of neutral nickel analyzed by Litzén et al. [21], most of the $3d^9nl$ configurations were found to be best described in the J_1l (otherwise known as *JK* or J_cK) coupling scheme, in which the total angular momentum of the core ($3d^9$ in this case) is combined with the orbital momentum of the valence electron to produce the *K* quantum number, which is then combined with the spin of the valence electron to obtain the final total angular momentum *J*. In Ni I, Litzén et al. [21] found the lowest $[3d^9]nl$ configurations, 4s, 4p, and 5p, to be better described in the *LS* coupling scheme, which is also the best for the $3d^84s^2$ and other $3d^8nln'l'$ configurations. In Cu II, our findings are similar, except that the $3d^95p$ configuration is better described by J_1l coupling. The J_1l purity of the $3d^9nl$ configurations increases from 65% for $3d^96p$ to 100% for $3d^9ns$, nh ($n \geq 6$), and ng ($n \geq 6$), and generally increases with increasing *n* and *l*. This general trend is disrupted in $3d^96p$. This configuration strongly interacts with $3d^84s4p$, unlike $3d^95p$, for which this interaction is somewhat weaker, resulting in the average J_1l purity of 75%. For the $3d^84s4p$ configuration, in agreement with the previous analysis by Roth [54], we found that the best description is obtained by first combining the quantum numbers of the 4s and 4p electrons with each other to produce an intermediate *LSJ* term ($^3P^\circ$ or $^1P^\circ$) of the 4s4p valence shell, and then combine it with *LSJ* of the $3d^8$ core. Although the average purity of $3d^84s4p$ in this coupling scheme is rather high, 73%, many of its levels are strongly mixed with other configurations, rendering assignment of single-configuration labels arbitrary. Several levels attributed to this configuration have a leading percentage of the composition smaller than 30%. In a few cases, such as the levels at 139331.149 cm^{-1} (J = 3) and 139710.491 cm^{-1} (J = 2), we assigned labels corresponding to the second leading term in the composition, in order to preserve uniqueness of the labels within the *J* manifolds. The labels assigned to the levels do not fully describe their physical nature; percentage compositions given in Table A2 are somewhat better in this regard.

The high J_1l purity of the $3d^9ng$ configurations and very small values of parameters describing interactions between $3d^9$ and ng shells allows for accurate prediction of the levels that were missing in Ross's levels list. This permitted us to find some of these levels from the lines left unclassified in his analysis.

For each level, Table A2 gives a number of observed lines on which this level is based. Most of the levels are based on more than two observed lines. However, 71 of them are derived from only one or two observed lines. Most of them are firmly identified, because the observed lines are the strongest predicted ones, and the level positions are confirmed by trends along series. Five of such levels are marked as questionable, because the corresponding series are strongly perturbed, or because the observed lines are not the strongest predicted to occur from these levels. The absence of predicted stronger lines could be explained by different registration sensitivity in exposures used in different spectral regions.

We identified the previously missing $3d^84s^2\,^1S_0$ level based on one line observed at 1807.8535 Å in two exposures in our VUV spectra and at 1807.84 Å by Shenstone [1]. We assigned this line to the strongest predicted transition from this level, terminating at $3d^94p\,^1P^\circ{}_1$. Although this line is observed to be about 20 times stronger than predicted, this was the only unidentified line within the interval of ±3000 cm^{-1}, and it places the $3d^84s^2\,^1S_0$ level within 25 cm^{-1} of its predicted position. Therefore, we believe that this identification is correct.

Five odd-parity levels are given in Table A2 with revised designations. The level with J = 3 at 121524.8509 cm^{-1} was previously labeled by Sugar and Musgrove [17] as $3d^95p\,^3D^\circ$ with a 53% contribution of this term in the composition, as given by Roth [54]. Although our calculation yields the same leading term in the composition of this level, the total contribution from the $3d^84s4p$ configuration is calculated to be greater than from $3d^95p$. Therefore, we designated this level by the second largest component, $3d^8(^3F)4s4p(^3P^\circ)$ $^1F^\circ$. A marginally larger contribution of the $3d^95p$ $^3D^\circ$ term was found in the J = 3 level at 121079.1501 cm^{-1}, which we attributed to this configuration. The J_1l-coupling designation gives a better description of this level, the leading term being 54% of $3d^9(^2D_{5/2})5p$ $^2[5/2]^\circ$. In effect, the identifications of these two J = 3 levels at 121079.1501 cm^{-1} and 121524.8509 cm^{-1} have been interchanged. This revision is supported by better agreement of observed and calculated line intensities.

The J = 4 level at 134742.863 cm^{-1} was labeled by Sugar and Musgrove [17] as $3d^96p$ $^3F^\circ$, and its composition was given as 49% of this term and 39% of $3d^8(^1G)4s4p(^3P^\circ)$ $^3F^\circ$. We found the leading term to be 50% of $3d^8(^3F)4s4p(^1P^\circ)$, which we use as a revised label. We note that Sugar and Musgrove gave the same configuration and term label to another J = 4 level at 139395.786 cm^{-1} with almost the same percentage composition. We found for the latter level 59% contribution from $3d^96p$ $^3F^\circ$, but designated it in J_1l-coupling as $3d^9(^2D_{5/2})6p$ $^2[7/2]^\circ$.

Another J = 4 level at 137938.904 cm^{-1} was designated by Sugar and Musgrove [17] as $3d^8(^3F)4s4p(^1P^\circ)$ $^3F^\circ$ with 53% of this term in the composition. We found the leading terms to be 49% of $3d^8(^1G)4s4p(^3P^\circ)$ $^3F^\circ$, 39% of $3d^96p$ $^3F^\circ$, and only 5% of the term given as label by Sugar and Musgrove. We changed the label to the leading term indicated by our calculation.

The J = 3 level at 139331.149 cm^{-1} was designated by Sugar and Musgrove [17] as $3d^96p$ $^3D^\circ$ with 56% of this term in the composition. In our calculation, the leading *LS* term is found to be 46% of $3d^96p$ $^1F^\circ$. However, in J_1l-coupling adopted for the $3d^96p$ configuration, the leading term is 32% of $3d^9(^2D_{5/2})6p$ $^2[7/2]^\circ$, which is used to label another level having 62% of this term. Thus, we labeled the level at 139331.149 cm^{-1} by the second leading *LS* term of its composition, $3d^8(^1G)4s4p(^3P^\circ)$ $^3F^\circ$.

4. Ionization Energy

Ross [2] derived the value for the ionization limit by extrapolation of quantum defects along 12 series of the type $3d^9nl$ (two *ns*, nine *nd*, and one *nf*, with *n* ranging from 4 to 10 for the *ns* J = 3 series, 4 to 9 for the *ns* J = 1 series, and 4 to 8 for the rest). Although Ross used *LS* designations for all the levels involved, the series he chose are almost pure in J_1l coupling, resulting in smooth behavior of quantum defects along the series. Using the RITZPL computer code by Sansonetti [55], we repeated Ross's derivation with our more accurate level values and obtained slightly different values for the series limits. In the derivation of the limit, we added two more 5-member $3d^9nd$ series, in which we found some new identifications. The weighted average of the limits obtained closely agrees with the average value adopted by Ross. In addition, series limits obtained from six three-member series with an exact fit of the polarization formula (Sansonetti [56]) produced values in close agreement with this average. Thus, we adopted the value for the ionization limit given by Ross [2], 163669.2(5) cm^{-1} or 20.292 39(6) eV, which is now confirmed.

Table A2 includes predicted energies for 89 $[3d^9]nd$, *nf*, *ng*, and *nh* levels for which no reliable identification could be found in the observed spectra. These levels are not necessarily those that produce the strongest predicted lines. Rather, they are those that are easier to accurately predict, because the corresponding series have relatively small perturbations. In this derivation, we used predicted positions for centers of gravity of groups of closely located levels along series, using the RITZPL (quantum-defect formula) and POLAR (polarization-formula) extrapolation codes by Sansonetti [55,56], which we combined with fine-structure intervals predicted by the LSF (described in Section 3). Uncertainties assigned to these levels in Table A2 are weighted means of the "observed-calculated" residuals of the LSF, calculated separately for known levels with J_1l purities of ≥99%, (97 to 98) %, and (92 to 96) %. Some of the remaining unclassified lines observed by Ross [2,16], Shenstone [1], and by us may be due to these levels. However, we could not reliably identify them because of lack of observed Ritz combinations.

5. Line Intensities

Procedures used in the present work to adjust observed line intensities to a common scale were described in detail by Kramida [57,58]. In this derivation, line intensities observed in each experiment are roughly modeled as described by Boltzmann level populations with certain effective excitation temperature T_{eff}. This temperature is found from a Boltzmann plot built using calculated transition probabilities. The ratios of predicted Boltzmann-population intensities are plotted against wavelengths to derive the response function of the registration equipment, which is then removed from the observed intensities to obtain an improved fit for T_{eff} from the Boltzmann plot. For photographic registration, an additional correction of non-linearity of intensity registration with exposure is deduced from a plot of ratios of calculated and observed intensities versus observed intensity. This procedure is repeated iteratively until convergence is achieved. Then it is easy to scale the reduced intensities observed in experiments with different T_{eff} and different sensitivity to a common value of T_{eff}. In this way, we combined intensities observed in our 22 VUV and 3 FTS spectrograms listed in Table 1 with those reported by Shenstone [1], Kaufman and Ward [15], and Ross [2,16] and reduced them to the scale derived from Ross's observed intensities to produce the average values given in Table A1. This scale corresponds to T_{eff} = 1.5 eV, which is about average for all experiments included in the analysis. For different experiments, the fitted value of T_{eff} varied between 0.6 eV in some of our VUV exposures to 11 eV in observations of Kaufman and Ward [15]. It should be noted that in most experiments analyzed here populations of highly excited levels were enhanced by resonant transfer from excited levels of helium or other rare gases used in the discharges. This population transfer is the mechanism leading to population inversion and producing laser action observed in hollow-cathode discharges. Thus, it should be expected and is indeed observed that level populations deviate from the Boltzmann distribution by a factor of three on average and strongly vary depending on discharge conditions. Line intensities are given in Table A1 in terms of total energy flux under the line contour.

As noted in Section 2, only E1-allowed lines were observed in laboratory discharges. Thus, the intensity scale adopted for E1-forbidden lines is different from the allowed lines. It is based on relative intensities observed in nebulae spectra by Thackeray [3], Aller et al. [4], and McKenna et al. [5], modified in such a way that the relative intensities of lines from a common upper level are consistent for different observations.

6. Transition Probabilities

Transition probabilities of Cu II were reported in 42 papers, the full list of which can be found in the NIST Atomic Probability Bibliographic Database (Kramida and Fuhr [59]). In addition, we have calculated them using Cowan's codes [52] using our LSF parameters described in Section 3. To assess the uncertainties of all available data sets, we used evaluation procedures described by Kramida [57].

The initial reference data set was constructed of 41 A-values determined by Ortiz et al. [50] from branching fractions measured using laser-induced breakdown spectroscopy and lifetimes calculated using the Cowan suite of codes [52], modified by inclusion of core-polarization effects. The high accuracy of theoretical lifetimes used in that work was confirmed by excellent agreement with accurate measurements made by Pinnington et al. [51] and Cederquist et al. [43], as well as with other advanced calculations.

This initial selection was expanded by inclusion of theoretical A-values from Dong and Fritzsche [45] obtained in a large-scale multiconfiguration Dirac-Fock calculation accounting for relaxation effects in the orbitals. Uncertainties of these theoretical data were estimated by comparing the results obtained by those authors in the Babushkin (length) and Coulomb (velocity) gauges. This comparison indicated that the stronger lines with line strength $S > 10^{-6}$ a.u. were calculated with uncertainties in A-values of about 12% on average, while for weaker lines the average uncertainty was much larger, 60%. These conclusions are supported by good agreement (less than 10%) of calculated lifetimes with measurements. For 18 transitions the A-values reported by Dong and

Fritzsche [45] were normalized to lifetimes of the upper levels measured by Pinnington et al. [51] and by Cederquist et al. [43].

For three transitions representing the sole allowed decay channels of their upper levels, the lifetimes measured by Pinnington et al. [51] and by Cederquist et al. [43] were directly converted into *A*-values and also added to the reference data set. Further expansion of the reference data set was provided by results of Crespo López-Urrutia et al. [44] obtained by combining their own measurements of emission branching fractions with radiative lifetimes measured in other studies. From this work, the results obtained using the lifetimes reliably measured by Cederquist et al. [43] and by Kono and Hattori [48] were used without adjustments. However, for several transitions we re-normalized their results to lifetimes more accurately measured by Pinnington et al. [51]. Uncertainties of Crespo López-Urrutia et al. [44] were estimated by combining in quadrature systematic contributions from the uncertainties of the lifetimes and statistical uncertainties, which we estimated to vary from less than 5% for the strongest lines to 90% for the weakest lines, which we assumed to have a signal-to-noise ratio about 1.

Kono and Hattori [48] measured radiative lifetimes of some of the $3d^94p$ and $3d^95s$ levels by a delayed coincidence technique and combined them with branching fractions, some of which were measured in a specially designed discharge tube, and some of which were obtained in a single-configuration LSF-adjusted intermediate-coupling calculation. These authors did not report separately their measured branching fractions, which impedes assessment of their results. Instead, they used all of their data, both experimental and theoretical, together, and adjusted the resulting *A*-values to satisfy the *J*-file sum rule. We renormalized several of their reported *A*-values to radiative lifetimes measured more accurately by Brown et al. [42], Pinnington et al. [51], and Cederquist et al. [43] and estimated the uncertainties by a method similar to the one described above in evaluation of results of Crespo López-Urrutia et al. [44].

Assessment of measurements made by Neger and Jäger [49] and by Hefferlin et al. [47] (hereafter referred to as N88 and H71) presented the greatest difficulties, since they grossly disagree with each other. N88 reported relative transition probabilities (*A*-values) of seven Cu II lines measured by using an axial discharge type of an exploding copper wire. The rather vague description of these measurements does not include the temperature value used in the data reduction. However, Neger and Jäger compare their results with those reported by Lux [60], obtained with similar equipment, and note that Lux determined the plasma temperature to be 26500(3500) K (2.3(3) eV) from a Boltzmann plot using several Cu I lines with known transition probabilities. The relative *A*-values reported by N88 (using the line at 4555.92 Å as reference) have uncertainties ranging from 18% to 25% and perfectly agree with those of Lux [60]. Thus, we assumed that they used Lux's temperature value in their data reduction and restored the observed line intensities from the reported line ratios and this temperature. Two of the lines reported in N88 are in common with more accurate measurements of Ortiz et al. [50].

Hefferlin et al. [47] (H71) determined relative intensities of 11 Cu II lines from photoelectric radiance measurements on the Burnout V experimental magnetic-confinement fusion reactor at the Oak Ridge National Laboratory. To derive the relative log(*gf*) values from these measurements, they assumed the electron temperature to be 67 eV ($\pm$50%). Neger and Jäger [49] blamed this large uncertainty in the temperature value for the discrepancy of the H71 results with theirs. We note that Cu^+ should be completely ionized at this high temperature. Thus, an obvious path would be to dismiss the results of H71 as erroneous. However, their reported relative log(*gf*) values appear to be internally consistent and agree well with other independent estimates. Thus, we restored the observed line intensities from them, using the 67 eV temperature as reported in H71, and attempted to reconcile the two sets of relative intensities (from H71 and N88) by using the following procedure.

As a first step, to determine initial estimates of the excitation temperatures in H71 and N88, we used Boltzmann plots with reference *A*-values for the lines at 4043.48 Å and 4227.94 Å from Ortiz et al. [50], and for the line at 4909.73 Å from Cederquist et al. [43], while for all other lines we used our Cowan-code calculations. As a second step, using the temperatures derived from these

Boltzmann plots, we determined the adjusted A-values from the H71 and N88 observed line intensities. Then we replaced the Cowan-code A-values in the reference set with logarithmic means of thus obtained H71 and N88 adjusted A-values and found adjusted temperature values from the Boltzmann plots. The second step was repeated iteratively until reasonable convergence, i.e., until the change in the temperature values between iterations decreased below 0.001 eV. The final Boltzmann plots for H71 and N88 are shown in Figure 4a,b, respectively.

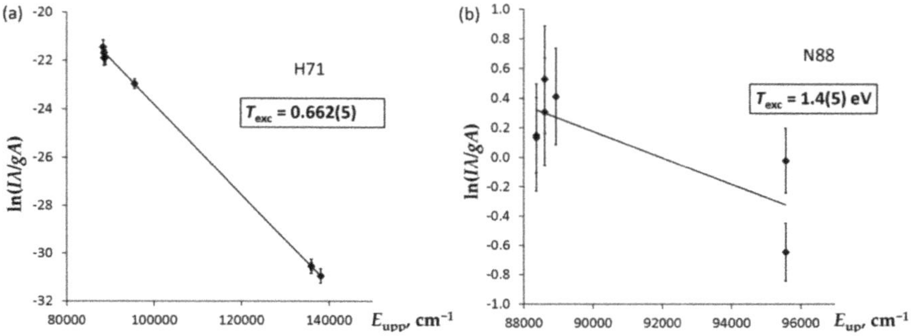

Figure 4. Boltzmann plots for relative intensities of observed lines: (**a**) Measurements of Hefferlin et al. [47] (H71); (**b**) Measurements of Neger and Jäger [49] (N88). The A-values used in these plots are the final adopted values given in Table A1. The error bars are combined uncertainties of A-values and relative intensities (see text).

The excitation temperatures determined for the H71 and N88 data sets are 0.662(5) eV and 1.4(5) eV, respectively. The latter is somewhat lower than the Lux [60] value used in N88, 2.3(3) eV. The final adjusted A-values deduced from the H71 and N88 data are compared with the adopted A-values in Figure 5. In this plot, the reference A-values used for the two leftmost lines (4043.48 Å and 4227.94 Å) and for the rightmost line at 4909.73 Å are from Ortiz et al. [50] and Cederquist et al. [43], while the rest of the lines use the mean of H71 and N88. The error bars are combined uncertainties of the measured line ratios and reference A-values. This plot shows that the adjusted A-values from H71 and N88 are consistent with each other, as well as with the adopted reference A-values.

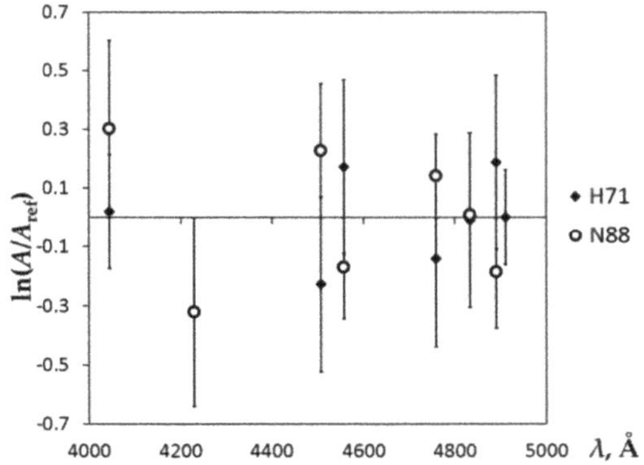

Figure 5. Comparison of adjusted A-values deduced from the line ratios reported by Hefferlin et al. [47] (H71) and by Neger and Jäger [49] (N88). The error bars are combined uncertainties of the measured line ratios and adopted reference A-values.

With the reference data set constructed from data described above, assessment of uncertainties in the theoretical study by Biémont et al. [41] was relatively easy to do by plotting the ratio of the calculated values to the reference ones against line strength S and calculating standard deviations for different ranges of S. This procedure resulted in estimated uncertainties of 9% for $S \geq 8.5$ atomic units (a.u.), 22% for S = (0.2 to 8.5) a.u., and 120% for $S < 0.2$ a.u.

Assessment procedures described above resulted in 210 reference A-values with uncertainties less than 25% covering a wide range of line strengths from 0.06 a.u. to 155 a.u., after which it became possible to evaluate the uncertainties of our LSF calculations. Comparison of our calculated A-values with the reference ones is illustrated in Figure 6.

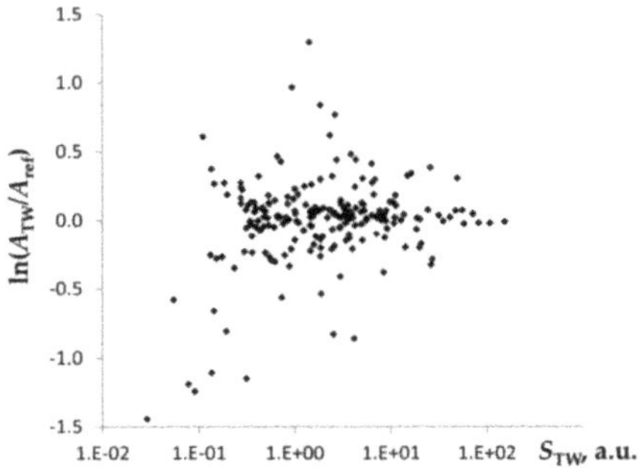

Figure 6. Comparison of transition probabilities calculated in this work (A_{TW}) with reference values (A_{ref}, see text). Natural logarithm of the ratio A_{TW}/A_{ref} is plotted against our calculated line strength, S_{TW} (in atomic units).

This plot displays a typical behavior of theoretical transition probabilities. Namely, the strongest lines exhibit the smallest discrepancies from the reference values, while for weaker lines the magnitude of discrepancies grows. A similar comparison was made between A-values produced in our two LSFs (the final and preliminary ones, described in Section 3). From these comparisons, we estimated the uncertainties of our A-values to be $\leq$10% for $S > 200$ a.u., $\leq$15% for S = (50 to 200) a.u., $\leq$30% for S = (4.3 to 50) a.u., and $\geq$50% for smaller S. We included 79 of our best calculated A-values in Table A1.

Studies of E1-forbidden transition probabilities of Cu II are very scarce. Garstang [46] calculated A-values for several M1 and E2 transitions from the first excited configuration, $3d^94s$, down to lower-lying levels of the same configuration and to the ground state, $3d^{10}\,{}^1S_0$. He used the pseudo-relativistic Hartree-Fock method with superposition of configurations and LSF-adjusted Slater parameters, similar to the one implemented in Cowan's codes [52] used in the present work, but limited by inclusion of only three low-lying configurations of even parity. Beck [61] made an ab initio restricted non-relativistic multiconfiguration Hartree-Fock calculation for the $3d^{10}\,{}^1S_0$–$3d^94s$ ${}^{1,3}D_2$ E2 transition and obtained $A = 2.33$ s^{-1}, somewhat larger than the sum of Garstang's values for transitions from ${}^1D_2$, 1.7 s^{-1}, and ${}^3D_2$, 0.12 s^{-1}. Prior [37] measured the A-value for the $3d^{10}\,{}^1S_0$–$3d^94s$ ${}^1D_2$ transition at 3806.3 Å to be 1.60(24) s^{-1}, in excellent agreement with Garstang's semiempirical value. The most recent multiconfiguration Dirac-Hartree-Fock calculation of Andersson et al. [40] gave $A = 1.937$ s^{-1} for this transition. However, their calculated wavelength is 3724.7 Å, significantly shorter than experimental. Adjustment to the experimental wavelength yields $A_{adj} = 1.74$ s^{-1}, in good agreement with Prior [37] and Garstang [46]. A similar adjustment of the value of Andersson et al. [40] for the $3d^{10}\,{}^1S_0$–$3d^94s$ ${}^3D_2$ transition at 4375.8 Å (which Andersson et al. [40] calculated to be at 4275.8 Å) gives $A_{adj} = 0.093$ s^{-1}, about 30% lower than Garstang's result. For these two E2 transitions, our A-values calculated with Cowan's codes are too high by a factor of about 1.7. We assume that our A-value for the predicted $3d^94s$ ${}^3D_3$–${}^3D_1$ E2 transition at 48318 Å, 9×10^{-8} s^{-1}, has a similar low accuracy. The $3d^{10}\ ^1S_0$–$3d^94s\ ^1D_2$ transition at 3806.3 Å was observed by Thackeray [3] and by McKenna et al. [5] in emission spectra of nebulae.

It should be noted that, in addition to the two E2 transitions mentioned above, Andersson et al. [40] give calculated *A*-values for the magnetic-octupole (M3) and hyperfine-induced $3d^{10}\ ^1S_0$–$3d^94s\ ^3D_3$ transitions at 4559 Å. The latter are different for various HFS components of the two isotopes (^{63}Cu and ^{65}Cu) and are typically on the order of a few times 10^{-9} s^{-1}, three orders of magnitude greater than for the M3 transition. This transition, as well as the $3d^{10}\ ^1S_0$–$3d^94s\ ^3D_2$ E2 transition at 4375.8 Å, was observed by Aller et al. [4] in the spectrum of the RR Telescopii nebula.

A-values for M1 transitions are relatively easier to calculate than E1 and E2 transitions, since their calculation does not involve radial integrals, but only amplitudes of eigenvector components. We compared our calculated *A*-values for M1 transitions with those of Garstang [46] and found that all of them agree to better than 15%. We included in Table A1 our *A*-value for the $3d^{10}\ ^1S_0$–$3d^94s\ ^3D_1$ M1 transition at 4165.7 Å, which was not given by Garstang [46]. It turned out to be extremely small, 2.1×10^{-12} s^{-1}. Our calculation indicates that there is no E2 contribution to the total decay rate of the $3d^94s\ ^3D_1$ level. This makes the identification of the line observed at this wavelength by McKenna et al. [5] with this transition questionable. Possible explanations for this observation could be (1) our calculated M1 *A*-value is greatly underestimated; (2) there is a considerable contribution of E2 transition to this line, which for unknown reason is computed as negligibly small by Cowan's codes; (3) hyperfine-induced transition significantly increases the total radiative rate; (4) the population of the $3d^94s\ ^3D_1$ level in RR Telescopii is many orders of magnitude greater than that of $^3D_2$ and $^1D_2$; or (5) the identification is incorrect, and the observed line is possibly due to some other species. As explained above, the first explanation is rather unlikely. To confirm or disprove the second and third explanations, more extensive atomic calculations are needed, while checking the fourth one requires population-kinetics modeling.

7. Conclusions

The present work provides a comprehensive list of all observed, classified, and predicted spectral lines of Cu II, which includes 2557 transitions with wavelengths from 675 Å to 10.9 μm. Over 600 of them were measured in this work using grating and Fourier-transform spectrometers. Experimental wavelengths of 2443 transitions were used in a least-squares level optimization procedure to produce optimized values for 379 energy levels, of which 29 are newly identified, and nine are revised. The previous analysis of Ross [2,16] is largely confirmed. However, our extended base for the levels optimization results in more dependable level values and Ritz wavelengths that can be used as secondary standards in the vacuum ultraviolet region. An improved theoretical interpretation of the energy levels was made by a parametric least-squares fitting with Cowan's atomic codes, which included all experimental levels. The fitted Slater parameters were used to calculate radiative transition probabilities for electric-dipole, magnetic-dipole, and electric-quadrupole transitions, which were critically evaluated together with other published data to construct a list of recommended *A*-values for 555 transitions.

Acknowledgments: This work was partially supported by the National Aeronautics and Space Administration (NASA) under interagency agreement NNH10AN38I.

Author Contributions: All authors contributed equally to this work.

Conflicts of Interest: The authors declare no conflict of interest.

Appendix A

Table A1. Spectral lines of Cu II.

λ_{obs} [a] (Å)	σ_{obs} [b] (cm^{-1})	λ_{Ritz} [c] (Å)	$\Delta\lambda_{obs\text{-}Ritz}$ (Å)	I_{obs} [d] (arb. u.)	Char [e]	Lower Level		Upper Level		A (s^{-1})	Acc [f]	Line Ref. [g]	TP Ref. [g]	Notes [h]
675.5994(20)	148016.7(4)	675.60192(8)	−0.0025	73000		d^{10}	$^1S_0$	5f	$(3/2)^2[3/2]^\circ_1$			TW		
677.675(3)	147563.3(6)	677.67816(8)	−0.003	29000		d^{10}	$^1S_0$	7p	$(5/2)^2[3/2]^\circ_1$			TW		
685.1377(20)	145956.1(4)	685.14058(8)	−0.0029	240000		d^{10}	$^1S_0$	5f	$(5/2)^2[3/2]^\circ_1$			TW		
685.3922(20)	145901.9(4)	685.39671(8)	−0.0045	45000		d^{10}	$^1S_0$	5f	$(5/2)^2[1/2]^\circ_1$			TW		
709.3098(20)	140982.1(4)	709.31287(9)	−0.0031	250000		d^{10}	$^1S_0$	6p	$(3/2)^2[1/2]^\circ_1$			TW		
718.1766(20)	139241.5(4)	718.17860(9)	−0.0020	310000		d^{10}	$^1S_0$	6p	$(5/2)^2[3/2]^\circ_1$			TW		
724.4881(20)	138028.5(4)	724.48867(9)	−0.0006	490000		d^{10}	$^1S_0$	4f	$(3/2)^2[3/2]^\circ_1$			TW		
735.5215(11)	135957.95(20)	735.52023(9)	0.0013	780000		d^{10}	$^1S_0$	4f	$(5/2)^2[3/2]^\circ_1$			TW		
736.0331(8)	135863.46(16)	736.03185(9)	0.0012	870000		d^{10}	$^1S_0$	4f	$(5/2)^2[1/2]^\circ_1$			TW		
763.2658(20)	131016.0(3)	763.2692(3)	−0.0034	140000		4s	$^3D_3$	sp	$(^3P)^1P^\circ\ ^3P^\circ_2$			TWn		
768.4821(20)	130126.6(3)	768.48565(19)	−0.0035	150000		4s	$^3D_3$	8p	$(5/2)^2[3/2]^\circ_2$			TWn		
776.484(3)	128785.6(5)	776.4877(3)	−0.004	58000		4s	$^3D_1$	sp	$(^3P)^1P^\circ\ ^3P^\circ_0$			TWn		
777.739(3)	128577.8(5)	777.74333(3)	−0.004	56000		4s	$^3D_2$	8p	$(5/2)^2[3/2]^\circ_1$			TW		
779.2932(20)	128321.4(3)	779.29473(5)	−0.0015	590000		4s	$^3D_3$	sp	$(^1D)^1P^\circ\ ^1D^\circ_2$			TW		
779.328(3)	128315.6(5)	779.33660(4)	−0.008	480000		4s	$^3D_2$	6f	$(5/2)^2[1/2]^\circ_1$			TWn		
784.9098(20)	127403.2(3)	784.91229(5)	−0.0025	69000		4s	$^3D_2$	sp	$(^1D)^1P^\circ\ ^1D^\circ_2$			TW		
786.392(3)	127163.1(5)	786.39222(4)	−0.000	21000	*	4s	$^3D_1$	6f	$(5/2)^2[1/2]^\circ_1$			TWn		
786.392(3)	127163.1(5)	786.39795(4)	−0.006	21000	*	4s	$^1D_2$	6f	$(3/2)^2[3/2]^\circ_2$			TW		
787.907(3)	126918.5(5)	787.9056(3)	0.001	33000		4s	$^3D_2$	7p	$(3/2)^2[3/2]^\circ_1$			TWn		
788.012(3)	126901.6(5)	788.01778(20)	−0.006	45000		4s	$^1D_2$	sp	$(^1D)^1P^\circ\ ^1P^\circ_1$			TWn		
789.3937(20)	126679.5(3)	789.3933(3)	0.0004	100000	*	4s	$^1D_2$	sp	$(^3P)^1P^\circ\ ^3P^\circ_2$			TWn		
789.3937(20)	126679.5(3)	789.3977(3)	−0.0040	100000	*	4s	$^3D_2$	7p	$(3/2)^2[5/2]^\circ_2$			TWn		
789.6575(20)	126637.2(3)	789.65867(5)	−0.0011	59000		4s	$^3D_2$	7p	$(3/2)^2[1/2]^\circ_1$			TWn		
794.9719(20)	125790.6(3)	794.97430(20)	−0.0024	56000		4s	$^1D_2$	8p	$(5/2)^2[3/2]^\circ_2$			TWn		
797.4530(20)	125399.2(3)	797.45499(4)	−0.0020	560000		4s	$^3D_3$	7p	$(5/2)^2[3/2]^\circ_2$			TW		
797.6193(20)	125373.1(3)	797.6218(5)	−0.0025	35000		4s	$^3D_1$	7p	$(3/2)^2[1/2]^\circ_0$			TWn		
798.977(3)	125160.0(5)	798.97928(3)	−0.002	23000		4s	$^1D_2$	8p	$(5/2)^2[3/2]^\circ_1$			TW		
800.6592(20)	124897.1(3)	800.66084(5)	−0.0017	26000		4s	$^1D_2$	6f	$(5/2)^2[1/2]^\circ_1$			TWn		
801.8229(20)	124715.8(3)	801.82460(4)	−0.0017	84000		4s	$^3D_2$	7p	$(5/2)^2[3/2]^\circ_1$			TW		
803.3370(20)	124480.8(3)	803.33841(4)	−0.0014	77000		4s	$^3D_2$	7p	$(5/2)^2[3/2]^\circ_2$			TW		
806.5454(20)	123985.6(3)	806.54699(5)	−0.0016	180000		4s	$^1D_2$	sp	$(^1D)^1P^\circ\ ^1D^\circ_2$			TW		
809.2942(20)	123564.5(3)	809.29524(4)	−0.0011	17000		4s	$^3D_1$	7p	$(5/2)^2[3/2]^\circ_1$			TW		
809.706(3)	123501.6(5)	809.7079(3)	−0.002	12000		4s	$^1D_2$	7p	$(3/2)^2[3/2]^\circ_1$			TWn		
809.9630(20)	123462.4(3)	809.9649(3)	−0.0019	45000		4s	$^1D_2$	7p	$(3/2)^2[3/2]^\circ_2$			TWn		
810.9987(20)	123304.8(3)	810.99829(11)	0.0004	520000		d^{10}	$^1S_0$	5p	$(3/2)^2[3/2]^\circ_1$			TW		
810.6365(20)	123359.8(3)	810.6361(3)	0.0005	86000		4s	$^1D_2$	7p	$(3/2)^2[5/2]^\circ_3$			TWn		
811.2843(20)	123261.4(3)	811.2839(4)	0.0004	110000		4s	$^1D_2$	7p	$(3/2)^2[5/2]^\circ_2$			TWn		
811.5559(20)	123220.1(3)	811.55946(5)	−0.0036	13000		4s	$^1D_2$	7p	$(3/2)^2[1/2]^\circ_1$			TWn		

Table A1. *Cont.*

λ_{obs} [a] (Å)	σ_{obs} [b] (cm^{-1})	λ_{Ritz} [c] (Å)	$\Delta\lambda_{obs-Ritz}$ (Å)	I_{obs} [d] (arb. u.)	Char [e]	Lower Level		Upper Level		A (s^{-1})	Acc [f]	Line Ref. [g]	TP Ref. [g]	Notes [h]
813.882(3)	122867.9(4)	813.88328(11)	−0.001	760000		d^{10}	$^1S_0$	5p	$(3/2)^2[1/2]^\circ_1$			TW		
823.059(3)	121497.9(4)	823.05584(4)	0.004	2000		4s	$^1D_2$	7p	$(5/2)^2[5/2]^\circ_3$			TW		
823.8361(18)	121383.4(3)	823.83758(6)	−0.0015	90000		4s	$^1D_2$	7p	$(5/2)^2[5/2]^\circ_2$			TW		
824.4127(20)	121298.5(3)	824.41507(4)	−0.0024	6100		4s	$^1D_2$	7p	$(5/2)^2[3/2]^\circ_1$			TW		
824.6630(19)	121261.7(3)	824.6649(4)	−0.0019	90000		4s	$^1D_2$	7p	$(5/2)^2[7/2]^\circ_3$			TWn		
826.014(3)	121063.3(5)	826.01546(5)	−0.001	4400		4s	$^1D_2$	7p	$(5/2)^2[3/2]^\circ_2$			TW		
826.9946(22)	120919.8(3)	826.99598(12)	−0.0014	1100000		d^{10}	$^1S_0$	5p	$(5/2)^2[3/2]^\circ_1$			TW		
829.360(5)	120575.0(8)	829.35106(14)	0.008	7400		4s	$^3D_2$	sp	$(^1G)^3P^\circ\ ^3G^\circ_3$			TWn		
836.042(20)	119611(3)	836.02765(3)	0.014	27000		4s	$^3D_3$	6p	$(3/2)^2[3/2]^\circ_2$			S36c		
841.136(3)	118886.8(4)	841.13445(3)	0.002	30000		4s	$^3D_2$	6p	$(3/2)^2[3/2]^\circ_1$			TW		
842.4973(24)	118694.7(3)	842.49630(3)	0.0010	66000		4s	$^3D_2$	6p	$(3/2)^2[3/2]^\circ_2$			TW		
844.6125(21)	118397.5(3)	844.61287(4)	−0.0003	88000		4s	$^3D_2$	6p	$(3/2)^2[5/2]^\circ_2$			TW		
844.9124(21)	118355.5(3)	844.91208(4)	0.0003	410000		4s	$^3D_2$	6p	$(3/2)^2[5/2]^\circ_3$			TW		
848.8062(20)	117812.5(3)	848.80734(3)	−0.0011	920000		4s	$^3D_3$	6p	$(5/2)^2[5/2]^\circ_3$			TW		
849.3599(19)	117735.7(3)	849.35931(3)	0.0006	240000		4s	$^3D_1$	6p	$(3/2)^2[3/2]^\circ_1$			TW		
850.7481(19)	117543.6(3)	850.74794(3)	0.0001	140000		4s	$^3D_1$	6p	$(3/2)^2[3/2]^\circ_2$			TW		
851.3012(16)	117467.24(23)	851.30253(3)	−0.0013	1100000		4s	$^3D_3$	6p	$(5/2)^2[7/2]^\circ_4$			TW		
851.776(3)	117401.8(4)	851.77122(3)	0.004	21000		4s	$^3D_3$	sp	$(^1G)^3P^\circ\ ^3F^\circ_3$			TW		
852.9074(21)	117246.0(3)	852.90622(4)	0.0012	370000		4s	$^3D_1$	6p	$(3/2)^2[5/2]^\circ_2$			TW		
853.5444(24)	117158.5(3)	853.54260(15)	0.0018	45000		4s	$^1D_2$	sp	$(^1G)^3P^\circ\ ^3G^\circ_3$			TWn		
855.4770(19)	116893.9(3)	855.476052(24)	0.0009	270000		4s	$^3D_2$	6p	$(5/2)^2[5/2]^\circ_3$			TW		
855.6994(21)	116863.5(3)	855.70010(3)	−0.0007	720000		4s	$^3D_2$	sp	$(^3F)^1P^\circ\ ^3F^\circ_2$			TW		
858.4844(20)	116484.4(3)	858.486771(24)	−0.0024	990000		4s	$^3D_2$	sp	$(^1G)^3P^\circ\ ^3F^\circ_3$			TW		
858.5639(20)	116473.6(3)	858.566433(20)	−0.0025	890000		4s	$^3D_3$	6p	$(5/2)^2[7/2]^\circ_3$			TW		
859.1522(24)	116393.8(3)	859.15072(3)	0.0014	14000		4s	$^3D_2$	6p	$(5/2)^2[3/2]^\circ_1$			TW		
860.7214(24)	116181.6(3)	860.72158(4)	−0.0001	22000		4s	$^3D_2$	6p	$(5/2)^2[5/2]^\circ_2$			TW		
861.9932(11)	116010.19(14)	861.99338(6)	−0.0002	1900000		4s	$^3D_3$	sp	$(^1G)^3P^\circ\ ^3F^\circ_4$			TW		
862.8221(20)	115898.7(3)	862.82245(4)	−0.0003	78000		4s	$^3D_2$	6p	$(5/2)^2[3/2]^\circ_2$			TW		
864.163(4)	115718.9(5)	864.15435(3)	0.009	130000		4s	$^3D_3$	sp	$(^3P)^3P^\circ\ ^1D^\circ_2$			TW		
864.214(3)	115712.0(4)	864.21371(3)	0.001	670000		4s	$^3D_1$	sp	$(^3F)^1P^\circ\ ^3F^\circ_2$			TW		
865.3907(8)	115554.74(11)	865.389986(23)	0.0007	1600000		4s	$^3D_2$	6p	$(5/2)^2[7/2]^\circ_3$			K66		
866.4430(20)	115414.4(3)	866.442524(13)	0.0005	64000		4s	$^3D_2$	4f	$(3/2)^2[7/2]^\circ_3$			TW		
867.7321(21)	115242.9(3)	867.73348(3)	−0.0014	320000		4s	$^3D_1$	6p	$(5/2)^2[3/2]^\circ_1$			TW		
869.0634(18)	115066.40(24)	869.06394(3)	−0.0005	250000		4s	$^3D_2$	sp	$(^3F)^1P^\circ\ ^3D^\circ_1$			TW		
869.3338(24)	115030.6(3)	869.33592(4)	−0.0021	1200000		4s	$^3D_1$	6p	$(5/2)^2[5/2]^\circ_2$			TW		
870.5390(18)	114871.37(24)	870.538673(24)	0.0003	240000		4s	$^3D_3$	sp	$(^3F)^1P^\circ\ ^3D^\circ_2$			TW		
871.0674(18)	114801.68(24)	871.06737(3)	0.0000	130000		4s	$^3D_2$	sp	$(^3P)^3P^\circ\ ^1D^\circ_2$			TW		
873.2629(21)	114513.0(3)	873.262654(21)	0.0003	310000		4s	$^3D_3$	sp	$(^3F)^1P^\circ\ ^3F^\circ_3$			TW		
876.7226(20)	114061.2(3)	876.722436(13)	0.0002	830000		4s	$^3D_3$	4f	$(5/2)^2[7/2]^\circ_3$			TW		

Table A1. *Cont.*

λ_{obs} [a] (Å)	σ_{obs} [b] (cm^{-1})	λ_{Ritz} [c] (Å)	$\Delta\lambda_{obs\text{-}Ritz}$ (Å)	I_{obs} [d] (arb. u.)	Char [e]	Lower Level		Upper Level		A (s^{-1})	Acc [f]	Line Ref. [g]	TP Ref. [g]	Notes [h]
876.973(5)	114028.6(7)	876.962109(24)	0.011	420000		4s	$^3D_1$	4f	$(3/2)^2[3/2]^\circ_1$			TW		
877.0107(24)	114023.7(3)	877.01184(4)	−0.0012	890000		4s	$^3D_3$	sp	$(^3P)^3P^\circ\,^5S^\circ_2$			TW		
877.5531(20)	113953.2(3)	877.554632(23)	−0.0015	820000		4s	$^3D_2$	sp	$(^3F)^1P^\circ\,^3D^\circ_2$			TW		
877.8469(21)	113915.1(3)	877.84692(3)	0.0000	610000		4s	$^3D_1$	sp	$(^3F)^1P^\circ\,^3D^\circ_1$			TW		
878.6984(8)	113804.69(10)	878.69833(3)	0.0001	1500000		4s	$^3D_3$	sp	$(^3F)^1P^\circ\,^3D^\circ_3$			K66		
879.8897(24)	113650.6(3)	879.89110(3)	−0.0014	21000		4s	$^3D_1$	sp	$(^3P)^3P^\circ\,^1D^\circ_2$			TW		
880.3225(24)	113594.7(3)	880.322766(20)	−0.0003	38000		4s	$^3D_2$	sp	$(^3F)^1P^\circ\,^3F^\circ_3$			TW		
881.473(3)	113446.4(4)	881.47711(3)	−0.004	3100		4s	$^1D_2$	sp	$(^3F)^1P^\circ\,^3F^\circ_2$			TW		
883.2800(20)	113214.4(3)	883.27981(3)	0.0002	41000		4s	$^3D_1$	sp	$(^1G)^3P^\circ\,^3F^\circ_2$			TW,S36r		
883.8404(20)	113142.6(3)	883.838830(10)	0.0016	55000		4s	$^3D_2$	4f	$(5/2)^2[7/2]^\circ_3$			TW		
884.1340(21)	113105.0(3)	884.13295(4)	0.0011	290000		4s	$^3D_2$	sp	$(^3P)^3P^\circ\,^5S^\circ_2$			TW		
884.4346(18)	113066.58(23)	884.43449(3)	0.0001	420000		4s	$^1D_2$	sp	$(^1G)^3P^\circ\,^3F^\circ_3$			TW		
884.8262(20)	113016.5(3)	884.826019(13)	0.0002	34000		4s	$^3D_2$	4f	$(5/2)^2[1/2]^\circ_1$			TW		
885.8463(20)	112886.4(3)	885.846966(24)	−0.0006	770000		4s	$^3D_2$	sp	$(^3F)^1P^\circ\,^3D^\circ_3$			TW		
886.4163(24)	112813.8(3)	886.413782(20)	0.0025	25000		4s	$^3D_3$	sp	$(^3F)^1P^\circ\,^3F^\circ_4$			TW		
886.5113(20)	112801.7(3)	886.510950(24)	0.0004	280000		4s	$^3D_1$	sp	$(^3F)^1P^\circ\,^3D^\circ_2$			TW		
886.9435(8)	112746.75(10)	886.94322(3)	0.0003	1300000		4s	$^3D_3$	sp	$(^3P)^3P^\circ\,^3D^\circ_2$			K66		
890.5677(6)	112287.93(8)	890.56671(3)	0.0010	1300000		4s	$^3D_2$	sp	$(^3P)^3P^\circ\,^3P^\circ_1$			K66		
891.7636(24)	112137.3(3)	891.76309(3)	0.0005	23000		4s	$^1D_2$	6p	$(5/2)^2[7/2]^\circ_3$			TW		
892.4154(6)	112055.44(8)	892.41418(3)	0.0012	1100000		4s	$^3D_3$	sp	$(^3P)^3P^\circ\,^3D^\circ_3$			K66		
893.634(5)	111902.7(7)	893.624985(19)	0.009	1100000		4s	$^3D_1$	4f	$(5/2)^2[1/2]^\circ_0$			TW		
893.6787(6)	111897.04(8)	893.67746(3)	0.0012	940000		4s	$^3D_3$	sp	$(^3P)^3P^\circ\,^3P^\circ_2$			K66		
894.2260(20)	111828.56(25)	894.227192(24)	−0.0012	860000		4s	$^3D_2$	sp	$(^3P)^3P^\circ\,^3D^\circ_2$			TW		
896.7583(15)	111512.77(19)	896.758608(23)	−0.0003	1000000		4s	$^3D_2$	sp	$(^3P)^3P^\circ\,^3D^\circ_1$			TW		
896.9741(20)	111485.94(25)	896.97600(3)	−0.0019	870000		4s	$^3D_1$	sp	$(^3P)^3P^\circ\,^3P^\circ_0$			TW		
897.7927(18)	111384.29(22)	897.79300(3)	−0.0003	410000		4s	$^1D_2$	sp	$(^3P)^3P^\circ\,^1D^\circ_2$			TW		
899.7904(18)	111136.99(22)	899.78867(3)	0.0018	910000	*	4s	$^3D_2$	sp	$(^3P)^3P^\circ\,^3D^\circ_3$			TW		
899.7904(18)	111136.99(22)	899.79200(3)	−0.0016	910000	*	4s	$^3D_1$	sp	$(^3P)^3P^\circ\,^3P^\circ_1$			TW		
901.0757(16)	110978.47(20)	901.07654(7)	−0.0009	660000	*	4s	$^1D_2$	sp	$(^3P)^3P^\circ\,^1P^\circ_1$			TW,S36r		
901.0757(16)	110978.47(20)	901.07292(3)	0.0027	660000	*	4s	$^3D_2$	sp	$(^3P)^3P^\circ\,^3P^\circ_2$			TW		
901.322(3)	110948.2(4)	901.32128(3)	0.000	25000		4s	$^1D_2$	sp	$(^1G)^3P^\circ\,^3F^\circ_2$			TWn		
903.522(6)	110678.0(7)	903.52887(3)	−0.007	16000		4s	$^3D_1$	sp	$(^3P)^3P^\circ\,^3D^\circ_2$			S36c		
906.1130(20)	110361.52(24)	906.11330(3)	−0.0003	670000		4s	$^3D_1$	sp	$(^3P)^3P^\circ\,^3D^\circ_1$			TW		
910.5183(21)	109827.56(25)	910.51831(3)	−0.0000	240000		4s	$^3D_1$	sp	$(^3P)^3P^\circ\,^3P^\circ_2$			TW		
911.630	109693.6	911.629867(16)		15000	:	4s	$^1D_2$	4f	$(5/2)^2[3/2]^\circ_1$			S36c		
911.679	109687.7	911.67901(4)		15000	:	4s	$^1D_2$	sp	$(^3P)^3P^\circ\,^5S^\circ_2$			S36c		
912.025	109646.2	912.024525(12)		7400	:	4s	$^1D_2$	4f	$(5/2)^2[3/2]^\circ_2$			S36c		
912.4142(24)	109599.3(3)	912.415956(16)	−0.0018	15000		4s	$^1D_2$	4f	$(5/2)^2[1/2]^\circ_1$			TW		
913.47(4)	109472(5)	913.50160(3)	−0.03	7300		4s	$^1D_2$	sp	$(^3F)^1P^\circ\,^3D^\circ_3$			S36c		
914.2128(6)	109383.72(7)	914.21305(11)	−0.0003	830000		4s	$^3D_3$	sp	$(^3P)^3P^\circ\,^5D^\circ_4$			K66		

Table A1. *Cont.*

λ_{obs} [a] (Å)	σ_{obs} [b] (cm^{-1})	λ_{Ritz} [c] (Å)	$\Delta\lambda_{obs\text{-}Ritz}$ (Å)	I_{obs} [d] (arb. u.)	Char [e]	Lower Level		Upper Level		A (s^{-1})	Acc [f]	Line Ref. [g]	TP Ref. [g]	Notes [h]
917.3051(20)	109014.98(23)	917.30556(8)	−0.0004	380000		4s	$^3D_3$	sp	$(^3P)^3P^\circ\ ^5D^\circ_2$			TW		
922.0188(8)	108457.66(9)	922.01869(14)	0.0001	520000		4s	$^3D_3$	sp	$(^1D)^3P^\circ\ ^3P^\circ_2$			K66		
922.4141(16)	108411.18(19)	922.41586(3)	−0.0018	350000		4s	$^1D_2$	sp	$(^3P)^3P^\circ\ ^3D^\circ_2$			TW		
924.2381(8)	108197.23(9)	924.23840(7)	−0.0003	550000		4s	$^3D_2$	sp	$(^3P)^3P^\circ\ ^5D^\circ_3$			K66		
925.119(10)	108094.2(12)	925.09896(8)	0.020	510000	*	4s	$^3D_2$	sp	$(^3P)^3P^\circ\ ^5D^\circ_2$			TW		
925.119(10)	108094.2(12)	925.10963(3)	0.009	510000	*	4s	$^1D_2$	sp	$(^3P)^3P^\circ\ ^3D^\circ_1$			TW		
925.119(10)	108094.2(12)	925.12612(7)	−0.007	510000	*	4s	$^3D_2$	sp	$(^3P)^3P^\circ\ ^5D^\circ_1$			TW		
929.702(3)	107561.4(3)	929.70175(3)	0.000	18000		4s	$^1D_2$	sp	$(^3P)^3P^\circ\ ^3P^\circ_2$			TW		
929.8940(20)	107539.14(23)	929.89272(14)	0.0013	18000		4s	$^3D_2$	sp	$(^1D)^3P^\circ\ ^3P^\circ_2$			TW		
932.9398(8)	107188.05(9)	932.93989(10)	−0.0001	590000		4s	$^3D_3$	sp	$(^1D)^3P^\circ\ ^3D^\circ_3$	1.7e+08	C	K66	B00	
934.9761(18)	106954.61(21)	934.97102(14)	0.0050	280000	?	4s	$^3D_1$	sp	$(^3P)^3P^\circ\ ^5D^\circ_0$			TWn		
935.0564(14)	106945.42(17)	935.05755(8)	−0.0012	380000		4s	$^3D_1$	sp	$(^3P)^3P^\circ\ ^5D^\circ_2$			TW		
		935.08529(7)			m	4s	$^3D_1$	sp	$(^3P)^3P^\circ\ ^5D^\circ_1$			S36c		
935.2307(22)	106925.49(25)	935.23232(7)	−0.0016	450000		4s	$^3D_3$	sp	$(^1D)^3P^\circ\ ^3D^\circ_2$			TW		
935.3408(18)	106912.90(20)	935.34317(5)	−0.0023	360000		4s	$^3D_2$	sp	$(^1D)^3P^\circ\ ^3P^\circ_1$			TW		
935.892(3)	106849.9(3)	935.89753(15)	−0.006	620000		4s	$^3D_3$	sp	$(^1D)^3P^\circ\ ^3F^\circ_4$	1.12e+08	C	TW	B00	
937.815(6)	106630.8(7)	937.81726(7)	−0.002	48000		4s	$^3D_3$	sp	$(^1D)^3P^\circ\ ^3F^\circ_3$			S36c		
939.5218(15)	106437.12(16)	939.52288(6)	−0.0010	190000		4s	$^3D_3$	sp	$(^1D)^3P^\circ\ ^3F^\circ_2$			TW		
943.3333(18)	106007.07(20)	943.33466(7)	−0.0014	580000		4s	$^3D_2$	sp	$(^1D)^3P^\circ\ ^3D^\circ_2$	1.4e+08	C	TW	B00	
945.5223(17)	105761.65(19)	945.52474(6)	−0.0025	510000		4s	$^3D_1$	sp	$(^1D)^3P^\circ\ ^3P^\circ_1$			TW		
945.8771(21)	105721.98(24)	945.87663(7)	0.0005	400000		4s	$^3D_2$	sp	$(^1D)^3P^\circ\ ^3D^\circ_1$			TW		
945.9614(21)	105712.56(24)	945.96464(7)	−0.0033	460000		4s	$^3D_2$	sp	$(^1D)^3P^\circ\ ^3F^\circ_3$	1.1e+08	C	TW	B00	
947.6989(17)	105518.74(19)	947.70005(6)	−0.0011	14000		4s	$^3D_2$	sp	$(^1D)^3P^\circ\ ^3F^\circ_2$			TW		
951.4068(15)	105107.51(16)	951.40774(8)	−0.0010	24000		4s	$^3D_1$	sp	$(^1D)^3P^\circ\ ^3P^\circ_0$			TW		
954.3815(15)	104779.91(16)	954.38279(7)	−0.0013	270000		4s	$^1D_2$	sp	$(^3P)^3P^\circ\ ^5D^\circ_3$			TW		
955.306(5)	104678.5(6)	955.30043(8)	0.005	9900		4s	$^1D_2$	sp	$(^3P)^3P^\circ\ ^5D^\circ_2$			TW		
955.320(5)	104677.0(6)	955.32939(7)	−0.009	9900		4s	$^1D_2$	sp	$(^3P)^3P^\circ\ ^5D^\circ_1$			TW		
956.2871(17)	104571.11(19)	956.29010(7)	−0.0030	300000		4s	$^3D_1$	sp	$(^1D)^3P^\circ\ ^3D^\circ_1$	1.8e+08	C	TW	B00	
958.1503(17)	104367.76(19)	958.15393(6)	−0.0037	380000		4s	$^3D_1$	sp	$(^1D)^3P^\circ\ ^3F^\circ_2$	1.6e+08	C	TW	B00	
960.4115(17)	104122.04(19)	960.41317(15)	−0.0017	270000		4s	$^1D_2$	sp	$(^1D)^3P^\circ\ ^3P^\circ_2$			TW		
966.2283(21)	103495.21(23)	966.22839(6)	−0.0001	11000		4s	$^1D_2$	sp	$(^1D)^3P^\circ\ ^3P^\circ_1$			TW		
967.8711(21)	103319.54(23)	967.87255(7)	−0.0015	4800		4s	$^3D_3$	sp	$(^3P)^3P^\circ\ ^5P^\circ_2$			TW		
968.0392(17)	103301.60(18)	968.04176(16)	−0.0025	250000		4s	$^3D_3$	sp	$(^3P)^3P^\circ\ ^5P^\circ_3$			TW		
972.2703(19)	102852.06(20)	972.26872(10)	0.0016	24000		4s	$^1D_2$	sp	$(^1D)^3P^\circ\ ^3D^\circ_3$			TW		
973.509(10)	102721.2(11)	973.49928(20)	0.010	18000		4s	$^3D_2$	sp	$(^3P)^3P^\circ\ ^5P^\circ_1$			S36c		
974.7581(15)	102589.56(15)	974.75876(8)	−0.0007	200000		4s	$^1D_2$	sp	$(^1D)^3P^\circ\ ^3D^\circ_2$			TW		
976.5524(15)	102401.06(15)	976.55293(7)	−0.0005	110000		4s	$^3D_2$	sp	$(^3P)^3P^\circ\ ^5P^\circ_2$			TW		
976.7238(17)	102383.09(18)	976.72519(17)	−0.0014	35000		4s	$^3D_2$	sp	$(^3P)^3P^\circ\ ^5P^\circ_3$			TW		
977.5662(13)	102294.86(14)	977.56714(7)	−0.0009	190000		4s	$^1D_2$	sp	$(^1D)^3P^\circ\ ^3F^\circ_3$			TW		
979.4200(15)	102101.25(16)	979.42055(6)	−0.0006	12000		4s	$^1D_2$	sp	$(^1D)^3P^\circ\ ^3F^\circ_2$			TW		
984.02(4)	101624(4)	983.979887(19)	0.04	9100		4s	$^3D_3$	5p	$(3/2)^2[3/2]^\circ_2$			S36c		
984.5331(15)	101570.98(15)	984.53339(21)	−0.0003	88000		4s	$^3D_1$	sp	$(^3P)^3P^\circ\ ^5P^\circ_1$			TW		

Table A1. *Cont.*

λ_{obs} [a] (Å)	σ_{obs} [b] (cm^{-1})	λ_{Ritz} [c] (Å)	$\Delta\lambda_{obs-Ritz}$ (Å)	I_{obs} [d] (arb. u.)	Char [e]	Lower Level	Upper Level	A (s^{-1})	Acc [f]	Line Ref. [g]	TP Ref. [g]	Notes [h]
987.6568(12)	101249.75(12)	987.65677(8)	−0.0000	61000		4s $^3D_1$	sp $(^3P)^3P^o\,{}^5P^o_2$			TW		
989.2364(12)	101088.07(12)	989.236270(17)	0.0002	30000		4s $^3D_3$	5p $(3/2)^2[5/2]^o_3$			TW		
992.9530(13)	100709.70(13)	992.952929(16)	0.0001	210000		4s $^3D_2$	5p $(3/2)^2[3/2]^o_2$			TW		
998.3067(14)	100169.62(14)	998.305876(14)	0.0008	40000		4s $^3D_2$	5p $(3/2)^2[5/2]^o_3$			TW		
999.797(3)	100020.3(3)	999.793812(15)	0.003	7600		4s $^3D_2$	5p $(3/2)^2[1/2]^o_1$			TW		
1001.0129(19)	99898.81(19)	1001.012710(14)	0.0002	22000		4s $^3D_2$	5p $(3/2)^2[5/2]^o_2$			TW		
1004.0540(17)	99596.24(17)	1004.055195(18)	−0.0012	330000		4s $^3D_3$	sp $(^3F)^3P^o\,{}^1F^o_3$			TW		
1006.980(6)	99306.8(6)	1006.983925(18)	−0.004	13000		4s $^3D_1$	5p $(3/2)^2[3/2]^o_1$			S36c		
1008.5674(17)	99150.54(16)	1008.568623(17)	−0.0012	290000		4s $^3D_3$	5p $(5/2)^2[5/2]^o_3$			TW		
1008.7274(17)	99134.81(16)	1008.728151(15)	−0.0008	310000		4s $^3D_2$	sp $(^3F)^3P^o\,{}^1D^o_2$			TW		
1010.2680(17)	98983.64(16)	1010.26867(8)	−0.0007	320000		4s $^1D_2$	sp $(^3P)^3P^o\,{}^5P^o_2$			TW		
1010.4520(18)	98965.61(18)	1010.45303(18)	−0.0011	48000		4s $^1D_2$	sp $(^3P)^3P^o\,{}^5P^o_3$			TW		
1010.640(3)	98947.2(3)	1010.639186(19)	0.001	18000		4s $^3D_3$	5p $(5/2)^2[5/2]^o_2$			TW		
1011.436	98869.4	1011.435606(17)		29000	:	4s $^3D_1$	5p $(3/2)^2[1/2]^o_1$			S36c		
1012.5956(18)	98756.11(18)	1012.596906(18)	−0.0013	290000		4s $^3D_3$	5p $(5/2)^2[7/2]^o_3$			TW		
1012.6813(18)	98747.75(18)	1012.683073(16)	−0.0017	50000		4s $^3D_1$	5p $(3/2)^2[5/2]^o_2$			TW		
1013.382(20)	98679.5(19)	1013.399849(15)	−0.018	15000		4s $^3D_2$	sp $(^3F)^3P^o\,{}^1F^o_3$			S36c		
1017.9976(18)	98232.06(18)	1017.997872(13)	−0.0002	70000		4s $^3D_2$	5p $(5/2)^2[5/2]^o_3$			TW		
1018.0628(19)	98225.77(18)	1018.064039(22)	−0.0013	180000		4s $^3D_1$	5p $(3/2)^2[1/2]^o_0$			TW		
1018.7031(24)	98164.03(23)	1018.707024(17)	−0.0039	550000		4s $^3D_3$	5p $(5/2)^2[3/2]^o_2$			TW		
1019.6525(18)	98072.62(17)	1019.654307(13)	−0.0018	240000		4s $^3D_2$	5p $(5/2)^2[3/2]^o_1$			TW		
1020.1070(18)	98028.93(18)	1020.107372(16)	−0.0004	280000		4s $^3D_2$	5p $(5/2)^2[5/2]^o_2$			TW		
1022.102	97837.6	1022.101982(13)		91000	:	4s $^3D_2$	5p $(5/2)^2[7/2]^o_3$			S36c		
1027.825(3)	97292.8(3)	1027.830823(19)	−0.006	1800000		4s $^1D_2$	5p $(3/2)^2[3/2]^o_2$	1.7e+08	C	TW	B00	
1028.3251(18)	97245.51(17)	1028.327701(13)	−0.0026	670000		4s $^3D_2$	5p $(5/2)^2[3/2]^o_2$			TW		
1029.7493(18)	97111.01(17)	1029.750492(24)	−0.0012	290000		4s $^3D_3$	sp $(^3F)^3P^o\,{}^3F^o_2$			TW		
1030.2612(18)	97062.76(17)	1030.26297(7)	−0.0017	500000		4s $^3D_3$	sp $(^3F)^3P^o\,{}^1G^o_4$			TW		
1031.7650(18)	96921.30(17)	1031.766015(15)	−0.0010	240000		4s $^3D_1$	5p $(5/2)^2[3/2]^o_1$			TW		
1033.5668(18)	96752.33(17)	1033.567510(18)	−0.0007	200000		4s $^1D_2$	5p $(3/2)^2[5/2]^o_3$			TW		
1035.161(3)	96603.3(3)	1035.162498(19)	−0.001	85000		4s $^1D_2$	5p $(3/2)^2[1/2]^o_1$			TW		
1036.465(3)	96481.8(3)	1036.469217(18)	−0.004	1800000		4s $^1D_2$	5p $(3/2)^2[5/2]^o_2$	1.01e+08	C	TW	B00	
1039.341(3)	96214.8(3)	1039.34743(3)	−0.006	2100000		4s $^3D_3$	sp $(^3F)^3P^o\,{}^3F^o_3$	2.0e+08	C	TW	B00	
1039.579(3)	96192.8(3)	1039.581895(21)	−0.003	1900000		4s $^3D_2$	sp $(^3F)^3P^o\,{}^3F^o_2$	2.0e+08	C	TW	B00	
1044.5185(8)	95737.89(7)	1044.51849(6)	0.0000	2300000		4s $^3D_3$	sp $(^3F)^3P^o\,{}^3F^o_4$	2.3e+08	C	K66	B00	
1044.7437(8)	95717.26(7)	1044.743170(19)	0.0005	2200000		4s $^1D_2$	sp $(^3F)^3P^o\,{}^1D^o_2$	4.8e+08	C	K66	B00	
1049.3642(18)	95295.80(17)	1049.363833(22)	0.0004	770000		4s $^3D_2$	sp $(^3F)^3P^o\,{}^3F^o_3$	1.11e+08	C	TW	B00	
1049.7551(6)	95260.31(5)	1049.755242(19)	−0.0001	2100000		4s $^1D_2$	sp $(^3F)^3P^o\,{}^1F^o_3$	1.04e+08	C	K66	B00	
1050.1523(18)	95224.28(17)	1050.15339(3)	−0.0010	530000		4s $^3D_2$	sp $(^3F)^3P^o\,{}^3D^o_1$	1.4e+08	C	TW	B00	
1050.4013(18)	95201.71(17)	1050.40243(3)	−0.0012	520000		4s $^3D_3$	sp $(^3F)^3P^o\,{}^3D^o_2$	8.2e+07	C	TW	B00	
1052.1742(18)	95041.30(16)	1052.174562(23)	−0.0004	1000000		4s $^3D_1$	sp $(^3F)^3P^o\,{}^3F^o_2$	7.3e+07	C	TW	B00	
1054.6907(6)	94814.53(5)	1054.689891(17)	0.0008	2800000		4s $^1D_2$	5p $(5/2)^2[5/2]^o_3$	2.1e+08	C	K66	B00	
1055.792(3)	94715.6(3)	1055.79644(4)	−0.005	2500000		4s $^3D_3$	sp $(^3F)^3P^o\,{}^3G^o_3$	2.1e+08	C	TW	B00	

Table A1. *Cont.*

λ_{obs} [a] (Å)	σ_{obs} [b] (cm^{-1})	λ_{Ritz} [c] (Å)	$\Delta\lambda_{obs\text{-}Ritz}$ (Å)	I_{obs} [d] (arb. u.)	Char [e]	Lower Level		Upper Level		A (s^{-1})	Acc [f]	Line Ref. [g]	TP Ref. [g]	Notes [h]
1056.9549(6)	94611.42(5)	1056.954368(20)	0.0005	2300000		4s	$^1D_2$	5p	$(5/2)^2[5/2]^{\circ}_2$	4.0e+08	C	K66	B00	
1058.798(3)	94446.7(3)	1058.79854(4)	−0.000	1800000		4s	$^3D_3$	sp	$(^3F)^3P^{\circ}\ ^3D^{\circ}_3$	1.9e+08	C	TW	B00	
1059.0963(8)	94420.12(7)	1059.095825(18)	0.0005	1400000		4s	$^1D_2$	5p	$(5/2)^2[7/2]^{\circ}_3$			K66		
1060.631(3)	94283.5(3)	1060.63409(3)	−0.003	1700000		4s	$^3D_2$	sp	$(^3F)^3P^{\circ}\ ^3D^{\circ}_2$	3.0e+08	C	TW	B00	
1063.0027(18)	94073.14(16)	1063.00505(3)	−0.0024	1400000		4s	$^3D_1$	sp	$(^3F)^3P^{\circ}\ ^3D^{\circ}_1$	4.2e+08	C	TW	B00	
1065.7837(18)	93827.67(16)	1065.781839(17)	0.0019	560000		4s	$^1D_2$	5p	$(5/2)^2[3/2]^{\circ}_2$			TW		
1066.1356(18)	93796.70(16)	1066.13397(4)	0.0017	590000		4s	$^3D_2$	sp	$(^3F)^3P^{\circ}\ ^3G^{\circ}_3$			TW		
1069.1944(18)	93528.36(16)	1069.19523(4)	−0.0008	1000000		4s	$^3D_2$	sp	$(^3F)^3P^{\circ}\ ^3D^{\circ}_3$	1.03e+08	C	TW	B00	
1070.3134(18)	93430.58(16)	1070.31087(6)	0.0025	380000		4s	$^3D_3$	sp	$(^3F)^3P^{\circ}\ ^3G^{\circ}_4$			TW		
1073.7444(18)	93132.04(16)	1073.74516(3)	−0.0008	720000		4s	$^3D_1$	sp	$(^3F)^3P^{\circ}\ ^3D^{\circ}_2$	1.5e+08	C	TW	B00	
1077.889(20)	92773.9(17)	1077.87559(3)	0.013	20000		4s	$^1D_2$	sp	$(^3F)^3P^{\circ}\ ^3F^{\circ}_2$			S36c		
1086.1134(18)	92071.42(15)	1086.10984(8)	0.0036	72000		4s	$^3D_3$	sp	$(^3F)^3P^{\circ}\ ^5F^{\circ}_3$			TW		
1088.405(5)	91877.6(4)	1088.39510(3)	0.010	270000		4s	$^1D_2$	sp	$(^3F)^3P^{\circ}\ ^3F^{\circ}_3$			TW		
1089.237(6)	91807.4(5)	1089.24451(3)	−0.007	39000		4s	$^1D_2$	sp	$(^3F)^3P^{\circ}\ ^3D^{\circ}_1$			S36c		
1091.293(3)	91634.43(24)	1091.29136(8)	0.001	40000		4s	$^3D_2$	sp	$(^3F)^3P^{\circ}\ ^5F^{\circ}_2$			TW		
1094.399(3)	91374.35(24)	1094.40214(8)	−0.003	440000		4s	$^3D_3$	sp	$(^3F)^3P^{\circ}\ ^5F^{\circ}_4$			TW		
1097.0512(18)	91153.45(15)	1097.05257(8)	−0.0014	320000		4s	$^3D_2$	sp	$(^3F)^3P^{\circ}\ ^5F^{\circ}_3$			TW		
1100.523(3)	90865.86(24)	1100.52419(3)	−0.001	25000		4s	$^1D_2$	sp	$(^3F)^3P^{\circ}\ ^3D^{\circ}_2$			TW		
1101.836(6)	90757.6(5)	1101.83633(13)	−0.000	8200		4s	$^3D_1$	sp	$(^3F)^3P^{\circ}\ ^5F^{\circ}_1$			S36c		
1105.1751(18)	90483.40(15)	1105.17629(8)	−0.0012	60000		4s	$^3D_1$	sp	$(^3F)^3P^{\circ}\ ^5F^{\circ}_2$			TW		
1106.447	90379.4	1106.44670(4)		23000	:	4s	$^1D_2$	sp	$(^3F)^3P^{\circ}\ ^3G^{\circ}_3$			S36c		
1109.744	90110.9	1109.74420(4)		7200	:	4s	$^1D_2$	sp	$(^3F)^3P^{\circ}\ ^3D^{\circ}_3$			S36c		
1111.753(6)	89948.0(5)	1111.75742(11)	−0.004	3000		4s	$^3D_3$	sp	$(^3F)^3P^{\circ}\ ^5G^{\circ}_3$			S36c		
1119.9480(18)	89289.86(15)	1119.94659(11)	0.0014	98000		4s	$^3D_3$	sp	$(^3F)^3P^{\circ}\ ^5G^{\circ}_4$			TW		
1123.2265(18)	89029.24(14)	1123.22580(11)	0.0007	21000		4s	$^3D_2$	sp	$(^3F)^3P^{\circ}\ ^5G^{\circ}_3$			TW		
1127.2516(18)	88711.34(14)			18000		4p	$^3F^{\circ}_2$	10s	$(3/2)^2[3/2]_1$			TWn		
1130.888(12)	88426.1(9)	1130.8852(3)	0.003	4500	*	4s	$^3D_1$	sp	$(^3F)^3P^{\circ}\ ^5G^{\circ}_2$			S36c		
1130.888(12)	88426.1(9)	1130.89779(11)	−0.010	4500	*	4p	$^3P^{\circ}_1$	9s	$(3/2)^2[3/2]_2$			S36c		
		1130.96216(16)			m	4p	$^3P^{\circ}_2$	8d	$(5/2)^2[3/2]_1$			S36cn		
1135.3657(18)	88077.35(14)	1135.34826(20)	0.0174	8200	?	4p	$^3F^{\circ}_3$	10s	$(5/2)^2[5/2]_2$			TW		X
1142.6393(18)	87516.68(14)	1142.64027(10)	−0.0009	73000		4s	$^3D_2$	sp	$(^3F)^3P^{\circ}\ ^5D^{\circ}_2$			TW		
1144.853(3)	87347.48(22)	1144.85531(8)	−0.003	120000		4s	$^3D_3$	sp	$(^3F)^3P^{\circ}\ ^5D^{\circ}_3$			TW		
1147.7628(18)	87126.02(14)			25000		4s	$^3D_1$	sp	$(^3F)^3P^{\circ}\ ^5D^{\circ}_1$			TW		
1157.0194(18)	86428.97(14)	1157.02043(8)	−0.0010	15000		4s	$^3D_2$	sp	$(^3F)^3P^{\circ}\ ^5D^{\circ}_3$			TW		
1157.8714(18)	86365.38(14)	1157.87172(10)	−0.0004	19000		4s	$^3D_1$	sp	$(^3F)^3P^{\circ}\ ^5D^{\circ}_2$			TW		
		1157.88056(11)			m	4p	$^3P^{\circ}_1$	9s	$(5/2)^2[5/2]_2$			TW		
1162.5991(18)	86014.17(13)	1162.60059(14)	−0.0015	6300		4s	$^3D_3$	sp	$(^3F)^3P^{\circ}\ ^5D^{\circ}_4$			TW		
1185.899	84324.2	1185.89903(5)		19000	:	4p	$^3P^{\circ}_2$	8s	$(5/2)^2[5/2]_3$			S36c		
1201.627(6)	83220.5(4)	1201.62566(8)	0.001	18000		4p	$^3F^{\circ}_3$	7d	$(5/2)^2[9/2]_4$			S36c		
1204.643(20)	83012.1(14)	1204.61568(7)	0.027	9100	*	4p	$^3F^{\circ}_4$	7d	$(5/2)^2[7/2]_4$			S36c		
1204.643(20)	83012.1(14)	1204.63538(5)	0.008	9100	*	4p	$^3F^{\circ}_4$	7d	$(5/2)^2[7/2]_3$			S36c		
1204.643(20)	83012.1(14)	1204.65288(9)	−0.010	9100	*	4s	$^1D_2$	sp	$(^3F)^3P^{\circ}\ ^5D^{\circ}_3$			S36c		
1205.180(12)	82975.2(8)	1205.14656(6)	0.034	4800	*	4p	$^3F^{\circ}_4$	7d	$(5/2)^2[5/2]_3$			S36c		

Table A1. *Cont.*

λ_{obs} [a] (Å)	σ_{obs} [b] (cm^{-1})	λ_{Ritz} [c] (Å)	$\Delta\lambda_{obs-Ritz}$ (Å)	I_{obs} [d] (arb. u.)	Char [e]	Lower Level		Upper Level		A (s^{-1})	Acc [f]	Line Ref. [g]	TP Ref. [g]	Notes [h]
1205.180(12)	82975.2(8)	1205.19424(8)	−0.014	4800	*	4p	$^1F^\circ_3$	7d	$(3/2)^2[5/2]_3$			S36c		
1205.180(12)	82975.2(8)	1205.2024(3)	−0.022	4800	*	4p	$^3D^\circ_2$	9s	$(3/2)^2[3/2]_1$			S36c		
1205.180(12)	82975.2(8)	1205.21874(7)	−0.039	4800	*	4p	$^3F^\circ_2$	8s	$(3/2)^2[3/2]_2$			S36cn		
1205.901(6)	82925.5(4)	1205.90271(8)	−0.002	18000		4p	$^3F^\circ_4$	7d	$(5/2)^2[9/2]_5$			S36c		
1206.771(6)	82865.8(4)	1206.76906(5)	0.002	4400		4p	$^3P^\circ_1$	8s	$(5/2)^2[5/2]_2$			S36c		
1214.540	82335.7	1214.53971(8)		9600	:	4p	$^3D^\circ_3$	9s	$(5/2)^2[5/2]_3$			S36c		
1214.554	82334.7	1214.55446(5)		8600	:	4p	$^3F^\circ_3$	8s	$(5/2)^2[5/2]_2$			S36c		
1219.334	82012.0	1219.33363(5)		8500	:	4p	$^3F^\circ_4$	8s	$(5/2)^2[5/2]_3$			S36c		
1235.93(5)	80911(3)	1235.87291(5)	0.06	4000		4p	$^3F^\circ_2$	8s	$(5/2)^2[5/2]_2$			S36c		
1240.026(6)	80643.5(4)	1240.02707(5)	−0.001	7400		4p	$^3P^\circ_1$	6d	$(3/2)^2[3/2]_2$			S36c		
1241.964	80517.6	1241.96398(3)		14000	:	4p	$^3P^\circ_2$	7s	$(3/2)^2[3/2]_2$			S36c		
1243.03(5)	80448(3)	1243.08557(6)	−0.05	7300		4p	$^3P^\circ_1$	6d	$(3/2)^2[1/2]_1$			S36c		
1248.796(3)	80077.12(17)	1248.79156(5)	0.005	26000		4p	$^3P^\circ_2$	6d	$(5/2)^2[5/2]_3$			TW		
1250.058(4)	79996.30(22)	1250.04822(6)	0.010	69000		4p	$^3P^\circ_2$	6d	$(5/2)^2[3/2]_2$			TW		
1253.185(3)	79796.67(17)	1253.18074(7)	0.004	25000		4p	$^3P^\circ_2$	6d	$(5/2)^2[1/2]_1$			TW		
1255.163(6)	79670.9(4)	1255.15693(6)	0.006	6800		4p	$^3P^\circ_0$	6d	$(3/2)^2[3/2]_1$			S36c		
1257.675(6)	79511.8(4)	1257.68320(7)	−0.008	6700		4p	$^3P^\circ_0$	6d	$(3/2)^2[1/2]_1$			S36c		
1261.214(6)	79288.7(4)	1261.21537(5)	−0.001	3500		4p	$^1D^\circ_2$	8s	$(5/2)^2[5/2]_2$			S36c		
1262.929(6)	79181.0(4)	1262.92481(9)	0.004	19000		4p	$^3P^\circ_1$	6d	$(5/2)^2[1/2]_0$			S36c		
1265.510(3)	79019.53(17)	1265.50624(3)	0.004	72000		4p	$^3P^\circ_1$	7s	$(3/2)^2[3/2]_2$			TW		
1266.313(3)	78969.42(17)	1266.30992(4)	0.003	27000		4p	$^3P^\circ_1$	7s	$(3/2)^2[3/2]_1$			TW		
1268.71(5)	78820(3)	1268.66846(5)	0.04	3300		4p	$^3D^\circ_3$	8s	$(5/2)^2[5/2]_3$			S36c		
1269.446(6)	78774.5(4)	1269.44626(6)	0.000	6200		4p	$^3F^\circ_2$	6d	$(3/2)^2[5/2]_2$			S36c		
1271.327(6)	78658.0(4)	1271.31771(4)	0.010	12000		4p	$^3P^\circ_1$	6d	$(5/2)^2[5/2]_2$			S36c		
1272.043(3)	78613.69(17)	1272.04165(6)	0.001	31000		4p	$^3F^\circ_2$	6d	$(3/2)^2[7/2]_3$			TW		
1273.705(6)	78511.1(4)	1273.70053(5)	0.005	11000		4p	$^3P^\circ_1$	6d	$(5/2)^2[3/2]_1$			S36c		
1274.071	78488.6	1274.07065(3)		17000	:	4p	$^3F^\circ_3$	7s	$(3/2)^2[3/2]_2$			S36c		
1274.465	78464.3	1274.46493(3)		16000	:	4p	$^3P^\circ_2$	7s	$(5/2)^2[5/2]_2$			S36c		
1275.5713(15)	78396.24(9)	1275.57159(3)	−0.0003	75000		4p	$^3P^\circ_2$	7s	$(5/2)^2[5/2]_3$			TW		
1279.948(20)	78128.2(12)	1279.96123(5)	−0.013	2700		4p	$^3F^\circ_3$	6d	$(5/2)^2[5/2]_2$			S36c		
1280.271(3)	78108.48(17)	1280.26800(4)	0.003	30000		4p	$^3F^\circ_3$	6d	$(5/2)^2[7/2]_3$			TW,S36r		
1281.094(3)	78058.29(17)	1281.09394(9)	−0.000	10000		s^2	$^3F_4$	7p	$(5/2)^2[5/2]^\circ_3$			TW		
1281.228(20)	78050.1(12)	1281.25682(6)	−0.029	16000		4p	$^3F^\circ_3$	6d	$(5/2)^2[5/2]_3$			S36c		
1281.462(3)	78035.88(17)	1281.46142(4)	0.000	44000		4p	$^3P^\circ_0$	7s	$(3/2)^2[3/2]_1$			TW		
1282.455(3)	77975.44(17)	1282.45466(5)	0.000	72000		4p	$^3F^\circ_3$	6d	$(5/2)^2[9/2]_4$			TW		
1283.824(6)	77892.3(4)	1283.82966(10)	−0.005	5300		s^2	$^3F_4$	7p	$(5/2)^2[7/2]^\circ_4$			S36c		
1284.872(3)	77828.74(16)	1284.87116(5)	0.001	36000		4p	$^3F^\circ_4$	6d	$(5/2)^2[7/2]_4$			TW		
1285.521(6)	77789.5(4)	1285.51841(5)	0.003	5400		4p	$^1F^\circ_3$	6d	$(3/2)^2[5/2]_3$			S36c		
1285.901(20)	77766.5(12)	1285.92203(6)	−0.021	5100		4p	$^3F^\circ_4$	6d	$(5/2)^2[5/2]_3$			S36c		
		1287.43101(6)			m	4p	$^1F^\circ_3$	6d	$(3/2)^2[7/2]_4$			TW		
1287.4690(8)	77671.77(5)	1287.46810(6)	0.0009	53000		4p	$^3F^\circ_4$	6d	$(5/2)^2[9/2]_5$			K66		
1297.550(4)	77068.31(21)	1297.54980(3)	0.000	4900		4p	$^3F^\circ_2$	7s	$(3/2)^2[3/2]_2$			TW		
1297.979(20)	77042.8(12)	1297.9922(9)	−0.013	4900		s^2	$^3F_2$	7p	$(3/2)^2[3/2]^\circ_1$			S36cn		
1298.3952(8)	77018.15(5)	1298.39471(4)	0.0005	60000		4p	$^3F^\circ_2$	7s	$(3/2)^2[3/2]_1$			K66		

Table A1. *Cont.*

λ_{obs} [a] (Å)	σ_{obs} [b] (cm^{-1})	λ_{Ritz} [c] (Å)	$\Delta\lambda_{obs\text{-}Ritz}$ (Å)	I_{obs} [d] (arb. u.)	Char [e]	Lower Level		Upper Level		A (s^{-1})	Acc [f]	Line Ref. [g]	TP Ref. [g]	Notes [h]
1298.917(12)	76987.2(7)	1298.90527(6)	0.012	4700		4p	$^1D^\circ_2$	6d	$(3/2)^2[7/2]_3$			S36c		
1299.2684(8)	76966.39(5)	1299.26777(3)	0.0006	44000		4p	$^3P^\circ_1$	7s	$(5/2)^2[5/2]_2$			K66		
1303.661(4)	76707.04(21)	1303.66001(5)	0.001	2900		4p	$^3F^\circ_2$	6d	$(5/2)^2[5/2]_2$			TW		
1303.978(3)	76688.38(16)	1303.97824(4)	0.000	6400		4p	$^3F^\circ_2$	6d	$(5/2)^2[7/2]_3$			TW		
1305.562(3)	76595.40(16)	1305.56082(6)	0.001	19000		4p	$^3D^\circ_3$	6d	$(3/2)^2[7/2]_4$			TW		
1308.2982(6)	76435.17(4)	1308.29687(3)	0.0013	99000		4p	$^3F^\circ_3$	7s	$(5/2)^2[5/2]_2$			K66		
1309.464(3)	76367.12(16)	1309.46310(3)	0.001	52000		4p	$^3F^\circ_3$	7s	$(5/2)^2[5/2]_3$			TW		
1311.76(4)	76233.5(23)	1311.79458(10)	−0.04	3900		s^2	$^3F_3$	7p	$(5/2)^2[5/2]^\circ_3$			S36c		
1314.1498(8)	76094.83(5)	1314.14936(4)	0.0004	50000		4p	$^1F^\circ_3$	7s	$(3/2)^2[3/2]_2$			K66		
1314.3371(6)	76083.98(3)	1314.33637(3)	0.0007	150000		4p	$^3F^\circ_4$	7s	$(5/2)^2[5/2]_3$			K66		
1320.687(3)	75718.15(16)	1320.68581(5)	0.002	30000		4p	$^1F^\circ_3$	6d	$(5/2)^2[7/2]_4$			TW		
1321.798(3)	75654.53(16)	1321.79611(6)	0.002	11000		4p	$^1F^\circ_3$	6d	$(5/2)^2[5/2]_3$			TW		
1322.628(6)	75607.0(3)	1322.63248(13)	−0.004	21000		4p	$^1P^\circ_1$	6d	$(3/2)^2[1/2]_0$			S36c		
1323.188(20)	75575.0(11)	1323.20408(6)	−0.016	9700		4p	$^1F^\circ_3$	6d	$(5/2)^2[3/2]_2$			S36c		
1323.812(20)	75539.4(11)	1323.79410(6)	0.018	21000		4p	$^3D^\circ_1$	6d	$(3/2)^2[5/2]_2$			S36c		
1325.272(20)	75456.2(11)	1325.24199(5)	0.030	3300		4p	$^3D^\circ_1$	6d	$(3/2)^2[3/2]_2$			S36c		
1325.515(3)	75442.39(15)	1325.51345(3)	0.001	8300		4p	$^1D^\circ_2$	7s	$(3/2)^2[3/2]_2$			TW		
1326.396(3)	75392.26(15)	1326.39519(4)	0.001	28000		4p	$^1D^\circ_2$	7s	$(3/2)^2[3/2]_1$			TW		
1328.415(3)	75277.66(15)	1328.41289(5)	0.002	12000		4p	$^3D^\circ_2$	6d	$(3/2)^2[5/2]_3$			TW		
1329.656(20)	75207.4(11)	1329.66943(6)	−0.014	3100		4p	$^3D^\circ_2$	6d	$(3/2)^2[3/2]_2$			S36c		
1331.892(3)	75081.17(15)	1331.89052(5)	0.001	14000		4p	$^1D^\circ_2$	6d	$(5/2)^2[5/2]_2$			TW		
1332.224(3)	75062.46(15)	1332.22268(4)	0.001	17000		4p	$^1D^\circ_2$	6d	$(5/2)^2[7/2]_3$			TW		
1333.0457(8)	75016.18(5)	1333.04501(3)	0.0007	58000		4p	$^3D^\circ_3$	7s	$(3/2)^2[3/2]_2$			K66		
		1333.06670(3)			m	4p	$^3F^\circ_2$	7s	$(5/2)^2[5/2]_2$			TW		
1334.56(4)	74931.2(22)	1334.50604(5)	0.05	5500	p	4p	$^1D^\circ_2$	6d	$(5/2)^2[3/2]_1$			S36c		X
1334.650(4)	74926.02(20)	1334.65445(7)	−0.005	5800		4p	$^1P^\circ_1$	6d	$(3/2)^2[3/2]_1$			TW		
1334.686(6)	74924.0(3)	1334.68792(17)	−0.002	1500		s^2	$^3F_2$	7p	$(5/2)^2[5/2]^\circ_2$			TW		
1337.55(4)	74763.8(22)	1337.51122(8)	0.03	1400		4p	$^1P^\circ_1$	6d	$(3/2)^2[1/2]_1$			S36c		
1339.46(4)	74656.7(22)	1339.49492(5)	−0.03	1300		4p	$^3D^\circ_3$	6d	$(5/2)^2[5/2]_2$			S36c		
1339.7734(23)	74639.49(13)	1339.77126(5)	0.0022	14000		4p	$^3D^\circ_3$	6d	$(5/2)^2[7/2]_4$			TW		
1340.9161(23)	74575.88(13)	1340.91390(6)	0.0022	6700		4p	$^3D^\circ_3$	6d	$(5/2)^2[5/2]_3$			TW		
1350.5963(20)	74041.37(11)	1350.59358(4)	0.0027	29000		4p	$^1F^\circ_3$	7s	$(5/2)^2[5/2]_2$			TW		
1351.8384(8)	73973.34(4)	1351.83646(3)	0.0019	53000		4p	$^1F^\circ_3$	7s	$(5/2)^2[5/2]_3$			K66		
1355.3066(8)	73784.04(4)	1355.30507(4)	0.0015	36000		4p	$^3D^\circ_1$	7s	$(3/2)^2[3/2]_1$			K66		
1358.7736(6)	73595.78(3)	1358.7729(3)	0.0007	15000000		d^{10}	$^1S_0$	4p	$^1P^\circ_1$	3.3e+08	C+	K66	B09	
1359.0107(8)	73582.94(4)	1359.00914(3)	0.0016	38000		4p	$^3D^\circ_2$	7s	$(3/2)^2[3/2]_2$			K66		
1359.9394(20)	73532.69(11)	1359.93602(4)	0.0034	11000		4p	$^3D^\circ_2$	7s	$(3/2)^2[3/2]_1$			TW		
1362.6004(6)	73389.09(3)	1362.59959(3)	0.0008	49000		4p	$^1D^\circ_2$	7s	$(5/2)^2[5/2]_2$			K66		
1363.506(2)	73340.36(13)	1363.50304(4)	0.003	12000		4p	$^1P^\circ_1$	7s	$(3/2)^2[3/2]_2$			TW		
1367.9508(6)	73102.04(3)	1367.9509(3)	−0.0001	5800000		d^{10}	$^1S_0$	4p	$^3D^\circ_1$	8.3e+07	C+	K66	D05	
1370.257(6)	72979.0(3)	1370.25181(6)	0.005	2700		4p	$^1P^\circ_1$	6d	$(5/2)^2[5/2]_2$			S36c		

Table A1. *Cont.*

λ_{obs} [a] (Å)	σ_{obs} [b] (cm^{-1})	λ_{Ritz} [c] (Å)	$\Delta\lambda_{obs-Ritz}$ (Å)	I_{obs} [d] (arb. u.)	Char [e]	Lower Level	Upper Level	A (s^{-1})	Acc [f]	Line Ref. [g]	TP Ref. [g]	Notes [h]
1370.561(3)	72962.83(18)	1370.55975(3)	0.001	1800		4p $^3D^\circ_3$	7s $(5/2)^2[5/2]_2$			TW		
1371.8400(6)	72894.80(3)	1371.83967(3)	0.0003	43000		4p $^3D^\circ_3$	7s $(5/2)^2[5/2]_3$			K66		
1375.522(20)	72699.7(11)	1375.50173(4)	0.020	6900		4p $^3P^\circ_2$	5d $(3/2)^2[5/2]_3$			S36c		
1393.129(2)	71780.88(13)	1393.12738(3)	0.001	18000		4p $^3D^\circ_1$	7s $(5/2)^2[5/2]_2$			TW		
1398.6428(16)	71497.88(8)	1398.64196(11)	0.0009	40000		s^2 $^3F_4$	6p $(3/2)^2[5/2]^\circ_3$			TW		
1399.3532(17)	71461.59(9)	1399.35262(3)	0.0005	15000		4p $^3D^\circ_2$	7s $(5/2)^2[5/2]_3$			TW		
1402.7785(8)	71287.09(4)	1402.77693(4)	0.0016	98000		4p $^1P^\circ_1$	7s $(5/2)^2[5/2]_2$			K66		
1403.985(3)	71225.85(16)	1403.99171(5)	−0.007	4900		4p $^3P^\circ_1$	5d $(3/2)^2[5/2]_2$			TW		
1407.1688(16)	71064.68(8)	1407.16862(5)	0.0001	60000		4p $^3P^\circ_1$	5d $(3/2)^2[3/2]_2$			TW		
1408.8131(23)	70981.74(12)	1408.81230(5)	0.0008	4900		4p $^3P^\circ_1$	5d $(3/2)^2[3/2]_1$			TW		
1414.4390(19)	70699.41(10)	1414.4371(4)	0.0019	26000		s^2 $^3F_2$	sp $(^1G)^3P^\circ\ ^3G^\circ_3$			TWn		
1414.8980(15)	70676.47(8)	1414.89768(6)	0.0003	49000		4p $^3P^\circ_1$	5d $(3/2)^2[1/2]_1$			TW		
1418.4250(14)	70500.73(7)	1418.42631(3)	−0.0014	140000		4p $^3P^\circ_2$	5d $(5/2)^2[5/2]_3$			TW		
1419.7465(20)	70435.11(10)	1419.74554(4)	0.0010	7500		4p $^3F^\circ_3$	5d $(3/2)^2[7/2]_4$			TW		
1421.3746(13)	70354.43(7)	1421.37346(5)	0.0011	35000		4p $^3P^\circ_2$	5d $(5/2)^2[3/2]_1$			TW		
1421.7587(6)	70335.42(3)	1421.75866(4)	0.0000	200000		4p $^3P^\circ_2$	5d $(5/2)^2[3/2]_2$			K66		
1427.5920(12)	70048.03(6)	1427.59106(6)	0.0009	58000		4p $^3P^\circ_0$	5d $(3/2)^2[3/2]_1$			TW		
1427.8283(8)	70036.43(4)	1427.82905(7)	−0.0007	180000		s^2 $^3F_4$	6p $(5/2)^2[5/2]^\circ_3$			K66		
1428.3572(8)	70010.50(4)	1428.35801(11)	−0.0008	190000		s^2 $^3F_3$	6p $(3/2)^2[3/2]^\circ_2$			K66		
1430.2425(8)	69918.21(4)	1430.24252(5)	−0.0000	240000		4p $^3P^\circ_2$	5d $(5/2)^2[1/2]_1$			K66		
1433.8415(12)	69742.72(6)	1433.84011(7)	0.0014	53000		4p $^3P^\circ_0$	5d $(3/2)^2[1/2]_1$			TW		
1434.452(3)	69713.01(13)	1434.45240(14)	−0.000	3600		s^2 $^3F_3$	6p $(3/2)^2[5/2]^\circ_2$			TWn		
1434.7712(12)	69697.52(6)	1434.76999(8)	0.0013	76000		4p $^3P^\circ_1$	5d $(5/2)^2[1/2]_0$			TW		
1434.9035(6)	69691.10(3)	1434.90377(9)	−0.0003	270000		s^2 $^3F_4$	6p $(5/2)^2[7/2]^\circ_4$			K66		
1435.3153(13)	69671.10(6)	1435.31565(13)	−0.0004	110000		s^2 $^3F_3$	6p $(3/2)^2[5/2]^\circ_3$			TW		
1436.2352(12)	69626.48(6)	1436.23585(7)	−0.0007	130000		s^2 $^3F_4$	sp $(^1G)^3P^\circ\ ^3F^\circ_3$			TW		
1442.1398(12)	69341.40(6)	1442.13845(4)	0.0014	66000		4p $^3P^\circ_2$	6s $(3/2)^2[3/2]_2$			TW		
1443.5423(12)	69274.04(6)	1443.54170(6)	0.0006	73000		4p $^3F^\circ_2$	5d $(3/2)^2[5/2]_2$			TW		
1444.1295(22)	69245.87(10)	1444.13017(4)	−0.0007	7700		4p $^3P^\circ_2$	6s $(3/2)^2[3/2]_1$			TW		
1445.9841(12)	69157.05(6)	1445.98335(4)	0.0007	97000		4p $^3P^\circ_1$	5d $(5/2)^2[5/2]_2$			TW		
1446.9008(22)	69113.24(10)	1446.90033(6)	0.0005	4600		4p $^3F^\circ_2$	5d $(3/2)^2[3/2]_2$			TW		
1448.6390(19)	69030.31(9)	1448.63819(6)	0.0008	6100		4p $^3F^\circ_2$	5d $(3/2)^2[3/2]_1$			TW		
1449.0578(8)	69010.36(4)	1449.05784(12)	−0.0000	180000		s^2 $^3F_2$	6p $(3/2)^2[3/2]^\circ_1$			K66		
1450.3032(12)	68951.10(6)	1450.30363(4)	−0.0004	170000		4p $^3F^\circ_2$	5d $(3/2)^2[7/2]_3$			TW		
1452.2950(12)	68856.53(6)	1452.29331(5)	0.0017	97000		4p $^3P^\circ_1$	5d $(5/2)^2[3/2]_1$			TW		
1452.670(20)	68838.7(9)	1452.69545(4)	−0.025	3600		4p $^3P^\circ_1$	5d $(5/2)^2[3/2]_2$			S36c		
1455.6643(13)	68697.16(6)	1455.66225(7)	0.0021	22000		s^2 $^3F_4$	6p $(5/2)^2[7/2]^\circ_3$			TW		
1457.1778(12)	68625.81(6)	1457.17554(5)	0.0023	55000		4p $^3F^\circ_3$	5d $(5/2)^2[5/2]_2$			TW		
1458.0022(6)	68587.00(3)	1458.00133(5)	0.0009	150000		4p $^3F^\circ_3$	5d $(5/2)^2[7/2]_3$			K66		
1459.4131(6)	68520.70(3)	1459.41216(15)	0.0009	220000		s^2 $^3F_2$	6p $(3/2)^2[5/2]^\circ_2$			K66		
1460.3076(13)	68478.72(6)	1460.30573(14)	0.0018	14000		s^2 $^3F_2$	6p $(3/2)^2[5/2]^\circ_3$			TW		
1460.4620(13)	68471.48(6)	1460.45917(4)	0.0028	17000		4p $^3F^\circ_3$	5d $(5/2)^2[5/2]_3$			TW		

Table A1. *Cont.*

λ_{obs} [a] (Å)	σ_{obs} [b] (cm^{-1})	λ_{Ritz} [c] (Å)	$\Delta\lambda_{obs-Ritz}$ (Å)	I_{obs} [d] (arb. u.)	Char [e]	Lower Level		Upper Level		A (s^{-1})	Acc [f]	Line Ref. [g]	TP Ref. [g]	Notes [h]
1461.5557(11)	68420.25(5)	1461.55369(5)	0.0020	60000		4p	$^3P^\circ_1$	5d	$(5/2)^2[1/2]_1$			TW		
1462.8534(18)	68359.55(9)	1462.85242(6)	0.0010	5100		s^2	$^3F_4$	4f	$(3/2)^2[9/2]^\circ_5$			TW		
1463.7512(6)	68317.62(3)	1463.75130(4)	−0.0001	350000		4p	$^3F^\circ_3$	5d	$(5/2)^2[9/2]_4$			K66		
1463.8367(6)	68313.63(3)	1463.83785(5)	−0.0012	200000		4p	$^3F^\circ_4$	5d	$(5/2)^2[7/2]_4$			K66		
1463.9947(12)	68306.26(6)	1463.99219(5)	0.0025	20000		4p	$^3F^\circ_3$	5d	$(5/2)^2[3/2]_2$			TW		
1464.1173(22)	68300.54(10)	1464.11637(6)	0.0009	4600		4p	$^1F^\circ_3$	5d	$(3/2)^2[5/2]_2$			TW		
1464.6035(11)	68277.86(5)	1464.60141(5)	0.0021	69000		4p	$^1F^\circ_3$	5d	$(3/2)^2[5/2]_3$			TW		
1465.0339(18)	68257.81(9)	1465.03633(14)	−0.0024	4300		s^2	$^3F_2$	6p	$(3/2)^2[1/2]^\circ_1$			TW		
1465.5404(6)	68234.22(3)	1465.54069(16)	−0.0003	170000		s^2	$^3F_4$	sp	$(^1G)^3P^\circ\ ^3F^\circ_4$			K66		
1466.0705(6)	68209.54(3)	1466.07026(9)	0.0002	280000		s^2	$^3F_3$	6p	$(5/2)^2[5/2]^\circ_3$			K66		
1466.5247(10)	68188.42(5)	1466.52373(4)	0.0010	79000		4p	$^3F^\circ_4$	5d	$(5/2)^2[5/2]_3$			K66		
1466.7305(12)	68178.85(6)	1466.72839(10)	0.0022	42000		s^2	$^3F_3$	sp	$(^3F)^1P^\circ\ ^3F^\circ_2$			TW		
1467.574(5)	68139.68(24)	1467.57153(6)	0.002	1400		4p	$^1F^\circ_3$	5d	$(3/2)^2[3/2]_2$			TW		
1469.6935(6)	68041.40(3)	1469.69291(5)	0.0006	150000		4p	$^1F^\circ_3$	5d	$(3/2)^2[7/2]_4$			K66		
1469.8457(12)	68034.35(5)	1469.84329(5)	0.0024	49000		4p	$^3F^\circ_4$	5d	$(5/2)^2[9/2]_4$			TW		
1470.6970(6)	67994.97(3)	1470.69711(4)	−0.0001	500000		4p	$^3F^\circ_4$	5d	$(5/2)^2[9/2]_5$			K66		
1471.072(5)	67977.65(24)	1471.07289(5)	−0.001	2700		4p	$^1F^\circ_3$	5d	$(3/2)^2[7/2]_3$			TW		
1472.3946(6)	67916.58(3)	1472.3950(4)	−0.0004	6800000		d^{10}	$^1S_0$	4p	$^3P^\circ_1$	8.e+06	E	K66	D05se	
1473.5295(8)	67864.27(4)	1473.53001(10)	−0.0005	100000		s^2	$^3F_3$	6p	$(5/2)^2[7/2]^\circ_4$			K66		
1473.9786(6)	67843.59(3)	1473.97844(4)	0.0002	190000		4p	$^3P^\circ_1$	6s	$(3/2)^2[3/2]_2$			K66		
1474.9348(6)	67799.61(3)	1474.93480(9)	−0.0000	230000		s^2	$^3F_3$	sp	$(^1G)^3P^\circ\ ^3F^\circ_3$			K66		
1475.4361(22)	67776.57(10)	1475.4383(9)	−0.0023	23000		s^2	$^1D_2$	sp	$(^1D)^1P^\circ\ ^1P^\circ_1$			TWn		
1476.0596(6)	67747.94(3)	1476.05914(4)	0.0005	150000		4p	$^3P^\circ_1$	6s	$(3/2)^2[3/2]_1$			K66		
1478.2385(13)	67648.08(6)	1478.23606(6)	0.0024	23000		4p	$^1D^\circ_2$	5d	$(3/2)^2[5/2]_2$			TW		
1481.5438(6)	67497.16(3)	1481.54375(12)	0.0000	130000		s^2	$^3F_3$	6p	$(5/2)^2[5/2]^\circ_2$			K66		
1484.178(5)	67377.37(24)	1484.17557(13)	0.002	1200		s^2	$^3F_4$	sp	$(^1G)^3P^\circ\ ^3H^\circ_5$			TW		
1485.3282(8)	67325.19(4)	1485.32773(4)	0.0005	110000		4p	$^1D^\circ_2$	5d	$(3/2)^2[7/2]_3$			K66		
1485.6143(13)	67312.22(6)	1485.60994(4)	0.0043	53000		4p	$^3F^\circ_3$	6s	$(3/2)^2[3/2]_2$			TW		
1485.6788(8)	67309.30(4)	1485.67761(4)	0.0012	110000		4p	$^3P^\circ_2$	6s	$(5/2)^2[5/2]_2$			K66		
1487.612(5)	67221.81(24)	1487.60927(6)	0.003	1400		4p	$^3D^\circ_3$	5d	$(3/2)^2[5/2]_2$			TW		
1487.7805(12)	67214.22(5)	1487.77920(12)	0.0013	68000		s^2	$^3F_3$	6p	$(5/2)^2[3/2]^\circ_2$			TW		
1487.9719(12)	67205.57(5)	1487.96988(5)	0.0020	70000		4p	$^3F^\circ_2$	5d	$(5/2)^2[5/2]_2$			TW		
1488.1123(12)	67199.23(5)	1488.11000(4)	0.0023	53000		4p	$^3D^\circ_3$	5d	$(3/2)^2[5/2]_3$			TW		
1488.2853(13)	67191.42(6)	1488.2853(12)	−0.0000	34000		s^2	$^1D_2$	8p	$(5/2)^2[5/2]^\circ_2$			TWn		
1488.6366(6)	67175.56(3)	1488.63692(4)	−0.0003	1100000		4p	$^3P^\circ_2$	6s	$(5/2)^2[5/2]_3$			K66		
1488.8319(12)	67166.75(6)	1488.83094(5)	0.0009	120000		4p	$^3F^\circ_2$	5d	$(5/2)^2[7/2]_3$			TW		
1491.178(5)	67061.07(24)	1491.17634(6)	0.002	1400		4p	$^3D^\circ_3$	5d	$(3/2)^2[3/2]_2$			TW		
1491.384(6)	67051.8(3)	1491.39392(4)	−0.010	1200		4p	$^3F^\circ_2$	5d	$(5/2)^2[5/2]_3$			TW		
1492.1535(10)	67017.23(4)	1492.15247(11)	0.0010	95000		s^2	$^3F_2$	6p	$(5/2)^2[5/2]^\circ_3$			K66		
1492.6819(10)	66993.51(4)	1492.68146(13)	0.0004	100000		4p	$^1P^\circ_1$	5d	$(3/2)^2[1/2]_0$			K66		

Table A1. *Cont.*

λ_{obs} [a] (Å)	σ_{obs} [b] (cm^{-1})	λ_{Ritz} [c] (Å)	$\Delta\lambda_{obs\text{-}Ritz}$ (Å)	I_{obs} [d] (arb. u.)	Char [e]	Lower Level	Upper Level	A (s^{-1})	Acc [f]	Line Ref. [g]	TP Ref. [g]	Notes [h]
1492.8346(6)	66986.66(3)	1492.83423(11)	0.0004	330000		s^2 $^3F_2$	sp $(^3F)^1P^\circ\ ^3F^\circ_2$	5.5e+08	D+	K66	TW	
1493.3675(6)	66962.75(3)	1493.36656(4)	0.0009	180000		4p $^3D^\circ_3$	5d $(3/2)^2[7/2]_4$			K66		
1494.661(6)	66904.8(3)	1494.65244(5)	0.008	49000		4p $^3F^\circ_2$	5d $(5/2)^2[3/2]_1$			S36c		
1494.7930(19)	66898.89(8)	1494.79138(4)	0.0016	6700		4p $^3D^\circ_3$	5d $(3/2)^2[7/2]_3$			TW		
1495.4296(6)	66870.42(3)	1495.42965(9)	−0.0001	200000		$s^{23}F$ 3	6p $(5/2)^2[7/2]^\circ_3$			K66		
1496.6862(6)	66814.27(3)	1496.68654(4)	−0.0003	260000		4p $^3P^\circ_0$	6s $(3/2)^2[3/2]_1$			K66		
1498.5779(14)	66729.93(6)	1498.57545(7)	0.0024	18000		s^2 $^3F_3$	4f $(3/2)^2[7/2]^\circ_3$			TW		
1499.5135(10)	66688.30(4)	1499.51299(7)	0.0005	79000		s^2 $^3F_3$	4f $(3/2)^2[7/2]^\circ_4$			K66		
1500.016(3)	66665.97(13)	1500.0148(7)	0.001	14000		s^2 $^1D_2$	8p $(5/2)^2[3/2]^\circ_2$			TWn		
1501.3359(12)	66607.35(5)	1501.33622(11)	−0.0003	110000		s^2 $^3F_2$	sp $(^1G)^3P^\circ\ ^3F^\circ_3$			TW		
1502.811(3)	66541.96(13)	1502.80934(8)	0.002	2600		s^2 $^3F_3$	4f $(3/2)^2[9/2]^\circ_4$			TW		
1503.3675(6)	66517.34(3)	1503.36800(13)	−0.0005	150000		s^2 $^3F_2$	6p $(5/2)^2[3/2]^\circ_1$			K66		
1504.7561(6)	66455.95(3)	1504.75691(4)	−0.0008	220000		s^2 $^3F_4$	4f $(5/2)^2[9/2]^\circ_5$			K66		
1505.3866(6)	66428.12(3)	1505.38756(4)	−0.0010	160000		s^2 $^3F_4$	4f $(5/2)^2[7/2]^\circ_4$			K66		
1505.8576(12)	66407.34(5)	1505.85714(18)	0.0005	60000		s^2 $^3F_3$	sp $(^1G)^3P^\circ\ ^3F^\circ_4$			TW		
1507.4705(22)	66336.29(10)	1507.4706(5)	−0.0000	9700		s^2 $^3P_2$	7f $(5/2)^2[5/2]^\circ_2$			TWn		
1507.6008(22)	66330.56(10)	1507.59910(4)	0.0017	3200		s^2 $^3F_4$	4f $(5/2)^2[5/2]^\circ_3$			TW		
1508.1835(6)	66304.93(3)	1508.18443(13)	−0.0009	170000		s^2 $^3F_2$	6p $(5/2)^2[5/2]^\circ_2$			K66		
1508.4845(22)	66291.70(10)	1508.4829(15)	0.0016	6400		s^2 $^3P_2$	7f $(5/2)^2[3/2]^\circ_1$			TWn		
1508.6313(6)	66285.25(3)	1508.63203(5)	−0.0007	150000		s^2 $^3F_4$	4f $(5/2)^2[7/2]^\circ_3$			K66		
1510.5051(6)	66203.02(3)	1510.50562(5)	−0.0005	210000		4p $^1F^\circ_3$	5d $(5/2)^2[7/2]_4$			K66		
1510.724(3)	66193.41(13)	1510.72666(5)	−0.002	8900		4p $^1F^\circ_3$	5d $(5/2)^2[7/2]_3$			TW		
1512.1738(12)	66129.96(5)	1512.17367(6)	0.0002	98000		s^2 $^3F_4$	sp $(^3F)^1P^\circ\ ^3G^\circ_4$			TW		
1512.4640(8)	66117.28(3)	1512.46442(11)	−0.0004	97000		s^2 $^3F_3$	sp $(^3P)^3P^\circ\ ^1D^\circ_2$			K66		
1512.5393(19)	66113.98(8)	1512.5391(6)	0.0003	18000		s^2 $^3P_1$	8p $(3/2)^2[1/2]^\circ_1$			TWn		
1513.0157(22)	66093.17(9)	1513.0151(5)	0.0005	9800		s^2 $^3P_1$	7f $(5/2)^2[5/2]^\circ_2$			TWn		
1513.3651(6)	66077.91(3)	1513.36565(4)	−0.0005	140000		4p $^1F^\circ_3$	5d $(5/2)^2[5/2]_3$			K66		
1513.9340(13)	66053.08(6)	1513.93287(14)	0.0011	46000		s^2 $^1D_2$	6f $(5/2)^2[7/2]^\circ_3$			TW		
1514.2339(8)	66040.00(3)	1514.23361(6)	0.0003	80000		4p $^3D^\circ_1$	5d $(3/2)^2[5/2]_2$			K66		
1514.3380(12)	66035.46(5)	1514.33770(11)	0.0003	110000		s^2 $^1D_2$	8p $(5/2)^2[3/2]^\circ_1$			TW		
1514.4921(6)	66028.74(3)	1514.49222(8)	−0.0001	700000		s^2 $^3F_4$	sp $(^3F)^1P^\circ\ ^3D^\circ_3$	4.8e+08	D+	K66	TW	
1514.6461(13)	66022.02(6)	1514.64663(14)	−0.0005	48000		s^2 $^3F_2$	6p $(5/2)^2[3/2]^\circ_2$			TW		
1514.805(5)	66015.12(23)	1514.80334(10)	0.001	13000		s^2 $^1D_2$	6f $(5/2)^2[5/2]^\circ_2$			TW		
1515.456(3)	65986.75(12)	1515.46044(10)	−0.005	8600		s^2 $^1D_2$	6f $(5/2)^2[3/2]^\circ_2$			TW		
1515.491(5)	65985.22(23)	1515.49438(15)	−0.004	4300		s^2 $^1D_2$	6f $(5/2)^2[3/2]^\circ_1$			TW		
1516.9018(13)	65923.84(6)	1516.90090(5)	0.0009	43000		4p $^1F^\circ_3$	5d $(5/2)^2[9/2]_4$			TW		
1517.1600(12)	65912.63(5)	1517.15960(5)	0.0004	81000		4p $^1F^\circ_3$	5d $(5/2)^2[3/2]_2$			TW		
1517.6309(8)	65892.17(3)	1517.63100(4)	−0.0001	130000		4p $^3F^\circ_2$	6s $(3/2)^2[3/2]_2$			K66		
1517.9297(8)	65879.20(3)	1517.92967(6)	0.0000	87000		4p $^3D^\circ_1$	5d $(3/2)^2[3/2]_2$			K66		
1519.4914(6)	65811.49(3)	1519.49170(4)	−0.0003	410000		4p $^3P^\circ_1$	6s $(5/2)^2[5/2]_2$			K66		
1519.8366(6)	65796.55(3)	1519.83686(4)	−0.0003	540000		4p $^3F^\circ_2$	6s $(3/2)^2[3/2]_1$			K66		

Table A1. *Cont.*

λ_{obs} [a] (Å)	σ_{obs} [b] (cm^{-1})	λ_{Ritz} [c] (Å)	$\Delta\lambda_{obs\text{-}Ritz}$ (Å)	I_{obs} [d] (arb. u.)	Char [e]	Lower Level	Upper Level	A (s^{-1})	Acc [f]	Line Ref. [g]	TP Ref. [g]	Notes [h]
		1519.84246(6)			m	4p $^3D^\circ_1$	5d $(3/2)^2[3/2]_1$			K66		
1519.923(3)	65792.81(13)	1519.9165(7)	0.007	11000	*	s^2 $^3P_0$	8p $(3/2)^2[1/2]^\circ_1$			TWn		
1519.923(3)	65792.81(13)	1519.9261(7)	−0.003	11000	*	s^2 $^3F_2$	sp $(^3P)^3P^\circ\ ^3S^\circ_1$			TWn		
1520.0166(12)	65788.75(5)	1520.01664(6)	−0.0000	53000		4p $^3D^\circ_2$	5d $(3/2)^2[5/2]_2$			TW		
1520.3899(12)	65772.60(5)	1520.38979(17)	0.0001	150000		s^2 $^1D_2$	6f $(5/2)^2[1/2]^\circ_1$			TWn		
1520.5397(6)	65766.12(3)	1520.53944(5)	0.0003	150000		4p $^3D^\circ_2$	5d $(3/2)^2[5/2]_3$			K66		
1521.4252(22)	65727.84(9)	1521.4270(16)	−0.0018	8000		s^2 $^3P_0$	7f $(5/2)^2[3/2]^\circ_1$			TWn		
1522.5070(19)	65681.14(8)	1522.50482(12)	0.0022	8900		s^2 $^3F_3$	sp $(^1G)^3P^\circ\ ^3F^\circ_2$			TWn		
1522.5767(6)	65678.14(3)	1522.57664(11)	0.0001	97000		s^2 $^3F_2$	6p $(5/2)^2[7/2]^\circ_3$			K66		
1523.7415(8)	65627.93(3)	1523.74102(6)	0.0005	76000		4p $^3D^\circ_2$	5d $(3/2)^2[3/2]_2$			K66		
1523.9231(15)	65620.11(6)	1523.9230(10)	0.0001	31000		s^2 $^3P_1$	8p $(3/2)^2[3/2]^\circ_1$			TWn		
1524.8604(8)	65579.77(3)	1524.85998(5)	0.0004	130000		4p $^1D^\circ_2$	5d $(5/2)^2[5/2]_2$			K66		
1525.656(20)	65545.6(9)	1525.63103(13)	0.025	92000		s^2 $^3F_3$	sp $(^3F)^1P^\circ\ ^3G^\circ_3$			S36c		
1525.6433(13)	65546.12(5)	1525.64065(7)	0.0026	73000		4p $^1P^\circ_1$	5d $(3/2)^2[5/2]_2$			TW		
1525.656(20)	65545.6(9)	1525.66850(6)	−0.013	97000		4p $^3D^\circ_2$	5d $(3/2)^2[3/2]_1$			S36c		
1525.7649(6)	65540.90(3)	1525.76429(5)	0.0006	150000		4p $^1D^\circ_2$	5d $(5/2)^2[7/2]_3$			K66		
1525.8388(8)	65537.72(3)	1525.83781(9)	0.0010	92000		s^2 $^3F_2$	4f $(3/2)^2[7/2]^\circ_3$			K66		
1526.932(5)	65490.79(21)	1526.92724(7)	0.005	32000		4p $^3D^\circ_1$	5d $(3/2)^2[1/2]_1$			TW		
1526.9944(7)	65488.12(3)	1526.9931(5)	0.0013	130000		s^2 $^3P_2$	sp $^3D^\circ_3$			K66n		
1527.8127(13)	65453.05(6)	1527.81239(9)	0.0003	42000		s^2 $^3F_2$	4f $(3/2)^2[5/2]^\circ_2$			TW		
1528.4583(22)	65425.40(9)	1528.45612(3)	0.0022	4700		4p $^1D^\circ_2$	5d $(5/2)^2[5/2]_3$			TW		
1528.785(20)	65411.4(9)	1528.89495(9)	−0.110	19000	?	s^2 $^3F_2$	4f $(3/2)^2[5/2]^\circ_3$			S36c		X
1529.3927(22)	65385.43(9)	1529.39267(7)	0.0000	4900		4p $^1P^\circ_1$	5d $(3/2)^2[3/2]_2$			TW		
1531.2874(14)	65304.53(6)	1531.28647(11)	0.0010	70000		s^2 $^3F_2$	4f $(3/2)^2[3/2]^\circ_1$			TW		
1531.3338(14)	65302.55(6)	1531.33448(7)	−0.0007	41000		4p $^1P^\circ_1$	5d $(3/2)^2[3/2]_1$			TW		
1531.4120(14)	65299.21(6)	1531.4121(10)	−0.0002	60000		s^2 $^3P_0$	8p $(3/2)^2[3/2]^\circ_1$			TWn		
1531.8555(6)	65280.31(3)	1531.85562(4)	−0.0001	700000		4p $^3F^\circ_3$	6s $(5/2)^2[5/2]_2$			K66		
		1531.87878(5)			m	4p $^1D^\circ_2$	5d $(5/2)^2[3/2]_1$			TW		
1532.128(3)	65268.70(13)	1532.13042(10)	−0.003	390000		s^2 $^3F_3$	sp $(^3F)^1P^\circ\ ^3D^\circ_2$	1.0e+09	D+	TW	TW	
1532.8083(15)	65239.73(6)	1532.80783(7)	0.0004	17000		4p $^3D^\circ_2$	5d $(3/2)^2[1/2]_1$			TW		
1533.9867(6)	65189.61(3)	1533.98624(12)	0.0005	170000		s^2 $^3F_2$	sp $(^3F)^1P^\circ\ ^3D^\circ_1$			K66		
1534.6282(13)	65162.37(6)	1534.62719(10)	0.0010	43000		s^2 $^3F_3$	sp $(^1G)^3P^\circ\ ^3H^\circ_4$			TW		
1535.0023(6)	65146.48(3)	1535.00194(4)	0.0004	350000		4p $^3F^\circ_3$	6s $(5/2)^2[5/2]_3$			K66		
1535.5242(6)	65124.34(3)	1535.52352(4)	0.0007	130000		4p $^3D^\circ_3$	5d $(5/2)^2[7/2]_4$			K66		
1536.193(5)	65096.00(21)	1536.19869(21)	−0.006	8200		s^2 $^3P_2$	6f $(3/2)^2[3/2]^\circ_1$			TW		
1536.9317(16)	65064.70(7)	1536.93333(14)	−0.0016	24000		s^2 $^3P_2$	6f $(3/2)^2[3/2]^\circ_2$			TW		
1537.3218(12)	65048.19(5)	1537.32130(14)	0.0005	120000		s^2 $^3P_2$	6f $(3/2)^2[5/2]^\circ_3$			TW		
1537.5581(6)	65038.19(3)	1537.55884(8)	−0.0007	860000		s^2 $^3F_4$	sp $(^3F)^1P^\circ\ ^3F^\circ_4$	1.4e+09	D+	K66	TW	
1538.4795(6)	64999.24(3)	1538.47917(3)	0.0003	96000		4p $^3D^\circ_3$	5d $(5/2)^2[5/2]_3$			K66		
1538.522(5)	64997.45(21)	1538.52706(8)	−0.005	40000		4p $^1P^\circ_1$	5d $(3/2)^2[1/2]_1$			TW		
1540.2392(6)	64924.98(3)	1540.23914(12)	0.0001	120000		s^2 $^3F_2$	sp $(^3P)^3P^\circ\ ^1D^\circ_2$			K66		
1540.3886(6)	64918.68(3)	1540.38850(4)	0.0001	290000		4p $^1F^\circ_3$	6s $(3/2)^2[3/2]_2$			K66		

Table A1. *Cont.*

λ_{obs} [a] (Å)	σ_{obs} [b] (cm^{-1})	λ_{Ritz} [c] (Å)	$\Delta\lambda_{obs-Ritz}$ (Å)	I_{obs} [d] (arb. u.)	Char [e]	Lower Level		Upper Level		A (s^{-1})	Acc [f]	Line Ref. [g]	TP Ref. [g]	Notes [h]
1540.5879(6)	64910.29(3)	1540.58814(9)	−0.0002	600000		s^2	$^3F_3$	sp	$(^3F)^1P^\circ\ ^3F^\circ_3$	9.e+08	D+	K66	TW	
1541.246(5)	64882.59(22)	1541.24593(13)	−0.000	5200		s^2	$^3P_1$	6f	$(3/2)^2[5/2]^\circ_2$			TW		
1541.7007(22)	64863.43(9)	1541.70280(5)	−0.0021	860000		4p	$^3F^\circ_4$	6s	$(5/2)^2[5/2]_3$			K66		
1541.7542(18)	64861.18(8)	1541.75580(18)	−0.0016	260000		s^2	$^1D_2$	sp	$(^1D)^1P^\circ\ ^1D^\circ_2$	1.4e+09	D+	K66	TW	
1541.9565(22)	64852.67(9)	1541.95701(21)	−0.0005	11000		s^2	$^3P_1$	6f	$(3/2)^2[3/2]^\circ_1$			TW		
1542.4007(13)	64834.00(6)	1542.40025(4)	0.0004	55000		4p	$^3D^\circ_3$	5d	$(5/2)^2[3/2]_2$			TW		
1542.6989(22)	64821.46(9)	1542.69717(14)	0.0017	10000		s^2	$^3P_1$	6f	$(3/2)^2[3/2]^\circ_2$			TW		
1543.1328(13)	64803.24(5)	1543.1327(9)	0.0001	64000		s^2	$^3P_2$	sp	$(^1D)^1P^\circ\ ^1P^\circ_1$			TWn		
1544.6764(6)	64738.48(3)	1544.67690(8)	−0.0005	430000		s^2	$^3F_3$	4f	$(5/2)^2[9/2]^\circ_4$			K66		
1547.9581(8)	64601.23(3)	1547.95800(7)	0.0001	59000		s^2	$^3F_3$	4f	$(5/2)^2[7/2]^\circ_4$			K66		
1548.4169(12)	64582.09(5)	1548.4165(12)	0.0004	200000		s^2	$^3P_2$	sp	$(^3P)^1P^\circ\ ^3P^\circ_2$	1.0e+09	D+	TWn	TW	
1548.942(3)	64560.22(12)	1548.9433(10)	−0.002	15000		s^2	$^3P_1$	sp	$(^1D)^1P^\circ\ ^1P^\circ_1$			TWn		
1549.56(7)	64534(3)	1549.62492(22)	−0.06	24000		s^2	$^3P_0$	6f	$(3/2)^2[3/2]^\circ_1$			S36nc		
1550.0971(17)	64512.09(7)	1550.09682(9)	0.0003	10000		4p	$^3D^\circ_1$	5d	$(5/2)^2[1/2]_0$			TW		
1550.2978(13)	64503.73(5)	1550.29648(7)	0.0014	29000		s^2	$^3F_3$	4f	$(5/2)^2[5/2]^\circ_3$			TW		
1550.6528(6)	64488.969(25)	1550.65296(13)	−0.0002	180000		s^2	$^3F_2$	sp	$(^1G)^3P^\circ\ ^3F^\circ_2$	7.3e+08	D+	K66n	TW	
1551.3886(6)	64458.383(25)	1551.38877(7)	−0.0002	250000		s^2	$^3F_3$	4f	$(5/2)^2[7/2]^\circ_3$			K66		
1552.297(5)	64420.66(22)	1552.29519(14)	0.002	1100		s^2	$^3F_3$	sp	$(^3P)^3P^\circ\ ^5S^\circ_2$			TW		
1552.6450(6)	64406.223(25)	1552.64631(10)	−0.0013	720000		s^2	$^3F_4$	sp	$(^3F)^1P^\circ\ ^3G^\circ_5$	1.9e+09	D+	K66	TW	
1553.3473(13)	64377.11(5)	1553.3472(13)	0.0001	96000		s^2	$^1D_2$	7p	$(3/2)^2[3/2]^\circ_1$			TWn		
1553.8961(6)	64354.367(25)	1553.89596(14)	0.0001	210000		s^2	$^3F_2$	sp	$(^3F)^1P^\circ\ ^3G^\circ_3$	1.4e+09	D+	K66	TW	
1554.2933(12)	64337.92(5)	1554.2934(11)	−0.0001	120000		s^2	$^1D_2$	7p	$(3/2)^2[3/2]^\circ_2$			TWn		
1555.1336(6)	64303.157(25)	1555.13425(8)	−0.0007	810000		s^2	$^3F_3$	sp	$(^3F)^1P^\circ\ ^3G^\circ_4$	1.3e+09	D+	K66	TW	
1555.7018(6)	64279.671(25)	1555.70295(9)	−0.0012	880000		s^2	$^3F_4$	sp	$(^3P)^3P^\circ\ ^3D^\circ_3$	6.8e+08	D+	K66	TW	
1556.0269(8)	64266.24(3)	1556.02548(4)	0.0014	40000		4p	$^1D^\circ_2$	6s	$(3/2)^2[3/2]_2$			K66		
1556.6836(21)	64239.13(9)	1556.6810(10)	0.0027	19000		s^2	$^3P_0$	sp	$(^1D)^1P^\circ\ ^1P^\circ_1$			TWn		
1556.7668(12)	64235.70(5)	1556.7668(12)	0.0001	180000		s^2	$^1D_2$	7p	$(3/2)^2[5/2]^\circ_3$			TWn		
1557.5867(6)	64201.884(25)	1557.58652(10)	0.0002	110000		s^2	$^3F_3$	sp	$(^3F)^1P^\circ\ ^3D^\circ_3$			K66		
1558.1590(13)	64178.30(6)	1558.1587(13)	0.0003	62000		s^2	$^3P_1$	sp	$(^3P)^1P^\circ\ ^3P^\circ_0$			TWn		
1558.3446(6)	64170.659(25)	1558.34444(4)	0.0002	180000		4p	$^1D^\circ_2$	6s	$(3/2)^2[3/2]_1$			K66		
1559.1587(14)	64137.15(6)	1559.1576(13)	0.0011	210000		s^2	$^1D_2$	7p	$(3/2)^2[5/2]^\circ_2$			TWn		
1563.108(3)	63975.10(12)	1563.1083(13)	−0.000	9400		s^2	$^3P_1$	8p	$(5/2)^2[5/2]^\circ_2$			TWn		
1563.1959(13)	63971.51(5)	1563.19353(5)	0.0024	22000		4p	$^3D^\circ_1$	5d	$(5/2)^2[5/2]_2$			TW		
1565.9244(6)	63860.043(24)	1565.92414(4)	0.0003	190000		4p	$^3F^\circ_2$	6s	$(5/2)^2[5/2]_2$			K66		
1566.4148(4)	63840.051(16)	1566.41460(4)	0.0002	220000		4p	$^3D^\circ_3$	6s	$(3/2)^2[3/2]_2$			K66		
1569.2124(8)	63726.24(3)	1569.21212(4)	0.0003	40000		4p	$^3F^\circ_2$	6s	$(5/2)^2[5/2]_3$			K66		
1569.4154(11)	63717.99(4)	1569.41524(11)	0.0001	42000		s^2	$^3F_2$	sp	$(^3F)^1P^\circ\ ^3F^\circ_3$			TW		
1570.0365(11)	63692.79(4)	1570.0367(8)	−0.0002	94000		s^2	$^3P_2$	8p	$(5/2)^2[3/2]^\circ_2$			TW,S36r		
1570.5722(12)	63671.06(5)	1570.57051(5)	0.0017	17000		4p	$^3D^\circ_1$	5d	$(5/2)^2[3/2]_1$			TW		
1573.1666(22)	63566.06(9)	1573.16667(4)	−0.0001	4000		4p	$^3D^\circ_2$	5d	$(5/2)^2[5/2]_3$			TW		
1575.3547(12)	63477.77(5)	1575.35310(6)	0.0016	34000		4p	$^1P^\circ_1$	5d	$(5/2)^2[5/2]_2$			TW		
1576.0523(12)	63449.67(5)	1576.0520(8)	0.0003	47000		s^2	$^3P_1$	8p	$(5/2)^2[3/2]^\circ_2$			TWn		
1577.2693(22)	63400.71(9)	1577.26680(5)	0.0025	4500		4p	$^3D^\circ_2$	5d	$(5/2)^2[3/2]_2$			TW		

Table A1. *Cont.*

λ_{obs} [a] (Å)	σ_{obs} [b] (cm^{-1})	λ_{Ritz} [c] (Å)	$\Delta\lambda_{obs-Ritz}$ (Å)	I_{obs} [d] (arb. u.)	Char [e]	Lower Level		Upper Level		A (s^{-1})	Acc [f]	Line Ref. [g]	TP Ref. [g]	Notes [h]
1579.4926(8)	63311.47(3)	1579.49148(9)	0.0011	64000		s^2	$^3F_2$	4f	$(5/2)^2[5/2]^\circ_3$			K66		
1580.0260(11)	63290.10(4)	1580.02469(9)	0.0013	43000		s^2	$^3F_2$	4f	$(5/2)^2[5/2]^\circ_2$			TW		
1580.6258(6)	63266.081(24)	1580.62531(9)	0.0005	71000		s^2	$^3F_2$	4f	$(5/2)^2[7/2]^\circ_3$			K66		
1581.4029(14)	63234.99(6)	1581.40630(5)	−0.0034	5300		4p	$^3D^\circ_1$	5d	$(5/2)^2[1/2]_1$			TW		
1581.420(12)	63234.3(5)	1581.41834(10)	0.001	2300		s^2	$^3F_2$	4f	$(5/2)^2[3/2]^\circ_1$			S36c		
1581.9962(6)	63211.277(24)	1581.99510(10)	0.0011	82000		s^2	$^3F_3$	sp	$(^3F)^1P^\circ\ ^3F^\circ_4$			K66		
1582.6074(20)	63186.86(8)	1582.60633(9)	0.0011	6600		s^2	$^3F_2$	4f	$(5/2)^2[3/2]^\circ_2$			TW		
1582.8471(11)	63177.30(4)	1582.84558(6)	0.0015	35000		4p	$^1P^\circ_1$	5d	$(5/2)^2[3/2]_1$			TW		
1583.6831(4)	63143.946(16)	1583.68224(10)	0.0009	160000		s^2	$^3F_3$	sp	$(^3P)^3P^\circ\ ^3D^\circ_2$			K66		
1586.276(5)	63040.74(21)	1586.27882(10)	−0.003	4100		s^2	$^3P_2$	6f	$(5/2)^2[5/2]^\circ_3$			TW		
1587.0607(22)	63009.56(9)	1587.05935(12)	0.0014	6100		s^2	$^3F_2$	sp	$(^3F)^1P^\circ\ ^3D^\circ_3$			TW		
1587.716(3)	62983.56(11)	1587.71486(5)	0.001	2600		4p	$^3D^\circ_2$	5d	$(5/2)^2[1/2]_1$			TW		
1590.1649(4)	62886.560(16)	1590.16460(4)	0.0003	110000		4p	$^1F^\circ_3$	6s	$(5/2)^2[5/2]_2$			K66		
1593.5559(4)	62752.741(16)	1593.55527(4)	0.0006	190000		4p	$^1F^\circ_3$	6s	$(5/2)^2[5/2]_3$			K66		
1596.749(6)	62627.26(24)	1596.74558(10)	0.003	5100		s^2	$^1D_2$	5f	$(3/2)^2[3/2]^\circ_1$			S36c		
1598.4025(4)	62562.465(16)	1598.40212(4)	0.0004	140000		4p	$^3D^\circ_1$	6s	$(3/2)^2[3/2]_1$			K66		
1601.211(3)	62452.75(12)	1601.20967(12)	0.001	2800		s^2	$^3F_3$	sp	$(^3P)^3P^\circ\ ^3D^\circ_3$			TW		
1602.2739(8)	62411.30(3)	1602.27264(12)	0.0013	28000		s^2	$^3F_2$	sp	$(^3P)^3P^\circ\ ^3P^\circ_1$			K66		
1602.3887(4)	62406.831(16)	1602.38797(4)	0.0007	130000		4p	$^3D^\circ_2$	6s	$(3/2)^2[3/2]_2$			K66		
1603.2280(23)	62374.16(9)	1603.22651(14)	0.0014	13000		s^2	$^1D_2$	7p	$(5/2)^2[5/2]^\circ_3$			TW		
1604.8475(4)	62311.216(16)	1604.84729(4)	0.0002	56000		4p	$^3D^\circ_2$	6s	$(3/2)^2[3/2]_1$			K66		
1605.2810(4)	62294.390(16)	1605.28112(12)	−0.0001	150000		s^2	$^3F_3$	sp	$(^3P)^3P^\circ\ ^3P^\circ_2$			K66		
1606.1963(17)	62258.89(6)	1606.19537(24)	0.0009	76000		s^2	$^1D_2$	7p	$(5/2)^2[5/2]^\circ_2$			TW		
1606.8338(4)	62234.190(15)	1606.83396(4)	−0.0002	170000		4p	$^1D^\circ_2$	6s	$(5/2)^2[5/2]_2$			K66		
1608.3931(17)	62173.85(6)	1608.39195(15)	0.0012	87000		s^2	$^1D_2$	7p	$(5/2)^2[3/2]^\circ_1$			TW		
1608.6395(4)	62164.332(15)	1608.63928(6)	0.0002	72000		4p	$^1P^\circ_1$	6s	$(3/2)^2[3/2]_2$			K66		
1609.3421(23)	62137.19(9)	1609.3430(15)	−0.0009	10000		s^2	$^1D_2$	7p	$(5/2)^2[7/2]^\circ_3$			TWn		
1610.2967(8)	62100.36(3)	1610.29617(4)	0.0005	21000		4p	$^1D^\circ_2$	6s	$(5/2)^2[5/2]_3$			K66		
1611.1185(20)	62068.68(8)	1611.11784(6)	0.0006	12000		4p	$^1P^\circ_1$	6s	$(3/2)^2[3/2]_1$			TW		
1614.159(5)	61951.76(21)	1614.16058(12)	−0.001	2100		s^2	$^3F_2$	sp	$(^3P)^3P^\circ\ ^3D^\circ_2$			TW		
1614.4945(18)	61938.89(7)	1614.49462(18)	−0.0002	75000		s^2	$^1D_2$	7p	$(5/2)^2[3/2]^\circ_2$			TW		
1617.9152(8)	61807.94(3)	1617.91504(4)	0.0002	24000		4p	$^3D^\circ_3$	6s	$(5/2)^2[5/2]_2$			K66		
1621.4262(4)	61674.099(15)	1621.42522(4)	0.0010	84000		4p	$^3D^\circ_3$	6s	$(5/2)^2[5/2]_3$			K66		
1622.4281(4)	61636.013(15)	1622.42766(12)	0.0004	80000		s^2	$^3F_2$	sp	$(^3P)^3P^\circ\ ^3D^\circ_1$			K66		
1623.1726(6)	61607.743(23)	1623.1731(3)	−0.0005	36000		s^2	$^3F_4$	sp	$(^3P)^3P^\circ\ ^5D^\circ_4$			K66		
1629.603(3)	61364.65(12)	1629.6017(12)	0.001	9500		s^2	$^3P_2$	7p	$(3/2)^2[3/2]^\circ_2$			TWn		
1630.2669(18)	61339.65(7)	1630.26799(21)	−0.0011	32000		s^2	$^3F_4$	sp	$(^3P)^3P^\circ\ ^5D^\circ_3$			TW		
1636.0706(23)	61122.06(9)	1636.06917(21)	0.0014	19000		s^2	$^3P_2$	7p	$(3/2)^2[1/2]^\circ_1$			TWn		
1636.6049(20)	61102.10(7)	1636.60469(14)	0.0002	11000		s^2	$^3F_2$	sp	$(^3P)^3P^\circ\ ^3P^\circ_2$			TW		
1645.6583(23)	60765.95(8)	1645.6575(22)	0.0008	12000		s^2	$^3P_1$	7p	$(3/2)^2[1/2]^\circ_0$			TWn		

Table A1. *Cont.*

λ_{obs} [a] (Å)	σ_{obs} [b] (cm^{-1})	λ_{Ritz} [c] (Å)	$\Delta\lambda_{obs\text{-}Ritz}$ (Å)	I_{obs} [d] (arb. u.)	Char [e]	Lower Level		Upper Level		A (s^{-1})	Acc [f]	Line Ref. [g]	TP Ref. [g]	Notes [h]
1649.4570(14)	60626.01(5)	1649.45739(4)	−0.0004	26000		4p	$^3D^\circ_1$	6s	$(5/2)^2[5/2]_2$			K66		
1656.322	60374.74	1656.32175(4)		27000	:	4p	$^3D^\circ_2$	6s	$(5/2)^2[5/2]_2$			S36c		
1660.0022(14)	60240.88(5)	1660.00075(4)	0.0014	18000		4p	$^3D^\circ_2$	6s	$(5/2)^2[5/2]_3$			K66		
1663.0029(14)	60132.19(5)	1663.00184(5)	0.0011	31000		4p	$^1P^\circ_1$	6s	$(5/2)^2[5/2]_2$			K66		
1672.7773(23)	59780.82(8)	1672.7755(4)	0.0018	8600		s^2	$^3F_3$	sp	$(^3P)^3P^\circ\ ^5D^\circ_4$			TW		
1680.27(4)	59514.1(14)	1680.31169(23)	−0.04	980		s^2	$^3F_3$	sp	$(^3P)^3P^\circ\ ^5D^\circ_3$			S36c		
1683.152(5)	59412.32(18)	1683.1596(3)	−0.007	34000	*	s^2	$^3F_4$	sp	$(^1D)^3P^\circ\ ^3D^\circ_3$	1.1e+08	C	TW	B00	
1683.152(5)	59412.32(18)	1683.1583(3)	−0.006	34000	*	s^2	$^3F_3$	sp	$(^3P)^3P^\circ\ ^5D^\circ_2$			TW		
1695.9031(23)	58965.63(8)	1695.90238(20)	0.0007	12000		s^2	$^3P_2$	7p	$(5/2)^2[3/2]^\circ_2$			TW		
1699.0958(20)	58854.83(7)	1699.0950(5)	0.0008	22000	*	s^2	$^3F_3$	sp	$(^1D)^3P^\circ\ ^3P^\circ_2$			TW		
1699.0958(20)	58854.83(7)	1699.10217(23)	−0.0064	22000	*	s^2	$^3F_4$	sp	$(^1D)^3P^\circ\ ^3F^\circ_3$			TW		
1702.9236(23)	58722.54(8)	1702.92295(21)	0.0006	8600		s^2	$^3P_1$	7p	$(5/2)^2[3/2]^\circ_2$			TW		
1714.655(20)	58320.8(7)	1714.6631(3)	−0.008	460		s^2	$^3F_2$	sp	$(^3P)^3P^\circ\ ^5D^\circ_3$			S36c		
1717.7188(22)	58216.75(7)	1717.7210(3)	−0.0023	11000		s^2	$^3F_2$	sp	$(^3P)^3P^\circ\ ^5D^\circ_1$			TW		
1734.2254(23)	57662.63(8)	1734.2267(5)	−0.0013	4200		s^2	$^3F_2$	sp	$(^1D)^3P^\circ\ ^3P^\circ_2$			TW		
1736.5531(22)	57585.34(7)	1736.5563(3)	−0.0032	8900		s^2	$^3F_3$	sp	$(^1D)^3P^\circ\ ^3D^\circ_3$			TW		
1744.506(20)	57322.8(7)	1744.5159(3)	−0.010	22000		s^2	$^3F_3$	sp	$(^1D)^3P^\circ\ ^3D^\circ_2$	5.9e+07	C	S36c	B00	
1753.2791(21)	57035.99(7)	1753.28069(22)	−0.0016	21000		s^2	$^3F_2$	sp	$(^1D)^3P^\circ\ ^3P^\circ_1$			TW		
1759.5030(21)	56834.23(7)	1759.50415(22)	−0.0012	3400		s^2	$^3F_3$	sp	$(^1D)^3P^\circ\ ^3F^\circ_2$			TW		
1781.573(3)	56130.16(10)	1781.5715(3)	0.002	4400		s^2	$^3F_2$	sp	$(^1D)^3P^\circ\ ^3D^\circ_2$			TW		
1790.6606(21)	55845.31(7)	1790.6599(3)	0.0007	33000		s^2	$^3F_2$	sp	$(^1D)^3P^\circ\ ^3D^\circ_1$	4.5e+07	C	TW	B00	
1800.9806(24)	55525.31(7)	1800.9789(6)	0.0018	11000		s^2	$^3F_4$	sp	$(^3P)^3P^\circ\ ^5P^\circ_3$			TW		
1807.8535(21)	55314.22(7)			160000		4p	$^1P^\circ_1$	s^2	$^1S_0$			TW,S36r		
1856.937(20)	53852.1(6)	1856.92890(17)	0.009	14000		s^2	$^1D_2$	6p	$(5/2)^2[3/2]^\circ_1$			S36c		
1861.57(4)	53718.2(12)	1861.6226(3)	−0.06	10000		s^2	$^3F_3$	sp	$(^3P)^3P^\circ\ ^5P^\circ_2$			S36c		
1874.166(5)	53357.06(14)	1874.16676(18)	−0.001	14000		s^2	$^1D_2$	6p	$(5/2)^2[3/2]^\circ_2$			TW		
1875.74638(18)	53312.112(5)	1875.74622(7)	0.00016	20000		s^2	$^3F_4$	5p	$(3/2)^2[5/2]^\circ_3$			F		
1882.253(3)	53127.82(8)	1882.2566(10)	−0.003	17000		s^2	$^1D_2$	sp	$(^3P)^3P^\circ\ ^3S^\circ_1$			TWn		
1903.858(5)	52524.93(15)	1903.86698(15)	−0.009	410		s^2	$^1D_2$	sp	$(^3F)^1P^\circ\ ^3D^\circ_1$			TW		
1920.6709(8)	52065.141(22)	1920.67133(7)	−0.0004	43000		4p	$^3P^\circ_2$	4d	$(3/2)^2[5/2]_3$	1.5e+07	C	F	B00	
1922.14229(17)	52025.285(5)	1922.14231(11)	−0.00003	94000		s^2	$^3F_3$	5p	$(3/2)^2[3/2]^\circ_2$			F		
1924.548(3)	51960.26(9)	1924.5458(22)	0.002	5200		s^2	$^1G_4$	7p	$(5/2)^2[7/2]^\circ_3$			TWn		
1928.46(4)	51854.9(11)	1928.4863(3)	−0.03	46000		s^2	$^1D_2$	sp	$(^3P)^3P^\circ\ ^1P^\circ_1$			S36cn		
		1929.60764(17)			m	s^2	$^1D_2$	sp	$(^1G)^3P^\circ\ ^3F^\circ_2$			S36cn		
1929.75131(6)	51820.1488(16)	1929.75129(5)	0.00002	340000		s^2	$^3F_4$	sp	$(^3F)^3P^\circ\ ^1F^\circ_3$	2.1e+07	C	F	B00	
1942.3030(7)	51485.272(19)	1942.30290(12)	0.0001	31000		s^2	$^3F_3$	5p	$(3/2)^2[5/2]^\circ_3$			F		
1944.59596(14)	51424.564(4)	1944.59601(6)	−0.00005	6000000		4s	$^3D_3$	4p	$^3D^\circ_2$	1.99e+07	C+	F	D05se	
1946.4928(3)	51374.451(7)	1946.49291(7)	−0.0001	130000		s^2	$^3F_4$	5p	$(5/2)^2[5/2]^\circ_3$			F		
1952.5792(18)	51214.31(5)	1952.57557(12)	0.0037	97000		s^2	$^3F_3$	5p	$(3/2)^2[5/2]^\circ_2$			F		
1957.5176(3)	51085.108(8)	1957.51767(7)	−0.0000	190000		s^2	$^3F_4$	5p	$(5/2)^2[7/2]^\circ_4$			F		
1968.008(3)	50812.79(8)	1968.01129(8)	−0.003	9700		4p	$^3P^\circ_2$	4d	$(3/2)^2[1/2]_1$	1.8e+07	C	TW	O07	

Table A1. *Cont.*

λ_{obs} [a] (Å)	σ_{obs} [b] (cm^{-1})	λ_{Ritz} [c] (Å)	$\Delta\lambda_{obs\text{-}Ritz}$ (Å)	I_{obs} [d] (arb. u.)	Char [e]	Lower Level		Upper Level		A (s^{-1})	Acc [f]	Line Ref. [g]	TP Ref. [g]	Notes [h]
1970.4936(6)	50748.707(15)	1970.49395(7)	−0.0004	760000		4s	$^{3}D_{2}$	4p	$^{1}P^{\circ}_{1}$	5.3e+06	C+	F	K82cor	
1974.467(3)	50646.57(8)	1974.46773(8)	−0.000	2800		s^2	$^{1}D_{2}$	4f	$(5/2)^2[5/2]^{\circ}_3$			TW		
1977.0266(3)	50581.009(6)	1977.02671(15)	−0.0001	80000		s^2	$^{3}F_{2}$	5p	$(3/2)^2[3/2]^{\circ}_1$			F		
1979.95578(4)	50506.1785(10)	1979.95579(3)	−0.00002	18000000		4s	$^{3}D_{2}$	4p	$^{3}D^{\circ}_{2}$	9.6e+07	C+	F	D05se	
1982.1503(20)	50450.26(5)	1982.14827(12)	0.0021	16000		s^2	$^{3}F_{3}$	sp	$(^3F)^3P^{\circ}\ ^1D^{\circ}_2$			F		
1984.763(3)	50383.85(8)	1984.76431(19)	−0.001	14000		s^2	$^{3}P_{2}$	6p	$(5/2)^2[3/2]^{\circ}_2$			TW		
1986.307(3)	50344.68(8)	1986.30794(14)	−0.001	8100		s^2	$^{1}D_{2}$	sp	$(^3F)^1P^{\circ}\ ^3D^{\circ}_3$			TW		
1989.85478(3)	50254.9236(9)	1989.85480(3)	−0.00001	7400000		4s	$^{3}D_{2}$	4p	$^{3}D^{\circ}_{1}$	9.8e+07	C+	F	D05	
1990.18000(19)	50246.711(5)	1990.18009(7)	−0.00009	670000		4p	$^{3}P^{\circ}_{1}$	4d	$(3/2)^2[3/2]_2$	1.3e+08	C	F	B00	
1993.8395(21)	50154.49(5)	1993.8395(12)	−0.0000	34000		s^2	$^{3}P_{2}$	sp	$(^3P)^3P^{\circ}\ ^3S^{\circ}_1$			TWn		
1994.257(3)	50143.98(8)	1994.25958(15)	−0.002	17000		s^2	$^{3}F_{2}$	5p	$(3/2)^2[1/2]^{\circ}_1$			TW		
1998.5575(16)	50036.09(4)	1998.55790(7)	−0.0004	34000		4p	$^{3}F^{\circ}_{3}$	4d	$(3/2)^2[5/2]_3$			F		
1999.5327(18)	50011.69(5)	1999.53361(8)	−0.0009	140000		4p	$^{3}P^{\circ}_{1}$	4d	$(3/2)^2[3/2]_1$	5.8e+07	C	TW	B00	
1999.69752(6)	49991.3640(16)	1999.69754(5)	−0.00002	20000000		4s	$^{3}D_{3}$	4p	$^{3}D^{\circ}_{3}$	2.13e+08	B	F	C94c	
2002.9033(20)	49911.36(5)	2002.9020(12)	0.0013	73000		s^2	$^{3}P_{1}$	sp	$(^3P)^3P^{\circ}\ ^3S^{\circ}_1$			TWn		
2009.32(15)	49752(4)	2009.28239(8)	0.04	6800		4p	$^{3}F^{\circ}_{4}$	4d	$(3/2)^2[5/2]_3$			S36c		
2012.98037(14)	49661.540(4)	2012.98041(6)	−0.00004	1800000		4p	$^{3}P^{\circ}_{2}$	4d	$(5/2)^2[5/2]_3$	2.5e+08	C	F	B00	
2015.5822(4)	49597.442(10)	2015.58222(8)	0.0000	1400000		4s	$^{3}D_{1}$	4p	$^{1}P^{\circ}_{1}$	2.1e+07	C+	F	K82cor	
2015.8660(24)	49590.46(6)	2015.8649(12)	0.0011	39000		s^2	$^{3}P_{0}$	sp	$(^3P)^3P^{\circ}\ ^3S^{\circ}_1$			TWn		
2016.8964(6)	49565.129(14)	2016.89673(6)	−0.0003	1700000		4s	$^{3}D_{3}$	4p	$^{1}D^{\circ}_{2}$	8.1e+06	C+	F	D05se	
2017.6096(16)	49547.61(4)	2017.60977(12)	−0.0002	160000		s^2	$^{3}F_{3}$	5p	$(5/2)^2[5/2]^{\circ}_3$			F		
2022.1913(16)	49435.37(4)	2022.19147(7)	−0.0002	170000		4p	$^{3}F^{\circ}_{3}$	4d	$(3/2)^2[7/2]_4$	1.14e+07	C+	F	O07	
2025.48773(12)	49354.924(3)	2025.48775(5)	−0.00001	8100000		4s	$^{3}D_{1}$	4p	$^{3}D^{\circ}_{2}$	6.8e+07	C+	F	D05se	
2025.919(3)	49344.43(8)	2025.91683(13)	0.002	25000		s^2	$^{3}F_{3}$	5p	$(5/2)^2[5/2]^{\circ}_2$			TW		
2027.1336(6)	49314.858(15)	2027.13416(9)	−0.0005	610000		4p	$^{3}P^{\circ}_{1}$	4d	$(3/2)^2[1/2]_1$	1.3e+08	C	F	O07se	
2029.471(3)	49258.06(8)	2029.47227(16)	−0.001	37000		s^2	$^{3}F_{2}$	sp	$(^3F)^3P^{\circ}\ ^1D^{\circ}_2$			TW		
2029.9491(7)	49246.469(17)	2029.94908(6)	0.0000	560000		4p	$^{3}P^{\circ}_{2}$	4d	$(5/2)^2[3/2]_1$	8.0e+07	B	F	O07	
2031.03579(16)	49220.123(4)	2031.03598(6)	−0.00020	2700000		4p	$^{3}P^{\circ}_{2}$	4d	$(5/2)^2[3/2]_2$	4.5e+08	B	F	O07	
2033.844(3)	49152.18(8)	2033.84242(8)	0.001	62000		4p	$^{3}F^{\circ}_{4}$	4d	$(3/2)^2[7/2]_4$	8.9e+06	C+	TW	O07	
2035.85321(6)	49103.6704(15)	2035.85327(4)	−0.00006	19000000		4s	$^{3}D_{1}$	4p	$^{3}D^{\circ}_{1}$	3.7e+08	C+	F	D05	
2036.9201(3)	49077.954(8)	2036.92001(9)	0.0001	730000		4p	$^{3}P^{\circ}_{0}$	4d	$(3/2)^2[3/2]_1$	1.9e+08	C	F	B00	
2037.12672(4)	49072.9776(9)	2037.12668(3)	0.00004	13000000		4s	$^{3}D_{2}$	4p	$^{3}D^{\circ}_{3}$	1.36e+08	B	F	C94c	
2043.8019(5)	48912.725(11)	2043.80148(7)	0.0004	21000000		4s	$^{3}D_{3}$	4p	$^{1}F^{\circ}_{3}$	1.42e+08	C+	R1c	D05se	
2047.67825(16)	48820.143(4)	2047.67825(11)	−0.00000	1300000		4p	$^{1}P^{\circ}_{1}$	4d	$(3/2)^2[1/2]_0$	5.32e+08	B+	F	B09	
2054.25259(21)	48663.922(5)	2054.25263(8)	−0.00003	810000		4p	$^{3}F^{\circ}_{2}$	4d	$(3/2)^2[5/2]_2$	1.4e+08	C	F	B00	
2054.4171(3)	48660.025(7)	2054.41739(8)	−0.0003	960000		4p	$^{3}P^{\circ}_{1}$	4d	$(5/2)^2[1/2]_0$	6.7e+08	C	F	B00	
2054.97836(4)	48646.7370(10)	2054.97836(3)	0.00001	18000000		4s	$^{3}D_{2}$	4p	$^{1}D^{\circ}_{2}$	1.65e+08	C+	F	D05se	
2058.611(3)	48560.91(8)	2058.61377(13)	−0.003	49000		s^2	$^{3}F_{3}$	5p	$(5/2)^2[3/2]^{\circ}_2$			TW		
2062.4201(4)	48471.230(8)	2062.42011(6)	−0.0000	1500000		4p	$^{3}P^{\circ}_{1}$	4d	$(5/2)^2[5/2]_2$	1.7e+08	C	F	B00	

Table A1. *Cont.*

λ_{obs} [a] (Å)	σ_{obs} [b] (cm^{-1})	λ_{Ritz} [c] (Å)	$\Delta\lambda_{obs\text{-}Ritz}$ (Å)	I_{obs} [d] (arb. u.)	Char [e]	Lower Level	Upper Level	A (s^{-1})	Acc [f]	Line Ref. [g]	TP Ref. [g]	Notes [h]
2066.2609(5)	48381.143(12)	2066.26109(10)	−0.0002	960000		4p $^3P^o_0$	4d $(3/2)^2[1/2]_1$	2.00e+08	B+	F	O07	
2067.368(3)	48355.23(7)	2067.36392(16)	0.004	35000		s^2 $^3F_2$	5p $(5/2)^2[5/2]^o_3$			TW		
2069.9356(13)	48295.27(3)	2069.93487(7)	0.0007	240000		4p $^3F^o_2$	4d $(3/2)^2[3/2]_2$	3.7e+07	C	F	B00	
2074.213(3)	48195.68(8)	2074.20995(16)	0.003	39000		s^2 $^3F_2$	5p $(5/2)^2[3/2]^o_1$			TW		
		2074.24066(15)			m	s^2 $^3P_1$	sp $(^3F)^1P^o\ ^3D^o_2$			TW		
2078.66147(9)	48092.5549(21)	2078.66153(6)	−0.00005	6700000		4p $^3P^o_2$	4d $(5/2)^2[1/2]_1$	6.3e+08	B+	F	O07	
2080.0593(13)	48060.24(3)	2080.05953(9)	−0.0002	210000		4p $^3F^o_2$	4d $(3/2)^2[3/2]_1$	4.9e+07	C	R1c	B00	
2082.920(20)	47994.2(5)	2082.91535(6)	0.005	56000		4s $^3D_2$	4p $^1F^o_3$	1.0e+06	E	S36c	D05se	
2084.3229(20)	47961.94(5)	2084.32369(22)	−0.0008	170000		s^2 $^3F_4$	sp $(^3F)^3P^o\ ^3F^o_4$			TW		
2085.2735(9)	47940.081(21)	2085.27476(6)	−0.0012	860000		4p $^3F^o_3$	4d $(5/2)^2[5/2]_2$	6.7e+07	C	R1c	B00	
2085.297(4)	47939.55(9)	2085.30956(7)	−0.013	2500000		4s $^3D_3$	4p $^3F^o_2$	4.1e+06	C+	TW	D05cor	
2087.91812(11)	47879.3666(25)	2087.91817(7)	−0.00005	3400000		4p $^3F^o_2$	4d $(3/2)^2[7/2]_3$	4.2e+08	B	F	B00	
2087.96979(24)	47878.182(5)	2087.96987(6)	−0.00008	3700000		4p $^3F^o_3$	4d $(5/2)^2[7/2]_3$	2.6e+08	C	F	B00	
2093.6366(5)	47748.606(11)	2093.63705(6)	−0.0004	2400000		4p $^3P^o_1$	4d $(5/2)^2[3/2]_1$	2.20e+08	B	R1c	O07	
2094.7925(9)	47722.261(21)	2094.79321(6)	−0.0007	290000		4p $^3P^o_1$	4d $(5/2)^2[3/2]_2$	2.21e+07	C+	R1c	O07	
2096.1891(14)	47690.47(3)	2096.19052(9)	−0.0014	120000		4p $^1F^o_3$	4d $(3/2)^2[5/2]_2$	1.6e+07	C	R1c	B00	
2098.3067(3)	47642.348(7)	2098.30677(8)	−0.0001	1200000		4p $^1F^o_3$	4d $(3/2)^2[5/2]_3$	1.01e+08	C	F	B00	
2098.3971(2)	47640.295(6)	2098.39724(8)	−0.0001	5000000		4p $^3F^o_4$	4d $(5/2)^2[7/2]_4$	2.09e+08	B	F	O07	
2098.7405(9)	47632.502(21)	2098.74108(6)	−0.0006	300000		4p $^3F^o_3$	4d $(5/2)^2[5/2]_3$	2.0e+07	C	R1c	B00	
2100.397(3)	47594.94(7)	2100.39318(8)	0.004	53000		4p $^3F^o_4$	4d $(5/2)^2[7/2]_3$			TW		
2104.79587(5)	47495.4827(11)	2104.79586(4)	0.00002	62000000		4s $^3D_1$	4p $^1D^o_2$	7.7e+07	C+	F	D05se	
2106.3816(11)	47459.73(3)	2106.3791(3)	0.0025	240000		s^2 $^3F_3$	sp $(^3F)^3P^o\ ^1G^o_4$			F		
2110.308(3)	47371.45(7)	2110.30833(15)	−0.001	90000		s^2 $^3P_2$	sp $(^3F)^1P^o\ ^3D^o_3$			TW		
2111.2934(4)	47349.333(10)	2111.29323(8)	0.0001	860000		4p $^3F^o_4$	4d $(5/2)^2[5/2]_3$	6.6e+07	C	R1c	B00	
2112.09930(19)	47331.268(4)	2112.09946(9)	−0.00016	55000000		4s $^1D_2$	4p $^1P^o_1$	3.5e+08	C+	F	D05se	
2112.5245(14)	47321.74(3)	2112.52195(8)	0.0026	110000		4p $^1F^o_3$	4d $(3/2)^2[3/2]_2$	1.2e+07	C	R1c	B00	
2117.30881(9)	47214.8259(20)	2117.30874(7)	0.00006	12000000		4p $^3F^o_3$	4d $(5/2)^2[9/2]_4$	7.2e+08	B	F	O07	
2118.3748(20)	47191.07(5)	2118.37484(7)	−0.0001	260000		4p $^3F^o_3$	4d $(5/2)^2[3/2]_2$	2.7e+07	B	TW	O07	
2122.97861(13)	47088.745(3)	2122.97858(6)	0.00003	71000000		4s $^1D_2$	4p $^3D^o_2$	2.2e+08	C+	F	D05se	
2125.10512(12)	47041.631(3)	2125.10507(7)	0.00004	3800000		4p $^1F^o_3$	4d $(3/2)^2[7/2]_4$	2.28e+08	B+	F	O07	
2125.2670(5)	47038.048(12)	2125.26686(8)	0.0001	610000		4p $^1D^o_2$	4d $(3/2)^2[5/2]_2$	6.6e+07	C	F	B00	
2126.04361(6)	47020.8676(14)	2126.04362(5)	−0.00001	74000000		4s $^3D_2$	4p $^3F^o_2$	1.41e+08	B	F	C94c	
2130.0848(5)	46931.669(10)	2130.08459(9)	0.0002	1000000		4p $^3F^o_4$	4d $(5/2)^2[9/2]_4$	4.4e+07	B+	R1c	O07	
2131.2548(6)	46905.908(13)	2131.25595(8)	−0.0011	190000		4p $^1F^o_3$	4d $(3/2)^2[7/2]_3$	1.3e+07	C	R1c	B00	
2134.3400(4)	46838.113(9)	2134.34030(8)	−0.0003	9500000		4p $^3F^o_4$	4d $(5/2)^2[9/2]_5$	7.7e+08	B	R1c	B00	
2135.399	46814.89	2135.39887(8)			:	4p $^3P^o_0$	4d $(5/2)^2[3/2]_1$	4.e+07	E		O07se	
2135.98008(7)	46802.1541(16)	2135.98004(6)	0.00004	140000000		4s $^3D_3$	4p $^3F^o_4$	4.59e+08	A+	F	P97	
2144.706(10)	46611.76(22)	2144.70358(8)	0.002	160000	*	4p $^3D^o_3$	4d $(3/2)^2[5/2]_2$	1.5e+07	C	R1c	B00	

Table A1. *Cont.*

λ_{obs} [a] (Å)	σ_{obs} [b] (cm^{-1})	λ_{Ritz} [c] (Å)	$\Delta\lambda_{obs\text{-}Ritz}$ (Å)	I_{obs} [d] (arb. u.)	Char [e]	Lower Level		Upper Level		A (s^{-1})	Acc [f]	Line Ref. [g]	TP Ref. [g]	Notes [h]
2144.706(10)	46611.76(22)	2144.72206(16)	−0.016	160000	*	s^2	$^3F_3$	sp	$(^3F)^3P^\circ\ ^3F^\circ_3$			R1c		
2145.4920(4)	46594.683(9)	2145.49206(7)	−0.0001	1200000		4p	$^3P^\circ_1$	4d	$(5/2)^2[1/2]_1$	1.14e+08	B	R1c	O07	
2146.9187(4)	46563.722(8)	2146.91894(7)	−0.0002	1100000		4p	$^3D^\circ_3$	4d	$(3/2)^2[5/2]_3$	1.04e+08	C	F	B00	
		2148.94587(6)			m	4p	$^3F^\circ_2$	4d	$(5/2)^2[5/2]_2$	1.12e+08	C	S36c	B00	
2148.9830(6)	46518.999(13)	2148.98283(8)	0.0002	59000000		4s	$^3D_3$	4p	$^3F^\circ_3$	8.8e+07	B	R1c	C94c	
2151.8083(4)	46457.927(9)	2151.80815(7)	0.0001	4200000		4p	$^3F^\circ_2$	4d	$(5/2)^2[7/2]_3$	2.59e+08	B	R1c	B00	
2152.910(20)	46434.2(4)	2152.90029(9)	0.010	23000		4p	$^1D^\circ_2$	4d	$(3/2)^2[3/2]_1$			S36c		
2158.4108(10)	46315.829(21)	2158.41130(19)	−0.0005	60000		s^2	$^3F_2$	sp	$(^3F)^3P^\circ\ ^3F^\circ_2$			R1c		
2161.31978(23)	46253.498(5)	2161.31990(7)	−0.00012	4200000		4p	$^1D^\circ_2$	4d	$(3/2)^2[7/2]_3$	3.1e+08	B	F	B00	
2161.7998(10)	46243.228(21)	2161.80255(7)	−0.0027	58000		4p	$^3D^\circ_3$	4d	$(3/2)^2[3/2]_2$			R1c		
2166.850(20)	46135.5(4)	2166.8678(3)	−0.018	21000		s^2	$^3F_3$	sp	$(^3F)^3P^\circ\ ^3F^\circ_4$			S36c		
2174.98119(9)	45963.0038(18)	2174.98126(6)	−0.00007	7500000		4p	$^3D^\circ_3$	4d	$(3/2)^2[7/2]_4$	4.9e+08	B	F	O07	
2179.41026(22)	45869.606(5)	2179.40994(6)	0.00033	72000000		4s	$^3D_1$	4p	$^3F^\circ_2$	2.15e+08	B	F	C94c	
2180.7508(5)	45841.413(11)			680000		s^2	$^3F_4$	sp	$(^3F)^3P^\circ\ ^3G^\circ_5$			R1c		
2181.4243(4)	45827.261(9)	2181.42461(7)	−0.0003	290000		4p	$^3D^\circ_3$	4d	$(3/2)^2[7/2]_3$	2.4e+07	C	R1c	B00	
2182.8585(5)	45797.154(11)	2182.85780(7)	0.0007	560000		4p	$^3F^\circ_2$	4d	$(5/2)^2[3/2]_1$	8.0e+07	C+	R1c	O07	
2189.3693(10)	45660.976(21)	2189.36930(8)	0.0000	240000		4p	$^3P^\circ_0$	4d	$(5/2)^2[1/2]_1$	1.3e+07	C	R1c	O07se	
2189.62967(9)	45655.5468(18)	2189.62971(6)	−0.00004	47000000		4s	$^1D_2$	4p	$^3D^\circ_3$	1.04e+08	B	F,R1c	C94c	
2190.500(20)	45637.4(4)	2190.5192(3)	−0.019	57000		s^2	$^1G_4$	6p	$(3/2)^2[5/2]^\circ_3$			S36c		
2192.26759(8)	45600.6159(16)	2192.26753(6)	0.00007	73000000		4s	$^3D_2$	4p	$^3F^\circ_3$	2.8e+08	B	F	C94c	
2195.68180(17)	45529.716(4)	2195.68192(7)	−0.00012	4200000		4p	$^1F^\circ_3$	4d	$(5/2)^2[7/2]_4$	4.2e+08	B	F	O07	
2197.8688(15)	45484.42(3)	2197.86726(7)	0.0016	71000		4p	$^1F^\circ_3$	4d	$(5/2)^2[7/2]_3$			R1c		
2200.300(20)	45434.2(4)	2200.31543(20)	−0.015	47000		s^2	$^3P_0$	sp	$(^3P)^3P^\circ\ ^3D^\circ_1$			S36c		
2200.5078(3)	45429.874(6)	2200.50822(8)	−0.0004	1600000		4p	$^3D^\circ_1$	4d	$(3/2)^2[5/2]_2$	2.1e+08	C	F	B00	
2201.000(20)	45419.7(4)	2201.02822(20)	−0.028	33000		s^2	$^3F_2$	sp	$(^3F)^3P^\circ\ ^3F^\circ_3$			S36c		
2209.8049(4)	45238.760(8)	2209.80507(7)	−0.0002	1900000		4p	$^1F^\circ_3$	4d	$(5/2)^2[5/2]_3$	2.1e+08	C	R1c	B00	
2210.26692(11)	45229.3048(22)	2210.26693(6)	−0.00001	56000000		4s	$^1D_2$	4p	$^1D^\circ_2$	1.53e+08	C+	F	D05se	
2212.7470(6)	45178.617(12)	2212.74730(8)	−0.0003	1100000		4p	$^3D^\circ_2$	4d	$(3/2)^2[5/2]_2$	1.4e+08	C	F	B00	
2215.1054(3)	45130.521(5)	2215.10550(8)	−0.0001	3800000		4p	$^3D^\circ_2$	4d	$(3/2)^2[5/2]_3$	4.9e+08	B	F	B00	
2218.10770(6)	45069.4401(12)	2218.10772(5)	−0.00002	66000000		4s	$^3D_2$	4p	$^3P^\circ_1$	3.5e+08	C+	F	D05se	
2218.5123(4)	45061.222(9)	2218.51209(8)	0.0002	1900000		4p	$^3D^\circ_1$	4d	$(3/2)^2[3/2]_2$	2.6e+08	C	R1c	B00	
2221.649(3)	44997.60(6)	2221.6477(8)	0.001	99000		s^2	$^1D_2$	sp	$(^1D)^3P^\circ\ ^3P^\circ_2$			R1c		
2224.6906(5)	44936.092(10)	2224.69062(11)	−0.0000	1300000		4p	$^1P^\circ_1$	4d	$(3/2)^2[5/2]_2$	1.3e+08	C	R1c	B00	
2226.7798(4)	44893.936(8)	2226.77999(6)	−0.0002	2200000		4p	$^1D^\circ_2$	4d	$(5/2)^2[5/2]_2$	2.9e+08	C	R1c	B00	
2228.86748(6)	44851.8904(11)	2228.86745(6)	0.00003	35000000		4s	$^3D_1$	4p	$^3P^\circ_0$	3.88e+08	B+	F	P97	
2229.8529(4)	44832.072(8)	2229.85346(6)	−0.0006	1800000		4p	$^1D^\circ_2$	4d	$(5/2)^2[7/2]_3$	2.1e+08	C	R1c	B00	
2230.1439(7)	44826.223(15)	2230.14604(9)	−0.0022	1900000		4p	$^3D^\circ_1$	4d	$(3/2)^2[3/2]_1$	1.9e+08	C	R1c	B00	
2230.3979(7)	44821.118(13)	2230.39901(9)	−0.0011	440000		4p	$^1F^\circ_3$	4d	$(5/2)^2[9/2]_4$	2.0e+07	C	R1c	O07se	

Table A1. *Cont.*

λ_{obs} [a] (Å)	σ_{obs} [b] (cm^{-1})	λ_{Ritz} [c] (Å)	$\Delta\lambda_{obs\text{-}Ritz}$ (Å)	I_{obs} [d] (arb. u.)	Char [e]	Lower Level		Upper Level		A (s^{-1})	Acc [f]	Line Ref. [g]	TP Ref. [g]	Notes [h]
2230.9512(6)	44810.003(12)	2230.95279(8)	−0.0016	1600000		4p	$^3D^\circ_2$	4d	$(3/2)^2[3/2]_2$	2.2e+08	C	R1c	B00	
2231.5817(4)	44797.343(8)	2231.58204(7)	−0.0003	1100000		4p	$^1F^\circ_3$	4d	$(5/2)^2[3/2]_2$	1.14e+08	B	R1c	O07	
2242.1424(11)	44586.363(21)	2242.14217(6)	0.0003	240000		4p	$^1D^\circ_2$	4d	$(5/2)^2[5/2]_3$	1.07e+07	C	R1c	B00	
2242.61775(10)	44576.9146(20)	2242.61764(7)	0.00011	54000000		4s	$^1D_2$	4p	$^1F^\circ_3$	2.5e+08	C+	F	D05se	
2242.718(3)	44574.93(5)	2242.71791(9)	−0.000	100000		4p	$^3D^\circ_2$	4d	$(3/2)^2[3/2]_1$	5.8e+07	C	TW	B00	
2243.0938(7)	44567.454(14)	2243.09395(10)	−0.0001	230000		4p	$^1P^\circ_1$	4d	$(3/2)^2[3/2]_2$	4.0e+07	C	R1c	B00	
2247.0017(5)	44489.953(11)	2247.00170(9)	−0.0000	66000000		4s	$^3D_3$	4p	$^3P^\circ_2$	3.3e+08	B	R1c	C94c	
2248.9666(4)	44451.086(8)	2248.96667(6)	−0.0001	1800000		4p	$^3D^\circ_3$	4d	$(5/2)^2[7/2]_4$	9.1e+07	B	R1c	O07	
2251.8564(5)	44394.047(9)	2251.85601(8)	0.0004	72000		4p	$^3D^\circ_2$	4d	$(3/2)^2[7/2]_3$	5.8e+06	C	R1c	B00	
2253.0326(16)	44370.87(3)	2253.0273(3)	0.0053	51000		s^2	$^1D_2$	sp	$(^1D)^3P^\circ\ ^3P^\circ_1$			R1c		
2254.9880(4)	44332.402(8)	2254.98778(11)	0.0002	620000		4p	$^1P^\circ_1$	4d	$(3/2)^2[3/2]_1$	1.9e+08	C	R1c	B00	
2263.2130(4)	44171.302(8)	2263.21318(6)	−0.0002	580000		4p	$^1D^\circ_2$	4d	$(5/2)^2[3/2]_1$	1.54e+08	B+	R1c	O07	
2263.7857(4)	44160.128(8)	2263.78578(6)	−0.0000	1700000		4p	$^3D^\circ_3$	4d	$(5/2)^2[5/2]_3$	1.8e+08	C	R1c	B00	
2264.5671(16)	44144.89(3)	2264.56423(6)	0.0029	50000		4p	$^1D^\circ_2$	4d	$(5/2)^2[3/2]_2$	1.4e+06	E	R1c	O07se	
2265.3643(4)	44129.358(8)	2265.36424(10)	0.0001	190000		4p	$^3D^\circ_1$	4d	$(3/2)^2[1/2]_1$	1.25e+08	B	R1c	O07	
2274.7407(6)	43947.476(11)	2274.74107(8)	−0.0004	150000		4p	$^3P^\circ_2$	5s	$(3/2)^2[3/2]_2$	4.e+06	E	R1c	C94	
2276.2577(4)	43918.191(8)	2276.25797(7)	−0.0003	16000000		4s	$^3D_1$	4p	$^3P^\circ_1$	6.3e+07	C+	R1c	D05se	
2278.3378(5)	43878.098(10)	2278.33737(10)	0.0004	220000		4p	$^3D^\circ_2$	4d	$(3/2)^2[1/2]_1$	1.32e+08	B+	R1c	O07	
2280.9423(4)	43827.999(8)	2280.9428(3)	−0.0005	260000		s^2	$^3F_3$	sp	$(^3F)^3P^\circ\ ^3G^\circ_4$			R1c		
2284.202(4)	43765.46(7)	2284.19835(22)	0.004	17000		s^2	$^1G_4$	sp	$(^1G)^3P^\circ\ ^3F^\circ_3$			R1c		
2286.6447(4)	43718.712(8)	2286.64494(6)	−0.0002	810000		4p	$^3D^\circ_3$	4d	$(5/2)^2[3/2]_2$	1.52e+08	B+	R1c	O07	
2289.4160(4)	43665.797(8)	2289.41619(8)	−0.0002	170000		4p	$^3P^\circ_2$	5s	$(3/2)^2[3/2]_1$	7.e+06	D	R1c	C94	
2290.1606(7)	43651.600(14)	2290.16119(24)	−0.0006	49000		s^2	$^3F_2$	sp	$(^3F)^3P^\circ\ ^3D^\circ_3$			R1c		
2291.0018(4)	43635.575(8)	2291.00110(12)	0.0007	550000		4p	$^1P^\circ_1$	4d	$(3/2)^2[1/2]_1$	1.76e+08	B+	R1c	O07	
2292.6896(6)	43603.454(11)	2292.69060(8)	−0.0010	19000		4s	$^1D_2$	4p	$^3F^\circ_2$	4.e+05	E	R1c	D05cor	
2292.9699(6)	43598.124(11)	2292.9700(3)	−0.0002	20000		s^2	$^3F_4$	sp	$(^3F)^3P^\circ\ ^5F^\circ_4$			R1c		
2294.3674(4)	43571.571(8)	2294.36758(7)	−0.0002	15000000		4s	$^3D_2$	4p	$^3P^\circ_2$	3.4e+07	C+	R1c	C94c	
2299.4885(5)	43474.542(9)	2299.48891(10)	−0.0004	140000		4p	$^3D^\circ_1$	4d	$(5/2)^2[1/2]_0$	1.07e+08	C	R1c	B00	
2309.5189(4)	43285.747(8)	2309.51890(6)	−0.0000	200000		4p	$^3D^\circ_1$	4d	$(5/2)^2[5/2]_2$	1.02e+07	C	R1c	B00	
2315.681(5)	43170.57(10)	2315.6826(4)	−0.002	17000		s^2	$^1D_2$	sp	$(^1D)^3P^\circ\ ^3F^\circ_3$			R1c		
2323.0039(5)	43034.495(9)	2323.00409(6)	−0.0002	200000		4p	$^3D^\circ_2$	4d	$(5/2)^2[5/2]_2$			R1c		
2323.9280(6)	43017.383(11)	2323.92802(7)	0.0000	86000		4p	$^1D^\circ_2$	4d	$(5/2)^2[1/2]_1$			R1c		
2325.9100(17)	42980.73(3)	2325.90832(12)	0.0017	25000		4p	$^1P^\circ_1$	4d	$(5/2)^2[1/2]_0$			R1c		
2327.5669(6)	42950.137(11)	2327.5672(4)	−0.0003	12000		s^2	$^3F_3$	sp	$(^3F)^3P^\circ\ ^5F^\circ_2$			R1c		
2333.743(3)	42836.49(6)	2333.75088(22)	−0.008	23000		s^2	$^1G_4$	6p	$(5/2)^2[7/2]^\circ_3$			R1c		
2336.1707(4)	42791.971(8)	2336.17057(10)	0.0001	230000		4p	$^1P^\circ_1$	4d	$(5/2)^2[5/2]_2$	4.9e+07	C	R1c	B00	
2339.7275(5)	42726.924(10)	2339.72732(6)	0.0002	68000		4p	$^3D^\circ_2$	4d	$(5/2)^2[5/2]_3$			R1c		
2341.3713(12)	42696.930(21)			17000		s^2	$^3F_4$	sp	$(^3F)^3P^\circ\ ^5F^\circ_5$			R1c		
2342.1723(17)	42682.33(3)	2342.1734(4)	−0.0011	55000		s^2	$^3P_2$	sp	$(^3P)^3P^\circ\ ^5D^\circ_3$			R1c		
2347.890(20)	42578.4(4)	2347.8850(4)	0.005	9000	*	s^2	$^3P_2$	sp	$(^3P)^3P^\circ\ ^5D^\circ_1$			S36c		

Table A1. *Cont.*

λ_{obs} [a] (Å)	σ_{obs} [b] (cm^{-1})	λ_{Ritz} [c] (Å)	$\Delta\lambda_{obs\text{-}Ritz}$ (Å)	I_{obs} [d] (arb. u.)	Char [e]	Lower Level		Upper Level		A (s^{-1})	Acc [f]	Line Ref. [g]	TP Ref. [g]	Notes [h]
2347.890(20)	42578.4(4)	2347.914(3)	−0.024	9000	*	5s	$(3/2)^2[3/2]_2$	sp	$(^3P)^1P^\circ\ ^3P^\circ_2$			S36cn		
2348.7330(4)	42563.115(8)	2348.73311(7)	−0.0001	120000		4p	$^3D^\circ_1$	4d	$(5/2)^2[3/2]_1$	2.8e+07	B	R1c	O07	
2350.1902(8)	42536.727(15)	2350.18820(6)	0.0020	14000		4p	$^3D^\circ_1$	4d	$(5/2)^2[3/2]_2$			R1c		
2352.2911(12)	42498.740(21)	2352.29447(19)	−0.0034	30000		s^2	$^1G_4$	4f	$(3/2)^2[9/2]^\circ_5$			R1c		
2353.9437(5)	42468.905(9)	2353.9434(4)	0.0003	27000		s^2	$^3F_3$	sp	$(^3F)^3P^\circ\ ^5F^\circ_3$			R1c		
2355.0143(6)	42449.600(11)	2355.01447(8)	−0.0001	610000		4p	$^3P^\circ_1$	5s	$(3/2)^2[3/2]_2$	2.5e+07	C+	R1c	K82	
2356.6402(5)	42420.316(9)	2356.64033(9)	−0.0001	1400000		4s	$^3D_1$	4p	$^3P^\circ_2$	3.0e+06	C+	R1c	C94c	
2360.6389(9)	42348.465(16)	2360.6391(9)	−0.0002	9200		s^2	$^3P_1$	sp	$(^3P)^3P^\circ\ ^5D^\circ_0$			R2nc		
2361.1901(12)	42338.579(21)	2361.1910(5)	−0.0008	37000		s^2	$^3P_1$	sp	$(^3P)^3P^\circ\ ^5D^\circ_2$			R1c		
2362.6809(9)	42311.867(16)	2362.68143(7)	−0.0005	26000		4p	$^3D^\circ_2$	4d	$(5/2)^2[3/2]_1$	6.3e+06	C+	R1c	O07	
2364.1538(6)	42285.508(10)	2364.15386(7)	−0.0000	57000		4p	$^3D^\circ_2$	4d	$(5/2)^2[3/2]_2$	3.4e+06	B	R1c	O07	
2366.980(3)	42235.03(5)	2366.9785(4)	0.001	8700		5s	$(5/2)^2[5/2]_3$	sp	$(^1D)^1P^\circ\ ^1D^\circ_2$			R1c		
2369.8893(5)	42183.179(9)	2369.88902(9)	0.0003	12000000		4s	$^1D_2$	4p	$^3F^\circ_3$	5.3e+07	B	R1c	C94c	
2370.7464(5)	42167.930(8)	2370.74698(8)	−0.0006	490000		4p	$^3P^\circ_1$	5s	$(3/2)^2[3/2]_1$	2.4e+07	C+	R1c	K82	
2376.3030(4)	42069.335(8)	2376.30277(10)	0.0003	330000		4p	$^1P^\circ_1$	4d	$(5/2)^2[3/2]_1$	1.61e+08	C+	R1c	O07se	
2377.7920(6)	42042.994(10)	2377.79223(10)	−0.0003	12000		4p	$^1P^\circ_1$	4d	$(5/2)^2[3/2]_2$	3.5e+06	D+	R1c	O07	
2378.4047(7)	42032.163(13)	2378.4063(6)	−0.0015	19000		s^2	$^3F_2$	sp	$(^3F)^3P^\circ\ ^5F^\circ_1$			R1c		
2378.8442(12)	42024.399(21)	2378.8439(9)	0.0002	220000		s^2	$^3P_2$	sp	$(^1D)^3P^\circ\ ^3P^\circ_2$			R1c		
2379.4048(12)	42014.499(21)	2379.4055(5)	−0.0007	53000		s^2	$^3P_0$	sp	$(^3P)^3P^\circ\ ^5D^\circ_1$			R1c		
2384.80(6)	41919.5(11)	2384.85893(9)	−0.06	49000		4p	$^3F^\circ_3$	5s	$(3/2)^2[3/2]_2$	6.e+05	E	S36c	K82cal	
2384.9439(4)	41916.926(8)	2384.94408(9)	−0.0002	100000		4p	$^3P^\circ_2$	5s	$(5/2)^2[5/2]_2$	7.e+06	E	R1c	K82cal	
2385.0951(18)	41914.27(3)	2385.0945(4)	0.0006	7400		5s	$(5/2)^2[5/2]_2$	sp	$(^1D)^1P^\circ\ ^1D^\circ_2$			R1c		
2392.6852(18)	41781.32(3)	2392.6858(9)	−0.0006	6900		s^2	$^3P_1$	sp	$(^1D)^3P^\circ\ ^3P^\circ_2$			R1c		
2393.2599(5)	41771.287(9)	2393.2602(4)	−0.0003	23000		s^2	$^3F_3$	sp	$(^3F)^3P^\circ\ ^5F^\circ_4$			R1c		
2394.0296(5)	41757.858(9)	2394.0296(4)	−0.0000	20000		s^2	$^3F_2$	sp	$(^3F)^3P^\circ\ ^5F^\circ_2$			R1c		
2400.1140(7)	41652.008(12)	2400.11404(9)	−0.0001	3700000		4s	$^1D_2$	4p	$^3P^\circ_1$	6.9e+06	C+	R1c	D05se	
2403.3373(5)	41596.149(8)	2403.33713(9)	0.0002	2200000		4p	$^3P^\circ_2$	5s	$(5/2)^2[5/2]_3$	1.06e+08	C+	F	C94	
2414.1881(5)	41409.205(9)	2414.18856(8)	−0.0004	36000		4p	$^3D^\circ_1$	4d	$(5/2)^2[1/2]_1$	2.e+06	E	R1c	O07cal	
2414.8568(5)	41397.740(9)	2414.8563(4)	0.0005	37000		s^2	$^3P_2$	sp	$(^1D)^3P^\circ\ ^3P^\circ_1$			R1c		
2421.9424(6)	41276.637(11)	2421.9425(4)	−0.0001	6700		s^2	$^3F_2$	sp	$(^3F)^3P^\circ\ ^5F^\circ_3$			R1c		
2424.4338(3)	41234.223(5)	2424.43430(9)	−0.0005	1100000		4p	$^3P^\circ_0$	5s	$(3/2)^2[3/2]_1$	5.0e+07	C+	F	C94	
2426.558(5)	41198.13(8)	2426.5614(9)	−0.003	10000		4d	$(5/2)^2[9/2]_5$	8f	$(5/2)^2[11/2]^\circ_6$			R1c		
2426.996(9)	41190.69(15)	2427.016(3)	−0.019	5000		5s	$(5/2)^2[5/2]_2$	7p	$(3/2)^2[5/2]^\circ_2$			R2nc		
2428.9274(5)	41157.944(8)	2428.92746(8)	−0.0000	38000		4p	$^3D^\circ_2$	4d	$(5/2)^2[1/2]_1$	1.93e+07	C+	R1c	O07	
2430.6772(13)	41128.318(21)	2430.67651(24)	0.0007	5600		s^2	$^1G_4$	sp	$(^1G)^3P^\circ\ ^3H^\circ_4$			R1c		
2432.420(3)	41098.86(5)	2432.4183(5)	0.001	4800		4d	$(5/2)^2[9/2]_4$	8f	$(5/2)^2[11/2]^\circ_5$			R1c		
2442.6646(5)	40926.494(9)			90000		s^2	$^3F_4$	sp	$(^3F)^3P^\circ\ ^5G^\circ_5$			R1c		
2443.3256(5)	40915.423(8)	2443.32555(11)	0.0001	44000		4p	$^1P^\circ_1$	4d	$(5/2)^2[1/2]_1$	2.00e+07	B	R1c	O07	
2448.2143(13)	40833.727(21)	2448.2131(4)	0.0012	46000		s^2	$^3P_0$	sp	$(^1D)^3P^\circ\ ^3P^\circ_1$			R1c		
2452.951(6)	40754.88(10)	2452.9576(7)	−0.006	3900		s^2	$^3P_2$	sp	$(^1D)^3P^\circ\ ^3D^\circ_3$			R1c		

Table A1. *Cont.*

λ_{obs} [a] (Å)	σ_{obs} [b] (cm^{-1})	λ_{Ritz} [c] (Å)	$\Delta\lambda_{obs-Ritz}$ (Å)	I_{obs} [d] (arb. u.)	Char [e]	Lower Level			Upper Level	A (s^{-1})	Acc [f]	Line Ref. [g]	TP Ref. [g]	Notes [h]
2456.008(4)	40704.16(7)	2455.99461(18)	0.013	7600		s^2	$^1G_4$	4f	$(5/2)^2[9/2]^\circ_4$			R1c		
2462.6140(19)	40594.98(3)	2462.61266(17)	0.0014	11000		s^2	$^1G_4$	4f	$(5/2)^2[9/2]^\circ_5$			R1c		
2464.307(5)	40567.10(8)	2464.30282(17)	0.004	3500		s^2	$^1G_4$	4f	$(5/2)^2[7/2]^\circ_4$			R1c		
2468.3475(8)	40500.690(13)	2468.3480(6)	−0.0005	33000		s^2	$^3P_1$	sp	$(^1D)^3P^\circ\ ^3P^\circ_0$			R1c		
2468.5000(5)	40498.187(8)	2468.50111(9)	−0.0011	170000		4p	$^3F^\circ_2$	5s	$(3/2)^2[3/2]_2$	1.30e+07	C+	R1c	K82	
2473.3333(4)	40419.053(7)	2473.33345(9)	−0.0002	570000		4p	$^3P^\circ_1$	5s	$(5/2)^2[5/2]_2$	5.9e+07	C+	R1c	C94	
2476.4439(7)	40368.287(11)	2476.44459(19)	−0.0007	15000		s^2	$^1G_4$	4f	$(5/2)^2[11/2]^\circ_5$			R1c		
2483.784(5)	40249.00(8)	2483.7876(5)	−0.003	2900		s^2	$^3P_1$	sp	$(^1D)^3P^\circ\ ^3D^\circ_2$			R1c		
2485.7919(5)	40216.489(8)	2485.79176(9)	0.0002	1000000		4p	$^3F^\circ_2$	5s	$(3/2)^2[3/2]_1$	1.42e+08	C+	R1c	C94	L
2489.6523(6)	40154.136(10)	2489.65236(11)	−0.0001	660000		4s	$^1D_2$	4p	$^3P^\circ_2$	1.0e+06	D+	R1c	C94c	
2499.001(8)	40003.94(12)	2499.0132(4)	−0.013	4900		s^2	$^3P_2$	sp	$(^1D)^3P^\circ\ ^3F^\circ_2$			R1c		
2501.4933(13)	39964.076(21)	2501.4947(5)	−0.0014	4800		s^2	$^3P_1$	sp	$(^1D)^3P^\circ\ ^3D^\circ_1$			R1c		
2506.2727(3)	39887.871(5)	2506.27258(10)	0.0001	1000000		4p	$^3F^\circ_3$	5s	$(5/2)^2[5/2]_2$	2.0e+08	C+	F	C94	L
2514.2919(10)	39760.659(16)	2514.2931(4)	−0.0012	4200		s^2	$^3P_1$	sp	$(^1D)^3P^\circ\ ^3F^\circ_2$			R1c		
2515.0822(7)	39748.166(11)	2515.0831(3)	−0.0008	8400		5s	$(5/2)^2[5/2]_3$	7p	$(5/2)^2[5/2]^\circ_3$			R1c		
2518.9484(5)	39687.164(8)	2518.9484(5)	−0.0001	34000		s^2	$^3F_3$	sp	$(^3F)^3P^\circ\ ^5G^\circ_4$			R1c		
2526.3277(14)	39571.246(21)	2526.3234(4)	0.0043	1900		s^2	$^3F_4$	sp	$(^3F)^3P^\circ\ ^5D^\circ_3$			R1c		
2526.5923(5)	39567.103(7)	2526.59232(10)	−0.0000	500000		4p	$^3F^\circ_3$	5s	$(5/2)^2[5/2]_3$	2.7e+07	C+	R1c	K82	
2529.30395(21)	39524.686(3)	2529.30385(10)	0.00010	940000		4p	$^1F^\circ_3$	5s	$(3/2)^2[3/2]_2$	1.06e+08	C+	F	C94	
2542.9345(8)	39312.840(12)	2542.9357(4)	−0.0012	3200		5s	$(5/2)^2[5/2]_3$	7p	$(5/2)^2[3/2]^\circ_2$			R1c		
2544.80509(20)	39283.945(3)	2544.80482(11)	0.00027	1200000		4p	$^3F^\circ_4$	5s	$(5/2)^2[5/2]_3$	1.94e+08	B	F	B00	
2553.3430(5)	39152.595(8)	2553.3430(5)	−0.0000	14000		s^2	$^3F_2$	sp	$(^3F)^3P^\circ\ ^5G^\circ_3$			R1c		
2556.3698(9)	39106.241(13)	2556.3691(3)	0.0007	2800		4d	$(5/2)^2[9/2]_5$	7f	$(5/2)^2[9/2]^\circ_5$			R1c		
2558.2130(5)	39078.067(8)	2558.2134(4)	−0.0004	5400		4d	$(5/2)^2[9/2]_5$	7f	$(5/2)^2[11/2]^\circ_6$			R1c		
2559.4301(14)	39059.485(21)	2559.4301(13)	−0.0000	2700		4d	$(5/2)^2[3/2]_2$	7f	$(5/2)^2[5/2]^\circ_2$			R2nc		
2559.7925(20)	39053.96(3)	2559.7938(18)	−0.0012	1300		4d	$(5/2)^2[3/2]_1$	8p	$(3/2)^2[1/2]^\circ_1$			R2nc		
2564.7257(6)	38978.841(9)	2564.7264(4)	−0.0007	5100		4d	$(5/2)^2[9/2]_4$	7f	$(5/2)^2[11/2]^\circ_5$			R1c		
2565.0459(7)	38973.975(10)	2565.0471(5)	−0.0011	2500		4d	$(3/2)^2[7/2]_3$	7f	$(3/2)^2[9/2]^\circ_4$			R1c		
2568.9061(20)	38915.41(3)	2568.9075(4)	−0.0014	2400		4d	$(5/2)^2[1/2]_1$	6f	$(3/2)^2[3/2]^\circ_2$			R1c		
2571.7551(6)	38872.306(9)	2571.75556(8)	−0.0004	220000		4p	$^1D^\circ_2$	5s	$(3/2)^2[3/2]_2$	1.10e+07	C+	R1c	K82	
2574.4124(7)	38832.185(11)	2574.4125(5)	−0.0001	2100		s^2	$^3F_3$	sp	$(^3F)^3P^\circ\ ^5D^\circ_2$			R1c		
2574.6371(7)	38828.796(11)	2574.6371(5)	−0.0000	4600		4d	$(3/2)^2[7/2]_4$	7f	$(3/2)^2[9/2]^\circ_5$			R1c		
2590.4012(9)	38592.515(14)	2590.3994(4)	0.0019	990		4d	$(5/2)^2[5/2]_3$	7f	$(5/2)^2[7/2]^\circ_4$			R1c		
2590.52825(17)	38590.6226(25)	2590.52826(8)	−0.00001	440000		4p	$^1D^\circ_2$	5s	$(3/2)^2[3/2]_1$	6.0e+07	C+	F	C94	L
2598.8126(6)	38467.613(9)	2598.81194(10)	0.0006	400000		4p	$^3F^\circ_2$	5s	$(5/2)^2[5/2]_2$	4.2e+07	C+	R1c	K82	L
2600.26992(14)	38446.0554(20)	2600.26973(8)	0.00018	500000		4p	$^3D^\circ_3$	5s	$(3/2)^2[3/2]_2$	1.01e+08	C+	F	C94	
2604.5253(21)	38383.24(3)	2604.5089(11)	0.0164	870	?	sp	$(^3F)^3P^\circ\ ^3F^\circ_3$	10s	$(5/2)^2[5/2]_2$			R1c		X
2606.5804(14)	38352.984(21)	2606.5819(5)	−0.0015	1700		4d	$(5/2)^2[7/2]_3$	7f	$(5/2)^2[7/2]^\circ_3$			R1c		
2606.8761(8)	38348.634(12)	2606.8759(4)	0.0001	3400		4d	$(5/2)^2[7/2]_3$	7f	$(5/2)^2[9/2]^\circ_4$			R1c		
2606.9973(10)	38346.850(15)	2606.9973(4)	0.0001	2500		4d	$(5/2)^2[7/2]_3$	7f	$(5/2)^2[7/2]^\circ_4$			R1c		
2607.041	38346.21	2607.0410(4)			:	4d	$(5/2)^2[7/2]_3$	7f	$(5/2)^2[5/2]^\circ_3$					
2609.9088(6)	38304.075(9)	2609.9094(3)	−0.0006	8400		4d	$(5/2)^2[7/2]_4$	7f	$(5/2)^2[9/2]^\circ_5$			R1c		
2610.7937(14)	38291.094(21)	2610.7940(5)	−0.0003	1600		4d	$(5/2)^2[5/2]_2$	7f	$(5/2)^2[7/2]^\circ_3$			R1c		

Table A1. *Cont.*

λ_{obs} [a] (Å)	σ_{obs} [b] (cm^{-1})	λ_{Ritz} [c] (Å)	$\Delta\lambda_{obs-Ritz}$ (Å)	I_{obs} [d] (arb. u.)	Char [e]	Lower Level		Upper Level		A (s^{-1})	Acc [f]	Line Ref. [g]	TP Ref. [g]	Notes [h]
2611.2544(8)	38284.338(11)	2611.2546(4)	−0.0002	810		4d	$(5/2)^2[5/2]_2$	7f	$(5/2)^2[5/2]^\circ_3$			R1c		
2614.4127(7)	38238.092(11)	2614.4126(7)	0.0001	15000		s^2	$^3F_4$	sp	$(^3F)^3P^\circ\ ^5D^\circ_4$			R1c		
2619.2104(5)	38168.054(8)	2619.21044(15)	−0.0000	8000		s^2	$^1D_2$	5p	$(3/2)^2[3/2]^\circ_2$			R1c		
2620.6656(5)	38146.862(7)	2620.66627(10)	−0.0007	56000		4p	$^3F^\circ_2$	5s	$(5/2)^2[5/2]_3$	1.3e+06	E	R1c	K82cal	
2636.6187(6)	37916.065(8)	2636.61965(16)	−0.0010	2200		s^2	$^1D_2$	5p	$(3/2)^2[3/2]^\circ_1$			R1c		
2648.6056(7)	37744.477(9)	2648.6059(4)	−0.0003	3000		s^2	$^3F_3$	sp	$(^3F)^3P^\circ\ ^5D^\circ_3$			R1c		
2655.9644(7)	37639.906(9)	2655.9642(5)	0.0002	840		s^2	$^3F_2$	sp	$(^3F)^3P^\circ\ ^5D^\circ_2$			R1c		
2666.2906(5)	37494.140(8)	2666.29047(11)	0.0001	230000		4p	$^1F^\circ_3$	5s	$(5/2)^2[5/2]_2$	2.0e+07	C+	R1c	K82	
2667.4229(15)	37478.224(21)	2667.4250(4)	−0.0021	990		5s	$(3/2)^2[3/2]_1$	7p	$(5/2)^2[3/2]^\circ_1$			R1c		
2676.0660(6)	37357.184(8)	2676.06654(15)	−0.0005	3100		s^2	$^1D_2$	5p	$(3/2)^2[5/2]^\circ_2$			R1c		
2681.4966(15)	37281.533(21)	2681.4957(7)	0.0008	2100		5s	$(3/2)^2[3/2]_2$	7p	$(5/2)^2[5/2]^\circ_2$			R1c		
2682.7484(8)	37264.137(11)	2682.75015(9)	−0.0017	440		4p	$^3D^\circ_1$	5s	$(3/2)^2[3/2]_2$	9.e+05	E	R1c	K82cal	
2689.2993(5)	37173.370(7)	2689.29932(11)	0.0000	410000		4p	$^1F^\circ_3$	5s	$(5/2)^2[5/2]_3$	5.9e+07	C+	R1c	C94	
2692.4978(15)	37129.213(21)	2692.4942(4)	0.0036	3200		4d	$(5/2)^2[7/2]_3$	6f	$(3/2)^2[9/2]^\circ_4$			R1c		
2700.96242(17)	37012.8599(24)	2700.96252(9)	−0.00011	520000		4p	$^3D^\circ_2$	5s	$(3/2)^2[3/2]_2$	1.02e+08	C+	F	C94	
2703.18424(13)	36982.4398(18)	2703.18448(8)	−0.00024	540000		4p	$^3D^\circ_1$	5s	$(3/2)^2[3/2]_1$	1.17e+08	C+	F	C94	
2704.5194(23)	36964.18(3)	2704.5157(16)	0.0037	1100		s^2	$^3P_1$	sp	$(^3P)^3P^\circ\ ^5P^\circ_1$			R1c		
2708.2695(8)	36913.003(11)	2708.2695(3)	−0.0000	7200		4d	$(5/2)^2[1/2]_1$	8p	$(5/2)^2[3/2]^\circ_1$			R1c		
2709.7592(6)	36892.711(8)	2709.7597(3)	−0.0006	1600		4d	$(5/2)^2[1/2]_1$	6f	$(5/2)^2[5/2]^\circ_2$			R1c		
2710.2454(16)	36886.093(21)	2710.2434(6)	0.0020	2700		s^2	$^3P_2$	sp	$(^3P)^3P^\circ\ ^5P^\circ_2$			R1c		
2710.6069(16)	36881.173(21)	2710.6071(7)	−0.0001	700		4d	$(5/2)^2[1/2]_0$	6f	$(3/2)^2[3/2]^\circ_1$			R1c		
2711.5767(23)	36867.98(3)	2711.5711(13)	0.0057	350		s^2	$^3P_2$	sp	$(^3P)^3P^\circ\ ^5P^\circ_3$			R2nc		
2711.8649(6)	36864.066(8)	2711.8639(3)	0.0010	18000		4d	$(5/2)^2[1/2]_1$	6f	$(5/2)^2[3/2]^\circ_2$			R1c		
2713.5078(5)	36841.748(7)	2713.50740(10)	0.0004	400000		4p	$^1D^\circ_2$	5s	$(5/2)^2[5/2]_2$	8.9e+07	C+	R1c	C94	
2715.4038(6)	36816.024(8)	2715.4039(6)	−0.0000	10000		4d	$(5/2)^2[1/2]_1$	6f	$(5/2)^2[1/2]^\circ_0$			R1c		
2718.7770(3)	36770.349(4)	2718.77770(14)	−0.0007	440000		4p	$^1P^\circ_1$	5s	$(3/2)^2[3/2]_2$	8.0e+07	C+	F	C94	
2721.6764(2)	36731.179(3)	2721.67628(9)	0.0002	300000		4p	$^3D^\circ_2$	5s	$(3/2)^2[3/2]_1$	3.9e+07	C+	F	K82	L
2727.6947(6)	36650.142(8)	2727.6948(5)	−0.0001	620		4d	$(5/2)^2[1/2]_1$	6f	$(5/2)^2[1/2]^\circ_1$			R2nc		
2728.204(7)	36643.30(10)	2728.2019(16)	0.002	310		s^2	$^3P_0$	sp	$(^3P)^3P^\circ\ ^5P^\circ_1$			R1c		
2731.9477(6)	36593.089(7)	2731.94820(16)	−0.0005	6300		s^2	$^1D_2$	sp	$(^3F)^3P^\circ\ ^1D^\circ_2$			R1c		
2737.3415(6)	36520.989(7)	2737.34194(10)	−0.0005	48000		4p	$^1D^\circ_2$	5s	$(5/2)^2[5/2]_3$	1.9e+06	D	R1c	C94	
2739.7662(6)	36488.669(8)	2739.76664(14)	−0.0004	86000		4p	$^1P^\circ_1$	5s	$(3/2)^2[3/2]_1$	4.0e+06	D	R1c	C94	
2745.2710(6)	36415.506(7)	2745.27056(10)	0.0005	140000		4p	$^3D^\circ_3$	5s	$(5/2)^2[5/2]_2$	1.40e+07	C+	R1c	C94	
2757.3283(7)	36256.276(9)	2757.3290(4)	−0.0007	770		4d	$(3/2)^2[1/2]_1$	6f	$(3/2)^2[5/2]^\circ_2$			R1c		
2759.6070(8)	36226.340(11)	2759.6065(7)	0.0005	760		4d	$(3/2)^2[1/2]_1$	6f	$(3/2)^2[3/2]^\circ_1$			R1c		
2762.481(20)	36188.6(3)	2762.4581(4)	0.023	1000	*	4d	$(3/2)^2[5/2]_3$	7f	$(5/2)^2[7/2]^\circ_4$			R1c		
2762.481(20)	36188.6(3)	2762.5072(5)	−0.026	1000	*	4d	$(3/2)^2[5/2]_3$	7f	$(5/2)^2[5/2]^\circ_3$			R1c		
2769.6690(6)	36094.738(7)	2769.66882(9)	0.0002	310000		4p	$^3D^\circ_3$	5s	$(5/2)^2[5/2]_3$	6.7e+07	C+	R1c	C94	
2788.2614(7)	35854.067(9)	2788.2619(3)	−0.0005	5700		4d	$(5/2)^2[9/2]_5$	6f	$(5/2)^2[9/2]^\circ_5$			R1c		
2789.2226(17)	35841.712(21)	2789.2218(12)	0.0009	210		5p	$(5/2)^2[7/2]^\circ_3$	10s	$(5/2)^2[5/2]_2$			R1c		
2791.7945(6)	35808.695(8)	2791.7947(4)	−0.0002	54000		4d	$(5/2)^2[9/2]_5$	6f	$(5/2)^2[11/2]^\circ_6$			R1c		
2792.224(11)	35803.19(14)	2792.2125(3)	0.012	420	*	4d	$(5/2)^2[9/2]_5$	6f	$(5/2)^2[11/2]^\circ_5$			R1c		
2792.224(11)	35803.19(14)	2792.2312(4)	−0.007	420	*	4d	$(5/2)^2[3/2]_2$	6f	$(5/2)^2[7/2]^\circ_3$			R1c		

Table A1. *Cont.*

λ_{obs} [a] (Å)	σ_{obs} [b] (cm^{-1})	λ_{Ritz} [c] (Å)	$\Delta\lambda_{obs-Ritz}$ (Å)	I_{obs} [d] (arb. u.)	Char [e]	Lower Level		Upper Level		A (s^{-1})	Acc [f]	Line Ref. [g]	TP Ref. [g]	Notes [h]
2793.6079(7)	35785.452(9)	2793.6091(3)	−0.0012	620		4d	$(5/2)^2[3/2]_2$	8p	$(5/2)^2[3/2]^{\circ}_1$			R1c		
2795.2979(6)	35763.818(8)	2795.2980(3)	−0.0001	8000		4d	$(5/2)^2[3/2]_2$	6f	$(5/2)^2[5/2]^{\circ}_3$			R1c		
2795.6571(7)	35759.223(9)	2795.6567(3)	0.0004	4600		4d	$(5/2)^2[9/2]_4$	6f	$(5/2)^2[9/2]^{\circ}_4$			R1c		
2795.8729(11)	35756.463(14)	2795.8732(3)	−0.0003	610		4d	$(5/2)^2[9/2]_4$	6f	$(5/2)^2[7/2]^{\circ}_4$			R1c		
2796.2625(7)	35751.482(9)	2796.2624(5)	0.0001	820		4d	$(3/2)^2[7/2]_3$	6f	$(3/2)^2[7/2]^{\circ}_3$			R1c		
2797.2549(6)	35738.798(8)	2797.2557(3)	−0.0008	7500		4d	$(5/2)^2[3/2]_1$	6f	$(5/2)^2[5/2]^{\circ}_2$			R1c		
2797.4336(6)	35736.516(8)	2797.4337(3)	−0.0001	8300		4d	$(5/2)^2[3/2]_2$	6f	$(5/2)^2[3/2]^{\circ}_2$			R1c		
2797.5493(12)	35735.038(16)	2797.5494(5)	−0.0001	410		4d	$(5/2)^2[3/2]_2$	6f	$(5/2)^2[3/2]^{\circ}_1$			R1c		
2798.8296(17)	35718.692(21)			200		5p	$(5/2)^2[7/2]^{\circ}_4$	10s	$(5/2)^2[5/2]_3$			R1c		
2799.5280(6)	35709.781(8)	2799.5291(3)	−0.0010	35000		4d	$(5/2)^2[9/2]_4$	6f	$(5/2)^2[11/2]^{\circ}_5$			R1c		
2799.6805(6)	35707.837(8)	2799.6812(4)	−0.0007	10000		4d	$(3/2)^2[7/2]_3$	6f	$(3/2)^2[9/2]^{\circ}_4$			R1c		
2801.0500(6)	35690.379(8)	2801.05008(16)	−0.0001	1000		s^2	$^1D_2$	5p	$(5/2)^2[5/2]^{\circ}_3$			R1c		
2801.3182(24)	35686.96(3)	2801.3220(17)	−0.0038	400		4d	$(3/2)^2[3/2]_2$	sp	$^3D^{\circ}_3$			R2nc		
2803.2711(17)	35662.102(21)	2803.2706(6)	0.0005	390		4d	$(5/2)^2[3/2]_1$	6f	$(5/2)^2[1/2]^{\circ}_0$			R1c		
2804.1888(17)	35650.432(21)	2804.1896(12)	−0.0008	200		5p	$(5/2)^2[5/2]^{\circ}_2$	10s	$(5/2)^2[5/2]_2$			R1c		
2807.1550(8)	35612.763(10)	2807.1556(5)	−0.0006	770		4d	$(3/2)^2[7/2]_4$	6f	$(3/2)^2[7/2]^{\circ}_4$			R1c		
2810.3655(17)	35572.082(21)	2810.3657(4)	−0.0002	190		4d	$(3/2)^2[7/2]_4$	6f	$(3/2)^2[9/2]^{\circ}_4$			R1c		
2810.8038(6)	35566.536(7)	2810.8039(4)	−0.0001	14000		4d	$(3/2)^2[7/2]_4$	6f	$(3/2)^2[9/2]^{\circ}_5$			R1c		
2813.6315(7)	35530.793(9)	2813.63100(16)	0.0005	2900		s^2	$^1D_2$	5p	$(5/2)^2[3/2]^{\circ}_1$			R1c		
2816.1978(9)	35498.417(11)	2816.1982(5)	−0.0004	3700		4d	$(3/2)^2[3/2]_1$	6f	$(3/2)^2[3/2]^{\circ}_2$			R1c		
2817.0843(7)	35487.247(9)	2817.08465(18)	−0.0004	370		s^2	$^1D_2$	5p	$(5/2)^2[5/2]^{\circ}_2$			R1c		
2828.6968(7)	35341.570(9)	2828.6970(3)	−0.0002	3500		4d	$(5/2)^2[5/2]_3$	6f	$(5/2)^2[9/2]^{\circ}_4$			R1c		
2828.9184(12)	35338.802(15)	2828.9187(3)	−0.0003	3500		4d	$(5/2)^2[5/2]_3$	6f	$(5/2)^2[7/2]^{\circ}_4$			R1c		
2830.2314(8)	35322.408(10)	2830.2323(3)	−0.0008	6100		4d	$(5/2)^2[5/2]_3$	6f	$(5/2)^2[5/2]^{\circ}_3$			R1c		
2832.4214(7)	35295.099(9)	2832.4217(3)	−0.0004	3400		4d	$(5/2)^2[5/2]_3$	6f	$(5/2)^2[3/2]^{\circ}_2$			R1c		
2833.053(2)	35287.23(3)	2833.0250(5)	0.028	170	?	sp	$(^3F)^3P^{\circ}\ ^3D^{\circ}_3$	7d	$(5/2)^2[3/2]_2$			R1c		X
2834.9699(17)	35263.372(21)	2834.9704(5)	−0.0005	510		4d	$(3/2)^2[3/2]_2$	6f	$(3/2)^2[3/2]^{\circ}_2$			R1c		
2836.2898(8)	35246.962(10)	2836.2911(5)	−0.0013	4600		4d	$(3/2)^2[3/2]_2$	6f	$(3/2)^2[5/2]^{\circ}_3$			R1c		
2836.6971(17)	35241.902(21)	2836.7201(6)	−0.0230	840	?	sp	$(^3F)^3P^{\circ}\ ^3F^{\circ}_2$	9s	$(5/2)^2[5/2]_2$			R1c		X
2837.3682(6)	35233.566(7)	2837.36814(10)	0.0001	150000		4p	$^3D^{\circ}_1$	5s	$(5/2)^2[5/2]_2$	2.3e+07	C+	R1c	C94	
2840.4918(6)	35194.823(8)	2840.49162(17)	0.0002	5800		s^2	$^3P_2$	5p	$(3/2)^2[3/2]^{\circ}_2$			R1c		
2846.8683(7)	35115.996(8)	2846.8693(5)	−0.0009	4300		4d	$(5/2)^2[7/2]_3$	6f	$(5/2)^2[7/2]^{\circ}_3$			R1c		
2848.4999(6)	35095.883(8)	2848.5005(3)	−0.0005	11000		4d	$(5/2)^2[7/2]_3$	6f	$(5/2)^2[9/2]^{\circ}_4$			R1c		
2848.7252(6)	35093.108(8)	2848.7252(3)	−0.0000	9100		4d	$(5/2)^2[7/2]_3$	6f	$(5/2)^2[7/2]^{\circ}_4$			R1c		
2849.9526(10)	35077.995(12)	2849.9500(3)	0.0026	160		4d	$(5/2)^2[7/2]_3$	6f	$(5/2)^2[5/2]^{\circ}_2$			R1c		
2851.8949(8)	35054.106(10)	2851.8944(5)	0.0005	4700		4d	$(5/2)^2[5/2]_2$	6f	$(5/2)^2[7/2]^{\circ}_3$			R1c		
2852.0764(7)	35051.875(8)	2852.0764(3)	0.0001	12000		4d	$(5/2)^2[7/2]_4$	6f	$(5/2)^2[9/2]^{\circ}_5$			R1c		
2852.1785(11)	35050.621(13)	2852.1793(3)	−0.0008	6300		4d	$(5/2)^2[7/2]_4$	6f	$(5/2)^2[9/2]^{\circ}_4$			R1c		
2852.4042(6)	35047.847(8)	2852.4046(3)	−0.0004	3900		4d	$(5/2)^2[7/2]_4$	6f	$(5/2)^2[7/2]^{\circ}_4$			R1c		
2853.7403(7)	35031.439(8)	2853.7401(3)	0.0002	630		4d	$(5/2)^2[7/2]_4$	6f	$(5/2)^2[5/2]^{\circ}_3$			R1c		
2854.9858(7)	35016.157(9)	2854.9860(3)	−0.0002	2100		4d	$(5/2)^2[5/2]_2$	6f	$(5/2)^2[5/2]^{\circ}_2$			R1c		

Table A1. *Cont.*

λ_{obs} [a] (Å)	σ_{obs} [b] (cm^{-1})	λ_{Ritz} [c] (Å)	$\Delta\lambda_{obs-Ritz}$ (Å)	I_{obs} [d] (arb. u.)	Char [e]	Lower Level	Upper Level	A (s^{-1})	Acc [f]	Line Ref. [g]	TP Ref. [g]	Notes [h]
		2855.0785(5)			m	4d $(3/2)^2[5/2]_3$	6f $(3/2)^2[7/2]^{\circ}_3$			R1nc		
2855.0930(17)	35014.842(21)	2855.0937(3)	−0.0007	310		4d $(5/2)^2[5/2]_2$	6f $(5/2)^2[5/2]^{\circ}_3$			R1c		
2855.3206(7)	35012.052(8)	2855.3215(5)	−0.0009	4700		4d $(3/2)^2[5/2]_3$	6f $(3/2)^2[7/2]^{\circ}_4$			R1c		
2856.2109(7)	35001.138(8)	2856.2100(4)	0.0010	310		4d $(5/2)^2[7/2]_4$	6f $(5/2)^2[11/2]^{\circ}_5$			R1c		
2857.4418(17)	34986.062(21)	2857.4425(5)	−0.0007	310		4d $(5/2)^2[5/2]_2$	6f $(5/2)^2[3/2]^{\circ}_1$			R1c		
2857.7484(6)	34982.309(8)	2857.74808(11)	0.0003	19000		4p $^3D^{\circ}_2$	5s $(5/2)^2[5/2]_2$	5.e+05	E	R1c	K82cal	
2859.0051(7)	34966.932(8)	2859.0054(5)	−0.0003	5300		4d $(3/2)^2[5/2]_2$	6f $(3/2)^2[7/2]^{\circ}_3$			R1c		
2859.9192(7)	34955.757(9)	2859.9187(4)	0.0005	2300		4d $(3/2)^2[5/2]_2$	6f $(3/2)^2[5/2]^{\circ}_2$			R1c		
2860.2493(9)	34951.723(11)	2860.24866(19)	0.0006	1100		s^2 $^3P_1$	5p $(3/2)^2[3/2]^{\circ}_2$			R1c		
2862.3233(7)	34926.399(8)	2862.3233(5)	−0.0000	3000		4d $(3/2)^2[5/2]_3$	6f $(3/2)^2[5/2]^{\circ}_3$			R1c		
2866.2706(17)	34878.302(21)	2866.2702(5)	0.0004	740		4d $(3/2)^2[5/2]_2$	6f $(3/2)^2[5/2]^{\circ}_3$			R1c		
2868.7914(8)	34847.656(9)	2868.7906(4)	0.0008	440		4d $(5/2)^2[1/2]_0$	8p $(5/2)^2[3/2]^{\circ}_1$			R1c		
2872.9461(7)	34797.263(8)	2872.9460(5)	0.0001	580		4d $(5/2)^2[1/2]_0$	6f $(5/2)^2[3/2]^{\circ}_1$			R1c		
2875.3341(7)	34768.365(9)	2875.3346(5)	−0.0005	570		5p $(5/2)^2[3/2]^{\circ}_2$	8d $(5/2)^2[5/2]_3$			R1c		
2877.6996(6)	34739.786(7)	2877.69898(16)	0.0007	170000		4p $^1P^{\circ}_1$	5s $(5/2)^2[5/2]_2$	2.3e+07	C+	R1c	C94	
2880.7003(7)	34703.600(9)	2880.69947(17)	0.0009	2000		s^2 $^1D_2$	5p $(5/2)^2[3/2]^{\circ}_2$			R1c		
2881.0216(8)	34699.730(9)	2881.02191(20)	−0.0003	1000		s^2 $^3P_1$	5p $(3/2)^2[3/2]^{\circ}_1$			R1c		
2883.1886(18)	34673.651(21)			2800		5p $(5/2)^2[3/2]^{\circ}_2$	8d $(5/2)^2[1/2]_1$			R1c		
2884.1954(6)	34661.548(8)	2884.19598(11)	−0.0006	41000		4p $^3D^{\circ}_2$	5s $(5/2)^2[5/2]_3$	1.20e+07	C+	R1c	C94	
2884.7560(7)	34654.812(9)	2884.75571(16)	0.0003	280		$s^{23}P$ $_2$	5p $(3/2)^2[5/2]^{\circ}_3$			R1c		
2897.2206(7)	34505.726(8)	2897.21944(17)	0.0011	1200		s^2 $^3P_2$	5p $(3/2)^2[1/2]^{\circ}_1$			R1c		
2907.9162(8)	34378.816(10)	2907.9158(3)	0.0004	1300		s^2 $^3P_0$	5p $(3/2)^2[3/2]^{\circ}_1$			R1c		
2908.863(3)	34367.63(3)	2908.8746(4)	−0.012	130		sp $(^3F)^3P^{\circ}$ $^3D^{\circ}_3$	8s $(5/2)^2[5/2]_3$			R1c		X
2912.4526(18)	34325.271(21)			250		5p $(5/2)^2[3/2]^{\circ}_1$	8d $(5/2)^2[1/2]_0$			R1c		
2917.7762(9)	34262.646(11)	2917.77629(20)	−0.0001	620		s^2 $^3P_1$	5p $(3/2)^2[1/2]^{\circ}_1$			R1c		
2923.2569(18)	34198.411(21)	2923.2586(8)	−0.0017	240		5p $(5/2)^2[7/2]^{\circ}_3$	8d $(5/2)^2[7/2]_3$			R1c		
2926.2499(7)	34163.434(8)	2926.2501(4)	−0.0002	240		5p $(5/2)^2[3/2]^{\circ}_2$	9s $(5/2)^2[5/2]_3$			R1c		
2927.2539(7)	34151.717(9)	2927.2535(5)	0.0003	2900		5p $(5/2)^2[7/2]^{\circ}_3$	8d $(5/2)^2[9/2]_4$			R1c		
2928.1916(18)	34140.781(21)	2928.18542(19)	0.0062	240		s^2 $^3P_1$	5p $(3/2)^2[5/2]^{\circ}_2$			R1c		
2931.789(3)	34098.89(4)	2931.7851(4)	0.004	230		sp $(^3F)^3P^{\circ}$ $^3G^{\circ}_3$	8s $(5/2)^2[5/2]_3$			R1c		
2932.2253(7)	34093.818(9)	2932.2254(6)	−0.0002	1200		5p $(5/2)^2[7/2]^{\circ}_4$	8d $(5/2)^2[7/2]_4$			R1c		
2933.565(6)	34078.25(7)	2933.5614(6)	0.003	230		sp $(^3F)^3P^{\circ}$ $^3F^{\circ}_4$	7d $(5/2)^2[7/2]_4$			R1c		
2936.9553(7)	34038.912(8)	2936.9555(7)	−0.0002	2900		5p $(5/2)^2[7/2]^{\circ}_4$	8d $(5/2)^2[9/2]_5$			R1c		
2939.7041(9)	34007.084(10)	2939.7036(8)	0.0005	1100		5p $(5/2)^2[5/2]^{\circ}_2$	8d $(5/2)^2[7/2]_3$			R1c		
2941.2137(18)	33989.631(21)	2941.2085(6)	0.0052	1100		sp $(^3F)^3P^{\circ}$ $^3F^{\circ}_4$	7d $(5/2)^2[9/2]_5$			R1c		
2942.623(5)	33973.35(6)	2942.6237(18)	−0.001	230		5p $(5/2)^2[3/2]^{\circ}_1$	8d $(5/2)^2[5/2]_2$			R1c		
2945.370(8)	33941.66(9)	2945.3640(3)	0.007	6100		s^2 $^3P_0$	5p $(3/2)^2[1/2]^{\circ}_1$			S36c		
2947.3037(11)	33919.401(12)	2947.3037(11)	0.0000	670		5p $(5/2)^2[3/2]^{\circ}_1$	8d $(5/2)^2[3/2]_1$			R1c		
2949.3463(18)	33895.911(21)	2949.3453(10)	0.0010	550		5p $(3/2)^2[5/2]^{\circ}_3$	9d $(5/2)^2[9/2]_4$			R1c		
2957.3211(19)	33804.511(21)	2957.3242(6)	−0.0032	1100		5p $(5/2)^2[5/2]^{\circ}_3$	8d $(5/2)^2[7/2]_4$			R1c		
2959.3285(19)	33781.581(21)	2959.3276(5)	0.0009	740		5p $(5/2)^2[5/2]^{\circ}_3$	8d $(5/2)^2[5/2]_3$			R1c		

Table A1. *Cont.*

λ_{obs} [a] (Å)	σ_{obs} [b] (cm^{-1})	λ_{Ritz} [c] (Å)	$\Delta\lambda_{obs-Ritz}$ (Å)	I_{obs} [d] (arb. u.)	Char [e]	Lower Level		Upper Level		A (s^{-1})	Acc [f]	Line Ref. [g]	TP Ref. [g]	Notes [h]
2968.744(3)	33674.44(3)	2968.7445(4)	−0.000	200		4d	$(3/2)^2[7/2]_3$	6f	$(5/2)^2[9/2]^\circ_4$			R1c		
2973.6495(7)	33618.897(8)	2973.64712(23)	0.0024	1300		s^2	$^3P_1$	5p	$(3/2)^2[1/2]^\circ_0$			R1c		
2975.271(5)	33600.57(6)	2975.2681(3)	0.003	300		sp	$(^3F)^3P^\circ\ ^3F^\circ_3$	7d	$(5/2)^2[7/2]_3$			R1c		
2975.6294(8)	33596.529(9)	2975.6286(7)	0.0008	1200		5p	$(5/2)^2[7/2]^\circ_3$	9s	$(5/2)^2[5/2]_2$			R1c		
2977.0081(19)	33580.971(21)	2977.0087(16)	−0.0006	1100		5p	$(3/2)^2[5/2]^\circ_2$	9s	$(3/2)^2[3/2]_1$			R1c		
2981.7867(12)	33527.156(14)	2981.7860(4)	0.0007	1000		5s	$(5/2)^2[5/2]_3$	6p	$(3/2)^2[3/2]^\circ_2$			R1c		
2981.945(3)	33525.38(3)	2981.9449(5)	−0.000	1400		sp	$(^3F)^3P^\circ\ ^3F^\circ_3$	7d	$(5/2)^2[9/2]_4$			R1c		
2983.7677(10)	33504.898(11)	2983.7658(3)	0.0018	1100		4d	$(5/2)^2[1/2]_1$	5f	$(3/2)^2[3/2]^\circ_1$			R1c		
2986.3345(8)	33476.101(9)	2986.3327(3)	0.0017	5700		4d	$(5/2)^2[1/2]_1$	5f	$(3/2)^2[3/2]^\circ_2$			R1c		
2987.2351(7)	33466.009(8)	2987.2353(5)	−0.0002	1400		5p	$(5/2)^2[7/2]^\circ_4$	9s	$(5/2)^2[5/2]_3$			R1c		
2992.6695(19)	33405.241(21)	2992.6698(7)	−0.0003	460		5p	$(5/2)^2[5/2]^\circ_2$	9s	$(5/2)^2[5/2]_2$			R1c		
2993.024(5)	33401.28(5)			460		5p	$(3/2)^2[3/2]^\circ_2$	8d	$(3/2)^2[5/2]_2$			R1c		
2993.2667(8)	33398.576(9)	2993.2675(4)	−0.0008	920		5s	$(5/2)^2[5/2]_2$	6p	$(3/2)^2[3/2]^\circ_1$			R1c		
2996.8412(19)	33358.741(21)	2996.8382(6)	0.0030	720		sp	$(^3F)^3P^\circ\ ^1F^\circ_3$	8d	$(5/2)^2[7/2]_4$			R1c		
2998.893(5)	33335.92(6)	2998.8956(6)	−0.003	180		sp	$(^3F)^3P^\circ\ ^1F^\circ_3$	8d	$(5/2)^2[5/2]_3$			R1c		
2999.8710(19)	33325.051(21)	2999.8701(8)	0.0009	530		5p	$(3/2)^2[5/2]^\circ_3$	9s	$(3/2)^2[3/2]_2$			R1c		
3004.058(5)	33278.60(6)	3004.0555(5)	0.003	260		4d	$(3/2)^2[3/2]_2$	6f	$(5/2)^2[7/2]^\circ_3$			R1c		
3010.5921(19)	33206.381(21)	3010.5910(4)	0.0011	1700		5s	$(5/2)^2[5/2]_2$	6p	$(3/2)^2[3/2]^\circ_2$			R1c		
3012.2768(19)	33187.811(21)	3012.2786(5)	−0.0018	170		5s	$(5/2)^2[5/2]_3$	6p	$(3/2)^2[5/2]^\circ_3$			R1c		
3013.2896(11)	33176.656(12)	3013.2888(5)	0.0008	170		5p	$(5/2)^2[5/2]^\circ_3$	9s	$(5/2)^2[5/2]_3$			R1c		
3014.5445(8)	33162.846(9)	3014.54413(18)	0.0004	6200		s^2	$^3P_2$	sp	$(^3F)^3P^\circ\ ^1F^\circ_3$			R1c		
3024.1816(19)	33057.171(21)	3024.1822(19)	−0.0005	230		5s	$(3/2)^2[3/2]_2$	sp	$(^1G)^3P^\circ\ ^3G^\circ_3$			R2nc		
3027.388(6)	33022.16(6)	3027.3945(17)	−0.006	76		5p	$(3/2)^2[3/2]^\circ_1$	9s	$(3/2)^2[3/2]_1$			R1c		
3037.6082(20)	32911.061(21)	3037.6081(19)	0.0001	140		sp	$(^3F)^3P^\circ\ ^1D^\circ_2$	8d	$(5/2)^2[5/2]_2$			R1c		
3037.801(3)	32908.97(3)	3037.8031(6)	−0.002	710		5s	$(5/2)^2[5/2]_2$	6p	$(3/2)^2[5/2]^\circ_2$			R1c		
3041.6787(8)	32867.019(9)	3041.6786(5)	0.0002	4500		5s	$(5/2)^2[5/2]_2$	6p	$(3/2)^2[5/2]^\circ_3$			R1c		
3042.8556(8)	32854.308(9)	3042.8551(5)	0.0005	340		5p	$(5/2)^2[3/2]^\circ_1$	7d	$(3/2)^2[3/2]_2$			R1c		
3049.2831(10)	32785.058(11)	3049.2836(8)	−0.0005	130		5p	$(3/2)^2[3/2]^\circ_2$	9s	$(3/2)^2[3/2]_2$			R1c		
3051.9472(20)	32756.441(21)	3051.9509(7)	−0.0038	130		sp	$(^3F)^3P^\circ\ ^1F^\circ_3$	9s	$(5/2)^2[5/2]_2$			R1c		
3052.1596(20)	32754.161(21)	3052.15798(24)	0.0016	130		s^2	$^1D_2$	sp	$(^3F)^3P^\circ\ ^3F^\circ_3$			R1c		
3053.5737(15)	32738.993(16)	3053.5735(6)	0.0003	190		5p	$(5/2)^2[3/2]^\circ_1$	7d	$(3/2)^2[1/2]_1$			R1c		
3055.6134(8)	32717.140(9)	3055.61251(18)	0.0009	500		s^2	$^3P_2$	5p	$(5/2)^2[5/2]^\circ_3$			R1c		
3056.8491(9)	32703.914(9)	3056.8484(3)	0.0007	310		sp	$(^3F)^3P^\circ\ ^3F^\circ_2$	7d	$(5/2)^2[7/2]_3$			R1c		
3059.8637(9)	32671.695(10)	3059.8632(5)	0.0005	420		5p	$(5/2)^2[5/2]^\circ_3$	7d	$(3/2)^2[7/2]_4$			R1c		
3062.2814(8)	32645.902(8)	3062.2814(5)	−0.0000	590		5s	$(5/2)^2[5/2]_2$	6p	$(3/2)^2[1/2]^\circ_1$			R1c		
3066.6018(12)	32599.910(13)	3066.5985(3)	0.0033	400		sp	$(^3F)^3P^\circ\ ^3F^\circ_3$	8s	$(5/2)^2[5/2]_3$			R1c		
3080.3202(20)	32454.730(21)	3080.3284(6)	−0.0082	400		sp	$(^3F)^3P^\circ\ ^5F^\circ_2$	7s	$(3/2)^2[3/2]_2$			R1c		
3081.4480(14)	32442.852(15)	3081.4527(3)	−0.0046	150		4d	$(5/2)^2[9/2]_4$	5f	$(3/2)^2[7/2]^\circ_4$			R1c		
3082.935(3)	32427.21(3)	3082.9348(3)	−0.000	97		4d	$(5/2)^2[3/2]_2$	5f	$(3/2)^2[5/2]^\circ_2$			R1c		
3083.3677(8)	32422.654(8)	3083.3669(3)	0.0009	970		4d	$(5/2)^2[3/2]_2$	5f	$(3/2)^2[5/2]^\circ_3$			R1c		
3085.4342(20)	32400.940(21)	3085.4421(3)	−0.0079	95		4d	$(5/2)^2[3/2]_1$	5f	$(3/2)^2[5/2]^\circ_2$			R1c		
3085.5818(20)	32399.390(21)	3085.6280(3)	−0.0462	94	?	4d	$(5/2)^2[9/2]_4$	5f	$(3/2)^2[5/2]^\circ_3$			R1c		X

Table A1. *Cont.*

λ_{obs} [a] (Å)	σ_{obs} [b] (cm^{-1})	λ_{Ritz} [c] (Å)	$\Delta\lambda_{obs-Ritz}$ (Å)	I_{obs} [d] (arb. u.)	Char [e]	Lower Level		Upper Level		A (s^{-1})	Acc [f]	Line Ref. [g]	TP Ref. [g]	Notes [h]
3088.7489(7)	32366.170(8)	3088.74858(24)	0.0003	550		4d	$(5/2)^2[9/2]_4$	5f	$(3/2)^2[9/2]°_5$			R1c		
3097.8651(13)	32270.929(13)	3097.86615(23)	−0.0011	410		s^2	$^3P_1$	5p	$(5/2)^2[5/2]°_2$			R1c		
3110.4745(10)	32140.112(10)	3110.4732(4)	0.0013	70		sp	$(^3F)^3P°\ ^3D°_3$	6d	$(3/2)^2[7/2]_4$			R1c		
3121.3959(8)	32027.662(8)	3121.3961(5)	−0.0002	150		4d	$(5/2)^2[9/2]_5$	7p	$(5/2)^2[7/2]°_4$			R1c		
3121.6428(7)	32025.129(7)	3121.6415(3)	0.0013	240		4d	$(5/2)^2[5/2]_3$	5f	$(3/2)^2[7/2]°_4$			R1c		
3121.8708(21)	32022.790(21)	3121.8736(3)	−0.0027	90		4d	$(5/2)^2[5/2]_3$	5f	$(3/2)^2[7/2]°_3$			R1c		
3124.7229(8)	31993.563(8)	3124.7222(3)	0.0006	440		s^2	$^3P_0$	5p	$(5/2)^2[3/2]°_1$			R1c		
3139.7885(9)	31840.054(9)	3139.7857(6)	0.0028	610	bl	sp	$(^3F)^3P°\ ^1D°_2$	7d	$(3/2)^2[5/2]_2$			R1c		
3140.4073(9)	31833.781(9)	3140.4064(6)	0.0008	230		sp	$(^3F)^3P°\ ^1D°_2$	7d	$(3/2)^2[5/2]_3$			R1c		
3146.0122(8)	31777.068(8)	3146.0119(3)	0.0003	480		4d	$(5/2)^2[7/2]_3$	5f	$(3/2)^2[7/2]°_3$			R1c		
3148.7869(8)	31749.067(8)	3148.7856(5)	0.0013	170		sp	$(^3F)^3P°\ ^1D°_2$	7d	$(3/2)^2[7/2]_3$			R1c		
3149.6815(21)	31740.050(21)	3149.6769(3)	0.0046	260	bl	4d	$(5/2)^2[7/2]_3$	5f	$(3/2)^2[5/2]°_2$			R1c		
3150.2634(9)	31734.187(9)	3150.2636(3)	−0.0002	88		4d	$(5/2)^2[7/2]_4$	5f	$(3/2)^2[7/2]°_4$			R1c		
3150.540(8)	31731.40(8)	3150.4999(3)	0.040	4600	?	4d	$(5/2)^2[7/2]_4$	5f	$(3/2)^2[7/2]°_3$			S36c		X
3150.6422(9)	31730.372(9)	3150.64124(19)	0.0010	3000		s^2	$^3P_2$	5p	$(5/2)^2[3/2]°_2$			R1c		
3151.0505(7)	31726.261(7)	3151.0506(3)	−0.0001	10000		5s	$(5/2)^2[5/2]_3$	6p	$(5/2)^2[5/2]°_3$			R1c		
3152.9000(7)	31707.651(7)	3152.8992(3)	0.0008	3500		4d	$(5/2)^2[7/2]_3$	5f	$(3/2)^2[9/2]°_4$			R1c		
3154.0935(8)	31695.653(8)	3154.0934(4)	0.0001	550		5s	$(5/2)^2[5/2]_3$	sp	$(^3F)^1P°\ ^3F°_2$			R1c		
3155.3760(9)	31682.771(9)	3155.3765(5)	−0.0005	380		4d	$(5/2)^2[5/2]_3$	7p	$(5/2)^2[5/2]°_3$			R1c		
3155.8323(21)	31678.190(21)	3155.8290(3)	0.0033	190		4d	$(5/2)^2[5/2]_2$	5f	$(3/2)^2[5/2]°_2$			R1c		
3156.2820(7)	31673.677(7)	3156.2817(3)	0.0003	1900		4d	$(5/2)^2[5/2]_2$	5f	$(3/2)^2[5/2]°_3$			R1c		
3157.8901(8)	31657.548(8)	3157.88937(23)	0.0007	1900		4d	$(5/2)^2[7/2]_4$	5f	$(3/2)^2[9/2]°_5$			R1c		
3158.6729(8)	31649.703(8)	3158.6734(5)	−0.0005	3700		5s	$(3/2)^2[3/2]_1$	6p	$(3/2)^2[3/2]°_1$			R1c		
3162.0434(7)	31615.968(7)	3162.0438(4)	−0.0004	2300		5p	$(5/2)^2[3/2]°_2$	7d	$(5/2)^2[5/2]_3$			R1c		
3163.6838(11)	31599.575(11)	3163.6842(4)	−0.0003	610		4d	$(5/2)^2[5/2]_2$	5f	$(3/2)^2[3/2]°_2$			R1c		
3166.5879(7)	31570.596(7)	3166.5873(5)	0.0006	3500		5p	$(5/2)^2[3/2]°_2$	7d	$(5/2)^2[3/2]_2$			R1c		
3170.6936(8)	31529.717(8)	3170.6935(7)	0.0001	540		5p	$(3/2)^2[1/2]°_0$	7d	$(3/2)^2[3/2]_1$			R1c		
3173.6092(9)	31500.752(9)	3173.6088(3)	0.0004	330		sp	$(^3F)^3P°\ ^3D°_2$	6d	$(3/2)^2[5/2]_3$			R1c		
3174.9680(12)	31487.271(11)	3174.96642(22)	0.0016	840		s^2	$^3P_1$	5p	$(5/2)^2[3/2]°_2$			R1c		
3176.3094(16)	31473.974(16)	3176.3108(3)	−0.0014	230		4d	$(5/2)^2[1/2]_1$	5f	$(5/2)^2[5/2]°_2$			R1c		
3177.7392(8)	31459.813(8)	3177.7397(7)	−0.0005	1700		5p	$(5/2)^2[3/2]°_2$	7d	$(5/2)^2[1/2]_1$			R1c		
3177.9692(8)	31457.536(8)	3177.9705(5)	−0.0012	2700		5s	$(3/2)^2[3/2]_1$	6p	$(3/2)^2[3/2]°_2$			R1c		
3179.3173(17)	31444.198(17)	3179.3163(4)	0.0010	230		4d	$(5/2)^2[1/2]_1$	5f	$(5/2)^2[3/2]°_1$			R1c		
3179.7846(8)	31439.577(8)	3179.7842(4)	0.0004	6400		4d	$(5/2)^2[1/2]_0$	5f	$(3/2)^2[3/2]°_1$			R1c		
3180.2932(9)	31434.550(9)	3180.2935(6)	−0.0003	470		5p	$(3/2)^2[1/2]°_0$	7d	$(3/2)^2[1/2]_1$			R1c		
3180.7930(21)	31429.610(21)	3180.7922(4)	0.0008	230		sp	$(^3F)^3P°\ ^3D°_2$	6d	$(3/2)^2[3/2]_2$			R1c		
3182.1717(7)	31415.994(7)	3182.1718(3)	−0.0002	11000		4d	$(5/2)^2[1/2]_1$	5f	$(5/2)^2[3/2]°_2$			R1c		
3184.6224(12)	31391.819(12)	3184.6233(5)	−0.0010	480		4d	$(5/2)^2[7/2]_4$	7p	$(5/2)^2[5/2]°_3$			R1c		
3184.8404(8)	31389.670(8)	3184.8409(3)	−0.0005	15000		4d	$(5/2)^2[1/2]_1$	5f	$(5/2)^2[1/2]°_1$			R1c		
3185.7249(7)	31380.955(7)	3185.7256(4)	−0.0006	4000		5s	$(5/2)^2[5/2]_3$	6p	$(5/2)^2[7/2]°_4$			R1c		
3186.0148(7)	31378.100(7)	3186.0153(4)	−0.0005	6400		4d	$(5/2)^2[1/2]_1$	5f	$(5/2)^2[1/2]°_0$			R1c		

Table A1. *Cont.*

λ_{obs} [a] (Å)	σ_{obs} [b] (cm^{-1})	λ_{Ritz} [c] (Å)	$\Delta\lambda_{obs\text{-}Ritz}$ (Å)	I_{obs} [d] (arb. u.)	Char [e]	Lower Level		Upper Level		A (s^{-1})	Acc [f]	Line Ref. [g]	TP Ref. [g]	Notes [h]
3186.3411(8)	31374.887(8)	3186.3415(4)	−0.0004	3000		5s	$(5/2)^2[5/2]_2$	sp	$(^3F)^1P^\circ\ ^3F^\circ_2$			R1c		
3187.0427(13)	31367.980(13)	3187.0386(5)	0.0041	1200		5s	$(3/2)^2[3/2]_2$	6p	$(3/2)^2[3/2]^\circ_1$			R1c		
3188.7231(22)	31351.450(21)	3188.7259(4)	−0.0028	490		sp	$(^3F)^3P^\circ\ ^3D^\circ_2$	6d	$(3/2)^2[7/2]_3$			R1c		
3192.3023(22)	31316.300(21)	3192.3011(3)	0.0012	250		5s	$(5/2)^2[5/2]_3$	sp	$(^1G)^3P^\circ\ ^3F^\circ_3$			R1c		
3198.1052(10)	31259.480(9)	3198.1052(8)	−0.0001	1000		5p	$(5/2)^2[3/2]^\circ_1$	7d	$(5/2)^2[1/2]_0$			R1c		
3204.5231(16)	31196.877(16)	3204.5259(6)	−0.0029	640		sp	$(^3F)^3P^\circ\ ^3G^\circ_4$	6d	$(5/2)^2[7/2]_3$			R1c		
3206.6848(13)	31175.847(12)	3206.6848(5)	−0.0000	1800		5s	$(3/2)^2[3/2]_2$	6p	$(3/2)^2[3/2]^\circ_2$			R1c		
3208.3026(16)	31160.127(16)	3208.3076(6)	−0.0050	1500		5s	$(3/2)^2[3/2]_1$	6p	$(3/2)^2[5/2]^\circ_2$			R1c		
3216.9896(14)	31075.987(14)	3216.9895(6)	0.0001	520		5p	$(3/2)^2[5/2]^\circ_2$	7d	$(3/2)^2[5/2]_2$			R1c		
3217.3123(22)	31072.870(21)	3217.3104(4)	0.0019	650		5p	$(5/2)^2[7/2]^\circ_3$	7d	$(5/2)^2[5/2]_2$			R1c		
3217.6410(17)	31069.696(17)	3217.6411(6)	−0.0001	390		5p	$(3/2)^2[5/2]^\circ_2$	7d	$(3/2)^2[5/2]_3$			R1rc		
3218.2664(15)	31063.658(15)	3218.2653(6)	0.0011	260		sp	$(^3F)^3P^\circ\ ^3G^\circ_4$	6d	$(5/2)^2[9/2]_4$			R1c		
3218.6278(16)	31060.170(16)	3218.6244(5)	0.0034	390		5p	$(5/2)^2[7/2]^\circ_3$	7d	$(5/2)^2[7/2]_4$			R1c		
3218.7642(8)	31058.854(8)	3218.7651(4)	−0.0009	2000		5p	$(5/2)^2[7/2]^\circ_3$	7d	$(5/2)^2[7/2]_3$			R1c		
3221.9702(22)	31027.950(21)	3221.9718(6)	−0.0016	260		5p	$(3/2)^2[5/2]^\circ_2$	7d	$(3/2)^2[3/2]_2$			R1c		
3225.3388(8)	30995.545(8)	3225.3392(3)	−0.0004	2100		5s	$(5/2)^2[5/2]_2$	sp	$(^1G)^3P^\circ\ ^3F^\circ_3$			R1c		
3226.3290(16)	30986.033(16)	3226.3301(12)	−0.0011	260		5p	$(3/2)^2[1/2]^\circ_1$	7d	$(3/2)^2[1/2]_0$			R1c		
3226.4395(9)	30984.972(9)	3226.4380(5)	0.0015	640		5p	$(3/2)^2[5/2]^\circ_2$	7d	$(3/2)^2[7/2]_3$			R1c		
3226.5812(10)	30983.611(9)	3226.5808(5)	0.0004	5200		5p	$(5/2)^2[7/2]^\circ_3$	7d	$(5/2)^2[9/2]_4$			R1c		
3229.5526(8)	30955.105(8)	3229.5524(5)	0.0003	970		5p	$(5/2)^2[7/2]^\circ_4$	7d	$(5/2)^2[7/2]_4$			R1c		
3233.3754(13)	30918.508(12)	3233.3722(4)	0.0032	1000		5p	$(5/2)^2[7/2]^\circ_4$	7d	$(5/2)^2[5/2]_3$			R1c		
3234.6710(14)	30906.125(13)	3234.6691(6)	0.0019	510		5p	$(3/2)^2[1/2]^\circ_1$	7d	$(3/2)^2[3/2]_2$			R1c		
3234.7336(10)	30905.527(9)	3234.7341(5)	−0.0005	930		5s	$(5/2)^2[5/2]_2$	6p	$(5/2)^2[3/2]^\circ_1$			R1c		
3237.1430(16)	30882.525(15)	3237.1420(7)	0.0009	250		sp	$(^3F)^3P^\circ\ ^5F^\circ_3$	7s	$(5/2)^2[5/2]_2$			R1c		
3237.2417(11)	30881.583(11)	3237.2413(5)	0.0004	1100		5p	$(5/2)^2[5/2]^\circ_2$	7d	$(5/2)^2[5/2]_2$			R1c		
		3237.5629(5)			m	5p	$(5/2)^2[7/2]^\circ_4$	7d	$(5/2)^2[9/2]_4$			R1nc		
3237.5736(12)	30878.417(11)	3237.5752(6)	−0.0016	1300		5s	$(3/2)^2[3/2]_2$	6p	$(3/2)^2[5/2]^\circ_2$			R1c		
3238.7160(10)	30867.526(10)	3238.7141(4)	0.0019	1900		5p	$(5/2)^2[5/2]^\circ_2$	7d	$(5/2)^2[7/2]_3$			R1c		
3238.8232(8)	30866.504(7)	3238.8228(6)	0.0005	5500		5p	$(5/2)^2[7/2]^\circ_4$	7d	$(5/2)^2[9/2]_5$			R1c		
3239.6499(13)	30858.628(12)	3239.6507(5)	−0.0008	250		sp	$(^3F)^3P^\circ\ ^1D^\circ_2$	8s	$(3/2)^2[3/2]_2$			R1c		
3241.8139(12)	30838.030(11)	3241.8139(5)	−0.0000	860		5p	$(5/2)^2[3/2]^\circ_1$	7d	$(5/2)^2[5/2]_2$			R1c		
3242.425(10)	30832.21(10)	3242.4129(5)	0.012	250	*	5p	$(5/2)^2[5/2]^\circ_2$	7d	$(5/2)^2[5/2]_3$			R1nc		
3242.425(10)	30832.21(10)	3242.4292(7)	−0.004	250	*	sp	$(^3F)^3P^\circ\ ^1D^\circ_2$	8s	$(3/2)^2[3/2]_1$			R1c		
3245.9402(14)	30798.829(13)	3245.9405(6)	−0.0003	1500		5p	$(3/2)^2[5/2]^\circ_3$	7d	$(3/2)^2[5/2]_3$			R1c		
3246.7898(22)	30790.770(21)	3246.7839(7)	0.0059	620		5p	$(3/2)^2[1/2]^\circ_1$	7d	$(3/2)^2[1/2]_1$			R1c		
3250.4652(8)	30755.955(8)	3250.4650(4)	0.0003	11000		4d	$(3/2)^2[1/2]_1$	5f	$(3/2)^2[3/2]^\circ_2$			R1c		
3250.8739(13)	30752.089(12)	3250.8733(7)	0.0006	1200		5p	$(5/2)^2[3/2]^\circ_1$	7d	$(5/2)^2[3/2]_1$			R1c		
3251.76(3)	30743.7(3)	3251.7620(6)	0.00	240	*	sp	$(^3F)^3P^\circ\ ^3G^\circ_4$	5g	$(5/2)^2[11/2]_5$			R1c		
3251.76(3)	30743.7(3)	3251.7913(5)	−0.03	240	*	5p	$(5/2)^2[3/2]^\circ_1$	7d	$(5/2)^2[3/2]_2$			R1c		
3252.7830(8)	30734.041(8)	3252.7838(6)	−0.0008	2400		5p	$(3/2)^2[5/2]^\circ_3$	7d	$(3/2)^2[7/2]_4$			R1c		
3253.1078(12)	30730.972(11)	3253.1070(5)	0.0008	1200		5s	$(5/2)^2[5/2]_3$	6p	$(5/2)^2[3/2]^\circ_2$			R1c		
3257.1234(16)	30693.086(15)	3257.1222(5)	0.0012	240		5s	$(5/2)^2[5/2]_2$	6p	$(5/2)^2[5/2]^\circ_2$			R1c		

Table A1. *Cont.*

λ_{obs} [a] (Å)	σ_{obs} [b] (cm^{-1})	λ_{Ritz} [c] (Å)	$\Delta\lambda_{obs\text{-}Ritz}$ (Å)	I_{obs} [d] (arb. u.)	Char [e]	Lower Level		Upper Level		A (s^{-1})	Acc [f]	Line Ref. [g]	TP Ref. [g]	Notes [h]
3260.0249(11)	30665.770(11)	3260.0253(5)	−0.0004	2400		5p	$(5/2)^2[5/2]^{\circ}_3$	7d	$(5/2)^2[7/2]_4$			R1c		
3260.1697(17)	30664.408(16)	3260.1697(4)	0.0000	240		5p	$(5/2)^2[5/2]^{\circ}_3$	7d	$(5/2)^2[7/2]_3$			R1c		
3261.6469(8)	30650.520(8)	3261.6476(3)	−0.0007	1800		5p	$(5/2)^2[3/2]^{\circ}_2$	8s	$(5/2)^2[5/2]_3$			R1c		
3263.5546(14)	30632.604(13)	3263.5530(7)	0.0017	350		5p	$(5/2)^2[3/2]^{\circ}_1$	7d	$(5/2)^2[1/2]_1$			R1c		
3263.9177(11)	30629.197(10)	3263.9177(4)	0.0000	3500		5p	$(5/2)^2[5/2]^{\circ}_3$	7d	$(5/2)^2[5/2]_3$			R1c		
3265.3948(14)	30615.342(13)	3265.3933(6)	0.0015	1100		5s	$(3/2)^2[3/2]_2$	6p	$(3/2)^2[1/2]^{\circ}_1$			R1c		
3268.0990(16)	30590.010(15)	3268.0988(7)	0.0002	230		5p	$(3/2)^2[1/2]^{\circ}_0$	8s	$(3/2)^2[3/2]_1$			R1c		
3268.1895(14)	30589.163(13)	3268.1879(5)	0.0016	450		5p	$(5/2)^2[5/2]^{\circ}_3$	7d	$(5/2)^2[9/2]_4$			R1c		
3268.7591(13)	30583.833(12)	3268.7589(5)	0.0002	1400		5p	$(5/2)^2[5/2]^{\circ}_3$	7d	$(5/2)^2[3/2]_2$			R1c		
3270.1104(16)	30571.195(15)	3270.1113(4)	−0.0008	1000		sp	$(^3F)^3P^{\circ}\ ^3D^{\circ}_1$	6d	$(3/2)^2[5/2]_2$			R1c		
3275.9027(13)	30517.143(12)	3275.9057(6)	−0.0031	2200		5p	$(3/2)^2[3/2]^{\circ}_1$	7d	$(3/2)^2[5/2]_2$			R1c		
3278.9653(23)	30488.640(21)	3278.9636(4)	0.0017	210		sp	$(^3F)^3P^{\circ}\ ^3D^{\circ}_1$	6d	$(3/2)^2[3/2]_2$			R1c		
3281.0757(23)	30469.030(21)	3281.0724(6)	0.0034	210		5p	$(3/2)^2[3/2]^{\circ}_1$	7d	$(3/2)^2[3/2]_2$			R1c		
3281.6964(8)	30463.268(8)	3281.69629(23)	0.0001	9700		4d	$(5/2)^2[9/2]_5$	5f	$(5/2)^2[9/2]^{\circ}_5$			R1c		
3282.0106(9)	30460.351(8)	3282.01019(24)	0.0005	1100		4d	$(5/2)^2[9/2]_5$	5f	$(5/2)^2[9/2]^{\circ}_4$			R1c		
3282.6022(11)	30454.862(10)	3282.6016(3)	0.0006	530		4d	$(5/2)^2[9/2]_5$	5f	$(5/2)^2[7/2]^{\circ}_4$			R1c		
3283.0958(16)	30450.283(15)	3283.0969(4)	−0.0011	210		sp	$(^3F)^3P^{\circ}\ ^3D^{\circ}_1$	6d	$(3/2)^2[3/2]_1$			R1c		
3283.2431(13)	30448.917(12)	3283.2433(7)	−0.0002	940		5p	$(3/2)^2[3/2]^{\circ}_1$	7d	$(3/2)^2[3/2]_1$			R1c		
3290.4174(4)	30382.530(4)	3290.41683(16)	0.0005	38000		4d	$(5/2)^2[9/2]_5$	5f	$(5/2)^2[11/2]^{\circ}_6$	5.9e+07	D+	F_Re	TW	
3291.0597(8)	30376.600(7)	3291.05952(22)	0.0002	1100		4d	$(5/2)^2[9/2]_5$	5f	$(5/2)^2[11/2]^{\circ}_5$			R1c		
3291.8090(16)	30369.686(15)	3291.8073(3)	0.0017	2000		4d	$(5/2)^2[9/2]_4$	5f	$(5/2)^2[9/2]^{\circ}_5$			R1c		
3292.1232(8)	30366.788(7)	3292.1232(3)	−0.0000	7700		4d	$(5/2)^2[9/2]_4$	5f	$(5/2)^2[9/2]^{\circ}_4$			R1c		
3292.7189(10)	30361.294(9)	3292.7183(3)	0.0007	3500		4d	$(5/2)^2[9/2]_4$	5f	$(5/2)^2[7/2]^{\circ}_4$			R1c		
3292.9965(13)	30358.735(12)	3292.9984(3)	−0.0020	1500		4d	$(5/2)^2[9/2]_4$	5f	$(5/2)^2[7/2]^{\circ}_3$			R1c		
3293.3326(9)	30355.637(8)	3293.3327(4)	−0.0001	2800		4d	$(3/2)^2[7/2]_3$	5f	$(3/2)^2[7/2]^{\circ}_3$			R1c		
3294.3354(8)	30346.397(7)	3294.3355(3)	−0.0002	3800		4d	$(5/2)^2[3/2]_2$	5f	$(5/2)^2[5/2]^{\circ}_2$			R1c		
3295.1019(8)	30339.338(7)	3295.1018(3)	0.0001	11000		4d	$(5/2)^2[3/2]_2$	5f	$(5/2)^2[5/2]^{\circ}_3$			R1c		
3297.1983(8)	30320.048(7)	3297.1986(3)	−0.0003	9300		4d	$(5/2)^2[3/2]_1$	5f	$(5/2)^2[5/2]^{\circ}_2$			R1c		
3297.3461(12)	30318.689(11)	3297.3492(4)	−0.0031	1200		4d	$(3/2)^2[7/2]_3$	5f	$(3/2)^2[5/2]^{\circ}_2$			R1c		
3297.5688(10)	30316.642(9)	3297.5686(4)	0.0002	1400		4d	$(5/2)^2[3/2]_2$	5f	$(5/2)^2[3/2]^{\circ}_1$			R1c		
3300.2124(14)	30292.358(13)	3300.2124(4)	0.0000	380		sp	$(^3F)^3P^{\circ}\ ^3G^{\circ}_3$	7s	$(3/2)^2[3/2]_2$			R1c		
3300.4370(9)	30290.296(9)	3300.4373(4)	−0.0003	6100		4d	$(5/2)^2[3/2]_1$	5f	$(5/2)^2[3/2]^{\circ}_1$			R1c		
3300.6409(9)	30288.425(9)	3300.6406(3)	0.0003	9100		4d	$(5/2)^2[3/2]_2$	5f	$(5/2)^2[3/2]^{\circ}_2$			R1c		
3300.8815(9)	30286.218(8)	3300.8808(3)	0.0006	13000		4d	$(3/2)^2[7/2]_3$	5f	$(3/2)^2[9/2]^{\circ}_4$	5.4e+07	D+	R1c	TW	
3301.2286(7)	30283.034(6)	3301.22844(23)	0.0001	25000		4d	$(5/2)^2[9/2]_4$	5f	$(5/2)^2[11/2]^{\circ}_5$	5.6e+07	D+	F_Re	TW	
3303.1817(12)	30265.128(11)	3303.1836(6)	−0.0018	1400		5p	$(3/2)^2[3/2]^{\circ}_2$	7d	$(3/2)^2[5/2]_2$			R1c		
3303.5132(21)	30262.091(20)	3303.5122(4)	0.0010	5800	*	4d	$(5/2)^2[3/2]_2$	5f	$(5/2)^2[1/2]^{\circ}_1$			R1c		
3303.5132(21)	30262.091(20)	3303.5147(3)	−0.0015	5800	*	4d	$(5/2)^2[3/2]_1$	5f	$(5/2)^2[3/2]^{\circ}_2$			R1c		
3303.8698(10)	30258.825(9)	3303.8706(6)	−0.0008	2300		5p	$(3/2)^2[3/2]^{\circ}_2$	7d	$(3/2)^2[5/2]_3$			R1c		
3306.3890(15)	30235.771(14)	3306.3913(4)	−0.0023	180		4d	$(5/2)^2[3/2]_1$	5f	$(5/2)^2[1/2]^{\circ}_1$			R1c		
3307.6576(11)	30224.175(10)	3307.6570(4)	0.0006	1200		4d	$(5/2)^2[3/2]_1$	5f	$(5/2)^2[1/2]^{\circ}_0$			R1c		

Table A1. *Cont.*

λ_{obs} [a] (Å)	σ_{obs} [b] (cm^{-1})	λ_{Ritz} [c] (Å)	$\Delta\lambda_{obs-Ritz}$ (Å)	I_{obs} [d] (arb. u.)	Char [e]	Lower Level	Upper Level	A (s^{-1})	Acc [f]	Line Ref. [g]	TP Ref. [g]	Notes [h]
3307.8727(13)	30222.210(12)	3307.8664(4)	0.0063	1200		4d $(3/2)^2[7/2]_4$	5f $(3/2)^2[7/2]^{\circ}_4$			R1c		X
3308.1052(14)	30220.086(13)	3308.1074(5)	−0.0023	1100		sp $(^3F)^3P^{\circ}\ ^1F^{\circ}_3$	7d $(5/2)^2[7/2]_4$			R1c		
3308.2526(15)	30218.739(14)	3308.2561(4)	−0.0035	180		sp $(^3F)^3P^{\circ}\ ^1F^{\circ}_3$	7d $(5/2)^2[7/2]_3$			R1c		
3308.4337(13)	30217.085(12)	3308.4367(6)	−0.0029	360		5p $(3/2)^2[3/2]^{\circ}_2$	7d $(3/2)^2[3/2]_2$			R1c		
3310.338(3)	30199.70(3)	3310.3389(4)	−0.001	89		sp $(^3F)^3P^{\circ}\ ^3D^{\circ}_3$	6d $(5/2)^2[5/2]_2$			R1c		
3312.0278(13)	30184.296(11)	3312.0276(4)	0.0001	550		sp $(^3F)^3P^{\circ}\ ^3D^{\circ}_3$	6d $(5/2)^2[7/2]_4$			R1c		
3312.1151(14)	30183.500(13)	3312.1155(5)	−0.0004	970		sp $(^3F)^3P^{\circ}\ ^1F^{\circ}_3$	7d $(5/2)^2[5/2]_3$			R1c		
3312.6771(10)	30178.380(9)	3312.6783(3)	−0.0012	610		4d $(3/2)^2[7/2]_4$	5f $(3/2)^2[5/2]^{\circ}_3$			R1c		
3315.7440(8)	30150.467(7)	3315.7431(3)	0.0010	1300		4d $(3/2)^2[7/2]_4$	5f $(3/2)^2[9/2]^{\circ}_4$			R1c		
3316.2756(8)	30145.634(7)	3316.2752(3)	0.0004	16000		4d $(3/2)^2[7/2]_4$	5f $(3/2)^2[9/2]^{\circ}_5$	5.6e+07	D+	R1c	TW	
3316.5116(16)	30143.489(15)	3316.5129(6)	−0.0013	260		sp $(^3F)^3P^{\circ}\ ^1F^{\circ}_3$	7d $(5/2)^2[9/2]_4$			R1c		
3317.1383(8)	30137.794(7)	3317.1386(4)	−0.0003	5500		4d $(3/2)^2[3/2]_1$	5f $(3/2)^2[5/2]^{\circ}_2$			R1c		
3318.3147(23)	30127.110(21)	3318.3232(12)	−0.0085	84		sp $(^3F)^3P^{\circ}\ ^5F^{\circ}_1$	7s $(5/2)^2[5/2]_2$			R1c		
3319.0203(11)	30120.706(10)	3319.0216(5)	−0.0013	840		sp $(^3F)^3P^{\circ}\ ^3D^{\circ}_3$	6d $(5/2)^2[5/2]_3$			R1c		
3321.1137(23)	30101.720(21)	3321.1114(7)	0.0023	500		5p $(3/2)^2[3/2]^{\circ}_2$	7d $(3/2)^2[1/2]_1$			R1c		
3321.5526(8)	30097.743(8)	3321.5528(3)	−0.0002	1400		5p $(5/2)^2[7/2]^{\circ}_3$	8s $(5/2)^2[5/2]_2$			R1c		
3321.7168(14)	30096.255(13)	3321.7165(8)	0.0003	410		4d $(3/2)^2[1/2]_1$	7p $(5/2)^2[3/2]^{\circ}_2$			R1c		
3322.6363(10)	30087.927(9)	3322.6351(4)	0.0012	2000		4d $(3/2)^2[3/2]_1$	5f $(3/2)^2[3/2]^{\circ}_1$			R1c		
3324.8295(9)	30068.080(8)	3324.8292(7)	0.0003	1200		5p $(3/2)^2[5/2]^{\circ}_2$	8s $(3/2)^2[3/2]_1$			R1c		
3325.0235(13)	30066.326(11)	3325.0210(4)	0.0025	720		5s $(5/2)^2[5/2]_2$	6p $(5/2)^2[7/2]^{\circ}_3$			R1c		
3325.8187(14)	30059.137(12)	3325.8184(4)	0.0003	3000		4d $(3/2)^2[3/2]_1$	5f $(3/2)^2[3/2]^{\circ}_2$			R1c		
3325.9236(13)	30058.189(11)	3325.9242(4)	−0.0006	400		5p $(5/2)^2[7/2]^{\circ}_3$	8s $(5/2)^2[5/2]_3$			R1c		
3327.9129(16)	30040.222(15)	3327.9161(5)	−0.0032	160		sp $(^3F)^3P^{\circ}\ ^3D^{\circ}_3$	6d $(5/2)^2[3/2]_2$			R1c		
3335.4075(14)	29972.725(12)	3335.4064(5)	0.0011	520		5p $(3/2)^2[1/2]^{\circ}_1$	8s $(3/2)^2[3/2]_2$			R1c		
3337.5951(12)	29953.080(11)	3337.5942(4)	0.0008	3000		5p $(5/2)^2[7/2]^{\circ}_4$	8s $(5/2)^2[5/2]_3$			R1c		
3338.0369(14)	29949.116(12)	3338.03567(24)	0.0012	9200		4d $(5/2)^2[5/2]_3$	5f $(5/2)^2[9/2]^{\circ}_4$			R1c		
3338.6475(9)	29943.638(8)	3338.6475(3)	0.0001	9900		4d $(5/2)^2[5/2]_3$	5f $(5/2)^2[7/2]^{\circ}_4$	2.7e+07	D+	R1c	TW	
3338.9360(9)	29941.051(8)	3338.9355(3)	0.0005	3600		4d $(5/2)^2[5/2]_3$	5f $(5/2)^2[7/2]^{\circ}_3$			R1c		
3339.0850(9)	29939.715(8)	3339.0845(4)	0.0006	3600		4d $(3/2)^2[3/2]_2$	5f $(3/2)^2[7/2]^{\circ}_3$			R1c		
3340.8308(13)	29924.070(11)	3340.8312(8)	−0.0003	720		5s $(5/2)^2[5/2]_3$	sp $(^1G)^3P^{\circ}\ ^3F^{\circ}_4$			R1c		
3341.7610(13)	29915.741(11)	3341.7607(5)	0.0003	570		sp $(^3F)^3P^{\circ}\ ^3G^{\circ}_3$	6d $(5/2)^2[7/2]_4$			R1c		
3342.8004(10)	29906.439(9)	3342.8003(4)	0.0001	710		5p $(5/2)^2[5/2]^{\circ}_2$	8s $(5/2)^2[5/2]_2$			R1c		
3342.9641(9)	29904.975(8)	3342.9640(3)	0.0001	1800		4d $(5/2)^2[5/2]_3$	5f $(5/2)^2[5/2]^{\circ}_2$			R1c		
3343.2140(24)	29902.740(21)	3343.2134(4)	0.0006	280		4d $(3/2)^2[3/2]_2$	5f $(3/2)^2[5/2]^{\circ}_2$			R1c		
3343.7214(9)	29898.202(8)	3343.7215(3)	−0.0001	14000		4d $(3/2)^2[3/2]_2$	5f $(3/2)^2[5/2]^{\circ}_3$			R1c		
3343.7515(9)	29897.933(8)	3343.7531(3)	−0.0016	14000		4d $(5/2)^2[5/2]_3$	5f $(5/2)^2[5/2]^{\circ}_3$			R1c		
3347.2268(18)	29866.892(16)	3347.2279(4)	−0.0011	140		5p $(5/2)^2[5/2]^{\circ}_2$	8s $(5/2)^2[5/2]_3$			R1c		
3347.6754(9)	29862.890(8)	3347.6762(3)	−0.0008	550		5p $(5/2)^2[3/2]^{\circ}_1$	8s $(5/2)^2[5/2]_2$			R1c		
3348.7955(14)	29852.902(12)	3348.7967(4)	−0.0012	540		4d $(3/2)^2[3/2]_2$	5f $(3/2)^2[3/2]^{\circ}_1$			R1c		
3348.8824(14)	29852.127(12)	3348.8809(5)	0.0015	620		sp $(^3F)^3P^{\circ}\ ^3G^{\circ}_3$	6d $(5/2)^2[5/2]_3$			R1c		
3349.4567(8)	29847.009(7)	3349.4569(3)	−0.0002	4200		4d $(5/2)^2[5/2]_3$	5f $(5/2)^2[3/2]^{\circ}_2$			R1c		
3352.0324(12)	29824.075(11)	3352.0304(4)	0.0020	6400		4d $(3/2)^2[3/2]_2$	5f $(3/2)^2[3/2]^{\circ}_2$			R1c		

Table A1. *Cont.*

λ_{obs} [a] (Å)	σ_{obs} [b] (cm^{-1})	λ_{Ritz} [c] (Å)	$\Delta\lambda_{obs-Ritz}$ (Å)	I_{obs} [d] (arb. u.)	Char [e]	Lower Level		Upper Level		A (s^{-1})	Acc [f]	Line Ref. [g]	TP Ref. [g]	Notes [h]
3352.079	29823.66	3352.0793(6)			:	5p	$(3/2)^2[5/2]^{\circ}_3$	8s	$(3/2)^2[3/2]_2$					
3354.0672(12)	29805.983(11)	3354.0677(3)	−0.0005	330		sp	$(^3F)^3P^{\circ}\ ^3D^{\circ}_2$	7s	$(3/2)^2[3/2]_2$			R1c		
3355.3107(13)	29794.937(11)	3355.31179(19)	−0.0011	130		5s	$(5/2)^2[5/2]_2$	4f	$(3/2)^2[5/2]^{\circ}_3$			R1c		
3356.8877(17)	29780.940(15)	3356.8867(3)	0.0010	64		s^2	$^3P_2$	sp	$(^3F)^3P^{\circ}\ ^3F^{\circ}_3$			R1c		
3357.4722(10)	29775.756(9)	3357.4732(5)	−0.0010	640		sp	$(^3F)^3P^{\circ}\ ^1D^{\circ}_2$	7d	$(5/2)^2[5/2]_2$			R1c		
3357.9368(14)	29771.636(12)	3357.9364(5)	0.0004	510		sp	$(^3F)^3P^{\circ}\ ^3G^{\circ}_3$	6d	$(5/2)^2[3/2]_2$			R1c		
3359.0587(9)	29761.693(8)	3359.0575(4)	0.0012	950		sp	$(^3F)^3P^{\circ}\ ^1D^{\circ}_2$	7d	$(5/2)^2[7/2]_3$			R1c		
3359.7217(9)	29755.820(8)	3359.7209(3)	0.0008	670		sp	$(^3F)^3P^{\circ}\ ^3D^{\circ}_2$	7s	$(3/2)^2[3/2]_1$			R1c		
3363.8287(17)	29719.491(15)	3363.8300(10)	−0.0013	310		4d	$(3/2)^2[3/2]_1$	7p	$(5/2)^2[5/2]^{\circ}_2$			R1c		
3365.4414(17)	29705.250(15)	3365.4436(7)	−0.0021	610		sp	$(^3F)^3P^{\circ}\ ^5D^{\circ}_3$	5d	$(3/2)^2[3/2]_2$			R1c		
3365.6475(8)	29703.431(7)	3365.6470(3)	0.0005	11000		4d	$(5/2)^2[7/2]_3$	5f	$(5/2)^2[9/2]^{\circ}_4$	2.9e+07	D+	R1c	TW	
3366.2696(10)	29697.942(9)	3366.2690(3)	0.0006	10000		4d	$(5/2)^2[7/2]_3$	5f	$(5/2)^2[7/2]^{\circ}_4$			R1c		
3366.5619(8)	29695.364(7)	3366.5618(3)	0.0001	6100		4d	$(5/2)^2[7/2]_3$	5f	$(5/2)^2[7/2]^{\circ}_3$			R1c		
3366.8560(24)	29692.770(21)	3366.8553(4)	0.0007	120		5s	$(5/2)^2[5/2]_2$	4f	$(3/2)^2[3/2]^{\circ}_1$			R1c		
3370.1508(11)	29663.742(10)	3370.1503(4)	0.0005	610		5p	$(5/2)^2[5/2]^{\circ}_3$	8s	$(5/2)^2[5/2]_3$			R1c		
3370.4529(7)	29661.083(6)	3370.45289(23)	−0.0000	22000		4d	$(5/2)^2[7/2]_4$	5f	$(5/2)^2[9/2]^{\circ}_5$	4.3e+07	D+	F_Re	TW	
3370.6573(11)	29659.285(10)	3370.6573(3)	−0.0001	1100		4d	$(5/2)^2[7/2]_3$	5f	$(5/2)^2[5/2]^{\circ}_2$			R1c		
3370.7846(8)	29658.165(7)	3370.7840(3)	0.0006	5900		4d	$(5/2)^2[7/2]_4$	5f	$(5/2)^2[9/2]^{\circ}_4$			R1c		
3371.4075(9)	29652.685(8)	3371.4079(3)	−0.0004	6200		4d	$(5/2)^2[7/2]_4$	5f	$(5/2)^2[7/2]^{\circ}_4$			R1c		
3371.7016(9)	29650.099(8)	3371.7016(3)	−0.0000	870		4d	$(5/2)^2[7/2]_4$	5f	$(5/2)^2[7/2]^{\circ}_3$			R1c		
3373.5914(8)	29633.490(7)	3373.5912(3)	0.0002	9700		4d	$(5/2)^2[5/2]_2$	5f	$(5/2)^2[7/2]^{\circ}_3$			R1c		
3374.4423(13)	29626.018(11)	3374.4427(4)	−0.0005	170		5s	$(3/2)^2[3/2]_1$	sp	$(^3F)^1P^{\circ}\ ^3F^{\circ}_2$			R1c		
3374.9515(8)	29621.548(7)	3374.9510(4)	0.0005	12000		4d	$(3/2)^2[5/2]_3$	5f	$(3/2)^2[7/2]^{\circ}_4$	4.6e+07	D+	R1c	TW	
3375.2221(12)	29619.173(11)	3375.2222(4)	−0.0000	1400		4d	$(3/2)^2[5/2]_3$	5f	$(3/2)^2[7/2]^{\circ}_3$			R1c		
3376.6139(8)	29606.965(7)	3376.6143(3)	−0.0004	1300		4d	$(5/2)^2[7/2]_4$	5f	$(5/2)^2[5/2]^{\circ}_3$			R1c		
3377.0834(17)	29602.849(15)	3377.0829(4)	0.0006	170		sp	$(^3F)^3P^{\circ}\ ^3F^{\circ}_2$	6d	$(3/2)^2[5/2]_2$			R1c		
3377.2601(14)	29601.300(12)	3377.2583(3)	0.0019	110		4d	$(5/2)^2[7/2]_3$	5f	$(5/2)^2[3/2]^{\circ}_2$			R1c		
3377.7037(8)	29597.412(7)	3377.7038(3)	−0.0001	5300		4d	$(5/2)^2[5/2]_2$	5f	$(5/2)^2[5/2]^{\circ}_2$			R1c		
3378.3846(15)	29591.448(13)	3378.3831(4)	0.0015	170		sp	$(^3F)^3P^{\circ}\ ^3F^{\circ}_2$	6d	$(3/2)^2[5/2]_3$			R1c		
3378.5094(8)	29590.354(7)	3378.5094(3)	0.0001	3500		4d	$(5/2)^2[5/2]_2$	5f	$(5/2)^2[5/2]^{\circ}_3$			R1c		
3379.4421(9)	29582.188(8)	3379.4410(4)	0.0011	500		4d	$(3/2)^2[5/2]_3$	5f	$(3/2)^2[5/2]^{\circ}_2$			R1c		
3379.9595(9)	29577.660(8)	3379.9602(3)	−0.0007	3800		4d	$(3/2)^2[5/2]_3$	5f	$(3/2)^2[5/2]^{\circ}_3$			R1c		
3380.3311(8)	29574.409(7)	3380.33017(22)	0.0009	1100		4d	$(5/2)^2[7/2]_4$	5f	$(5/2)^2[11/2]^{\circ}_5$			R1c		
3380.7117(8)	29571.079(7)	3380.7116(4)	0.0002	7600		4d	$(3/2)^2[5/2]_2$	5f	$(3/2)^2[7/2]^{\circ}_3$	3.4e+07	D+	R1c	TW	
3381.1021(11)	29567.665(10)	3381.1027(4)	−0.0006	1400		4d	$(5/2)^2[5/2]_2$	5f	$(5/2)^2[3/2]^{\circ}_1$			R1c		
3384.3322(8)	29539.446(7)	3384.3324(3)	−0.0002	1300		4d	$(5/2)^2[5/2]_2$	5f	$(5/2)^2[3/2]^{\circ}_2$			R1c		
3384.7698(14)	29535.627(12)	3384.7669(6)	0.0029	160		5p	$(3/2)^2[3/2]^{\circ}_1$	8s	$(3/2)^2[3/2]_2$			R1c		
3384.9450(9)	29534.098(8)	3384.9442(4)	0.0009	3600		4d	$(3/2)^2[5/2]_2$	5f	$(3/2)^2[5/2]^{\circ}_2$			R1c		
3385.4657(9)	29529.556(8)	3385.4650(3)	0.0006	2100		4d	$(3/2)^2[5/2]_2$	5f	$(3/2)^2[5/2]^{\circ}_3$			R1c		
3387.3535(17)	29513.099(15)	3387.3515(4)	0.0021	100		4d	$(5/2)^2[5/2]_2$	5f	$(5/2)^2[1/2]^{\circ}_1$			R1c		
3387.7986(17)	29509.222(15)	3387.8000(8)	−0.0015	210		5p	$(3/2)^2[3/2]^{\circ}_1$	8s	$(3/2)^2[3/2]_1$			R1c		

Table A1. *Cont.*

λ_{obs} [a] (Å)	σ_{obs} [b] (cm^{-1})	λ_{Ritz} [c] (Å)	$\Delta\lambda_{obs\text{-}Ritz}$ (Å)	I_{obs} [d] (arb. u.)	Char [e]	Lower Level		Upper Level		A (s^{-1})	Acc [f]	Line Ref. [g]	TP Ref. [g]	Notes [h]
3388.4491(14)	29503.557(12)	3388.4504(4)	−0.0013	420		4d	$(3/2)^2[5/2]_3$	5f	$(3/2)^2[3/2]^\circ_2$			R1c		
		3390.6470(10)			m	4d	$(3/2)^2[3/2]_2$	7p	$(5/2)^2[5/2]^\circ_2$			R1c		
3390.664(20)	29484.29(17)	3390.6678(4)	−0.004	260		4d	$(3/2)^2[5/2]_2$	5f	$(3/2)^2[3/2]^\circ_1$			R1c		
3390.9325(24)	29481.950(21)	3390.9337(4)	−0.0012	100		sp	$(^3F)^3P^\circ\ ^3F^\circ_2$	6d	$(3/2)^2[3/2]_1$			R1c		
3392.7462(24)	29466.190(21)	3392.7444(4)	0.0019	130		s^2	$^3P_1$	sp	$(^3F)^3P^\circ\ ^3D^\circ_1$			R1c		
3393.9761(18)	29455.513(16)	3393.9803(13)	−0.0042	150	*	sp	$(^3P)^3P^\circ\ ^5P^\circ_2$	7g	$(5/2)^2[7/2]_3$			R1c		X
3393.9761(18)	29455.513(16)	3393.9829(4)	−0.0069	150	*	4d	$(3/2)^2[5/2]_2$	5f	$(3/2)^2[3/2]^\circ_2$			R1c		X
3393.9909(15)	29455.384(13)	3393.9914(6)	−0.0004	150		sp	$(^3F)^3P^\circ\ ^3G^\circ_4$	7s	$(5/2)^2[5/2]_3$			R1c		
3395.2150(9)	29444.765(8)	3395.2154(4)	−0.0004	2000		sp	$(^3F)^3P^\circ\ ^3D^\circ_2$	6d	$(5/2)^2[5/2]_2$			R1c		
3395.5206(10)	29442.115(9)	3395.5191(4)	0.0015	450		sp	$(^3F)^3P^\circ\ ^3F^\circ_2$	6d	$(3/2)^2[7/2]_3$			R1c		
3397.3752(9)	29426.043(8)	3397.3754(3)	−0.0002	740		sp	$(^3F)^3P^\circ\ ^3D^\circ_2$	6d	$(5/2)^2[7/2]_3$			R1c		
3400.4535(9)	29399.406(8)	3400.4535(6)	0.0000	96		4d	$(3/2)^2[3/2]_2$	7p	$(5/2)^2[3/2]^\circ_1$			R1c		
3402.8301(9)	29378.873(7)	3402.8303(4)	−0.0002	1500		4d	$(5/2)^2[1/2]_0$	5f	$(5/2)^2[3/2]^\circ_1$			R1c		
3406.835(2)	29344.340(21)	3406.8351(4)	−0.000	140		5s	$(3/2)^2[3/2]_2$	sp	$(^3F)^1P^\circ\ ^3F^\circ_2$			R1c		
3409.1599(8)	29324.327(7)	3409.1597(4)	0.0001	1400		4d	$(5/2)^2[1/2]_0$	5f	$(5/2)^2[1/2]^\circ_1$			R1c		
3410.4564(17)	29313.180(15)	3410.4549(5)	0.0015	90		5s	$(5/2)^2[5/2]_2$	sp	$(^3P)^3P^\circ\ ^1D^\circ_2$			R1c		
3410.6738(13)	29311.311(11)	3410.6739(9)	−0.0000	90		5p	$(3/2)^2[1/2]^\circ_1$	7d	$(5/2)^2[1/2]_0$			R1c		
3412.2696(10)	29297.604(9)	3412.2691(4)	0.0005	130		sp	$(^3F)^3P^\circ\ ^3D^\circ_2$	6d	$(5/2)^2[3/2]_1$			R1c		
3413.7074(14)	29285.265(12)	3413.7080(5)	−0.0007	180		sp	$(^3F)^3P^\circ\ ^3D^\circ_2$	6d	$(5/2)^2[3/2]_2$			R1c		
3413.8952(9)	29283.654(8)	3413.8957(6)	−0.0006	220		5p	$(3/2)^2[3/2]^\circ_2$	8s	$(3/2)^2[3/2]_2$			R1c		
3416.9362(12)	29257.593(11)	3416.9349(4)	0.0012	170		sp	$(^3F)^3P^\circ\ ^1F^\circ_3$	8s	$(5/2)^2[5/2]_2$			R1c		
3421.5612(10)	29218.046(8)	3421.5612(4)	−0.0001	850		sp	$(^3F)^3P^\circ\ ^1F^\circ_3$	8s	$(5/2)^2[5/2]_3$			R1c		
3428.7658(10)	29156.654(8)	3428.7659(5)	−0.0001	330		5s	$(3/2)^2[3/2]_1$	6p	$(5/2)^2[3/2]^\circ_1$			R1c		
3430.1011(12)	29145.304(10)	3430.1015(4)	−0.0004	81		s^2	$^3P_0$	sp	$(^3F)^3P^\circ\ ^3D^\circ_1$			R1c		
3433.9380(9)	29112.740(8)	3433.9259(7)	0.0120	120	?	4d	$(3/2)^2[5/2]_3$	7p	$(5/2)^2[7/2]^\circ_4$			R1c		X
3437.1829(19)	29085.256(16)	3437.1784(5)	0.0045	78		sp	$(^3F)^3P^\circ\ ^3D^\circ_2$	6d	$(5/2)^2[1/2]_1$			R1c		
3444.1360(17)	29026.540(14)	3444.1368(7)	−0.0008	75		sp	$(^3F)^3P^\circ\ ^3D^\circ_1$	6d	$(5/2)^2[1/2]_0$			R1c		
3445.9053(16)	29011.637(13)	3445.9035(5)	0.0018	37		5p	$(3/2)^2[5/2]^\circ_2$	7d	$(5/2)^2[5/2]_2$			R1c		
3451.4534(9)	28965.003(8)	3451.4543(4)	−0.0009	360		5s	$(3/2)^2[3/2]_2$	sp	$(^1G)^3P^\circ\ ^3F^\circ_3$			R1c		
3453.9306(10)	28944.230(9)	3453.9308(6)	−0.0002	540		5s	$(3/2)^2[3/2]_1$	6p	$(5/2)^2[5/2]^\circ_2$			R1c		
3460.045(3)	28893.080(21)	3460.0456(7)	−0.000	350		sp	$(^3F)^3P^\circ\ ^3F^\circ_4$	6d	$(5/2)^2[7/2]_4$			R1c		
3460.4304(14)	28889.865(11)	3460.4309(5)	−0.0005	69		5p	$(3/2)^2[1/2]^\circ_1$	7d	$(5/2)^2[5/2]_2$			R1c		
3461.9510(15)	28877.176(12)	3461.9515(5)	−0.0005	100		5s	$(5/2)^2[5/2]_2$	sp	$(^1G)^3P^\circ\ ^3F^\circ_2$			R1nc		
3463.4085(10)	28865.024(9)	3463.4094(3)	−0.0009	140		sp	$(^3F)^3P^\circ\ ^3D^\circ_1$	7s	$(3/2)^2[3/2]_2$			R1c		
3467.679(3)	28829.480(21)	3467.6793(7)	−0.001	200		sp	$(^3F)^3P^\circ\ ^3F^\circ_4$	6d	$(5/2)^2[5/2]_3$			R1c		
3469.4366(10)	28814.873(9)	3469.4375(4)	−0.0009	1000		sp	$(^3F)^3P^\circ\ ^3D^\circ_1$	7s	$(3/2)^2[3/2]_1$			R1c		
3470.7569(14)	28803.912(11)	3470.7552(8)	0.0017	130		5p	$(3/2)^2[1/2]^\circ_1$	7d	$(5/2)^2[3/2]_1$			R1c		
3471.1558(10)	28800.602(8)	3471.1560(4)	−0.0003	260		sp	$(^3F)^3P^\circ\ ^1D^\circ_2$	8s	$(5/2)^2[5/2]_2$			R1c		
3472.0275(10)	28793.371(9)	3472.0278(3)	−0.0003	130		sp	$(^3F)^3P^\circ\ ^3F^\circ_3$	7s	$(3/2)^2[3/2]_2$			R1c		
3473.0008(10)	28785.302(8)	3473.0010(4)	−0.0002	330		5s	$(5/2)^2[5/2]_3$	sp	$(^3F)^1P^\circ\ ^3D^\circ_2$			R1c		

Table A1. *Cont.*

λ_{obs} [a] (Å)	σ_{obs} [b] (cm^{-1})	λ_{Ritz} [c] (Å)	$\Delta\lambda_{obs\text{-}Ritz}$ (Å)	I_{obs} [d] (arb. u.)	Char [e]	Lower Level		Upper Level		A (s^{-1})	Acc [f]	Line Ref. [g]	TP Ref. [g]	Notes [h]
3476.4708(17)	28756.571(14)	3476.4703(7)	0.0005	97		sp	$(^3F)^3P^\circ\ ^3F^\circ_4$	6d	$(5/2)^2[9/2]_4$			R1c		
3478.950(3)	28736.080(21)	3478.9487(7)	0.001	640		sp	$(^3F)^3P^\circ\ ^3F^\circ_4$	6d	$(5/2)^2[9/2]_5$			R1c		
3483.8304(9)	28695.825(8)	3483.8297(4)	0.0006	310		4d	$(3/2)^2[1/2]_1$	5f	$(5/2)^2[3/2]^\circ_2$			R1c		
3487.0291(9)	28669.502(7)	3487.0290(4)	0.0001	470		4d	$(3/2)^2[1/2]_1$	5f	$(5/2)^2[1/2]^\circ_1$			R1c		
3488.0212(16)	28661.348(13)	3488.0222(6)	−0.0010	31		5s	$(3/2)^2[3/2]_1$	6p	$(5/2)^2[3/2]^\circ_2$			R1c		
3488.4370(9)	28657.932(7)	3488.4369(5)	0.0001	150		4d	$(3/2)^2[1/2]_1$	5f	$(5/2)^2[1/2]^\circ_0$			R1c		
3501.612(3)	28550.110(21)	3501.6122(4)	−0.000	59		5p	$(5/2)^2[3/2]^\circ_2$	6d	$(3/2)^2[5/2]_2$			R1c		
3503.0098(9)	28538.716(8)	3503.0101(3)	−0.0003	150		5p	$(5/2)^2[3/2]^\circ_2$	6d	$(3/2)^2[5/2]_3$			R1c		
3504.6643(15)	28525.244(12)	3504.6642(4)	0.0001	59		s^2	$^3P_1$	sp	$(^3F)^3P^\circ\ ^3D^\circ_2$			R1c		
3506.8362(11)	28507.578(9)	3506.8359(4)	0.0003	330		sp	$(^3F)^3P^\circ\ ^3D^\circ_3$	7s	$(5/2)^2[5/2]_2$			R1c		
3507.3014(10)	28503.797(8)	3507.3011(4)	0.0003	58		sp	$(^3F)^3P^\circ\ ^3D^\circ_1$	6d	$(5/2)^2[5/2]_2$			R1c		
3515.2292(11)	28439.515(9)	3515.2302(4)	−0.0011	850		sp	$(^3F)^3P^\circ\ ^3D^\circ_3$	7s	$(5/2)^2[5/2]_3$			R1c		
3516.1395(14)	28432.152(11)	3516.1396(4)	−0.0001	110		sp	$(^3F)^3P^\circ\ ^3F^\circ_3$	6d	$(5/2)^2[5/2]_2$			R1c		
3516.7786(9)	28426.985(8)	3516.7792(3)	−0.0006	280		5s	$(5/2)^2[5/2]_3$	sp	$(^3F)^1P^\circ\ ^3F^\circ_3$			R1c		
3518.0440(10)	28416.761(8)	3518.0449(4)	−0.0009	230		sp	$(^3F)^3P^\circ\ ^3F^\circ_3$	6d	$(5/2)^2[7/2]_4$			R1c		
3518.4565(9)	28413.429(7)	3518.4562(3)	0.0003	420		sp	$(^3F)^3P^\circ\ ^3F^\circ_3$	6d	$(5/2)^2[7/2]_3$			R1c		
3522.643(3)	28379.660(21)	3522.6429(6)	0.000	84		5s	$(3/2)^2[3/2]_2$	6p	$(5/2)^2[3/2]^\circ_2$			R1c		
3525.5026(16)	28356.644(13)	3525.5023(5)	0.0003	55		sp	$(^3F)^3P^\circ\ ^3D^\circ_1$	6d	$(5/2)^2[3/2]_1$			R1c		
3525.9383(15)	28353.140(12)	3525.9370(4)	0.0013	140		sp	$(^3F)^3P^\circ\ ^3F^\circ_3$	6d	$(5/2)^2[5/2]_3$			R1c		
3534.6784(20)	28283.034(16)	3534.6781(9)	0.0002	27		5p	$(5/2)^2[3/2]^\circ_1$	6d	$(3/2)^2[1/2]_0$			R1c		
3534.809(5)	28281.99(4)	3534.8076(3)	0.002	410	*	4d	$(3/2)^2[7/2]_3$	5f	$(5/2)^2[9/2]^\circ_4$			R1c		
3534.809(5)	28281.99(4)	3534.8127(5)	−0.004	410	*	s^2	$^3P_2$	sp	$(^3F)^3P^\circ\ ^3G^\circ_3$			R1c		
3535.0258(14)	28280.254(11)	3535.0262(4)	−0.0003	1100		sp	$(^3F)^3P^\circ\ ^3F^\circ_3$	6d	$(5/2)^2[9/2]_4$			R1c		
3535.4942(10)	28276.508(8)	3535.4937(3)	0.0005	220		4d	$(3/2)^2[7/2]_3$	5f	$(5/2)^2[7/2]^\circ_4$			R1c		
3535.8163(12)	28273.932(9)	3535.8167(3)	−0.0004	190		4d	$(3/2)^2[7/2]_3$	5f	$(5/2)^2[7/2]^\circ_3$			R1c		
3535.9757(10)	28272.657(8)	3535.9767(5)	−0.0009	140		sp	$(^3F)^3P^\circ\ ^3F^\circ_3$	6d	$(5/2)^2[3/2]_2$			R1c		
3540.1850(15)	28239.042(12)	3540.1870(5)	−0.0020	110		sp	$(^3F)^3P^\circ\ ^3G^\circ_3$	7s	$(5/2)^2[5/2]_2$			R1c		
3540.3356(19)	28237.841(15)	3540.3346(4)	0.0010	110		4d	$(3/2)^2[7/2]_3$	5f	$(5/2)^2[5/2]^\circ_2$			R1c		
3541.2200(20)	28230.789(16)	3541.2196(4)	0.0004	27		4d	$(3/2)^2[7/2]_3$	5f	$(5/2)^2[5/2]^\circ_3$			R1c		
3546.7568(21)	28186.719(17)	3546.7551(5)	0.0017	27		5p	$(3/2)^2[3/2]^\circ_2$	7d	$(5/2)^2[7/2]_3$			R1c		
3548.7423(11)	28170.949(9)	3548.7419(4)	0.0004	1400		sp	$(^3F)^3P^\circ\ ^3G^\circ_3$	7s	$(5/2)^2[5/2]_3$			R1c		
3551.4887(13)	28149.165(10)	3551.4887(3)	0.0001	530		4d	$(3/2)^2[7/2]_4$	5f	$(5/2)^2[9/2]^\circ_5$			R1c		
3551.8563(10)	28146.252(8)	3551.8563(3)	−0.0000	210		4d	$(3/2)^2[7/2]_4$	5f	$(5/2)^2[9/2]^\circ_4$			R1c		
3552.5495(12)	28140.760(9)	3552.5490(3)	0.0005	110		4d	$(3/2)^2[7/2]_4$	5f	$(5/2)^2[7/2]^\circ_4$			R1c		
3552.8735(15)	28138.194(12)	3552.8751(3)	−0.0016	53		4d	$(3/2)^2[7/2]_4$	5f	$(5/2)^2[7/2]^\circ_3$			R1c		
3555.4287(21)	28117.972(17)	3555.43262(15)	−0.0039	53		5s	$(5/2)^2[5/2]_3$	4f	$(5/2)^2[7/2]^\circ_4$			R1c		
3556.915(5)	28106.22(4)	3556.9167(3)	−0.001	79	*	5s	$(5/2)^2[5/2]_2$	sp	$(^3F)^1P^\circ\ ^3F^\circ_3$			R1c		
3556.915(5)	28106.22(4)	3556.9231(7)	−0.008	79	*	5p	$(3/2)^2[3/2]^\circ_2$	7d	$(5/2)^2[3/2]_2$			R1nc		
3558.332(4)	28095.03(3)	3558.3303(4)	0.002	52		4d	$(3/2)^2[7/2]_4$	5f	$(5/2)^2[5/2]^\circ_3$			R1c		
3562.4574(10)	28062.497(8)	3562.4572(3)	0.0002	100		4d	$(3/2)^2[7/2]_4$	5f	$(5/2)^2[11/2]^\circ_5$			R1c		
3563.1578(10)	28056.981(8)	3563.1580(4)	−0.0002	78		4d	$(3/2)^2[3/2]_1$	5f	$(5/2)^2[5/2]^\circ_2$			R1c		
3565.7620(21)	28036.491(17)	3565.7601(4)	0.0018	52		5p	$(3/2)^2[5/2]^\circ_2$	8s	$(5/2)^2[5/2]_2$			R1c		
3565.8472(10)	28035.821(8)	3565.8496(4)	−0.0024	210		5s	$(3/2)^2[3/2]_2$	6p	$(5/2)^2[7/2]^\circ_3$			R1c		
3566.9405(11)	28027.228(9)	3566.9405(5)	−0.0000	52		4d	$(3/2)^2[3/2]_1$	5f	$(5/2)^2[3/2]^\circ_1$			R1c		

Table A1. *Cont.*

λ_{obs} [a] (Å)	σ_{obs} [b] (cm^{-1})	λ_{Ritz} [c] (Å)	$\Delta\lambda_{obs\text{-}Ritz}$ (Å)	I_{obs} [d] (arb. u.)	Char [e]	Lower Level	Upper Level	A (s^{-1})	Acc [f]	Line Ref. [g]	TP Ref. [g]	Notes [h]
3567.801(3)	28020.470(21)	3567.79746(15)	0.003	73		5s $(5/2)^2[5/2]_3$	4f $(5/2)^2[5/2]^\circ_3$			R1c		
3568.7006(18)	28013.405(14)	3568.7005(4)	0.0001	78		s^2 $^3P_2$	sp $(^3F)^3P^\circ\ ^3D^\circ_3$			R1c		
3573.5908(14)	27975.072(11)	3573.58964(18)	0.0011	520		5s $(5/2)^2[5/2]_3$	4f $(5/2)^2[7/2]^\circ_3$			R1c		
3575.373(4)	27961.13(3)	3575.3666(24)	0.006	26	*	sp $(^1D)^3P^\circ\ ^3F^\circ_2$	9s $(3/2)^2[3/2]_1$			R1nc		
3575.373(4)	27961.13(3)	3575.3746(5)	−0.002	26	*	4d $(3/2)^2[3/2]_1$	5f $(5/2)^2[1/2]^\circ_0$			R1c		
3577.5874(19)	27943.821(15)	3577.5764(4)	0.0110	51	?	5s $(3/2)^2[3/2]_1$	4f $(3/2)^2[3/2]^\circ_1$			R1c		X
3578.4051(17)	27937.436(13)	3578.4043(6)	0.0008	130		5s $(5/2)^2[5/2]_3$	sp $(^3P)^3P^\circ\ ^5S^\circ_2$			R1c		
3581.3166(13)	27914.724(10)	3581.3179(4)	−0.0013	26		5p $(3/2)^2[1/2]^\circ_1$	8s $(5/2)^2[5/2]_2$			R1c		
3583.6322(15)	27896.687(11)	3583.6328(3)	−0.0005	77		sp $(^3F)^3P^\circ\ ^3F^\circ_2$	7s $(3/2)^2[3/2]_2$			R1c		
3588.6079(9)	27858.009(7)	3588.6074(3)	0.0005	200		4d $(3/2)^2[3/2]_2$	5f $(5/2)^2[7/2]^\circ_3$			R1c		
3590.0862(10)	27846.538(8)	3590.0870(3)	−0.0008	650		sp $(^3F)^3P^\circ\ ^3F^\circ_2$	7s $(3/2)^2[3/2]_1$			R1c		
3592.1150(19)	27830.811(15)	3592.1139(5)	0.0011	51		5p $(5/2)^2[7/2]^\circ_3$	6d $(3/2)^2[7/2]_4$			R1c		
3592.3528(10)	27828.969(8)	3592.3522(5)	0.0005	770		5s $(3/2)^2[3/2]_1$	sp $(^3F)^1P^\circ\ ^3D^\circ_1$			R1c		
3593.2610(10)	27821.935(8)	3593.2613(4)	−0.0002	100		4d $(3/2)^2[3/2]_2$	5f $(5/2)^2[5/2]^\circ_2$			R1c		
3594.1722(18)	27814.882(14)	3594.1729(4)	−0.0008	130		4d $(3/2)^2[3/2]_2$	5f $(5/2)^2[5/2]^\circ_3$			R1c		
3597.100(3)	27792.240(21)	3597.1080(5)	−0.008	51		4d $(3/2)^2[3/2]_2$	5f $(5/2)^2[3/2]^\circ_1$			R1c		
3600.4380(11)	27766.477(8)	3600.4382(4)	−0.0002	76		5p $(5/2)^2[5/2]^\circ_2$	6d $(3/2)^2[5/2]_2$			R1c		
3600.7635(10)	27763.967(7)	3600.7638(4)	−0.0002	130		4d $(3/2)^2[3/2]_2$	5f $(5/2)^2[3/2]^\circ_2$			R1c		
3601.9187(19)	27755.063(15)	3601.9162(4)	0.0025	200		5p $(5/2)^2[5/2]^\circ_2$	6d $(3/2)^2[5/2]_3$			R1c		
3602.2331(11)	27752.641(8)	3602.2324(3)	0.0007	1200		sp $(^3F)^3P^\circ\ ^3D^\circ_2$	7s $(5/2)^2[5/2]_2$			R1c		
3606.6576(11)	27718.596(8)	3606.6576(4)	0.0000	1900		5s $(5/2)^2[5/2]_3$	sp $(^3F)^1P^\circ\ ^3D^\circ_3$			R1c		
3611.0901(10)	27684.573(7)	3611.0902(3)	−0.0001	370		sp $(^3F)^3P^\circ\ ^3D^\circ_2$	7s $(5/2)^2[5/2]_3$			R1c		
3611.172	27683.94	3611.1721(4)			:	5p $(5/2)^2[5/2]^\circ_2$	6d $(3/2)^2[3/2]_2$					
3615.0429(20)	27654.303(16)	3615.04192(19)	0.0010	25		5s $(5/2)^2[5/2]_2$	4f $(5/2)^2[7/2]^\circ_3$			R1c		
3616.8636(10)	27640.382(7)	3616.8630(4)	0.0007	890		5p $(5/2)^2[3/2]^\circ_1$	6d $(3/2)^2[3/2]_2$			R1c		
3621.4008(19)	27605.753(15)	3621.4013(5)	−0.0005	26		5p $(5/2)^2[5/2]^\circ_2$	6d $(3/2)^2[7/2]_3$			R1c		
3621.8942(11)	27601.992(8)	3621.8927(4)	0.0016	77		5p $(5/2)^2[3/2]^\circ_1$	6d $(3/2)^2[3/2]_1$			R1c		
3626.350(3)	27568.080(21)	3626.3111(9)	0.039	77	?	sp $(^3F)^3P^\circ\ ^1G^\circ_4$	6d $(5/2)^2[7/2]_4$			R1c		X
3626.973(3)	27563.340(21)	3626.9734(4)	−0.000	51		5p $(5/2)^2[5/2]^\circ_3$	6d $(3/2)^2[5/2]_2$			R1c		
3628.4741(10)	27551.940(8)	3628.4733(4)	0.0008	130		5p $(5/2)^2[5/2]^\circ_3$	6d $(3/2)^2[5/2]_3$			R1c		
3629.0907(16)	27547.259(12)	3629.0858(5)	0.0049	51		5s $(3/2)^2[3/2]_2$	sp $(^3F)^1P^\circ\ ^3D^\circ_1$			R1c		
3629.3186(19)	27545.529(15)	3629.3176(3)	0.0011	26		4d $(3/2)^2[5/2]_3$	5f $(5/2)^2[9/2]^\circ_4$			R1c		
3630.0412(11)	27540.046(8)	3630.0408(3)	0.0004	100		4d $(3/2)^2[5/2]_3$	5f $(5/2)^2[7/2]^\circ_4$			R1c		
3630.6454(10)	27535.463(7)	3630.6450(4)	0.0004	380		sp $(^3F)^3P^\circ\ ^3F^\circ_2$	6d $(5/2)^2[5/2]_2$			R1c		
3631.6198(22)	27528.075(17)	3631.61939(23)	0.0004	51		5s $(5/2)^2[5/2]_2$	4f $(5/2)^2[1/2]^\circ_1$			R1c		
3633.1159(10)	27516.740(8)	3633.1151(4)	0.0008	520		sp $(^3F)^3P^\circ\ ^3F^\circ_2$	6d $(5/2)^2[7/2]_3$			R1c		
3635.1554(17)	27501.302(13)	3635.1443(4)	0.0111	210	bl	4d $(3/2)^2[5/2]_3$	5f $(5/2)^2[5/2]^\circ_2$			R1c		X
3636.0783(17)	27494.322(13)	3636.0773(4)	0.0009	77		4d $(3/2)^2[5/2]_3$	5f $(5/2)^2[5/2]^\circ_3$			R1c		
3637.867(3)	27480.800(21)	3637.8663(4)	0.001	26		5p $(5/2)^2[5/2]^\circ_3$	6d $(3/2)^2[3/2]_2$			R1c		
3641.0912(20)	27456.470(15)	3641.0918(5)	−0.0007	52		sp $(^3F)^3P^\circ\ ^3F^\circ_2$	6d $(5/2)^2[5/2]_3$			R1c		

Table A1. *Cont.*

λ_{obs} [a] (Å)	σ_{obs} [b] (cm^{-1})	λ_{Ritz} [c] (Å)	$\Delta\lambda_{obs\text{-}Ritz}$ (Å)	I_{obs} [d] (arb. u.)	Char [e]	Lower Level		Upper Level		A (s^{-1})	Acc [f]	Line Ref. [g]	TP Ref. [g]	Notes [h]
3641.7928(11)	27451.180(8)	3641.7903(4)	0.0025	1000		s^2	$^1G_4$	5p	$(3/2)^2[5/2]^\circ_3$			R1c		
3642.4505(14)	27446.224(11)	3642.4488(4)	0.0017	77		4d	$(3/2)^2[5/2]_2$	5f	$(5/2)^2[5/2]^\circ_3$			R1c		
3642.8238(22)	27443.411(17)	3642.8229(4)	0.0009	26		4d	$(3/2)^2[5/2]_3$	5f	$(5/2)^2[3/2]^\circ_2$			R1c		
3643.0146(11)	27441.974(8)	3643.0150(5)	−0.0004	770		5p	$(5/2)^2[3/2]^\circ_1$	6d	$(3/2)^2[1/2]_1$			R1c		
3643.7566(10)	27436.386(7)	3643.7571(5)	−0.0005	1300		5p	$(5/2)^2[5/2]^\circ_3$	6d	$(3/2)^2[7/2]_4$			R1c		
3647.081(4)	27411.38(3)	3647.0800(10)	0.001	160		sp	$(^3F)^3P^\circ\ ^1G^\circ_4$	6d	$(5/2)^2[9/2]_5$			R1c		
3648.2463(20)	27402.622(15)	3648.2475(5)	−0.0011	52		5p	$(5/2)^2[5/2]^\circ_3$	6d	$(3/2)^2[7/2]_3$			R1c		
3648.8875(15)	27397.807(11)	3648.8850(4)	0.0025	390		5s	$(5/2)^2[5/2]_2$	sp	$(^3F)^1P^\circ\ ^3D^\circ_3$			R1c		
3649.2205(22)	27395.307(17)	3649.2180(4)	0.0025	52		4d	$(3/2)^2[5/2]_2$	5f	$(5/2)^2[3/2]^\circ_2$			R1c		
3651.800(3)	27375.960(21)	3651.7990(5)	0.001	100		sp	$(^3F)^3P^\circ\ ^3F^\circ_2$	6d	$(5/2)^2[3/2]_2$			R1c		
3664.2913(10)	27282.636(8)	3664.2900(5)	0.0014	130		5s	$(3/2)^2[3/2]_2$	sp	$(^3P)^3P^\circ\ ^1D^\circ_2$			R1c		
3678.6688(20)	27176.009(15)	3678.6702(6)	−0.0014	80		sp	$(^3F)^3P^\circ\ ^3F^\circ_2$	6d	$(5/2)^2[1/2]_1$			R1c		
3682.4238(14)	27148.298(11)	3682.4254(7)	−0.0016	3900		sp	$(^3F)^3P^\circ\ ^3F^\circ_4$	7s	$(5/2)^2[5/2]_3$			R1c		
3686.5552(10)	27117.875(8)	3686.5539(4)	0.0013	16000		4p	$^3F^\circ_3$	s^2	$^1G_4$	9.7e+05	C+	R1c	O07	
3693.3279(12)	27068.148(9)	3693.3271(10)	0.0009	81		4d	$(3/2)^2[1/2]_0$	7p	$(3/2)^2[1/2]^\circ_1$			R2nc		
3703.9305(11)	26990.667(8)	3703.9285(5)	0.0020	680		sp	$(^3F)^3P^\circ\ ^1F^\circ_3$	6d	$(3/2)^2[7/2]_4$			R1c		
3724.1764(11)	26843.940(8)	3724.1770(3)	−0.0006	390		5p	$(5/2)^2[3/2]^\circ_2$	7s	$(3/2)^2[3/2]_2$			R1c		
3725.4537(17)	26834.737(12)	3725.4544(4)	−0.0007	110		4p	$^3F^\circ_4$	s^2	$^1G_4$			R1c		
3728.6574(10)	26811.681(7)	3728.6569(4)	0.0005	280		sp	$(^3F)^3P^\circ\ ^3D^\circ_1$	7s	$(5/2)^2[5/2]_2$			R1c		
3730.3433(10)	26799.564(7)	3730.3437(5)	−0.0005	250		5s	$(5/2)^2[5/2]_2$	sp	$(^3P)^3P^\circ\ ^3P^\circ_1$			R1c		
3731.1462(13)	26793.797(9)	3731.1479(3)	−0.0018	170		5p	$(5/2)^2[3/2]^\circ_2$	7s	$(3/2)^2[3/2]_1$			R1c		
3738.6488(10)	26740.029(7)	3738.6478(3)	0.0010	1700		sp	$(^3F)^3P^\circ\ ^3F^\circ_3$	7s	$(5/2)^2[5/2]_2$			R1c		
3740.3281(11)	26728.024(8)	3740.3278(4)	0.0003	280		5s	$(5/2)^2[5/2]_3$	sp	$(^3F)^1P^\circ\ ^3F^\circ_4$			R1c		
3742.0597(10)	26715.656(7)	3742.0592(4)	0.0006	430		5s	$(3/2)^2[3/2]_1$	sp	$(^3F)^1P^\circ\ ^3D^\circ_2$			R1c		
3748.1909(10)	26671.956(7)	3748.1901(3)	0.0008	1800		sp	$(^3F)^3P^\circ\ ^3F^\circ_3$	7s	$(5/2)^2[5/2]_3$			R1c		
3749.782(10)	26660.64(7)	3749.7756(4)	0.006	400	*	5s	$(5/2)^2[5/2]_3$	sp	$(^3P)^3P^\circ\ ^3D^\circ_2$			R1c		
3749.782(10)	26660.64(7)	3749.7826(5)	−0.001	400	*	sp	$(^3F)^3P^\circ\ ^1D^\circ_2$	6d	$(3/2)^2[5/2]_2$			R1c		
3751.3859(10)	26649.241(7)	3751.3858(4)	0.0001	310		sp	$(^3F)^3P^\circ\ ^1D^\circ_2$	6d	$(3/2)^2[5/2]_3$			R1c		
3761.4270(14)	26578.103(10)	3761.4269(5)	0.0001	150		sp	$(^3F)^3P^\circ\ ^1D^\circ_2$	6d	$(3/2)^2[3/2]_2$			R1c		
3766.8662(14)	26539.726(10)	3766.8669(5)	−0.0007	230		sp	$(^3F)^3P^\circ\ ^1D^\circ_2$	6d	$(3/2)^2[3/2]_1$			R1c		
3772.5256(10)	26499.913(7)	3772.5262(5)	−0.0006	480		sp	$(^3F)^3P^\circ\ ^1D^\circ_2$	6d	$(3/2)^2[7/2]_3$			R1c		
3774.9743(20)	26482.724(14)	3774.9751(4)	−0.0008	59		5p	$(5/2)^2[3/2]^\circ_2$	6d	$(5/2)^2[5/2]_2$			R1c		
3777.6444(10)	26464.006(7)	3777.6455(3)	−0.0011	150		5p	$(5/2)^2[3/2]^\circ_2$	6d	$(5/2)^2[7/2]_3$			R1c		
3781.9355(12)	26433.980(9)	3781.9350(4)	0.0005	290		5s	$(3/2)^2[3/2]_2$	sp	$(^3F)^1P^\circ\ ^3D^\circ_2$			R1c		
3786.2696(10)	26403.721(7)	3786.2702(5)	−0.0006	3200		5p	$(5/2)^2[3/2]^\circ_2$	6d	$(5/2)^2[5/2]_3$			R1c		
3795.4413(11)	26339.919(8)	3795.4419(4)	−0.0006	450		5s	$(5/2)^2[5/2]_2$	sp	$(^3P)^3P^\circ\ ^3D^\circ_2$			R1c		
3796.0675(12)	26335.574(8)	3796.0687(4)	−0.0012	580		5p	$(5/2)^2[3/2]^\circ_2$	6d	$(5/2)^2[3/2]_1$			R1c		
3796.1694(12)	26334.867(9)	3796.1694(11)	−0.0000	820		5p	$(3/2)^2[1/2]^\circ_1$	6d	$(3/2)^2[1/2]_0$			R1c		
3797.8489(10)	26323.221(7)	3797.8496(5)	−0.0007	4200		5p	$(5/2)^2[3/2]^\circ_2$	6d	$(5/2)^2[3/2]_2$			R1c		
3801.5563(10)	26297.551(7)	3801.5556(5)	0.0007	670		5p	$(3/2)^2[1/2]^\circ_0$	6d	$(3/2)^2[3/2]_1$			R1c		
3806.27(5)	26265.0(3)	3806.333(3)	−0.06	7400		d^{10}	$^1S_0$	4s	$^1D_2$	1.58e+00	C+	T53,M97,P84	P84	X,E2
3818.8787(11)	26178.268(7)	3818.8793(8)	−0.0006	2000		5p	$(5/2)^2[3/2]^\circ_1$	6d	$(5/2)^2[1/2]_0$			R1c		

Table A1. *Cont.*

λ_{obs} [a] (Å)	σ_{obs} [b] (cm^{-1})	λ_{Ritz} [c] (Å)	$\Delta\lambda_{obs\text{-}Ritz}$ (Å)	I_{obs} [d] (arb. u.)	Char [e]	Lower Level		Upper Level		A (s^{-1})	Acc [f]	Line Ref. [g]	TP Ref. [g]	Notes [h]
3824.8323(11)	26137.521(8)	3824.8321(6)	0.0002	1000		5p	$(3/2)^2[1/2]^\circ_0$	6d	$(3/2)^2[1/2]_1$			R1c		
3826.9209(10)	26123.256(7)	3826.9216(6)	−0.0007	3000		5p	$(5/2)^2[3/2]^\circ_2$	6d	$(5/2)^2[1/2]_1$			R1c		
3836.1646(11)	26060.310(7)	3836.1655(3)	−0.0010	1800		5p	$(5/2)^2[5/2]^\circ_2$	7s	$(3/2)^2[3/2]_2$			R1c		
3841.4818(11)	26024.240(7)	3841.4819(4)	−0.0001	590		5s	$(5/2)^2[5/2]_2$	sp	$(^3P)^3P^\circ\ ^3D^\circ_1$			R1c		
3842.5889(12)	26016.742(8)	3842.5882(3)	0.0007	2000		5p	$(5/2)^2[3/2]^\circ_1$	7s	$(3/2)^2[3/2]_2$			R1c		
3843.5628(13)	26010.150(9)	3843.5624(3)	0.0004	150		5p	$(5/2)^2[5/2]^\circ_2$	7s	$(3/2)^2[3/2]_1$			R1c		
3849.5824(13)	25969.479(9)	3849.5810(5)	0.0014	1700		5s	$(5/2)^2[5/2]_3$	sp	$(^3P)^3P^\circ\ ^3D^\circ_3$			R1c		
3850.0092(15)	25966.600(10)	3850.0099(3)	−0.0007	1600		5p	$(5/2)^2[3/2]^\circ_1$	7s	$(3/2)^2[3/2]_1$			R1c		
3851.1050(11)	25959.212(7)	3851.1020(4)	0.0030	890		s^2	$^1G_4$	sp	$(^3F)^3P^\circ\ ^1F^\circ_3$			R1c		
3859.41(6)	25903.4(4)	3859.4330(7)	−0.02	180		4d	$(5/2)^2[3/2]_2$	6p	$(3/2)^2[3/2]^\circ_2$			S36c		
3861.3421(11)	25890.391(8)	3861.3424(4)	−0.0004	1300		5p	$(5/2)^2[7/2]^\circ_3$	6d	$(5/2)^2[5/2]_2$			R1c		
3862.1245(11)	25885.146(8)	3862.1243(4)	0.0002	1200		5p	$(3/2)^2[5/2]^\circ_2$	6d	$(3/2)^2[5/2]_3$			R1c		
3863.6403(11)	25874.991(8)	3863.6404(4)	−0.0001	1200		5p	$(5/2)^2[7/2]^\circ_3$	6d	$(5/2)^2[7/2]_4$			R1c		
3864.1370(11)	25871.665(7)	3864.1365(4)	0.0005	3200		5p	$(5/2)^2[7/2]^\circ_3$	6d	$(5/2)^2[7/2]_3$			R1c		
3866.3047(11)	25857.160(7)	3866.3035(3)	0.0013	1200		5p	$(5/2)^2[5/2]^\circ_3$	7s	$(3/2)^2[3/2]_2$			R1c		
3868.3711(11)	25843.348(7)	3868.3707(3)	0.0004	1100		sp	$(^3F)^3P^\circ\ ^3F^\circ_2$	7s	$(5/2)^2[5/2]_2$			R1c		
3871.334(3)	25823.570(21)	3871.3321(10)	0.002	210		sp	$(^3F)^3P^\circ\ ^1G^\circ_4$	7s	$(5/2)^2[5/2]_3$			R1c		
3872.7645(15)	25814.031(10)	3872.7678(5)	−0.0033	550		5p	$(3/2)^2[5/2]^\circ_2$	6d	$(3/2)^2[3/2]_2$			R1c		
3873.2075(11)	25811.078(7)	3873.2058(6)	0.0017	940		5s	$(5/2)^2[5/2]_3$	sp	$(^3P)^3P^\circ\ ^3P^\circ_2$			R1c		
3878.5873(12)	25775.278(8)	3878.5875(3)	−0.0002	860		sp	$(^3F)^3P^\circ\ ^3F^\circ_2$	7s	$(5/2)^2[5/2]_3$			R1c		
3878.667	25774.75	3878.6670(5)			:	5p	$(3/2)^2[1/2]^\circ_1$	6d	$(3/2)^2[5/2]_2$					
3879.3966(11)	25769.901(7)	3879.3976(4)	−0.0010	3800		5p	$(5/2)^2[7/2]^\circ_4$	6d	$(5/2)^2[7/2]_4$			R1c		
3879.8973(21)	25766.576(14)	3879.8977(4)	−0.0005	140		5p	$(5/2)^2[7/2]^\circ_4$	6d	$(5/2)^2[7/2]_3$			R1c		
3884.1312(9)	25738.489(6)	3884.1313(5)	−0.0001	8100		5p	$(5/2)^2[7/2]^\circ_3$	6d	$(5/2)^2[9/2]_4$	3.1e+07	D+	F_Re	TW	
3884.5339(11)	25735.821(7)	3884.5350(5)	−0.0010	3800		5p	$(3/2)^2[5/2]^\circ_2$	6d	$(3/2)^2[7/2]_3$	3.1e+07	D+	R1c	TW	
3885.2865(17)	25730.836(11)	3885.2789(5)	0.0076	85	bl	5p	$(5/2)^2[7/2]^\circ_3$	6d	$(5/2)^2[3/2]_2$			R1c		X
3890.0864(14)	25699.088(9)	3890.0865(4)	−0.0001	2200		5p	$(5/2)^2[5/2]^\circ_2$	6d	$(5/2)^2[5/2]_2$			R1c		
3891.1277(12)	25692.211(8)	3891.1267(5)	0.0010	1500		5p	$(3/2)^2[1/2]^\circ_1$	6d	$(3/2)^2[3/2]_2$			R1c		
3892.9237(11)	25680.358(7)	3892.9223(4)	0.0014	3000		5p	$(5/2)^2[5/2]^\circ_2$	6d	$(5/2)^2[7/2]_3$			R1c		
3894.6198(21)	25669.175(14)	3894.61915(19)	0.0006	280		5s	$(3/2)^2[3/2]_2$	4f	$(5/2)^2[5/2]^\circ_3$			R1c		
3896.6915(11)	25655.527(7)	3896.6912(4)	0.0003	1900		5p	$(5/2)^2[3/2]^\circ_1$	6d	$(5/2)^2[5/2]_2$			R1c		
3896.9491(20)	25653.832(13)	3896.9487(5)	0.0004	280		5p	$(3/2)^2[1/2]^\circ_1$	6d	$(3/2)^2[3/2]_1$			R1c		
3897.7292(20)	25648.698(13)	3897.7262(5)	0.0030	2500		5s	$(5/2)^2[5/2]_2$	sp	$(^3P)^3P^\circ\ ^3D^\circ_3$			R1c		
3900.0557(11)	25633.397(7)	3900.0565(5)	−0.0008	1100		5p	$(5/2)^2[7/2]^\circ_4$	6d	$(5/2)^2[9/2]_4$			R1c		
3901.2315(12)	25625.672(8)	3901.2316(5)	−0.0000	280		5p	$(3/2)^2[5/2]^\circ_3$	6d	$(3/2)^2[5/2]_2$			R1c		
3902.0837(24)	25620.076(16)	3902.0821(5)	0.0016	140		5p	$(5/2)^2[5/2]^\circ_2$	6d	$(5/2)^2[5/2]_3$			R1c		
3902.9667(12)	25614.280(8)	3902.9668(4)	−0.0002	2500		5p	$(3/2)^2[5/2]^\circ_3$	6d	$(3/2)^2[5/2]_3$			R1c		
3903.1761(7)	25612.905(5)	3903.1759(5)	0.0002	8200		5p	$(5/2)^2[7/2]^\circ_4$	6d	$(5/2)^2[9/2]_5$	3.5e+07	D+	F_Re	TW	
3905.1121(14)	25600.208(9)	3905.1133(6)	−0.0012	830		4d	$(3/2)^2[1/2]_0$	5f	$(3/2)^2[3/2]^\circ_1$			R1c		
3907.2743(24)	25586.042(16)	3907.2616(8)	0.0127	82	?	5s	$(3/2)^2[3/2]_2$	sp	$(^3P)^3P^\circ\ ^5S^\circ_2$			R1c		X

Table A1. *Cont.*

λ_{obs} [a] (Å)	σ_{obs} [b] (cm^{-1})	λ_{Ritz} [c] (Å)	$\Delta\lambda_{obs-Ritz}$ (Å)	I_{obs} [d] (arb. u.)	Char [e]	Lower Level		Upper Level		A (s^{-1})	Acc [f]	Line Ref. [g]	TP Ref. [g]	Notes [h]
3912.4911(12)	25551.927(8)	3912.4900(5)	0.0011	690		5p	$(5/2)^2[5/2]^\circ_2$	6d	$(5/2)^2[3/2]_1$			R1c		
3913.8355(14)	25543.150(9)	3913.8369(5)	−0.0014	160		5p	$(3/2)^2[5/2]^\circ_3$	6d	$(3/2)^2[3/2]_2$			R1c		
3914.311(3)	25540.050(21)	3914.3081(9)	0.002	110	bl	4d	$(5/2)^2[9/2]_4$	6p	$(3/2)^2[5/2]^\circ_3$			R1c		
3918.3817(12)	25513.515(8)	3918.3792(4)	0.0025	400		s^2	$^1G_4$	5p	$(5/2)^2[5/2]^\circ_3$			R1c		
3919.1725(11)	25508.366(7)	3919.1710(4)	0.0016	1000		5p	$(5/2)^2[3/2]^\circ_1$	6d	$(5/2)^2[3/2]_1$			R1c		
3920.6546(11)	25498.724(7)	3920.6561(5)	−0.0015	4000		5p	$(3/2)^2[5/2]^\circ_3$	6d	$(3/2)^2[7/2]_4$	3.3e+07	D+	R1c	TW	
3921.071(10)	25496.02(7)	3921.0693(5)	0.002	130	*	5p	$(5/2)^2[3/2]^\circ_1$	6d	$(5/2)^2[3/2]_2$			R1c		
3921.071(10)	25496.02(7)	3921.0811(4)	−0.010	130	*	5p	$(5/2)^2[5/2]^\circ_3$	6d	$(5/2)^2[5/2]_2$			R1c		
3921.4136(15)	25493.789(10)	3921.4117(6)	0.0019	450		5p	$(3/2)^2[1/2]^\circ_1$	6d	$(3/2)^2[1/2]_1$			R1c		
3923.4504(11)	25480.555(7)	3923.4506(5)	−0.0003	2100		5p	$(5/2)^2[5/2]^\circ_3$	6d	$(5/2)^2[7/2]_4$			R1c		
3923.9630(12)	25477.226(8)	3923.9622(4)	0.0008	400		5p	$(5/2)^2[5/2]^\circ_3$	6d	$(5/2)^2[7/2]_3$			R1c		
3925.8560(11)	25464.942(7)	3925.8554(5)	0.0006	390		5p	$(3/2)^2[5/2]^\circ_3$	6d	$(3/2)^2[7/2]_3$			R1c		
3933.2684(11)	25416.953(7)	3933.2688(5)	−0.0004	2000		5p	$(5/2)^2[5/2]^\circ_3$	6d	$(5/2)^2[5/2]_3$			R1c		
3934.1202(11)	25411.450(7)	3934.1175(3)	0.0027	470		sp	$(^3F)^3P^\circ\ ^1F^\circ_3$	7s	$(3/2)^2[3/2]_2$			R1c		
3935.9626(14)	25399.555(9)	3935.9560(6)	0.0067	380	bl	5s	$(3/2)^2[3/2]_1$	sp	$(^3P)^3P^\circ\ ^3P^\circ_0$			R1c		X
3940.9702(12)	25367.282(8)	3940.9707(5)	−0.0004	300		5s	$(3/2)^2[3/2]_2$	sp	$(^3F)^1P^\circ\ ^3D^\circ_3$			R1c		
3944.5822(12)	25344.054(8)	3944.5826(5)	−0.0004	490		5p	$(5/2)^2[5/2]^\circ_3$	6d	$(5/2)^2[9/2]_4$			R1c		
3945.5770(12)	25337.664(8)	3945.5767(5)	0.0004	2300		5p	$(3/2)^2[3/2]^\circ_1$	6d	$(3/2)^2[5/2]_2$			R1c		
3945.7664(13)	25336.448(9)	3945.7662(5)	0.0002	1500		5p	$(5/2)^2[5/2]^\circ_3$	6d	$(5/2)^2[3/2]_2$			R1c		
3952.0661(11)	25296.062(7)	3952.0660(6)	0.0001	980		5p	$(5/2)^2[3/2]^\circ_1$	6d	$(5/2)^2[1/2]_1$			R1c		
3958.4734(13)	25255.118(8)	3958.4707(5)	0.0027	840		5p	$(3/2)^2[3/2]^\circ_1$	6d	$(3/2)^2[3/2]_2$			R1c		
3974.3504(22)	25154.229(14)	3974.3528(7)	−0.0024	91		4d	$(5/2)^2[5/2]_2$	6p	$(3/2)^2[3/2]^\circ_2$			R1c		
3975.5387(13)	25146.711(8)	3975.5380(9)	0.0007	90		4d	$(3/2)^2[1/2]_0$	7p	$(5/2)^2[3/2]^\circ_1$			R1c		
3985.2145(11)	25085.657(7)	3985.2136(5)	0.0009	700		5p	$(3/2)^2[3/2]^\circ_2$	6d	$(3/2)^2[5/2]_2$			R1c		
3987.0237(11)	25074.274(7)	3987.0244(5)	−0.0007	2300		5p	$(3/2)^2[3/2]^\circ_2$	6d	$(3/2)^2[5/2]_3$			R1c		
3990.7807(13)	25050.670(8)	3990.7773(5)	0.0033	960		5s	$(3/2)^2[3/2]_1$	sp	$(^3P)^3P^\circ\ ^3P^\circ_1$			R1c		
3993.3023(12)	25034.852(8)	3993.3022(5)	0.0001	2600		sp	$(^3F)^3P^\circ\ ^1F^\circ_3$	6d	$(5/2)^2[7/2]_4$			R1c		
3998.3675(12)	25003.138(8)	3998.3684(5)	−0.0009	860		5p	$(3/2)^2[3/2]^\circ_2$	6d	$(3/2)^2[3/2]_2$			R1c		
4003.4759(12)	24971.235(8)	4003.4736(5)	0.0023	1600		sp	$(^3F)^3P^\circ\ ^1F^\circ_3$	6d	$(5/2)^2[5/2]_3$			R1c		
4004.5155(14)	24964.752(9)	4004.5160(6)	−0.0004	400		5p	$(3/2)^2[3/2]^\circ_2$	6d	$(3/2)^2[3/2]_1$			R1c		
4006.1651(12)	24954.473(7)	4006.1666(3)	−0.0015	1500		sp	$(^3F)^3P^\circ\ ^1D^\circ_2$	7s	$(3/2)^2[3/2]_2$			R1c		
4014.2360(12)	24904.301(7)	4014.2342(4)	0.0018	870		sp	$(^3F)^3P^\circ\ ^1D^\circ_2$	7s	$(3/2)^2[3/2]_1$			R1c		
4015.1963(12)	24898.345(8)	4015.1954(5)	0.0009	860		sp	$(^3F)^3P^\circ\ ^1F^\circ_3$	6d	$(5/2)^2[9/2]_4$			R1c		
4016.4254(14)	24890.726(9)	4016.4218(6)	0.0036	1500		sp	$(^3F)^3P^\circ\ ^1F^\circ_3$	6d	$(5/2)^2[3/2]_2$			R1c		
4030.3508(21)	24804.727(13)	4030.3525(7)	−0.0018	330		5p	$(3/2)^2[3/2]^\circ_2$	6d	$(3/2)^2[1/2]_1$			R1c		
4032.6469(12)	24790.604(7)	4032.6470(3)	−0.0001	1500		5p	$(5/2)^2[3/2]^\circ_2$	7s	$(5/2)^2[5/2]_2$			R1c		
4042.549(3)	24729.880(21)	4042.5473(7)	0.002	67		4d	$(5/2)^2[1/2]_1$	6p	$(5/2)^2[3/2]^\circ_1$			R1c		
4043.4879(9)	24724.139(6)	4043.4858(5)	0.0021	27000		4p	$^1F^\circ_3$	s^2	$^1G_4$	1.14e+06	C+	F_Re	O07	
4043.7515(4)	24722.527(3)	4043.75122(21)	0.0003	17000		5p	$(5/2)^2[3/2]^\circ_2$	7s	$(5/2)^2[5/2]_3$			F_Re		
4053.6529(21)	24662.142(13)	4053.6521(4)	0.0008	3100		5p	$(3/2)^2[1/2]^\circ_0$	7s	$(3/2)^2[3/2]_1$			R1c		
4065.0094(12)	24593.244(7)	4065.0090(4)	0.0004	1600		sp	$(^3F)^3P^\circ\ ^1D^\circ_2$	6d	$(5/2)^2[5/2]_2$			R1c		

Table A1. *Cont.*

λ_{obs} [a] (Å)	σ_{obs} [b] (cm^{-1})	λ_{Ritz} [c] (Å)	$\Delta\lambda_{obs-Ritz}$ (Å)	I_{obs} [d] (arb. u.)	Char [e]	Lower Level		Upper Level		A (s^{-1})	Acc [f]	Line Ref. [g]	TP Ref. [g]	Notes [h]
4065.3723(22)	24591.049(13)	4065.3730(5)	−0.0007	230		5s	$(3/2)^2[3/2]_1$	sp	$(^3P)^3P^\circ\ ^3D^\circ_2$			R1c		
4068.1058(12)	24574.526(7)	4068.1056(4)	0.0001	2000		sp	$(^3F)^3P^\circ\ ^1D^\circ_2$	6d	$(5/2)^2[7/2]_3$			R1c		
4089.4787(13)	24446.094(8)	4089.4788(5)	−0.0001	480		sp	$(^3F)^3P^\circ\ ^1D^\circ_2$	6d	$(5/2)^2[3/2]_1$			R1c		
4112.4816(13)	24309.359(8)	4112.4801(5)	0.0015	820		5s	$(3/2)^2[3/2]_2$	sp	$(^3P)^3P^\circ\ ^3D^\circ_2$			R1c		
4118.2403(24)	24275.367(14)	4118.2398(5)	0.0005	57		5s	$(3/2)^2[3/2]_1$	sp	$(^3P)^3P^\circ\ ^3D^\circ_1$			R1c		
4125.171(4)	24234.580(21)	4125.1717(8)	−0.000	11		4d	$(5/2)^2[1/2]_1$	6p	$(5/2)^2[3/2]^\circ_2$			R1c		
4125.9352(17)	24230.094(10)	4125.9352(9)	0.0001	65		5p	$(3/2)^2[1/2]^\circ_1$	6d	$(5/2)^2[1/2]_0$			R1c		
4131.3610(4)	24198.2729(25)	4131.3610(3)	0.0000	16000		5p	$(5/2)^2[7/2]^\circ_3$	7s	$(5/2)^2[5/2]_2$			F_Re		
4132.7116(21)	24190.365(12)	4132.7109(3)	0.0007	1000		5p	$(3/2)^2[5/2]^\circ_2$	7s	$(3/2)^2[3/2]_2$			R1c		
4141.2965(5)	24140.219(3)	4141.2968(4)	−0.0003	9700		5p	$(3/2)^2[5/2]^\circ_2$	7s	$(3/2)^2[3/2]_1$			F_Re		
4143.0170(13)	24130.195(7)	4143.01624(20)	0.0007	2100		5p	$(5/2)^2[7/2]^\circ_3$	7s	$(5/2)^2[5/2]_3$			R1c		
4147.8136(20)	24102.291(11)	4147.8131(6)	0.0004	190		4d	$(5/2)^2[3/2]_2$	6p	$(5/2)^2[5/2]^\circ_3$			R1c		
4151.904(3)	24078.544(15)	4151.9059(6)	−0.002	47		4d	$(5/2)^2[9/2]_4$	6p	$(5/2)^2[5/2]^\circ_3$			R1c		
4153.6234(13)	24068.579(7)	4153.6236(3)	−0.0003	3600		5p	$(3/2)^2[1/2]^\circ_1$	7s	$(3/2)^2[3/2]_2$			R1c		
4154.642(4)	24062.680(21)	4154.642(4)	−0.000	9		s^2	$^3P_2$	sp	$(^3F)^3P^\circ\ ^5G^\circ_2$			R1c		
4161.13984(15)	24025.1034(9)	4161.13989(14)	−0.00005	17000		5p	$(5/2)^2[7/2]^\circ_4$	7s	$(5/2)^2[5/2]_3$			F_Re		
4162.2967(14)	24018.426(8)	4162.2967(4)	0.0000	3000		5p	$(3/2)^2[1/2]^\circ_1$	7s	$(3/2)^2[3/2]_1$			R1c		
4164.2826(10)	24006.972(6)	4164.2827(3)	−0.0001	8700		5p	$(5/2)^2[5/2]^\circ_2$	7s	$(5/2)^2[5/2]_2$			F_Re		
4165.67(15)	23999.0(9)	4165.775(3)	−0.10	3		d^{10}	$^1S_0$	4s	$^3D_1$	2.1e−12	C+	M97	TW	X,M1
4166.5884(23)	23993.687(13)	4166.5874(5)	0.0010	350		5s	$(3/2)^2[3/2]_2$	sp	$(^3P)^3P^\circ\ ^3D^\circ_1$			R1c		
4171.8513(5)	23963.419(3)	4171.8521(3)	−0.0007	10000		5p	$(5/2)^2[3/2]^\circ_1$	7s	$(5/2)^2[5/2]_2$			F_Re		
4176.1248(12)	23938.897(7)	4176.1247(3)	0.0001	1400		5p	$(5/2)^2[5/2]^\circ_2$	7s	$(5/2)^2[5/2]_3$			R1c		
4179.5117(4)	23919.4987(21)	4179.5113(3)	0.0003	12000		5p	$(3/2)^2[5/2]^\circ_3$	7s	$(3/2)^2[3/2]_2$			F_Re		
4185.149(4)	23887.280(21)	4185.1555(11)	−0.006	8		s^2	$^1D_2$	sp	$(^3F)^3P^\circ\ ^5D^\circ_3$			R1c		
4195.3590(16)	23829.148(9)	4195.3583(5)	0.0007	210		5p	$(3/2)^2[5/2]^\circ_2$	6d	$(5/2)^2[5/2]_2$			R1c		
4195.7735(13)	23826.794(8)	4195.7740(7)	−0.0005	340		4d	$(5/2)^2[9/2]_5$	6p	$(5/2)^2[7/2]^\circ_4$			R1c		
4198.6561(13)	23810.436(8)	4198.6568(4)	−0.0007	240		5p	$(3/2)^2[5/2]^\circ_2$	6d	$(5/2)^2[7/2]_3$			R1c		
4199.435(4)	23806.020(21)	4199.4464(9)	−0.011	110		4d	$(3/2)^2[3/2]_1$	6p	$(3/2)^2[3/2]^\circ_1$			R1c		
4201.735(3)	23792.990(17)	4201.7309(10)	0.004	15		5p	$(3/2)^2[3/2]^\circ_1$	6d	$(5/2)^2[1/2]_0$			R1c		
4201.888(4)	23792.120(21)	4201.8877(16)	0.001	22		sp	$(^3P)^3P^\circ\ ^5D^\circ_3$	8d	$(5/2)^2[9/2]_4$			R1c		
4207.672(3)	23759.417(16)	4207.5870(9)	0.085	15	?	sp	$(^3F)^3P^\circ\ ^3G^\circ_4$	5d	$(3/2)^2[5/2]_3$			R1c		X
4209.321(4)	23750.110(21)	4209.3213(10)	−0.000	29		4d	$(3/2)^2[1/2]_1$	6p	$(3/2)^2[1/2]^\circ_1$			R1c		
4211.8649(8)	23735.766(5)	4211.86561(23)	−0.0007	11000		5p	$(5/2)^2[5/2]^\circ_3$	7s	$(5/2)^2[5/2]_3$			F_Re		
4212.315(3)	23733.232(16)	4212.3160(7)	−0.001	14		4d	$(5/2)^2[9/2]_4$	6p	$(5/2)^2[7/2]^\circ_4$			R1c		
4216.9124(13)	23707.356(7)	4216.9115(5)	0.0008	640		5p	$(3/2)^2[1/2]^\circ_1$	6d	$(5/2)^2[5/2]_2$			R1c		
4219.38(20)	23693.5(11)	4219.5840(6)	−0.20	340		4d	$(5/2)^2[3/2]_2$	sp	$(^1G)^3P^\circ\ ^3F^\circ_3$			S36c		
4221.427(2)	23682.000(14)	4221.4276(5)	−0.000	69		5p	$(3/2)^2[5/2]^\circ_2$	6d	$(5/2)^2[3/2]_1$			R1c		
4223.8195(14)	23668.588(8)	4223.8197(6)	−0.0002	100		4d	$(5/2)^2[9/2]_4$	sp	$(^1G)^3P^\circ\ ^3F^\circ_3$			R1c		
4224.344(4)	23665.650(21)	4224.3455(3)	−0.002	14		4d	$(5/2)^2[1/2]_1$	4f	$(3/2)^2[5/2]^\circ_2$			R1c		
4225.1956(14)	23660.880(8)	4225.1967(6)	−0.0012	140		4d	$(5/2)^2[5/2]_3$	6p	$(5/2)^2[5/2]^\circ_3$			R1c		
4227.9422(14)	23645.509(8)	4227.9397(5)	0.0025	7900		4p	$^3D^\circ_3$	s^2	$^1G_4$	5.1e+05	B	R1c	O07	

Table A1. *Cont.*

λ_{obs} [a] (Å)	σ_{obs} [b] (cm^{-1})	λ_{Ritz} [c] (Å)	$\Delta\lambda_{obs-Ritz}$ (Å)	I_{obs} [d] (arb. u.)	Char [e]	Lower Level	Upper Level	A (s^{-1})	Acc [f]	Line Ref. [g]	TP Ref. [g]	Notes [h]
4230.4486(14)	23631.500(8)	4230.4495(4)	−0.0009	4000		5p $(3/2)^2[3/2]^{\circ}_1$	7s $(3/2)^2[3/2]_2$			R1c		
4232.8346(15)	23618.180(8)	4232.8362(6)	−0.0016	130		5s $(3/2)^2[3/2]_2$	sp $(^3P)^3P^{\circ}\ ^3D^{\circ}_3$			R1c		
4239.4445(20)	23581.356(11)	4239.4468(4)	−0.0023	7000		5p $(3/2)^2[3/2]^{\circ}_1$	7s $(3/2)^2[3/2]_1$			F_Re		
4240.410(3)	23575.985(14)	4240.4120(8)	−0.002	64		4d $(5/2)^2[3/2]_1$	6p $(5/2)^2[3/2]^{\circ}_1$			R1c		
4241.31(4)	23570.97(22)	4241.271(3)	0.04	48	*	sp $(^3P)^3P^{\circ}\ ^5D^{\circ}_4$	8d $(5/2)^2[7/2]_4$			R1nc		
4241.31(4)	23570.97(22)	4241.3236(9)	−0.01	48	*	4d $(3/2)^2[3/2]_2$	6p $(3/2)^2[3/2]^{\circ}_1$			R1c		
4243.2516(15)	23560.199(8)	4243.2501(5)	0.0015	320		5p $(3/2)^2[1/2]^{\circ}_1$	6d $(5/2)^2[3/2]_1$			R1c		
4246.9860(20)	23539.483(11)	4246.9715(4)	0.0145	390	*	5p $(3/2)^2[5/2]^{\circ}_3$	6d $(5/2)^2[7/2]_3$			R1nc		X
4246.9860(20)	23539.483(11)	4246.9847(7)	0.0013	390	*	4d $(3/2)^2[1/2]_0$	5f $(5/2)^2[3/2]^{\circ}_1$			R1c		
4251.0194(15)	23517.149(8)	4251.0200(6)	−0.0006	300		4d $(5/2)^2[1/2]_1$	4f $(3/2)^2[3/2]^{\circ}_1$			R1c		
4254.630(4)	23497.190(21)	4254.6275(11)	0.003	12		4d $(3/2)^2[7/2]_3$	6p $(3/2)^2[5/2]^{\circ}_2$			R2nc		
4255.6348(14)	23491.644(8)	4255.6343(4)	0.0006	1100		4d $(5/2)^2[1/2]_1$	4f $(3/2)^2[3/2]^{\circ}_2$			R1c		
4256.846(4)	23484.960(21)	4256.8484(7)	−0.002	30		4d $(3/2)^2[1/2]_0$	5f $(5/2)^2[1/2]^{\circ}_1$			R1c		
4267.611(4)	23425.720(21)	4267.6128(13)	−0.002	120		s^2 $^1G_4$	sp $(^3F)^3P^{\circ}\ ^1G^{\circ}_4$			R1c		
4268.103(4)	23423.020(21)	4268.1012(12)	0.002	18		sp $(^3F)^3P^{\circ}\ ^5F^{\circ}_4$	5d $(5/2)^2[9/2]_5$			R1c		
4276.0484(9)	23379.499(5)	4276.0496(4)	−0.0012	8700		5p $(3/2)^2[3/2]^{\circ}_2$	7s $(3/2)^2[3/2]_2$			F_Re		
4277.805(4)	23369.900(21)	4277.8016(6)	0.003	150		4d $(5/2)^2[7/2]_4$	6p $(5/2)^2[5/2]^{\circ}_3$			R1c		
4279.9621(15)	23358.120(8)	4279.9594(3)	0.0028	5100		sp $(^3F)^3P^{\circ}\ ^1F^{\circ}_3$	7s $(5/2)^2[5/2]_2$			R1c		
4280.845(4)	23353.300(21)	4280.8437(6)	0.002	6		4d $(5/2)^2[5/2]_2$	6p $(5/2)^2[5/2]^{\circ}_3$			R1c		
4281.838(3)	23347.889(15)	4281.8368(8)	0.001	29		5p $(3/2)^2[1/2]^{\circ}_1$	6d $(5/2)^2[1/2]_1$			R1c		
4285.2433(13)	23329.334(7)	4285.2420(4)	0.0012	2000		5p $(3/2)^2[3/2]^{\circ}_2$	7s $(3/2)^2[3/2]_1$			R1c		
4286.465(3)	23322.683(16)	4286.4615(7)	0.004	11		4d $(5/2)^2[5/2]_2$	sp $(^3F)^1P^{\circ}\ ^3F^{\circ}_2$			R1c		
4287.13(6)	23319.1(3)	4287.0450(10)	0.08	180		4d $(3/2)^2[7/2]_4$	6p $(3/2)^2[5/2]^{\circ}_3$			S36c		
4287.779(3)	23315.537(17)	4287.7744(7)	0.005	11		4d $(5/2)^2[5/2]_3$	6p $(5/2)^2[7/2]^{\circ}_4$			R1c		
4291.084(3)	23297.580(16)	4291.0824(24)	0.002	640		5s $(5/2)^2[5/2]_3$	sp $(^3P)^3P^{\circ}\ ^5D^{\circ}_4$			R1c		
4292.4705(14)	23290.055(8)	4292.4694(3)	0.0011	9000		sp $(^3F)^3P^{\circ}\ ^1F^{\circ}_3$	7s $(5/2)^2[5/2]_3$			R1c		
4296.123(3)	23270.256(14)	4296.1187(5)	0.004	17		5p $(3/2)^2[3/2]^{\circ}_1$	6d $(5/2)^2[5/2]_2$			R1c		
4299.696(3)	23250.920(16)	4299.6946(6)	0.001	11		4d $(5/2)^2[5/2]_3$	sp $(^1G)^3P^{\circ}\ ^3F^{\circ}_3$			R1c		
4308.710(3)	23202.274(16)	4308.7114(9)	−0.001	28		4d $(3/2)^2[5/2]_2$	6p $(3/2)^2[3/2]^{\circ}_1$			R1c		
4320.7611(19)	23137.564(10)	4320.7616(7)	−0.0005	270		4d $(5/2)^2[1/2]_1$	sp $(^3P)^3P^{\circ}\ ^1D^{\circ}_2$			R1c		
4335.639(3)	23058.167(16)	4335.6353(9)	0.004	17		4d $(3/2)^2[5/2]_3$	6p $(3/2)^2[3/2]^{\circ}_2$			R1c		
4341.039(3)	23029.484(15)	4341.0414(14)	−0.002	17		5s $(5/2)^2[5/2]_3$	sp $(^3P)^3P^{\circ}\ ^5D^{\circ}_3$			R1c		
4343.152(4)	23018.280(21)	4343.1532(5)	−0.001	11		5p $(3/2)^2[3/2]^{\circ}_2$	6d $(5/2)^2[5/2]_2$			R1c		
4344.700(4)	23010.080(21)	4344.6972(9)	0.003	6		4d $(3/2)^2[5/2]_2$	6p $(3/2)^2[3/2]^{\circ}_2$			R1c		
4346.687(4)	22999.560(21)	4346.6883(5)	−0.001	240		5p $(3/2)^2[3/2]^{\circ}_2$	6d $(5/2)^2[7/2]_3$			R1c		
4357.339(4)	22943.340(21)	4357.3345(6)	0.004	6		4d $(5/2)^2[5/2]_2$	sp $(^1G)^3P^{\circ}\ ^3F^{\circ}_3$			R1c		
4360.102(3)	22928.797(14)	4360.0973(17)	0.005	47		5s $(5/2)^2[5/2]_3$	sp $(^3P)^3P^{\circ}\ ^5D^{\circ}_2$			R1c		
4365.3705(14)	22901.127(7)	4365.3697(3)	0.0008	7300		sp $(^3F)^3P^{\circ}\ ^1D^{\circ}_2$	7s $(5/2)^2[5/2]_2$			R1c		
4373.460(3)	22858.770(15)	4373.4591(7)	0.001	18		5p $(3/2)^2[3/2]^{\circ}_2$	6d $(5/2)^2[3/2]_2$			R1c		
4375.8(5)	22847(3)	4375.690(3)	0.1	2000		d^{10} $^1S_0$	4s $^3D_2$	9.e−02	D+	A73,P84	A08adj	X,E2
4378.3840(14)	22833.061(7)	4378.3848(3)	−0.0007	1500		sp $(^3F)^3P^{\circ}\ ^1D^{\circ}_2$	7s $(5/2)^2[5/2]_3$			R1c		
4396.423(3)	22739.375(15)	4396.4196(6)	0.004	13		4d $(5/2)^2[9/2]_4$	6p $(5/2)^2[7/2]^{\circ}_3$			R1c		

Table A1. *Cont.*

λ_{obs} [a] (Å)	σ_{obs} [b] (cm^{-1})	λ_{Ritz} [c] (Å)	$\Delta\lambda_{obs-Ritz}$ (Å)	I_{obs} [d] (arb. u.)	Char [e]	Lower Level		Upper Level		A (s^{-1})	Acc [f]	Line Ref. [g]	TP Ref. [g]	Notes [h]
4402.363(3)	22708.694(16)	4402.3616(15)	0.002	33		5s	$(5/2)^2[5/2]_2$	sp	$(^3P)^3P^\circ\,^5D^\circ_3$			R1c		
4410.941(3)	22664.535(17)	4410.9371(9)	0.004	7		4d	$(5/2)^2[1/2]_0$	6p	$(5/2)^2[3/2]^\circ_1$			R1c		
4419.088(4)	22622.750(21)	4419.0825(3)	0.006	7		4d	$(5/2)^2[3/2]_2$	4f	$(3/2)^2[7/2]^\circ_3$			R1c		
4422.583(3)	22604.871(16)	4422.5816(15)	0.002	15		5s	$(5/2)^2[5/2]_2$	sp	$(^3P)^3P^\circ\,^5D^\circ_1$			R1c		
4435.693(3)	22538.066(15)	4435.6906(4)	0.002	80		4d	$(5/2)^2[3/2]_2$	4f	$(3/2)^2[5/2]^\circ_2$			R1c		
4440.8836(15)	22511.721(8)	4440.8827(4)	0.0009	520		4d	$(5/2)^2[3/2]_1$	4f	$(3/2)^2[5/2]^\circ_2$			R1c		
4441.709(14)	22507.54(7)	4441.7193(7)	−0.011	180		sp	$(^3F)^3P^\circ\,^3D^\circ_3$	5d	$(3/2)^2[7/2]_4$			R1c		
4444.8314(14)	22491.727(7)	4444.8307(3)	0.0007	1200		4d	$(5/2)^2[3/2]_2$	4f	$(3/2)^2[5/2]^\circ_3$			R1c		
4462.6902(16)	22401.721(8)	4462.6875(5)	0.0027	1600		4d	$(5/2)^2[9/2]_4$	4f	$(3/2)^2[9/2]^\circ_5$			R1c		
4485.280	22288.90	4485.2800(6)			:	4d	$(5/2)^2[1/2]_1$	sp	$(^3F)^1P^\circ\,^3D^\circ_2$			W93		
4495.356(3)	22238.942(15)	4495.3577(8)	−0.002	38		sp	$(^3F)^3P^\circ\,^3G^\circ_3$	5d	$(3/2)^2[7/2]_4$			R1c		
4505.9982(8)	22186.418(4)	4505.9996(4)	−0.0013	11000		4p	$^3P^\circ_2$	s^2	$^3P_1$	6.3e+05	C	F_Re	H71,N88	L
4515.5194(15)	22139.637(7)	4515.5195(3)	−0.0001	960		4d	$(5/2)^2[5/2]_3$	4f	$(3/2)^2[7/2]^\circ_4$			R1c		
4516.0492(15)	22137.040(7)	4516.0499(4)	−0.0007	2700		5p	$(3/2)^2[5/2]^\circ_2$	7s	$(5/2)^2[5/2]_2$			R1c		
4529.983(3)	22068.951(15)	4529.9803(3)	0.002	70		5p	$(3/2)^2[5/2]^\circ_2$	7s	$(5/2)^2[5/2]_3$			R1c		
4533.814(4)	22050.300(21)	4533.8122(3)	0.002	54		4d	$(5/2)^2[5/2]_3$	4f	$(3/2)^2[5/2]^\circ_3$			R1c		
4541.0325(15)	22015.251(7)	4541.0337(4)	−0.0013	6900		5p	$(3/2)^2[1/2]^\circ_1$	7s	$(5/2)^2[5/2]_2$			R1c		
4542.15(4)	22009.82(19)	4542.1167(8)	0.04	390	*	4d	$(5/2)^2[3/2]_2$	sp	$(^3P)^3P^\circ\,^1D^\circ_2$			R1c		
4542.15(4)	22009.82(19)	4542.1719(10)	−0.02	390	*	4d	$(3/2)^2[1/2]_1$	6p	$(5/2)^2[3/2]^\circ_1$			R1c		
4546.433(3)	21989.099(12)	4546.4353(6)	−0.002	410		sp	$(^3F)^3P^\circ\,^3D^\circ_2$	5d	$(3/2)^2[5/2]_3$			R1c		
4551.807(3)	21963.141(13)	4551.8069(8)	−0.000	260		4d	$(3/2)^2[7/2]_3$	sp	$(^3F)^1P^\circ\,^3F^\circ_2$			R1c		
4555.9211(5)	21943.307(3)	4555.9193(4)	0.0018	18000		4p	$^3P^\circ_2$	s^2	$^3P_2$	7.1e+05	C	F_Re	H71,N88	L
4557.5077(15)	21935.668(7)	4557.5079(4)	−0.0002	1100		4d	$(5/2)^2[7/2]_3$	4f	$(3/2)^2[7/2]^\circ_3$			R1c		
4558.7(5)	21929.9(24)	4558.949(4)	−0.2	3000		d^{10}	$^1S_0$	4s	$^3D_3$			A73		X,HF
4575.180(7)	21850.94(3)	4575.1748(4)	0.005	550	*	4d	$(5/2)^2[7/2]_3$	4f	$(3/2)^2[5/2]^\circ_2$			R1c		
		4575.185(3)			m	sp	$(^3P)^3P^\circ\,^3D^\circ_2$	10s	$(5/2)^2[5/2]_2$			R1c		
4575.180(7)	21850.94(3)	4575.1871(7)	−0.007	550	*	sp	$(^3F)^3P^\circ\,^3D^\circ_2$	5d	$(3/2)^2[3/2]_2$			R1c		
4575.6529(15)	21848.681(7)	4575.6532(3)	−0.0003	850		4d	$(5/2)^2[7/2]_4$	4f	$(3/2)^2[7/2]^\circ_4$			R1c		
4584.899(3)	21804.623(12)	4584.8994(3)	−0.001	60		4d	$(5/2)^2[7/2]_3$	4f	$(3/2)^2[5/2]^\circ_3$			R1c		
4586.274(3)	21798.084(13)	4586.2719(3)	0.002	300		5p	$(3/2)^2[5/2]^\circ_3$	7s	$(5/2)^2[5/2]_3$			R1c		
4588.1661(15)	21789.095(7)	4588.1667(4)	−0.0006	620		4d	$(5/2)^2[5/2]_2$	4f	$(3/2)^2[5/2]^\circ_2$			R1c		
4589.601(3)	21782.285(12)	4589.6035(8)	−0.003	220		4d	$(3/2)^2[3/2]_1$	sp	$(^3F)^1P^\circ\,^3F^\circ_2$			R1c		
4594.330(3)	21759.863(13)	4594.3638(14)	−0.034	200	?	sp	$(^3F)^3P^\circ\,^5F^\circ_3$	6s	$(3/2)^2[3/2]_2$			R1c		X
4594.436(3)	21759.358(15)	4594.4374(3)	−0.001	130		4d	$(5/2)^2[7/2]_4$	4f	$(3/2)^2[5/2]^\circ_3$			R1c		
4596.9056(15)	21747.671(7)	4596.9063(5)	−0.0007	4800		4d	$(5/2)^2[7/2]_3$	4f	$(3/2)^2[9/2]^\circ_4$			R1c		
4597.9473(15)	21742.744(7)	4597.9466(3)	0.0007	3200		4d	$(5/2)^2[5/2]_2$	4f	$(3/2)^2[5/2]^\circ_3$			R1c		
4606.5041(21)	21702.356(10)	4606.4944(5)	0.0097	730	bl	4d	$(5/2)^2[7/2]_4$	4f	$(3/2)^2[9/2]^\circ_4$			R1c		X
4608.4661(16)	21693.117(7)	4608.4662(5)	−0.0001	3100		4d	$(5/2)^2[7/2]_4$	4f	$(3/2)^2[9/2]^\circ_5$			R1c		
4609.416(3)	21688.646(14)	4609.3996(6)	0.016	560		sp	$(^3F)^3P^\circ\,^3D^\circ_2$	5d	$(3/2)^2[7/2]_3$			R1c		X
4619.659(3)	21640.559(15)	4619.6507(6)	0.008	82		4d	$(5/2)^2[5/2]_2$	4f	$(3/2)^2[3/2]^\circ_1$			R1c		
4625.097(3)	21615.114(13)	4625.1004(5)	−0.003	2100		4d	$(5/2)^2[5/2]_2$	4f	$(3/2)^2[3/2]^\circ_2$			R1c		

Table A1. *Cont.*

λ_{obs} [a] (Å)	σ_{obs} [b] (cm^{-1})	λ_{Ritz} [c] (Å)	$\Delta\lambda_{obs\text{-}Ritz}$ (Å)	I_{obs} [d] (arb. u.)	Char [e]	Lower Level		Upper Level		A (s^{-1})	Acc [f]	Line Ref. [g]	TP Ref. [g]	Notes [h]
4631.801(3)	21583.829(16)	4631.8076(7)	−0.007	90		4d	$(3/2)^2[7/2]_3$	sp	$(^1G)^3P^\circ\,^3F^\circ_3$			R1c		
4635.260(3)	21567.722(12)	4635.2599(16)	0.000	230		4d	$(5/2)^2[7/2]_4$	sp	$(^1G)^3P^\circ\,^3F^\circ_4$			R1c		
4639.5854(16)	21547.616(8)	4639.5835(9)	0.0019	1100		4d	$(5/2)^2[3/2]_1$	sp	$(^1G)^3P^\circ\,^3F^\circ_2$			R1nc		
4647.012(5)	21513.180(21)	4647.0121(12)	−0.000	950		4d	$(5/2)^2[9/2]_5$	sp	$(^1G)^3P^\circ\,^3H^\circ_5$			R1c		
4649.2705(16)	21502.730(7)	4649.27084(23)	−0.0003	6700		4d	$(5/2)^2[1/2]_1$	4f	$(5/2)^2[5/2]^\circ_2$			R1c		
4660.3044(17)	21451.820(8)	4660.3070(7)	−0.0026	5600		4d	$(5/2)^2[1/2]_0$	4f	$(3/2)^2[3/2]^\circ_1$			R1c		
4661.3627(16)	21446.950(7)	4661.3620(3)	0.0007	7100		4d	$(5/2)^2[1/2]_1$	4f	$(5/2)^2[3/2]^\circ_1$			R1c		
4662.654(3)	21441.010(12)	4662.6475(11)	0.007	13000		4d	$(5/2)^2[1/2]_1$	sp	$(^3P)^3P^\circ\,^5S^\circ_2$			R1c		
4667.315(2)	21419.599(11)	4667.3119(12)	0.003	11000		4d	$(5/2)^2[9/2]_4$	sp	$(^1G)^3P^\circ\,^3H^\circ_5$			R1c		
4671.7016(2)	21399.4867(11)	4671.70176(20)	−0.0002	29000		4d	$(5/2)^2[1/2]_1$	4f	$(5/2)^2[3/2]^\circ_2$	5.6e+07	D+	F_Re	TW	
4673.5772(5)	21390.8989(25)	4673.5774(5)	−0.0002	20000		4d	$(5/2)^2[1/2]_1$	4f	$(5/2)^2[1/2]^\circ_0$	1.6e+08	D+	F_Re	TW	L
4681.9938(5)	21352.4459(22)	4681.9935(3)	0.0003	36000		4d	$(5/2)^2[1/2]_1$	4f	$(5/2)^2[1/2]^\circ_1$	1.1e+08	D+	F_Re	TW	L
4687.7662(18)	21326.153(8)	4687.7648(4)	0.0014	3000		5p	$(3/2)^2[3/2]^\circ_2$	7s	$(5/2)^2[5/2]_2$			R1c		
4702.125(3)	21261.032(13)	4702.1291(9)	−0.004	550		4d	$(5/2)^2[5/2]_2$	sp	$(^3P)^3P^\circ\,^1D^\circ_2$			R1c		
4737.909(4)	21100.456(16)	4737.9023(11)	0.006	170		4d	$(3/2)^2[3/2]_1$	6p	$(5/2)^2[5/2]^\circ_2$			R1c		
4744.588(3)	21070.751(11)	4744.5892(8)	−0.001	520		sp	$(^3F)^3P^\circ\,^3D^\circ_1$	5d	$(3/2)^2[5/2]_2$			R1c		
4753.4684(16)	21031.389(7)	4753.4691(7)	−0.0007	2100		4d	$(5/2)^2[9/2]_4$	sp	$(^1G)^3P^\circ\,^3H^\circ_4$			R1c		
4758.4334(9)	21009.445(4)	4758.4334(7)	0.0000	15000		4p	$^3P^\circ_1$	s^2	$^3P_0$	1.3e+06	C	F_Re	H71,N88	
4765.913(3)	20976.472(14)	4765.9115(6)	0.002	62		sp	$(^3F)^3P^\circ\,^3F^\circ_3$	5d	$(3/2)^2[5/2]_3$			R1c		
4766.7392(16)	20972.837(7)	4766.7394(9)	−0.0002	4500		4d	$(5/2)^2[1/2]_1$	sp	$(^3P)^3P^\circ\,^3P^\circ_0$			R1c		
4772.9651(17)	20945.481(7)	4772.9656(5)	−0.0005	590		4d	$(3/2)^2[1/2]_1$	4f	$(3/2)^2[5/2]^\circ_2$			R1c		
4781.0855(17)	20909.907(8)	4781.0769(8)	0.0086	670		sp	$(^3F)^3P^\circ\,^3D^\circ_1$	5d	$(3/2)^2[3/2]_2$			R1c		X
4792.709(3)	20859.194(14)	4792.7117(21)	−0.002	430		5s	$(3/2)^2[3/2]_1$	sp	$(^3P)^3P^\circ\,^5D^\circ_2$			R1c		
4797.306(3)	20839.208(14)	4797.3080(18)	−0.002	150		5s	$(5/2)^2[5/2]_3$	sp	$(^1D)^3P^\circ\,^3D^\circ_2$			R1c		
4800.112(3)	20827.027(11)	4800.1106(8)	0.001	590		sp	$(^3F)^3P^\circ\,^3D^\circ_1$	5d	$(3/2)^2[3/2]_1$			R1c		
4800.584(3)	20824.980(14)	4800.5808(9)	0.003	370		4d	$(5/2)^2[5/2]_2$	sp	$(^1G)^3P^\circ\,^3F^\circ_2$			R1nc		
4805.6557(19)	20803.001(8)	4805.6527(6)	0.0030	3600		4d	$(5/2)^2[3/2]_2$	sp	$(^3F)^1P^\circ\,^3F^\circ_3$			R1c		
4807.0463(17)	20796.983(7)	4807.0463(8)	0.0000	7000		4d	$(3/2)^2[1/2]_1$	4f	$(3/2)^2[3/2]^\circ_1$			R1c		
4811.1487(21)	20779.250(9)	4811.1475(6)	0.0012	1000		4d	$(5/2)^2[9/2]_4$	sp	$(^3F)^1P^\circ\,^3F^\circ_3$			R1c		
4812.9476(5)	20771.4831(22)	4812.9474(4)	0.0003	36000		4d	$(3/2)^2[1/2]_1$	4f	$(3/2)^2[3/2]^\circ_2$	1.3e+08	D+	F_Re	TW	
4814.868(4)	20763.199(16)	4814.868(4)	0.000	310		5s	$(5/2)^2[5/2]_3$	sp	$(^1D)^3P^\circ\,^3F^\circ_4$			R1c		
4820.269(3)	20739.934(15)	4820.2672(6)	0.002	160		sp	$(^3F)^3P^\circ\,^3F^\circ_3$	5d	$(3/2)^2[7/2]_4$			R1c		
4829.3349(17)	20701.001(7)	4829.3340(3)	0.0009	1100		4d	$(5/2)^2[9/2]_5$	4f	$(5/2)^2[9/2]^\circ_4$			R1c		
4832.2454(17)	20688.533(7)	4832.2439(5)	0.0014	14000		4p	$^3P^\circ_1$	s^2	$^3P_1$	2.5e+05	C	R1c	H71,N88	
4833.761(3)	20682.046(12)	4833.7603(9)	0.001	1000		4d	$(3/2)^2[1/2]_1$	sp	$(^3F)^1P^\circ\,^3D^\circ_1$			R1c		
4834.673(3)	20678.146(12)	4834.6703(18)	0.002	440		5s	$(3/2)^2[3/2]_2$	sp	$(^3P)^3P^\circ\,^5D^\circ_3$			R1c		
4836.799(2)	20669.054(11)	4836.7982(8)	0.001	790		sp	$(^3F)^3P^\circ\,^3D^\circ_3$	5d	$(5/2)^2[7/2]_4$			R1c		
4837.434(3)	20666.341(15)	4837.4350(20)	−0.001	360		4d	$(5/2)^2[1/2]_0$	sp	$(^3P)^3P^\circ\,^1P^\circ_1$			R2nc		
4840.1833(17)	20654.604(7)	4840.1829(8)	0.0005	1100		4d	$(3/2)^2[7/2]_3$	6p	$(5/2)^2[7/2]^\circ_3$			R1c		
4841.832(3)	20647.573(15)	4841.8320(24)	−0.000	91	?	4f	$(5/2)^2[9/2]^\circ_5$	8g	$(5/2)^2[11/2]_6$			R2nc		
4847.3822(17)	20623.930(7)	4847.3823(8)	−0.0000	1500		4d	$(5/2)^2[1/2]_1$	sp	$(^3P)^3P^\circ\,^3P^\circ_1$			R1c		
4851.2625(11)	20607.434(5)	4851.2617(4)	0.0008	12000		4d	$(5/2)^2[9/2]_4$	4f	$(5/2)^2[9/2]^\circ_4$	2.9e+07	D+	F_Re	TW	

Table A1. *Cont.*

λ_{obs} [a] (Å)	σ_{obs} [b] (cm^{-1})	λ_{Ritz} [c] (Å)	$\Delta\lambda_{obs\text{-}Ritz}$ (Å)	I_{obs} [d] (arb. u.)	Char [e]	Lower Level		Upper Level		A (s^{-1})	Acc [f]	Line Ref. [g]	TP Ref. [g]	Notes [h]
4854.98733(18)	20591.6237(7)	4854.98743(17)	−0.00011	34000		4d	$(5/2)^2[9/2]_5$	4f	$(5/2)^2[9/2]^\circ_5$	3.5e+07	D+	F_Re	TW	L
4859.068(7)	20574.33(3)	4859.0673(19)	0.001	97		5s	$(3/2)^2[3/2]_2$	sp	$(^3P)^3P^\circ\ ^5D^\circ_1$			R1c		
4861.5612(18)	20563.780(8)	4861.56049(20)	0.0008	3600		4d	$(5/2)^2[9/2]_5$	4f	$(5/2)^2[7/2]^\circ_4$			R1c		
4866.254(3)	20543.951(12)	4866.2550(8)	−0.001	700		sp	$(^3F)^3P^\circ\ ^3D^\circ_3$	5d	$(5/2)^2[5/2]_3$			R1c		
4873.3056(17)	20514.223(7)	4873.3035(5)	0.0021	9900		4d	$(3/2)^2[7/2]_3$	4f	$(3/2)^2[7/2]^\circ_3$	2.4e+07	D+	F_Re	TW	
4877.151(3)	20498.050(13)	4877.1492(3)	0.002	100		4d	$(5/2)^2[9/2]_4$	4f	$(5/2)^2[9/2]^\circ_5$			R1c		
4883.2352(17)	20472.510(7)	4883.2351(4)	0.0001	2900		4d	$(3/2)^2[7/2]_3$	4f	$(3/2)^2[7/2]^\circ_4$			R1c		
4883.7832(17)	20470.213(7)	4883.7825(3)	0.0007	2900		4d	$(5/2)^2[9/2]_4$	4f	$(5/2)^2[7/2]^\circ_4$			R1c		
4889.7005(17)	20445.441(7)	4889.6998(5)	0.0008	12000		4p	$^3P^\circ_1$	s^2	$^3P_2$	1.9e+05	C	R1c	H71,N88	
4893.506(3)	20429.543(11)	4893.5089(5)	−0.003	770		4d	$(3/2)^2[7/2]_3$	4f	$(3/2)^2[5/2]^\circ_2$			R1c		
4896.414(3)	20417.408(14)	4896.4163(10)	−0.002	1900		4d	$(3/2)^2[1/2]_1$	sp	$(^3P)^3P^\circ\ ^1D^\circ_2$			R1c		
4900.472(3)	20400.501(11)	4900.4712(9)	0.001	920		sp	$(^3F)^3P^\circ\ ^3G^\circ_3$	5d	$(5/2)^2[7/2]_4$			R1c		
4901.42635(21)	20396.5291(9)	4901.42634(15)	0.00002	26000		4d	$(5/2)^2[3/2]_2$	4f	$(5/2)^2[5/2]^\circ_3$	6.2e+07	D+	F_Re	TW	
4904.534(4)	20383.605(16)	4904.5146(21)	0.020	110	?	4f	$(5/2)^2[3/2]^\circ_1$	9s	$(3/2)^2[3/2]_2$			R1c		X
4905.768(3)	20378.478(11)	4905.7667(5)	0.001	580		4d	$(3/2)^2[7/2]_4$	4f	$(3/2)^2[7/2]^\circ_3$			R1c		
4906.5663(3)	20375.1629(11)	4906.56612(16)	0.0001	21000		4d	$(5/2)^2[3/2]_2$	4f	$(5/2)^2[5/2]^\circ_2$	4.4e+07	D+	F_Re	TW	
4907.1427(24)	20372.770(10)	4907.1424(4)	0.0002	1000		4d	$(5/2)^2[9/2]_4$	4f	$(5/2)^2[5/2]^\circ_3$			R1c		
4908.2819(24)	20368.041(10)	4908.2838(8)	−0.0019	970		4d	$(5/2)^2[7/2]_3$	sp	$(^1G)^3P^\circ\ ^3H^\circ_4$			F_Re		
4909.0397(13)	20364.897(5)	4909.0390(5)	0.0007	6100		4d	$(5/2)^2[9/2]_5$	4f	$(5/2)^2[11/2]^\circ_5$			F_Re		
4909.73351(2)	20362.01913(10)	4909.733510(20)	0.00000	160000		4d	$(5/2)^2[9/2]_5$	4f	$(5/2)^2[11/2]^\circ_6$	2.04e+08	B+	F_Re	C84	L
4912.3645(5)	20351.1138(21)	4912.3644(3)	0.0001	17000		4d	$(5/2)^2[3/2]_2$	4f	$(5/2)^2[7/2]^\circ_3$			F_Re		
4912.91989(16)	20348.8131(7)	4912.91987(14)	0.00002	29000		4d	$(5/2)^2[3/2]_1$	4f	$(5/2)^2[5/2]^\circ_2$	6.0e+07	D+	F_Re	TW	
4915.8321(17)	20336.758(7)	4915.8312(4)	0.0009	11000		4d	$(3/2)^2[7/2]_4$	4f	$(3/2)^2[7/2]^\circ_4$	2.5e+07	D+	F_Re	TW	
4918.1079(21)	20327.348(9)	4918.1061(4)	0.0019	2400		4d	$(5/2)^2[9/2]_4$	4f	$(5/2)^2[7/2]^\circ_3$			R1c		
4918.3779(5)	20326.2320(20)	4918.3778(4)	0.0000	54000		4d	$(3/2)^2[7/2]_3$	4f	$(3/2)^2[9/2]^\circ_4$	2.9e+08	C	F_Re	H71	
4920.0348(12)	20319.387(5)	4920.0345(4)	0.0003	6800		4d	$(5/2)^2[3/2]_2$	4f	$(5/2)^2[3/2]^\circ_1$			F_Re		
4921.4627(21)	20313.492(9)	4921.4666(12)	−0.0039	5900		4d	$(5/2)^2[3/2]_2$	sp	$(^3P)^3P^\circ\ ^5S^\circ_2$			F_Re		
4926.4228(5)	20293.0397(22)	4926.4232(3)	−0.0004	26000		4d	$(5/2)^2[3/2]_1$	4f	$(5/2)^2[3/2]^\circ_1$	9.e+07	D+	F_Re	TW	
4926.811(4)	20291.440(17)	4926.8186(15)	−0.007	740		sp	$(^3F)^3P^\circ\ ^5F^\circ_4$	6s	$(5/2)^2[5/2]_3$			R1c		
4927.860(3)	20287.121(14)	4927.8590(12)	0.001	830		4d	$(5/2)^2[3/2]_1$	sp	$(^3P)^3P^\circ\ ^5S^\circ_2$			R1c		
4930.713(3)	20275.383(11)	4930.7110(8)	0.002	620		sp	$(^3F)^3P^\circ\ ^3G^\circ_3$	5d	$(5/2)^2[5/2]_3$			R1c		
4931.55496(20)	20271.9213(8)	4931.55505(16)	−0.00009	28000		4d	$(5/2)^2[3/2]_2$	4f	$(5/2)^2[3/2]^\circ_2$	7.1e+07	D+	F_Re	TW	
4931.6982(5)	20271.3325(20)	4931.6981(4)	0.0001	140000		4d	$(5/2)^2[9/2]_4$	4f	$(5/2)^2[11/2]^\circ_5$	1.9e+08	C	F_Re	H71	L
4933.0777(24)	20265.664(10)	4933.0745(5)	0.0032	1100		4d	$(5/2)^2[9/2]_5$	sp	$(^3F)^1P^\circ\ ^3G^\circ_4$			R1c		
4937.22029(23)	20248.6601(9)	4937.22031(22)	−0.00002	26000		4d	$(3/2)^2[3/2]_1$	4f	$(3/2)^2[5/2]^\circ_2$	1.1e+08	D+	F_Re	TW	
4937.516(3)	20247.449(11)	4937.5188(4)	−0.003	1100		4d	$(3/2)^2[7/2]_4$	4f	$(3/2)^2[5/2]^\circ_3$			R1c		
4937.9740(4)	20245.5693(16)	4937.97372(21)	0.0003	16000		4d	$(5/2)^2[3/2]_1$	4f	$(5/2)^2[3/2]^\circ_2$	2.8e+07	D+	F_Re	TW	
4939.648(3)	20238.709(11)	4939.6526(8)	−0.005	1200		4d	$(3/2)^2[3/2]_2$	6p	$(5/2)^2[7/2]^\circ_3$			R1c		
4940.0695(17)	20236.982(7)	4940.0693(6)	0.0002	4700		4d	$(5/2)^2[3/2]_1$	4f	$(5/2)^2[1/2]^\circ_0$			R1c		
4940.912(8)	20233.53(3)	4940.9093(20)	0.003	130		5s	$(5/2)^2[5/2]_2$	sp	$(^1D)^3P^\circ\ ^3D^\circ_1$			R1c		
4943.0240(7)	20224.886(3)	4943.0250(4)	−0.0010	20000		4d	$(5/2)^2[3/2]_2$	4f	$(5/2)^2[1/2]^\circ_1$	4.0e+07	D+	F_Re	TW	
4949.4750(18)	20198.526(7)	4949.4735(4)	0.0015	3700		4d	$(5/2)^2[3/2]_1$	4f	$(5/2)^2[1/2]^\circ_1$			R1c		
4951.4463(18)	20190.484(7)	4951.4464(6)	−0.0000	4200		4d	$(3/2)^2[7/2]_4$	4f	$(3/2)^2[9/2]^\circ_4$			F_Re		

Table A1. *Cont.*

λ_{obs} [a] (Å)	σ_{obs} [b] (cm^{-1})	λ_{Ritz} [c] (Å)	$\Delta\lambda_{obs\text{-}Ritz}$ (Å)	I_{obs} [d] (arb. u.)	Char [e]	Lower Level	Upper Level	A (s^{-1})	Acc [f]	Line Ref. [g]	TP Ref. [g]	Notes [h]
4951.6184(14)	20189.783(6)	4951.6201(3)	−0.0016	12000		4d $(5/2)^2[5/2]_3$	4f $(5/2)^2[9/2]^{\circ}_4$			F_Re		
4953.7244(5)	20181.1995(20)	4953.7246(4)	−0.0002	82000		4d $(3/2)^2[7/2]_4$	4f $(3/2)^2[9/2]^{\circ}_5$	3.1e+08	C	F_Re	H71	
4955.9564(18)	20172.111(7)	4955.9566(5)	−0.0002	6300		4d $(5/2)^2[9/2]_4$	sp $(^3F)^1P^{\circ}\ ^3G^{\circ}_4$			R1c		
4969.8062(18)	20115.896(7)	4969.8046(6)	0.0016	4400		4d $(5/2)^2[7/2]_3$	sp $(^3F)^1P^{\circ}\ ^3F^{\circ}_3$	1.1e+07	D+	R1c	TW	
4971.222(4)	20110.167(15)	4971.2261(9)	−0.004	280		sp $(^3F)^3P^{\circ}\ ^3G^{\circ}_3$	5d $(5/2)^2[3/2]_2$			R1c		
4973.136(3)	20102.429(12)	4973.1394(7)	−0.004	280		sp $(^3F)^3P^{\circ}\ ^3F^{\circ}_2$	5d $(3/2)^2[5/2]_2$			R1c		
4973.6975(19)	20100.158(8)	4973.6959(8)	0.0016	9000		4d $(3/2)^2[3/2]_1$	4f $(3/2)^2[3/2]^{\circ}_1$	3.6e+07	D+	F_Re	TW	
4974.1542(15)	20098.313(6)	4974.1533(5)	0.0008	9700		4d $(3/2)^2[3/2]_2$	4f $(3/2)^2[7/2]^{\circ}_3$	3.0e+07	D+	F_Re	TW	
4975.064(3)	20094.637(12)	4975.0659(8)	−0.002	560		4d $(5/2)^2[3/2]_2$	sp $(^3F)^1P^{\circ}\ ^3D^{\circ}_3$			R1c		
4980.0153(19)	20074.659(8)	4980.0135(6)	0.0018	11000		4d $(3/2)^2[3/2]_1$	4f $(3/2)^2[3/2]^{\circ}_2$			R1c		
4981.0125(23)	20070.640(9)	4981.0133(6)	−0.0008	670		4d $(5/2)^2[7/2]_4$	sp $(^3F)^1P^{\circ}\ ^3F^{\circ}_3$			R1c		
4985.1421(23)	20054.014(9)	4985.1381(6)	0.0040	5100		4d $(5/2)^2[5/2]_2$	sp $(^3F)^1P^{\circ}\ ^3F^{\circ}_3$	2.1e+07	D+	R1c	TW	
4985.50499(7)	20052.5541(3)	4985.50498(7)	0.00001	70000		4d $(5/2)^2[5/2]_3$	4f $(5/2)^2[7/2]^{\circ}_4$	9.7e+07	C+	F_Re	TW	
4988.327(4)	20041.212(15)	4988.326(4)	0.000	430		5d $(5/2)^2[9/2]_5$	8f $(5/2)^2[11/2]^{\circ}_6$	5.7e+06	D+	R1c	TW	
4995.2054(22)	20013.614(9)	4995.2055(5)	−0.0001	800		4d $(3/2)^2[3/2]_2$	4f $(3/2)^2[5/2]^{\circ}_2$			R1c		
4999.584(3)	19996.088(12)	4999.5826(20)	0.001	440		5d $(5/2)^2[9/2]_4$	8f $(5/2)^2[11/2]^{\circ}_5$			R1c		
5002.294(3)	19985.255(12)	5002.2998(10)	−0.006	440		4d $(3/2)^2[3/2]_1$	sp $(^3F)^1P^{\circ}\ ^3D^{\circ}_1$			R1c		
5006.79983(15)	19967.2680(6)	5006.79978(15)	0.00005	46000		4d $(3/2)^2[3/2]_2$	4f $(3/2)^2[5/2]^{\circ}_3$	1.20e+08	C+	F_Re	TW	
5009.8505(3)	19955.1095(10)	5009.85058(17)	−0.0001	35000		4d $(5/2)^2[5/2]_3$	4f $(5/2)^2[5/2]^{\circ}_3$	5.4e+07	D+	F_Re	TW	
5012.6197(3)	19944.0855(12)	5012.6199(3)	−0.0002	37000		4d $(5/2)^2[7/2]_3$	4f $(5/2)^2[9/2]^{\circ}_4$	9.6e+07	C+	F_Re	TW	L
5015.2190(8)	19933.749(3)	5015.22038(21)	−0.0014	11000		4d $(5/2)^2[5/2]_3$	4f $(5/2)^2[5/2]^{\circ}_2$			F_Re		
5020.126(3)	19914.266(10)	5020.1256(5)	0.000	2000		4p $^3F^{\circ}_3$	s^2 $^3P_2$			R1c		
5021.2785(4)	19909.6939(15)	5021.27849(23)	0.0000	32000		4d $(5/2)^2[5/2]_3$	4f $(5/2)^2[7/2]^{\circ}_3$	3.2e+07	E	F_Re	TW	L
5022.599(3)	19904.459(11)	5022.6044(7)	−0.005	760		sp $(^3F)^3P^{\circ}\ ^3D^{\circ}_2$	5d $(5/2)^2[7/2]_3$			R1c		
5024.0251(16)	19898.810(6)	5024.0227(3)	0.0024	8700		4d $(5/2)^2[7/2]_4$	4f $(5/2)^2[9/2]^{\circ}_4$			F_Re		
5030.789(3)	19872.057(11)	5030.7892(12)	−0.000	1400		4d $(5/2)^2[5/2]_3$	sp $(^3P)^3P^{\circ}\ ^5S^{\circ}_2$			R1c		
5031.299(3)	19870.040(10)	5031.2968(8)	0.003	1400		4d $(3/2)^2[5/2]_2$	6p $(5/2)^2[7/2]^{\circ}_3$			R1c		
5032.544(3)	19865.125(11)	5032.5461(8)	−0.002	1100		4d $(3/2)^2[3/2]_2$	4f $(3/2)^2[3/2]^{\circ}_1$			R1c		
5034.171(3)	19858.705(13)	5034.1729(8)	−0.002	460		sp $(^3F)^3P^{\circ}\ ^3F^{\circ}_2$	5d $(3/2)^2[3/2]_1$			R1c		
5036.7302(21)	19848.616(8)	5036.7320(7)	−0.0019	540		4d $(5/2)^2[1/2]_1$	sp $(^3P)^3P^{\circ}\ ^3D^{\circ}_1$			R1c		
5039.0109(13)	19839.632(5)	5039.0141(6)	−0.0032	17000		4d $(3/2)^2[3/2]_2$	4f $(3/2)^2[3/2]^{\circ}_2$	3.9e+07	D+	F_Re	TW	
5041.3310(9)	19830.502(4)	5041.33123(23)	−0.0003	14000		4d $(5/2)^2[5/2]_3$	4f $(5/2)^2[3/2]^{\circ}_2$	1.6e+07	D+	F_Re	TW	
5042.093(3)	19827.505(11)	5042.0348(18)	0.058	310	?	sp $(^3F)^3P^{\circ}\ ^1G^{\circ}_4$	5d $(3/2)^2[7/2]_3$			R1c		X
5044.2776(21)	19818.918(8)	5044.2768(10)	0.0008	940		4d $(5/2)^2[3/2]_1$	sp $(^3P)^3P^{\circ}\ ^3P^{\circ}_0$			R1c		
5047.34772(23)	19806.8629(9)	5047.34770(18)	0.00002	27000		4d $(5/2)^2[7/2]_3$	4f $(5/2)^2[7/2]^{\circ}_4$			F_Re		
5051.79210(4)	19789.43784(14)	5051.79209(4)	0.00000	120000		4d $(5/2)^2[7/2]_4$	4f $(5/2)^2[9/2]^{\circ}_5$	1.55e+08	C+	F_Re	TW	L
5054.344(3)	19779.447(10)	5054.3487(6)	−0.005	950		sp $(^3F)^3P^{\circ}\ ^3F^{\circ}_2$	5d $(3/2)^2[7/2]_3$			R1c		
5054.7758(20)	19777.757(8)	5054.7728(5)	0.0030	1300		4d $(3/2)^2[5/2]_3$	4f $(3/2)^2[7/2]^{\circ}_3$			R1c		
5058.90906(16)	19761.5981(6)	5058.90923(14)	−0.00017	48000		4d $(5/2)^2[7/2]_4$	4f $(5/2)^2[7/2]^{\circ}_4$	4.8e+07	D+	F_Re	TW	
5059.418(5)	19759.611(21)	5059.4131(13)	0.005	160		6s $(5/2)^2[5/2]_2$	6f $(3/2)^2[5/2]^{\circ}_2$			R1c		
5060.6437(9)	19754.825(4)	5060.6415(5)	0.0022	16000		4p $^3P^{\circ}_0$	s^2 $^3P_1$	4.1e+05	C	F_Re	H71	L
5065.45858(8)	19736.0471(3)	5065.45861(8)	−0.00002	70000		4d $(3/2)^2[5/2]_3$	4f $(3/2)^2[7/2]^{\circ}_4$	1.61e+08	C+	F_Re	TW	

Table A1. *Cont.*

λ_{obs} [a] (Å)	σ_{obs} [b] (cm^{-1})	λ_{Ritz} [c] (Å)	$\Delta\lambda_{obs-Ritz}$ (Å)	I_{obs} [d] (arb. u.)	Char [e]	Lower Level		Upper Level		A (s^{-1})	Acc [f]	Line Ref. [g]	TP Ref. [g]	Notes [h]
5067.09416(17)	19729.6767(7)	5067.09423(17)	−0.00008	46000		4d	$(3/2)^2[5/2]_2$	4f	$(3/2)^2[7/2]^\circ_3$	1.23e+08	C+	F_Re	TW	
5069.431(3)	19720.584(11)	5069.4315(10)	−0.001	1600		4d	$(3/2)^2[3/2]_1$	sp	$(^3P)^3P^\circ\ ^1D^\circ_2$			R1c		
5072.3034(6)	19709.4147(23)	5072.30253(22)	0.0008	32000		4d	$(5/2)^2[7/2]_3$	4f	$(5/2)^2[5/2]^\circ_3$			F_Re		
5076.5136(20)	19693.069(8)	5076.5143(5)	−0.0007	830		4d	$(3/2)^2[5/2]_3$	4f	$(3/2)^2[5/2]^\circ_2$			R1c		
5077.8070(18)	19688.053(7)	5077.8071(3)	−0.0001	6300		4d	$(5/2)^2[7/2]_3$	4f	$(5/2)^2[5/2]^\circ_2$			R1c		
5081.099(4)	19675.296(14)	5081.1053(16)	−0.006	330		5s	$(3/2)^2[3/2]_1$	sp	$(^1D)^3P^\circ\ ^3P^\circ_1$			R1c		
5082.525(4)	19669.778(14)	5082.5262(15)	−0.001	160		5p	$(5/2)^2[3/2]^\circ_1$	5d	$(3/2)^2[1/2]_0$			R1c		
5083.9795(12)	19664.150(5)	5083.97879(22)	0.0007	21000		4d	$(5/2)^2[7/2]_4$	4f	$(5/2)^2[5/2]^\circ_3$			F_Re		
5084.0173(5)	19664.0034(18)	5084.0175(3)	−0.0002	14000		4d	$(5/2)^2[7/2]_3$	4f	$(5/2)^2[7/2]^\circ_3$			F_Re,R1r		
5088.27603(10)	19647.5455(4)	5088.27603(9)	0.00000	57000		4d	$(5/2)^2[5/2]_2$	4f	$(5/2)^2[5/2]^\circ_3$			F_Re		
5088.4882(8)	19646.726(3)	5088.4896(4)	−0.0013	25000		4d	$(3/2)^2[5/2]_3$	4f	$(3/2)^2[5/2]^\circ_3$	4.8e+07	D+	F_Re	TW	
5088.9426(11)	19644.972(4)	5088.9421(5)	0.0005	19000		4d	$(3/2)^2[5/2]_2$	4f	$(3/2)^2[5/2]^\circ_2$	5.2e+07	D+	F_Re	TW	
5089.492(5)	19642.851(21)	5089.4972(7)	−0.005	330		sp	$(^3F)^3P^\circ\ ^3D^\circ_2$	5d	$(5/2)^2[3/2]_1$			R1c		
5093.81535(17)	19626.1799(7)	5093.81536(14)	−0.00001	41000		4d	$(5/2)^2[5/2]_2$	4f	$(5/2)^2[5/2]^\circ_2$	4.9e+07	D+	F_Re	TW	
5094.442(3)	19623.766(11)	5094.4408(7)	0.001	330		sp	$(^3F)^3P^\circ\ ^3D^\circ_2$	5d	$(5/2)^2[3/2]_2$			R1c		
5095.7489(19)	19618.733(7)	5095.7478(3)	0.0011	3100		4d	$(5/2)^2[7/2]_4$	4f	$(5/2)^2[7/2]^\circ_3$			R1c		
5100.0629(10)	19602.138(4)	5100.0650(3)	−0.0021	12000		4d	$(5/2)^2[5/2]_2$	4f	$(5/2)^2[7/2]^\circ_3$			F_Re		
5100.9790(21)	19598.618(8)	5100.9762(5)	0.0028	9900		4d	$(3/2)^2[5/2]_2$	4f	$(3/2)^2[5/2]^\circ_3$	1.6e+07	D+	R1c	TW	
5103.283(3)	19589.770(13)	5103.2831(6)	−0.000	830		4d	$(3/2)^2[5/2]_3$	4f	$(3/2)^2[9/2]^\circ_4$			R1c		
5104.573(3)	19584.818(11)	5104.5755(3)	−0.002	830		4d	$(5/2)^2[7/2]_3$	4f	$(5/2)^2[3/2]^\circ_2$			R1c		
5108.3335(19)	19570.402(7)	5108.3328(4)	0.0007	7300		4d	$(5/2)^2[5/2]_2$	4f	$(5/2)^2[3/2]^\circ_1$			R1c		
5109.825(3)	19564.691(13)	5109.8270(21)	−0.002	500		sp	$(^1D)^3P^\circ\ ^3D^\circ_1$	5g	$(3/2)^2[5/2]_2$			R1c		
5109.880(6)	19564.481(21)	5109.8767(13)	0.003	330		4d	$(5/2)^2[5/2]_2$	sp	$(^3P)^3P^\circ\ ^5S^\circ_2$			R1c		
5110.3417(19)	19562.712(7)	5110.3409(5)	0.0008	1700		4d	$(5/2)^2[7/2]_4$	4f	$(5/2)^2[11/2]^\circ_5$			R1c		
5115.535(3)	19542.852(13)	5115.5386(11)	−0.004	170		sp	$(^3F)^3P^\circ\ ^3D^\circ_1$	5d	$(5/2)^2[1/2]_0$			R1c		
5120.7535(4)	19522.9360(15)	5120.75319(19)	0.0003	20000		4d	$(5/2)^2[5/2]_2$	4f	$(5/2)^2[3/2]^\circ_2$	2.0e+07	D+	F_Re	TW	
5121.765(3)	19519.080(13)	5121.7672(7)	−0.002	1000		4d	$(3/2)^2[5/2]_3$	4f	$(3/2)^2[3/2]^\circ_2$			R1c		
5124.4745(7)	19508.760(3)	5124.4753(5)	−0.0007	30000		4d	$(5/2)^2[7/2]_3$	sp	$(^3F)^1P^\circ\ ^3G^\circ_4$			F_Re		
5127.682(3)	19496.556(13)	5127.7027(9)	−0.020	570		4d	$(3/2)^2[5/2]_2$	4f	$(3/2)^2[3/2]^\circ_1$			R1c		X
5127.7428(21)	19496.326(8)	5127.7327(8)	0.0101	630	?	4d	$(5/2)^2[3/2]_2$	sp	$(^3P)^3P^\circ\ ^3P^\circ_1$			R1c		X
5130.583(3)	19485.533(12)	5130.5829(10)	0.000	920		4d	$(3/2)^2[3/2]_2$	sp	$(^3P)^3P^\circ\ ^1D^\circ_2$			R1c		
5133.123(3)	19475.890(11)	5133.1211(4)	0.002	500		4d	$(5/2)^2[5/2]_2$	4f	$(5/2)^2[1/2]^\circ_1$			R1c		
5134.6720(24)	19470.016(9)	5134.6725(9)	−0.0005	170		4d	$(5/2)^2[3/2]_1$	sp	$(^3P)^3P^\circ\ ^3P^\circ_1$			R1c		
5136.3939(19)	19463.489(7)	5136.3932(5)	0.0007	1300		4d	$(5/2)^2[7/2]_4$	sp	$(^3F)^1P^\circ\ ^3G^\circ_4$			R1c		
5138.613(4)	19455.085(17)	5138.6113(20)	0.001	340		4d	$(3/2)^2[5/2]_3$	sp	$(^1G)^3P^\circ\ ^3F^\circ_4$			R1c		
5151.206(2)	19407.522(9)	5151.2073(8)	−0.001	2500		4d	$(5/2)^2[7/2]_3$	sp	$(^3F)^1P^\circ\ ^3D^\circ_3$			R1c		
5157.255(4)	19384.761(14)	5157.2580(9)	−0.003	340		sp	$(^3F)^3P^\circ\ ^3D^\circ_3$	6s	$(3/2)^2[3/2]_2$			R1c		
5158.0914(6)	19381.6168(21)	5158.0916(4)	−0.0002	13000		4d	$(5/2)^2[1/2]_0$	4f	$(5/2)^2[3/2]^\circ_1$	2.4e+07	D+	F_Re	TW	
5159.103(4)	19377.817(17)	5159.0971(13)	0.006	2300		sp	$(^3F)^3P^\circ\ ^3F^\circ_4$	5d	$(5/2)^2[7/2]_4$			R1c		
5163.252(3)	19362.244(11)	5163.2501(8)	0.002	330		4d	$(5/2)^2[7/2]_4$	sp	$(^3F)^1P^\circ\ ^3D^\circ_3$			R1c		
5167.6923(22)	19345.609(8)	5167.6825(8)	0.0098	11000	?	4d	$(5/2)^2[5/2]_2$	sp	$(^3F)^1P^\circ\ ^3D^\circ_3$			R1c		X
5171.644(3)	19330.828(12)	5171.6406(13)	0.003	570		4d	$(3/2)^2[7/2]_3$	sp	$(^3F)^1P^\circ\ ^3G^\circ_3$			R1c		
5175.9651(23)	19314.689(9)	5175.9637(23)	0.0014	1400	*	4d	$(3/2)^2[3/2]_1$	sp	$(^3P)^3P^\circ\ ^1P^\circ_1$			R1nc		
5175.9651(23)	19314.689(9)	5175.9653(10)	−0.0002	1400	*	4d	$(5/2)^2[1/2]_1$	sp	$(^3P)^3P^\circ\ ^3P^\circ_2$			R1c		
5183.3664(7)	19287.110(3)	5183.3664(5)	−0.0000	19000		4d	$(5/2)^2[1/2]_0$	4f	$(5/2)^2[1/2]^\circ_1$	2.4e+07	D+	F_Re	TW	
5184.052(3)	19284.561(11)	5184.0520(11)	−0.000	340		4d	$(3/2)^2[3/2]_1$	sp	$(^1G)^3P^\circ\ ^3P^\circ_2$			R1nc		

Table A1. *Cont.*

λ_{obs} [a] (Å)	σ_{obs} [b] (cm^{-1})	λ_{Ritz} [c] (Å)	$\Delta\lambda_{obs\text{-}Ritz}$ (Å)	I_{obs} [d] (arb. u.)	Char [e]	Lower Level		Upper Level		A (s^{-1})	Acc [f]	Line Ref. [g]	TP Ref. [g]	Notes [h]
5192.638(6)	19252.671(21)	5192.6239(13)	0.015	240		sp	$(^3F)^3P^\circ\ ^3F^\circ_4$	5d	$(5/2)^2[5/2]_3$			R1c		
5194.349(4)	19246.331(16)	5194.3399(17)	0.009	170		sp	$(^3F)^3P^\circ\ ^5F^\circ_2$	6s	$(5/2)^2[5/2]_2$			R1c		
5205.112(4)	19206.536(17)	5205.1059(8)	0.006	500		sp	$(^3F)^3P^\circ\ ^3D^\circ_2$	5d	$(5/2)^2[1/2]_1$			R1c		
5207.1357(22)	19199.070(8)	5207.1340(15)	0.0018	29000		4d	$(3/2)^2[7/2]_4$	sp	$(^1G)^3P^\circ\ ^3H^\circ_5$			R1c		
5229.518(3)	19116.900(10)	5229.5194(11)	−0.002	3200		4d	$(3/2)^2[5/2]_2$	sp	$(^3P)^3P^\circ\ ^1D^\circ_2$			R1c		
5229.709(6)	19116.201(21)	5229.7107(10)	−0.002	170		sp	$(^3F)^3P^\circ\ ^3G^\circ_3$	6s	$(3/2)^2[3/2]_2$			R1c		
5234.471(23)	19098.81(8)	5234.429(3)	0.042	170	*	sp	$(^3P)^3P^\circ\ ^1P^\circ_1$	9s	$(3/2)^2[3/2]_2$			R1nc		
5234.471(23)	19098.81(8)	5234.4943(14)	−0.023	170	*	sp	$(^3F)^3P^\circ\ ^3F^\circ_4$	5d	$(5/2)^2[9/2]_4$			R1c		
5239.5473(24)	19080.307(9)	5239.5490(8)	−0.0017	330		4d	$(5/2)^2[9/2]_4$	sp	$(^3F)^1P^\circ\ ^3F^\circ_4$			R1c		
5245.340(3)	19059.237(12)	5245.3423(13)	−0.003	19000		sp	$(^3F)^3P^\circ\ ^3F^\circ_4$	5d	$(5/2)^2[9/2]_5$			R1c		
5247.109(4)	19052.811(16)	5247.1155(9)	−0.007	160		4d	$(3/2)^2[7/2]_3$	sp	$(^3F)^1P^\circ\ ^3D^\circ_2$			R1c		
5247.976(4)	19049.661(16)	5247.9750(7)	0.001	160		5p	$(5/2)^2[3/2]^\circ_2$	5d	$(3/2)^2[5/2]_2$			R1c		
5251.545(3)	19036.718(9)	5251.5445(8)	0.000	330		4d	$(5/2)^2[3/2]_2$	sp	$(^3P)^3P^\circ\ ^3D^\circ_2$			R1c		
5254.2141(20)	19027.046(7)	5254.2139(6)	0.0002	350		5p	$(5/2)^2[3/2]^\circ_2$	5d	$(3/2)^2[5/2]_3$			R1c		
5258.8249(21)	19010.364(8)	5258.8237(8)	0.0011	580		4d	$(5/2)^2[3/2]_1$	sp	$(^3P)^3P^\circ\ ^3D^\circ_2$			R1c		
5261.0461(22)	19002.338(8)	5261.0443(8)	0.0018	330		sp	$(^3F)^3P^\circ\ ^3D^\circ_1$	5d	$(5/2)^2[5/2]_2$			R1c		
5269.9892(14)	18970.091(5)	5269.9904(6)	−0.0012	23000		4p	$^3P^\circ_2$	s^2	$^1D_2$			F_Re		
5276.5244(15)	18946.597(5)	5276.5241(9)	0.0003	16000		4d	$(3/2)^2[7/2]_3$	sp	$(^1G)^3P^\circ\ ^3H^\circ_4$			F_Re		
5280.9533(23)	18930.707(8)	5280.9563(7)	−0.0030	480		sp	$(^3F)^3P^\circ\ ^3F^\circ_3$	5d	$(5/2)^2[5/2]_2$			R1c		
5285.360(4)	18914.923(13)	5285.3600(13)	0.000	480		4d	$(3/2)^2[3/2]_2$	sp	$(^3F)^1P^\circ\ ^3G^\circ_3$			R1c		
5289.1103(21)	18901.512(8)	5289.1096(6)	0.0007	800		sp	$(^3F)^3P^\circ\ ^3F^\circ_3$	5d	$(5/2)^2[7/2]_4$			R1c		
5291.8196(20)	18891.835(7)	5291.8216(7)	−0.0020	1600		sp	$(^3F)^3P^\circ\ ^3F^\circ_3$	5d	$(5/2)^2[7/2]_3$			R1c		
5297.4020(23)	18871.927(8)	5297.4047(9)	−0.0028	480		4d	$(3/2)^2[3/2]_1$	sp	$(^3F)^1P^\circ\ ^3D^\circ_2$			R1c		
5310.771(3)	18824.421(11)	5310.7694(24)	0.001	310		4f	$(5/2)^2[1/2]^\circ_1$	7g	$(5/2)^2[3/2]_2$			R1c		
5314.6020(22)	18810.851(8)	5314.6025(9)	−0.0005	480		4d	$(3/2)^2[7/2]_4$	sp	$(^1G)^3P^\circ\ ^3H^\circ_4$			R1c		
5315.989(4)	18805.942(15)	5315.9868(7)	0.003	160		5p	$(5/2)^2[3/2]^\circ_2$	5d	$(3/2)^2[3/2]_1$			R1c		
5321.666(3)	18785.883(12)	5321.666(3)	0.000	310		4f	$(5/2)^2[1/2]^\circ_0$	7g	$(5/2)^2[3/2]_1$			R1c		
5321.840(3)	18785.267(11)	5321.8352(21)	0.005	310	*	4f	$(5/2)^2[3/2]^\circ_2$	7g	$(5/2)^2[5/2]_2$			R1nc		
5321.840(3)	18785.267(11)	5321.8386(23)	0.002	310	*	4f	$(5/2)^2[3/2]^\circ_2$	7g	$(5/2)^2[5/2]_3$			R1c		
5324.070(4)	18777.398(13)	5324.0734(24)	−0.003	310		4f	$(5/2)^2[3/2]^\circ_2$	7g	$(5/2)^2[3/2]_2$			R1c		
5324.3542(21)	18776.397(7)	5324.3531(6)	0.0011	1500		sp	$(^3F)^3P^\circ\ ^3F^\circ_3$	5d	$(5/2)^2[5/2]_3$			R1c		
5325.319(3)	18772.994(12)	5325.3132(19)	0.006	310		4f	$(5/2)^2[11/2]^\circ_5$	7g	$(5/2)^2[11/2]_5$			R1c		
5327.9841(23)	18763.605(8)			920		4f	$(5/2)^2[11/2]^\circ_6$	7g	$(5/2)^2[13/2]_7$	1.1e+07	D+	R1c	TW	
5328.8109(22)	18760.694(8)			1600		4f	$(5/2)^2[11/2]^\circ_5$	7g	$(5/2)^2[13/2]_6$	1.1e+07	D+	R1c	TW	
5328.963(4)	18760.157(13)			460		4f	$(3/2)^2[3/2]^\circ_2$	7g	$(3/2)^2[5/2]_3$	9.e+06	D+	R1c	TW	
5331.63(4)	18750.78(14)	5331.6093(19)	0.02	150	*	4f	$(5/2)^2[7/2]^\circ_4$	8d	$(5/2)^2[7/2]_4$			R1nc		X
5331.63(4)	18750.78(14)	5331.674(3)	−0.05	150	*	5s	$(3/2)^2[3/2]_2$	sp	$(^1D)^3P^\circ\ ^3D^\circ_3$			R1c		X
5335.317(3)	18737.818(12)	5335.3169(21)	−0.000	300		4f	$(5/2)^2[3/2]^\circ_1$	7g	$(5/2)^2[5/2]_2$			R2nc		
5335.518(3)	18737.112(12)	5335.5179(6)	−0.000	300		4p	$^3F^\circ_2$	s^2	$^3P_1$			R1c		
5336.203(3)	18734.707(12)			150		4f	$(3/2)^2[3/2]^\circ_1$	7g	$(3/2)^2[5/2]_2$			R1c		
5337.580(13)	18729.87(5)	5337.5664(20)	0.013	150	*	4f	$(5/2)^2[3/2]^\circ_1$	7g	$(5/2)^2[3/2]_2$			R1c		

Table A1. *Cont.*

λ_{obs} [a] (Å)	σ_{obs} [b] (cm^{-1})	λ_{Ritz} [c] (Å)	$\Delta\lambda_{obs\text{-}Ritz}$ (Å)	I_{obs} [d] (arb. u.)	Char [e]	Lower Level		Upper Level		A (s^{-1})	Acc [f]	Line Ref. [g]	TP Ref. [g]	Notes [h]
5337.580(13)	18729.87(5)	5337.593(3)	−0.013	150	*	4f	$(5/2)^2[3/2]^\circ_1$	7g	$(5/2)^2[3/2]_1$			R1c		
5338.460(3)	18726.784(10)	5338.4584(6)	0.002	300		4d	$(3/2)^2[1/2]_1$	4f	$(5/2)^2[3/2]^\circ_1$			R1c		
5340.098(3)	18721.041(11)	5340.0973(8)	0.001	1300		4d	$(5/2)^2[3/2]_2$	sp	$(^3P)^3P^\circ\,^3D^\circ_1$			R1c		
5340.878(3)	18718.305(12)	5340.8750(20)	0.003	300		4f	$(5/2)^2[7/2]^\circ_3$	7g	$(5/2)^2[9/2]_4$			R1c		
5342.197(3)	18713.684(12)	5342.1987(24)	−0.001	150		4f	$(5/2)^2[7/2]^\circ_3$	7g	$(5/2)^2[7/2]_3$			R1c		
		5344.316(4)			m	4f	$(3/2)^2[9/2]^\circ_4$	7g	$(3/2)^2[9/2]_4$			R1nc		
5344.366(4)	18706.091(16)	5344.3657(21)	0.000	150	*	4f	$(5/2)^2[7/2]^\circ_3$	7g	$(5/2)^2[5/2]_2$			R1nc		
5344.366(4)	18706.091(16)	5344.3692(24)	−0.003	150	*	4f	$(5/2)^2[7/2]^\circ_3$	7g	$(5/2)^2[5/2]_3$			R1c		
5345.160(3)	18703.312(10)	5345.154(3)	0.006	420	*	4f	$(3/2)^2[9/2]^\circ_5$	7g	$(3/2)^2[11/2]_5$			R1c		
5345.160(3)	18703.312(10)			420	*	4f	$(3/2)^2[9/2]^\circ_5$	7g	$(3/2)^2[11/2]_6$	1.1e+07	D+	R1nc	TW	
5345.576(2)	18701.857(9)	5345.5723(9)	0.003	300		sp	$(^3F)^3P^\circ\,^3D^\circ_1$	5d	$(5/2)^2[3/2]_1$			R1c		
5347.6243(22)	18694.693(8)	5347.6243(8)	−0.0000	1500		4d	$(5/2)^2[3/2]_1$	sp	$(^3P)^3P^\circ\,^3D^\circ_1$			R1c		
5347.684(3)	18694.484(11)	5347.6891(8)	−0.005	750		4d	$(3/2)^2[7/2]_3$	sp	$(^3F)^1P^\circ\,^3F^\circ_3$			R1c		
5347.809(3)	18694.047(12)	5347.809(3)	0.000	420		4f	$(3/2)^2[9/2]^\circ_4$	7g	$(3/2)^2[11/2]_5$	1.1e+07	D+	R1c	TW	
5349.076(4)	18689.619(13)	5349.0730(24)	0.003	150		4f	$(5/2)^2[5/2]^\circ_2$	7g	$(5/2)^2[7/2]_3$			R1c		
5351.245(4)	18682.043(14)	5351.2456(21)	−0.000	150	*	4f	$(5/2)^2[5/2]^\circ_2$	7g	$(5/2)^2[5/2]_2$			R1nc		
5351.245(4)	18682.043(14)	5351.2491(24)	−0.004	150	*	4f	$(5/2)^2[5/2]^\circ_2$	7g	$(5/2)^2[5/2]_3$			R1c		
5351.579(4)	18680.878(12)	5351.5804(12)	−0.001	300		4d	$(3/2)^2[5/2]_2$	sp	$(^1G)^3P^\circ\,^3F^\circ_2$			R1nc		
5352.0268(21)	18679.315(7)	5352.0245(5)	0.0023	6000		4d	$(3/2)^2[1/2]_1$	4f	$(5/2)^2[3/2]^\circ_2$			R1c		
5353.863(4)	18672.908(12)	5353.8649(20)	−0.002	300		4f	$(5/2)^2[5/2]^\circ_3$	7g	$(5/2)^2[9/2]_4$			R1c		
5354.4850(20)	18670.740(7)	5354.4863(8)	−0.0013	4500		4d	$(3/2)^2[1/2]_1$	4f	$(5/2)^2[1/2]^\circ_0$			R1c		
5355.186(3)	18668.294(11)	5355.1870(24)	−0.001	410		4f	$(5/2)^2[5/2]^\circ_3$	7g	$(5/2)^2[7/2]_4$	7.4e+06	D+	R1c	TW	
5356.8109(21)	18662.633(7)	5356.8091(8)	0.0018	2100		4d	$(5/2)^2[5/2]_3$	sp	$(^3F)^1P^\circ\,^3F^\circ_4$			R1c		
5357.374(4)	18660.670(13)	5357.3727(21)	0.002	300	*	4f	$(5/2)^2[5/2]^\circ_3$	7g	$(5/2)^2[5/2]_2$			R1nc		
5357.374(4)	18660.670(13)	5357.3761(24)	−0.002	300	*	4f	$(5/2)^2[5/2]^\circ_3$	7g	$(5/2)^2[5/2]_3$			R1c		
5361.907(4)	18644.894(12)	5361.914(4)	−0.007	150	*	4f	$(3/2)^2[5/2]^\circ_3$	7g	$(3/2)^2[7/2]_3$			R1c		
5361.907(4)	18644.894(12)			150	*	4f	$(3/2)^2[5/2]^\circ_3$	7g	$(3/2)^2[7/2]_4$	9.e+06	D+	R1nc	TW	
5365.5363(20)	18632.284(7)	5365.5363(6)	−0.0000	5500		4d	$(3/2)^2[1/2]_1$	4f	$(5/2)^2[1/2]^\circ_1$			R1c		
5366.243(3)	18629.831(11)	5366.2476(7)	−0.005	440		sp	$(^3F)^3P^\circ\,^3D^\circ_2$	6s	$(3/2)^2[3/2]_2$			R1c		
5368.3825(21)	18622.406(7)	5368.3839(7)	−0.0013	7400		sp	$(^3F)^3P^\circ\,^3F^\circ_3$	5d	$(5/2)^2[9/2]_4$			R1c		
5371.6264(21)	18611.160(7)	5371.6264(7)	0.0001	880		sp	$(^3F)^3P^\circ\,^3F^\circ_3$	5d	$(5/2)^2[3/2]_2$			R1c		
5375.276(4)	18598.524(12)	5375.276(4)	0.000	290		4f	$(3/2)^2[5/2]^\circ_2$	7g	$(3/2)^2[7/2]_3$	8.1e+06	D+	R1c	TW	
5381.9479(23)	18575.468(8)	5381.9511(21)	−0.0032	430	*	4f	$(5/2)^2[7/2]^\circ_4$	7g	$(5/2)^2[9/2]_4$			R1c		
5381.9479(23)	18575.468(8)	5381.9485(20)	−0.0006	430	*	4f	$(5/2)^2[7/2]^\circ_4$	7g	$(5/2)^2[9/2]_5$	8.2e+06	D+	R1nc	TW	
5382.329(4)	18574.152(15)	5382.3347(19)	−0.006	140		4f	$(5/2)^2[7/2]^\circ_4$	7g	$(5/2)^2[11/2]_5$			R1c		
5383.289(4)	18570.842(12)	5383.2871(24)	0.001	140		4f	$(5/2)^2[7/2]^\circ_4$	7g	$(5/2)^2[7/2]_4$			R1c		
5384.818(3)	18565.569(11)	5384.8222(16)	−0.005	430		4d	$(3/2)^2[1/2]_0$	6p	$(3/2)^2[1/2]^\circ_1$			R1c		
5386.435(4)	18559.995(14)	5386.437(4)	−0.003	420	*	4f	$(3/2)^2[7/2]^\circ_4$	7g	$(3/2)^2[9/2]_4$			R1c		
5386.435(4)	18559.995(14)			420	*	4f	$(3/2)^2[7/2]^\circ_4$	7g	$(3/2)^2[9/2]_5$	1.0e+07	D+	R1nc	TW	
5386.8405(22)	18558.597(8)	5386.8419(10)	−0.0014	720		4d	$(5/2)^2[1/2]_0$	sp	$(^3P)^3P^\circ\,^3P^\circ_1$			R1c		
5390.029(4)	18547.619(13)	5390.0296(21)	−0.001	140	*	4f	$(5/2)^2[9/2]^\circ_5$	7g	$(5/2)^2[9/2]_4$			R1c		
5390.029(4)	18547.619(13)	5390.0270(20)	0.002	140	*	4f	$(5/2)^2[9/2]^\circ_5$	7g	$(5/2)^2[9/2]_5$			R1nc		

Table A1. *Cont.*

λ_{obs} [a] (Å)	σ_{obs} [b] (cm^{-1})	λ_{Ritz} [c] (Å)	$\Delta\lambda_{obs\text{-}Ritz}$ (Å)	I_{obs} [d] (arb. u.)	Char [e]	Lower Level		Upper Level		A (s^{-1})	Acc [f]	Line Ref. [g]	TP Ref. [g]	Notes [h]
5390.4163(22)	18546.286(8)	5390.4169(13)	−0.0006	4800		4d	$(3/2)^2[5/2]_2$	sp	$(^3F)^1P^\circ\ ^3G^\circ_3$			R1c		
5391.6966(22)	18541.882(8)	5391.6987(10)	−0.0021	1700		4d	$(5/2)^2[9/2]_5$	sp	$(^3F)^1P^\circ\ ^3G^\circ_5$			R1c		
5393.9372(21)	18534.180(7)	5393.9372(7)	0.0001	2600		sp	$(^3F)^3P^\circ\ ^3D^\circ_2$	6s	$(3/2)^2[3/2]_1$			R1c		
5397.2943(21)	18522.652(7)	5397.2952(5)	−0.0008	580		4d	$(3/2)^2[7/2]_3$	4f	$(5/2)^2[9/2]^\circ_4$			R1c		
5397.37	18522.38	5397.3730(17)			:	sp	$(^3F)^1P^\circ\ ^3D^\circ_3$	9s	$(5/2)^2[5/2]_3$					
5398.573(4)	18518.266(12)	5398.573(4)	−0.000	140		4f	$(3/2)^2[7/2]^\circ_3$	7g	$(3/2)^2[9/2]_4$	9.e+06	D+	R1c	TW	
5403.714(4)	18500.647(13)	5403.7093(9)	0.005	140		5p	$(5/2)^2[3/2]^\circ_2$	5d	$(3/2)^2[1/2]_1$			R1c		
5405.652(3)	18494.013(11)	5405.6515(6)	0.001	280		4p	$^3F^\circ_2$	s^2	$^3P_2$			R1c		
5407.447(3)	18487.877(11)	5407.4452(23)	0.001	280		5s	$(3/2)^2[3/2]_2$	sp	$(^1D)^3P^\circ\ ^3D^\circ_2$			R1c		
5408.383(4)	18484.675(12)	5408.3840(24)	−0.001	280		5s	$(3/2)^2[3/2]_1$	sp	$(^1D)^3P^\circ\ ^3D^\circ_1$			R1c		
5422.002(4)	18438.246(12)	5422.0053(21)	−0.003	140		4f	$(5/2)^2[9/2]^\circ_4$	7g	$(5/2)^2[9/2]_4$			R1c		
5422.392(3)	18436.920(11)	5422.3947(19)	−0.003	270		4f	$(5/2)^2[9/2]^\circ_4$	7g	$(5/2)^2[11/2]_5$	7.2e+06	D+	R1c	TW	
5428.274(3)	18416.944(11)	5428.2724(9)	0.001	140		4d	$(5/2)^2[7/2]_3$	sp	$(^3F)^1P^\circ\ ^3F^\circ_4$			R1c		
5437.1438(21)	18386.899(7)	5437.1431(5)	0.0006	2100		4d	$(3/2)^2[7/2]_4$	4f	$(5/2)^2[9/2]^\circ_4$			R1c		
5437.5787(23)	18385.428(8)	5437.5790(4)	−0.0002	6700		4d	$(3/2)^2[7/2]_3$	4f	$(5/2)^2[7/2]^\circ_4$			R1c		
5441.6461(22)	18371.686(7)	5441.6471(9)	−0.0011	2100		4d	$(5/2)^2[7/2]_4$	sp	$(^3F)^1P^\circ\ ^3F^\circ_4$			R1c		
5448.1914(23)	18349.615(8)	5448.1938(9)	−0.0024	260		4d	$(5/2)^2[7/2]_3$	sp	$(^3P)^3P^\circ\ ^3D^\circ_2$			R1c		
5449.409(4)	18345.516(12)	5449.4072(10)	0.002	2700		4d	$(5/2)^2[3/2]_2$	sp	$(^3P)^3P^\circ\ ^3D^\circ_3$			R1c		
5458.0942(22)	18316.323(7)	5458.0941(9)	0.0001	390		4d	$(3/2)^2[5/2]_3$	sp	$(^3F)^1P^\circ\ ^3D^\circ_2$			R1c		
5466.57(6)	18287.94(20)	5466.5527(5)	0.01	1600	*	4d	$(3/2)^2[7/2]_3$	4f	$(5/2)^2[5/2]^\circ_3$			R1c		
5466.57(6)	18287.94(20)	5466.6267(9)	−0.06	1600	*	4d	$(5/2)^2[5/2]_2$	sp	$(^3P)^3P^\circ\ ^3D^\circ_2$			R1c		
5468.563(3)	18281.260(9)	5468.5633(19)	−0.000	510		5s	$(3/2)^2[3/2]_1$	sp	$(^1D)^3P^\circ\ ^3F^\circ_2$			R1c		
5469.376(3)	18278.541(10)	5469.3734(8)	0.003	1300		4d	$(3/2)^2[3/2]_2$	sp	$(^3F)^1P^\circ\ ^3F^\circ_3$			R1c		
5469.6826(21)	18277.518(7)	5469.6819(4)	0.0006	4400		4d	$(3/2)^2[7/2]_4$	4f	$(5/2)^2[9/2]^\circ_5$			R1c		
5472.465(3)	18268.225(11)	5472.4631(10)	0.002	130		4d	$(3/2)^2[5/2]_2$	sp	$(^3F)^1P^\circ\ ^3D^\circ_2$			R1c		
5472.9475(24)	18266.614(8)	5472.9467(5)	0.0007	740		4d	$(3/2)^2[7/2]_3$	4f	$(5/2)^2[5/2]^\circ_2$			R1c		
5473.01	18266.42	5473.006(2)			:	sp	$(^3F)^1P^\circ\ ^3F^\circ_3$	7g	$(5/2)^2[9/2]_4$					
5473.119(4)	18266.040(12)	5473.1211(8)	−0.002	920		5p	$(5/2)^2[5/2]^\circ_2$	5d	$(3/2)^2[5/2]_2$			R1c		
		5473.1623(16)			m	4f	$(5/2)^2[7/2]^\circ_3$	9s	$(5/2)^2[5/2]_3$			R1nc		
5477.127(3)	18252.676(11)	5477.1275(13)	−0.001	130		4d	$(3/2)^2[1/2]_1$	sp	$(^3P)^3P^\circ\ ^3P^\circ_0$			R1c		
5478.0268(23)	18249.677(8)	5478.0262(4)	0.0006	3700		4d	$(3/2)^2[7/2]_4$	4f	$(5/2)^2[7/2]^\circ_4$			R1c		
5479.9089(22)	18243.409(7)	5479.9071(7)	0.0018	3700		5p	$(5/2)^2[5/2]^\circ_2$	5d	$(3/2)^2[5/2]_3$			R1c		
5480.1613(24)	18242.569(8)	5480.1619(5)	−0.0006	1900		4d	$(3/2)^2[7/2]_3$	4f	$(5/2)^2[7/2]^\circ_3$			R1c		
5482.524(3)	18234.708(10)	5482.5262(15)	−0.002	790		sp	$(^3F)^3P^\circ\ ^3G^\circ_4$	6s	$(5/2)^2[5/2]_3$			R1c		
5486.205(3)	18222.473(11)	5486.2037(8)	0.001	120		5p	$(5/2)^2[3/2]^\circ_1$	5d	$(3/2)^2[5/2]_2$			R1c		
5493.5268(22)	18198.186(7)	5493.5289(6)	−0.0021	240		5p	$(5/2)^2[7/2]^\circ_3$	5d	$(3/2)^2[7/2]_4$			R1c		
5495.0454(23)	18193.157(8)	5495.0446(23)	0.0008	430		5s	$(3/2)^2[3/2]_2$	sp	$(^1D)^3P^\circ\ ^3F^\circ_3$			R1c		
5496.868(3)	18187.126(11)	5496.8685(11)	−0.001	510		4d	$(5/2)^2[3/2]_2$	sp	$(^3P)^3P^\circ\ ^3P^\circ_2$			R1c		
5504.021(4)	18163.488(12)	5504.0559(5)	−0.035	120	?	4d	$(3/2)^2[7/2]_3$	4f	$(5/2)^2[3/2]^\circ_2$			R1c		X
5504.844(3)	18160.774(11)	5504.8442(11)	−0.000	240		4d	$(5/2)^2[3/2]_1$	sp	$(^3P)^3P^\circ\ ^3P^\circ_2$			R1c		
5507.428(4)	18152.254(12)	5507.4337(4)	−0.006	240		4d	$(3/2)^2[7/2]_4$	4f	$(5/2)^2[5/2]^\circ_3$			R1c		
5512.8649(22)	18134.351(7)	5512.8650(6)	−0.0001	240		5p	$(5/2)^2[7/2]^\circ_3$	5d	$(3/2)^2[7/2]_3$			R1c		

Table A1. *Cont.*

λ_{obs} [a] (Å)	σ_{obs} [b] (cm^{-1})	λ_{Ritz} [c] (Å)	$\Delta\lambda_{obs-Ritz}$ (Å)	I_{obs} [d] (arb. u.)	Char [e]	Lower Level		Upper Level		A (s^{-1})	Acc [f]	Line Ref. [g]	TP Ref. [g]	Notes [h]
5521.249(3)	18106.815(9)	5521.2475(5)	0.001	230		4d	$(3/2)^2[7/2]_4$	4f	$(5/2)^2[7/2]^\circ_3$			R1c		
5521.729(3)	18105.239(9)	5521.7315(8)	−0.002	1100		5p	$(5/2)^2[5/2]^\circ_2$	5d	$(3/2)^2[3/2]_2$			R1c		
5525.438(3)	18093.088(9)	5525.4391(6)	−0.001	230		5p	$(5/2)^2[7/2]^\circ_4$	5d	$(3/2)^2[7/2]_4$			R1c		
5525.53	18092.78	5525.532(6)			:	sp	$(^3F)^1P^\circ\ ^3D^\circ_2$	8d	$(5/2)^2[5/2]_2$					
5527.1968(22)	18087.330(7)	5527.1993(7)	−0.0025	2200		4d	$(3/2)^2[7/2]_3$	sp	$(^3F)^1P^\circ\ ^3G^\circ_4$			R1c		
5527.6813(22)	18085.744(7)	5527.6804(5)	0.0009	690		4d	$(3/2)^2[3/2]_1$	4f	$(5/2)^2[5/2]^\circ_2$			R1c		
5534.668(3)	18062.913(10)	5534.6726(8)	−0.004	230		5p	$(5/2)^2[5/2]^\circ_3$	5d	$(3/2)^2[5/2]_2$			R1c		
5535.0494(22)	18061.669(7)	5535.0478(8)	0.0016	4100		5p	$(5/2)^2[3/2]^\circ_1$	5d	$(3/2)^2[3/2]_2$			R1c		
5537.670(20)	18053.12(7)	5537.6662(21)	0.004	340		sp	$(^3F)^3P^\circ\ ^1G^\circ_4$	5d	$(5/2)^2[7/2]_4$			R1c		
5538.3836(22)	18050.796(7)	5538.3835(6)	0.0001	880		4d	$(3/2)^2[7/2]_4$	4f	$(5/2)^2[11/2]^\circ_5$			R1c		
5541.6120(22)	18040.280(7)	5541.6123(6)	−0.0003	1300		5p	$(5/2)^2[5/2]^\circ_3$	5d	$(3/2)^2[5/2]_3$			R1c		
5543.5368(22)	18034.016(7)	5543.5385(7)	−0.0016	1900		sp	$(^3F)^3P^\circ\ ^3F^\circ_2$	5d	$(5/2)^2[5/2]_2$			R1c		
5544.7815(22)	18029.968(7)	5544.7803(6)	0.0012	220		4d	$(3/2)^2[3/2]_1$	4f	$(5/2)^2[3/2]^\circ_1$			R1c		
5555.5122(22)	17995.143(7)	5555.5123(7)	−0.0001	5500		sp	$(^3F)^3P^\circ\ ^3F^\circ_2$	5d	$(5/2)^2[7/2]_3$			R1c		
5558.314(3)	17986.072(11)	5558.3108(10)	0.003	110		4d	$(3/2)^2[7/2]_3$	sp	$(^3F)^1P^\circ\ ^3D^\circ_3$			R1c		
5559.417(4)	17982.504(11)	5559.4167(5)	0.000	110		4d	$(3/2)^2[3/2]_1$	4f	$(5/2)^2[3/2]^\circ_2$			R1c		
5560.5761(24)	17978.755(8)	5560.5739(8)	0.0022	610		5p	$(5/2)^2[3/2]^\circ_1$	5d	$(3/2)^2[3/2]_1$			R1c		
5562.074(3)	17973.913(11)	5562.0730(8)	0.001	110		4d	$(3/2)^2[3/2]_1$	4f	$(5/2)^2[1/2]^\circ_0$			R1c		
5562.6460(23)	17972.065(7)	5562.6478(9)	−0.0018	770		4d	$(5/2)^2[5/2]_2$	sp	$(^3P)^3P^\circ\ ^3D^\circ_1$			R1c		
5567.002(3)	17958.004(9)	5567.0018(8)	−0.000	210		4d	$(3/2)^2[5/2]_3$	sp	$(^3F)^1P^\circ\ ^3F^\circ_3$			R1c		
5569.670(3)	17949.401(11)	5569.6723(16)	−0.002	210		5d	$(5/2)^2[9/2]_5$	7f	$(5/2)^2[9/2]^\circ_5$			R1c		
5571.6406(24)	17943.052(8)	5571.6415(7)	−0.0009	210		5p	$(5/2)^2[5/2]^\circ_2$	5d	$(3/2)^2[7/2]_3$			R1c		
5578.4356(23)	17921.196(7)	5578.4337(17)	0.0019	520		5d	$(5/2)^2[9/2]_5$	7f	$(5/2)^2[11/2]^\circ_6$	1.0e+07	D+	R1c	TW	
5579.431(4)	17917.999(12)	5579.4269(20)	0.004	210		5d	$(5/2)^2[3/2]_2$	7f	$(5/2)^2[5/2]^\circ_3$			R1c		
5581.9475(24)	17909.921(8)	5581.9507(9)	−0.0032	310		4d	$(3/2)^2[5/2]_2$	sp	$(^3F)^1P^\circ\ ^3F^\circ_3$			R1c		
5582.173(4)	17909.196(11)	5582.1725(18)	0.001	100		5d	$(5/2)^2[9/2]_4$	7f	$(5/2)^2[9/2]^\circ_4$			R1c		
5583.864(3)	17903.773(11)	5583.8665(11)	−0.002	210		4d	$(3/2)^2[1/2]_1$	sp	$(^3P)^3P^\circ\ ^3P^\circ_1$			R1c		
5584.346(3)	17902.230(11)	5584.3413(22)	0.004	480		5d	$(3/2)^2[7/2]_3$	7f	$(3/2)^2[9/2]^\circ_4$	9.e+06	D+	R1c	TW	
5591.382(3)	17879.700(9)	5591.3777(6)	0.005	210		sp	$(^3F)^3P^\circ\ ^3F^\circ_2$	5d	$(5/2)^2[5/2]_3$			R1c		
5592.5298(24)	17876.032(8)	5592.5272(19)	0.0026	510		5d	$(5/2)^2[9/2]_4$	7f	$(5/2)^2[11/2]^\circ_5$	9.e+06	D+	R1c	TW	
5593.7697(13)	17872.070(4)	5593.7708(4)	−0.0011	5400		4d	$(3/2)^2[3/2]_2$	4f	$(5/2)^2[5/2]^\circ_3$			F_Re		
5600.4662(23)	17850.701(7)	5600.4661(5)	0.0001	500		4d	$(3/2)^2[3/2]_2$	4f	$(5/2)^2[5/2]^\circ_2$			R1c		
5600.581(4)	17850.334(11)	5600.5810(10)	0.000	300		4d	$(3/2)^2[7/2]_4$	sp	$(^3F)^1P^\circ\ ^3D^\circ_3$			R1c		
5607.282(3)	17829.001(9)	5607.2823(22)	0.000	300		5d	$(3/2)^2[7/2]_4$	7f	$(3/2)^2[9/2]^\circ_5$	8.e+06	D+	R1c	TW	
5608.027(3)	17826.635(8)	5608.0216(5)	0.005	200		4d	$(3/2)^2[3/2]_2$	4f	$(5/2)^2[7/2]^\circ_3$			R1c		
5615.2350(23)	17803.751(7)	5615.2379(6)	−0.0030	5500		5p	$(5/2)^2[5/2]^\circ_3$	5d	$(3/2)^2[7/2]_4$			R1c		
5618.016(3)	17794.938(11)	5618.0200(6)	−0.004	98		4d	$(3/2)^2[3/2]_2$	4f	$(5/2)^2[3/2]^\circ_1$			R1c		
5619.885(7)	17789.021(21)	5619.8874(16)	−0.003	190		4d	$(3/2)^2[3/2]_2$	sp	$(^3P)^3P^\circ\ ^5S^\circ_2$			R1c		
5620.7814(23)	17786.183(7)	5620.7805(5)	0.0008	850		4d	$(3/2)^2[5/2]_3$	4f	$(5/2)^2[9/2]^\circ_4$			R1c		
5621.7027(24)	17783.268(8)	5621.7019(10)	0.0009	530		4d	$(5/2)^2[1/2]_0$	sp	$(^3P)^3P^\circ\ ^3D^\circ_1$			R1c		
5629.240(3)	17759.456(11)	5629.218(3)	0.022	190	?	5d	$(5/2)^2[5/2]_3$	7f	$(5/2)^2[7/2]^\circ_3$			R1c		X
5630.589(4)	17755.202(11)	5630.5900(18)	−0.001	95		5d	$(5/2)^2[5/2]_3$	7f	$(5/2)^2[9/2]^\circ_4$			R1c		

Table A1. *Cont.*

λ_{obs} [a] (Å)	σ_{obs} [b] (cm^{-1})	λ_{Ritz} [c] (Å)	$\Delta\lambda_{obs\text{-}Ritz}$ (Å)	I_{obs} [d] (arb. u.)	Char [e]	Lower Level	Upper Level	A (s^{-1})	Acc [f]	Line Ref. [g]	TP Ref. [g]	Notes [h]
5631.153(3)	17753.425(11)	5631.1561(18)	−0.003	190		5d $(5/2)^2[5/2]_3$	7f $(5/2)^2[7/2]^{\circ}_4$			R1c		
5631.358(3)	17752.777(11)	5631.3600(20)	−0.002	95		5d $(5/2)^2[5/2]_3$	7f $(5/2)^2[5/2]^{\circ}_3$			R1c		
5633.0442(23)	17747.464(7)	5633.0461(5)	−0.0020	3400		4d $(3/2)^2[3/2]_2$	4f $(5/2)^2[3/2]^{\circ}_2$			R1c		
5633.602(3)	17745.707(10)	5633.6032(12)	−0.001	950		4d $(5/2)^2[5/2]_3$	sp $(^3P)^3P^{\circ}\,^3P^{\circ}_2$			R1c		
5635.5118(23)	17739.693(7)	5635.5130(11)	−0.0012	3000		4d $(5/2)^2[7/2]_4$	sp $(^3F)^1P^{\circ}\,^3G^{\circ}_5$			R1c		
5637.162(4)	17734.501(13)	5637.1546(21)	0.007	190		sp $(^3F)^3P^{\circ}\,^1G^{\circ}_4$	5d $(5/2)^2[9/2]_5$			R1c		
5637.468(3)	17733.536(8)	5637.4686(8)	−0.000	1100		sp $(^3F)^3P^{\circ}\,^3F^{\circ}_2$	5d $(5/2)^2[3/2]_1$			R1c		
5641.2645(23)	17721.603(7)	5641.2645(18)	−0.0000	2300		5p $(3/2)^2[1/2]^{\circ}_1$	5d $(3/2)^2[1/2]_0$			R1c		
5643.5338(24)	17714.477(8)	5643.5346(7)	−0.0008	280		sp $(^3F)^3P^{\circ}\,^3F^{\circ}_2$	5d $(5/2)^2[3/2]_2$			R1c		
5648.0170(24)	17700.416(8)	5648.0161(6)	0.0009	430		4d $(3/2)^2[3/2]_2$	4f $(5/2)^2[1/2]^{\circ}_1$			R1c		
5651.708(4)	17688.857(11)	5651.7093(9)	−0.001	370		sp $(^3F)^3P^{\circ}\,^3D^{\circ}_1$	6s $(3/2)^2[3/2]_2$			R1c		
5656.632(3)	17673.459(10)	5656.6271(10)	0.005	7300		5p $(5/2)^2[3/2]^{\circ}_1$	5d $(3/2)^2[1/2]_1$			R1c		
5664.4833(23)	17648.963(7)	5664.4829(5)	0.0004	2900		4d $(3/2)^2[5/2]_3$	4f $(5/2)^2[7/2]^{\circ}_4$			R1c		
5667.434(3)	17639.775(11)	5667.4344(19)	−0.001	180		5d $(5/2)^2[7/2]_3$	7f $(5/2)^2[9/2]^{\circ}_4$	5.8e+06	D+	R1c	TW	
5668.007(3)	17637.991(10)	5668.0080(18)	−0.001	180		5d $(5/2)^2[7/2]_3$	7f $(5/2)^2[7/2]^{\circ}_4$			R1c		
5670.329(3)	17630.769(10)	5670.3245(16)	0.004	350		5d $(5/2)^2[7/2]_4$	7f $(5/2)^2[9/2]^{\circ}_5$	7.4e+06	D+	R1c	TW	
5674.694(4)	17617.207(12)	5674.6947(7)	−0.001	180	*	sp $(^3F)^3P^{\circ}\,^3F^{\circ}_3$	6s $(3/2)^2[3/2]_2$			R1c		
5674.694(4)	17617.207(12)	5674.6968(9)	−0.003	180	*	sp $(^3F)^3P^{\circ}\,^1F^{\circ}_3$	5d $(3/2)^2[5/2]_2$			R1nc		
5676.005(3)	17613.137(10)	5676.0028(11)	0.002	1300		4d $(5/2)^2[7/2]_4$	sp $(^3P)^3P^{\circ}\,^3D^{\circ}_3$			R1c		
5678.556(3)	17605.224(11)	5678.554(3)	0.002	87		5d $(5/2)^2[5/2]_2$	7f $(5/2)^2[7/2]^{\circ}_3$			R1c		
5681.353(3)	17596.559(11)	5681.3597(11)	−0.007	87		4d $(5/2)^2[5/2]_2$	sp $(^3P)^3P^{\circ}\,^3D^{\circ}_3$			R1c		
5681.998(3)	17594.559(9)	5681.9923(7)	0.006	430		sp $(^3F)^3P^{\circ}\,^1F^{\circ}_3$	5d $(3/2)^2[5/2]_3$			R1c		
5682.433(3)	17593.214(9)	5682.4316(9)	0.001	2100		sp $(^3F)^3P^{\circ}\,^3D^{\circ}_1$	6s $(3/2)^2[3/2]_1$			R1c		
5689.83(4)	17570.34(12)	5689.8869(11)	−0.06	170		4d $(3/2)^2[3/2]_2$	sp $(^3F)^1P^{\circ}\,^3D^{\circ}_3$			S36c		
5694.526(3)	17555.852(10)	5694.5253(14)	0.001	170		4d $(3/2)^2[3/2]_1$	sp $(^3P)^3P^{\circ}\,^3P^{\circ}_0$			R1c		
5695.931(3)	17551.521(10)	5695.9321(5)	−0.001	340		4d $(3/2)^2[5/2]_3$	4f $(5/2)^2[5/2]^{\circ}_3$			R1c		
5706.012(4)	17520.514(12)	5706.0055(7)	0.006	450		4p $^1F^{\circ}_3$	s^2 $^3P_2$			R1c		
5710.7111(24)	17506.097(7)	5710.7089(5)	0.0022	250		4d $(3/2)^2[5/2]_3$	4f $(5/2)^2[7/2]^{\circ}_3$			R1c		
5711.586(4)	17503.416(11)	5711.5823(5)	0.003	250		4d $(3/2)^2[5/2]_2$	4f $(5/2)^2[5/2]^{\circ}_3$			R1c		
5721.7840(24)	17472.220(7)	5721.7856(7)	−0.0015	4100		4p $^3P^{\circ}_1$	s^2 $^1D_2$			R1c		
5726.442(3)	17458.009(11)	5726.4406(6)	0.001	160		4d $(3/2)^2[5/2]_2$	4f $(5/2)^2[7/2]^{\circ}_3$			R1c		
5731.004(4)	17444.110(12)	5731.0007(11)	0.004	160		4d $(3/2)^2[1/2]_1$	sp $(^3P)^3P^{\circ}\,^3D^{\circ}_2$			R1c		
5736.659(4)	17426.915(11)	5736.6603(5)	−0.001	160		4d $(3/2)^2[5/2]_3$	4f $(5/2)^2[3/2]^{\circ}_2$			R1c		
5752.536(4)	17378.819(11)	5752.5355(6)	0.000	580		4d $(3/2)^2[5/2]_2$	4f $(5/2)^2[3/2]^{\circ}_2$			R1c		
5759.015(4)	17359.266(11)	5759.0170(20)	−0.002	150		sp $(^3F)^1P^{\circ}\,^3G^{\circ}_5$	6g $(5/2)^2[11/2]_6$			R1c		
5759.4219(24)	17358.040(7)	5759.4213(7)	0.0005	6300		sp $(^3F)^3P^{\circ}\,^1F^{\circ}_3$	5d $(3/2)^2[7/2]_4$			R1c		
5761.220(3)	17352.622(8)	5761.2165(10)	0.004	2000		sp $(^3F)^3P^{\circ}\,^3D^{\circ}_3$	6s $(5/2)^2[5/2]_2$			R1c		
5761.8052(24)	17350.860(7)	5761.8056(7)	−0.0003	2200		4d $(3/2)^2[5/2]_3$	sp $(^3F)^1P^{\circ}\,^3G^{\circ}_4$			R1c		
5768.143(4)	17331.796(11)	5768.1482(7)	−0.005	150		4d $(3/2)^2[5/2]_2$	4f $(5/2)^2[1/2]^{\circ}_1$			R1c		
5779.656(4)	17297.270(12)	5779.6592(8)	−0.003	360		sp $(^3F)^3P^{\circ}\,^3F^{\circ}_2$	5d $(5/2)^2[1/2]_1$			R1c		
5783.920(4)	17284.519(11)	5783.9194(20)	0.001	6400		5p $(3/2)^2[3/2]^{\circ}_1$	5d $(3/2)^2[1/2]_0$			R1c		
5801.130(7)	17233.242(21)	5801.116(3)	0.014	140		5s $(5/2)^2[5/2]_3$	sp $(^3P)^3P^{\circ}\,^5P^{\circ}_2$			R1c		

Table A1. *Cont.*

λ_{obs} [a] (Å)	σ_{obs} [b] (cm^{-1})	λ_{Ritz} [c] (Å)	$\Delta\lambda_{obs\text{-}Ritz}$ (Å)	I_{obs} [d] (arb. u.)	Char [e]	Lower Level	Upper Level	A (s^{-1})	Acc [f]	Line Ref. [g]	TP Ref. [g]	Notes [h]
5805.989(3)	17218.822(8)	5805.9869(11)	0.002	4700		sp $(^3F)^3P^\circ\ ^3D^\circ_3$	6s $(5/2)^2[5/2]_3$			R1c		
5825.823(2)	17160.200(7)	5825.8247(9)	−0.002	9100		sp $(^3F)^3P^\circ\ ^1D^\circ_2$	5d $(3/2)^2[5/2]_2$	1.1e+07	D+	R1c	TW	
5833.5146(24)	17137.573(7)	5833.5141(7)	0.0005	6400		sp $(^3F)^3P^\circ\ ^1D^\circ_2$	5d $(3/2)^2[5/2]_3$	7.9e+06	D+	R1c	TW	
5836.619(4)	17128.458(13)	5836.6234(11)	−0.004	130		4d $(3/2)^2[1/2]_1$	sp $(^3P)^3P^\circ\ ^3D^\circ_1$			R1c		
5842.495(3)	17111.233(8)	5842.4918(7)	0.003	1900		4p $^1D^\circ_2$	s^2 $^3P_1$			R1c		
5851.787(4)	17084.062(11)	5851.7811(12)	0.006	930		sp $(^3F)^3P^\circ\ ^3G^\circ_3$	6s $(5/2)^2[5/2]_2$			R1c		
5858.54(10)	17064.4(3)	5858.353(5)	0.19	9100		sp $(^3F)^1P^\circ\ ^3F^\circ_2$	7g $(3/2)^2[7/2]_3$			S36cn		X
5890.443(4)	16971.949(11)	5890.4592(12)	−0.016	120		4d $(3/2)^2[3/2]_2$	sp $(^3P)^3P^\circ\ ^3P^\circ_1$			R1c		X
5897.971(3)	16950.287(7)	5897.9758(12)	−0.005	23000		sp $(^3F)^3P^\circ\ ^3G^\circ_3$	6s $(5/2)^2[5/2]_3$			R1c		
5901.188(3)	16941.047(10)	5901.1910(8)	−0.003	1700		4p $^3F^\circ_3$	s^2 $^1D_2$			R1c		
5909.753(3)	16916.494(9)	5909.7579(9)	−0.005	1500		sp $(^3F)^3P^\circ\ ^1D^\circ_2$	5d $(3/2)^2[3/2]_1$			R1c		
5910.813(4)	16913.460(11)	5910.8150(20)	−0.002	57		sp $(^3F)^1P^\circ\ ^3F^\circ_4$	7d $(5/2)^2[9/2]_5$			R1c		
5926.691(3)	16868.147(8)	5926.6916(6)	−0.000	1300		4p $^1D^\circ_2$	s^2 $^3P_2$			R1c		
5929.634(3)	16859.777(10)	5929.6380(11)	−0.004	110		4d $(3/2)^2[7/2]_4$	sp $(^3F)^1P^\circ\ ^3F^\circ_4$			R1c		
5937.576(2)	16837.224(7)	5937.5815(7)	−0.005	2000		sp $(^3F)^3P^\circ\ ^1D^\circ_2$	5d $(3/2)^2[7/2]_3$	1.0e+07	D+	R1c	TW	
5939.775(7)	16830.992(21)	5939.7773(20)	−0.002	54		sp $(^1G)^3P^\circ\ ^3H^\circ_4$	6g $(3/2)^2[11/2]_5$			R1c		
5941.1951(4)	16826.9685(10)	5941.1951(3)	−0.0001	31000		5p $(5/2)^2[3/2]^\circ_2$	5d $(5/2)^2[5/2]_3$	3.6e+07	D+	F_Re	TW	
5941.826(4)	16825.183(11)	5941.8291(17)	−0.003	1600		4d $(3/2)^2[1/2]_0$	6p $(5/2)^2[3/2]^\circ_1$			R1c		
5979.015(3)	16720.532(8)	5979.0193(8)	−0.004	2800		sp $(^3F)^3P^\circ\ ^3F^\circ_2$	6s $(3/2)^2[3/2]_2$			R1c		
5988.3141(17)	16694.567(5)	5988.3135(14)	0.0006	7600		5p $(5/2)^2[3/2]^\circ_1$	5d $(5/2)^2[1/2]_0$	9.e+07	D+	F_Re	TW	
5993.259(3)	16680.793(7)	5993.2605(8)	−0.001	3200		5p $(5/2)^2[3/2]^\circ_2$	5d $(5/2)^2[3/2]_1$			R1c		
5995.5902(19)	16674.307(5)	5995.5876(10)	0.0026	4600		5p $(3/2)^2[1/2]^\circ_0$	5d $(3/2)^2[3/2]_1$	3.0e+07	D+	F_Re	TW	
6000.1169(8)	16661.7272(22)	6000.1168(6)	0.0001	38000		5p $(5/2)^2[3/2]^\circ_2$	5d $(5/2)^2[3/2]_2$	7.5e+07	D+	F_Re	TW	
6002.304(4)	16655.656(10)	6002.305(3)	−0.000	98		6s $(5/2)^2[5/2]_3$	sp $(^1D)^1P^\circ\ ^1D^\circ_2$			R1c		
6013.411(3)	16624.893(7)	6013.4138(8)	−0.003	3200		sp $(^3F)^3P^\circ\ ^3F^\circ_2$	6s $(3/2)^2[3/2]_1$			R1c		
6018.369(5)	16611.196(13)	6018.3711(12)	−0.002	480		sp $(^3F)^3P^\circ\ ^1D^\circ_2$	5d $(3/2)^2[1/2]_1$			R1c		
6023.263(3)	16597.700(7)	6023.2636(9)	−0.000	3900		sp $(^3F)^3P^\circ\ ^3D^\circ_2$	6s $(5/2)^2[5/2]_2$			R1c		
6050.917(4)	16521.845(11)	6050.916(3)	0.001	91		6s $(5/2)^2[5/2]_2$	sp $(^1D)^1P^\circ\ ^1D^\circ_2$			R1c		
6071.492(4)	16465.858(10)	6071.4896(22)	0.002	130		sp $(^3F)^1P^\circ\ ^3G^\circ_3$	6g $(3/2)^2[9/2]_4$			R1c		
6072.217(3)	16463.891(7)	6072.2167(9)	0.001	4200		sp $(^3F)^3P^\circ\ ^3D^\circ_2$	6s $(5/2)^2[5/2]_3$			R1c		
6080.3334(19)	16441.915(5)	6080.3370(7)	−0.0036	11000		4p $^3D^\circ_3$	s^2 $^3P_2$			F_Re		
6083.086(4)	16434.475(10)	6083.0848(22)	0.001	130		6p $(5/2)^2[7/2]^\circ_3$	8d $(5/2)^2[9/2]_4$	6.9e+06	D+	R1c	TW	
6085.021(4)	16429.250(11)	6085.021(3)	−0.000	87		4d $(5/2)^2[1/2]_1$	sp $(^3P)^3P^\circ\ ^5D^\circ_1$			R1c		
6097.328(3)	16396.088(8)	6097.3235(10)	0.004	4800		5p $(3/2)^2[5/2]^\circ_2$	5d $(3/2)^2[5/2]_2$	1.9e+07	D+	R1c	TW	
6099.989(3)	16388.936(7)	6099.9896(7)	−0.001	3700		5p $(5/2)^2[7/2]^\circ_3$	5d $(5/2)^2[5/2]_2$			R1c		
6105.747(3)	16373.480(7)	6105.7469(8)	0.000	3600		5p $(3/2)^2[5/2]^\circ_2$	5d $(3/2)^2[5/2]_3$	9.e+06	D+	R1c	TW	
6107.4055(21)	16369.034(6)	6107.4083(12)	−0.0028	5100		5p $(3/2)^2[1/2]^\circ_0$	5d $(3/2)^2[1/2]_1$	3.3e+07	D+	F_Re	TW	
6110.872(3)	16359.748(7)	6110.8706(7)	0.002	3700		5p $(5/2)^2[7/2]^\circ_3$	5d $(5/2)^2[7/2]_4$			R1c		
6114.4926(14)	16350.061(4)	6114.4911(7)	0.0016	15000		5p $(5/2)^2[7/2]^\circ_3$	5d $(5/2)^2[7/2]_3$	3.4e+07	D+	F_Re	TW	
6150.3818(9)	16254.6548(23)	6150.3813(6)	0.0005	16000		5p $(5/2)^2[7/2]^\circ_4$	5d $(5/2)^2[7/2]_4$	3.6e+07	D+	F_Re	TW	
6154.2215(8)	16244.5134(21)	6154.2211(7)	0.0004	25000		5p $(5/2)^2[3/2]^\circ_2$	5d $(5/2)^2[1/2]_1$	1.0e+08	D+	F_Re	TW	
6157.716(4)	16235.296(10)	6157.7151(10)	0.000	4700		5p $(3/2)^2[5/2]^\circ_2$	5d $(3/2)^2[3/2]_2$			R1c		
6157.956(8)	16234.662(21)	6157.9650(6)	−0.009	530		5p $(5/2)^2[7/2]^\circ_3$	5d $(5/2)^2[5/2]_3$			R1c		

Table A1. *Cont.*

λ_{obs} [a] (Å)	σ_{obs} [b] (cm^{-1})	λ_{Ritz} [c] (Å)	$\Delta\lambda_{obs\text{-}Ritz}$ (Å)	I_{obs} [d] (arb. u.)	Char [e]	Lower Level		Upper Level		A (s^{-1})	Acc [f]	Line Ref. [g]	TP Ref. [g]	Notes [h]
6160.573(4)	16227.765(11)	6160.5707(14)	0.003	320		4d	$(3/2)^2[7/2]_4$	sp	$(^3F)^1P^\circ\ ^3G^\circ_5$			R1c		
6172.033(3)	16197.634(7)	6172.0339(8)	−0.000	16000		5p	$(5/2)^2[5/2]^\circ_2$	5d	$(5/2)^2[5/2]_2$	3.4e+07	D+	F_Re	TW	
6174.296(5)	16191.698(12)	6174.2920(12)	0.004	190		4d	$(3/2)^2[5/2]_3$	sp	$(^3P)^3P^\circ\ ^3D^\circ_2$			R1c		
6186.8770(14)	16158.772(4)	6186.8803(8)	−0.0033	18000		5p	$(5/2)^2[5/2]^\circ_2$	5d	$(5/2)^2[7/2]_3$	3.6e+07	D+	F_Re	TW	
6188.6763(11)	16154.075(3)	6188.6761(7)	0.0002	12000		5p	$(5/2)^2[3/2]^\circ_1$	5d	$(5/2)^2[5/2]_2$	3.1e+07	D+	F_Re	TW	
6189.317(4)	16152.402(11)	6189.3237(10)	−0.007	760		5p	$(3/2)^2[5/2]^\circ_2$	5d	$(3/2)^2[3/2]_1$			R1c		
6192.684(4)	16143.620(11)	6192.6855(12)	−0.001	76		4d	$(3/2)^2[5/2]_2$	sp	$(^3P)^3P^\circ\ ^3D^\circ_2$			R1c		
6198.091(3)	16129.537(7)	6198.0890(5)	0.002	5000		5p	$(5/2)^2[7/2]^\circ_4$	5d	$(5/2)^2[5/2]_3$	9.e+06	D+	R1c	TW	
6199.751(3)	16125.219(8)	6199.7460(10)	0.005	1700		5p	$(3/2)^2[5/2]^\circ_3$	5d	$(3/2)^2[5/2]_2$			R1c		
6203.715(4)	16114.915(11)	6203.712(3)	0.004	37		6p	$(5/2)^2[3/2]^\circ_2$	8d	$(5/2)^2[5/2]_3$			R1c		
6204.2569(19)	16113.507(5)	6204.2577(10)	−0.0008	11000		5p	$(3/2)^2[1/2]^\circ_1$	5d	$(3/2)^2[3/2]_2$	4.9e+07	D+	F_Re	TW	
6208.4527(19)	16102.618(5)	6208.4549(8)	−0.0022	9900		5p	$(3/2)^2[5/2]^\circ_3$	5d	$(3/2)^2[5/2]_3$	2.9e+07	D+	F_Re	TW	
6208.988(4)	16101.230(10)	6208.9891(14)	−0.001	75		4d	$(3/2)^2[7/2]_4$	sp	$(^3P)^3P^\circ\ ^3D^\circ_3$			R1c		
6214.542(4)	16086.839(11)			150		6p	$(5/2)^2[3/2]^\circ_2$	8d	$(5/2)^2[3/2]_2$			R1c		
6216.9386(8)	16080.6383(21)	6216.9385(6)	0.0001	39000		5p	$(5/2)^2[7/2]^\circ_3$	5d	$(5/2)^2[9/2]_4$	9.4e+07	C+	F_Re	TW	
6219.8492(9)	16073.1134(22)	6219.8488(7)	0.0005	24000		5p	$(3/2)^2[5/2]^\circ_2$	5d	$(3/2)^2[7/2]_3$	9.4e+07	C+	F_Re	TW	
6221.288(4)	16069.396(11)	6221.2875(7)	0.001	740		5p	$(5/2)^2[7/2]^\circ_3$	5d	$(5/2)^2[3/2]_2$			R1c		
6231.396(3)	16043.331(8)	6231.3934(6)	0.002	290		5p	$(5/2)^2[5/2]^\circ_2$	5d	$(5/2)^2[5/2]_3$			R1c		
6236.344(3)	16030.601(7)	6236.3471(10)	−0.003	1200		5p	$(3/2)^2[1/2]^\circ_1$	5d	$(3/2)^2[3/2]_1$			R1c		
6250.417(3)	15994.508(8)	6250.4216(8)	−0.005	720		5p	$(5/2)^2[5/2]^\circ_3$	5d	$(5/2)^2[5/2]_2$			R1c		
6257.838(3)	15975.542(7)	6257.8373(7)	0.000	3400		5p	$(5/2)^2[7/2]^\circ_4$	5d	$(5/2)^2[9/2]_4$	4.1e+06	D+	F_Re	TW	
6261.8477(13)	15965.311(3)	6261.8464(6)	0.0013	19000		5p	$(5/2)^2[5/2]^\circ_3$	5d	$(5/2)^2[7/2]_4$	3.7e+07	D+	F_Re	TW	
6265.650(3)	15955.623(7)	6265.6480(8)	0.002	2800		5p	$(5/2)^2[5/2]^\circ_3$	5d	$(5/2)^2[7/2]_3$			R1c		
6273.34762(8)	15936.04484(21)	6273.34763(8)	−0.00000	47000		5p	$(5/2)^2[7/2]^\circ_4$	5d	$(5/2)^2[9/2]_5$	1.06e+08	C+	F_Re	TW	
6276.660(5)	15927.636(14)	6276.6713(19)	−0.012	14000		sp	$(^3F)^3P^\circ\ ^3F^\circ_4$	6s	$(5/2)^2[5/2]_3$			R1c		
6288.695(3)	15897.154(7)	6288.6936(9)	0.001	6600		5p	$(5/2)^2[5/2]^\circ_2$	5d	$(5/2)^2[3/2]_1$	1.7e+07	D+	R1c	TW	
6296.239(4)	15878.107(11)	6296.2430(8)	−0.004	34		5p	$(5/2)^2[5/2]^\circ_2$	5d	$(5/2)^2[3/2]_2$			R1c		
6299.864(5)	15868.969(11)	6299.810(3)	0.054	34	?	sp	$(^3P)^3P^\circ\ ^1D^\circ_2$	6g	$(3/2)^2[5/2]_3$			R1c		X
6301.0135(8)	15866.0749(21)	6301.0137(7)	−0.0001	27000		5p	$(3/2)^2[5/2]^\circ_3$	5d	$(3/2)^2[7/2]_4$	1.0e+08	C+	F_Re	TW	
6305.9712(10)	15853.601(3)	6305.9718(8)	−0.0006	12000		5p	$(5/2)^2[3/2]^\circ_1$	5d	$(5/2)^2[3/2]_1$	5.7e+07	D+	F_Re	TW	
6311.3006(21)	15840.214(5)	6311.3059(5)	−0.0053	18000		5p	$(5/2)^2[5/2]^\circ_3$	5d	$(5/2)^2[5/2]_3$	3.5e+07	D+	F_Re	TW	
6312.491(3)	15837.227(7)	6312.4915(11)	−0.000	14000		5p	$(3/2)^2[3/2]^\circ_1$	5d	$(3/2)^2[5/2]_2$	5.8e+07	D+	R1c	TW	
6313.564(3)	15834.535(7)	6313.5627(7)	0.001	1000		5p	$(5/2)^2[3/2]^\circ_1$	5d	$(5/2)^2[3/2]_2$			R1c		
6317.788(3)	15823.950(7)	6317.7831(12)	0.004	1300		4p	$^3D^\circ_1$	s^2	$^3P_0$			R1c		
6318.944(5)	15821.053(13)	6318.9422(15)	0.002	33		4d	$(3/2)^2[3/2]_2$	sp	$(^3P)^3P^\circ\ ^3D^\circ_3$			R1c		
6326.465(3)	15802.245(7)	6326.4649(8)	0.000	2500		5p	$(3/2)^2[5/2]^\circ_3$	5d	$(3/2)^2[7/2]_3$	5.0e+06	D+	R1c	TW	
6357.414(3)	15725.318(7)	6357.4193(13)	−0.005	16000		5p	$(3/2)^2[1/2]^\circ_1$	5d	$(3/2)^2[1/2]_1$	5.5e+07	D+	F_Re	TW	
6363.566(5)	15710.115(11)	6363.568(4)	−0.001	32		6p	$(3/2)^2[5/2]^\circ_3$	9d	$(5/2)^2[9/2]_4$			R1c		
6373.267(3)	15686.203(8)	6373.2678(7)	−0.001	6900		5p	$(5/2)^2[5/2]^\circ_3$	5d	$(5/2)^2[9/2]_4$			R1c		
6377.248(7)	15676.410(17)	6377.2432(12)	0.005	13000		5p	$(3/2)^2[3/2]^\circ_1$	5d	$(3/2)^2[3/2]_2$			R1c		
6377.842(3)	15674.951(7)	6377.8383(7)	0.004	19000		5p	$(5/2)^2[5/2]^\circ_3$	5d	$(5/2)^2[3/2]_2$	1.9e+07	D+	F_Re	TW	
6380.758(4)	15667.788(9)	6380.7645(7)	−0.007	1300		5p	$(5/2)^2[3/2]^\circ_2$	6s	$(3/2)^2[3/2]_2$			R1c		
6385.263(3)	15656.734(7)	6385.2618(10)	0.001	1200		sp	$(^3F)^3P^\circ\ ^3D^\circ_1$	6s	$(5/2)^2[5/2]_2$			R1c		

Table A1. *Cont.*

λ_{obs} [a] (Å)	σ_{obs} [b] (cm^{-1})	λ_{Ritz} [c] (Å)	$\Delta\lambda_{obs-Ritz}$ (Å)	I_{obs} [d] (arb. u.)	Char [e]	Lower Level		Upper Level		A (s^{-1})	Acc [f]	Line Ref. [g]	TP Ref. [g]	Notes [h]
6393.074(6)	15637.604(15)	6393.0792(20)	−0.005	130	*	sp	$(^3F)^1P^\circ\ ^3G^\circ_4$	6g	$(5/2)^2[9/2]_4$			R1c		
6393.074(6)	15637.604(15)	6393.0735(21)	0.001	130	*	sp	$(^3F)^1P^\circ\ ^3G^\circ_4$	6g	$(5/2)^2[9/2]_5$			R1nc		
6393.957(3)	15635.446(8)	6393.9587(13)	−0.002	2100		sp	$(^3F)^1P^\circ\ ^3G^\circ_4$	6g	$(5/2)^2[11/2]_5$	3.4e+06	D+	R1c	TW	
6403.384(3)	15612.428(8)	6403.3840(15)	−0.000	9600		4d	$(3/2)^2[1/2]_0$	4f	$(3/2)^2[3/2]^\circ_1$	2.7e+07	D+	R1c	TW	
6411.142(4)	15593.535(10)	6411.1518(12)	−0.010	10000		5p	$(3/2)^2[3/2]^\circ_1$	5d	$(3/2)^2[3/2]_1$	5.8e+07	D+	F_Re	TW	
6414.569(3)	15585.204(7)	6414.5612(11)	0.008	12000		5p	$(3/2)^2[3/2]^\circ_2$	5d	$(3/2)^2[5/2]_2$	1.4e+07	D+	R1c	TW	
6414.625(3)	15585.069(7)	6414.6165(8)	0.008	16000		sp	$(^3F)^3P^\circ\ ^3F^\circ_3$	6s	$(5/2)^2[5/2]_2$			R1c		
6418.160(4)	15576.483(10)	6418.1566(24)	0.004	2100		4f	$(5/2)^2[1/2]^\circ_1$	6g	$(5/2)^2[3/2]_2$	1.5e+07	D+	R1c	TW	
		6418.192(3)			m	4f	$(5/2)^2[1/2]^\circ_1$	6g	$(5/2)^2[3/2]_1$			R1nc		
6419.941(9)	15572.162(21)	6419.9515(7)	−0.010	62		5p	$(5/2)^2[3/2]^\circ_2$	6s	$(3/2)^2[3/2]_1$			R1c		
6423.8842(11)	15562.604(3)	6423.8846(8)	−0.0003	19000		5p	$(3/2)^2[3/2]^\circ_2$	5d	$(3/2)^2[5/2]_3$	5.6e+07	C+	F_Re	TW	
6427.416(9)	15554.052(21)	6427.411(3)	0.005	31		4f	$(5/2)^2[3/2]^\circ_2$	6g	$(5/2)^2[7/2]_3$			R1c		
6432.416(3)	15541.962(7)	6432.415(3)	0.001	5600		4f	$(5/2)^2[3/2]^\circ_2$	6g	$(5/2)^2[5/2]_3$	1.5e+07	D+	R1c	TW	
6433.597(4)	15539.109(10)	6433.5944(22)	0.003	920		4f	$(5/2)^2[11/2]^\circ_6$	6g	$(5/2)^2[11/2]_6$	2.7e+06	D+	R1c	TW	
6434.081(5)	15537.941(12)	6434.075(3)	0.006	610		4f	$(5/2)^2[1/2]^\circ_0$	6g	$(5/2)^2[3/2]_1$			R1c		
6434.799(3)	15536.208(8)	6434.7944(13)	0.004	610		4f	$(5/2)^2[11/2]^\circ_5$	6g	$(5/2)^2[11/2]_5$			R1c		
6437.596(3)	15529.458(8)	6437.5974(24)	−0.002	460		4f	$(5/2)^2[3/2]^\circ_2$	6g	$(5/2)^2[3/2]_2$	6.7e+06	D+	R1c	TW	
6441.202(5)	15520.764(12)	6441.1920(9)	0.010	1200		4p	$^3F^\circ_2$	s^2	$^1D_2$			R1c		
6441.677(3)	15519.618(8)	6441.6790(8)	−0.002	13000		sp	$(^3F)^3P^\circ\ ^1F^\circ_3$	5d	$(5/2)^2[7/2]_4$			R1c		
6441.7371(11)	15519.474(3)			10000		4f	$(5/2)^2[11/2]^\circ_6$	6g	$(5/2)^2[13/2]_7$	2.4e+07	D+	F_Re	TW	
6442.964(3)	15516.519(7)	6442.964(3)	−0.000	7600		4f	$(5/2)^2[11/2]^\circ_5$	6g	$(5/2)^2[13/2]_6$	2.3e+07	D+	R1c	TW	
6443.587(3)	15515.019(8)	6443.587(3)	0.000	4900		4f	$(3/2)^2[3/2]^\circ_2$	6g	$(3/2)^2[5/2]_3$	2.0e+07	D+	R1c	TW	
6445.692(20)	15509.95(5)	6445.673(3)	0.020	1400	*	6p	$(5/2)^2[3/2]^\circ_2$	9s	$(5/2)^2[5/2]_3$			R1nc		
6445.692(20)	15509.95(5)	6445.7022(9)	−0.010	1400	*	sp	$(^3F)^3P^\circ\ ^1F^\circ_3$	5d	$(5/2)^2[7/2]_3$			R1c		
6447.627(5)	15505.296(13)	6447.632(3)	−0.005	61		sp	$(^1G)^3P^\circ\ ^3F^\circ_3$	8d	$(5/2)^2[9/2]_4$			R1c		
6448.558(3)	15503.059(8)	6448.5593(8)	−0.002	20000		4p	$^3D^\circ_1$	s^2	$^3P_1$			R1c		
6449.63(3)	15500.49(7)	6449.6175(15)	0.01	30	*	4d	$(3/2)^2[5/2]_3$	sp	$(^3P)^3P^\circ\ ^3D^\circ_3$			R1c		
6449.63(3)	15500.49(7)	6449.655(3)	−0.03	30	*	sp	$(^3P)^3P^\circ\ ^5S^\circ_2$	6g	$(5/2)^2[5/2]_2$			R1nc		
6449.63(3)	15500.49(7)	6449.659(4)	−0.03	30	*	sp	$(^3P)^3P^\circ\ ^5S^\circ_2$	6g	$(5/2)^2[5/2]_3$			R1nc		
6450.874(3)	15497.493(8)	6450.8738(17)	−0.000	610		4d	$(3/2)^2[1/2]_0$	sp	$(^3F)^1P^\circ\ ^3D^\circ_1$			R1c		
6452.116(3)	15494.508(7)	6452.1161(24)	0.000	1300		4f	$(5/2)^2[3/2]^\circ_1$	6g	$(5/2)^2[5/2]_2$	1.4e+07	D+	R1c	TW	
6454.161(3)	15489.601(8)			910		4f	$(3/2)^2[3/2]^\circ_1$	6g	$(3/2)^2[5/2]_2$			R1c		
6454.894(9)	15487.842(21)	6454.895(3)	−0.001	30		6p	$(5/2)^2[7/2]^\circ_4$	8d	$(5/2)^2[7/2]_4$			R1rc		
6456.119(13)	15484.90(3)	6456.140(9)	−0.021	30		5s	$(3/2)^2[3/2]_1$	sp	$(^3P)^3P^\circ\ ^5P^\circ_1$			R1c		
6457.187(3)	15482.342(7)	6457.1848(19)	0.002	2600		4f	$(5/2)^2[7/2]^\circ_3$	6g	$(5/2)^2[9/2]_4$			R1c		
6457.368(4)	15481.908(10)	6457.371(3)	−0.004	1600		4f	$(5/2)^2[3/2]^\circ_1$	6g	$(5/2)^2[3/2]_1$			R1c		
6458.263(9)	15479.762(21)	6458.259(3)	0.004	61		4f	$(3/2)^2[9/2]^\circ_5$	6g	$(3/2)^2[9/2]_5$			R1c		
6460.304(3)	15474.870(8)	6460.304(3)	0.001	1000		4f	$(5/2)^2[7/2]^\circ_3$	6g	$(5/2)^2[7/2]_3$			R1c		
6462.140(3)	15470.475(8)	6462.1413(20)	−0.002	60		4f	$(3/2)^2[9/2]^\circ_4$	6g	$(3/2)^2[9/2]_4$			R1c		
6462.70(10)	15469.13(24)	6462.6086(21)	0.09	1000		4f	$(3/2)^2[7/2]^\circ_3$	7d	$(3/2)^2[7/2]_3$			S36c		
6465.342(15)	15462.81(4)	6465.3543(24)	−0.012	30	*	4f	$(5/2)^2[7/2]^\circ_3$	6g	$(5/2)^2[5/2]_2$			R1c		
6465.342(15)	15462.81(4)	6465.359(3)	−0.017	30	*	4f	$(5/2)^2[7/2]^\circ_3$	6g	$(5/2)^2[5/2]_3$			R1c		
6466.245(3)	15460.653(7)			5600		4f	$(3/2)^2[9/2]^\circ_5$	6g	$(3/2)^2[11/2]_6$	2.3e+07	D+	R1c	TW	

Table A1. *Cont.*

λ_{obs} [a] (Å)	σ_{obs} [b] (cm^{-1})	λ_{Ritz} [c] (Å)	$\Delta\lambda_{obs\text{-}Ritz}$ (Å)	I_{obs} [d] (arb. u.)	Char [e]	Lower Level		Upper Level		A (s^{-1})	Acc [f]	Line Ref. [g]	TP Ref. [g]	Notes [h]
6470.1386(19)	15451.350(5)	6470.1383(19)	0.0003	8600		4f	$(3/2)^2[9/2]^\circ_4$	6g	$(3/2)^2[11/2]_5$	2.3e+07	D+	F_Re	TW	
6470.160(3)	15451.298(7)	6470.1668(8)	−0.006	15000		sp	$(^3F)^3P^\circ\ ^3F^\circ_3$	6s	$(5/2)^2[5/2]_3$			F_Re		
6475.426(4)	15438.734(9)	6475.4257(24)	−0.000	1000		4f	$(5/2)^2[5/2]^\circ_2$	6g	$(5/2)^2[5/2]_2$	7.5e+06	D+	R1c	TW	
6476.184(4)	15436.925(9)	6476.1819(19)	0.003	1000		4f	$(5/2)^2[5/2]^\circ_3$	6g	$(5/2)^2[9/2]_4$			R1c		
6477.868(9)	15432.912(21)	6477.860(3)	0.008	120		6p	$(5/2)^2[7/2]^\circ_4$	8d	$(5/2)^2[9/2]_5$			R1c		
6479.316(3)	15429.463(7)	6479.3164(21)	−0.000	4100		4f	$(5/2)^2[5/2]^\circ_3$	6g	$(5/2)^2[7/2]_4$	1.5e+07	D+	R1c	TW	
6481.436(3)	15424.418(7)	6481.4350(12)	0.001	15000		5p	$(3/2)^2[3/2]^\circ_2$	5d	$(3/2)^2[3/2]_2$	3.4e+07	D+	R1c	TW	
		6484.404(3)			m	4f	$(5/2)^2[5/2]^\circ_3$	6g	$(5/2)^2[5/2]_3$	6.5e+06	D+	R1nc	TW	
6484.417(3)	15417.326(8)	6484.4174(9)	−0.000	7800		5p	$(5/2)^2[3/2]^\circ_1$	5d	$(5/2)^2[1/2]_1$	1.4e+07	D+	F_Re	TW	
6488.816(3)	15406.875(7)	6488.814(3)	0.002	2700		4f	$(3/2)^2[5/2]^\circ_3$	6g	$(3/2)^2[7/2]_4$	1.9e+07	D+	R1c	TW	
6494.029(3)	15394.507(6)	6494.0320(6)	−0.003	18000		sp	$(^3F)^3P^\circ\ ^1F^\circ_3$	5d	$(5/2)^2[5/2]_3$			F_Re		
6497.040(9)	15387.372(21)	6497.041(3)	−0.001	89		4f	$(3/2)^2[5/2]^\circ_3$	6g	$(3/2)^2[5/2]_3$			R1c		
6508.401(3)	15360.514(7)	6508.400(3)	0.001	2700		4f	$(3/2)^2[5/2]^\circ_2$	6g	$(3/2)^2[7/2]_3$	1.7e+07	D+	R1c	TW	
6516.458(3)	15341.522(8)	6516.4637(12)	−0.006	6600		5p	$(3/2)^2[3/2]^\circ_2$	5d	$(3/2)^2[3/2]_1$			R1c		
6517.316(3)	15339.501(7)	6517.3166(21)	−0.000	6600		4f	$(5/2)^2[7/2]^\circ_4$	6g	$(5/2)^2[9/2]_5$	1.7e+07	D+	R1c	TW	
6518.236(4)	15337.337(9)	6518.2366(12)	−0.001	150		4f	$(5/2)^2[7/2]^\circ_4$	6g	$(5/2)^2[11/2]_5$			R1c		
6520.497(3)	15332.019(8)	6520.4970(21)	−0.000	1000		4f	$(5/2)^2[7/2]^\circ_4$	6g	$(5/2)^2[7/2]_4$	5.7e+06	D+	R1c	TW	
6521.270(6)	15330.200(13)	6521.2756(14)	−0.005	480		4p	$^1P^\circ_1$	s^2	$^3P_0$			R1c		
6523.820(3)	15324.209(7)	6523.820(3)	−0.000	4500		4f	$(3/2)^2[7/2]^\circ_4$	6g	$(3/2)^2[9/2]_5$	2.1e+07	D+	R1c	TW	
6525.641(9)	15319.932(21)	6525.650(3)	−0.008	29		4f	$(5/2)^2[7/2]^\circ_4$	6g	$(5/2)^2[5/2]_3$			R1c		
6526.646(6)	15317.574(13)	6526.655(3)	−0.009	88		4f	$(3/2)^2[7/2]^\circ_4$	6g	$(3/2)^2[7/2]_4$			R1c		
6529.167(3)	15311.658(8)	6529.1668(21)	0.001	1000		4f	$(5/2)^2[9/2]^\circ_5$	6g	$(5/2)^2[9/2]_5$	4.6e+06	D+	R1c	TW	
6530.082(3)	15309.514(7)	6530.0829(22)	−0.001	13000		4f	$(5/2)^2[9/2]^\circ_5$	6g	$(5/2)^2[11/2]_6$	2.0e+07	D+	R1c	TW	
6536.679(9)	15294.062(21)	6536.6969(17)	−0.017	29		4d	$(3/2)^2[5/2]_2$	sp	$(^3P)^3P^\circ\ ^3P^\circ_2$			R1c		
6541.637(3)	15282.471(7)	6541.6369(20)	0.000	3400		4f	$(3/2)^2[7/2]^\circ_3$	6g	$(3/2)^2[9/2]_4$	1.9e+07	D+	R1c	TW	
6544.485(6)	15275.822(14)	6544.480(3)	0.004	58	*	4f	$(3/2)^2[7/2]^\circ_3$	6g	$(3/2)^2[7/2]_4$			R1nc		
6544.485(6)	15275.822(14)	6544.489(3)	−0.004	58	*	4f	$(3/2)^2[7/2]^\circ_3$	6g	$(3/2)^2[7/2]_3$			R1c		
6550.304(5)	15262.251(12)	6550.3097(9)	−0.006	150		5p	$(3/2)^2[3/2]^\circ_2$	5d	$(3/2)^2[7/2]_3$			R1c		
6551.286(3)	15259.964(8)	6551.2873(8)	−0.002	1900		4p	$^3D^\circ_1$	s^2	$^3P_2$			R1c		
6554.791(6)	15251.804(13)	6554.7923(9)	−0.002	5100		4p	$^3D^\circ_2$	s^2	$^3P_1$			R1c		
6556.193(6)	15248.542(14)	6556.206(3)	−0.013	430		4d	$(5/2)^2[1/2]_1$	sp	$(^1D)^3P^\circ\ ^3P^\circ_1$			R1c		
6559.657(4)	15240.489(9)	6559.6523(8)	0.005	13000		sp	$(^3F)^3P^\circ\ ^1F^\circ_3$	5d	$(5/2)^2[9/2]_4$	4.9e+06	D+	R1c	TW	
6564.492(3)	15229.264(8)	6564.4941(8)	−0.002	10000		sp	$(^3F)^3P^\circ\ ^1F^\circ_3$	5d	$(5/2)^2[3/2]_2$			R1c		
6567.949(6)	15221.249(13)	6567.9621(7)	−0.013	290		5s	$(5/2)^2[5/2]_2$	5p	$(3/2)^2[3/2]^\circ_2$	5.1e+05	C	R1c	B00	
6576.147(6)	15202.273(13)	6576.1510(20)	−0.004	1000		4f	$(5/2)^2[9/2]^\circ_4$	6g	$(5/2)^2[9/2]_4$	4.3e+06	D+	R1c	TW	
6577.0812(15)	15200.114(3)	6577.0816(12)	−0.0003	5700		4f	$(5/2)^2[9/2]^\circ_4$	6g	$(5/2)^2[11/2]_5$	1.5e+07	D+	F_Re	TW	
6579.376(9)	15194.812(21)	6579.386(3)	−0.010	57		4f	$(5/2)^2[9/2]^\circ_4$	6g	$(5/2)^2[7/2]_3$			R1c		
6592.896(9)	15163.652(21)	6592.900(3)	−0.004	120		5s	$(3/2)^2[3/2]_1$	sp	$(^3P)^3P^\circ\ ^5P^\circ_2$			R1c		
6602.096(9)	15142.522(21)	6602.094(3)	0.003	57		6p	$(5/2)^2[5/2]^\circ_3$	8d	$(5/2)^2[7/2]_4$			R1c		
6603.527(9)	15139.242(21)	6603.523(4)	0.003	29		6p	$(3/2)^2[5/2]^\circ_3$	9s	$(3/2)^2[3/2]_2$			R1c		
6612.087(9)	15119.642(21)	6612.086(3)	0.001	57		6p	$(5/2)^2[5/2]^\circ_3$	8d	$(5/2)^2[5/2]_3$			R1c		
6624.291(3)	15091.787(7)	6624.2890(9)	0.002	8200		sp	$(^3F)^3P^\circ\ ^1D^\circ_2$	5d	$(5/2)^2[5/2]_2$	2.2e+07	D+	R1c	TW	

Table A1. *Cont.*

λ_{obs} [a] (Å)	σ_{obs} [b] (cm^{-1})	λ_{Ritz} [c] (Å)	$\Delta\lambda_{obs\text{-}Ritz}$ (Å)	I_{obs} [d] (arb. u.)	Char [e]	Lower Level		Upper Level		A (s^{-1})	Acc [f]	Line Ref. [g]	TP Ref. [g]	Notes [h]
6626.243(3)	15087.341(8)	6626.2420(20)	0.001	710		5d	$(5/2)^2[1/2]_1$	8p	$(5/2)^2[3/2]^{\circ}_1$			R1c		
6631.476(3)	15075.436(7)	6631.4733(7)	0.003	3300		5p	$(5/2)^2[7/2]^{\circ}_3$	6s	$(3/2)^2[3/2]_2$			R1c		
6635.173(7)	15067.037(17)	6635.1692(18)	0.003	85		5d	$(5/2)^2[1/2]_1$	6f	$(5/2)^2[5/2]^{\circ}_2$			R1c		
6641.395(3)	15052.920(7)	6641.3938(9)	0.002	8000		sp	$(^3F)^3P^{\circ}\ ^1D^{\circ}_2$	5d	$(5/2)^2[7/2]_3$	2.6e+07	D+	R1c	TW	
6647.796(4)	15038.427(10)	6647.7986(18)	−0.003	850		5d	$(5/2)^2[1/2]_1$	6f	$(5/2)^2[3/2]^{\circ}_2$			R1c		
6648.764(5)	15036.237(11)	6648.7717(14)	−0.008	7300		5p	$(3/2)^2[3/2]^{\circ}_2$	5d	$(3/2)^2[1/2]_1$	1.1e+07	D+	R1c	TW	
6651.323(4)	15030.453(10)	6651.3254(21)	−0.003	700		sp	$(^3F)^1P^{\circ}\ ^3F^{\circ}_3$	6g	$(5/2)^2[9/2]_4$	3.8e+06	D+	R1c	TW	
6654.635(9)	15022.972(21)	6654.635(3)	−0.000	56		sp	$(^3F)^1P^{\circ}\ ^3F^{\circ}_3$	6g	$(5/2)^2[7/2]_3$			R1c		
6660.961(3)	15008.705(7)	6660.9606(8)	0.000	9200		4p	$^3D^{\circ}_2$	s^2	$^3P_2$			R1c		
6663.950(9)	15001.972(21)	6663.9466(6)	0.004	560		5s	$(5/2)^2[5/2]_3$	5p	$(3/2)^2[5/2]^{\circ}_3$			R1c		
6692.710(6)	14937.507(14)	6692.7144(7)	−0.005	700		sp	$(^3F)^3P^{\circ}\ ^1D^{\circ}_2$	5d	$(5/2)^2[5/2]_3$			R1c		
6711.983(6)	14894.615(12)	6711.980(3)	0.003	56		5d	$(3/2)^2[1/2]_1$	6f	$(3/2)^2[5/2]^{\circ}_2$			R1c		
6716.700(3)	14884.155(7)	6716.7062(8)	−0.007	1300		5p	$(5/2)^2[5/2]^{\circ}_2$	6s	$(3/2)^2[3/2]_2$			R1c		
6717.687(6)	14881.968(12)	6717.687(4)	−0.001	280		5s	$(3/2)^2[3/2]_2$	sp	$(^3P)^3P^{\circ}\ ^5P^{\circ}_2$			R1c		
6725.481(10)	14864.722(21)	6725.490(4)	−0.010	28		5d	$(3/2)^2[1/2]_1$	6f	$(3/2)^2[3/2]^{\circ}_1$			R2c		
6725.834(10)	14863.942(21)	6725.850(8)	−0.016	28		5s	$(3/2)^2[3/2]_2$	sp	$(^3P)^3P^{\circ}\ ^5P^{\circ}_3$			R1nc		
6736.397(5)	14840.633(10)	6736.4201(7)	−0.023	6300		5p	$(5/2)^2[3/2]^{\circ}_1$	6s	$(3/2)^2[3/2]_2$			R1c		X
6739.599(3)	14833.582(7)	6739.598(3)	0.002	140		5d	$(3/2)^2[1/2]_1$	6f	$(3/2)^2[3/2]^{\circ}_2$	7.4e+06	D+	R1c	TW	
6743.740(7)	14824.474(16)	6743.737(3)	0.003	56		5d	$(5/2)^2[1/2]_1$	6f	$(5/2)^2[1/2]^{\circ}_1$			R2nc		
6758.855(4)	14791.323(8)	6758.8577(10)	−0.003	6300		sp	$(^3F)^3P^{\circ}\ ^1D^{\circ}_2$	5d	$(5/2)^2[3/2]_1$	1.0e+07	D+	R1c	TW	
6760.142(5)	14788.506(11)	6760.1422(8)	0.000	56		5p	$(5/2)^2[5/2]^{\circ}_2$	6s	$(3/2)^2[3/2]_1$			R1c		
6765.794(10)	14776.152(21)	6765.7884(19)	0.006	28		sp	$(^1G)^3P^{\circ}\ ^3H^{\circ}_4$	6g	$(5/2)^2[11/2]_5$			R1c		
6767.561(7)	14772.294(16)	6767.5789(9)	−0.018	28		sp	$(^3F)^3P^{\circ}\ ^1D^{\circ}_2$	5d	$(5/2)^2[3/2]_2$			R1c		
6770.361(3)	14766.186(7)	6770.3604(10)	0.000	3100		4p	$^1P^{\circ}_1$	s^2	$^3P_2$			R1c		
6779.440(7)	14746.410(15)	6779.4453(18)	−0.005	28		5p	$(3/2)^2[1/2]^{\circ}_1$	5d	$(5/2)^2[1/2]_0$			R1c		
6780.114(4)	14744.945(9)	6780.1122(7)	0.002	4900		5p	$(5/2)^2[3/2]^{\circ}_1$	6s	$(3/2)^2[3/2]_1$			R1c		
6786.482(5)	14731.110(11)	6786.4808(7)	0.001	84		5s	$(5/2)^2[5/2]_3$	5p	$(3/2)^2[5/2]^{\circ}_2$			R1c		
6802.138(3)	14697.204(7)	6802.1387(19)	−0.001	220		5d	$(5/2)^2[9/2]_5$	6f	$(5/2)^2[9/2]^{\circ}_5$	3.5e+06	D+	R1c	TW	
6806.215(3)	14688.401(7)	6806.2136(9)	0.001	4100		sp	$(^3F)^3P^{\circ}\ ^3F^{\circ}_2$	6s	$(5/2)^2[5/2]_2$			R1c		
6809.6453(23)	14681.001(5)	6809.6436(8)	0.0016	5200		5p	$(5/2)^2[5/2]^{\circ}_3$	6s	$(3/2)^2[3/2]_2$			F_Re		
6814.690(14)	14670.13(3)	6814.6913(22)	−0.001	28		5d	$(5/2)^2[3/2]_2$	8p	$(5/2)^2[3/2]^{\circ}_1$			R1c		
6821.060(5)	14656.434(11)	6821.0571(19)	0.002	140		5d	$(5/2)^2[9/2]_4$	6f	$(5/2)^2[9/2]^{\circ}_4$	3.2e+06	D+	R1c	TW	
6822.342(8)	14653.679(17)	6822.3460(19)	−0.004	28		5d	$(5/2)^2[9/2]_4$	6f	$(5/2)^2[7/2]^{\circ}_4$			R1c		
6823.201(3)	14651.833(7)	6823.2010(23)	0.000	6400		5d	$(5/2)^2[9/2]_5$	6f	$(5/2)^2[11/2]^{\circ}_6$	1.8e+07	D+	R1c	TW	
6824.750(4)	14648.508(8)	6824.7493(19)	0.001	700		5d	$(5/2)^2[3/2]_2$	6f	$(5/2)^2[5/2]^{\circ}_3$	7.2e+06	D+	R1c	TW	
6830.531(3)	14636.112(7)	6830.5296(22)	0.001	1300		5d	$(3/2)^2[7/2]_3$	6f	$(3/2)^2[9/2]^{\circ}_4$	1.8e+07	D+	R1c	TW	
6833.026(4)	14630.766(9)	6833.0244(19)	0.002	560		5d	$(5/2)^2[3/2]_1$	6f	$(5/2)^2[5/2]^{\circ}_2$	8.1e+06	D+	R1c	TW	
6837.491(4)	14621.212(8)	6837.4936(18)	−0.002	720		5d	$(5/2)^2[3/2]_2$	6f	$(5/2)^2[3/2]^{\circ}_2$	9.e+06	D+	R1c	TW	
6844.156(3)	14606.974(7)	6844.1542(20)	0.002	3200		5d	$(5/2)^2[9/2]_4$	6f	$(5/2)^2[11/2]^{\circ}_5$	1.8e+07	D+	R1c	TW	
6846.104(20)	14602.82(4)	6846.065(3)	0.039	880		sp	$(^3F)^3P^{\circ}\ ^1G^{\circ}_4$	6s	$(5/2)^2[5/2]_3$			R1c		
6847.112(7)	14600.669(14)	6847.112(3)	−0.000	170		5d	$(5/2)^2[3/2]_1$	6f	$(5/2)^2[3/2]^{\circ}_1$			R1c		
6849.991(6)	14594.532(13)	6849.988(4)	0.003	56		4d	$(5/2)^2[1/2]_1$	sp	$(^1D)^3P^{\circ}\ ^3P^{\circ}_0$			R1c		
6852.428(7)	14589.342(14)	6852.432(3)	−0.004	84		5d	$(3/2)^2[3/2]_1$	6f	$(3/2)^2[5/2]^{\circ}_2$			R1c		

Table A1. *Cont.*

λ_{obs} [a] (Å)	σ_{obs} [b] (cm^{-1})	λ_{Ritz} [c] (Å)	$\Delta\lambda_{obs-Ritz}$ (Å)	I_{obs} [d] (arb. u.)	Char [e]	Lower Level	Upper Level	A (s^{-1})	Acc [f]	Line Ref. [g]	TP Ref. [g]	Notes [h]
6863.061(3)	14566.739(7)	6863.0599(23)	0.001	2100		5d $(3/2)^2[7/2]_4$	6f $(3/2)^2[9/2]°_5$	1.8e+07	D+	R1c	TW	
6867.88(20)	14556.5(4)	6868.184(4)	−0.30	470		4d $(5/2)^2[5/2]_2$	sp $(^3P)^3P°\ ^5D°_2$			S36c		
6868.790(3)	14554.589(7)	6868.7864(9)	0.004	2900		sp $(^3F)^3P°\ ^3F°_2$	6s $(5/2)^2[5/2]_3$			R1c		
6872.230(3)	14547.303(7)	6872.2293(10)	0.001	3200		4p $^1F°_3$	s^2 $^1D_2$			R1c		
6879.403(3)	14532.136(7)	6879.4043(7)	−0.001	2500		5s $(5/2)^2[5/2]_2$	5p $(3/2)^2[1/2]°_1$	5.5e+06	C	R1c	B00	
6893.496(6)	14502.427(12)	6893.4896(18)	0.006	220		5d $(5/2)^2[5/2]_3$	6f $(5/2)^2[9/2]°_4$			R1c		
6894.810(4)	14499.663(9)	6894.8061(18)	0.004	700		5d $(5/2)^2[5/2]_3$	6f $(5/2)^2[7/2]°_4$	1.0e+07	D+	R1c	TW	
6902.617(4)	14483.263(9)	6902.6139(19)	0.003	560		5d $(5/2)^2[5/2]_3$	6f $(5/2)^2[5/2]°_3$	8.0e+06	D+	R1c	TW	
6915.653(10)	14455.962(21)	6915.6509(18)	0.002	56		5d $(5/2)^2[5/2]_3$	6f $(5/2)^2[3/2]°_2$			R1c		
6928.602(6)	14428.946(12)	6928.592(3)	0.009	340		5d $(3/2)^2[3/2]_2$	6f $(3/2)^2[5/2]°_3$	1.1e+07	D+	R1c	TW	
6937.552(4)	14410.331(8)	6937.5472(7)	0.005	2100		5s $(5/2)^2[5/2]_2$	5p $(3/2)^2[5/2]°_2$	3.9e+06	C	R1c	B00	
6939.099(7)	14407.119(14)	6939.098(3)	0.001	280		5d $(5/2)^2[7/2]_3$	6f $(5/2)^2[7/2]°_3$	4.8e+06	D+	R1c	TW	
6948.795(7)	14387.015(15)	6948.7964(20)	−0.001	2000		5d $(5/2)^2[7/2]_3$	6f $(5/2)^2[9/2]°_4$	1.1e+07	D+	R1c	TW	
6950.139(6)	14384.234(13)	6950.1341(20)	0.005	1700		5d $(5/2)^2[7/2]_3$	6f $(5/2)^2[7/2]°_4$	3.1e+06	D+	R1c	TW	
6952.870(4)	14378.583(7)	6952.8664(20)	0.004	4200		5d $(5/2)^2[7/2]_4$	6f $(5/2)^2[9/2]°_5$	1.4e+07	D+	R1c	TW	
6953.480(10)	14377.322(21)	6953.4782(20)	0.002	110		5d $(5/2)^2[7/2]_4$	6f $(5/2)^2[9/2]°_4$			R1c		
6953.930(6)	14376.391(13)	6953.914(3)	0.016	2100		5d $(3/2)^2[5/2]_3$	6f $(3/2)^2[7/2]°_4$	1.5e+07	D+	R1c	TW	
6954.820(10)	14374.552(21)	6954.8176(19)	0.002	230		5d $(5/2)^2[7/2]_4$	6f $(5/2)^2[7/2]°_4$	4.4e+06	D+	R1c	TW	
6957.884(10)	14368.222(20)	6957.869(3)	0.015	1800	*	5d $(5/2)^2[5/2]_2$	6f $(5/2)^2[7/2]°_3$	1.1e+07	D+	R1c	TW	
6957.884(10)	14368.222(20)	6957.876(4)	0.008	1800	*	sp $(^1G)^3P°\ ^3H°_5$	6g $(5/2)^2[13/2]_6$			R1c		
6957.884(10)	14368.222(20)	6957.876(3)	0.008	1800	*	5d $(3/2)^2[5/2]_3$	6f $(3/2)^2[5/2]°_2$			R1c		
6963.433(10)	14356.772(21)	6963.427(3)	0.006	230		5d $(3/2)^2[5/2]_2$	6f $(3/2)^2[7/2]°_3$	1.3e+07	D+	R1c	TW	
6968.846(10)	14345.622(21)	6968.847(3)	−0.002	57		5d $(3/2)^2[5/2]_2$	6f $(3/2)^2[5/2]°_2$			R1c		
6976.305(10)	14330.282(21)	6976.2990(20)	0.006	230		5d $(5/2)^2[5/2]_2$	6f $(5/2)^2[5/2]°_2$	5.6e+06	D+	R1c	TW	
6977.571(4)	14327.683(8)	6977.5641(10)	0.007	4300		5p $(3/2)^2[5/2]°_2$	5d $(5/2)^2[5/2]_2$			R1c		
6986.548(8)	14309.273(17)	6986.5269(20)	0.021	2100		5p $(3/2)^2[3/2]°_1$	5d $(5/2)^2[1/2]_0$			R1c		
6996.556(5)	14288.805(9)	6996.5446(10)	0.012	1400		5p $(3/2)^2[5/2]°_2$	5d $(5/2)^2[7/2]_3$			R1c		
7022.859(4)	14235.289(7)	7022.8519(9)	0.007	2800		sp $(^3F)^3P°\ ^1F°_3$	6s $(3/2)^2[3/2]_2$			R1c		
7037.388(4)	14205.899(8)	7037.3853(10)	0.003	2900		5p $(3/2)^2[1/2]°_1$	5d $(5/2)^2[5/2]_2$			R1c		
7083.772(6)	14112.882(12)	7083.7743(13)	−0.003	730		5p $(3/2)^2[1/2]°_0$	5d $(5/2)^2[1/2]_1$			R1c		
7123.112(11)	14034.939(21)	7123.106(3)	0.006	59	*	6s $(5/2)^2[5/2]_2$	7p $(5/2)^2[5/2]°_3$			R1c		
7123.112(11)	14034.939(21)	7123.1061(21)	0.006	59	*	sp $(^3F)^1P°\ ^3G°_5$	5g $(3/2)^2[11/2]_6$			R1c		
7123.112(11)	14034.939(21)	7123.1212(21)	−0.010	59	*	sp $(^3F)^1P°\ ^3G°_5$	5g $(3/2)^2[11/2]_5$			R1c		
7127.032(8)	14027.218(15)	7127.0305(11)	0.002	150		5p $(3/2)^2[5/2]°_2$	5d $(5/2)^2[3/2]_1$			R1c		
7139.649(5)	14002.431(10)	7139.653(3)	−0.004	1500		6s $(5/2)^2[5/2]_3$	7p $(5/2)^2[7/2]°_4$			R1c		
7157.757(5)	13967.006(10)	7157.7517(8)	0.006	1500		5s $(5/2)^2[5/2]_3$	sp $(^3F)^3P°\ ^1D°_2$	5.2e+05	C	R1c	B00	
7182.106(7)	13919.655(14)	7182.104(5)	0.002	60		6s $(5/2)^2[5/2]_2$	7p $(5/2)^2[5/2]°_2$			R1c		
7189.458(4)	13905.422(7)	7189.4534(12)	0.004	1300		5p $(3/2)^2[1/2]°_1$	5d $(5/2)^2[3/2]_1$			R1c		
7194.900(3)	13894.903(6)	7194.8933(10)	0.007	6800		4p $^1D°_2$	s^2 $^1D_2$			F_Re		
7255.7937(20)	13778.293(4)	7255.7899(9)	0.0038	6400		sp $(^3F)^3P°\ ^1D°_2$	6s $(3/2)^2[3/2]_2$	6.5e+06	D+	F_Re	TW	
7271.507(6)	13748.519(11)	7271.4991(10)	0.008	460		5p $(3/2)^2[5/2]°_3$	5d $(5/2)^2[9/2]_4$			R1c		
7306.511(4)	13682.653(7)	7306.5042(9)	0.007	3700		sp $(^3F)^3P°\ ^1D°_2$	6s $(3/2)^2[3/2]_1$			R1c		
7326.007(4)	13646.242(7)	7326.0039(8)	0.003	6900		5s $(5/2)^2[5/2]_2$	sp $(^3F)^3P°\ ^1D°_2$	1.80e+07	B	R1c	B00	

Table A1. *Cont.*

λ_{obs} [a] (Å)	σ_{obs} [b] (cm^{-1})	λ_{Ritz} [c] (Å)	$\Delta\lambda_{obs-Ritz}$ (Å)	I_{obs} [d] (arb. u.)	Char [e]	Lower Level		Upper Level		A (s^{-1})	Acc [f]	Line Ref. [g]	TP Ref. [g]	Notes [h]
7331.6940(19)	13635.656(4)	7331.6941(7)	−0.0001	5100		5p	$(5/2)^2[3/2]^\circ_2$	6s	$(5/2)^2[5/2]_2$			F_Re		
7382.275(3)	13542.229(6)	7382.2720(14)	0.003	3500		4d	$(3/2)^2[1/2]_0$	4f	$(5/2)^2[3/2]^\circ_1$	1.6e+07	D+	F_Re	TW	
7396.156(6)	13516.815(11)	7396.1523(12)	0.003	1900		5p	$(3/2)^2[3/2]^\circ_2$	5d	$(5/2)^2[5/2]_2$			R1c		
7399.8836(16)	13510.005(3)	7399.8787(7)	0.0049	14000		5s	$(5/2)^2[5/2]_3$	sp	$(^3F)^3P^\circ\ ^1F^\circ_3$	2.57e+07	B	F_Re	B00	
7404.3561(11)	13501.8442(20)	7404.3532(6)	0.0029	55000		5p	$(5/2)^2[3/2]^\circ_2$	6s	$(5/2)^2[5/2]_3$	2.0e+07	D+	F_Re	TW	L
7417.483(5)	13477.950(9)	7417.4819(12)	0.001	2400		5p	$(3/2)^2[3/2]^\circ_2$	5d	$(5/2)^2[7/2]_3$			R1c		
7420.556(4)	13472.368(7)	7420.5668(9)	−0.010	4100		5s	$(3/2)^2[3/2]_1$	5p	$(3/2)^2[3/2]^\circ_2$	3.1e+06	C	R1c	B00	
7422.602(5)	13468.656(9)	7422.5904(11)	0.011	790		4p	$^3D^\circ_3$	s^2	$^1D_2$			R1c		
7434.155(4)	13447.725(7)	7434.1525(15)	0.002	3700		4d	$(3/2)^2[1/2]_0$	4f	$(5/2)^2[1/2]^\circ_1$	1.3e+07	D+	R1c	TW	
7438.1504(13)	13440.5007(24)	7438.1505(10)	−0.0001	11000		5p	$(3/2)^2[1/2]^\circ_0$	6s	$(3/2)^2[3/2]_1$	1.0e+07	D+	F_Re	TW	
7481.554(8)	13362.528(15)	7481.5552(10)	−0.001	190		5p	$(3/2)^2[3/2]^\circ_2$	5d	$(5/2)^2[5/2]_3$			R1c		
7493.289(8)	13341.602(14)	7493.2883(20)	0.000	130		6p	$(5/2)^2[7/2]^\circ_3$	7d	$(5/2)^2[7/2]_3$	3.0e+06	D+	R1c	TW	
7562.0141(17)	13220.350(3)	7562.0167(9)	−0.0026	16000		5s	$(3/2)^2[3/2]_1$	5p	$(3/2)^2[3/2]^\circ_1$	3.8e+07	B	F_Re	B00	
7564.324(8)	13216.314(14)	7564.3057(14)	0.018	64		5p	$(3/2)^2[3/2]^\circ_2$	5d	$(5/2)^2[3/2]_1$			R1c		
7575.242(7)	13197.266(12)	7575.2309(12)	0.011	220		5p	$(3/2)^2[3/2]^\circ_2$	5d	$(5/2)^2[3/2]_2$			R1c		
7579.023(3)	13190.681(5)	7579.0283(10)	−0.005	20000		5s	$(3/2)^2[3/2]_2$	5p	$(3/2)^2[3/2]^\circ_2$	4.1e+07	B	F_Re	B00	
7579.850(5)	13189.242(9)	7579.8494(9)	0.001	8400		5s	$(5/2)^2[5/2]_2$	sp	$(^3F)^3P^\circ\ ^1F^\circ_3$	1.15e+07	B	F_Re	B00	
7583.273(8)	13183.289(13)	7583.339(4)	−0.066	64	?	4d	$(5/2)^2[1/2]_0$	sp	$(^1D)^3P^\circ\ ^3P^\circ_1$			R1c		X
7652.3326(7)	13064.3147(12)	7652.3337(5)	−0.0011	26000		5s	$(5/2)^2[5/2]_3$	5p	$(5/2)^2[5/2]^\circ_3$	2.35e+07	B	F_Re	B00	
7664.6451(12)	13043.3283(20)	7664.6465(7)	−0.0014	52000		5p	$(5/2)^2[7/2]^\circ_3$	6s	$(5/2)^2[5/2]_2$	3.5e+07	D+	F_Re	TW	L
7681.787(4)	13014.223(7)	7681.7968(10)	−0.010	570		5p	$(3/2)^2[5/2]^\circ_2$	6s	$(3/2)^2[3/2]_2$			R1c		
7712.418(6)	12962.535(9)	7712.421(3)	−0.003	640		6p	$(5/2)^2[3/2]^\circ_2$	7d	$(5/2)^2[5/2]_3$	3.7e+06	D+	R1c	TW	
7726.637(4)	12938.681(6)	7726.6439(10)	−0.007	9000		5s	$(3/2)^2[3/2]_2$	5p	$(3/2)^2[3/2]^\circ_1$	1.70e+07	B	F_Re	B00	
7738.6656(12)	12918.5693(21)	7738.6644(9)	0.0012	30000		5p	$(3/2)^2[5/2]^\circ_2$	6s	$(3/2)^2[3/2]_1$	4.5e+07	D+	F_Re	TW	L
7739.499(8)	12917.179(14)	7739.505(4)	−0.006	1600		6p	$(5/2)^2[3/2]^\circ_2$	7d	$(5/2)^2[3/2]_2$	7.1e+06	D+	R1c	TW	
7744.0889(18)	12909.522(3)	7744.0905(8)	−0.0016	7800		5p	$(5/2)^2[7/2]^\circ_3$	6s	$(5/2)^2[5/2]_3$			F_Re		
7754.3688(20)	12892.408(3)	7754.3653(10)	0.0035	11000		5p	$(3/2)^2[1/2]^\circ_1$	6s	$(3/2)^2[3/2]_2$	9.e+06	D+	F_Re	TW	
7766.516(9)	12872.244(16)	7766.511(7)	0.006	32		6p	$(3/2)^2[1/2]^\circ_1$	7d	$(3/2)^2[1/2]_0$			R1c		
7773.196(7)	12861.182(12)	7773.1992(9)	−0.003	3200		5s	$(5/2)^2[5/2]_3$	5p	$(5/2)^2[5/2]^\circ_2$	1.08e+06	C	R1c	B00	
7778.7353(13)	12852.0236(22)	7778.7347(8)	0.0006	25000		5p	$(5/2)^2[5/2]^\circ_2$	6s	$(5/2)^2[5/2]_2$	1.3e+07	D+	F_Re	TW	L
7805.1886(13)	12808.4659(21)	7805.1878(7)	0.0008	26000		5p	$(5/2)^2[3/2]^\circ_1$	6s	$(5/2)^2[5/2]_2$	1.6e+07	D+	F_Re	TW	L
7807.6534(12)	12804.4224(20)	7807.6526(7)	0.0007	82000		5p	$(5/2)^2[7/2]^\circ_4$	6s	$(5/2)^2[5/2]_3$	3.6e+07	C+	F_Re	TW	L
7812.3182(19)	12796.777(3)	7812.3164(10)	0.0018	10000		5p	$(3/2)^2[1/2]^\circ_1$	6s	$(3/2)^2[3/2]_1$	1.2e+07	D+	F_Re	TW	
7820.577(4)	12783.264(6)	7820.5768(10)	−0.000	7300		5s	$(3/2)^2[3/2]_1$	5p	$(3/2)^2[1/2]^\circ_1$	1.51e+07	B	F_Re	B00	
7825.6528(12)	12774.9718(20)	7825.6530(7)	−0.0002	59000		5s	$(5/2)^2[5/2]_3$	5p	$(5/2)^2[7/2]^\circ_4$	5.2e+07	B	F_Re	B00	L
7845.0792(12)	12743.3378(20)	7845.0795(9)	−0.0003	24000		5p	$(3/2)^2[5/2]^\circ_3$	6s	$(3/2)^2[3/2]_2$	3.8e+07	D+	F_Re	TW	L
7853.984(9)	12728.890(15)	7853.980(4)	0.004	190		6p	$(5/2)^2[5/2]^\circ_2$	7d	$(5/2)^2[5/2]_2$			R1c		
7860.576(5)	12718.216(7)	7860.5740(10)	0.002	5700		5p	$(5/2)^2[5/2]^\circ_2$	6s	$(5/2)^2[5/2]_3$			R1c		
7862.652(9)	12714.857(14)	7862.654(3)	−0.002	250		6p	$(5/2)^2[5/2]^\circ_2$	7d	$(5/2)^2[7/2]_3$	2.9e+06	D+	R1c	TW	
7886.076(9)	12677.091(15)	7886.097(5)	−0.021	61		6p	$(3/2)^2[1/2]^\circ_1$	7d	$(3/2)^2[1/2]_1$			R1c		
7890.568(4)	12669.874(6)	7890.5666(8)	0.001	3500		5s	$(5/2)^2[5/2]_3$	5p	$(5/2)^2[7/2]^\circ_3$	5.8e+06	B	F_Re	B00	
7895.8050(13)	12661.4700(22)	7895.8039(8)	0.0011	21000		5s	$(3/2)^2[3/2]_1$	5p	$(3/2)^2[5/2]^\circ_2$	4.3e+07	B	F_Re	B00	L
7902.5482(13)	12650.6660(21)	7902.5499(8)	−0.0017	25000		5s	$(3/2)^2[3/2]_2$	5p	$(3/2)^2[5/2]^\circ_3$	4.9e+07	B	F_Re	B00	L

Table A1. *Cont.*

λ_{obs} [a] (Å)	σ_{obs} [b] (cm^{-1})	λ_{Ritz} [c] (Å)	$\Delta\lambda_{obs\text{-}Ritz}$ (Å)	I_{obs} [d] (arb. u.)	Char [e]	Lower Level		Upper Level		A (s^{-1})	Acc [f]	Line Ref. [g]	TP Ref. [g]	Notes [h]
7907.351(10)	12642.983(16)	7907.364(5)	−0.014	61		6p	$(5/2)^2[5/2]^\circ_2$	7d	$(5/2)^2[3/2]_1$			R1c		
7944.4368(6)	12583.9633(10)	7944.4365(5)	0.0003	17000		5s	$(5/2)^2[5/2]_2$	5p	$(5/2)^2[3/2]^\circ_1$	4.7e+07	B	F_Re	B00	L
7967.067(5)	12548.220(8)	7967.065(4)	0.002	2100		6p	$(3/2)^2[5/2]^\circ_3$	7d	$(3/2)^2[7/2]_4$	1.0e+07	D+	R1c	TW	
7972.031(5)	12540.406(7)	7972.0306(10)	0.000	7800		5s	$(5/2)^2[5/2]_2$	5p	$(5/2)^2[5/2]^\circ_2$	2.9e+07	B	R1c	B00	
7981.574(9)	12525.413(15)	7981.4451(23)	0.128	59	?	4f	$(5/2)^2[7/2]^\circ_3$	6d	$(3/2)^2[7/2]_4$			R1c		X
7988.1598(13)	12515.0857(20)	7988.1620(7)	−0.0022	47000		5p	$(5/2)^2[5/2]^\circ_3$	6s	$(5/2)^2[5/2]_3$	1.7e+07	D+	F_Re	TW	L
7996.793(3)	12501.575(5)	7996.7854(10)	0.007	9000		5s	$(3/2)^2[3/2]_2$	5p	$(3/2)^2[1/2]^\circ_1$	3.2e+07	B	F_Re	B00	
8006.515(5)	12486.394(8)	8006.525(4)	−0.009	440		6p	$(3/2)^2[5/2]^\circ_2$	7d	$(3/2)^2[7/2]_3$	5.9e+06	D+	R2nc	TW	
8026.479(3)	12455.338(5)	8026.4828(12)	−0.004	7100		5p	$(3/2)^2[3/2]^\circ_1$	6s	$(3/2)^2[3/2]_2$	8.2e+06	D+	F_Re	TW	
8042.489(7)	12430.544(11)	8042.494(4)	−0.005	230		6p	$(5/2)^2[3/2]^\circ_1$	7d	$(5/2)^2[3/2]_1$			R1c		
		8053.200(6)			m	6p	$(3/2)^2[5/2]^\circ_2$	7d	$(3/2)^2[1/2]_1$			R1c		
8053.339(16)	12413.796(24)	8053.358(4)	−0.019	230		sp	$(^1G)^3P^\circ\ ^3F^\circ_3$	7d	$(5/2)^2[7/2]_4$	2.3e+06	D+	R1c	TW	
8054.227(7)	12412.428(11)	8054.239(3)	−0.012	110		sp	$(^1G)^3P^\circ\ ^3F^\circ_3$	7d	$(5/2)^2[7/2]_3$			R1c		
8074.999(7)	12380.498(11)	8074.999(3)	0.000	420		6p	$(5/2)^2[7/2]^\circ_3$	8s	$(5/2)^2[5/2]_2$	3.5e+06	D+	R1c	TW	
8075.456(7)	12379.797(11)	8075.4576(10)	−0.001	2300		5s	$(3/2)^2[3/2]_2$	5p	$(3/2)^2[5/2]^\circ_2$	2.8e+05	C	R1c	B00	
8088.5846(15)	12359.7038(23)	8088.5888(11)	−0.0042	13000		5p	$(3/2)^2[3/2]^\circ_1$	6s	$(3/2)^2[3/2]_1$	1.6e+07	D+	F_Re	TW	
8095.5267(14)	12349.1051(21)	8095.5268(8)	−0.0001	26000		5s	$(5/2)^2[5/2]_2$	5p	$(5/2)^2[7/2]^\circ_3$	3.7e+07	B	F_Re	B00	
8098.519(7)	12344.542(11)	8098.5396(14)	−0.020	550		sp	$(^3F)^1P^\circ\ ^3G^\circ_4$	5g	$(3/2)^2[9/2]_5$			R1c		
8100.868(7)	12340.963(11)	8100.883(3)	−0.015	83		6p	$(5/2)^2[7/2]^\circ_3$	8s	$(5/2)^2[5/2]_3$			R1c		
8103.352(7)	12337.180(11)	8103.353(4)	−0.001	550		sp	$(^1G)^3P^\circ\ ^3F^\circ_3$	7d	$(5/2)^2[9/2]_4$	2.7e+06	D+	R1c	TW	
8154.003(6)	12260.545(9)	8154.012(4)	−0.009	2000		6p	$(5/2)^2[7/2]^\circ_4$	7d	$(5/2)^2[9/2]_5$	5.4e+06	D+	R1c	TW	
8192.221(6)	12203.347(9)	8192.2333(13)	−0.012	10700		5p	$(3/2)^2[3/2]^\circ_2$	6s	$(3/2)^2[3/2]_2$	2.0e+07	D+	F_Re	TW	
8192.330(3)	12203.185(4)	8192.3279(10)	0.002	3300		sp	$(^3F)^3P^\circ\ ^1F^\circ_3$	6s	$(5/2)^2[5/2]_2$	5.9e+06	D+	F_Re	TW	
8201.898(7)	12188.949(11)	8201.908(4)	−0.010	52		6p	$(3/2)^2[3/2]^\circ_2$	7d	$(3/2)^2[7/2]_3$			R1c		
8235.272(4)	12139.553(6)	8235.2788(14)	−0.007	5600		5s	$(3/2)^2[3/2]_1$	5p	$(3/2)^2[1/2]^\circ_0$	4.6e+07	B	F_Re	B00	
8256.944(6)	12107.691(9)	8256.9412(13)	0.002	3000		5p	$(3/2)^2[3/2]^\circ_2$	6s	$(3/2)^2[3/2]_1$			F_Re		
8277.5525(14)	12077.5460(21)	8277.5527(8)	−0.0003	20000		5s	$(5/2)^2[5/2]_3$	5p	$(5/2)^2[3/2]^\circ_2$	4.3e+07	B	F_Re	B00	
8283.1521(14)	12069.3813(21)	8283.1520(9)	0.0000	23000		sp	$(^3F)^3P^\circ\ ^1F^\circ_3$	6s	$(5/2)^2[5/2]_3$			F_Re		
8298.461(7)	12047.116(11)	8298.465(3)	−0.004	72		sp	$(^3F)^1P^\circ\ ^3F^\circ_2$	7d	$(5/2)^2[5/2]_2$			R1c		
8306.475(5)	12035.493(7)	8306.4866(14)	−0.011	890		4p	$^3D^\circ_2$	s^2	$^1D_2$			R1c		
8308.148(7)	12033.070(11)	8308.150(3)	−0.002	120		sp	$(^3F)^1P^\circ\ ^3F^\circ_2$	7d	$(5/2)^2[7/2]_3$			R1c		
8328.395(6)	12003.817(8)	8328.393(3)	0.001	470		6p	$(5/2)^2[5/2]^\circ_3$	7d	$(5/2)^2[7/2]_4$	3.1e+06	D+	R1c	TW	
8333.074(7)	11997.077(10)	8333.072(4)	0.001	470		6p	$(5/2)^2[3/2]^\circ_2$	8s	$(5/2)^2[5/2]_3$	2.3e+06	D+	R1c	TW	
8333.617(8)	11996.294(12)	8333.6418(20)	−0.024	580		sp	$(^3F)^1P^\circ\ ^3G^\circ_5$	5g	$(5/2)^2[9/2]_5$			R1c		
8336.333(5)	11992.386(8)	8336.3323(20)	0.001	3500		sp	$(^3F)^1P^\circ\ ^3G^\circ_5$	5g	$(5/2)^2[11/2]_6$	1.4e+06	D+	R1c	TW	
8353.837(7)	11967.259(10)	8353.843(3)	−0.006	230		6p	$(5/2)^2[5/2]^\circ_3$	7d	$(5/2)^2[5/2]_3$	3.1e+06	D+	R1c	TW	
8381.868(7)	11927.237(11)	8381.872(4)	−0.004	45		6p	$(5/2)^2[5/2]^\circ_3$	7d	$(5/2)^2[9/2]_4$			R1c		
8385.624(7)	11921.895(11)	8385.629(4)	−0.005	67		6p	$(5/2)^2[5/2]^\circ_3$	7d	$(5/2)^2[3/2]_2$			R1c		
8394.566(8)	11909.195(11)	8394.5745(12)	−0.008	110		4f	$(5/2)^2[9/2]^\circ_4$	5g	$(3/2)^2[9/2]_5$			R1c		
8402.897(7)	11897.388(9)	8402.9047(11)	−0.007	1600		5s	$(3/2)^2[3/2]_1$	sp	$(^3F)^3P^\circ\ ^1D^\circ_2$	3.9e+06	C	R1c	B00	
8450.062(8)	11830.982(11)	8450.078(3)	−0.016	63		sp	$(^3F)^1P^\circ\ ^3D^\circ_2$	6d	$(3/2)^2[5/2]_3$			R1c		
8477.298(5)	11792.971(7)	8477.3077(17)	−0.009	3900		4p	$^1P^\circ_1$	s^2	^{1}D textsubscript2			R1c		

Table A1. *Cont.*

λ_{obs} [a] (Å)	σ_{obs} [b] (cm^{-1})	λ_{Ritz} [c] (Å)	$\Delta\lambda_{obs\text{-}Ritz}$ (Å)	I_{obs} [d] (arb. u.)	Char [e]	Lower Level		Upper Level		A (s^{-1})	Acc [f]	Line Ref. [g]	TP Ref. [g]	Notes [h]
8503.394(5)	11756.780(7)	8503.3976(9)	−0.004	3600		5s	$(5/2)^2[5/2]_2$	5p	$(5/2)^2[3/2]^o_2$	1.6e+06	C	R1c	B00	
8511.0626(20)	11746.187(3)	8511.0647(10)	−0.0021	11000		sp	$(^3F)^3P^o\ ^1D^o_2$	6s	$(5/2)^2[5/2]_2$	1.0e+07	D+	F_Re	TW	
8535.196(11)	11712.975(15)	8535.2020(23)	−0.006	77		sp	$(^3P)^3P^o\ ^3D^o_1$	5g	$(5/2)^2[5/2]_2$			R1c		
8574.093(11)	11659.838(15)	8574.083(3)	0.010	110		sp	$(^3F)^1P^o\ ^3F^o_4$	6d	$(5/2)^2[9/2]_5$			R1c		
8606.681(10)	11615.690(13)	8606.6732(12)	0.008	690		5s	$(3/2)^2[3/2]_2$	sp	$(^3F)^3P^o\ ^1D^o_2$	7.9e+06	B	R1c	B00	
8609.132(5)	11612.384(7)	8609.1359(12)	−0.004	1100		sp	$(^3F)^3P^o\ ^1D^o_2$	6s	$(5/2)^2[5/2]_3$			R1c		
8729.820(9)	11451.845(12)	8729.822(3)	−0.002	160		sp	$(^1G)^3P^o\ ^3H^o_4$	5g	$(3/2)^2[11/2]_5$			R1c		
8730.233(7)	11451.303(9)	8730.232(3)	0.002	110		sp	$(^1G)^3P^o\ ^3F^o_3$	8s	$(5/2)^2[5/2]_2$			R1c		
8797.094(7)	11364.269(9)	8797.0968(22)	−0.002	380		sp	$(^3F)^1P^o\ ^3F^o_4$	5g	$(5/2)^2[9/2]_5$	1.6e+06	D+	R1c	TW	
8800.120(8)	11360.362(10)	8800.122(3)	−0.002	74		sp	$(^3F)^1P^o\ ^3F^o_4$	5g	$(5/2)^2[11/2]_5$			R1c		
8810.397(8)	11347.110(10)	8810.397(4)	0.000	290		6p	$(5/2)^2[7/2]^o_4$	8s	$(5/2)^2[5/2]_3$	2.4e+06	D+	R1c	TW	
8937.241(8)	11186.064(10)	8937.281(3)	−0.039	13		5d	$(5/2)^2[5/2]_3$	5f	$(3/2)^2[7/2]^o_4$			R1c		X
8959.159(9)	11158.698(11)	8959.1604(12)	−0.001	1300		5s	$(3/2)^2[3/2]_2$	sp	$(^3F)^3P^o\ ^1F^o_3$	1.6e+05	C	R1c	B00	
8967.150(9)	11148.754(11)	8967.143(4)	0.007	250		sp	$(^1G)^3P^o\ ^3F^o_2$	6d	$(3/2)^2[1/2]_1$			R1rc		
8991.446(9)	11118.629(11)	8991.447(7)	−0.001	12		sp	$(^3P)^3P^o\ ^1P^o_1$	6d	$(3/2)^2[1/2]_1$			R2nc		
9001.419(8)	11106.311(9)	9001.428(5)	−0.009	25		6p	$(3/2)^2[3/2]^o_1$	8s	$(3/2)^2[3/2]_2$			R1c		
9005.757(8)	11100.961(9)	9005.754(4)	0.003	4300		sp	$(^3F)^1P^o\ ^3G^o_3$	5g	$(3/2)^2[9/2]_4$	2.5e+06	D+	R1c	TW	
9015.187(9)	11089.349(11)	9015.144(4)	0.043	310		sp	$(^3F)^1P^o\ ^3G^o_3$	5g	$(3/2)^2[7/2]_4$			R1rc		X
9029.347(8)	11071.959(9)	9029.343(4)	0.003	300		sp	$(^3F)^1P^o\ ^3F^o_2$	8s	$(5/2)^2[5/2]_2$			R1c		
9036.147(8)	11063.627(9)	9036.132(4)	0.015	600		sp	$(^1G)^3P^o\ ^3H^o_5$	5g	$(3/2)^2[11/2]_6$	9.e+05	D+	R1rc	TW	
9067.974(8)	11024.796(10)	9067.9717(16)	0.002	180		5s	$(5/2)^2[5/2]_3$	sp	$(^3F)^3P^o\ ^3F^o_2$	3.9e+05	C	R1c	B00	
9086.933(9)	11001.793(11)	9086.928(4)	0.005	1200		6p	$(5/2)^2[5/2]^o_3$	8s	$(5/2)^2[5/2]_3$	1.7e+06	D+	R1c	TW	
9089.413(9)	10998.792(11)	9089.4094(24)	0.003	46		5d	$(5/2)^2[7/2]_3$	5f	$(3/2)^2[9/2]^o_4$			R1c		
9093.704(9)	10993.602(11)	9093.632(4)	0.072	12		sp	$(^3P)^3P^o\ ^1D^o_2$	6d	$(3/2)^2[5/2]_2$			R1c		X
9096.064(9)	10990.749(11)	9096.054(6)	0.011	81		5d	$(5/2)^2[1/2]_1$	7p	$(5/2)^2[3/2]^o_2$			R1c		
9097.529(9)	10988.980(11)	9097.525(3)	0.004	460		sp	$(^3P)^3P^o\ ^3P^o_2$	7s	$(5/2)^2[5/2]_3$			R1c		
9098.493(9)	10987.815(11)	9098.489(3)	0.005	69		5d	$(5/2)^2[5/2]_2$	5f	$(3/2)^2[5/2]^o_3$			R1c		
9101.459(12)	10984.235(15)	9101.4285(22)	0.030	23		5d	$(5/2)^2[7/2]_4$	5f	$(3/2)^2[9/2]^o_5$			R1c		
9103.234(6)	10982.093(8)	9103.2366(12)	−0.003	2600		5p	$(3/2)^2[5/2]^o_2$	6s	$(5/2)^2[5/2]_2$			R1c		
9107.895(9)	10976.473(11)	9107.878(6)	0.016	230		5s	$(5/2)^2[5/2]_3$	sp	$(^3F)^3P^o\ ^1G^o_4$	9.1e+04	C	R1c	B00	
9115.700(9)	10967.075(11)	9115.643(10)	0.057	46	?	5f	$(5/2)^2[11/2]^o_5$	9d	$(5/2)^2[9/2]_4$			R1c		X
9125.941(9)	10954.767(11)	9125.932(4)	0.009	340		sp	$(^1G)^3P^o\ ^3F^o_2$	5g	$(3/2)^2[7/2]_3$			R2nc		
9154.181(18)	10920.973(21)	9154.160(4)	0.021	170		sp	$(^1G)^3P^o\ ^3F^o_2$	5g	$(3/2)^2[5/2]_3$			R2c		
9158.409(9)	10915.931(11)	9158.393(4)	0.016	450		sp	$(^3P)^3P^o\ ^3P^o_1$	5g	$(5/2)^2[3/2]_1$			R1c		
9179.306(12)	10891.081(14)	9179.319(7)	−0.014	610		5d	$(3/2)^2[1/2]_1$	7p	$(3/2)^2[1/2]^o_1$			R2nc		
9205.3214(22)	10860.301(3)	9205.3242(12)	−0.0028	5800		5p	$(3/2)^2[1/2]^o_1$	6s	$(5/2)^2[5/2]_2$			F_Re		
9219.491(10)	10843.610(12)	9219.466(5)	0.025	32		5d	$(5/2)^2[5/2]_3$	7p	$(5/2)^2[5/2]^o_3$			R1c		
9226.742(10)	10835.088(12)	9226.7371(12)	0.005	730		5s	$(3/2)^2[3/2]_1$	5p	$(5/2)^2[3/2]^o_1$	1.3e+06	C	R1c	B00	
9230.567(11)	10830.599(13)	9230.577(3)	−0.010	370		sp	$(^3P)^3P^o\ ^3D^o_3$	7s	$(5/2)^2[5/2]_3$			R1c		
9234.272(8)	10826.253(10)	9234.258(3)	0.014	1300		sp	$(^3F)^1P^o\ ^3D^o_3$	6d	$(5/2)^2[7/2]_4$			R1c		
9277.399(12)	10775.926(14)	9277.400(9)	−0.000	11	?	5f	$(5/2)^2[9/2]^o_5$	8g	$(5/2)^2[11/2]_6$	4.6e+06	D+	R1nc	TW	

Table A1. *Cont.*

λ_{obs} [a] (Å)	σ_{obs} [b] (cm^{-1})	λ_{Ritz} [c] (Å)	$\Delta\lambda_{obs-Ritz}$ (Å)	I_{obs} [d] (arb. u.)	Char [e]	Lower Level		Upper Level		A (s^{-1})	Acc [f]	Line Ref. [g]	TP Ref. [g]	Notes [h]
9296.634(12)	10753.631(14)	9296.580(6)	0.054	21		sp	$(^1D)^3P^{\circ}\ ^3F^{\circ}_2$	5d	$(3/2)^2[5/2]_3$			R1c		X
9312.387(13)	10735.440(15)	9312.450(9)	−0.064	21		5f	$(5/2)^2[9/2]^{\circ}_5$	7g	$(3/2)^2[11/2]_5$			R1c		X
9317.944(9)	10729.037(10)	9317.943(4)	0.002	260		sp	$(^3F)^1P^{\circ}\ ^3D^{\circ}_1$	6d	$(3/2)^2[5/2]_2$			R1c		
9324.302(9)	10721.722(10)	9324.3128(20)	−0.011	21		sp	$(^3F)^1P^{\circ}\ ^3G^{\circ}_4$	6d	$(5/2)^2[7/2]_3$			R1c		
9331.893(6)	10713.000(7)	9331.8965(12)	−0.004	2400		5s	$(3/2)^2[3/2]_2$	5p	$(5/2)^2[5/2]^{\circ}_3$	1.2e+06	C	R1c	B00	
9339.711(8)	10704.032(9)	9339.7151(17)	−0.004	360		5s	$(5/2)^2[5/2]_2$	sp	$(^3F)^3P^{\circ}\ ^3F^{\circ}_2$	6.8e+05	C	R1c	B00	
9364.118(14)	10676.133(16)	9364.046(7)	0.072	10		4d	$(3/2)^2[7/2]_4$	sp	$(^1D)^3P^{\circ}\ ^3F^{\circ}_3$			R1c		X
9374.435(19)	10664.383(21)	9374.4412(23)	−0.006	10		4f	$(5/2)^2[3/2]^{\circ}_2$	6d	$(5/2)^2[5/2]_2$			R1c		
9391.049(12)	10645.517(14)	9390.9259(21)	0.123	21	?	4f	$(5/2)^2[3/2]^{\circ}_2$	6d	$(5/2)^2[7/2]_3$			R1c		X
9415.060(9)	10618.368(10)	9415.044(6)	0.016	21		4d	$(3/2)^2[7/2]_3$	sp	$(^1D)^3P^{\circ}\ ^3F^{\circ}_2$			R1c		
9424.164(12)	10608.110(14)	9424.151(4)	0.013	100		sp	$(^3F)^1P^{\circ}\ ^3D^{\circ}_1$	6d	$(3/2)^2[3/2]_1$			R1c		
9444.862(10)	10584.863(11)	9444.858(10)	0.005	10		4d	$(3/2)^2[5/2]_2$	sp	$(^1D)^3P^{\circ}\ ^3D^{\circ}_3$			R1c		
9451.495(11)	10577.435(12)	9451.5130(13)	−0.018	1400		5p	$(3/2)^2[5/2]^{\circ}_3$	6s	$(5/2)^2[5/2]_3$			R1c		
9458.336(9)	10569.785(10)	9458.335(3)	0.000	72		4f	$(5/2)^2[7/2]^{\circ}_3$	6d	$(5/2)^2[7/2]_4$			R1c		
9459.889(10)	10568.049(11)	9459.895(4)	−0.005	31		sp	$(^3F)^1P^{\circ}\ ^3G^{\circ}_4$	6d	$(5/2)^2[9/2]_5$			R1c		
9460.791(9)	10567.042(10)	9460.790(5)	0.001	1000		sp	$(^3P)^3P^{\circ}\ ^3P^{\circ}_0$	5g	$(5/2)^2[3/2]_1$			R1c		
9461.294(10)	10566.480(11)	9461.3089(22)	−0.015	10		4f	$(5/2)^2[7/2]^{\circ}_3$	6d	$(5/2)^2[7/2]_3$			R1c		
9473.011(12)	10553.411(13)	9473.0050(12)	0.006	1300		5s	$(3/2)^2[3/2]_2$	5p	$(5/2)^2[3/2]^{\circ}_1$	1.13e+06	C	R1c	B00	
9474.119(10)	10552.176(11)	9474.106(5)	0.013	10		5d	$(5/2)^2[7/2]_4$	7p	$(5/2)^2[7/2]^{\circ}_4$			R1c		
9485.256(10)	10539.787(11)	9485.2726(23)	−0.017	21		4f	$(5/2)^2[5/2]^{\circ}_3$	6d	$(5/2)^2[5/2]_2$			R1c		
9492.663(19)	10531.563(21)	9492.653(4)	0.010	10		4f	$(3/2)^2[3/2]^{\circ}_1$	6d	$(3/2)^2[3/2]_2$			R1c		
9493.808(21)	10530.293(23)	9493.987(16)	−0.179	10	?	sp	$(^3P)^3P^{\circ}\ ^5P^{\circ}_3$	6s	$(3/2)^2[3/2]_2$			R2c		X
9500.248(10)	10523.154(11)	9500.262(3)	−0.014	52		sp	$(^3P)^3P^{\circ}\ ^3D^{\circ}_1$	7s	$(5/2)^2[5/2]_2$			R1c		
9502.093(10)	10521.111(11)	9502.1498(21)	−0.057	10	?	4f	$(5/2)^2[5/2]^{\circ}_3$	6d	$(5/2)^2[7/2]_3$			R1c		X
9504.184(9)	10518.796(10)	9504.220(4)	−0.035	360		sp	$(^3P)^3P^{\circ}\ ^1D^{\circ}_2$	5g	$(3/2)^2[7/2]_3$			R1c		X
9505.598(10)	10517.232(11)	9505.605(3)	−0.008	10		4f	$(5/2)^2[3/2]^{\circ}_2$	6d	$(5/2)^2[3/2]_1$			R1c		
9510.299(19)	10512.033(21)	9510.298(7)	0.001	42		sp	$(^3P)^3P^{\circ}\ ^5P^{\circ}_2$	6s	$(3/2)^2[3/2]_2$			R1c		
9512.262(8)	10509.864(9)	9512.2653(15)	−0.004	2900		5s	$(3/2)^2[3/2]_2$	5p	$(5/2)^2[5/2]^{\circ}_2$	1.3e+06	C	R1c	B00	
9515.588(11)	10506.190(12)	9515.595(3)	−0.007	210		4f	$(5/2)^2[7/2]^{\circ}_3$	6d	$(5/2)^2[5/2]_3$			R1c		
9527.379(10)	10493.188(11)	9527.376(4)	0.003	31		4f	$(3/2)^2[3/2]^{\circ}_1$	6d	$(3/2)^2[3/2]_1$			R1c		
9530.897(10)	10489.315(11)	9530.905(3)	−0.008	160		4f	$(5/2)^2[11/2]^{\circ}_5$	6d	$(5/2)^2[9/2]_4$			R1c		
9546.928(10)	10471.701(11)	9546.928(3)	−0.000	210		4f	$(5/2)^2[11/2]^{\circ}_6$	6d	$(5/2)^2[9/2]_5$			R1c		
9548.691(10)	10469.768(11)	9548.702(3)	−0.011	21		4f	$(5/2)^2[3/2]^{\circ}_1$	6d	$(5/2)^2[3/2]_1$			R1c		
9565.600(10)	10451.261(11)	9565.623(4)	−0.024	53		4f	$(3/2)^2[9/2]^{\circ}_5$	6d	$(3/2)^2[7/2]_4$			R1c		
9585.611(10)	10429.442(11)	9585.628(3)	−0.017	32		4f	$(3/2)^2[5/2]^{\circ}_3$	6d	$(3/2)^2[3/2]_2$			R1c		
9597.666(20)	10416.343(21)	9597.614(7)	0.052	11		sp	$(^3P)^3P^{\circ}\ ^5P^{\circ}_2$	6s	$(3/2)^2[3/2]_1$			R1c		
9599.817(10)	10414.009(11)	9599.843(3)	−0.026	11		4f	$(5/2)^2[5/2]^{\circ}_2$	6d	$(5/2)^2[3/2]_1$			R1c		
9602.377(10)	10411.232(11)	9602.374(3)	0.003	22		4f	$(3/2)^2[7/2]^{\circ}_4$	6d	$(3/2)^2[5/2]_3$			R1c		
9605.15(9)	10408.22(10)	9605.075(6)	0.08	54	*	5d	$(5/2)^2[5/2]_3$	7p	$(5/2)^2[3/2]^{\circ}_2$			R1c		
9605.15(9)	10408.22(10)	9605.193(4)	−0.04	54	*	4f	$(3/2)^2[9/2]^{\circ}_4$	6d	$(3/2)^2[7/2]_3$			R1c		
9610.896(10)	10402.004(11)	9610.884(5)	0.012	430		5d	$(5/2)^2[1/2]_0$	5f	$(3/2)^2[3/2]^{\circ}_1$	5.9e+06	D+	R1c	TW	
9613.596(20)	10399.083(21)	9613.594(3)	0.002	65		4f	$(5/2)^2[9/2]^{\circ}_5$	6d	$(5/2)^2[7/2]_4$			R1c		
9630.420(10)	10380.916(11)	9630.428(3)	−0.009	11		4f	$(3/2)^2[7/2]^{\circ}_3$	6d	$(3/2)^2[5/2]_2$			R1c		

Table A1. *Cont.*

λ_{obs} [a] (Å)	σ_{obs} [b] (cm^{-1})	λ_{Ritz} [c] (Å)	$\Delta\lambda_{obs-Ritz}$ (Å)	I_{obs} [d] (arb. u.)	Char [e]	Lower Level		Upper Level		A (s^{-1})	Acc [f]	Line Ref. [g]	TP Ref. [g]	Notes [h]
9630.969(11)	10380.324(11)	9631.023(3)	−0.054	11		4f	$(5/2)^2[5/2]^\circ_3$	6d	$(5/2)^2[3/2]_2$			R1c		X
9637.149(11)	10373.668(12)	9637.148(3)	0.000	1100		sp	$(^3F)^1P^\circ\ ^3D^\circ_3$	5g	$(5/2)^2[9/2]_4$			R1c		
9646.751(10)	10363.342(11)	9646.770(3)	−0.019	22		4f	$(5/2)^2[7/2]^\circ_4$	6d	$(5/2)^2[5/2]_3$			R1c		
9649.361(10)	10360.539(11)	9649.378(3)	−0.017	110	*	sp	$(^3F)^1P^\circ\ ^3D^\circ_3$	5g	$(5/2)^2[7/2]_4$			R1c		X
9649.361(10)	10360.539(11)	9649.399(3)	−0.038	110	*	sp	$(^3F)^1P^\circ\ ^3D^\circ_3$	5g	$(5/2)^2[7/2]_3$			R1nc		X
9651.107(10)	10358.665(11)	9651.115(4)	−0.008	22		4f	$(3/2)^2[3/2]^\circ_2$	6d	$(3/2)^2[1/2]_1$			R1c		
9657.363(11)	10351.954(11)	9657.369(4)	−0.005	22		4f	$(5/2)^2[1/2]^\circ_1$	6d	$(5/2)^2[1/2]_1$			R1c		
9664.119(10)	10344.717(11)	9664.140(3)	−0.020	11		4f	$(3/2)^2[5/2]^\circ_2$	6d	$(3/2)^2[3/2]_1$			R1c		
9670.167(10)	10338.248(11)	9670.172(8)	−0.006	11		6d	$(5/2)^2[9/2]_4$	8f	$(5/2)^2[11/2]^\circ_5$	3.5e+06	D+	R1c	TW	
9688.602(8)	10318.577(8)	9688.6189(13)	−0.017	1600		5s	$(3/2)^2[3/2]_2$	5p	$(5/2)^2[7/2]^\circ_3$	2.5e+05	C	R1c	B00	
9715.855(20)	10289.633(21)	9715.789(3)	0.066	12		4f	$(5/2)^2[9/2]^\circ_4$	6d	$(5/2)^2[7/2]_4$			R1c		X
9718.927(10)	10286.381(11)	9718.9263(24)	0.000	23		4f	$(5/2)^2[9/2]^\circ_4$	6d	$(5/2)^2[7/2]_3$			R1c		
9732.121(21)	10272.435(22)	9732.1007(17)	0.021	1100	*	sp	$(^3F)^1P^\circ\ ^3G^\circ_4$	5g	$(5/2)^2[9/2]_5$			R1nc		
9732.148(13)	10272.407(13)	9732.1264(21)	0.022	1100	*	sp	$(^3F)^1P^\circ\ ^3G^\circ_4$	5g	$(5/2)^2[9/2]_4$	2.1e+06	D+	F_Re	TW	
9734.473(10)	10269.953(11)	9734.488(3)	−0.015	24		4f	$(5/2)^2[7/2]^\circ_4$	6d	$(5/2)^2[9/2]_5$			R1c		
9735.802(4)	10268.551(4)	9735.8031(20)	−0.001	6700		sp	$(^3F)^1P^\circ\ ^3G^\circ_4$	5g	$(5/2)^2[11/2]_5$	9.e+06	D+	F_Re	TW	
		9781.688(5)			m	sp	$(^3F)^1P^\circ\ ^3D^\circ_1$	5g	$(3/2)^2[5/2]_2$			R1c		
9781.844(13)	10220.219(13)	9781.880(4)	−0.037	120		4f	$(3/2)^2[7/2]^\circ_3$	6d	$(3/2)^2[7/2]_3$			R1c		X
9792.465(10)	10209.134(11)	9792.4709(19)	−0.006	50		4f	$(5/2)^2[1/2]^\circ_1$	5g	$(5/2)^2[5/2]_2$			R1c		
9794.085(10)	10207.445(10)	9794.067(3)	0.018	38		sp	$(^3P)^3P^\circ\ ^3D^\circ_2$	7s	$(5/2)^2[5/2]_2$			R1c		
9813.2038(22)	10187.5583(23)	9813.2047(20)	−0.0009	8600		4f	$(5/2)^2[1/2]^\circ_1$	5g	$(5/2)^2[3/2]_2$	4.1e+07	C+	F_Re	TW	
9813.334(8)	10187.423(8)	9813.325(4)	0.009	2600		4f	$(5/2)^2[1/2]^\circ_1$	5g	$(5/2)^2[3/2]_1$	1.7e+07	D+	R1c	TW	
9817.398(9)	10183.206(9)	9817.3974(12)	0.001	240		4f	$(5/2)^2[3/2]^\circ_2$	5g	$(5/2)^2[7/2]_3$	1.8e+06	D+	R1c	TW	
9824.237(10)	10176.117(11)	9824.2394(9)	−0.002	100		4f	$(5/2)^2[11/2]^\circ_6$	5g	$(5/2)^2[9/2]_5$			R1c		
9827.041(12)	10173.213(12)	9827.0477(21)	−0.006	78		4f	$(5/2)^2[11/2]^\circ_5$	5g	$(5/2)^2[9/2]_4$			R1c		
9827.982(4)	10172.240(4)	9827.9787(7)	0.003	3200		4f	$(5/2)^2[11/2]^\circ_6$	5g	$(5/2)^2[11/2]_6$	7.5e+06	D+	F_Re	TW	
9828.973(10)	10171.214(10)	9828.9743(16)	−0.001	710		5p	$(3/2)^2[3/2]^\circ_2$	6s	$(5/2)^2[5/2]_2$			R1c		
9830.795(4)	10169.329(4)	9830.7965(20)	−0.001	2700		4f	$(5/2)^2[11/2]^\circ_5$	5g	$(5/2)^2[11/2]_5$	7.7e+06	D+	F_Re	TW	
9835.299(10)	10164.672(11)	9835.299(3)	−0.001	100		4f	$(3/2)^2[3/2]^\circ_2$	5g	$(3/2)^2[7/2]_3$			R1c		
9837.8361(21)	10162.0505(22)	9837.8372(17)	−0.0011	10000		4f	$(5/2)^2[3/2]^\circ_2$	5g	$(5/2)^2[5/2]_3$	4.4e+07	C+	F_Re	TW	
9850.495(5)	10148.991(6)	9850.504(4)	−0.010	2400		4f	$(5/2)^2[1/2]^\circ_0$	5g	$(5/2)^2[3/2]_1$	2.8e+07	D+	F_Re	TW	
9858.723(7)	10140.521(8)	9858.7254(24)	−0.002	2500		4f	$(5/2)^2[3/2]^\circ_2$	5g	$(5/2)^2[3/2]_2$	1.9e+07	D+	R1c	TW	
9861.28027(17)	10137.89135(18)			21000		4f	$(5/2)^2[11/2]^\circ_6$	5g	$(5/2)^2[13/2]_7$	6.7e+07	B	F_Re	TW	
9864.13643(18)	10134.95593(19)			19000		4f	$(5/2)^2[11/2]^\circ_5$	5g	$(5/2)^2[13/2]_6$	6.6e+07	B	F_Re	TW	
9868.0934(21)	10130.8920(22)	9868.0940(21)	−0.0006	9100		4f	$(3/2)^2[3/2]^\circ_2$	5g	$(3/2)^2[5/2]_3$	5.6e+07	C+	F_Re	TW	
9870.816(16)	10128.098(17)	9870.8012(20)	0.014	97		5s	$(5/2)^2[5/2]_3$	sp	$(^3F)^3P^\circ\ ^3F^\circ_3$	5.4e+05	C	R1c	B00	
9878.465(8)	10120.255(9)	9878.193(5)	0.273	280	?	sp	$(^3P)^3P^\circ\ ^5S^\circ_2$	5g	$(5/2)^2[5/2]_2$			R1c		X
9881.464(3)	10117.184(3)	9881.4639(15)	0.000	8400		4f	$(5/2)^2[7/2]^\circ_3$	5g	$(5/2)^2[9/2]_4$	2.1e+07	D	F_Re	TW	
9883.730(9)	10114.864(9)	9883.7091(24)	0.021	560		4f	$(3/2)^2[9/2]^\circ_5$	5g	$(3/2)^2[9/2]_5$	5.4e+06	D+	R1c	TW	
9883.9672(10)	10114.6216(10)	9883.9674(10)	−0.0001	5100		4f	$(5/2)^2[3/2]^\circ_1$	5g	$(5/2)^2[5/2]_2$	3.9e+07	C+	F_Re	TW	
9884.996(11)	10113.569(11)	9884.989(4)	0.007	28		6p	$(5/2)^2[7/2]^\circ_3$	6d	$(3/2)^2[7/2]_4$			R1c		
9892.939(4)	10105.448(4)	9892.940(4)	−0.000	4800		4f	$(3/2)^2[3/2]^\circ_1$	5g	$(3/2)^2[5/2]_2$	3.1e+07	D	F_Re	TW	
9894.337(7)	10104.021(7)	9894.3443(14)	−0.007	5300		4f	$(5/2)^2[7/2]^\circ_3$	5g	$(5/2)^2[7/2]_3$	1.0e+07	D+	R1c	TW	

Table A1. *Cont.*

λ_{obs} [a] (Å)	σ_{obs} [b] (cm^{-1})	λ_{Ritz} [c] (Å)	$\Delta\lambda_{obs-Ritz}$ (Å)	I_{obs} [d] (arb. u.)	Char [e]	Lower Level		Upper Level		A (s^{-1})	Acc [f]	Line Ref. [g]	TP Ref. [g]	Notes [h]
9899.567(11)	10098.683(11)	9899.414(6)	0.153	57	?	sp	$(^3P)^3P^\circ\ ^5S^\circ_2$	5g	$(5/2)^2[3/2]_1$			R1c		X
9905.136(23)	10093.005(24)	9905.091(3)	0.045	1800	*	4f	$(5/2)^2[3/2]^\circ_1$	5g	$(5/2)^2[3/2]_2$	3.7e+06	D+	F_Re	TW	
9905.18(8)	10092.97(8)	9905.214(4)	−0.04	1800	*	4f	$(5/2)^2[3/2]^\circ_1$	5g	$(5/2)^2[3/2]_1$	1.5e+07	D+	R1c	TW	
9911.988(12)	10086.028(12)	9911.986(3)	0.002	58		sp	$(^3F)^1P^\circ\ ^3D^\circ_2$	7s	$(3/2)^2[3/2]_1$			R1c		
9915.090(21)	10082.873(22)	9915.0672(18)	0.022	630	*	4f	$(5/2)^2[7/2]^\circ_3$	5g	$(5/2)^2[5/2]_2$			R1c		
9915.090(21)	10082.873(22)	9915.1061(20)	−0.017	630	*	4f	$(5/2)^2[7/2]^\circ_3$	5g	$(5/2)^2[5/2]_3$			R1c		
9916.4171(3)	10081.5233(3)	9916.4172(3)	−0.0001	15000		4f	$(3/2)^2[9/2]^\circ_5$	5g	$(3/2)^2[11/2]_6$	6.6e+07	B	F_Re	TW	
9917.9509(10)	10079.9642(10)	9917.9510(10)	−0.0001	9800		4f	$(5/2)^2[5/2]^\circ_2$	5g	$(5/2)^2[7/2]_3$	3.8e+07	C+	F_Re	TW	
9925.5883(4)	10072.2080(4)	9925.5883(4)	−0.0000	13000		4f	$(3/2)^2[9/2]^\circ_4$	5g	$(3/2)^2[11/2]_5$	6.5e+07	B	F_Re	TW	
9926.0209(15)	10071.7690(16)	9926.0211(13)	−0.0002	4900		4f	$(5/2)^2[5/2]^\circ_3$	5g	$(5/2)^2[9/2]_4$			F_Re		
9938.775(4)	10058.844(4)	9938.7730(16)	0.002	5300		4f	$(5/2)^2[5/2]^\circ_2$	5g	$(5/2)^2[5/2]_2$	2.1e+07	C+	F_Re	TW	
9938.9960(6)	10058.6207(6)	9938.9960(6)	0.0000	11000		4f	$(5/2)^2[5/2]^\circ_3$	5g	$(5/2)^2[7/2]_4$	4.4e+07	C+	F_Re	TW	
9948.895(11)	10048.612(11)	9948.8880(21)	0.007	61		4f	$(3/2)^2[5/2]^\circ_3$	5g	$(3/2)^2[9/2]_4$			R1c		
9959.970(3)	10037.439(3)	9959.9679(18)	0.002	5800		4f	$(5/2)^2[5/2]^\circ_3$	5g	$(5/2)^2[5/2]_3$	1.9e+07	C+	F_Re	TW	
9960.3485(8)	10037.0575(8)	9960.3485(8)	−0.0000	11000		4f	$(3/2)^2[5/2]^\circ_3$	5g	$(3/2)^2[7/2]_4$	5.6e+07	B	F_Re	TW	
9981.399(12)	10015.890(12)	9981.379(3)	0.020	130		4f	$(5/2)^2[5/2]^\circ_3$	5g	$(5/2)^2[3/2]_2$	1.9e+06	D+	R1c	TW	
9994.027(11)	10003.234(11)	9994.019(3)	0.008	900		4f	$(3/2)^2[5/2]^\circ_3$	5g	$(3/2)^2[5/2]_3$	9.e+06	D+	R1c	TW	
10006.588(3)	9990.678(3)	10006.5881(24)	−0.001	8400		4f	$(3/2)^2[5/2]^\circ_2$	5g	$(3/2)^2[7/2]_3$	4.9e+07	C+	F_Re	TW	
10022.9672(3)	9974.3510(3)	10022.9672(3)	−0.0000	20000		4f	$(5/2)^2[7/2]^\circ_4$	5g	$(5/2)^2[9/2]_5$	4.9e+07	B	F_Re	TW	
10026.894(4)	9970.445(4)	10026.8943(17)	−0.001	1500		4f	$(5/2)^2[7/2]^\circ_4$	5g	$(5/2)^2[11/2]_5$			F_Re		
10036.224(3)	9961.176(3)	10036.2243(9)	−0.000	5700		4f	$(5/2)^2[7/2]^\circ_4$	5g	$(5/2)^2[7/2]_4$	1.64e+07	C+	F_Re	TW	
10038.0920(5)	9959.3223(5)	10038.0919(5)	0.0001	14000		4f	$(3/2)^2[7/2]^\circ_4$	5g	$(3/2)^2[9/2]_5$	6.0e+07	B	F_Re	TW	
10040.499(13)	9956.935(13)	10040.481(5)	0.017	560		4f	$(3/2)^2[5/2]^\circ_2$	5g	$(3/2)^2[5/2]_2$	9.e+06	D+	R1c	TW	
10049.773(12)	9947.746(12)	10049.7881(19)	−0.015	1200		4f	$(3/2)^2[7/2]^\circ_4$	5g	$(3/2)^2[7/2]_4$	9.e+06	D+	R1c	TW	
10051.0231(20)	9946.5092(20)	10051.0218(6)	0.0014	5100		4f	$(5/2)^2[9/2]^\circ_5$	5g	$(5/2)^2[9/2]_5$	1.32e+07	C+	F_Re	TW	
10054.93572(22)	9942.63880(21)	10054.93573(22)	−0.00001	24000		4f	$(5/2)^2[9/2]^\circ_5$	5g	$(5/2)^2[11/2]_6$	5.8e+07	B	F_Re	TW	
10057.609(11)	9939.996(11)	10057.6090(20)	−0.000	180		4f	$(5/2)^2[7/2]^\circ_4$	5g	$(5/2)^2[5/2]_3$	1.7e+06	D+	R1c	TW	
10080.3499(10)	9917.5719(10)	10080.3499(10)	−0.0000	11000		4f	$(3/2)^2[7/2]^\circ_3$	5g	$(3/2)^2[9/2]_4$	5.5e+07	B	F_Re	TW	
10092.14(3)	9905.99(3)	10092.1155(21)	0.02	990	*	4f	$(3/2)^2[7/2]^\circ_3$	5g	$(3/2)^2[7/2]_4$	1.4e+06	D+	R1nc	TW	
10092.14(3)	9905.99(3)	10092.151(3)	−0.01	990	*	4f	$(3/2)^2[7/2]^\circ_3$	5g	$(3/2)^2[7/2]_3$	7.3e+06	D+	R1c	TW	
10162.808(8)	9837.103(8)	10162.8099(18)	−0.001	3800		4f	$(5/2)^2[9/2]^\circ_4$	5g	$(5/2)^2[9/2]_4$	1.25e+07	C+	F_Re	TW	
10166.8202(21)	9833.2217(20)	10166.8192(16)	0.0010	18000		4f	$(5/2)^2[9/2]^\circ_4$	5g	$(5/2)^2[11/2]_5$	4.4e+07	B	F_Re	TW	
10176.433(12)	9823.933(11)	10176.4346(17)	−0.002	1700		4f	$(5/2)^2[9/2]^\circ_4$	5g	$(5/2)^2[7/2]_3$	1.4e+06	D+	R1c	TW	
10193.660(10)	9807.331(10)	10193.6484(21)	0.012	1100		5s	$(5/2)^2[5/2]_2$	sp	$(^3F)^3P^\circ\ ^3F^\circ_3$			R1c		
10225.073(10)	9777.201(10)	10225.077(4)	−0.003	1700		6p	$(5/2)^2[7/2]^\circ_3$	5g	$(3/2)^2[9/2]_4$			R1c		
10237.22(3)	9765.60(3)	10237.183(4)	0.03	120	*	6p	$(5/2)^2[7/2]^\circ_3$	5g	$(3/2)^2[7/2]_4$			R1nc		
10237.22(3)	9765.60(3)	10237.219(4)	−0.00	120	*	6p	$(5/2)^2[7/2]^\circ_3$	5g	$(3/2)^2[7/2]_3$			R1c		
10247.043(10)	9756.239(9)	10247.038(4)	0.005	1300		sp	$(^3F)^1P^\circ\ ^3D^\circ_2$	6d	$(5/2)^2[7/2]_3$	1.9e+06	D+	R1c	TW	
10278.686(16)	9726.204(15)	10278.6604(14)	0.026	540		5s	$(3/2)^2[3/2]_2$	5p	$(5/2)^2[3/2]^\circ_2$			R1c		
10297.40(3)	9708.53(3)	10297.147(5)	0.25	56	?	sp	$(^1G)^3P^\circ\ ^3H^\circ_4$	6d	$(5/2)^2[9/2]_5$			R1c		X
10343.469(10)	9665.287(9)	10343.472(3)	−0.003	6100		sp	$(^3F)^1P^\circ\ ^3F^\circ_3$	5g	$(5/2)^2[9/2]_4$	1.17e+07	C+	R1c	TW	
10357.603(10)	9652.098(9)	10357.586(3)	0.017	780		sp	$(^3F)^1P^\circ\ ^3F^\circ_3$	5g	$(5/2)^2[7/2]_3$	4.8e+06	D+	R1c	TW	

Table A1. *Cont.*

λ_{obs} [a] (Å)	σ_{obs} [b] (cm^{-1})	λ_{Ritz} [c] (Å)	$\Delta\lambda_{obs\text{-}Ritz}$ (Å)	I_{obs} [d] (arb. u.)	Char [e]	Lower Level		Upper Level		A (s^{-1})	Acc [f]	Line Ref. [g]	TP Ref. [g]	Notes [h]
10380.323(21)	9630.972(20)	10380.297(3)	0.026	130	*	sp	$(^3F)^1P^{\circ}\ ^3F^{\circ}_3$	5g	$(5/2)^2[5/2]_2$			R1nc		
10380.323(21)	9630.972(20)	10380.340(3)	−0.016	130	*	sp	$(^3F)^1P^{\circ}\ ^3F^{\circ}_3$	5g	$(5/2)^2[5/2]_3$			R1c		
10424.302(8)	9590.341(7)	10424.309(3)	−0.007	3700		5d	$(5/2)^2[1/2]_1$	5f	$(5/2)^2[3/2]^{\circ}_2$	1.2e+07	D+	F_Re	TW	
10453.012(7)	9564.000(7)	10453.004(4)	0.008	4800		5d	$(5/2)^2[1/2]_1$	5f	$(5/2)^2[1/2]^{\circ}_1$	2.7e+07	D+	F_Re	TW	
10465.664(10)	9552.438(10)	10465.665(4)	−0.002	3800		5d	$(5/2)^2[1/2]_1$	5f	$(5/2)^2[1/2]^{\circ}_0$	3.6e+07	D+	F_Re	TW	
10575.703(11)	9453.046(10)	10575.690(6)	0.013	180		6p	$(5/2)^2[5/2]^{\circ}_2$	6d	$(3/2)^2[7/2]_3$	3.0e+06	D+	R1c	TW	
10609.294(11)	9423.116(10)	10609.300(5)	−0.006	240		5d	$(3/2)^2[1/2]_1$	5f	$(3/2)^2[3/2]^{\circ}_1$	1.0e+07	D+	R1c	TW	
10624.895(12)	9409.280(11)	10624.901(4)	−0.007	96		sp	$(^1G)^3P^{\circ}\ ^3H^{\circ}_4$	5g	$(5/2)^2[11/2]_5$			R1c		
10641.797(11)	9394.335(9)	10641.822(5)	−0.025	3600		5d	$(3/2)^2[1/2]_1$	5f	$(3/2)^2[3/2]^{\circ}_2$	3.0e+07	D+	F_Re	TW	
10742.332(8)	9306.416(7)	10742.3392(23)	−0.007	4800		5d	$(5/2)^2[9/2]_5$	5f	$(5/2)^2[9/2]^{\circ}_5$	7.7e+06	D+	F_Re	TW	
10745.703(12)	9303.497(11)	10745.703(3)	−0.001	110		5d	$(5/2)^2[9/2]_5$	5f	$(5/2)^2[9/2]^{\circ}_4$			R1c		
10752.044(14)	9298.010(12)	10752.046(3)	−0.002	53		5d	$(5/2)^2[9/2]_5$	5f	$(5/2)^2[7/2]^{\circ}_4$			R1c		
10768.359(14)	9283.923(12)	10768.360(4)	−0.001	210		5d	$(3/2)^2[7/2]_3$	5f	$(3/2)^2[7/2]^{\circ}_3$	5.7e+06	D+	R1c	TW	
10787.828(10)	9267.168(9)	10787.827(3)	0.001	210		5d	$(5/2)^2[3/2]_2$	5f	$(5/2)^2[7/2]^{\circ}_3$	1.9e+06	D+	R1c	TW	
10791.529(10)	9263.990(9)	10791.519(3)	0.010	1400		5d	$(5/2)^2[9/2]_4$	5f	$(5/2)^2[9/2]^{\circ}_4$	7.e+06	D+	R1c	TW	
10797.923(12)	9258.504(11)	10797.915(3)	0.008	160		5d	$(5/2)^2[9/2]_4$	5f	$(5/2)^2[7/2]^{\circ}_4$			R1c		
10800.96(3)	9255.90(3)	10800.929(3)	0.03	110	*	5d	$(5/2)^2[9/2]_4$	5f	$(5/2)^2[7/2]^{\circ}_3$			R1c		
10800.985(12)	9255.879(10)	10800.972(7)	0.014	110	*	sp	$(^3F)^3P^{\circ}\ ^5D^{\circ}_3$	4d	$(3/2)^2[5/2]_2$			F_Re		
10836.3452(6)	9225.6765(5)	10836.3453(6)	−0.0001	8300		5d	$(5/2)^2[9/2]_5$	5f	$(5/2)^2[11/2]^{\circ}_6$	4.2e+07	D+	F_Re	TW	
10849.473(3)	9214.514(3)	10849.477(3)	−0.004	6300		5d	$(3/2)^2[7/2]_3$	5f	$(3/2)^2[9/2]^{\circ}_4$	4.1e+07	D+	F_Re	TW	
10852.407(10)	9212.022(9)	10852.401(4)	0.007	2100		5d	$(5/2)^2[3/2]_1$	5f	$(5/2)^2[5/2]^{\circ}_2$	1.8e+07	D+	R1c	TW	
10865.050(17)	9201.303(15)	10865.010(5)	0.039	100		5d	$(5/2)^2[3/2]_2$	5f	$(5/2)^2[3/2]^{\circ}_1$			R1c		
10885.074(16)	9184.376(14)	10885.066(5)	0.008	200	*	sp	$(^1G)^3P^{\circ}\ ^3F^{\circ}_3$	6d	$(3/2)^2[7/2]_4$	2.1e+06	D+	R1c	TW	
10885.074(16)	9184.376(14)	10885.064(8)	0.010	200	*	sp	$(^3F)^1P^{\circ}\ ^3D^{\circ}_1$	6d	$(5/2)^2[1/2]_0$			R1c		
10889.9718(12)	9180.2457(11)	10889.9720(12)	−0.0003	6800		5d	$(5/2)^2[9/2]_4$	5f	$(5/2)^2[11/2]^{\circ}_5$	4.2e+07	D+	F_Re	TW	
10904.886(9)	9167.691(7)	10904.902(5)	−0.016	2600		5d	$(3/2)^2[3/2]_1$	5f	$(3/2)^2[5/2]^{\circ}_2$	2.8e+07	D+	F_Re	TW	
10925.167(13)	9150.672(11)	10925.155(4)	0.012	97		5d	$(3/2)^2[7/2]_4$	5f	$(3/2)^2[9/2]^{\circ}_4$	1.3e+06	D+	R1c	TW	
10929.791(10)	9146.800(9)	10929.799(4)	−0.008	480		5d	$(5/2)^2[3/2]_2$	5f	$(5/2)^2[1/2]^{\circ}_1$	1.2e+07	D+	R1c	TW	
10930.9327(21)	9145.8452(17)	10930.9342(19)	−0.0015	5300		5d	$(3/2)^2[7/2]_4$	5f	$(3/2)^2[9/2]^{\circ}_5$	4.2e+07	D+	F_Re	TW	
10959.809(11)	9121.748(9)	10959.819(5)	−0.010	4600		5d	$(3/2)^2[3/2]_2$	5f	$(3/2)^2[7/2]^{\circ}_3$	3.0e+06	D+	R1c	TW	
10964.502(11)	9117.844(9)	10964.527(5)	−0.025	3200		5d	$(3/2)^2[3/2]_1$	5f	$(3/2)^2[3/2]^{\circ}_1$	1.0e+07	D+	R1c	TW	
10966.532(11)	9116.156(9)	10966.524(5)	0.008	900		5d	$(5/2)^2[3/2]_1$	5f	$(5/2)^2[1/2]^{\circ}_0$	7.e+06	D+	R1c	TW	
10973.938(5)	9110.004(4)	10973.944(3)	−0.006	3100		5d	$(5/2)^2[5/2]_3$	5f	$(5/2)^2[9/2]^{\circ}_4$	5.5e+06	D+	F_Re	TW	
10980.553(5)	9104.516(4)	10980.559(3)	−0.006	4700		5d	$(5/2)^2[5/2]_3$	5f	$(5/2)^2[7/2]^{\circ}_4$	2.0e+07	D+	F_Re	TW	
10983.677(10)	9101.926(8)	10983.675(3)	0.002	1600		5d	$(5/2)^2[5/2]_3$	5f	$(5/2)^2[7/2]^{\circ}_3$	3.9e+06	D+	R1c	TW	
11008.410(12)	9081.477(10)	11008.408(4)	0.002	3300		sp	$(^3F)^1P^{\circ}\ ^3D^{\circ}_3$	7s	$(5/2)^2[5/2]_3$			R1c		
11009.932(6)	9080.221(5)	11009.932(4)	−0.000	3200		5d	$(3/2)^2[3/2]_2$	5f	$(3/2)^2[5/2]^{\circ}_3$	2.9e+07	D+	F_Re	TW	
11027.380(13)	9065.854(11)	11027.388(3)	−0.008	190		5d	$(5/2)^2[5/2]_3$	5f	$(5/2)^2[5/2]^{\circ}_2$	1.6e+06	D+	R1c	TW	
11035.977(10)	9058.792(8)	11035.978(4)	−0.002	2700		5d	$(5/2)^2[5/2]_3$	5f	$(5/2)^2[5/2]^{\circ}_3$	1.6e+07	D+	R1c	TW	
11098.345(11)	9007.885(9)	11098.352(3)	−0.006	150		5d	$(5/2)^2[5/2]_3$	5f	$(5/2)^2[3/2]^{\circ}_2$	4.5e+06	D+	R1c	TW	
11100.523(11)	9006.118(9)	11100.529(5)	−0.006	150		5d	$(3/2)^2[3/2]_2$	5f	$(3/2)^2[3/2]^{\circ}_2$	1.1e+07	D+	R1c	TW	
11114.771(5)	8994.573(4)	11114.773(3)	−0.001	2500		5d	$(5/2)^2[7/2]_3$	5f	$(5/2)^2[9/2]^{\circ}_4$	2.4e+07	D+	F_Re	TW	

Table A1. *Cont.*

λ_{obs} [a] (Å)	σ_{obs} [b] (cm^{-1})	λ_{Ritz} [c] (Å)	$\Delta\lambda_{obs\text{-}Ritz}$ (Å)	I_{obs} [d] (arb. u.)	Char [e]	Lower Level		Upper Level		A (s^{-1})	Acc [f]	Line Ref. [g]	TP Ref. [g]	Notes [h]
11121.559(10)	8989.083(8)	11121.558(3)	0.001	3300		5d	$(5/2)^2[7/2]_3$	5f	$(5/2)^2[7/2]^\circ_4$	9.e+06	E	R1c	TW	
11123.1488(20)	8987.7987(16)	11123.1487(18)	0.0001	6400		5d	$(5/2)^2[7/2]_4$	5f	$(5/2)^2[9/2]^\circ_5$	3.3e+07	D+	F_Re	TW	
11124.761(11)	8986.496(9)	11124.755(4)	0.006	2400		5d	$(5/2)^2[7/2]_3$	5f	$(5/2)^2[7/2]^\circ_3$	1.2e+07	D+	R1c	TW	
11125.451(5)	8985.939(4)	11125.454(4)	−0.003	3400		5d	$(3/2)^2[5/2]_3$	5f	$(3/2)^2[7/2]^\circ_4$	3.5e+07	D+	F_Re	TW	
11126.754(11)	8984.886(9)	11126.756(3)	−0.001	2100		5d	$(5/2)^2[7/2]_4$	5f	$(5/2)^2[9/2]^\circ_4$	4.5e+06	D+	R1c	TW	
11133.556(12)	8979.397(9)	11133.556(3)	0.000	2300		5d	$(5/2)^2[7/2]_4$	5f	$(5/2)^2[7/2]^\circ_4$	9.e+06	D+	R1c	TW	
11141.872(21)	8972.695(17)	11141.845(5)	0.027	490		sp	$(^3F)^1P^\circ\ ^3D^\circ_1$	7s	$(3/2)^2[3/2]_1$			R1c		
11156.494(10)	8960.935(8)	11156.492(5)	0.002	3400		5d	$(3/2)^2[5/2]_2$	5f	$(3/2)^2[7/2]^\circ_3$	3.2e+07	D+	R1c	TW	
11173.078(11)	8947.635(8)	11173.081(4)	−0.003	2200		5d	$(5/2)^2[5/2]_2$	5f	$(5/2)^2[7/2]^\circ_3$	2.2e+07	D+	F_Re	TW	
11180.070(13)	8942.039(11)	11180.072(4)	−0.002	800		5d	$(3/2)^2[5/2]_3$	5f	$(3/2)^2[5/2]^\circ_3$	1.0e+07	D+	R1c	TW	
11190.552(13)	8933.663(11)	11190.534(4)	0.018	94		5d	$(5/2)^2[7/2]_4$	5f	$(5/2)^2[5/2]^\circ_3$	2.0e+06	D+	R1c	TW	
11202.751(12)	8923.935(9)	11202.717(5)	0.034	350		5d	$(3/2)^2[5/2]_2$	5f	$(3/2)^2[5/2]^\circ_2$	1.1e+07	D+	R1c	TW	
11208.490(13)	8919.366(11)	11208.425(5)	0.065	84		5d	$(3/2)^2[5/2]_2$	5f	$(3/2)^2[5/2]^\circ_3$	1.2e+06	D+	R1c	TW	X
11218.324(11)	8911.547(9)	11218.317(4)	0.007	1200		5d	$(5/2)^2[5/2]_2$	5f	$(5/2)^2[5/2]^\circ_2$	1.6e+07	D+	R1c	TW	
11227.209(11)	8904.495(9)	11227.208(4)	0.000	1100		5d	$(5/2)^2[5/2]_2$	5f	$(5/2)^2[5/2]^\circ_3$	5.4e+06	D+	R1c	TW	
23063.7	4335.82	23063.701(9)			:	4s	$^3D_3$	4s	$^1D_2$	2.0e−01	C+		TW,G64	M1
29261.7	3417.43	29261.733(10)			:	4s	$^3D_2$	4s	$^1D_2$	1.55e−02	C+		TW,G64	M1
44127.2	2266.18	44127.15(3)			:	4s	$^3D_1$	4s	$^1D_2$	2.7e−02	C+		TW,G64	M1
48317.6	2069.639	48317.60(4)			:	4s	$^3D_3$	4s	$^3D_1$	9.e−08	E		TW	E2
86861.8	1151.254	86861.79(7)			:	4s	$^3D_2$	4s	$^3D_1$	5.6e−02	C+		TW,G64	M1
108887	918.385	108886.80(16)			:	4s	$^3D_3$	4s	$^3D_2$	1.8e−02	C+		TW,G64	M1

[a] Observed wavelength between 2000 Å and 20000 Å is given in standard air; outside of this region, it is in vacuum. The standard uncertainty in the last decimal place is given in parentheses after the value. Conversion from air to vacuum was made using the five-parameter formula from Peck and Reeder [38];
[b] Observed wavenumber in vacuum;
[c] Ritz wavelength and its uncertainty were obtained in the least-squares level optimization procedure using the LOPT code [39]. For lines that alone determine one of the energy levels of the transition, this column is blank;
[d] Observed intensities from different experiments have been normalized to a uniform scale (see text). They are proportional to the energy flux under the line profile and have uncertainties of a factor of three on average. Intensities of parity-forbidden transitions are given on a different scale, since most of them were observed only in nebulas;
[e] Line character code: bl—blended line; p—perturbed by a close line; *—the given intensity value is shared by two or more transitions; m—masked by another strong line (no wavelength measurement available);—the value given in the observed wavelength column is a rounded Ritz wavelength (no wavelength measurement available); ?—questionable identification;
[f] Transition probability accuracy code: A+—transition probability uncertainty is likely ≤2%; B+—≤7%; B—≤7%; C+—≤18%; C—≤25%; D—≤50%; E—>50%;
[g] Key to observed wavelength and transition probability references: A73—Aller et al. [4]; A08—Andersson et al. [40]; B00—Biémont et al. [41]; B09—Brown et al. [42]; C84—Cederquist et al. [43]; C94—Crespo López-Urrutia et al. [44]; D05—Dong and Fritzsche [45]; G64—Garstang [46]; H71—Hefferlin et al. [47]; K66—Kaufman and Ward [15]; K82—Kono and Hattori [48]; M97—McKenna et al. [5]; N88—Neger and Jäger [49]; O07—Ortiz et al. [50]; P84—Prior [37]; P97—Pinnington et al. [51]; R1—Ross [2]; R2—Ross [16]; S36—Shenstone [1]; T53—Thackeray [3]; W93—Wagatsuma and Hirokawa [36]; F—this work, FTS measurements with Cu/Ge/Pt/Ar hollow cathode; F_Re—this work, measurements with Cu/Re/Ar hollow cathode; TW—this work, grating measurements. Lower-case letters after the reference have the following meaning: c—corrected in this work; n—new identification; r—revised identification; cal—calculated *A*-value; se—*A*-value was semiempirically adjusted by ratio of observed and calculated lifetime;
[h] Notes: X—excluded from level-optimization procedure; L—lasing line; M1—magnetic-dipole transition; E2—electric-quadrupole transition; HF—hyperfine-induced transition.

Table A2. Energy levels of Cu II.

Label [a]		Configuration	Term	J	Level [b], cm^{-1}	Unc. [c], cm^{-1}	Landé g [d]	Leading Percentages [e]	Note [f]	N_{lines} [g]
d^{10}	^1S	$3d^{10}$	^1S	0	0.000	0.017		97% + 2% 4d ^1S		15
4s	^3D	$3d^94s$	^3D	3	21928.7326	0.0014	1.32	98%		59
4s	^3D	$3d^94s$	^3D	2	22847.1176	–	1.16	89% + 9% 4s ^1D		77
4s	^3D	$3d^94s$	^3D	1	23998.3718	0.0009	0.48	98%		48
4s	^1D	$3d^94s$	^1D	2	26264.5502	0.0012	1.00	89% + 9% 4s ^3D		64
4p	$^3P^o$	$3d^94p$	$^3P^o$	2	66418.6849	0.0014	1.49	96% + 2% 4p $^3D^o$		31
4p	$^3P^o$	$3d^94p$	$^3P^o$	1	67916.5572	0.0011	1.49	95% + 2% 4p $^1P^o$		42
4p	$^3F^o$	$3d^94p$	$^3F^o$	3	68447.7349	0.0013	1.06	62% + 34% 4p $^1F^o$		32
4p	$^3F^o$	$3d^94p$	$^3F^o$	4	68730.8876	0.0017	1.23	98%		23
4p	$^3P^o$	$3d^94p$	$^3P^o$	0	68850.2628	0.0014		98%		12
s^2	^3F	$3d^84s^2$	^3F	4	69704.7015	0.0019		95% + 4% p^2 $(^3F)^1S$ ^3F		32
4p	$^3F^o$	$3d^94p$	$^3F^o$	2	69867.9849	0.0011	0.67	88% + 6% 4p $^3D^o$		37
4p	$^1F^o$	$3d^94p$	$^1F^o$	3	70841.4669	0.0014		49% + 31% 4p $^3D^o$ + 17% 4p $^3F^o$		40
4p	$^1D^o$	$3d^94p$	$^1D^o$	2	71493.8548	0.0007	1.08	54% + 35% 4p $^3D^o$ + 9% 4p $^3F^o$		36
s^2	^3F	$3d^84s^2$	^3F	3	71531.542	0.003		95% + 4% p^2 $(^3F)^1S$ ^3F		57
4p	$^3D^o$	$3d^94p$	$^3D^o$	3	71920.0961	0.0008		64% + 19% 4p $^3F^o$ + 15% 4p $^1F^o$		36
s^2	^3F	$3d^84s^2$	^3F	2	72723.817	0.004		94% + 4% p^2 $(^3F)^1S$ ^3F		52
4p	$^3D^o$	$3d^94p$	$^3D^o$	1	73102.0408	0.0007	0.47	92% + 3% 4p $^1P^o$		31
4p	$^3D^o$	$3d^94p$	$^3D^o$	2	73353.2957	0.0008	0.99	54% + 41% 4p $^1D^o$		39
4p	$^1P^o$	$3d^94p$	$^1P^o$	1	73595.8143	0.0018	1.04	93% + 4% 4p $^3D^o$		36
s^2	^1D	$3d^84s^2$	^1D	2	85388.772	0.003		73% + 21% s^2 ^3P		49
s^2	^3P	$3d^84s^2$	^3P	2	88362.001	0.003		74% + 21% s^2 ^1D		44
s^2	^3P	$3d^84s^2$	^3P	1	88605.096	0.004		95% + 4% p^2 $(^3P)^1S$ ^3P		38
s^2	^3P	$3d^84s^2$	^3P	0	88926.002	0.003		95% + 4% p^2 $(^3P)^1S$ ^3P		17
s^2	^1G	$3d^84s^2$	^1G	4	95565.619	0.003	0.98	95% + 4% p^2 $(^1G)^1S$ ^1G		18
sp	$(^3F)^3P^o$ $^5D^o$	$3d^8(^3F)4s4p(^3P^o)$	$^5D^o$	4	107942.795	0.010		93% + 4% sp $(^3P)^3P^o$ $^5D^o$		2
5s	$(5/2)^2[5/2]$	$3d^9(^2D_{5/2})5s$	$^2[5/2]$	3	108014.8372	0.0012		100%		48
5s	$(5/2)^2[5/2]$	$3d^9(^2D_{5/2})5s$	$^2[5/2]$	2	108335.6078	0.0012		98% + 2% 5s $(3/2)^2[3/2]$		48
sp	$(^3F)^3P^o$ $^5D^o$	$3d^8(^3F)4s4p(^3P^o)$	$^5D^o$	3	109276.015	0.006		91% + 4% sp $(^3P)^3P^o$ $^5D^o$		8
5s	$(3/2)^2[3/2]$	$3d^9(^2D_{3/2})5s$	$^2[3/2]$	1	110084.4773	0.0011		100%		34
sp	$(^3F)^3P^o$ $^5D^o$	$3d^8(^3F)4s4p(^3P^o)$	$^5D^o$	2	110363.725	0.008		91% + 5% sp $(^3P)^3P^o$ $^5D^o$		4
5s	$(3/2)^2[3/2]$	$3d^9(^2D_{3/2})5s$	$^2[3/2]$	2	110366.1542	0.0011		98% + 2% 5s $(5/2)^2[5/2]$		47
sp	$(^3F)^3P^o$ $^5G^o$	$3d^8(^3F)4s4p(^3P^o)$	$^5G^o$	5	110631.196	0.009		82% + 13% sp $(^3F)^3P^o$ $^5F^o$		1
sp	$(^3F)^3P^o$ $^5D^o$	$3d^8(^3F)4s4p(^3P^o)$	$^5D^o$	1	111124.39	0.14		93% + 5% sp $(^3P)^3P^o$ $^5D^o$		1
sp	$(^3F)^3P^o$ $^5G^o$	$3d^8(^3F)4s4p(^3P^o)$	$^5G^o$	4	111218.705	0.008		83% + 10% sp $(^3F)^3P^o$ $^5F^o$ + 5% sp $(^3F)^3P^o$ $^3G^o$		2
sp	$(^3F)^3P^o$ $^5G^o$	$3d^8(^3F)4s4p(^3P^o)$	$^5G^o$	3	111876.412	0.008		89% + 7% sp $(^3F)^3P^o$ $^5F^o$		3
sp	$(^3F)^3P^o$ $^5F^o$	$3d^8(^3F)4s4p(^3P^o)$	$^5F^o$	5	112401.632	0.022		86% + 12% sp $(^3F)^3P^o$ $^5G^o$		1
sp	$(^3F)^3P^o$ $^5G^o$	$3d^8(^3F)4s4p(^3P^o)$	$^5G^o$	2	112424.679	0.023		95% + 3% sp $(^3F)^3P^o$ $^5F^o$		2
sp	$(^3F)^3P^o$ $^5F^o$	$3d^8(^3F)4s4p(^3P^o)$	$^5F^o$	4	113302.823	0.006		84% + 9% sp $(^3F)^3P^o$ $^5G^o$		5
sp	$(^3F)^3P^o$ $^5F^o$	$3d^8(^3F)4s4p(^3P^o)$	$^5F^o$	3	114000.452	0.007		86% + 7% sp $(^3F)^3P^o$ $^5G^o$		5

Table A2. *Cont.*

Label [a]	Configuration	Term	J	Level [b], cm^{-1}	Unc. [c], cm^{-1}	Landé g [d]	Leading Percentages [e]	Note [f]	N_{lines} [g]
sp $(^3F)^3P^\circ\ ^5F^\circ$	$3d^8(^3F)4s4p(^3P^\circ)$	$^5F^\circ$	2	114481.674	0.006		92% + 3% sp $(^3F)^3P^\circ\ ^5G^\circ$		6
4d $(5/2)^2[1/2]$	$3d^9(^2D_{5/2})4d$	$^2[1/2]$	1	114511.2386	0.0012		91% + 7% 4d $(3/2)^2[1/2]$ + 1% 4d2		39
sp $(^3F)^3P^\circ\ ^5F^\circ$	$3d^8(^3F)4s4p(^3P^\circ)$	$^5F^\circ$	1	114755.953	0.011		97%		3
sp $(^3F)^3P^\circ\ ^3G^\circ$	$3d^8(^3F)4s4p(^3P^\circ)$	$^3G^\circ$	4	115359.532	0.005		71% + 20% sp $(^3F)^3P^\circ\ ^1G^\circ$ + 6% sp $(^3F)^3P^\circ\ ^5G^\circ$		7
sp $(^3F)^3P^\circ\ ^3G^\circ$	$3d^8(^3F)4s4p(^3P^\circ)$	$^3G^\circ$	5	115546.114	0.011		93% + 6% sp $(^3F)^3P^\circ\ ^5G^\circ$		1
4d $(5/2)^2[9/2]$	$3d^9(^2D_{5/2})4d$	$^2[9/2]$	5	115568.99497	0.0012		99%		22
4d $(5/2)^2[3/2]$	$3d^9(^2D_{5/2})4d$	$^2[3/2]$	2	115638.8036	0.0011		93% + 4% 4d $(5/2)^2[5/2]$ + 2% 4d $(3/2)^2[3/2]$		42
4d $(5/2)^2[9/2]$	$3d^9(^2D_{5/2})4d$	$^2[9/2]$	4	115662.5622	0.0014		97% + 1% 4d $(5/2)^2[7/2]$		32
4d $(5/2)^2[3/2]$	$3d^9(^2D_{5/2})4d$	$^2[3/2]$	1	115665.1539	0.0011		97% + 1% 4d $(3/2)^2[1/2]$ + 1% 4d $(3/2)^2[3/2]$		30
4d $(5/2)^2[5/2]$	$3d^9(^2D_{5/2})4d$	$^2[5/2]$	3	116080.2237	0.0010		97% + 1% 4d $(5/2)^2[7/2]$ + 1% 4d $(3/2)^2[5/2]$		35
4d $(5/2)^2[7/2]$	$3d^9(^2D_{5/2})4d$	$^2[7/2]$	3	116325.9148	0.0011		93% + 4% 4d $(3/2)^2[7/2]$ + 1% 4d $(5/2)^2[5/2]$		37
4d $(5/2)^2[7/2]$	$3d^9(^2D_{5/2})4d$	$^2[7/2]$	4	116371.18040	0.0011		95% + 2% 4d $(3/2)^2[7/2]$ + 1% 4d $(5/2)^2[9/2]$		35
sp $(^3F)^3P^\circ\ ^3D^\circ$	$3d^8(^3F)4s4p(^3P^\circ)$	$^3D^\circ$	3	116375.406	0.003		44% + 35% sp $(^3F)^3P^\circ\ ^3G^\circ$ + 13% sp $(^3F)^3P^\circ\ ^3F^\circ$		17
4d $(5/2)^2[5/2]$	$3d^9(^2D_{5/2})4d$	$^2[5/2]$	2	116387.7873	0.0010		92% + 3% 4d $(3/2)^2[3/2]$ + 3% 4d $(5/2)^2[3/2]$		43
4d $(5/2)^2[1/2]$	$3d^9(^2D_{5/2})4d$	$^2[1/2]$	0	116576.5758	0.0018		53% + 45% 4d $(3/2)^2[1/2]$		16
sp $(^3F)^3P^\circ\ ^3G^\circ$	$3d^8(^3F)4s4p(^3P^\circ)$	$^3G^\circ$	3	116643.960	0.003		57% + 36% sp $(^3F)^3P^\circ\ ^3D^\circ$		17
sp $(^3F)^3P^\circ\ ^3D^\circ$	$3d^8(^3F)4s4p(^3P^\circ)$	$^3D^\circ$	2	117130.340	0.003		75% + 10% sp $(^3F)^3P^\circ\ ^3F^\circ$		27
4d $(3/2)^2[1/2]$	$3d^9(^2D_{3/2})4d$	$^2[1/2]$	1	117231.4014	0.0019		91% + 7% 4d $(5/2)^2[1/2]$		28
sp $(^3F)^3P^\circ\ ^3F^\circ$	$3d^8(^3F)4s4p(^3P^\circ)$	$^3F^\circ$	4	117666.626	0.005		88% + 3% sp $(^3F)^3P^\circ\ ^5F^\circ$		15
4d $(3/2)^2[7/2]$	$3d^9(^2D_{3/2})4d$	$^2[7/2]$	3	117747.3504	0.0015		95% + 4% 4d $(5/2)^2[7/2]$		37
4d $(3/2)^2[7/2]$	$3d^9(^2D_{3/2})4d$	$^2[7/2]$	4	117883.0985	0.0013		96% + 2% 4d $(5/2)^2[7/2]$		35
4d $(3/2)^2[3/2]$	$3d^9(^2D_{3/2})4d$	$^2[3/2]$	1	117928.2197	0.0018		98%		31
sp $(^3F)^3P^\circ\ ^3D^\circ$	$3d^8(^3F)4s4p(^3P^\circ)$	$^3D^\circ$	1	118071.302	0.003		87% + 6% sp $(^1D)^3P^\circ\ ^3D^\circ$		22
sp $(^3F)^3P^\circ\ ^3F^\circ$	$3d^8(^3F)4s4p(^3P^\circ)$	$^3F^\circ$	3	118142.950	0.003		63% + 14% sp $(^3F)^3P^\circ\ ^1F^\circ$ + 10% sp $(^3F)^3P^\circ\ ^3D^\circ$		32
4d $(3/2)^2[3/2]$	$3d^9(^2D_{3/2})4d$	$^2[3/2]$	2	118163.2663	0.0015		82% + 12% 4d $(3/2)^2[5/2]$ + 2% 4d $(5/2)^2[5/2]$		42
4d $(3/2)^2[5/2]$	$3d^9(^2D_{3/2})4d$	$^2[5/2]$	3	118483.8135	0.0015		98% + 1% 4d $(5/2)^2[5/2]$		37
4d $(3/2)^2[5/2]$	$3d^9(^2D_{3/2})4d$	$^2[5/2]$	2	118531.9058	0.0016		86% + 12% 4d $(3/2)^2[3/2]$ + 1% 4d $(5/2)^2[3/2]$		35
sp $(^3F)^3P^\circ\ ^1G^\circ$	$3d^8(^3F)4s4p(^3P^\circ)$	$^1G^\circ$	4	118991.330	0.007		75% + 20% sp $(^3F)^3P^\circ\ ^3G^\circ$		9
sp $(^3F)^3P^\circ\ ^3F^\circ$	$3d^8(^3F)4s4p(^3P^\circ)$	$^3F^\circ$	2	119039.6355	0.0019		82% + 8% sp $(^3F)^3P^\circ\ ^3D^\circ$		34
5p $(5/2)^2[3/2]^\circ$	$3d^9(^2D_{5/2})5p$	$^2[3/2]^\circ$	2	120092.3828	0.0012		96% + 2% sp + 1% 5p $(3/2)^2[3/2]^\circ$		40
5p $(5/2)^2[7/2]^\circ$	$3d^9(^2D_{5/2})5p$	$^2[7/2]^\circ$	3	120684.7128	0.0013		86% + 13% sp		33
5p $(5/2)^2[7/2]^\circ$	$3d^9(^2D_{5/2})5p$	$^2[7/2]^\circ$	4	120789.80865	0.0013		95% + 4% sp		20
5p $(5/2)^2[5/2]^\circ$	$3d^9(^2D_{5/2})5p$	$^2[5/2]^\circ$	2	120876.0141	0.0015		51% + 41% sp + 3% 5p $(3/2)^2[3/2]^\circ$		41
5p $(5/2)^2[3/2]^\circ$	$3d^9(^2D_{5/2})5p$	$^2[3/2]^\circ$	1	120919.5715	0.0012		83% + 15% 5p $(3/2)^2[1/2]^\circ$ + 1% sp		43
5p $(5/2)^2[5/2]^\circ$	$3d^9(^2D_{5/2})5p$	$^2[5/2]^\circ$	3	121079.1501	0.0012		54% + 35% sp + 3% 5p $(3/2)^2[5/2]^\circ$		45
sp $(^3F)^3P^\circ\ ^1F^\circ$	$3d^8(^3F)4s4p(^3P^\circ)$	$^1F^\circ$	3	121524.8509	0.0014		39% + 43% 5p $^3D^\circ$ + 6% sp $(^3F)^3P^\circ\ ^3F^\circ$	cd	37
sp $(^3F)^3P^\circ\ ^1D^\circ$	$3d^8(^3F)4s4p(^3P^\circ)$	$^1D^\circ$	2	121981.8546	0.0015		36% + 32% 5p $^3D^\circ$ + 22% 5p $^3F^\circ$		44
5p $(3/2)^2[1/2]^\circ$	$3d^9(^2D_{3/2})5p$	$^2[1/2]^\circ$	0	122224.0199	0.0020		99%		13
4d $(3/2)^2[1/2]$	$3d^9(^2D_{3/2})4d$	$^2[1/2]$	0	122415.957	0.003		45% + 34% 4d $(5/2)^2[1/2]$ + 10% 5d $(5/2)^2[1/2]$		12
5p $(3/2)^2[5/2]^\circ$	$3d^9(^2D_{3/2})5p$	$^2[5/2]^\circ$	2	122745.9491	0.0013		85% + 4% 5p $(3/2)^2[3/2]^\circ$ + 4% 5p $(5/2)^2[5/2]^\circ$		39

Table A2. *Cont.*

	Label [a]	Configuration	Term	J	Level [b], cm^{-1}	Unc. [c], cm^{-1}	Landé g [d]	Leading Percentages [e]	Note [f]	N_{lines} [g]
5p	$(3/2)^2[1/2]^\circ$	$3d^9(^2D_{3/2})5p$	$^2[1/2]^\circ$	1	122867.7407	0.0015		75% + 14% 5p $(5/2)^2[3/2]^\circ$ + 9% 5p $(3/2)^2[3/2]^\circ$		39
5p	$(3/2)^2[5/2]^\circ$	$3d^9(^2D_{3/2})5p$	$^2[5/2]^\circ$	3	123016.8175	0.0014		91% + 7% sp + 1% 5p $(5/2)^2[7/2]^\circ$		27
5p	$(3/2)^2[3/2]^\circ$	$3d^9(^2D_{3/2})5p$	$^2[3/2]^\circ$	1	123304.823	0.003		86% + 9% 5p $(3/2)^2[1/2]^\circ$ + 4% sp		27
5p	$(3/2)^2[3/2]^\circ$	$3d^9(^2D_{3/2})5p$	$^2[3/2]^\circ$	2	123556.8261	0.0016		76% + 13% sp + 3% 5p $(5/2)^2[5/2]^\circ$		43
sp	$(^3P)^3P^\circ\ ^5P^\circ$	$3d^8(^3P)4s4p(^3P^\circ)$	$^5P^\circ$	3	125230.061	0.017		88% + 8% sp $(^1D)^3P^\circ\ ^3D^\circ$		6
sp	$(^3P)^3P^\circ\ ^5P^\circ$	$3d^8(^3P)4s4p(^3P^\circ)$	$^5P^\circ$	2	125248.121	0.008		88% + 3% sp $(^1D)^3P^\circ\ ^3P^\circ$		11
sp	$(^3P)^3P^\circ\ ^5P^\circ$	$3d^8(^3P)4s4p(^3P^\circ)$	$^5P^\circ$	1	125569.33	0.04		94% + 2% sp $(^1D)^3P^\circ\ ^3P^\circ$		5
sp	$(^1D)^3P^\circ\ ^3F^\circ$	$3d^8(^1D)4s4p(^3P^\circ)$	$^3F^\circ$	2	128365.736	0.006		69% + 15% sp $(^1D)^3P^\circ\ ^3D^\circ$		10
sp	$(^1D)^3P^\circ\ ^3F^\circ$	$3d^8(^1D)4s4p(^3P^\circ)$	$^3F^\circ$	3	128559.314	0.008		68% + 11% sp $(^3P)^3P^\circ\ ^5D^\circ$ + 9% sp $(^1D)^3P^\circ\ ^3D^\circ$		6
sp	$(^1D)^3P^\circ\ ^3D^\circ$	$3d^8(^1D)4s4p(^3P^\circ)$	$^3D^\circ$	1	128569.150	0.008		66% + 9% sp $(^1D)^3P^\circ\ ^3P^\circ$ + 9% sp $(^3P)^3P^\circ\ ^3P^\circ$		7
sp	$(^1D)^3P^\circ\ ^3F^\circ$	$3d^8(^1D)4s4p(^3P^\circ)$	$^3F^\circ$	4	128778.037	0.017		64% + 28% sp $(^3P)^3P^\circ\ ^5D^\circ$		2
sp	$(^1D)^3P^\circ\ ^3D^\circ$	$3d^8(^1D)4s4p(^3P^\circ)$	$^3D^\circ$	2	128854.036	0.008		59% + 19% sp $(^1D)^3P^\circ\ ^3F^\circ$		8
s^2	^1S	$3d^84s^2$	^1S	0	128910.03	0.06		94% + 5% $p^2\ (^1S)^1S\ ^1S$	N	1
sp	$(^1D)^3P^\circ\ ^3P^\circ$	$3d^8(^1D)4s4p(^3P^\circ)$	$^3P^\circ$	0	129105.778	0.009		63% + 32% sp $(^3P)^3P^\circ\ ^3P^\circ$		3
sp	$(^1D)^3P^\circ\ ^3D^\circ$	$3d^8(^1D)4s4p(^3P^\circ)$	$^3D^\circ$	3	129116.774	0.011		69% + 10% sp $(^1D)^3P^\circ\ ^3F^\circ$ + 7% sp $(^3P)^3P^\circ\ ^5P^\circ$		6
sp	$(^1D)^3P^\circ\ ^3P^\circ$	$3d^8(^1D)4s4p(^3P^\circ)$	$^3P^\circ$	1	129759.750	0.006		56% + 19% sp $(^1D)^3P^\circ\ ^3D^\circ$ + 18% sp $(^3P)^3P^\circ\ ^3P^\circ$		9
sp	$(^1D)^3P^\circ\ ^3P^\circ$	$3d^8(^1D)4s4p(^3P^\circ)$	$^3P^\circ$	2	130386.404	0.016		74% + 12% sp $(^3P)^3P^\circ\ ^3P^\circ$ + 8% sp $(^1D)^3P^\circ\ ^3D^\circ$		8
sp	$(^3P)^3P^\circ\ ^5D^\circ$	$3d^8(^3P)4s4p(^3P^\circ)$	$^5D^\circ$	1	130940.488	0.008		91% + 5% sp $(^3F)^3P^\circ\ ^5D^\circ$		8
sp	$(^3P)^3P^\circ\ ^5D^\circ$	$3d^8(^3P)4s4p(^3P^\circ)$	$^5D^\circ$	2	130943.661	0.009		88% + 5% sp $(^3F)^3P^\circ\ ^5D^\circ$		9
sp	$(^3P)^3P^\circ\ ^5D^\circ$	$3d^8(^3P)4s4p(^3P^\circ)$	$^5D^\circ$	0	130953.558	0.016		91% + 6% sp $(^3F)^3P^\circ\ ^5D^\circ$	N	2
sp	$(^3P)^3P^\circ\ ^5D^\circ$	$3d^8(^3P)4s4p(^3P^\circ)$	$^5D^\circ$	3	131044.310	0.008		81% + 11% sp $(^1D)^3P^\circ\ ^3F^\circ$		10
sp	$(^3P)^3P^\circ\ ^5D^\circ$	$3d^8(^3P)4s4p(^3P^\circ)$	$^5D^\circ$	4	131312.426	0.013		66% + 25% sp $(^1D)^3P^\circ\ ^3F^\circ$ + 5% sp $(^1G)^3P^\circ\ ^3F^\circ$		5
6s	$(5/2)^2[5/2]$	$3d^9(^2D_{5/2})6s$	$^2[5/2]$	3	133594.2323	0.0013		100%		27
6s	$(5/2)^2[5/2]$	$3d^9(^2D_{5/2})6s$	$^2[5/2]$	2	133728.0387	0.0013		100%		29
sp	$(^3P)^3P^\circ\ ^3P^\circ$	$3d^8(^3P)4s4p(^3P^\circ)$	$^3P^\circ$	2	133825.927	0.004		55% + 22% sp $(^3P)^3P^\circ\ ^3D^\circ$ + 9% sp $(^1D)^3P^\circ\ ^3P^\circ$		13
sp	$(^3P)^3P^\circ\ ^3D^\circ$	$3d^8(^3P)4s4p(^3P^\circ)$	$^3D^\circ$	3	133984.325	0.004		47% + 31% sp $(^3F)^1P^\circ\ ^3D^\circ$ + 6% sp $(^1D)^3P^\circ\ ^3D^\circ$		14
sp	$(^3F)^1P^\circ\ ^3G^\circ$	$3d^8(^3F)4s4p(^1P^\circ)$	$^3G^\circ$	5	134110.870	0.004		93% + 2% 4f $^3G^\circ$		9
sp	$(^3P)^3P^\circ\ ^3D^\circ$	$3d^8(^3P)4s4p(^3P^\circ)$	$^3D^\circ$	1	134359.847	0.003		52% + 24% sp $(^3P)^3P^\circ\ ^3P^\circ$ + 10% sp $(^1D)^3P^\circ\ ^3P^\circ$		16
sp	$(^3P)^3P^\circ\ ^3D^\circ$	$3d^8(^3P)4s4p(^3P^\circ)$	$^3D^\circ$	2	134675.522	0.003		50% + 22% sp $(^3P)^3P^\circ\ ^3P^\circ$ + 7% sp $(^3F)^1P^\circ\ ^3D^\circ$		18
sp	$(^3F)^1P^\circ\ ^3F^\circ$	$3d^8(^3F)4s4p(^1P^\circ)$	$^3F^\circ$	4	134742.863	0.003		50% + 31% sp $(^3F)^1P^\circ\ ^3G^\circ$ + 7% sp $(^1G)^3P^\circ\ ^3F^\circ$	cd	13
sp	$(^3P)^3P^\circ\ ^3P^\circ$	$3d^8(^3P)4s4p(^3P^\circ)$	$^3P^\circ$	1	135135.168	0.003		38% + 32% sp $(^3P)^3P^\circ\ ^3D^\circ$ + 14% sp $(^1D)^3P^\circ\ ^3P^\circ$		10
sp	$(^3P)^3P^\circ\ ^3P^\circ$	$3d^8(^3P)4s4p(^3P^\circ)$	$^3P^\circ$	0	135484.075	0.004		60% + 31% sp $(^1D)^3P^\circ\ ^3P^\circ$ + 5% 4f $^3P^\circ$		6
6s	$(3/2)^2[3/2]$	$3d^9(^2D_{3/2})6s$	$^2[3/2]$	1	135664.5204	0.0015		100%		21
sp	$(^3F)^1P^\circ\ ^3D^\circ$	$3d^8(^3F)4s4p(^1P^\circ)$	$^3D^\circ$	3	135733.433	0.003		26% sp $(^3P)^3P^\circ\ ^3D^\circ$ + 14% sp $(^3F)^1P^\circ\ ^3F^\circ$ + 13% sp $(^3F)^1P^\circ$ $^3D^\circ$		20
6s	$(3/2)^2[3/2]$	$3d^9(^2D_{3/2})6s$	$^2[3/2]$	2	135760.1548	0.0016		99%		27
sp	$(^3F)^1P^\circ\ ^3G^\circ$	$3d^8(^3F)4s4p(^1P^\circ)$	$^3G^\circ$	4	135834.6720	0.0019		49% + 22% sp $(^3F)^1P^\circ\ ^3F^\circ$ + 9% 4f $^3G^\circ$		17
4f	$(5/2)^2[1/2]^\circ$	$3d^9(^2D_{5/2})4f$	$^2[1/2]^\circ$	1	135863.6857	0.0016		97% + 2% sp + 1% 4f $(5/2)^2[3/2]^\circ$		19
4f	$(5/2)^2[1/2]^\circ$	$3d^9(^2D_{5/2})4f$	$^2[1/2]^\circ$	0	135902.1365	0.0023		94% + 4% sp		8
4f	$(5/2)^2[3/2]^\circ$	$3d^9(^2D_{5/2})4f$	$^2[3/2]^\circ$	2	135910.7245	0.0011		94% + 4% 4f $(5/2)^2[5/2]^\circ$ + 2% sp		23

Table A2. *Cont.*

	Label [a]	Configuration	Term	J	Level [b], cm^{-1}	Unc. [c], cm^{-1}	Landé g [d]	Leading Percentages [e]	Note [f]	N_{lines} [g]
4f	$(5/2)^2[11/2]^\circ$	$3d^9(^2D_{5/2})4f$	$^2[11/2]^\circ$	6	135931.01412	0.0012		99% + 1% sp		5
4f	$(5/2)^2[11/2]^\circ$	$3d^9(^2D_{5/2})4f$	$^2[11/2]^\circ$	5	135933.89499	0.0019		98% + 1% sp		11
sp	$(^3P)^3P^\circ\ ^5S^\circ$	$3d^8(^3P)4s4p(^3P^\circ)$	$^5S^\circ$	2	135952.279	0.005		83% + 4% 4f $^1D^\circ$		12
4f	$(5/2)^2[3/2]^\circ$	$3d^9(^2D_{5/2})4f$	$^2[3/2]^\circ$	1	135958.1919	0.0016		97% + 2% sp + 1% 4f $(5/2)^2[1/2]^\circ$		20
4f	$(5/2)^2[7/2]^\circ$	$3d^9(^2D_{5/2})4f$	$^2[7/2]^\circ$	3	135989.9176	0.0012		41% + 37% sp + 11% 4f $(5/2)^2[5/2]^\circ$		33
4f	$(5/2)^2[5/2]^\circ$	$3d^9(^2D_{5/2})4f$	$^2[5/2]^\circ$	2	136013.9671	0.0011		84% + 9% sp + 5% 4f $(5/2)^2[3/2]^\circ$		17
4f	$(5/2)^2[5/2]^\circ$	$3d^9(^2D_{5/2})4f$	$^2[5/2]^\circ$	3	136035.3328	0.0010		86% + 11% 4f $(5/2)^2[7/2]^\circ$ + 3% sp		28
4f	$(5/2)^2[7/2]^\circ$	$3d^9(^2D_{5/2})4f$	$^2[7/2]^\circ$	4	136132.77781	0.0010		88% + 8% 4f $(5/2)^2[9/2]^\circ$ + 3% sp		26
4f	$(5/2)^2[9/2]^\circ$	$3d^9(^2D_{5/2})4f$	$^2[9/2]^\circ$	5	136160.61825	0.0011		98% + 2% sp		14
4f	$(5/2)^2[9/2]^\circ$	$3d^9(^2D_{5/2})4f$	$^2[9/2]^\circ$	4	136269.9996	0.0014		75% + 14% sp + 9% 4f $(5/2)^2[7/2]^\circ$		20
5d	$(5/2)^2[1/2]$	$3d^9(^2D_{5/2})5d$	$^2[1/2]$	1	136336.8971	0.0020		98% + 1% 5d $(3/2)^2[1/2]$		17
sp	$(^3F)^1P^\circ\ ^3F^\circ$	$3d^8(^3F)4s4p(^1P^\circ)$	$^3F^\circ$	3	136441.817	0.003		25% + 25% sp $(^3F)^1P^\circ\ ^3G^\circ$ + 11% sp $(^1G)^3P^\circ\ ^3F^\circ$		21
sp	$(^1G)^3P^\circ\ ^3H^\circ$	$3d^8(^1G)4s4p(^3P^\circ)$	$^3H^\circ$	4	136693.948	0.003		97%		10
5d	$(5/2)^2[9/2]$	$3d^9(^2D_{5/2})5d$	$^2[9/2]$	5	136725.85349	0.0013		100%		14
5d	$(5/2)^2[3/2]$	$3d^9(^2D_{5/2})5d$	$^2[3/2]$	2	136754.1104	0.0018		97% + 2% 5d $(5/2)^2[5/2]$		25
5d	$(5/2)^2[9/2]$	$3d^9(^2D_{5/2})5d$	$^2[9/2]$	4	136765.3514	0.0018		99%		20
5d	$(5/2)^2[3/2]$	$3d^9(^2D_{5/2})5d$	$^2[3/2]$	1	136773.1713	0.0021		99%		19
sp	$(^3F)^1P^\circ\ ^3D^\circ$	$3d^8(^3F)4s4p(^1P^\circ)$	$^3D^\circ$	2	136800.137	0.003		41% + 18% sp $(^3P)^3P^\circ\ ^1D^\circ$ + 14% 6p		14
5d	$(5/2)^2[5/2]$	$3d^9(^2D_{5/2})5d$	$^2[5/2]$	3	136919.3511	0.0014		99% + 1% 5d $(5/2)^2[7/2]$		36
5d	$(5/2)^2[7/2]$	$3d^9(^2D_{5/2})5d$	$^2[7/2]$	3	137034.778	0.003		99% + 1% 5d $(5/2)^2[5/2]$ + 1% 5d $(3/2)^2[7/2]$		23
5d	$(5/2)^2[7/2]$	$3d^9(^2D_{5/2})5d$	$^2[7/2]$	4	137044.4647	0.0017		99%		22
5d	$(5/2)^2[5/2]$	$3d^9(^2D_{5/2})5d$	$^2[5/2]$	2	137073.6465	0.0019		97% + 2% 5d $(5/2)^2[3/2]$		24
sp	$(^3F)^1P^\circ\ ^3G^\circ$	$3d^8(^3F)4s4p(^1P^\circ)$	$^3G^\circ$	3	137078.190	0.005		52% + 19% sp $(^3F)^1P^\circ\ ^3F^\circ$ + 14% sp $(^1G)^3P^\circ\ ^3F^\circ$		7
sp	$(^1G)^3P^\circ\ ^3H^\circ$	$3d^8(^1G)4s4p(^3P^\circ)$	$^3H^\circ$	5	137082.175	0.005		96% + 2% 4f 1H°		6
sp	$(^1G)^3P^\circ\ ^3F^\circ$	$3d^8(^1G)4s4p(^3P^\circ)$	$^3F^\circ$	2	137212.779	0.004		36% + 36% sp $(^3F)^1P^\circ\ ^3F^\circ$ + 14% sp $(^3P)^3P^\circ\ ^1D^\circ$	RJ	12
sp	$(^3P)^3P^\circ\ ^1P^\circ$	$3d^8(^3P)4s4p(^3P^\circ)$	$^1P^\circ$	1	137242.914	0.008		82% + 5% sp $(^3P)^3P^\circ\ ^3P^\circ$	N	6
5d	$(5/2)^2[1/2]$	$3d^9(^2D_{5/2})5d$	$^2[1/2]$	0	137614.140	0.004		60% + 37% 5d $(3/2)^2[1/2]$ + 1% 6d $(5/2)^2[1/2]$		7
sp	$(^3P)^3P^\circ\ ^1D^\circ$	$3d^8(^3P)4s4p(^3P^\circ)$	$^1D^\circ$	2	137648.800	0.004		50% + 9% sp $(^3P)^3P^\circ\ ^3D^\circ$ + 9% sp $(^3F)^1P^\circ\ ^3D^\circ$		15
sp	$(^3F)^1P^\circ\ ^3D^\circ$	$3d^8(^3F)4s4p(^1P^\circ)$	$^3D^\circ$	1	137913.450	0.004		28% + 16% 4f $^3D^\circ$ + 16% 4f		13
sp	$(^1G)^3P^\circ\ ^3F^\circ$	$3d^8(^1G)4s4p(^3P^\circ)$	$^3F^\circ$	4	137938.904	0.007		49% + 39% 6p $^3F^\circ$ + 5% sp $(^3F)^1P^\circ\ ^3F^\circ$	cd	6
4f	$(3/2)^2[3/2]^\circ$	$3d^9(^2D_{3/2})4f$	$^2[3/2]^\circ$	2	138002.8856	0.0023		94% + 3% sp + 2% 4f $(3/2)^2[5/2]^\circ$		10
4f	$(3/2)^2[3/2]^\circ$	$3d^9(^2D_{3/2})4f$	$^2[3/2]^\circ$	1	138028.384	0.003		59% + 26% sp + 10% 6p		14
4f	$(3/2)^2[9/2]^\circ$	$3d^9(^2D_{3/2})4f$	$^2[9/2]^\circ$	5	138064.2971	0.0021		99% + 1% sp		10
4f	$(3/2)^2[9/2]^\circ$	$3d^9(^2D_{3/2})4f$	$^2[9/2]^\circ$	4	138073.5826	0.0022		98% + 1% sp		10
4f	$(3/2)^2[5/2]^\circ$	$3d^9(^2D_{3/2})4f$	$^2[5/2]^\circ$	3	138130.5345	0.0015		97% + 1% 4f $(3/2)^2[7/2]^\circ$		17
4f	$(3/2)^2[5/2]^\circ$	$3d^9(^2D_{3/2})4f$	$^2[5/2]^\circ$	2	138176.8797	0.0018		89% + 7% sp + 1% 4f $(3/2)^2[3/2]^\circ$		17
4f	$(3/2)^2[7/2]^\circ$	$3d^9(^2D_{3/2})4f$	$^2[7/2]^\circ$	4	138219.8605	0.0015		98%		12
4f	$(3/2)^2[7/2]^\circ$	$3d^9(^2D_{3/2})4f$	$^2[7/2]^\circ$	3	138261.5822	0.0017		93% + 1% sp + 1% 4f $(3/2)^2[5/2]^\circ$		19
6p	$(5/2)^2[7/2]^\circ$	$3d^9(^2D_{5/2})6p$	$^2[7/2]^\circ$	3	138401.956	0.003		62% + 29% sp + 2% 6p $(5/2)^2[5/2]^\circ$		21
sp	$(^3P)^3P^\circ\ ^3S^\circ$	$3d^8(^3P)4s4p(^3P^\circ)$	$^3S^\circ$	1	138516.49	0.03		97%	N	5

Table A2. *Cont.*

	Label [a]	Configuration	Term	J	Level [b], cm^{-1}	Unc. [c], cm^{-1}	Landé g [d]	Leading Percentages [e]	Note [f]	N_{lines} [g]
5d	$(3/2)^2[1/2]$	$3d^9(^2D_{3/2})5d$	$^2[1/2]$	1	138593.046	0.003		98% + 1% 5d $(5/2)^2[1/2]$		17
6p	$(5/2)^2[3/2]^\circ$	$3d^9(^2D_{5/2})6p$	$^2[3/2]^\circ$	2	138745.817	0.005		91% + 4% sp + 3% 6p $(5/2)^2[5/2]^\circ$		14
5d	$(3/2)^2[7/2]$	$3d^9(^2D_{3/2})5d$	$^2[7/2]$	3	138819.0637	0.0019		99% + 1% 5d $(5/2)^2[7/2]$		15
5d	$(3/2)^2[7/2]$	$3d^9(^2D_{3/2})5d$	$^2[7/2]$	4	138882.8921	0.0019		99%		15
5d	$(3/2)^2[3/2]$	$3d^9(^2D_{3/2})5d$	$^2[3/2]$	1	138898.334	0.003		99%		18
5d	$(3/2)^2[3/2]$	$3d^9(^2D_{3/2})5d$	$^2[3/2]$	2	138981.246	0.003		97% + 2% 5d $(3/2)^2[5/2]$ + 1% 5d $(5/2)^2[3/2]$		19
6p	$(5/2)^2[5/2]^\circ$	$3d^9(^2D_{5/2})6p$	$^2[5/2]^\circ$	2	139028.705	0.005		41% + 40% sp + 11% 6p $(3/2)^2[5/2]^\circ$		11
5d	$(3/2)^2[5/2]$	$3d^9(^2D_{3/2})5d$	$^2[5/2]$	3	139119.4295	0.0020		100%		18
5d	$(3/2)^2[5/2]$	$3d^9(^2D_{3/2})5d$	$^2[5/2]$	2	139142.049	0.003		97% + 2% 5d $(3/2)^2[3/2]$		24
6p	$(5/2)^2[3/2]^\circ$	$3d^9(^2D_{5/2})6p$	$^2[3/2]^\circ$	1	139241.130	0.005		73% + 7% 6p $(3/2)^2[3/2]^\circ$ + 4% 6p $(3/2)^2[1/2]^\circ$		13
sp	$(^1G)^3P^\circ\ ^3F^\circ$	$3d^8(^1G)4s4p(^3P^\circ)$	$^3F^\circ$	3	139331.149	0.003		21% + 46% 6p $^1F^\circ$ + 15% 6p $^3D^\circ$	cd	21
6p	$(5/2)^2[7/2]^\circ$	$3d^9(^2D_{5/2})6p$	$^2[7/2]^\circ$	4	139395.786	0.004		59% + 37% sp		11
sp	$(^3F)^1P^\circ\ ^3F^\circ$	$3d^8(^3F)4s4p(^1P^\circ)$	$^3F^\circ$	2	139710.491	0.004		19% + 42% 6p $^1D^\circ$ + 15% sp $(^1G)^3P^\circ\ ^3F^\circ$		15
6p	$(5/2)^2[5/2]^\circ$	$3d^9(^2D_{5/2})6p$	$^2[5/2]^\circ$	3	139741.097	0.003		59% + 31% sp + 4% 6p $(5/2)^2[7/2]^\circ$		18
5d	$(3/2)^2[1/2]$	$3d^9(^2D_{3/2})5d$	$^2[1/2]$	0	140589.344	0.006		47% + 20% 6d $(5/2)^2[1/2]$ + 17% 5d $(5/2)^2[1/2]$		4
6p	$(3/2)^2[1/2]^\circ$	$3d^9(^2D_{3/2})6p$	$^2[1/2]^\circ$	1	140981.510	0.005		94% + 3% 6p $(5/2)^2[3/2]^\circ$ + 2% sp		8
6p	$(3/2)^2[5/2]^\circ$	$3d^9(^2D_{3/2})6p$	$^2[5/2]^\circ$	3	141202.628	0.006		89% + 8% sp		12
6p	$(3/2)^2[5/2]^\circ$	$3d^9(^2D_{3/2})6p$	$^2[5/2]^\circ$	2	141244.556	0.006		62% + 22% sp + 13% 6p $(3/2)^2[3/2]^\circ$	ci	9
6p	$(3/2)^2[3/2]^\circ$	$3d^9(^2D_{3/2})6p$	$^2[3/2]^\circ$	2	141542.001	0.005		73% + 13% 6p $(3/2)^2[5/2]^\circ$ + 10% sp		13
6p	$(3/2)^2[3/2]^\circ$	$3d^9(^2D_{3/2})6p$	$^2[3/2]^\circ$	1	141734.175	0.005		79% + 16% sp + 1% 5p	ci	10
sp	$(^1G)^3P^\circ\ ^3G^\circ$	$3d^8(^1G)4s4p(^3P^\circ)$	$^3G^\circ$	3	143423.319	0.020		99%	N	4
7s	$(5/2)^2[5/2]$	$3d^9(^2D_{5/2})7s$	$^2[5/2]$	3	144814.9118	0.0014		100%		26
7s	$(5/2)^2[5/2]$	$3d^9(^2D_{5/2})7s$	$^2[5/2]$	2	144882.9859	0.0016		100%		26
5f	$(5/2)^2[1/2]^\circ$	$3d^9(^2D_{5/2})5f$	$^2[1/2]^\circ$	0	145889.334	0.004		100%		6
5f	$(5/2)^2[1/2]^\circ$	$3d^9(^2D_{5/2})5f$	$^2[1/2]^\circ$	1	145900.904	0.003		84% + 16% 5f $(5/2)^2[3/2]^\circ$		10
5f	$(5/2)^2[3/2]^\circ$	$3d^9(^2D_{5/2})5f$	$^2[3/2]^\circ$	2	145927.231	0.003		100%		13
5f	$(5/2)^2[11/2]^\circ$	$3d^9(^2D_{5/2})5f$	$^2[11/2]^\circ$	5	145945.5969	0.0019		100%		5
5f	$(5/2)^2[11/2]^\circ$	$3d^9(^2D_{5/2})5f$	$^2[11/2]^\circ$	6	145951.5299	0.0014		100%		2
5f	$(5/2)^2[3/2]^\circ$	$3d^9(^2D_{5/2})5f$	$^2[3/2]^\circ$	1	145955.447	0.004		84% + 16% 5f $(5/2)^2[1/2]^\circ$		10
5f	$(5/2)^2[5/2]^\circ$	$3d^9(^2D_{5/2})5f$	$^2[5/2]^\circ$	3	145978.142	0.003		98% + 2% 5f $(5/2)^2[7/2]^\circ$		12
5f	$(5/2)^2[5/2]^\circ$	$3d^9(^2D_{5/2})5f$	$^2[5/2]^\circ$	2	145985.199	0.003		100%		12
5f	$(5/2)^2[7/2]^\circ$	$3d^9(^2D_{5/2})5f$	$^2[7/2]^\circ$	3	146021.279	0.003		98% + 2% 5f $(5/2)^2[5/2]^\circ$		13
5f	$(5/2)^2[7/2]^\circ$	$3d^9(^2D_{5/2})5f$	$^2[7/2]^\circ$	4	146023.862	0.003		86% + 14% 5f $(5/2)^2[9/2]^\circ$		13
5f	$(5/2)^2[9/2]^\circ$	$3d^9(^2D_{5/2})5f$	$^2[9/2]^\circ$	4	146029.350	0.003		86% + 14% 5f $(5/2)^2[7/2]^\circ$		13
5f	$(5/2)^2[9/2]^\circ$	$3d^9(^2D_{5/2})5f$	$^2[9/2]^\circ$	5	146032.2635	0.0021		100%		7
5g	$(5/2)^2[3/2]$	$3d^9(^2D_{5/2})5g$	$^2[3/2]$	1	146051.118	0.004		100%		5
5g	$(5/2)^2[3/2]$	$3d^9(^2D_{5/2})5g$	$^2[3/2]$	2	146051.2431	0.003		100%		4
5g	$(5/2)^2[13/2]$	$3d^9(^2D_{5/2})5g$	$^2[13/2]$	6	146068.85092	0.0019		100%		1
5g	$(5/2)^2[13/2]$	$3d^9(^2D_{5/2})5g$	$^2[13/2]$	7	146068.90547	0.0012		100%		1
5g	$(5/2)^2[5/2]$	$3d^9(^2D_{5/2})5g$	$^2[5/2]$	3	146072.7739	0.0020		100%		5
5g	$(5/2)^2[5/2]$	$3d^9(^2D_{5/2})5g$	$^2[5/2]$	2	146072.8134	0.0018		100%		6

Table A2. *Cont.*

	Label [a]	Configuration	Term	J	Level [b], cm^{-1}	Unc. [c], cm^{-1}	Landé g [d]	Leading Percentages [e]	Note [f]	N_{lines} [g]
5g	$(5/2)^2[7/2]$	$3d^9(^2D_{5/2})5g$	$^2[7/2]$	3	146093.9312	0.0014		100%		5
5g	$(5/2)^2[7/2]$	$3d^9(^2D_{5/2})5g$	$^2[7/2]$	4	146093.9535	0.0012		100%		2
5g	$(5/2)^2[11/2]$	$3d^9(^2D_{5/2})5g$	$^2[11/2]$	5	146103.2223	0.0018		100%		7
5g	$(5/2)^2[11/2]$	$3d^9(^2D_{5/2})5g$	$^2[11/2]$	6	146103.25704	0.0011		100%		3
5g	$(5/2)^2[9/2]$	$3d^9(^2D_{5/2})5g$	$^2[9/2]$	4	146107.1016	0.0016		100%		7
5g	$(5/2)^2[9/2]$	$3d^9(^2D_{5/2})5g$	$^2[9/2]$	5	146107.1288	0.0010		100%		6
6d	$(5/2)^2[1/2]$	$3d^9(^2D_{5/2})6d$	$^2[1/2]$	1	146215.634	0.004		98% + 1% 6d $(5/2)^2[3/2]$		7
6d	$(5/2)^2[9/2]$	$3d^9(^2D_{5/2})6d$	$^2[9/2]$	5	146402.715	0.004		100%		8
6d	$(5/2)^2[3/2]$	$3d^9(^2D_{5/2})6d$	$^2[3/2]$	2	146415.599	0.004		97% + 2% 6d $(5/2)^2[5/2]$		12
6d	$(5/2)^2[9/2]$	$3d^9(^2D_{5/2})6d$	$^2[9/2]$	4	146423.201	0.003		100%		10
6d	$(5/2)^2[3/2]$	$3d^9(^2D_{5/2})6d$	$^2[3/2]$	1	146427.948	0.003		99% + 1% 6d $(5/2)^2[1/2]$		12
6d	$(5/2)^2[5/2]$	$3d^9(^2D_{5/2})6d$	$^2[5/2]$	3	146496.100	0.003		99% + 1% 6d $(5/2)^2[7/2]$		16
6d	$(5/2)^2[7/2]$	$3d^9(^2D_{5/2})6d$	$^2[7/2]$	3	146556.381	0.003		99% + 1% 6d $(5/2)^2[5/2]$		19
6d	$(5/2)^2[7/2]$	$3d^9(^2D_{5/2})6d$	$^2[7/2]$	4	146559.703	0.003		100%		14
6d	$(5/2)^2[5/2]$	$3d^9(^2D_{5/2})6d$	$^2[5/2]$	2	146575.101	0.003		97% + 2% 6d $(5/2)^2[3/2]$ + 1% 7s $(3/2)^2[3/2]$		23
7s	$(3/2)^2[3/2]$	$3d^9(^2D_{3/2})7s$	$^2[3/2]$	1	146886.1667	0.0021		100%		20
7s	$(3/2)^2[3/2]$	$3d^9(^2D_{3/2})7s$	$^2[3/2]$	2	146936.3180	0.0018		99%		24
6d	$(5/2)^2[1/2]$	$3d^9(^2D_{5/2})6d$	$^2[1/2]$	0	147097.835	0.005		60% + 29% 6d $(3/2)^2[1/2]$ + 4% 7d $(5/2)^2[1/2]$		6
7p	$(5/2)^2[3/2]°$	$3d^9(^2D_{5/2})7p$	$^2[3/2]°$	2	147327.659	0.007		94% + 4% sp + 1% 7p $(5/2)^2[5/2]°$		10
7p	$(5/2)^2[7/2]°$	$3d^9(^2D_{5/2})7p$	$^2[7/2]°$	3	147525.93	0.06		98% + 1% sp	R	3
7p	$(5/2)^2[3/2]°$	$3d^9(^2D_{5/2})7p$	$^2[3/2]°$	1	147562.672	0.006		94% + 3% sp + 1% 5f $(3/2)^2[3/2]°$		8
7p	$(5/2)^2[7/2]°$	$3d^9(^2D_{5/2})7p$	$^2[7/2]°$	4	147596.655	0.006		100%		4
7p	$(5/2)^2[5/2]°$	$3d^9(^2D_{5/2})7p$	$^2[5/2]°$	2	147647.699	0.009		94% + 3% sp + 2% 7p $(5/2)^2[3/2]°$		7
7p	$(5/2)^2[5/2]°$	$3d^9(^2D_{5/2})7p$	$^2[5/2]°$	3	147762.990	0.005		98%		9
5f	$(3/2)^2[3/2]°$	$3d^9(^2D_{3/2})5f$	$^2[3/2]°$	2	147987.359	0.004		99%		8
5f	$(3/2)^2[3/2]°$	$3d^9(^2D_{3/2})5f$	$^2[3/2]°$	1	148016.157	0.004		99% + 1% 7p $(5/2)^2[3/2]°$		11
5f	$(3/2)^2[9/2]°$	$3d^9(^2D_{3/2})5f$	$^2[9/2]°$	5	148028.7360	0.0023		100%		5
5f	$(3/2)^2[9/2]°$	$3d^9(^2D_{3/2})5f$	$^2[9/2]°$	4	148033.574	0.003		100%		6
5f	$(3/2)^2[5/2]°$	$3d^9(^2D_{3/2})5f$	$^2[5/2]°$	3	148061.467	0.003		97% + 2% 5f $(3/2)^2[7/2]°$		9
5f	$(3/2)^2[5/2]°$	$3d^9(^2D_{3/2})5f$	$^2[5/2]°$	2	148066.011	0.003		99%		11
5f	$(3/2)^2[7/2]°$	$3d^9(^2D_{3/2})5f$	$^2[7/2]°$	3	148102.986	0.003		97% + 3% 5f $(3/2)^2[5/2]°$		9
5f	$(3/2)^2[7/2]°$	$3d^9(^2D_{3/2})5f$	$^2[7/2]°$	4	148105.366	0.003		100%		5
5g	$(3/2)^2[5/2]$	$3d^9(^2D_{3/2})5g$	$^2[5/2]$	3	148133.777	0.003		100%		3
5g	$(3/2)^2[5/2]$	$3d^9(^2D_{3/2})5g$	$^2[5/2]$	2	148133.832	0.005		100%		3
5g	$(3/2)^2[11/2]$	$3d^9(^2D_{3/2})5g$	$^2[11/2]$	5	148145.7906	0.0022		100%		3
5g	$(3/2)^2[11/2]$	$3d^9(^2D_{3/2})5g$	$^2[11/2]$	6	148145.8203	0.0021		100%		3
5g	$(3/2)^2[7/2]$	$3d^9(^2D_{3/2})5g$	$^2[7/2]$	3	148167.557	0.003		100%		5
5g	$(3/2)^2[7/2]$	$3d^9(^2D_{3/2})5g$	$^2[7/2]$	4	148167.5920	0.0017		100%		4
5g	$(3/2)^2[9/2]$	$3d^9(^2D_{3/2})5g$	$^2[9/2]$	4	148179.1541	0.0019		100%		4
5g	$(3/2)^2[9/2]$	$3d^9(^2D_{3/2})5g$	$^2[9/2]$	5	148179.1829	0.0016		100%		4

Table A2. *Cont.*

	Label [a]	Configuration	Term	J	Level [b], cm^{-1}	Unc. [c], cm^{-1}	Landé g [d]	Leading Percentages [e]	Note [f]	N_{lines} [g]
6d	$(3/2)^2[1/2]$	$3d^9(^2D_{3/2})6d$	$^2[1/2]$	1	148361.542	0.004		99%		10
6d	$(3/2)^2[7/2]$	$3d^9(^2D_{3/2})6d$	$^2[7/2]$	3	148481.763	0.003		100%		11
6d	$(3/2)^2[7/2]$	$3d^9(^2D_{3/2})6d$	$^2[7/2]$	4	148515.532	0.004		100%		9
6d	$(3/2)^2[3/2]$	$3d^9(^2D_{3/2})6d$	$^2[3/2]$	1	148521.575	0.003		100%		12
6d	$(3/2)^2[3/2]$	$3d^9(^2D_{3/2})6d$	$^2[3/2]$	2	148559.958	0.003		99% + 1% 6d $(3/2)^2[5/2]$		15
6d	$(3/2)^2[5/2]$	$3d^9(^2D_{3/2})6d$	$^2[5/2]$	3	148631.096	0.003		100%		13
6d	$(3/2)^2[5/2]$	$3d^9(^2D_{3/2})6d$	$^2[5/2]$	2	148642.489	0.003		99% + 1% 6d $(3/2)^2[3/2]$		13
6d	$(3/2)^2[1/2]$	$3d^9(^2D_{3/2})6d$	$^2[1/2]$	0	149202.607	0.007		51% + 28% 7d $(5/2)^2[1/2]$ + 6% 6d $(5/2)^2[1/2]$		3
7p	$(3/2)^2[1/2]^\circ$	$3d^9(^2D_{3/2})7p$	$^2[1/2]^\circ$	0	149371.08	0.08		94% + 5% sp	N	2
7p	$(3/2)^2[1/2]^\circ$	$3d^9(^2D_{3/2})7p$	$^2[1/2]^\circ$	1	149484.111	0.008		95% + 4% sp + 1% 7p $(5/2)^2[3/2]^\circ$	N	5
7p	$(3/2)^2[5/2]^\circ$	$3d^9(^2D_{3/2})7p$	$^2[5/2]^\circ$	2	149525.97	0.05		64% + 26% 7p $(3/2)^2[3/2]^\circ$ + 8% sp	N	4
7p	$(3/2)^2[5/2]^\circ$	$3d^9(^2D_{3/2})7p$	$^2[5/2]^\circ$	3	149624.47	0.05		92% + 6% sp	N	2
7p	$(3/2)^2[3/2]^\circ$	$3d^9(^2D_{3/2})7p$	$^2[3/2]^\circ$	2	149726.69	0.05		63% + 33% 7p $(3/2)^2[5/2]^\circ$ + 3% sp	N	3
7p	$(3/2)^2[3/2]^\circ$	$3d^9(^2D_{3/2})7p$	$^2[3/2]^\circ$	1	149765.88	0.05		95% + 4% sp	N	4
sp	$(^1D)^1P^\circ\ ^1D^\circ$	$3d^8(^1D)4s4p(^1P^\circ)$	$^1D^\circ$	2	150249.887	0.008		38% + 28% sp $(^3P)^1P^\circ\ ^3P^\circ$ + 11% 7p $^1D^\circ$		8
8s	$(5/2)^2[5/2]$	$3d^9(^2D_{5/2})8s$	$^2[5/2]$	3	150742.896	0.003		100%		13
8s	$(5/2)^2[5/2]$	$3d^9(^2D_{5/2})8s$	$^2[5/2]$	2	150782.454	0.003		100%		13
6f	$(5/2)^2[1/2]^\circ$	$3d^9(^2D_{5/2})6f$	$^2[1/2]^\circ$	1	151161.379	0.007		63% + 21% sp + 5% 8p $(5/2)^2[3/2]^\circ$	N	6
6f	$(5/2)^2[1/2]^\circ$	$3d^9(^2D_{5/2})6f$	$^2[1/2]^\circ$	0	151327.262	0.008		98% + 1% sp		2
6f	$(5/2)^2[11/2]^\circ$	$3d^9(^2D_{5/2})6f$	$^2[11/2]^\circ$	5	151372.330	0.004		100%		4
6f	$(5/2)^2[3/2]^\circ$	$3d^9(^2D_{5/2})6f$	$^2[3/2]^\circ$	1	151373.840	0.006		96% + 3% 6f $(5/2)^2[1/2]^\circ$		5
6f	$(5/2)^2[3/2]^\circ$	$3d^9(^2D_{5/2})6f$	$^2[3/2]^\circ$	2	151375.318	0.004		52% + 46% 6f $(5/2)^2[5/2]^\circ$		7
6f	$(5/2)^2[11/2]^\circ$	$3d^9(^2D_{5/2})6f$	$^2[11/2]^\circ$	6	151377.688	0.005		100%		2
6f	$(5/2)^2[5/2]^\circ$	$3d^9(^2D_{5/2})6f$	$^2[5/2]^\circ$	3	151402.621	0.004		71% + 28% 6f $(5/2)^2[7/2]^\circ$		7
6f	$(5/2)^2[5/2]^\circ$	$3d^9(^2D_{5/2})6f$	$^2[5/2]^\circ$	2	151403.942	0.004		53% + 44% 6f $(5/2)^2[3/2]^\circ$		8
6f	$(5/2)^2[7/2]^\circ$	$3d^9(^2D_{5/2})6f$	$^2[7/2]^\circ$	4	151419.022	0.004		90% + 10% 6f $(5/2)^2[9/2]^\circ$		8
6f	$(5/2)^2[9/2]^\circ$	$3d^9(^2D_{5/2})6f$	$^2[9/2]^\circ$	4	151421.791	0.004		90% + 10% 6f $(5/2)^2[7/2]^\circ$		9
6f	$(5/2)^2[9/2]^\circ$	$3d^9(^2D_{5/2})6f$	$^2[9/2]^\circ$	5	151423.056	0.004		100%		4
8p	$(5/2)^2[3/2]^\circ$	$3d^9(^2D_{5/2})8p$	$^2[3/2]^\circ$	1	151424.241	0.004		33% 6f $(5/2)^2[1/2]^\circ$ + 36% sp + 20% 8p $(5/2)^2[3/2]^\circ$		8
6g	$(5/2)^2[3/2]$	$3d^9(^2D_{5/2})6g$	$^2[3/2]$	1	151440.091	0.008		100%		2
6g	$(5/2)^2[3/2]$	$3d^9(^2D_{5/2})6g$	$^2[3/2]$	2	151440.178	0.006		100%		2
6f	$(5/2)^2[7/2]^\circ$	$3d^9(^2D_{5/2})6f$	$^2[7/2]^\circ$	3	151441.899	0.006		70% + 27% 6f $(5/2)^2[5/2]^\circ$ + 2% sp		7
6g	$(5/2)^2[13/2]$	$3d^9(^2D_{5/2})6g$	$^2[13/2]$	6	151450.414	0.007		100%		2
6g	$(5/2)^2[13/2]$	$3d^9(^2D_{5/2})6g$	$^2[13/2]$	7	151450.488	0.003		100%		1
6g	$(5/2)^2[5/2]$	$3d^9(^2D_{5/2})6g$	$^2[5/2]$	3	151452.690	0.007		100%		4
6g	$(5/2)^2[5/2]$	$3d^9(^2D_{5/2})6g$	$^2[5/2]$	2	151452.701	0.006		100%		4
6h	$(5/2)^2[5/2]^\circ$	$3d^9(^2D_{5/2})6h$	$^2[5/2]^\circ$	2	[151458.3]	6		100%	pf	0
6h	$(5/2)^2[5/2]^\circ$	$3d^9(^2D_{5/2})6h$	$^2[5/2]^\circ$	3	[151458.4]	6		100%	pf	0
6h	$(5/2)^2[15/2]^\circ$	$3d^9(^2D_{5/2})6h$	$^2[15/2]^\circ$	7	[151461.7]	6		100%	pf	0
6h	$(5/2)^2[15/2]^\circ$	$3d^9(^2D_{5/2})6h$	$^2[15/2]^\circ$	8	[151461.8]	6		100%	pf	0
6h	$(5/2)^2[7/2]^\circ$	$3d^9(^2D_{5/2})6h$	$^2[7/2]^\circ$	3	[151463.9]	6		100%	pf	0

Table A2. *Cont.*

	Label [a]	Configuration	Term	*J*	Level [b], cm^{-1}	Unc. [c], cm^{-1}	Landé *g* [d]	Leading Percentages [e]	Note [f]	N_{lines} [g]
6h	$(5/2)^2[7/2]^\circ$	$3d^9(^2D_{5/2})6h$	$^2[7/2]^\circ$	4	[151464.0]	6		100%	pf	0
6g	$(5/2)^2[7/2]$	$3d^9(^2D_{5/2})6g$	$^2[7/2]$	3	151464.789	0.006		100%		4
6g	$(5/2)^2[7/2]$	$3d^9(^2D_{5/2})6g$	$^2[7/2]$	4	151464.796	0.005		100%		2
6h	$(5/2)^2[9/2]^\circ$	$3d^9(^2D_{5/2})6h$	$^2[9/2]^\circ$	4	[151468.8]	6		100%	pf	0
6h	$(5/2)^2[9/2]^\circ$	$3d^9(^2D_{5/2})6h$	$^2[9/2]^\circ$	5	[151468.9]	6		100%	pf	0
6h	$(5/2)^2[13/2]^\circ$	$3d^9(^2D_{5/2})6h$	$^2[13/2]^\circ$	6	[151469.8]	6		100%	pf	0
6h	$(5/2)^2[13/2]^\circ$	$3d^9(^2D_{5/2})6h$	$^2[13/2]^\circ$	7	[151470.0]	6		100%	pf	0
6g	$(5/2)^2[11/2]$	$3d^9(^2D_{5/2})6g$	$^2[11/2]$	5	151470.113	0.003		100%		5
6g	$(5/2)^2[11/2]$	$3d^9(^2D_{5/2})6g$	$^2[11/2]$	6	151470.130	0.005		100%		3
6h	$(5/2)^2[11/2]^\circ$	$3d^9(^2D_{5/2})6h$	$^2[11/2]^\circ$	5	[151471.4]	6		100%	pf	0
6h	$(5/2)^2[11/2]^\circ$	$3d^9(^2D_{5/2})6h$	$^2[11/2]^\circ$	6	[151471.5]	6		100%	pf	0
6g	$(5/2)^2[9/2]$	$3d^9(^2D_{5/2})6g$	$^2[9/2]$	4	151472.264	0.005		100%		5
6g	$(5/2)^2[9/2]$	$3d^9(^2D_{5/2})6g$	$^2[9/2]$	5	151472.278	0.005		100%		3
7d	$(5/2)^2[1/2]$	$3d^9(^2D_{5/2})7d$	$^2[1/2]$	1	151552.191	0.007		98% + 1% 7d $(5/2)^2[3/2]$		2
7d	$(5/2)^2[9/2]$	$3d^9(^2D_{5/2})7d$	$^2[9/2]$	5	151656.317	0.005		100%		5
7d	$(5/2)^2[3/2]$	$3d^9(^2D_{5/2})7d$	$^2[3/2]$	2	151662.985	0.005		96% + 4% 7d $(5/2)^2[5/2]$		6
7d	$(5/2)^2[9/2]$	$3d^9(^2D_{5/2})7d$	$^2[9/2]$	4	151668.328	0.005		99%		7
7d	$(5/2)^2[3/2]$	$3d^9(^2D_{5/2})7d$	$^2[3/2]$	1	151671.666	0.006		99% + 1% 7d $(5/2)^2[1/2]$		4
7d	$(5/2)^2[5/2]$	$3d^9(^2D_{5/2})7d$	$^2[5/2]$	3	151708.347	0.004		99% + 1% 7d $(5/2)^2[7/2]$		8
7d	$(5/2)^2[7/2]$	$3d^9(^2D_{5/2})7d$	$^2[7/2]$	3	151743.558	0.003		99% + 1% 7d $(5/2)^2[5/2]$		13
7d	$(5/2)^2[7/2]$	$3d^9(^2D_{5/2})7d$	$^2[7/2]$	4	151744.916	0.005		99%		8
7d	$(5/2)^2[5/2]$	$3d^9(^2D_{5/2})7d$	$^2[5/2]$	2	151757.601	0.004		96% + 4% 7d $(5/2)^2[3/2]$		8
8p	$(5/2)^2[3/2]^\circ$	$3d^9(^2D_{5/2})8p$	$^2[3/2]^\circ$	2	152054.78	0.03		53% + 21% 8p $(5/2)^2[5/2]^\circ$ + 20% sp	N	5
7d	$(5/2)^2[1/2]$	$3d^9(^2D_{5/2})7d$	$^2[1/2]$	0	152179.051	0.007		52% + 28% 7d $(3/2)^2[1/2]$ + 9% 8d $(5/2)^2[1/2]$		2
8p	$(5/2)^2[5/2]^\circ$	$3d^9(^2D_{5/2})8p$	$^2[5/2]^\circ$	2	152580.19	0.05		65% + 27% 8p $(5/2)^2[3/2]^\circ$ + 7% sp	N	2
sp	$(^3P)^1P^\circ\ ^3P^\circ$	$3d^8(^3P)4s4p(^1P^\circ)$	$^3P^\circ$	0	152783.41	0.05		77% + 9% 8p $^3P^\circ$ + 5% 7p $^3P^\circ$	N	2
8s	$(3/2)^2[3/2]$	$3d^9(^2D_{3/2})8s$	$^2[3/2]$	1	152814.032	0.007		100%		4
8s	$(3/2)^2[3/2]$	$3d^9(^2D_{3/2})8s$	$^2[3/2]$	2	152840.475	0.005		100%		6
sp	$(^3P)^1P^\circ\ ^3P^\circ$	$3d^8(^3P)4s4p(^1P^\circ)$	$^3P^\circ$	2	152944.11	0.05		39% + 19% sp $(^1D)^1P^\circ\ ^1D^\circ$ + 16% 8p $^3D^\circ$	N	4
sp	$(^1D)^1P^\circ\ ^1P^\circ$	$3d^8(^1D)4s4p(^1P^\circ)$	$^1P^\circ$	1	153165.24	0.04		22% + 18% 6f $^3D^\circ$ + 17% 6f $^1P^\circ$	N	5
6f	$(3/2)^2[5/2]^\circ$	$3d^9(^2D_{3/2})6f$	$^2[5/2]^\circ$	3	153410.212	0.006		89% + 8% sp + 1% 6f $(3/2)^2[7/2]^\circ$		5
6f	$(3/2)^2[3/2]^\circ$	$3d^9(^2D_{3/2})6f$	$^2[3/2]^\circ$	2	153426.632	0.006		52% + 42% 6f $(3/2)^2[5/2]^\circ$ + 5% sp		7
6f	$(3/2)^2[9/2]^\circ$	$3d^9(^2D_{3/2})6f$	$^2[9/2]^\circ$	5	153449.633	0.005		100%		2
6f	$(3/2)^2[9/2]^\circ$	$3d^9(^2D_{3/2})6f$	$^2[9/2]^\circ$	4	153455.178	0.005		100%		4
6f	$(3/2)^2[3/2]^\circ$	$3d^9(^2D_{3/2})6f$	$^2[3/2]^\circ$	1	153457.747	0.009		55% + 23% sp + 15% 8p $(3/2)^2[1/2]^\circ$		6
6f	$(3/2)^2[5/2]^\circ$	$3d^9(^2D_{3/2})6f$	$^2[5/2]^\circ$	2	153487.668	0.005		56% + 37% 6f $(3/2)^2[3/2]^\circ$ + 5% sp		8
6f	$(3/2)^2[7/2]^\circ$	$3d^9(^2D_{3/2})6f$	$^2[7/2]^\circ$	4	153495.854	0.006		100%		3
6f	$(3/2)^2[7/2]^\circ$	$3d^9(^2D_{3/2})6f$	$^2[7/2]^\circ$	3	153498.834	0.006		98% + 2% 6f $(3/2)^2[5/2]^\circ$		3
6g	$(3/2)^2[5/2]$	$3d^9(^2D_{3/2})6g$	$^2[5/2]$	3	153517.905	0.007		100%		2
6g	$(3/2)^2[5/2]$	$3d^9(^2D_{3/2})6g$	$^2[5/2]$	2	153517.985	0.008		100%		1
6g	$(3/2)^2[11/2]$	$3d^9(^2D_{3/2})6g$	$^2[11/2]$	5	153524.933	0.005		100%		2

Table A2. *Cont.*

	Label [a]	Configuration	Term	J	Level [b], cm^{-1}	Unc. [c], cm^{-1}	Landé g [d]	Leading Percentages [e]	Note [f]	N_{lines} [g]
6g	(3/2)^2[11/2]	3d^9(^{2}D$_{3/2}$)6g	^{2}[11/2]	6	153524.950	0.007		100%		1
6g	(3/2)^2[7/2]	3d^9(^{2}D$_{3/2}$)6g	^{2}[7/2]	3	153537.395	0.007		100%		2
6g	(3/2)^2[7/2]	3d^9(^{2}D$_{3/2}$)6g	^{2}[7/2]	4	153537.414	0.006		100%		3
6h	(3/2)^2[7/2]°	3d^9(^{2}D$_{3/2}$)6h	^{2}[7/2]°	3	[153543.7]	6		100%	pf	0
6h	(3/2)^2[7/2]°	3d^9(^{2}D$_{3/2}$)6h	^{2}[7/2]°	4	[153543.8]	6		100%	pf	0
6g	(3/2)^2[9/2]	3d^9(^{2}D$_{3/2}$)6g	^{2}[9/2]	4	153544.054	0.005		100%		3
6g	(3/2)^2[9/2]	3d^9(^{2}D$_{3/2}$)6g	^{2}[9/2]	5	153544.069	0.007		100%		2
6h	(3/2)^2[13/2]°	3d^9(^{2}D$_{3/2}$)6h	^{2}[13/2]°	6	[153546.0]	6		100%	pf	0
6h	(3/2)^2[13/2]°	3d^9(^{2}D$_{3/2}$)6h	^{2}[13/2]°	7	[153546.2]	6		100%	pf	0
6h	(3/2)^2[9/2]°	3d^9(^{2}D$_{3/2}$)6h	^{2}[9/2]°	4	[153552.0]	6		100%	pf	0
6h	(3/2)^2[9/2]°	3d^9(^{2}D$_{3/2}$)6h	^{2}[9/2]°	5	[153552.1]	6		100%	pf	0
6h	(3/2)^2[11/2]°	3d^9(^{2}D$_{3/2}$)6h	^{2}[11/2]°	5	[153554.3]	6		100%	pf	0
6h	(3/2)^2[11/2]°	3d^9(^{2}D$_{3/2}$)6h	^{2}[11/2]°	6	[153554.4]	6		100%	pf	0
7d	(3/2)^2[1/2]	3d^9(^{2}D$_{3/2}$)7d	^{2}[1/2]	1	153658.567	0.006		99%		5
7d	(3/2)^2[7/2]	3d^9(^{2}D$_{3/2}$)7d	^{2}[7/2]	3	153730.935	0.005		100%		5
7d	(3/2)^2[7/2]	3d^9(^{2}D$_{3/2}$)7d	^{2}[7/2]	4	153750.851	0.005		100%		3
7d	(3/2)^2[3/2]	3d^9(^{2}D$_{3/2}$)7d	^{2}[3/2]	1	153753.738	0.007		100%		2
7d	(3/2)^2[3/2]	3d^9(^{2}D$_{3/2}$)7d	^{2}[3/2]	2	153773.884	0.006		97% + 3% 7d (3/2)^2[5/2]		5
7d	(3/2)^2[5/2]	3d^9(^{2}D$_{3/2}$)7d	^{2}[5/2]	3	153815.644	0.006		100%		5
7d	(3/2)^2[5/2]	3d^9(^{2}D$_{3/2}$)7d	^{2}[5/2]	2	153821.937	0.006		97% + 3% 7d (3/2)^2[3/2]		4
sp	(^{3}P)^{1}P° ^{3}D°	3d^8(^{3}P)4s4p(^{1}P°)	^{3}D°	3	153850.18	0.04		62% + 9% sp (^{1}D)^{1}P° ^{1}F° + 8% 8p ^{3}F°	N	2
7d	(3/2)^2[1/2]	3d^9(^{2}D$_{3/2}$)7d	^{2}[1/2]	0	153853.763	0.011		49% + 44% 8d (5/2)^2[1/2] + 2% 8d (3/2)^2[1/2]		2
8p	(3/2)^2[3/2]°	3d^9(^{2}D$_{3/2}$)8p	^{2}[3/2]°	1	154225.21	0.04		38% + 44% sp + 6% 7f (5/2)^2[3/2]°	N	2
9s	(5/2)^2[5/2]	3d^9(^{2}D$_{5/2}$)9s	^{2}[5/2]	3	154255.815	0.005		100%		4
9s	(5/2)^2[5/2]	3d^9(^{2}D$_{5/2}$)9s	^{2}[5/2]	2	154281.251	0.008		100%		4
7f	(5/2)^2[11/2]°	3d^9(^{2}D$_{5/2}$)7f	^{2}[11/2]°	5	154641.392	0.006		99%		2
7f	(5/2)^2[11/2]°	3d^9(^{2}D$_{5/2}$)7f	^{2}[11/2]°	6	154647.056	0.005		100%		2
7f	(5/2)^2[3/2]°	3d^9(^{2}D$_{5/2}$)7f	^{2}[3/2]°	1	154653.77?	0.07		51% + 25% 8p (3/2)^2[3/2]° + 12% 8p (3/2)^2[1/2]°	N	2
7f	(5/2)^2[5/2]°	3d^9(^{2}D$_{5/2}$)7f	^{2}[5/2]°	3	154672.123	0.006		62% + 37% 7f (5/2)^2[7/2]°		4
7f	(5/2)^2[7/2]°	3d^9(^{2}D$_{5/2}$)7f	^{2}[7/2]°	4	154672.766	0.006		91% + 9% 7f (5/2)^2[9/2]°		5
7f	(5/2)^2[9/2]°	3d^9(^{2}D$_{5/2}$)7f	^{2}[9/2]°	4	154674.551	0.006		91% + 9% 7f (5/2)^2[7/2]°		4
7f	(5/2)^2[9/2]°	3d^9(^{2}D$_{5/2}$)7f	^{2}[9/2]°	5	154675.247	0.005		100%		4
7f	(5/2)^2[7/2]°	3d^9(^{2}D$_{5/2}$)7f	^{2}[7/2]°	3	154678.877	0.008		62% + 36% 7f (5/2)^2[5/2]° + 1% sp		3
7g	(5/2)^2[3/2]	3d^9(^{2}D$_{5/2}$)7g	^{2}[3/2]	1	154688.020	0.011		100%		2
7g	(5/2)^2[3/2]	3d^9(^{2}D$_{5/2}$)7g	^{2}[3/2]	2	154688.112	0.009		100%		3
7g	(5/2)^2[13/2]	3d^9(^{2}D$_{5/2}$)7g	^{2}[13/2]	6	154694.589	0.008		100%		1
7g	(5/2)^2[13/2]	3d^9(^{2}D$_{5/2}$)7g	^{2}[13/2]	7	154694.619	0.008		100%		1
7g	(5/2)^2[5/2]	3d^9(^{2}D$_{5/2}$)7g	^{2}[5/2]	3	154695.997	0.008		100%		4
7g	(5/2)^2[5/2]	3d^9(^{2}D$_{5/2}$)7g	^{2}[5/2]	2	154696.009	0.007		100%	N	5
7f	(5/2)^2[5/2]°	3d^9(^{2}D$_{5/2}$)7f	^{2}[5/2]°	2	154698.288	0.020		39% + 28% 8p (3/2)^2[3/2]° + 15% 7f (5/2)^2[3/2]°	N	3
7h	(5/2)^2[5/2]°	3d^9(^{2}D$_{5/2}$)7h	^{2}[5/2]°	2	[154702.4]	2.0		100%	pf	0

Table A2. *Cont.*

	Label [a]	Configuration	Term	J	Level [b], cm^{-1}	Unc. [c], cm^{-1}	Landé g [d]	Leading Percentages [e]	Note [f]	N_{lines} [g]
7h	$(5/2)^2[5/2]°$	$3d^9(^2D_{5/2})7h$	$^2[5/2]°$	3	[154702.5]	2.0		100%	pf	0
7g	$(5/2)^2[7/2]$	$3d^9(^2D_{5/2})7g$	$^2[7/2]$	3	154703.597	0.008		100%		2
7g	$(5/2)^2[7/2]$	$3d^9(^2D_{5/2})7g$	$^2[7/2]$	4	154703.625	0.008		100%		2
7h	$(5/2)^2[15/2]°$	$3d^9(^2D_{5/2})7h$	$^2[15/2]°$	7	[154704.6]	2.0		100%	pf	0
7h	$(5/2)^2[15/2]°$	$3d^9(^2D_{5/2})7h$	$^2[15/2]°$	8	[154704.6]	2.0		100%	pf	0
7h	$(5/2)^2[7/2]°$	$3d^9(^2D_{5/2})7h$	$^2[7/2]°$	3	[154706.0]	2.0		100%	pf	0
7h	$(5/2)^2[7/2]°$	$3d^9(^2D_{5/2})7h$	$^2[7/2]°$	4	[154706.1]	2.0		100%	pf	0
7g	$(5/2)^2[11/2]$	$3d^9(^2D_{5/2})7g$	$^2[11/2]$	5	154706.911	0.007		100%		3
7g	$(5/2)^2[9/2]$	$3d^9(^2D_{5/2})7g$	$^2[9/2]$	4	154708.235	0.007		100%		5
7g	$(5/2)^2[9/2]$	$3d^9(^2D_{5/2})7g$	$^2[9/2]$	5	154708.244	0.007		100%	N	2
7h	$(5/2)^2[9/2]°$	$3d^9(^2D_{5/2})7h$	$^2[9/2]°$	4	[154709.1]	2.0		100%	pf	0
7h	$(5/2)^2[9/2]°$	$3d^9(^2D_{5/2})7h$	$^2[9/2]°$	5	[154709.1]	2.0		100%	pf	0
7h	$(5/2)^2[13/2]°$	$3d^9(^2D_{5/2})7h$	$^2[13/2]°$	6	[154709.7]	2.0		100%	pf	0
7h	$(5/2)^2[13/2]°$	$3d^9(^2D_{5/2})7h$	$^2[13/2]°$	7	[154709.8]	2.0		100%	pf	0
7h	$(5/2)^2[11/2]°$	$3d^9(^2D_{5/2})7h$	$^2[11/2]°$	5	[154710.7]	2.0		100%	pf	0
7h	$(5/2)^2[11/2]°$	$3d^9(^2D_{5/2})7h$	$^2[11/2]°$	6	[154710.8]	2.0		100%	pf	0
8p	$(3/2)^2[1/2]°$	$3d^9(^2D_{3/2})8p$	$^2[1/2]°$	1	154719.09	0.03		29% + 32% 8p $(3/2)^2[3/2]°$ + 19% sp	N	3
8d	$(5/2)^2[1/2]$	$3d^9(^2D_{5/2})8d$	$^2[1/2]$	1	154766.034	0.022		98% + 1% 8d $(5/2)^2[3/2]$		1
8d	$(5/2)^2[9/2]$	$3d^9(^2D_{5/2})8d$	$^2[9/2]$	5	154828.718	0.008		100%		2
8d	$(5/2)^2[3/2]$	$3d^9(^2D_{5/2})8d$	$^2[3/2]$	2	154832.657	0.011		95% + 5% 8d $(5/2)^2[5/2]$		1
8d	$(5/2)^2[9/2]$	$3d^9(^2D_{5/2})8d$	$^2[9/2]$	4	154836.434	0.006		100%		4
8d	$(5/2)^2[3/2]$	$3d^9(^2D_{5/2})8d$	$^2[3/2]$	1	154838.973?	0.013		98% + 1% 8d $(5/2)^2[1/2]$		2
8d	$(5/2)^2[5/2]$	$3d^9(^2D_{5/2})8d$	$^2[5/2]$	3	154860.741	0.006		98% + 1% 8d $(5/2)^2[7/2]$		5
8d	$(5/2)^2[7/2]$	$3d^9(^2D_{5/2})8d$	$^2[7/2]$	3	154883.104	0.009		98% + 1% 8d $(5/2)^2[5/2]$		2
8d	$(5/2)^2[7/2]$	$3d^9(^2D_{5/2})8d$	$^2[7/2]$	4	154883.625	0.007		100%		6
8d	$(5/2)^2[5/2]$	$3d^9(^2D_{5/2})8d$	$^2[5/2]$	2	154892.916	0.020		95% + 5% 8d $(5/2)^2[3/2]$		2
8d	$(5/2)^2[1/2]$	$3d^9(^2D_{5/2})8d$	$^2[1/2]$	0	155244.842?	0.021		32% + 29% 9d $(5/2)^2[1/2]$ + 20% 8d $(3/2)^2[1/2]$		1
9s	$(3/2)^2[3/2]$	$3d^9(^2D_{3/2})9s$	$^2[3/2]$	1	156326.913	0.018		100%		4
9s	$(3/2)^2[3/2]$	$3d^9(^2D_{3/2})9s$	$^2[3/2]$	2	156341.878	0.009		99% + 1% 10s $(5/2)^2[5/2]$		5
10s	$(5/2)^2[5/2]$	$3d^9(^2D_{5/2})10s$	$^2[5/2]$	3	156508.501	0.022		100%		1
10s	$(5/2)^2[5/2]$	$3d^9(^2D_{5/2})10s$	$^2[5/2]$	2	156526.436	0.015		99% + 1% 9s $(3/2)^2[3/2]$		2
7f	$(3/2)^2[9/2]°$	$3d^9(^2D_{3/2})7f$	$^2[9/2]°$	5	156711.894	0.007		91% + 6% 8f $(5/2)^2[11/2]°$ + 3% 8f $(5/2)^2[9/2]°$		2
7f	$(3/2)^2[9/2]°$	$3d^9(^2D_{3/2})7f$	$^2[9/2]°$	4	156721.308	0.007		97% + 3% 8f $(5/2)^2[9/2]°$		2
7f	$(3/2)^2[7/2]°$	$3d^9(^2D_{3/2})7f$	$^2[7/2]°$	4	[156754.9]	7		99% + 1% 8f $(5/2)^2[7/2]°$	sf	0
8f	$(5/2)^2[11/2]°$	$3d^9(^2D_{5/2})8f$	$^2[11/2]°$	5	156761.443	0.008		94% + 6% 7f $(3/2)^2[9/2]°$		3
7g	$(3/2)^2[5/2]$	$3d^9(^2D_{3/2})7g$	$^2[5/2]$	3	156763.044	0.014		99% + 1% 8g $(5/2)^2[5/2]$		1
7g	$(3/2)^2[5/2]$	$3d^9(^2D_{3/2})7g$	$^2[5/2]$	2	156763.090	0.011		99% + 1% 8g $(5/2)^2[5/2]$		1
8f	$(5/2)^2[11/2]°$	$3d^9(^2D_{5/2})8f$	$^2[11/2]°$	6	156767.066	0.016		100%		2
7g	$(3/2)^2[11/2]$	$3d^9(^2D_{3/2})7g$	$^2[11/2]$	6	156767.609	0.011		99% + 1% 8g $(5/2)^2[11/2]$	N	1
7g	$(3/2)^2[11/2]$	$3d^9(^2D_{3/2})7g$	$^2[11/2]$	5	156767.630	0.011		99% + 1% 8g $(5/2)^2[11/2]$		2
7g	$(3/2)^2[7/2]$	$3d^9(^2D_{3/2})7g$	$^2[7/2]$	3	156775.405	0.014		99% + 1% 8g $(5/2)^2[7/2]$		2

Table A2. *Cont.*

Label [a]	Configuration	Term	*J*	Level [b], cm^{-1}	Unc. [c], cm^{-1}	Landé *g* [d]	Leading Percentages [e]	Note [f]	N_{lines} [g]
7g (3/2)^2[7/2]	3d^9(^{2}D$_{3/2}$)7g	^{2}[7/2]	4	156775.430	0.014		99% + 1% 8g (5/2)^2[7/2]	N	1
8f (5/2)^2[7/2]°	3d^9(^{2}D$_{5/2}$)8f	^{2}[7/2]°	4	[156779.7]	7		94% + 5% 8f (5/2)^2[9/2]° + 1% 7f (3/2)^2[7/2]°	sf	0
7g (3/2)^2[9/2]	3d^9(^{2}D$_{3/2}$)7g	^{2}[9/2]	4	156779.847	0.014		100%		2
7g (3/2)^2[9/2]	3d^9(^{2}D$_{3/2}$)7g	^{2}[9/2]	5	156779.855	0.014		100%	N	1
8f (5/2)^2[9/2]°	3d^9(^{2}D$_{5/2}$)8f	^{2}[9/2]°	4	[156781.7]	7		92% + 5% 8f (5/2)^2[7/2]° + 2% 7f (3/2)^2[9/2]°	sf	0
8f (5/2)^2[9/2]°	3d^9(^{2}D$_{5/2}$)8f	^{2}[9/2]°	5	[156782.4]	7		97% + 3% 7f (3/2)^2[9/2]°	sf	0
8g (5/2)^2[3/2]	3d^9(^{2}D$_{5/2}$)8g	^{2}[3/2]	2	[156797.3]	2.0		100%	pf	0
8g (5/2)^2[7/2]	3d^9(^{2}D$_{5/2}$)8g	^{2}[7/2]	3	[156805.7]	2.0		99% + 1% 7g (3/2)^2[7/2]	pf	0
8g (5/2)^2[9/2]	3d^9(^{2}D$_{5/2}$)8g	^{2}[9/2]	5	[156807.8]	2.0		100%	pf	0
8g (5/2)^2[11/2]	3d^9(^{2}D$_{5/2}$)8g	^{2}[11/2]	6	156808.189?	0.010		99% + 1% 7g (3/2)^2[11/2]	N	2
9d (5/2)^2[9/2]	3d^9(^{2}D$_{5/2}$)9d	^{2}[9/2]	5	[156888.5]	2.0		100%	sf	0
9d (5/2)^2[9/2]	3d^9(^{2}D$_{5/2}$)9d	^{2}[9/2]	4	156912.740	0.011		51% + 44% 8d (3/2)^2[7/2] + 5% 9d (5/2)^2[7/2]		2
8d (3/2)^2[5/2]	3d^9(^{2}D$_{3/2}$)8d	^{2}[5/2]	2	156958.11?	0.06		42% + 38% 9d (5/2)^2[5/2] + 16% 8d (3/2)^2[3/2]		1
9g (5/2)^2[3/2]	3d^9(^{2}D$_{5/2}$)9g	^{2}[3/2]	2	[158241.1]	2.0		100%	pf	0
9g (5/2)^2[7/2]	3d^9(^{2}D$_{5/2}$)9g	^{2}[7/2]	3	[158246.8]	2.0		100%	pf	0
10d (5/2)^2[1/2]	3d^9(^{2}D$_{5/2}$)10d	^{2}[1/2]	1	[158285.3]	4		98% + 2% 10d (5/2)^2[3/2]	sf	0
10d (5/2)^2[3/2]	3d^9(^{2}D$_{5/2}$)10d	^{2}[3/2]	2	[158306.0]	5		95% + 4% 10d (5/2)^2[5/2]	sf	0
10d (5/2)^2[9/2]	3d^9(^{2}D$_{5/2}$)10d	^{2}[9/2]	5	[158306.9]	2.0		100%	sf	0
10d (5/2)^2[9/2]	3d^9(^{2}D$_{5/2}$)10d	^{2}[9/2]	4	[158308.9]	2.0		100%	sf	0
10d (5/2)^2[3/2]	3d^9(^{2}D$_{5/2}$)10d	^{2}[3/2]	1	[158310.4]	4		98% + 2% 10d (5/2)^2[1/2]	sf	0
10d (5/2)^2[5/2]	3d^9(^{2}D$_{5/2}$)10d	^{2}[5/2]	3	[158323.3]	4		98% + 2% 10d (5/2)^2[7/2]	sf	0
10d (5/2)^2[5/2]	3d^9(^{2}D$_{5/2}$)10d	^{2}[5/2]	2	[158333.4]	5		95% + 4% 10d (5/2)^2[3/2]	sf	0
10d (5/2)^2[7/2]	3d^9(^{2}D$_{5/2}$)10d	^{2}[7/2]	3	[158334.1]	4		98% + 2% 10d (5/2)^2[5/2]	sf	0
10d (5/2)^2[7/2]	3d^9(^{2}D$_{5/2}$)10d	^{2}[7/2]	4	[158334.2]	2.0		100%	sf	0
10s (3/2)^2[3/2]	3d^9(^{2}D$_{3/2}$)10s	^{2}[3/2]	1	158579.33	0.14		100%	N	1
8f (3/2)^2[9/2]°	3d^9(^{2}D$_{3/2}$)8f	^{2}[9/2]°	4	[158851.1]	8		100%	sf	0
8f (3/2)^2[7/2]°	3d^9(^{2}D$_{3/2}$)8f	^{2}[7/2]°	4	[158864.4]	8		100%	sf	0
8g (3/2)^2[5/2]	3d^9(^{2}D$_{3/2}$)8g	^{2}[5/2]	3	[158869.3]	2.0		100%	pf	0
8g (3/2)^2[11/2]	3d^9(^{2}D$_{3/2}$)8g	^{2}[11/2]	6	[158871.7]	2.0		100%	pf	0
8g (3/2)^2[7/2]	3d^9(^{2}D$_{3/2}$)8g	^{2}[7/2]	4	[158875.8]	2.0		100%	pf	0
8h (3/2)^2[7/2]°	3d^9(^{2}D$_{3/2}$)8h	^{2}[7/2]°	3	[158876.9]	2.0		100%	pf	0
8h (3/2)^2[7/2]°	3d^9(^{2}D$_{3/2}$)8h	^{2}[7/2]°	4	[158877.0]	2.0		100%	pf	0
8h (3/2)^2[13/2]°	3d^9(^{2}D$_{3/2}$)8h	^{2}[13/2]°	6	[158877.9]	2.0		100%	pf	0
8g (3/2)^2[9/2]	3d^9(^{2}D$_{3/2}$)8g	^{2}[9/2]	5	[158878.0]	2.0		100%	pf	0
8h (3/2)^2[13/2]°	3d^9(^{2}D$_{3/2}$)8h	^{2}[13/2]°	7	[158878.0]	2.0		100%	pf	0
8h (3/2)^2[9/2]°	3d^9(^{2}D$_{3/2}$)8h	^{2}[9/2]°	4	[158880.4]	2.0		100%	pf	0
8h (3/2)^2[9/2]°	3d^9(^{2}D$_{3/2}$)8h	^{2}[9/2]°	5	[158880.5]	2.0		100%	pf	0
8h (3/2)^2[11/2]°	3d^9(^{2}D$_{3/2}$)8h	^{2}[11/2]°	5	[158881.4]	2.0		100%	pf	0
8h (3/2)^2[11/2]°	3d^9(^{2}D$_{3/2}$)8h	^{2}[11/2]°	6	[158881.5]	2.0		100%	pf	0
9d (3/2)^2[1/2]	3d^9(^{2}D$_{3/2}$)9d	^{2}[1/2]	1	[158935.9]	4		99%	sf	0

Table A2. *Cont.*

Label [a]	Configuration	Term	J	Level [b], cm^{-1}	Unc. [c], cm^{-1}	Landé g [d]	Leading Percentages [e]	Note [f]	N_{lines} [g]
9d (3/2)^2[7/2]	3d^9(^{2}D$_{3/2}$)9d	^{2}[7/2]	3	[158961.1]	2.0		100%	sf	0
9d (3/2)^2[7/2]	3d^9(^{2}D$_{3/2}$)9d	^{2}[7/2]	4	[158966.7]	2.0		100%	sf	0
9d (3/2)^2[3/2]	3d^9(^{2}D$_{3/2}$)9d	^{2}[3/2]	1	[158970.8]	2.0		100%	sf	0
9d (3/2)^2[3/2]	3d^9(^{2}D$_{3/2}$)9d	^{2}[3/2]	2	[158977.0]	5		96% + 4% 9d (3/2)^2[5/2]	sf	0
9d (3/2)^2[5/2]	3d^9(^{2}D$_{3/2}$)9d	^{2}[5/2]	3	[158999.4]	2.0		100%	sf	0
9d (3/2)^2[5/2]	3d^9(^{2}D$_{3/2}$)9d	^{2}[5/2]	2	[159000.4]	5		96% + 4% 9d (3/2)^2[3/2]	sf	0
10g (5/2)^2[3/2]	3d^9(^{2}D$_{5/2}$)10g	^{2}[3/2]	2	[159273.3]	2.0		100%	pf	0
10g (5/2)^2[7/2]	3d^9(^{2}D$_{5/2}$)10g	^{2}[7/2]	3	[159277.4]	2.0		100%	pf	0
9g (3/2)^2[5/2]	3d^9(^{2}D$_{3/2}$)9g	^{2}[5/2]	3	[160312.5]	2.0		100%	pf	0
9g (3/2)^2[11/2]	3d^9(^{2}D$_{3/2}$)9g	^{2}[11/2]	6	[160314.1]	2.0		100%	pf	0
9g (3/2)^2[7/2]	3d^9(^{2}D$_{3/2}$)9g	^{2}[7/2]	4	[160317.0]	2.0		100%	pf	0
9g (3/2)^2[9/2]	3d^9(^{2}D$_{3/2}$)9g	^{2}[9/2]	5	[160318.5]	2.0		100%	pf	0
10d (3/2)^2[1/2]	3d^9(^{2}D$_{3/2}$)10d	^{2}[1/2]	1	[160361.5]	4		99%	sf	0
10d (3/2)^2[7/2]	3d^9(^{2}D$_{3/2}$)10d	^{2}[7/2]	3	[160378.4]	2.0		100%	sf	0
10d (3/2)^2[7/2]	3d^9(^{2}D$_{3/2}$)10d	^{2}[7/2]	4	[160382.2]	2.0		100%	sf	0
10d (3/2)^2[3/2]	3d^9(^{2}D$_{3/2}$)10d	^{2}[3/2]	1	[160385.0]	2.0		100%	sf	0
10d (3/2)^2[3/2]	3d^9(^{2}D$_{3/2}$)10d	^{2}[3/2]	2	[160388.7]	5		96% + 3% 10d (3/2)^2[5/2]	sf	0
10d (3/2)^2[5/2]	3d^9(^{2}D$_{3/2}$)10d	^{2}[5/2]	3	[160404.5]	2.0		100%	sf	0
10d (3/2)^2[5/2]	3d^9(^{2}D$_{3/2}$)10d	^{2}[5/2]	2	[160405.1]	5		96% + 3% 10d (3/2)^2[3/2]	sf	0
10g (3/2)^2[5/2]	3d^9(^{2}D$_{3/2}$)10g	^{2}[5/2]	3	[161344.3]	2.0		100%	pf	0
10g (3/2)^2[11/2]	3d^9(^{2}D$_{3/2}$)10g	^{2}[11/2]	6	[161345.4]	2.0		100%	pf	0
10g (3/2)^2[7/2]	3d^9(^{2}D$_{3/2}$)10g	^{2}[7/2]	4	[161347.6]	2.0		100%	pf	0
10g (3/2)^2[9/2]	3d^9(^{2}D$_{3/2}$)10g	^{2}[9/2]	5	[161348.6]	2.0		100%	pf	0
	Cu III (3d^9 ^{2}D$_{5/2}$)	Limit		163669.2	0.5				

[a] Label used in the column of Leading Percentages;

[b] Level values were obtained in the least-squares optimization procedure using the LOPT code [39] (see text), except the following: (1) Values in square brackets were obtained using extrapolations along level series (see column "Notes"); (2) The ionization limit is quoted from Ross [2] (see text). A question mark after the value indicates an uncertain identification;

[c] Uncertainties (one standard deviation) are specified for separations from the 3d^{9}4s ^{3}D$_2$ level (at 22847.1176 cm^{-1}). To determine uncertainties relative to the ground level, the given values should be combined in quadrature with the uncertainty of the ground level, 0.017 cm^{-1};

[d] Experimental Landé g-factors are quoted from Sugar and Musgrove [17];

[e] The three leading contributions to the eigenvector are given, if their rounded value is $\geq$1%. The first percentage refers to the configuration and term given in the second and third columns, unless otherwise specified after the percentage value. For the 3d^9nl levels designated in the J_1l (a.k.a. JK) coupling scheme, the percentage value of the 3d^{8}4s4p configuration is the sum of percentage contributions of all terms of this configuration;

[f] Key to the notes: N—newly identified level; R—revised level value (new identification); RJ—revised J value (was 1 in Sugar and Musgrove [17] and Ross [2]); ci—the two previously known levels, for which the identifications have been interchanged; cd—previously known levels, for which the configuration and/or term designations have been revised; sf, pf—level values found by extrapolation using the Ritz quantum-defect or polarization formulas, respectively (see text in Section 4), in combination with the least-squares parametric fitting (see text);

[g] Number of connecting lines included in the level optimization procedure for this level.

Table 3. Least-squares fitting parameters for Cu II.

Configuration	Parameter [a]	LSF [b]	STD [c]	Group [d]	HFR [e]	LSF/HF
Even parity						
$3d^{10}$	E_{av}	5775.8	52		0.0	
$3d^94s$	E_{av}	26843.5	21		9367.0	2.8658
	$\zeta(3d)$	819.9	2.0	1	814.7	1.0064
	$G^2(3d,4s)$	8519.5	132	2	9945.5	0.8566
$3d^94d$	E_{av}	119071.8	12		98893.7	1.2040
	$\zeta(3d)$	827.0	2.0	1	821.7	1.0064
	$\zeta(4d)$	13.7	fixed		13.7	1.0000
	$F^2(3d,4d)$	3237.5	96		4085.1	0.7925
	$F^4(3d,4d)$	1109.0	fixed		1386.3	0.8000
	$G^0(3d,4d)$	1426.8	47		1440.0	0.9908
	$G^2(3d,4d)$	705.6	152		1276.5	0.5528
	$G^4(3d,4d)$	538.0	228		890.6	0.6041
$3d^95s$	E_{av}	109078.5	20		92286.1	1.1820
	$\zeta(3d)$	826.4	2.0	1	821.1	1.0064
	$G^2(3d,5s)$	1537.1	24	2	1794.4	0.8566
$3d^95d$	E_{av}	137813.1	10		120253.4	1.1460
	$\zeta(3d)$	827.6	2.0	1	822.3	1.0064
	$\zeta(5d)$	6.1	fixed		6.1	1.0000
	$F^2(3d,5d)$	1905.2	73	5	1567.9	1.2152
	$F^4(3d,5d)$	463.5	fixed		579.4	0.8000
	$G^0(3d,5d)$	455.2	20	8	616.8	0.7380
	$G^2(3d,5d)$	412.6	18	8	559.0	0.7380
	$G^4(3d,5d)$	290.1	13	8	393.0	0.7380
$3d^95g$	E_{av}	146914.9	9		129486.8	1.1346
	$\zeta(3d)$	828.0	2.0	1	822.7	1.0064
	$F^2(3d,5g)$	133.4	fixed		166.7	0.8000
	$F^4(3d,5g)$	5.2	fixed		6.5	0.8000
$3d^96s$	E_{av}	134521.0	20		117385.5	1.1460
	$\zeta(3d)$	827.4	2.0	1	822.1	1.0064
	$G^2(3d,6s)$	564.8	9	2	659.4	0.8566
$3d^96d$	E_{av}	147347.9	10		129891.5	1.1344
	$\zeta(3d)$	827.8	2.0	1	822.5	1.0064
	$\zeta(6d)$	3.2	fixed		3.2	1.0000
	$F^2(3d,6d)$	952.5	36	5	783.9	1.2152
	$F^4(3d,6d)$	240.4	fixed		300.5	0.8000
	$G^0(3d,6d)$	182.2	17	9	320.8	0.5680
	$G^2(3d,6d)$	166.8	15	9	293.6	0.5680
	$G^4(3d,6d)$	117.6	11	9	207.0	0.5680
$3d^96g$	E_{av}	152289.1	9		134856.3	1.1293
	$\zeta(3d)$	828.0	2.0	1	822.7	1.0064
	$F^2(3d,6g)$	77.6	fixed		97.0	0.8000
	$F^4(3d,6g)$	3.9	fixed		4.9	0.8000
$3d^97s$	E_{av}	145697.1	20		128417.0	1.1346
	$\zeta(3d)$	827.7	2.0	1	822.4	1.0064
	$G^2(3d,7s)$	273.2	4	2	318.9	0.8566
$3d^97d$	E_{av}	152539.1	10		135114.4	1.1290
	$\zeta(3d)$	827.9	2.0	1	822.6	1.0064
	$\zeta(7d)$	1.9	fixed		1.9	1.0000
	$F^2(3d,7d)$	546.1	21	5	449.4	1.2152
	$F^4(3d,7d)$	140.5	fixed		175.6	0.8000
	$G^0(3d,7d)$	149.8	fixed		187.2	0.8000
	$G^2(3d,7d)$	166.8	fixed		172.4	0.8000
	$G^4(3d,7d)$	117.6	fixed		121.8	0.8000
$3d^97g$	E_{av}	155529.5	10		138096.9	1.1262
	$\zeta(3d)$	828.0	2.0	1	822.7	1.0064
	$F^2(3d,7g)$	49.0	fixed		61.3	0.8000
	$F^4(3d,7g)$	2.7	fixed		3.4	0.8000

Table 3. *Cont.*

Configuration	Parameter [a]	LSF [b]	STD [c]	Group [d]	HFR [e]	LSF/HF
Even parity						
$3d^98s$	E_{av}	151604.1	20		134260.1	1.1292
	$\zeta(3d)$	827.8	2.0	1	822.5	1.0064
	$G^2(3d,8s)$	153.2	2.0	2	178.9	0.8566
$3d^98d$	E_{av}	155694.0	14		138268.4	1.1260
	$\zeta(3d)$	827.9	2.0	1	822.6	1.0064
	$\zeta(8d)$	1.2	fixed		1.2	1.0000
	$F^2(3d,8d)$	342.5	13	5	281.9	1.2152
	$F^4(3d,8d)$	89.1	fixed		111.4	0.8000
	$G^0(3d,8d)$	95.0	fixed		118.8	0.8000
	$G^2(3d,8d)$	87.7	fixed		109.6	0.8000
	$G^4(3d,8d)$	62.0	fixed		77.5	0.8000
$3d^98g$	E_{av}	157632.2	17		140200.6	1.1243
	$\zeta(3d)$	828.0	2.0	1	822.7	1.0064
	$F^2(3d,8g)$	32.9	fixed		41.1	0.8000
	$F^4(3d,8g)$	2.0	fixed		2.5	0.8000
$3d^99s$	E_{av}	155106.1	20		137730.5	1.1262
	$\zeta(3d)$	827.9	2.0	1	822.6	1.0064
	$G^2(3d,9s)$	94.7	1.0	2	110.5	0.8566
$3d^99d$	E_{av}	157752.1	14		140319.6	1.1242
	$\zeta(3d)$	828.0	2.0	1	822.7	1.0064
	$\zeta(9d)$	0.8	fixed		0.8	1.0000
	$F^2(3d,9d)$	229.1	9	5	188.5	1.2152
	$F^4(3d,9d)$	60.0	fixed		75.0	0.8000
	$G^0(3d,9d)$	64.0	fixed		80.0	0.8000
	$G^2(3d,9d)$	59.2	fixed		74.0	0.8000
	$G^4(3d,9d)$	41.8	fixed		52.2	0.8000
$3d^99g$	E_{av}	159073.9	17		141643.1	1.1231
	$\zeta(3d)$	828.0	2.0	1	822.7	1.0064
	$F^2(3d,9g)$	23.1	fixed		28.9	0.8000
	$F^4(3d,9g)$	1.4	fixed		1.7	0.8000
$3d^910s$	E_{av}	157351.8	23		139959.4	1.1243
	$\zeta(3d)$	827.9	2.0	1	822.6	1.0064
	$G^2(3d,10s)$	62.5	1.0	2	73.0	0.8566
$3d^910d$	E_{av}	159159.7	10		141729.3	1.1230
	$\zeta(3d)$	828.0	2.0	1	822.7	1.0064
	$\zeta(10d)$	0.6	fixed		0.6	1.0000
	$F^2(3d,10d)$	160.7	6	5	132.3	1.2152
	$F^4(3d,10d)$	42.3	fixed		52.9	0.8000
	$G^0(3d,10d)$	45.1	fixed		56.4	0.8000
	$G^2(3d,10d)$	41.8	fixed		52.2	0.8000
	$G^4(3d,10d)$	29.5	fixed		36.9	0.8000
$3d^910g$	E_{av}	160104.8	17		142674.4	1.1222
	$\zeta(3d)$	828.0	2.0	1	822.7	1.0064
	$F^2(3d,10g)$	16.8	fixed		21.0	0.8000
	$F^4(3d,10g)$	1.1	fixed		1.4	0.8000
$3d^84s^2$	E_{av}	88372.4	16		62522.0	1.4135
	$F^2(3d,3d)$	91560.3	114	3	109696.0	0.8347
	$F^4(3d,3d)$	58255.9	103	4	68373.1	0.8520
	$\alpha(3d)$	93.8	3	7	0.0	
	$\zeta(3d)$	892.2	3	1	886.5	1.0064
$3d^84p^2$	E_{av}	188427.0	fixed		162579.7	1.1590
	$F^2(3d,3d)$	92340.0	115	3	110630.1	0.8347
	$F^4(3d,3d)$	58793.9	104	4	69004.6	0.8520
	$\alpha(3d)$	93.8	3	7	0.0	
	$F^2(4p,4p)$	28518.8	fixed		35648.5	0.8000
	$\zeta(3d)$	898.5	3	1	892.8	1.0064
	$\zeta(4p)$	619.0	fixed		619.0	1.0000
	$F^2(3d,4p)$	12983.5	fixed		16229.4	0.8000
	$G^1(3d,4p)$	4628.9	fixed		5786.1	0.8000
	$G^3(3d,4p)$	3922.3	fixed		4902.9	0.8000

Table 3. *Cont.*

Configuration	Parameter [a]	LSF [b]	STD [c]	Group [d]	HFR [e]	LSF/HF
Even parity						
$3d^84s4d$	E_{av}	186352.1	fixed		160504.8	1.1610
	F^2(3d,3d)	92375.3	115	3	110672.4	0.8347
	F^4(3d,3d)	58816.8	104	4	69031.5	0.8520
	α(3d)	93.8	3	7	0.0	
	ζ(3d)	898.6	3	1	892.9	1.0064
	F^2(3d,4d)	5461.6	208	5	4494.5	1.2152
	G^2(3d,4s)	9256.3	144	2	10805.6	0.8566
$3d^84s5s$	E_{av}	178348.4	fixed		152501.1	1.1695
	F^2(3d,3d)	92277.5	115	3	110555.2	0.8347
	F^4(3d,3d)	58748.9	104	4	68951.7	0.8520
	α(3d)	93.8	3	7	0.0	
	ζ(3d)	897.8	3	1	892.1	1.0064
	G^2(3d,4s)	9552.4	148	2	11151.3	0.8566
	G^2(3d,5s)	1270.3	20	2	1482.9	0.8566
$3d^84d^2$	E_{av}	307518.0	fixed		281670.7	1.0918
	F^2(3d,3d)	93279.6	116	3	111755.9	0.8347
	F^4(3d,3d)	59441.2	105	4	69764.2	0.8520
	α(3d)	93.8	3	7	0.0	
	ζ(3d)	906.0	3	1	900.2	1.0064
	F^2(3d,4d)	7552.1	288	5	6214.9	1.2152
$3d^84s5d$	E_{av}	209861.7	fixed		184014.4	1.1405
	F^2(3d,3d)	92407.3	115	3	110710.8	0.8347
	F^4(3d,3d)	58838.5	104	4	69056.9	0.8520
	α(3d)	93.8	3	7	0.0	
	ζ(3d)	899.0	3	1	893.3	1.0064
	F^2(3d,5d)	1997.4	76	5	1643.7	1.2152
	G^2(3d,4s)	9429.0	147	2	11007.3	0.8566
Configuration interaction						
$3d^94d$-$3d^95d$	$R_d{}^0$(3d4d,3d5d)	147.8	7	6	135.3	1.0921
	$R_d{}^2$(3d4d,3d5d)	2400.9	108	6	2198.4	1.0921
	$R_d{}^4$(3d4d,3d5d)	959.9	43	6	879.0	1.0921
	$R_e{}^0$(3d4d,3d5d)	1028.7	46	6	942.0	1.0921
	$R_e{}^2$(3d4d,3d5d)	921.6	41	6	843.9	1.0921
	$R_e{}^4$(3d4d,3d5d)	645.4	29	6	591.0	1.0921
$3d^94d$-$3d^96d$	$R_d{}^0$(3d4d,3d6d)	106.6	5	6	97.6	1.0921
	$R_d{}^2$(3d4d,3d6d)	1631.1	73	6	1493.6	1.0921
	$R_d{}^4$(3d4d,3d6d)	684.3	31	6	626.6	1.0921
	$R_e{}^0$(3d4d,3d6d)	741.5	33	6	679.0	1.0921
	$R_e{}^2$(3d4d,3d6d)	667.2	30	6	610.9	1.0921
	$R_e{}^4$(3d4d,3d6d)	467.9	21	6	428.4	1.0921
$3d^94d$-$3d^97d$	$R_d{}^0$(3d4d,3d7d)	81.5	4	6	74.6	1.0921
	$R_d{}^2$(3d4d,3d7d)	1212.2	54	6	1110.0	1.0921
	$R_d{}^4$(3d4d,3d7d)	520.2	23	6	476.3	1.0921
	$R_e{}^0$(3d4d,3d7d)	566.5	25	6	518.7	1.0921
	$R_e{}^2$(3d4d,3d7d)	510.8	23	6	467.7	1.0921
	$R_e{}^4$(3d4d,3d7d)	358.5	16	6	328.3	1.0921
$3d^94d$-$3d^98d$	$R_d{}^0$(3d4d,3d8d)	64.9	3	6	59.4	1.0921
	$R_d{}^2$(3d4d,3d8d)	950.0	43	6	869.9	1.0921
	$R_d{}^4$(3d4d,3d8d)	412.8	18	6	378.0	1.0921
	$R_e{}^0$(3d4d,3d8d)	451.0	20	6	413.0	1.0921
	$R_e{}^2$(3d4d,3d8d)	407.2	18	6	372.9	1.0921
	$R_e{}^4$(3d4d,3d8d)	286.0	13	6	261.9	1.0921
$3d^94d$-$3d^99d$	$R_d{}^0$(3d4d,3d9d)	53.2	2.0	6	48.7	1.0921
	$R_d{}^2$(3d4d,3d9d)	771.9	35	6	706.8	1.0921
	$R_d{}^4$(3d4d,3d9d)	338.0	15	6	309.5	1.0921
	$R_e{}^0$(3d4d,3d9d)	370.1	17	6	338.9	1.0921
	$R_e{}^2$(3d4d,3d9d)	334.4	15	6	306.2	1.0921
	$R_e{}^4$(3d4d,3d9d)	234.9	11	6	215.1	1.0921

Table 3. *Cont.*

Configuration	Parameter [a]	LSF [b]	STD [c]	Group [d]	HFR [e]	LSF/HF
Configuration interaction						
$3d^94d$-$3d^910d$	$R_d{}^0$(3d4d,3d10d)	44.8	2.0	6	41.0	1.0921
	$R_d{}^2$(3d4d,3d10d)	643.8	29	6	589.5	1.0921
	$R_d{}^4$(3d4d,3d10d)	283.4	13	6	259.5	1.0921
	$R_e{}^0$(3d4d,3d10d)	310.7	14	6	284.5	1.0921
	$R_e{}^2$(3d4d,3d10d)	281.0	13	6	257.3	1.0921
	$R_e{}^4$(3d4d,3d10d)	197.5	9	6	180.8	1.0921
$3d^95d$-$3d^96d$	$R_d{}^2$(3d5d,3d6d)	1147.0	51	6	1050.3	1.0921
	$R_d{}^4$(3d5d,3d6d)	454.3	20	6	416.0	1.0921
	$R_e{}^0$(3d5d,3d6d)	485.8	22	6	444.8	1.0921
	$R_e{}^2$(3d5d,3d6d)	442.4	20	6	405.1	1.0921
	$R_e{}^4$(3d5d,3d6d)	311.5	14	6	285.2	1.0921
$3d^95d$-$3d^97d$	$R_d{}^2$(3d5d,3d7d)	850.8	38	6	779.1	1.0921
	$R_d{}^4$(3d5d,3d7d)	346.3	16	6	317.1	1.0921
	$R_e{}^0$(3d5d,3d7d)	371.2	17	6	339.9	1.0921
	$R_e{}^2$(3d5d,3d7d)	339.0	15	6	310.4	1.0921
	$R_e{}^4$(3d5d,3d7d)	238.7	11	6	218.6	1.0921
$3d^95d$-$3d^98d$	$R_d{}^2$(3d5d,3d8d)	666.6	30	6	610.4	1.0921
	$R_d{}^4$(3d5d,3d8d)	275.3	12	6	252.1	1.0921
	$R_e{}^0$(3d5d,3d8d)	295.5	13	6	270.6	1.0921
	$R_e{}^2$(3d5d,3d8d)	270.3	12	6	247.5	1.0921
	$R_e{}^4$(3d5d,3d8d)	190.6	9	6	174.5	1.0921
$3d^95d$-$3d^99d$	$R_d{}^2$(3d5d,3d9d)	541.3	24	6	495.7	1.0921
	$R_d{}^4$(3d5d,3d9d)	225.8	10	6	206.8	1.0921
	$R_e{}^0$(3d5d,3d9d)	242.4	11	6	222.0	1.0921
	$R_e{}^2$(3d5d,3d9d)	222.0	10	6	203.3	1.0921
	$R_e{}^4$(3d5d,3d9d)	156.6	7	6	143.4	1.0921
$3d^95d$-$3d^910d$	$R_d{}^2$(3d5d,3d10d)	451.6	20	6	413.5	1.0921
	$R_d{}^4$(3d5d,3d10d)	189.4	8	6	173.4	1.0921
	$R_e{}^0$(3d5d,3d10d)	203.7	9	6	186.5	1.0921
	$R_e{}^2$(3d5d,3d10d)	186.4	8	6	170.7	1.0921
	$R_e{}^4$(3d5d,3d10d)	131.6	6	6	120.5	1.0921
$3d^96d$-$3d^97d$	$R_d{}^2$(3d6d,3d7d)	631.0	28	6	577.8	1.0921
	$R_d{}^4$(3d6d,3d7d)	250.7	11	6	229.6	1.0921
	$R_e{}^0$(3d6d,3d7d)	267.7	12	6	245.1	1.0921
	$R_e{}^2$(3d6d,3d7d)	245.7	11	6	225.0	1.0921
	$R_e{}^4$(3d6d,3d7d)	173.4	8	6	158.8	1.0921
$3d^96d$-$3d^98d$	$R_d{}^2$(3d6d,3d8d)	494.0	22	6	452.3	1.0921
	$R_d{}^4$(3d6d,3d8d)	199.6	9	6	182.8	1.0921
	$R_e{}^0$(3d6d,3d8d)	213.3	10	6	195.3	1.0921
	$R_e{}^2$(3d6d,3d8d)	196.0	9	6	179.5	1.0921
	$R_e{}^4$(3d6d,3d8d)	138.5	6	6	126.8	1.0921
$3d^96d$-$3d^99d$	$R_d{}^2$(3d6d,3d9d)	401.0	18	6	367.2	1.0921
	$R_d{}^4$(3d6d,3d9d)	163.7	7	6	149.9	1.0921
	$R_e{}^0$(3d6d,3d9d)	175.1	8	6	160.3	1.0921
	$R_e{}^2$(3d6d,3d9d)	161.0	7	6	147.4	1.0921
	$R_e{}^4$(3d6d,3d9d)	113.7	5	6	104.1	1.0921
$3d^96d$-$3d^910d$	$R_d{}^2$(3d6d,3d10d)	334.6	15	6	306.4	1.0921
	$R_d{}^4$(3d6d,3d10d)	137.3	6	6	125.7	1.0921
	$R_e{}^0$(3d6d,3d10d)	146.9	7	6	134.5	1.0921
	$R_e{}^2$(3d6d,3d10d)	135.3	6	6	123.9	1.0921
	$R_e{}^4$(3d6d,3d10d)	95.6	4	6	87.5	1.0921
$3d^97d$-$3d^98d$	$R_d{}^2$(3d7d,3d8d)	382.7	17	6	350.4	1.0921
	$R_d{}^4$(3d7d,3d8d)	152.8	7	6	139.9	1.0921
	$R_e{}^0$(3d7d,3d8d)	163.0	7	6	149.3	1.0921
	$R_e{}^2$(3d7d,3d8d)	150.2	7	6	137.5	1.0921
	$R_e{}^4$(3d7d,3d8d)	106.0	5	6	97.1	1.0921

Table 3. *Cont.*

Configuration	Parameter [a]	LSF [b]	STD [c]	Group [d]	HFR [e]	LSF/HF
Configuration interaction						
$3d^97d$-$3d^99d$	R_d^2(3d7d,3d9d)	310.7	14	6	284.5	1.0921
	R_d^4(3d7d,3d9d)	125.4	6	6	114.8	1.0921
	R_e^0(3d7d,3d9d)	133.7	6	6	122.4	1.0921
	R_e^2(3d7d,3d9d)	123.4	6	6	113.0	1.0921
	R_e^4(3d7d,3d9d)	87.3	4	6	79.9	1.0921
$3d^97d$-$3d^910d$	R_d^2(3d7d,3d10d)	259.2	12	6	237.3	1.0921
	R_d^4(3d7d,3d10d)	105.3	5	6	96.4	1.0921
	R_e^0(3d7d,3d10d)	112.3	5	6	102.8	1.0921
	R_e^2(3d7d,3d10d)	103.6	5	6	94.9	1.0921
	R_e^4(3d7d,3d10d)	73.3	3	6	67.1	1.0921
$3d^98d$-$3d^99d$	R_d^2(3d8d,3d9d)	249.3	11	6	228.3	1.0921
	R_d^4(3d8d,3d9d)	99.9	4	6	91.5	1.0921
	R_e^0(3d8d,3d9d)	106.5	5	6	97.5	1.0921
	R_e^2(3d8d,3d9d)	98.4	4	6	90.1	1.0921
	R_e^4(3d8d,3d9d)	69.6	3	6	63.7	1.0921
$3d^98d$-$3d^910d$	R_d^2(3d8d,3d10d)	207.8	9	6	190.3	1.0921
	R_d^4(3d8d,3d10d)	83.9	4	6	76.8	1.0921
	R_e^0(3d8d,3d10d)	89.4	4	6	81.9	1.0921
	R_e^2(3d8d,3d10d)	82.8	4	6	75.8	1.0921
	R_e^4(3d8d,3d10d)	58.5	3	6	53.6	1.0921
$3d^99d$-$3d^910d$	R_d^2(3d9d,3d10d)	171.3	8	6	156.9	1.0921
	R_d^4(3d9d,3d10d)	68.8	3	6	63.0	1.0921
	R_e^0(3d9d,3d10d)	73.3	3	6	67.1	1.0921
	R_e^2(3d9d,3d10d)	67.9	3	6	62.2	1.0921
	R_e^4(3d9d,3d10d)	48.1	2.0	6	44.0	1.0921
Odd parity						
$3d^94p$	E_{av}	73281.0	22		53995.6	1.3572
	ζ(3d)	828.8	5	1	818.3	1.0128
	ζ(4p)	538.0	20	9	444.2	1.2112
	F^2(3d,4p)	13098.9	152		13901.7	0.9423
	G^1(3d,4p)	4368.5	65		5338.7	0.8183
	G^3(3d,4p)	3615.9	363		4299.1	0.8411
$3d^94f$	E_{av}	136895.6	17		119289.7	1.1476
	ζ(3d)	833.2	5	1	822.7	1.0128
$3d^95p$	E_{av}	121911.6	24		104871.2	1.1625
	ζ(3d)	832.0	5	1	821.5	1.0128
	ζ(5p)	159.8	6	9	131.9	1.2112
	F^2(3d,5p)	3512.5	21	2	3548.4	0.9899
	G^1(3d,5p)	1262.1	94	3	1287.6	0.9802
	G^3(3d,5p)	1174.4	306	10	1102.4	1.0653
$3d^95f$	E_{av}	146833.4	16		129308.6	1.1355
	ζ(3d)	833.2	5	1	822.7	1.0128
$3d^96p$	E_{av}	139748.8	29		122652.3	1.1394
	ζ(3d)	832.7	5	1	822.2	1.0128
	ζ(6p)	69.9	3	9	57.7	1.2112
	F^2(3d,6p)	1443.6	9	2	1458.4	0.9899
	G^1(3d,6p)	521.8	39	3	532.4	0.9802
	G^3(3d,6p)	493.4	129	10	463.1	1.0653
$3d^96f$	E_{av}	152224.4	17		134753.4	1.1297
	ζ(3d)	833.2	5	1	822.7	1.0128
$3d^96h$	E_{av}	152299.5	16		134875.8	1.1292
	ζ(3d)	833.2	5	1	822.7	1.0128
$3d^97p$	E_{av}	148438.8	22		131122.8	1.1321
	ζ(3d)	833.0	5	1	822.5	1.0128
	ζ(7p)	36.8	1.0	9	30.4	1.2112
	F^2(3d,7p)	738.7	4	2	746.3	0.9899
	G^1(3d,7p)	268.6	20	3	274.0	0.9802
	G^3(3d,7p)	255.8	67	10	240.1	1.0653

Table 3. *Cont.*

Configuration	Parameter [a]	LSF [b]	STD [c]	Group [d]	HFR [e]	LSF/HF
Odd parity						
$3d^97f$	E_{av}	155498.8	22		138032.9	1.1265
	$\zeta(3d)$	833.2	5	1	822.7	1.0128
$3d^97h$	E_{av}	155540.7	21		138110.8	1.1262
	$\zeta(3d)$	833.2	5	1	822.7	1.0128
$3d^98p$	E_{av}	153145.3	54		135833.1	1.1275
	$\zeta(3d)$	833.1	5	1	822.6	1.0128
	$\zeta(8p)$	21.8	1.0	9	18.0	1.2112
	$F^2(3d,8p)$	429.6	3	2	434.0	0.9899
	$G^1(3d,8p)$	156.8	12	3	160.0	0.9802
	$G^3(3d,8p)$	150.0	39	10	140.8	1.0653
$3d^98f$	E_{av}	157610.4	28		140157.8	1.1245
	$\zeta(3d)$	833.2	5	1	822.7	1.0128
$3d^98h$	E_{av}	157629.5	26		140210.6	1.1242
	$\zeta(3d)$	833.2	5	1	822.7	1.0128
$3d^84s4p$	E_{av}	132837.3	12		104548.6	1.2706
	$F^2(3d,3d)$	92783.8	119	4	110131.5	0.8425
	$F^4(3d,3d)$	59745.4	207	5	68667.5	0.8701
	$\alpha(3d)$	84.8	3	8	0.0	
	$\zeta(3d)$	900.8	5	1	889.4	1.0128
	$\zeta(4p)$	732.4	27	9	604.7	1.2112
	$F^2(3d,4p)$	15857.7	94	2	16019.7	0.9899
	$G^2(3d,4s)$	8059.3	145	7	10125.0	0.7960
	$G^1(3d,4p)$	5532.7	82	11	5751.7	0.9619
	$G^3(3d,4p)$	4703.6	208	12	4854.4	0.9689
	$G^1(4s,4p)$	37321.3	52	6	47726.6	0.7820
$3d^84s4f$	E_{av}	211778.8	fixed		183501.6	1.1541
	$F^2(3d,3d)$	93301.4	119	4	110745.9	0.8425
	$F^4(3d,3d)$	60104.6	208	5	69080.2	0.8701
	$\alpha(3d)$	84.8	3	8	0.0	
	$\zeta(3d)$	905.0	5	1	893.6	1.0128
	$G^2(3d,4s)$	8816.3	158	7	11076.0	0.7960
$3d^84p4d$	E_{av}	244758.6	fixed		216481.4	1.1306
	$F^2(3d,3d)$	93638.6	120	4	111146.1	0.8425
	$F^4(3d,3d)$	60341.3	209	5	69352.4	0.8701
	$\alpha(3d)$	84.8	3	8	0.0	
	$\zeta(3d)$	907.5	5	1	896.1	1.0128
	$\zeta(4p)$	873.4	33	9	721.1	1.2112
	$F^2(3d,4p)$	17746.3	105	2	17927.6	0.9899
	$G^1(3d,4p)$	6222.8	92	11	6469.1	0.9619
	$G^3(3d,4p)$	5367.4	237	12	5539.4	0.9689
$3d^84s5p$	E_{av}	195045.6	fixed		166768.4	1.1696
	$F^2(3d,3d)$	93179.2	119	4	110600.7	0.8425
	$F^4(3d,3d)$	60019.6	208	5	68982.6	0.8701
	$\alpha(3d)$	84.8	3	8	0.0	
	$\zeta(3d)$	903.9	5	1	892.5	1.0128
	$\zeta(5p)$	182.9	7	9	151.0	1.2112
	$F^2(3d,5p)$	3559.6	21	2	3596.0	0.9899
	$G^2(3d,4s)$	8773.2	158	7	11021.9	0.7960
	$G^1(3d,5p)$	1184.8	89	3	1208.8	0.9802
	$G^3(3d,5p)$	1145.1	298	10	1074.9	1.0653
	$G^1(4s,5p)$	4181.2	6	6	5346.9	0.7820

Table 3. *Cont.*

Configuration	Parameter [a]	LSF [b]	STD [c]	Group [d]	HFR [e]	LSF/HF
Odd parity						
$3d^84s5f$	E_{av}	222119.6	fixed		193842.4	1.1459
	$F^2(3d,3d)$	93300.7	119	4	110745.0	0.8425
	$F^4(3d,3d)$	60104.1	208	5	69079.8	0.8701
	$\alpha(3d)$	84.8	3	8	0.0	
	$\zeta(3d)$	905.0	5	1	893.6	1.0128
	$G^2(3d,4s)$	8827.6	159	7	11090.3	0.7960

[a] All omitted single-configuration parameters were fixed at HFR values scaled by a factor of 0.80 for the direct and exchange electrostatic parameters F^k and G^k, and 1.0 for spin-orbit parameters ζ. All omitted configuration-interaction parameters were fixed at HFR values scaled by a factor of 0.94 in both parities;
[b] Parameter values determined in the least-squares fitting procedure (see Section 3);
[c] Standard deviation of the least-squares fitting;
[d] Parameters within each numbered group were linked together in the LSF procedure, so that the ratios to ab initio HFR values were the same for each parameter in the group;
[e] The ab initio Hartree-Fock-Relativistic parameter values as computed by Cowan's codes [52]. In this calculation, we included both relativistic and Breit corrections (in Cowan's codes, the latter affect only the average energies of configurations) and used the scaling factor of 1.0 for the exchange contribution.

References

1. Shenstone, A.G. The first spark spectrum of copper. *Philos. Trans. R. Soc. Lond. Ser. A* **1936**, *235*, 195–243. [CrossRef]
2. Ross, C.B., Jr. Vacuum Ultraviolet Standards in the Spectrum of Cu II. Ph.D. Thesis, Purdue University, West Lafayette, IN, USA, 1969.
3. Thackeray, A.D. Identifications in the spectra of Eta Carinae and RR Telescopii. *Mon. Not. R. Astron. Soc.* **1953**, *113*, 211–236. [CrossRef]
4. Aller, L.H.; Polidan, R.S.; Rhodes, E.J., Jr.; Wares, G.W. The spectrum of RR Telescopii in 1968. *Astrophys. Space Sci.* **1973**, *20*, 93–110. [CrossRef]
5. McKenna, F.C.; Keenan, F.P.; Hambly, N.C.; Allende Prieto, C.; Rolleston, W.R.J.; Aller, L.H.; Feibelman, W.A. The optical spectral line list of RR Telescopii. *Astrophys. J. Suppl. Ser.* **1997**, *109*, 225–239. [CrossRef]
6. Wallerstein, G.; Gilroy, K.K.; Zethson, T.; Johansson, S.; Hamann, F. Line identifications in the spectrum of η Carinae as observed in 1990–1991 with CCD detectors. *Publ. Astron. Soc. Pac.* **2001**, *113*, 1210–1214. [CrossRef]
7. Jaschek, J.; Jaschek, M. *The Behavior of Chemical Elements in Stars*; Cambridge University Press: Cambridge, UK, 1995.
8. Danezis, E.; Theodossiou, E. The UV Spectrum of the Be Star 88 Herculis. *Astrophys. Space Sci.* **1990**, *174*, 49–90. [CrossRef]
9. Samain, D. A High Spectral Resolution Atlas of the Balloon Ultraviolet Spectrum of the Sun: 1950–2000 Å. *Astron. Astrophys. Suppl. Ser.* **1995**, *113*, 237–255.
10. McNeil, J.R.; Collins, G.J.; Persson, K.B.; Franzen, D.L. CW laser oscillation in Cu II. *Appl. Phys. Lett.* **1975**, *27*, 595–598. [CrossRef]
11. McNeil, J.R.; Collins, G.J.; Persson, K.B.; Franzen, D.L. Ultraviolet laser action from Cu II in the 2500-Å region. *Appl. Phys. Lett.* **1976**, *28*, 207–209. [CrossRef]
12. Jain, K. New UV and IR transitions in gold, copper, and cadmium hollow cathode lasers. *IEEE J. Quantum Electron.* **1980**, *16*, 387–391. [CrossRef]
13. Zinchenko, S.P.; Ivanov, I.G. Pulsed hollow-cathode ion lasers: Pumping and lasing parameters. *Quantum Electron.* **2012**, *42*, 518–523. [CrossRef]
14. Reader, J.; Meissner, K.W.; Andrew, K.L. Improved Cu II standard wavelengths in the vacuum ultraviolet. *J. Opt. Soc. Am.* **1960**, *50*, 221–227. [CrossRef]
15. Kaufman, V.; Ward, J.F. Measurement and calculation of Cu II, Ge II, Si II, and C I vacuum-ultraviolet lines. *J. Opt. Soc. Am.* **1966**, *56*, 1591–1597. [CrossRef]

16. Ross, C.B. *Wavelengths and Energy Levels of Singly Ionized Copper, Cu II*; University California Report LA-4498; Los Alamos Scientific Lab: Los Alamos, NM, USA, 1970.
17. Sugar, J.; Musgrove, A. Energy levels of copper, Cu I through Cu XXIX. *J. Phys. Chem. Ref. Data* **1990**, *19*, 527–616. [CrossRef]
18. Sansonetti, J.E.; Martin, W.C. Handbook of basic atomic spectroscopic data. *J. Phys. Chem. Ref. Data* **2005**, *34*, 1559–2259. [CrossRef]
19. Kramida, A.; Ralchenko, Y.; Reader, J.; NIST ASD Team. *NIST Atomic Spectra Database*; version 5.4; National Institute of Standards and Technology: Gaithersburg, MD, USA, 2016. Available online: http://physics.nist.gov/asd (accessed on 15 February 2017).
20. Nave, G.; Sansonetti, C.J. Reference wavelengths in the spectra of Fe, Ge, and Pt in the region near 1935 Å. *J. Opt. Soc. Am. B* **2004**, *21*, 442–453. [CrossRef]
21. Litzén, U.; Brault, J.W.; Thorne, A.P. Spectrum and term system of neutral nickel, Ni I. *Phys. Scr.* **1993**, *47*, 628–673. [CrossRef]
22. Wiese, W.L.; Martin, G.A. Transition Probabilities. In *Wavelengths and Transition Probabilities for Atoms and Atomic Ions*; National Standard Reference Data Series NSRDS-68; National Bureau of Standards: Gaithersburg, DC, USA, 1980; Part II; pp. 359–406.
23. Coursey, J.S.; Schwab, D.J.; Tsai, J.J.; Dragoset, R.A. *Atomic Weights and Isotopic Compositions*; version 4.0; National Institute of Standards and Technology: Gaithersburg, MD, USA, 2015. Available online: http://physics.nist.gov/Comp (accessed on 15 February 2017).
24. Elbel, M.; Fischer, W. Zur Isotopieverschiebung im Kupfer I- und II-Spektrum. *Z. Phys.* **1961**, *165*, 151–170. [CrossRef]
25. Elbel, M.; Fischer, W.; Hartmann, M. Hyperfeinstruktur und Isotopieverschiebung im Kupfer II-Spektrum. *Z. Phys.* **1963**, *176*, 288–292. [CrossRef]
26. Reader, J.; Davis, S.P. Promethium 147 hyperfine structure under high resolution. *J. Opt. Soc. Am.* **1963**, *53*, 431–435. [CrossRef]
27. Danzmann, K.; Günther, M.; Fisher, J.; Kock, M.; Kühne, M. High current hollow cathode as a radiometric transfer standard source for the extreme vacuum ultraviolet. *Appl. Opt.* **1988**, *27*, 4947–4951. [CrossRef] [PubMed]
28. Nave, G.; Griesmann, U.; Brault, J.W.; Abrams, M.C. XGREMLIN: Interferograms and Spectra from Fourier Transform Spectrometers Analysis. Astrophysics Source Code Library, record ascl:1511.004, 2015. Available online: https://github.com/gnave/Xgremlin (accessed on 9 December 2015).
29. Brault, J.W. High precision Fourier transform spectrometry: The critical role of phase corrections. *Microchim. Acta* **1987**, *93*, 215–227. [CrossRef]
30. Kaufman, V.; Andrew, K.L. Germanium vacuum ultraviolet Ritz standards. *J. Opt. Soc. Am.* **1962**, *52*, 1223–1237. [CrossRef]
31. Sansonetti, C.J.; Veza, D. Doppler-free measurement of the 546 nm line of mercury. *J. Phys. B* **2010**, *43*, 205003. [CrossRef]
32. Nave, G.; Sansonetti, C.J. Wavelengths of the $3d^6(^5D)4s\ a^6D$–$3d^5(^6S)4s4p\ y^6P$ multiplet of Fe II (UV 8). *J. Opt. Soc. Am. B* **2011**, *28*, 737–745. [CrossRef]
33. Whaling, W.; Anderson, W.H.C.; Carle, M.T.; Brault, J.W.; Zarem, H.A. Argon ion linelist and level energies in the hollow-cathode discharge. *J. Quant. Spectrosc. Radiat. Transf.* **1995**, *53*, 1–22. [CrossRef]
34. Kaufman, V. Wavelengths, energy levels, and pressure shifts in mercury 198. *J. Opt. Soc. Am.* **1962**, *52*, 866–870. [CrossRef]
35. Kramida, A. Re-optimized energy levels and Ritz wavelengths of ^{198}Hg I. *J. Res. Natl. Inst. Stand. Technol.* **2011**, *116*, 599–619. [CrossRef]
36. Wagatsuma, K.; Hirokawa, K. Observation of singly-ionized copper emission lines from a Grimm-type glow discharge plasma with argon-helium gas mixtures in a visible wavelength region. *Spectrochim. Acta B* **1993**, *48*, 1039–1044. [CrossRef]
37. Prior, M.H. Radiative decay rates of metastable Ar III and Cu II ions. *Phys. Rev. A* **1984**, *30*, 3051–3056. [CrossRef]
38. Peck, E.R.; Reeder, K. Dispersion of air. *J. Opt. Soc. Am.* **1972**, *62*, 958–962. [CrossRef]
39. Kramida, A.E. The program LOPT for least-squares optimization of energy levels. *Comput. Phys. Commun.* **2011**, *182*, 419–434. [CrossRef]

40. Andersson, M.; Yao, K.; Hutton, R.; Zou, Y.; Chen, C.Y.; Brage, T. Hyperfine-state-dependent lifetimes along the Ni-like isoelectronic sequence. *Phys. Rev. A* **2008**, *77*, 042509. [CrossRef]
41. Biémont, E.; Pinnington, E.H.; Quinet, P.; Zeippen, C.J. Core-polarization effects in Cu II. *Phys. Scr.* **2000**, *61*, 567–580. [CrossRef]
42. Brown, M.S.; Federman, S.R.; Irving, R.E.; Cheng, S.; Curtis, L.J. Lifetimes and oscillator strengths for ultraviolet transitions in singly ionized copper. *Astrophys. J.* **2009**, *702*, 880–883. [CrossRef]
43. Cederquist, H.; Mannervik, S.; Kisielinski, M.; Forsberg, P.; Martinson, I.; Curtis, L.J.; Ramanujam, P.S. Lifetimes of some excited levels in Cu I and Cu II. *Phys. Scr.* **1984**, *T8*, 104–106. [CrossRef]
44. Crespo López-Urrutia, J.R.; Kenner, B.; Neger, T.; Jäger, H. Absolute transition probabilities of Cu II lines. *J. Quant. Spectrosc. Radiat. Transf.* **1994**, 52, 111–114. [CrossRef]
45. Dong, C.Z.; Fritzsche, S. Relativistic, relaxation, and correlation effects in spectra of Cu II. *Phys. Rev. A* **2005**, *72*, 012507. [CrossRef]
46. Garstang, R.H. Transition probabilities of forbidden lines. *J. Res. Natl. Bur. Stand. Sect. A* **1964**, *68*, 61–73. [CrossRef]
47. Hefferlin, R.; Kuhlman, H.; Penz, J.; Wheeler, D. Approximate relative log gf for green lines of Cu^+. *Bull. Am. Phys. Soc.* **1971**, *16*, 106.
48. Kono, A.; Hattori, S. Lifetimes and transition probabilities in Cu II. *J. Opt. Soc. Am.* **1982**, *72*, 601–605. [CrossRef]
49. Neger, T.; Jäger, H.Z. Transition probabilities of Cu II lines. *Z. Naturforsch. A* **1988**, *43*, 507–508. [CrossRef]
50. Ortiz, M.; Mayo, R.; Biémont, É.; Quinet, P.; Malcheva, G.; Blagoev, K. Radiative parameters for some transitions arising from the $3d^94d$ and $3d^84s^2$ electronic configurations in Cu II spectrum. *J. Phys. B* **2007**, *40*, 167–176. [CrossRef]
51. Pinnington, E.H.; Rieger, G.; Kernahan, J.A.; Biémont, E. Beam-laser measurements and relativistic Hartree-Fock calculations of the lifetimes of the $3d^94p$ levels in Cu II. *Can. J. Phys.* **1997**, *75*, 1–9. [CrossRef]
52. Cowan, R.D. *The Theory of Atomic Structure and Spectra*; University California Press: Berkeley, CA, USA, 1981.
53. Azarov, V.I. Formal approach to the solution of the complex-spectra identification problem. 2. Implementation. *Phys. Scr.* **1993**, *48*, 656–667. [CrossRef]
54. Roth, C. Odd configurations in singly-ionized copper. *J. Res. Natl. Bur. Stand. Sect. A* **1969**, *73*, 599–609. [CrossRef]
55. Sansonetti, C.J.; National Institute of Standards and Technology, Gaithersburg, MD, USA. Fortran computer code RITZPL. Personal communication, 2005.
56. Sansonetti, C.J.; National Institute of Standards and Technology, Gaithersburg, MD, USA. Fortran computer code POLAR. Personal communication, 2005.
57. Kramida, A. Critical evaluation of data on atomic energy levels, wavelengths, and transition probabilities. *Fusion Sci. Technol.* **2013**, *63*, 313–323.
58. Kramida, A. Critically evaluated energy levels and spectral lines of singly ionized indium (In II). *J. Res. Natl. Inst. Stand. Technol.* **2013**, *118*, 52–104. [CrossRef] [PubMed]
59. Kramida, A.; Fuhr, J.R. *NIST Atomic Transition Probability Bibliographic Database*; version 9.0; National Institute of Standards and Technology: Gaithersburg, MD, USA, 2010. Available online: http://physics.nist.gov/Fvalbib (accessed on 15 February 2017).
60. Lux, B. Untersuchungen über den Axialdurchschlag bei der Elektrischen Explosion von Kupferdrähten. Ph.D. Thesis, Universität Kiel, Kiel, Germany, 1973.
61. Beck, D.R. Many-electron effects in and operator forms for electron quadrupole transition probabilities. *Phys. Rev. A* **1981**, *23*, 159–171. [CrossRef]

atoms

MDPI

Article

The Third Spectrum of Indium: In III

Swapnil and Tauheed Ahmad *

Physics Department, Aligarh Muslim University, Aligarh 202002, India; swapnilamu@gmail.com
* Correspondence: ahmadtauheed@rediffmail.com; Tel.: +91-9837-404-077

Academic Editor: Joseph Reader
Received: 1 February 2017; Accepted: 5 June 2017; Published: 13 June 2017

Abstract: The present investigation reports on the extended study of the third spectrum of indium (In III). This spectrum was previously analyzed in many articles, but, nevertheless, this study represents a significant extension of the previous analyses. The main new contribution is connected to the observation of transitions involving core-excited configurations. Previous data are critically evaluated and in some cases are corrected. The spectra were recorded on 3-m as well as on 10.7-m normal incidence spectrographs using a triggered spark source. Theoretical calculations were made with Cowan's code. The analysis results in the identifications of 70 spectral lines and determination of 24 new energy levels. In addition, the manuscript represents a compilation of all presently available data on In III.

Keywords: spectra; ionized atoms; wavelengths; energy levels; ionization energies

1. Introduction

The third spectrum of indium (In III) belongs to the Ag I isoelectronic sequence with the ground state [Kr] $4d^{10}5s\ ^2S_{1/2}$. The outer electronic excitation gives rise to the [Kr] $4d^{10}n\ell$ ($n \geq 5$, for $\ell \leq 2$; $n \geq \ell + 1$ otherwise) type of configurations with a simple doublet structure, while core excitation involving the configurations such as $4d^95s$ (5p + 4f), $4d^95s^2$ and $4d^95p^2$ makes a complex three-electron system having both doublet and quartet terms.

Several authors studied the In III spectrum, and it is appropriate to summarize their work briefly. The first work on the third spectrum of indium was done by Rao et al. [1], followed by Lang [2], Douglas [3] and Nodwell [4]. Rao et al. [1] identified 12 lines in the wavelength region 2983–5918 Å and established 13 levels belonging to the $4d^{10}$(5s, 6s, 7s, 5p, 6p, 5d, 6d, 4f, 5f and 5g) configurations. However, only six of those levels could be verified by later workers [2–4]. Nodwell [4] studied the indium spectrum in more detail. He recorded the indium spectra on a 2-m vacuum grating spectrograph and identified 56 lines of In III in the wavelength region 685–6198 Å. He established 27 energy levels including six doubtful. This work is listed in the Atomic Energy Levels (AEL) compilation [5]. Bhatia [6] investigated the In III spectrum more comprehensively using a 3-m normal incidence vacuum spectrograph in the range 340–2300 Å with a 1200 lines/mm grating giving a reciprocal dispersion of 2.775 Å/mm and a prism spectrograph in the region 2300 Å to 9500 Å with a disruptive electrodeless discharge. He revised and extended the earlier analysis and established the levels of the $4d^{10}ns$ (n = 5–12), $4d^{10}n$p (n = 5–9), $4d^{10}n$d (n = 5–9), $4d^{10}nf$ (n = 4–7), $4d^{10}ng$ (n = 5–9), $4d^{10}nh$ (n = 6–9), $4d^95s^2$, and $4d^95s5p$ configurations. Kaufman et al. [7] studied the core-excited transition array $4d^{10}5s$–$4d^95s5p$ in the isoelectronic sequence from In III to Te VI. The spectra were recorded on 10.7-m normal and grazing incidence spectrographs using a sliding spark source. Out of 23 possible levels of the $4d^95s5p$ configuration, they reported only 10 that can combine with the ground level $4d^{10}5s\ ^2S_{1/2}$. Kilbane et al. [8] studied photoabsorption spectra of In II–IV with a dual laser plasma (DLP) technique. They reported the $4d^{10}5s$–{$4d^95snp$ (n = 6–11) + $4d^95snf$ (n = 4–11)} transition array. They could not observe the $4d^{10}5s$–$4d^95s5p$ transitions as they lie beyond the region of their investigation. Recently, Ryabtsev et al. [9] added a new configuration $4d^95p^2$ to the In III-Te VI sequence and observed the $4d^{10}5p$–$4d^95p^2$ transition array in the range 250–600 Å using a 6.65-m normal incidence spectrograph equipped with a 1200 lines/mm grating giving a reciprocal linear dispersion of 1.25 Å/mm. They were able to determine only 13 levels out of 28 levels of the $4d^95p^2$ configuration. Skočić et al. [10] studied Stark shifts of some prominent lines of In III (6s–6p, 6p–6d, and 4f–5d).

As mentioned above, a number of publications on In III appeared in the literature [1–11]. Among these, Bhatia's [6] analysis was the most comprehensive and contained a large number of one-electron configurations. However, after careful examination of these results, a number of irregularities were noticed in Bhatia's results, for example, many lines classified did not match the In III characteristics on our recorded spectra and 17 reported lines have incorrect conversion between wavenumbers and wavelengths. Moreover, the levels of the $4d^95s5p$ configuration reported by Kaufman et al. [7] and the levels of $4d^95p^2$ configuration established by Ryabtsev et al. [9] are still incomplete. These facts prompted us to re-investigate the In III spectrum in detail. A Grotrian energy level diagram of In III is illustrated in Figure 1 showing the basic configurations and possible transition between them.

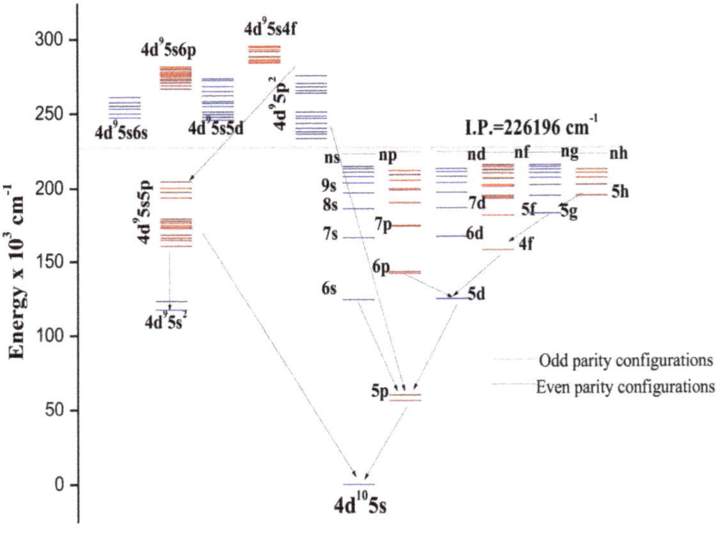

Figure 1. Grotrian diagram of In III. "I.P." denotes the ionization potential (see Section 6). Arrows denote the observed transition arrays.

2. Experiment Detail

The spectra were recorded at two different places. A 3-m vacuum spectrograph equipped with 2400 lines per mm holographic grating was employed at Antigonish laboratory in Nova Scotia, Canada with a triggered spark source to cover the wavelength region 350–2080 Å. This spectrograph gives the first order inverse dispersion of 1.385 Å/mm. For ionization separation of the spectral lines, either the charging potential of the source was varied or an inductance with a varying number of turns was inserted in series in the circuit. The charging unit was a 14.5 μF low inductance fast charging capacitor and the charging potential was varied between 2 and 6 kV. Y.N. Joshi of St. Francis Xavier University, Antigonish (Canada) provided the indium spectra that were recorded on the 10.7-m normal incidence vacuum spectrograph of the National Institute of Standards and Technology (NIST) also using a triggered spark source. The NIST spectrograph was equipped with 1200 lines/mm grating with an inverse dispersion 0.78 Å/mm. The spectrograms were measured either on an *Abbec* comparator at Aligarh or on a semi-automatic Grant's comparator in Antigonish, Canada. Known standard lines of oxygen, carbon, aluminium and silicon [11] were used as internal standards for the calibration of wavelengths. We estimated our measurements uncertainty for sharp and unblended lines to be within ±0.006 Å for wavelength below 900 Å and ±0.008 Å above that.

3. Theoretical Calculations

The ab initio calculations were performed by the Hartree–Fock method with relativistic corrections using Cowan code [12] with superposition of configurations including $4d^{10}ns$ (n = 5–12), $4d^{10}nd$ (n = 5–9), $4d^{10}ng$ (n = 5–9), $4d^9(5s^2 + 5p^2)$, $4d^95s(5d + 6s)$ configurations for the even parity system and $4d^{10}np$ (n = 5–9), $4d^{10}nf$ (n = 4–7), $4d^{10}nh$ (n = 6–9), $4d^95snp$ (n = 5–11), $4d^95snf$ (n = 4–12), $4d^85s^25p$ for the odd parity matrix involving a total of 52 configurations in our calculations. The initial scaling of the Slater energy parameters was kept at 100% of the Hartree–Fock values for E_{av} and ζ_{nl}, 85% for F^k, and 80% for the G^k as well as R^k integrals. These parameters were more refined at a later stage as least squares fitted parametric calculations were performed. The main output from these programs includes the values of energy levels, wavelengths, weighted transition rates and weighted oscillator strengths. The transition probability of lines depends on the line strength and is greatly affected by the cancellation factor [13], is also calculated by Cowan's code programs [12]. The Hartree–Fock (HFR) and least-squares-fitted (LSF) energy parameters used in the present calculations are given in Table 1 along with their scaling factor (ratio of the LSF value to the HFR value) of the parameters. The standard deviations for the even and odd parity systems are 172 cm^{-1} and 216 cm^{-1}, respectively.

Table 1. Least Square Fitted (LSF) Energy Parameters (in cm^{-1}) for In III.

Configuration	Parameters [a]	LSF	STD [#]	Group [b]	HFR	LSF/HFR
Even Parity						
5s	E_{av}	1525.6	247		1560.9	0.9774
6s	E_{av}	126,947.0	245		124,323.4	1.0211
7s	E_{av}	169,472.1	245		166,231.3	1.0195
8s	E_{av}	189,397.9	245		185,887.2	1.0189
9s	E_{av}	200,378.4	245		196,742.7	1.0185
10s	E_{av}	207,041.1	139	1	203,380.9	1.0180
11s	E_{av}	211,473.3	142	1	207,734.7	1.0180
12s	E_{av}	214,537.1	144	1	210,744.4	1.0180
13s	E_{av}	216,749.1	145	1	212,917.3	1.0180
14s	E_{av}	218,395.3	147	1	214,534.4	1.0180
5d	E_{av}	128,785.8	179		124,706.2	1.0327
	$\zeta(5d)$	149.6	125	3	120.5	1.2415
6d	E_{av}	170,730.6	174		167,143.9	1.0215
	$\zeta(6d)$	65.7	55	3	52.9	1.2420
7d	E_{av}	190,146.5	174		186,483.2	1.0196
	$\zeta(7d)$	35.0	29	3	28.2	1.2411
8d	E_{av}	200,844.9	174		197,127.9	1.0189
	$\zeta(8d)$	20.9	17	3	16.8	1.2440
9d	E_{av}	207,385.2	174	2	203,639.2	1.0184
	$\zeta(9d)$	13.4	11	3	10.8	1.2407
10d	E_{av}	211,739.0	177	2	207,914.3	1.0184
	$\zeta(10d)$	9.1	8	3	7.3	1.2466
11d	E_{av}	214,754.5	180	2	210,875.3	1.0184
	$\zeta(11d)$	6.5	5	3	5.2	1.2500
12d	E_{av}	216,933.8	182	2	213,015.3	1.0184
	$\zeta(12d)$	4.7	4	3	3.8	1.2368
5g	E_{av}	186,530.4	173		182,689.5	1.0210
	$\zeta(5d)$	0.3	Fixed		0.3	1.0000
6g	E_{av}	198,656.5	173		194,809.9	1.0197
	$\zeta(5d)$	0.2	Fixed		0.2	1.0000

Table 1. *Cont.*

Configuration	Parameters [a]	LSF	STD [#]	Group [b]	HFR	LSF/HFR
7g	E_{av}	205,968.6	173		202,130.8	1.0190
	ζ(5d)	0.1	Fixed		0.1	1.0000
8g	E_{av}	210,713.4	173		206,885.7	1.0185
	ζ(5d)	0.1	Fixed		0.1	1.0000
9g	E_{av}	213,967.7	173	4	210,144.3	1.0182
	ζ(5d)	0.0	Fixed		0.0	
10g	E_{av}	216,340.0	175	4	212,474.2	1.0182
	ζ(5d)	0.0	Fixed		0.0	
11g	E_{av}	218,091.5	177	4	214,194.4	1.0182
	ζ(11g)	0.0	Fixed		0.0	
12g	E_{av}	219,428.0	178	4	215,507.0	1.0182
	ζ(12g)	0.0	Fixed		0.0	
$4d^95s^2$	E_{av}	122,546.3	196		124,206.8	0.9866
	ζ(4d)	2827.6	53	5	2706.2	1.0449
$4d^95p^2$	E_{av}	248,495.0	551		246,591.5	1.0077
	F^2(5p,5p)	27,409.1	4453		39,408.9	0.6955
	ζ(4d)	2855.5	54	5	2732.9	1.0449
	ζ(5p)	3648.0	138		2988.1	1.2208
	F^2(4d,5p)	20,103.9	530		24,204.1	0.8306
	G^1(4d,5p)	6453.4	Fixed		7592.2	0.8500
	G^3(4d,5p)	5725.3	Fixed		6735.6	0.8500
$4d^95s5d$ *	E_{av}	262,301.3	2850		253,437.2	1.0350
$4d^95s6s$ *	E_{av}	252,606.4	Fixed		252,606.4	1.0000
	σ [#]	172				
Odd Parity						
5p	E_{av}	60,352.9	42	1	59,151.7	1.0203
	ζ(5p)	2671.6	242	2	2505.1	1.0665
6p	E_{av}	145,688.0	103	1	142,788.3	1.0203
	ζ(6p)	863.2	78	2	809.4	1.0665
7p	E_{av}	178,073.7	125	1	174,529.4	1.0203
	ζ(7p)	396.3	36	2	371.6	1.0665
8p	E_{av}	194,213.5	137	1	190,348.0	1.0203
	ζ(8p)	216.2	20	2	202.7	1.0666
9p	E_{av}	203,466.4	143	1	199,416.7	1.0203
	ζ(9p)	131.0	12	2	122.8	1.0668
10p	E_{av}	209,274.8	147	1	205,109.5	1.0203
	ζ(10p)	85.4	8	2	80.1	1.0662
11p	E_{av}	213,160.3	150	1	208,917.6	1.0203
	ζ(11p)	58.8	5	2	55.1	1.0672
12p	E_{av}	215,887.1	152	1	211,590.2	1.0203
	ζ(12p)	42.1	4	2	39.5	1.0658
4f	E_{av}	162,121.7	202		158,107.9	1.0254
	ζ(4f)	1.2	Fixed		1.2	1.0000
5f	E_{av}	185,069.7	206		181,299.2	1.0208
	ζ(5f)	0.8	Fixed		0.8	1.0000
6f	E_{av}	191,442.6	406	3	193,937.2	0.9871
	ζ(6f)	0.5	Fixed		0.5	1.0000
7f	E_{av}	198,958.4	422	3	201,550.9	0.9871
	ζ(7f)	0.3	Fixed		0.3	1.0000
8f	E_{av}	203,826.5	432	3	206,482.5	0.9871
	ζ(8f)	0.2	Fixed		0.2	1.0000
9f	E_{av}	207,155.3	439	3	209,854.6	0.9871
	ζ(9f)	0.2	Fixed		0.2	1.0000
10f	E_{av}	209,529.7	445	3	212,260.0	0.9871
	ζ(10f)	0.1	Fixed		0.1	1.0000
11f	E_{av}	211,278.0	448	3	214,031.1	0.9871
	ζ(11f)	0.1	Fixed		0.1	1.0000
12f	E_{av}	212,609.0	451	3	215,379.4	0.9871
	ζ(12f)	0.1	Fixed		0.1	1.0000
6h	E_{av}	198,520.8	109	4	194,930.4	1.0184
	ζ(6h)	0.1	Fixed		0.1	1.0000
7h	E_{av}	205,935.4	113	4	202,210.9	1.0184
	ζ(7h)	0.1	Fixed		0.1	1.0000

Table 1. *Cont.*

Configuration	Parameters [a]	LSF	STD [#]	Group [b]	HFR	LSF/HFR
8h	E_{av}	210,752.2	116	4	206,940.6	1.0184
	ζ(8h)	0.1	Fixed		0.1	1.0000
9h	E_{av}	214,053.7	118	4	210,182.4	1.0184
	ζ(9h)	0.0	Fixed		0.0	
10h	E_{av}	216,415.1	119	4	212,501.1	1.0184
	ζ(10h)	0.0	Fixed		0.0	
$4d^95s5p$	E_{av}	179,088.7	71		177,339.0	1.0099
	ζ(4d)	2747.9	79		2718.8	1.0107
	ζ(5p)	3591.3	175		2994.7	1.1992
	F^2(4d,5p)	196,94.2	839		24,199.1	0.8138
	G^2(4d,5s)	12,656.6	1434		13,315.8	0.9505
	G^1(4d,5p)	7671.9	563	5	7655.9	1.0021
	G^3(4d,5p)	6794.6	499	5	6780.4	1.0021
	G^1(5s,5p)	33,087.5	270		48,386.8	0.6838
$4d^95s6p$	E_{av}	275,446.9	127		272,620.0	1.0104
	σ [#]	216				

[a] All configuration-interaction parameters R^k for even and odd parity configurations were fixed at 80% of the Hartree–Fock value. [b] Parameters in each numbered group were linked together with their ratio fixed at the Hartree–Fock level. [#] σ and STD are the standard deviations of the fit for the levels and parameters, respectively. * Only E_{av} of unobserved interacting configurations are given.

4. Spectrum Analysis

The initial approach of the analysis was to identify In III lines with correct ionization character. A computer code FIND3 [14] was useful in the analysis to search for new levels. A total of 91 levels have been established, of which 24 are new; they are assembled in Table 2 along with least squares fitted values and LS percentage composition. Two hundred fifty-one lines have been classified in In III and they are given in Table 3 along with their transition probabilities. In the present analysis, apart from the one- electron spectrum $4d^{10}n\ell$, the configurations involving inner-shell excitation, such as $4d^95s$ (5p + 4f), $4d^95s^2$ and $4d^95p^2$ have also been studied extensively. The following sections describe them in detail.

Table 2. Optimized energy levels of in III.

				LS Compositions [d]								
J	Energy [a] cm^{-1}	Unc [b]	ΔEo-c [c] cm^{-1}		1st Component			2nd Component		3rd Component	No. of Lines [e]	Lev. Ref. [f]
Even Parity												
0.5	0.00	0.3	0	99	$4d^{10}5s$	^2S					18	B *
2.5	115,572.19	0.25	71	97	$4d^95s^2$	^2D	3	$4d^95p^2(^1S)$	^2D		9	B *
1.5	122,419.73	0.22	−74	95	$4d^95s^2$	^2D	3	$4d^95p^2(^1S)$	^2D		16	B *
0.5	126,879.89	0.24	0	100	$4d^{10}6s$	^2S					7	B *
1.5	128,458.36	0.23	6	97	$4d^{10}5d$	^2D	2	$4d^95s^2$	^2D		14	B *
2.5	128,748.33	0.25	−6	99	$4d^{10}5d$	^2D					10	B *
0.5	169,434.59	0.25	0	100	$4d^{10}7s$	^2S					3	B *
1.5	170,535.76	0.24	−8	100	$4d^{10}6d$	^2D					4	B *
2.5	170,718.81	0.3	8	100	$4d^{10}6d$	^2D					2	B *
3.5	186,527.40	0.3	0	100	$4d^{10}5g$	^2G					4	B *
4.5	186,528.26	0.3	−1	100	$4d^{10}5g$	^2G					4	B *
0.5	189,374.5	0.3	1	100	$4d^{10}8s$	^2S					4	B *
1.5	190,038.8	0.3	−4	100	$4d^{10}7d$	^2D					4	B *
2.5	190,136.3	0.4	4	100	$4d^{10}7d$	^2D					2	B *
4.5	198,654.0	0.8	0	100	$4d^{10}6g$	^2G					1	B *
3.5	198,654.3	0.4	0	100	$4d^{10}6g$	^2G					2	B *
0.5	200,362.77	0.3	0	100	$4d^{10}9s$	^2S					4	B *
1.5	200,778.32	0.23	−2	100	$4d^{10}8d$	^2D					5	B *
2.5	200,836.01	0.24	2	100	$4d^{10}8d$	^2D					3	B *
3.5	205,966.56	0.3	1	100	$4d^{10}7g$	^2G					3	B *
4.5	205,966.76	0.3	0	100	$4d^{10}7g$	^2G					2	B *
0.5	207,068.43	0.3	38	100	$4d^{10}10s$	^2S					4	B *
1.5	207,338.8	0.4	−3	100	$4d^{10}9d$	^2D					3	B *
2.5	207,379.7	0.4	3	100	$4d^{10}9d$	^2D					2	B *
4.5	210,710.88	0.3	−1	100	$4d^{10}8g$	^2G					1	B *
3.5	210,713.04	0.3	1	100	$4d^{10}8g$	^2G					3	B *
0.5	211,462.1	0.3	−3	100	$4d^{10}11s$	^2S					4	B *
1.5	(211,708.6)			100	$4d^{10}10d$	^2D						
2.5	(211,732.5)			100	$4d^{10}10d$	^2D						
4.5	213,966.18	0.3	−1	100	$4d^{10}9g$	^2G					2	B *
3.5	213,966.94	0.4	1	100	$4d^{10}9g$	^2G					1	B *
0.5	214,497.7	1.4	−33	100	$4d^{10}12s$	^2S					2	B *
1.5	(214,732.2)			100	$4d^{10}11d$	^2D						
2.5	(214,749.3)			100	$4d^{10}11d$	^2D						
3.5	(216,339)			100	$4d^{10}10g$	^2G						
4.5	(216,339.2)			100	$4d^{10}10g$	^2G						
0.5	(216,744.7)			100	$4d^{10}13s$	^2S						
1.5	(216,917.2)			100	$4d^{10}12d$	^2D						

Table 2. *Cont.*

J	Energy [a] cm^{-1}	Unc [b]	$\Delta Eo\text{-}c$ [c] cm^{-1}	*LS* Compositions [d] 1st Component			2nd Component			3rd Component			No. of Lines [e]	Lev. Ref. [f]
2.5	(216,929.7)			100	$4d^{10}12d$	^2D								
3.5	(218,090.7)			100	$4d^{10}11g$	^2G								
4.5	(218,090.8)			100	$4d^{10}11g$	^2G								
0.5	(218,391.9)			100	$4d^{10}14s$	^2S								
3.5	(219,427.4)			100	$4d^{10}12g$	^2G								
4.5	(219,427.4)			100	$4d^{10}12g$	^2G								
2.5	235,451.0	0.4	248	55	$4d^95p^2(^3P)$	^4D	11	$4d^95p^2(^3P)$	^2D	9	$4d^95p^2(^1D)$	^2D	3	TW
3.5	236,170.2	0.4	−577	84	$4d^95p^2(^3P)$	^4D	9	$4d^95p^2(^3P)$	^4F				3	TW
4.5	(236,964.5)			68	$4d^95p^2(^1D)$	^2G	19	$4d^95s(^3D)5d$	^2G	9	$4d^95p^2(^3P)$	^4F		
1.5	237,145.6	1.3	−104	51	$4d^95p^2(^1D)$	^2P	14	$4d^95s(^3D)5d$	^2P	10	$4d^95p^2(^1D)$	^2D	2	R+TW
0.5	237,201.7	2	343	52	$4d^95p^2(^1D)$	^2S	21	$4d^95p^2(^1D)$	^2P	9	$4d^95s(^3D)5d$	^2S	1	R *
1.5	238,830.9	0.4	207	66	$4d^95p^2(^3P)$	^4D	14	$4d^95p^2(^3P)$	^2D	8	$4d^95p^2(^1D)$	^2P	3	R *
2.5	239,739.3	0.4	−71	37	$4d^95p^2(^1D)$	^2D	11	$4d^95s(^3D)5d$	^2D	11	$4d^95p^2(^3P)$	^2F	3	TW
3.5	240,891.5	0.5	−191	29	$4d^95p^2(^1D)$	^2F	20	$4d^95p^2(^3P)$	^4F	20	$4d^95p^2(^1D)$	^2G	2	TW
2.5	241,544.42	0.3	73	49	$4d^95p^2(^3P)$	^2F	19	$4d^95p^2(^3P)$	^4F	14	$4d^95p^2(^3P)$	^4P	7	TW
0.5	241,683.8	0.8	−448	90	$4d^95p^2(^3P)$	^4D	5	$4d^95p^2(^1D)$	^2S				3	TW
3.5	(243,413.3)			43	$4d^95p^2(^3P)$	^4F	25	$4d^95p^2(^1D)$	^2G	11	$4d^95p^2(^3P)$	^4D		
0.5	243,423.3	2	−224	45	$4d^95p^2(^1D)$	^2P	15	$4d^95p^2(^1D)$	^2S	14	$4d^95s(^3D)5d$	^2P	1	R *
4.5	244,559.9	0.4	−49	91	$4d^95p^2(^3P)$	^4F	6	$4d^95p^2(^1D)$	^2G				3	TW
3.5	(245,494.4)			33	$4d^95p^2(^1D)$	^2F	24	$4d^95p^2(^1D)$	^2G	16	$4d^95p^2(^3P)$	^2F		
1.5	245,704.30	0.5	152	44	$4d^95p^2(^3P)$	^2D	19	$4d^95p^2(^3P)$	^4F	17	$4d^95p^2(^1D)$	^2D	3	R *
1.5	246,474.9	0.4	−11	67	$4d^95p^2(^3P)$	^4F	12	$4d^95p^2(^3P)$	^4P	7	$4d^95p^2(^3P)$	^4D	4	R *
2.5	246,503.2	0.3	64	65	$4d^95p^2(^3P)$	^4P	16	$4d^95p^2(^3P)$	^4D	5	$4d^95p^2(^1D)$	2F	5	R *
2.5	247,098.2	0.5	−468	42	$4d^95p^2(^3P)$	^4F	25	$4d^95p^2(^1D)$	^2F	16	$4d^95p^2(^3P)$	2F	2	R *
3.5	(247,109.4)			100	$4d^95s(^3D)6s$	^4D								
1.5	247,551.6	0.5	250	28	$4d^95p^2(^1D)$	^2D	36	$4d^95P^2(^3P)$	^2D	11	$4d^95s(^3D)5d$	^2D	3	TW
2.5	(248,375.7)			73	$4d^95s(^3D)6s$	^4D	15	$4d^95s(^3D)6s$	^2D	12	$4d^95s(^1D)6s$	^2D		
1.5	(249,769.2)			46	$4d^95s(^3D)6s$	^4D	33	$4d^95s(^1D)6s$	^2D	21	$4d^95s(^3D)6s$	^2D		
2.5	250,202.6	0.7	403	53	$4d^95p^2(^3P)$	^2D	12	$4d^95p^2(^1D)$	^2D	11	$4d^95p^2(^1D)$	^2F	2	R *
1.5	250,491.4	0.7	−73	34	$4d^95p^2(^3P)$	^4P	25	$4d^95p^2(^3P)$	^2P	10	$4d^95p^2(^3P)$	^4D	2	R *
0.5	251,157.4	0.4	40	78	$4d^95p^2(^3P)$	^2P	8	$4d^95p^2(^3P)$	^4P				4	R *
2.5	251,829.7	0.4	−221	18	$4d^95p^2(^1D)$	^2F	23	$4d^95p^2(^3P)$	^2D	15	$4d^95p^2(^3P)$	^4F	4	R *
2.5	(253,116.3)			83	$4d^95s(^3D)6s$	^2D	10	$4d^95s(^3D)6s$	^4D	6	$4d^95s(^1D)6s$	^2D		
3.5	253,457.6	0.5	556	61	$4d^95p^2(^3P)$	^2F	28	$4d^95p^2(^3P)$	^4F				3	TW
1.5	253,703.7	0.6	162	54	$4d^95p^2(^3P)$	^2P	37	$4d^95p^2(^3P)$	^4P				3	R *
0.5	254,006.1	0.4	−58	83	$4d^95p^2(^3P)$	^4P	14	$4d^95p^2(^3P)$	^2P				4	TW
Odd Parity														
0.5	57,184.0	0.3	367	100	$4d^{10}5p$	$^2P^O$							22	B *
1.5	61,527.0	base	731	100	$4d^{10}5p$	$^2P^O$							30	B *
0.5	144,589.32	0.24	−10	100	$4d^{10}6p$	$^2P^O$							8	B *

Table 2. *Cont.*

J	Energy [a] cm^{-1}	Unc [b]	ΔEo-c [c] cm^{-1}	LS Compositions [d]	1st Component			2nd Component			3rd Component		No. of Lines [e]	Lev. Ref. [f]
1.5	145,926.21	0.23	46	100	$4d^{10}6p$	$^2P^O$							14	B *
2.5	161,974.14	0.3	14	99	$4d^{10}4f$	$^2F^O$							7	B *
3.5	161,982.00	0.3	−13	99	$4d^{10}4f$	$^2F^O$							6	B *
2.5	163,890.3	0.4	534	90	$4d^95s(^3D)5p$	$^4P^O$	8	$4d^95s(^3D)5p$	$^4D^O$				5	TW
3.5	167,308.1	0.4	31	74	$4d^95s(^3D)5p$	$^4F^O$	12	$4d^95s(^3D)5p$	$^4D^O$	8	$4d^95s(^3D)5p$	$^2F^O$	4	TW
1.5	167,339.24	0.3	−126	77	$4d^95s(^3D)5p$	$^4P^O$	12	$4d^95s(^3D)5p$	$^4D^O$	6	$4d^95s(^1D)5p$	$^2P^O$	6	B *
2.5	167,465.9	0.4	−214	62	$4d^95s(^3D)5p$	$^4F^O$	20	$4d^95s(^1D)5p$	$^2F^O$	9	$4d^95s(^3D)5p$	$^2F^O$	3	TW
4.5	168,947.6	0.5	−340	100	$4d^95s(^3D)5p$	$^4F^O$							3	TW
1.5	170,813.7	0.5	−143	86	$4d^95s(^3D)5p$	$^4F^O$	7	$4d^95s(^3D)5p$	$^4P^O$				4	K+TW
0.5	171,315.7	1	−96	87	$4d^95s(^3D)5p$	$^4P^O$	7	$4d^95s(^3D)5p$	$^4D^O$	5	$4d^95s(^1D)5p$	$^2P^O$	2	K+TW
3.5	174,043.59	0.4	−153	82	$4d^95s(^3D)5p$	$^4D^O$	9	$4d^95s(^1D)5p$	$^2F^O$	5	$4d^95s(^3D)5p$	$^4F^O$	6	TW
2.5	174,496.6	0.5	238	42	$4d^95s(^3D)5p$	$^4D^O$	32	$4d^95s(^1D)5p$	$^2F^O$	9	$4d^95s(^3D)5p$	$^2F^O$	3	TW
1.5	175,539.35	0.3	201	38	$4d^95s(^1D)5p$	$^2D^O$	23	$4d^95s(^3D)5p$	$^4D^O$	17	$4d^95s(^3D)5p$	$^2D^O$	5	K *
2.5	176,090.19	0.4	−281	23	$4d^95s(^1D)5p$	$^2F^O$	30	$4d^95s(^3D)5p$	$^4F^O$	17	$4d^95s(^3D)5p$	$^4D^O$	5	TW
0.5	177,263.38	0.3	−190	94	$4d^{10}7p$	$^2P^O$	2	$4d^95s(^1D)5p$	$^2P^O$				4	B *
1.5	177,867.74	0.23	−172	92	$4d^{10}7p$	$^2P^O$	4	$4d^95s(^1D)5p$	$^2P^O$				7	B *
0.5	178,187.72	0.3	121	79	$4d^95s(^3D)5p$	$^4D^O$	11	$4d^95s(^3D)5p$	$^4P^O$	10	$4d^95s(^1D)5p$	$^2P^O$	6	K *
1.5	178,616.85	0.3	−213	61	$4d^95s(^1D)5p$	$^2P^O$	22	$4d^95s(^3D)5p$	$^4D^O$	7	$4d^{10}7p$	$^2P^O$	5	K *
0.5	179,320.9	0.3	4	63	$4d^95s(^1D)5p$	$^2P^O$	18	$4d^95s(^3D)5p$	$^2P^O$	12	$4d^95s(^3D)5p$	$^4D^O$	2	K *
3.5	180,060.3	0.6	23	47	$4d^95s(^1D)5p$	$^2F^O$	28	$4d^95s(^3D)5p$	$^2F^O$	20	$4d^95s(^3D)5p$	$^4F^O$	2	TW
1.5	180,943.95	0.3	216	37	$4d^95s(^3D)5p$	$^4D^O$	25	$4d^95s(^1D)5p$	$^2D^O$	14	$4d^95s(^1D)5p$	$^2P^O$	7	K *
2.5	182,399.28	0.4	231	40	$4d^95s(^1D)5p$	$^2D^O$	28	$4d^95s(^3D)5p$	$^4D^O$	26	$4d^95s(^3D)5p$	$^2D^O$	5	TW
3.5	184,895.95	0.3	−11	96	$4d^{10}5f$	$^2F^O$	4	$4d^95s(^3D)5p$	$^2F^O$				5	B *
2.5	185,024.81	0.3	9	99	$4d^{10}5f$	$^2F^O$							6	B *
3.5	191,104.1	0.9	404	26	$4d^95s(^3D)5p$	$^2F^O$	58	$4d^{10}6f$	$^2F^O$	14	$4d^95s(^1D)5p$	$^2F^O$	1	TW
2.5	(191,336.8)			98	$4d^{10}6f$	$^2F^O$	2	$4d^95s(^3D)5p$	$^2F^O$					
1.5	191,509.1	0.3	−55	78	$4d^95s(^3D)5p$	$^2P^O$	9	$4d^{10}8p$	$^2P^O$	8	$4d^95s(^1D)5p$	$^2P^O$	3	K *
3.5	(192,475.5)			42	$4d^{10}6f$	$^2F^O$	32	$4d^95s(^3D)5p$	$^2F^O$	23	$4d^95s(^1D)5p$	$^2F^O$		
0.5	193,938.4	0.4	141	96	$4d^{10}8p$	$^2P^O$	3	$4d^95s(^3D)5p$	$^2P^O$				1	B *
1.5	194,333.3	0.3	−223	91	$4d^{10}8p$	$^2P^O$	8	$4d^95s(^3D)5p$	$^2P^O$				5	B *
2.5	194,902.6	0.4	−57	43	$4d^95s(^3D)5p$	$^2D^O$	25	$4d^95s(^3D)5p$	$^2F^O$	23	$4d^95s(^1D)5p$	$^2D^O$	3	B *
0.5	198,382.2	0.4	464	73	$4d^95s(^3D)5p$	$^2P^O$	18	$4d^95s(^1D)5p$	$^2P^O$				3	B *
2.5	(198,489.7)			54	$4d^{10}7f$	$^2F^O$	19	$4d^95s(^3D)5p$	$^2F^O$	10	$4d^95s(^3D)5p$	$^2D^O$		
4.5	(198,518.6)			100	$4d^{10}6h$	$^2H^O$								
5.5	(198,519.4)			100	$4d^{10}6h$	$^2H^O$								
2.5	198,799.3	0.4	−782	28	$4d^95s(^3D)5p$	$^2F^O$	45	$4d^{10}7f$	$^2F^O$	11	$4d^95s(^1D)5p$	$^2F^O$	3	TW
3.5	(199,029)			99	$4d^{10}7f$	$^2F^O$								
1.5	202,135.0	0.5	28	64	$4d^95s(^3D)5p$	$^2D^O$	30	$4d^95s(^1D)5p$	$^2D^O$				3	B *
0.5	(203,388.5)			99	$4d^{10}9p$	$^2P^O$								
1.5	(203,556.3)			99	$4d^{10}9p$	$^2P^O$								

Table 2. *Cont.*

J	Energy [a] cm^{-1}	Unc [b]	ΔE_{o-c} [c] cm^{-1}	*LS* Compositions [d]	1st Component			2nd Component			3rd Component		No. of Lines [e]	Lev. Ref. [f]
3.5	(203,854.4)			100	$4d^{10}8f$	$^2F^O$								
2.5	(203,879.1)			99	$4d^{10}8f$	$^2F^O$								
4.5	206,012.3	0.4	79	100	$4d^{10}7h$	$^2H^O$							2	B *
5.5	206,012.78	0.5	79	100	$4d^{10}7h$	$^2H^O$							1	B *
3.5	(207,170.2)			100	$4d^{10}9f$	$^2F^O$								
2.5	(207,177.7)			100	$4d^{10}9f$	$^2F^O$								
0.5	(209,199.1)			100	$4d^{10}10p$	$^2P^O$								
1.5	(209,320.8)			100	$4d^{10}10p$	$^2P^O$								
3.5	(209,539)			100	$4d^{10}10f$	$^2F^O$								B *
2.5	(209,542.4)			100	$4d^{10}10f$	$^2F^O$								
4.5	210,743.3	0.5	−9	100	$4d^{10}8h$	$^2H^O$							2	B *
5.5	210,743.8	0.7	−6	100	$4d^{10}8h$	$^2H^O$							1	B *
3.5	(211,284.2)			100	$4d^{10}11f$	$^2F^O$								
2.5	(211,286.1)			100	$4d^{10}11f$	$^2F^O$								
3.5	(212,613.4)			100	$4d^{10}12f$	$^2F^O$								
2.5	(212,614.6)			100	$4d^{10}12f$	$^2F^O$								
0.5	(213,103.9)			100	$4d^{10}11p$	$^2P^O$								
1.5	(213,189.4)			100	$4d^{10}11p$	$^2P^O$								
4.5	213,987.3	0.6	−65	100	$4d^{10}9h$	$^2H^O$							2	B *
5.5	213,987.75	0.8	−64	100	$4d^{10}9h$	$^2H^O$							1	B *
0.5	(215,845.4)			100	$4d^{10}12p$	$^2P^O$								
1.5	(215,907.2)			100	$4d^{10}12p$	$^2P^O$								
4.5	(216,413.2)			100	$4d^{10}10h$	$^2H^O$								
5.5	(216,413.4)			100	$4d^{10}10h$	$^2H^O$								
1.5	271,244.30		−184	60	$4d^95s(^3D)6p$	$^4P^O$	18	$4d^95s(^3D)6p\,^4D^O$	13	$4D^95s(^1D)6p$	$^2P^O$			Ki
0.5	277,535.30		−3	62	$4d^95s(^3D)6p$	$^4P^O$	18	$4d^95s(^3D)6p\,^2P^O$	12	$4d^95s(^1D)6p$	$^2P^O$			Ki
0.5	278,100.20		−121	73	$4d^95s(^3D)6p$	$^4D^O$	23	$4d^95s(^3D)6p\,^2P^O$						Ki
1.5	278,906.10		50	34	$4d^95s(^3D)6p$	$^4D^O$	22	$4d^95s(^3D)6p\,^4P^O$	20	$4d^95s(^3D)6p$	$^2D^O$			Ki
1.5	279,955.20		258	55	$4d^95s(^1D)6p$	$^2P^O$	21	$4d^95s(^3D)6p\,^4D^O$	11	$4d^95s(^1D)6p$	$^2D^O$			Ki
1.5	286,407.80		−76	75	$4d^95s(^3D)4f$	$^4P^O$	23	$4d^95s(^3D)4f\,^4D^O$						Ki
1.5	288,020.60		−178	34	$4d^95s(^3D)4f$	$^2D^O$	32	$4d^95s(^3D)4f\,^4F^O$	18	$4d^95s(^3D)4f$	$^4D^O$			Ki
0.5	288,423.30		−303	73	$4d^95s(^3D)4f$	$^4D^O$	17	$4d^95s(^3D)4f\,^2P^O$	6	$4d^95s(^1D)4f$	$^2P^O$			Ki
0.5	289,472.80		−265	66	$4d^95s(^3D)4f$	$^2P^O$	27	$4d^95s(^1D)4f\,^2P^O$	5	$4d^95s(^3D)4f$	$^4D^O$			Ki
1.5	290,682.60		117	34	$4d^95s(^3D)4f$	$^2P^O$	28	$4d^95s(^3D)4f\,^4F^O$	26	$4d^95s(^1D)4f$	$^2D^O$			Ki
1.5	295,198.90		26	31	$4d^95s(^3D)4f$	$^4F^O$	29	$4d^95s(^3D)4f\,^2P^O$	19	$4d^95s(^3D)4f$	$^2D^O$			Ki
1.5	296,892.10		69	73	$4d^95s(^1D)4f$	$^2P^O$	10	$4d^95s(^3D)4f\,^2D^O$	10	$4d^95s(^3D)4f$	$^4D^O$			Ki
1.5	297,860.40		246	62	$4d^95s(^1D)4f$	$^2D^O$	24	$4d^95s(^3D)4f\,^2P^O$	9	$4d^95s(^3D)4f$	$^4F^O$			Ki
0.5	297,860.50		368	66	$4d^95s(^1D)4f$	$^2P^O$	16	$4d^95s(^3D)4f\,^4D^O$	14	$4d^95s(^3D)4f$	$^2P^O$			Ki
1.5	305,280.50		2	38	$4d^95s(^3D)7p$	$^4P^O$	37	$4d^95s(^3D)7p\,^2P^O$	15	$4d^95s(^3D)7P$	$^4D^O$			Ki
1.5	312,862.60		244	43	$4d^95s(^3D)5f$	$^2D^O$	27	$4d^95s(^3D)5f\,^4F^O$	12	$4d^95s(^3D)5f$	$^2P^O$			Ki
0.5	313,588.90		272	59	$4d^95s(^3D)5f$	$^4D^O$	30	$4d^95s(^3D)5f\,^2P^O$						Ki

Table 2. *Cont.*

J	Energy [a] cm^{-1}	Unc [b]	ΔEo-c [c] cm^{-1}	*LS* Compositions [d]	1st Component			2nd Component		3rd Component		No. of Lines [e]	Lev. Ref. [f]
1.5	314,636.90		−317	29	$4d^95s(^3D)5f$	$^2P^O$	26	$4d^95s(^3D)5f\,^4F^O$	23	$4d^95s(^1D)5f$	$^2D^O$		Ki
1.5	321,653.00		116	70	$4d^95s(^1D)5f$	$^2P^O$	9	$4d^95s(^3D)5f\,^2D^O$	9	$4d^95s(^3D)5f$	$^4D^O$		Ki
0.5	321,653.40		−352	61	$4d^95s(^1D)5f$	$^2P^O$	15	$4d^95s(^3D)5f\,^4D^O$	10	$4d^95s(^3D)5f$	$^2P^O$		Ki
1.5	325,283.70		−311	48	$4d^95s(^3D)6f$	$^2D^O$	22	$4d^95s(^3D)6f\,^4F^O$	14	$4d^95s(^3D)6f$	$^2P^O$		Ki
0.5	326,170.10		−153	39	$4d^95s(^3D)6f$	$^4D^O$	36	$4d^95s(^3D)6f\,^2P^O$	7	$4d^95s(^3D)8P$	$^2P^O$		Ki
1.5	328,106.90		348	26	$4d^95s(^3D)6f$	$^2P^O$	23	$4d^95s(^3D)6f\,^4F^O$	18	$4d^95s(^1D)6f$	$^2D^O$		Ki
1.5	332,461.90		162	28	$4d^95s(^3D)6f$	$^2P^O$	23	$4d^95s(^3D)6f\,^4F^O$	19	$4d^95s(^3D)6f$	$^2D^O$		Ki
1.5	333,429.00		42	46	$4d^95s(^3D)7f$	$^2D^O$	17	$4d^95s(^3D)7f\,^4F^O$	16	$4d^95s(^3D)7f$	$^2P^O$		Ki
0.5	333,429.80		−63	59	$4d^95s(^3D)7f$	$^2P^O$	23	$4d^95s(^3D)7f\,^4D^O$	10	$4d^95s(^1D)6f$	$^2P^O$		Ki

[a] Energy values are optimized from observed wavelengths using the least squares level optimization code LOPT [15]. Values enclosed in parentheses correspond to unobserved energy levels found from the parametric least squares fitting. [b] Uncertainties resulting from the level optimization procedure are given on the level of one standard deviation. They correspond to uncertainties of level separations from $4d^{10}5p\ ^2P_{3/2}$. To determine uncertainties of excitation energies from the ground level, the given values should be combined in quadrature with the uncertainty of the ground level, 0.3 cm^{-1}. If this column is blank, the level value was not included in the level optimization. [c] Differences between observed energies and those calculated in the parametric least squares fitting. [d] Only three leading *LS* components are given. [e] Number of observed lines determining the level in the optimization procedure LOPT [15]. [f] Reference to the level source as B, K, R, K+TW, R+TW and TW stand for Bhatia [6], Kaufman et al. [7], Ryabtsev et al. [9], previous value [7] has been revised, previous value [9] has been revised, and this work ; * stands for levels from [6,7,9] re-optimized in this work. Ki stands for Kilbane et al. [8] level values, which have not been included in the level optimization.

Table 3. List of classified lines in In III spectrum.

I_{obs} [a]	ch [b]	λ_{obs} [c] Å	σ_{obs} cm^{-1}	λ_{Ritz} [d] Å	$\Delta\lambda_{O\text{-}Ritz}$ [e] Å	Classification [f]					E_{low} cm^{-1}	E_{upp} cm^{-1}	gA [h] S^{-1}		Lin. Ref [i]
50		494.715(8)	202,137	494.7189(13)	−0.004	$4d^{10}5s$	$(^1S)^2S_{0.5}$	-	$4d^95s5p$	$(^3D)^2D_{1.5}$	0.0	202,135.0	1.42E+08	#	TW
200		504.080(6)	198,381.2	504.0775(12)	0.003	$4d^{10}5s$	$(^1S)^2S_{0.5}$	-	$4d^95s5p$	$(^3D)^2P_{0.5}$	0.0	198,382.2	5.08E+09		TW
50		508.066(6)	196,824.8	508.0730(12)	−0.007	$4d^{10}5p$	$(^1S)^2P_{0.5}$	-	$4d^95p^2$	$(^3P)^4P_{0.5}$	57,184.0	254,006.1	8.17E+08		TW
100		508.846(6)	196,523.1	508.8548(15)	−0.009	$4d^{10}5p$	$(^1S)^2P_{0.5}$	-	$4d^95p^2$	$(^3P)^2P_{1.5}$	57,184.0	253,703.7	4.60E+08	#	R
20		514.583(8)	194,332	514.5798(11)	0.003	$4d^{10}5s$	$(^1S)^2S_{0.5}$	-	$4d^{10}8p$	$(^1S)^2P_{1.5}$	0.0	194,333.3	3.30E+09		B
750		515.532(6)	193,974.4	515.5346(12)	−0.003	$4d^{10}5p$	$(^1S)^2P_{0.5}$	-	$4d^95p^2$	$(^3P)^2P_{0.5}$	57,184.0	251,157.4	3.94E+09		R
45		519.544(6)	192,476.5	519.5369(11)	0.007	$4d^{10}5p$	$(^1S)^2P_{1.5}$	-	$4d^95p^2$	$(^3P)^4P_{0.5}$	61,527.0	254,006.1	2.39E+08		TW
620		520.357(6)	192,175.8	520.3544(15)	0.003	$4d^{10}5p$	$(^1S)^2P_{1.5}$	-	$4d^95p^2$	$(^3P)^2P_{1.5}$	61,527.0	253,703.7	8.31E+09		R
120		522.166(6)	191,510.0	522.1684(11)	−0.002	$4d^{10}5s$	$(^1S)^2S_{0.5}$	-	$4d^95s5p$	$(^3D)^2P_{1.5}$	0.0	191,509.1	1.07E+10		K
300		525.300(6)	190,367.4	525.2995(14)	0.001	$4d^{10}5p$	$(^1S)^2P_{0.5}$	-	$4d^95p^2$	$(^1D)^2D_{1.5}$	57,184.0	247,551.6	1.85E+10		R
250		525.482(6)	190,301.5	525.4786(12)	0.003	$4d^{10}5p$	$(^1S)^2P_{1.5}$	-	$4d^95p^2$	$(^1D)^2F_{2.5}$	61,527.0	251,829.7	5.20E+09		R

Table 3. *Cont.*

I_{obs} [a]	ch [b]	λ_{obs} [c] Å	σ_{obs} cm^{-1}	λ_{Ritz} [d] Å	$\Delta\lambda_{O\text{-}Ritz}$ [e] Å	Classification [f]					E_{low} cm^{-1}	E_{upp} cm^{-1}	gA [h] S^{-1}		Lin. Ref [i]
540		527.348(6)	189,628.1	527.3416(12)	0.006	$4d^{10}5p$	$(^1S)^2P_{1.5}$	-	$4d^95p^2$	$(^3P)^2P_{0.5}$	61,527.0	251,157.4	3.46E+09		R
230		528.287(6)	189,291.0	528.2874(12)	0.000	$4d^{10}5p$	$(^1S)^2P_{0.5}$	-	$4d^95p^2$	$(^3P)^4F_{1.5}$	57,184.0	246,474.9	1.15E+06	#	R
630		529.200(6)	188,964.5	529.2002(19)	0.000	$4d^{10}5p$	$(^1S)^2P_{1.5}$	-	$4d^95p^2$	$(^3P)^4P_{1.5}$	61,527.0	250,491.4	7.53E+09		R
450		530.000(6)	188,679.2	530.0102(20)	−0.010	$4d^{10}5p$	$(^1S)^2P_{1.5}$	-	$4d^95p^2$	$(^3P)^2D_{2.5}$	61,527.0	250,202.6	2.73E+10		R
460		530.448(6)	188,519.9	530.4469(14)	0.001	$4d^{10}5p$	$(^1S)^2P_{0.5}$	-	$4d^95p^2$	$(^3P)^2D_{1.5}$	57,184.0	245,704.3	4.21E+09		R
300		540.613(6)	184,975.2	540.6101(10)	0.003	$4d^{10}5p$	$(^1S)^2P_{1.5}$	-	$4d^95p^2$	$(^3P)^4P_{2.5}$	61,527.0	246,503.2	4.37E+09		R
200		540.678(6)	184,953.0	540.6928(11)	−0.015	$4d^{10}5p$	$(^1S)^2P_{1.5}$	-	$4d^95p^2$	$(^3P)^4F_{1.5}$	61,527.0	246,474.9	6.21E+08		R
480		549.764(6)	181,896.2	549.764(6)	0.000	$4d^{10}5p$	$(^1S)^2P_{1.5}$	-	$4d^95p^2$	$(^1D)^2P_{0.5}$	61,527.0	243,423.3	2.42E+09		R
570		550.518(6)	181,647.1	550.5186(13)	−0.001	$4d^{10}5p$	$(^1S)^2P_{0.5}$	-	$4d^95p^2$	$(^3P)^4D_{1.5}$	57,184.0	238,830.9	3.66E+09		R
25		552.660(6)	180,943.1	552.6573(11)	0.003	$4d^{10}5s$	$(^1S)^2S_{0.5}$	-	$4d^95s5p$	$(^3D)^4D_{1.5}$	0.0	180,943.95	4.73E+08		K
15		555.069(6)	180,157.8	555.0720(24)	−0.003	$4d^{10}5p$	$(^1S)^2P_{1.5}$	-	$4d^95p^2$	$(^3P)^4D_{0.5}$	61,527.0	241,683.8	3.56E+08		TW
20		555.501(6)	180,017.7	555.501(6)	0.000	$4d^{10}5p$	$(^1S)^2P_{0.5}$	-	$4d^95p^2$	$(^1D)^2S_{0.5}$	57,184.0	237,201.7	4.56E+09		TW
390		555.669(6)	179,963.3	555.674(4)	−0.005	$4d^{10}5p$	$(^1S)^2P_{0.5}$	-	$4d^95p^2$	$(^1D)^2P_{1.5}$	57,184.0	237,145.6	1.08E+09		R
150		557.662(6)	179,320.1	557.6595(13)	0.003	$4d^{10}5s$	$(^1S)^2S_{0.5}$	-	$4d^95s5p$	$(^1D)^2P_{0.5}$	0.0	179,320.9	1.47E+09		K
150		559.857(6)	178,617.0	559.8576(11)	−0.001	$4d^{10}5s$	$(^1S)^2S_{0.5}$	-	$4d^95s5p$	$(^1D)^2P_{1.5}$	0.0	178,616.85	1.27E+10		K
130		561.210(6)	178,186.4	561.2059(11)	0.004	$4d^{10}5s$	$(^1S)^2S_{0.5}$	-	$4d^95s5p$	$(^3D)^4D_{0.5}$	0.0	178,187.72	5.47E+09		K
200		562.214(6)	177,868.2	562.2155(11)	−0.001	$4d^{10}5s$	$(^1S)^2S_{0.5}$	-	$4^{10}7p$	$(^1S)^2P_{1.5}$	0.0	177,867.74	5.56E+09		B, TW
160		564.131(6)	177,263.8	564.1323(11)	−0.001	$4d^{10}5s$	$(^1S)^2S_{0.5}$	-	$4^{10}7p$	$(^1S)^2P_{0.5}$	0.0	177,263.38	3.77E+09		B, TW
480		569.421(6)	175,617.0	569.416(4)	0.005	$4d^{10}5p$	$(^1S)^2P_{1.5}$	-	$4d^95p^2$	$(^1D)^2P_{1.5}$	61,527.0	237,145.6	1.72E+09		R
80		569.677(6)	175,538.1	569.6728(12)	0.004	$4d^{10}5s$	$(^1S)^2S_{0.5}$	-	$4d^95s5p$	$(^1D)^2D_{1.5}$	0.0	175,539.35	7.80E+08		K
80		583.723(6)	171,314.1	583.718(3)	0.005	$4d^{10}5s$	$(^1S)^2S_{0.5}$	-	$4d^95s5p$	$(^3D)^4P_{0.5}$	0.0	171,315.7	9.39E+08		TW
50		585.440(6)	170,811.7	585.4331(18)	0.007	$4d^{10}5s$	$(^1S)^2S_{0.5}$	-	$4d^95s5p$	$(^3D)^4F_{1.5}$	0.0	170,813.7	2.95E+07		K
100		597.596(6)	167,337.1	597.5885(13)	0.008	$4d^{10}5s$	$(^1S)^2S_{0.5}$	-	$4d^95s5p$	$(^3D)^4P_{1.5}$	0.0	167,339.24	1.18E+09		B, TW
30		635.672(8)	157,313.8	635.673(6)	−0.001	$4d^{10}5p$	$(^1S)^2P_{0.5}$	-	$4d^{10}12s$	$(^1S)^2S_{0.5}$	57,184.0	214,497.7	3.71E+07		B
32		648.185(8)	154,276.9	648.1801(15)	0.005	$4d^{10}5p$	$(^1S)^2P_{0.5}$	-	$4d^{10}11s$	$(^1S)^2S_{0.5}$	57,184.0	211,462.1	5.21E+07		B
33		653.721(8)	152,970.5	653.720(6)	0.001	$4d^{10}5p$	$(^1S)^2P_{1.5}$	-	$4d^{10}12s$	$(^1S)^2S_{0.5}$	61,527.0	214,497.7	6.79E+07		B
41		665.979(8)	150,154.9	665.9794(17)	0.000	$4d^{10}5p$	$(^1S)^2P_{0.5}$	-	$4d^{10}9d$	$(^1S)^2D_{1.5}$	57,184.0	207,338.8	9.52E+07		B
40		666.963(8)	149,933.4	666.9552(15)	0.008	$4d^{10}5p$	$(^1S)^2P_{1.5}$	-	$4d^{10}11s$	$(^1S)^2S_{0.5}$	61,527.0	211,462.1	9.55E+07		B
40		667.177(8)	149,885.3	667.1807(15)	−0.004	$4d^{10}5p$	$(^1S)^2P_{0.5}$	-	$4d^{10}10s$	$(^1S)^2S_{0.5}$	57,184.0	207,068.43	7.66E+07		B
80		685.273(6)	145,927.2	685.2779(16)	−0.005	$4d^{10}5s$	$(^1S)^2S_{0.5}$	-	$4d^{10}6p$	$(^1S)^2P_{1.5}$	0.0	145,926.21	4.63E+07		B, TW
41		685.612(8)	145,855.1	685.6232(17)	−0.011	$4d^{10}5p$	$(^1S)^2P_{1.5}$	-	$4d^{10}9d$	$(^1S)^2D_{2.5}$	61,527.0	207,379.7	1.68E+08		B
67		685.815(8)	145,811.9	685.8156(17)	−0.001	$4d^{10}5p$	$(^1S)^2P_{1.5}$	-	$4d^{10}9d$	$(^1S)^2D_{1.5}$	61,527.0	207,338.8	1.88E+07		B
48		687.076(8)	145,544.3	687.0896(15)	−0.014	$4d^{10}5p$	$(^1S)^2P_{1.5}$	-	$4d^{10}10s$	$(^1S)^2S_{0.5}$	61,527.0	207,068.43	1.41E+08		B
200		691.610(6)	144,590.2	691.6140(17)	−0.004	$4d^{10}5s$	$(^1S)^2S_{0.5}$	-	$4d^{10}6p$	$(^1S)^2P_{0.5}$	0.0	144,589.32	2.10E+07		B, TW
50		696.399(8)	143,595.8	696.4064(13)	−0.007	$4d^{10}5p$	$(^1S)^2P_{0.5}$	-	$4d^{10}8d$	$(^1S)^2D_{1.5}$	57,184.0	200,778.32	1.79E+08		B
65		698.422(8)	143,179.9	698.4276(15)	−0.006	$4d^{10}5p$	$(^1S)^2P_{0.5}$	-	$4d^{10}9s$	$(^1S)^2S_{0.5}$	57,184.0	200,362.77	1.20E+08		B
68		717.834(8)	139,308.0	717.8287(12)	0.005	$4d^{10}5p$	$(^1S)^2P_{1.5}$	-	$4d^{10}8d$	$(^1S)^2D_{2.5}$	61,527.0	200,836.01	3.09E+08		B
60		718.135(8)	139,249.6	718.1260(12)	0.009	$4d^{10}5p$	$(^1S)^2P_{1.5}$	-	$4d^{10}8d$	$(^1S)^2D_{1.5}$	61,527.0	200,778.32	3.45E+07		B

Table 3. *Cont.*

I_{obs} [a]	ch [b]	λ_{obs} [c] Å	σ_{obs} cm^{-1}	λ_{Ritz} [d] Å	$\Delta\lambda_{O\text{-}Ritz}$ [e] Å	Classification [f]					E_{low} cm^{-1}	E_{upp} cm^{-1}	gA [h] S^{-1}		Lin. Ref [i]
70		720.281(8)	138,834.7	720.2755(15)	0.006	4d^{10}5p	(^{1}S)^{2}P$_{1.5}$	-	4d^{10}9s	(^{1}S)^{2}S$_{0.5}$	61,527.0	200,362.77	2.20E+08		B
15		752.699(6)	132,855.2	752.7014(21)	−0.002	4d^{10}5p	(^{1}S)^{2}P$_{0.5}$	-	4d^{10}7d	(^{1}S)^{2}D$_{1.5}$	57,184.0	190,038.8	3.91E+08		B, TW
20		756.484(6)	132,190.5	756.4840(21)	0.000	4d^{10}5p	(^{1}S)^{2}P$_{0.5}$	-	4d^{10}8s	(^{1}S)^{2}S$_{0.5}$	57,184.0	189,374.5	2.04E+08		B, TW
10		777.547(6)	128,609.6	777.549(3)	−0.002	4d^{10}5p	(^{1}S)^{2}P$_{1.5}$	-	4d^{10}7d	(^{1}S)^{2}D$_{2.5}$	61,527.0	190,136.3	6.60E+08		B, TW
65		778.142(8)	128,511.2	778.1387(21)	0.003	4d^{10}5p	(^{1}S)^{2}P$_{1.5}$	-	4d^{10}7d	(^{1}S)^{2}D$_{1.5}$	61,527.0	190,038.8	7.34E+07		B
30		782.187(6)	127,846.7	782.1819(20)	0.005	4d^{10}5p	(^{1}S)^{2}P$_{1.5}$	-	4d^{10}8s	(^{1}S)^{2}S$_{0.5}$	61,527.0	189,374.5	3.73E+08		B, TW
180		882.207(6)	113,352.1	882.2095(22)	−0.003	4d^{10}5p	(^{1}S)^{2}P$_{0.5}$	-	4d^{10}6d	(^{1}S)^{2}D$_{1.5}$	57,184.0	170,535.76	1.12E+09		B, TW
150		890.870(6)	112,249.8	890.8639(23)	0.006	4d^{10}5p	(^{1}S)^{2}P$_{0.5}$	-	4d^{10}7s	(^{1}S)^{2}S$_{0.5}$	57,184.0	169,434.59	4.00E+08		B, TW
120		915.824(6)	109,191.3	915.8196(21)	0.004	4d^{10}5p	(^{1}S)^{2}P$_{1.5}$	-	4d^{10}6d	(^{1}S)^{2}D$_{2.5}$	61,527.0	170,718.81	1.84E+09		B, TW
25		917.355(6)	109,009.1	917.3575(20)	−0.002	4d^{10}5p	(^{1}S)^{2}P$_{1.5}$	-	4d^{10}6d	(^{1}S)^{2}D$_{1.5}$	61,527.0	170,535.76	2.04E+08		B, TW
120		926.723(8)	107,907.1	926.7189(21)	0.004	4d^{10}5p	(^{1}S)^{2}P$_{1.5}$	-	4d^{10}7s	(^{1}S)^{2}S$_{0.5}$	61,527.0	169,434.59	7.20E+08		B
200		1153.839(8)	86,667.2	1153.844(5)	−0.005	d^{9}5s5p	(^{3}D)^{4}P$_{1.5}$	-	4d^{9}5p^2	(^{3}P)^{4}P$_{0.5}$	167,339.24	254,006.1	3.60E+08	#	TW
6		1162.895(8)	85,992.3	1162.903(6)	−0.008	d^{9}5s5p	(^{3}D)^{4}F$_{2.5}$	-	4d^{9}5p^2	(^{3}P)^{2}F$_{3.5}$	167,465.9	253,457.6	3.59E+09		TW
50		1201.523(8)	83,227.7	1201.532(5)	−0.009	4d^{9}5s^2	(^{2}D)^{2}D$_{2.5}$	-	4d^{9}5s5p	(^{3}D)^{2}F$_{2.5}$	115,572.19	198,799.3	2.25E+09		TW
80		1210.468(8)	826,12.7	1210.465(5)	0.003	d^{9}5s5p	(^{3}D)^{4}P$_{2.5}$	-	4d^{9}5p^2	(^{3}P)^{4}P$_{2.5}$	163,890.3	246,503.2	2.49E+09		TW
300	w	1254.458(16)	79,715.7	1254.465(7)	−0.007	4d^{9}5s^2	(^{2}D)^{2}D$_{1.5}$	-	4d^{9}5s5p	(^{3}D)^{2}D$_{1.5}$	122,419.73	202,135.0	7.59E+09		B, TW
300	w	1260.567(16)	79,329.4	1260.551(6)	0.016	4d^{9}5s^2	(^{2}D)^{2}D$_{2.5}$	-	4d^{9}5s5p	(^{3}D)^{2}D$_{2.5}$	115,572.19	194,902.6	1.49E+10		TW
200		(1263.152)	79,167.0	1263.201(5)		4d^{9}5s5p	(^{3}D)^{4}P$_{1.5}$	-	4d^{9}5p^2	(^{3}P)^{4}P$_{2.5}$	167,339.24	246,503.2	3.92E+09		TW
100		(1263.594)	79,139.3	1263.653(6)		4d^{9}5s5p	(^{3}D)^{4}P$_{1.5}$	-	4d^{9}5p^2	(^{3}P)^{4}F$_{1.5}$	167,339.24	246,474.9	3.06E+08		TW
15		1285.588(8)	77,785.4	1285.577(6)	0.011	4d^{9}5s5p	(^{3}D)^{4}D$_{3.5}$	-	4d^{9}5p^2	(^{1}D)^{2}F$_{2.5}$	174,043.59	251,829.7	4.50E+06	#	TW
20		1287.752(8)	77,654.7	1287.762(5)	−0.010	4d^{9}5s5p	(^{3}D)^{4}P$_{2.5}$	-	4d^{9}5p^2	(^{3}P)^{2}F$_{2.5}$	163,890.3	241,544.42	1.29E+09		TW
150		1294.468(8)	77,251.8	1294.468(5)	0.000	4d^{9}5s5p	(^{3}D)^{4}F$_{3.5}$	-	4d^{9}5p^2	(^{3}P)^{4}F$_{4.5}$	167,308.1	244,559.9	5.06E+09		TW
250	w	1309.269(16)	76,378.5	1309.251(6)	0.018	4d^{9}5s^2	(^{2}D)^{2}D$_{1.5}$	-	4d^{9}5s5p	(^{3}D)^{2}F$_{2.5}$	122,419.73	198,799.3	8.07E+09		TW
15		1315.880(8)	75,994.8	1315.880(8)	0.000	4d^{9}5s5p	(^{3}D)^{4}D$_{2.5}$	-	4d^{9}5p^2	(^{3}P)^{4}P$_{1.5}$	174,496.6	250,491.4	1.81E+09		TW
280	w	1316.430(16)	75,963.0	1316.440(7)	−0.010	4d^{9}5s^2	(^{2}D)^{2}D$_{1.5}$	-	4d^{9}5s5p	(^{3}D)^{2}P$_{0.5}$	122,419.73	198,382.2	3.01E+09		TW
70		1318.399(8)	75,849.6	1318.409(5)	−0.010	4d^{9}5s5p	(^{3}D)^{4}P$_{2.5}$	-	4d^{9}5p^2	(^{1}D)^{2}D$_{2.5}$	163,890.3	239,739.3	1.16E+09		TW
60		1318.946(8)	75,818.1	1318.941(6)	0.005	4d^{9}5s5p	(^{3}D)^{4}D$_{0.5}$	-	4d^{9}5p^2	(^{3}P)^{4}P$_{0.5}$	178,187.72	254,006.1	1.12E+08	#	TW
100		1320.314(8)	75,739.6	1320.315(6)	−0.001	4d^{9}5s5p	(^{1}D)^{2}F$_{2.5}$	-	4d^{9}5p^2	(^{1}D)^{2}F$_{2.5}$	176,090.19	251,829.7	3.21E+09		TW
150		1321.683(8)	75,661.1	1321.681(7)	0.002	4d^{9}5s5p	(^{3}D)^{4}F$_{1.5}$	-	4d^{9}5p^2	(^{3}P)^{4}F$_{1.5}$	170,813.7	246,474.9	4.82E+09		TW
200		1322.526(8)	75,612.9	1322.536(6)	−0.010	4d^{9}5s5p	(^{3}D)^{4}F$_{4.5}$	-	4d^{9}5p^2	(^{3}P)^{4}F$_{4.5}$	168,947.6	244,559.9	1.78E+10		TW
250	w	1323.944(16)	75,531.9	1323.944(16)	0.000	4d^{9}5s^2	(^{2}D)^{2}D$_{2.5}$	-	4d^{9}5s5p	(^{3}D)^{2}F$_{3.5}$	115,572.19	191,104.1	1.75E+10		TW
80		1349.914(8)	74,078.8	1349.919(6)	−0.005	4d^{9}5s5p	(^{3}D)^{4}F$_{2.5}$	-	4d^{9}5p^2	(^{3}P)^{2}F$_{2.5}$	167,465.9	241,544.42	6.37E+09		TW
60		1357.284(8)	73,676.5	1357.282(7)	0.002	4d^{10}5d	(^{1}S)^{2}D$_{1.5}$	-	4d^{9}5s5p	(^{3}D)^{2}D$_{1.5}$	128,458.36	202,135.0	4.03E+09		TW
100		1358.998(8)	73,583.6	1359.002(7)	−0.004	4d^{9}5s5p	(^{3}D)^{4}F$_{3.5}$	-	4d^{9}5p^2	(^{1}D)^{2}F$_{3.5}$	167,308.1	240,891.5	1.66E+09		TW
200		1362.445(8)	73,397.5	1362.448(8)	−0.003	4d^{9}5s5p	(^{1}D)^{2}F$_{3.5}$	-	4d^{9}5p^2	(^{3}P)^{2}F$_{3.5}$	180,060.3	253,457.6	1.10E+10		TW
15		1370.424(8)	72,970.1	1370.432(7)	−0.008	4d^{9}5s5p	(^{3}D)^{4}D$_{0.5}$	-	4d^{9}5p^2	(^{3}P)^{2}P$_{0.5}$	178,187.72	251,157.4	6.79E+08		TW
8	f	1378.569(16)	72,539.0	1378.539(8)	0.030	4d^{9}5s5p	(^{1}D)^{2}P$_{1.5}$	-	4d^{9}5p^2	(^{3}P)^{2}P$_{0.5}$	178,616.85	251,157.4	4.19E+08		TW
160		1380.066(8)	72,460.3	1380.079(5)	−0.013	4d^{9}5s5p	(^{3}D)^{4}D$_{3.5}$	-	4d^{9}5p^2	(^{3}P)^{4}P$_{2.5}$	174,043.59	246,503.2	6.43E+09		TW
160		1380.638(8)	72,430.3	1380.621(6)	0.017	4d^{9}5s5p	(^{3}D)^{4}F$_{3.5}$	-	4d^{9}5p^2	(^{1}D)^{2}D$_{2.5}$	167,308.1	239,739.3	1.06E+09		TW
180		1383.510(8)	72,279.9	1383.510(6)	0.000	4d^{9}5s5p	(^{3}D)^{4}P$_{2.5}$	-	4d^{9}5p^2	(^{3}P)^{4}D$_{3.5}$	163,890.3	236,170.2	1.06E+10		TW
40		1389.976(8)	71,943.7	1389.972(7)	0.004	4d^{9}5s5p	(^{3}D)^{4}F$_{4.5}$	-	4d^{9}5p^2	(^{1}D)^{2}F$_{3.5}$	168,947.6	240,891.5	4.36E+08		TW
8	f	1390.554(16)	71,913.8	1390.558(5)	−0.004	4d^{9}5s^2	(^{2}D)^{2}D$_{1.5}$	-	4d^{10}8p	(^{1}S)^{2}P$_{1.5}$	122,419.73	194,333.3	1.89E+07	#	TW
20		1397.429(8)	71,560.0	1397.415(6)	0.014	4d^{9}5s5p	(^{3}D)^{4}P$_{2.5}$	-	4d^{9}5p^2	(^{3}P)^{4}D$_{2.5}$	163,890.3	235,451.0	2.21E+09		TW

Table 3. *Cont.*

I_{obs} [a]	ch [b]	λ_{obs} [c] Å	σ_{obs} cm^{-1}	λ_{Ritz} [d] Å	$\Delta\lambda_{O\text{-}Ritz}$ [e] Å	Classification [f]					E_{low} cm^{-1}	E_{upp} cm^{-1}	gA [h] s^{-1}		Lin. Ref [i]
200		1398.755(8)	71,492.1	1398.765(7)	−0.010	4d^{9}5s5p	(^{3}D)^{4}P$_{1.5}$	-	4d^{9}5p^2	(^{3}P)^{4}D$_{1.5}$	167,339.24	238,830.9	1.78E+09		TW
10		1399.355(8)	71,461.5	1399.357(7)	−0.002	4d^{9}5s5p	(^{1}D)^{2}F$_{2.5}$	-	4d^{9}5p^2	(^{1}D)^{2}D$_{1.5}$	176,090.19	247,551.6	2.30E+09		TW
100		1401.254(8)	71,364.6	1401.247(7)	0.007	4d^{9}5s5p	(^{3}D)^{4}F$_{2.5}$	-	4d^{9}5p^2	(^{3}P)^{4}D$_{1.5}$	167,465.9	238,830.9	4.84E+09		TW
100		1402.439(8)	71,304.3	1402.438(8)	0.001	4d^{9}5s5p	(^{1}D)^{2}D$_{2.5}$	-	4d^{9}5p^2	(^{3}P)^{2}P$_{1.5}$	182,399.28	253,703.7	6.34E+09		TW
280	w	1403.017(16)	71,275.0	1403.029(6)	−0.012	4d^{10}5p	(^{1}S)^{2}P$_{0.5}$	-	4d^{10}5d	(^{1}S)^{2}D$_{1.5}$	57,184.0	128,458.36	4.26E+09		B, TW
150		1404.342(8)	71,207.7	1404.343(7)	−0.001	4d^{9}5s5p	(^{3}D)^{4}D$_{2.5}$	-	4d^{9}5p^2	(^{3}P)^{2}D$_{1.5}$	174,496.6	245,704.3	3.40E+09		TW
200		1407.302(8)	71,058.0	1407.295(7)	0.007	4d^{9}5s5p	(^{1}D)^{2}D$_{2.5}$	-	4d^{9}5p^2	(^{3}P)^{2}F$_{3.5}$	182,399.28	253,457.6	4.50E+09		TW
120		1408.292(8)	71,008.0	1408.292(7)	0.000	4d^{9}5s5p	(^{1}D)^{2}F$_{2.5}$	-	4d^{9}5p^2	(^{3}P)^{4}F$_{2.5}$	176,090.1,9	247,098.2	3.16E+09		TW
2	f	1411.028(16)	70,870.3	1411.032(14)	−0.004	4d^{9}5s5p	(^{3}D)^{4}F$_{1.5}$	-	4d^{9}5p^2	(^{3}P)^{4}D$_{0.5}$	170,813.7	241,683.8	3.00E+09		TW
2	f	1413.821(16)	70,730.3	1413.813(9)	0.008	4d^{9}5s5p	(^{3}D)^{4}F$_{1.5}$	-	4d^{9}5p^2	(^{3}P)^{2}F$_{2.5}$	170,813.7	241,544.42	1.55E+09		TW
150		1418.119(8)	70,515.9	1418.112(6)	0.007	4d^{9}5s5p	(^{3}D)^{4}D$_{3.5}$	-	4d^{9}5p^2	(^{3}P)^{4}F$_{4.5}$	174,043.59	244,559.9	5.34E+09		TW
5	f	1420.204(16)	70,412.4	1420.192(7)	0.012	4d^{9}5s5p	(^{1}D)^{2}F$_{2.5}$	-	4d^{9}5p^2	(^{3}P)^{4}P$_{2.5}$	176,090.19	246,503.2	4.82E+08	#	TW
4	f	1421.105(16)	70,367.8	1421.098(15)	0.007	4d^{9}5s5p	(^{3}D)^{4}P$_{0.5}$	-	4d^{9}5p^2	(^{3}P)^{4}D$_{0.5}$	171,315.7	241,683.8	1.38E+09		TW
6		1421.649(8)	70,340.9	1421.647(6)	0.002	4d^{10}5d	(^{1}S)^{2}D$_{1.5}$	-	4d^{9}5s5p	(^{3}D)^{2}F$_{2.5}$	128,458.36	198,799.3	4.94E+09		TW
100		1425.676(8)	70,142.2	1425.673(8)	0.003	4d^{9}5s5p	(^{1}D)^{2}F$_{3.5}$	-	4d^{9}5p^2	(^{3}P)^{2}D$_{2.5}$	180,060.3	250,202.6	4.78E+09		TW
35		1430.130(8)	69,923.7	1430.127(7)	0.003	4d^{10}5d	(^{1}S)^{2}D$_{1.5}$	-	4d^{9}5s5p	(^{3}D)^{2}P$_{0.5}$	128,458.36	198,382.2	1.38E+09		TW
280	w	1434.800(16)	69,696.1	1434.805(6)	−0.005	4d^{10}5p	(^{1}S)^{2}P$_{0.5}$	-	4d^{10}6s	(^{1}S)^{2}S$_{0.5}$	57,184.0	126,879.89	1.08E+09		B, TW
35	bl	1439.854(16)	69,451.5	1439.830(4)	0.024	4d^{9}5s^2	(^{2}D)^{2}D$_{2.5}$	–	4d^{10}5f	(^{1}S)^{2}F$_{2.5}$	115,572.19	185,024.81	7.33E+06		TW
100		1440.281(8)	69,430.9	1440.291(7)	−0.010	4d^{9}5s5p	(^{1}D)^{2}D$_{2.5}$	-	4d^{9}5p^2	(^{1}D)^{2}F$_{2.5}$	182,399.28	251,829.7	2.33E+09		TW
200	bl	1442.512(16)	69,323.5	1442.507(5)	0.005	4d^{9}5s^2	(^{2}D)^{2}D$_{2.5}$	-	4d^{10}5f	(^{1}S)^{2}F$_{3.5}$	115,572.19	184,895.95	2.52E+08		TW
110		1447.387(8)	69,090.0	1447.401(6)	−0.014	4d^{9}5s^2	(^{2}D)^{2}D$_{1.5}$	-	4d^{9}5s5p	(^{3}D)^{2}P$_{1.5}$	122,419.73	191,509.1	1.09E+08	#	B
100		1467.495(8)	68,143.3	1467.504(6)	−0.009	4d^{9}5s5p	(^{3}D)^{4}F$_{3.5}$	-	4d^{9}5p^2	(^{3}P)^{4}D$_{2.5}$	167,308.1	235,451.0	6.54E+09		TW
120		1468.172(8)	68,111.9	1468.175(6)	−0.003	4d^{9}5s5p	(^{3}D)^{4}P$_{1.5}$	-	4d^{9}5p^2	(^{3}P)^{4}D$_{2.5}$	167,339.24	235,451.0	3.47E+09		TW
40		1481.468(8)	67,500.6	1481.463(5)	0.005	4d^{9}5s5p	(^{3}D)^{4}D$_{3.5}$	-	4d^{9}5p^2	(^{3}P)^{2}F$_{2.5}$	174,043.59	241,544.42	1.54E+09		TW
8	f	1482.483(16)	67,454.4	1482.505(6)	−0.022	4d^{10}6s	(^{1}S)^{2}S$_{0.5}$	-	4d^{10}8p	(^{1}S)^{2}P$_{1.5}$	126,879.89	194,333.3	8.27E+06		B
300	w	1487.623(23)	67,221.3	1487.595(10)	0.028	4d^{9}5s5p	(^{3}D)^{4}F$_{4.5}$	-	4d^{9}5p^2	(^{3}P)^{4}D$_{3.5}$	168,947.6	236,170.2	1.06E+10		B, TW
300	w	1487.623(23)	67,221.3	1487.623(5)	0.000	4d^{10}5p	(^{1}S)^{2}P$_{1.5}$	-	4d^{10}5d	(^{1}S)^{2}D$_{2.5}$	61,527.0	128,748.33	8.44E+09		TW
15		1491.235(8)	67,058.5	1491.235(8)	0.000	4d^{10}6s	(^{1}S)^{2}S$_{0.5}$	-	4d^{10}8p	(^{1}S)^{2}P$_{0.5}$	126,879.89	193,938.4	5.84E+05		B
5	f	1491.474(16)	67,047.8	1491.473(10)	0.001	4d^{9}5s5p	(^{3}D)^{4}D$_{2.5}$	-	4d^{9}5p^2	(^{3}P)^{2}F$_{2.5}$	174,496.6	241,544.42	2.20E+09		TW
200	w	1494.066(16)	66,931.4	1494.068(5)	−0.002	4d^{10}5p	(^{1}S)^{2}P$_{1.5}$	-	4d^{10}5d	(^{1}S)^{2}D$_{1.5}$	61,527.0	128,458.36	7.21E+08		B, TW
5		1495.389(8)	66,872.2	1495.377(6)	0.012	4d^{10}6p	(^{1}S)^{2}P$_{0.5}$	-	4d^{10}11s	(^{1}S)^{2}S$_{0.5}$	144,589.32	211,462.1	2.09E+07		B
20		1505.020(8)	66,444.3	1505.021(7)	−0.001	4d^{10}5d	(^{1}S)^{2}D$_{1.5}$	-	4d^{9}5s5p	(^{3}D)^{2}D$_{2.5}$	128,458.36	194,902.6	6.10E+08		TW
40	*	1511.615(23)	66,154.4	1511.619(11)	−0.004	4d^{9}5s5p	(^{3}D)^{4}D$_{1.5}$	-	4d^{9}5p^2	(^{3}P)^{4}F$_{2.5}$	180,943.95	247,098.2	1.23E+09		TW
40	*	1511.615(23)	66,154.4	1511.618(8)	−0.003	4d^{10}5d	(^{1}S)^{2}D$_{2.5}$	-	4d^{9}5s5p	(^{3}D)^{2}D$_{2.5}$	128,748.33	194,902.6	1.91E+08		TW
35		1515.040(8)	66,004.9	1515.035(6)	0.005	4d^{9}5s5p	(^{1}D)^{2}D$_{1.5}$	-	4d^{9}5p^2	(^{3}P)^{2}F$_{2.5}$	175,539.35	241,544.42	2.10E+09		TW
10		1518.024(8)	65,875.1	1518.028(5)	−0.004	4d^{10}5d	(^{1}S)^{2}D$_{1.5}$	-	4d^{10}8p	(^{1}S)^{2}P$_{1.5}$	128,458.36	194,333.3	2.33E+07	#	B
60		1522.164(8)	65,695.9	1522.169(6)	−0.005	4d^{9}5s5p	(^{3}D)^{4}D$_{3.5}$	-	4d^{9}5p^2	(^{1}D)^{2}D$_{2.5}$	174,043.59	239,739.3	7.16E+08		TW
25		1524.750(8)	65,584.5	1524.740(6)	0.010	4d^{10}5d	(^{1}S)^{2}D$_{2.5}$	-	4d^{10}8p	(^{1}S)^{2}P$_{1.5}$	128,748.33	194,333.3	7.42E+07		B
10		1525.344(8)	65,559.0	1525.338(6)	0.006	4d^{9}5s5p	(^{3}D)^{4}D$_{1.5}$	-	4d^{9}5p^2	(^{3}P)^{4}P$_{2.5}$	180,943.95	246,503.2	1.21E+08	#	TW

Table 3. *Cont.*

I_{obs} [a]	ch [b]	λ_{obs} [c] Å	σ_{obs} cm^{-1}	λ_{Ritz} [d] Å	$\Delta\lambda_{O-Ritz}$ [e] Å	Classification [f]					E_{low} cm^{-1}	E_{upp} cm^{-1}	gA [h] s^{-1}		Lin. Ref [i]
14		1525.869(8)	65,536.4	1525.881(6)	−0.012	4d^{10}6p	(^{1}S)^{2}P$_{1.5}$	-	4d^{10}11s	(^{1}S)^{2}S$_{0.5}$	145,926.21	211,462.1	3.95E+07		B
100		1526.000(8)	65,530.8	1525.996(7)	0.004	4d^{9}5s5p	(^{3}D)^{4}D$_{1.5}$	-	4d^{9}5p^2	(^{3}P)^{4}F$_{1.5}$	180,943.95	246,474.9	7.67E+08		TW
40		1527.784(8)	65,454.3	1527.785(6)	−0.001	4d^{9}5s5p	(^{1}D)^{2}F$_{2.5}$	-	4d^{9}5p^2	(^{3}P)^{2}F$_{2.5}$	176,090.19	241,544.42	7.27E+07	#	TW
52		1529.704(8)	65,372.1	1529.713(5)	−0.009	4d^{9}5s^2	(^{2}D)^{2}D$_{2.5}$	-	4d^{9}5s5p	(^{3}D)^{4}D$_{1.5}$	115,572.19	180,943.95	7.42E+06	#	B
260	w	1530.169(16)	65,352.3	1530.154(6)	0.015	4d^{10}5p	(^{1}S)^{2}P$_{1.5}$	-	4d^{10}6s	(^{1}S)^{2}S$_{0.5}$	61,527.0	126,879.89	1.82E+09		B, TW
180	w	1532.926(16)	65,234.7	1532.902(6)	0.024	4d^{10}5p	(^{1}S)^{2}P$_{0.5}$	-	4d^{9}5s^2	(^{2}D)^{2}D$_{1.5}$	57,184.0	122,419.73	2.79E+09		B, TW
5	f	1534.868(16)	65,152.2	1534.865(10)	0.003	4d^{9}5s5p	(^{1}D)^{2}D$_{2.5}$	-	4d^{9}5p^2	(^{1}D)^{2}D$_{1.5}$	182,399.28	247,551.6	1.00E+08	#	TW
48		1547.300(8)	64,628.7	1547.288(6)	0.012	4d^{10}6s	(^{1}S)^{2}S$_{0.5}$	-	4d^{9}5s5p	(^{3}D)^{2}P$_{1.5}$	126,879.89	191,509.1	1.64E+07		B
25		1579.653(8)	63,305.0	1579.654(7)	−0.001	4d^{9}5s5p	(^{1}D)^{2}D$_{2.5}$	-	4d^{9}5p^2	(^{3}P)^{2}D$_{1.5}$	182,399.28	245,704.3	4.09E+07	#	TW
30		(1593.384)	62,759.5	1593.352(8)		4d^{10}5d	(^{1}S)^{2}D$_{2.5}$	-	4d^{9}5s5p	(^{3}D)^{2}P$_{1.5}$	128,748.33	191,509.1	4.16E+06	#	B
32		(1593.592)	62,751.3	1593.639(8)		4d^{10}6p	(^{1}S)^{2}P$_{0.5}$	-	4d^{10}9d	(^{1}S)^{2}D$_{1.5}$	144,589.32	207,338.8	8.02E+07		B
78		1597.329(8)	62,604.5	1597.314(4)	0.015	4d^{9}5s^2	(^{2}D)^{2}D$_{1.5}$	-	4d^{10}5f	(^{1}S)^{2}F$_{2.5}$	122,419.73	185,024.81	1.92E+08		B
50		1600.535(8)	62,479.1	1600.535(6)	0.000	4d^{10}6p	(^{1}S)^{2}P$_{0.5}$	-	4d^{10}10s	(^{1}S)^{2}S$_{0.5}$	144,589.32	207,068.43	3.15E+07		B
60	w	(1605.211)	62,297.1	1605.251(6)		4d^{9}5s^2	(^{2}D)^{2}D$_{2.5}$	-	4^{10}7p	(^{1}S)^{2}P$_{1.5}$	115,572.19	177,867.74	7.79E+05	#	B
10		1609.613(8)	62,126.7	1609.616(7)	−0.003	4d^{9}5s5p	(^{3}D)^{4}D$_{3.5}$	-	4d^{9}5p^2	(^{3}P)^{4}D$_{3.5}$	174,043.59	236,170.2	3.86E+08		TW
400	w	1625.301(16)	61,527.1	1625.303(9)	−0.002	4d^{10}5s	(^{1}S)^{2}S$_{0.5}$	-	4d^{10}5p	(^{1}S)^{2}P$_{1.5}$	0.0	61,527.0	3.34E+09		B, TW
60		1627.249(8)	61,453.4	1627.247(8)	0.002	4d^{10}6p	(^{1}S)^{2}P$_{1.5}$	-	4d^{10}9d	(^{1}S)^{2}D$_{2.5}$	145,926.21	207,379.7	1.37E+08		B
49		1628.330(8)	61,412.6	1628.331(8)	−0.001	4d^{10}6p	(^{1}S)^{2}P$_{1.5}$	-	4d^{10}9d	(^{1}S)^{2}D$_{1.5}$	145,926.21	207,338.8	1.52E+07		B
35		1635.534(8)	61,142.1	1635.531(6)	0.003	4d^{10}6p	(^{1}S)^{2}P$_{1.5}$	-	4d^{10}10s	(^{1}S)^{2}S$_{0.5}$	145,926.21	207,068.43	5.93E+07		B
180	w	1642.237(16)	60,892.6	1642.232(6)	0.005	4d^{10}5p	(^{1}S)^{2}P$_{1.5}$	-	4d^{9}5s^2	(2D)^{2}D$_{1.5}$	61,527.0	122,419.73	4.70E+08		B, TW
65		1667.581(8)	59,967.1	1667.579(5)	0.002	4d^{9}5s^2	(^{2}D)^{2}D$_{2.5}$	-	4d^{9}5s5p	(^{1}D)^{2}D$_{1.5}$	115,572.19	175,539.35	2.38E+07	#	B
60		(1708.662)	58,525.3	1708.694(5)		4d^{9}5s^2	(^{2}D)^{2}D$_{1.5}$	-	4d^{9}5s5p	(^{3}D)^{4}D$_{1.5}$	122,419.73	180,943.95	2.48E+06	#	B
400	w	1748.728(16)	57,184.4	1748.741(11)	−0.013	4d^{10}5s	(^{1}S)^{2}S$_{0.5}$	-	4d^{10}5p	(^{1}S)^{2}P$_{0.5}$	0.0	57,184.0	1.38E+09		B, TW
14		1757.432(8)	56,901.2	1757.433(8)	−0.001	4d^{9}5s^2	(^{2}D)^{2}D$_{1.5}$	-	4d^{9}5s5p	(^{1}D)^{2}P$_{0.5}$	122,419.73	179,320.9	6.87E+06	#	B
68		1767.840(8)	56,566.2	1767.832(5)	0.008	4d^{10}5d	(^{1}S)^{2}D$_{1.5}$	-	4d^{10}5f	(^{1}S)^{2}F$_{2.5}$	128,458.36	185,024.81	6.40E+07		B
53		1776.943(8)	56,276.4	1776.941(5)	0.002	4d^{10}5d	(^{1}S)^{2}D$_{2.5}$	-	4d^{10}5f	(^{1}S)^{2}F$_{2.5}$	128,748.33	185,024.81	1.31E+07		B
40		1779.457(8)	56,196.9	1779.451(5)	0.006	4d^{9}5s^2	(^{2}D)^{2}D$_{1.5}$	-	4d^{9}5s5p	(^{1}D)^{2}P$_{1.5}$	122,419.73	178,616.85	5.00E+06	#	B
65		1779.704(8)	56,189.1	1779.708(5)	−0.004	4d^{10}6p	(^{1}S)^{2}P$_{0.5}$	-	4d^{10}8d	(^{1}S)^{2}D$_{1.5}$	144,589.32	200,778.32	1.44E+08		B
70		1781.020(8)	56,147.6	1781.019(5)	0.001	4d^{10}5d	(^{1}S)^{2}D$_{2.5}$	-	4d^{10}5f	(^{1}S)^{2}F$_{3.5}$	128,748.33	184,895.95	2.40E+08		B
50		1792.962(8)	55,773.63	1792.968(6)	−0.006	4d^{10}6p	(^{1}S)^{2}P$_{0.5}$	-	4d^{10}9s	(^{1}S)^{2}S$_{0.5}$	144,589.32	200,362.77	5.11E+07		B
50		1793.146(8)	55,767.91	1793.143(5)	0.003	4d^{9}5s^2	(^{2}D)^{2}D$_{1.5}$	-	4d^{9}5s5p	(^{3}D)^{4}D$_{0.5}$	122,419.73	178,187.72	2.70E+05	#	B
19		1803.500(8)	55,447.74	1803.491(5)	0.009	4d^{9}5s^2	(^{2}D)^{2}D$_{1.5}$	-	4^{10}7p	(^{1}S)^{2}P$_{1.5}$	122,419.73	177,867.74	9.39E+05	#	B
60		1821.158(8)	54,910.12	1821.169(5)	−0.011	4d^{10}6p	(^{1}S)^{2}P$_{1.5}$	-	4d^{10}8d	(^{1}S)^{2}D$_{2.5}$	145,926.21	200,836.01	2.44E+08		B
41		1823.097(8)	54,851.72	1823.084(5)	0.013	4d^{10}6p	(^{1}S)^{2}P$_{1.5}$	-	4d^{10}8d	(^{1}S)^{2}D$_{1.5}$	145,926.21	200,778.32	2.70E+07		B
29		1823.363(8)	54,843.71	1823.365(6)	−0.002	4d^{9}5s^2	(^{2}D)^{2}D$_{1.5}$	-	4^{10}7p	(^{1}S)^{2}P$_{0.5}$	122,419.73	177,263.38	8.19E+06	#	B
60		1837.006(8)	54,436.40	1837.001(6)	0.005	4d^{10}6p	(^{1}S)^{2}P$_{1.5}$	-	4d^{10}9s	(^{1}S)^{2}S$_{0.5}$	145,926.21	200,362.77	9.55E+07		B
80	w	1850.280(16)	54,045.9	1850.303(8)	−0.023	4d^{10}5p	(^{1}S)^{2}P$_{1.5}$	-	4d^{9}5s^2	(2D)^{2}D$_{2.5}$	61,527.0	115,572.19	2.28E+08		B, TW
70		1882.547(8)	53,119.52	1882.544(6)	0.003	4d^{9}5s^2	(^{2}D)^{2}D$_{1.5}$	-	4d^{9}5s5p	(^{1}D)^{2}D$_{1.5}$	122,419.73	175,539.35	1.09E+07	#	B
13		1905.284(8)	52,485.61	1905.285(6)	−0.001	4d^{10}5d	(^{1}S)^{2}D$_{1.5}$	-	4d^{9}5s5p	(^{3}D)^{4}D$_{1.5}$	128,458.36	180,943.95	2.37E+05	#	B

Table 3. *Cont.*

I_{obs} [a]	ch [b]	λ_{obs} [c] Å	σ_{obs} cm^{-1}	λ_{Ritz} [d] Å	$\Delta\lambda_{O\text{-}Ritz}$ [e] Å	Classification [f]					E_{low} cm^{-1}	E_{upp} cm^{-1}	gA [h] s^{-1}		Lin. Ref [i]
19		1915.881(8)	52,195.31	1915.870(6)	0.011	4d^{10}5d	(^{1}S)^{2}D$_{2.5}$	-	4d^{9}5s5p	(^{3}D)^{4}D$_{1.5}$	128,748.33	180,943.95	1.26E+05	#	B
50		1923.343(10)	51,992.8	1923.343(10)	0.000	4d^{10}4f	(^{1}S)^{2}F$_{2.5}$	-	4d^{10}9g	(^{1}S)^{2}G$_{3.5}$	161,974.14	213,966.94	2.19E+08	sh	
46		1923.654(10)	51,984.4	1923.662(7)	−0.008	4d^{10}4f	(^{1}S)^{2}F$_{3.5}$	-	4d^{10}9g	(^{1}S)^{2}G$_{4.5}$	161,982.00	213,966.18	2.84E+08	sh	
75		1931.728(8)	51,767.12	1931.731(7)	−0.003	4d^{9}5s^2	(^{2}D)^{2}D$_{2.5}$	-	4d^{9}5s5p	(^{3}D)^{4}P$_{1.5}$	115,572.19	167,339.24	1.33E+07	#	B
58		(1932.89)	51,736.00	1932.854(8)		4d^{10}6s	(^{1}S)^{2}S$_{0.5}$	-	4d^{9}5s5p	(^{1}D)^{2}P$_{1.5}$	126,879.89	178,616.85	3.74E+06		B
28		1949.021(8)	51,307.81	1949.020(6)	0.001	4d^{10}6s	(^{1}S)^{2}S$_{0.5}$	-	4d^{9}5s5p	(^{3}D)^{4}D$_{0.5}$	126,879.89	178,187.72	2.11E+06		B
38		1961.245(8)	50,988.02	1961.252(6)	−0.007	4d^{10}6s	(^{1}S)^{2}S$_{0.5}$	-	4d^{10}7p	(^{1}S)^{2}P$_{1.5}$	126,879.89	177,867.74	3.29E+05	#	B
75		(1965.976)	50,865.32	1966.083(11)		4d^{10}5d	(^{1}S)^{2}D$_{1.5}$	-	4d^{9}5s5p	(^{1}D)^{2}P$_{0.5}$	128,458.36	179,320.9	1.19E+05	#	B
30		1984.780(8)	50,383.42	1984.777(6)	0.003	4d^{10}6s	(^{1}S)^{2}S$_{0.5}$	-	4d^{10}7p	(^{1}S)^{2}P$_{0.5}$	126,879.89	177,263.38	1.38E+05	#	B
10		1993.680(8)	50,158.50	1993.680(6)	0.000	4d^{10}5d	(^{1}S)^{2}D$_{1.5}$	-	4d^{9}5s5p	(^{1}D)^{2}P$_{1.5}$	128,458.36	178,616.85	4.67E+05	#	B
66		2004.620(20)	49,868.6	2004.624(8)	−0.004	4d^{10}5d	(^{1}S)^{2}D$_{2.5}$	-	4d^{9}5s5p	(^{1}D)^{2}P$_{1.5}$	128,748.33	178,616.85	3.64E+07		B
36		2010.200(20)	49,730.2	2010.235(8)	−0.035	4d^{10}5d	(^{1}S)^{2}D$_{1.5}$	-	4d^{9}5s5p	(^{3}D)^{4}D$_{0.5}$	128,458.36	178,187.72	2.68E+07	#	B
63		2023.260(20)	49,409.3	2023.255(7)	0.005	4d^{10}5d	(^{1}S)^{2}D$_{1.5}$	-	4d^{10}7p	(^{1}S)^{2}P$_{1.5}$	128,458.36	177,867.74	8.57E+06	#	B
76		2035.190(20)	49,119.7	2035.201(8)	−0.011	4d^{10}5d	(^{1}S)^{2}D$_{2.5}$	-	4d^{10}7p	(^{1}S)^{2}P$_{1.5}$	128,748.33	177,867.74	5.75E+07		B
76		2048.310(20)	48,805.1	2048.313(8)	−0.003	4d^{10}5d	(^{1}S)^{2}D$_{1.5}$	-	4d^{10}7p	(^{1}S)^{2}P$_{0.5}$	128,458.36	177,263.38	1.84E+07	#	B
74		2051.070(20)	48,739.4	2051.092(8)	−0.022	4d^{10}4f	(^{1}S)^{2}F$_{2.5}$	-	4d^{10}8g	(^{1}S)^{2}G$_{3.5}$	161,974.14	210,713.04	3.39E+08		B
78		2051.410(20)	48,731.3	2051.423(9)	−0.013	4d^{10}4f	(^{1}S)^{2}F$_{3.5}$	-	4d^{10}8g	(^{1}S)^{2}G$_{3.5}$	161,982.00	210,713.04	1.25E+07		B
10		2136.480(20)	46,791.2	2136.488(9)	−0.008	4d^{10}5d	(^{1}S)^{2}D$_{2.5}$	-	4d^{9}5s5p	(^{1}D)^{2}D$_{1.5}$	128,748.33	175,539.35	3.77E+05	#	B
81		(2154.04)	46,409.8	2154.039(11)		4d^{9}5s^2	(^{2}D)^{2}D$_{2.5}$	-	4d^{10}4f	(^{1}S)^{2}F$_{3.5}$	115,572.19	161,982.00	1.48E+08		B
80		2154.400(20)	46,402.0	2154.404(10)	−0.004	4d^{9}5s^2	(^{2}D)^{2}D$_{2.5}$	-	4d^{10}4f	(^{1}S)^{2}F$_{2.5}$	115,572.19	161,974.14	8.14E+06		B
77		2199.550(20)	45,449.7	2199.558(13)	−0.008	4d^{10}6p	(^{1}S)^{2}P$_{0.5}$	-	4d^{10}7d	(^{1}S)^{2}D$_{1.5}$	144,589.32	190,038.8	3.15E+08		B
42		(2201.47)	45,410.0	2201.686(19)		4d^{10}4f	(^{1}S)^{2}F$_{2.5}$	-	4d^{10}9d	(^{1}S)^{2}D$_{2.5}$	161,974.14	207,379.7	2.98E+05		B
10		(2203.54)	45,367.4	2203.671(18)		4d^{10}4f	(^{1}S)^{2}F$_{2.5}$	-	4d^{10}9d	(^{1}S)^{2}D$_{1.5}$	161,974.14	207,338.8	4.11E+06		B
45		2225.480(20)	44,920.2	2225.512(11)	−0.032	4d^{9}5s^2	(^{2}D)^{2}D$_{1.5}$	-	4d^{9}5s5p	(^{3}D)^{4}P$_{1.5}$	122,419.73	167,339.24	8.27E+05	#	B
75		2232.170(20)	44,785.5	2232.188(13)	−0.018	4d^{10}6p	(^{1}S)^{2}P$_{0.5}$		4d^{10}8s	(^{1}S)^{2}S$_{0.5}$	144,589.32	189,374.5	9.32E+07		B
77		2261.230(20)	44,210.0	2261.227(19)	0.003	4d^{10}6p	(^{1}S)^{2}P$_{1.5}$	-	4d^{10}7d	(^{1}S)^{2}D$_{2.5}$	145,926.21	190,136.3	5.25E+08		B
74		2266.230(20)	44,112.5	2266.226(14)	0.004	4d^{10}6p	(^{1}S)^{2}P$_{1.5}$	-	4d^{10}7d	(^{1}S)^{2}D$_{1.5}$	145,926.21	190,038.8	5.80E+07		B
78		2272.370(20)	43,993.3	2272.417(9)	−0.047	4d^{10}4f	(^{1}S)^{2}F$_{2.5}$	-	4d^{10}7g	(^{1}S)^{2}G$_{3.5}$	161,974.14	205,966.56	5.74E+08		B
80		2272.810(20)	43,984.8	2272.812(13)	−0.002	4d^{10}4f	(^{1}S)^{2}F$_{3.5}$	-	4d^{10}7g	(^{1}S)^{2}G$_{4.5}$	161,982.00	205,966.76	7.44E+08		B
81		2300.890(20)	43,448.1	2300.878(14)	0.012	4d^{10}6p	(^{1}S)^{2}P$_{1.5}$	-	4d^{10}8s	(^{1}S)^{2}S$_{0.5}$	145,926.21	189,374.5	1.72E+08		B
74		2527.380(20)	39,554.8	2527.403(11)	−0.023	4d^{9}5s^2	(^{2}D)^{2}D$_{1.5}$	-	4d^{10}4f	(^{1}S)^{2}F$_{2.5}$	122,419.73	161,974.14	1.45E+09		B
10		(2572.42)	38,862.3	2572.446(15)		4d^{10}4f	(^{1}S)^{2}F$_{2.5}$	-	4d^{10}8d	(^{1}S)^{2}D$_{2.5}$	161,974.14	200,836.01	4.91E+05		B
30		(2572.94)	38,854.4	2572.966(17)		4d^{10}4f	(^{1}S)^{2}F$_{3.5}$	-	4d^{10}8d	(^{1}S)^{2}D$_{2.5}$	161,982.00	200,836.01	9.82E+06		B
22		(2576.15)	38,806.0	2576.270(15)		4d^{10}4f	(^{1}S)^{2}F$_{2.5}$	-	4d^{10}8d	(^{1}S)^{2}D$_{1.5}$	161,974.14	200,778.32	6.77E+06		B
85		2725.460(20)	36,680.2	2725.462(19)	−0.002	4d^{10}4f	(^{1}S)^{2}F$_{2.5}$	-	4d^{10}6g	(^{1}S)^{2}G$_{3.5}$	161,974.14	198,654.3	1.13E+09		B
86	*	2726.07(6)	36,672.0	2726.046(23)	0.02	4d^{10}4f	(^{1}S)^{2}F$_{3.5}$	-	4d^{10}6g	(^{1}S)^{2}G$_{3.5}$	161,982.00	198,654.3	4.18E+07		B
86	*	2726.07(6)	36,672.0	2726.07(6)	0.000	4d^{10}4f	(^{1}S)^{2}F$_{3.5}$	-	4d^{10}6g	(^{1}S)^{2}G$_{4.5}$	161,982.00	198,654.0	1.46E+09		B
26		(2923.41)	34,196.62	2923.23(3)		4d^{10}7p	(^{1}S)^{2}P$_{0.5}$	-	4d^{10}11s	(^{1}S)^{2}S$_{0.5}$	177,263.38	211,462.1	8.81E+06		B
80		2982.800(20)	33,515.77	2982.799(14)	0.001	4d^{10}5d	(^{1}S)^{2}D$_{1.5}$	-	4d^{10}4f	(^{1}S)^{2}F$_{2.5}$	128,458.36	161,974.14	2.16E+09		S

Table 3. *Cont.*

I_{obs} [a]	ch [b]	λ_{obs} [c] Å	σ_{obs} cm^{-1}	λ_{Ritz} [d] Å	$\Delta\lambda_{O\text{-}Ritz}$ [e] Å	Classification [f]					E_{low} cm^{-1}	E_{upp} cm^{-1}	gA [h] s^{-1}	Lin. Ref [i]
82		3008.080(20)	33,234.11	3008.120(15)	−0.040	$4d^{10}5d$	$(^1S)^2D_{2.5}$	-	$4d^{10}4f$	$(^1S)^2F_{3.5}$	128,748.33	161,982.00	4.91E+09	S
77		(3008.76)	33,226.60	3008.832(17)		$4d^{10}5d$	$(^1S)^2D_{2.5}$	-	$4d^{10}4f$	$(^1S)^2F_{2.5}$	128,748.33	161,974.14	2.44E+08	B
45		(3293.56)	30,353.54	3293.51(3)		$4d^95s^2$	$(^2D)^2D_{2.5}$	-	$4d^{10}6p$	$(^1S)^2P_{1.5}$	115,572.19	145,926.21	4.41E+06	B
21		3438.970(20)	29,070.14	3438.960(18)	0.010	$4d^{10}5f$	$(^1S)^2F_{3.5}$	-	$4d^{10}9g$	$(^1S)^2G_{4.5}$	184,895.95	213,966.18	2.06E+08	B
37		(3551.03)	28,152.80	3550.84(6)		$4d^{10}4f$	$(^1S)^2F_{3.5}$	-	$4d^{10}7d$	$(^1S)^2D_{2.5}$	161,982.00	190,136.3	1.84E+07	B
28		(3562.35)	28,063.34	3562.18(5)		$4d^{10}4f$	$(^1S)^2F_{2.5}$	-	$4d^{10}7d$	$(^1S)^2D_{1.5}$	161,974.14	190,038.8	1.27E+07	B
30	*	3640.69(10)	27,459.5	3640.69(10)	0.00	$4d^{10}5g$	$(^1S)^2G_{4.5}$	-	$4d^{10}9h$	$(^1S)^2H_{5.5}$	186,528.26	213,987.75	1.37E+08	B
30	*	3640.69(10)	27,459.5	3640.64(8)	0.05	$4d^{10}5g$	$(^1S)^2G_{3.5}$	-	$4d^{10}9h$	$(^1S)^2H_{4.5}$	186,527.40	213,987.3	1.11E+08	B
30	*	3640.69(10)	27,459.5	3640.75(8)	−0.06	$4d^{10}5g$	$(^1S)^2G_{4.5}$	-	$4d^{10}9h$	$(^1S)^2H_{4.5}$	186,528.26	213,987.3	2.53E+06	B
91		3853.010(20)	25,946.38	3853.001(17)	0.009	$4d^{10}6p$	$(^1S)^2P_{0.5}$	-	$4d^{10}6d$	$(^1S)^2D_{1.5}$	144,589.32	170,535.76	1.15E+09	B
65		3872.630(20)	25,814.93	3872.630(20)	0.000	$4d^{10}5f$	$(^1S)^2F_{3.5}$	-	$4d^{10}8g$	$(^1S)^2G_{4.5}$	184,895.95	210,710.88	3.11E+08	B
45		3891.740(20)	25,688.17	3891.731(19)	0.009	$4d^{10}5f$	$(^1S)^2F_{2.5}$	-	$4d^{10}8g$	$(^1S)^2G_{3.5}$	185,024.81	210,713.04	2.42E+08	B
86		(4023.82)	24,844.98	4023.77(3)		$4d^{10}6p$	$(^1S)^2P_{0.5}$	-	$4d^{10}7s$	$(^1S)^2S_{0.5}$	144,589.32	169,434.59	2.29E+08	B
90		4032.320(20)	24,792.61	4032.322(20)	−0.002	$4d^{10}6p$	$(^1S)^2P_{1.5}$	-	$4d^{10}6d$	$(^1S)^2D_{2.5}$	145,926.21	170,718.81	1.81E+09	S
88		4062.310(20)	24,609.59	4062.316(17)	−0.006	$4d^{10}6p$	$(^1S)^2P_{1.5}$	-	$4d^{10}6d$	$(^1S)^2D_{1.5}$	145,926.21	170,535.76	1.97E+08	B
81		4071.640(20)	24,553.20	4071.629(19)	0.011	$4d^{10}4f$	$(^1S)^2F_{2.5}$	-	$4d^{10}5g$	$(^1S)^2G_{3.5}$	161,974.14	186,527.40	2.96E+09	B
92		4072.780(20)	24,546.32	4072.790(19)	−0.010	$4d^{10}4f$	$(^1S)^2F_{3.5}$	-	$4d^{10}5g$	$(^1S)^2G_{4.5}$	161,982.00	186,528.26	1.10E+08	B
40	*	4128.42(10)	24,215.5	4128.42(10)	0.00	$4d^{10}5g$	$(^1S)^2G_{4.5}$	-	$4d^{10}8h$	$(^1S)^2H_{5.5}$	186,528.26	210,743.8	2.43E+08	B
40	*	4128.42(10)	24,215.5	4128.35(8)	0.07	$4d^{10}5g$	$(^1S)^2G_{3.5}$	-	$4d^{10}8h$	$(^1S)^2H_{4.5}$	186,527.40	210,743.3	1.98E+08	B
40	*	4128.42(10)	24,215.5	4128.50(8)	−0.08	$4d^{10}5g$	$(^1S)^2G_{4.5}$	-	$4d^{10}8h$	$(^1S)^2H_{4.5}$	186,528.26	210,743.3	4.50E+06	B
38		(4233.56)	23,614.13	4233.50(6)		$4d^{10}6d$	$(^1S)^2D_{2.5}$	-	$4d^{10}8p$	$(^1S)^2P_{1.5}$	170,718.81	194,333.3	7.87E+06	B
64		(4250.94)	23,517.59	4251.42(4)		$4d^{10}7p$	$(^1S)^2P_{0.5}$	-	$4d^{10}8d$	$(^1S)^2D_{1.5}$	177,263.38	200,778.32	7.93E+07	B
88		4252.600(20)	23,508.41	4252.605(20)	−0.005	$4d^{10}6p$	$(^1S)^2P_{1.5}$	-	$4d^{10}7s$	$(^1S)^2S_{0.5}$	145,926.21	169,434.59	3.93E+08	B
80		(4252.91)	23,506.69	4252.95(4)		$4d^95s^2$	$(^2D)^2D_{1.5}$	-	$4d^{10}6p$	$(^1S)^2P_{1.5}$	122,419.73	145,926.21	1.51E+07	B
40		(4328.03)	23,098.71	4327.90(5)		$4d^{10}7p$	$(^1S)^2P_{0.5}$	-	$4d^{10}9s$	$(^1S)^2S_{0.5}$	177,263.38	200,362.77	2.32E+07	B
12		4352.620(20)	22,968.21	4352.609(19)	0.011	$4d^{10}7p$	$(^1S)^2P_{1.5}$	-	$4d^{10}8d$	$(^1S)^2D_{2.5}$	177,867.74	200,836.01	1.48E+08	B
2		4363.560(20)	22,910.63	4363.569(19)	−0.009	$4d^{10}7p$	$(^1S)^2P_{1.5}$	-	$4d^{10}8d$	$(^1S)^2D_{1.5}$	177,867.74	200,778.32	1.65E+07	B
50		(4444.36)	22,494.11	4444.18(4)		$4d^{10}7p$	$(^1S)^2P_{1.5}$	-	$4d^{10}9s$	$(^1S)^2S_{0.5}$	177,867.74	200,362.77	4.83E+07	B
22		(4479.97)	22,315.32	4480.24(8)		$4d^{10}5f$	$(^1S)^2F_{2.5}$	-	$4d^{10}9d$	$(^1S)^2D_{1.5}$	185,024.81	207,338.8	5.53E+06	B
87		(4509.78)	22,167.81	4509.42(4)		$4d^95s^2$	$(^2D)^2D_{1.5}$	-	$4d^{10}6p$	$(^1S)^2P_{0.5}$	122,419.73	144,589.32	6.28E+07	B
73	*	4744.58(6)	21,070.8	4744.62(4)	−0.04	$4d^{10}5f$	$(^1S)^2F_{3.5}$	-	$4d^{10}7g$	$(^1S)^2G_{3.5}$	184,895.95	205,966.56	1.44E+07	B
73	*	4744.58(6)	21,070.8	4744.58(5)	0.00	$4d^{10}5f$	$(^1S)^2F_{3.5}$	-	$4d^{10}7g$	$(^1S)^2G_{4.5}$	184,895.95	205,966.76	5.05E+08	B
63		4773.830(20)	20,941.69	4773.815(19)	0.015	$4d^{10}5f$	$(^1S)^2F_{2.5}$	-	$4d^{10}7g$	$(^1S)^2G_{3.5}$	185,024.81	205,966.56	3.93E+08	B
44	*	5130.85(10)	19,484.5	5130.85(10)	0.00	$4d^{10}5g$	$(^1S)^2G_{4.5}$	-	$4d^{10}7h$	$(^1S)^2H_{5.5}$	186,528.26	206,012.78	5.20E+08	B
44	*	5130.85(10)	19,484.5	5130.75(8)	0.10	$4d^{10}5g$	$(^1S)^2G_{3.5}$	-	$4d^{10}7h$	$(^1S)^2H_{4.5}$	186,527.40	206,012.3	4.23E+08	B
44	*	5130.85(10)	19,484.5	5130.98(8)	−0.13	$4d^{10}5g$	$(^1S)^2G_{4.5}$	-	$4d^{10}7h$	$(^1S)^2H_{4.5}$	186,528.26	206,012.3	9.62E+06	B

Table 3. *Cont.*

I_{obs} [a]	ch [b]	λ_{obs} [c] Å	σ_{obs} cm^{-1}	λ_{Ritz} [d] Å	$\Delta\lambda_{O\text{-}Ritz}$ [e] Å	Classification [f]					E_{low} cm^{-1}	E_{upp} cm^{-1}	gA [h] S^{-1}	Lin. Ref [i]
72		(5248.77)	19,046.78	5248.90(6)		$4d^{10}6s$	$(^1S)^2S_{0.5}$	-	$4d^{10}6p$	$(^1S)^2P_{1.5}$	126,879.89	145,926.21	4.79E+08	S
70		(5644.96)	17,710.00	5645.14(7)		$4d^{10}6s$	$(^1S)^2S_{0.5}$	-	$4d^{10}6p$	$(^1S)^2P_{0.5}$	126,879.89	144,589.32	1.96E+08	B
76		(5722.71)	17,469.39	5723.22(7)		$4d^{10}5d$	$(^1S)^2D_{1.5}$	-	$4d^{10}6p$	$(^1S)^2P_{1.5}$	128,458.36	145,926.21	2.50E+07	B
70		(5819.41)	17,179.11	5819.83(8)		$4d^{10}5d$	$(^1S)^2D_{2.5}$	-	$4d^{10}6p$	$(^1S)^2P_{1.5}$	128,748.33	145,926.21	3.39E+08	B
40		(6197.72)	16,130.50	6197.54(9)		$4d^{10}5d$	$(^1S)^2D_{1.5}$	-	$4d^{10}6p$	$(^1S)^2P_{0.5}$	128,458.36	144,589.32	1.02E+08	B
10		(6520.50)	15,332.01	6519.8(5)		$4d^{10}6g$	$(^1S)^2G_{4.5}$	-	$4d^{10}9h$	$(^1S)^2H_{5.5}$	198,654.0	213,987.75	1.49E+08	B

[a] Observed relative intensities on an arbitrary scale (1–400) for the blackening of the lines on the photographic plates. Response functions of the instruments were not taken into account. [b] Character of the observed line encoded as follows: w-wide line; f-faint line; sh- shaded line; *- intensity shared by two or more transitions. [c] Observed and Ritz wavelengths are given in standard air for wavenumber σ between 5000 cm^{-1} and 50,000 cm^{-1} and in vacuum ouside of this range. The uncertainty (standard deviation) in the last digit is given in parentheses for both λ_{obs} and $\lambda_{Ritz,}$ (λ) denotes values not included in the level optimization. [d] Ritz wavelengths and their uncertainties were determined in the least-squares level optimization procedure LOPT [15]. [e] Difference between observed and Ritz wavelength. If this column is blank, the line was excluded from the level optimization because its observed wavelength deviates from the Ritz value by more than our given uncertainty. [f] Classification specifies the lower and upper levels of the transition. [h] Weighted transition probability values ($g = 2J_{upper} + 1$ is statistical weight of the upper level). If marked as # then the given gA values are too unreliable for the transitions whose cancellation factor |CF| < 0.10 in our calculations with Cowan's code [12]. [i] Reference to the source: B—Bhatia et al. [6]; B, TW—Wavelength from this work; K—Kaufman et al. [7]; R—Ryabtsev et al. [9]; S-Skočić et al. [10]; TW—this work.

4.1. The $4d^{10}5s$– $[4d^{10}np]$ Transition Array

The resonance transitions $4d^{10}5s$–$4d^{10}5p$ were first reported by Rao [1], and confirmed by all other workers [2–5]. We observed these two lines in our indium spectra with high intensity. They were the main reference in establishing the In III ionization characteristics. Bhatia [6] reported the levels of $4d^{10}np$ (n = 5–9). We agreed with Bhatia's analysis only up to $4d^{10}8p$. The $4d^{10}5s$–$4d^{10}9p$ transitions could not be seen in our spectra. The reported level value of $4d^{10}9p\ ^2P_{3/2,1/2}$ at 201,180.3 cm^{-1} did not fit in our least squares fitted parametric calculations. Our predicted values were found to be at 203,388.5 cm^{-1} and 203,556.3 cm^{-1} for $^2P_{3/2}$ and $^2P_{1/2}$, respectively. A plot of the energy differences between observed and Hartree–Fock (HF) calculated values of $4d^{10}np$ (n = 5–9) $^2P_{3/2}$ series is shown in Figure 2, and it is evident from this figure that the reported value for $4d^{10}9p$ levels shows an irregular behavior. Therefore, this reported level seems to be doubtful. We did not find any alternative value as $4d^{10}9p$ transitions were too weak to be observed on our plates.

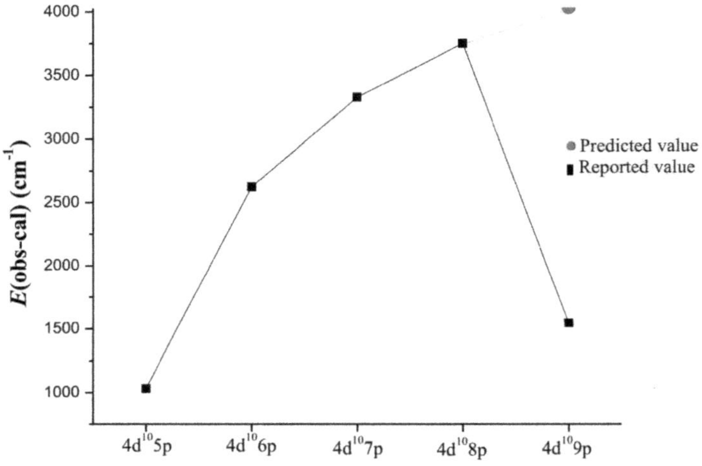

Figure 2. A plot of the observed and calculated energy difference in $4d^{10}np$ series of In III.

4.2. The $4d^{10}np$–$[4d^{10}\{ns + nd\} + 4d^9 5s^2]$ Transition Array

The second excitation, $4d^{10}5p$–$[4d^{10}(6s + 5d) + 4d^9 5s^2]$ transitions, is also observed to be quite strong. In the $4d^{10}ns$ series, we observed transitions $4d^{10}5p$–$4d^{10}\ ns$ (n = 6–8) and $4d^{10}6p$–$4d^{10}ns$ (n = 9–12), and, in the $4d^{10}nd$ series, three transitions are possible between each of the $4d^{10}np$–$4d^{10}nd$ configurations out of which two transitions, namely $^2P_{1/2}$–$^2D_{3/2}$ and $^2P_{3/2}$–$^2D_{5/2}$, were observed to be quite strong, while the third transition, $^2P_{3/2}$–$^2D_{3/2}$, was predicted to be weak in the series. All these three transitions were observed in $4d^{10}$[5p–nd (n = 5–7)]. Thus, we confirmed the levels of the $4d^{10}ns$ (n = 6–12) and $4d^{10}nd$ (n = 5–7) configurations. The transitions from $4d^{10}nd$ (n = 8, 9) to $4d^{10}5p$ were not observed on our plates. However, these transitions were reported by Bhatia [6]. We examined these levels and found their scaling factor to be quite regular. Secondly, a similar plot as in Figure 2 with the average energy difference between the calculated and observed values shows a regular behavior for the $4d^{10}ns$ and $4d^{10}nd$ series (Figure 3). Although we could not confirm the levels of the $4d^{10}nd$ (n = 8, 9) configurations, on the basis of their regularity, we included them in Table 2 for the sake of completeness.

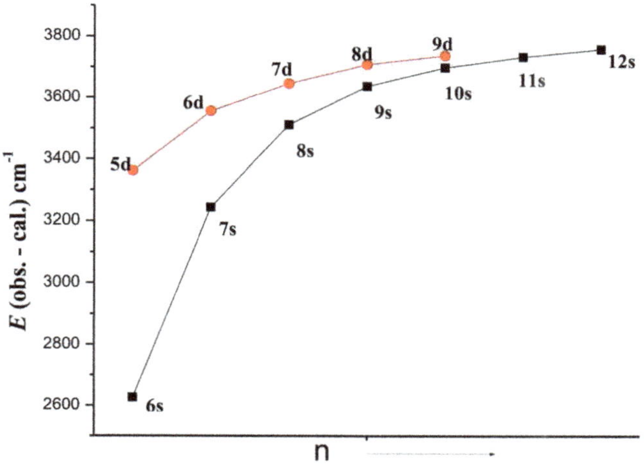

Figure 3. A plot *E*(obs-cal) for the $4d^{10}ns$ and $4d^{10}nd$ series in In III.

The other configuration $4d^95s^2$ in even parity system has two inverted ^2D levels having the same energy range as the $4d^{10}5d$ ^2D levels. Both ^2D levels of these two configurations interact with each other. As a result of this interaction, $4d^{10}5p$–$4d^95s^2$ transitions are observed. Further confirmation of these two levels was made by the observed transitions from the levels of the $4d^95s5p$ configuration that will be discussed later.

4.3. The $4d^{10}(nf + ng + nh)$ Configurations

The $4d^{10}5d$–$4d^{10}4f$ transitions lie beyond our wavelength region of investigation (above 2080 Å); therefore, we could not confirm them experimentally in the present work. However, these levels were well established by Nodwell [4] along with levels of the $4d^{10}ng$ (n = 5–7) series by observing transitions from $4d^{10}4f$. The repeated appearance of the $4d^{10}4f$ $^2F_{5/2,7/2}$ interval in transitions from the $4d^{10}$(5g, 6g and 7g) $^2G_{7/2,9/2}$ levels confirms the correctness of the $4d^{10}4f$ levels. The latter were compiled in AEL [5] and were later confirmed by Bhatia [6]. The $4d^{10}5d$–$4d^{10}5f$ transitions lie in our wavelength region. We observed a pair of lines from $4d^{10}5d$ $^2D_{3/2}$ and $^2D_{5/2}$, and two transitions from $4d^95s^2$ $^2D_{5/2,3/2}$, thus confirming $4d^{10}5f$ $^2F_{5/2}$. The other level $4d^{10}5f$ $^2F_{7/2}$ is also expected to gives two transitions, one from $4d^{10}5d$ $^2D_{5/2}$ and the other from $4d^95s^2$ $^2D_{5/2}$; both were in fact found. Furthermore, the level positions agree well with theoretical prediction of an inverted doublet. The $4d^{10}6f$ $^2F_{5/2,7/2}$ levels are strongly mixed with the $4d^95s5p$ $^2F_{5/2,7/2}$ levels. Bhatia [6] reported only the $4d^{10}6f$ $^2F_{5/2}$ level at 198,499.3 cm^{-1}, but our least squares fitted calculation predicted at 191,337 cm^{-1}. This large deviation does not seem to be right. Bhatia [6] reported unresolved $4d^{10}7f$ levels, but we did not find his identified lines on our line list. Therefore, his $4d^{10}6f$ and $4d^{10}7f$ levels could not be confirmed.

Neither the $4d^{10}4f$–$4d^{10}ng$ (n = 5–7) nor $4d^{10}5g$–$4d^{10}nh$ (n = 7–9) transitions lie in our wavelength region. Therefore, they could not be confirmed in the present work. However, we have compared Bhatia's experimental results [6] with theoretical calculations for the known spectra in the isoelectronic sequence from Ag I–Sn IV [11], and they appear to be regular. The $4d^{10}4f$–(8g + 9g) transitions do lie in our wavelength region, but they are too weak to be verified. However, we have included them in our LSF calculations for the sake of completeness.

4.4. The $4d^9 5s5p$ Configuration

This configuration arises due to core excitation of the ground level configuration $4d^{10}5s$. A number of levels from this configuration were reported by Bhatia [6]. Kaufman et al. [7] revised three levels of this configuration by observing transitions from the ground level $4d^{10}5s$ $^2S_{1/2}$, thus connecting only $J = 1/2$ and $3/2$ levels. The remaining levels of Bhatia (with $J = 5/2$, $7/2$ and $9/2$) still remain to be verified. In the present investigation, we agreed with six levels of Kaufman et al. [7] but revised four levels. The ionization separation on our recorded spectrum in this wavelength region was quite clear, thus new levels could be found with full confidence. The level $^2P_{1/2}$ reported by Bhatia [6] at 199,561.2 cm^{-1} was revised by Kaufman et al. [7] to a new position at 197,081 cm^{-1}. The line (507.406 Å) used by Kaufman et al. [7] for this transition actually belongs to O III (507.391 Å) [11] and the line used by Bhatia was not found on our spectrograms. We found an unclassified In III line with moderate intensity at 504.080 Å that has been assigned to this transition, yielding the level value at 198,382.2 cm^{-1} that also fits well in the least squares calculations.

Kaufman et al. [7] had revised another $J = 1/2$ level of Bhatia [6] and re-designated it as a $J = 3/2$ level at 170,888 cm^{-1} based on Bhatia's line list as they did not observe the corresponding lines. We also could not find the lines associated with this level in our line list. Therefore, this level was rejected. According to our analysis, we found that the lowest $J = 1/2$ level reported by Kaufman et al. [7] at 170,812 cm^{-1} is in fact a $J = 3/2$ level and the replacement for the lowest $J = 1/2$ level is found at 171,315.7 cm^{-1}. The lowest $J = 3/2$ level of this configuration reported by Kaufman et al. [7] at 167,079 cm^{-1} is in fact based on an In IV line (598.526 Å) [16,17]. However, Bhatia [6] had reported this level at 167,339.1 cm^{-1}, which was based on a correct In III line at (597.589 Å), and we agree with this identification. Moreover, it also gives two transitions from the recently found $4d^9 5p^2$ configuration [9] that confirm the identification of this level.

The highest $J = 3/2$ level was not found by Kaufman et al. [7] because calculations predict a weak transition to the ground level. However, Bhatia [6] had reported this level at 202,132.3 cm^{-1}. We found two strong lines with correct In III ionization characteristics, which we classified as transitions from $4d^9 5s^2$ levels to the level in question. Thus, we confirmed Bhatia's level value. Table 4 shows the summary of the $J = 1/2$ and $3/2$ levels of the $4d^9 5s5p$ configuration given by previous researchers [6,7] and the present analysis.

Table 4. Energy level values ($J = 1/2$ & $3/2$ Levels) of $4d^9 5s5p$ Configuration.

Configuration ($4d^9 5s5p$)	Previous Work		This Work	
	Bhatia [6]	Kaufman et al. [7]		
$(^3D)^4P_{1/2}$	170,888.3	170,812	171,315.7	Revised
$(^1D)^2P_{1/2}$	178,187.5	178,187	178,187.72	Verified
$(^3D)^4D_{1/2}$	179,321.0	179,321	179,320.9	Verified
$(^3D)^2P_{1/2}$	199,561.2	197,081	198,382.2	Revised
$(^3D)^4P_{3/2}$	167,339.1	167,079	167,339.24	Revised
$(^3D)^4F_{3/2}$	170,918.9	170,888	170,813.7	Revised
$(^1D)^2D_{3/2}$	175,538.7	175,538	175,539.35	Verified
$(^1D)^2P_{3/2}$	178,616.5	178,616	178,616.85	Verified
$(^3D)^4D_{3/2}$	180,945.0	180,945	180,943.95	Verified
$(^3D)^2P_{3/2}$	191,509.2	191,508	191,509.1	Verified
$(^3D)^2D_{3/2}$	202,132.3	-	202,134.5	Verified

The remaining 12 levels of this configuration with higher J values (5/2–9/2) were considered next. These levels have only been reported by Bhatia [6] through the transitions from $4d^9 5s^2$. We found lines corresponding to transitions from the $J = 5/2$ level at 194,902.6 cm^{-1} and confirmed only this level in Bhatia's list. We were successful in locating 10 remaining levels of $J = 5/2$ and $7/2$ from transitions to $4d^9 5s^2$ and $4d^9 5p^2$ levels. The level with the highest J value (9/2) does not connect to any other known configuration except $4d^9 5p^2$, which was partially known. We extended that configuration to include $J = 7/2$ levels. This paved the way for the establishment of the $J = 9/2$ level. We found three transitions placing the $J = 9/2$ level at 168,947.6 cm^{-1}. All 23 levels of $4d^95s5p$ configuration are now known experimentally.

4.5. The $4d^95s$ (nf + np) Configurations

These are the configurations that arise due to the core excitation. The $4d^95s4f$ configuration has a large energy spread and contains 39 levels. Since the ground configuration contains only the $^2S_{1/2}$ level, only $J = 1/2$ and $3/2$ levels of the $4d^95s4f$ configuration can decay to the ground configuration. Kilbane et al. [8] have studied the $4d^95snf$ (n = 4–12) and $4d^95snp$ (n = 6–11) configurations using a photoabsorption technique. They reported 10 levels of $4d^95s4f$ and seven levels of $4d^95s6p$ belonging to $J = 1/2$ and $3/2$. In our spectra, these transitions lie in the shorter wavelength region, where reflectivity of the grating falls considerably in the normal incidence setting. Therefore, these transitions appeared with very weak intensity on our spectrograms. Secondly, a large number of In V [18] and In VI [19] transitions overlap in this region. Therefore, it was very difficult to identify confidently In III lines of this array. Moreover, these levels lie above the ionization limit and consequently have a very small population. Therefore, these levels could not be located in the present work. However, we performed least squares fitted parametric calculations to provide a precise prediction of the remaining levels of the $4d^95snf$ (n = 4–7) and $4d^95snp$ (n = 6–7) configurations based on the identification made in reference [8].

4.6. The $4d^95p^2$ Configuration

The first attempt to study the low-lying autoionizing configuration $4d^95p^2$ in the sequence In III–Te VI was made by Ryabtsev et al. [9], connecting this configuration with $4d^{10}5p$. It is important to note that all the levels of this configuration lie above the ionization limit. It was difficult to arrange experimental conditions providing for a reasonable population above the ionization limit. Certainly it was advantageous to identify the broad lines due to continuum effect, but only the strongest transitions could be observed. Not many pairs connecting to both $4d^{10}5p$ $^2P_{1/2,3/2}$ were found to confirm these levels. However, the lines used to locate these levels have a definite In III characteristic and show continuum broadening effect. Out of 28 levels of $4d^95p^2$, only 13 levels with $J = 1/2, 3/2$ and $5/2$ were reported by Ryabtsev et al. [9]. We should point out that two levels (^{1}D) $^2S_{1/2}$ and (^{1}D) $^2P_{3/2}$ were reported by Ryabtsev et al. [9] with the same energy level values. They were based on the double classification of the same pair of lines (555.669 Å and 569.421 Å). We agreed with assignments of these lines to (^{1}D) $^2P_{3/2}$ giving the level value at 237,145.6 cm^{-1} as both transitions are predicted to be of the comparable intensity. However, the (^{1}D) $^2S_{1/2}$ level is predicted to have one strong and one weak transition, and we found one unclassified line on our plate at 555.501 Å, which we used to establish this level at 237,201.7 cm^{-1}. Several levels have also been confirmed through transitions to the $4d^95s5p$ configuration. The higher J values of $4d^95p^2$ configuration ($J = 7/2$ and $9/2$) could only be established through transitions from $4d^95s5p$. We were successful in establishing three $J = 7/2$ and one $J = 9/2$ levels. One $J = 9/2$ and two $J = 7/2$ levels remain unknown. The study of the $4d^95p^2$ and $4d^95s5p$ configurations together complemented each other. The other even parity configuration $4d^95s5d$ lies above the ionization limit and partially overlaps the $4d^95p^2$ configuration. It has also been incorporated in the least squares fitted parametric calculation to interpret the results.

5. Optimization of the Energy Levels

The transition wavelengths observed for this spectrum were used to derive the energy level values. For this purpose, a least-squares level optimization code LOPT [15] was used. The essential factors for the level optimization procedure are the correct identification of the spectral lines and estimation of their uncertainties. The wavelength uncertainty is determined by the combined effect of the statistical deviation of the line position measured on the comparator and systematic uncertainty of reference wavelengths used in the fitting. Ryabtsev et al. [9] reported the uncertainty of autoionized lines to be ±0.006 Å. Our wavelength accuracy for sharp and unblended lines is estimated to be within ±0.006 Å and ±0.008 Å below and above 900 Å. We estimated the uncertainty of Bhatia's lines to be ±0.008 Å for lines below 2000 Å with the comparison of our measurement and Kaufman et al. [7] for sharp and unblended lines. Bhatia mentioned in his paper that the prism lines are not accurate to more than 0.01 Å. However, he gave wavelengths above 2000 Å with only two places after the decimal point implying that the uncertainty is at least 0.02 Å or higher. In our level optimization with Bhatia's lines [6], we noticed several lines showing a deviation around 0.22 Å for the region 2000–4000 Å from their Ritz values. The deviation increases up to 0.8 Å for the longer-wavelength region 4000–6500 Å. We, therefore, did not use these lines with large deviation in the level optimization. All of the lines used in the optimization of the level values were given an estimated uncertainty to find the final optimized energy level values with an estimated uncertainty for each level. Since the level $4d^{10}5p$ (^1S) $^2P_{3/2}$ connects with the largest number of observed transitions, it was adopted as the base level, hence all the level uncertainties in Table 2 are given with respect to this level. All the given uncertainties are taken to be at the level of one standard deviation.

6. Ionization Potential

Since more than one series with three members are known in In III, its ionization potential can be determined with good accuracy. The value of ionization potential of In III given in AEL [5] at 226,100 cm^{-1} was derived by Catalan and Rico [20] by comparison of the third spectra from Y to In. Bhatia [6] improved the value of ionization potential by using $4d^{10}ng$ (n = 5–9) and $4d^{10}nh$ (n = 6–9) in frames of the polarization theory [21]. He calculated the In III limit at 226,191 cm^{-1}; this value is listed in the NIST Atomic Spectra Database [11]. We have calculated the ionization potential from two series, ns (n = 5–12) and ng (n = 5–9) using the Ritz quantum defect extrapolation method with the aid of the RITZPL code [22]. However, the non-penetrating ($4d^{10}ng$) series is certainly expected to give more accurate value. The value of IP obtained using the three-parameter extended Ritz formula [22] for the $4d^{10}ns$ (n = 5–12) series is 226,196.58 cm^{-1}, while the values obtained by fitting the two-parameter extended Ritz formula for the two $4d^{10}ng$ $^2G_{7/2,9/2}$ (n = 5–9) series are 226,197.00 cm^{-1} and 226,195.08 cm^{-1}, respectively. The limits calculated by the POLAR code [22] for the ng (n = 5–9) $^2G_{7/2,\,9/2}$ series were found to be 226,197.28 and 226,195.35cm^{-1}, respectively. The adopted value is the average of these calculations at 226,196.3 cm^{-1} ± 1.0 cm^{-1} (28.0448 ± 0.0001 eV) differing by 5 cm^{-1} from Bhatia's value.

7. Conclusions

A total of 91 energy levels have been established, among which three levels are revised and 21 are new. All of these levels were based on the identification of 218 spectral transitions, 70 being new. The results were interpreted using Cowan's codes and the least square fitted parametric theory. The optimized energy levels and their calculated values are given in Table 2 along with the level uncertainty, *LS*-percentage compositions and number of connecting transitions. All of the classified transitions are given in Table 3 along with their weighted transition probabilities (*gA*) obtained with least squares fitted energy parameters. This table also contains the Ritz wavelengths of all transitions with their uncertainties obtained by using the level optimization code (LOPT).

Author Contributions: All authors contributed equally to this work.

Conflicts of Interest: The authors declare no conflict of interest.

References

1. Rao, K.R. On the Spectra of Doubly-Ionised Gallium and Indium. *Proc. Phys. Soc.* **1927**, *39*, 150–160. [CrossRef]
2. Lang, R.J. On the Spectra of Zn II, Cd II, In III and Sn IV. *Proc. Natl. Acad. Sci. USA* **1929**, *15*, 414–418. [CrossRef] [PubMed]

3. Archer, D.H. The Spectra of Indium. Master's Thesis, University of British Columbia, Vancouve, BC, Canada, 1948.
4. Nodwell, R.A. A Study of Spark Spectrum of Indium. Ph.D. Thesis, University of British Columbia, Vancouve, BC, Canada, 1956.
5. Moore, C.E. *Atomic Energy Levels, National Bureau of Standards Circular 467*; US Govt. Printing Office: Washington, DC, USA, 1958; Volume III.
6. Bhatia, K.S. Spectrum of Doubly Ionised Indium. *J. Phys. B At. Mol. Phys.* **1978**, *11*, 2421–2434. [CrossRef]
7. Kaufman, V.; Sugar, J.; VanKleef, T.A.M.; Joshi, Y.N. Resonance transition $4d^{10}5s$-$4d^{9}5s5p$ in the Ag I sequence of In III, Sn IV, Sb V, and Te VI. *J. Opt. Soc. Am. B* **1985**, *2*, 426–429. [CrossRef]
8. Kilbane, D.; Mosnier, J.-P.; Kennedy, E.T.; Costello, J.T.; van Kampen, P. 4d Photoabsorption Spectra of Indium (In II–In IV). *J. Opt. Soc. Am. B Opt. Phys.* **2006**, *39*, 773–782. [CrossRef]
9. Ryabtsev, A.; Churilov, S.S.; Kononov, É.Y. $4d^{9}5p^{2}$ Configuration in the Spectra of In III–Te VI. *Opt. Spectrosc.* **2007**, *102*, 354–362. [CrossRef]
10. Skočić, M.; Burger, M.; Bukvić, S.; Djeniže, S. Line intensity and broadening in the In III spectrum. *J. Phys. B At. Mol. Opt. Phys.* **2012**, *45*, 225701. [CrossRef]
11. Kramida, A.; Ralchenko, Y.; Reader, J.; NIST ASD Team. NIST Atomic Spectra Database, v. 5.4, National Institute of Standards and Technology, Gaithersburg, MD, USA. 2016. Available online: http://physics.nist.gov/ASD (access on 7 June 2017).
12. Cowan, R.D. *The Theory of Atomic Structure and Spectra*; University California Press: Berkeley, CA, USA, 1981.
13. Kramida, A. Critically Evaluated Energy Levels and Spectral Lines of Singly Ionized Indium (In II). *J. Res. Natl. Inst. Tech.* **2013**, *118*, 52–104. [CrossRef] [PubMed]
14. van het Hof, G.J. *A Computer Program—FIND3, for Searching the Levels*; Zeeman Lab: Amsterdam, The Netherland, 1994.
15. Kramida, A.E. The program LOPT for least-squares optimization of energy levels. *Comput. Phys. Commun.* **2011**, *182*, 419–434. [CrossRef]
16. Ryabtsev, A.N.; Kononov, E.Y. High Lying Configurations in the Spectrum of Three Times Ionized Indium (In IV). *J. Quant. Spectrosc. Radiat. Transf.* **2016**, *168*, 89–101. [CrossRef]
17. Swapnil; Tauheed, A. Revised and Extended Analysis of the Fourth Spectrum of Indium: In IV. *J. Quant. Spectrosc. Radiat. Transf.* **2013**, *129*, 31–47. [CrossRef]
18. Joshi, Y.N.; VanKleef, T.A.M.; Kushawaha, V.S. The Fifth Spectrum of Indium: In V. *Can. J. Phys.* **1976**, *54*, 889–894. [CrossRef]
19. Joshi, Y.N.; VanKleef, T.A.M. Sixth Spectrum of Indium: In VI. *J. Opt. Soc. Am.* **1982**, *72*, 259–267. [CrossRef]
20. Catalán, M.A.; Rico, F.R. Series y potenciales de ionizacion en los espectros III de los elementos del grupo del paladio. *An. Fis. Quim. Ser. A* **1957**, *53*, 85.
21. Edlen, B. Wavelength measurements in the vacuum ultra-violet. *Rep. Prog. Phys.* **1963**, *26*, 181. [CrossRef]
22. Sansonetti, C.J. (National Institute of Standards and Technology, Gaithersburg, ML, USA). Computer Programs RITZPL and POLAR. Private Communication, 2005.

atoms

Article

Identification and Plasma Diagnostics Study of Extreme Ultraviolet Transitions in Highly Charged Yttrium

Roshani Silwal [1,2,*], Endre Takacs [1,2], Joan M. Dreiling [2], John D. Gillaspy [2,3] and Yuri Ralchenko [2]

1 Department of Physics and Astronomy, Clemson University, Clemson, SC 29634, USA; etakacs@clemson.edu
2 National Institute of Standards and Technology, Gaithersburg, MD 20899, USA; joan.dreiling@nist.gov (J.M.D.); jgillasp@nsf.gov (J.D.G.); yuri.ralchenko@nist.gov (Y.R.)
3 National Science Foundation, Arlington, VA 22230, USA
* Correspondence: rsilwal@clemson.edu

Academic Editor: Joseph Reader
Received: 17 July 2017; Accepted: 12 September 2017; Published: 18 September 2017

Abstract: Extreme ultraviolet spectra of the L-shell ions of highly charged yttrium (Y^{26+}–Y^{36+}) were observed in the electron beam ion trap of the National Institute of Standards and Technology using a flat-field grazing-incidence spectrometer in the wavelength range of 4 nm-20 nm. The electron beam energy was systematically varied from 2.3 keV–6.0 keV to selectively produce different ionization stages. Fifty-nine spectral lines corresponding to $\Delta n = 0$ transitions within the $n = 2$ and $n = 3$ shells have been identified using detailed collisional-radiative (CR) modeling of the non-Maxwellian plasma. The uncertainties of the wavelength determinations ranged between 0.0004 nm and 0.0020 nm. Li-like resonance lines, $2s$–$2p_{1/2}$ and $2s$–$2p_{3/2}$, and the Na-like D lines, $3s$–$3p_{1/2}$ and $3s$–$3p_{3/2}$, have been measured and compared with previous measurements and calculations. Forbidden magnetic dipole (M1) transitions were identified and analyzed for their potential applicability in plasma diagnostics using large-scale CR calculations including approximately 1.5 million transitions. Several line ratios were found to show strong dependence on electron density and, hence, may be implemented in the diagnostics of hot plasmas, in particular in fusion devices.

Keywords: highly charged ions; yttrium; spectroscopy; extreme ultraviolet; Li-like; Na-like; magnetic dipole; plasma diagnostics; electron beam ion trap; non-Maxwellian plasma

1. Introduction

Multi-electron ions are under intense theoretical study as state-of-the-art calculations rival highly accurate measurements sensitive to higher order terms of quantum electrodynamics (QED) corrections to atomic energy levels [1]. While elements with a high-Z atomic number have these effects amplified, ions in the medium-Z region have special importance because they allow for more accurate experiments and provide constraints to theoretical trends. In the past few years, the electron beam ion trap (EBIT) research program at the National Institute of Standards and Technology (NIST) has reported accurate measurements in the extreme ultraviolet (EUV) region that focus on systematic observations of transitions in L-shell, M-shell and N-shell ions [2–14]. The work reported here extends these results to a range of previously unobserved transitions of a fifth row element, yttrium.

Yttrium was chosen for the current investigation because of its relevance as a possible diagnostic impurity in tokamak fusion plasmas. For instance, together with strontium, zirconium, niobium and molybdenum, yttrium has been injected into the Texas Experimental Tokamak (TEXT) [15,16], the Joint European Torus (JET) tokamak [17] and the Princeton tokamaks [18–20] and has also been observed in laser-produced linear plasmas [16,21]. L-shell ions of high-Z elements, especially Be-like to Ne-like [22–26], and a few M-shell ions such as Na-like, Mg-like and Al-like [15,16,27–30] were used to diagnose these hot plasmas for decades. The elemental abundance of yttrium in stars also makes it astronomically important. Its relevance in nuclear astrophysics, weak interaction physics and nuclear structure physics has been discussed [31–34].

Among the various transitions in these elements, special interest is devoted to forbidden transitions that originate from long-lived metastable energy levels. The importance of the forbidden transitions in medium-Z and high-Z elements has been demonstrated by different researchers for astrophysical [35] and fusion [18,36] plasmas. For example, charge states near closed shells include potentially useful forbidden transitions such as those between the $2s^22p^5$–$2s2p^6$ configurations of F-like ions and the magnetic dipole transition $2s^22p^5\ ^2P_{3/2}$–$^2P_{1/2}$ in the same ion. These have been extensively investigated in earlier studies [37–40].

There have been a few EUV measurements of highly charged yttrium over the past couple of decades. Alexander et al. observed the EUV spectra of Y IX–XIII in the wavelength range of 4.5 nm–35 nm using vacuum spark [41]. Ekberg et al. performed a series of measurements for the identification of transitions in Si-like Y XXVI [42], Al-like Y XXVII [43] and Mg-like Y XXVIII [21,28] in the EUV spectra emitted from line-focus laser-produced plasmas as part of the X-ray laser research program. Reader et al. have reported observations of F-like Y XXXI [37], Mg-like Y XXVIII [44] and Na-like Y XXVIV [16,45] using laser-produced and tokamak plasmas in a series of systematic spectroscopic studies. Similar experiments for moderate charge states also reported observations of multiply-charged yttrium spectra (Y II–XI); see, e.g., [46–50]. Despite these experiments, the second row isoelectronic sequences of yttrium have largely been unexplored in the EUV region to date.

In this paper, we report the systematic study and identification of atomic spectral lines of the L-shell charge states of yttrium ranging from Li-like to Ne-like ions (Y XXX–Y XXXVII) created and trapped in the NIST EBIT [51,52]. We also present the most pronounced spectral lines of the Na-like, Mg-like and Al-like yttrium charge states, as these can provide benchmark experimental results for precise multi-electron atomic theory calculations. Na-like D1 and D2 lines originate from quasi-hydrogenic ions and have been used as a probe of QED contributions due to their high intensities and the available precise ab initio calculations [6,53–55]. We report the first data for the wavelengths of the Na-like D1 and D2 yttrium lines measured with an EBIT to provide accurate experimental results that complement the previously reported measurements of Reader et al. [45] in laser-produced and tokamak plasmas.

In addition to the spectral analysis, we also discuss the forbidden magnetic dipole (M1) transitions of highly charged yttrium ions that are potentially important for plasma diagnostics. The spectroscopy of forbidden magnetic dipole lines can help deduce important plasma parameters such as the density and temperature of plasmas. These parameters are obtained in practice from intensity ratios of various atomic spectral lines rather than direct measurements, which are difficult or even impossible in fusion, laboratory and astrophysical plasmas [56]. The availability of accurate collisional-radiative models makes this technique a reliable tool for plasma diagnostics [57–59].

M1 transition probabilities strongly depend on the spectroscopic charge Z_{sp}, and for highly ionized ions, these transitions become prominent. At low electron densities, the radiative decay rates are substantially larger than collisional depopulation from both metastable and allowed excited levels. At higher densities, however, the metastable levels decay both by collision and radiation, whereas allowed transitions still take place mostly by radiative decay. This makes the ratio of the allowed versus forbidden transitions dependent on the electron density.

The following sections describe the experimental method, the theoretical calculations that aided in line identifications, the list of the observed transitions and their uncertainties and a discussion of the diagnostic capabilities of some of the M1 transitions.

2. Experiment

The NIST EBIT and a multi-cathode metal vapor vacuum arc ion source (MEVVA) were used to produce highly charged yttrium ions, and the ion spectra were recorded with a custom-made EUV flat field grazing incidence spectrometer [60]. Both the MEVVA and the NIST EBIT are discussed in detail elsewhere [51,61], but we will now briefly review the most important details.

The MEVVA, which produces singly-charged ions by sparking a high voltage across metal cathodes, is located $\approx$ 2 m above the central trapping region of the EBIT. The ions are created at a potential of about 10 kV above ground and are accelerated towards the center of the EBIT through several electrodes at lower voltages. The trapping region, consisting of the drift tubes (upper, middle and lower), is floated on top of the voltage of the cylindrically-shaped shield electrode. To capture the ions in the trap, the shield electrode voltage is very briefly (on the order of 10^{-3} s) switched to a potential of about 9.6 kV, and the middle drift tube voltage is simultaneously raised by an additional 0.4 kV. Then, precisely at the arrival of the ions, the middle drift tube is pulsed down to the shield voltage in order to trap and confine the plasma in the trap. During the entire timing sequence, the lower and upper drift tubes are kept at constant potentials with respect to the shield (0.5 kV and 0.26 kV, respectively) to create axial trapping. Radial confinement is accomplished by a combination of the axial 2.7 T magnetic field and the space charge of the intense electron beam, which is directed through the drift tubes to further ionize the ions. The electron beam energy can then be set as required for the experiment by adjusting the shield electrode voltage. The beam energy in the EBIT is determined by the voltage difference between the electron gun and the middle drift tube, taking into account the space charge of the electron beam in the interaction region [62]. The latter depends on the density of the electron beam and therefore scales with the beam energy and current in addition to the ion cloud neutralization factor, which is generally difficult to quantify. In our experiment, the modeling of the observed spectral line intensities showed that the space charge correction was approximately 150 eV. Electron beam currents were varied between 66 mA and 147 mA during the measurement. To control the charge-state distribution of the yttrium ions, the energy of the electron beam was systematically varied from 2.3 keV–6.0 keV.

The flat field EUV grazing incidence spectrometer [60] is equipped with a liquid-nitrogen-cooled charged-coupled device (CCD) detector with 2048 $\times$ 512 active pixels of 13.5 μm $\times$ 13.5 μm size each. The spectrometer consists of a gold-coated spherical focusing mirror that focuses light radiated from the EBIT plasma onto a slit, followed by a gold-coated concave reflection grating with a groove spacing of approximately 1200 lines mm^{-1}. The instrument has a resolving power of $\lambda/d\lambda \approx 400$. The 2D images recorded by the CCD were hardware collapsed along the vertical axis, so that the resulting image was a 1D (2048 $\times$ 1) spectrum. Ten 60 s frames of yttrium spectra were collected in a set, giving a total acquisition time of 600 s for each energy. The spectra were filtered of cosmic rays using a program that identifies outlier intensities among different frames within the same set. If the intensity of a channel in a certain frame is five or more Poisson standard deviations away from the mean of all of the frames, it is replaced by the average value of the other frames before the frames are summed together to form the overall spectrum.

3. Wavelength Calibration

Spectra emitted by yttrium ions were acquired in the approximate wavelength region of 4 nm to 20 nm. Wavelength calibration was performed using highly charged neon lines (Ne V–VIII), xenon lines (Xe XLI–XLII), barium lines (Ba XLIII–XLIV), oxygen lines (O V–VI) and iron lines (Fe XXIII–XXIV) [5,7,11,63,64], as described in this section. Neon and carbon dioxide gases were injected into the EBIT as neutral atoms from the gas injection setup described by Fahy et al. [2], with the injection pressure normally on the order of 10^{-3} Pa. Iron ions were loaded from the MEVVA ion source. Small amounts of barium and xenon ions are always present in the trap as heavy ion contaminants from the electron gun and the ion pumps. In order to prevent long-term accumulation of these ions, the EBIT trap was emptied and reloaded every 10 s.

The calibration lines were fitted using unweighted Gaussian profiles, and the locations of the peaks were noted in terms of the channel (pixel) numbers corresponding to the respective lines. The literature-recommended wavelengths [5,11,64] were plotted as a function of channel number weighted with the uncertainty in these wavelengths. A third order polynomial from the fit was used to convert the uncertainties in channel number to the uncertainties in wavelength. The statistical uncertainties of the calibration lines were then determined from the quadrature sum of these uncertainties with the adopted wavelength uncertainties from the literature.

The final calibration function was a third order polynomial that describes the wavelength versus channel as a fit weighted by the inverse square of the total uncertainties of the lines. The latter was calculated as the quadrature sum of the overall statistical uncertainty and the systematic uncertainty. The systematic uncertainty was estimated to be 0.0006 nm by requiring the reduced chi-square of the fit to be 1 according to the standard statistical procedure [65]. Systematic uncertainty may arise from several factors during the experiment such as small device vibrations or uneven pixel response. The residual of the literature values of the calibration lines with respect to their calibrated wavelength provided an assessment of the quality of the calibration. Including their uncertainties, 95% of the residual should lie within two standard deviation (σ) of their mean (μ): $P(\mu - 2\sigma \leq x \leq \mu + 2\sigma) \approx 0.9545$. Figure 1 shows the calibration data points and 95% confidence band of the fit.

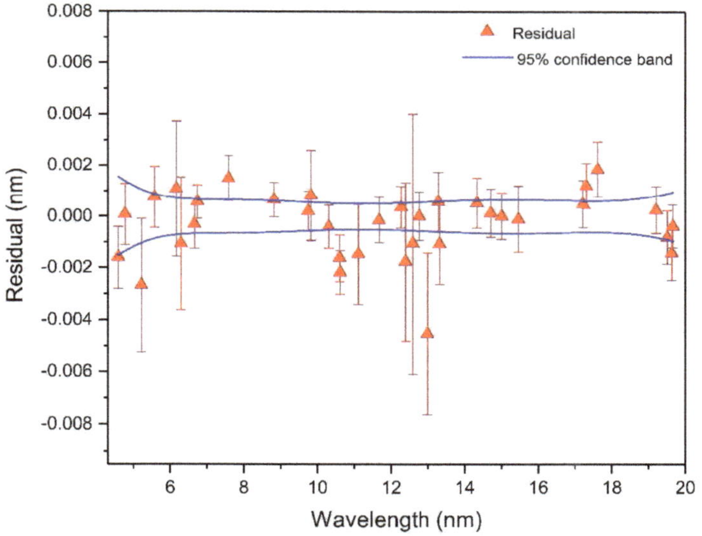

Figure 1. Residual of the adopted wavelength of the calibration lines with respect to their calibrated wavelength. The individual uncertainties of the data points are as described in the text. The solid (blue) line corresponds to the 95% confidence band of the calibration fit.

In calculating the overall calibration uncertainty contributing to the total uncertainty at a given wavelength for the identified yttrium lines, we have used the 95% confidence band at the position of the line. The calibration uncertainties reported are equal to the vertical width of the confidence band divided by four (equivalent to one standard deviation). The calibration uncertainty calculated from the 68.3% confidence band corresponding to one standard deviation gives comparable results, as expected.

4. Theoretical Modeling

The spectral modeling for the non-Maxwellian EBIT plasma was performed with the collisional-radiative code NOMAD [66] that has been extensively used in EBIT spectroscopy. The yttrium plasma was assumed to be in the steady state, optically thin and uniform with electron density of 10^{11} cm^{-3}. The electron beam energy distribution was modeled by a Gaussian function with the full width at half maximum of 40 eV.

A detailed collisional-radiative (CR) model would generally require a large amount of atomic data, such as energy levels, wavelengths, transition probabilities and cross-sections. For the present analysis, we make use of the Flexible Atomic Code (FAC) [67], which is based on a fully-relativistic model potential and can consistently generate all required data. In total, our CR model included 13 ionization stages from Si-like to He-like ions of yttrium, about 5000 atomic levels and nearly 1.5 million transitions describing spontaneous radiative decays, electron-impact ionization and excitation, as well as radiative recombination. NOMAD also takes into account the charge exchange of ions with neutral atoms present in the trap, which shifts the ionization balance to lower charge states. Within the model, the density of neutral atoms is a free parameter and adjusted such that the theoretical and experimental spectra closely agree. The neutral densities obtained from the spectra of the current experiment are consistent with previous values under similar EBIT conditions.

Another adjustable parameter (although less important due to the lower sensitivity of the results to its variations) is the space charge correction to the electron beam energy as described in the experimental section. A generally good match between the observed and calculated line intensity ratios was obtained with a 150 eV correction to the values calculated from the applied voltages.

The CR model used these data to build and solve a system of rate equations to determine level populations and line intensities for EBIT plasmas of given electron energies. With this approach, NOMAD was used to simulate the yttrium emission as the electron beam energy was systematically changed during the experiment. The calculated spectra were convoluted with the spectrometer energy resolution and corrected for the efficiency of the grazing incidence instrument to obtain the theoretical result.

5. Line Identification

Yttrium spectra were taken as a function of the calibrated wavelength and fitted with unweighted Gaussians to determine the line positions. The uncertainty associated with the identified lines was then calculated from the quadrature sum of the uncertainty of the line fit that corresponds to the statistical uncertainty, the calibration uncertainty, the systematic uncertainty (estimated using the calibration data as discussed above) and uncertainty assigned for a possible small systematic line asymmetry (discussed below), which might be due to line blends or instrument asymmetries. In order to reach the desired ionization stages, the beam energy was systematically varied from 2.3 keV–6.0 keV. By matching theoretical and experimental spectra, we were able to conveniently identify most of the yttrium lines, as shown in Figure 2.

Some of the yttrium lines were also observed in the second and third orders of diffraction, in addition to the first order. Second order and third order yttrium spectra were plotted simultaneously as a visual aid to better identify the observed lines. They were obtained by dividing the line intensities of the first order experimental spectra by 2.5 and eight and multiplying the wavelength by two and three, respectively. Since we observed the same yttrium lines at several different beam energies, our reported wavelengths are the weighted averages of the positions of these lines using the formula for the best combined estimate of N measurements of the same quantity, $x_{CE} = \frac{\sum w_i \times x_i}{\sum w_i}$, x_i being the line position at different energies. The weight w_i is given by $w_i = \frac{1}{s_i^2}$, where s_i is the total uncertainty corresponding to each measurement. A few lines were blended with unresolved features, making it difficult to precisely determine their positions. In such cases, the spectra at energies that gave the cleanest and strongest signals were solely used. As a test for unanticipated systematics, the difference in the individual wavelengths at different energies with their weighted average was calculated. This difference was binned to get a histogram that represents a normal distribution about their mean, which should be zero. The distribution was fitted with a Gaussian function, and the mean value of 0.0003 nm was assigned to be the uncertainty due to unknown line asymmetries. As mentioned earlier, this uncertainty was added in quadrature to the rest of the uncertainties to get the total line uncertainty.

For the lines that we observed in second and third order, the wavelengths and the corresponding uncertainties were divided by two and three, respectively, and the weighted average was calculated accordingly. The total uncertainty for each of the identified lines was computed using the error propagation method $s = \sqrt{\frac{1}{\Sigma \frac{1}{s_i^2}}}$.

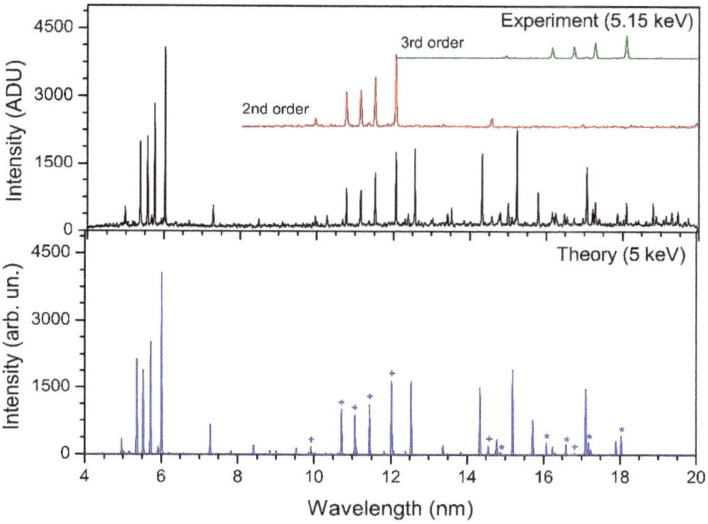

Figure 2. Comparison of the experimental spectrum (**top**) of yttrium with the theoretical spectrum (**bottom**). The intensity is given in analog to digital units (ADU) for the measured spectra. The second and third order spectra for the experimental data are also shown (red and green insets, respectively). The theoretical spectrum includes the second (+) and third (∗) order spectra and is calculated at an energy of 5 keV for the electron beam energy of 5.15 keV, to account for the space-charge correction in the experiment.

The most prominent observed lines were unambiguously identified through comparison with theory. For instance, the Be-like lines $2s^2\ ^1S_0$–$2s2p\ ^1P_1$ and $2s^2\ ^1S_0$–$2s2p\ ^3P_1$ were identified at 6.0322(5) nm and 15.2336(7) nm experimentally compared to the calculated values of 6.0098 nm and 15.1907 nm. The measured wavelengths of the Li-like, B-like, C-like and N-like yttrium lines are within 1.3% of our theoretical values, sufficient for line identification purposes. We note that for electron beam energies below 3.75 keV, where M-shell yttrium ion charge states become prominent, a more accurate relativistic many-body perturbation theory (RMBPT) calculation had to be invoked to match theoretical and experimental data. These calculations were performed by Safronova et al. [68] for the lines of the Ne-like, Na-like, Mg-like and Al-like charge states of yttrium lines.

In order to help with the identification of lines that were close in wavelength, we considered the evolution of charge states with electron beam energy by plotting the line intensities as a function of the beam energy. The ionization energies of the different charge states determine the minimum beam energy required for the emergence of a particular charge state. For instance, the ionization energy of Y XXXIV is 4.299 keV [64]; hence, a beam energy of 4.299 keV or higher is required to observe spectral lines from Be-like Y XXXV. This gives an idea of the range of charge states one is supposed to observe at a particular beam energy. In addition, lines emitted from the same ionization state usually depend in a similar way on the beam energy. These qualitative dependencies aided the line identification as illustrated in Figure 3. In order to verify these qualitative assumptions, we have used our detailed CR model calculations to make a final assignment based on the line intensity dependences.

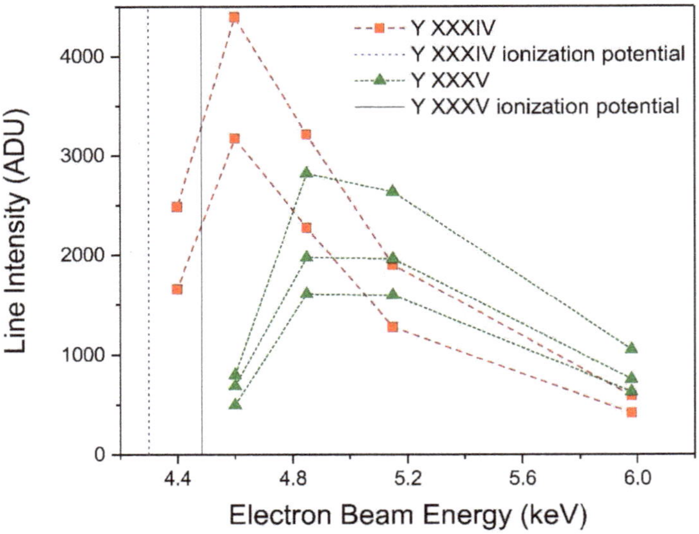

Figure 3. Line intensity plotted as a function of electron beam energy for three Y XXXV lines (green triangles) and two Y XXXIV lines (red squares). The solid (black) and dotted (blue) vertical lines depict the ionization potential of Y XXXV and Y XXXIV, respectively.

6. Results and Discussions

Table 1 presents the yttrium lines identified in our experiment for the charge state range between Y XXVII and Y XXXVII. We focused on lines that originate from $\Delta n = 0$ transitions within the $n = 2$ and $n = 3$ principal quantum number states. Most of the lines are electric dipole (E1) transitions, while a few lines are magnetic dipole (M1) transitions. All M1 lines correspond to transitions within the $2s^22p^m$ ground state configurations of different charge states ($m = 1, 2, 3, 4, 5$), with the exception of the line at 16.4817(7) nm, which originates from within the excited configuration $2p^53s$ of Ne-like yttrium ion. The energy levels within the ground configuration are close and result in longer wavelength forbidden lines, while the energy levels contributing to the allowed (E1) transitions are further separated and give rise to shorter wavelength lines. Level notations are taken from FAC and are given in *jj*-coupling. The plus sign stands for the j value of $l + 1/2$, and the minus sign represents the j value of $l - 1/2$, where l is the orbital angular momentum. As an example, the line at 5.9329(4) nm connects the $2s^22p^3$ upper level with a total angular momentum of $J = 5/2$ to the $2s2p^4$ of $J = 3/2$. It should be noted that in FAC, the subshells that couple to zero angular momentum are omitted in the notation. As an illustration of this, the $2p^5$ configuration is noted as $2p_+^3$ when both electrons on the $2p_-$ subshell are present and couple to zero joint angular momentum. Yttrium spectra recorded at different energies are shown in Figures 4 and 5, with the identified lines labeled by their isoelectronic sequence.

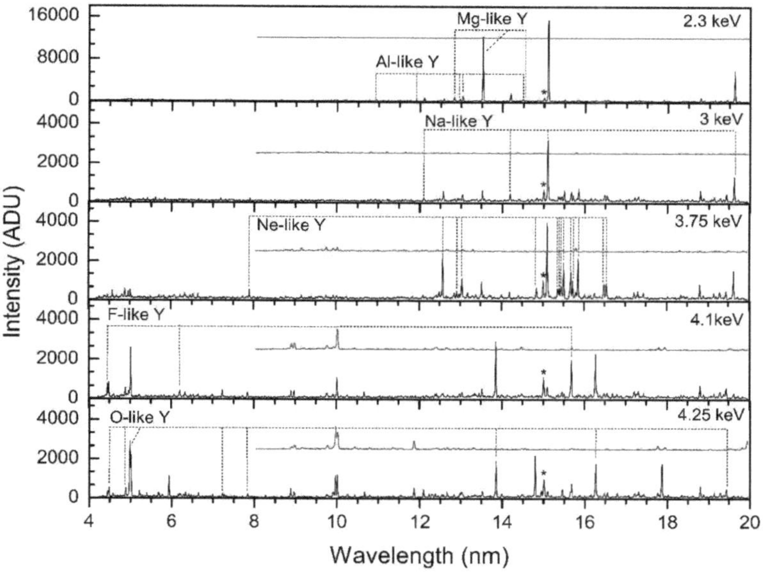

Figure 4. Yttrium spectra at beam energies from 2.30 keV–4.25 keV. The black and red (shifted) spectra correspond to the first and second order Y spectra, respectively. The * marks the impurity coming from oxygen at 15.0099(5) nm.

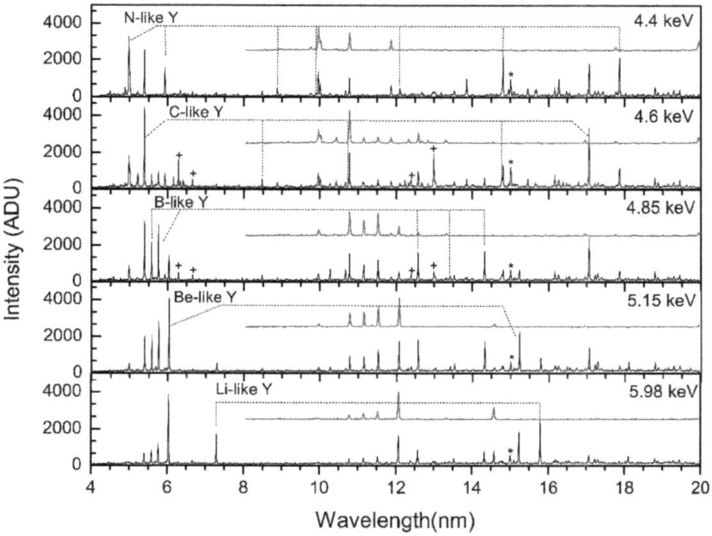

Figure 5. Yttrium spectra at beam energies from 4.40 keV–5.98 keV. The black and red (shifted) spectra correspond to the first and second order Y spectra, respectively. The + marks the Na-like and Mg-like xenon impurities at beam energies of 4.6 keV and 4.85 keV and the * marks the impurity coming from oxygen at 15.0099(5) nm.

Table 1. The table below presents the list of the yttrium lines identified Y^{26+}–Y^{36+}. Previous measurements and calculations are reported as well. Isoelectronic sequence is abbreviated as Seq., configuration is abbreviated as Config., and the level number is abbreviated as No.

Ion Charge	Seq.	Type	Lower Level			Upper Level			Experimental Wavelength (nm)		Theoretical Wavelength (nm)	
			No.	Config.	$Term_J$	No.	Config.	$Term_J$	This Work	Previous Work	This Work	Previous Work
36	Li	E1	1	$2s$	$2s_+$	2	$2p$	$2p_+$	7.2874(6)		7.2771	7.2893 [69] 7.2887(1) [70] 7.2892 [71] 7.2890(1) [72] 7.2888 [73]
36	Li	E1	1	$2s$	$2s_+$	3	$2p$	$2p_-$	15.7862(9)		15.7139	15.7878 [69] 15.7868(5) [70] 15.7865 [71] 15.7874(4) [72] 15.7867 [73]
35	Be	E1	1	$2s^2$	$(2s_+^2)_0$	6	$2s2p$	$(2s_+,2p_+)_1$	6.0322(5)		6.0098	6.0337(20) ^{f} [24] 6.0283 [74]
35	Be	E1	1	$2s^2$	$(2s_+^2)_0$	3	$2s2p$	$(2s_+,2p_-)_1$	15.2336(7)		15.1907	15.2345(20) ^{f} [24] 15.2302 [74]
34	B	E1	1	$2p$	$2p_-$	7	$2s2p^2$	$((2s_+,2p_-)_1,2p_+)_{1/2}$	5.5768(6)		5.5310	5.5771 ^{f} [25]
34	B	E1	1	$2p$	$2p_-$	6	$2s2p^2$	$((2s_+,2p_-)_1,2p_+)_{3/2}$	5.7623(6)		5.7254	5.7629 ^{f} [25]
34	B	E1	1	$2p$	$2p_-$	3	$2s2p^2$	$2s_+$	12.5693(6)		12.5372	
34	B	E1	2	$2p$	$2p_+$	5	$2s2p^2$	$((2s_+,2p_-)_1,2p_+)_{5/2}$	13.4185(8)		13.3700	
34	B	M1	1	$2p$	$2p_-$	2	$2p$	$2p_+$	14.3234(5)		14.3363	14.321 [75] 14.322 ^{f} [25]
33	C	E1	1	$2p^2$	$(2p_-^2)_0$	7	$2s2p^3$	$(2s_+,2p_+)_1$	5.3878(5)		5.3571	
33	C	E1	3	$2p^2$	$(2p_-,2p_+)_2$	7	$2s2p^3$	$(2s_+,2p_+)_1$	8.4792(19) ^{b}		8.4071	
33	C	E1	2	$2p^2$	$(2p_-,2p_+)_1$	5	$2s2p^3$	$(2s_+,2p_+)_2$	11.1236(9) ^{b}		11.1132	
33	C	M1	1	$2p^2$	$(2p_-^2)_0$	3	$2p^2$	$(2p_-,2p_+)_2$	14.7700(10)		14.7668	
33	C	M1	1	$2p^2$	$(2p_-^2)_0$	2	$2p^2$	$(2p_-,2p_+)_1$	17.0632(7)		17.1036	17.0625 [76] 17.0558 [76]
32	N	E1	1	$2p^3$	$2p_+$	8	$2s2p^4$	$(2s_+,(2p_+^2)_2)_{3/2}$	4.9858(6)		4.9593	

Table 1. *Cont.*

Ion Charge	Seq.	Type	Lower Level			Upper Level			Experimental Wavelength (nm)		Theoretical Wavelength (nm)	
			No.	Config.	$Term_J$	No.	Config.	$Term_J$	This Work	Previous Work	This Work	Previous Work
32	N	E1	1	$2p^3$	$2p_+$	6	$2s2p^4$	$(2s_+,(2p_+^2)_2)_{5/2}$	5.9329(4)		5.9151	
32	N	E1	2	$2p^3$	$(2p_-,(2p_+^2)_2)_{3/2}$	6	$2s2p^4$	$(2s_+,(2p_+^2)_2)_{5/2}$	8.8822(7)		8.8358	
32	N	E1	3	$2p^3$	$(2p_-,(2p_+^2)_2)_{5/2}$	6	$2s2p^4$	$(2s_+,(2p_+^2)_2)_{5/2}$	9.9054(10)		9.8612	
32	N	M1	1	$2p^3$	$2p_+$	4	$2p^3$	$(2p_-,(2p_+^2)_0)_{1/2}$	12.0926(6)		12.0717	
32	N	M1	1	$2p^3$	$2p_+$	3	$2p^3$	$(2p_-,(2p_+^2)_2)_{5/2}$	14.8036(5)		14.7819	
32	N	M1	1	$2p^3$	$2p_+$	2	$2p^3$	$(2p_-,(2p_+^2)_2)_{3/2}$	17.8665(6)		17.8947	
31	O	E1	1	$2p^4$	$(2p_+^2)_2$	7	$2s2p^5$	$(2s_+,(2p_+^3)_{3/2})_1$	4.4854(8)	4.4857(15) [77]	4.4567	
31	O	E1	2	$2p^4$	$(2p_+^2)_0$	7	$2s2p^5$	$(2s_+,(2p_+^3)_{3/2})_1$	4.8871(12)	4.8882(15) [77]	4.8569	
31	O	E1	1	$2p^4$	$(2p_+^2)_2$	6	$2s2p^5$	$(2s_+,(2p_+^3)_{3/2})_2$	5.0103(5)	5.0085(15) [77]	4.9828	
31	O	E1	3	$2p^4$	$(2p_-,(2p_+^3)_{3/2})_1$	6	$2s2p^5$	$(2s_+,(2p_+^3)_{3/2})_2$	7.2352(8)	7.2356(15) [77]	7.1754	
31	O	E1	4	$2p^4$	$(2p_-,(2p_+^3)_{3/2})_2$	6	$2s2p^5$	$(2s_+,(2p_+^3)_{3/2})_2$	7.8430(8)		7.7848	
31	O	M1	1	$2p^4$	$(2p_+^2)_2$	4	$2p^4$	$(2p_-,(2p_+^3)_{3/2})_2$	13.8581(6)		13.8442	13.89(2) [77]
31	O	M1	1	$2p^4$	$(2p_+^2)_2$	3	$2p^4$	$(2p_-,(2p_+^3)_{3/2})_1$	16.2725(9)		16.307	16.28(2) [77]
31	O	E1	16	$2p^33p$	$(2p_+,3p_+)_3$	32	$2p^33d$	$(2p_+,3d_+)_4$	19.4383(8)		19.4639	
30	F	E1	1	$2p^5$	$(2p_+^3)_{3/2}$	3	$2s$	$2s_+$	4.4500(7)	4.4496(15) [37]	4.417	4.4083 [26] 4.4486 ^{f} [26] 4.4492 [39]
30	F	E1	2	$2p^5$	$2p_-$	3	$2s$	$2s_+$	6.2115(14) ^{b}	6.2107(15) [37]	6.1454	6.1299 [26] 6.2109 ^{f} [26]
30	F	M1	1	$2p^5$	$(2p_+^3)_{3/2}$	2	$2p^5$	$2p_-$	15.6801(11)		15.7043	15.681(12) ^{f} [37] 15.654(5) ^{f} [78] 15.678 ^{f} [26] 15.678(12) [79] 15.71 [38] 15.6826 [39]
												15.685 ^{f} [80]

Table 1. *Cont.*

Ion Charge	Seq.	Type	Lower Level			Upper Level			Experimental Wavelength (nm)		Theoretical Wavelength (nm)	
			No.	Config.	$Term_J$	No.	Config.	$Term_J$	This Work	Previous Work	This Work	Previous Work
29	Ne	E1	3	$2p^5 3s$	$((2p_+^3)_{3/2}), 3s_+)_1$	20	$2p^5 3p$	$(2p_-, 3p_-)_0$	7.8983(8)		7.9003 [a]	7.914 [f] [81]
29	Ne	E1	3	$2p^5 3s$	$((2p_+^3)_{3/2}, 3s_+)_1$	11	$2p^5 3p$	$((2p_+^3)_{3/2}, 3p_+)_0$	12.5743(7)		12.5696 [a]	12.576 [f] [81]
29	Ne	E1	12	$2p^5 3p$	$(2p_-, 3p_-)_1$	24	$2p^5 3d$	$(2p_-, 3d_-)_2$	12.9238(8)		12.9267 [a]	
29	Ne	E1	5	$2p^5 3p$	$((2p_+^3)_{3/2}, 3p_-)_2$	15	$2p^5 3d$	$((2p_+^3)_{3/2}, 3d_-)_3$	13.0471(8)		13.0550 [a]	
29	Ne	E1	2	$2p^5 3s$	$((2p_+^3)_{3/2}, 3s_+)_2$	10	$2p^5 3p$	$((2p_+^3)_{3/2}, 3p_+)_2$	14.8480(7)		14.8389 [a]	
29	Ne	E1	10	$2p^5 3p$	$((2p_+^3)_{3/2}, 3p_+)_2$	22	$2p^5 3d$	$((2p_+^3)_{3/2}, 3d_+)_3$	15.3559(10)		15.3587 [a]	
29	Ne	E1	19	$2p^5 3p$	$(2p_-, 3p_+)_2$	26	$2p^5 3d$	$(2p_-, 3d_+)_3$	15.3945(10)		15.3972 [a]	
29	Ne	E1	18	$2p^5 3p$	$(2p_-, 3p_+)_1$	25	$2p^5 3d$	$(2p_-, 3d_+)_2$	15.4387(10)		15.4444 [a]	
29	Ne	E1	3	$2p^5 3s$	$((2p_+^3)_{3/2}, 3s_+)_1$	10	$2p^5 3p$	$((2p_+^3)_{3/2}, 3p+)_2$	15.4902(18) [b]	15.497(15) [82] 15.50 [83]	15.4882 [a]	15.503 [f] [81] 15.50 [84]
29	Ne	E1	9	$2p^5 3s$	$(2p_-, 3s_+)_1$	20	$2p^5 3p$	$(2p_-, 3p_-)_0$	15.5024(8) [b]		15.4904 [a]	15.498 [f] [81]
29	Ne	E1	6	$2p^5 3p$	$((2p_+^3)_{3/2}, 3p_+)_3$	17	$2p^5 3d$	$((2p_+^3)_{3/2}, 3d_+)_4$	15.6711(10)		15.6769 [a]	
29	Ne	E1	9	$2p^5 3s$	$(2p_-, 3s_+)_1$	19	$2p^5 3p$	$(2p_-, 3p_+)_2$	15.7208(7)	15.714(15) [82] 15.71 [83]	15.7085 [a]	15.723 [f] [81] 15.71 [84]
29	Ne	E1	2	$2p^5 3s$	$((2p_+^3)_{3/2}, 3s_+)_2$	6	$2p^5 3p$	$((2p_+^3)_{3/2}, 3p_+)_3$	15.8537(7)		15.8455 [a]	
29	Ne	M1	3	$2p^5 3s$	$((2p_+^3)_{3/2}, 3s_+)_1$	8	$2p^5 3s$	$(2p_-, 3s_+)_0$	16.4817(7)		16.4843 [a]	
29	Ne	E1	3	$2p^5 3s$	$((2p_+^3)_{3/2}, 3s_+)_1$	7	$2p^5 3p$	$((2p_+^3)_{3/2}, 3p_+)_1$	16.5411(8)	16.537(15) [82]	16.5488 [a]	16.542 [f] [81] 16.463 [85] 16.484 [85]
28	Na	E1	2	$3p$	$3p_-$	4	$3d$	$3d_-$	12.0979(8)	12.098(20) [16]	12.1353	12.09248 [86] 12.0993(7) [f] [45]
28	Na	E1	3	$3p$	$3p_+$	5	$3d$	$3d_+$	14.1938(7)	14.1938(6) [16]	14.2458	14.1873 [86] 14.1959(7) [f] [45]

Table 1. *Cont.*

Ion Charge	Seq.	Type	Lower Level			Upper Level			Experimental Wavelength (nm)		Theoretical Wavelength (nm)	
			No.	Config.	Term_J	No.	Config.	Term_J	This Work	Previous Work	This Work	Previous Work
28	Na	E1	1	$3s^2$	$3s_+$	3	$3p$	$3p_+$	15.1037(5)	15.1035(10) [16]	15.0542	15.1038 [55] 15.10402(40) [7] 15.1033 ^{f} [29] 15.1038(7) ^{f} [45] 15.0658 [86]
28	Na	E1	1	$3s^2$	$3s_+$	2	$3p$	$3p_-$	19.6212(7)	19.6215(10) [16]	19.5175	19.6199 [55] 19.6209(7) [7] 19.6219 ^{f} [29] 19.6213(7) [45]
27	Mg	E1	5	$3s3p$	$(3s_+,3p_+)_1$	14	$3s3d$	$(3s_+,3d_+)_2$	12.8333(9)	12.8352(10) [21] 12.8349(5) [15] 12.8301(15) [44]	12.7875	
27	Mg	E1	1	$3s^2$	$(3s_+^2)_0$	5	$3s3p$	$(3s_+,3p_+)_1$	13.5276(5)	13.5279(10) [21] 13.5283(5) [15] 13.5216(15) [44]	13.4437	13.5276 [87] 13.5213 [88]
27	Mg	E1	3	3s3p	$(3s_+,3p_-)_1$	7	$3p^2$	$(3p_-,3p_+)_2$	14.5650(20) ^{b}	14.5603(10) [21]	14.528	
26	Al	E1	1	$3p$	$3p_-$	10	$3s3p^2$	$(3s_+,(3p_+^2)_0)_{3/2}$	10.9388(15) ^{w,b}	10.9391(20) [30] 10.9413(10) [43]	10.8578	
26	Al	E1	2	$3p$	$3p_+$	11	$3s3d$	$3d_-$	11.9072(12) ^{w,b}	11.9131(20) [30] 11.9110(10) [43]	11.8248	
26	Al	E1	2	$3p$	$3p_+$	10	$3s3p^2$	$(3s_+,(3p_+^2)_2)_3$	12.9717(11) ^{w}	12.9729(20) [30] 12.9745(10) [43]	12.852	
26	Al	E1	1	$3p$	$3p_-$	8	$3s3p^2$	$((3s_+,3p_-)_1),3p_+)_{1/2}$	13.0401(8)	13.0416(20) [30] 13.0417(10) [43]	12.9265	
26	Al	E1	1	$3p$	$3p_-$	6	$3s3p^2$	$((3s_+,3p_-)_1),3p_+)_{3/2}$	14.4883(8) ^{w}	14.4914(20) [30] 14.4910(10) [43]	14.4675	

^{a} Wavelength from Safronova et al.'s calculation [68]; ^{b} blended with other line feature; ^{w} weak lines; and ^{f} fitted values.

A total of 59 spectral lines were identified in this work from the Li-like to the Al-like isoelectronic sequences. Of these lines, 38 are new and 21 correspond to previously measured transitions in O-like, F-like, Na-like, Mg-like and Al-like charge states [15,16,21,24–26,29,30,37,43–45,77]. The previously measured transitions are also listed in Table 1 together with their currently measured wavelengths. We observed an O VI line at 15.0099(5) nm in all of the spectra due to impurities in the trap. At 4.60 keV and 4.85 keV drift tube voltages, we also observed impurity lines due to xenon, including a Xe XLII line at 15.0116(7) nm blended with the above-mentioned O VI line. Mg-like Xe XLIII lines were observed at 6.2903(6) nm and 12.9969(9) nm wavelengths, and two Na-like Xe XLIV lines were found at 6.6628(7) nm and 12.3939(7) nm. These lines are listed in the NIST Atomic Spectra Database [7,64] at 6.288(3) nm, 12.993(3) nm, 6.6628(5) nm and 12.394(1) nm, respectively.

Two Li-like yttrium lines were identified at 7.2874(6) nm/170.134(15) eV and 15.7862(9) nm/ 78.5395(44) eV, corresponding to the $(2s_+)$–$(2p_+)$ and $(2s_+)$–$(2p_-)$ electric dipole transitions, respectively. The Li-like isoelectronic sequence has been extensively studied both theoretically and experimentally due to its simple electronic structure. Highly accurate ab initio calculations agree with precise experimental results at the high-Z end of the isoelectronic sequence [1,89]. Although recent results are sensitive to higher order QED terms, further developments are expected, especially in the moderately high-Z region where experiments can provide accurate data due to the wavelength range available to grazing incidence EUV spectrometers. Our current relative uncertainty of 57×10^{-6} for the wavelength of the $(2s_+)$–$(2p_-)$ transition shows good agreement with previous high precision calculations [70–72]; however, the $(2s_+)$–$(2p_+)$ theoretical results [70,72] are slightly outside our relative uncertainty of 82×10^{-6} as shown in Figure 6. Upon close examination, a small feature of unidentified origin was found in the low wavelength wings of both Li-like lines. They were taken into account with the inclusion of a second small Gaussian peak in the fits. The reported results and uncertainties reflect the inclusion of these features, and we therefore believe that they are not responsible for the slight disagreement between the theoretical values and our results for the $(2s_+)$–$(2p_+)$ line.

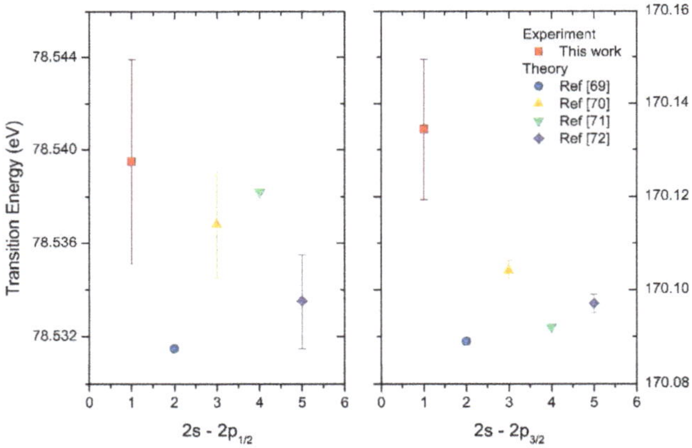

Figure 6. Comparison of measured and calculated Li-like yttrium lines.

The two electric dipole Be-like yttrium lines in the measured spectra at 6.0322(5) nm and 15.2336(7) nm correspond to the transitions of $(2s_+^2)_0$–$(2s_+, 2p_+)_1$ and $(2s_+^2)_0$–$(2s_+, 2p_-)_1$, respectively. Be-like ions are quasi two-electron systems. Therefore, calculations at the level of the precision of the measurement are difficult. Denne et al. [24] predicted these wavelengths in yttrium by fitting the difference between the theoretical and experimental wave numbers and then extrapolating to attain the fine-structure separation. Their predictions of 6.0337(20) nm and 15.2345(20) nm had uncertainties much larger than our measurements. Thus, we provide a considerable increase in the accuracy of the wavelengths.

A similar approach was used by Mrynas et al. [25] for the predictions of the wavelengths of B-like yttrium $(2p_-)$–$((2s_+,2p_-)_1,2p_+)_{1/2}$, $(2p_-)$–$((2s_+,2p_-)_1,2p_+)_{3/2}$ and $(2p_-)$–$(2p_+)$ lines. The obtained respective values were 5.5771 nm, 5.7629 nm and 14.322 nm, but no uncertainties were provided for the fitted wavelength predictions. These results are in a generally good agreement with our observed values of 5.5768(6) nm, 5.7623(6) nm and 14.3234(5) nm. Beyond these three transitions, two additional lines have been identified for B-like yttrium, as shown in Table 1. Out of the five reported transitions, four are E1, and one is an M1 transition.

With an increasing number of electrons, the electronic structure of open shell ions becomes more difficult for theory. However, the experimental wavelength determinations are as accurate as for their simpler structure counterparts. Accurate wavelength results in these ions can provide guidance for further theoretical work for the better understanding of the electron-electron interactions in these systems. Here, we report E1 and M1 transitions for both C-like and N-like yttrium ions.

Behring et al. [77] observed O-like yttrium transitions by irradiating a solid yttrium target with 24 frequency-tripled laser beams. We identified six E1 transitions and two M1 transitions in the same system and provide wavelength values for these in Table 1. The observations of Behring et al. are consistent with our measurements with the exception of the M1 transition $(2p_+^2)_2$–$(2p_-,(2p_+^3)_{3/2})_2$. Our measurement of 13.8581(6) nm for this M1 line is at a shorter wavelength than their predicted wavelength of 13.89(2) nm. The M1 transition at 16.2725(9) nm agrees with their predicted wavelength of 16.28(2) nm [77].

Wavelengths of F-like yttrium lines were measured by Reader et al. [37] using laser produced plasmas. The $(2p_+^3)_{3/2}$–$(2s_+)$ and $(2p_-)$–$(2s_+)$ transitions in these ions were measured to be 4.4496(15) nm and 6.2107(15) nm, respectively. Our results for the same transitions indicated 4.4500(7) nm and 6.2115(14) nm wavelengths and are in good agreement with the previously observed values. Calculations by Feldman et al. [26] reported values of 4.4083 nm and 6.1299 nm that are further away from these measurements than our FAC calculated wavelengths values of 4.417 nm and 6.1454 nm.

The M1 transition $(2p_+^3)_{3/2}$–$(2p_-)$ in F-like Y is interesting due to its potential for plasma diagnostics [37,38]. Reader et al. [37] predicted the wavelength by comparing the observed fine-structure intervals with Dirac–Fock calculations and obtained 15.681(12) nm. This is in agreement with our measured value of 15.6801(11) nm.

The ground state of Ne-like ions is a closed shell. However, the low lying excited states have interesting features that have been exploited in many experiments and observations [90]. The level structure has been investigated for use in the diagnosis of astrophysical and laboratory plasmas [91,92] and has been used in soft X-ray laser schemes [93]. We report fourteen E1 and one M1 transitions in Ne-like Y. The Ne-like yttrium lines were identified using the theoretical values from highly accurate RMBPT calculations [68].

In our spectra, we were able to identify four E1 transitions of Na-like yttrium ions. The two most prominent ones are the well-known Na-like D_1 $(3s_+)$–$(3p_+)$ and D_2 $(3s_+)$–$(3p_-)$ lines. Our measured wavelength values of 15.1037(5) nm and 19.6212(7) nm, respectively, lie within the uncertainty of the 15.1035(10) nm and 19.6215(10) nm measurements by Reader et al. in tokamak plasmas and laser-produced plasmas [16]. Seely et al. [29] reported calculated values of 15.0961 nm and 15.0310 nm for the D_1 line and 19.6047 nm and 19.4851 nm for the D_2 line with and without QED corrections, respectively. This illustrates the importance of QED corrections at this level of experimental accuracy. Their fitted values of 15.1033 nm and 19.6219 nm for these transitions are within our experimental uncertainty. Our measured wavelengths also agree well with Blundell's calculated values of 15.10402(40) nm and 19.6209(7) nm for the D_1 and D_2 lines [7]. Gillaspy et al. [7] have pointed out that the accuracy of the measurements in medium-Z to high-Z systems is sensitive to the finite nuclear size correction in the otherwise calculable QED terms. This illustrates the importance of these transitions in studies at the interface of atomic and nuclear physics.

Our goal in these studies was the identification of lines in the L-shell ion states of yttrium in the EUV region. The few M-shell charge states we report here appeared at the low energy end of our systematic scans. The highest isoelectronic sequences investigated here were those for Mg and Al. In Mg-like yttrium, we report the observation of three E1 lines that have been previously measured by Ekberg et al. [21], Sugar et al. [15] and Reader et al. [44]. Similarly, the five Al-like yttrium E1 lines that we identified were previously measured by Ekberg et al. [43] and Sugar et al. [30]. The slight disagreement with some of the previous measurements might be due to weak lines and blends with other line features.

7. Diagnostically Important M1 Transitions

Among the 59 identified yttrium lines listed in Table 1, 10 lines are due to forbidden M1 transitions. States that decay via M1 transitions have a different dependence on collisional depopulation from states with E1 transitions. Thus, the corresponding intensity ratios of M1 to E1 lines are sensitive to the electron densities and temperatures, thereby making them potential candidates for plasma diagnostics. To analyze the feasibility of this, calculations with the CR modeling code NOMAD were performed for Maxwellian electron energy distribution plasmas with electron densities ranging from 10^{12} cm^{-3}–10^{20} cm^{-3} and temperatures from 1500 eV–6000 eV, which provide the largest abundance of the ions.

The most sensitive intensity ratios for the spectral lines of Y^{30+}–Y^{33+} ions are presented in Figure 7. The figure provides examples for sensitivity to the electron density (n_e) in the range of 10^{12} cm^{-3}–10^{19} cm^{-3}. At low densities, radiative rates for both forbidden and allowed transitions are much stronger than the collisional rates, and therefore, no dependence on electron density arises. However, at higher densities, collisional quenching dominates radiative decay for forbidden lines, and the line intensity ratios become sensitive to n_e.

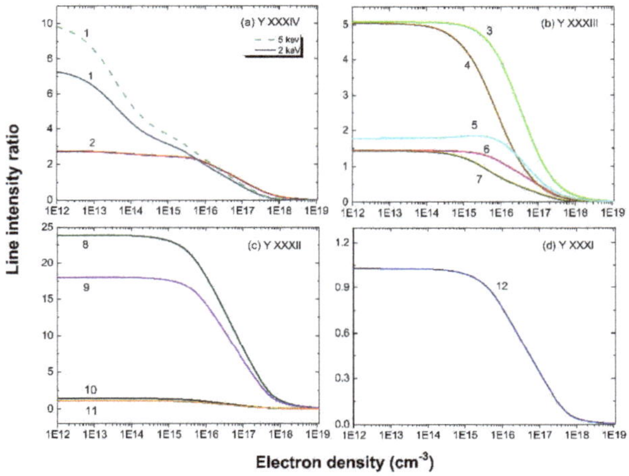

Figure 7. Density-sensitive line ratios for (**a**) C-like Y XXXIV, (**b**) N-like Y XXXIII, (**c**) O-like Y XXXII and (**d**) F-like Y XXXI. The number labels correspond to the line ratios: (1) 17.0632(7) nm/11.1236(9) nm at 5-keV electron energy (dash) and 2-keV electron energy (solid), (2) 17.0632(7) nm/8.4792(19) nm, (3) 17.8665(6) nm/9.9054(10) nm, (4) 14.8036(5) nm/9.9054(10) nm, (5) 12.0926(6) nm/9.9054(10) nm, (6) 17.8665(6) nm/8.8822(7) nm, (7) 14.8036(5) nm/8.8822(7) nm, (8) 13.8581(6) nm/19.4383(8) nm, (9) 16.2725(9) nm/19.4383(8) nm, (10) 16.2725(9) nm/7.2352(8) nm, (11) 13.8581(6) nm/7.2352(8) nm and (12) 15.6801(11) nm/6.21115(14) nm.

The five C-like yttrium lines listed in Table 1 include an M1 transition at a 17.0632(7) nm wavelength. The ratio of this M1 line to the E1 line at 8.4792(19) nm varies by a factor of 45 or less in the electron density range of 10^{15} cm^{-3}–10^{18} cm^{-3}. The line intensity ratio I(17.0632(7))/I(11.1236(9)) varies by more than two orders of magnitude for the same range of electron density. This line ratio also shows dependence on the electron temperature at lower densities as illustrated in Figure 7a. The line ratio of the M1 line at 17.0632(7) nm to another line at the FAC wavelength of 12.3869 nm shows a similar dependence on electron density and temperature because both of the E1 transitions at 11.1236(9) nm and 12.3869 nm arise from the same upper level decaying to the second and third energy levels, respectively. The line at 12.3869 nm, which is about half of the intensity of line at 11.1236(9) nm, could not be resolved due to strong blend with a Na-like Xe line at 12.3939(7) nm. However, at other plasma conditions with no Xe, this line should be easily resolved.

Among the seven identified N-like yttrium lines, four of the lines arise from E1 transitions, and three arise from M1 transitions. The ratios of the M1/E1 intensity show a dependence on n_e between 10^{15} cm^{-3} and 10^{18} cm^{-3} of a maximum of two orders of magnitude. According to our FAC calculations, the transition probability of the M1 line at 14.8036(5) nm is 6.75×10^5 s^{-1} compared to the transition probability of 3.9×10^6 s^{-1} for the M1 line at 12.0926(6) nm and 3.7×10^6 s^{-1} for the M1 line at 17.8665(6) nm. This explains why the ratios I(14.8036(5))/I(9.9054(10)) and I(14.8036(5))/I(8.8822(7)) start decreasing at densities of 10^{15} cm^{-3}, whereas the intensity ratios of the M1 line at 12.0926(6) nm to the E1 lines and the M1 line at 17.8665(6) nm to the E1 lines only start to fall off at densities of 10^{16} cm^{-3} and higher. These transitions are illustrated in the Grotrian diagram shown in Figure 8.

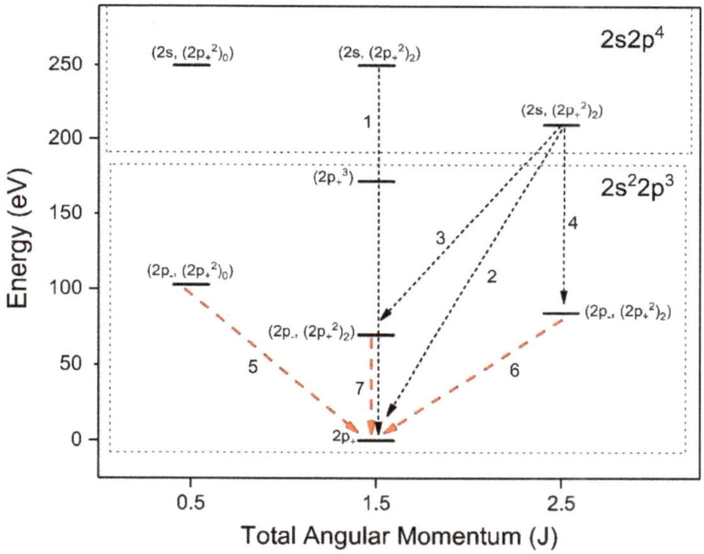

Figure 8. Partial Grotrian diagram of the ground state and first lowest excited configurations of N-like yttrium. The number labels in increasing order from 1–7 correspond to the lines at 4.9858(6) nm, 5.9329(4) nm, 8.8822(7) nm, 9.9054(10) nm, 12.0926(6) nm, 14.8036(5) nm and 17.8665(6) nm, respectively. Only three of the energy levels for the $2s2p^4$ configuration are shown. The dashed red lines correspond to magnetic dipole (M1) transitions, and the dotted black lines correspond to electric dipole (E1) transitions.

A closer look at the the density dependence of M1/E1 ratios in N-like ions gives an insight into the population scheme of allowed and metastable upper levels. For instance, let us take the population and depopulation channels of the metastable level in the ground state configuration $2p^3$ that is the upper level of the 17.8665(6) nm M1 transition and a $2s2p^4$ excited state level that is the origin of three E1 transitions (5.9329(4) nm, 8.8822(7) nm and 9.9054(10) nm wavelengths). At a temperature of 1500 eV, the metastable level $2p^3$ at a lower density of 10^{12} cm^{-3} is depopulated by radiative decay with 99.96% probability. At a higher density of 10^{16} cm^{-3}, the depopulation is 73% by collisional excitation to higher levels, 3% by collisional deexcitation to a lower ground state level, and only a 21% probability remains for radiative decay. At an even more elevated electron density of 10^{18} cm^{-3}, the level is depopulated mostly by collisional excitation with nearly 94% probability, leaving 4% to collisional deexcitation, 1% to radiative recombination and a negligible probability (0.27%) to radiation. The $2s2p^4$ excited level is depopulated 100% by radiative decay at 10^{12} cm^{-3} and 10^{16} cm^{-3}, and the probability only slightly lowers to 99.6% at 10^{18} cm^{-3}. This means that the ratio of the intensity of the $2s2p^4$ E1 transitions to that of the $2p^3$ M1 transition shows a strong variation with the electron density.

Out of the eight O-like yttrium lines, we observed two that originate from M1 transitions at 13.8581(6) nm and 16.2725(9) nm. The intensity ratios between the M1 and E1 transitions vary by more than an order of magnitude in the density ranges we investigated. For instance, the intensity ratio I(13.8581(6))/I(19.4383(8)) changes by a factor of 67 for densities ranging from 10^{15} cm^{-3}–10^{18} cm^{-3}.

For the one M1 and two E1 lines in F-like Y, an order of magnitude variation is seen for the intensity ratio of M1 line at 15.6801(11) nm to the E1 line at 6.2115(14) nm.

8. Conclusions

New and previously-measured EUV lines in L-shell ions along with transitions in a few M-shell ions of highly charged yttrium were observed. The measurements were performed with an electron beam ion trap, and spectral lines were recorded in the wavelength region of 4 nm–20 nm. The experimental uncertainties were combinations of statistical and systematic uncertainties that included sources with calibration origins and uncertainties from unresolved blends. The total uncertainties ranged between 0.0004 nm and 0.0020 nm. Line identifications were inferred from comparisons with spectra simulated from the collisional-radiative model NOMAD based on a non-Maxwellian distribution designed for EBIT-like environments. For Ne-like Y ions, a better agreement between theory and experiment was found using relativistic many-body perturbation theory (RMBPT) [68]. Several of the identified forbidden M1 transitions were found to be potentially useful for density diagnostics of laboratory, fusion and astrophysical plasmas.

Acknowledgments: This work was funded by the Measurement Science and Engineering (MSE) Research Grant Programs of the Department of Commerce at the National Institute of Standards and Technology (NIST). We would like to thank the Department of Physics and Astronomy at Clemson University for their support. We are grateful to M. Safronova for providing the theory for Ne-like Y ions. J.M.D. acknowledges support from a National Research Council Associateship award at NIST.

Author Contributions: E.T. and J.D.G. contributed to the design and construction of the apparatus. R.S., E.T., Y.R. and J.M.D. conceived of and designed the experiments. R.S., J.M.D., E.T. and J.D.G. performed the experiments. Y.R. performed the calculations. R.S. analyzed the data. All authors contributed to writing the paper.

Conflicts of Interest: The authors declare no conflict of interest.

References

1. Sapirstein, J.; Cheng, K.T. Tests of Quantum Electrodynamics with EBIT. *Can. J. Phys.* **2008**, *86*, 25–31.
2. Fahy, K.; Sokell, E.; O'Sullivan, G.; Aguilar, A.; Pomeroy, J.M.; Tan, J.N.; Gillaspy, J.D. Extreme-Ultraviolet Spectroscopy of Highly Charged Xenon Ions Created Using an Electron-Beam Ion Trap. *Phys. Rev. A* **2007**, *75*, 032520.
3. Podpaly, Y.A.; Gillaspy, J.D.; Reader, J.; Ralchenko, Y. EUV Measurements of Kr XXI–Kr XXXIV and the Effect of a Magnetic-Dipole Line on Allowed Transitions. *J. Phys. B At. Mol. Opt. Phys.* **2014**, *47*, 095702.
4. Podpaly, Y.A.; Gillaspy, J.D.; Reader, J.; Ralchenko, Y. Measurements and Identifications of Extreme Ultraviolet Spectra of Highly-Charged Sm and Er. *J. Phys. B At. Mol. Opt. Phys.* **2015**, *48*, 025002.

5. Reader, J.; Gillaspy, J.D.; Osin, D.; Ralchenko, Y. Extreme Ultraviolet Spectra and Analysis of $\Delta n = 0$ Transitions in Highly Charged Barium. *J. Phys. B At. Mol. Opt. Phys.* **2014**, *47*, 145003.
6. Gillaspy, J.D.; Draganić, I.N.; Ralchenko, Y.; Reader, J.; Tan, J.N.; Pomeroy, J.M.; Brewer, S.M. Measurement of the D-Line Doublet in High-Z Highly Charged Sodiumlike Ions. *Phys. Rev. A* **2009**, *80*, 010501.
7. Gillaspy, J.D.; Osin, D.; Ralchenko, Y.; Reader, J.; Blundell, S.A. Transition Energies of the *D* lines in Na-like ions. *Phys. Rev. A* **2013**, *87*, 062503.
8. Kilbane, D.; O'Sullivan, G.; Podpaly, Y.A.; Gillaspy, J.D.; Reader, J.; Ralchenko, Y. EUV Spectra of Rb-like to Ni-like Dysprosium Ions in an Electron Beam Ion Trap. *Eur. Phys. J. D* **2014**, *68*, 222.
9. Kilbane, D.; O'Sullivan, G.; Gillaspy, J.D.; Ralchenko, Y.; Reader, J. EUV Spectra of Rb-like to Cu-like Gadolinium Ions in an Electron-Beam Ion Trap. *Phys. Rev. A* **2012**, *86*, 042503.
10. Kilbane, D.; Gillaspy, J.D.; Ralchenko, Y.; Reader, J.; O'Sullivan, G. Extreme Ultraviolet Spectra from *N*-Shell Ions of Gd, Dy and W. *Phys. Scr.* **2013**, *T156*, 014012.
11. Osin, D.; Reader, J.; Gillaspy, J.D.; Ralchenko, Y. Extreme Ultraviolet Spectra of Highly Charged Xenon Observed with an Electron Beam Ion Trap. *J. Phys. B At. Mol. Opt. Phys.* **2012**, *45*, 245001.
12. Osin, D.; Gillaspy, J.D.; Reader, J.; Ralchenko, Y. EUV Magnetic-Dipole Lines from Highly-Charged High-Z Ions with an Open 3D Shell. *Eur. Phys. J. D* **2012**, *66*, 286.
13. Ralchenko, Y.; Reader, J.; Pomeroy, J.M.; Tan, J.N.; Gillaspy, J.D. Spectra of W^{39+}–W^{47+} in the 12–20 nm Region Observed with an EBIT Light Source. *J. Phys. B At. Mol. Opt. Phys.* **2007**, *40*, 3861–3875.
14. Ralchenko, Y.; Draganic, I.N.; Tan, J.N.; Gillaspy, J.D.; Pomeroy, J.M.; Reader, J.; Feldman, U.; Holland, G.E. EUV Spectra of Highly-Charged Ions W^{54+}–W^{63+} Relevant to ITER Diagnostics. *J. Phys. B At. Mol. Opt. Phys.* **2008**, *41*, 021003.
15. Sugar, J.; Kaufman, V.; Indelicato, P.; Rowan, W.L. Analysis of Magnesiumlike Spectra from Cu XVIII to Mo XXXI. *J. Opt. Soc. Am. B* **1989**, *6*, 1437–1443.
16. Reader, J.; Kaufman, V.; Sugar, J.; Ekberg, J.O.; Feldman, U.; Brown, C.M.; Seely, J.F.; Rowan, W.L. 3*s*–3*p*, 3*p*–3*d*, and 3*d*–4*f* Transitions of Sodiumlike Ions. *J. Opt. Soc. Am. B* **1987**, *4*, 1821–1828.
17. Jupén, C.; Denne, B.; Martinson, I. Transitions in Al-like, Mg-like and Na-like Kr and Mo, Observed in the JET Tokamak. *Phys. Scr.* **1990**, *41*, 669–674.
18. Hinnov, E. Highly Ionized Atoms in Tokamak Discharges. *Phys. Rev. A* **1976**, *14*, 1533–1541.
19. Hinnov, E.; Boody, F.; Cohen, S.; Feldman, U.; Hosea, J.; Sato, K.; Schwob, J.L.; Suckewer, S.; Wouters, A. Spectrum Lines of Highly Ionized, Zinc, Germanium, Zirconium, Molybdenum, and Silver Injected into Princeton Large Torus and Tokamak Fusion Test Reactor Tokamak Discharges. *J. Opt. Soc. Am. B* **1986**, *3*, 1288–1294.
20. Suckewer, S.; Hinnov, E.; Cohen, S.; Finkenthal, M.; Sato, K. Identification of Magnetic Dipole Lines above 2000 Å in Several Highly Ionized Mo and Zr Ions on the PLT Tokamak. *Phys. Rev. A* **1982**, *26*, 1161–1163.
21. Ekberg, J.O.; Feldman, U.; Seely, J.F.; Brown, C.M. Transitions and Energy Levels in Mg-like Ge XXI–Zr XXIX Observed in Laser-Produced Linear Plasmas. *Phys. Scr.* **1989**, *40*, 643–651.
22. Nilsen, J.; Beiersdorfer, P.; Widmann, K.; Decaux, V.; Elliott, S.R. Energies of Neon-like $n = 4$ to $n = 2$ Resonance Lines. *Phys. Scr.* **1996**, *54*, 183–187.
23. Fischer, C.F. Multiconfiguration Hartree-Fock Breit-Pauli Results for $2p_{1/2}$–$2p_{3/2}$ Transitions in the Boron Sequence. *J. Phys. B At. Mol. Phys.* **1983**, *16*, 157–165.
24. Denne, B.; Magyar, G.; Jacquinot, J. Berylliumlike Mo XXXIX and Lithiumlike Mo XL Observed in the Joint European Torus Tokamak. *Phys. Rev. A* **1989**, *40*, 3702–3705.
25. Myrnäs, R.; Jupén, C.; Miecznik, G.; Martinson, I.; Denne-Hinnov, B. Transitions in Boronlike Ni XXIV, Ge XXVIII, Kr XXXII and Mo XXXVIII and Fluorinelike Zr XXXII and Mo XXXIV, Observed in the JET Tokamak. *Phys. Scr.* **1994**, *49*, 429–435.
26. Feldman, U.; Ekberg, J.O.; Seely, J.F.; Brown, C.M.; Kania, D.R.; MacGowan, B.J.; Keane, C.J.; Behring, W.E. Transitions of the Type 2*s*–2*p* in Highly Charged Flourinelike and Oxygenlike Mo, Cd, In, and Sn. *J. Opt. Soc. Am. B* **1991**, *8*, 531–537.
27. Curtis, L.J.; Ramanujam, P.S. Isoelectronic Wavelength Predictions for Magnetic-Dipole, Electric-Quadrupole, and Intercombination Transitions in the Mg Sequence. *J. Opt. Soc. Am.* **1983**, *73*, 979–984.
28. Ekberg, J.O.; Feldman, U.; Seely, J.F.; Brown, C.M.; MacGowan, B.J.; Kania, D.R.; Keane, C.J. Analysis of Magnesiumlike Spectra from Mo XXXI to Cs XLIV. *Phys. Scr.* **1991**, *43*, 19–32.

29. Seely, J.F.; Wagner, R.A. QED Contributions to the $3s$–$3p$ Transitions in Highly Charged Na-like Ions. *Phys. Rev. A* **1990**, *41*, 5246–5249.
30. Sugar, J.; Kaufman, V.; Rowan, W.L. Aluminiumlike Spectra of Copper Through Molybdenum. *J. Opt. Soc. Am. B* **1988**, *5*, 2183–2189.
31. Zhao, G.; Magain, P. Abundances of Neutron Capture Elements in Metal-Poor Dwarfs I. Yttrium and Zirconium. *Astron. Astrophys.* **1991**, *244*, 425–432.
32. Redfors, A.; Cowley, C.R. Elemental Abundances of Yttrium and Zirconium in the Mercury-Manganese Stars ϕ Herculis, κ Cancri and ι Coronae Borealis. *Astron. Astrophys.* **1993**, *271*, 273–275.
33. Kessler, T.; Moore, I.D.; Kudryavtsev, Y.; Peräjärvi, K.; Popov, A.; Ronkanen, P.; Sonoda, T.; Tordoff, B.; Wendt, K.D.A.; Äystö, J. Off-line Studies of the Laser Ionization of Yttrium at the IGISOL Facility. *Nucl. Instrum. Methods Phys. Res. Sect. B* **2008**, *266*, 681–700.
34. Wahlgren, G.M.; Carpenter, K.G.; Norris, R.P. Heavy Elements and Cool Stars. *AIP Conf. Proc. Astron.* **2008**, doi:10.1063/1.3099261.
35. Osterbrock, D.E. *Astrophysics of Gaseous Nebulae*; W. H. Freeman & Co Ltd.: San Francisco, CA, USA, 1974.
36. Denne, B.; Hinnov, E. Spectral Lines of Highly-Ionized Atoms for the Diagnostics of Fusion Plasmas. *Phys. Scr.* **1987**, *35*, 811–818.
37. Reader, J. $2s^22p^5$–$2s2p^6$ Transitions in the Flourinelike Ions Sr^{29+} and Y^{30+}. *Phys. Rev. A* **1982**, *26*, 501–503.
38. Aggarwal, K.M.; Keenan, F.P. Radiative Rates for E1, E2, M1, and M2 Transitions in F-like ions with $37 \leq Z \leq 53$. *At. Data Nucl. Data Tables* **2016**, *109–110*, 205–338.
39. Jönsson, P.; Alkauskas, A.; Gaigalas, G. Energies and E1, M1, E2 Transition Rates for States of the $2s^22p^5$ and $2s2p^6$ Configurations in Fluorine-like Ions Between Si VI and W LXVI. *At. Data Nucl. Data Tables* **2013**, *99*, 431–446.
40. Khatri, I.; Goyal, A.; Aggarwal, S.; Singh, A.K.; Mohan, M. Extreme Ultraviolet and Soft X-ray Spectral Lines in Rb XXIX. *Chin. Phys. B* **2016**, *25*, 033201.
41. Alexander, E.; Even-Zohar, M.; Fraenkel, B.S.; Goldsmith, S. Classification of Transitions in the EUV Spectra of Y IX–XIII, Zr X–XIV, Nb XI–XV, and Mo XII–XVI. *J. Opt. Soc. Am.* **1971**, *61*, 508–514.
42. Ekberg, J.O.; Jupén, C.; Brown, C.M.; Feldman, U.; Seely, J.F. Classification of Resonance Transitions in Ge XIX, Se XXI, Sr XXV, Y XXVI and Zr XXVll. *Phys. Scr.* **1992**, *46*, 120–126.
43. Ekberg, J.O.; Redfors, A.; Brown, C.M.; Feldman, U.; Seely, J.F. Transitions and Energy Levels in Al-like Ge XX, Se XXII, Sr XXVI, Y XXVII and Zr XXVIII. *Phys. Scr.* **1991**, *44*, 539–547.
44. Reader, J. $3s^2$–$3s3p$ and $3s3p$–$3s3d$ Transitions in Magnesiumlike Ions from Sr^{26+} to Rh^{33+}. *J. Opt. Soc. Am.* **1983**, *73*, 796–799.
45. Reader, J.; Ekberg, J.O.; Feldman, U.; Brown, C.M.; Seely, J.F. Spectra and Energy Levels of Sodiumlike Ions from Y^{28+} to Sn^{39+}. *J. Opt. Soc. Am. B* **1990**, *7*, 1176–1181.
46. Ateqad, N.; Chaghtai, M.S.Z.; Rahimullah, K. Addition to the Analysis of Y VII, VIII and Mo X. *J. Phys. B At. Mol. Phys.* **1984**, *17*, 4617–4622.
47. Litzén, U.; Hansson, A. Additions to the Spectra and Energy Levels of the Zinc-like Ions Y X–Cd XIX. *Phys. Scr.* **1989**, *40*, 468–471.
48. Nilsson, A.E.; Johansson, S.; Kurucz, R.L. The Spectrum of Singly Ionized Yttrium, Y II. *Phys. Scr.* **1991**, *44*, 226–257.
49. Epstein, G.L.; Reader, J. Spectrum and Energy Levels of Triply Ionized Yttrium (Y IV). *J. Opt. Soc. Am.* **1982**, *72*, 476–492.
50. Reader, J.; Acquista, N. Spectrum and Energy Levels of Ten-Times Ionized Yttrium (Y XI). *J. Opt. Soc. Am.* **1979**, *69*, 1285–1288.
51. Gillaspy, J.D. First Results from the EBIT at NIST. *Phys. Scr.* **1997**, *T71*, 99.
52. Gillaspy, J.D.; Aglitskiy, Y.; Bell, E.W.; Brown, C.M.; Chandler, C.T.; Deslattes, R.D.; Feldman, U.; Hudson, L.T.; Laming, J.M.; Meyer, E.S.; et al. Overview of the Electron Beam Ion Trap Program at NIST. *Phys. Scr.* **1995**, *T59*, 392–395.
53. Gillaspy, J.D. Testing QED in Sodium-like Gold and Xenon: Using Atomic Spectroscopy and an EBIT to Probe the Quantum Vacuum. *J. Instrum.* **2010**, *5*, C10005.
54. Gillaspy, J.D. Precision Spectroscopy of Trapped Highly Charged Heavy Elements: Pushing the Limits of Theory and Experiment. *Phys. Scr.* **2014**, *89*, 114004.

55. Sapirstein, J.; Cheng, K.T. S-Matrix Calculations of Energy Levels of Sodiumlike Ions. *Phys. Rev. A* **2015**, *91*, 062508.
56. Griem, H.R. *Principles of Plasma Spectroscopy*; Cambridge University Press: Cambridge, UK, 1997.
57. Ralchenko, Y.; Gillaspy, J.D.; Reader, J.; Osin, D.; Curry, J.J.; Podpaly, Y.A. Magnetic-Dipole Lines in $3d^n$ Ions of High-Z Elements: Identification, Diagnostic Potential and Dielectronic Resonances. *Phys. Scr.* **2013**, *T156*, 014082.
58. Ralchenko, Y.; Draganić, I.N.; Osin, D.; Gillaspy, J.D.; Reader, J. Spectroscopy of Diagnostically Important Magnetic-Dipole Lines in Highly Charged $3d^n$ Ions of Tungsten. *Phys. Rev. A* **2011**, *83*, 032517.
59. Ralchenko, Yu. Density Dependence of the Forbidden Lines in Ni-like Tungsten. *J. Phys. B At. Mol. Opt. Phys.* **2007**, *40*, F175–F180.
60. Blagojević, B.; Le Bigot, E.-O.; Fahy, K.; Aguilar, A.; Makonyi, K.; Takács, E.; Tan, J.N.; Pomeory, J.M.; Burnett, J.H.; Gillaspy, J.D.; et al. A High Efficiency Ultrahigh Vacuum Compatible Flat Field Spectrometer for Extreme Ultraviolet Wavelengths. *Rev. Sci. Instrum.* **2005**, *76*, 083102.
61. Holland, G.E.; Boyer, C.N.; Seely, J.F.; Tan, J.N.; Pomeroy, J.M.; Gillaspy, J.D. Low Jitter Metal Vapor Vacuum Arc Ion Source for Electron Beam Ion Trap Injections. *Rev. Sci. Instrum.* **2005**, *76*, 073304.
62. Gillaspy, J.D. Highly Charged Ions. *J. Phys. B At. Mol. Opt. Phys.* **2001**, *34*, R93–R130.
63. Kramida, A.; Brown, C.M.; Feldman, U.; Reader, J. Extension and New Level Optimization of the Ne IV Spectrum. *Phys. Scr.* **2012**, *85*, 025303.
64. Ralchenko, Y.; Kramida, A.; Reader, J.; The NIST ASD Team. NIST Atomic Spectra Database (Version 5), 2011. Available online: http://physics.nist.gov/asd (accessed on 15 January 2017).
65. Hughes, I.G.; Hase, T.P.A. *Measurements and Their Uncertainties*; Oxford University Press: Oxford, UK, 2010.
66. Ralchenko, Y.; Maron, Y. Accelerated Recombination Due to Resonant Deexcitation of Metastable States. *J. Quant. Spectosc. Radiat. Transf.* **2001**, *71*, 609–621.
67. Gu, M.F. The Flexible Atomic Code. *Can. J. Phys.* **2008**, *86*, 675–689.
68. Safronova, U.I.; Cowan, T.E.; Safronova, M.S. Relativistic Many-Body Calculations of Electric-Dipole Lifetimes, Transition Rates and Oscillator Strengths for $2l^{-1}$–$3l'$ States in Ne-like Ions. *J. Phys. B At. Mol. Opt. Phys.* **2005**, *38*, 2741–2763.
69. Kim,Y.-K.; Baik, D.H.; Indelicato, P.; Desclaux, J.P. Resonance Transition Energies of Li-, Na-, and Cu-like Ions. *Phys. Rev. A* **1991**, *44*, 148–166.
70. Kozhedub, Y.S.; Volotka, A.V.; Artemyev, A.N.; Glazov, D.A.; Plunien, G.; Shabaev, V.M.; Tupitsyn, I.I.; Stöhlker, T. Relativistic Recoil, Electron-Correlation, and QED Effects on the $2p_j$–$2s$ Transition Energies in Li-like Ions. *Phys. Rev. A* **2010**, *81*, 042513.
71. Blundell, S.A. Calculations of the Screened Self-Energy and Vacuum Polarization in Li-like, Na-like, and Cu-like Ions. *Phys. Rev. A* **1993**, *47*, 1790–1803.
72. Sapirstein, J.; Cheng, K.T. S-Matrix Calculations of Energy Levels of the Lithium Isoelectronic Sequence. *Phys. Rev. A* **2011**, *83*, 012504.
73. Seely, J.F. QED Contributions to the $2p$–$2s$ Transitions in Highly Charged Li-like Ions. *Phys. Rev. A* **1989**, *39*, 3682–3685.
74. Verdebout, S.; Nazé, C.; Jönsson, P.; Rynkun, P.; Godefroid, M.; Gaigalas, G. Hyperfine Structures and Landé g_J- Factors for $n = 2$ States in Beryllium-, Boron-, Carbon-, and Nitrogen-like Ions from Relativisitc Configuration Interaction Calculations. *At. Data Nucl. Data Tables* **2014**, *100*, 1111–1155.
75. Huang, K.-N.; Kim, Y.-K.; Cheng, K.T.; Desclaux, J.P. Correlation and Relativistic Effects in Spin-Orbit Splitting. *Phys. Rev. Lett.* **1982**, *48*, 1245.
76. Liu, H.; Jiang, G.; Hu, F.; Wang, C.-K.; Wang, Z.-B.; Yang, J.-M. Intercombination Transitions of the Carbon-like Isoelectronic Sequence. *Chin. Phys. B* **2013**, *22*, 073202.
77. Behring, W.E.; Brown, C.M.; Feldman, U.; Seely, J.F.; Reader, J.; Richardson, M.C. Transitions of the Type $2s$–$2p$ in Oxygenlike Y, Zr, and Nb. *J. Opt. Soc. Am. B* **1986**, *3*, 1113–1115.
78. Edlén, B. The 2_p Interval of $2s^2 2p^5$ and $2s^2 2p$. *Opt. Pura Apl.* **1977**, *10*, 123–129.
79. Kaufman, V.; Sugar, J. Forbidden Lines in $ns^2 np^k$ Ground Configurations and $nsnp$ Excited Configurations of Beryllium through Molybdenum Atoms and Ions. *J. Phys. Chem. Ref. Data* **1986**, *15*, 321–426.
80. Curtis, L.J.; Ramanujam, P.S. Ground-State Fine Sturcture for the B and F Isoelectronic Sequence Using the Extended Regular Doublet Law. *Phys. Rev. A* **1982**, *26*, 3672–3675.

81. Nilsen, J.; Scofield, J.H. Wavelengths of Neon-like $3p \rightarrow 3s$ X-ray Laser Transitions. *Phys. Scr.* **1994**, *49*, 588–591.
82. Shimkaveg, G.M.; Carter, M.R.; Walling, R.S.; Ticehurst, J.M.; Mrowka, S.; Trebes, J.E.; MacGowan, B.J.; Dasilva, L.B.; Matthews, D.L.; London, R.A.; et al. X-ray Laser Coherence Experiments in Neon-like Yttrium. In Proceedings of the International Conference on Lasers '91, San Diego, CA, USA, 9–13 December 1992; pp. 84–92.
83. Matthews, D.L.; Hagelstein, P.L.; Rosen, M.D.; Eckart, M.J.; Ceglio, N.M.; Hazi, A.U.; Medecki, H.; Macgowan, B.J.; Trebes, J.E.; Whitten, B.L.; et al. Demonstration of a Soft X-Ray Amplifier. *Phys. Rev. Lett.* **1985**, *54*, 110.
84. Cogordan, J.A.; Lunell, S. Energies of $2p^5 3s$, 3p, and 3d Levels of Neon-like Ions from Relativistic MCDF Calculations, $20 \leq Z \leq 54$. *Phys. Scr.* **1985**, 33, 406–411.
85. Biersdorfer, P.; Obst, M.; Safronova, U.I. Radiative Decay Probabilities of the $(2s^2 2p^5_{1/2} 3s_{1/2})_{J=0}$ Level in Neonlike Ions. *Phys. Rev. A* **2011**, *83*, 012514.
86. Fontes, C.J.; Zhang, H.L. Relativistic Distorted-Wave Collision Strengths for $\Delta n = 0$ Transitions in the 67 Li-like, F-like and Na-like Ions with $26 \leq Z \leq 92$. *At. Data Nucl. Data Tables* **2017**, *113*, 293–315.
87. Santana, J.A. Relativistic MR-MP Energy Levels: Low-Lying States in the Mg Isoelectronic Sequence. *At. Data Nucl. Data Tables* **2016**, *111–112*, 87–186.
88. Safronova, U.I.; Jonson, W.R.; Berry, H.G. Excitation Energies and Transition Rates in Magensiumlike Ions. *Phys. Rev. A* **2000**, *61*, 052503.
89. Beiersdorfer, P. Precision Energy-Level Measurements and QED of Highly Charged Ions. *Can. J. Phys.* **2009**, *87*, 9–14.
90. Andersson, M.; Grumer, J.; Brage, T.; Zou, Y.-M.; Hutton, R. Analysis of the Competition Between Forbidden and Hyperfine-Induced Transitions in Ne-like Ions. *Phys. Rev. A* **2016**, *93*, 032506.
91. Beiersdorfer, P.; von Goeler, S.; Bitter, M.; Thorn, D.B. Measurement of the $3D \rightarrow 2p$ Resonance to Intercombination Line-Intensity Ratio in Neonlike Fe XVII, Ge XXIII, and Se XXV. *Phys. Rev. A* **2001**, *64*, 032705.
92. Mauche, C.W.; Liedahl, D.A.; Fournier, K.B. First Application of the Fe XVII $I(17.10\ \text{Å})/I(17.05\ \text{Å})$ Line Ratio to Constrain the Plasma Density of a Cosmic X-Ray Source. *Astrophys. J.* **2001**, *560*, 992–996.
93. Tallents, G.J. The Physics of Soft X-ray Lasers Pumped by Electron Collisions in Laser Plasmas. *J. Phys. D* **2003**, *36*, R259–R276.

atoms

MDPI

Article

Spectrum of Singly Charged Uranium (U II) : Theoretical Interpretation of Energy Levels, Partition Function and Classified Ultraviolet Lines

Ali Meftah [1,2], Mourad Sabri [1], Jean-François Wyart [2,3] and Wan-Ü Lydia Tchang-Brillet [2,4,*]

1 Laboratoire de Physique et Chimie Quantique, Université Mouloud Mammeri, BP 17 RP, 15000 Tizi-Ouzou, Algeria; ali.meftah@obspm.fr (A.M.); mouradsabri48@yahoo.fr (M.S.)
2 LERMA, Observatoire de Paris, PSL Research University, CNRS, F-92195 Meudon, France; jean-francois.wyart@u-psud.fr
3 Laboratoire Aimé Cotton, CNRS UMR9188, Université Paris-Sud, ENS Cachan, Université Paris-Saclay, Bâtiment 505, F-91405 Orsay Cedex, France
4 Sorbonne Université, UPMC Université Paris 06, LERMA, F-75005 Paris, France
* Correspondence: lydia.tchang-brillet@obspm.fr; Tel.: +33-145-077-576

Academic Editor: Joseph Reader
Received: 23 March 2017; Accepted: 16 June 2017; Published: 26 June 2017

Abstract: In an attempt to improve U II analysis, the lowest configurations of both parities have been interpreted by means of the Racah-Slater parametric method, using Cowan codes. In the odd parity, including the ground state, 253 levels of the interacting configurations $5f^37s^2 + 5f^36d7s + 5f^36d^2 + 5f^47p + 5f^5$ are interpreted by 24 free parameters and 64 constrained ones, with a root mean square (*rms*) deviation of 60 cm^{-1}. In the even parity, the four known configurations $5f^47s, 5f^46d, 5f^26d^27s, 5f^26d7s^2$ and the unknown $5f^26d^3$ form a basis for interpreting 125 levels with a *rms* deviation of 84 cm^{-1}. Due to perturbations, the theoretical description of the higher configurations $5f^37s7p + 5f^36d7p$ remains unsatisfactory. The known and predicted levels of U II are used for a determination of the partition function. The parametric study led us to a re-investigation of high resolution ultraviolet spectrum of uranium recorded at the Meudon Observatory in the late eighties, of which the analysis was unachieved. In the course of the present study, a number of 451 lines of U II has been classified in the region 2344 –2955 Å. One new level has been established as $5f^36d7p\ (^4I)^6K(J = 5.5)$ at 39113.98 ±0.1 cm^{-1}.

Keywords: uranium; actinide ions; emission spectrum; energy levels; energy parameters; partition function

1. Introduction

The spectroscopy of uranium is of interest in many respects. Being the element with the highest atomic number (Z = 92) naturally available, the nuclear decay of ^{238}U provides a tool for the evaluation of the age of the Universe [1]. In the astrophysical plasma models, the ionized uranium (U II) transition at 3859.572 Å ($5f^36d7s\ ^6L_{11/2} - 5f^36d7p\ ^6M_{13/2}$) is used for the diagnostics. Not only are specific radiative data for this transition needed, but so are partition functions that depend on energy levels relative to the ground level $5f^37s^2\ ^4I_{9/2}$. Therefore a comprehensive picture of the level scheme in ionized uranium is desired. For U II, as for other complex spectra of heavy elements, the interpretation of the observed emission lines does not allow a complete determination of the energy level scheme. Nevertheless, by application of the Racah-Slater parametric method, the energy parameters adjusted against known experimental energy values E_{exp} should lead to improved predictions of energies E_{th} for the levels left undetermined. The relevance of the methods that were used with success in lower-Z elements [2] was confirmed in the cases of several higher-Z elements, including thorium [3,4], despite the fact that the non-relativistic perturbative model was primarily unsatisfactory for these heavy systems. Due to the limited computer capacities in the early times, the previous calculations on U II [5,6] had to neglect configuration interaction (CI) effects or to use truncated bases of configurations. Therefore, the necessary limitation of core configurations $5f^3$ and $5f^4$ to their lowest LS terms impaired the calculated energies and wave functions for the $5f^3ll'$ and $5f^4l$ levels, consequence of the large spin-orbit interactions and the intermediate coupling conditions.

The critical compilation of energy levels and spectra of actinides published in 1992 by Blaise and Wyart [3] provided preliminary tables of energy levels of both parities in U II. The compilation of U II was based on emission data from Steinhaus et al. [7] and on new FTS measurements by Palmer et al. [8]. In particular, energy values of the lowest levels of configurations involving $5f, 6d, 7s, 7p$ electrons were reported, thus updating the previous estimates by Brewer [9]. The list of experimental energy levels in U II was further extended by Blaise et al. [6]. Although the earlier calculations [5,6] usefully supported the search for energy levels, it is worth taking advantage of the present possibilities of Cowan codes [10,11] implemented on modern computers for improving the interpretation of the level scheme in U II, and more generally in actinides. This is the main purpose of the present work.

On the experimental side, a set of uranium emission spectra in the ultraviolet range (1000–3000 Å) recorded in the late eighties at the Meudon Observatory was available at the beginning of the present work. The original aim of these recordings, involving one of the present authors (JFW), was to support the critical compilation of the U III spectrum in Blaise and Wyart [3] by supplementing the Fourier Transform measurements [12] in the range (2000 Å– 4 μ) with data in the shorter wavelength range. However the spectrograms had been only partly measured and had never been completely analyzed, leaving most of the experimental material unpublished. Only improvements for U III were reported in an EGAS conference [13] and in the compilation[3]. The analysis of the unknown spectrum of U IV was also planned but never initiated. With the recent publication of IR data on uranium [14], these unused ultraviolet data represent an opportunity for a new step in a comprehensive description of ionized uranium emission spectra.

2. Available Experimental Data

The available experimental spectra in the wavelength range 1000–3000 Å were emitted by a vacuum triggered spark source with uranium electrode and recorded on photographic plates using the high-resolution 10.7 m normal incidence vacuum ultraviolet spectrograph of the Meudon Observatory. The spectrograph is equipped with a 3600 lines/mm holographic concave grating, leading to a linear dispersion of 0.26 Å/mm on the plates. At the time of the experiment, only partial measurement of the plates was carried out on a semi-automatic comparator (microdensitometer). Wavelength calibration was insured by external reference lines in a superimposed spectrum from a iron Penning discharge source. In addition to U III lines, the spectrograms contain known lines from U V [15] and U VI [16], and a number of unidentified lines. Among the last ones, many likely belong to the unknown U IV spectrum. Although the discharge conditions were favorable for producing more than doubly charged ions, many sharp lines were present at the long wavelength end, which we presumed to belong to U II. In the present work, more complete measurements have been resumed for the wavelength range 2250–2955 Å, by digitizing the spectral plates using a flatbed scanner. The plates were scanned simultaneously with a precision ruler with markings every 1 mm, allowing interpolation between markings for correction of possible distortions, as described in [17]. Then the positions of lines were determined by superimposing two symmetrical profiles of the line displayed by a "homemade" software that mimics the rotating prism set-up of the comparator [18]. For wavelength calibration, internal standards were preferred. These were chosen among the U III wavelengths from Fourier Transform Spectrometry (FTS) [12] and the U II Ritz wavelengths calculated from level energies determined by FTS [6]. For the wavelength range shorter than 2350 Å, some U V wavelengths [15] were used. The uncertainty of the wavelength measurements varies between ± 0.001 and ± 0.003 Å.

Figure 1 shows a section of the triggered spark spectrum between 2863–2875Å. The shape of the lines, from relatively sharp (for U II) to hazy (for U IV) may be attributed to Doppler broadening as higher charged ions are produced in hotter part of the sparks. Identified U II lines are numbered. The numbers and corresponding wavelengths can be found in Table 9.

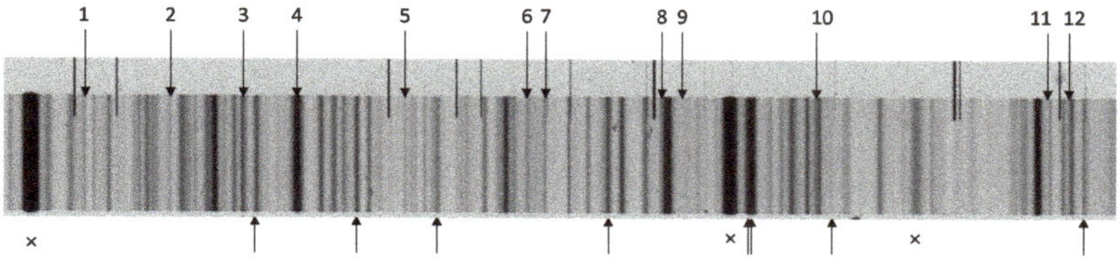

Figure 1. Section of a vacuum triggered spark spectrum (2863–2875 Å). Downward arrows: the U II lines identified in the present work, their numbers and corresponding wavelengths can be found in Table 9; Upward arrows: U III lines from [12]; x: Unidentified lines, likely from U IV. The superimposed iron spectrum is visible above the uranium spectrum but not used in the present work.

3. Theoretical Interpretation of the Energy Levels

Figure 2 shows a diagram of U II configurations of both parities included in the present study. The energy levels spread as predicted by ab initio calculations in the Relativistic Hartree-Fock mode (Cf text below).

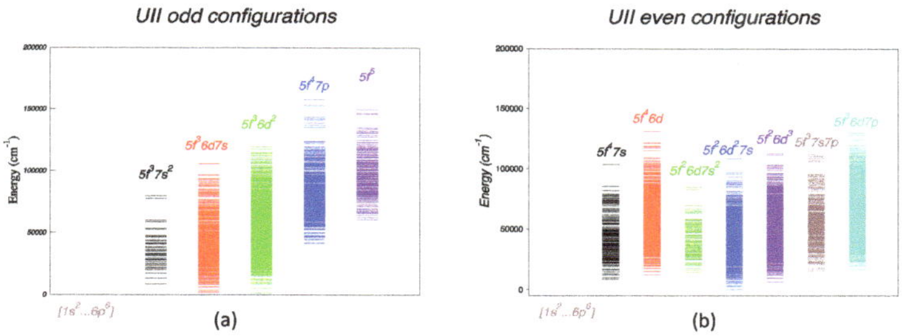

Figure 2. Energy levels of U II belonging to the configurations included in the present study as predicted by ab initio HFR calculations. (**a**) Odd parity configurations. (**b**) Even parity configurations.

We started our work from the energy values tabulated in the final publication on U II [6]. Table 1 recalls the account of various observables available in [6] used for checking the validity of theoretical calculations. The lowest levels of configurations with $5f, 6d, 7s$ and $7p$ electrons, as determined in [6], may be supplemented by predictions for unknown configurations given in [3] according to Brewer's work [9]. This guided our choice of the bases of interacting electronic configurations.

Table 1. Summary of experimental data available for the parametric interpretation of U II levels. N_{tot}: total number of levels; N_{ZE}: number of levels with Landé factor measured by Zeeman effect, N_{IS}: number of levels with measured isotope shift; N_{ident}: number of levels with empirical identification. The number of classified lines per level was not given in [6].

	Odd Parity Levels	Even Parity Levels
N_{tot}	354	809
N_{ZE}	137	355
N_{IS}	114	401
N_{ident}	109	113

3.1. Odd Parity Levels

The present work benefited from the similarities between lanthanides and actinides elements. In the periodic table of elements, neodymium occupies the same position in the lanthanide row as does uranium in the actinide row. Similarities between the two spectra do exist, although configurations built on the $5f^3$ core are much lower relative to the core $5f^4$ in U II than are those built on $4f^3$ relative to $4f^4$ in Nd II.

In Nd II, the basis set of overlapping odd configurations $4f^35d6s + 4f^35d^2 + 4f^36s^2 + 4f^46p + 4f^5$ was adopted for the parametric interpretation of 596 energy levels of these configurations [19,20] by means of the Cowan's codes [10,11]. It had been proven to be adequate by a *root mean square* (*rms*) deviation as small as 53 cm^{-1}. Since in U II the corresponding configurations (with principal quantum numbers increased by one) are also the lowest ones in the parity, we used the same basis set for resuming the calculation of odd parity levels, including the lowest odd parity levels listed in [6]. However, in the case of U II, the $5f^5$ configuration is unknown but is involved by electrostatic interaction with the $5f^36d^2$ configuration. The next higher unknown configuration $5f^26d7s7p$ was not included in the basis. Since its lowest level is expected at 38000 $\pm$ 5000 cm^{-1} above the ground state, according to Brewer's estimates [9], it should not overlap the other odd levels included in the calculation, although CI repulsion effects could be present.

In the first step of the calculation, the *RCN* and *RCN*2 codes were used in the Relativistic Hartree-Fock (*HFR*) mode. The electrostatic and spin-orbit radial integrals were then scaled with factors obtained as averages from earlier actinide calculations [4] and helped for generating the set of parameters for the first diagonalization by the *RCG* code. Appropriate corrections on the average energy E_{av} parameters were made for establishing a fair correspondence between calculated and experimental energies and Landé factors [6] for the low levels of f^3ds, f^3d^2 and f^3s^2. Then the iterative Least Squares Fit (LSF) of the energy parameters was performed by means of the *RCE* code, minimizing the *rms* deviation $\Delta E = \sqrt{\sum_i (E_i^{exp} - E_i^{th})^2/(N_i - N_p)}$, where N_i and N_p are respectively the number of experimental energies and the number of free parameters, with a few dozens of experimental energies to start. Constraints on the parameters were applied for preventing uncontrolled divergence problems. Step by step, the number of levels in the fit was increased up to 253 with a final *rms* deviation of 60 cm^{-1}. Final values for 22 free parameters and 64 constrained ones are reported in Table 2. The electrostatic and spin-orbit integrals are listed with their fitted values and their *HFR* values from which the scaling factors $SF(P) = P_{fit}/P_{HFR}$ are derived. In addition to the explicit CI effects, second order CI effects of distant configurations have been taken into account by using effective parameters. These are α, β and γ for the $5f^n$ core configurations and *Slater forbidden* parameters for $5f^nnl$ (enabled by a specific option of the *RCG* code). Their initial values were chosen semi-empirically by comparison with earlier works [4].

The comparison of experimental and calculated levels is given in Table 3, which is ordered by increasing theoretical energies E_{th}. One may notice that leading *LS* components of eigenfunctions often represent a small part of the total wave functions. As an example, the eigenfunction of the level at E_{exp} = 8379.697 cm^{-1} has three leading components representing respectively only 14, 11 and 10 percent of the total wavefunction. However, the calculation seems correct, as shown by the small deviations for both energies and Landé factors: E_{exp}–E_{th} = 22 cm^{-1} and g_{exp}– g_{th} = 0.002. At higher energies, in the bulk of calculated levels, it occurs that leading *LS* components become as small as 3% only. Considering that the configuration sharing is more meaningful than tiny term components, we have summed the squared amplitudes in the wave functions separately for the 5 configurations. The dominant configuration and its percentage are respectively reported in the last two columns of the table.

Below 20,000 cm^{-1}, it was possible to establish a reliable correspondence between experimental and theoretical energies for the LSF procedure, which was generally supported by agreement between g_{th} and g_{exp} Landé factors, when availabie [6]. However, a few exceptions have been observed. As an example, the two J = 7.5 levels at 8394.362 and 8521.922 cm^{-1} were previously [6] assigned respectively as $f^3d^2\ ^6M_{15/2}$ and $f^3ds\ ^6K_{15/2}$, based on the empirical identification of Landé factors and isotope shifts. In the present work, the initial HFR step predicted these two levels in an inverted order of energies at respectively 8536 and 8438 cm^{-1}, with strongly mixed eigenfunctions. Furthermore, the LSF following this inverted order led to smaller deviations, although physically unsatisfactory. Therefore this order has been adopted in Table 3. Above 20,000 cm^{-1} Landé factors are missing for a majority of levels and the quantum number J is reported as ambiguous for some of them. Since the correspondence between calculated and experimental energies becomes uncertain as energy increases, identifications above 25,726 cm^{-1} are not reported here.

At an intermediate step of the parametric fitting, the previously assigned J value, $J = 11/2$, of the level E_{exp} = 13,695.737 cm^{-1} raised questions. It was found that on one hand, no other level with $J = 11/2$ was predicted between the two experimental levels at 13,270.612 and 13,961.850 cm^{-1}, and on the other hand, one $J = 9/2$ level was missing between E_{exp} = 13,450.490 and 14,265.976 cm^{-1}. The possibility for correcting the J-value for E_{exp} = 13,695.737 was thus examined. Indeed, while a $J = 11/2$ attribution can be justified by only one unique transition with a $J = 13/2$ even level, a $J = 9/2$ value is supported by 16 lines of [7] and by two unidentified lines from the infrared line list of [14] that fit transitions with $J = 7/2$ even levels. Table 4 collects the transitions supporting the present assignation of a $J = 9/2$ for this level. Similar ambiguity for some other levels led us to be cautious and to avoid inclusion of too many E_{exp} values in the LSF fitting process with no other reason but a small ΔE value.

Table 2. Fitted parameters (in cm^{-1}) for odd parity configurations of U II compared with HFR radial integrals. The scaling factors $SF(P) = P_{fit}/P_{HFR}$ (dimensionless) are replaced by $\Delta E = E_{fit} - E_{HFR}$ for E_{av} average energies (in cm^{-1}). Constraints on some parameters are in the second column under 'Cstr.' (denoted 'f' for fixed parameters or 'rn', which link parameters of the same 'rn' to vary in a constant ratio). The *HFR* values of E_{av} parameters are relative to the ground state configuration $5f^37s^2$ taken as zero value.

		$5f^37s^2$				$5f^36d7s$				$5f^36d^2$			
Param. *P*	**Cstr.**	P_{fit}	**Unc.**	P_{HFR}	$\Delta E/SF$	P_{fit}	**Unc.**	P_{HFR}	$\Delta E/SF$	P_{fit}	**Unc.**	P_{HFR}	$\Delta E/SF$
E_{av}		22711	54	0	22711	30141	35	4716	25425	41875	29	15459	26916
$F^2(ff)$	r1	47923	244	70159	0.683	47163	241	69047	0.683	46426	237	67969	0.683
$F^4(ff)$	r2	31974	437	45448	0.704	31411	429	44648	0.704	30868	422	43877	0.704
$F^6(ff)$	r3	21971	563	33210	0.662	21568	552	32601	0.662	21180	542	32015	0.662
α	r4	36.3	1			36.3	1			36.3	1		
β	f	−600				−600				−600			
γ	f	1500				1500				1500			
$F^2(dd)$										23885	401	33922	0.704
$F^4(dd)$										12305	782	22307	0.551
${\alpha_d}^2$	f									10			
ζ_f	r5	1732.4	4	1868.5	0.927	1705.7	4	1839.8	0.927	1681.2	4	1813.4	0.927
ζ_d	r6					1531.8	14	1793.1	0.854	1410.4	13	1650.9	0.854
$F^1(fd)$	r7					880	202			880	202		
$F^2(fd)$	r8					19605	205	28026	0.700	18724	196	26766	0.700
$F^4(fd)$	r9					13940	363	15151	0.920	13251	345	14404	0.920
$G^1(fd)$	r10					11304	81	18156	0.623	10995	78	17660	0.623
$G^2(fd)$	r11					833	295			833	295		
$G^3(fd)$	r12					11699	227	13054	0.896	11236	218	12535	0.896
$G^4(fd)$	r13					2858	355			2858	355		
$G^5(fd)$	r14					8054	353	9682	0.832	7696	337	9250	0.832
$G^3(fs)$						2583	88	4198	0.615				
$G^2(ds)$						13609	247	21081	0.646				

Table 2. *Cont.*

	$5f^47p$					$5f^5$				
Param. P	**Cstr.**	P_{fit}	**Unc.**	P_{HFR}	$\Delta E/SF$	**Cstr.**	P_{fit}	**Unc.**	P_{HFR}	$\Delta E/SF$
E_{av}		60158	196	42943	17215	f	61000		58456	2544
$F^2(ff)$	r1	43973	224	64376	0.683	f	38003		55478	0.686
$F^4(ff)$	r2	29072	397	41323	0.704	f	24513		35118	0.691
$F^6(ff)$	r3	19894	509	30071	0.662	f	17151		25410	0.675
α	r4	37.7	1				36.5	1		
β	f	−600					−600			
γ	f	1500					1500			
ζ_f	r5	1550.4	4	1672.2	0.927		1333.5	3	1437.9	0.927
ζ_p	f	3325.8	209	3293.7	1.01					
$F^1(fp)$	f	100								
$F^2(fp)$	f	7215		8016	0.900					
$G^2(fp)$	f	2058		2058	1.000					
$G^4(fp)$	f	1800		1800	1.000					
Configuration Interaction										
$5f^37s^2 - 5f^36d7s$										
$R^2(fs,fd)$	r15	−6614	71	−10104	0.65					
$R^3(fs,df)$	r15	−1442	15	−2204	0.65					
$5f^3s^2 - 5f^3d^2$										
$R^2(ss,dd)$	r15	14670	157	22412	0.65					
$5f^37s^2 - 5f^5$										
$R^3(ss,ff)$	r15	4516	48	6900	0.65					
$5f^36d7s - 5f^36d^2$										
$R^2(fs,fd)$	r15	−6605	71	−9955	0.65					
$R^3(fs,df)$	r15	−1500	16	−2284	0.65					
$R^2(ds,dd)$	r15	−16091	172	−24584	0.65					
$5f^36d7s - 5f^47p$										
$R^1(ds,fp)$	r15	−9384	101	−14337	0.65					
$R^3(ds,pf)$	r15	−2805	30	−4286	0.65					
$5f^36d7s - 5f^5$										
$R^3(ds,ff)$	r15	−3507	38	−5357	0.65					
$5f^36d^2 - 5f^47p$										
$R^1(dd,fp)$	r15	5580	60	8526	0.66					
$R^3(dd,fp)$	r15	2490	27	3805	0.66					
$5f^36d^2 - 5f^5$										
$R^1(dd,ff)$	r15	15269	164	23326	0.66					
$R^3(dd,ff)$	r15	10171	109	15537	0.66					
$R^5(dd,ff)$	r15	7325	78	11191	0.66					
$5f^47p - 5f^5$										
$R^2(fp,ff)$	r15	−2887	31	−4410	0.66					
$R^4(fp,ff)$	r15	−2501	27	−3821	0.66					

Table 3. Energy levels of U II, odd parity. Comparison of experimental energies and Landé factors with values calculated from the parameter set of Table 2. $\Delta E = E^{exp} - E^{th}$. The percentage of the leading term (notations from Cowan codes) and its configuration are specified by columns 7–9. The dominant configuration and its percentage are reported in the last two columns.

J	E^{exp} (cm^{-1})	E^{th} (cm^{-1})	ΔE (cm^{-1})	g_L^{th}	g_L^{exp}	% 1st *comp*	*Conf*	*Term*	*Main conf*	%
4.5	0.000	−56.8	56	0.756	0.765	77	f3s2	(4I)4I	f3s2	91.7
5.5	289.041	224.4	64	0.656	0.655	77	f3ds	(4I)6L	f3ds	99.9
4.5	914.765	930.2	−15	0.604	0.605	71	f3ds	(4I)6K	f3ds	96.2
6.5	1749.123	1715.5	33	0.864	0.865	45	f3ds	(4I)6L	f3ds	93.8
5.5	2294.696	2320.7	−26	0.868	0.865	47	f3ds	(4I)6K	f3ds	97.5
5.5	4420.871	4406.4	14	0.971	0.97	89	f3s2	(4I)4I	f3s2	93.6
6.5	4585.434	4577.9	7	0.793	0.785	28	f3d2	(4I)6M	f3d2	51.8
2.5	4706.273	4674.8	31	0.477	0.480	33	f3ds	(4I)6H	f3ds	96.8
7.5	5259.653	5247.0	12	1.007	1.015	68	f3ds	(4I)6L	f3ds	97.6
3.5	5401.503	5352.6	48	0.767	0.690	26	f3ds	(4I)6I	f3ds	98.0
6.5	5526.750	5549.1	−22	1.019	1.020	70	f3ds	(4I)6K	f3ds	98.3
3.5	5667.331	5695.5	−28	0.657	0.735	51	f3ds	(4I)6I	f3ds	96.9
5.5	5790.641	5827.3	−36	0.851	0.860	39	f3ds	(4I)6K	f3ds	95.5
6.5	6283.431	6392.9	−109	0.785	0.790	39	f3d2	(4I)6M	f3ds	54.5
4.5	6445.035	6471.2	−26	0.835	0.840	43	f3ds	(4I)6I	f3ds	97.6
0.5		6999.4		2.398		20	f3ds	(4F)4Pa	f3ds	97.8
1.5	7017.172	7096.0	−78	0.612	0.620	58	f3s2	(4F)4F	f3s2	90.8
4.5	7166.632	7259.7	−93	0.945	0.940	20	f3ds	(4I)6H	f3ds	94.9
5.5	7598.353	7626.3	−27	0.971	0.980	18	f3ds	(4I)4Ia	f3ds	98.0
3.5	7547.374	7629.2	−81	0.802	0.790	21	f3ds	(4I)4Ha	f3ds	84.8
6.5	8276.733	8248.0	28	1.093	1.090	84	f3s2	(4I)4I	f3s2	89.9
4.5	8379.697	8357.5	22	0.841	0.840	14	f3ds	(4I)6I	f3ds	76.1
1.5	8400.125	8426.1	−25	0.086	0.150	68	f3ds	(4I)6G	f3ds	97.7
2.5	8430.185	8432.6	−2	0.719	0.720	38	f3ds	(4I)6G	f3ds	94.3
7.5	8394.362	8437.9	−43	1.052	0.960	55	f3ds	(4I)6K	f3ds	74.2
5.5	8510.866	8446.9	63	0.854	0.860	11	f3ds	(4I)4Kb	f3ds	78.7
7.5	8521.922	8535.6	−13	0.968	1.060	41	f3d2	(4I)6M	f3d2	57.1
6.5	8755.640	8767.9	−12	1.042	1.040	14	f3ds	(4I)4Lb	f3ds	92.2
8.5	8853.748	8815.5	38	1.105	1.105	83	f3ds	(4I)6L	f3ds	98.6
3.5	9075.732	9020.5	55	0.873	0.870	15	f3ds	(4I)6H	f3ds	68.6
2.5	9344.625	9250.9	93	0.751	0.79	25	f3s2	(4G)4G	f3s2	47.1
4.5	9241.971	9254.9	−12	1.023	1.015	12	f3ds	(4I)6H	f3ds	77.6
5.5	9553.187	9584.8	−31	1.053	1.060	56	f3ds	(4I)6I	f3ds	96.4
6.5	9626.113	9637.3	−11	0.946	0.950	39	f3ds	(4I)4Kb	f3ds	83.9
4.5	9690.665	9707.7	−17	0.991	0.995	10	f3s2	(2H)2H2	f3ds	60.9
1.5	9881.618	9911.4	−29	0.272		51	f3ds	(4F)6G	f3ds	98.3
3.5	9933.226	9916.5	16	0.823	0.82	27	f3ds	(4I)4Hb	f3ds	88.1
4.5	9882.726	9967.6	−84	0.878	0.875	16	f3s2	(4I)4Ib	f3ds	43.9
2.5	10285.072	10178.1	106	0.454	0.42	35	f3ds	(4F)6H	f3ds	93.1
7.5	10198.312	10250.6	−52	0.968	0.960	44	f3ds	(4I)4Lb	f3ds	79.5
2.5	10366.253	10437.7	−71	0.922		58	f3s2	(4F)4F	f3s2	68.1
3.5	10444.432	10437.7	6	0.878	0.865	12	f3ds	(4F)4Hb	f3ds	74.7
1.5	-	10643.1		1.731		30	f3ds	(4F)6D	f3ds	97.8
5.5	10740.958	10688.6	52	0.690	0.685	68	f3d2	(4I)6L	f3d2	84.1
2.5	10732.087	10867.1	−135	0.953		29	f3ds	(4F)6G	f3ds	92.0
2.5	11350.714	11227.9	122	1.254		17	f3ds	(4F)6D	f3ds	95.3
1.5	-	11230.7		1.617		58	f3s2	(4S)4S	f3s2	91.6
3.5	11363.537	11289.2	74	1.033		13	f3d2	(4I)6Ia	f3ds	69.2
8.5	11382.321	11330.6	51	1.179	1.185	75	f3ds	(4I)6K	f3ds	96.4
4.5	11544.672	11426.0	118	0.673	0.690	45	f3d2	(4I)6Ka	f3d2	61.1
3.5	-	11571.9		0.994		23	f3s2	(4F)4F	f3s2	52.5
7.5	11708.483	11664.6	43	1.175	1.175	77	f3s2	(4I)4I	f3s2	92.2

Table 3. *Cont.*

J	E^{exp} (cm^{-1})	E^{th} (cm^{-1})	ΔE (cm^{-1})	g_L^{th}	g_L^{exp}	% 1st*comp*	*Conf*	*Term*	*Main conf*	%
3.5	11707.835	11743.7	−35	0.781	0.705	37	f3ds	(4G)6I	f3ds	60.0
5.5	11784.953	11809.2	−24	1.097		36	f3ds	(4I)6H	f3ds	97.0
6.5	11813.450	11833.5	−20	1.008	1.09	26	f3ds	(4I)6I	f3ds	81.0
6.5	11787.315	11841.2	−53	1.030	0.940	37	f3ds	(4I)6I	f3ds	83.9
4.5	11797.343	11854.3	−56	1.005		16	f3ds	(4I)6G	f3ds	96.3
0.5	-	11902.1		1.361		22	f3ds	(4F)2P	f3ds	95.6
2.5	12112.402	12095.7	16	0.957		10	f3ds	(4F)6G	f3ds	94.6
4.5	12055.788	12117.5	−61	0.976		18	f3ds	(4G)6I	f3ds	93.8
8.5	12033.378	12121.2	−87	1.014	1.005	78	f3d2	(4I)6M	f3d2	90.9
3.5	12092.319	12161.5	−69	0.823		22	f3ds	(4F)6H	f3ds	86.6
9.5	12350.555	12255.1	95	1.176	1.200	89	f3ds	(4I)6L	f3ds	99.0
1.5	-	12392.5		0.562		19	f3ds	(4G)4Fa	f3ds	90.4
5.5	12530.613	12519.6	10	0.987	1.000	12	f3d2	(4I)6Ka	f3ds	74.9
6.5	12629.355	12578.1	51	1.044	1.095	23	f3d2	(4I)6L	f3ds	65.8
3.5	12627.826	12582.6	45	0.965		24	f3s2	(4G)4G	f3ds	49.3
7.5	12660.559	12591.2	69	1.072	1.015	8	f3ds	(4I)4Lb	f3ds	88.1
5.5	12638.060	12690.0	−51	0.997		14	f3ds	(4I)4Ib	f3ds	78.1
2.5	-	12691.9		0.713		33	f3s2	(4G)4G	f3ds	47.4
4.5	12687.308	12718.0	−30	0.916		13	f3d2	(4I)6Ia	f3d2	49.9
1.5	-	12891.3		0.905		8	f3ds	(4F)4Db	f3ds	78.5
7.5	13015.838	12958.6	57	1.070	1.125	30	f3ds	(4I)4Kb	f3ds	96.0
3.5	13183.793	13117.1	66	0.982		22	f3ds	(4F)6G	f3ds	70.5
2.5	-	13125.8		0.997		20	f3ds	(4F)6G	f3ds	83.5
5.5	13089.590	13133.0	−43	0.941	0.940	34	f3d2	(4I)6Ka	f3d2	53.3
4.5	13275.365	13247.5	27	0.956		15	f3ds	(4I)6H	f3ds	78.9
5.5	13270.612	13272.1	−1	1.134		20	f3ds	(4I)6G	f3ds	92.7
6.5	13344.198	13275.1	69	0.905	0.910	35	f3d2	(4I)6L	f3d2	60.6
3.5	13450.362	13423.3	27	0.979		19	f3ds	(4I)6G	f3ds	80.0
4.5	13450.490	13449.0	1	1.014		15	f3ds	(4G)6I	f3ds	69.0
7.5	13503.319	13524.3	−20	1.101	1.13	19	f3ds	(4I)6I	f3ds	87.2
4.5	13695.737	13683.4	12	1.058		21	f3ds	(4F)6H	f3ds	85.3
3.5	13733.500	13730.8	2	0.934		22	f3ds	(4F)6G	f3ds	64.9
2.5	-	13751.0		0.638		17	f3d2	(4I)6Ha	f3ds	61.8
0.5	-	13817.0		0.009		51	f3ds	(4F)6F	f3ds	94.3
6.5	13975.278	13932.4	42	1.021		19	f3ds	(4I)2K	f3ds	81.3
2.5	13967.812	13944.8	23	0.734		18	f3ds	(4I)4Ga	f3ds	87.9
5.5	13961.850	13980.2	−18	0.972		14	f3ds	(4I)4Ib	f3ds	63.9
3.5	14107.329	14042.4	64	0.829		16	f3ds	(4F)4Hb	f3ds	84.4
9.5	-	14049.9		1.232		72	f3ds	(4I)6K	f3ds	97.9
1.5	-	14142.4		1.176		34	f3ds	(4F)6F	f3ds	88.8
8.5	14177.723	14148.0	29	1.072	1.085	60	f3ds	(4I)4Lb	f3ds	86.1
4.5	14265.976	14230.0	35	1.110		17	f3ds	(4I)4Hb	f3ds	81.3
4.5	14366.889	14399.2	−32	1.078		8	f3ds	(4I)6H	f3ds	90.1
1.5	-	14477.7		1.043		14	f3ds	(4S)6D	f3ds	65.8
6.5	-	14496.9		1.142		16	f3ds	(4I)4Ia	f3ds	96.8
6.5	14599.600	14557.6	42	0.934		17	f3d2	(4I)6L	f3d2	61.8
4.5	14654.181	14628.3	25	1.057		12	f3ds	(4F)6G	f3ds	75.8
2.5	-	14660.8		0.777		17	f3ds	(4F)4Gb	f3ds	71.4
7.5	14724.776	14738.1	−13	1.170	1.170	34	f3ds	(4I)6I	f3ds	96.5
5.5	14709.266	14771.1	−61	0.925	0.920	10	f3ds	(4I)4Ka	f3ds	49.4
3.5	14900.134	14873.4	26	1.010		7	f3ds	(4G)6G	f3ds	84.7
0.5	-	14905.3		1.636		32	f3ds	(4S)6D	f3ds	92.3
4.5	-	14981.0		1.177		44	f3s2	(4F)4F	f3s2	71.3
6.5	14991.377	14993.8	−2	1.180		17	f3ds	(4F)6H	f3ds	88.6
2.5	-	15013.4		1.198		13	f3ds	(4F)6P	f3ds	89.2

Table 3. *Cont.*

J	E^{exp} (cm^{-1})	E^{th} (cm^{-1})	ΔE (cm^{-1})	g_L^{th}	g_L^{exp}	% 1st comp	*Conf*	*Term*	*Main conf*	%
3.5	15147.878	15080.0	67	0.989		9	f3ds	(4F)4Gb	f3ds	79.0
1.5	-	15205.3		1.007		10	f3ds	(4G)6G	f3ds	66.4
5.5	15234.383	15205.3	29	1.058		7	f3ds	(4F)6H	f3ds	78.7
2.5	-	15280.5		1.147		33	f3ds	(4F)6F	f3ds	71.8
1.5	-	15308.6		1.985		55	f3ds	(4F)6P	f3ds	93.4
5.5	15330.434	15334.3	−3	1.123		14	f3ds	(4I)6G	f3ds	82.6
4.5	15413.346	15397.6	15	0.984		14	f3ds	(4G)6I	f3ds	52.3
0.5	-	15411.9		0.611		25	f3ds	(4S)4Db	f3ds	81.9
3.5	15587.280	15423.3	164	1.031		9	f3ds	(4I)4Ga	f3ds	63.2
5.5	15430.900	15483.0	−52	1.073		16	f3s2	(2H)2H2	f3ds	54.4
9.5	15534.868	15604.8	−69	1.093	1.095	89	f3d2	(4I)6M	f3ds	97.5
7.5	15692.655	15633.9	58	1.005		62	f3d2	(4I)6L	f3d2	92.4
10.5	-	15655.2		1.230		91	f3ds	(4I)6L	f3ds	99.7
8.5	15767.762	15683.0	84	1.200		36	f3ds	(4I)6I	f3ds	96.2
3.5	-	15689.4		1.136		11	f3ds	(4F)6P	f3ds	82.3
6.5	15717.452	15736.5	−19	1.026	1.04	35	f3d2	(4I)6Ka	f3d2	69.0
5.5	15863.755	15811.3	52	1.027		14	f3d2	(4I)6Ia	f3d2	51.9
4.5	15916.166	15879.5	36	0.971		28	f3d2	(4I)6Kb	f3ds	55.2
2.5	-	15883.4		0.628		10	f3d2	(4I)6Ha	f3d2	49.2
6.5	15884.560	15900.5	−15	1.072	1.150	12	f3ds	(4I)4Ka	f3ds	72.3
7.5	16156.487	16066.8	89	1.150		19	f3ds	(4I)6H	f3ds	82.9
2.5	16213.945	16073.4	140	1.010		7	f3ds	(4G)2F	f3ds	84.5
0.5	-	16099.2	0	2.345		32	f3ds	(4F)6D	f3ds	93.9
4.5	16063.244	16101.2	−37	0.889		21	f3d2	(4I)6Kb	f3ds	44.9
5.5	16003.163	16120.1	−116	1.091		20	f3ds	(4G)6I	f3ds	75.8
3.5	-	16140.4		1.052		15	f3ds	(4G)6H	f3ds	75.0
1.5	-	16176.8		1.130		9	f3ds	(4F)4Pa	f3ds	85.5
2.5	16336.514	16281.8	54	1.161		7	f3ds	(4F)6P	f3ds	74.4
3.5	16239.757	16285.9	−46	1.230		14	f3ds	(4F)6P	f3ds	83.5
4.5	16338.719	16290.5	48	1.040		6	f3ds	(4I)4Ia	f3ds	87.9
8.5	-	16419.3		1.098		22	f3ds	(4I)4La	f3ds	78.7
3.5	16376.820	16443.1	−66	1.048		9	f3ds	(4I)4Gb	f3ds	63.7
6.5	16532.589	16466.5	66	1.043		14	f3ds	(4I)2I	f3ds	77.6
1.5	-	16471.3		1.008		12	f3ds	(4F)2D	f3ds	86.2
2.5	16473.747	16481.7	−7	0.952		13	f3ds	(4F)6F	f3ds	65.1
5.5	16397.828	16490.5	−92	1.045		13	f3s2	(2H)2H2	f3ds	51.3
1.5	-	16528.6		0.844		10	f3s2	(4D)4D	f3ds	49.5
4.5	16514.265	16528.9	−14	0.986		16	f3s2	(4G)4G	f3ds	51.5
6.5	16618.369	16650.2	−31	0.987		60	f3s2	(2K)2K	f3s2	68.2
2.5	-	16660.2		1.339		22	f3ds	(4F)6P	f3ds	80.3
3.5	16672.399	16723.4	−51	1.019		7	f3ds	(4G)6G	f3ds	69.2
7.5	16690.212	16731.2	−40	0.986		14	f3ds	(2K)4M	f3ds	51.9
4.5	16819.694	16758.4	61	1.033		11	f3ds	(4F)6G	f3ds	68.2
5.5	16699.409	16782.6	−83	1.114		16	f3ds	(4I)6H	f3ds	78.5
2.5	16838.258	16808.1	30	0.982		10	f3ds	(4G)6H	f3ds	83.1
0.5	-	16830.2		0.230		14	f3d2	(4I)6F	f3d2	71.1
2.5	-	16867.1		1.060		8	f3ds	(4G)6H	f3ds	58.8
3.5	16857.015	16868.5	−11	0.838		10	f3d2	(4F)6I	f3ds	54.8
5.5	16758.024	16899.5	−141	1.026		12	f3ds	(4I)4Ka	f3ds	65.5
8.5	16982.510	16958.9	23	1.143		22	f3ds	(4I)4Kb	f3ds	91.3
4.5	-	16964.3		1.026		11	f3ds	(4F)4Hb	f3ds	64.7
6.5	16990.271	17005.7	−15	1.118		12	f3ds	(4I)4Ib	f3ds	59.7
1.5	-	17006.8		0.859		9	f3d2	(4I)6F	f3ds	55.0
2.5	17008.229	17102.2	−94	1.154		10	f3ds	(4F)6D	f3ds	84.7
3.5	-	17117.2		0.920		8	f3ds	(4G)4Hb	f3ds	60.7

Table 3. *Cont.*

J	E^{exp} (cm^{-1})	E^{th} (cm^{-1})	ΔE (cm^{-1})	g_L^{th}	g_L^{exp}	% 1st *comp*	*Conf*	*Term*	*Main conf*	%
5.5	17205.372	17170.4	34	1.120		28	f3ds	(4F)6G	f3ds	76.8
1.5	-	17192.9		0.957		17	f3ds	(4F)2P	f3ds	82.7
0.5	-	17217.5		0.499		30	f3s2	(2P)2P	f3s2	53.9
4.5	17200.903	17280.6	−79	0.979	0.970	10	f3ds	(4G)4Ib	f3d2	50.5
7.5	17259.216	17301.3	−42	1.118		12	f3ds	(4I)4Ib	f3ds	79.9
2.5	-	17320.0		0.619		18	f3ds	(4G)6H	f3ds	48.2
3.5	17381.930	17363.1	18	1.184		17	f3ds	(4F)6D	f3ds	84.2
1.5	-	17364.5		0.679		7	f3ds	(4F)4Fa	f3ds	77.1
4.5	17388.631	17392.7	−4	1.010		13	f3ds	(4I)4Hb	f3ds	65.2
6.5	17461.883	17423.5	38	1.155		18	f3ds	(4I)6G	f3ds	84.6
3.5	-	17481.4		1.163		15	f3ds	(4S)6D	f3ds	67.6
7.5	17556.776	17523.7	33	1.055		11	f3d2	(4I)6L	f3ds	68.6
1.5	-	17579.4		0.960		15	f3d2	(4I)6F	f3ds	57.4
2.5	17621.175	17607.3	13	0.941		5	f3d2	(4I)6Hb	f3ds	57.1
3.5	17560.922	17619.6	−58	0.991		10	f3ds	(4G)4Gb	f3ds	72.9
5.5	17755.028	17785.9	−30	1.050		11	f3ds	(2H)4K2	f3ds	84.0
6.5	17775.960	17808.2	−32	1.046		15	f3d2	(4I)6Ka	f3d2	57.7
0.5	-	17812.2		0.076		13	f3d2	(4I)6F	f3ds	51.0
3.5	17823.434	17820.9	2	0.950		12	f3d2	(4F)6I	f3d2	51.4
2.5	-	17857.4		0.987		5	f3s2	(4D)4D	f3d2	44.1
5.5	17922.769	17934.8	−12	0.973		11	f3d2	(4I)4Ka	f3d2	65.0
6.5	17888.312	17935.3	−47	1.157		21	f3ds	(4F)6H	f3ds	90.7
3.5	-	17959.5		0.963		12	f3d2	(4I)6Ha	f3ds	48.7
4.5	18009.838	17965.3	44	1.119		12	f3ds	(4F)6F	f3ds	71.6
9.5	-	17995.9		1.146		68	f3ds	(4I)4Lb	f3ds	81.7
8.5	18032.564	18015.8	16	1.044		25	f3d2	(4I)6L	f3d2	92.4
3.5	18041.437	18095.3	−53	0.913		9	f3d2	(4F)6I	f3ds	51.0
2.5	18178.854	18128.3	50	1.074		5	f3s2	(4D)4D	f3ds	54.7
5.5	18102.958	18150.0	−47	1.117		7	f3ds	(4F)6F	f3ds	79.1
4.5	-	18182.7		1.081		8	f3d2	(4I)6Ha	f3ds	70.3
3.5	-	18232.9		0.974		7	f3d2	(4I)6Ha	f3d2	40.4
3.5	18291.412	18326.3	−34	0.893		19	f3d2	(4I)6Ib	f3d2	63.0
1.5	-	18385.9		0.921		8	f3ds	(4F)6F	f3ds	59.8
4.5	-	18424.3		1.061		8	f3ds	(4F)4Hb	f3ds	68.1
0.5	-	18436.4		0.102		30	f3ds	(4G)6F	f3ds	59.7
1.5	-	18451.3		0.966		10	f3ds	(4S)6D	f3ds	54.7
2.5	18594.848	18459.0	135	0.995		4	f3d2	(4I)6Ha	f3d2	49.9
7.5	-	18486.3		1.112		7	f3ds	(4I)4Ia	f3ds	50.0
4.5	18526.283	18503.5	22	0.990		10	f3d2	(4I)6Ia	f3d2	54.0
3.5	-	18505.9		0.955		6	f3d2	(4I)6Ia	f3ds	50.6
7.5	18539.154	18523.8	15	1.098		10	f3ds	(4I)4Ia	f3ds	50.6
6.5	18451.979	18576.6	−124	1.116		40	f3ds	(4G)6I	f3ds	89.1
0.5	-	18623.1		1.461		35	f3ds	(4F)4Pb	f3ds	79.0
1.5	-	18664.9		1.001		11	f3ds	(4F)4Pb	f3ds	57.2
2.5	18852.922	18728.5	124	1.145		5	f3ds	(4S)6D	f3ds	72.2
5.5	18754.949	18740.0	14	1.013		12	f3d2	(4I)6Kb	f3d2	60.4
6.5	18788.827	18792.5	−3	1.042		14	f3d2	(4I)4La	f3d2	67.4
8.5	-	18796.5		1.153		29	f3ds	(2H)4K2	f3ds	90.2
3.5	18796.998	18857.5	−60	0.841		15	f3d2	(4I)6Hb	f3d2	61.4
5.5	18892.289	18858.6	33	0.969		24	f3d2	(4I)6Kb	f3d2	61.0
2.5	-	18913.1		1.008		5	f3d2	(4F)6Ga	f3ds	60.9
7.5	19017.870	18939.8	78	1.161		16	f3ds	(4I)4Ka	f3ds	81.9
10.5	-	18945.7		1.157		92	f3d2	(4I)6M	f3d2	100.
0.5	-	18964.1		0.717		10	f3ds	(4F)4Db	f3ds	72.7
4.5	19004.689	18967.7	36	1.024		5	f3ds	(4F)4Gb	f3ds	65.0
2.5	-	19032.8		0.700		14	f3d2	(4I)6Hb	f3ds	52.2

Table 3. *Cont.*

J	E^{exp} (cm^{-1})	E^{th} (cm^{-1})	ΔE (cm^{-1})	g_L^{th}	g_L^{exp}	% 1st comp	Conf	Term	Main conf	%
1.5	-	19059.2		1.033		16	f3ds	(4G)6F	f3ds	59.4
1.5	-	19112.9		1.073		13	f3ds	(4F)4Fb	f3ds	76.4
4.5	19147.489	19141.1	6	1.179		16	f3ds	(4F)6F	f3ds	70.5
5.5	19129.394	19157.3	−27	1.019		11	f3ds	(4I)4Gb	f3ds	62.9
8.5	19242.364	19215.9	26	1.042		48	f3d2	(4I)6L	f3d2	89.0
4.5	19237.394	19219.6	17	1.174		12	f3s2	(4G)4G	f3ds	66.2
6.5	19276.989	19249.2	27	1.166		8	f3ds	(4I)6G	f3ds	67.6
0.5	-	19274.8		0.018		13	f3ds	(4F)4Da	f3ds	62.7
4.5	19375.292	19300.7	74	1.031		11	f3d2	(4G)6K	f3ds	58.0
3.5	19316.823	19365.5	−48	1.211		16	f3ds	(4F)6P	f3ds	78.0
1.5	-	19372.2		0.983		12	f3d2	(4I)6G	f3ds	49.7
5.5	19473.367	19473.3		0.982		13	f3d2	(4I)6Kb	f3d2	64.7
3.5	-	19490.5		1.086		11	f3ds	(4F)4Fb	f3d2	50.5
7.5	19510.817	19551.1	−40	1.045		8	f3ds	(2K)4K	f3ds	71.7
7.5	-	19592.9		1.049		23	f3s2	(2K)2K	f3s2	47.0
4.5	19627.056	19598.3	28	1.093		9	f3d2	(4G)6K	f3ds	64.9
3.5	19570.868	19600.8	−29	1.017		5	f3ds	(4S)4Db	f3d2	56.0
2.5	19469.535	19630.2	−160	1.085		9	f3ds	(4G)6F	f3ds	68.6
1.5	-	19743.7		0.654		26	f3d2	(4I)6G	f3d2	48.0
7.5	19748.316	19758.5	−10	1.078		20	f3d2	(4I)6Ka	f3d2	46.6
2.5	-	19760.1		1.182		9	f3s2	(2D)2D1	f3ds	59.1
6.5	19694.329	19779.6	−85	1.023		9	f3ds	(2K)4K	f3ds	61.3
5.5	19733.338	19797.2	−63	1.079		13	f3ds	(4G)6H	f3ds	68.9
4.5	19869.609	19841.7	27	1.053		7	f3d2	(4I)6Hb	f3d2	52.8
3.5	-	19846.5		0.967		7	f3ds	(4S)4Db	f3ds	68.2
9.5	-	19865.8		1.141		23	f3ds	(4I)4La	f3ds	81.4
6.5	-	19922.7		1.035		15	f3d2	(4I)6Kb	f3d2	56.3
4.5	20018.685	19953.9	64	0.910		28	f3d2	(4G)6K	f3d2	52.5
5.5	-	19960.6		1.063		16	f3ds	(2K)4K	f3ds	79.3
2.5	-	20001.7		0.970		10	f3d2	(4I)6G	f3d2	51.3
1.5	-	20038.3		0.807		8	f3d2	(4I)6G	f3ds	50.0
3.5	20084.775	20039.4	45	0.978		6	f3ds	(4F)4Gb	f3d2	45.1
0.5	-	20067.8		1.423		16	f3ds	(4G)6D	f3ds	58.5
6.5	20055.098	20097.5	−42	1.129		19	f3ds	(4F)6G	f3ds	66.5
5.5	-	20103.3		1.080		11	f3ds	(4G)6G	f3ds	73.9
3.5	20263.434	20107.4	156	1.038		11	f3d2	(4F)6H	f3d2	54.8
7.5	20148.474	20124.4	24	0.985		18	f3d2	(4I)4Ma	f3d2	51.4
8.5	-	20130.9		1.093		26	f3d2	(4I)6Ka	f3d2	83.3
4.5	20033.114	20149.0	−115	1.128		13	f3d2	(4I)6Ha	f3ds	47.2
2.5	-	20196.5		1.138		8	f3ds	(4G)6F	f3ds	49.0
1.5	-	20202.8		1.050		10	f3ds	(4D)6F	f3ds	67.7
6.5	-	20217.0		1.067		9	f3d2	(4I)4Lb	f3ds	66.3
5.5	-	20249.0		1.138		16	f3s2	(2H)2H1	f3ds	47.4
0.5	-	20290.5		0.067		17	f3ds	(4G)6F	f3ds	58.5
4.5	20340.780	20306.0	34	1.083		13	f3d2	(4F)6I	f3d2	51.7
2.5	-	20356.6		1.013		16	f3s2	(2D)2D1	f3ds	59.5
4.5	-	20458.9		0.949		7	f3ds	(2H)4I2	f3ds	52.5
6.5	-	20499.6		1.108		23	f3ds	(4I)4Hb	f3ds	86.7
3.5	20514.235	20526.7	−12	1.030		9	f3ds	(4G)6H	f3ds	68.5
4.5	20500.810	20528.4	−27	1.006		8	f3d2	(4F)6I	f3ds	61.3
2.5	-	20537.7		1.034		6	f3ds	(2G)4G1	f3ds	54.5
5.5	-	20541.7		1.078		9	f3ds	(4G)4Ib	f3ds	51.7
3.5	-	20595.8		1.055		7	f3ds	(4F)6F	f3ds	58.1
1.5	-	20632.0		1.171		7	f3ds	(4G)6F	f3ds	64.4
5.5	-	20634.6		1.069		8	f3ds	(2H)4I2	f3ds	67.9
4.5	-	20658.4		1.023		11	f3d2	(4I)6Hb	f3ds	49.6

Table 3. *Cont.*

J	E^{exp} (cm^{-1})	E^{th} (cm^{-1})	ΔE (cm^{-1})	g_L^{th}	g_L^{exp}	% 1st *comp*	*Conf*	*Term*	*Main conf*	%
7.5	-	20663.4		1.098		21	f3ds	(4G)6I	f3ds	71.6
8.5	-	20670.8		1.068		14	f3ds	(2K)4M	f3d2	50.5
2.5	-	20704.1		1.122		19	f3ds	(4G)6F	f3ds	65.7
7.5	-	20771.0		1.031		11	f3d2	(4I)4Ma	f3ds	52.3
6.5	-	20779.4		1.008		10	f3ds	(2K)4L	f3ds	61.0
3.5	-	20780.2		1.054		7	f3d2	(4F)6H	f3d2	52.7
0.5	-	20796.7		1.267		16	f3ds	(4G)6D	f3ds	65.5
4.5	20899.429	20808.9	90	1.065		8	f3ds	(4G)6G	f3ds	61.3
2.5	-	20851.8		0.970		26	f3s2	(2D)2D1	f3s2	33.0
5.5	-	20951.3		1.110		15	f3d2	(4I)6Ib	f3d2	55.1
1.5	-	20961.9		0.786		8	f3ds	(2P)4F	f3ds	58.5
3.5	20890.426	20965.3	−74	1.130		9	f3ds	(4F)4Fb	f3ds	50.1
5.5	-	21022.3		1.084		7	f3ds	(2H> 2I2	f3ds	59.9
7.5	20946.239	21023.1	−76	1.008		23	f3d2	(4I)4Mb	f3d2	79.4
4.5	21103.432	21118.7	−15	1.078		7	f3d2	(4F)6H	f3ds	50.6
1.5	-	21159.6		0.906		11	f3ds	(4G)6F	f3ds	54.0
3.5	21107.881	21165.2	−57	0.870		14	f3d2	(4G)6I	f3d2	49.3
7.5	-	21215.3		1.111		13	f3ds	(4F)6H	f3ds	84.4
4.5	-	21232.1		1.062		14	f3ds	(4G)4Hb	f3ds	55.3
6.5	-	21247.7		1.005		6	f3ds	(2H)2K2	f3d2	52.8
2.5	-	21256.4		1.004		6	f3d2	(4I)6G	f3d2	51.8
1.5	-	21288.7		1.046		5	f3ds	(4S)6D	f3ds	71.5
4.5	21387.040	21309.6	77	0.940		9	f3d2	(4G)6I	f3d2	74.1
5.5	-	21309.5		1.129		13	f3ds	(4F)6F	f3ds	62.0
2.5	-	21350.9		1.045		10	f3ds	(4G)6F	f3ds	57.4
0.5	-	21372.9		1.981		13	f3ds	(4D)6D	f3ds	64.7
3.5	-	21415.6		1.018		9	f3ds	(4F)2G	f3ds	70.7
5.5	-	21494.8		1.104		4	f3ds	(4F)6G	f3ds	66.0
1.5	-	21502.1		0.977		6	f3ds	(4G)4Fb	f3ds	64.9
8.5	-	21531.9		1.072		12	f3ds	(2K)4M	f3d2	58.7
2.5	-	21553.7		1.045		7	f3ds	(4D)6F	f3ds	65.9
8.5	-	21582.8		1.058		24	f3ds	(4I)4Kb	f3ds	78.3
6.5	21645.939	21597.3	48	1.152		12	f3d2	(4I)6Ha	f3ds	51.0
0.5	-	21604.6		0.518		11	f3s2	(2P)2P	f3ds	44.8
9.5	-	21618.8		1.159		73	f3d2	(4I)6L	f3d2	99.9
3.5	-	21651.4		1.144		11	f3ds	(4G)4Gb	f3ds	67.0
4.5	-	21656.8		1.177		21	f3ds	(2H)4F2	f3ds	82.7
5.5	-	21662.8		1.075		6	f3d2	(4I)6Hb	f3d2	65.7
2.5	-	21718.6		1.114		7	f3d2	(4F)6Ga	f3ds	51.7
0.5	-	21721.7		0.048		14	f3d2	(4S)6F	f3d2	58.4
4.5	21793.334	21750.4	42	1.067		10	f3d2	(4F)6Ga	f3d2	65.1
7.5	-	21768.7		1.100		9	f3ds	(2H)2K2	f3ds	92.0
6.5	-	21803.0		1.117		7	f3ds	(4I)4Ha	f3ds	75.2
5.5	-	21819.9		1.081		7	f3d2	(4I)6Ha	f3ds	51.0
4.5	-	21821.1		1.070		7	f3d2	(4G)6I	f3ds	52.6
3.5	-	21855.1		1.096		6	f3ds	(2H> 2F2	f3ds	63.9
6.5	21858.433	21863.2	−4	1.015		31	f3d2	(4I)6Kb	f3d2	80.9
2.5	-	21873.3		1.090		7	f3ds	(4F)4Pb	f3ds	70.7
5.5	-	21886.1		1.088		7	f3ds	(4F)4Gb	f3ds	53.7
2.5	-	21942.1		1.178		8	f3d2	(4F)6P	f3d2	70.4
1.5	-	21944.2		0.872		8	f3d2	(4S)6F	f3ds	44.7
4.5	-	21981.2		1.074		8	f3d2	(4F)6H	f3ds	65.1
3.5	-	22046.5		1.089		6	f3d2	(4F)6Ga	f3ds	61.1
1.5	-	22059.7		1.278		11	f3ds	(4G)6D	f3ds	58.5
2.5	-	22070.4		0.903		6	f3ds	(4G)4Gb	f3ds	60.4
5.5	-	22089.4		1.062		24	f3s2	(2I)2I	f3d2	33.5

Table 3. *Cont.*

J	E^{exp} (cm^{-1})	E^{th} (cm^{-1})	ΔE (cm^{-1})	g_L^{th}	g_L^{exp}	% 1st *comp*	*Conf*	*Term*	*Main conf*	%
5.5	-	22123.0		1.023		14	f3d2	(4G)6K	f3d2	62.3
1.5	-	22127.9		1.207		88	f3d2	(4I)6M	f3d2	100.
8.5	-	22150.9		1.120		16	f3ds	(4I)6I	f3ds	84.2
3.5	-	22152.9		1.206		18	f3ds	(4G)6F	f3ds	74.8
3.5	-	22197.0		1.101		8	f3d2	(4I)6F	f3d2	59.6
4.5	22230.453	22197.0	33	1.114		7	f3s2	(2G)2G1	f3d2	41.3
6.5	-	22225.0		1.093		4	f3ds	(4G)6G	f3d2	51.3
2.5	-	22226.7		1.125		8	f3ds	(2H)4F2	f3ds	70.4
4.5	-	22262.3		0.943		8	f3d2	(4G)6I	f3ds	49.9
1.5	-	22269.6		1.121		6	f3ds	(4D)6D	f3d2	50.6
6.5	-	22273.6		1.086		10	f3ds	(4F)4Hb	f3ds	64.9
4.5	22305.305	22289.0	16	1.090		15	f3ds	(4I)2G	f3ds	55.5
7.5	-	22300.7		1.062		15	f3ds	(4I)4Ib	f3ds	58.6
3.5	-	22350.7		0.985		9	f3d2	(4I)4Hc	f3d2	61.8
4.5	-	22398.8		1.054		8	f3s2	(2G)2G1	f3ds	42.7
1.5	-	22425.5		1.053		9	f3ds	(4F)4Fb	f3ds	62.9
3.5	-	22449.8		1.049		7	f3ds	(4G)6G	f3d2	50.1
5.5	-	22470.4		1.019		17	f3d2	(4G)6K	f3d2	78.5
5.5	22567.167	22555.9	11	1.080		11	f3d2	(4G)6K	f3d2	50.3
7.5	-	22582.5		1.220		23	f3ds	(4G)6H	f3ds	68.3
2.5	-	22589.1		0.927		6	f3d2	(4I)4Gb	f3d2	58.8
4.5	-	22604.6		1.115		5	f3ds	(4I)4Ga	f3ds	49.4
0.5	-	22620.1		0.894		11	f3ds	(4D)6D	f3ds	60.9
3.5	-	22645.0		1.020		5	f3ds	(4I)4Ga	f3ds	51.8
6.5	-	22692.1		1.223		23	f3ds	(4G)6G	f3ds	80.8
2.5	-	22693.8		1.021		6	f3ds	(2D)4F2	f3ds	61.9
5.5	-	22708.2		1.185		6	f3ds	(4G)6G	f3ds	53.5
3.5	-	22729.6		1.131		7	f3ds	(4D)6D	f3ds	58.6
5.5	22813.792	22788.4	25	1.049		6	f3ds	(2K)4I	f3ds	52.7
3.5	-	22793.4		1.215		8	f3s2	(4D)4D	f3ds	68.0
7.5	-	22819.3		1.048		12	f3d2	(4I)6Kb	f3d2	64.3
6.5	-	22820.8		1.050		7	f3ds	(4G)4Ib	f3ds	55.7
0.5	-	22833.5		1.834		20	f3ds	(2D)2S1	f3ds	80.9
1.5	-	22833.8		1.287		9	f3d2	(4F)6P	f3d2	59.8
5.5	-	22851.9		1.043		7	f3d2	(4F)6I	f3d2	61.3
3.5	-	22857.0		1.182		7	f3s2	(4D)4D	f3ds	58.4
8.5	-	22888.2		1.135		34	f3d2	(4I)6Ka	f3d2	82.6
1.5	-	22942.6		1.091		7	f3ds	(4G)6D	f3ds	72.5
3.5	-	22963.0		1.051		8	f3ds	(2H> 2F2	f3ds	64.5
2.5	-	22972.1		1.012		5	f3ds	(4G)4Fb	f3ds	62.2
1.5	-	22991.6		0.806		8	f3d2	(4F)6Ga	f3ds	51.1
4.5	-	23006.5		1.047		3	f3d2	(4I)6Ib	f3d2	46.1
2.5	-	23031.6		1.181		4	f3ds	(2H> 2F2	f3ds	53.7
2.5	-	23062.2		1.204		14	f3ds	(4G)6D	f3ds	66.3
4.5	23106.350	23094.0	12	1.023		7	f3d2	(4G)6I	f3d2	63.8
6.5	23013.222	23100.7	−87	1.066		9	f3ds	(4G)6H	f3d2	53.5
8.5	-	23125.4		1.173		34	f3ds	(4G)6I	f3ds	84.5
5.5	23043.896	23135.7	−91	1.111		11	f3s2	(4G)4G	f3ds	49.8
0.5	-	23170.1		0.636		11	f3d2	(4F)6Fa	f3d2	53.0
5.5	-	23170.0		1.131		9	f3ds	(2H)4G2	f3ds	76.9
1.5	-	23225.8		1.043		5	f3ds	(2D)4P1	f3ds	54.4
3.5	-	23259.0		1.082		6	f3d2	(4F)6Fb	f3d2	51.4
7.5	23371.611	23282.8	88	1.047		26	f3d2	(4I)6Kb	f3d2	82.3
2.5	-	23293.3		0.964		5	f3d2	(4F)6Fa	f3ds	48.9
9.5	-	23318.7		1.196		47	f3d2	(4I)6Ka	f3d2	99.9

Table 3. *Cont.*

J	E^{exp} (cm^{-1})	E^{th} (cm^{-1})	ΔE (cm^{-1})	g_L^{th}	g_L^{exp}	% $1^{st}comp$	*Conf*	*Term*	*Main conf*	%
6.5	-	23329.5		1.092		7	f3d2	(4F)6H	f3d2	69.3
2.5	-	23330.0		1.174		6	f3ds	(2D)4P1	f3ds	73.2
1.5	-	23364.1		1.369		14	f3d2	(4F)6P	f3d2	49.4
0.5	-	23375.3		1.338		23	f3ds	(2D)2S1	f3ds	78.6
4.5	-	23397.0		1.097		11	f3ds	(4F)4Fb	f3ds	60.5
3.5	-	23461.6		1.152		6	f3d2	(4F)6P	f3d2	54.8
4.5	-	23474.8		1.020		13	f3d2	(4I)6G	f3d2	69.7
8.5	-	23507.1		1.032		28	f3d2	(4I)4Ma	f3d2	79.9
6.5	-	23534.5		1.060		5	f3ds	(2K)4L	f3ds	53.0
2.5	-	23557.4		1.091		6	f3d2	(4I)6F	f3d2	45.6
5.5	23528.305	23572.0	−43	1.079		6	f3d2	(4F)6H	f3d2	59.3
3.5	-	23585.2		1.047		4	f3ds	(2H)4G2	f3ds	54.3
5.5	-	23594.8		1.146		9	f3ds	(2H)4H2	f3ds	56.3
3.5	-	23601.2		1.214		9	f3ds	(4G)6D	f3ds	51.7
1.5	-	23648.2		0.829		9	f3ds	(2D)4F1	f3ds	62.1
7.5	-	23651.9		1.084		15	f3ds	(2H)4K2	f3ds	75.8
5.5	-	23671.9		1.137		6	f3d2	(4F)6Ga	f3d2	58.2
0.5	-	23688.3		0.487		15	f3d2	(4F)6Fa	f3ds	49.3
4.5	-	23706.2		1.100		8	f3d2	(4I)6G	f3d2	58.8
3.5	-	23726.6		1.131		3	f3ds	(4D)6F	f3ds	56.0
9.5	-	23730.0		1.088		41	f3ds	(2K)4M	f3ds	99.9
6.5	-	23735.2		1.093		8	f3d2	(4F)6I	f3ds	62.3
4.5	-	23760.6		1.112		10	f3d2	(4I)6F	f3d2	51.8
0.5	-	23774.7		0.513		14	f3s2	(4D)4D	f3ds	46.4
2.5	-	23811.4		0.973		9	f3d2	(4S)6F	f3d2	64.0
1.5	-	23813.8		0.906		15	f3d2	(4F)6Fa	f3d2	47.4
5.5	-	23814.8		1.095		5	f3s2	(4G)4G	f3d2	65.2
6.5	-	23819.7		1.094		5	f3ds	(4G)4Ib	f3d2	51.7
7.5	-	23823.9		1.037		21	f3d2	(4I)4Mb	f3d2	89.3
2.5	-	23844.1		1.166		5	f3ds	(2D)4P1	f3ds	62.3
6.5	-	23916.1		1.130		11	f3d2	(4I)6Ib	f3d2	65.2
2.5	-	23926.4		1.092		6	f3ds	(4F)4Ga	f3ds	75.7
1.5	-	23940.0		0.948		6	f3ds	(4D)6G	f3ds	65.3
3.5	-	23945.5		1.031		4	f3d2	(4G)6I	f3d2	52.2
4.5	23924.333	23956.4	−32	1.100		5	f3ds	(2K)4I	f3ds	70.9
0.5	-	23978.6		1.410		10	f3d2	(4S)6F	f3d2	66.6
8.5	-	23985.8		0.995		13	f3d2	(4I)4N	f3d2	54.4
5.5	-	24014.6		1.150		11	f3d2	(4F)6H	f3ds	49.7
5.5	-	24061.4		1.115		8	f3ds	(4G)6F	f3ds	70.7
7.5	-	24091.8		1.123		10	f3d2	(4I)6Ha	f3d2	64.6
4.5	-	24098.6		1.129		7	f3d2	(4I)6G	f3ds	50.5
1.5	-	24104.3		0.872		10	f3d2	(4F)6Gb	f3ds	43.9
5.5	-	24163.8		1.111		8	f3ds	(4G)4Gb	f3ds	54.0
3.5	-	24199.1		0.952		6	f3d2	(4I)4Hc	f3d2	62.1
0.5	-	24202.1		0.767		9	f3ds	(2D)4P1	f3ds	61.4
2.5	-	24214.8		1.139		5	f3ds	(4F)4Db	f3ds	62.7
3.5	-	24215.6		1.249		8	f3ds	(4G)6F	f3ds	48.6
4.5	-	24232.9		1.084		11	f3d2	(4F)6Ga	f3d2	58.8
6.5	-	24278.4		1.060		7	f3d2	(4G)6K	f3d2	64.9
9.5	-	24314.3		1.013		50	f3d2	(4I)4N	f3d2	99.9
2.5	-	24319.2		0.969		8	f3d2	(4G)6Ha	f3d2	57.6
2.5	-	24359.1		1.202		6	f3d2	(4F)6P	f3ds	48.6
1.5	-	24369.2		1.084		6	f3ds	(2D)4P1	f3ds	53.5
4.5	-	24377.2		1.100		4	f3ds	(4F)2H	f3d2	63.1
5.5	-	24383.2		1.098		7	f3ds	(2K)4I	f3ds	65.1
6.5	-	24388.0		1.063		10	f3ds	(2K)4K	f3ds	57.9

Table 3. *Cont.*

J	E^{exp} (cm^{-1})	E^{th} (cm^{-1})	ΔE (cm^{-1})	g_L^{th}	g_L^{exp}	% 1st *comp*	*Conf*	*Term*	*Main conf*	%
3.5	-	24403.0		1.168		8	f3ds	(4D)6G	f3ds	51.2
10.5	-	24433.6		1.209		71	f3d2	(4I)6L	f3d2	99.9
7.5	-	24441.5		1.077		12	f3ds	(2K)4K	f3ds	53.7
1.5	-	24488.1		0.673		20	f3d2	(4F)6Gb	f3d2	57.1
1.5	-	24505.5		1.121		5	f3ds	(2D)4P1	f3ds	65.9
4.5	24446.491	24505.8	−59	1.070		4	f3d2	(4I)2Ha	f3d2	47.6
3.5	-	24548.7		1.089		5	f3ds	(4D)4Gb	f3ds	53.8
2.5	-	24570.0		1.060		6	f3d2	(4I)4Ga	f3d2	59.3
1.5	-	24602.0		1.162		9	f3ds	(4D)6D	f3ds	63.0
8.5	-	24605.5		1.069		19	f3d2	(4I)4Mb	f3d2	58.0
4.5	-	24707.3		0.959		4	f3d2	(4I)4Ib	f3ds	46.1
5.5	24771.463	24720.3	51	1.096		7	f3d2	(4G)6I	f3ds	58.6
6.5	-	24735.1		1.056		36	f3d2	(4G)6K	f3d2	91.9
2.5	-	24740.2		1.161		11	f3ds	(4G)4Db	f3d2	49.4
7.5	-	24754.0		1.125		8	f3d2	(4I)6Ha	f3d2	52.1
3.5	-	24790.7		1.124		4	f3ds	(4D)6D	f3d2	55.0
6.5	-	24812.7		1.058		7	f3ds	(4G)4Hb	f3d2	53.6
0.5	-	24840.6		0.546		9	f3d2	(4I)6F	f3d2	55.9
7.5	-	24851.5		1.093		10	f3ds	(2K)4L	f3ds	53.1
3.5	-	24861.8		1.247		9	f3ds	(4G)6D	f3ds	60.8
2.5	-	24877.5		1.259		6	f3ds	(4D)6D	f3ds	53.9
1.5	-	24889.3		0.985		3	f3d2	(4F)6Fa	f3d2	54.6
5.5	-	24920.7		1.094		5	f3ds	(4G)6F	f3d2	58.8
4.5	24888.132	24945.0	−56	1.161		9	f3ds	(4G)6F	f3ds	50.8
2.5	-	24964.3		1.135		5	f3ds	(2P)4F	f3ds	52.9
6.5	-	24986.3		1.073		5	f3d2	(4G)6K	f3ds	50.2
4.5	24984.711	25008.7	−23	1.023		3	f3ds	(2H)4I1	f3ds	57.1
8.5	-	25008.2		1.077		14	f3d2	(4I)6Ia	f3d2	52.8
7.5	-	25019.1		1.182		11	f3d2	(4F)6H	f3d2	78.6
3.5	-	25045.7		1.077		4	f3ds	(2P)2Fa	f3d2	54.7
5.5	25111.924	25063.7	48	1.170		6	f3ds	(4F)4Gb	f3ds	56.2
3.5	-	25082.7		1.126		5	f3ds	(4G)6G	f3d2	49.5
1.5	-	25084.7		1.215		11	f3d2	(4F)6Da	f3d2	56.5
6.5	-	25130.0		1.110		10	f3ds	(4G)6H	f3ds	50.0
4.5	-	25154.2		1.173		7	f3ds	(2H> 2G2	f3ds	51.7
2.5	-	25158.7		1.017		8	f3ds	(2H)2F2	f3d2	49.9
2.5	-	25179.5		1.213		5	f3d2	(4F)6Fa	f3d2	51.9
1.5	-	25219.2		0.913		7	f3d2	(4I)4Fb	f3ds	51.9
4.5	-	25236.5		1.036		9	f3s2	(2H)2H1	f3ds	48.1
5.5	25245.183	25237.4	7	1.021		25	f3d2	(4G)6I	f3d2	71.9
8.5	-	25302.6		1.037		23	f3d2	(4I)4Mb	f3d2	54.4
2.5	-	25320.0		1.094		8	f3d2	(4S)6F	f3d2	49.8
4.5	-	25325.8		1.152		6	f3ds	(4F)4Fb	f3ds	58.6
6.5	-	25376.5		1.091		8	f3s2	(2I)2I	f3d2	68.6
0.5	-	25379.7		1.245		11	f3ds	(4D)6D	f3ds	67.7
3.5	-	25396.3		1.018		5	f3ds	(4F)2G	f3d2	47.5
6.5	-	25411.7		1.092		9	f3s2	(2I)2I	f3d2	50.4
3.5	-	25453.7		1.070		4	f3d2	(4I)2G	f3d2	47.5
4.5	25439.103	25463.6	−24	1.076		5	f3d2	(4S)6F	f3d2	58.6
5.5	25423.490	25476.5	−53	0.977		7	f3ds	(2I)4K	f3d2	54.5
1.5	-	25482.2		1.014		8	f3ds	(2D)4S1	f3ds	53.0
3.5	-	25516.3		1.045		4	f3d2	(4G)6I	f3d2	56.6
0.5	-	25521.6		0.983		6	f3ds	(2D> 2P1	f3ds	53.1
5.5	25514.656	25537.8	−23	1.082		10	f3ds	(4G)2I	f3ds	49.4
4.5	-	25545.5		1.169		12	f3d2	(4S)6F	f3d2	58.7
6.5	-	25558.5		1.096		7	f3ds	(4G)4Ia	f3ds	63.1

Table 3. *Cont.*

J	E^{exp} (cm^{-1})	E^{th} (cm^{-1})	ΔE (cm^{-1})	g_L^{th}	g_L^{exp}	% 1stcomp	Conf	Term	Main conf	%
2.5	-	25582.3		1.134		9	f3d2	(4F)6Fa	f3ds	53.6
2.5	-	25603.8		0.986		6	f3ds	(4D)4Fb	f3d2	48.7
6.5	-	25647.8		1.088		42	f3s2	(2I)2I	f2s2	44.7
4.5	-	25665.7		1.137		4	f3ds	(4G)6F	f3d2	57.5
1.5	-	25666.4		1.063		7	f3ds	(4D)6G	f3d2	49.1
7.5	-	25671.0		1.079		7	f3ds	(2L)4M	f3ds	67.3
5.5	25726.260	25729.7	−3	1.101		4	f3d2	(4I)6G	f3ds	60.1
1.5	-	25754.9		1.030		6	f3d2	(4F)6Gb	f3d2	64.2

Table 4. Transitions of the U II odd parity level at 13695.737 cm^{-1} with J = 4.5 to even levels of J = 3.5.

wl_{Ritz} in Air (Å)	wn_{Ritz} (cm^{-1})	wn_{exp} (cm^{-1})	E_{even} (cm^{-1})	J	E_{odd} (cm^{-1})	J	Int	*Ref*
14539.464	6875.953	6875.950	20571.690	3.5	13695.737	4.5	18.15	[14]
13112.082	7624.469	7624.468	21320.206	3.5	13695.737	4.5	9.22	[14]
4566.3396	21893.243	21893.239	35588.980	3.5	13695.737	4.5	142	[7]
4452.1493	22454.758	22454.757	36150.495	3.5	13695.737	4.5	112	[7]
4354.3144	22959.274	22959.276	36655.011	3.5	13695.737	4.5	135	[7]
4343.3456	23017.256	23017.274	36712.993	3.5	13695.737	4.5	139	[7]
4299.6016	23251.430	23251.429	36947.167	3.5	13695.737	4.5	131	[7]
4103.8897	24360.253	24360.293	38055.990	3.5	13695.737	4.5	236	[7]
4058.3724	24633.463	24633.459	38329.200	3.5	13695.737	4.5	269	[7]
4036.9830	24763.977	24763.973	38459.714	3.5	13695.737	4.5	138	[7]
4026.7550	24826.878	24826.899	38522.615	3.5	13695.737	4.5	151	[7]
3990.6927	25051.223	25051.233	38746.960	3.5	13695.737	4.5	126	[7]
3965.9476	25207.524	25207.546	38903.261	3.5	13695.737	4.5	133	[7]
3891.9441	25686.821	25686.815	39382.558	3.5	13695.737	4.5	171	[7]
3765.4517	26549.697	26549.706	40245.434	3.5	13695.737	4.5	166	[7]
3734.8311	26767.359	26767.361	40463.096	3.5	13695.737	4.5	157	[7]
3728.6081	26812.033	26812.083	40507.770	3.5	13695.737	4.5	141	[7]
3697.7707	27035.628	27035.603	40731.365	3.5	13695.737	4.5	217	[7]

3.2. Even Parity Levels

Similarly to the odd parity study, the *RCN* and *RCN2* codes were used in the Relativistic Hartree-Fock (*HFR*) mode. Considering the large CI interaction integrals within the group $5f^2(6d+7s)^3$, the previously undetermined configuration $5f^26d^3$ was added to the four lowest configurations $5f^47s, 5f^46d, 5f^26d^27s, 5f^26d7s^2$. Appropriate scaling of Slater and spin-orbit integrals and corrections on the average energy parameters were applied for preparing the initial input data of the *RCG* code and of the LSF in the *RCE* code. In the final cycle of optimization, 125 levels and 22 free parameters led to a *rms* deviation of 84 cm^{-1}, i.e., which is less satisfactory than in the odd parity. One of these levels, given at E_{exp} = 22917.453 cm^{-1} without any label in [6] has been identified as the lowest level of the $5f^26d^3$ configuration, slightly above the error bars of Brewer's predictions [9]. It is seen that the scaling factors of fitted parameters reported in Table 5 are not very different from those obtained in the opposite parity (Table 2). With regard to the unachieved status of the parametric interpretation in the even parity, only the dominant configuration and the first component of the eigenfunctions are given in Table 6, together with the energies and Landé factors calculated in the final LSF.

Attempts to interpret $5f^37s7p + 5f^36d7p$ with the same method of parametric fitting could not go beyond the optimization of the average energy E_{av} and spin-orbit ζ_{5f} parameters. In Table 7 energy parameters adopted for $5f^37s7p + 5f^36d7p$ are reported and they lead to the calculated energies in Table 8. The empirical attribution of E_{exp} levels to configurations derived from isotope shifts and transition intensities in [6] are not fully supported by the present calculations. There were more

$5f^37s7p$ labels in Table 2 of [6] than predicted from the present work (Cf Table 8). The quantitative evaluation of the CI effects within the whole group $5f^4(7s+6d)+5f^2(6d+7s)^3+5f^37s7p+5f^36d7p$ has been attempted but has failed.

Table 5. Fitted parameters (in cm^{-1}) for even parity configurations of U II with $5f^2$ and $5f^4$ cores compared with HFR radial integrals. The scaling factors are $SF(P)=P_{fit}/P_{HFR}$ (dimensionless). They are replaced by $\Delta E = E_{fit} - E_{HFR}$ for E_{av} average energies (in cm^{-1}). Constraints on some parameters are in the 'Cstr' columns (denoted 'f' for fixed parameters or 'rn', which link parameters of the same 'rn' to vary in a constant ratio). The HFR values of E_{av} parameters are relative to the lowest odd configuration $5f^37s^2$ taken as zero value.

	$5f^47s$					**$5f^46d$**				
Param. P	P_{fit}	Cstr.	Unc.	P_{HFR}	$\Delta E/SF$	P_{fit}	Cstr.	Unc.	P_{HFR}	$\Delta E/SF$
E_{av}	32165		112	15872	16293	46815		153	29044	17771
$F^2(ff)$	42100	r1	471	63821	0.660	41167	r1	461	62408	0.660
$F^4(ff)$	25599	r2	810	40934	0.625	24977	r2	790	39939	0.625
$F^6(ff)$	18480	r3	814	29779	0.621	18016	r3	794	29030	0.621
α	19.5	r4	2			19.5	r4	2		
β	−600	f				−600	f			
γ	1600	f				1600	f			
ζ_f	1557	r5	9	1661	0.938	1529	r5	9	1631	0.938
ζ_d						1145	r6	19	1369	0.836
$F^1(fd)$						509	f			
$F^2(fd)$						20390	r8	520	24906	0.819
$F^4(fd)$						14468	r9	776	13477	1.074
$G^1(fd)$						11163	r10	248	18109	0.616
$G^2(fd)$						1524	f	197		
$G^3(fd)$						13293	r11	511	12342	1.077
$G^4(fd)$						2691	f			
$G^5(fd)$						7527	r12	923	8963	0.840
$G^3(fs)$	2132		113	4561	0.467					
	$5f^26d7s^2$					**$5f^26d^27s$**				
Param. P	P_{fit}	Cstr.	Unc.	P_{HFR}	$\Delta E/SF$	P_{fit}	Cstr.	Unc.	P_{HFR}	$\Delta E/SF$
E_{av}	39115		305	11841	27274	43050		208	12973	30077
$F^2(ff)$	49005	r1	548	74289	0.660	48413	r1	542	73381	0.660
$F^4(ff)$	30284	r2	958	48424	0.625	29870	r2	945	47763	0.625
$F^6(ff)$	22025	r3	971	35490	0.621	21710	r3	957	34984	0.621
α	19.5	r4	2			19.5	r4	2		
β	−600	f				−600	f			
γ	1600	f				1600	f			
$F^2(dd)$						23302	r13	1546	37756	0.617
$F^4(dd)$						14997	r16	3057	25080	0.598
$\alpha(dd)$						10	f			
ζ_f	1908	r5	11	2036	0.937	1884	r5	11	2010	0.937
ζ_d	1889	r6	32	2259	0.931	1760	r6		2104	0.837
$F^1(fd)$	509	f				509	f			
$F^2(fd)$	25399	r8	647	31025	0.819	24437	r8	623	29850	0.819
$F^4(fd)$	18040	r9	968	16803	1.074	17283	r9	927	16099	1.074
$G^1(fd)$	11285	r10	250	18306	0.616	11030	r10	245	17892	0.616
$G^2(fd)$	1524	f				1524	f			
$G^3(fd)$	14821	r11	570	13760	1.077	14324	r11	551	13299	1.077
$G^4(fd)$	2691	f				2691	f			
$G^5(fd)$	8727	r12	1071	10389	0.840	8394	r12	1030	9996	0.840
$G^3(fs)$	1578	f		2450	0.644	1885	r6	100	4033	0.467
$G^2(ds)$						12984		597	20874	0.622

Table 5. *Cont.*

			$5f^26d^3$		
Param. P	P_{fit}	Cstr.	Unc.	P_{HFR}	$\Delta E/SF$
E_{av}	52093	401	20882	31211	
$F^2(ff)$	47830	r1	535	74289	0.660
$F^4(ff)$	29460	r2	932	44131	0.625
$F^6(ff)$	21410	r3	943	34500	0.621
α	19.5	r4	2		
β	−600	f			
γ	1600	f			
$F^2(dd)$	22416	r13	1488	36319	0.617
$F^4(dd)$	14359	r16	2926	24013	0.598
$\alpha(dd)$	10	f			
ζ_f	1860	r5	11	1986	0.937
ζ_d	1635	r6	27	1956	0.836
$F^1(fd)$	509	f			
$F^2(fd)$	25399	r8	647	31025	0.819
$F^4(fd)$	18040	r9	968	16803	1.074
$G^1(fd)$	11285	r10	250	18306	0.616
$G^2(fd)$	1524	f			
$G^3(fd)$	13799	r11	570	12811	1.077
$G^4(fd)$	2691	f			
$G^5(fd)$	8050	r12	988	9587	0.840
Configuration Interaction					
$5f^47s - 5f^46d$					
$R^2(fs,fd)$	−6216	r14	357	−10574	0.588
$R^3(fs,df)$	−1875	r14	108	−3189	0.588
$5f^47s - 5f^26d7s^2$					
$R^3(ff,ds)$	−2310	r14	133	−3930	0.588
$5f^47s - 5f^26d^27s$					
$R^1(ff,dd)$	10962	r15	214	23117	0.474
$R^3(ff,dd)$	7708	r15	150	16225	0.474
$R^5(ff,dd)$	5672	r15	111	11961	0.474
$5f^46d - 5f^26d7s^2$					
$R^3(ff,ss)$	3556	r14	204	6050	0.588
$5f^46d - 5f^26d^27s$					
$R^3(ff,ds)$	−2385	r14	137	−4056	0.588
$5f^46d - 5f^26d^3$					
$R^1(ff,dd)$	10762	r15	210	22694	0.474
$R^3(ff,dd)$	7466	r15	146	15745	0.474
$R^5(ff,dd)$	5465	r15	107	11525	0.474
$5f^26d7s^2 - 5f^26d^27s$					
$R^2(fs,fd)$	−5657	r14	324	−9621	0.588
$R^3(fs,df)$	−908	r14	52	−1545	0.588
$R^2(ds,dd)$	−14874	r14	853	−25302	0.588
$5f^26d7s^2 - 5f^26d^3$					
$R^2(ss,dd)$	13119	r14	752	22316	0.588
$5f^26d^27s - 5f^26d^3$					
$R^2(fs,fd)$	−5558	r14	319	−9454	0.588
$R^4(fs,df)$	−938	r14	54	−1596	0.588
$R^2(ds,dd)$	−14638	r14	840	−24900	0.588

Table 6. Energy levels of U II, even parity with $5f^2$ and $5f^4$ parent configurations. Comparison of experimental energies and Landé factors with values calculated from the parameter set of Table 5. $\Delta E = E^{exp} - E^{th}$. The percentage, the configuration and the *LS* name of the leading component in the corresponding configuration are given in the last three columns.

J	E^{exp} (cm^{-1})	E^{th} (cm^{-1})	ΔE (cm^{-1})	g_L^{th}	g_L^{exp}	% 1stcomp	Conf	Term
3.5	4663.803	4647	16	0.500	0.490	71	f4s	(5I)6I
4.5	5716.449	5564	152	0.830	0.830	40	f4s	(5I)6I
5.5	8347.690	8327	19	1.030	1.040	62	f4s	(5I)6I
4.5	8423.418	8423		0.797	0.790	41	f4s	(5I)4I
6.5	10740.265	10772	−31	1.142	1.145	72	f4s	(5I)6I
1.5	10987.204	10954	32	0.690	0.645	24	f4s	(5F)6F
2.5	11252.337	11138	114	1.175		22	f4s	(5F)6F
5.5	11389.469	11419	−30	0.961	0.970	59	f4s	(5I)4I
0.5		12254		−0.516		70	f4s	(5F)6F
5.5	12513.881	12493	19	0.676	0.680	61	f4s	(5I)6L
3.5	12804.950	12821	−16	0.922		12	f4s	(3G)4G2
7.5	12862.155	12880	−17	1.206	1.22	69	f4s	(5I)6I
4.5	13023.114	12905	118	1.134		11	f4s	(3G)4G2
1.5	13006.990	13044	−37	0.701		34	f4s	(5F)6F
5.5	13783.030	13733	49	0.695	0.685	56	f2d2s	(3H)6L
2.5	13758.142	13807	−49	0.931		30	f4s	(5G)6G
6.5	13865.969	13875	−9	1.068	1.10	61	f4s	(5I)4I
3.5	14018.821	13979	39	1.260		46	f4s	(5F)6F
1.5		14204		0.370		34	f4s	(5F)4F
2.5	14239.503	14439	−199	1.517		34	f4s	(5S)6S
8.5	14796.725	14742	54	1.245		61	f4s	(5I)6I
3.5	14767.466	14759	8	1.044		30	f4s	(5G)6G
2.5	14848.575	14955	−107	1.004		26	f4s	(5F)4F
2.5	15087.785	15088		0.951		23	f4s	(5G)4G
6.5	15392.416	15353	38	0.871	0.880	72	f4d	(5I)6L
3.5	15679.555	15734	−55	0.589	0.615	20	f2d2s	(3H)6Ia
3.5	15812.498	15857	−44	0.588	0.590	17	f2d2s	(5I)6I
1.5	15888.905	15870	18	1.529		42	f4s	(5S)4S
7.5	15992.765	15937	55	1.129	1.20	53	f4s	(5I)4I
6.5	15962.320	15959	3	0.903	0.900	46	f2d2s	(3H)6L
4.5	16211.704	16356	−145	0.663	0.615	53	f4d	(5I)6K
5.5	16379.878	16364	14	1.283		24	f4s	(5F)6F
4.5	16804.920	16546	259	0.759	0.845	19	f2d2s	(3H)6K
4.5	16656.412	16711	−55	1.318		52	f4s	(5F)6F
5.5	16706.303	16913	−207	0.788	0.790	36	f4s	(3K)4K2
4.5	17225.885	17216	9	0.991		14	f4s	(3H)6K
5.5	17434.363	17438	−4	0.795	0.800	20	f2ds2	(3K)4K2
6.5	17380.868	17463	−82	0.973		31	f4s	(3K)4K2
4.5	17392.211	17604	−212	0.860	0.785	16	f2d2s	(3H)6K
3.5		17683		1.115		49	f4s	(5F)4F
2.5		18060		0.680		17	f4d	(5I)4G
7.5	18136.366	18062	73	1.004	1.005	76	f4d	(5I)6L
4.5	18084.435	18154	−69	0.971		19	f4d	(5G)6G
4.5	18200.092	18334	−134	0.836	0.780	27	f2ds2	(3H)4I
3.5		18599		0.980		25	f4s	(5G)4G
4.5	18536.705	18600	−63	0.967		26	f4d	(5I)6I
2.5		18675		0.575		10	f4d	(5I)6H
5.5	18827.008	18694	133	0.908	0.945	36	f2d2s	(3H)6K
7.5	18656.355	18699	−42	1.053		21	f4s	(3L)4L
0.5		18737		2.635		39	f4s	(5D)6D
6.5	18617.807	18791	−173	0.908		42	f4s	(3L)4L
5.5	18654.316	18850	−196	0.874	0.880	70	f4d	(5I)6K
2.5		19047		0.751		12	f4s	(3G)4G2

Table 6. *Cont.*

J	E^{exp} (cm^{-1})	E^{th} (cm^{-1})	ΔE (cm^{-1})	g_L^{th}	g_L^{exp}	% 1st *comp*	*Conf*	*Term*
5.5	19097.594	19096	1	1.330		42	f4s	(5F)6F
2.5		19134		0.933		10	f4s	(5F)6F
1.5		19159		0.448		17	f2d2s	(3F)6Ga
3.5		19246		1.001		13	f4s	(3G)4G2
2.5	19395.168	19330	64	0.779		7	f2d2s	(3H)6Ha
2.5		19354		1.001		10	f4s	(3H)6Ha
1.5		19412		0.996		17	f4s	(5F)6F
3.5	19517.729	19546	−28	0.821	0.815	15	f4d	(5I)6I
7.5	19743.511	19756	−12	1.017	1.000	67	f2d2s	(3H)6L
1.5		19796		0.191		48	f4d	(5I)6G
2.5		19863		0.909		12	f2d2s	(3H)6Ha
5.5	19840.514	19899	−58	0.947		8	f4d	(5I)6K
6.5	19977.100	19935	41	0.969	0.960	32	f2d2s	(3H)6L
3.5	19971.328	19977	−5	0.857	0.860	11	f4d	(5I)4H
8.5	20230.479	20127	102	1.099		20	f4s	(5I)6I
5.5	20353.992	20310	43	1.029	1.015	29	f2d2s	(3H)6Ia
4.5		20365		1.208		50	f4s	(5F)4F
7.5	20425.567	20445	−20	0.975		32	f4s	(3M)4M
3.5	20571.690	20474	97	0.947	0.935	8	f4d	(3H)6Ha
0.5		20496		1.065		26	f4s	(3P)4P2
1.5		20530		1.274		20	f4s	(5D)6D
8.5	20739.844	20612	127	1.095	1.11	74	f4d	(5I)6L
2.5	20678.779	20672	6	1.066		8	f4s	(1D)2D3
4.5	20635.272	20721	−86	0.914	0.945	13	f2d2s	(3H)6Ia
6.5	20702.037	20789	−87	1.034	0.990	40	f4d	(5I)6K
1.5		20828		1.079		12	f4s	(3P)4P2
6.5	20934.186	20858	76	1.265		45	f4s	(5G)6G
3.5	20961.720	20901	60	0.877	0.855	11	f4d	(5I)6H
2.5		20917		0.750		14	f4d	(5I)6H
5.5	20742.878	20940	−197	1.012		29	f4d	(5I)6I
5.5	20932.139	21050	−118	1.173		25	f4s	(5G)6G
1.5		21053		1.534		31	f4s	(5D)6D
4.5	21154.557	21066	88	1.061	1.010	13	f4d	(5I)4H
4.5	21053.528	21089	−35	1.215		21	f4s	(3F)4F4
3.5	21207.738	21190	17	1.303	1.150	19	f4s	(5D)6D
3.5	21320.206	21514	−194	0.822	0.835	14	f4d	(3H)6Ia
5.5	21691.517	21532	159	0.961	0.975	15	f2d2s	(3H)4I
4.5	21555.275	21619	−63	0.915	1.025	9	f2d2s	(3H)4Ic
6.5	21710.768	21641	68	0.917	0.915	31	f2d2s	(3H)4Lb
3.5		21650		1.062		9	f4s	(3F)2F4
2.5		21719		0.862		15	f4d	(5F)6G
4.5		21720		1.001		26	f4s	(5G)4G
1.5		21728		0.478		12	f4s	(3F)4F3
0.5		21778		0.852		34	f4s	(5D)4D
0.5		21942		1.493		33	f4s	(3P)4P2
2.5		21953		1.186		15	f4s	(3D)2D1
3.5	21860.051	21954	−94	0.718	0.67	16	f2d2s	(3H)6Ia
3.5	22158.070	22053	104	0.910		11	f4d	(5I)6H
5.5	22157.162	22058	98	1.171		39	f4s	(5G)4G
6.5	21975.590	22058	−82	1.030	1.03	29	f4d	(5I)6K
2.5		22142		1.003		8	f4s	(3F)4F3
1.5		22153		0.300		25	f4d	(5F)6G
4.5	22165.179	22197	−32	1.007	0.895	12	f4d	(5F)6H
3.5	22250.398	22216	34	0.863	0.885	12	f2d2s	(3F)6I
4.5	22429.865	22303	126	0.874	0.935	13	f4s	(3I)4I1
5.5	22389.574	22326	62	0.992	1.040	6	f2d2s	(3H)6K

Table 6. *Cont.*

J	E^{exp} (cm^{-1})	E^{th} (cm^{-1})	ΔE (cm^{-1})	g_L^{th}	g_L^{exp}	% 1st *comp*	*Conf*	*Term*
6.5	22615.319	22534	80	0.986	0.995	28	f2d2s	(3H)4K
0.5		22613		1.207		14	f2ds2	(3F)2P
5.5	22764.904	22625	139	1.030	0.980	17	f4s	(1H)2H1
4.5	22642.478	22634	8	0.936	0.875	8	f2d2s	(3I)4I1
2.5		22696		1.168		14	f2d2s	(5D)6D
3.5	22815.123	22740	74	0.786		26	f2ds2	(3F)4H
7.5		22776		1.032		40	f4d	(5I)6K
3.5	22960.667	22891	69	0.997	0.945	9	f2d2s	(3H)6Ha
4.5	22868.033	22902	−34	0.943	0.980	9	f2d2s	(5I)6H
5.5	22917.453	22942	−25	0.759	0.860	38	f2d3	(3H)6L
6.5	23107.566	22945	161	1.120	1.060	29	f2d2s	(3H)6Ia
2.5	23029.458	23039	−9	0.988		17	f4d	(5I)6G
9.5		23076		1.160		71	f4d	(5I)6L
1.5		23104		1.446		15	f2d2s	(3H)6D
5.5	23241.365	23121	119	0.968	0.96	17	ds2	(3H)4I *
2.5		23148		1.070		13	f2d2s	(5D)6D
4.5	23241.033	23168	72	0.959	1.050	6	f2d2s	(5I)6I
3.5	23257.613	23205	52	0.597		21	f4d	(5G)6I
6.5	23234.820	23223	11	1.024	1.090	29	f4d	(5I)6I
2.5	23353.601	23264	89	0.779		20	f2d2s	(3H)6Ha
7.5	23262.359	23350	−87	1.102	1.070	24	f2d2s	(3H)6K
3.5		23412		0.960		13	f4d	(5I)6G
0.5		23428		2.116		14	f2d2s	(3H)6D
8.5		23441		1.107		73	f2d2s	(3H)6L
6.5		23492		1.202		24	f4s	(3H)4H3
4.5		23501		1.073		12	f4s	(3G)4G2
5.5		23628		1.033		11	f4d	(5I)4H
3.5		23644		0.849		11	f2d2s	(3H)6Ib
1.5	23673.649	23648	25	1.276		25	f4d	(5D)4D
6.5	23635.919	23712	−77	0.986	0.920	18	f2d2s	(3H)4Lb
2.5	23700.946	23739	−38	0.868		17	f2d2s	(3F)6H
5.5		23792		0.923		22	f2ds2	(3H)4I
4.5	23817.508	23802	15	0.958	0.870	11	f2d2s	(3H)6K
0.5		23827		1.989		9	f4d	(5S)6D
2.5	23905.877	23828	77	1.085		9	f4d	(3F)6H
3.5	23803.252	23831	−27	0.991		7	f4d	(3F)4G
7.5	24071.418	23927	143	1.023		41	f4s	(3L)4L
1.5		23943		0.969		7	f2d2s	(3H)6Ga
4.5		23962		0.947		8	f2d2s	(3H)6K
3.5	23895.471	24064	−169	0.969	0.735	10	f4s	(3G)2G2
6.5	24159.696	24072	86	0.922	0.965	31	f4s	(3L)4L
8.5		24074		1.069		42	f4s	(3L)4L
5.5	24010.467	24077	−66	0.934	0.975	18	f4d	(5I)4K
7.5	24247.529	24122	124	1.113		22	f4d	(5I)6I
4.5	24220.675	24158	62	1.094		7	f4d	(5F)6G
5.5		24168		0.989		7	f2d2s	(3H)2H3
2.5		24213		1.052		11	f4d	(5F)6F
3.5	24209.303	24243	−34	1.086		12	f2d2s	(3F)6Ga
1.5		24292		0.768		17	f2d2s	(5F)6F
2.5		24299		1.004		17	f4d	(5F)6G
6.5		24375		1.024		9	f4d	(5I)4K
7.5	24423.656	24381	42	0.995		38	f4d	(3L)2L
5.5		24432		1.037		12	f4s	(5G)4G
4.5		24440		1.016		12	f2d2s	(3F)6I
3.5		24491		1.262		17	f4s	(5D)6D
1.5		24501		0.628		15	f2d2s	(3H)6Ga
4.5		24593		1.023		13	f4d	(3H)6K

Table 6. *Cont.*

J	E^{exp} (cm^{-1})	E^{th} (cm^{-1})	ΔE (cm^{-1})	g_L^{th}	g_L^{exp}	% 1st comp	Conf	Term
0.5		24650		0.273		29	f2d2s	(3F)6Fa
7.5		24675		0.974		17	f4d	(3K)4M2
8.5		24686		1.084		30	f4d	(5I)6K
0.5		24712		−0.082		43	f4d	(5F)6F
3.5	24709.449	24720	−10	0.943		16	f2d2s	(3H)6Ib
5.5		24722		1.007		11	f4d	(5I)6H
1.5		24746		1.386		15	f4d	(5F)6D
4.5	24684.135	24802	−117	0.981	0.935	7	f2d2s	(3H)6K
2.5		24840		0.805		12	f2d2s	(3F)4G
9.5		24845		1.114		46	f4s	(3L)4L
5.5	24857.570	24893	−35	1.068		7	f2d2s	(3H)4Ga
4.5		24928		0.975		11	f2d2s	(3H)6K
3.5	24862.698	24946	−83	0.993		10	f2d2s	(3H)6Ib
6.5		24977		1.111		18	f4s	(3I)4I1
4.5		24981		1.159		16	f4s	(3F)4F2
2.5		24984		0.830		8	f2d2s	(3H)4Gc
8.5	25053.005	25075	−22	1.063		43	f4s	(3M)4M
3.5		25132		1.140		9	f4d	(3F)4F3
2.5		25247		0.917		7	f4s	(3F)2F3
6.5		25248		1.000		14	f2d2s	(3H)6L
3.5		25294		1.056		8	f4d	(5I)6G
5.5	25356.972	25334	22	0.997	1.020	9	f4d	(3H)4Kb
4.5		25343		1.003		9	f2d2s	(3H)4H
3.5		25346		1.020		10	f4d	(3H)6Ha
4.5		25424		0.904		23	f2d2s	(3H)6Ib
1.5		25434		0.868		7	f4d	(5F)6F
8.5		25458		1.157		22	f4d	(5I)6I
0.5		25477		1.513		12	f4d	(5G)4D
7.5	25399.465	25518	−119	0.986		32	f4s	(3M)4M
3.5		25532		0.989		9	f4s	(3F)2F2
4.5		25537		0.950		7	f2d2s	(3H)4I
1.5	25582.631	25561	21	1.305		14	f4d	(5S)6D
2.5		25564		0.793		9	f4d	(5G)6H
8.5		25575		1.040		44	f4s	(3M)2M
2.5		25628		0.942		7	f2d2s	(3H)6Ga
5.5	25626.941	25635	−8	1.038		8	f4s	(5I)6H
3.5		25637		1.038		8	f2d2s	(5F)6F
10.5		25657		1.215		72	f4d	(5I)6L
1.5		25669		0.638		15	f2d2s	(3H)4F
6.5		25714		1.088		14	f4d	(3H)6L
7.5	25667.906	25733	−65	1.164	1.100	29	f4s	(3I)4I1
5.5		25746		0.978		20	f2d2s	(3H)4Kb
2.5		25748		1.061		6	f2d2s	(3F)6Ga
3.5		25784		1.051		13	f2d2s	(3H)6Ga
7.5		25784		1.078		15	f2d2s	(3H)6L
2.5		25875		1.115		9	f4d	(5F)6F
6.5		25892		0.998		20	f2d2s	(3H)6L
5.5		25894		1.083		22	f2d2s	(3H)6Ha
1.5		25981		0.913		21	f2d2s	(3F)6Fa
0.5		26012		1.555		14	f4d	(5S)6D
4.5		26038		1.116		7	f2d2s	(5F)6G
6.5		26058		1.176		12	f4d	(5I)6G
2.5		26094		1.261		16	f4s	(3P)4P2
0.5		26143		0.461		29	f4s	(3D)4D1
5.5	26158.897	26164	−5	1.010		18	f2d2s	(3F)6I
1.5		26166		1.167		18	f4s	(3P)4P2
3.5		26246		0.996		7	f2d2s	(3H)4Hb

Table 6. *Cont.*

J	E^{exp} (cm^{-1})	E^{th} (cm^{-1})	ΔE (cm^{-1})	g_L^{th}	g_L^{exp}	% $1^{st}comp$	*Conf*	*Term*
5.5		26321		1.134		8	f4d	(3G)4G2
4.5		26343		1.008		9	f2d2s	(3H)6Ib
2.5		26364		1.209		16	f4s	(5D)4D
6.5		26375		1.035		18	f4s	(3I)2I1
8.5		26386		1.162		13	f2d2s	(3H)4Ka
1.5		26397		1.140		12	f4s	(3P)2P2
3.5		26446		1.063		4	f4d	(3H)4H2
4.5		26457		1.042		7	f4d	(5F)6G
2.5		26470		0.912		5	f4s	(3H)6D
7.5	26527.106	26493	33	1.095	1.075	17	f4d	(3H)4Lb
1.5		26521		1.308		10	f4d	(5F)6P
5.5		26544		1.093		9	f2d2s	(3H)4Kb
4.5		26569		1.066		7	f4d	(5D)6D
3.5		26623		1.160		6	f4d	(5S)6D
5.5	26628.496	26633	−5	1.065	1.155	19	f4d	(3H)6K
9.5		26641		1.176		41	f4d	(5I)6K
0.5		26642		0.493		8	f2d2s	(5G)6F
8.5		26703		1.050		30	f4d	(5I)6K
4.5		26717		1.094		7	f4s	(1G)2G4
0.5		26793		0.457		14	f4d	(5F)4D
2.5		26801		1.049		18	f2d2s	(3F)6Fa
3.5		26811		0.980		7	f2.d	(3H)4Hh
6.5		26842		1.015		12	f4d	(3H)4I
4.5		26856		1.076		6	f2d2s	(3G)2G2
5.5	26989.437	26863	125	1.103	1.095	13	f4d	(3H)6K
7.5		26868		1.061		34	f4s	(3H)6K
5.5		26903		1.059		12	f4d	(3H)6K
1.5		26918		0.766		6	f2d2s	(5F)6G
3.5		26919		1.111		14	f4s	(3F)4F4
7.5	26931.699	26961	−29	1.058		18	f4d	(5I)4L
2.5		26966		1.098		4	f2d2s	(3H)6D
3.5		26974		1.143		5	f4d	(5D)4D
4.5		26984		1.056		9	f4d	(5F)6F
5.5		27019		1.135		9	f4d	(5I)6G
6.5		27037		0.995		14	f2d2s	(3H)4Ka
1.5		27069		0.490		17	f2d2s	(3F)6Gb
2.5		27161		1.031		5	f4d	(5F)6G
6.5		27170		1.006		6	f4d	(3H)4Ka
4.5		27177		1.117		9	f4d	(5S)6D
3.5		27192		0.990		5	f4d	(5D)4D
4.5		27287		1.014		13	f2d2s	(3F)6H
5.5		27295		1.094		6	f4d	(5I)6G
3.5		27356		1.029		5	f2d2s	(3H)4H2
1.5		27358		1.067		9	f2d2s	(3H)6F
3.5		27382		1.005		4	f2d2s	(3F)6H
2.5		27385		0.982		7	f4d	(5S)6D
4.5		27390		1.060		5	f2d2s	(3G)2G2
2.5		27434		0.998		9	f4d	(3F)2F4
6.5		27439		1.033		17	f2d2s	(3H)4I
4.5		27484		1.089		9	f2d2s	(3F)4H

Table 6. *Cont.*

J	E^{exp} (cm^{-1})	E^{th} (cm^{-1})	ΔE (cm^{-1})	g_L^{th}	g_L^{exp}	% $1^{st}comp$	Conf	Term
6.5		27508		1.098		9	f2d2s	(3H)6Ia
1.5		27520		0.933		6	f4d	(3F)6P
9.5		27553		1.178		89	f2d2s	(3H)6L
0.5		27574		1.729		17	f2d2s	(3F)4P
7.5		27597		1.181		15	f4d	(5I)6H
5.5		27618		1.033		9	f2d2s	(3H)6Ib
3.5		27629		1.046		6	f4d	(5F)6D
4.5		27659		0.944		9	f2d2s	(3H)6Ib
2.5		27661		0.922		7	f2d2s	(3H)6Hb
3.5		27737		0.930		9	f2d2s	(3H)4Hb
7.5	27695.597	27788	−92	1.073	1.090	9	f4d	(3H)4Lb

*: The level at 23241.365 cm^{-1} has a leading component belonging to the $5f^26d7s^2$ configuration but the dominant configuration is $5f^26d^27s$.

Table 7. Adopted parameters (in cm^{-1}) for even parity configurations $5f^37s7p$ and $5f^36d7p$ of U II compared with HFR radial integrals. The scaling factors are $SF(P) = P_{fit}/P_{HFR}$. Constraints on some parameters (denoted 'rn' in the 'Unc' columns of standard errors) link parameters of the same 'rn' to vary in a constant ratio. The HFR values of E_{av} parameters are relative to the lowest even configuration $5f^47s$ taken as zero value.

	$5f^37s7p$					$5f^36d7p$				
Param. P	P_{fit}	**Cstr.**	**Unc.**	P_{HFR}	$\Delta E/SF$	P_{fit}	**Cstr.**	**Unc.**	P_{HFR}	$\Delta E/SF$
E_{av}	54614		598	10338	44276	61291		312	17718	43573
$F^2(ff)$	48327	f		70448	0.686	47577	f		69355	0.686
$F^4(ff)$	45655	f		45655	0.691	31003	f		44867	0.691
$F^6(ff)$	22523	f		33367	0.675	22117	f		32766	0.675
α	36.5	f				36.5	f			
β	−600	f				−600	f			
γ	1500	f				1500	f			
ζ_f	1809	r1	102	1875	0.965	1781	r1	100	1846	0.965
ζ_d						1624	f		1902	0.854
ζ_p	5118	f		4490	1.14	4232	f		3713	1.14
$F^2(fp)$	7201	f		9001	0.80	6461	f		8077	0.80
$F^2(fd)$						20160	f		29007	0.695
$F^4(fd)$						15386	f		15765	0.976
$F^2(dp)$						13971	f		17464	0.80
$G^1(fd)$						11676	f		18742	0.623
$G^3(fd)$						12558	f		13562	0.926
$G^5(fd)$						8692	f		10084	0.862
$G^2(fp)$	2205	f		2205		1552	f		1939	0.8
$G^4(fp)$	1978	f		1978		1382	f		1727	0.8
$G^3(fs)$	2618	f		4328	0.605					
$G^1(dp)$						6543	f		10904	0.6
$G^3(dp)$						4812	f		8019	0.6
$G^1(sp)$	23515	f		26415	0.89					
Configuration Interaction										
$5f^37s7p - 5f^36d7p$										
$R^2(fs,fd)$	−6710	f	−10167	0.66						
$R^3(fs,df)$	−1442	f	−2185	0.66						
$R^2(sp,dp)$	−11708	f	−17742	0.66						
$R^1(sp,pd)$	−10971	f	−16622	0.66						

Table 8. Energy levels for even parity configurations $5f^37s7p$ and $5f^36d7p$ of U II. $\Delta E = E^{exp} - E^{th}$. The percentage, the configuration and the *LS* name of the leading component in the corresponding configuration are given in the last three columns.

J	E^{exp} (cm^{-1})	E^{th} (cm^{-1})	ΔE (cm^{-1})	g_L^{th}	g_L^{exp}	% 1st*comp*	*Conf*	*Term*
4.5	23315.092	22981	333	0.649	0.875	56	$5f^37s7p$	(4I)6I
5.5	24608.168	24608		0.874	0.910	30	$5f^37s7p$	(4I)6I
3.5	24342.199	24688	−345	0.577	0.760	53	$5f^37s7p$	(4I)6I
4.5	25437.162	25193	243	0.887	0.930	15	$5f^37s7p$	(4I)6I
6.5	26191.312	25937	253	0.753	0.890	49	$5f^36d7p$	(4I)6M
5.5		26810		0.733		33	$5f^36d7p$	(4I)6La
5.5		27873		0.834		25	$5f^37s7p$	(4I)6K
4.5		28301		0.729		30	$5f^36d7p$	(4I)4If
5.5	28154.447	28532	−378	0.862	0.890	18	$5f^37s7p$	(4I)6Lb
6.5		28989		1.033		48	$5f^37s7p$	(4I)6K
5.5		29614		0.970		18	$5f^37s7p$	(4I)6K
4.5		29689		0.865		42	$5f^37s7p$	(4I)6I
2.5		30187		0.362		65	$5f^37s7p$	(4I)6H
3.5		30321		0.723		31	$5f^37s7p$	(4I)6H
1.5		30387		0.375		46	$5f^37s7p$	(4F)6G
7.5	30341.673	30527	−185	0.910	1.010	59	$5f^36d7p$	(4I)6M
4.5		30725		0.874		19	$5f^37s7p$	(4I)6H
2.5		31004		0.787		8	$5f^36d7p$	(4I)4Ga
6.5		31106		0.919		36	$5f^36d7p$	(4I)6La
5.5		31210		0.965		15	$5f^37s7p$	(4I)4Ka
3.5		31231		0.705		14	$5f^36d7p$	(4I)6Ia
6.5		31719		1.005		21	$5f^37s7p$	(4I)6I
0.5		31781		0.118		28	$5f^37s7p$	(4F)6F
1.5		32017		1.030		12	$5f^37s7p$	(4F)6F
2.5		32113		0.862		34	$5f^37s7p$	(4F)6G
6.5	32535.021	32250	283	0.940	0.990	29	$5f^36d7p$	(4I)6Lb
5.5		32326		0.930		35	$5f^36d7p$	(4I)6Kb
6.5		32464		1.093		21	$5f^37s7p$	(4I)6H
4.5		32642		0.693		40	$5f^36d7p$	(4I)6Ka
6.5		32856		0.840		27	$5f^36d7p$	(4I)4Ld
7.5		32894		1.139		47	$5f^37s7p$	(4I)6K
3.5		33045		0.673		31	$5f^36d7p$	(4I)4Hf
5.5		33215		0.799		25	$5f^36d7p$	(4I)6Lb
4.5		33289		0.888		11	$5f^37s7p$	(2H)4I2
0.5		33509		0.499		12	$5f^36d7p$	(4F)6F
4.5		33651		0.809		21	$5f^36d7p$	(4I)6Kb
6.5		33806		1.036		21	$5f^37s7p$	(4I)4Ka
2.5		33859		1.057		10	$5f^37s7p$	(4F)6G
5.5		33985		1.026		22	$5f^37s7p$	(4I)4Ia
5.5	34207.000	34347		0.837		15	$5f^36d7p$	(4I)6La
3.5		34351		0.951		22	$5f^37s7p$	(4F)6G
6.5		34409		0.936		11	$5f^36d7p$	(4I)4Lc
5.5		34427		1.048		12	$5f^37s7p$	(4I)6I
2.5		34433		0.540		26	$5f^37s7p$	(4G)6H
1.5		34435		0.866		14	$5f^37s7p$	(4G)6G
3.5		34466		0.698		25	$5f^36d7p$	(4I)6I
4.5		34496		0.920		11	$5f^36d7p$	(4I)6Ia
7.5		34524		0.959		17	$5f^36d7p$	(4I)4Mb
4.5		34560		0.990		14	$5f^37s7p$	(4I)6H
3.5		34761		0.935		17	$5f^37s7p$	(4F)6G
1.5		34812		1.605		42	$5f^37s7p$	(4S)6P
8.5	34632.367	34911	−279	1.024	1.085	60	$5f^36d7p$	(4I)6M

Table 8. *Cont.*

J	E^{exp} (cm^{-1})	E^{th} (cm^{-1})	ΔE (cm^{-1})	g_L^{th}	g_L^{exp}	% $1^{st}comp$	Conf	Term
2.5		35093		0.837		17	$5f^37s7p$	(4F)4Ga
3.5		35149		0.908		9	$5f^37s7p$	(4I)6H
7.5		35301		1.085		17	$5f^36d7p$	(4I)6I
4.5		35338		0.895		10	$5f^36d7p$	(4I)6K
2.5		35498		1.074		17	$5f^37s7p$	(4F)6F
6.5		35514		0.990		31	$5f^36d7p$	(4I)6Kb
7.5		35529		1.051		27	$5f^36d7p$	(4I)6La
3.5		35552		0.873		12	$5f^36d7p$	(4I)6H
3.5		35666		0.834		6	$5f^36d7p$	(4I)6H
4.5		35696		0.811		13	$5f^36d7p$	(4I)6H
5.5		35744		0.966		19	$5f^37s7p$	(4I)2I
4.5		35780		0.867		11	$5f^36d7p$	(4I)4Ia
5.5		35804		0.901		14	$5f^36d7p$	(4I)4Ke
6.5		35809		1.064		30	$5f^37s7p$	(4I)2K
7.5		35892		1.196		34	$5f^37s7p$	(4I)6H
3.5		35906		1.009		9	$5f^36d7p$	(4F)6G
5.5		35985		0.961		13	$5f^36d7p$	(4I)6Ka
1.5		36046		0.867		14	$5f^37s7p$	(4F)6F
3.5		36064		1.089		10	$5f^37s7p$	(4F)6D
4.5		36137		0.865		10	$5f^36d7p$	(4I)6Ib
2.5		36149		1.444		33	$5f^37s7p$	(4S)6P
8.5		36364		1.210		31	$5f^37s7p$	(4I)6I
7.5		36389		1.059		29	$5f^36d7p$	(4I)6Lb
6.5		36410		1.019		14	$5f^36d7p$	(4I)6I
2.5		36417		0.607		17	$5f^36d7p$	(4F)6Ha
1.5		36462		1.200		14	$5f^37s7p$	(4S)2P
5.5		36516		0.934		39	$5f^36d7p$	(4I)6Ka
0.5		36680		1.812		15	$5f^36d7p$	(4S)4Pa
3.5		36721		0.935		9	$5f^37s7p$	(4F)6I
4.5		36849		0.904		18	$5f^36d7p$	(4I)4Hf
2.5		36946		1.020		5	$5f^36d7p$	(4F)4Fa
4.5		37004		1.163		34	$5f^37s7p$	(4F)6G
3.5		37008		0.949		10	$5f^36d7p$	(4I)6Ia
1.5		37053		0.995		12	$5f^36d7p$	(4I)6Ga
0.5		37112		2.063		27	$5f^37s7p$	(4F)6D
2.5		37313		0.864		6	$5f^36d7p$	(4I)6Ga
3.5		37369		0.935		16	$5f^37s7p$	(4G)6H
2.5		37477		0.897		13	$5f^36d7p$	(4G)6H
1.5		37537		0.695		14	$5f^36d7p$	(4F)4Ff
7.5		37620		1.065		13	$5f^37s7p$	(4I)4Ka
6.5		37789		1.127		26	$5f^37s7p$	(4I)4Ha
3.5		37822		0.986		8	$5f^37s7p$	(4F)4Ga
1.5		37855		1.172		15	$5f^36d7p$	(4F)6F
5.5		37879		0.929		11	$5f^36d7p$	(4I)6Kb
7.5		37917		0.996		18	$5f^36d7p$	(4I)4Mb
2.5		37960		0.861		9	$5f^36d7p$	(4G)6H
8.5	37308.326	37991	−683	1.062	1.070	18	$5f^36d7p$	(4I)6M
4.5		38040		0.895		18	$5f^36d7p$	(4I)6I
3.5		38063		0.821		10	$5f^36d7p$	(4F)6I
1.5		38080		0.578		23	$5f^36d7p$	(4G)6G
6.5		38156		0.931		15	$5f^36d7p$	(4I)4Ld
5.5		38183		0.896		13	$5f^36d7p$	(4I)6K
4.5		38205		0.863		12	$5f^36d7p$	(4I)6I
2.5		38217		0.816		9	$5f^36d7p$	(4I)6Hb
0.5		38228		2.117		26	$5f^37s7p$	(4F)6D

Table 8. *Cont.*

J	E^{exp} (cm^{-1})	E^{th} (cm^{-1})	ΔE (cm^{-1})	g_L^{th}	g_L^{exp}	% 1st comp	Conf	Term
3.5		38436		0.941		9	$5f^36d7p$	(4F)4Hd
2.5		38452		0.879		17	$5f^37s7p$	(4G)2F
7.5		38497		1.029		25	$5f^36d7p$	(4I)4Lc
2.5		38554		0.760		11	$5f^36d7p$	(4I)6Gb
4.5		38630		1.028		10	$5f^36d7p$	(4G)6H
6.5		38721		0.974		15	$5f^36d7p$	(4I)6Kb
1.5		38794		1.134		8	$5f^36d7p$	(4F)6D
5.5		38809		0.916		22	$5f^36d7p$	(4I)6K
8.5		38835		1.187		29	$5f^37s7p$	(4I)6K
2.5		38916		1.074		12	$5f^36d7p$	(4F)6F
4.5		38931		1.064		7	$5f^36d7p$	(4I)6I
4.5		38975		0.830		20	$5f^36d7p$	(4G)6K
3.5		38993		0.936		8	$5f^37s7p$	(4I)4Ha
2.5		39000		0.891		14	$5f^37s7p$	(4G)6G
1.5		39006		0.407		30	$5f^36d7p$	(4I)6Gb
9.5	39809.365	39050	758	1.111	1.105	48	$5f^36d7p$	(4I)6M
5.5		39065		1.092		5	$5f^36d7p$	(4I)6Kb
6.5		39122		1.038		14	$5f^36d7p$	(4I)6Ka
4.5		39161		1.136		14	$5f^36d7p$	(4F)6D

3.3. Partition Function

To get an idea of how semi-empirical parametric calculations could influence the value of the partition function $Q(T) = \sum_i (2J_i + 1)exp(-E_i/k_BT)$ (k_B : Boltzmann constant), we made an estimation of the partition function of U II for a typical stellar temperature. The temperature chosen is 4825 K ($k_BT = 3353.54$ cm^{-1}), which is the temperature quoted by Cayrel et al. [1] for a metal-poor star showing the U II line at 3859.57 Å in its spectrum.

Since experimental levels are incompletely determined, a partition function calculated with only known experimental energies would be underestimated. Therefore we calculated the partition function with all the available experimental energies supplemented by the final least squares fitted energies when experimental ones are missing. In the expression of the partition function we included all the levels below 46 000 cm^{-1} of both parities. The result is : $Q_{exp/LSF}(T) = 122.99$, which is the best value possible in the present case. When the partition function is calculated with the same number of levels, but with all the fitted energies, the result is: Q_{LSF}(T) = 120.99, which agrees with $Q_{exp/LSF}(T)$ within 2%. When *ab initio* HFR energy values are used, we have Q_{HFR}(T) = 89.19, which is 26% smaller. Consequently, in absence of complete experimental level energies, the energies calculated from fitted parameters provide a realistic estimation of the partition function.

3.4. Transition Probabilities

The parametric calculations provide gA values for transition probabilities (g: upper level statistical weight; A: Einstein coefficient of spontaneous emission) between calculated levels. Extensive comparison with experimental transition probabilities is not possible because of the scarcity of measurements. Furthermore, because of the strongly mixed wave functions, weak transitions are sensitive to small changes of energy parameters and may not be reliable for comparison. Nevertheless, it is interesting to consider the line at 3859.6 Å, which is strong and used as cosmochronometer [1]. Chen and Borzileri [21] measured the gA value for this line and found 2.8×10^8 s^{-1}, to be compared with previous measurement 1.1×10^8 s^{-1} by Corliss [22]. Nilsson et al. [23] derived branching ratios from relative intensities measured in FTS spectra and combined with radiative lifetime of the upper level at 26191 cm^{-1} to find a g_lf value of 0.856 for the oscillator strength weighted by the lower level degeneracy. The corresponding gA value (Equation (1) of [23]) is 3.8×10^8 s^{-1} in agreement with the value of 3.5 × 10^8 s^{-1} calculated by Kurucz [24]. Our calculations lead to gA = 1.53 × 10^9 s^{-1}, four times larger, but they confirm the order of magnitude. However, the parametric study for the high even levels of $5f^37s7p + 5f^36d7p$ is still unachieved, since treated without all the interacting even configurations. Its results should be taken with caution.

4. Classified Lines of U II in the Ultraviolet

On our spectrograms described in Section 2, some lines were relatively sharp and were likely emitted by singly charged uranium ions. For identification of U II lines, we searched experimental wave numbers matching the Ritz wave numbers calculated from the energy differences of known U II energy levels reported in [6], even when the level was not assigned with quantum numbers. The maximum uncertainty of the wavelength measurements is estimated to be ±0.003 Å . Thus the corresponding uncertainty on wave numbers should be less than ±0.05 cm^{-1}. To take into account any possible perturbations in the spark spectrum, we chose a tolerance of ±0.1 cm^{-1} for a criterion of identification. Indeed, according to [14], the level energies in [6], therefore the Ritz wave numbers, have negligible uncertainties of about ±0.01 cm^{-1}. Table 9 lists the 451 lines between 2344 and 2955 Å identified as U II transitions, with calculated Ritz wavelengths, experimental wavelengths, deviations exp-Ritz and line intensities, together with the corresponding upper and lower levels. One line has triple identification and 24 lines have double identification. These concern mostly lines with two deviations of opposite signs. Otherwise, the line with the smallest deviation is retained. No gA values were available here for confirmation of identifications since the even levels involved in these transitions have only experimental energy values but no quantum numbers assigned except the J values.

Search of new levels of $5f^36d7p$ close to the predicted energies of Table 8 was attempted using the possible U II lines left unidentified. Unfortunately, only one chain of transitions supported by calculated transition probabilities could be found leading to a level $5f^36d7p$ $(^4I)^6K$ with $J = 5.5$ at 39113.98 ± 0.1 cm^{-1}. Table 10 lists the six transitions that establish this level.

Table 9. Ultraviolet transitions of U II emitted from a vacuum spark source. wl_{Ritz} : Ritz wavelength calculated with experimental energies from [6]; wl_{exp}: experimental wavelength ; $\Delta wl = wl_{exp} - wl_{Ritz}$; wn_{exp}: experimental wavenumbers; $\Delta wn = wn_{exp} - (E_{even} - E_{odd})$.

wl_{Ritz} in Air (Å)	wl_{exp} in Air (Å)	*Int*	*Note*	wn_{exp} (cm^{-1})	Δwl (Å)	Δwn (cm^{-1})	E_{odd} (cm^{-1})	J_{odd}	E_{even} (cm^{-1})	J_{even}
2343.5696	2343.5707	49		42656.867	0.0012	−0.021	0.000	4.5	42656.888	5.5
2348.8952	2348.8968	73		42560.149	0.0016	−0.029	289.041	5.5	42849.219	5.5
2390.9748	2390.9742	10		41811.216	−0.0006	0.011	0.000	4.5	41811.205	5.5
2401.1302	2401.1330	125		41634.329	0.0029	−0.050	289.041	5.5	41923.420	6.5
2423.7052	2423.7102	52		41246.529	0.0051	−0.086	0.000	4.5	41246.615	4.5
2427.0021	2427.0022	43		41190.593	0.0001	−0.001	914.765	4.5	42105.359	5.5
2448.0954	2448.0942	97		40835.731	−0.0012	0.020	289.041	5.5	41124.752	6.5
2448.9324	2448.9265	7		40821.854	−0.0059	0.099	0.000	4.5	40821.755	4.5
2471.0901	2471.0868	55		40455.794	−0.0033	0.054	2294.696	5.5	42750.436	6.5
2477.1835	2477.1824	30		40356.253	−0.0010	0.017	1749.123	6.5	42105.359	5.5
2478.6816	2478.6852	37	as	40331.791	0.0036	−0.059	914.765	4.5	41246.615	4.5
2481.1377	2481.1412	24		40291.868	0.0034	−0.056	289.041	5.5	40580.965	5.5
2484.0042	2484.0095	16		40245.347	0.0054	−0.087	0.000	4.5	40245.434	3.5
2484.6702	2484.6667	41		40234.703	−0.0035	0.057	0.000	4.5	40234.646	5.5
2490.2907	2490.2899	9		40143.856	−0.0009	0.014	914.765	4.5	41058.607	5.5
2491.4292	2491.4330	8		40125.442	0.0037	−0.060	1749.123	6.5	41874.625	6.5
2506.8037	2506.8012	25		39879.462	−0.0025	0.039	914.765	4.5	40794.188	4.5
2512.5746	2512.5784	13		39787.777	0.0038	−0.060	0.000	4.5	39787.837	5.5
2514.7696	2514.7686	19	LA	39753.125	−0.0010	0.016	4420.871	5.5	44173.980	6.5
2518.9755	2518.9760	37		39686.730	0.0005	−0.008	0.000	4.5	39686.738	5.5
2533.2401	2533.2370	42		39463.326	−0.0031	0.049	0.000	4.5	39463.277	5.5
2537.6966	2537.6954	163		39393.998	−0.0012	0.018	289.041	5.5	39683.021	5.5
2538.4329	2538.4355	85		39382.517	0.0026	−0.041	0.000	4.5	39382.558	3.5
2538.7351	2538.7384	33		39377.819	0.0033	−0.051	289.041	5.5	39666.911	4.5
2539.1756	2539.1760	96	as	39371.032	0.0004	−0.006	289.041	5.5	39660.079	6.5
2540.7030	2540.7065	39	c	39347.316	0.0034	−0.053	1749.123	6.5	41096.492	6.5
2541.3669	2541.3655	39	c	39337.112	−0.0014	0.022	0.000	4.5	39337.090	4.5
2554.3761	2554.3725	134	as	39136.819	−0.0035	0.054	1749.123	6.5	40885.888	6.5

Table 9. *Cont.*

wl_{Ritz} in Air (Å)	wl_{exp} in Air (Å)	*Int*	*Note*	wn_{exp} (cm^{-1})	Δwl (Å)	Δwn (cm^{-1})	E_{odd} (cm^{-1})	J_{odd}	E_{even} (cm^{-1})	J_{even}
2556.1928	2556.1946	82	LA	39108.922	0.0018	−0.028	0.000	4.5	39108.950	5.5
2560.1798	2560.1743	20		39048.133	−0.0055	0.084	289.041	5.5	39337.090	4.5
2560.3421	2560.3457	29		39045.519	0.0037	−0.056	0.000	4.5	39045.575	4.5
2561.7992	2561.7934	46		39023.455	−0.0058	0.089	914.765	4.5	39938.131	5.5
2565.4072	2565.4130	104	LA	38968.400	0.0058	−0.088	0.000	4.5	38968.488	5.5
2567.2954	2567.2987	174	as	38939.778	0.0032	−0.049	2294.696	5.5	41234.523	6.5
2567.9515	2567.9578	11		38929.783	0.0063	−0.095	0.000	4.5	38929.878	4.5
2568.9777	2568.9783	26	LA,as	38914.318	0.0006	−0.009	5259.653	7.5	44173.980	6.5
2569.7085	2569.7095	45	LA	38903.246	0.0010	−0.015	0.000	4.5	38903.261	3.5
2575.2266	2575.2291	23	LA	38819.872	0.0025	−0.037	289.041	5.5	39108.950	5.5
2577.3205	2577.3219	20		38788.354	0.0015	−0.022	0.000	4.5	38788.376	4.5
2578.7860	2578.7797	155		38766.427	−0.0063	0.095	1749.123	6.5	40515.455	5.5
2579.5692	2579.5681	29		38754.579	−0.0011	0.017	0.000	4.5	38754.562	4.5
2584.4158	2584.4163	42	LA	38681.882	0.0005	−0.008	0.000	4.5	38681.890	3.5
2584.9012	2584.9028	10	c	38674.602	0.0016	−0.024	0.000	4.5	38674.626	4.5
2586.1972	2586.1965	44		38655.256	−0.0007	0.010	4420.871	5.5	43076.117	4.5
2591.2483	2591.2450	107	as	38579.949	−0.0034	0.050	0.000	4.5	38579.899	4.5
2592.5704	2592.5690	60		38560.247	−0.0014	0.021	0.000	4.5	38560.226	3.5
2593.5699	2593.5698	30		38545.372	−0.0001	0.002	0.000	4.5	38545.370	5.5
2601.4681	2601.4695	72		38428.328	0.0014	−0.020	4420.871	5.5	42849.219	5.5
2604.2985	2604.2993	48	p	38386.572	0.0008	−0.012	0.000	4.5	38386.584	3.5
2606.7253	2606.7266	63		38350.837	0.0013	−0.019	0.000	4.5	38350.856	4.5
2607.3014	2607.3069	65		38342.302	0.0055	−0.081	289.041	5.5	38631.424	5.5
2608.1733	2608.1800	28	p	38329.466	0.0067	−0.099	4420.871	5.5	42750.436	6.5
2609.2426	2609.2457	23		38313.811	0.0031	−0.046	0.000	4.5	38313.857	5.5
2609.8933	2609.8900	253		38304.353	−0.0033	0.049	914.765	4.5	39219.069	5.5
2612.4565	2612.4555	11		38266.741	−0.0010	0.015	0.000	4.5	38266.726	4.5
2613.9584	2613.9578	11		38244.748	−0.0005	0.008	0.000	4.5	38244.740	3.5
2615.9468	2615.9422	24		38215.737	−0.0046	0.067	0.000	4.5	38215.670	4.5
2616.0690	2616.0679	35		38213.901	−0.0012	0.017	0.000	4.5	38213.884	5.5
2620.8611	2620.8670	19		38143.931	0.0059	−0.086	914.765	4.5	39058.782	5.5
2621.4511	2621.4463	21		38135.502	−0.0048	0.070	1749.123	6.5	39884.555	6.5
2623.5499	2623.5514	17		38104.907	0.0015	−0.022	4420.871	5.5	42525.800	5.5
2624.9155	2624.9101	24		38085.183	−0.0054	0.078	4420.871	5.5	42505.976	6.5
2625.2536	2625.2508	21		38080.241	−0.0028	0.041	0.000	4.5	38080.200	5.5
2628.9275	2628.9276	41		38026.982	0.0001	−0.002	0.000	4.5	38026.984	4.5
2632.6555	2632.6570	32		37973.118	0.0015	−0.021	0.000	4.5	37973.139	5.5
2632.9771	2632.9786	42		37968.480	0.0015	−0.021	0.000	4.5	37968.501	4.5
2634.3223	2634.3286	35		37949.022	0.0063	−0.091	2294.696	5.5	40243.809	6.5
2635.1207	2635.1213	26		37937.607	0.0006	−0.008	1749.123	6.5	39686.738	5.5
2635.3792	2635.3781	69	as	37933.914	−0.0011	0.016	1749.123	6.5	39683.021	5.5
2635.5278	2635.5306	102		37931.719	0.0028	−0.041	0.000	4.5	37931.760	4.5
2637.6935	2637.6967	15		37900.569	0.0032	−0.046	4420.871	5.5	42321.486	6.5
2639.5742	2639.5720	20	p	37873.643	−0.0022	0.032	914.765	4.5	38788.376	4.5
2639.8350	2639.8351	26	p	37869.868	0.0001	−0.001	0.000	4.5	37869.869	5.5
2641.5456	2641.5488	20		37845.305	0.0031	−0.045	0.000	4.5	37845.350	3.5
2641.9333	2641.9291	8	p	37839.856	−0.0041	0.059	914.765	4.5	38754.562	4.5
2644.1238	2644.1281	20		37808.391	0.0043	−0.062	0.000	4.5	37808.453	5.5
2645.4716	2645.4749	78	LA	37789.143	0.0033	−0.047	0.000	4.5	37789.190	4.5
2648.7844	2648.7857	9		37741.914	0.0013	−0.018	0.000	4.5	37741.932	4.5
2649.0644	2649.0686	68		37737.884	0.0041	−0.059	289.041	5.5	38026.984	4.5
2650.7354	2650.7392	2		37714.100	0.0038	−0.054	1749.123	6.5	39463.277	5.5
2652.1885	2652.1883	29		37693.494	−0.0003	0.004	1749.123	6.5	39442.613	7.5
2652.8221	2652.8233	83		37684.471	0.0012	−0.017	4420.871	5.5	42105.359	5.5
2656.5889	2656.5910	172		37631.029	0.0021	−0.030	914.765	4.5	38545.824	3.5
2660.1401	2660.1369	20		37580.873	−0.0032	0.045	289.041	5.5	37869.869	5.5
2662.8483	2662.8539	21		37542.527	0.0057	−0.080	4420.871	5.5	41963.478	4.5
2663.2920	2663.2908	4		37536.369	−0.0012	0.017	5526.750	6.5	43063.102	6.5
2664.1581	2664.1519	11	p	37524.237	−0.0062	0.088	4420.871	5.5	41945.020	6.5
2664.4580	2664.4578	23	p	37519.928	−0.0002	0.003	4585.434	6.5	42105.359	5.5
2665.6926	2665.6909	36		37502.572	−0.0016	0.023	4420.871	5.5	41923.420	6.5
2665.8632	2665.8615	6	LA	37500.173	−0.0017	0.024	289.041	5.5	37789.190	4.5
2666.5295	2666.5315	14		37490.754	0.0021	−0.029	5259.653	7.5	42750.436	6.5
2667.8790	2667.8833	25	b,	37471.758	0.0043	−0.061	914.765	4.5	38386.584	3.5
2668.0123	2668.0134	7		37469.931	0.0011	−0.015	1749.123	6.5	39219.069	5.5
2669.1658	2669.1602	28		37453.832	−0.0056	0.078	4420.871	5.5	41874.625	6.5
2669.2273	2669.2230	28		37452.951	−0.0043	0.060	289.041	5.5	37741.932	4.5
2670.5030	2670.5089	49		37434.916	0.0059	−0.083	2294.696	5.5	39729.695	6.5
2672.2712	2672.2736	11		37410.195	0.0024	−0.034	289.041	5.5	37699.270	5.5
2672.4852	2672.4830	182		37407.265	−0.0022	0.031	5259.653	7.5	42666.887	7.5

Table 9. *Cont.*

wl_{Ritz} in Air (Å)	wl_{exp} in Air (Å)	*Int*	*Note*	wn_{exp} (cm^{-1})	Δwl (Å)	Δwn (cm^{-1})	E_{odd} (cm^{-1})	J_{odd}	E_{even} (cm^{-1})	J_{even}
2672.7077	2672.7030	72		37404.185	−0.0047	0.066	0.000	4.5	37404.119	5.5
2675.1142	2675.1087	37		37370.552	−0.0054	0.076	289.041	5.5	37659.517	4.5
2675.8767	2675.8790	35	LA	37359.794	0.0024	−0.033	1749.123	6.5	39108.950	5.5
2676.4154	2676.4115	187	LA	37352.362	−0.0039	0.055	6283.431	6.5	43635.738	6.5
2676.6836	2676.6849	11		37348.546	0.0014	−0.019	1749.123	6.5	39097.688	6.5
2677.4419	2677.4355	23	p	37338.076	−0.0065	0.090	4585.434	6.5	41923.420	6.5
2678.7931	2678.7924	72	p	37319.163	−0.0007	0.010	4585.434	6.5	41904.587	5.5
2683.2766	2683.2723	55		37256.861	−0.0043	0.060	1749.123	6.5	39005.924	5.5
2683.4216	2683.4147	4	c	37254.884	−0.0070	0.097	289.041	5.5	37543.828	4.5
2683.4719	2683.4724	12	c	37254.083	0.0004	−0.006	289.041	5.5	37543.130	5.5
2684.0314	2684.0318	24	as	37246.318	0.0004	−0.005	5259.653	7.5	42505.976	6.5
2685.9761	2685.9705	10	LA,p	37219.443	−0.0056	0.078	1749.123	6.5	38968.488	5.5
2689.1080	2689.1152	13	D	37175.918	0.0072	−0.099	0.000	4.5	37176.017	4.5
2689.1195	2689.1152	13	D	37175.918	−0.0043	0.059	289.041	5.5	37464.900	6.5
2689.6460	2689.6496	8		37168.531	0.0036	−0.050	2294.696	5.5	39463.277	5.5
2691.0334	2691.0336	219	p	37149.415	0.0003	−0.004	4420.871	5.5	41570.290	5.5
2693.7311	2693.7346	55	as	37112.170	0.0036	−0.049	914.765	4.5	38026.984	4.5
2697.3932	2697.3884	20		37061.899	−0.0048	0.066	5259.653	7.5	42321.486	6.5
2697.9173	2697.9129	24	p	37054.694	−0.0044	0.060	1749.123	6.5	38803.757	7.5
2698.4845	2698.4782	148		37046.932	−0.0063	0.087	4585.434	6.5	41632.279	6.5
2700.2512	2700.2548	12	as	37022.562	0.0036	−0.049	4420.871	5.5	41443.482	6.5
2705.7866	2705.7917	13		36946.806	0.0051	−0.070	1749.123	6.5	38695.999	5.5
2706.9739	2706.9728	63		36930.686	−0.0012	0.016	8276.733	6.5	45207.403	7.5
2708.9821	2708.9885	66		36903.206	0.0064	−0.087	4585.434	6.5	41488.727	5.5
2709.5050	2709.5064	30	LA,p	36896.153	0.0013	−0.018	4420.871	5.5	41317.042	5.5
2709.5564	2709.5498	77		36895.562	−0.0066	0.090	0.000	4.5	36895.472	4.5
2711.1029	2711.0998	13	LA,as	36874.468	−0.0032	0.043	914.765	4.5	37789.190	4.5
2711.7043	2711.7061	7		36866.223	0.0018	−0.024	5790.641	5.5	42656.888	5.5
2711.7820	2711.7807	22	b	36865.209	−0.0013	0.018	0.000	4.5	36865.191	5.5
2712.0582	2712.0588	8		36861.429	0.0005	−0.007	4420.871	5.5	41282.307	6.5
2714.5822	2714.5861	7		36827.115	0.0038	−0.052	914.765	4.5	37741.932	4.5
2715.5344	2715.5334	5		36814.267	−0.0010	0.013	2294.696	5.5	39108.950	5.5
2716.4220	2716.4269	9		36802.158	0.0049	−0.067	289.041	5.5	37091.266	4.5
2716.8633	2716.8693	9		36796.166	0.0060	−0.081	1749.123	6.5	38545.370	5.5
2718.0425	2718.0410	39		36780.304	−0.0015	0.020	1749.123	6.5	38529.407	7.5
2718.0444	2718.0410	39		36780.304	−0.0035	0.047	5259.653	7.5	42039.910	7.5
2719.3675	2719.3687	21		36762.350	0.0013	−0.017	1749.123	6.5	38511.490	5.5
2723.1625	2723.1671	139		36711.073	0.0046	−0.062	1749.123	6.5	38460.258	6.5
2725.0668	2725.0738	4	D,c	36685.386	0.0071	−0.095	8521.922	7.5	45207.403	7.5
2725.0752	2725.0738	4	D,c	36685.386	−0.0014	0.019	5259.653	7.5	41945.020	6.5
2725.2686	2725.2662	137	as	36682.796	−0.0024	0.032	6283.431	6.5	42966.195	7.5
2726.6810	2726.6841	27		36663.725	0.0031	−0.042	5259.653	7.5	41923.420	6.5
2727.7730	2727.7792	77	as	36649.007	0.0061	−0.082	4585.434	6.5	41234.523	6.5
2728.4664	2728.4701	7		36639.726	0.0037	−0.050	289.041	5.5	36928.817	4.5
2728.6183	2728.6233	8		36637.669	0.0050	−0.067	4420.871	5.5	41058.607	5.5
2733.9627	2733.9667	45		36566.071	0.0040	−0.054	289.041	5.5	36855.166	4.5
2734.0667	2734.0715	7		36564.670	0.0048	−0.064	1749.123	6.5	38313.857	5.5
2734.2730	2734.2762	285		36561.932	0.0032	−0.043	5401.503	3.5	41963.478	4.5
2735.5783	2735.5728	22	p	36544.602	−0.0055	0.073	4420.871	5.5	40965.400	6.5
2738.9799	2738.9848	35		36499.079	0.0049	−0.065	2294.696	5.5	38793.840	6.5
2739.3900	2739.3921	20	D	36493.652	0.0021	−0.028	2294.696	5.5	38788.376	4.5
2739.3906	2739.3921	20	D,LA	36493.652	0.0015	−0.020	289.041	5.5	36782.713	6.5
2740.6331	2740.6355	6		36477.099	0.0024	−0.032	289.041	5.5	36766.172	5.5
2740.8273	2740.8282	18		36474.534	0.0010	−0.013	289.041	5.5	36763.588	4.5
2740.9305	2740.9325	9		36473.146	0.0020	−0.027	4585.434	6.5	41058.607	5.5
2741.7458	2741.7506	24		36462.264	0.0047	−0.063	914.765	4.5	37377.092	4.5
2742.0571	2742.0563	15		36458.198	−0.0008	0.010	5259.653	7.5	41717.841	7.5
2744.4027	2744.3996	25		36427.069	−0.0032	0.042	914.765	4.5	37341.792	4.5
2745.0627	2745.0659	7		36418.227	0.0032	−0.043	5526.750	6.5	41945.020	6.5
2746.1590	2746.1557	12		36403.775	−0.0033	0.044	1749.123	6.5	38152.854	7.5
2746.6917	2746.6870	179		36396.733	−0.0048	0.063	5526.750	6.5	41923.420	6.5
2747.3598	2747.3547	23		36387.891	−0.0051	0.067	0.000	4.5	36387.824	3.5
2748.4450	2748.4475	7		36373.424	0.0025	−0.033	6283.431	6.5	42656.888	5.5
2748.5078	2748.5044	17		36372.671	−0.0034	0.045	5259.653	7.5	41632.279	6.5
2749.9421	2749.9398	37		36353.685	−0.0023	0.030	8853.748	8.5	45207.403	7.5
2750.3794	2750.3750	15		36347.933	−0.0044	0.058	5526.750	6.5	41874.625	6.5
2750.5536	2750.5520	7		36345.594	−0.0017	0.022	8276.733	6.5	44622.305	7.5

Table 9. *Cont.*

wl_{Ritz} in Air (Å)	wl_{exp} in Air (Å)	*Int*	*Note*	wn_{exp} (cm^{-1})	Δwl (Å)	Δwn (cm^{-1})	E_{odd} (cm^{-1})	J_{odd}	E_{even} (cm^{-1})	J_{even}
2751.2231	2751.2161	21		36336.820	−0.0070	0.092	2294.696	5.5	38631.424	5.5
2752.4357	2752.4349	13		36320.730	−0.0008	0.011	1749.123	6.5	38069.842	5.5
2754.1493	2754.1480	80		36298.143	−0.0013	0.017	914.765	4.5	37212.891	3.5
2756.8379	2756.8307	22		36262.821	−0.0072	0.095	289.041	5.5	36551.767	5.5
2756.9499	2756.9462	9	c	36261.301	−0.0037	0.049	914.765	4.5	37176.017	4.5
2757.5498	2757.5455	32		36253.420	−0.0043	0.056	0.000	4.5	36253.364	5.5
2758.9517	2758.9473	10	as	36235.000	−0.0043	0.057	914.765	4.5	37149.708	4.5
2759.7839	2759.7817	24		36224.045	−0.0022	0.029	1749.123	6.5	37973.139	5.5
2762.7151	2762.7167	12		36185.566	0.0017	−0.022	1749.123	6.5	37934.711	7.5
2762.8494	2762.8471	18		36183.858	−0.0022	0.029	5259.653	7.5	41443.482	6.5
2762.8797	2762.8769	38		36183.468	−0.0028	0.037	0.000	4.5	36183.431	5.5
2763.4090	2763.4131	29		36176.447	0.0041	−0.054	914.765	4.5	37091.266	4.5
2763.6889	2763.6843	14		36172.897	−0.0046	0.060	5790.641	5.5	41963.478	4.5
2764.2397	2764.2400	20	D	36165.625	0.0003	−0.004	5526.750	6.5	41692.379	5.5
2764.2449	2764.2400	20	D	36165.625	−0.0048	0.063	2294.696	5.5	38460.258	6.5
2764.6629	2764.6561	56		36160.182	−0.0067	0.088	4420.871	5.5	40580.965	5.5
2765.3970	2765.3989	49		36150.470	0.0019	−0.025	0.000	4.5	36150.495	3.5
2766.7528	2766.7518	13		36132.792	−0.0010	0.013	5790.641	5.5	41923.420	6.5
2766.8721	2766.8729	28	as	36131.211	0.0008	−0.010	0.000	4.5	36131.221	5.5
2767.6745	2767.6669	314		36120.845	−0.0076	0.099	1749.123	6.5	37869.869	5.5
2768.8587	2768.8516	37		36105.395	−0.0071	0.093	1749.123	6.5	37854.425	5.5
2770.0418	2770.0399	34		36089.906	−0.0019	0.025	0.000	4.5	36089.881	3.5
2770.7417	2770.7376	65		36080.819	−0.0041	0.054	6445.035	4.5	42525.800	5.5
2772.1759	2772.1743	9		36062.120	−0.0016	0.021	4420.871	5.5	40482.970	5.5
2772.3887	2772.3910	9		36059.300	0.0023	−0.030	1749.123	6.5	37808.453	5.5
2772.6325	2772.6313	38		36056.175	−0.0012	0.015	2294.696	5.5	38350.856	4.5
2773.6033	2773.6039	29		36043.532	0.0006	−0.008	5526.750	6.5	41570.290	5.5
2775.0145	2775.0118	6		36025.245	−0.0027	0.035	5526.750	6.5	41551.960	5.5
2775.2114	2775.2079	7		36022.699	−0.0035	0.045	5259.653	7.5	41282.307	6.5
2775.3724	2775.3675	5		36020.628	−0.0049	0.064	5790.641	5.5	41811.205	5.5
2775.8213	2775.8148	11		36014.828	−0.0066	0.085	4420.871	5.5	40435.614	6.5
2776.8169	2776.8205	7		36001.784	0.0036	−0.047	1749.123	6.5	37750.954	5.5
2778.4471	2778.4532	12	p	35980.629	0.0060	−0.078	914.765	4.5	36895.472	4.5
2779.4472	2779.4478	2	D	35967.757	0.0006	−0.008	1749.123	6.5	37716.888	7.5
2779.4508	2779.4478	2	D	35967.757	−0.0029	0.038	1749.123	6.5	37716.842	6.5
2780.0321	2780.0339	11	LA	35960.175	0.0018	−0.023	0.000	4.5	35960.198	4.5
2781.0310	2781.0317	9		35947.273	0.0007	−0.009	289.041	5.5	36236.323	4.5
2781.5634	2781.5684	21		35940.336	0.0050	−0.065	914.765	4.5	36855.166	4.5
2782.0684	2782.0730	5		35933.818	0.0046	−0.059	44357.295	4.5	8423.418	4.5
2783.2065	2783.2076	12		35919.174	0.0011	−0.014	2294.696	5.5	38213.884	5.5
2783.2899	2783.2969	3		35918.021	0.0070	−0.090	0.000	4.5	35918.111	5.5
2783.3968	2783.4034	41	?	35916.647	0.0066	−0.085	5526.750	6.5	41443.482	6.5
2784.4497	2784.4498	16		35903.149	0.0001	−0.001	5526.750	6.5	41429.900	6.5
2784.5592	2784.5533	95	b	35901.815	−0.0060	0.077	5790.641	5.5	41692.379	5.5
2784.6660	2784.6669	11		35900.350	0.0009	−0.011	289.041	5.5	36189.402	5.5
2784.9076	2784.9081	4		35897.241	0.0005	−0.006	8276.733	6.5	44173.980	6.5
2788.5251	2788.5246	11		35850.685	−0.0005	0.007	0.000	4.5	35850.678	5.5
2789.2284	2789.2236	79		35841.700	−0.0048	0.062	5790.641	5.5	41632.279	6.5
2791.2531	2791.2572	3		35815.592	0.0041	−0.052	5259.653	7.5	41075.297	8.5
2793.9333	2793.9363	47		35781.248	0.0030	−0.039	289.041	5.5	36070.328	4.5
2794.0612	2794.0576	8	D,as	35779.695	−0.0036	0.046	5790.641	5.5	41570.290	5.5
2794.0636	2794.0576	8	D,as	35779.695	−0.0060	0.077	8394.362	7.5	44173.980	6.5
2794.9276	2794.9215	37		35768.636	−0.0062	0.079	8853.748	8.5	44622.305	7.5
2795.2282	2795.2335	41		35764.643	0.0053	−0.068	914.765	4.5	36679.476	3.5
2796.2329	2796.2312	5		35751.882	−0.0017	0.022	0.000	4.5	35751.860	5.5
2797.1416	2797.1451	19		35740.201	0.0035	−0.045	914.765	4.5	36655.011	3.5
2800.1004	2800.0975	38		35702.521	−0.0029	0.037	5526.750	6.5	41229.234	7.5
2803.8298	2803.8262	40		35655.042	−0.0036	0.046	1749.123	6.5	37404.119	5.5
2805.2406	2805.2421	14	D	35637.045	0.0015	−0.019	2294.696	5.5	37931.760	4.5
2805.2455	2805.2421	14	D	35637.045	−0.0034	0.043	914.765	4.5	36551.767	5.5
2806.4938	2806.4936	11		35621.158	−0.0002	0.002	6283.431	6.5	41904.587	5.5
2807.1192	2807.1167	69	as	35613.251	−0.0025	0.032	289.041	5.5	35902.260	4.5
2809.0118	2809.0131	34		35589.208	0.0013	−0.017	1749.123	6.5	37338.348	7.5
2809.6382	2809.6426	29		35581.234	0.0044	−0.056	9626.113	6.5	45207.403	7.5
2809.9856	2809.9791	15	p	35576.974	−0.0066	0.083	289.041	5.5	35865.932	4.5
2813.0414	2813.0421	17		35538.240	0.0007	−0.009	4706.273	2.5	40244.522	1.5
2813.5474	2813.5491	7		35531.835	0.0017	−0.022	5526.750	6.5	41058.607	5.5
2814.7037	2814.7027	5		35517.273	−0.0010	0.013	4420.871	5.5	39938.131	5.5
2817.9580	2817.9578	37		35476.246	−0.0002	0.002	914.765	4.5	36391.009	3.5

Table 9. *Cont.*

wl_{Ritz} in Air (Å)	wl_{exp} in Air (Å)	*Int*	*Note*	wn_{exp} (cm^{-1})	Δwl (Å)	Δwn (cm^{-1})	E_{odd} (cm^{-1})	J_{odd}	E_{even} (cm^{-1})	J_{even}
2819.0247	2819.0286	7		35462.770	0.0039	−0.049	289.041	5.5	35751.860	5.5
2819.2845	2819.2778	8		35459.636	−0.0067	0.084	6445.035	4.5	41904.587	5.5
2820.2640	2820.2637	6		35447.240	−0.0003	0.004	2294.696	5.5	37741.932	4.5
2820.5036	2820.4993	13	as	35444.279	−0.0043	0.054	4585.434	6.5	40029.659	6.5
2821.1209	2821.1202	103		35436.482	−0.0006	0.008	914.765	4.5	36351.239	3.5
2826.2615	2826.2551	73	as	35372.099	−0.0064	0.080	0.000	4.5	35372.019	5.5
2826.6653	2826.6723	15		35366.879	0.0070	−0.087	4420.871	5.5	39787.837	5.5
2826.7289	2826.7250	25		35366.219	−0.0039	0.049	6445.035	4.5	41811.205	5.5
2828.9347	2828.9339	37		35338.608	−0.0007	0.009	914.765	4.5	36253.364	5.5
2829.2940	2829.2865	132		35334.205	−0.0075	0.094	5790.641	5.5	41124.752	6.5
2829.3965	2829.4002	53	p	35332.784	0.0038	−0.047	1749.123	6.5	37081.954	7.5
2830.0727	2830.0759	10		35324.349	0.0032	−0.040	0.000	4.5	35324.389	5.5
2831.5587	2831.5603	26	p	35305.831	0.0016	−0.020	5790.641	5.5	41096.492	6.5
2832.0616	2832.0605	44		35299.599	−0.0011	0.014	5259.653	7.5	40559.238	8.5
2832.0988	2832.0965	14		35299.150	−0.0023	0.029	4585.434	6.5	39884.555	6.5
2833.8199	2833.8173	29		35277.715	−0.0026	0.032	2294.696	5.5	37572.379	6.5
2834.0646	2834.0590	57	as	35274.707	−0.0056	0.070	914.765	4.5	36189.402	5.5
2834.5444	2834.5520	13		35268.572	0.0076	−0.094	914.765	4.5	36183.431	5.5
2834.5554	2834.5520	13		35268.572	−0.0035	0.043	6283.431	6.5	41551.960	5.5
2835.5690	2835.5682	10		35255.932	−0.0008	0.010	289.041	5.5	35544.963	4.5
2836.9143	2836.9164	19		35239.182	0.0021	−0.026	4420.871	5.5	39660.079	6.5
2837.1943	2837.1947	18		35235.725	0.0004	−0.005	914.765	4.5	36150.495	3.5
2837.3302	2837.3253	27		35234.103	−0.0049	0.061	0.000	4.5	35234.042	5.5
2839.7944	2839.8005	27		35203.393	0.0061	−0.075	914.765	4.5	36118.233	4.5
2839.8803	2839.8864	34	D	35202.328	0.0061	−0.075	4585.434	6.5	39787.837	5.5
2839.8925	2839.8864	34	D	35202.328	−0.0061	0.076	289.041	5.5	35491.293	5.5
2840.4603	2840.4659	16		35195.146	0.0056	−0.069	1749.123	6.5	36944.338	5.5
2842.4803	2842.4828	54		35170.173	0.0025	−0.031	2294.696	5.5	37464.900	6.5
2842.8541	2842.8515	26		35165.612	−0.0027	0.033	40882.028	5.5	5716.449	4.5
2845.5385	2845.5356	6		35132.445	−0.0028	0.035	1749.123	6.5	36881.533	6.5
2845.9556	2845.9514	9	as	35127.312	−0.0042	0.052	914.765	4.5	36042.025	4.5
2846.1491	2846.1474	11	D	35124.893	−0.0017	0.021	8510.866	5.5	43635.738	6.5
2846.1497	2846.1474	11	D	35124.893	−0.0023	0.028	8276.733	6.5	43401.598	7.5
2846.3700	2846.3683	16		35122.167	−0.0017	0.021	7166.632	4.5	42288.778	5.5
2849.9862	2849.9799	7		35077.659	−0.0063	0.077	5259.653	7.5	40337.235	8.5
2852.7458	2852.7524	9		35043.573	0.0065	−0.080	5526.750	6.5	40570.403	6.5
2853.4221	2853.4200	348		35035.374	−0.0021	0.026	289.041	5.5	35324.389	5.5
2853.5636	2853.5639	12	LA,D	35033.607	0.0003	−0.004	6283.431	6.5	41317.042	5.5
2853.5653	2853.5639	12	LA,D	35033.607	−0.0014	0.017	1749.123	6.5	36782.713	6.5
2854.9132	2854.9129	4		35017.053	−0.0003	0.004	1749.123	6.5	36766.172	5.5
2855.7135	2855.7198	135		35007.158	0.0064	−0.078	8394.362	7.5	43401.598	7.5
2856.0308	2856.0297	182		35003.360	−0.0011	0.014	914.765	4.5	35918.111	5.5
2856.2831	2856.2829	67		35000.257	−0.0002	0.003	2294.696	5.5	37294.950	6.5
2856.6147	2856.6188	13	as	34996.141	0.0042	−0.051	9626.113	6.5	44622.305	7.5
2857.2259	2857.2328	3		34988.621	0.0069	−0.084	5526.750	6.5	40515.455	5.5
2858.9077	2858.9146	124		34968.038	0.0069	−0.085	0.000	4.5	34968.123	5.5
2859.8812	2859.8868	30		34956.151	0.0056	−0.069	5526.750	6.5	40482.970	5.5
2860.4656	2860.4578	141	p	34949.178	−0.0079	0.096	0.000	4.5	34949.082	3.5
2860.7997	2860.8038	48		34944.950	0.0042	−0.051	289.041	5.5	35234.042	5.5
2862.2375	2862.2426	37		34927.384	0.0052	−0.063	7598.353	5.5	42525.800	5.5
2862.4075	2862.4069	69		34925.379	−0.0006	0.007	289.041	5.5	35214.413	4.5
2862.6160	2862.6160	53	LA	34922.828	0.0000	0.000	4585.434	6.5	39508.262	7.5
2863.5279	2863.5306	29	#1	34911.674	0.0027	−0.033	5259.653	7.5	40171.360	8.5
2863.8629	2863.8701	4	c	34907.535	0.0072	−0.088	7598.353	5.5	42505.976	6.5
2864.4082	2864.4097	34	#2	34900.960	0.0015	−0.018	2294.696	5.5	37195.674	6.5
2865.1378	2865.1415	113	#3	34892.046	0.0036	−0.044	4420.871	5.5	39312.961	6.5
2865.6808	2865.6847	558	#4, LA	34885.431	0.0039	−0.048	0.000	4.5	34885.479	5.5
2866.1576	2866.1603	85		34879.643	0.0027	−0.033	8521.922	7.5	43401.598	7.5
2866.7879	2866.7887	24	#5	34871.997	0.0008	−0.010	6445.035	4.5	41317.042	5.5
2868.0074	2868.0134	63	#6	34857.106	0.0060	−0.073	4585.434	6.5	39442.613	7.5
2868.1857	2868.1865	47	#7	34855.003	0.0007	−0.009	2294.696	5.5	37149.708	4.5
2869.3858	2869.3803	62	#8, p	34840.505	−0.0054	0.066	5667.331	3.5	40507.770	3.5
2869.6612	2869.6595	22	#9	34837.116	−0.0017	0.021	914.765	4.5	35751.860	5.5
2870.9740	2870.9721	116	#10	34821.188	−0.0019	0.023	289.041	5.5	35110.206	4.5
2872.5019	2872.5038	25	c	34802.620	0.0020	−0.024	1749.123	6.5	36551.767	5.5
2872.5897	2872.5899	2		34801.577	0.0002	−0.003	6445.035	4.5	41246.615	4.5
2873.0033	2873.0085	44		34796.507	0.0052	−0.063	2294.696	5.5	37091.266	4.5
2873.2953	2873.2939	50	#11	34793.050	−0.0014	0.017	5526.750	6.5	40319.783	5.5
2873.5191	2873.5141	146	#12	34790.384	−0.0050	0.060	5790.641	5.5	40580.965	5.5
2874.0820	2874.0812	86		34783.519	−0.0007	0.009	1749.123	6.5	36532.633	7.5
2874.4694	2874.4633	21		34778.896	−0.0061	0.074	0.000	4.5	34778.822	5.5

Table 9. *Cont.*

wl_{Ritz} in Air (Å)	wl_{exp} in Air (Å)	*Int*	*Note*	wn_{exp} (cm^{-1})	Δwl (Å)	Δwn (cm^{-1})	E_{odd} (cm^{-1})	J_{odd}	E_{even} (cm^{-1})	J_{even}
2874.7708	2874.7635	148		34775.264	−0.0073	0.088	6283.431	6.5	41058.607	5.5
2875.1857	2875.1898	89	T,p	34770.108	0.0041	−0.049	289.041	5.5	35059.198	6.5
2875.1866	2875.1898	89	T,p	34770.108	0.0031	−0.038	8510.866	5.5	43281.012	5.5
2875.1952	2875.1898	89	T,p	34770.108	−0.0055	0.066	914.765	4.5	35684.807	5.5
2876.5188	2876.5190	55		34754.041	0.0002	−0.003	1749.123	6.5	36503.167	5.5
2877.5677	2877.5722	41	LA	34741.325	0.0045	−0.054	0.000	4.5	34741.379	4.5
2877.7302	2877.7226	194		34739.509	−0.0075	0.091	2294.696	5.5	37034.114	5.5
2878.9404	2878.9408	30		34724.810	0.0003	−0.004	5790.641	5.5	40515.455	5.5
2879.0798	2879.0806	118		34723.123	0.0008	−0.010	7598.353	5.5	42321.486	6.5
2879.9719	2879.9729	15	p	34712.365	0.0010	−0.012	43060.067	5.5	8347.690	5.5
2881.7318	2881.7380	18		34691.103	0.0062	−0.075	2294.696	5.5	36985.874	6.5
2881.7944	2881.7925	23		34690.447	−0.0018	0.022	7598.353	5.5	42288.778	5.5
2881.8744	2881.8796	16		34689.399	0.0052	−0.063	8276.733	6.5	42966.195	7.5
2881.9893	2881.9962	28	as	34687.995	0.0070	−0.084	4420.871	5.5	39108.950	5.5
2882.5843	2882.5784	127		34680.989	−0.0059	0.071	0.000	4.5	34680.918	4.5
2882.7370	2882.7381	233		34679.068	0.0012	−0.014	289.041	5.5	34968.123	5.5
2882.9252	2882.9279	165		34676.785	0.0027	−0.032	4420.871	5.5	39097.688	6.5
2885.1866	2885.1884	45		34649.620	0.0018	−0.022	2294.696	5.5	36944.338	5.5
2885.3330	2885.3323	10	c	34647.893	−0.0007	0.009	5667.331	3.5	40315.215	3.5
2885.5754	2885.5836	7		34644.875	0.0082	−0.098	5790.641	5.5	40435.614	6.5
2885.6088	2885.6149	33		34644.499	0.0062	−0.074	7166.632	4.5	41811.205	5.5
2886.0312	2886.0284	32		34639.535	−0.0027	0.033	914.765	4.5	35554.267	3.5
2886.1638	2886.1640	29		34637.908	0.0002	−0.003	4420.871	5.5	39058.782	5.5
2886.4456	2886.4514	28	LA	34634.459	0.0058	−0.070	289.041	5.5	34923.570	6.5
2886.4796	2886.4792	37	D	34634.126	−0.0004	0.005	2294.696	5.5	36928.817	4.5
2886.4824	2886.4792	37	D	34634.126	−0.0033	0.039	0.000	4.5	34634.087	4.5
2886.9223	2886.9234	53		34628.796	0.0012	−0.014	7166.632	4.5	41795.442	3.5
2887.0060	2887.0111	7		34627.745	0.0051	−0.061	40344.255	5.5	5716.449	4.5
2887.2481	2887.2529	165		34624.845	0.0048	−0.057	5259.653	7.5	39884.555	6.5
2887.5908	2887.5910	70	LA	34620.790	0.0003	−0.003	9553.187	5.5	44173.980	6.5
2888.2558	2888.2565	96	LA	34612.813	0.0008	−0.009	0.000	4.5	34612.822	4.5
2888.7371	2888.7399	83		34607.021	0.0028	−0.034	4420.871	5.5	39027.926	6.5
2889.1209	2889.1232	39		34602.430	0.0023	−0.027	6283.431	6.5	40885.888	6.5
2889.2613	2889.2607	29		34600.783	−0.0006	0.007	2294.696	5.5	36895.472	4.5
2889.6236	2889.6258	198	LA	34596.412	0.0022	−0.026	289.041	5.5	34885.479	5.5
2890.4257	2890.4214	43		34586.889	−0.0043	0.052	2294.696	5.5	36881.533	6.5
2891.6259	2891.6265	6	as	34572.478	0.0007	−0.008	8276.733	6.5	42849.219	5.5
2891.6805	2891.6847	19		34571.783	0.0042	−0.050	8394.362	7.5	42966.195	7.5
2891.7924	2891.7926	66	D	34570.493	0.0002	−0.002	2294.696	5.5	36865.191	5.5
2891.8001	2891.7926	66	D	34570.493	−0.0075	0.090	42918.093	5.5	8347.690	5.5
2894.8391	2894.8403	45		34534.101	0.0012	−0.014	1749.123	6.5	36283.238	5.5
2895.5407	2895.5443	47		34525.704	0.0036	−0.043	7166.632	4.5	41692.379	5.5
2895.6254	2895.6257	34		34524.734	0.0003	−0.004	1749.123	6.5	36273.861	6.5
2895.7279	2895.7315	4		34523.472	0.0037	−0.044	4585.434	6.5	39108.950	5.5
2896.0761	2896.0695	46		34519.443	−0.0065	0.078	5526.750	6.5	40046.115	5.5
2896.6728	2896.6710	717	as,IV	34512.275	−0.0018	0.021	4585.434	6.5	39097.688	6.5
2897.4573	2897.4562	47		34502.923	−0.0012	0.014	5526.750	6.5	40029.659	6.5
2898.1286	2898.1254	144	as	34494.955	−0.0032	0.038	5259.653	7.5	39754.570	8.5
2898.3689	2898.3677	51		34492.072	−0.0013	0.015	289.041	5.5	34781.098	4.5
2898.5602	2898.5686	146		34489.681	0.0084	−0.100	289.041	5.5	34778.822	5.5
2898.7085	2898.7080	30		34488.022	−0.0004	0.005	2294.696	5.5	36782.713	6.5
2898.9194	2898.9231	174		34485.463	0.0037	−0.044	42833.197	5.5	8347.690	5.5
2899.9121	2899.9080	19		34473.751	−0.0040	0.048	8276.733	6.5	42750.436	6.5
2899.9419	2899.9443	21		34473.320	0.0024	−0.028	4585.434	6.5	39058.782	5.5
2900.2638	2900.2714	46	D,c	34469.432	0.0076	−0.090	8379.697	4.5	42849.219	5.5
2900.2656	2900.2714	46	D,c	34469.432	0.0058	−0.069	5259.653	7.5	39729.154	7.5
2900.3168	2900.3173	23	c	34468.886	0.0005	−0.006	2294.696	5.5	36763.588	4.5
2900.7128	2900.7164	27	p	34464.144	0.0036	−0.043	4706.273	2.5	39170.460	2.5
2902.3902	2902.3976	39	p	34444.185	0.0074	−0.088	8521.922	7.5	42966.195	7.5
2902.4127	2902.4100	29	p	34444.037	−0.0027	0.032	5790.641	5.5	40234.646	5.5
2902.8069	2902.8073	19	LA	34439.323	0.0004	−0.005	0.000	4.5	34439.328	5.5
2903.7855	2903.7816	121		34427.768	−0.0039	0.046	42775.412	5.5	8347.690	5.5
2904.5041	2904.5065	63	LA	34419.176	0.0024	−0.028	289.041	5.5	34708.245	6.5
2905.8166	2905.8110	46		34403.724	−0.0056	0.066	7166.632	4.5	41570.290	5.5
2906.0896	2906.0976	81		34400.331	0.0080	−0.095	5259.653	7.5	39660.079	6.5
2906.9576	2906.9550	70		34390.184	−0.0025	0.030	8276.733	6.5	42666.887	7.5
2907.0539	2907.0554	41		34388.997	0.0014	−0.017	289.041	5.5	34678.055	6.5
2908.0936	2908.0988	14		34376.658	0.0052	−0.062	6445.035	4.5	40821.755	4.5
2908.4109	2908.4115	89		34372.962	0.0006	−0.007	4420.871	5.5	38793.840	6.5
2909.0748	2909.0674	24		34365.212	−0.0074	0.087	7598.353	5.5	41963.478	4.5

Table 9. *Cont.*

wl_{Ritz} in Air (Å)	wl_{exp} in Air (Å)	*Int*	*Note*	wn_{exp} (cm^{-1})	Δwl (Å)	Δwn (cm^{-1})	E_{odd} (cm^{-1})	J_{odd}	E_{even} (cm^{-1})	J_{even}
2909.6946	2909.6885	63		34357.877	−0.0061	0.072	5526.750	6.5	39884.555	6.5
2910.4278	2910.4339	63		34349.081	0.0061	−0.072	6445.035	4.5	40794.188	4.5
2910.5243	2910.5222	10		34348.039	−0.0021	0.025	5259.653	7.5	39607.667	6.5
2910.6385	2910.6395	32		34346.655	0.0010	−0.012	7598.353	5.5	41945.020	6.5
2911.7385	2911.7334	17		34333.752	−0.0052	0.061	4420.871	5.5	38754.562	4.5
2912.5792	2912.5779	40	LA	34323.796	−0.0013	0.015	289.041	5.5	34612.822	4.5
2913.9646	2913.9679	26		34307.424	0.0032	−0.038	8755.640	6.5	43063.102	6.5
2914.2520	2914.2493	70	D	34304.111	−0.0027	0.032	289.041	5.5	34593.120	5.5
2914.2561	2914.2493	70	D	34304.111	−0.0068	0.080	45044.296	7.5	10740.265	6.5
2914.6285	2914.6235	106		34299.707	−0.0050	0.059	914.765	4.5	35214.413	4.5
2914.7246	2914.7324	82	p	34298.425	0.0078	−0.092	0.000	4.5	34298.517	5.5
2915.4993	2915.4945	97	as	34289.459	−0.0048	0.056	8394.362	7.5	42683.765	8.5
2915.5822	2915.5801	38		34288.453	−0.0021	0.025	1749.123	6.5	36037.551	7.5
2916.7057	2916.7052	57	D	34275.226	−0.0005	0.006	5526.750	6.5	39801.970	6.5
2916.7136	2916.7052	57	D	34275.226	−0.0083	0.098	4420.871	5.5	38695.999	5.5
2916.9351	2916.9366	37		34272.507	0.0015	−0.018	8394.362	7.5	42666.887	7.5
2917.5409	2917.5390	207		34265.431	−0.0020	0.023	5401.503	3.5	39666.911	4.5
2917.9089	2917.9034	26		34261.152	−0.0055	0.065	5526.750	6.5	39787.837	5.5
2938.9919	2938.9908	114		34015.338	−0.0010	0.012	1749.123	6.5	35764.449	7.5
2940.0800	2940.0783	18	p	34002.757	−0.0017	0.020	1749.123	6.5	35751.860	5.5
2940.2834	2940.2880	274	p	34000.331	0.0047	−0.054	9075.732	3.5	43076.117	4.5
2940.4294	2940.4290	46		33998.701	−0.0004	0.005	5401.503	3.5	39400.199	4.5
2941.3079	2941.3068	30		33988.555	−0.0011	0.013	2294.696	5.5	36283.238	5.5
2941.6963	2941.6897	32		33984.131	−0.0067	0.077	8521.922	7.5	42505.976	6.5
2941.9164	2941.9176	213	LA	33981.498	0.0012	−0.014	5526.750	6.5	39508.262	7.5
2942.1196	2942.1204	74		33979.156	0.0008	−0.009	2294.696	5.5	36273.861	6.5
2942.4224	2942.4278	7	D	33975.606	0.0054	−0.062	10198.312	7.5	44173.980	6.5
2942.4268	2942.4278	7	D,LA	33975.606	0.0010	−0.011	4706.273	2.5	38681.890	3.5
2942.7456	2942.7519	100		33971.864	0.0063	−0.073	7598.353	5.5	41570.290	5.5
2942.8515	2942.8516	49	LA	33970.713	0.0001	−0.001	914.765	4.5	34885.479	5.5
2943.8954	2943.8964	160		33958.656	0.0010	−0.012	2294.696	5.5	36253.364	5.5
2944.3342	2944.3300	12	as	33953.655	−0.0042	0.048	7598.353	5.5	41551.960	5.5
2944.5416	2944.5456	34		33951.169	0.0040	−0.046	6283.431	6.5	40234.646	5.5
2945.5252	2945.5297	7		33939.830	0.0045	−0.052	5259.653	7.5	39199.535	7.5
2945.5971	2945.5947	10		33939.081	−0.0023	0.027	5790.641	5.5	39729.695	6.5
2945.8164	2945.8129	13		33936.567	−0.0035	0.040	5526.750	6.5	39463.277	5.5
2945.8896	2945.8919	96	D,LA	33935.658	0.0023	−0.026	1749.123	6.5	35684.807	5.5
2945.8980	2945.8919	96	D	33935.658	−0.0062	0.071	5401.503	3.5	39337.090	4.5
2946.2771	2946.2831	13		33931.152	0.0060	−0.069	4585.434	6.5	38516.655	6.5
2946.6046	2946.6073	32	p	33927.418	0.0027	−0.031	5526.750	6.5	39454.199	6.5
2946.6329	2946.6252	42	p	33927.212	−0.0076	0.088	8394.362	7.5	42321.486	6.5
2947.5118	2947.5119	51		33917.006	0.0001	−0.001	2294.696	5.5	36211.703	6.5
2948.0897	2948.0907	161		33910.347	0.0010	−0.012	289.041	5.5	34199.400	5.5
2949.4511	2949.4501	79	p	33894.718	−0.0010	0.012	2294.696	5.5	36189.402	5.5
2949.6008	2949.6057	38	as	33892.930	0.0049	−0.056	4420.871	5.5	38313.857	5.5
2949.6888	2949.6890	33		33891.973	0.0002	−0.002	7166.632	4.5	41058.607	5.5
2949.8281	2949.8363	23		33890.280	0.0082	−0.094	7598.353	5.5	41488.727	5.5
2949.9708	2949.9735	14		33888.704	0.0027	−0.031	2294.696	5.5	36183.431	5.5
2950.5021	2950.5005	15		33882.651	−0.0017	0.019	4585.434	6.5	38468.066	7.5
2950.5656	2950.5655	26		33881.905	−0.0002	0.002	1749.123	6.5	35631.026	7.5
2951.0563	2951.0646	84	D	33876.174	0.0084	−0.096	5790.641	5.5	39666.911	4.5
2951.0724	2951.0646	84	D	33876.174	−0.0078	0.089	0.000	4.5	33876.085	5.5
2951.9221	2951.9153	23	p	33866.412	−0.0069	0.079	914.765	4.5	34781.098	4.5
2953.0016	2953.0029	105		33853.938	0.0013	−0.015	4706.273	2.5	38560.226	3.5
2953.5797	2953.5870	60		33847.243	0.0073	−0.084	5667.331	3.5	39514.658	2.5
2953.7715	2953.7756	188	p	33845.082	0.0041	−0.047	7598.353	5.5	41443.482	6.5
2954.5230	2954.5233	326	p	33836.521	0.0003	−0.004	2294.696	5.5	36131.221	5.5
2954.6867	2954.6825	143		33834.698	−0.0042	0.048	5259.653	7.5	39094.303	7.5

LA: line already assigned as U II transition in [7], as: asymmetrical line, c: complex line shape, p: line resolved on the plate, but perturbed by a close line, b: broad line, ?: line given by [12] as U III without classification, IV : this line could be blended with a strong U IV line, D: line with double identification, T: line with triple identification, #n: line number in Figure 1.

Table 10. Transitions establishing the newly determined even parity level $5f^36d7p$ $(^4I)^6K(J = 5.5)$ of the U^+ ion at 39113.98 ± 0.1 cm^{-1}. In $\log(g_lf)$, f is the absorption oscillator strength and g_l, the statistical weight of the lower level. gA is the upper level statistical weight g multiplied by the Einstein coefficient of spontaneous emission. CF is the cancellation factor defined by Equation (14.107), p432 in [10].

E_{th} (cm^{-1})	J	*Odd Level*	wn_{th} (cm^{-1})	*log(g*$_l$*f)*	*gA* (s^{-1})	*CF*	E_{exp} (cm^{-1})	wn_{exp} (cm^{-1})	λ_{exp} in Air (Å)	Int_{exp} (arb.)
−56.8	4.5	$5f^37s^2$ (4I) 4I	38865.8	0.263	1.847E+09	0.63	0. 000	39114.382	2555.8378	545
224.4	5.5	$5f^36d7s$ (4I) 6L	38584.6	−1.518	3.014E+07	0.03	289.041	38824.783	2574.9033	4
1715.5	6.5	$5f^36d7s$ (4I) 6L	37093.5	0.181	1.391E+09	−0.27	1749.123	37364.867	2675.5157	228
2320.7	5.5	$5f^36d7s$ (4I) 6K	36488.3	−0.252	4.974E+08	−0.21	2294.696	36819.292	2715.1628	36
4406.4	5.5	$5f^37s^2$ (4I) 4I	34402.6	−0.877	1.048E+08	0.20	4420.871	34692.797	2881.5973	14
4577.9	6.5	$5f^36d7s$ (4I) 6L	34231.1	−0.160	5.404E+08	−0.16	4585.434	34528.617	2895.3001	33

5. Conclusions

The lowest energy levels of the singly ionized uranium are interpreted following the Racah-Slater parametric method by means of Cowan codes. In the odd parity, the number of interpreted levels is about ten times larger than the number of free parameters. The relatively small *rms* deviation of the energies and the deviations between g_L^{th} and g_L^{exp} Landé factors for many levels show that the present model is robust. Some experimental level energies, although supported by the high accuracy of the observed FTS wave numbers, could not be attributed unambiguously to a theoretical level energy. The limitations of the present theoretical description are even more obvious in the even parity with larger *rms* deviations on the energies for both groups of configurations studied. After 70 years of investigations, the spectrum of U II still deserves further experimental studies for removing uncertain interpretations. The main difficulties are due to the ambiguities on the J values of levels, the determination of which would need a more complete study of Zeeman effect. Furthermore, the description of the strongly mixed CI wave functions could only be confirmed by the value of the Landé factor. By remembering the sentence *Levels without known g values are less certain because of the possibilities of fortuitous coincidences* written in [6], we do consider that the present calculations are satisfactory in spite the uninterpreted levels. A theoretical interpretation of the core configurations $5f^4$ and $5f^3(6d + 7s)$ of U III is presently under way for a better knowledge of appropriate scaling factors of the *HFR* radial integrals to be used in U II. An estimate of the partition function shows that level energies from parametric fit are preferable for its calculation. On the experimental side, a list of 451 ultraviolet spectral lines from high resolution vacuum spark spectra identified as U II transitions is reported, as well as six other transitions establishing a new energy level in the even parity configuration $5f^36d7p$.

Acknowledgments: The photographic spectrograms were recorded between 1986 and 1988 with technical assistance of Françoise Launay and Maurice Benharrous. Christophe Blaess is acknowledged for digitizing the spectrograms. The financial support of the French CNRS – PNPS national program is acknowledged. This work is part of the Plas@Par LabEx project managed by the ANR (ANR-11-IDEX-0004-02). AM and MS wish to acknowledge supports from Université Mouloud Mammeri, Tizi-Ouzou, Algeria and from the project CNEPRU D00520110032, Algeria.

Author Contributions: These authors contributed equally to this work.

Conflicts of Interest: The authors declare no conflict of interest.

References

1. Cayrel, R.; Hill, V.; Beers, T.; Barbuy, B.; Spite, M.; Spite, F.; Plez, B.; Andersen, J.; Bonifacio, P.; Francois, P.; et al. Measurement of stellar age from uranium decay. *Nature* **2001**, *409*, 691–692.
2. Judd, B. R. Complex atomic spectra. *Rep. Prog. Phys.* **1985**, *48*, 907–954.
3. Blaise, J.; Wyart, J.-F. *Selected Constants Energy Levels and Atomic Spectra of Actinides*; Centre National de la Recherche Scientifique: Paris, France, 1992; Volume 20.

4. Wyart, J.-F.; Blaise, J.; Worden, E.F. Studies of electronic configurations in the emission spectra of lanthanides and actinides: application to the interpretation of Es I and Es II, predictions for Fm I. *J. Sol. State Chem.* **2005**, *178*, 589–602
5. Guyon, F.; Blaise, J.; Wyart, J.-F. Etude paramétrique des configurations impaires profondes dans les spectres de l'uranium UI et UII. *J. Phys.* **1974**, *35*, 929–933.
6. Blaise, J.; Wyart, J.-F.; Vergès, J.; Engleman, R., Jr.; Palmer, B.A.; Radziemski, L.J. Energy levels and isotope shifts for singly ionized uranium (U II). *J. Opt. Soc. Am. B* **1994**, *11*, 1897–1929.
7. Steinhaus, D.W.; Radziemski, L.J., Jr.; Cowan, R.D.; Blaise, J.; Guelachvili, G.; Ben Osman, Z.; Vergès, J. *Present Status of the Analyses of the First and Second Spectra of Uranium (U I and U II) as Derived from Measurements of Optical Spectra*; LASL Report LA-4501; Los Alamos Scientific Lab., N. Mex.: Los Alamos, NM, USA, 1971.
8. Palmer, B.A.; Keller, R.A.; Engleman, R., Jr. *An Atlas of Uranium Emission Intensities in A Hollow Cathode Discharge*; LASL Informal Report LA-8251-MS,UC-34a; Los Alamos Scientific Lab., N. Mex.: Los Alamos, NM, USA, 1980.
9. Brewer, L. Energies of the electronic configurations of the singly, doubly and triply ionized lanthanides and actinides. *J. Opt. Soc. Am.* **1971**, *12*, 1666–1682.
10. Cowan, R.D. *The Theory of Atomic Structure and Spectra*; University of California Press: Berkeley, CA, USA, 1981.
11. Kramida, A. PC Version of Cowan Codes. Available online: http://das101.isan.troitsk.ru (accessed on 21 August 2012).
12. Palmer, B.A.; Engleman, R., Jr. Wavelengths and energy levels of doubly ionized uranium obtained using a Fourier Transform spectrometer. *J. Opt. Soc. Am. B* **1984**, *1*, 609–625.
13. Blaise, J.; Wyart, J.-F.; Palmer, B.A.; Engleman, R., Jr.; Launay, F. Analysis of the spectrum of doubly ionized Uranium (U III). In *19th EGAS, Dublin: European Group for Atomic Spectroscopy: 14–17 July 1987: Abstracts*; European Physical Society: Mulhouse, France, 1987; pp. A3–08.
14. Redman, S.L.; Lawler, J.E.; Nave, G.; Ramsey, L.W.; Mahadevan, S. The infrared spectrum of Uranium Hollow athode Lamps from 850nm to 4000nm. *Astrophys. J. Supp. Ser.* **2011**, *195*, 24.
15. Wyart, J.-F.; Kaufman, V.; Sugar, J. Analysis of the Spectrum of Four-Times-Ionized Uranium (U5). *Phys. Scr.* **1980**, *22*, 389–396.
16. Kaufman, V.; Radziemski, L.F., Jr. The sixth spectrum of Uranium (UVI). *J. Opt. Soc. Am.* **1976**, *66*, 599–600.
17. Meftah, A.; Wyart, J.-F.; Tchang-Brillet, W.-Ü.L.; Blaess, C.; Champion, N. Spectrum and energy levels of the Yb^{4+} free ion (Yb V) *Phys. Scr.* **2013**, *88*, 045305.
18. Tomkins, F.S.; Fred, M. A photoelectric setting device for a spectrum plate comparator. *J. Opt. Soc. Am.* **1951**, *41*, 641.
19. Wyart, J.-F. Theoretical interpretation of the Nd II spectrum: Odd parity energy levels. *Phys. Scr.* **2010**, *82*, 035302.
20. Wyart, J.-F. On the interpretation of complex atomic spectra by means of the parametric Racah-Slater method and Cowan codes. *Can. J. Phys.* **2011**, *89*, 451.
21. Chen, H.-L.; Borzileri, C. Laser induced fluorescence studies of U II produced by photoionization of uranium. *J. Chem. Phys.* **1981**, *74*, 6063–6069.
22. Corliss, C.H. Oscillator strengths for lines of ionized uranium (U II). *J. Res. Nat. Bur. Stand. Sect. A* **1976**, *80*, 429.
23. Nilsson, H.; Ivarsson, S.; Johansson, S.; Lundberg, H. Experimental oscillator strengths in U II of cosmological interest. *Astron. Astrophys.* **2002**, *381*, 1090–1093.
24. Kurucz, R.L. Available online: http://kurucz.harvard.edu/linelists/gfnew/ (accessed on 31 March 2017).

atoms

MDPI

Article

The Role of the Hyperfine Structure for the Determination of Improved Level Energies of Ta II, Pr II and La II

Laurentius Windholz

Institute of Experimental Physics, Graz University of Technology, Petersgasse 16, A-8010 Graz, Austria; windholz@tugraz.at

Academic Editor: Joseph Reader
Received: 19 January 2017; Accepted: 21 February 2017; Published: 28 February 2017

Abstract: For the determination of improved energy levels of ionic spectra of elements with large values of nuclear magnetic dipole moment (and eventually large values of nuclear quadrupole moments), it is necessary to determine the center of gravity of spectral lines from resolved hyperfine structure patterns appearing in highly resolved spectra. This is demonstrated on spectral lines of Ta II, Pr II and La II. Blend situations (different transitions with accidentally nearly the same wave number difference between the combining levels) must also be considered.

Keywords: energy levels; Ta II; Pr II; La II

1. Introduction

The laser spectroscopy group at Graz University of Technology has been concerned since 1990 with investigations of the hyperfine (hf) structure of several elements. The spectra of tantalum, praseodymuim and lanthanum were investigated most intensely. As a source of free atoms, a hollow cathode lamp was used in which a low-pressure plasma of the treated element was generated by cathode sputtering. For starting the discharge, a noble gas (argon or neon) at a typical pressure of 0.5 mbar was used. This source of free atoms and ions was investigated by tunable laser light (band width ca. 1 MHz) by scanning the laser frequency across the selected wavelength range. Either laser-induced fluorescence light or the change of the discharge impedance (optogalvanic detection) was observed. Details of the experimental arrangement can be found in various publications, e.g., [1–3].

In this paper, spectra of Ta, Pr and La are treated. These elements have in their natural abundance either only one dominant isotope (Ta, La, see Table 1) or are isotopically pure (Pr). Their nuclear magnetic dipole moment μ is large enough to cause hyperfine splitting of the spectral lines larger than the Doppler width in the spectra. Thus, in most cases, the observed hf structure can be used as valuable help for the classification of the spectral lines. The isotope composition and nuclear moments can be found in Table 1. For Pr and La, the quadrupole moment is quite small and can be neglected for most of the energy levels.

Table 1. Isotope composition and nuclear moments of the investigated elements (natural abundance). In the spectra of Ta and La, we observed only the dominant isotopes ^{181}Ta and ^{139}La.

Element	Z	Isotope	Natural Abundance %	Lifetime (Years)	Nuclear Spin Quantum Number I	Magnetic Moment μ (μ_N)	Electric Quadrupole Moment Q (10^{-28} m^2)
Ta	73	180	0.012	1.2×10^{15}	9	+4.825(11)	+4.95(2)
Ta	73	181	99.998	stable	7/2	+2.3705(7)	+3.28(6)
Pr	59	141	100	stable	5/2	+4.2754(5)	−0.059(4)
La	57	138	0.09	1.05×10^{11}	5	+3.713646(7)	+0.45(2)
La	57	139	99.91	stable	7/2	+2.7830455(9)	+0.20(1)

* Based on uncorrected proton moment, 2.79277564 nm. Values of μ and Q from [4].

While at the early stage of the investigations, the hf constants of already known energy levels were determined, and it turned out later that the list of energy levels given in literature [5,6] is far from being complete. Thus, the focus was directed to the finding of new energy levels in order to explain spectral lines that could not be classified as transitions between known energy levels. An overview of how previously unknown energy levels can be found is given in Ref. [7].

In order to get accurate start wavelengths for laser spectroscopic investigations, spectra with high resolution and high wavelength precision are needed. These requirements can be fulfilled by means of Fourier-transform (FT) spectroscopy. Several co-operations led to the availability of spectra of Ta [8], Pr [9], and La [10]. In these spectra, much more lines can be found than listed in commonly used wavelength tables [11]. The spectra were taken with a resolution between 0.03 and 0.05 cm^{-1} and carefully wavelength calibrated using Ar II lines [12].

For strong lines which were classified and for which the hf constants of the combining levels were known, one finds that the center of gravity (cg) wavelengths determined from the FT spectra usually differ from wavelengths calculated from the known level energies. The conversion from wave numbers to standard air wavelengths and back was performed using the formula given by Reeder and Peck [13] for the refractive index of air.

Thus, exploiting the low uncertainty of the cg wavelengths determined from the FT spectra, improved level energies were determined. This was made step by step, beginning with the upper levels combining with the ground level. From these upper levels, transitions to lower levels were searched and energies of low-lying levels were corrected, and so on. Finally, the level energies were determined by a global fit procedure.

Since each spectrum contains several tenths of thousands of spectral lines, and since the treated elements have several hundreds or thousands energy levels, one can imagine that this procedure is very time-consuming. Thus, first the spectra of the first ions, Ta II, Pr II and La II, were used to perform a final determination of level energies. For most of the levels, an uncertainty of the level energy below 0.01 cm^{-1} was achieved.

For the determination of the hf constants of the levels involved in an investigated transition (hf resolved spectra either from an FT spectrum or a laser spectroscopic scan), we used a software called "Fitter" which was very helpful [14].

It is clear that a suitable computer program is needed to manage such huge numbers of lines and the extended FT spectra. Thus, a program called "Elements" was developed (for descriptions, see Refs. [7,15]). One can select a certain wavelength and then go from one line to the next. The corresponding part of the FT spectrum is automatically shown. For classified lines, the combining levels and the hf pattern is shown in graphical form. For unclassified lines, classification suggestions (transitions between known energy levels within a selected wave number deviation) are also shown. A part of the FT spectrum can be copied easily to a simulation window where it can be compared with such a suggestion, for which the hf pattern is graphically shown. If no agreement between the pattern from the FT spectrum and any suggestion can be found, one has to assume that a previously unknown energy level is involved in the structure.

In the following sections, peculiarities of the investigated spectra are discussed in more details.

2. Ta II

Energy levels of Ta II are listed in the famous tables of Moore [5]. The given data are based mainly on works of Kiess [16], who published separately a collection of Ta II energy levels including the classified lines. Concerning the hf structure, a relatively low number of publications can be found. In 1952, Brown and Tamboulian [17] determined for the first time the nuclear moments of ^{181}Ta investigating the hf structure of 7 Ta II lines. In 1987, Engleman Jr. [18] determined the hf constants of several Ta II levels and improved the energy values, but the results were presented only at a symposium but never published. Eriksson et al. [19] investigated in 2002 the Ta II spectrum with respect to applications in astrophysics. Laser spectroscopic determinations of the hf constants of Ta II were performed by Messnarz and Guthöhrlein [20,21] at approximately the same time. During work on his thesis, Messnarz discovered some new energy levels of Ta II and could correct some incorrect classifications. Zilio and Pickering [22] investigated an FT spectrum of Ta II and published hf constants of several levels.

The FT spectrum taken by J. Pickering at Imperial College London and spectra taken at the Kitt Peak Observatory by Engleman Jr. were used later in Graz [8,23,24]. In these papers, the discovery of new energy levels of Ta I and the determination of hf constants of already known Ta I levels are reported. The papers are part of a series of works published in Zeitschrift für Physik, the succeeding European Journal of Physics and later in Physica Scripta, entitled "Investigation of the hyperfine structure of Ta I lines, Part I to X". As can be seen from the list of authors, a strong collaboration between the group in Graz and the group of G. H. Guthöhrlein in Hamburg took place.

In the early stage of the investigations, it was very helpful to have a tool for distinguishing Ta I and Ta II spectral lines. This could be done with the help of photographic spectra, taken with a classical spectrograph, using for one trace of the spectrum a direct current (DC) hollow cathode lamp and, for a second trace, a discharge with pulsed excitation. In the pulsed discharge, the ionic lines are much more intense, allowing a clear distinguishing between Ta I and Ta II lines. One example of such spectra—in comparison with the FT spectra—is given in Ref. [25]. Another example where two spectral lines, one belonging to Ta I and the second to Ta II, are located side by side (Figure 1).

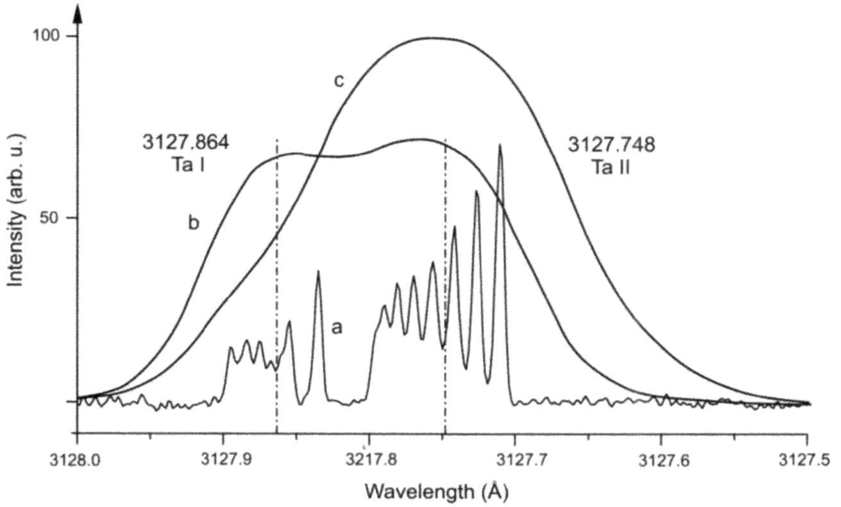

Figure 1. Comparison of a part of a highly resolved Fourier-transform (FT) spectrum with photographic spectra. For a description, see text.

In Figure 1, trace a shows the FT spectrum, full width at half maximum (FWHM) ca. 2.2 GHz. The light source was a hollow cathode lamp, operated by DC. The hf patterns of the lines are well resolved. The cg positions are marked with vertical dash-dotted lines. Traces b and c show photographic spectra, digitized from a photo plate generated by means of an Ebert-mounted grating spectrograph, focal length 2 m, 7th order (dispersion 0.72 Å/mm). The resolution is ca. 25 GHz (0.08 Å). In trace b, a hollow cathode lamp operated by DC was used, while, in trace c, the discharge was pulsed. This pulsed operation enhanced significantly the intensity of the ionic line compared to the atomic one. Thus, this spectral line on the photo plate causes difficulties in finding the cg wavelengths, since width and line center position strongly depend on the ratio of the intensities of the two lines and thus from the discharge conditions. Nevertheless, the two different photographic spectra made it easy to identify the structure at 3127.748 Å as an ionic line. The classification of the lines (and also of the lines in the subsequent figures) is given in Table A1 (see Appendix A).

In the FT spectra from Kitt Peak Observatory, an electrodeless microwave discharge was used as the light source. In these spectra, one can find a large number of unclassified lines. Some of them showed well resolved hf patterns. Analyzing these lines, a previously unknown system of high lying even parity ionic states with energies above 72,000 cm^{-1} could be found, while the highest previously known even level is located at 40,900 cm^{-1}. First, results were published (together with the description of the classification program "Elements") in 2002 [15], all new energy levels in Ref. [26].

For finding accurate cg wavelengths of lines, it is very important to take into account their hf pattern. Figure 2 shows an example where, in the FT spectrum, practically only a single peak is visible. However, treating the peak wavelength as cg is not correct. Components with small intensities, but located far from the highest peak, shift the cg wavelength to the middle of the low-frequency wing of the peak.

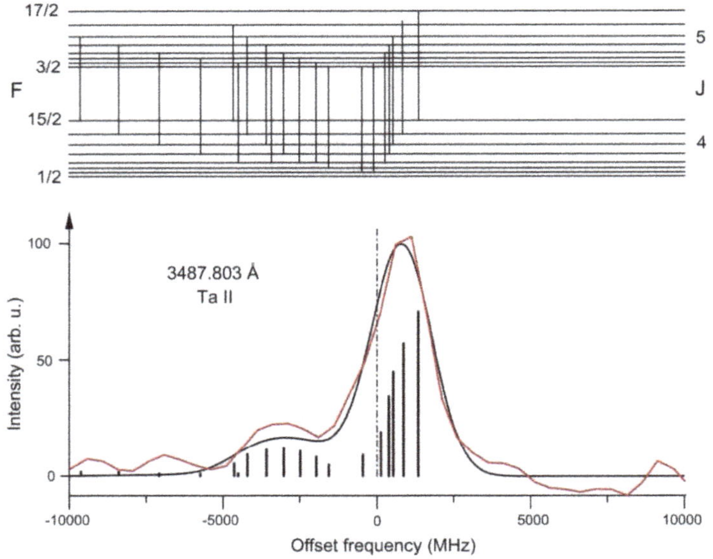

Figure 2. Example of a Ta II spectral line where the center of gravity (cg) wavelength is different from the single large peak appearing in the FT spectrum. Red line: FT spectrum, black line: simulation, full width at half maximum (FWHM) 2.2 GHz. Shown is also the hyperfine (hf) level scheme, the transitions and the components (theoretical intensity ratios). The components on the left side of the high peak (built by overlapping hf components for which $\Delta F = \Delta J$) are only barely visible in the FT spectrum. The cg is shifted against the peak by 0.8 GHz (0.05 Å). The cg wavelength is marked by a vertical line (chain-dotted).

During the procedure of line classification and cg wavelength determination, one is quite often confronted with blend situations. As an example, Figure 3 shows a blend situation of two Ta II lines. With known hf constants of the four involved energy levels, it is possible to decompose the observed structure into two overlapping lines and to determine their cg wavelengths (in this case differing by 0.029 Å).

In the Ta FT spectra, ranging from 2120 Å to 46,000 Å, one can find a number of 12,200 spectral lines, from which as many lines as possible were classified. Around 1000 of them are Ar I or Ar II lines, which can be used for wavelength calibration. Ca. 3000 lines belong to the Ta II spectrum. Despite of all efforts, roughly 2000 lines are still not classified.

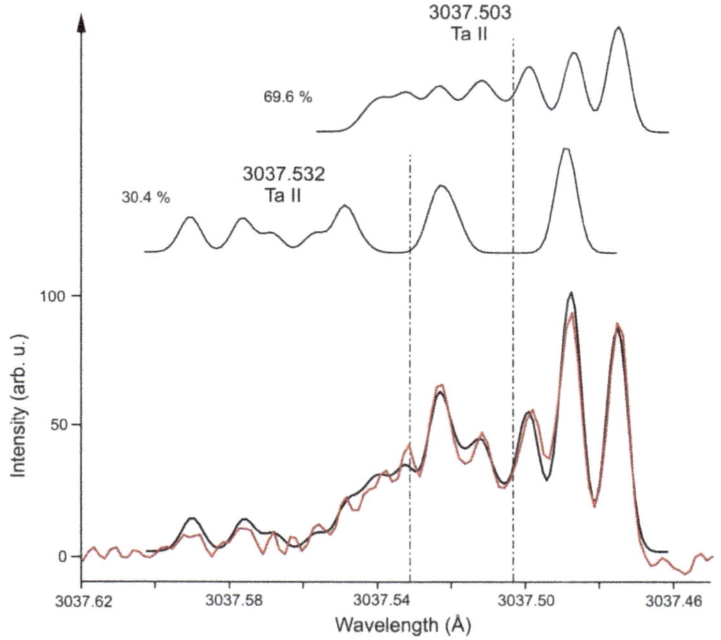

Figure 3. A blend situation of two Ta II lines. Red line: FT spectrum, black line: simulation, FWHM 2.2 GHz. In the upper part, normalized hf spectra of both lines are shown. Both profiles, added with the percentage given in the figure, gave the simulated sum profile, which describes the observed structure quite well. Thus, both cg wavelengths can be determined with high accuracy from the FT spectrum.

A systematic investigation of Ta II lines in the FT spectra available in Graz was performed in the PhD work of Uddin [27]. The determination of the hf constants of Ta II levels, either from laser spectroscopic records of Ta II lines or from the hf patterns of Ta II lines in the FT spectrum, made it possible to determine quite accurate values of cg wavelengths. From the classified Ta II lines, lines with good signal-to-noise ratio (SNR) were selected in order to re-calculate the level energies. Using the obtained vacuum wave numbers, a transition matrix was built up, and, in a global least squares fit, values for the level energies were calculated, together with their statistical uncertainties. The result was published recently [25].

The even parity system of Ta II was investigated theoretically—based on the results given in ref. [25]—by Stachowska et al. [28] performing a semi-empirical analysis, which confirmed also the new high lying levels above 72,000 cm^{-1} (Figure 4). For this analysis, an important point was to exclude energy levels given in literature but in reality not existing. This is sometimes very difficult, especially if it could not be verified under which assumptions (which lines) the corresponding level was introduced.

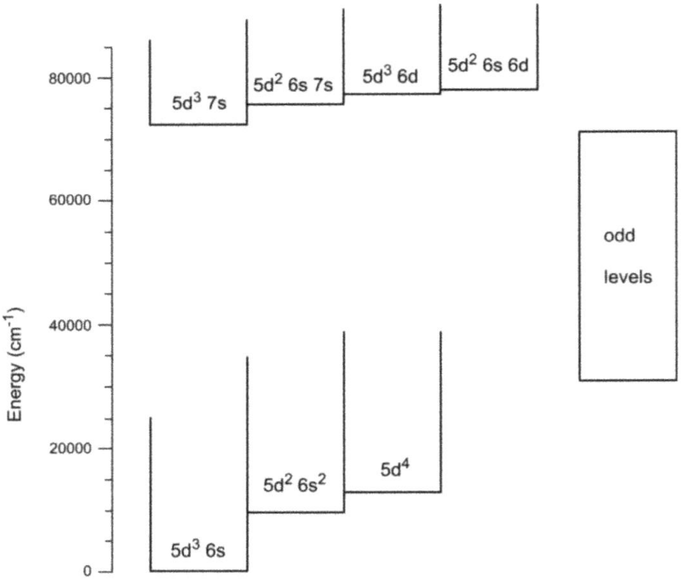

Figure 4. Simplified level scheme of Ta II. The theoretical analysis showed that the new even parity energy levels above 72,000 cm^{-1} belong to the configurations shown in the picture.

3. Pr II

First investigations of the hf structure of Pr II lines were performed in 1929 by White [29]. In 1941, Rosen et al. [30] investigated Zeeman patterns of Pr II lines and determined Landé factors and J values of 74 energy levels. Further progress in finding fine structure levels was made by Blaise et al. [31] in 1973. All available data on Pr energy levels (and of all other atoms of the lanthanide group) were collected by Martin et al. in 1978 [6]. The work of Blaise was continued by Ginibre [32–35], who achieved remarkable progress in the classification of Pr I and Pr II lines. She discovered a large number of previously unknown energy levels and determined their hf constants. In 2001, Ivarsson et al. [36] improved the wavelength accuracy of some Pr II lines of astrophysical relevance. Investigations of the hf structure and search for new energy levels were performed also by the groups in Hamburg [37] and Graz. First results were published in a common paper [38], but still many of the results achieved in Hamburg are still not published.

The available FT spectra of Pr (3260–9880 Å) were taken by members of the group in Graz using an FT spectrometer in Hannover (group of Prof. Tiemann) and a hollow cathode lamp brought from Graz to Hannover. A first analysis of the spectra revealed more than 9000 previously unknown spectral lines of Pr I and Pr II, from which ca. 1200 could be classified as transitions between already known energy levels. During this first examination, 24 previously unknown energy levels were also discovered [9]. Later, these FT spectra were very helpful for laser spectroscopic investigations since the excitation wavelength could be set precisely to interesting peaks in the FT spectrum.

In between the list of spectral lines in the FT spectrum ca. 30,000 lines are contained, among them only 200 Ar lines and 650 Pr II lines. All other lines belong to the spectrum of neutral Pr (Pr I). In some spectral regions, the number of lines is so big that nearly no wavelength can be found at which the Pr plasma does not emit light. For example, the region around 5800 Å is shown in Figure 5.

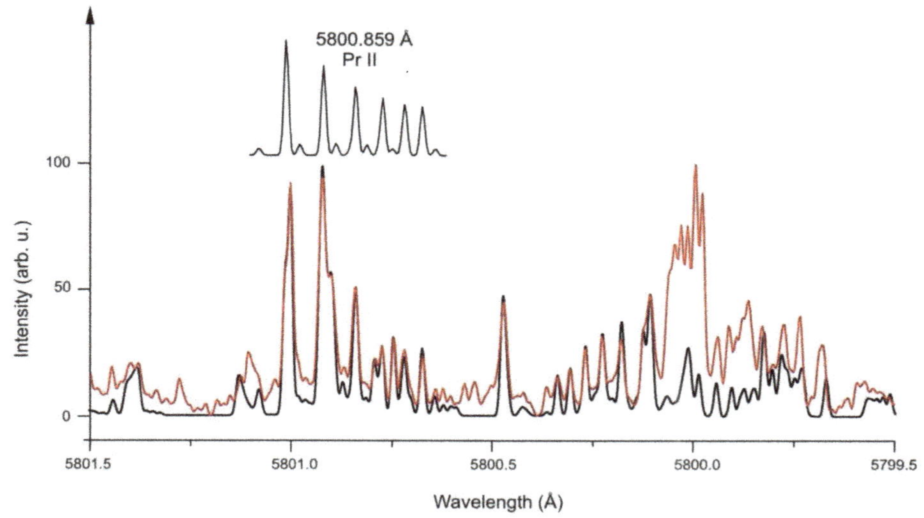

Figure 5. Part of an FT spectrum of Pr, using a direct current (DC) hollow cathode as light source (red line). As can be seen, there is nearly no wavelength at which the Pr plasma emits no light. The black curve shows a simulation taking into account all classified lines (FWHM 1.2 GHz). As can be seen (especially around 5800.0 Å), there are some quite dominant structures that are still not interpreted as transitions between known energy levels. Among the spectral lines, there is also one Pr II line (hf pattern shown in the upper part of the figure).

An overview concerning the already published newly discovered Pr I energy levels can be found in Ref. [39] and references therein.

In contrary to Ta II, even and odd levels of Pr II are not separated by a large energy difference. Thus, for the determination of improved energy values, one first has to build up two transition matrixes separately. These two could then connected by only one line, 4048.132. Even though this line had a low SNR of only 6, it must be used. A simplified level scheme is shown in Figure 6.

The FT spectra were also used to improve the accuracy of Pr II level energies. As can be seen from Figure 5, it may sometimes be tricky to find Pr II lines among the manifold of Pr I lines. Fortunately, in the blue and near infrared region, the identification becomes easier. Nevertheless, blend situations are quite frequently observed. As an example, in Figure 7, a blend of three lines is shown. In a more noisy spectrum or a spectrum with less resolution, one would notice only the dominant peak of the Pr I line at 5161.717 Å. Decomposing the observed pattern also allows for a precise determination of the cg wavelength of the involved Pr II line.

In Figure 8, a blend of two Pr II lines having nearly the same cg wavelength is shown. The wavelength difference is only 0.0025 Å (0.454 GHz or 0.015 cm^{-1}). The observed structure is well reproduced by adding the profiles of the two Pr II lines.

The wavelength of selected Pr II lines were determined carefully and a global fit of the level energies was performed. The results are published in Ref. [40].

Beside laser spectroscopic experiments, in which the plasma of a hollow cathode discharge was used as source of free Pr atoms, highly precise investigations of the hyperfine structure of Pr II lines were also made using collinear laser-ion beam spectroscopy (CLIBS) [41]. Such experiments were also performed earlier by Rivest et al. [42]. During the last time, CLIBS investigations were performed in the presence of a magnetic field in order to re-determine the Landé g_J-factors of the involved levels [43,44].

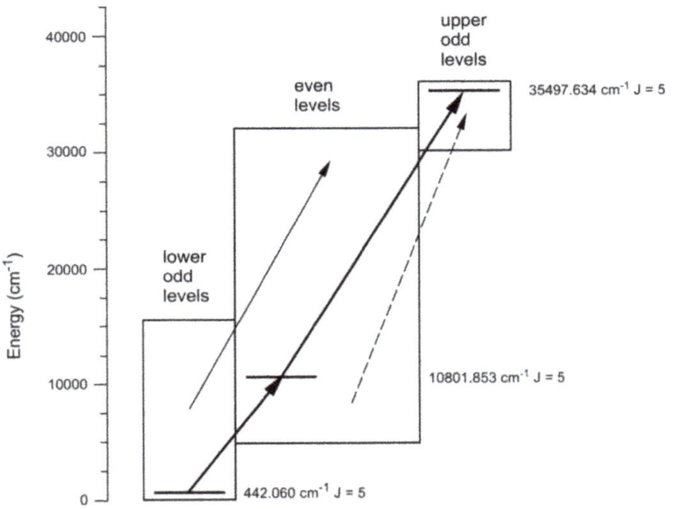

Figure 6. Simplified level scheme of Pr II. All transitions from lower odd levels to high-lying even levels are symbolized by the thin full arrow, all from the even levels to the upper odd levels by the dashed arrow. Only one ladder (bold arrows) could be found, which allowed for combining the two sub-systems of transitions.

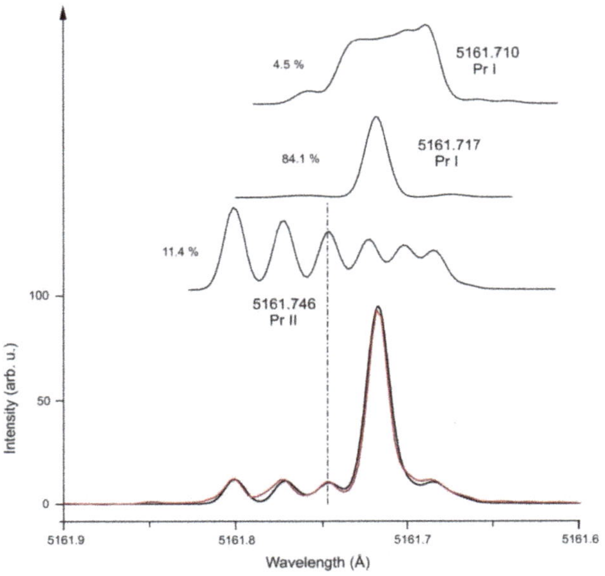

Figure 7. Blend of a Pr II line and two Pr I lines. Red line: FT spectrum, black line: simulation, FWHM 1.6 GHz). Knowing the hf constants of the involved levels, the cg wavelengths of all lines can be determined. In a less resolved spectrum, only the strong peak of the Pr I line at 5161.717 Å would be noticeable.

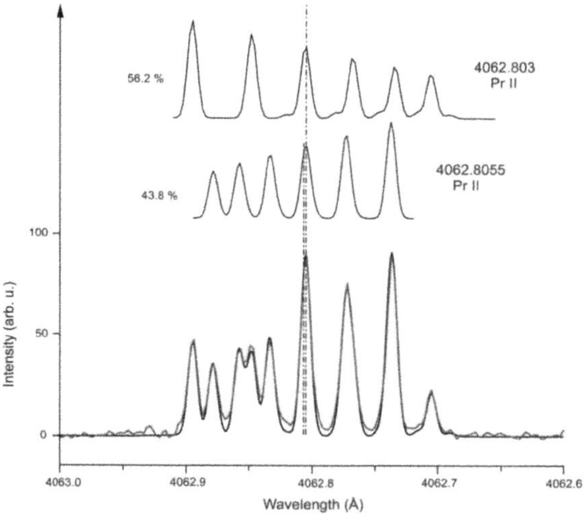

Figure 8. A blend situation of two Pr II lines with nearly the same cg wavelength. Red line: FT spectrum, black line: simulation, FWHM 1.6 GHz. In the upper part, normalized hf spectra of both lines are shown. Both profiles, added with the percentage given in the figure, gave the simulated sum profile (black), which describes the observed structure quite well. Thus, both cg wavelengths can be determined with high accuracy from the FT spectrum.

4. La II

Energy levels of La are also listed in Ref. [6]. Precise values of the hf structure constants of low lying metastable levels were obtained by a CLIBS technique by Höhle et al. [45] in 1982. These and other data were the basis of a theoretical interpretation of the hf structure of La II by Bauche et al. [46]. Later, CLIBS methods were also used by Li [47], Li [48], and Liang [49]. Some hf constants were determined by Lawler et al. [50] using an FT spectrum. Laser spectroscopic investigations of La II lines were made by Furmann et al. [51,52].

The FT spectra available in Graz (3225–16,600 Å) were taken in the group of Ferber (Laser Centre, University of Latvia, Riga, Latvia) with support of Kröger (Hochschule für Technik und Wirtschaft, Berlin, Germany). The near infrared part was analyzed in co-operation with the group of Basar (Physics Department, Istanbul University, Istanbul, Turkey). This was the first investigation of a La spectrum at wavelengths higher than 10,600 Å (investigated spectral range 8330–16,600 Å) [10]. The spectra were calibrated carefully using Ar II spectral lines. This allowed, together with the knowledge of the hf constants of the involved levels, a precise determination of transition wavelengths. From these values, improved energy values were determined. The results are submitted for publication [53]. Laser spectroscopy was performed in Graz mainly on atomic lines of La. These investigations are supported from the theoretical side by the group of Dembczyński (Institute of Materials Research and Quantum Engineering, Poznań University of Technology, Poznań, Poland).

As in the case of Pr, in the La spectrum blend situations are also observed quite frequently, despite the fact that the number of lines appearing in the spectra is lower (ca. 10,500). Figure 9 shows a typical blend situation.

Figure 10 shows an example for a La II line with widely split hf components. The line at 4151.957 Å is split into two groups of components. Since all components are well resolved, at such line, an independent determination of the hf constants of both combining levels is possible.

Additionally, Zeeman patterns of La II lines were investigated using a hollow cathode discharge in the presence of a magnetic field [54].

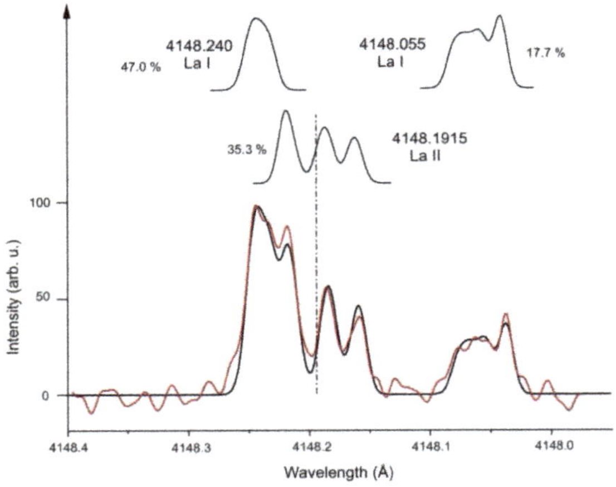

Figure 9. Blend of a La II and a La I line. Close to the blended lines another La I line appears. Red line: FT spectrum, black line: simulation, FWHM 2.4 GHz. Despite the fact that the signal-to noise ratio (SNR) is small (only 16 for the highest peak), the transitions can be clearly identified and their cg wavelength can be determined with good accuracy.

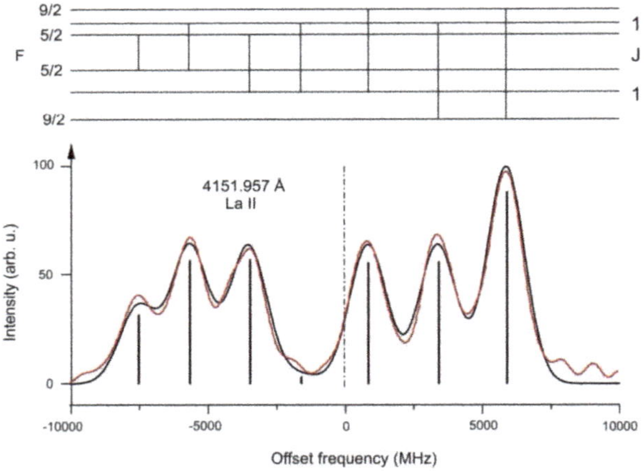

Figure 10. Example of a La II spectral line. Red line: FT spectrum, black line: simulation, FWHM 1.6 GHz. Also shown is the hf level scheme, the transitions and the components (theoretical intensity ratios).

5. Conclusions

Improved level energies of Ta II, Pr II, and La II were published. This paper is concerned with some peculiarities that had to be taken into account for an improvement of the energy values.

Acknowledgments: The author would like to thank all persons who contributed to the work described in the present article. Special thanks are devoted to G.H. Guthöhrlein, Universität der Bundeswehr Hamburg (Germany), for a very fruitful co-operation for more than 25 years. The work was theoretically supported by the group of J. Dembczyński, University of Technology, Poznań (Poland). Special thanks are devoted to W. Ernst, the present head (since 2002) of the Institute of Experimental Physics, Graz University of Technology, Graz, Austria, for allowing me to keep a room and a laser spectroscopy lab after my retirement (2014).

Conflicts of Interest: The author declares no conflict of interest.

Appendix A

Table A1. Classification of the lines shown in the figures and data of the involved energy levels. Cols.: columns, Sp.: spectrum, tw: this work.

	Line			Upper level					Lower Level					References to Cols.		
Fig. No.	**Sp.**	**Wavelength (Å)**	**SNR**	**Energy (cm^{-1})**	**J**	**P**	**A (MHz)**	**B(MHz)**	**Energy (cm^{-1})**	**J**	**P**	**A (MHz)**	**B(MHz)**	**5,10**	**8,9**	**13,14**
1	2	3	4	5	6	7	8	9	10	11	12	13	14	15	16	17
1	Ta II	3127.748	86	41708.994	5	o	914(5)	350(100)	9746.376	4	e	303(10)	1680(300)	[25]	[25]	[25]
1	Ta I	3127.864	30	31961.442	5/2	o	1243(3)	740(10)	0	3/2	e	509.084(1)	−1012.238(8)	[18]	tw	[55]
2	Ta II	3487.803	21	54048.682	5	o	650(10)	1200(200)	25385.546	4	e	730(20)	0(200)	[25]	[25]	[25]
3	Ta II	3037.503	39	39743.636	4	o	955(5)	−1200(100)	6831.437	3	e	360(10)	970(200)	[25]	[25]	[25]
3	Ta II	3037.532	17	46387.287	2	o	1802(15)	−130(50)	13475.416	1	e	−480(4)	724(20)	[25]	[25]	[21]
5	Pr II	5800.859	16	17676.112	5	e	805	-	442.060	5	o	1910.3(21)	-	[40]	[34]	[42]
7	Pr I	5161.710	15	33932.700	13/2	o	737(1)	-	14564.673	13/2	e	577(2)	-	[38]	[38]	[37]
7	Pr I	5161.717	280	29899.954	17/2	o	529(1)	-	10531.951	17/2	e	546(3)	-	[38]	[38]	[38]
7	Pr II	5161.746	38	23261.402	5	e	581.9(3)	−19(5)	3893.46	6	o	902.1	-	[40]	[41]	[34]
8	Pr II	4062.803	55	28009.828	7	e	556.5(7)	−34(25)	3403.226	6	o	−146.5(4)	-	[40]	[41]	[42]
8	Pr II	4062.8055	34	27604.990	6	e	597.0(5)	1(32)	2998.412	7	o	1435.2(16)	-	[40]	[41]	[42]
9	La I	4148.055	6	33820.316	1/2	o	−232.7(60)	-	9719.429	3/2	e	−655.138	−33.249	tw	[56]	[57]
9	La II	4148.1915	12	51524.005	2	e	−220(3)	-	27423.911	1	o	886.9(15)	−18.9(48)	[53]	[53]	[49]
9	La I	4148.240	16	38903.885	5/2	e	181(3)	-	14804.067	5/2	o	335.01(74)	23.64(95)	[58]	[58]	[59]
10	La II	4151.957	24	25973.360	1	o	547.3(30)	27(7)	1895.128	1	e	−1128.1(0.9)	49.8(65)	[53]	[52]	[45]

References

1. Windholz, L.; Gamper, B.; Binder, T. Variation of the observed widths of La I lines with the energy of the upper excited levels, demonstrated on previously unknown energy levels. *Spectr. Anal. Rev.* **2016**, *4*, 23–40. [CrossRef]
2. Siddiqui, I.; Khan, S.; Windholz, L. Experimental investigation of the hyperfine spectra of Pr I—Lines: Discovery of new fine structure energy levels of Pr I using LIF spectroscopy with medium angular momentum quantum number between 7/2 and 13/2. *Eur. Phys. J. D* **2016**, *70*, 44. [CrossRef]
3. Głowacki, P.; Uddin, Z.; Guthöhrlein, G.H.; Windholz, L.; Dembczyński, J. A study of the hyperfine structure of Ta I lines based on Fourier transform spectra and laser-induced fluorescence. *Phys. Scr.* **2009**, *80*, 025301. [CrossRef]
4. Stone, N.J. Table of nuclear magnetic dipole and electric quadrupole moments. *Atom. Data Nucl. Data* **2003**, *90*, 75–176. [CrossRef]
5. Moore, C.E. *Atomic Energy Levels, Vol. III*; Circular of the National Bureau of Standards 467; U.S. Government Printing Office: Washington, DC, USA, 1958.
6. Martin, W.C.; Zalubas, R.; Hagan, L. *Atomic Energy Levels—The Rare Earth Elements*; National Bureau of Standards: Washington, DC, USA, 1978.
7. Windholz, L. Finding of previously unknown energy levels using Fourier-transform and laser spectroscopy. *Phys. Scr.* **2016**, *91*, 114003. [CrossRef]
8. Messnarz, D.; Jaritz, N.; Arcimowicz, B.; Zilio, V.O.; Engleman, R., Jr.; Pickering, J.C.; Jäger, H.; Guthöhrlein, G.H.; Windholz, L. Investigation of the hyperfine structure of Ta I lines (VII). *Phys. Scr.* **2003**, *68*, 170–191. [CrossRef]
9. Gamper, B.; Uddin, Z.; Jagangir, M.; Allard, O.; Knöckel, H.; Tiemann, E.; Windholz, L. Investigation of the hyperfine structure of Pr I and Pr II lines based on highly resolved Fourier transform spectra. *J. Phys. B At. Mol. Opt.* **2011**, *44*, 045003. [CrossRef]
10. Güzelçimen, F.; Gö, B.; Tamanis, M.; Kruzins, A.; Ferber, R.; Windholz, L.; Kröger, S. High-resolution Fourier transform spectroscopy of lanthanum in Ar discharge in the near infrared. *Astrophys. Suppl. Ser.* **2013**, *218*, 18. [CrossRef]
11. Harrison, G.R. *Wavelength Tables*; Massachusetts Institute of Technology, The MIT Press: Cambridge, MA, USA, 1969.
12. Learner, R.C.M.; Thorne, A.P. Wavelength calibration of Fourier-transform emission spectra with applications to Fe I. *J. Opt. Soc. Am. B* **1988**, *5*, 2045–2059. [CrossRef]
13. Peck, E.R.; Reeder, K. Dispersion of air. *J. Opt. Soc. Am.* **1972**, *62*, 958–962. [CrossRef]
14. *Program package "Fitter"*; developed by Guthöhrlein, G.H., Helmut-Schmidt-Universität; Universität der Bundeswehr Hamburg: Hamburg, Germany, 1998.
15. Windholz, L.; Guthöhrlein, G.H. Classification of spectral lines by means of their hyperfine structure. Application to Ta I and Ta II levels. *Phys. Scr.* **2003**, *2003*, 55–60. [CrossRef]
16. Kiess, C.C. Description and analysis of the second spectrum of tantalum, Ta II. *J. Res. Natl. Bur. Stand.* **1962**, *66A*, 111–161. [CrossRef]
17. Brown, B.M.; Tomboulian, D.H. The nuclear moments of Ta181. *Phys. Rev.* **1952**, *88*, 1158–1162. [CrossRef]
18. Engleman, R., Jr. Improved Analysis of Ta I and Ta II. In Proceedings of the Symposium on Atomic Spectroscopy and Highly-Ionized Atoms, IL, USA, 16 August 1987.
19. Eriksson, M.; Litzén, U.; Wahlgren, G.M.; Leckrone, D.S. Spectral data for Ta II with application to the tantalum abundance in χ Lupi. *Phys. Scr.* **2002**, *65*, 480–489. [CrossRef]
20. Messnarz, D. Laserspektroskopische und Parametrische Analyse der Fein- und Hyperfeinstruktur des Tantal-Atoms und Tantal-Ions. Ph.D. Thesis, Wissenschaft & Technik Verlag, Berlin, Germany, 2011.
21. Messnarz, D.; Guthöhrlein, G.H. Laserspectroscopic Investigation of the hyperfine structure of Ta II—Lines. *Phys. Scr.* **2003**, *67*, 59–63. [CrossRef]
22. Zilio, V.O.; Pickering, J.C. Measurements of hyperfine structure in Ta II. *Mon. Not. R. Astron. Soc.* **2002**, *334*, 48–52. [CrossRef]
23. Jaritz, N.; Guthöhrlein, G.H.; Windholz, L.; Messnarz, D.; Engleman, R., Jr.; Pickering, J.C.; Jäger, H. Investigation of the hyperfine structure of Ta I lines (VIII). *Phys. Scr.* **2004**, *69*, 441–450. [CrossRef]

24. Jaritz, N.; Windholz, L.; Messnarz, D.; Jäger, H.; Engleman, R., Jr. Investigation of the hyperfine structure of Ta I lines (IX). *Phys. Scr.* **2005**, *71*, 611–620. [CrossRef]
25. Windholz, L.; Arcimowicz, B.; Uddin, Z. Revised energy levels and hyperfine structure constants of Ta II. *J. Quant. Spectrosc. Radiat. Transf.* **2016**, *176*, 97–121. [CrossRef]
26. Uddin, Z.; Windholz, L. New levels of Ta II with energies higher than 72,000 cm^{-1}. *J. Quant. Spectrosc. Radiat. Transf.* **2014**, *149*, 204–210. [CrossRef]
27. Uddin, Z. Hyperfine Structure Studies of Tantalum and Praseodymium. Ph.D Thesis, Graz University of Technology, Graz, Austria, 2006.
28. Stachowska, E.; Dembczyński, J.; Windholz, L.; Ruczkowski, J.; Elantkowska, M. Extended analysis of the system of even configurations of Ta II. *Atom. Data Nucl. Data* **2017**, *113*, 350–360. [CrossRef]
29. White, H.E. Hyperfine structure in singly ionized praseodymium. *Phys. Rev.* **1929**, *34*, 1397–1403. [CrossRef]
30. Rosen, N.; Harrison, G.R.; McNally, R., Jr. Zeeman effect data and preliminary classification of the spark spectrum of praseodymiun—Pr II. *Phys. Rev.* **1941**, *60*, 722–730. [CrossRef]
31. Blaise, J.; Verges, J.; Wyart, J.-F.; Camus, P.; Zalubas, R.J. Opt. *Soc. Am.* **1973**, *63*, 1315.
32. Ginibre, A. Fine and hyperfine structures in the configurations $4f^25d6s^2$ and $4f^25d^26s$ of neutral praseodymium. *Phys. Scr.* **1981**, *23*, 260–267. [CrossRef]
33. Ginibre-Emery, A. Classification et Étude Paramétrique Des Spectres Complexes Á L'Aide De L'Interprétation Des Structures Hyperfines: Spectres I et II du Praséodyme. Ph.D Thesis, Université de Paris-Sud, Centre d'Orsay, France, 1988.
34. Ginibre, A. Fine and hyperfine structures of singly ionized praseodymium: I. energy levels, hyperfine structures and Zeeman effect, classified lines. *Phys. Scr.* **1989**, *39*, 694–709. [CrossRef]
35. Ginibre, A. Fine and hyperfine structures of singly ionized praseodymium: II. Parametric interpretation of fine and hyperfine structures for the even levels of singly ionised praseodymium. *Phys. Scr.* **1989**, *39*, 710–721. [CrossRef]
36. Ivarsson, S.; Litzén, U.; Wahlgren, G.M. Accurate wavelengths, oscillatror strengths and hyperfine structure in slected praseodymium lines of astrophysical interest. *Phys. Scr.* **2001**, *64*, 455–461. [CrossRef]
37. Guthöhrlein, G.H. (Helmut-Schmidt-Universität, Universität der Bundeswehr Hamburg, Hamburg, Germany). Private communication, 2005.
38. Zaheer, U.; Driss, E.B.; Gamper, B.; Khan, S.; Siddiqui, I.; Guthöhrlein, G.H.; Windholz, L. Laser spectroscopic investigations of praseodymium I transitions: New energy levels. *Adv. Opt. Technol.* **2012**, *2012*, 639126.
39. Khan, S.; Siddiqui, I.; Iqbal, S.T.; Uddin, Z.; Guthöhrlein, G.H.; Windholz, L. Experimental investigation of the hyperfine structure of neutral praseodymium spectral lines and discovery of new energy levels. *Int. J. Chem.* **2017**, *9*, 7–29. [CrossRef]
40. Akhtar, N.; Windholz, L. Improved energy levels and wavelengths of Pr II from a high-resolution Fourier transform spectrum. *J. Phys. B At. Mol. Opt.* **2012**, *45*, 095001. [CrossRef]
41. Akhtar, N.; Anjum, N.; Hühnermann, H.; Windholz, L. A study of hyperfine transitions of singly ionized praseodymium ($^{141}Pr^+$) using collinear laser ion beam spectroscopy. *Eur. Phys. J. D* **2012**, *66*, 264. [CrossRef]
42. Rivest, R.C.; Izawa, M.R.; Rosner, S.D.; Scholl, T.J.; Wu, G.; Hol, R.A. Laser spectroscopic measurements of hyperfine structure in Pr II. *Can. J. Phys.* **2002**, *80*, 557–562. [CrossRef]
43. Werbowy, S.; Kwela, J.; Anjum, N.; Hühnermann, H.; Windholz, L. Zeeman effect of hyperfine-resolved spectral lines of singly ionized praesodymium using collinear laser-ion-beam spectroscopy. *Phys. Rev. A* **2014**, *90*, 032515. [CrossRef]
44. Werbowy, S.; Windholz, L. Revised Lande g_J-factors of some 141Pr II levels using collinear laser ion beam spectroscopy. *J. Quant. Spectrosc. Radiat. Transf.* **2016**, *187*, 267–273. [CrossRef]
45. Höhle, C.; Hühnermann, H.; Wagner, H. Measurements of the hyperfine structure constants of all the $5d^2$ and $5d6s$ levels in ^{139}La II. *Z. Phys. A Hadron Nucl.* **1982**, *304*, 279–283.
46. Bauche, J.; Wyart, J.-F. Interpretation of the hyperfine structures of the low even configurations of lanthanum II. *Z. Phys. A Hadron Nucl.* **1982**, *304*, 285–292. [CrossRef]
47. Li, M.; Ma, H.; Chen, M.; Chen, Z.; Lu, F.; Tang, J.; Yang, F. Hyperfine structure measurements in the lines 576.91 nm, 597.11 nm and 612.61 nm of La II. *Phys. Scr.* **2000**, *61*, 449–451.
48. Li, G.W.; Zhang, X.M.; Lu, F.Q.; Peng, X.J.; Yang, F.J. Hyperfine structure measurement of La II by collinear fast ion beam laser spectroscopy. *Jpn. J. Appl. Phys.* **2001**, *40*, 2508–2510. [CrossRef]

49. Liang, M.H. Hyperfine structure of singly ionized lanthanum and praseodymium. *Chin. Phys.* **2002**, *11*, 905–909. [CrossRef]
50. Lawler, J.E.; Bonvallet, G.; Sneden, C. Experimental radiative lifetimes, branching fractions, and oscillator strengths for La II and a new determination of the solar lanthanum abundance. *Astrophys. J.* **2001**, *556*, 452–460. [CrossRef]
51. Furmann, B.; Elantkowska, M.; Stefańska, D.; Ruczkowski, J.; Dembczyński, J. Hyperfine structure in La II even configuration levels. *J. Phys. B At. Mol. Opt.* **2008**, *41*, 235002. [CrossRef]
52. Furmann, B.; Ruczkowski, J.; Stefańska, D.; Elantkowska, M.; Dembczyński, J. Hyperfine structure in La II odd configuration levels. *J. Phys. B At. Mol. Opt.* **2008**, *41*, 215004. [CrossRef]
53. Gücelçimen, F.; Tonka, M.; Uddin, Z.; Bhatti, N.A.; Başar, G.; Windholz, L.; Kröger, S. Improved energy levels and wavelengths of La II from a high-resolution Fourier transform spectrum. *J. Quant. Spectrosc. Radiat. Transf.* **2017**, submitted for publication.
54. Werbowy, S.; Güney, C.; Windholz, L. Studies of Landé g_j-factors of singly ionized lanthanum by laser-induced fluorescence spectroscopy. *J. Quant. Spectrosc. Radiat. Transf.* **2016**, *179*, 33–39. [CrossRef]
55. Büttgenbach, S.; Meisel, G. Hyperfine structure measurements in the ground states $^4F_{3/2}$, $^4F_{5/2}$ and $^4F_{7/2}$ of Ta^{181} with the atomic beam magnetic resonance method. *Z. Phys.* **1971**, *244*, 149–162. [CrossRef]
56. Furmann, B.; Stefańska, D.; Dembczyński, J. Hyperfine structure analysis odd configurations levels in neutral lanthanum: I. Experimental. *Phys. Scr.* **2007**, *76*, 264. [CrossRef]
57. Childs, W.J.; Nielsen, U. Hyperfine structure of the $(5d+6s)^3$ configuration of ^{139}La I: New measurements and ab initio multiconfigurational Dirac-Fock calculations. *Phys. Rev. A* **1988**, *37*, 6–15. [CrossRef]
58. Güzelçimen, F.; Siddiqui, I.; Başar, G.; Kröger, S.; Windholz, L. New energy levels and hyperfine structure measurements of neutral lanthanum by laser-induced fluorescence spectroscopy. *J. Phys. B At. Mol. Opt.* **2012**, *45*, 135005. [CrossRef]
59. Jin, W.G.; Endo, T.; Uematsu, H.; Minowa, T.; Katsuragawa, H. Diode-laser hyperfine-structure spectroscopy of ^{138,139}La. *Phys. Rev. A* **2001**, *63*, 064501. [CrossRef]

atoms

MDPI

Article

Hyperfine Structure and Isotope Shifts in Dy II

Dylan F. Del Papa, Richard A. Holt and S. David Rosner *

Department of Physics & Astronomy, University of Western Ontario, London, ON N6A 3K7, Canada; ddelpapa@uwo.ca (D.F.D.P.); rholt@uwo.ca (R.A.H.)

* Correspondence: rosner@uwo.ca

Academic Editor: Joseph Reader

Received: 8 December 2016; Accepted: 13 January 2017; Published: 20 January 2017

Abstract: Using fast-ion-beam laser-fluorescence spectroscopy (FIBLAS), we have measured the hyperfine structure (hfs) of 14 levels and an additional four transitions in Dy II and the isotope shifts (IS) of 12 transitions in the wavelength range of 422–460 nm. These are the first precision measurements of this kind in Dy II. Along with hfs and IS, new undocumented transitions were discovered within 3 GHz of the targeted transitions. These atomic data are essential for astrophysical studies of chemical abundances, allowing correction for saturation and the effects of blended lines. Lanthanide abundances are important in diffusion modeling of stellar interiors, and in the mechanisms and history of nucleosynthesis in the universe. Hfs and IS also play an important role in the classification of energy levels, and provide a benchmark for theoretical atomic structure calculations.

Keywords: laser spectroscopy; ionized atoms; hyperfine structure; isotope shifts

1. Introduction

Accurate elemental abundances in stars and the interstellar medium are essential for understanding stellar formation, evolution, and structure. Deriving abundances from stellar spectra requires accurate atomic data: tabulated spectra of several ionization states (usually I–III), oscillator strengths, hyperfine structure (hfs), and isotope shifts (IS). Even though hfs and IS are too small to be resolved in most astrophysical spectra because of Doppler and rotational broadening, they contribute to the overall profile of a spectral line; extracting the elemental abundance from a line often requires a knowledge of the hfs to correct for saturation [1], and ignoring the underlying structure arising from hfs and IS can lead to errors as large as two to three orders of magnitude [2]. In addition, there can be errors in the determination of rotational, microturbulent and macroturbulent velocities from Fourier analysis of line profiles, and unknown shifts in wavelengths of saturated lines [3]. The lanthanides are particularly important because they provide a contiguous sequence of observable elements in which patterns of abundance may be observed and related to nucleosynthesis and chemical fractionation [4]. In this paper, we present new measurements of hfs and IS in Dy II, one of the lanthanide elements that is used to understand these processes.

The main mechanisms of neutron capture by which elements beyond ^{56}Fe are formed are the *r*-process and *s*-process, named by whether the rate of neutron capture is *r*apid or *s*low compared to the rate of β-decay [5]. The *s*-process is believed to take place in thermally pulsing asymptotic giant branch (AGB) stars [6], whereas the *r*-process requires the high neutron density characteristic of supernovas. Comparison of the elemental composition of a star with theoretical *r*- or *s*-process models gives information about the nature of these processes and about the environment in which the star formed. The observed ratio of the abundances of a pair of unstable and stable *r*-process elements, e.g., [Th]/[Eu], can yield the age of the star when compared with the theoretical production ratio [7]. There are several recent studies of abundances in very metal-poor stars, which are believed to have formed from the debris of the very first generation of initially metal-free massive stars (Population III). These studies shed light on the earliest nucleosynthesis of heavy elements in the universe and how they were incorporated into later stars [8,9]. CH stars, which are deficient in Fe and enhanced in C and *s*-process elements, obtained their heavy elements by mass transfer from now-unseen white dwarf binary companions when the latter were in the AGB stage of evolution. Thus the composition of CH stars reveals information about AGB synthesis of *s*-process elements [10].

Accurate abundance information is also required in the study of diffusion processes within stars, where chemical fractionation occurs due to a combination of gravity, convection and radiation pressure. The chemically peculiar (CP) stars exhibit typically large lanthanide abundance excesses relative to solar and meteoritic values, and the abundance pattern expected from nucleosynthesis may be greatly perturbed [4]. One of the most exciting recent developments in stellar astrophysics has been the rapid expansion of the field of asteroseismology [11], in which the observation of natural, resonant oscillations provides information about stellar interiors and evolution. These observations can be photometric or velocity measurements from Doppler shifts. In the latter case, it is essential to be able to model spectral line shapes accurately, requiring good atomic data.

Extensive spectroscopy of Dy II has been done principally by Conway and Worden [12], Wyart [13–15], and more recently by Nave and Griesmann [16], who pointed out the complexity of lanthanide spectra due to the large number of terms arising from the unfilled 4*f* shell and the fact that 4*f*, 5*d*, and 6*s* electrons all have similar binding energies. The NIST Atomic Spectra Database [17] lists 576 levels of which a significant number are only partly classified as to principal configuration and term, and configuration mixing is common. One of the tools that has been used to aid in level classification is IS, since the 'field shift' part of the IS (the dominant contribution in heavy atoms) depends on the electron density at the nucleus and therefore is a function of the electronic states involved in a transition. Early measurements of IS were carried out by Pacheva and Abadjieva [18]. Aufmuth [19] measured IS for Dy isotopes 162–164 in 29 lines of Dy II in order to check the mixing of three configurations calculated by Wyart [13–15]. Ahmad et al. [20] measured the IS of 62 spectral lines, three of which overlap with our measurements. The only previous report of hfs is by Murakawa and Kamei [21].

2. Experimental Setup

The FIBLAS method [22] we use is very well suited to the measurement of the detailed structure within an atomic transition because of the high spectral resolution (full-width-half-maximum linewidths of ~100 MHz), the high wavelength selectivity of a narrow-band laser, and the ability to mass-select the targeted ions in the fast beam. The transition linewidths are a combination of the natural width (here 20–70 MHz), and Doppler broadening from the ion source, significantly narrowed for fast ions by 'kinematic compression' [23]. The hfs and IS of an optical transition are often completely resolved, a goal difficult to achieve in other techniques, such as Fourier transform spectroscopy.

In our apparatus, Dy^+ ions are created using a Penning sputter ion source, engineered by us to yield a small energy spread of a few eV in the extracted beam [24,25]. Another important feature is the large population of metastable ions in the beam with energies up to ~22,000 cm^{-1}, greatly enhancing the number of transitions we can observe. The use of sputtering rather than thermal evaporation makes the source universal for any solid conducting element in the periodic table; this is a great advantage for the lanthanides and many other refractory metals of astrophysical importance. A discharge in the Ne support gas is easily struck at anode-cathode potentials of several hundred volts, with the discharge current regulated at 60 mA. Our experience with lanthanides is that there is an initial period of conditioning (up to 2 h), during which the potential drops to ~200–300 V, and the extracted ion current, as measured in a downstream Faraday cup, rises sharply to ~200 nA. We attribute this behavior to sputter-cleaning of the surfaces of the lanthanide cathode and anticathode to remove oxide layers, which form easily in air.

After initial acceleration to energies of 10–12 keV, the ion beam is focused by an einzel lens, steered, and mass-selected using a Wien filter. Removal of the large Ne^+ current from the beam reduces space-charge spreading and significantly lowers the collision-induced background fluorescence. The mass resolution of the Wien filter was selected by adjustment of its electric field to reject the Ne ions, while transmitting nearly equally the seven stable isotopes of Dy (see Table 1) to allow IS measurement. The mass-selected beam is then horizontally deflected by 5° to make it collinear with the laser beam. After further focusing and steering, the ions are accelerated by 478 eV into a 'Doppler tuning' region; this energy boost ensures the laser-induced fluorescence (LIF) is mainly confined to this region, from which it is transmitted out of the vacuum chamber by an array of optical fibers to a photomultiplier equipped with a short-pass filter. The filter allows rejection of the scattered laser light while transmitting most of the LIF, since the lower (metastable) energy states of the laser-induced transitions we observed ranged from ~4300 cm^{-1} to 22,000 cm^{-1}.

Table 1. Properties of the Dy isotopes. *I*, *μ*, and *Q* are the nuclear spin, magnetic dipole moment, and electric quadrupole moment, respectively.

Isotope	Mass (u) [a]	Abundance (%) [b]	*I* [c]	*μ* (nm) [c]	*Q* (b) [c]
^{156}Dy	155.924278	0.06	0		
^{158}Dy	157.924405	0.10	0		
^{160}Dy	159.925194	2.34	0		
^{161}Dy	160.926930	18.91	5/2	−0.480 (3)	+2.51 (2)
^{162}Dy	161.926795	25.51	0		
^{163}Dy	162.928728	24.90	5/2	+0.673 (4)	+2.318 (6)
^{164}Dy	163.929171	28.18	0		

[a] Reference [26]; [b] Reference [27]; [c] Reference [28].

The laser system is a single-mode ring dye laser operating with Stilbene 420 dye, whose usable output range is ~420 nm to 460 nm. It is pumped with the all-lines UV output of an Ar-ion laser, and is frequency stabilized to a few MHz. The dye laser wavelength is monitored to ~1 part in 10^7 by a traveling-mirror Michelson interferometer [29] using a polarization-stabilized HeNe laser as a reference. The laser beam is loosely focused to a waist located approximately in the Doppler-tuning region. Before starting the laser scan across the hfs/IS components of a transition, the laser frequency is manually tuned to the absorption line of ^{162}Dy, which is the even isotope (i.e., without hfs) closest to the mean mass of the stable isotopes. The overlap of the laser and ion beams is optimized by adjusting steering and focusing of the ion beam. This results in a scan where the peak intensities of the even isotopes approximately match their standard abundances, facilitating subsequent analysis of the spectrum. During the scan, a computer steps the laser frequency over the entire spectrum of the line (<20 GHz) in intervals of ~1/10th of a linewidth with a dwell time of typically 250 ms. The computer simultaneously records the LIF signal along with the ion beam current, laser beam power, and a calibration signal that corrects for the nonlinearity (~2%) of the laser scan. This calibration signal is obtained using a set of markers generated by a plane-parallel Fabry-Perot interferometer with a free spectral range of 665.980 MHz, which is used to convert channel number into laser frequency difference.

Because the Dy II hfs is complicated by the existence of an unusually large electric quadrupole interaction, we introduced some redundancy to the data in order to allow checks on the robustness of the analysis. This was done by taking spectra with the laser beam both parallel (P) and antiparallel (A) to the ion beam, and by changing the ion-beam energy. Both of these measures change the large Doppler shift for each isotope, resulting in significant changes in the appearance of the measured spectrum. This can result in a P (or A) spectrum that is much less congested than its counterpart, greatly facilitating the analysis (see below).

3. Data and Analysis

A typical spectrum with fully resolved hfs and IS is shown in Figure 1, in which the 'stick figure' below the data shows the fitted peak locations. At the left of the figure, a number of peaks belonging to another, weaker transition can be seen. The spectra in which such blends occurred are noted in the IS table below by asterisks. The relative intensities of the extra transitions ranged from 1% to 15%, except in the case of the 443.100 nm line, for which two nearly equal-intensity transitions were seen, precluding analysis. The peaks corresponding to ^{156}Dy and ^{158}Dy were too small to be seen above the noise in our spectra and were not included in the analysis.

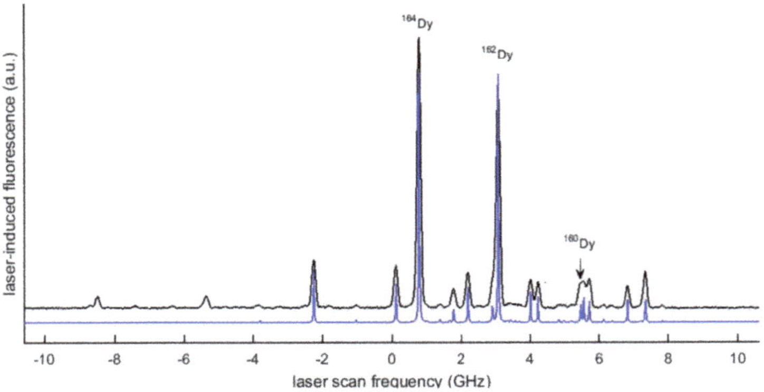

Figure 1. Laser-induced fluorescence spectrum of the transition $4f^{10}(^5I_6)6s_{1/2}\ (6,{}^1/_2)_{13/2}$–$4f^{10}(^5I)6p°_{13/2}$ at 438.430 nm in Dy II. The laser beam and ion beam were in parallel geometry. The observed spectrum (upper curve) is a single scan of 1024 channels at a dwell time of 400 ms per channel. The photomultiplier signal current for the ^{164}Dy peak was ~80 nA on a background of ~10 nA due to collisionally-induced light. The observed FWHM linewidth was 98.8 MHz, which includes the 10.6 MHz natural width. The lower 'stick figure' (blue online) shows the positions and amplitudes of the fitted components, and has been displaced vertically for clarity. The hfs components of ^{161}Dy and ^{163}Dy are not individually annotated. It is important to understand that the separations between peaks of different isotopes arise from a combination of IS and relative Doppler shift. Note that the components of a second, partially blended, transition are visible on the left side.

Figure 2a,b show another transition observed in both P and A modes to demonstrate the 'decongestion' that occurs in one of these modes.

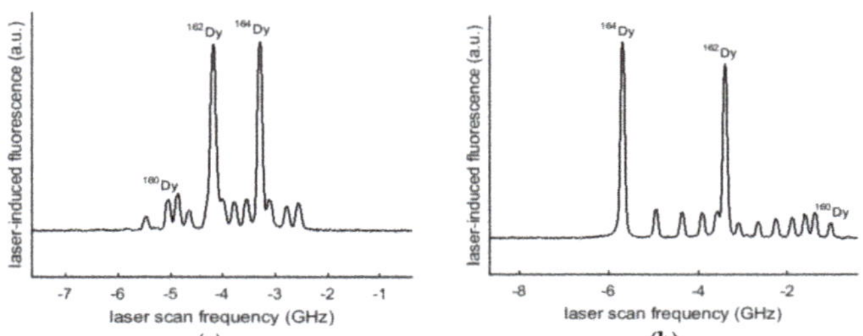

Figure 2. Laser-induced fluorescence spectrum of the transition $4f^9(^6H°)5d\ (^7I°)6s\ ^8I_{17/2}°$–$4f^9(^6H°)5d\ (^7H°)6p\ _{15/2}$ at 436.135 nm in Dy II, showing the advantage of viewing the same transition in two different laser beam/ion beam geometries: (**a**) Anti-parallel; (**b**) Parallel.

It is important to note that in the large separation $\Delta\nu_\ell$ any pair of peaks corresponding to two different isotopes shown in these spectra is predominantly a *differential Doppler shift* arising from the slightly different velocity of each isotope. In the A (P) mode, driving a transition whose rest-frame frequency is ν_0 requires a laser frequency ν_ℓ given by

$$\nu_\ell = \nu_0 \left(\frac{1 \mp \beta}{1 \pm \beta} \right)^{1/2} \tag{1}$$

where $\beta = \left(2eV/Mc^2\right)^{1/2}$ for a singly-charged ion of mass M accelerated from rest through a potential difference V. Thus, the ion-rest-frame peak separation $\Delta\nu_0$ (which is the IS within a sign) for a pair of masses M and M' is related to $\Delta\nu_\ell$ by

$$\Delta\nu_\ell = \Delta\nu_0 \left(\frac{1 \mp \beta'}{1 \pm \beta'} \right)^{1/2} + \nu_0 \left[\left(\frac{1 \mp \beta'}{1 \pm \beta'} \right)^{1/2} - \left(\frac{1 \mp \beta}{1 \pm \beta} \right)^{1/2} \right] \tag{2}$$

(This formula may also be applied to the separation of a pair of hf peaks of the same mass by setting $\beta' = \beta$.) Creating the model spectrum to be fit to the data thus requires knowledge of the kinetic energy of the fast ions, which depends on the potential in the ion source, the extraction voltage, and the potential in the Doppler-tuning region. It is possible to determine the beam energy directly by measuring the Doppler-shifted wavenumber of a given spectral line in both A and P modes; a simple calculation with Equation (1) yields the velocity of the isotope corresponding to that spectral line as well as the transition wavenumber in the ion rest frame. Such measurements have shown that the beam energy is, within ~1 eV, just the sum of the anode-cathode potential difference V_{ac} (the power-supply potential difference V_{supply} minus the 58.5 V drop across a 1 kΩ stabilizing ballast resistor) and the 10 kV or 12 kV extraction voltage, minus the potential in the central region of the Doppler-tuning region (−478 V), determined from its applied voltage using a numerical solution of Laplace's equation. This implies that the plasma region in which the ions are created is at the potential of the anode. In order to account for the Doppler shifts in a spectrum, we thus needed to monitor only V_{supply} during the few minutes of a scan. Typically, V_{supply} varied by <1 V in the current-regulated plasma, and only such scans were used as data.

The model used in the least-squares fit makes use of the standard formula [30] for the hyperfine contribution to the energy of a level with a nuclear spin I, an electronic angular momentum J, and a total angular momentum F, containing magnetic-dipole and electric-quadrupole terms, with parameters A and B, respectively:

$$\nu_{\text{hfs}}(F) = \frac{1}{2}AK + \frac{1}{2}B\frac{3K(K+1) - 4I(I+1)J(J+1)}{2I(2I-1)2J(2J-1)} \tag{3}$$

where $K = F(F+1) - I(I+1) - J(J+1)$. The amplitudes of the model peaks are constrained by the standard abundances of the isotopes, and, also, for the odd isotopes displaying hfs, by standard angular momentum recoupling coefficients:

$$a(F, F') \propto (2I+1)^{-1}(2F+1)(2F'+1) \begin{Bmatrix} F' & F & 1 \\ J & J' & I \end{Bmatrix}^2 \tag{4}$$

where the primes (non-primes) indicate the upper (lower) level of a transition. A further constraint was the use of known ratios (see Table 1) of the nuclear magnetic moments and electric quadrupole moments of the odd isotopes ^{161}Dy and ^{163}Dy, so that the hfs of only one isotope needed to be fit. We are thus neglecting the hyperfine anomaly, which is expected to be small compared to our measurement uncertainty. The experimental lineshapes combine natural broadening with asymmetric Doppler broadening, arising mainly from the unknown potential distribution in the source plasma; we interpret the 'tail' of slower-energy ions as those created in the 'cathode fall' region. In the course of data analysis, we observed that an asymmetric Lorentzian profile improved the consistency of the data. We note that the peak frequency *differences* determining the IS and hfs constants are insensitive to the detailed lineshape model used.

In addition to the four observable IS (measured with ^{164}Dy as the reference) and four hfs constants for a given transition, other adjustable parameters in the fit included the overall frequency offset, overall amplitude, background, linewidth, asymmetry, and laser power (to account for saturation). The large electric quadrupole contribution to the hfs created splittings very different from the standard 'flag' pattern obtained when the magnetic dipole interaction dominates. Without the help of pattern recognition, it was necessary to use trial and error along with global fitting algorithms, searching across up to six parameters to find a fit. Another issue was that the pattern of a spectrum was dependent mainly on the differences B'–B and A'–A, with only subtle changes in the fit spectrum resulting from varying the *individual* values of the constants. Thus, as an intermediate measure, the search space could be reduced by fixing the constants associated with either the lower or upper energy level. A very important factor in breaking these parameter correlations is the ability to measure the relatively weak satellite peaks ($\Delta F \neq \Delta J$) in the hyperfine spectra, since the frequencies of these peaks depend algebraically on the four hfs constants rather differently than for the principal peaks ($\Delta F = \Delta J$). We also dealt with these correlations experimentally by repeating measurements of most transitions at different beam energies (10 keV and 12 keV) and different relative beam orientations (A and P modes). In four cases, the satellite peaks were not visible, with the result that the associated hfs constants were not individually well-determined, but are rather to be regarded as a set of 'effective' parameters that reproduce the principal peaks of the observed spectrum very well so as to be of practical use in modeling astrophysical spectra.

The resulting hfs constants for six lower and eight upper levels are given in Table 2, while Table 3 lists the effective hfs parameters for the four transitions referred to above. For the levels listed in Table 2, we were able to use several transitions to determine the hfs constants in each case, providing important 'cross-checks'; however, the levels in Table 3 were only accessible using the transitions listed.

Table 2. Hyperfine structure constants of levels of ^{161}Dy II and ^{163}Dy II derived from transitions where satellite peaks are well fit. A and B are the magnetic dipole and electric quadrupole constants, respectively. J is the total angular momentum. Energies of odd-parity levels are italicized.

Configuration [a]	Term [a]	J [a]	Energy [a] (cm^{-1})	A (^{161}Dy) (MHz)	B (^{161}Dy) (MHz)	A (^{163}Dy) (MHz)	B (^{163}Dy) (MHz)
$4f^{10}(^5I_7)6s_{1/2}$	(7, 1/2)	15/2	4341.104	−251.89 (94)	1045 (48)	352.6 (1.3)	1104 (46)
$4f^{10}(^5I_6)6s_{1/2}$	(6, 1/2)	13/2	7485.117	93.22 (16)	−883.4 (7.1)	−130.48 (22)	−933.1 (7.5)
$4f^{10}(^5F_5)6s_{1/2}$	(5, 1/2)	11/2	13,338.27	−272.49 (23)	−818 (10)	381.44 (32)	−864 (11)
$4f^9(^6H^\circ)5d(^7K^\circ)6s$	$^8K^\circ$	19/2	*17,606.65*	−144.40 (30)	4081 (17)	202.13 (41)	4310 (18)
$4f^9(^6H^\circ)5d(^7K^\circ)6s$	$^6K^\circ$	19/2	*19,571.75*	−69.51 (35)	3861 (24)	97.29 (49)	4078 (25)
$4f^9(^6H^\circ)5d^2(^3F)$	°	11/2	*20,517.39*	−96.98 (37)	2076 (17)	135.76 (52)	2193 (18)
$4f^9(^6H^\circ)5d^2(^3P)$	°	15/2	*27,435.132*	−97.23 (12)	252.8 (7.1)	136.10 (17)	267.1 (7.5)
$4f^9(^6H^\circ)5d^2(^3F)$	°	13/2	*28,019.70*	−114.76 (56)	1104 (37)	159.67 (78)	1166 (39)
$4f^{1\circ}(^5I)6p$	°	13/2	*30,287.36*	278.44 (18)	−973.9 (7.3)	−389.76 (25)	−1028.6 (7.7)
	°	9/2	*36,466.34*	−110.97 (24)	801 (14)	155.34 (33)	846 (15)
$4f^9(^6H^\circ)6s6p(^3P^\circ)$		17/2	40,455.73	−165.08 (32)	3851 (17)	231.08 (44)	4067 (18)
$4f^9(^6H^\circ)5d(^7H^\circ)6p$		17/2	41,583.90	−102.69 (31)	1826 (18)	143.75 (43)	1929 (19)
$4f^9(^6H^\circ)5d(^7H^\circ)6p$		13/2	42,289.33	−96.92 (45)	2773 (38)	135.67 (63)	2929 (40)
$4f^9(^6H^\circ)5d(^7H^\circ)6p$		19/2	42,478.98	−113.16 (42)	3099 (24)	158.40 (59)	3273 (26)

[a] Reference [17].

Table 4 presents the IS for 12 transitions, using the standard sign convention that $IS \equiv \nu_{M'} - \nu_M$ where $M' > M$. The uncertainties in the data arise from the curve-fitting procedure, the residual non-linearity of the laser scan, and small drifts in V_{ac}. Equation (2) can be used to calculate the sensitivity of the measured separation $\Delta\nu_\ell$ of two spectral peaks to the accuracy of the ion-beam energy eV, and to drifts in that energy *during a scan*. If the drift is zero, $d(\Delta\nu_\ell)/dV$ is completely negligible for two hyperfine (hf) peaks of the same mass, and is ~0.15 MHz/V for peaks of the isotope pair (160, 164).

Since we know V to ~1 V, this effect is negligible compared to fitting errors. If the drift is non-zero, the sensitivity for any pair of peaks is ~12 MHz/V for typical values of ν_0 and the beam energy. For the hf spectrum of a given isotope, the effect of a drift is taken into account by the fitting error since the many hf peaks are fit with only four hf parameters. For the same reason any residual non-linearity in the frequency scale is also subsumed in the fitting error. However, drift in V can contribute an error to the IS data, just as non-linearity can. Accordingly, we have added a contribution to the error budget of the IS of 10 MHz to account for both of these effects. We present evidence (see below) that this estimate is conservative.

Table 3. Hyperfine structure constants of transitions of ^{161}Dy II and ^{163}Dy II where satellite peaks are not well fit. A and B are the magnetic dipole and electric quadrupole constants, respectively. J is the total angular momentum. Energies of odd-parity levels are italicized.

λ_{air} [a] (nm)	Configuration [b]	Term [b]	J [b]	Energy [b] (cm^{-1})	A (^{161}Dy) (MHz)	B (^{161}Dy) (MHz)	A (^{163}Dy) (MHz)	B (^{163}Dy) (MHz)
424.846	$4f^9(^6H^\circ)5d(^7H^\circ)6s$	$^8H^\circ$	11/2	*14,347.205*	−63.33 (91)	1905 (37)	88.7 (1.3)	2012 (39)
	$4f^9(^6H^\circ)6s6p(^3P^\circ)$		13/2	37,878.55	−26.0 (1.4)	1552 (31)	36.4 (2.0)	1639 (32)
436.135	$4f^9(^6H^\circ)5d(^7I^\circ)6s$	$^8I^\circ$	17/2	*14,895.06*	−94.86 (79)	2106 (60)	132.8 (1.1)	2224 (64)
	$4f^9(^6H^\circ)5d(^7H^\circ)6p$		15/2	37,817.31	−65.15 (91)	1598 (59)	91.2 (1.3)	1688 (62)
451.851	$4f^9(^6H^\circ)5d^2(^3F)$	$^8K^\circ$	21/2	*22,031.98*	11.69 (73)	−1613 (50)	−16.4 (1.0)	−1704 (53)
	$4f^9(^6H^\circ)5d(^7I^\circ)6p$	^8K	21/2	44,156.98	−7.72 (73)	−503 (50)	10.8 (1.0)	−531 (53)
457.385	$4f^9(^6H^\circ)5d^2(^3F)$	°	17/2	*20,884.42*	68.32 (87)	−1358 (63)	−95.6 (1.2)	−1434 (66)
	$4f^9(^6H^\circ)5d(^7H^\circ)6p$		17/2	42,741.69	44.11 (88)	−1206 (63)	−61.7 (1.2)	−1274 (66)

[a] Reference [31]. [b] Reference [17].

Table 4. Isotope shifts in Dy II, denoted by mass pairs: (M, M'). The signs are determined by the convention that $IS \equiv \nu_{M'} - \nu_M$ where $M' > M$. J_{lo}, E_{lo} and J_{up}, E_{up} are the total angular momentum and energy of the lower and upper levels of a transition, respectively.

λ_{air} [a] (nm)	J_{lo} [a]	E_{lo} [a] (cm^{-1})	J_{up} [a]	E_{up} [a] (cm^{-1})	(160, 164) (MHz)	(161, 164) (MHz)	(162, 164) (MHz)	(163, 164) (MHz)
422.203 *	15/2	4341.10	13/2	*28,019.70*	−610 (11)	−427 (10)	−306 (10)	−129 (10)
424.846 *	11/2	*14,347.21*	13/2	37,878.55	−960 (11)	−842 (11)	−456 (11)	−326 (10)
432.253 *	11/2	13,338.27	9/2	*36,466.34*	−1760 (11)	−1453 (10)	−857 (10)	−533 (10)
432.891	15/2	4341.10	15/2	*27,435.12*	−1204 (10)	−945 (10)	−591 (10)	−331 (10)
436.135 *	17/2	*14,895.06*	15/2	37,817.31	−1429 (10)	−1237 (11)	−694 (10)	−463 (11)
436.421	19/2	*19,571.75*	19/2	42,478.98	−1812 (10)	−1550 (10)	−870 (10)	−581 (10)
437.531	19/2	*17,606.65*	17/2	40,455.73	−108 (10)	−74 (10)	−52 (10)	−23 (10)
438.430 *	13/2	7485.12	13/2	30,287.36	−1457 (11)	−1175 (10)	−707 (10)	−408 (10)
451.851 *	21/2	*22,031.98*	21/2	44,156.98	−363 (10)	−334 (11)	−173 (10)	−133 (10)
454.167	19/2	*19,571.75*	17/2	41,583.90	−1759 (10)	−1506 (11)	−847 (10)	−565 (10)
457.385	17/2	*20,884.42*	17/2	42,741.69	−790 (11)	−700 (11)	−378 (10)	−272 (10)
459.178	11/2	*20,517.39*	13/2	42,289.33	−709 (11)	−628 (10)	−339 (10)	−244 (10)

[a] Reference [31]. * These transitions contained unidentified blends whose relative intensities were <15%. A spectrum at 443.100 nm contained a blend of two transitions of comparable relative intensity and could not be analyzed.

4. Discussion and Conclusions

We have measured hfs parameters in 14 levels of Dy II, and effective hfs parameters in a further four transitions. There is no previous hfs data for these levels for comparison, but the very large magnitude of the electric quadrupole constants is consistent with measurements in Dy I [32]. Comparison with hfs measurements in isoelectronic ^{159}Tb I is not useful as the electron configurations of the levels studied [33] are different from those in the present work, and also because the magnetic dipole and electric quadrupole moments are quite different from those of ^{161}Dy and ^{163}Dy.

Of the 12 transition ISs we measured, we can compare three with the data of Ahmad et al. [20] made with a Fabry-Perot spectrometer viewing a hollow-cathode discharge. As shown in Table 5, the agreement is excellent, and our results are more precise by an order of magnitude. None of the transitions measured by Aufmuth [19] overlap with our data.

Table 5. Comparison of IS measurements with previous work.

λ_{air} (nm)	Mass Pair	IS (MHz)	
		This Work	Ahmad [20]
432.891	(160, 164)	−1204 (10)	−1268 (90)
436.421	(160, 164)	−1812 (10)	−1820 (90)
437.531	(160, 164)	−108 (10)	~0 (90)

Another check on our IS values is through the conventional King plot analysis [29] (Chapter 6). As a reference transition in the King plots, we used the 597.4 nm transition in Dy I, for which ISs have been measured [34,35] at a high accuracy (~2 MHz). Both fits were excellent straight lines (see Figure 3). That linearity reflects almost entirely the quality of *our* data since the errors we have ascribed to our IS measurements are about five times greater than those for the reference transition. This suggests that the uncertainty of 10 MHz that we have attached to our IS measurements to account for source-voltage drift and residual nonlinearity is conservative.

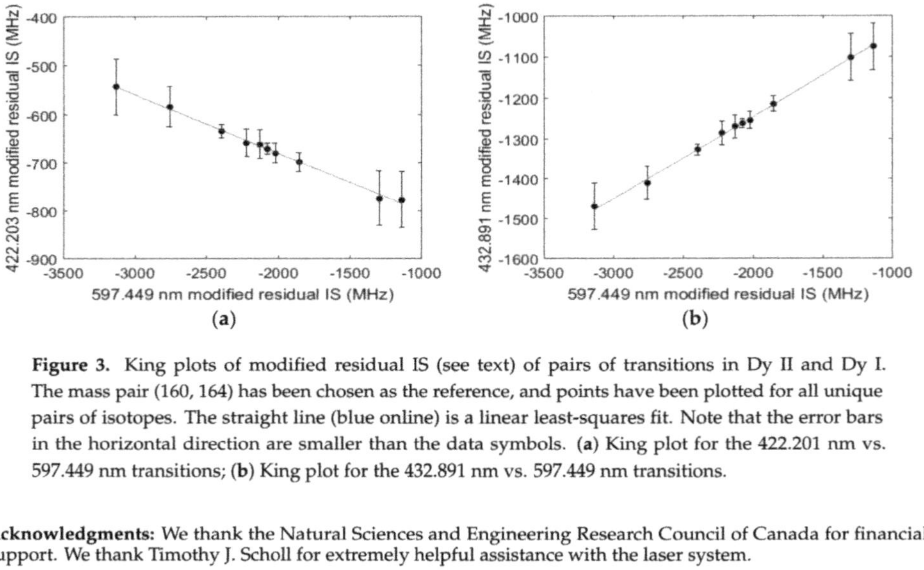

Figure 3. King plots of modified residual IS (see text) of pairs of transitions in Dy II and Dy I. The mass pair (160, 164) has been chosen as the reference, and points have been plotted for all unique pairs of isotopes. The straight line (blue online) is a linear least-squares fit. Note that the error bars in the horizontal direction are smaller than the data symbols. (**a**) King plot for the 422.201 nm vs. 597.449 nm transitions; (**b**) King plot for the 432.891 nm vs. 597.449 nm transitions.

Acknowledgments: We thank the Natural Sciences and Engineering Research Council of Canada for financial support. We thank Timothy J. Scholl for extremely helpful assistance with the laser system.

Author Contributions: All authors contributed equally to acquiring and analyzing data and writing the manuscript.

Conflicts of Interest: The authors declare no conflict of interest.

References

1. Abt, A. Hyperfine structure in the solar spectrum. *Astrophys. J.* **1952**, *115*, 199–205. [CrossRef]
2. Jomaron, C.M.; Dworetsky, M.M.; Allen, C.S. Manganese abundances in mercury-manganese stars. *Mon. Not. R. Astron. Soc.* **1999**, *303*, 555–564. [CrossRef]
3. Kurucz, R.L. Atomic data for interpreting stellar spectra: Isotopic and hyperfine data. *Phys. Scr.* **1993**, *T47*, 110–117. [CrossRef]
4. Cowley, C.R. Lanthanide rare earths in stellar spectra with emphasis on chemically peculiar stars. *Phys. Scr.* **1984**, *T8*, 28–38. [CrossRef]
5. Sneden, C.; Cowan, J.J.; Gallino, R. Neutron-capture elements in the early Galaxy. *Ann. Rev. Astron. Astrophys.* **2008**, *46*, 241–288. [CrossRef]

6. Aoki, W.; Ryan, S.G.; Norris, J.E.; Beers, T.C.; Ando, H.; Iwamoto, N.; Kajino, T.; Mathews, G.J.; Fujimoto, M.Y. Neutron Capture Elements in s-Process-Rich, Very Metal-Poor Stars. *Astrophys. J.* **2001**, *561*, 346–363. [CrossRef]
7. Christlieb, N.; Beers, T.C.; Barklem, P.S.; Bessell, M.; Hill, V.; Holmberg, J.; Horn, A.J.; Marsteller, B.; Mashonkina, L.; Qian, Y.-Z.; et al. The Hamburg/ESO R-process Enhanced Star survey (HERES). I. Project description, and discovery of two stars with strong enhancements of neutron-capture elements. *Astron. Astrophys.* **2004**, *428*, 1027–1037. [CrossRef]
8. François, P.; Depagne, E.; Hill, V.; Spite, M.; Spite, F.; Plez, B.; Beers, T.C.; Andersen, J.; James, G.; Barbuy, B.; et al. First stars. VIII. Enrichment of the neutron-capture elements in the early Galaxy. *Astron. Astrophys.* **2007**, *476*, 935–950. [CrossRef]
9. Burris, D.L.; Pilachowski, C.A.; Armandroff, T.E.; Sneden, C.; Cowan, J.; Roe, H. Neutron-capture elements in the early Galaxy: Insights from a large sample of metal-poor giants. *Astrophys. J.* **2000**, *544*, 302–319. [CrossRef]
10. Karinkuzhi, D.; Goswami, A. Chemical analysis of CH stars—II. Atmospheric parameters and elemental abundances. *Mon. Not. R. Astron. Soc.* **2015**, *446*, 2348–2362. [CrossRef]
11. Chaplin, W.J.; Miglio, A. Asteroseismology of Solar-Type and Red-Giant stars. *Ann. Rev. Astron. Astrophys.* **2013**, *51*, 353–392. [CrossRef]
12. Conway, J.G.; Worden, E.F. Preliminary level analysis of the first and second spectra of dysprosium, Dy I and Dy II. *J. Opt. Soc. Am.* **1971**, *61*, 704–726. [CrossRef]
13. Wyart, J.F. Interpretation du spectre de Dy II. I. Etude des configurations impaires. *Physica* **1972**, *61*, 182–190. [CrossRef]
14. Wyart, J.F. Interpretation du spectre de Dy II. II. Etude des configurations $4f^{9}6s6p$ ET $4f^{9}5d6p$. *Physica* **1972**, *61*, 191–199. [CrossRef]
15. Wyart, J.F. Interprétation du spectre de Dy II. III. Etude des configurations $4f^{10}6s$ ET $4f^{10}5d$. *Physica B+C* **1976**, *83*, 361–366. [CrossRef]
16. Nave, G.; Griesmann, U. New energy levels and classifications of spectral lines from neutral and singly-ionized dysprosium (Dy I and Dy II). *Phys. Scr.* **2000**, *62*, 463–473. [CrossRef]
17. Kramida, A.; Ralchenko, Y.; Reader, J.; NIST ASD Team. *NIST Atomic Spectra Database*; version 5.4; National Institute of Standards and Technology: Gaithersburg, MD, USA, 2016. Available online: http://physics.nist.gov/asd (accessed on 9 January 2017).
18. Pacheva, Y.; Abadjieva, L. ИЗОТОПНО ОТМЕСТВАНЕ В СПЕКТЪРА НА ДИСПРОЗИЯ. *Bull. de l'Institut de Physique et de Recherche Atomique (Bulgaria)* **1968**, *17*, 87–93. (In Bulgarian)
19. Aufmuth, P. Isotope shift and configuration mixing in dysprosium II. *Z. Phys. A* **1978**, *286*, 235–241. [CrossRef]
20. Ahmad, S.A.; Venugopalan, A.; Saksena, G.D. Electronic configuration mixing and isotope shifts in singly ionized dysprosium. *Spectrochim. Acta* **1983**, *38B*, 1115–1124. [CrossRef]
21. Murakawa, K.; Kamei, T. Hyperfine structure of the spectra of dysprosium, cobalt, vanadium, manganese, and lanthanum. *Phys. Rev.* **1953**, *92*, 325–327. [CrossRef]
22. Demtröder, W. *Laser Spectroscopy*, 5th ed.; Springer: Berlin, Germany, 2015; Volume 2, pp. 208–211.
23. Kaufman, S.L. High-resolution laser spectroscopy in fast beams. *Opt. Commun.* **1976**, *17*, 309–312. [CrossRef]
24. Nouri, Z.; Li, R.; Holt, R.A.; Rosner, S.D. A Penning sputter ion source with very low energy spread. *Nucl. Instrum. Methods Phys. Res. A* **2010**, *614*, 174–178. [CrossRef]
25. Nouri, Z.; Li, R.; Holt, R.A.; Rosner, S.D. Corrigendum to "A Penning sputter ion source with very low energy spread" [Nuclear Instr. and Meth. A 614 (2010) 174–178]. *Nucl. Instrum. Methods Phys. Res. A* **2010**, *621*, 717. [CrossRef]
26. Audi, G.; Wapstra, A.H.; Thibault, C. The AME2003 atomic mass evaluation. (II). Tables, graphs and references. *Nucl. Phys. A* **2003**, *729*, 337–676. [CrossRef]
27. De Bièvre, P.; Taylor, P.D.P. Table of the isotopic compositions of the elements. *Int. J. Mass Spectrom. Ion Proc.* **1993**, *123*, 149–166. [CrossRef]
28. Holden, N.E. Table of the isotopes. In *CRC Handbook of Chemistry and Physics*, 97th ed.; CRC Press: Boca Raton, FL, USA, 2016; p. 263. Available online: http://hbcponline.com/faces/documents/11_02/11_02_0263.xhtml (accessed on 17 January 2017).
29. Hall, J.L.; Lee, S.A. Interferometric real-time display of CW dye laser wavelength with sub-Doppler accuracy. *Appl. Phys. Lett.* **1976**, *29*, 367–369. [CrossRef]

30. Woodgate, G.K. *Elementary Atomic Structure*, 2nd ed.; Clarendon: Oxford, UK, 1980; p. 184.
31. Kurucz, R.L.; Bell, B. *Atomic Line Data*; Kurucz CD-ROM No. 23; Cambridge, Mass: Smithsonian Astrophysical Observatory: Cambridge, MA, USA, 1995.
32. Childs, W.J.; Crosswhite, H.; Goodman, L.S.; Pfeufer, V. Hyperfine structure of $4f^N\ 6s^2$ configurations in ^{159}Tb, ^{161,163}Dy, and ^{169}Tm. *J. Opt. Soc. Am. B* **1984**, *1*, 22–29. [CrossRef]
33. Childs, W.J. Hyperfine Structure of Many Atomic Levels of Tb^{159} and the Tb^{159} Nuclear Electric-Quadrupole Moment. *Phys. Rev. A* **1970**, *2*, 316–336. [CrossRef]
34. Dekker, J.W.M.; Klinkenberg, P.F.A.; Langkemper, J.F. Optical isotope shifts and nuclear deformation in dysprosium. *Physica* **1968**, *39*, 393–412. [CrossRef]
35. Wakasugi, M.; Horiguchi, T.; Jin, W.G.; Sakata, H.; Yoshizawa, Y. Changes of the Nuclear Charge Distribution of Nd, Sm, Gd and Dy from Optical Isotope Shifts. *J. Phys. Soc. Jpn.* **1990**, *59*, 2700–2713. [CrossRef]

atoms

MDPI

Article

Combining Multiconfiguration and Perturbation Methods: Perturbative Estimates of Core–Core Electron Correlation Contributions to Excitation Energies in Mg-Like Iron

Stefan Gustafsson [1], Per Jönsson [1,*], Charlotte Froese Fischer [2] and Ian Grant [3,4]

1 Materials Science and Applied Mathematics, Malmö University, SE-205 06 Malmö, Sweden; stefan.gustafsson@mah.se
2 Department of Computer Science, University of British Columbia, Vancouver, BC V6T 1Z4, Canada; charlotte.f.fischer@comcast.net
3 Mathematical Institute, University of Oxford, Woodstock Road, Oxford OX2 6GG, UK; iangrant15@btinternet.com
4 Department of Applied Mathematics and Theoretical Physics, Centre for Mathematical Sciences, Wilberforce Road, Cambridge CB3 0WA, UK
* Correspondence: per.jonsson@mah.se; Tel.: +46-40-66-57251

Academic Editor: Joseph Reader
Received: 25 November 2016; Accepted: 6 January 2017; Published: 12 January 2017

Abstract: Large configuration interaction (CI) calculations can be performed if part of the interaction is treated perturbatively. To evaluate the combined CI and perturbative method, we compute excitation energies for the $3l3l'$, $3l4l'$ and $3s5l$ states in Mg-like iron. Starting from a CI calculation including valence and core–valence correlation effects, it is found that the perturbative inclusion of core–core electron correlation halves the mean relative differences between calculated and observed excitation energies. The effect of the core–core electron correlation is largest for the more excited states. The final relative differences between calculated and observed excitation energies is 0.023%, which is small enough for the calculated energies to be of direct use in line identifications in astrophysical and laboratory spectra.

Keywords: excitation energies; multiconfiguration Dirac–Hartree–Fock; configuration interaction

1. Introduction

Transitions from highly charged ions are observed in the spectra of astrophysical sources as well as in Tokamak and laser-produced plasmas, and they are routinely used for diagnostic purposes [1]. Often, transitions between configurations in the same complex are used, but transitions from higher lying configurations are also important (see, e.g., [2] for a discussion of the higher lying states in the case of Mg-like iron). Transition energies are available from experiments for many ions and collected in various data bases [3], but large amounts of data are still lacking. Although experimental work is aided by a new generation of light sources such as EBITs [4], spectral identifications are still a difficult and time-consuming task. A way forward is provided by theoretical transition energies that support line identification and render consistency checks for experimental level designations.

Much work has been done to improve both multiconfiguration methods and perturbative methods, each with their strengths and weaknesses, in order to provide theoretical transition energies of spectroscopic accuracy, i.e., transition energies with uncertainties of the same order as the ones obtained from experiments and observations using Chandra, Hinode or other space based missions in the X-ray and EUV spectral ranges [5–8]. Further advancements for complex systems with several electrons outside a closed atomic core calls for a combination of multiconfiguration and perturbative methods [9] and also for methods based on new principles [10,11].

In this paper, we describe how the multiconfiguration Dirac–Hartree–Fock (MCDHF) and relativistic configuration interaction (CI) methods can be modified to include perturbative corrections that account for core–core electron correlation. Taking Mg-like iron as an example, we show how the corrections improve excitation energies for the more highly excited states.

2. Relativistic Multiconfiguration Methods

2.1. Multiconfiguration Dirac–Hartree–Fock and Configuration Interaction

In the MCDHF method [12,13], as implemented in the GRASP2K program package [14], the wave function $\Psi(\gamma PJM_J)$ for a state labeled γPJM_J, where J and M_J are the angular quantum numbers and P is the parity, is expanded in antisymmetrized and coupled configuration state functions (CSFs)

$$\Psi(\gamma PJM_J) = \sum_{j=1}^{M} c_j \Phi(\gamma_j PJM_J). \tag{1}$$

The labels $\{\gamma_j\}$ denote other appropriate information of the configuration state functions, such as orbital occupancy and coupling scheme. The CSFs are built from products of one-electron orbitals, having the general form

$$\psi_{n\kappa,m}(\mathbf{r}) = \frac{1}{r}\begin{pmatrix} P_{n\kappa}(r)\chi_{\kappa,m}(\theta,\varphi) \\ \imath Q_{n\kappa}(r)\chi_{-\kappa,m}(\theta,\varphi) \end{pmatrix}, \tag{2}$$

where $\chi_{\pm\kappa,m}(\theta,\varphi)$ are 2-component spin-orbit functions. The radial functions $\{P_{n\kappa}(r), Q_{n\kappa}(r)\}$ are numerically represented on a grid.

Wave functions for a number of targeted states are determined simultaneously in the extended optimal level (EOL) scheme. Given initial estimates of the radial functions, the energies E and expansion coefficients $\mathbf{c} = (c_1, \ldots, c_M)^t$ for the targeted states are obtained as solutions to the configuration interaction (CI) problem

$$\mathbf{Hc} = E\mathbf{c}, \tag{3}$$

where $\mathbf{H}$ is the CI matrix of dimension $M \times M$ with elements

$$H_{ij} = \langle \Phi(\gamma_i PJM_J)|H|\Phi(\gamma_j PJM_J)\rangle. \tag{4}$$

In relativistic calculations, the Hamiltonian H is often taken as the Dirac–Coulomb Hamiltonian. Once the expansion coefficients have been determined, the radial functions are improved by solving a set of differential equations results from applying the variational principle on a weighted energy functional of the targeted states together with additional terms needed to preserve orthonormality of the orbitals. The CI problem and the solution of the differential equations are iterated until the radial orbitals and the energy are converged to a specified tolerance.

The MCDHF calculations are often followed by CI calculations where terms representing the transverse photon interaction are added to the Dirac-Coulomb Hamiltonian and the vacuum polarization effects are taken into account by including the Uehling potential. Electron self-energies are calculated with the screened hydrogenic formula [12,15]. Due to the relative simplicity of the CI method, often much larger expansions are included in the final CI calculations compared to the MCDHF calculations.

2.2. Large Expansions and Perturbative Corrections

The number of CSFs in the wave function expansions depend on the shell structure of the ionic system as well as the model for electron correlation (to be discussed in Section 3). For accurate calculations, a large number of CSFs are required, leading to very large matrices. To handle these large matrices, the CSFs can a priori be divided into two groups. The first group, P, with m elements ($m \ll M$) contains CSFs that account for the major parts of the wave functions. The second group, Q, with $M - m$ elements contains CSFs that represent minor corrections. Allowing interaction between CSFs in group P, interaction between CSFs in group P and Q and diagonal interactions between CSFs in Q gives a matrix

$$\begin{pmatrix} H^{(PP)} & H^{(PQ)} \\ H^{(QP)} & H^{(QQ)} \end{pmatrix}, \tag{5}$$

where $H_{ij}^{(QQ)} = \delta_{ij} E_i^Q$. The restriction of $H^{(QQ)}$ to diagonal elements results in a huge reduction in the total number of matrix elements and corresponding computational time. The assumptions of the approximation and the connections to the method of deflation in numerical analysis are discussed in [13]. This form of the CI matrix, which has been available in the non-relativistic and relativistic multiconfiguration codes for a long time [16,17], yields energies that are similar to the ones obtained by applying second-order perturbation theory (PT) corrections to the energies of the smaller $m \times m$ matrix. The method is therefore referred to here as CI combined with second-order Brillouin–Wigner perturbation theory [18]. Note, however, that the CI method with restrictions on the interactions gives, in contrast to ordinary perturbative methods, wave functions that can be directly used to evaluate expectation values such as transition rates.

3. Calculations

Calculations were performed for states belonging to the $3s^2$, $3p^2$, $3s3d$, $3d^2$, $3s4s$, $3s4d$, $3p4p$, $3p4f$, $3d4s$, $3d4d$, $3s5s$, $3s5d$, $3s5g$ even configurations and the $3s3p$, $3p3d$, $3s4p$, $3p4s$, $3s4f$, $3p4d$, $3d4p$, $3d4f$, $3s5p$, $3s5f$ odd configurations of Mg-like iron. For $3d4f$, only states below the $3p5s$ configuration were included. The above configurations define the multireference (MR) for the even and odd parities, respectively. Following the procedure in [19], an initial MCDHF calculation for all even and odd reference states was done in the EOL scheme. The initial calculation was followed by separate calculations in the EOL scheme for the even and odd parity states. The MCDHF calculations for the even states were based on CSF expansions obtained by allowing single (S) and double (D) substitutions of orbitals in the even MR configurations to an increasing active set of orbitals. In a similar way, the calculations for the odd states were based on CSF expansions obtained by allowing single (S) and double (D) substitutions of orbitals in the odd MR configurations to an increasing active set of orbitals. To prevent the CSF expansions from growing unmanageably large and in order to obtain orbitals that are spatially localized in the valence and core–valence region, at most, single substitutions were allowed from the $2s^22p^6$ core. The $1s^2$ shell was always closed. The active sets of orbitals for the even and odd parity states were extended by layers to include orbitals with quantum numbers up to $n = 8$ and $l = 6$, at which point the excitation energies are well converged.

To investigate the effects of electron correlation, three sets of CI calculations were done. In the first set of CI calculations, one calculation was done for the even states and one calculation for the odd states, the SD substitutions were only allowed from the valence shells of the MR, and the CSFs account for valence–valence correlation. In the second set of calculations, SD substitutions were such that there was at most one substitution from the $2s^22p^6$ core, and the CSFs account for valence–valence and core–valence correlation. In the final set of calculations, all SD substitutions were allowed, and the CSFs account for valence–valence, core–valence and core–core correlation. When all substitutions are allowed, the number of CSFs grows very large. For this reason, we apply CI with second-order perturbation corrections. The CSFs describing valence–valence and core–valence effects (SD substitutions with at most one substitution from the $2s^22p^6$ core) were included in group P, whereas the CSFs accounting for core–core correlation (D substitutions from $2s^2 2p^6$) were included in group Q and treated in second-order perturbation theory. The number of CSFs for the different CI calculations are given in Table 1.

Table 1. Number of CSFs for the even and odd parity expansions for the different sets of CI calculations. VV are the expansions accounting for valence–valence correlation, VV+CV are the expansions accounting for valence–valence and core–valence correlation and VV+CV+CC are the expansions accounting for valence-valence, core–valence and core–core correlation.

	VV	VV+CV	VV+CV+CC
even	2738	644,342	5,624,158
odd	2728	630,502	6,214,393

4. Results

The excitation energies from the different CI calculations, along with observed energies from the NIST database [3], are displayed in Table 2. From the table, we see that states belonging to $3l3l'$, with the exception of $3s3p\ ^3P_{0,1,2}$, are too high for the valence–valence correlation calculation. The states belonging to $3l4l'$ and $3s5l$, on the other hand, are too low. When including also the core–valence correlation, the states belonging to $3l3l'$ go down in energy and approach the observed excitation energies. The states belonging to $3l4l'$ and $3s5l$ go up and are now too high. Including also the core–core correlation results in a rather small energy change for the states belonging to $3l3l'$. The main effect of the core–core correlation is to lower the energies of the states belonging to $3l4l'$ and $3s5l$, bringing them in very good agreement with observations. The labeling of levels is normally done by looking at the quantum designation of the leading component in the CSF expansion [20]. There are two levels (67 and 69) with $3p4d\ ^3D_3$ as the leading component in the corresponding CSF expansion. To distinguish these levels, we added subscripts A and B to the labels of the dominant component. In a similar way, subscripts A and B were added to distinguish levels 78 and 80, both with $3p4f\ ^3F_3$ as the leading component.

Table 2 indicates that there are a few states that are either misidentified or assigned with a label that is inconsistent with the labels of the current calculation. The observed energy for $3p4f\ ^3D_2$ (level 84) is 2417 cm^{-1} too low compared to the calculated value and the observed energy for $3s5s\ ^3S_1$ (level 92) is 33,948 cm^{-1} too high. There seem to be no other computed energy levels that match the observed energies. The observed energy for $3s5p\ ^1P_1^o$ (level 100) is 3733 cm^{-1} too low. The observed energy matches the computed energy of $3s5p\ ^3P_1^o$ (level 97), and, thus, it seems like an inconsistency in the labeling. Finally, $3s5f\ ^1F_3^o$ (level 117) is 101,545 cm^{-1} too high and there is no other computed energy level that matches. Removing the energy outliers above, the mean relative energy differences are, respectively, 0.217%, 0.051%, 0.023% for the valence, the valence and core–valence and the valence, core–valence and core–core calculations. The energy differences are mainly due to higher-order electron correlation effects that have not been accounted for in the calculations. At the same time, one should bear in mind that the observed excitation energies are also associated with uncertainties as reflected in the limited number of valid digits displayed in the NIST tables.

In Table 3, the excitation energies obtained by including core–core correlation in the CI calculations are compared with energies from calculations by Landi [2] using the FAC code and with energies by Aggarwal et al. [21] using CIV3 in the Breit–Pauli approximation. The uncertainties of the excitation energies for the latter calculations are substantially larger. The calculations by Landi support the conclusion that some of the levels in the NIST database are misidentified. One may note that Landi gives levels 78 and 80 the labels $3p4f\ ^3F_3$ and $3p4f\ ^1F_3$, respectively, whereas Aggarwal et al. reverse the labels. This illustrates that labeling is dependent on the calculation and that the labeling process is far from straightforward [20].

Table 2. Comparison of calculated and observed excitation energies in Mg-like iron (Fe XV). E_{VV} are energies from CI calculations that account for valence–valence correlation. E_{VV+CV} are energies from CI calculations that account for valence–valence and core–valence electron correlation. $E_{VV+CV+CC}$ are energies that account for valence–valence and core–valence electron correlation and where core–core electron correlation effects have been included perturbatively. E_{NIST} are observed energies from the NIST database ([3]). ΔE are energy differences with respect to E_{NIST}. All energies are in cm^{-1}.

No.	Level	E_{VV}	ΔE	E_{VV+CV}	ΔE	$E_{VV+CV+CC}$	ΔE	E_{NIST}
1	$3s^2\ ^1S_0$	0	0	0	0	0	0	0
2	$3s3p\ ^3P^o_0$	233,087	−755	233,828	−14	233,928	86	233,842
3	$3s3p\ ^3P^o_1$	238,936	−724	239,668	8	239,741	81	239,660
4	$3s3p\ ^3P^o_2$	253,017	−803	253,829	9	253,773	−47	253,820
5	$3s3p\ ^1P^o_1$	354,941	3030	352,169	258	352,091	180	351,911
6	$3p^2\ ^3P_0$	556,594	2070	554,643	119	554,895	371	554,524
7	$3p^2\ ^1D_2$	559,900	300	559,834	234	559,661	61	559,600
8	$3p^2\ ^3P_1$	566,524	1922	564,663	61	564,674	72	56,4602
9	$3p^2\ ^3P_2$	583,327	1524	581,933	130	581,870	67	581,803
10	$3p^2\ ^1S_0$	662,999	3372	660,269	642	660,229	602	659,627
11	$3s3d\ ^3D_1$	680,522	1750	678,954	182	678,329	−443	678,772
12	$3s3d\ ^3D_2$	681,520	1735	679,986	201	679,381	−404	679,785
13	$3s3d\ ^3D_3$	683,080	1664	681,603	187	680,952	−464	681,416
14	$3s3d\ ^1D_2$	766,690	4597	762,729	636	762,176	83	762,093
15	$3p3d\ ^3F^o_2$	929,158	917	928,565	324	928,086	−155	928,241
16	$3p3d\ ^3F^o_3$	938,885	759	938,469	343	938,068	−58	938,126
17	$3p3d\ ^1D^o_2$	950,226	1713	948,768	255	948,383	−130	948,513
18	$3p3d\ ^3F^o_4$	950,300	642	949,990	332	949,451	−207	949,658
19	$3p3d\ ^3D^o_1$	986,221	3353	983,077	209	982,740	−128	982,868
20	$3p3d\ ^3P^o_2$	986,499	2985	983,765	251	983,350	−164	983,514
21	$3p3d\ ^3D^o_3$	998,324	3472	995,088	236	994,712	−140	994,852
22	$3p3d\ ^3P^o_0$	998,597	2708	996,218	329	995,835	−54	995,889
23	$3p3d\ ^3P^o_1$	999,166	2923	996,547	304	996,127	−116	996,243
24	$3p3d\ ^3D^o_2$	999,755	3132	996,892	269	996,449	−174	996,623
25	$3p3d\ ^1F^o_3$	1,066,906	4391	1,063,163	648	1,062,704	189	1,062,515
26	$3p3d\ ^1P^o_1$	1,078,913	4026	1,075,795	908	1,075,306	419	1,074,887
27	$3d^2\ ^3F_2$	1,373,374	3043	1,370,858	527	1,369,758	−573	1,370,331
28	$3d^2\ ^3F_3$	1,374,983	2948	1,372,527	492	1,371,407	−628	1,372,035
29	$3d^2\ ^3F_4$	1,376,965	2909	1,374,580	524	1,373,475	−581	1,374,056
30	$3d^2\ ^1D_2$	1,405,702	3110	1,403,474	882	1,402,237	−355	1,402,592
31	$3d^2\ ^3P_0$	1,409,066		1,406,328		1,405,381		
32	$3d^2\ ^3P_1$	1,409,639		1,406,926		1,405,672		
33	$3d^2\ ^1G_4$	1,409,702	2644	1,407,974	916	1,406,831	−227	1,407,058
34	$3d^2\ ^3P_2$	1,411,053	3280	1,408,467	694	1,407,210	−563	1,407,773
35	$3d^2\ ^1S_0$	1,489,913	2859	1,488,993	1939	1,487,460	406	1,487,054
36	$3s4s\ ^3S_1$	1,761,471	−2229	1,764,876	1176	17,63,699	−1	1,763,700
37	$3s4s\ ^1S_0$	1,785,265	−1735	1,788,455	1455	1,787,322	322	1,787,000
38	$3s4p\ ^3P^o_0$	1,880,014		1,883,187		1,882,236		
39	$3s4p\ ^3P^o_1$	1,880,440		1,883,595		1,882,588		
40	$3s4p\ ^3P^o_2$	1,887,508		1,890,703		1,889,632		
41	$3s4p\ ^1P^o_1$	1,887,872	−2098	1,891,051	1081	1,890,042	72	1,889,970
42	$3s4d\ ^3D_1$	2,029,659	−1651	2,032,907	1597	2,031,683	373	2,031,310
43	$3s4d\ ^3D_2$	2,030,413	−1607	2,033,653	1633	2,032,413	393	2,032,020

Table 2. *Cont.*

No.	Level	E_{VV}	ΔE	E_{VV+CV}	ΔE	$E_{VV+CV+CC}$	ΔE	E_{NIST}
44	$3s4d\ ^3D_3$	2,031,636	−1544	2,034,880	1700	2,033,623	443	2,033,180
45	$3s4d\ ^1D_2$	2,032,991	−2289	2,036,318	1038	2,035,053	−227	2,035,280
46	$3p4s\ ^3P^o_0$	2,051,314		2,053,909		2,053,031		
47	$3p4s\ ^3P^o_1$	2,054,922		2,057,446		2,056,493		
48	$3p4s\ ^3P^o_2$	2,071,700		2,074,376		2,073,372		
49	$3p4s\ ^1P^o_1$	2,085,097		2,087,237		2,086,235		
50	$3s4f\ ^3F^o_2$	2,105,597	−2923	2,109,821	1301	2,108,281	−239	2,108,520
51	$3s4f\ ^3F^o_3$	2,105,804	−2816	2,110,029	1409	2,108,503	−117	2,108,620
52	$3s4f\ ^3F^o_4$	2,106,098	−2782	2,110,327	1447	2,108,798	−82	2,108,880
53	$3s4f\ ^1F^o_3$	2,120,519	−2631	2,124,654	1504	2,123,180	30	2,123,150
54	$3p4p\ ^1P_1$	2,152,851		2,155,266		2,154,244		
55	$3p4p\ ^3D_1$	2,167,018		2,169,386		2,168,341		
56	$3p4p\ ^3D_2$	2,168,756		2,171,070		2,170,006		
57	$3p4p\ ^3P_0$	2,173,624		2,175,566		2,174,583		
58	$3p4p\ ^3P_1$	2,181,779		2,183,914		2,182,831		
59	$3p4p\ ^3D_3$	2,184,022		2,186,457		2,185,350		
60	$3p4p\ ^3P_2$	2,189,341		2,191,385		2,190,270		
61	$3p4p\ ^3S_1$	2,192,119		2,194,460		2,193,367		
62	$3p4p\ ^1D_2$	2,206,894		2,208,893		2,207,746		
63	$3p4p\ ^1S_0$	2,235,724		2,237,406		2,236,314		
64	$3p4d\ ^3D^o_1$	2,311,660		2,314,071		2,313,090		
65	$3p4d\ ^1D^o_2$	2,311,989		2,314,331		2,313,312		
66	$3p4d\ ^3D^o_2$	2,312,449		2,314,882		2,313,865		
67	$3p4d\ ^3D^o_{3A}$	2,313,908		2,316,401		2,315,387		
68	$3p4d\ ^3F^o_2$	2,329,261		2,331,722		2,330,678		
69	$3p4d\ ^3D^o_{3B}$	2,330,539		2,333,084		2,332,039		
70	$3p4d\ ^3F^o_4$	2,337,384		2,339,922		2,338,857		
71	$3p4d\ ^1F^o_3$	2,337,651		2,340,302		2,339,278		
72	$3p4d\ ^3P^o_2$	2,341,803		2,344,120		2,343,033		
73	$3p4d\ ^3P^o_1$	2,342,778		2,345,091		2,344,049		
74	$3p4d\ ^3P^o_0$	2,346,915		2,349,198		2,348,199		
75	$3p4d\ ^1P^o_1$	2,350,169		2,352,543		2,351,513		
76	$3p4f\ ^3G_3$	2,377,507	−2653	2,381,283	1123	2,379,714	−446	2,380,160
77	$3p4f\ ^3G_4$	2,384,217	−2483	2,387,976	1276	2,386,434	−266	2,386,700
78	$3p4f\ ^3F_{3A}$	2,384,435		2,388,118		2,386,537		
79	$3p4f\ ^3F_2$	2,388,049	−2051	2,391,670	1570	2,390,091	−9	2,390,100
80	$3p4f\ ^3F_{3B}$	2,397,860		2,401,630		2,400,029		
81	$3p4f\ ^3G_5$	2,399,542	−2558	2,403,453	1353	2,401,876	−224	2,402,100
82	$3p4f\ ^3F_4$	2,400,524	−1576	2,404,286	2186	2,402,697	597	2,402,100
83	$3p4f\ ^3D_3$	2,411,680	−1320	2,415,368	2368	2,413,758	758	2,413,000
84	$3p4f\ ^3D_2$	2,414,633	333	2,418,319	4019	2,416,717	2417	2,414,300
85	$3p4f\ ^3D_1$	2,417,852	−2248	2,421,557	1457	2,419,975	−125	2,420,100
86	$3p4f\ ^1G_4$	2,426,828	−1872	2,430,497	1797	2,429,063	363	2,428,700
87	$3p4f\ ^1D_2$	2,433,430	−2570	2,437,039	1039	2,435,534	−466	2,436,000
88	$3d4s\ ^3D_1$	2,458,614		2,460,640		2,458,997		
89	$3d4s\ ^3D_2$	2,459,450		2,461,503		2,459,846		
90	$3d4s\ ^3D_3$	2,461,283		2,463,415		2,461,742		
91	$3d4s\ ^1D_2$	2,468,780		2,470,737		2,469,163		
92	$3s5s\ ^3S_1$	2,507,700	−37,100	2,512,036	−32,764	2,510,852	−33,948	2,544,800

Table 2. *Cont.*

No.	Level	E_{VV}	ΔE	E_{VV+CV}	ΔE	$E_{VV+CV+CC}$	ΔE	E_{NIST}
93	$3s5s\ ^1S_0$	2,516,613		2,520,681		2,519,752		
94	$3d4p\ ^1D^o_2$	2,561,358		2,563,408		2,561,899		
95	$3d4p\ ^3D^o_1$	2,564,069		2,567,301		2,565,949		
96	$3s5p\ ^3P^o_0$	2,564,472		2,568,582		2,567,624		
97	$3s5p\ ^3P^o_1$	2,565,848		2,568,791		2,567,639		
98	$3d4p\ ^3D^o_2$	2,567,134		2,569,092		2,567,703		
99	$3d4p\ ^3D^o_3$	2,568,154		2,571,175		2,569,693		
100	$3s5p\ ^1P^o_1$	2,568,200	1200	2,571,834	4834	2,570,733	3733	2,567,000
101	$3s5p\ ^3P^o_2$	2,569,213		2,572,157		2,570,743		
102	$3d4p\ ^3F^o_2$	2,570,296		2,572,316		2,571,126		
103	$3d4p\ ^3F^o_3$	2,573,116		2,575,101		2,573,592		
104	$3d4p\ ^3F^o_4$	2,576,139		2,578,374		2,576,829		
105	$3d4p\ ^3P^o_1$	2,583,286		2,585,242		2,583,862		
106	$3d4p\ ^3P^o_2$	2,583,400		2,585,407		2,583,960		
107	$3d4p\ ^3P^o_0$	2,583,734		2,585,658		2,584,322		
108	$3d4p\ ^1F^o_3$	2,592,868		2,594,519		2,593,236		
109	$3d4p\ ^1P^o_1$	2,603,279		2,605,145		2,604,533		
110	$3s5d\ ^3D_1$	2,637,190	−2910	2,641,400	1300	2,640,247	147	2,640,100
111	$3s5d\ ^3D_2$	2,637,419	−2481	2,641,630	1730	2,640,442	542	2,639,900
112	$3s5d\ ^3D_3$	2,637,852	−2448	2,642,072	1772	2,640,870	570	2,640,300
113	$3s5d\ ^1D_2$	2,639,773		2,643,981		2,642,888		
114	$3s5f\ ^3F^o_2$	2,672,676	−3724	2,677,360	960	2,675,889	−511	2,676,400
115	$3s5f\ ^3F^o_3$	2,672,770	−3630	2,677,455	1055	2,675,988	−412	2,676,400
116	$3s5f\ ^3F^o_4$	2,672,907	−3693	2,677,594	994	2,676,123	−477	2,676,600
117	$3s5f\ ^1F^o_3$	2,678,041	−104,659	2,682,597	−100,103	2,681,155	−101,545	2,782,700
118	$3s5g\ ^3G_3$	2,682,487		2,687,368		2,685,680		
119	$3s5g\ ^3G_4$	2,682,654		2,687,556		2,685,877		
120	$3s5g\ ^3G_5$	2,682,855		2,687,777		2,686,099		
121	$3s5g\ ^1G_4$	2,685,580		2,690,506		2,688,841		
122	$3d4d\ ^1F_3$	2,699,116		2,701,602		2,699,874		
123	$3d4d\ ^3D_1$	2,703,542		2,705,972		2,704,354		
124	$3d4d\ ^3D_2$	2,704,742		2,707,218		2,705,580		
125	$3d4d\ ^3D_3$	2,706,116		2,708,636		2,706,964		
126	$3d4d\ ^3G_3$	2,707,934		2,710,522		2,708,828		
127	$3d4d\ ^1P_1$	2,709,315		2,711,813		2,710,163		
128	$3d4d\ ^3G_4$	2,709,360		2,711,928		2,710,264		
129	$3d4d\ ^3G_5$	2,711,220		2,713,878		2,712,174		
130	$3d4d\ ^3S_1$	2,720,698		2,723,175		2,721,783		
131	$3d4d\ ^3F_2$	2,726,309		2,728,092		2,726,350		
132	$3d4d\ ^3F_3$	2,727,568		2,729,398		2,727,634		
133	$3d4d\ ^3F_4$	2,729,029		2,730,908		2,729,156		
134	$3d4d\ ^1D_2$	2,741,839		2,743,862		2,742,627		
135	$3d4d\ ^3P_0$	2,744,213		2,746,022		2,744,706		
136	$3d4d\ ^3P_1$	2,744,807		2,746,626		2,745,163		
137	$3d4d\ ^3P_2$	2,745,935		2,747,809		2,746,300		
138	$3d4d\ ^1G_4$	2,748,985		2,751,121		2,749,474		
139	$3d4f\ ^3H^o_4$	2,765,833		2,770,098		2,768,443		
140	$3d4f\ ^1G^o_4$	2,767,533		2,771,821		2,770,030		
141	$3d4f\ ^3H^o_5$	2,767,692		2,771,943		2,770,434		
142	$3d4d\ ^1S_0$	2,775,538		2,779,275		2,777,362		

Table 2. *Cont.*

No.	Level	E_{VV}	ΔE	E_{VV+CV}	ΔE	$E_{VV+CV+CC}$	ΔE	E_{NIST}
143	$3d4f\ ^3F^o_2$	2,776,151		2,779,298		2,778,011		
144	$3d4f\ ^3F^o_3$	2,776,264		2,779,933		2,778,867		
145	$3d4f\ ^3F^o_4$	2,776,981		2,780,796		2,780,729		
146	$3d4f\ ^1D^o_2$	2,786,768		2,790,305		2,788,248		

Table 3. Comparison of calculated and observed excitation energies in Mg-like iron (Fe XV). $E_{VV+CV+CC}$ are energies that account for valence–valence and core–valence electron correlation and where core–core electron correlation effects have been included perturbatively. E_{FAC} are energies by Landi [2] using the FAC code. E_{CIV3} are energies by Aggarwal et al. [21] using the CIV3 code. E_{NIST} are observed energies from the NIST database ([3]). ΔE are energy differences with respect to E_{NIST}. All energies are in cm^{-1}.

No.	Level	$E_{VV+CV+VV}$	ΔE	E_{FAC}	ΔE	E_{CIV3}	ΔE	E_{NIST}
1	$3s^2\ ^1S_0$	0	0	0	0	0	0	0
2	$3s3p\ ^3P^o_0$	233,928	86	233,068	−774	235,013	1171	233,842
3	$3s3p\ ^3P^o_1$	239,741	81	238,900	−760	240,511	851	239,660
4	$3s3p\ ^3P^o_2$	253,773	−47	252,917	−903	253,548	−272	253,820
5	$3s3p\ ^1P^o_1$	352,091	180	356,126	4215	356,262	4351	351,911
6	$3p^2\ ^3P_0$	554,895	371	556,994	2470	560,275	5751	554,524
7	$3p^2\ ^1D_2$	559,661	61	560,266	666	563,216	3616	559,600
8	$3p^2\ ^3P_1$	564,674	72	566,832	2230	569,295	4693	564,602
9	$3p^2\ ^3P_2$	581,870	67	583,564	1761	584,856	3053	581,803
10	$3p^2\ ^1S_0$	660,229	602	665,768	6141	665,260	5633	659,627
11	$3s3d\ ^3D_1$	678,329	−443	680,146	1374	687,680	8908	678,772
12	$3s3d\ ^3D_2$	679,381	−404	681,129	1344	688,733	8948	6797,85
13	$3s3d\ ^3D_3$	680,952	−464	682,667	1251	690,401	8985	681,416
14	$3s3d\ ^1D_2$	762,176	83	769,369	7276	774,295	12,202	762,093
15	$3p3d\ ^3F^o_2$	928,086	−155	928,786	545	938,265	10,024	928,241
16	$3p3d\ ^3F^o_3$	938,068	−58	938,555	429	947,307	9181	938,126
17	$3p3d\ ^1D^o_2$	948,383	−130	949,447	934	958,402	9889	948,513
18	$3p3d\ ^3F^o_4$	949,451	−207	949,927	269	957,820	8162	949,658
19	$3p3d\ ^3D^o_1$	982,740	−128	986,082	3214	995,526	12,658	982,868
20	$3p3d\ ^3P^o_2$	983,350	−164	986,407	2893	995,767	12,253	983,514
21	$3p3d\ ^3D^o_3$	994,712	−140	997,944	3092	1,007,026	12,174	994,852
22	$3p3d\ ^3P^o_0$	995,835	−54	998,762	2873	1,006,708	10,819	995,889
23	$3p3d\ ^3P^o_1$	996,127	−116	999,173	2930	1,007,366	11,123	996,243
24	$3p3d\ ^3D^o_2$	996,449	−174	999,578	2955	1,008,124	11,501	996,623
25	$3p3d\ ^1F^o_3$	1,062,704	189	1,070,794	8279	1,077,456	14,941	1,062,515
26	$3p3d\ ^1P^o_1$	1,075,306	419	1,083,826	8939	1,089,691	14,804	1,074,887
27	$3d^2\ ^3F_2$	1,369,758	−573	1,372,400	2069	1,388,111	17,780	1,370,331
28	$3d^2\ ^3F_3$	1,371,407	−628	1,373,988	1953	1,389,834	17,799	1,372,035
29	$3d^2\ ^3F_4$	1,373,475	−581	1,375,938	1882	1,391,941	17,885	1,374,056
30	$3d^2\ ^1D_2$	1,402,237	−355	1,407,428	4836	1,421,702	19,110	1,402,592
31	$3d^2\ ^3P_0$	1,405,381		1,409,507		1,424,577		
32	$3d^2\ ^3P_1$	1,405,672		1,410,109		1,425,246		
33	$3d^2\ ^1G_4$	1,406,831	−227	1,412,127	5069	1,425,872	18,814	1,407,058
34	$3d^2\ ^3P_2$	1,407,210	−563	1,411,643	3870	1,426,815	19,042	1,407,773
35	$3d^2\ ^1S_0$	1,487,460	406	1,498,668	11,614	1,508,954	21,900	1,487,054
36	$3s4s\ ^3S_1$	1,763,699	−1	1,760,910	−2790	1,764,005	305	1,763,700

Table 3. *Cont.*

No.	Level	$E_{VV+CV+VV}$	ΔE	E_{FAC}	ΔE	E_{CIV3}	ΔE	E_{NIST}
37	$3s4s\ ^1S_0$	1,787,322	322	1,786,052	−948	1,787,950	950	1,787,000
38	$3s4p\ ^3P^o_0$	1,882,236		1,880,319		1,883,685		
39	$3s4p\ ^3P^o_1$	1,882,588		1,880,746		1,884,091		
40	$3s4p\ ^3P^o_2$	1,889,632		1,887,756		1,890,313		
41	$3s4p\ ^1P^o_1$	1,890,042	72	1,888,124	−1846	1,890,631	661	1,889,970
42	$3s4d\ ^3D_1$	2,031,683	373	2,029,563	−1747	2,034,124	2814	2,031,310
43	$3s4d\ ^3D_2$	2,032,413	393	2,030,328	−1692	2,034,848	2828	2,0320,20
44	$3s4d\ ^3D_3$	2,033,623	443	2,031,544	−1636	2,036,055	2875	2,033,180
45	$3s4d\ ^1D_2$	2,035,053	−227	2,033,212	−2068	2,037,569	2289	2,035,280
46	$3p4s\ ^3P^o_0$	2,053,031		2,051,778		2,055,797		
47	$3p4s\ ^3P^o_1$	2,056,493		2,055,514		2,059,308		
48	$3p4s\ ^3P^o_2$	2,073,372		2,072,083		2,074,452		
49	$3p4s\ ^1P^o_1$	2,086,235		2,086,607		2,088,795		
50	$3s4f\ ^3F^o_2$	2,108,281	−239	2,107,228	−1292	2,110,073	1553	2,108,520
51	$3s4f\ ^3F^o_3$	2,108,503	−117	2,107,423	−1197	2,110,281	1661	2,108,620
52	$3s4f\ ^3F^o_4$	2,108,798	−82	2,107,701	−1179	2,110,567	1687	2,108,880
53	$3s4f\ ^1F^o_3$	2,123,180	30	2,124,054	904	2,125,886	2736	2,123,150
54	$3p4p\ ^1P_1$	2,154,244		2,167,343		2,158,599		
55	$3p4p\ ^3D_1$	2,168,341		2,153,046		2,171,635		
56	$3p4p\ ^3D_2$	2,170,006		2,169,173		2,173,578		
57	$3p4p\ ^3P_0$	2,174,583		2,175,103		2,178,812		
58	$3p4p\ ^3P_1$	2,182,831		2,182,790		2,185,901		
59	$3p4p\ ^3D_3$	2,185,350		2,184,242		2,187,229		
60	$3p4p\ ^3P_2$	2,190,270		2,190,674		2,193,265		
61	$3p4p\ ^3S_1$	2,193,367		2,192,597		2,195,756		
62	$3p4p\ ^1D_2$	2,207,746		2,209,221		2,211,163		
63	$3p4p\ ^1S_0$	2,236,314		2,239,314		2,241,187		
64	$3p4d\ ^3D^o_1$	2,313,090		2,311,999		2,318,014		
65	$3p4d\ ^1D^o_2$	2,313,312		2,312,326		2,318,179		
66	$3p4d\ ^3D^o_2$	2,313,865		2,312,835		2,318,826		
67	$3p4d\ ^3D^o_{3A}$	2,315,387		23,144,663		2,320,538		
68	$3p4d\ ^3F^o_2$	2,330,678		2,329,647		2,334,178		
69	$3p4d\ ^3D^o_{3B}$	2,332,039		2,331,0213		2,335,726		
70	$3p4d\ ^3F^o_4$	2,338,857		2,338,064		2,342,277		
71	$3p4d\ ^1F^o_3$	2,339,278		2,338,703		2,343,517		
72	$3p4d\ ^3P^o_2$	2,343,033		2,342,598		2,347,544		
73	$3p4d\ ^3P^o_1$	2,344,049		2,343,850		2,348,795		
74	$3p4d\ ^3P^o_0$	2,348,199		2,347,823		2,352,406		
75	$3p4d\ ^1P^o_1$	2,351,513		2,351,661		2,356,773		
76	$3p4f\ ^3G_3$	2,379,714	−446	2,379,430	−730	2,384,306	4146	2,380,160
77	$3p4f\ ^3G_4$	2,386,434	−266	2,386,688	−12	2,391,198	4498	2,386,700
78	$3p4f\ ^3F_{3A}$	2,386,537		2,386,430		2,390,473		
79	$3p4f\ ^3F_2$	2,390,091	−9	2,390,112	12	2,393,842	3742	2,390,100
80	$3p4f\ ^3F_{3B}$	2,400,029		2,399,796		2,402,786		
81	$3p4f\ ^3G_5$	2,401,876	−224	2,401,746	−354	2,405,617	3517	2,402,100
82	$3p4f\ ^3F_4$	2,402,697	597	2,402,507	407	2,405,496	3396	2,402,100
83	$3p4f\ ^3D_3$	2,413,758	758	2,414,120	1120	2,417,151	4151	2,413,000
84	$3p4f\ ^3D_2$	2,416,717	2417	2,417,276	2976	2,420,124	5824	2,414,300
85	$3p4f\ ^3D_1$	2,419,975	−125	2,420,512	412	2,423,219	3119	2,420,100
86	$3p4f\ ^1G_4$	2,429,063	363	2,432,908	4208	2,435,828	7128	2,428,700

Table 3. *Cont.*

No.	Level	$E_{VV+CV+VV}$	ΔE	E_{FAC}	ΔE	E_{CIV3}	ΔE	E_{NIST}
87	$3p4f\ ^1D_2$	2,435,534	−466	2,438,982	2982	2,440,239	4239	2,436,000
88	$3d4s\ ^3D_1$	2,458,997		2,458,814		2,468,047		
89	$3d4s\ ^3D_2$	2,459,846		2,459,675		2,468,969		
90	$3d4s\ ^3D_3$	2,461,742		2,461,461		2,470,911		
91	$3d4s\ ^1D_2$	2,469,163		2,470,364		2,479,437		
92	$3s5s\ ^3S_1$	2,510,852	−33,948	2,507,572	−37,228			2,544,800
93	$3s5s\ ^1S_0$	2,519,752		2,517,043				
94	$3d4p\ ^1D^o_2$	2,561,899		2,561,169		2,571,814		
95	$3d4p\ ^3D^o_1$	2,565,949		2,566,041		2,576,851		
96	$3s5p\ ^3P^o_0$	2,567,624		2,564,597				
97	$3s5p\ ^3P^o_1$	2,567,639		2,564,254				
98	$3d4p\ ^3D^o_2$	2,567,703		2,567,341		2,577,905		
99	$3d4p\ ^3D^o_3$	2,569,693		2,569,518		2,583,117		
100	$3s5p\ ^1P^o_1$	2,570,733	3733	2,568,358	1358			2,567,000
101	$3s5p\ ^3P^o_2$	2,570,743		2,568,240				
102	$3d4p\ ^3F^o_2$	2,571,126		2,570,526		2,580,319		
103	$3d4p\ ^3F^o_3$	2,573,592		2,573,370		2,579,847		
104	$3d4p\ ^3F^o_4$	2,576,829		2,576,531		2,586,036		
105	$3d4p\ ^3P^o_1$	2,583,862		2,584,287		2,593,158		
106	$3d4p\ ^3P^o_2$	2,583,960		2,584,326		2,593,586		
107	$3d4p\ ^3P^o_0$	2,584,322		2,584,699		2,593,641		
108	$3d4p\ ^1F^o_3$	2,593,236		2,596,425		2,604,571		
109	$3d4p\ ^1P^o_1$	2,604,533		2,607,817		2,610,870		
110	$3s5d\ ^3D_1$	2,640,247	147	2,637,143	−2957			2,640,100
111	$3s5d\ ^3D_2$	2,640,442	542	2,637,376	−2524			2,639,900
112	$3s5d\ ^3D_3$	2,640,870	570	2,637,804	−2496			2,640,300
113	$3s5d\ ^1D_2$	2,642,888		2,640,084	0			
114	$3s5f\ ^3F^o_2$	2,675,889	−511	2,673,354	−3046			2,676,400
115	$3s5f\ ^3F^o_3$	2,675,988	−412	2,673,444	−2956			2,676,400
116	$3s5f\ ^3F^o_4$	2,676,123	−477	2,673,575	−3025			2,676,600
117	$3s5f\ ^1F^o_3$	2,681,155	−101,545	2,679,558	−103,142			2,782,700
118	$3s5g\ ^3G_3$	2,685,680		2,683,089				
119	$3s5g\ ^3G_4$	2,685,877		2,683,272				
120	$3s5g\ ^3G_5$	2,686,099		2,683,494				
121	$3s5g\ ^1G_4$	2,688,841		2,686,809				
122	$3d4d\ ^1F_3$	2,699,874		2,697,717		2,710,391		
123	$3d4d\ ^3D_1$	2,704,354		2,702,464		2,714,967		
124	$3d4d\ ^3D_2$	2,705,580		2,703,625		2,716,229		
125	$3d4d\ ^3D_3$	2,706,964		2,705,001		2,717,578		
126	$3d4d\ ^3G_3$	2,708,828		2,707,726		2,717,919		
127	$3d4d\ ^1P_1$	2,710,163		2,708,170		2,721,079		
128	$3d4d\ ^3G_4$	2,710,264		2,709,064		2,719,345		
129	$3d4d\ ^3G_5$	2,712,174		2,710,955		2,721,463		
130	$3d4d\ ^3S_1$	2,721,783		2,720,286		2,732,634		
131	$3d4d\ ^3F_2$	2,726,350		2,726,401		2,738,407		
132	$3d4d\ ^3F_3$	2,727,634		2,727,604		2,739,745		
133	$3d4d\ ^3F_4$	2,729,156		2,729,075		2,741,293		
134	$3d4d\ ^1D_2$	2,742,627		2,743,889		2,755,547		
135	$3d4d\ ^3P_0$	2,744,706		2,745,181		2,757,907		
136	$3d4d\ ^3P_1$	2,745,163		2,745,727		2,758,477		

Table 3. *Cont.*

No.	Level	$E_{VV+CV+VV}$	ΔE	E_{FAC}	ΔE	E_{CIV3}	ΔE	E_{NIST}
137	$3d4d\ ^3P_2$	2,746,300		2,747,024		2,759,619		
138	$3d4d\ ^1G_4$	2,749,474		2,752,675		2,761,254		
139	$3d4f\ ^3H^o_4$	2,768,443		2,766,350		2,778,483		
140	$3d4f\ ^1G^o_4$	2,770,030		2,768,154		2,780,096		
141	$3d4f\ ^3H^o_5$	2,770,434		2,768,448		2,780,831		
142	$3d4d\ ^1S_0$	2,777,362		2,781,322		2,792,233		
143	$3d4f\ ^3F^o_2$	2,778,011		2,775,995		2,787,305		
144	$3d4f\ ^3F^o_3$	2,778,867		2,776,790		2,787,964		
145	$3d4f\ ^3F^o_4$	2,780,729		2,777,446		2,788,842		
146	$3d4f\ ^1D^o_2$	2,788,248		2,787,354		2,798,312		

5. Conclusions

CI with restrictions on the interactions (CI combined with second-order Brillouin–Wigner perturbation theory) makes it possible to handle large CSF expansions. The calculations including core–core correlation take around 20 h with 10 nodes on a cluster and bring the computed and observed excitation energies into very good agreement. To improve the computed excitation energies, the orbital set would need to be further extended leading to even larger matrices. The combined CI and perturbation method can be applied to include core–valence correlation in systems with many valence electrons and calculations. Calculations including valence–valence correlation and where core–valence correlation is treated perturbatively are in progress for P-, S-, and Cl-like systems.

Acknowledgments: Per Jönsson gratefully acknowledges support from the Swedish Research Council under contract 2015-04842.

Author Contributions: All authors contributed equally to the work.

Conflicts of Interest: The authors declare no conflicts of interest.

References

1. Young, P.R.; Del Zanna, G.; Mason, H.E.; Dere, K.P.; Landi, E.; Landini, M.; Doschek, G.A.; Brown, C.M.; Culhane, L.; Harra, L.K.; et al. EUV Emission Lines and Diagnostics Observed with Hinode/EIS. *Publ. Astron. Soc. Jpn.* **2007**, *59*, S857–S864.
2. Landi, E. Atomic data and spectral line intensities for Fe XV. *At. Data Nucl. Data Tables* **2011**, *97*, 587–647.
3. Kramida, A.; Ralchenko, Y.; Reader, J.; NIST ASD Team. *Atomic Spectra Database (ver. 5.2)*; National Institute of Standards and Technology: Gaithersburg, MD, USA, 2014. Available online: http://physics.nist.gov/asd (accessed on 28 december 2014).
4. Brown, G.V.; Beiersdorfer, P.; Utter, S.B.; Boyce, K.R.; Gendreau, K.C.; Kelley, R.; Porter, F.S.; Gygax, J. Measurements of Atomic Parameters of Highly Charged Ions for Interpreting Astrophysical Spectra. *Phys. Scr.* **2001**, doi:10.1238/Physica.Topical.092a00130.
5. Jönsson, P.; Bengtsson, P.; Ekman, J.; Gustafsson, S.; Karlsson, L.B.; Gaigalas, G.; Froese Fischer, C.; Kato, D.; Murakami, I; Sakaue, H.A.; et al. Relativistic CI calculations of spectroscopic data for the $2p^6$ and $2p^5 3l$ configurations in Ne-like ions between Mg III and Kr XXVII. *At. Data Nucl. Data Tables* **2014**, *100*, 1–154.
6. Wang, K.; Guo, X.L.; Li, S.; Si, R.; Dang, W.; Chen, Z.B.; Jönsson, P.; Hutton, R.; Chen, C.Y.; Yan, J. Calculations with spectroscopic accuracy: Energies and transition rates in the nitrogen isoelectronic sequence from Ar XII to Zn XXIV. *Astrophys. J. Suppl. Ser.* **2016**, *223*, 33.
7. Vilkas, M.J.; Ishikawa, Y. High-accuracy calculations of term energies and lifetimes of silicon-like ions with nuclear charges $Z = 24 - 30$. *J. Phys. B At. Mol. Opt. Phys.* **2004**, *37*, 1803–1816.
8. Gu, M.F. Energies of $1s^2 2l^q$ $(1 \leq q \leq 8)$ states for $Z \leq 60$ with a combined configuration interaction and many-body perturbation theory approach. *At. Data Nucl. Data Tables* **2005**, *89*, 267–293.

9. Kozlov, M.G.; Porsev, S.G.; Safronova, M.S.; Tupitsyn, I.I. CI-MBPT: A package of programs for relativistic atomic calculations based on a method combining configuration interaction and many-body perturbation theory. *Comput. Phys. Commun.* **2015**, *195*, 199–213.
10. Verdebout, S.; Rynkun, P.; Jönsson, P.; Gaigalas, G.; Froese Fischer, C.; Godefroid, M. A partitioned correlation function interaction approach for describing electron correlation in atoms. *J. Phys. B At. Mol. Opt. Phys.* **2013**, *46*, 085003.
11. Dzuba, V.A.; Berengut, J.; Harabati, C.; Flambaum, V.V. Combining configuration interaction with perturbation theory for atoms with large number of valence electrons. **2016**, arXiv:1611.00425v1.
12. Grant, I.P. *Relativistic Quantum Theory of Atoms and Molecules*; Springer: New York, NY, USA, 2007.
13. Froese Fischer, C.; Godefroid, M.; Brage, T.; Jönsson, P.; Gaigalas, G. Advanced multiconfiguration methods for complex atoms: Part I—Energies and wave functions. *J. Phys. B At. Mol. Opt. Phys.* **2016**, *49*, 182004.
14. Jönsson, P.; Gaigalas, G.; Bieroń, J.; Froese Fischer, C.; Grant, I.P. New Version: Grasp2K relativistic atomic structure package. *Comput. Phys. Commun.* **2013**, *184*, 2197–2203.
15. McKenzie, B.J.; Grant, I.P.; Norrington, P.H. A program to calculate transverse Breit and QED corrections to energylevels in a multiconfiguration Dirac-Fock environment. *Comput. Phys. Commun.* **1980**, *21*, 233–246.
16. Froese Fischer, C. The MCHF atomic-structure package. *Comput. Phys. Commun.* **1991**, *64*, 369–398.
17. Parpia, F.A.; Froese Fischer, C.; Grant, I.P. GRASP92: A package for large-scale relativistic atomic structure calculations. *Comput. Phys. Commun.* **1996**, *94*, 249–271.
18. Kotochigova, S.; Kirby, K.P.; Tupitsyn, I. Ab initio fully relativistic calculations of x-ray spectra of highly charged ions. *Phys. Rev. A* **2007**, *76*, 052513.
19. Gustafsson, S.; Jönsson, P.; Froese Fischer, C.; Grant, I.P. MCDHF and RCI calculations of energy levels, lifetimes and transition rates for $3l3l'$, $3l4l'$ and $3s5l$ states in Ca IX—As XXII and Kr XXV. *Astron. Astrophys.* **2017**, *579*, A76.
20. Gaigalas, G.; Froese Fischer, C.; Rynkun, P.; Jönsson, P. JJ2LSJ transformation and unique labeling for energy levels. *Atoms* **2016**, submitted.
21. Aggarwal, K.M.; Tayal, V.; Gupta, G.P.; Keenan, F.P. Energy levels and radiative rates for transitions in Mg-like iron, cobalt and nickel. *At. Data Nucl. Data Tables* **2007**, *93*, 615–710.

atoms

MDPI

Article

JJ2LSJ Transformation and Unique Labeling for Energy Levels

Gediminas Gaigalas [1,*], Charlotte Froese Fischer [2], Pavel Rynkun [1] and Per Jönsson [3]

1 Institute of Theoretical Physics and Astronomy, Vilnius University, Saulėtekio av. 3, LT-10222 Vilnius, Lithuania; pavel.rynkun@tfai.vu.lt
2 Department of Computer Science, University of British Columbia, Vancouver, BC V6T 1Z4, Canada; cff@cs.ubc.ca
3 Materials Science and Applied Mathematics, Malmö University, SE-205 06 Malmö, Sweden; per.jonsson@mah.se
* Correspondence: gediminas.gaigalas@tfai.vu.lt

Academic Editor: Joseph Reader
Received: 21 December 2016; Accepted:19 January 2017; Published: 27 January 2017

Abstract: The JJ2LSJ program, which is important not only for the GRASP2K package but for the atom theory in general, is presented. The program performs the transformation of atomic state functions (ASFs) from a *jj*-coupled CSF basis into an *LSJ*-coupled CSF basis. In addition, the program implements a procedure that assigns a unique label to all energy levels. Examples of how to use the JJ2LSJ program are given. Several cases are presented where there is a unique labeling problem.

Keywords: energy levels; *LSJ*-coupling; *jj*-coupling; JJ2LSJ transformation; unique label

1. Introduction

In principle, any valid coupling scheme can be used to represent the wave function in atomic structure calculations. Levels of an energy spectrum are identified and labeled with the help of sets of quantum numbers describing the coupling scheme used for the wave function. However, these quantum numbers are exact only for the cases of pure coupling. In a calculations of energy spectra one has to start with the coupling scheme closest to reality [1]. The most frequently used coupling schemes in atomic theory are the *LSJ* and *jj*. In atomic spectroscopy, the standard *LSJ* notation of the levels is frequently applied for classifying the low-lying level structures of atoms or ions.

Calculations may be performed in the relativistic (*jj*-coupling) scheme in order to get more accurate data that include relativistic effects. Thus, after a multiconfiguration Dirac-Hartree-Fock (MCDHF) or relativistic configuration interaction (RCI) [2] calculation the transformation to *LSJ*-coupling is needed. The JJ2LSJ code in GRASP2K [3] does this by applying a unitary transformation to the relativistic configuration state function (CSF) basis set which preserves orthonormality. The unitary transformation selected is the coupling transformation that changes the order of coupling from *jj* to *LSJ*, a transformation that does not involve the radial factor, only the spin-angular factor.

An energy level is normally assigned the label of the leading CSF in the wave function expansion. For many systems, two or more wave functions have the same leading CSFs giving rise to non-unique labels for the energy levels. We have such a situation for Si-like ions [4] and some other systems [5]. The new JJ2LSJ program implements a procedure that resolves these problems, assigning a unique label to all energy levels.

2. Theory

2.1. Transformation from jj- to LSJ-Coupling

Each nonrelativistic nl-orbital (except for ns) is associated with two relativistic orbitals $l_{\pm} \equiv j = l \pm 1/2$. In the transformation of the spin-angular factor $|l^{w}\alpha LS\rangle$ into a jj-coupled angular basis, two subshell states, one with $l_{-} \equiv j = l - 1/2$ and another one with $l_{+} \equiv j = l + 1/2$, may occur in the expansion. This shell-splitting

$$|l^{w}\alpha\nu LS\rangle \longrightarrow \left(|l_{-}^{w_1}\,\nu_1 J_1\rangle,\ |l_{+}^{w_2}\,\nu_2 J_2\rangle\right)\,, \tag{1}$$

obviously conserves the number of electrons, provided ($w = w_1 + w_2$), with $w_1(\max) = 2l$ and $w_2(\max) = 2(l+1)$. Making use of this notation, the transformation between the subshell states in LSJ- and jj-coupling can be written as

$$|l^{w}\alpha\nu LSJ\rangle = \sum_{\nu_1 J_1 \nu_2 J_2 w_1} |(l_{-}^{w_1}\nu_1 J_1,\ l_{+}^{(w-w_1)}\nu_2 J_2)\ J\rangle\langle(l_{-}^{w_1}\nu_1 J_1,\ l_{+}^{(w-w_1)}\nu_2 J_2)\ J|l^{w}\alpha\nu LSJ\rangle, \tag{2}$$

$$|(l_{-}^{w_1}\nu_1 J_1,\ l_{+}^{(w-w_1)}\nu_2 J_2)\ J\rangle = \sum_{\alpha\nu LS} |l^{w}\alpha\nu LSJ\rangle\langle l^{w}\alpha\nu LSJ|l(l_{-}^{w_1}\nu_1 J_1,\ l_{+}^{(w-w_1)}\nu_2 J_2)\ J\rangle, \tag{3}$$

which, in both cases, includes a summation over all the quantum numbers (except of n, l_{-}, and l_{+}). Here, $|(l_{-}^{w_1}\nu_1 J_1,\ l_{+}^{w_2}\nu_2 J_2)\ J\rangle$ is a coupled angular state with well-defined total angular momentum J which is built from the corresponding jj-coupled subshell states with $j_1 = l_{-} = l - \frac{1}{2}$, $j_2 = l_{+} = l + \frac{1}{2}$ and the total subshell angular momenta J_1 and J_2 , respectively.

An explicit expression for the coupling transformation coefficients

$$\langle(l_{-}^{w_1}\nu_1 J_1,\ l_{+}^{(w-w_1)}\nu_2 J_2)\ J|l^{w}\alpha\nu LSJ\rangle = \langle l^{w}\alpha\nu LSJ|(l_{-}^{w_1}\nu_1 J_1,\ l_{+}^{(w-w_1)}\nu_2 J_2)\ J\rangle \tag{4}$$

in (2) and (3) can be obtained only if we take the construction of the subshell states of w equivalent electrons from their corresponding *parent states* with $w-1$ electrons into account. In general, however, the *recursive* definition of the subshell states, out of their parent states, also leads to a recursive generation of the transformation matrices (4). These transformation coefficients can be chosen *real*: they occur very frequently as the *building blocks* in the transformation of all symmetry functions. The expressions and values of these coefficients are published in [6,7].

These transformation matrices, which are applied internally by the program JJ2LSJ, are consistent with the definition of the coefficients of fractional parentage [8,9] and with the phase system used in the [10]. So the program presented in the paper supports transformation from jj- to LSJ-coupling if ASF (which needs transformation) was created using the approach [7–10]. Otherwise the program may perform the transformation incorrectly.

2.2. Unique Labeling

An energy level is often given the label of the leading CSF in the wave function expansion. But it sometimes happens that two wave functions have the same largest CSF in LSJ- or jj-coupling, and then classification in energy spectra is not unique. The simplest way to have a unique identification of an energy level would be use a position number (POS) and symmetry J. But to get the energy spectra with unique labels in LSJ-coupling we should re-classify levels. For that purpose JJ2LSJ transformation with the unique labeling option can be used. To obtain unique labels the algorithm proposed in [11,12] is used: for a given set of wave functions with the same J and parity, the CSF with largest expansion coefficient is used as the label for the function containing this largest component. Once a label is assigned, the corresponding CSF is removed from consideration in the determination of the next label. In such a way we will get energy levels with unique labels. In this process, cases where one CSF is dominant (defines more than 50 % of the wave function composition) that CSF will give the label for the corresponding energy level, but when the composition is spread over a number of CSFs, and none particularly large, the label is defined by the algorithm. Thus labeling is done by blocks of levels, each of the same J and parity. The first step is to order the levels by energy and assign the POS (position) identifier with the lowest having POS = 1, the second POS = 2, etc. and then proceed with determining the label.

In the Section 4 we will present a few examples where wave functions have the same dominant term and where the unique labeling algorithm is needed.

3. The JJ2LSJ Program

JJ2LSJ program is intended to perform the transformation of ASFs from a jj-coupled CSF basis into an LSJ-coupled CSF basis. This program is written in FORTRAN90 and is included in the GRASP2K package [3]. It uses the same libraries as other programs in GRASP2K. The program is based on the earlier published LSJ program [13], but modified for speed up. The new program transforms only the most important components of large expansions. In addition, the new program provides an option to choose unique labeling versus labeling by the leading CSF in the wave function expansion.

For running the JJ2LSJ program we need several input files generated with the GRASP2K package: the CSFs list file (name.c) and the mixing coefficients file after MCDHF (name.m) or after RCI calculations (name.cm). The example below shows the execution of the JJ2LSJ program for the odd states of Si-like Sr (Sr XXV) ion, built on a CSF basis containing orbitals with principal quantum numbers up to $n = 7$ and expansion coefficients from RCI calculations. In the following example, the unique labeling option is chosen and the program is run in default mode. It should be remembered that the contribution to the wave function composition from a particular CSF is the square of the expansion coefficient. Thus a CSF with an expansion coefficient of 0.10 contributes 1% to the wave function composition.

```
>>jj2lsj

 jj2lsj: Transformation of ASFs from a jj-coupled CSF basis
         into an LSJ-coupled CSF basis  (Fortran 95 version)
         (C) Copyright by   G. Gaigalas and Ch. F. Fischer,
         (2017).
 Input files: name.c, name.(c)m
 Ouput files: name.lsj.lbl,
  (optional)  name.lsj.c, name.lsj.j,
              name.uni.lsj.lbl, name.uni.lsj.sum

 Name of state
>>odd7
 Loading Configuration Symmetry List File ...
 There are 49 relativistic subshells;
 There are 4420742 relativistic CSFs;
  ... load complete;

 Mixing coefficients from a CI calc.?
>>y
 Do you need a unique labeling? (y/n)
>>y
    nelec  =             14
    ncftot =        4420742
    nw     =             49
    nblock =              5

   block     ncf      nev    2j+1  parity
       1  190132        2       1      -1
       2  703411        7       3      -1
       3 1095473        8       5      -1
```

```
     4 1276414        4        7       -1
     5 1155312        1        9       -1
 Default settings?  (y/n)
>>y
 Maximum % of omitted composition is    1.000
 Below  5.0E-03 the eigenvector component is to be neglected for calculating
 Below  1.0E-03 the eigenvector composition is to be neglected for printing


 . . . . . . . . . . . . . . . . . . . . . . . . . . . . . . . . . . . . .
 Under investigation is the block:       1            The number of eigenvectors:  2
 The number of CSF (in jj-coupling):      190132       The number of CSF (in LS-coupling):        184
 Weights of major contributors to ASF in jj-coupling:

  Level  J Parity       CSF contributions

    1    0   -       0.89670 of       2    0.08762 of       1    0.00642 of       7    0.00250 of      13
                     0.00204 of       9
              Total sum over  weight (in jj) is:  0.99907840477285059

 Definition of leading CSF:

           2) 2s ( 2)   2p-( 2)   2p ( 4)   3s ( 1)   3p-( 1)   3p ( 2)
                                                1/2       1/2
                                                     1/2                  0


 Weights of major contributors to ASF in LS-coupling:

  Level  J Parity       CSF contributions

    1    0   -       0.89670 of       2    0.08762 of       1    0.00821 of       4    0.00441 of      11

              Total sum over  weight (in LSJ) is:  0.99862429376844597

 Definition of leading CSF:

         1)     2s( 2)    2p( 6)    3s( 2)    3p( 1)    3d( 1)
                   1S0       1S0       1S0       2P1       2D1       1S        1S        2P        3P        0
         2)     2s( 2)    2p( 6)    3s( 1)    3p( 3)
                   1S0       1S0       2S1       2P1       1S        2S        3P         0
         4)     2s( 2)    2p( 6)    3s( 1)    3p( 1)    3d( 2)
                   1S0       1S0       2S1       2P1       1S0       1S        2S        3P        3P        0
        11)     2s( 2)    2p( 6)    3p( 3)    3d( 1)
                   1S0       1S0       2P1       2D1       1S        2P        3P         0

 . . . . . . . . . . . . . . . . . . . . . .
 The new level is under investigation.
 Weights of major contributors to ASF in jj-coupling:

  Level  J Parity       CSF contributions

    2    0   -       0.88679 of       1    0.08724 of       2    0.00640 of       4    0.00505 of      13
                     0.00354 of      10
              Total sum over  weight (in jj) is:  0.99887286591520896

 Definition of leading CSF:

           1) 2s ( 2)   2p-( 2)   2p ( 4)   3s ( 2)   3p ( 1)   3d-( 1)
                                                          3/2       3/2
                                                               3/2        0
```

```
Weights of major contributors to ASF in LS-coupling:

 Level  J Parity       CSF contributions

   2    0   -        0.88679 of      1   0.08724 of      2   0.00852 of     11   0.00670 of      3
                     0.00537 of      8   0.00251 of      6
              Total sum over  weight (in LSJ) is:  0.99726815958280934

Definition of leading CSF:

        1)     2s( 2)    2p( 6)    3s( 2)    3p( 1)    3d( 1)
                 1S0       1S0       1S0       2P1       2D1       1S        1S        2P        3P         0
        2)     2s( 2)    2p( 6)    3s( 1)    3p( 3)
                 1S0       1S0       2S1       2P1       1S        2S        3P         0
        3)     2s( 2)    2p( 6)    3s( 1)    3p( 1)    3d( 2)
                 1S0       1S0       2S1       2P1       3P2       1S        2S        1P        3P         0
        6)     2s( 2)    2p( 6)    3s( 1)    3p( 1)    3d( 2)
                 1S0       1S0       2S1       2P1       3P2       1S        2S        3P        3P         0
        8)     2s( 2)    2p( 6)    3s( 1)    3p( 1)    3d( 2)
                 1S0       1S0       2S1       2P1       1D2       1S        2S        3P        3P         0
       11)     2s( 2)    2p( 6)    3p( 3)    3d( 1)
                 1S0       1S0       2P1       2D1       1S        2P        3P         0

. . . . . . . . . . . . . . . . . . . . . . . . . . . . . . . . . . . .

     ...........

jj2lsj: Execution complete.
```

The program, in default mode, produces the `name.lsj.lbl` file in which, for each ASF, the position, J, parity, total energy (in hartrees), and percentage of the wave function compositions are provided, followed by a list of expansion coefficients, their squares (compositions), and the CSF in LSJ-coupling. The example for `odd7.lsj.lbl` is given below. The label of the ASF is given by the the notation of the first line. So the level with total energy `-2794.938367562` is labeled `2s(2).2p(6).3s(2).3p_2P.3d_3F`.

```
Output file odd7.lsj.lbl
---------------------------------------------------------------------------
Pos   J   Parity       Energy Total       Comp. of ASF
 1    0     -          -2795.294072330        99.862%
        -0.94694267    0.89670042   2s(2).2p(6).3s_2S.3p(3)2P1_3P
        -0.29599906    0.08761544   2s(2).2p(6).3s(2).3p_2P.3d_3P
        -0.09062556    0.00821299   2s(2).2p(6).3s_2S.3p_3P.3d(2)1S0_3P
        -0.06641103    0.00441042   2s(2).2p(6).3p(3)2P1_2P.3d_3P
 2    0     -          -2793.959522946        99.727%
        -0.94169405    0.88678769   2s(2).2p(6).3s(2).3p_2P.3d_3P
         0.29537085    0.08724394   2s(2).2p(6).3s_2S.3p(3)2P1_3P
        -0.09229288    0.00851798   2s(2).2p(6).3p(3)2P1_2P.3d_3P
         0.08186357    0.00670164   2s(2).2p(6).3s_2S.3p_1P.3d(2)3P2_3P
        -0.07330494    0.00537361   2s(2).2p(6).3s_2S.3p_3P.3d(2)1D2_3P
        -0.05007114    0.00250712   2s(2).2p(6).3s_2S.3p_3P.3d(2)3P2_3P

    ...........

 4    2     -          -2794.938367562        99.746%
         0.61826021    0.38224569   2s(2).2p(6).3s(2).3p_2P.3d_3F
         0.54278368    0.29461412   2s(2).2p(6).3s_2S.3p(3)2P1_3P
         0.37809496    0.14295580   2s(2).2p(6).3s(2).3p_2P.3d_1D
         0.29469334    0.08684416   2s(2).2p(6).3s_2S.3p(3)2D3_3D
         0.17060037    0.02910449   2s(2).2p(6).3s(2).3p_2P.3d_3P
        -0.16453825    0.02707283   2s(2).2p(6).3s_2S.3p(3)4S3_5S
         0.12507106    0.01564277   2s(2).2p(6).3s_2S.3p(3)2D3_1D
        -0.06574377    0.00432224   2s(2).2p(6).3s(2).3p_2P.3d_3D
         0.06041988    0.00365056   2s(2).2p(6).3p(3)2P1_2P.3d_3F
```

```
     0.05069300     0.00256978   2s(2).2p(6).3s_2S.3p_3P.3d(2)1S0_3P
     0.03868154     0.00149626   2s(2).2p(6).3p(3)2P1_2P.3d_3P
     0.03576268     0.00127897   2s(2).2p(6).3p(3)2P1_2P.3d_1D
    -0.03187628     0.00101610   2s(2).2p(6).3s_2S.3p_1P.3d(2)3F2_3F

  ...........
-------------------------------------------------------------------------
```

When the unique labeling option is chosen, the `name.uni.lsj.lbl` and `name.uni.lsj.sum` files are produced. The format and information in the `name.uni.lsj.lbl` is the same as in `name.lsj.lbl`, but all the levels have unique labels. Please note that the third level, with total energy `-2794.938367562`, and a smaller largest component, was relabeled as `2s(2).2p(6).3s_2S.3p(3)2P1_3P` since the ^3F label had already been assigned.

```
Output file odd7.uni.lsj.lbl
-------------------------------------------------------------------------
Pos   J   Parity         Energy Total       Comp. of ASF
 1    0     -            -2795.294072330      99.862%
     -0.94694267     0.89670042   2s(2).2p(6).3s_2S.3p(3)2P1_3P
     -0.29599906     0.08761544   2s(2).2p(6).3s(2).3p_2P.3d_3P
     -0.09062556     0.00821299   2s(2).2p(6).3s_2S.3p_3P.3d(2)1S0_3P
     -0.06641103     0.00441042   2s(2).2p(6).3p(3)2P1_2P.3d_3P
 2    0     -            -2793.959522946      99.727%
     -0.94169405     0.88678769   2s(2).2p(6).3s(2).3p_2P.3d_3P
      0.29537085     0.08724394   2s(2).2p(6).3s_2S.3p(3)2P1_3P
     -0.09229288     0.00851798   2s(2).2p(6).3p(3)2P1_2P.3d_3P
      0.08186357     0.00670164   2s(2).2p(6).3s_2S.3p_1P.3d(2)3P2_3P
     -0.07330494     0.00537361   2s(2).2p(6).3s_2S.3p_3P.3d(2)1D2_3P
     -0.05007114     0.00250712   2s(2).2p(6).3s_2S.3p_3P.3d(2)3P2_3P

  ...........

 4    2     -            -2794.938367562      99.746%
      0.54278368     0.29461412   2s(2).2p(6).3s_2S.3p(3)2P1_3P
      0.61826021     0.38224569   2s(2).2p(6).3s(2).3p_2P.3d_3F
      0.37809496     0.14295580   2s(2).2p(6).3s(2).3p_2P.3d_1D
      0.29469334     0.08684416   2s(2).2p(6).3s_2S.3p(3)2D3_3D
      0.17060037     0.02910449   2s(2).2p(6).3s(2).3p_2P.3d_3P
     -0.16453825     0.02707283   2s(2).2p(6).3s_2S.3p(3)4S3_5S
      0.12507106     0.01564277   2s(2).2p(6).3s_2S.3p(3)2D3_1D
     -0.06574377     0.00432224   2s(2).2p(6).3s(2).3p_2P.3d_3D
      0.06041988     0.00365056   2s(2).2p(6).3p(3)2P1_2P.3d_3F
      0.05069300     0.00256978   2s(2).2p(6).3s_2S.3p_3P.3d(2)1S0_3P
      0.03868154     0.00149626   2s(2).2p(6).3p(3)2P1_2P.3d_3P
      0.03576268     0.00127897   2s(2).2p(6).3p(3)2P1_2P.3d_1D
     -0.03187628     0.00101610   2s(2).2p(6).3s_2S.3p_1P.3d(2)3F2_3F

  ...........
-------------------------------------------------------------------------
```

Below is the `odd7.uni.lsj.sum` file that provides the information – J, position, composition, serial number, and identification – for each ASF, where the serial number is the number of CSFs used in determining the composition. If the serial number of the composition is equal 1 the level is identified with the largest expansion coefficient, if 2, ..., etc., as in example below for levels with $J = 2$ Pos = 4 and Pos = 8, the levels are relabeled.

```
 Output file odd7.uni.lsj.sum
---------------------------------------------------------------------------
           Composition  Serial No.         Coupling
                        of compos.
 J =     0
-----------------------------------------------
Pos   1   0.896700420   1   2s(2).2p(6).3s_2S.3p(3)2P1_3P
Pos   2   0.886787690   1   2s(2).2p(6).3s(2).3p_2P.3d_3P
-----------------------------------------------

   ..........

           Composition  Serial No.         Coupling
                        of compos.
 J =     2
-----------------------------------------------
Pos   1   0.816598140   1   2s(2).2p(6).3s_2S.3p(3)4S3_5S
Pos   2   0.623434900   1   2s(2).2p(6).3s_2S.3p(3)2D3_3D
Pos   6   0.495079260   1   2s(2).2p(6).3s(2).3p_2P.3d_3P
Pos   5   0.477147080   1   2s(2).2p(6).3s(2).3p_2P.3d_3F
Pos   7   0.400735720   1   2s(2).2p(6).3s(2).3p_2P.3d_3D
Pos   4   0.294614120   2   2s(2).2p(6).3s_2S.3p(3)2P1_3P
Pos   3   0.283483530   1   2s(2).2p(6).3s_2S.3p(3)2D3_1D
Pos   8   0.137996860   3   2s(2).2p(6).3s(2).3p_2P.3d_1D
-----------------------------------------------

   ..........
---------------------------------------------------------------------------
```

The program can also be used in non-default mode. The typical run proceeds as follows:

```
  Default settings? (y/n)
>>n
  All levels (Y/N)
>>y
  Maximum % of omitted composition
>>0.5
  What is the value below which an eigenvector component is to be neglected
in the determination of the LSJ expansion: should be smaller than: 0.00500
>>0.003
  What is the value below which an eigenvector composition is to be neglected
  for printing?
>>0.0005
  Do you need the output file *.lsj.c? (y/n)
>>y
  Do you need the output file *.lsj.j? (y/n)
>>y
```

The non-default mode is useful in several cases:

(1) The present code allows the user, through the first parameter (`0.5`), to select the maximum percentage of the ASF composition that can be omitted. Given this information and with the help of the second parameter (`0.003`), it is easy to derive the largest small coefficient in the CSF expansion that may be included. However, with many components of about the same size, smaller values may be needed to meet the original objective. In this implementation, the user specifies the CSFs that can be omitted. The remaining CSFs define the basis that is to be transformed. By transforming this basis in decreasing order of importance, the desired percentage of the wave function can be transformed. A third parameter (`0.0005`) controls the printing of expansion coefficients in the LSJ basis and their contribution to the composition of the wave function. The default is to transform at least 99% of the wave function composition and print components in LSJ that contribute more than 0.1% to the composition. The cut-off for the jj-expansion has the value of 0.005, whereas the cut-off for printing is 0.001.

(2) In particular, the user may request a complete transformation, with a resulting list of CSFs in LSJ-coupling in `name.lsj.c` and their expansion coefficients in `name.lsj.j`. The two files have the same format as in ATSP2K [14]. Complete expansions are feasible only for small expansions. In this case the first and second parameter should be `0`.

(3) The non-default option should be used if we choose a unique labeling option, but the program will not give the unique identification for all levels. In this case we need to transform a larger amount of ASF with larger number of expansion coefficients. It can be done with help of the first and second parameter.

4. Results

In the recent calculations of energy spectra for Sr XXV [4] two pairs of odd levels with $J = 2$ had the same label and were separated by adding subscripts 'a' and 'b'. In the NIST database [15] for two ($3s\,3p^3(^2_1P)\,^3P^\circ_2$ and the $3s^2\,3p\,3d\,^3F^\circ_2$) of these levels there is no data and the $3s^2\,3p\,3d\,^1D^\circ_2$ level is not identified.

Running the `JJ2LSJ` program for Sr XXV levels 13 and 25 are relabeled in the Table 1. Table 1 gives also the labels from [4]. As we see level 13 had the same label as level 14, and 25 was labeled as 17. In Table 1 also the compositions in LSJ- coupling are given. In the Table 2 transition data of E1, M1, M2 transitions for relabeled levels are presented.

Table 1. Energy levels in cm^{-1} and LSJ-composition for Si-like Sr. In the original data levels 13 and 14 had the same label and subscripts 'a' and 'b' were introduced to separate the levels. Using the `JJ2LSJ` program levels 13 and 14 are now assigned unique labels.

No.	Level [4]	Level (Relabeled)	LSJ-Composition	$n = 7$
1	$3s^2\,3p^2(^3_2P)\,^3P_0$	$3s^2\,3p^2(^3_2P)\,^3P_0$	$0.82 + 0.16\;3s^2\,3p^2(^1_0S)\,^1S$	0
2	$3s^2\,3p^2(^3_2P)\,^3P_1$	$3s^2\,3p^2(^3_2P)\,^3P_1$	0.98	92 950
3	$3s^2\,3p^2(^3_2P)\,^3P_2$	$3s^2\,3p^2(^3_2P)\,^3P_2$	$0.54 + 0.44\;3s^2\,3p^2(^1_2D)\,^1D$	122 240
4	$3s^2\,3p^2(^1_2D)\,^1D_2$	$3s^2\,3p^2(^1_2D)\,^1D_2$	$0.54 + 0.44\;3s^2\,3p^2(^3_2P)\,^3P$	239 120
5	$3s^2\,3p^2(^1_0S)\,^1S_0$	$3s^2\,3p^2(^1_0S)\,^1S_0$	$0.81 + 0.16\;3s^2\,3p^2(^3_2P)\,^3P$	313 384
6	$3s\,3p^3(^4_3S)\,^5S^\circ_2$	$3s\,3p^3(^4_3S)\,^5S^\circ_2$	$0.82 + 0.14\;3s\,3p^3(^2_1P)\,^3P^\circ + 0.02\;3s\,3p^3(^2_3D)\,^3D^\circ$	537 112
7	$3s\,3p^3(^2_3D)\,^3D^\circ_1$	$3s\,3p^3(^2_3D)\,^3D^\circ_1$	$0.64 + 0.17\;3s\,3p^3(^2_1P)\,^3P^\circ + 0.09\;3s^2\,3p\,3d\,^3D^\circ$	648 560
8	$3s\,3p^3(^2_3D)\,^3D^\circ_2$	$3s\,3p^3(^2_3D)\,^3D^\circ_2$	$0.62 + 0.14\;3s\,3p^3(^2_1P)\,^3P^\circ + 0.11\;3s\,3p^3(^4_3S)\,^5S^\circ$	665 804
9	$3s\,3p^3(^2_3D)\,^3D^\circ_3$	$3s\,3p^3(^2_3D)\,^3D^\circ_3$	$0.88 + 0.10\;3s^2\,3p\,3d\,^3D^\circ$	700 704
10	$3s\,3p^3(^2_1P)\,^3P^\circ_0$	$3s\,3p^3(^2_1P)\,^3P^\circ_0$	$0.90 + 0.09\;3s^2\,3p\,3d\,^3P^\circ$	767 747
11	$3s\,3p^3(^2_3D)\,^1D^\circ_2$	$3s\,3p^3(^2_3D)\,^1D^\circ_2$	$0.28 + 0.23\;3s^2\,3p\,3d\,^1D^\circ + 0.17\;3s\,3p^3(^2_1P)\,^3P^\circ$	772 281
12	$3s\,3p^3(^2_1P)\,^3P^\circ_1$	$3s\,3p^3(^2_1P)\,^3P^\circ_1$	$0.66 + 0.16\;3s\,3p^3(^2_3D)\,^3D^\circ + 0.07\;3s^2\,3p\,3d\,^3P^\circ$	779 904
13	$3s^2\,3p\,3d\,^3F^\circ_{2\,a}$	$3s\,3p^3(^2_1P)\,^3P^\circ_2$	$0.29 + 0.38\;3s^2\,3p\,3d\,^3F^\circ + 0.14\;3s^2\,3p\,3d\,^1D^\circ$	845 815
14	$3s^2\,3p\,3d\,^3F^\circ_{2\,b}$	$3s^2\,3p\,3d\,^3F^\circ_2$	$0.48 + 0.21\;3s\,3p^3(^2_3D)\,^1D^\circ + 0.13\;3s\,3p^3(^2_1P)\,^3P^\circ$	882 783
15	$3s\,3p^3(^4_3S)\,^3S^\circ_1$	$3s\,3p^3(^4_3S)\,^3S^\circ_1$	$0.55 + 0.32\;3s\,3p^3(^2_1P)\,^1P^\circ + 0.04\;3s\,3p^3(^2_1P)\,^3P^\circ$	884 118
16	$3s^2\,3p\,3d\,^3F^\circ_3$	$3s^2\,3p\,3d\,^3F^\circ_3$	$0.89 + 0.04\;3s^2\,3p\,3d\,^3D^\circ + 0.03\;3s^2\,3p\,3d\,^1F^\circ$	906 676
17	$3s^2\,3p\,3d\,^3P^\circ_{2\,a}$	$3s^2\,3p\,3d\,^3P^\circ_2$	$0.50 + 0.20\;3s^2\,3p\,3d\,^3D^\circ + 0.16\;3s\,3p^3(^2_3D)\,^1D^\circ$	967 289
18	$3s^2\,3p\,3d\,^3D^\circ_1$	$3s^2\,3p\,3d\,^3D^\circ_1$	$0.42 + 0.26\;3s^2\,3p\,3d\,^3P^\circ + 0.16\;3s^2\,3p\,3d\,^1P^\circ$	971 435
19	$3s^2\,3p\,3d\,^3F^\circ_4$	$3s^2\,3p\,3d\,^3F^\circ_4$	0.98	989 440
20	$3s\,3p^3(^2_1P)\,^1P^\circ_1$	$3s\,3p^3(^2_1P)\,^1P^\circ_1$	$0.42 + 0.31\;3s\,3p^3(^4_3S)\,^3S^\circ + 0.17\;3s^2\,3p\,3d\,^3D^\circ$	1 007 850
21	$3s^2\,3p\,3d\,^3D^\circ_2$	$3s^2\,3p\,3d\,^3D^\circ_2$	$0.40 + 0.28\;3s^2\,3p\,3d\,^1D^\circ + 0.19\;3s\,3p^3(^2_3D)\,^1D^\circ$	1 052 275
22	$3s^2\,3p\,3d\,^3P^\circ_0$	$3s^2\,3p\,3d\,^3P^\circ_0$	$0.89 + 0.09\;3s\,3p^3(^2_1P)\,^3P^\circ$	1 060 646
23	$3s^2\,3p\,3d\,^3D^\circ_3$	$3s^2\,3p\,3d\,^3D^\circ_3$	$0.72 + 0.10\;3s^2\,3p\,3d\,^1F^\circ + 0.08\;3s\,3p^3(^2_3D)\,^3D^\circ$	1 064 175
24	$3s^2\,3p\,3d\,^3P^\circ_1$	$3s^2\,3p\,3d\,^3P^\circ_1$	$0.58 + 0.18\;3s^2\,3p\,3d\,^3D^\circ + 0.09\;3s\,3p^3(^2_1P)\,^3P^\circ$	1 075 683
25	$3s^2\,3p\,3d\,^3P^\circ_{2\,b}$	$3s^2\,3p\,3d\,^1D^\circ_2$	$0.14 + 0.37\;3s^2\,3p\,3d\,^3P^\circ + 0.23\;3s^2\,3p\,3d\,^3D^\circ$	1 091 772
26	$3s^2\,3p\,3d\,^1F^\circ_3$	$3s^2\,3p\,3d\,^1F^\circ_3$	$0.84 + 0.11\;3s^2\,3p\,3d\,^3D^\circ$	1 147 847
27	$3s^2\,3p\,3d\,^1P^\circ_1$	$3s^2\,3p\,3d\,^1P^\circ_1$	$0.73 + 0.12\;3s\,3p^3(^2_1P)\,^1P^\circ + 0.07\;3s^2\,3p\,3d\,^3D^\circ$	1 184 571

Table 2. Transition data for Si-like Sr where each level has been assigned a unique label.

Upper	Lower	EM	ΔE (cm^{-1})	λ (Å)	A (s^{-1})	gf	dT
$3s\,3p^3(^2_1P)\,^3P^o_2$	$3s^2\,3p^2(^3_2P)\,^3P_0$	M2	845815	118.23	9.680E+00	1.014E-10	
$3s\,3p^3(^2_1P)\,^3P^o_2$	$3s^2\,3p^2(^3_2P)\,^3P_1$	E1	752864	132.83	1.712E+07	2.264E-04	0.014
$3s\,3p^3(^2_1P)\,^3P^o_2$	$3s^2\,3p^2(^3_2P)\,^3P_1$	M2	752864	132.83	4.646E+00	6.144E-11	
$3s\,3p^3(^2_1P)\,^3P^o_2$	$3s^2\,3p^2(^3_2P)\,^3P_2$	M2	723575	138.20	1.952E+01	2.795E-10	
$3s\,3p^3(^2_1P)\,^3P^o_2$	$3s^2\,3p^2(^3_2P)\,^3P_2$	E1	723575	138.20	4.775E+09	6.837E-02	0.015
$3s\,3p^3(^2_1P)\,^3P^o_2$	$3s^2\,3p^2(^1_2D)\,^1D_2$	E1	606694	164.83	1.828E+09	3.723E-02	0.021
$3s\,3p^3(^2_1P)\,^3P^o_2$	$3s^2\,3p^2(^1_2D)\,^1D_2$	M2	606694	164.83	1.179E+02	2.402E-09	
$3s\,3p^3(^2_1P)\,^3P^o_2$	$3s^2\,3p^2(^1_0S)\,^1S_0$	M2	532430	187.82	3.162E+01	8.361E-10	
$3s^2\,3p\,3d\,^1P^o_1$	$3s\,3p^3(^2_1P)\,^3P^o_2$	M1	338756	295.20	2.315E+03	9.072E-08	
$3s\,3p^3(^2_1P)\,^3P^o_2$	$3s\,3p^3(^4_3S)\,^5S^o_2$	M1	308702	323.94	1.150E+04	9.043E-07	
$3s^2\,3p\,3d\,^1F^o_3$	$3s\,3p^3(^2_1P)\,^3P^o_2$	M1	302032	331.09	2.777E+03	3.195E-07	
$3s^2\,3p\,3d\,^1D^o_2$	$3s\,3p^3(^2_1P)\,^3P^o_2$	M1	245957	406.57	3.507E+03	4.346E-07	
$3s^2\,3p\,3d\,^3P^o_1$	$3s\,3p^3(^2_1P)\,^3P^o_2$	M1	229868	435.03	1.962E+03	1.670E-07	
$3s^2\,3p\,3d\,^3D^o_3$	$3s\,3p^3(^2_1P)\,^3P^o_2$	M1	218360	457.96	6.377E+03	1.403E-06	
$3s^2\,3p\,3d\,^3D^o_2$	$3s\,3p^3(^2_1P)\,^3P^o_2$	M1	206460	484.35	2.293E+03	4.032E-07	
$3s\,3p^3(^2_1P)\,^3P^o_2$	$3s\,3p^3(^2_3D)\,^3D^o_1$	M1	197254	506.96	1.069E+03	2.060E-07	
$3s\,3p^3(^2_1P)\,^3P^o_2$	$3s\,3p^3(^2_3D)\,^3D^o_2$	M1	180010	555.52	1.380E+04	3.193E-06	
$3s\,3p^3(^2_1P)\,^1P^o_1$	$3s\,3p^3(^2_1P)\,^3P^o_2$	M1	162035	617.15	1.312E+03	2.248E-07	
$3s\,3p^3(^2_1P)\,^3P^o_2$	$3s\,3p^3(^2_3D)\,^3D^o_3$	M1	145110	689.13	5.267E+03	1.875E-06	
$3s^2\,3p\,3d\,^3D^o_1$	$3s\,3p^3(^2_1P)\,^3P^o_2$	M1	125620	796.05	3.721E+02	1.061E-07	
$3s^2\,3p\,3d\,^3P^o_2$	$3s\,3p^3(^2_1P)\,^3P^o_2$	M1	121474	823.22	1.112E+02	5.648E-08	
$3s\,3p^3(^2_1P)\,^3P^o_2$	$3s\,3p^3(^2_3D)\,^1D^o_2$	M1	73533	1359.92	3.568E+03	4.947E-06	
$3s\,3p^3(^2_1P)\,^3P^o_2$	$3s\,3p^3(^2_1P)\,^3P^o_1$	M1	65910	1517.20	1.725E+03	2.976E-06	
$3s^2\,3p\,3d\,^3F^o_3$	$3s\,3p^3(^2_1P)\,^3P^o_2$	M1	60861	1643.09	2.184E+03	6.189E-06	
$3s\,3p^3(^4_3S)\,^3S^o_1$	$3s\,3p^3(^2_1P)\,^3P^o_2$	M1	38303	2610.71	6.515E+00	1.997E-08	
$3s^2\,3p\,3d\,^3F^o_2$	$3s\,3p^3(^2_1P)\,^3P^o_2$	M1	36968	2705.00	5.987E+02	3.284E-06	
$3s^2\,3p\,3d\,^1D^o_2$	$3s^2\,3p^2(^3_2P)\,^3P_0$	M2	1091772	91.59	9.719E+00	6.112E-11	
$3s^2\,3p\,3d\,^1D^o_2$	$3s^2\,3p^2(^3_2P)\,^3P_1$	M2	998822	100.12	2.608E+00	1.960E-11	
$3s^2\,3p\,3d\,^1D^o_2$	$3s^2\,3p^2(^3_2P)\,^3P_1$	E1	998822	100.12	5.171E+09	3.886E-02	0.002
$3s^2\,3p\,3d\,^1D^o_2$	$3s^2\,3p^2(^3_2P)\,^3P_2$	E1	969532	103.14	8.118E+09	6.474E-02	0.012
$3s^2\,3p\,3d\,^1D^o_2$	$3s^2\,3p^2(^3_2P)\,^3P_2$	M2	969532	103.14	1.596E+02	1.273E-09	
$3s^2\,3p\,3d\,^1D^o_2$	$3s^2\,3p^2(^1_2D)\,^1D_2$	M2	852652	117.28	5.942E+01	6.127E-10	
$3s^2\,3p\,3d\,^1D^o_2$	$3s^2\,3p^2(^1_2D)\,^1D_2$	E1	852652	117.28	1.036E+11	1.068E+00	0.003
$3s^2\,3p\,3d\,^1D^o_2$	$3s^2\,3p^2(^1_0S)\,^1S_0$	M2	778388	128.47	6.258E+01	7.742E-10	
$3s^2\,3p\,3d\,^1D^o_2$	$3s\,3p^3(^4_3S)\,^5S^o_2$	M1	554660	180.29	4.087E+03	9.957E-08	
$3s^2\,3p\,3d\,^1D^o_2$	$3s\,3p^3(^2_3D)\,^3D^o_1$	M1	443212	225.63	1.341E+02	5.116E-09	
$3s^2\,3p\,3d\,^1D^o_2$	$3s\,3p^3(^2_3D)\,^3D^o_2$	M1	425968	234.76	6.591E+03	2.723E-07	
$3s^2\,3p\,3d\,^1D^o_2$	$3s\,3p^3(^2_3D)\,^3D^o_3$	M1	391068	255.71	2.693E+03	1.320E-07	
$3s^2\,3p\,3d\,^1D^o_2$	$3s\,3p^3(^2_3D)\,^1D^o_2$	M1	319491	313.00	4.115E+01	3.022E-09	
$3s^2\,3p\,3d\,^1D^o_2$	$3s\,3p^3(^2_1P)\,^3P^o_1$	M1	311868	320.65	2.231E+03	1.719E-07	
$3s^2\,3p\,3d\,^1D^o_2$	$3s^2\,3p\,3d\,^3F^o_2$	M1	208989	478.49	1.141E+03	1.958E-07	
$3s^2\,3p\,3d\,^1D^o_2$	$3s\,3p^3(^4_3S)\,^3S^o_1$	M1	207654	481.57	1.285E+02	2.234E-08	
$3s^2\,3p\,3d\,^1D^o_2$	$3s^2\,3p\,3d\,^3F^o_3$	M1	185096	540.26	3.272E+03	7.160E-07	
$3s^2\,3p\,3d\,^1D^o_2$	$3s^2\,3p\,3d\,^3P^o_2$	M1	124483	803.32	1.272E+04	6.151E-06	
$3s^2\,3p\,3d\,^1D^o_2$	$3s^2\,3p\,3d\,^3P^o_2$	M1	124483	803.32	1.272E+04	6.151E-06	
$3s^2\,3p\,3d\,^1D^o_2$	$3s^2\,3p\,3d\,^3D^o_1$	M1	120337	831.00	5.753E+02	2.978E-07	
$3s^2\,3p\,3d\,^1P^o_1$	$3s^2\,3p\,3d\,^1D^o_2$	M1	92798	1077.60	2.314E+02	1.208E-07	
$3s^2\,3p\,3d\,^1D^o_2$	$3s\,3p^3(^2_1P)\,^1P^o_1$	M1	83922	1191.58	7.850E+02	8.355E-07	
$3s^2\,3p\,3d\,^1F^o_3$	$3s^2\,3p\,3d\,^1D^o_2$	M1	56074	1783.34	1.104E+02	3.686E-07	
$3s^2\,3p\,3d\,^1D^o_2$	$3s^2\,3p\,3d\,^3D^o_2$	M1	39497	2531.81	3.077E+01	1.478E-07	
$3s^2\,3p\,3d\,^1D^o_2$	$3s^2\,3p\,3d\,^3D^o_3$	M1	27597	3623.57	1.171E+02	1.153E-06	
$3s^2\,3p\,3d\,^1D^o_2$	$3s^2\,3p\,3d\,^3P^o_1$	M1	16089	6215.18	4.670E+01	1.352E-06	

Another example for which problems with unique labels occur is P-like W. Calculations using the MCDHF and RCI methods show that there are many levels with the same labels [16].

Table 3 presents the part of energy spectra with unique labels and LSJ-composition. The levels which were relabeled are marked with grey color.

Table 3. LSJ-composition and energy levels in cm^{-1} for P-like W from relativistic configuration interaction (RCI) calculations. Levels that are assigned new labels using the JJ2LSJ program are marked with grey background.

No.	Level	LSJ-Composition	$E(RCI)$
1	$3s^2\,3p^3({}^2_3D)\,{}^2D^{\circ}_{3/2}$	0.27 + 0.48 $3s^2\,3p^3({}^2_1P)\,{}^2P^{\circ}$ + 0.25 $3s^2\,3p^3({}^4_3S)\,{}^4S^{\circ}$	0
2	$3s^2\,3p^2({}^3_2P)\,{}^3P\,3d\,{}^4F_{3/2}$	0.34 + 0.31 $3s^2\,3p^2({}^1_0S)\,{}^1S\,3d\,{}^2D$ + 0.11 $3s^2\,3p^2({}^3_2P)\,{}^3P\,3d\,{}^4D$	1 853 012
3	$3s^2\,3p^2({}^1_2D)\,{}^1D\,3d\,{}^2F_{5/2}$	0.002 + 0.30 $3s^2\,3p^2({}^1_0S)\,{}^1S\,3d\,{}^2D$ + 0.20 $3s^2\,3p^2({}^3_2P)\,{}^3P\,3d\,{}^4D$	2 613 799
4	$3s^2\,3p^3({}^4_3S)\,{}^4S^{\circ}_{3/2}$	0.55 + 0.44 $3s^2\,3p^3({}^2_3D)\,{}^2D^{\circ}$	2 752 643
5	$3s^2\,3p^3({}^2_3D)\,{}^2D^{\circ}_{5/2}$	0.99	2 847 490
6	$3s^2\,3p^3({}^2_1P)\,{}^2P^{\circ}_{1/2}$	0.99	2 971 667
7	$3s\,3p^4({}^3_2P)\,{}^4P_{5/2}$	0.66 + 0.27 $3s\,3p^4({}^1_2D)\,{}^2D$ + 0.02 $3s^2\,3p^2({}^1_2D)\,{}^1D\,3d\,{}^2D$	4 157 536
8	$3s\,3p^4({}^1_2D)\,{}^2D_{3/2}$	0.24 + 0.33 $3s\,3p^4({}^3_2P)\,{}^2P$ + 0.11 $3s\,3p^4({}^3_2P)\,{}^4P$	4 398 405
9	$3s\,3p^4({}^1_0S)\,{}^2S_{1/2}$	0.54 + 0.24 $3s\,3p^4({}^3_2P)\,{}^4P$ + 0.07 $3s\,3p^4({}^3_2P)\,{}^2P$	4 413 592
10	$3s^2\,3p^2({}^3_2P)\,{}^3P\,3d\,{}^4F_{5/2}$	0.47 + 0.29 $3s^2\,3p^2({}^1_2D)\,{}^1D\,3d\,{}^2F$ + 0.16 $3s^2\,3p^2({}^3_2P)\,{}^3P\,3d\,{}^2F$	4 584 963
11	$3s^2\,3p^2({}^3_2P)\,{}^3P\,3d\,{}^4D_{1/2}$	0.80 + 0.13 $3s^2\,3p^2({}^3_2P)\,{}^3P\,3d\,{}^2P$ + 0.04 $3s^2\,3p^2({}^3_2P)\,{}^3P\,3d\,{}^4P$	4 611 635
12	$3s^2\,3p^2({}^3_2P)\,{}^3P\,3d\,{}^4D_{3/2}$	0.28 + 0.33 $3s^2\,3p^2({}^3_2P)\,{}^3P\,3d\,{}^4F$ + 0.12 $3s^2\,3p^2({}^3_2P)\,{}^3P\,3d\,{}^2P$	4 613 253
13	$3s^2\,3p^2({}^1_2D)\,{}^1D\,3d\,{}^2G_{7/2}$	0.52 + 0.18 $3s^2\,3p^2({}^3_2P)\,{}^3P\,3d\,{}^4F$ + 0.14 $3s^2\,3p^2({}^3_2P)\,{}^3P\,3d\,{}^2F$	4 683 926
14	$3s^2\,3p^2({}^3_2P)\,{}^3P\,3d\,{}^2D_{5/2}$	0.30 + 0.24 $3s^2\,3p^2({}^1_2D)\,{}^1D\,3d\,{}^2F$ + 0.12 $3s^2\,3p^2({}^3_2P)\,{}^3P\,3d\,{}^4P$	4 922 718
15	$3s^2\,3p^2({}^1_2D)\,{}^1D\,3d\,{}^2P_{1/2}$	0.35 + 0.31 $3s^2\,3p^2({}^3_2P)\,{}^3P\,3d\,{}^4P$ + 0.18 $3s^2\,3p^2({}^1_2D)\,{}^1D\,3d\,{}^2S$	5 004 842
16	$3s^2\,3p^2({}^1_2D)\,{}^1D\,3d\,{}^2D_{3/2}$	0.27 + 0.22 $3s^2\,3p^2({}^3_2P)\,{}^3P\,3d\,{}^4P$ + 0.15 $3s^2\,3p^2({}^1_2D)\,{}^1D\,3d\,{}^2P$	5 009 796
17	$3s^2\,3p^2({}^3_2P)\,{}^3P\,3d\,{}^4D_{7/2}$	0.47 + 0.37 $3s^2\,3p^2({}^3_2P)\,{}^3P\,3d\,{}^4F$ + 0.07 $3s^2\,3p^2({}^3_2P)\,{}^3P\,3d\,{}^2F$	5 242 340
18	$3s^2\,3p^2({}^3_2P)\,{}^3P\,3d\,{}^2P_{3/2}$	0.25 + 0.23 $3s^2\,3p^2({}^3_2P)\,{}^3P\,3d\,{}^4P$ + 0.20 $3s^2\,3p^2({}^1_2D)\,{}^1D\,3d\,{}^2P$	5 342 886
19	$3s^2\,3p^2({}^1_2D)\,{}^1D\,3d\,{}^2G_{9/2}$	0.62 + 0.37 $3s^2\,3p^2({}^3_2P)\,{}^3P\,3d\,{}^4F$	5 346 296
20	$3s^2\,3p^2({}^3_2P)\,{}^3P\,3d\,{}^2F_{5/2}$	0.33 + 0.25 $3s^2\,3p^2({}^1_2D)\,{}^1D\,3d\,{}^2D$ + 0.22 $3s^2\,3p^2({}^3_2P)\,{}^3P\,3d\,{}^4D$	5 371 289
21	$3s^2\,3p^2({}^3_2P)\,{}^3P\,3d\,{}^4P_{5/2}$	0.35 + 0.27 $3s^2\,3p^2({}^1_2D)\,{}^1D\,3d\,{}^2D$ + 0.14 $3s^2\,3p^2({}^3_2P)\,{}^3P\,3d\,{}^2D$	5 534 143
22	$3s^2\,3p^2({}^1_2D)\,{}^1D\,3d\,{}^2F_{7/2}$	0.44 + 0.18 $3s^2\,3p^2({}^3_2P)\,{}^3P\,3d\,{}^2F$ + 0.17 $3s^2\,3p^2({}^3_2P)\,{}^3P\,3d\,{}^4F$	5 544 364
23	$3s^2\,3p^2({}^3_2P)\,{}^3P\,3d\,{}^2D_{3/2}$	0.46 + 0.22 $3s^2\,3p^2({}^1_2D)\,{}^1D\,3d\,{}^2P$ + 0.14 $3s^2\,3p^2({}^1_2D)\,{}^1D\,3d\,{}^2D$	5 614 260
24	$3s^2\,3p^2({}^1_2D)\,{}^1D\,3d\,{}^2S_{1/2}$	0.36 + 0.31 $3s^2\,3p^2({}^3_2P)\,{}^3P\,3d\,{}^2P$ + 0.23 $3s^2\,3p^2({}^1_2D)\,{}^1D\,3d\,{}^2P$	5 645 215
25	$3s^2\,3p^3({}^2_1P)\,{}^2P^{\circ}_{3/2}$	0.51 + 0.28 $3s^2\,3p^3({}^2_3D)\,{}^2D^{\circ}$ + 0.20 $3s^2\,3p^3({}^4_3S)\,{}^4S^{\circ}$	5 738 961
26	$3s\,3p^3({}^4_3S)\,{}^5S\,3d\,{}^6D^{\circ}_{5/2}$	0.19 + 0.20 $3s^2\,3p^2({}^3_2P)\,{}^3P\,3d\,{}^4F^{\circ}$ + 0.11 $3s\,3p^3({}^2_1P)\,{}^3P\,3d\,{}^4F^{\circ}$	5 975 676
27	$3s\,3p^3({}^2_1P)\,{}^3P\,3d\,{}^4D^{\circ}_{3/2}$	0.20 + 0.23 $3s\,3p^3({}^4_3S)\,{}^5S\,3d\,{}^6D^{\circ}$ + 0.14 $3s\,3p^3({}^2_1P)\,{}^3P\,3d\,{}^4P^{\circ}$	5 993 157
28	$3s\,3p^3({}^2_3D)\,{}^3D\,3d\,{}^4P^{\circ}_{1/2}$	0.06 + 0.39 $3s\,3p^3({}^2_1P)\,{}^3P\,3d\,{}^4P^{\circ}$ + 0.29 $3s\,3p^3({}^4_3S)\,{}^5S\,3d\,{}^6D^{\circ}$	6 011 091
29	$3s\,3p^3({}^2_1P)\,{}^3P\,3d\,{}^4F^{\circ}_{7/2}$	0.23 + 0.17 $3s\,3p^3({}^2_1P)\,{}^3P\,3d\,{}^2F^{\circ}$ + 0.16 $3s\,3p^3({}^4_3S)\,{}^5S\,3d\,{}^6D^{\circ}$	6 061 088
30	$3s\,3p^3({}^2_3D)\,{}^3D\,3d\,{}^4F^{\circ}_{3/2}$	0.12 + 0.20 $3s\,3p^3({}^2_1P)\,{}^1P\,3d\,{}^2D^{\circ}$ + 0.11 $3s\,3p^3({}^2_1P)\,{}^3P\,3d\,{}^4P^{\circ}$	6 210 048
31	$3s\,3p^3({}^2_3D)\,{}^3D\,3d\,{}^4G^{\circ}_{5/2}$	0.20 + 0.21 $3s\,3p^3({}^2_1P)\,{}^1P\,3d\,{}^2F^{\circ}$ + 0.11 $3s^2\,3p\,3d^2({}^3_2F)\,{}^4G^{\circ}$	6 299 645
32	$3s\,3p^3({}^2_3D)\,{}^3D\,3d\,{}^4D^{\circ}_{1/2}$	0.11 + 0.25 $3s\,3p^3({}^2_1P)\,{}^1P\,3d\,{}^2P^{\circ}$ + 0.17 $3s\,3p^3({}^4_3S)\,{}^3S\,3d\,{}^4D^{\circ}$	6 335 321
33	$3s^2\,3p\,3d^2({}^3_2F)\,{}^4G^{\circ}_{5/2}$	0.42 + 0.15 $3s^2\,3p\,3d^2({}^1_2D)\,{}^2F^{\circ}$ + 0.13 $3s^2\,3p\,3d^2({}^3_2F)\,{}^2F^{\circ}$	6 551 091
34	$3s\,3p^3({}^2_3D)\,{}^3D\,3d\,{}^4G^{\circ}_{9/2}$	0.10 + 0.47 $3s\,3p^3({}^2_1P)\,{}^3P\,3d\,{}^4F^{\circ}$ + 0.30 $3s\,3p^3({}^4_3S)\,{}^5S\,3d\,{}^6D^{\circ}$	6 636 519
35	$3s\,3p^3({}^2_3D)\,{}^1D\,3d\,{}^2D^{\circ}_{5/2}$	0.05 + 0.20 $3s\,3p^3({}^2_1P)\,{}^3P\,3d\,{}^4P^{\circ}$ + 0.16 $3s\,3p^3({}^2_1P)\,{}^3P\,3d\,{}^2D^{\circ}$	6 771 988
36	$3s\,3p^3({}^4_3S)\,{}^5S\,3d\,{}^6D^{\circ}_{7/2}$	0.14 + 0.23 $3s\,3p^3({}^2_1P)\,{}^3P\,3d\,{}^2F^{\circ}$ + 0.19 $3s\,3p^3({}^2_1P)\,{}^3P\,3d\,{}^4D^{\circ}$	6 832 810
37	$3s\,3p^3({}^2_3D)\,{}^3D\,3d\,{}^2D^{\circ}_{3/2}$	0.07 + 0.19 $3s\,3p^3({}^2_1P)\,{}^3P\,3d\,{}^2D^{\circ}$ + 0.17 $3s\,3p^3({}^4_3S)\,{}^5S\,3d\,{}^4D^{\circ}$	6 845 169
38	$3s^2\,3p\,3d^2({}^3_2P)\,{}^4D^{\circ}_{1/2}$	0.36 + 0.26 $3s^2\,3p\,3d^2({}^1_0S)\,{}^2P^{\circ}$ + 0.12 $3s^2\,3p\,3d^2({}^3_2P)\,{}^2P^{\circ}$	6 891 509
39	$3s^2\,3p\,3d^2({}^3_2F)\,{}^4F^{\circ}_{3/2}$	0.30 + 0.25 $3s^2\,3p\,3d^2({}^3_2F)\,{}^2D^{\circ}$ + 0.07 $3s^2\,3p\,3d^2({}^3_2F)\,{}^4D^{\circ}$	6 929 835
40	$3s\,3p^3({}^2_1P)\,{}^1P\,3d\,{}^2F^{\circ}_{7/2}$	0.25 + 0.16 $3s\,3p^3({}^4_3S)\,{}^3S\,3d\,{}^4D^{\circ}$ + 0.11 $3s\,3p^3({}^2_3D)\,{}^3D\,3d\,{}^4G^{\circ}$	6 971 178
41	$3s\,3p^4({}^3_2P)\,{}^4P_{3/2}$	0.70 + 0.16 $3s\,3p^4({}^1_2D)\,{}^2D$ + 0.03 $3s^2\,3p^2({}^3_2P)\,{}^3P\,3d\,{}^4D$	6 998 090
42	$3s\,3p^3({}^2_1P)\,{}^3P\,3d\,{}^2P^{\circ}_{1/2}$	0.36 + 0.27 $3s\,3p^3({}^4_3S)\,{}^5S\,3d\,{}^4D^{\circ}$ + 0.07 $3s\,3p^3({}^2_1P)\,{}^3P\,3d\,{}^4D^{\circ}$	7 042 278
43	$3s\,3p^3({}^2_1P)\,{}^1P\,3d\,{}^2D^{\circ}_{5/2}$	0.17 + 0.12 $3s\,3p^3({}^2_1P)\,{}^3P\,3d\,{}^2F^{\circ}$ + 0.09 $3s\,3p^3({}^2_3D)\,{}^3D\,3d\,{}^4F^{\circ}$	7 084 012
44	$3s\,3p^3({}^2_1P)\,{}^1P\,3d\,{}^2P^{\circ}_{3/2}$	0.18 + 0.13 $3s\,3p^3({}^2_3D)\,{}^3D\,3d\,{}^4P^{\circ}$ + 0.11 $3s\,3p^3({}^4_3S)\,{}^3S\,3d\,{}^2D^{\circ}$	7 100 823
45	$3s\,3p^4({}^1_2D)\,{}^2D_{5/2}$	0.63 + 0.24 $3s\,3p^4({}^3_2P)\,{}^4P$ + 0.04 $3s^2\,3p^2({}^3_2P)\,{}^3P\,3d\,{}^4D$	7 144 999
46	$3s^2\,3p\,3d^2({}^3_2F)\,{}^4G^{\circ}_{7/2}$	0.55 + 0.14 $3s^2\,3p\,3d^2({}^3_2F)\,{}^2G^{\circ}$ + 0.13 $3s^2\,3p\,3d^2({}^3_2F)\,{}^4F^{\circ}$	7 224 945

5. Conclusions

In this paper, a new version of the JJ2LSJ program, consistent with the approach described in [7–10], is presented. The program performs the transformation of ASFs from a jj-to LSJ-coupling and provides the option to assign all level unique labels. Examples of the program use and explanations of possible options are given. In the paper, a few cases (Si-like Sr and P-like W) where the problem with unique labeling in energy spectra occur, are discussed and new labels are assigned.

The program is freely distributed. It may be obtained from the corresponding author.

Acknowledgments: Computations were performed on resources at the High Performance Computing Center "HPC Sauletekis" in Vilnius University Faculty of Physics.

Author Contributions: Gediminas Gaigalas developed the theory and created algorithms, performed programing work. Charlotte Froese Fischer created algorithms, performed programing work. Pavel Rynkun and Per Jonsson performed the calculations, tested the program. All authors wrote the paper.

Conflicts of Interest: The authors declare no conflict of interest.

References

1. Rudzikas, Z.B. *Theoretical Atomic Spectroscopy*; Cambridge University Press: Cambridge, UK, 2007.
2. Fischer, C.F.; Godefroid, M.R.; Brage, T.; Jönsson, P.; Gaigalas, G. Advanced multiconfiguration methods for complex atoms: I. Energies and wave functions. *J. Phys. B At. Mol. Opt. Phys.* **2016**, *49*, 182004.
3. Jönsson, P.; Gaigalas, G.; Bieroń, J.; Fischer, C.F.; Grant, I.P. New version: Grasp2K relativistic atomic structure package. *Comput. Phys. Commun.* **2013**, *184*, 2197–2203.
4. Jönsson, P.; Radžiūtė, L.; Gaigalas, G.; Godefroid, M.R.; Marques, J.P.; Brage, T.; Fischer, C.F.; Grant, I.P. Accurate multiconfiguration calculations of energy levels, lifetimes, and transition rates for the silicon isoelectronic sequence Ti IX - Ge XIX, Sr XXV, Zr XXVII, Mo XXIX. *Astron. Astrophys.* **2016**, *585*, A26.
5. Gaigalas, G.; Rynkun, P.; Fischer, C.F. Lifetimes of $4p^5 4d$ levels in highly ionized atoms. *Phys. Rev. A* **2015**, *91*, 022509.
6. Gaigalas, G.; Žalandauskas, T.; Rudzikas, Z. Analytical expressions for special cases of LS-jj transformation matrices for a shell of equivalent electrons. *Lith. J. Phys.* **2001**, *41*, 226–231.
7. Gaigalas, G.; Žalandauskas, T.; Rudzikas, Z. LS-jj transformation matrices for a shell of equivalent electrons. *At. Data Nucl. Data Tables* **2003**, *84*, 99–190.
8. Gaigalas, G.; Rudzikas, Z.; Fischer, C.F. Reduced coefficients (subcoefficients) of fractional parentage for $p-$, $d-$, and $f-$ shells. *At. Data Nucl. Data Tables* **1998**, *70*, 1–39.
9. Gaigalas, G.; Fritzsche, S.; Rudzikas, Z. Reduced coefficients of fractional parentage and matrix elements of the tensor $W^{(k_q k_j)}$ in jj-coupling. *At. Data Nucl. Data Tables* **2000**, *76*, 235–269.
10. Gaigalas, G.; Rudzikas, Z.; Fischer, C.F. An efficient approach for spin - angular integrations in atomic structure calculations. *J. Phys. B At. Mol. Opt. Phys.* **1997**, *30*, 3747–3771.
11. Fischer, C.F.; Tachiev, G. Breit-Pauli energy levels, lifetimes, and transition probabilities for the beryllium-like to neon-like sequences. *At. Data Nucl. Data Tables* **2004**, *87*, 1–184.
12. Fischer, C.F.; Gaigalas, G. Multiconfiguration Dirac-Hartree-Fock energy levels and transition probabilities for W XXXVIII. *Phys. Rev. A* **2012**, *85*, 042501.
13. Gaigalas, G.; Žalandauskas, T.; Fritzsche, S. Spectroscopic LSJ notation for atomic levels obtained from relativistic calculations. *Comput. Phys. Commun.* **2004**, *157*, 239–253.
14. Fischer, C.F.; Tachiev, G.; Gaigalas, G.; Godefroid, M.R. An MCHF atomic-structure package for large-scale calculations. *Comput. Phys. Commun.* **2007**, *176*, 559–579.
15. Kramida, A.E.; Ralchenko, Yu.; Reader, J.; NIST ASD Team. NIST Atomic Spectra Database (ver. 5.3), [Online]. National Institute of Standards and Technology: Gaithersburg, MD, USA, 2015. Available online: http://physics.nist.gov/asd (accessed on 2 December 2016).
16. Gaigalas, G.; Jönsson, P.; Rynkun, P. MCDHF and RCI calculations for P-like ions. **2016**, in preparation.

atoms

MDPI

Article

Core Effects on Transition Energies for $3d^k$ Configurations in Tungsten Ions

Charlotte Froese Fischer [1,*,†], Gediminas Gaigalas [2,†] and Per Jönsson [3,†]

1 Department of Computer Science, University of British Columbia, Vancouver V6T 1Z4, BC, Canada
2 Vilnius University, Institute of Theoretical Physics and Astronomy, Saulėtekio av. 3, LT-10222 Vilnius, Lithuania; gediminas.gaigalas@tfai.vu.lt
3 Materials Science and Applied Mathematics, Malmö University, SE-205 06 Malmö, Sweden; per.jonsson@mah.se
* Correspondence: cff@cs.ubc.ca
† These authors contributed equally to this work.

Academic Editor: Joseph Reader
Received: 20 December 2016; Accepted: 26 January 2017; Published: 8 February 2017

Abstract: All energy levels of the $3d^k$, $k = 1,2,\ldots,8,9$, configurations for tungsten ions, computed using the GRASP2K fully relativistic code based on the variational multiconfiguration Dirac–Hartree–Fock method, are reported. Included in the calculations are valence correlation where all $3s, 3p, 3d$ orbitals are considered to be valence orbitals, as well as core–valence and core–core effects from the $2s, 2p$ subshells. Results are compared with other recent theory and with levels obtained from the wavelengths of lines observed in the experimental spectra. It is shown that the core correlation effects considerably reduce the disagreement with levels linked directly to observed wavelengths, but may differ significantly from the NIST levels, where an unknown shift of the levels could not be determined from experimental wavelengths. For low values of k, levels were in good agreement with relativistic many-body perturbation levels, but for $2 < k < 8$, the present results were in better agreement with observation.

Keywords: core correlation effects; energy levels; multiconfiguration Dirac-Hartree-Fock; tungsten ions

1. Introduction

Because of their importance for the ITER project [1], spectra of tungsten ions have recently received much attention over a wide range of wavelengths. Of special interest are the NIST EBIT experiments reported by Ralchenko et al. [2], who studied tungsten ions with the ground states $3d$, $3d^2, \ldots, 3d^8$, and $3d^9$. Detailed collisional-radiative modelling was undertaken to identify the measured spectral lines. For the modelling they relied on energy levels, radiative transition probabilities, and electron-impact collisional cross-sections obtained using the relativistic Flexible Atomic Code (FAC) [3]. They found that many of the strong lines arose from magnetic dipole (M1) transitions. These lines were located in a narrow range of wavelengths, mostly well isolated with line ratios that could infer plasma properties, and were sensitive to electron densities. All these features make the M1 lines useful for plasma diagnostics. The measured observed wavelengths for M1 transitions and the FAC energy levels were analyzed by Kramida [4] for spectra for these ions, and form the basis for the energy levels included in the Atomic Spectra Database (ASD) [5].

At the same time, highly charged ions are of special interest for theory in that both correlation and relativistic effects are interrelated, and additional quantum electrodynamic (QED) corrections are needed for accurate results. Quinet [6] reports an extensive summary of a large variety of theoretical energy levels and forbidden transitions for all levels of $3d^k$ ground configurations, and compared their energy levels with the NIST energies. Included among the various methods were results that he obtained using the GRASP code developed by Norrington [7]. Most of the correlation included in the calculation was valence correlation restricted to the n = 3 complex. More recently, Guo et al. [8] computed energy levels, wavelengths, and transition probabilities for the same configurations for a number of ions, including tungsten. The theoretical basis for their work was the relativistic many-body perturbation theory (RMBPT) as described in [9], but small corrections for finite nuclear size, nuclear recoil, vacuum polarization, and self-energy correction were also included using standard procedures such as those in GRASP2K [10]. All basis orbitals were determined from the same central field, and all three types of correlation—valence–valence (VV), core–valence (CV), and core–core (CC)—where the core consists of the the full $1s, 2s, 2p$ core were included . Statistically, their energy levels were in much better agreement with NIST values than those of Quinet [6].

The purpose of the present work was to evaluate the accuracy of energy levels obtained from variational multconfiguration Dirac–Hartree–Fock methods as implemented in the GRASP2K code [10]. Included are all three correlation types as in the RMBPT calculation—except for the $1s^2$ core, that will be assumed to be inactive.

2. Multiconfiguration Dirac–Hartree–Fock (MCDHF) and Configuration Interaction Methods

In the MCDHF method [11,12], as implemented in the GRASP2K program package [10], the wave function $\Psi(\gamma PJM_J)$ for a state labeled γPJM_J, where J and M_J are the angular quantum numbers and P is the parity, is expanded in antisymmetrized and coupled configuration state functions (CSFs)

$$\Psi(\gamma PJM_J) = \sum_{j=1}^{M} c_j \Phi(\gamma_j PJM_J). \tag{1}$$

The labels $\{\gamma_j\}$ denote other appropriate information about the CSFs, such as orbital occupancy and coupling of the subshells. The CSFs are built from products of one-electron orbitals, having the general form

$$\psi_{n\kappa,m}(\mathbf{r}) = \frac{1}{r} \begin{pmatrix} P_{n\kappa}(r)\chi_{\kappa,m}(\theta,\varphi) \\ \imath Q_{n\kappa}(r)\chi_{-\kappa,m}(\theta,\varphi) \end{pmatrix}, \tag{2}$$

where $\chi_{\pm\kappa,m}(\theta,\varphi)$ are two-component spin–orbit functions. The radial functions $\{P_{n\kappa}(r), Q_{n\kappa}(r)\}$ are represented numerically on a grid.

Wave functions for a number of targeted states are determined simultaneously in the extended optimal level (EOL) scheme. Given initial estimates of the radial functions, the energies E and expansion coefficients $\mathbf{c} = (c_1, \ldots, c_M)^t$ for the targeted states are obtained as solutions to the configuration interaction (CI) problem

$$\mathbf{Hc} = E\mathbf{c}, \tag{3}$$

where $\mathbf{H}$ is the CI matrix of dimension $M \times M$ with elements

$$H_{ij} = \langle \Phi(\gamma_i PJM_J)|H|\Phi(\gamma_j PJM_J)\rangle. \tag{4}$$

Radial functions are solutions of systems of differential equations that define a stationary state of an energy functional for a wave function expansion.

Two types of expansions may be used. In the past, both usually were the same, but for large calculations, there are advantages to relaxing this restraint. The first is the expansion that determines the radial functions using the RMCDHF program of the GRASP2K package. For occupied orbitals, optimized radial functions can be obtained by applying the variational principal of an energy expression. However, when correlation orbitals are to be determined, the most effective orbitals are those that are in the same region of space as the occupied orbitals for a given type of correlation, as has been shown in partitioned configuration interaction (PCFI) studies [13]. In this work, we consider two regions: the $3s, 3p, 3d$ region for valence–valence (VV) correlation and the $2s, 2p$ region for core–valence (CV) and core–core (CC) correlations.

The second is an expansion for the relativistic configuration interaction (RCI) program that determines the wavefunction and its associated energy for a given Hamiltonian and based on a given orbital basis. In the present work, the Hamiltonian for RCI was the Dirac–Coulomb Hamiltonian (DC) plus the transverse photon interaction (DCB), the vacuum polarization effects as accounted for by the Uehling potential, and electron self-energies as calculated with the screened hydrogenic formula [12,14], namely the DCBQ Hamiltonian. The RCI program is relatively simple to parallelize efficiently [15,16] using message passing. As a result, much larger expansions are possible for RCI calculations than RMCDHF ones that build the orbital basis. Present calculations were done with forty-eight (48) processors for the larger cases.

The computational procedure was essentially the same for all ions. The first step was to perform Dirac–Hartree–Fock (DHF) calculations (in the EOL approximation) for all states associated with the $3s^2 3p^6 3d^k$ configuration. This calculation determined the $1s, 2s, 2p$ orbitals for all subsequent calculations. Then, sequentially, orbital sets of increasing size, with maximum principal quantum numbers $n = 3, 4, 5$, were determined from expansions that defined valence–valence correlation expansions. The latter were obtained from single- and double-excitations from the valence shells to those of the orbital set. Since the $3d$ shell is unfilled, excitations such as $3s^2 \rightarrow 3d^2$ are allowed and increase the generalized occupation number for the $3d$ orbitals but decrease those of $3s$. Variational methods determined the new orbitals introduced at each stage using the Dirac–Coulomb Hamiltonian. The $n = 6$ orbitals were targeted for core correlation effects. They were obtained from calculations that included CV correlation from the $n = 2$ shell where one orbital from the active core (either $2s$ or $2p$) and one $3s, 3p$, or $3d$ orbital were excited, as well as CC, where two $n = 2$ orbitals were excited. At the same time, excitations from $3s, 3p$ subshells were limited to single excitations for $3s$ or $3p$, thereby contracting the $n = 6$ orbitals to overlap more strongly with the $n = 2$ orbitals and reducing the size of the expansions. For the configurations $3d^k$, $k = 3, 4, 5, 6, 7$, the expansions were still exceedingly large and additional restrictions on interactions were imposed that define the energy functional. First, what might be considered a zero-order approximation was obtained that consisted of the CSFs of the $n = 5$ VV expansion that accounted for 99.9 percent of the normalized expansion. All other terms of the $n = 6$ expansions were treated as first-order corrections. In deriving the energy expression that determines the radial factors of the $n = 6$ orbitals, it was assumed that the interaction between CSFs of the first-order corrections could be neglected. This procedure optimizes the interaction of the $n = 6$ orbitals with the zero-order wave function, and has the effect of contracting the core–valence orbitals.

Each of these four orbital sets were then used in relativistic configuration interaction (RCI) calculations that included VV, CV, and CC correlation effects (excluding the $1s$ shell) for the three Hamiltonians—DC, DCB, and DCBQ. Again, for the cases where $k = 3, 4, 5, 6, 7$, the RCI calculations were performed under the assumption that interactions between CSF of the first-order correction could be ignored.

Table 1 summarizes the size of various expansions for the different $3d^k$ configurations, whereas Table 2 shows how the mean radii of the $n = 6$ orbitals are contracted relative to the valence correlation orbitals. Note that the size increases rapidly as the number of electrons (or holes) increases from one to five, as well as the number of J values and levels. The number of CSFs defining 99.9% of the wave function composition is relatively small. Increasing this percentage to 99.99% would include some higher order corrections. As for mean radii, it should be noted the the $3d$ orbitals (in non-relativistic notation) have a mean radius closer to the core than either $3s$ or $3p$. Listed in Table 2 are typical values for the $3d^5$ configuration. The mean radii are also depicted graphically in Figure 1. Correlation increases the generalized orbital occupation number of the $3d$ orbitals, but decreases those of all other occupied orbitals. The $n = 4$ and $n = 5$ orbitals have mean radii similar to those of the valence orbitals, whereas the $n = 6$ orbitals that are used to represent CC and CV correlation have mean radii either similar to $n = 2$ orbitals or between $n = 2$ and $n = 3$, as in CV correlation.

Table 1. Table showing the size (M) of the $n = 6$ relativistic configuration interaction (RCI) expansions and the size of the zero-order space (m) for the different tungsten ions.

J	M	m	J	M	m
$3d$			$3d^9$		
3/2	103 104	-	3/2	152 230	-
5/2	130 021	-	5/2	193 718	-
$3d^2$			$3d^8$		
0	109 376	-	0	138 241	-
1	306 873	-	1	388 664	-
2	453 546	-	2	576 194	-
3	526 871	-	3	672 708	-
4	529 065	-	4	679 881	-
$3d^3$			$3d^7$		
1/2	508 854	514	1/2	584 675	734
3/2	934 941	1 056	3/2	1 075 476	1 564
5/2	1 217 067	1 062	5/2	1 402 693	1 563
7/2	1 328 694	668	7/2	1 535 467	1 020
9/2	1 281 840	737	9/2	1 486 446	1 055
11/2	2216460	277	11/2	1 300 160	353
$3d^4$			$3d^6$		
0	433 540	925	0	462 613	1 113
1	1 228 917	1 070	1	1 311 786	1 244
2	1 840 515	1 688	2	1 965 798	2 071
3	2 187 525	1 375	3	2 338 660	1 738
4	2 261 243	1 624	4	2 420 366	1 921
5	2 095 354	632	5	2 246 438	761
6	1 771 535	572	6	1 902 774	659
$3d^5$					
1/2	1 022 700	1 119			
3/2	1 888 910	1 688			
5/2	2 480 422	2 352			
7/2	2 741 429	1 857			
9/2	2 687 207	1 306			
11/2	2 387 571	910			
13/2	1 943 915	329			

Table 2. Mean radii in a.u. of orbitals for the $3d^5$ configuration and their generalized occupation number w.

nl	$\langle nl\|r\|nl\rangle$	w
$1s$	1.83433D-02	2.00000
$2s$	7.64525D-02	1.99992
$2p_-$	6.33222D-02	1.99986
$2p$	7.10859D-02	3.99969
$3s$	1.91692D-01	1.99940
$3p_-$	1.81324D-01	1.99853
$3p$	1.93743D-01	3.99577
$3d_-$	1.67488D-01	2.00137
$3d$	1.71346D-01	3.00266
$4s$	2.04509D-01	1.24D-04
$4p_-$	1.89988D-01	1.45D-04
$4p$	2.01490D-01	2.94D-04
$4d_-$	1.71036D-01	1.73D-04
$4d$	1.70979D-01	2.82D-04
$4f_-$	1.94058D-01	5.94D-04
$4f$	1.97398D-01	8.24D-04
$5s$	2.03090D-01	1.93D-05
$5p_-$	1.95387D-01	2.23D-05
$5p$	1.97508D-01	4.08D-05
$5d_-$	2.12303D-01	2.88D-05
$5d$	2.17420D-01	4.47D-05
$5f_-$	1.86560D-01	1.30D-05
$5f$	1.85984D-01	2.01D-05
$5g_-$	1.97882D-01	3.38D-05
$5g$	2.00859D-01	5.11D-05
$6s$	1.31230D-01	6.77D-06
$6p_-$	1.19574D-01	8.04D-06
$6p$	1.20726D-01	1.40D-05
$6d_-$	1.18546D-01	1.71D-05
$6d$	1.24725D-01	2.58D-05
$6f_-$	8.84520D-02	7.35D-06
$6f$	9.29611D-02	1.10D-05
$6g_-$	7.72823D-02	2.26D-06
$6g$	7.88248D-02	3.31D-06
$6h_-$	1.62256D-01	2.42D-06
$6h$	8.04121D-02	7.65D-07

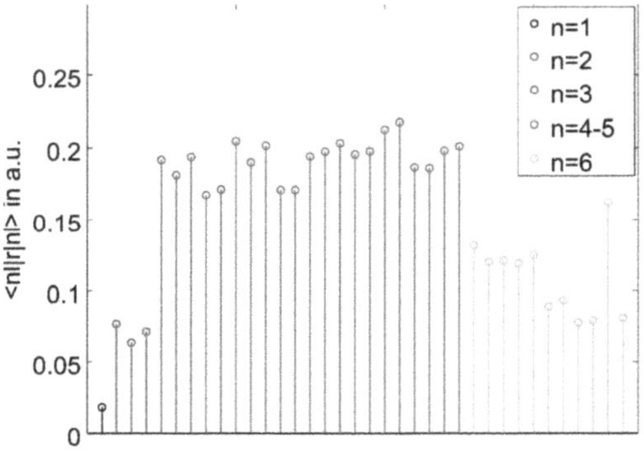

Figure 1. Plot of the mean radii of orbitals of the $3d^5$ configuration in the order listed in Table 2.

3. Results and Their Comparison

Table 3 reports some of the results for all levels of the $3d^k$ configurations of tungsten ions from RCI calculations for the DCBQ Hamiltonian. The classification of energy levels are presented in the LSJ- and jj-couplings. A set of three quantum numbers L, S, and seniority ν allows a one-to-one classification of $3d^k$ ($k = 3,4,5,6,7$) energy levels in LSJ-coupling. These quantum numbers are presented in Table 3 as ${}^{(2S+1)}L^{\nu}$. The $n = 5$ results include only VV correlation, whereas $n = 6$ include all three correlation effects. The next column is the energy levels as reported by NIST [5]. Included here are the different types of results. Energies with no square brackets are directly related to observed wavelengths—often these are in the lower portion of the spectrum. Then, there are levels that may be linked to an observed wavelength but the shift of the energy levels relative to the ground state is not known from experiment. These levels include a $+x$ or $+y$ in the table. Thus, the difference between two levels with the same $+x$ is known accurately, but not the levels themselves. Taking these factors into account, it is clear that the inclusion of core effects has reduced the discrepancy with NIST values by about a factor of 1/2. In the next column, the values found by Quinet [6] are generally like the VV results. From a general theoretical point of view, the the RMBPT results of Guo et al. [8] should be the most accurate. In the case of $3d^2$, RMBPT results have also been reported by Safronova and Safronova [17], and are reported in the last column. These results are not as accurate as those of Guo et al. In these tables, all energies are reported in the units of 1000 cm^{-1}.

Table 3. Energy level results for $3d, 3d^2, \ldots, 3d^8, 3d^9$ ground configuration of tungsten ions. Shown is a unique label in LSJ- and jj-notation, the J value, the present n = 5 result for valence–valence (VV) correlation, and n = 6 result for all three types of correlation, the Atomic Spectra Database (ASD) value [5], the Quinet value [6], the Guo et al. RMBPT$_g$ value [8], and the Safronova & Safronova RMBPT$_s$ value [17]. All energy levels are reported in 1000 cm^{-1}.

Label			J	Present Work		ASD	GRASP	RMBPT$_g$	RMBPT$_s$
LSJ-	jj-Couplings			$n=5$	$n=6$				
W^{55+} (K-like)									
$3d\ ^2D$	$3d_-$	(3/2,0)	3/2	0.00	0.00	0.00	0.00	0.00	
$3d\ ^2D$	$3d_+$	(0,5/2)	5/2	625.23	626.17	626.49	624.7	626.56	
W^{54+} (Ca-like)									
$3d^2\ ^3F$	$3d_-^2$	(2,0)	2	0.00	0.00	0.00	0.00	0.00	0.00
$3d^2\ ^3P$	$3d_-^2$	(0,0)	0	186.42	186.23	[188]	186.9	184.86	187.11
$3d^2\ ^3F$	$3d_-3d_+$	(3/2,5/2)	3	584.05	584.75	585.48	583.5	585.80	582.85
$3d^2\ ^3P$	$3d_-3d_+$	(3/2,5/2)	2	667.45	667.96	668.49	667.6	668.00	666.21
$3d^2\ ^3P$	$3d_-3d_+$	(3/2,5/2)	1	706.35	706.75	709.46+x	707.1	706.78	705.41
$3d^2\ ^1G$	$3d_-3d_+$	(3/2,5/2)	4	695.68	696.10	[697]	697.1	696.74	693.81
$3d^2\ ^3F$	$3d_+^2$	(0,4)	4	1234.31	1235.57	[1234]	1234.1	1237.00	1231.64
$3d^2\ ^3P$	$3d_+^2$	(0,2)	2	1298.91	1300.18	[1299]	1298.6	1300.28	1296.73
$3d^2\ ^1S$	$3d_+^2$	(0,0)	0	1492.04	1493.71	[1493]	1491.0	1491.18	1491.54
W^{53+} (Sc-like)									
$3d^3\ ^4F^3$	$3d_-^3$	(3/2,0)	3/2	0.00	0.00	0.00	0.00	0.00	
$3d^3\ ^4F^3$	$3d_-^2 3d_+$	(2,5/2)	5/2	528.39	529.07	530.03	528.2	530.51	
$3d^3\ ^4P^3$	$3d_-^2 3d_+$	(2,5/2)	3/2	579.43	579.99	580.86	579.9	580.86	
$3d^3\ ^2G^3$	$3d_-^2 3d_+$	(2,5/2)	7/2	610.41	610.86	[610]	611.7	611.86	
$3d^3\ ^4P^3$	$3d_-^2 3d_+$	(2,5/2)	1/2	622.72	623.22	623.95	623.6	623.53	
$3d^3\ ^2H^3$	$3d_-^2 3d_+$	(2,5/2)	9/2	609.94	610.32	[610]+x	612.0	611.62	
$3d^3\ ^2D^1$	$3d_-^2 3d_+$	(0,5/2)	5/2	811.84	812.07	812.22	814.2	811.77	
$3d^3\ ^4F^3$	$3d_-3d_+^2$	(3/2,4)	7/2	1127.31	1128.60	[1126]	1127.1	1130.58	
$3d^3\ ^4F^3$	$3d_-3d_+^2$	(3/2,4)	9/2	1164.81	1165.99	[1164]	1165.7	1168.15	

Table 3. *Cont.*

Label			J	Present Work		ASD	GRASP	RMBPT$_g$	RMBPT$_s$
LSJ-	jj-Couplings			$n=5$	$n=6$				
$3d^3\,{}^4P^3$	$3d_-3d_+^2$	(3/2,2)	3/2	1206.41	1207.73	[1206]	1206.2	1208.34	
$3d^3\,{}^2P^3$	$3d_-3d_+^2$	(3/2,2)	1/2	1230.34	1231.58	[1230]	1230.5	1232.08	
$3d^3\,{}^2D^3$	$3d_-3d_+^2$	(3/2,4)	5/2	1243.67	1244.61	[1244]	1245.0	1245.39	
$3d^3\,{}^2H^3$	$3d_-3d_+^2$	(3/2,4)	11/2	1242.38	1243.30	1243.51+x	1245.2	1245.42	
$3d^3\,{}^2F^3$	$3d_-3d_+^2$	(3/2,2)	5/2	1314.58	1315.54	[1315]	1316.4	1315.84	
$3d^3\,{}^2F^3$	$3d_-3d_+^2$	(3/2,2)	7/2	1318.68	1319.55	[1320]	1321.5	1320.10	
$3d^3\,{}^2D^1$	$3d_-3d_+^2$	(3/2,0)	3/2	1479.96	1481.26	[1482]	1481.3	1479.89	
$3d^3\,{}^2G^3$	$3d_+^3$	(0,9/2)	9/2	1762.93	1764.86		1762.9	1767.02	
$3d^3\,{}^2P^3$	$3d_+^3$	(0,3/2)	3/2	1876.44	1878.32		1877.0	1878.54	
$3d^3\,{}^2D^1$	$3d_+^3$	(0,5/2)	5/2	1958.00	1960.12		1957.9	1959.56	
W^{52+} (Ti-like)									
$3d^4\,{}^3P^2$	$3d_-^4$	(0,0)	0	0.00	0.00	0.00	0.00	0.00	
$3d^4\,{}^5D^4$	$3d_-^3\,3d_+$	(3/2,5/2)	1	515.87	516.51	517.63	516.0	518.08	
$3d^4\,{}^3H^4$	$3d_-^3\,3d_+$	(3/2,5/2)	4	613.24	613.54	[613]+y	615.6	614.79	
$3d^4\,{}^5D^4$	$3d_-^3\,3d_+$	(3/2,5/2)	2	637.98	638.39	[638]+x	639.9	639.34	
$3d^4\,{}^3F^2$	$3d_-^3\,3d_+$	(3/2,5/2)	3	665.84	666.09	665.5621+x	668.6	667.04	
$3d^4\,{}^5D^4$	$3d_-^2\,3d_+^2$	(2,2)	0	1101.86	1103.18	[1100]	1101.6	1104.66	
$3d^4\,{}^5D^4$	$3d_-^2\,3d_+^2$	(2,4)	2	1106.82	1107.98	1109.69	1107.6	1110.02	
$3d^4\,{}^3H^4$	$3d_-^2\,3d_+^2$	(2,4)	4	1125.54	1126.59	1127.27+y	1127.3	1129.11	
$3d^4\,{}^5D^4$	$3d_-^2\,3d_+^2$	(2,4)	3	1142.02	1143.02	[1141]	1144.0	1145.19	
$3d^4\,{}^3H^4$	$3d_-^2\,3d_+^2$	(2,4)	5	1172.24	1173.06	1173.35+y	1175.7	1175.60	
$3d^4\,{}^3D^4$	$3d_-^2\,3d_+^2$	(2,2)	1	1213.52	1214.54	[1213]	1215.4	1215.64	
$3d^4\,{}^1I^4$	$3d_-^2\,3d_+^2$	(2,4)	6	1195.60	1196.31	[1195]	1200.00	1199.02	
$3d^4\,{}^3F^4$	$3d_-^2\,3d_+^2$	(2,2)	3	1239.13	1239.92	[1240]	1242.5	1240.99	
$3d^4\,{}^3G^4$	$3d_-^2\,3d_+^2$	(2,2)	4	1242.41	1243.17	[1243]	1245.7	1244.47	
$3d^4\,{}^3F^4$	$3d_-^2\,3d_+^2$	(2,2)	2	1257.75	1258.62	[1258]	1260.6	1259.43	
$3d^4\,{}^3F^2$	$3d_-^2\,3d_+^2$	(2,0)	2	1359.28	1360.44	[1361]	1361.1	1360.35	
$3d^4\,{}^3F^2$	$3d_-^2\,3d_+^2$	(0,4)	4	1403.66	1404.22	1403.95+x	1408.6	1405.11	
$3d^4\,{}^1D^2$	$3d_-^2\,3d_+^2$	(0,2)	2	1505.68	1506.35	[1509]	1510.3	1505.82	
$3d^4\,{}^3P^4$	$3d_-^2\,3d_+^2$	(0,0)	0	1633.13	1634.15	[1637]	1636.5	1632.74	
$3d^4\,{}^5D^4$	$3d_-3d_+^3$	(3/2,9/2)	4	1714.26	1715.10		1715.3	1718.50	
$3d^4\,{}^3F^4$	$3d_-3d_+^3$	(3/2,9/2)	3	1725.24	1727.04		1725.9	1729.15	
$3d^4\,{}^3D^4$	$3d_-3d_+^3$	(3/2,3/2)	1	1766.70	1768.58		1767.1	1769.70	
$3d^4\,{}^3G^4$	$3d_-3d_+^3$	(3/2,9/2)	5	1773.76	1775.28		1776.4	1777.84	
$3d^4\,{}^3H^4$	$3d_-3d_+^3$	(3/2,9/2)	6	1778.76	1780.21		1782.4	1783.28	
$3d^4\,{}^3F^4$	$3d_-3d_+^3$	(3/2,3/2)	2	1841.18	1842.98		1842.9	1843.90	
$3d^4\,{}^3D^4$	$3d_-3d_+^3$	(3/2,3/2)	3	1857.70	1859.24		1860.2	1860.18	
$3d^4\,{}^1S^4$	$3d_-3d_+^3$	(3/2,3/2)	0	1922.88	1924.06		1925.8	1923.37	
$3d^4\,{}^3P^2$	$3d_-3d_+^3$	(3/2,5/2)	1	1983.87	1985.44		1987.2	1985.44	
$3d^4\,{}^3F^2$	$3d_-3d_+^3$	(3/2,5/2)	3	1979.96	1981.50		1983.6	1981.91	
$3d^4\,{}^1G^2$	$3d_-3d_+^3$	(3/2,5/2)	4	1985.00	1986.57		1988.7	1987.02	
$3d^4\,{}^1D^4$	$3d_-3d_+^3$	(3/2,5/2)	2	2018.63	2020.04		2022.8	2019.68	
$3d^4\,{}^3F^2$	$3d_+^4$	(0,4)	4	2376.23	2378.86		2376.1	2380.51	
$3d^4\,{}^1D^2$	$3d_+^4$	(0,2)	2	2460.51	2463.08		2461.4	2463.56	
$3d^4\,{}^3P^2$	$3d_+^4$	(0,0)	0	2662.74	2665.52		2663.5	2663.60	
W^{51+} (V-like)									
$3d^5\,{}^4P^3$	$3d_-^4\,3d_+$	(0,5/2)	5/2	0.00	0.00	0.00	0.00	0.00	
$3d^5\,{}^6S^5$	$3d_-^3\,3d_+^2$	(3/2,4)	5/2	469.71	470.75	71.63	469.1	472.03	

Table 3. *Cont.*

Label			J	Present Work		ASD	GRASP	RMBPT$_g$	RMBPT$_s$
LSJ-	jj-Couplings			$n=5$	$n=6$				
$3d^5\ ^4G^5$	$3d_-^3 3d_+^2$	(3/2,4)	7/2	564.98	565.80	66.25	566.2	566.41	
$3d^5\ ^4D^5$	$3d_-^3 3d_+^2$	(3/2,2)	3/2	579.61	580.50	80.89	579.8	580.44	
$3d^5\ ^2H^3$	$3d_-^3 3d_+^2$	(3/2,4)	11/2	576.03	576.78	[577]+x	578.5	577.80	
$3d^5\ ^2G^5$	$3d_-^3 3d_+^2$	(3/2,4)	9/2	620.92	621.61	[623]	623.7	622.20	
$3d^5\ ^4D^5$	$3d_-^3 3d_+^2$	(3/2,2)	5/2	650.71	651.45	[652]	652.8	651.27	
$3d^5\ ^4P^3$	$3d_-^3 3d_+^2$	(3/2,2)	1/2	679.60	680.38	[681]	680.8	679.83	
$3d^5\ ^2F^5$	$3d_-^3 3d_+^2$	(3/2,2)	7/2	687.73	688.28	88.18	690.9	687.90	
$3d^5\ ^2D^1$	$3d_-^3 3d_+^2$	(3/2,0)	3/2	823.99	824.95	[827]	825.5	823.60	
$3d^5\ ^6S^5$	$3d_-^2 3d_+^3$	(2,9/2)	5/2	1025.98	1027.97	[1015]	1024.9	1029.11	
$3d^5\ ^4D^5$	$3d_-^2 3d_+^3$	(2,9/2)	7/2	1096.84	1098.61	[1097]	1097.9	1099.59	
$3d^5\ ^4G^5$	$3d_-^2 3d_+^3$	(2,9/2)	11/2	1100.79	1102.51	1103.43	1103.0	1104.04	
$3d^5\ ^4G^5$	$3d_-^2 3d_+^3$	(2,9/2)	9/2	1116.98	1118.70	[1118]	1118.8	1119.70	
$3d^5\ ^4D^5$	$3d_-^2 3d_+^3$	(2,3/2)	1/2	1155.66	1157.40		1156.6	1157.55	
$3d^5\ ^4P^3$	$3d_-^2 3d_+^3$	(2,5/2)	3/2	1164.73	1166.79		1163.6	1166.64	
$3d^5\ ^2I^5$	$3d_-^2 3d_+^3$	(2,9/2)	13/2	1142.15	1143.78	[1143]	1145.6	1145.272	
$3d^5\ ^2F^5$	$3d_-^2 3d_+^3$	(2,3/2)	5/2	1174.89	1176.61		1176.3	1176.63	
$3d^5\ ^2H^3$	$3d_-^2 3d_+^3$	(2,5/2)	9/2	1217.34	1219.21		1218.4	1219.39	
$3d^5\ ^2G^5$	$3d_-^2 3d_+^3$	(2,3/2)	7/2	1237.88	1239.44		1240.9	1239.13	
$3d^5\ ^4F^3$	$3d_-^2 3d_+^3$	(2,5/2)	5/2	1254.59	1256.46		1255.8	1256.02	
$3d^5\ ^2D^5$	$3d_-^2 3d_+^3$	(2,3/2)	3/2	1259.49	1260.94		1262.1	1259.77	
$3d^5\ ^4P^3$	$3d_-^2 3d_+^3$	(2,5/2)	1/2	1308.19	1309.93		1309.8	1308.63	
$3d^5\ ^2G^3$	$3d_-^2 3d_+^3$	(2,5/2)	7/2	1307.82	1309.62		1309.9	1308.84	
$3d^5\ ^2G^3$	$3d_-^2 3d_+^3$	(0,9/2)	9/2	1379.66	1381.18		1383.8	1380.57	
$3d^5\ ^2P^3$	$3d_-^2 3d_+^3$	(0,3/2)	3/2	1504.94	1506.22		1510.4	1504.14	
$3d^5\ ^2D^1$	$3d_-^2 3d_+^3$	(0,5/2)	5/2	1533.17	1534.71		1537.4	1532.74	
$3d^5\ ^4P^3$	$3d_- 3d_+^4$	(3/2,4)	5/2	1660.92	1663.98		1658.7	1664.07	
$3d^5\ ^4F^3$	$3d_- 3d_+^4$	(3/2,4)	7/2	1733.68	1736.62		1733.1	1736.60	
$3d^5\ ^4D^5$	$3d_- 3d_+^4$	(3/2,2)	3/2	1759.25	1762.30		1758.1	1761.85	
$3d^5\ ^2H^3$	$3d_- 3d_+^4$	(3/2,4)	11/2	1746.45	1749.34		1747.2	1749.91	
$3d^5\ ^2G^5$	$3d_- 3d_+^4$	(3/2,4)	9/2	1806.21	1808.95		1807.7	1808.86	
$3d^5\ ^2D^3$	$3d_- 3d_+^4$	(3/2,2)	5/2	1843.82	1846.49		1844.6	1845.21	
$3d^5\ ^2G^3$	$3d_- 3d_+^4$	(3/2,2)	7/2	1871.70	1874.38		1874.1	1873.74	
$3d^5\ ^2P^3$	$3d_- 3d_+^4$	(3/2,2)	1/2	1933.91	1936.39		1937.0	1934.46	
$3d^5\ ^2D^1$	$3d_- 3d_+^4$	(3/2,0)	3/2	2063.04	2065.78		2065.5	2062.96	
$3d^5\ ^2D^1$	$3d_+^5$	(0,5/2)	5/2	2362.48	2366.70		2359.4	2365.33	
W^{50+} (Cr-like)									
$3d^6\ ^5D^4$	$3d_-^4 3d_+^2$	(0,4)	4	0.00	0.00	0.00	0.00	0.00	
$3d^6\ ^3D^4$	$3d_-^4 3d_+^2$	(0,2)	2	62.74	62.71	62.38	62.6	61.56	
$3d^6\ ^3P^2$	$3d_-^4 3d_+^2$	(0,0)	0	207.31	207.66	[208]+x	205.9	205.74	
$3d^6\ ^5D^4$	$3d_-^3 3d_+^3$	(3/2,9/2)	3	506.28	507.09	508.03	505.2	507.80	
$3d^6\ ^5D^4$	$3d_-^3 3d_+^3$	(3/2,9/2)	4	518.36	519.02	519.78	518.0	519.83	
$3d^6\ ^5D^4$	$3d_-^3 3d_+^3$	(3/2,3/2)	1	545.62	546.54	[545]	543.8	546.53	
$3d^6\ ^3G^4$	$3d_-^3 3d_+^3$	(3/2,9/2)	5	582.70	583.09	583.67	584.2	583.74	
$3d^6\ ^3H^4$	$3d_-^3 3d_+^3$	(3/2,9/2)	6	582.40	582.70	[583]	584.3	583.61	
$3d^6\ ^3F^4$	$3d_-^3 3d_+^3$	(3/2,3/2)	2	637.99	638.51	[639]	638.1	637.59	
$3d^6\ ^3D^4$	$3d_-^3 3d_+^3$	(3/2,3/2)	3	649.76	650.29	650.91	650.6	649.82	
$3d^6\ ^3P^4$	$3d_-^3 3d_+^3$	(3/2,3/2)	0	725.01	725.35	[729]	727.9	723.98	
$3d^6\ ^3P^2$	$3d_-^3 3d_+^3$	(3/2,5/2)	1	767.07	767.54	768.98+x	769.3	766.38	

Table 3. *Cont.*

Label			J	Present Work		ASD	GRASP	RMBPT$_g$	RMBPT$_s$
LSJ-	jj-Couplings			$n=5$	$n=6$				
$3d^6\ ^3D^4$	$3d_-^3 3d_+^3$	(3/2,5/2)	2	766.25	766.84	766.95	767.6	765.69	
$3d^6\ ^1G^2$	$3d_-^3 3d_+^3$	(3/2,5/2)	4	760.65	761.12	761.21	762.5	760.28	
$3d^6\ ^3F^2$	$3d_-^3 3d_+^3$	(3/2,5/2)	3	782.18	782.54	782.53	785.0	781.26	
$3d^6\ ^5D^4$	$3d_-^2 3d_+^4$	(2,4)	2	1058.57	1060.19		1055.6	1060.64	
$3d^6\ ^5D^4$	$3d_-^2 3d_+^4$	(2,2)	0	1083.07	1084.88		1079.6	1085.16	
$3d^6\ ^3H^4$	$3d_-^2 3d_+^4$	(2,4)	4	1108.16	1109.55		1106.9	1110.13	
$3d^6\ ^5D^4$	$3d_-^2 3d_+^4$	(2,4)	3	1135.23	1136.57		1134.6	1136.84	
$3d^6\ ^3H^4$	$3d_-^2 3d_+^4$	(2,4)	5	1142.11	1143.32		1142.4	1144.11	
$3d^6\ ^1I^4$	$3d_-^2 3d_+^4$	(2,4)	6	1169.18	1170.23		1170.5	1171.16	
$3d^6\ ^3F^4$	$3d_-^2 3d_+^4$	(2,2)	3	1196.79	1198.08		1197.0	1198.01	
$3d^6\ ^3D^4$	$3d_-^2 3d_+^4$	(2,2)	1	1217.26	1218.50		1217.8	1217.79	
$3d^6\ ^1G^4$	$3d_-^2 3d_+^4$	(2,2)	4	1232.82	1233.95		1234.1	1233.73	
$3d^6\ ^3F^4$	$3d_-^2 3d_+^4$	(2,2)	2	1243.66	1244.79		1244.0	1243.75	
$3d^6\ ^3F^2$	$3d_-^2 3d_+^4$	(2,0)	2	1336.95	1338.38		1336.9	1336.97	
$3d^6\ ^3F^2$	$3d_-^2 3d_+^4$	(0,4)	4	1374.79	1375.77		1376.9	1375.03	
$3d^6\ ^1D^2$	$3d_-^2 3d_+^4$	(0,2)	2	1518.97	1519.86		1523.2	1517.58	
$3d^6\ ^1S^0$	$3d_-^2 3d_+^4$	(0,0)	0	1660.58	1661.58		1664.9	1658.28	
$3d^6\ ^3P^2$	$3d_- 3d_+^5$	(3/2,5/2)	1	1663.26	1665.83		1657.7	1665.57	
$3d^6\ ^1G^2$	$3d_- 3d_+^5$	(3/2,5/2)	4	1764.33	1766.52		1762.0	1766.29	
$3d^6\ ^3P^2$	$3d_- 3d_+^5$	(3/2,5/2)	2	1813.76	1815.87		1811.7	1814.75	
$3d^6\ ^3F^2$	$3d_- 3d_+^5$	(3/2,5/2)	3	1831.23	1833.30		1830.3	1832.64	
$3d^6\ ^3P^2$	$3d_+^6$	(0,0)	0	2321.86	2325.36		2314.1	2323.82	
W^{49+} (Mn-like)									
$3d^7\ ^4F^3$	$3d_-^4 3d_+^3$	(0,9/2)	9/2	0.00	0.00	0.00	0.00	0.00	
$3d^7\ ^2P^3$	$3d_-^4 3d_+^3$	(0,3/2)	3/2	101.71	101.64	[103]+x	102.1	100.13	
$3d^7\ ^2D^1$	$3d_-^4 3d_+^3$	(0,5/2)	5/2	158.95	159.10	158.75	158.7	157.62	
$3d^7\ ^4F^3$	$3d_-^3 3d_+^4$	(3/2,4)	7/2	527.98	528.88	529.66	526.1	529.08	
$3d^7\ ^4F^3$	$3d_-^3 3d_+^4$	(3/2,4)	9/2	583.50	584.16	584.59	583.1	584.18	
$3d^7\ ^4P^3$	$3d_-^3 3d_+^4$	(3/2,2)	3/2	607.96	608.87	[608]	606.6	608.30	
$3d^7\ ^4P^3$	$3d_-^3 3d_+^4$	(3/2,4)	5/2	624.97	625.72	628.02+x	624.9	625.41	
$3d^7\ ^4P^3$	$3d_-^3 3d_+^4$	(3/2,2)	1/2	635.89	636.62	638.62+x	635.1	635.45	
$3d^7\ ^2H^3$	$3d_-^3 3d_+^4$	(3/2,4)	11/2	650.16	650.58	650.70	651.8	650.55	
$3d^7\ ^2F^3$	$3d_-^3 3d_+^4$	(3/2,2)	7/2	705.20	705.71	705.92	706.4	704.86	
$3d^7\ ^2F^3$	$3d_-^3 3d_+^4$	(3/2,2)	5/2	742.86	743.30	[747]	745.4	742.07	
$3d^7\ ^2D^1$	$3d_-^3 3d_+^4$	(3/2,0)	3/2	888.41	889.03	[893]	890.8	886.67	
$3d^7\ ^4F^3$	$3d_-^2 3d_+^5$	(2,5/2)	5/2	1115.46	1117.19		1112.0	1116.93	
$3d^7\ ^4P^3$	$3d_-^2 3d_+^5$	(2,5/2)	3/2	1147.62	1149.25		1145.1	1148.65	
$3d^7\ ^2P^3$	$3d_-^2 3d_+^5$	(2,5/2)	1/2	1192.13	1193.65		1189.9	1192.53	
$3d^7\ ^2H^3$	$3d_-^2 3d_+^5$	(2,5/2)	9/2	1185.68	1187.07		1184.9	1186.89	
$3d^7\ ^2F^3$	$3d_-^2 3d_+^5$	(2,5/2)	7/2	1210.79	1212.16		1210.3	1211.44	
$3d^7\ ^2D^1$	$3d_-^2 3d_+^5$	(0,5/2)	5/2	1410.07	1411.30		1411.0	1409.49	
$3d^7\ ^2D^1$	$3d_- 3d_+^6$	(3/2,0)	3/2	1751.87	1754.44		1746.4	1753.15	
W^{48+} (Fe-like)									
$3d^8\ ^3F$	$3d_-^4 3d_+^4$	(0,4)	4	0.00	0.00	0.00	0.00	0.00	
$3d^8\ ^1D$	$3d_-^4 3d_+^4$	(0,2)	2	72.15	72.12	[73.4]+x	72.8	71.26	
$3d^8\ ^3P$	$3d_-^4 3d_+^4$	(0,0)	0	229.94	230.10	[233]	230.7	228.17	
$3d^8\ ^3F$	$3d_-^3 3d_+^5$	(3/2,5/2)	3	525.18	526.07	526.65	523.2	526.13	
$3d^8\ ^3P$	$3d_-^3 3d_+^5$	(3/2,5/2)	2	600.38	601.15	603.12+x	599.7	600.69	

Table 3. *Cont.*

Label LSJ-	Label jj-Couplings		J	Present Work $n=5$	Present Work $n=6$	ASD	GRASP	RMBPT$_g$	RMBPT$_s$
$3d^8\,^3P$	$3d_-^3 3d_+^5$	(3/2,5/2)	1	642.01	642.71	644.76+x	642.7	642.14	
$3d^8\,^1G$	$3d_-^3 3d_+^5$	(3/2,5/2)	4	643.89	644.43	644.70	645.0	644.03	
$3d^8\,^3F$	$3d_-^2 3d_+^6$	(2,2)	2	1106.91	1108.59	[1106]	1103.6	1108.17	
$3d^8\,^1S$	$3d_-^2 3d_+^6$	(0,0)	0	1304.16	1305.75	[1306]	1301.7	1304.07	
W^{47+} (Co-like)									
$3d^9\,^2D$	$3d_-^4 3d_+^5$	(0,5/2)	5/2	0.00	0.00	0.00	0.00	0.00	
$3d^9\,^2D$	$3d_-^3 3d_+^6$	(3/2,0)	3/2	537.21	538.04	538.59	535.6	538.05	

The uncertainties of NIST energy levels not based on observed wavelengths are estimated as being less than 5000 cm^{-1}, or 5.00 in our table. In order to better understand the importance of various effects in Table 4, we report the NIST energy levels that are based on observation and differences of various theories for only those levels where NIST values are accurate, although there may be an unknown shift.

Table 4. Difference from NIST energy levels derived from observation. Shown is the *LS* label, the *J* value, the present $n=5$ result for VV correlation, and $n=6$ result for all three types of correlation, the ASD value [5], the Quinet value [6], the Guo et al. RMBPT$_g$ value [8], and the Safranova & Safronova RMBPT$_s$ value [17]. All energy levels are reported in 1000 cm^{-1}.

Label	J	Present Work $n=5$	Present Work $n=6$	ASD	GRASP	RMBPT$_g$	RMBPT$_s$
W^{55+} (K-like)							
$3d\,^2D$	3/2	0.00	0.00	0.00	0.00	0.00	
$3d\,^2D$	5/2	1.25	0.32	626.49	2.49	−0.07	
W^{54+} (Ca-like)							
$3d^2\,^3F$	2	0.00	0.00	0.00	0.00	0.00	0.00
$3d^2\,^3F$	3	1.43	0.73	585.48	1.98	−0.32	2.63
$3d^2\,^3P$	2	1.04	0.53	668.49	0.89	0.49	2.28
$3d^2\,^3P$	1	3.11	2.71	709.46+x	2.36	2.68	4.05
W^{53+} (Sc-like)							
$3d^3\,^4F3$	3/2	0.00	0.00	0.00	0.00	0.00	
$3d^3\,^4F3$	5/2	1.64	0.96	530.03	1.83	−0.48	
$3d^3\,^4P3$	3/2	1.43	0.87	580.86	0.96	0.0	
$3d^3\,^4P3$	1/2	1.23	0.73	623.95	0.35	0.42	
$3d^3\,^2D1$	5/2	0.38	0.15	812.22	−1.98	0.45	
$3d^3\,^2H3$	11/2	1.13	0.21	1234.51+x	−1.69	0.45	
W^{52+} (Ti-like)							
$3d^4\,^3P2$	0	0.00	0.00	0.00	0.00	0.00	
$3d^4\,^5D4$	1	1.76	1.12	517.63	1.63	−0.45	
$3d^4\,^3F2$	3	−0.28	−0.53	665.5621+x	−3.04	−1.48	
$3d^4\,^5D4$	2	2.87	1.71	1109.69	2.09	−0.33	
$3d^4\,^3H4$	4	1.73	0.68	1127.27+y	−0.03	−1.84	
$3d^4\,^3H4$	5	1.11	0.29	1173.35+y	−2.35	−22.25	
$3d^4\,^3F2$	4	0.29	−0.27	1403.95+x	−4.65	−1.16	

Table 4. *Cont.*

Label	J	Present Work $n=5$	$n=6$	ASD	GRASP	RMBPT$_g$	RMBPT$_s$
W^{51+} (V-like)							
$3d^5\ {}^4P^3$	5/2	0.00	0.00	0.00	0.00	0.00	
$3d^5\ {}^6S^5$	5/2	1.92	0.88	471.63	2.53	−0.40	
$3d^5\ {}^4G^5$	7/2	1.27	0.45	566.25	0.05	−0.16	
$3d^5\ {}^4D^5$	3/2	1.28	0.39	580.89	1.09	0.45	
$3d^5\ {}^2F^5$	7/2	0.45	−0.10	688.18	−2.72	0.28	
$3d^5\ {}^4G^5$	11/2	2.64	0.92	1103.43	0.43	−0.61	
W^{50+} (Cr-like)							
$3d^6\ {}^5D^4$	4	0.00	0.00	0.00	0.00	0.00	
$3d^6\ {}^3D^4$	2	−0.36	−0.29	62.38	−0.22	0.82	
$3d^6\ {}^5D^4$	3	1.75	1.04	508.03	2.83	0.23	
$3d^6\ {}^5D^4$	4	1.41	0.74	519.78	1.78	0.05	
$3d^6\ {}^3G^4$	5	0.97	0.44	583.67	−0.53	−0.07	
$3d^6\ {}^3D^4$	3	1.15	0.56	650.91	0.31	1.09	
$3d^6\ {}^3P^2$	1	1.91	1.44	768.98+x	−0.32	2.60	
$3d^6\ {}^3D^4$	2	0.70	0.11	766.95	−0.65	1.26	
$3d^6\ {}^1G^2$	4	0.56	−0.04	761.21	−1.29	0.93	
$3d^6\ {}^3F^2$	3	0.35	−0.08	782.53	−2.47	1.27	
W^{49+} (Mn-like)							
$3d^7\ {}^4F^3$	9/2	0.00	0.00	0.00	0.00	0.00	
$3d^7\ {}^2D^1$	5/2	−0.20	−0.35	158.75	0.05	1.13	
$3d^7\ {}^4F^3$	7/2	1.68	0.78	529.66	3.56	0.58	
$3d^7\ {}^4F^3$	9/2	1.09	0.43	584.59	1.49	0.41	
$3d^7\ {}^4P^3$	5/2	3.05	2.30	628.02+x	3.12	2.61	
$3d^7\ {}^4P^3$	1/2	2.73	2.00	638.62+x	3.52	3.17	
$3d^7\ {}^2H^3$	11/2	0.54	0.12	650.70	−1.10	0.15	
$3d^7\ {}^2F^3$	7/2	0.72	0.21	705.92	−0.48	1.06	
W^{48+} (Fe-like)							
$3d^8\ {}^3F$	4	0.00			0.00	0.00	0.00
$3d^8\ {}^3F$	3	1.47	0.58	526.65	3.45	0.52	
$3d^8\ {}^3P$	2	2.74	1.97	603.12+x	3.42	2.43	
$3d^8\ {}^3P$	1	2.73	2.05	644.76+x	2.06	2.62	
W^{47+} (Co-like)							
$3d^9\ {}^2D$	5/2	0.00	0.00	0.00	0.00	0.00	
$3d^9\ {}^2D$	3/2	1.38	0.55	538.59	2.99	0.54	

Table 4 shows clearly that the uncertainties of the present $n = 6$ results are smaller by about a factor of a half when no shifts are indicated in the NIST value. For these levels, the $n = 6$ results statistically differ less than the Quinet values that are similar to the less accurate $n = 5$ values. The most accurate results for $3d$ and $3d^9$ are the RMCDHF$_g$ results, although for $3d^9$, the $n = 6$ are almost of the same accuracy. RMBPT$_g$ is the more accurate for $3d^2$, with $n = 6$ almost the same. For $3d^8$, the two lower levels, RMBPT$_g$ is the more accurate, whereas $n = 6$ is the more accurate for the two upper levels. A similar pattern seems to hold for other spectra. An interesting case is $3d^7\ {}^4P\ J = 5/2$ and $1/2$, where both levels have an unknown shift. An exact theoretical value and an exact NIST value (except for the shift) would have the same difference for the two levels. In the present case, the

$n = 6$ differences are more similar than the RMCDHF$_g$ differences. In fact, from this table, we can conclude that any NIST value for which the theoretical difference from NIST for both methods is more than 1.00 has a noticeable error. Thus, for example, the $^3P_1$ level of $3d^8$ with an energy level of 644.70 Kcm^{-1} suggests that the NIST values is not accurate to two decimal places.

The errors in different theoretical results are shown in Figure 2. Note the similarity in accuracy of the present $n = 6$ results and values reported by Guo et al. [8].

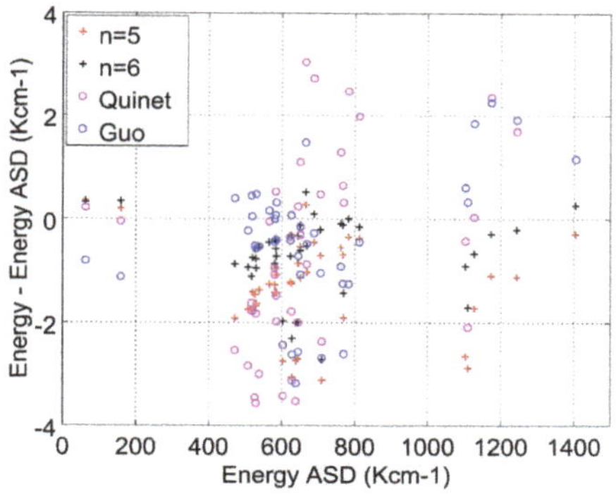

Figure 2. Plot comparing the accuracy of different theoretical methods.

The accuracy of theoretical energy levels are best evaluated by comparing theoretical wavelengths with wavelengths of observed lines in the spectrum. In Table 5, all wavelengths for M1 transitions between the $3d^k$ levels for the present $n = 5, 6$ results are compared with experimental results and other theory, when available. This table clearly shows the improvement in accuracy of $n = 6$ calculations over $n = 5$, as well as the GRASP results reported by Quinet [6], and in many cases the very close agreement with Guo et al. [8]. Two exceptions are the $3d^7\ ^4F^3 - 3d^7\ ^2F^3$ ($J = 9/2$ to $J = 7/2$) transition, for which the observed wavelength is 14.166(3) nm, the present $n = 6$ is 14.170 nm, and the Guo et al. value is 14.187 nm. Similarly, the $3d^8\ ^3F - 3d^8\ ^1G$ ($J = 4$ to $J = 4$) transition has an observed wavelength of 15.511(3) nm, whereas the present value is 15.518 nm and the Guo et al. value is 15.463 nm.

Table 5. Wavelengths from theory for observed M1 transitions compared with observed wavelengths (in nm). Included are some long wavelengths for transitions between close-lying levels.

Label and J for Lower		Label and J for Upper		Present Work $n = 5$	$n = 6$	Expt (Ref. [2])	GRASP	RMBPT$_g$	RCI$_g$
W^{55+} (K-like)									
$3d\ ^2D$	3/2	$3d\ ^2D$	5/2	15.994	15.970	15.962(3)	16.008	15.960	16.035
W^{54+} (Ca-like)									
$3d^2\ ^3F$	2	$3d^2\ ^3F$	3	17.122	17.101	17.080(3)	17.138	17.071	17.218
$3d^2\ ^3F$	2	$3d^2\ ^3P$	2	14.982	14.971	14.959(3)	14.980	14.970	14.924
$3d^2\ ^3F$	2	$3d^2\ ^3P$	1	14.157	14.149				
$3d^2\ ^3F$	2	$3d^2\ ^3P$	2	7.699	7.691				
$3d^2\ ^3P$	0	$3d^2\ ^3P$	1	19.233	19.211	19.177(3)	19.222	19.160	19.422
$3d^2\ ^3F$	3	$3d^2\ ^3P$	2	119.908	120.168				
$3d^2\ ^3F$	3	$3d^2\ ^1G$	4	89.580	89.805				
$3d^2\ ^3F$	3	$3d^2\ ^3F$	4	15.378	15.365				
$3d^2\ ^3F$	3	$3d^2\ ^3P$	2	13.989	13.977				

Table 5. *Cont.*

Label and J for Lower		Label and J for Upper		Present Work $n=5$	$n=6$	Expt (Ref. [2])	GRASP	RMBPT$_g$	RCI$_g$
$3d^2\ ^3P$	2	$3d^2\ ^3P$	1	257.054	257.793				
$3d^2\ ^3P$	2	$3d^2\ ^3P$	2	15.836	15.817				
$3d^2\ ^3P$	1	$3d^2\ ^3P$	2	16.876	16.851				
$3d^2\ ^3P$	1	$3d^2\ ^1S$	0	12.728	12.707				
$3d^2\ ^1G$	4	$3d^2\ ^3F$	4	18.566	18.537				
W^{53+} (Sc-like)									
$3d^3\ ^4F^3$	3/2	$3d^3\ ^4F^3$	5/2	18.925	18.901	18.867(3)	18.933	18.850	19.120
$3d^3\ ^4F^3$	3/2	$3d^3\ ^4P^3$	3/2	17.258	17.242	17.216(3)	17.243	17.216	17.315
$3d^3\ ^4F^3$	3/2	$3d^3\ ^4P^3$	1/2	16.059	16.046	16.027(3)	16.035	16.038	16.038
$3d^3\ ^4F^3$	3/2	$3d^3\ ^2D^1$	5/2	12.318	12.314	12.312(3)	12.282	12.319	12.225
$3d^3\ ^4F^3$	3/2	$3d^3\ ^4P^3$	3/2	8.289	8.280				
$3d^3\ ^4F^3$	3/2	$3d^3\ ^2P^3$	1/2	8.128	8.120				
$3d^3\ ^4F^3$	3/2	$3d^3\ ^2D^3$	5/2	8.041	8.035				
$3d^3\ ^4F^3$	3/2	$3d^3\ ^2F^3$	5/2	7.607	7.601				
$3d^3\ ^4F^3$	3/2	$3d^3\ ^2D^1$	3/2	6.757	6.751				
$3d^3\ ^4F^3$	3/2	$3d^3\ ^2P^3$	3/2	5.329	5.324				
$3d^3\ ^4F^3$	3/2	$3d^3\ ^2D^1$	5/2	5.107	5.102				
$3d^3\ ^4F^3$	5/2	$3d^3\ ^4P^3$	3/2	195.953	196.390				
$3d^3\ ^4F^3$	5/2	$3d^3\ ^2G^3$	7/2	121.923	122.264				
$3d^3\ ^4F^3$	5/2	$3d^3\ ^2D^1$	5/2	35.279	35.336				
$3d^3\ ^4F^3$	5/2	$3d^3\ ^4F^3$	7/2	16.697	16.680				
$3d^3\ ^4F^3$	5/2	$3d^3\ ^4P^3$	3/2	14.749	14.735				
$3d^3\ ^4F^3$	5/2	$3d^3\ ^2D^3$	5/2	13.981	13.975				
$3d^3\ ^4F^3$	5/2	$3d^3\ ^2F^3$	5/2	12.720	12.715				
$3d^3\ ^4F^3$	5/2	$3d^3\ ^2F^3$	7/2	12.654	12.651				
$3d^3\ ^4F^3$	5/2	$3d^3\ ^2D^1$	3/2	10.509	10.502				
$3d^3\ ^4F^3$	5/2	$3d^3\ ^2P^3$	3/2	7.418	7.412				
$3d^3\ ^4F^3$	5/2	$3d^3\ ^2D^1$	5/2	6.995	6.988				
$3d^3\ ^4P^3$	3/2	$3d^3\ ^4P^3$	1/2	230.965	231.304				
$3d^3\ ^4P^3$	3/2	$3d^3\ ^2D^1$	5/2	43.026	43.089				
$3d^3\ ^4P^3$	3/2	$3d^3\ ^4P^3$	3/2	15.949	15.930				
$3d^3\ ^4P^3$	3/2	$3d^3\ ^2P^3$	1/2	15.364	15.347				
$3d^3\ ^4P^3$	3/2	$3d^3\ ^2D^3$	5/2	15.055	15.046				
$3d^3\ ^4P^3$	3/2	$3d^3\ ^2F^3$	5/2	13.603	13.595				
$3d^3\ ^4P^3$	3/2	$3d^3\ ^2D^1$	3/2	11.105	11.095				
$3d^3\ ^4P^3$	3/2	$3d^3\ ^2P^3$	3/2	7.710	7.702				
$3d^3\ ^4P^3$	3/2	$3d^3\ ^2D^1$	5/2	7.254	7.246				
$3d^3\ ^2H^3$	9/2	$3d^3\ ^2G^3$	7/2	21322.871	18591.162				
$3d^3\ ^2H^3$	9/2	$3d^3\ ^4F^3$	7/2	19.329	19.295				
$3d^3\ ^2H^3$	9/2	$3d^3\ ^4F^3$	9/2	18.022	17.996				
$3d^3\ ^2H^3$	9/2	$3d^3\ ^2H^3$	11/2	15.812	15.798	15.785(3)	15.792	15.778	15.876
$3d^3\ ^2H^3$	9/2	$3d^3\ ^2F^3$	7/2	14.110	14.100				
$3d^3\ ^2H^3$	9/2	$3d^3\ ^2G^3$	9/2	8.673	8.661				
$3d^3\ ^2G^3$	7/2	$3d^3\ ^2D^1$	5/2	49.645	49.700				
$3d^3\ ^2G^3$	7/2	$3d^3\ ^4F^3$	7/2	19.346	19.315				
$3d^3\ ^2G^3$	7/2	$3d^3\ ^4F^3$	9/2	18.037	18.014				
$3d^3\ ^2G^3$	7/2	$3d^3\ ^2D^3$	5/2	15.791	15.779				
$3d^3\ ^2G^3$	7/2	$3d^3\ ^2F^3$	5/2	14.201	14.191				

Table 5. *Cont.*

Label and J for Lower		Label and J for Upper		Present Work $n = 5$	$n = 6$	Expt (Ref. [2])	GRASP	$RMBPT_g$	RCI_g
$3d^3\ ^2G^3$	7/2	$3d^3\ ^2F^3$	7/2	14.119	14.110				
$3d^3\ ^2G^3$	7/2	$3d^3\ ^2G^3$	9/2	8.677	8.665				
$3d^3\ ^2G^3$	7/2	$3d^3\ ^2D^1$	5/2	7.421	7.411				
$3d^3\ ^4P^3$	1/2	$3d^3\ ^4P^3$	3/2	17.132	17.108				
$3d^3\ ^4P^3$	1/2	$3d^3\ ^2P^3$	1/2	16.459	16.438				
$3d^3\ ^4P^3$	1/2	$3d^3\ ^2D^1$	3/2	11.665	11.654				
$3d^3\ ^4P^3$	1/2	$3d^3\ ^2P^3$	3/2	7.976	7.968				
$3d^3\ ^2D^1$	5/2	$3d^3\ ^4F^3$	7/2	31.700	31.592				
$3d^3\ ^2D^1$	5/2	$3d^3\ ^4P^3$	3/2	25.344	25.274				
$3d^3\ ^2D^1$	5/2	$3d^3\ ^2D^3$	5/2	23.157	23.119				
$3d^3\ ^2D^1$	5/2	$3d^3\ ^2F^3$	5/2	19.891	19.862				
$3d^3\ ^2D^1$	5/2	$3d^3\ ^2F^3$	7/2	19.730	19.705				
$3d^3\ ^2D^1$	5/2	$3d^3\ ^2D^1$	3/2	14.968	14.943				
$3d^3\ ^2D^1$	5/2	$3d^3\ ^2P^3$	3/2	9.393	9.379				
$3d^3\ ^2D^1$	5/2	$3d^3\ ^2D^1$	5/2	8.725	8.710				
$3d^3\ ^4F^3$	7/2	$3d^3\ ^4F^3$	9/2	266.613	267.448				
$3d^3\ ^4F^3$	7/2	$3d^3\ ^2D^3$	5/2	85.937	86.197				
$3d^3\ ^4F^3$	7/2	$3d^3\ ^2F^3$	5/2	53.397	53.492				
$3d^3\ ^4F^3$	7/2	$3d^3\ ^2F^3$	7/2	52.253	52.370				
$3d^3\ ^4F^3$	7/2	$3d^3\ ^2G^3$	9/2	15.733	15.717				
$3d^3\ ^4F^3$	7/2	$3d^3\ ^2D^1$	5/2	12.038	12.026				
$3d^3\ ^4F^3$	9/2	$3d^3\ ^2H^3$	11/2	128.917	129.345				
$3d^3\ ^4F^3$	9/2	$3d^3\ ^2F^3$	7/2	64.991	65.122				
$3d^3\ ^4F^3$	9/2	$3d^3\ ^2G^3$	9/2	16.719	16.698				
$3d^3\ ^4P^3$	3/2	$3d^3\ ^2P^3$	1/2	418.463	419.226				
$3d^3\ ^4P^3$	3/2	$3d^3\ ^2D^3$	5/2	268.399	271.110				
$3d^3\ ^4P^3$	3/2	$3d^3\ ^2F^3$	5/2	92.445	92.751				
$3d^3\ ^4P^3$	3/2	$3d^3\ ^2D^1$	3/2	36.557	36.559				
$3d^3\ ^4P^3$	3/2	$3d^3\ ^2P^3$	3/2	14.925	14.912				
$3d^3\ ^4P^3$	3/2	$3d^3\ ^2D^1$	5/2	13.305	13.291				
$3d^3\ ^2P^3$	1/2	$3d^3\ ^2D^1$	3/2	40.056	40.051				
$3d^3\ ^2P^3$	1/2	$3d^3\ ^2P^3$	3/2	15.477	15.462				
$3d^3\ ^2H^3$	11/2	$3d^3\ ^2G^3$	9/2	19.211	19.173				
$3d^3\ ^2D^3$	5/2	$3d^3\ ^2F^3$	5/2	141.016	140.984				
$3d^3\ ^2D^3$	5/2	$3d^3\ ^2F^3$	7/2	133.312	133.450				
$3d^3\ ^2D^3$	5/2	$3d^3\ ^2D^1$	3/2	42.321	42.257				
$3d^3\ ^2D^3$	5/2	$3d^3\ ^2P^3$	3/2	15.804	15.780				
$3d^3\ ^2D^3$	5/2	$3d^3\ ^2D^1$	5/2	13.999	13.976				
$3d^3\ ^2F^3$	5/2	$3d^3\ ^2F^3$	7/2	2440.155	2497.085				
$3d^3\ ^2F^3$	5/2	$3d^3\ ^2D^1$	3/2	60.469	60.343				
$3d^3\ ^2F^3$	5/2	$3d^3\ ^2P^3$	3/2	17.798	17.769				
$3d^3\ ^2F^3$	5/2	$3d^3\ ^2D^1$	5/2	15.542	15.514				
$3d^3\ ^2F^3$	7/2	$3d^3\ ^2G^3$	9/2	22.510	22.456				
$3d^3\ ^2F^3$	7/2	$3d^3\ ^2D^1$	5/2	15.642	15.611				
$3d^3\ ^2D^1$	3/2	$3d^3\ ^2P^3$	3/2	25.222	25.185				
$3d^3\ ^2D^1$	3/2	$3d^3\ ^2D^1$	5/2	20.919	20.883				
$3d^3\ ^2P^3$	3/2	$3d^3\ ^2D^1$	5/2	122.606	122.246				

Table 5. *Cont.*

Label and J for Lower		Label and J for Upper		Present Work $n = 5$	$n = 6$	Expt (Ref. [2])	GRASP	RMBPT$_g$	RCI$_g$
W^{52+} (Ti-like)									
$3d^4\,^3P^2$	0	$3d^4\,^5D^4$	1	19.385	19.361	19.319(3)	19.379	19.302	19.605
$3d^4\,^3P^2$	0	$3d^4\,^3D^4$	1	8.241	8.234				
$3d^4\,^3P^2$	0	$3d^4\,^3D^4$	1	5.660	5.654				
$3d^4\,^3P^2$	0	$3d^4\,^3P^2$	1	5.041	5.037				
$3d^4\,^5D^4$	1	$3d^4\,^5D^4$	2	81.888	82.053				
$3d^4\,^5D^4$	1	$3d^4\,^5D^4$	0	17.065	17.045				
$3d^4\,^5D^4$	1	$3d^4\,^5D^4$	2	16.922	16.907	16.890(3)	16.903	16.894	16.958
$3d^4\,^5D^4$	1	$3d^4\,^3D^4$	1	14.334	14.326				
$3d^4\,^5D^4$	1	$3d^4\,^3F^4$	2	13.479	13.475				
$3d^4\,^5D^4$	1	$3d^4\,^3F^2$	2	11.857	11.849				
$3d^4\,^5D^4$	1	$3d^4\,^1D^2$	2	10.103	10.103				
$3d^4\,^5D^4$	1	$3d^4\,^3P^4$	0	8.950	8.947				
$3d^4\,^5D^4$	1	$3d^4\,^3D^4$	1	7.995	7.987				
$3d^4\,^5D^4$	1	$3d^4\,^3F^4$	2	7.545	7.539				
$3d^4\,^5D^4$	1	$3d^4\,^1S^4$	0	7.107	7.105				
$3d^4\,^5D^4$	1	$3d^4\,^3P^2$	1	6.812	6.808				
$3d^4\,^5D^4$	1	$3d^4\,^1D^4$	2	6.654	6.651				
$3d^4\,^5D^4$	1	$3d^4\,^1D^2$	2	5.142	5.137				
$3d^4\,^5D^4$	1	$3d^4\,^3P^2$	0	4.658	4.653				
$3d^4\,^3H^4$	4	$3d^4\,^3F^2$	3	190.114	190.262				
$3d^4\,^3H^4$	4	$3d^4\,^3H^4$	4	19.520	19.491	19.445(3)	19.543	19.443	19.696
$3d^4\,^3H^4$	4	$3d^4\,^5D^4$	3	18.912	18.886				
$3d^4\,^3H^4$	4	$3d^4\,^3H^4$	5	17.889	17.872	17.846(3)	17.855	17.831	18.065
$3d^4\,^3H^4$	4	$3d^4\,^3F^4$	3	15.977	15.965				
$3d^4\,^3H^4$	4	$3d^4\,^3G^4$	4	15.894	15.882				
$3d^4\,^3H^4$	4	$3d^4\,^3F^2$	4	12.652	12.647				
$3d^4\,^3H^4$	4	$3d^4\,^5D^4$	4	9.083	9.071				
$3d^4\,^3H^4$	4	$3d^4\,^3F^4$	3	8.993	8.981				
$3d^4\,^3H^4$	4	$3d^4\,^3G^4$	5	8.617	8.608				
$3d^4\,^3H^4$	4	$3d^4\,^3D^4$	3	8.036	8.028				
$3d^4\,^3H^4$	4	$3d^4\,^3F^2$	3	7.317	7.310				
$3d^4\,^3H^4$	4	$3d^4\,^1G^2$	4	7.290	7.283				
$3d^4\,^3H^4$	4	$3d^4\,^3F^2$	4	5.672	5.665				
$3d^4\,^5D^4$	2	$3d^4\,^3F^2$	3	358.990	360.907				
$3d^4\,^5D^4$	2	$3d^4\,^5D^4$	2	21.329	21.295				
$3d^4\,^5D^4$	2	$3d^4\,^5D^4$	3	19.840	19.816				
$3d^4\,^5D^4$	2	$3d^4\,^3D^4$	1	17.375	17.356				
$3d^4\,^5D^4$	2	$3d^4\,^3F^4$	3	16.635	16.624				
$3d^4\,^5D^4$	2	$3d^4\,^3F^4$	2	16.135	16.123				
$3d^4\,^5D^4$	2	$3d^4\,^3F^2$	2	13.864	13.849				
$3d^4\,^5D^4$	2	$3d^4\,^1D^2$	2	11.525	11.521				
$3d^4\,^5D^4$	2	$3d^4\,^3F^4$	3	9.197	9.186				
$3d^4\,^5D^4$	2	$3d^4\,^3D^4$	1	8.860	8.848				
$3d^4\,^5D^4$	2	$3d^4\,^3F^4$	2	8.311	8.302				
$3d^4\,^5D^4$	2	$3d^4\,^3D^4$	3	8.199	8.191				
$3d^4\,^5D^4$	2	$3d^4\,^3F^2$	3	7.452	7.445				
$3d^4\,^5D^4$	2	$3d^4\,^3P^2$	1	7.430	7.424				
$3d^4\,^5D^4$	2	$3d^4\,^1D^4$	2	7.243	7.238				

Table 5. *Cont.*

Label and J for Lower		Label and J for Upper		Present Work $n=5$	$n=6$	Expt (Ref. [2])	GRASP	$RMBPT_g$	RCI_g
$3d^4\,^5D^4$	2	$3d^4\,^1D^2$	2	5.487	5.480				
$3d^4\,^3F^2$	3	$3d^4\,^5D^4$	2	22.677	22.630				
$3d^4\,^3F^2$	3	$3d^4\,^3H^4$	4	21.753	21.716				
$3d^4\,^3F^2$	3	$3d^4\,^5D^4$	3	21.001	20.968				
$3d^4\,^3F^2$	3	$3d^4\,^3F^4$	3	17.443	17.427				
$3d^4\,^3F^2$	3	$3d^4\,^3G^4$	4	17.344	17.329				
$3d^4\,^3F^2$	3	$3d^4\,^3F^4$	2	16.895	16.877				
$3d^4\,^3F^2$	3	$3d^4\,^3F^2$	2	14.421	14.402				
$3d^4\,^3F^2$	3	$3d^4\,^3F^2$	4	13.554	13.548	13.543(3)	13.513	13.549	13.495
$3d^4\,^3F^2$	3	$3d^4\,^1D^2$	2	11.907	11.901				
$3d^4\,^3F^2$	3	$3d^4\,^5D^4$	4	9.538	9.525				
$3d^4\,^3F^2$	3	$3d^4\,^3F^4$	3	9.439	9.426				
$3d^4\,^3F^2$	3	$3d^4\,^3F^4$	2	8.508	8.497				
$3d^4\,^3F^2$	3	$3d^4\,^3D^4$	3	8.390	8.381				
$3d^4\,^3F^2$	3	$3d^4\,^3F^2$	3	7.610	7.602				
$3d^4\,^3F^2$	3	$3d^4\,^1G^2$	4	7.581	7.573				
$3d^4\,^3F^2$	3	$3d^4\,^1D^4$	2	7.392	7.386				
$3d^4\,^3F^2$	3	$3d^4\,^3F^2$	4	5.847	5.839				
$3d^4\,^3F^2$	3	$3d^4\,^1D^2$	2	5.572	5.565				
$3d^4\,^5D^4$	0	$3d^4\,^3D^4$	1	89.555	89.798				
$3d^4\,^5D^4$	0	$3d^4\,^3D^4$	1	15.041	15.028				
$3d^4\,^5D^4$	0	$3d^4\,^3P^2$	1	11.338	11.334				
$3d^4\,^5D^4$	2	$3d^4\,^5D^4$	3	284.116	285.346				
$3d^4\,^5D^4$	2	$3d^4\,^3D^4$	1	93.723	93.840				
$3d^4\,^5D^4$	2	$3d^4\,^3F^4$	3	75.583	75.791				
$3d^4\,^5D^4$	2	$3d^4\,^3F^4$	2	66.257	66.382				
$3d^4\,^5D^4$	2	$3d^4\,^3F^2$	2	39.611	39.610				
$3d^4\,^5D^4$	2	$3d^4\,^1D^2$	2	25.072	25.102				
$3d^4\,^5D^4$	2	$3d^4\,^3F^4$	3	16.170	16.153				
$3d^4\,^5D^4$	2	$3d^4\,^3D^4$	1	15.154	15.138				
$3d^4\,^5D^4$	2	$3d^4\,^3F^4$	2	13.617	13.605				
$3d^4\,^5D^4$	2	$3d^4\,^3D^4$	3	13.318	13.311				
$3d^4\,^5D^4$	2	$3d^4\,^3F^2$	3	11.453	11.448				
$3d^4\,^5D^4$	2	$3d^4\,^3P^2$	1	11.402	11.396				
$3d^4\,^5D^4$	2	$3d^4\,^1D^4$	2	10.967	10.964				
$3d^4\,^5D^4$	2	$3d^4\,^1D^2$	2	7.387	7.380				
$3d^4\,^3H^4$	4	$3d^4\,^5D^4$	3	606.876	608.761				
$3d^4\,^3H^4$	4	$3d^4\,^3H^4$	5	214.114	215.206				
$3d^4\,^3H^4$	4	$3d^4\,^3F^4$	3	88.039	88.243				
$3d^4\,^3H^4$	4	$3d^4\,^3G^4$	4	85.561	85.781				
$3d^4\,^3H^4$	4	$3d^4\,^3F^2$	4	35.956	36.020				
$3d^4\,^3H^4$	4	$3d^4\,^5D^4$	4	16.986	16.966				
$3d^4\,^3H^4$	4	$3d^4\,^3F^4$	3	16.675	16.654				
$3d^4\,^3H^4$	4	$3d^4\,^3G^4$	5	15.427	15.416				
$3d^4\,^3H^4$	4	$3d^4\,^3D^4$	3	13.658	13.649				
$3d^4\,^3H^4$	4	$3d^4\,^3F^2$	3	11.704	11.697				
$3d^4\,^3H^4$	4	$3d^4\,^1G^2$	4	11.635	11.628				
$3d^4\,^3H^4$	4	$3d^4\,^3F^2$	4	7.996	7.986				
$3d^4\,^5D^4$	3	$3d^4\,^3F^4$	3	102.978	103.203				

Table 5. *Cont.*

Label and J for Lower		Label and J for Upper		Present Work $n = 5$	$n = 6$	Expt (Ref. [2])	GRASP	$RMBPT_g$	RCI_g
$3d^4\ ^5D^4$	3	$3d^4\ ^3G^4$	4	99.604	99.852				
$3d^4\ ^5D^4$	3	$3d^4\ ^3F^4$	2	86.407	86.507				
$3d^4\ ^5D^4$	3	$3d^4\ ^3F^2$	2	46.028	45.995				
$3d^4\ ^5D^4$	3	$3d^4\ ^3F^2$	4	38.221	38.285				
$3d^4\ ^5D^4$	3	$3d^4\ ^1D^2$	2	27.498	27.524				
$3d^4\ ^5D^4$	3	$3d^4\ ^5D^4$	4	17.475	17.453				
$3d^4\ ^5D^4$	3	$3d^4\ ^3F^4$	3	17.146	17.123				
$3d^4\ ^5D^4$	3	$3d^4\ ^3F^4$	2	14.303	14.287				
$3d^4\ ^5D^4$	3	$3d^4\ ^3D^4$	3	13.973	13.962				
$3d^4\ ^5D^4$	3	$3d^4\ ^3F^2$	3	11.934	11.926				
$3d^4\ ^5D^4$	3	$3d^4\ ^1G^2$	4	11.863	11.855				
$3d^4\ ^5D^4$	3	$3d^4\ ^1D^4$	2	11.408	11.402				
$3d^4\ ^5D^4$	3	$3d^4\ ^3F^2$	4	8.102	8.092				
$3d^4\ ^5D^4$	3	$3d^4\ ^1D^2$	2	7.584	7.575				
$3d^4\ ^3H^4$	5	$3d^4\ ^1I^4$	6	428.184	430.120				
$3d^4\ ^3H^4$	5	$3d^4\ ^3G^4$	4	142.509	142.637				
$3d^4\ ^3H^4$	5	$3d^4\ ^3F^2$	4	43.213	43.261				
$3d^4\ ^3H^4$	5	$3d^4\ ^5D^4$	4	18.450	18.418				
$3d^4\ ^3H^4$	5	$3d^4\ ^3G^4$	5	16.625	16.605				
$3d^4\ ^3H^4$	5	$3d^4\ ^3H^4$	6	16.488	16.470				
$3d^4\ ^3H^4$	5	$3d^4\ ^1G^2$	4	12.304	12.292				
$3d^4\ ^3H^4$	5	$3d^4\ ^3F^2$	4	8.306	8.293				
$3d^4\ ^1I^4$	6	$3d^4\ ^3G^4$	5	17.296	17.272				
$3d^4\ ^1I^4$	6	$3d^4\ ^3H^4$	6	17.148	17.126				
$3d^4\ ^3D^4$	1	$3d^4\ ^3F^4$	2	226.092	226.868				
$3d^4\ ^3D^4$	1	$3d^4\ ^3F^2$	2	68.606	68.541				
$3d^4\ ^3D^4$	1	$3d^4\ ^1D^2$	2	34.228	34.269				
$3d^4\ ^3D^4$	1	$3d^4\ ^3P^4$	0	23.832	23.832				
$3d^4\ ^3D^4$	1	$3d^4\ ^3D^4$	1	18.077	18.049				
$3d^4\ ^3D^4$	1	$3d^4\ ^3F^4$	2	15.932	15.912				
$3d^4\ ^3D^4$	1	$3d^4\ ^1S^4$	0	14.097	14.094				
$3d^4\ ^3D^4$	1	$3d^4\ ^3P^2$	1	12.981	12.972				
$3d^4\ ^3D^4$	1	$3d^4\ ^1D^4$	2	12.421	12.415				
$3d^4\ ^3D^4$	1	$3d^4\ ^1D^2$	2	8.019	8.009				
$3d^4\ ^3D^4$	1	$3d^4\ ^3P^2$	0	6.900	6.892				
$3d^4\ ^3F^4$	3	$3d^4\ ^3G^4$	4	3040.724	3074.775				
$3d^4\ ^3F^4$	3	$3d^4\ ^3F^4$	2	536.990	534.715				
$3d^4\ ^3F^4$	3	$3d^4\ ^3F^2$	2	83.228	82.973				
$3d^4\ ^3F^4$	3	$3d^4\ ^3F^2$	4	60.779	60.864				
$3d^4\ ^3F^4$	3	$3d^4\ ^1D^2$	2	37.516	37.533				
$3d^4\ ^3F^4$	3	$3d^4\ ^5D^4$	4	21.047	21.005				
$3d^4\ ^3F^4$	3	$3d^4\ ^3F^4$	3	20.571	20.529				
$3d^4\ ^3F^4$	3	$3d^4\ ^3F^4$	2	16.610	16.582				
$3d^4\ ^3F^4$	3	$3d^4\ ^3D^4$	3	16.166	16.147				
$3d^4\ ^3F^4$	3	$3d^4\ ^3F^2$	3	13.498	13.485				
$3d^4\ ^3F^4$	3	$3d^4\ ^1G^2$	4	13.407	13.393				
$3d^4\ ^3F^4$	3	$3d^4\ ^1D^4$	2	12.829	12.819				
$3d^4\ ^3F^4$	3	$3d^4\ ^3F^2$	4	8.794	8.780				
$3d^4\ ^3F^4$	3	$3d^4\ ^1D^2$	2	8.187	8.176				

Table 5. *Cont.*

Label and J for Lower		Label and J for Upper		Present Work $n = 5$	$n = 6$	Expt (Ref. [2])	GRASP	RMBPT$_g$	RCI$_g$
$3d^4\,{}^3G^4$	4	$3d^4\,{}^3F^2$	4	62.019	62.093				
$3d^4\,{}^3G^4$	4	$3d^4\,{}^5D^4$	4	21.194	21.149				
$3d^4\,{}^3G^4$	4	$3d^4\,{}^3F^4$	3	20.711	20.667				
$3d^4\,{}^3G^4$	4	$3d^4\,{}^3G^4$	5	18.820	18.793				
$3d^4\,{}^3G^4$	4	$3d^4\,{}^3D^4$	3	16.253	16.232				
$3d^4\,{}^3G^4$	4	$3d^4\,{}^3F^2$	3	13.559	13.544				
$3d^4\,{}^3G^4$	4	$3d^4\,{}^1G^2$	4	13.466	13.452				
$3d^4\,{}^3G^4$	4	$3d^4\,{}^3F^2$	4	8.820	8.805				
$3d^4\,{}^3F^4$	2	$3d^4\,{}^3F^2$	2	98.494	98.213				
$3d^4\,{}^3F^4$	2	$3d^4\,{}^1D^2$	2	40.334	40.367				
$3d^4\,{}^3F^4$	2	$3d^4\,{}^3F^4$	3	21.391	21.348				
$3d^4\,{}^3F^4$	2	$3d^4\,{}^3D^4$	1	19.648	19.609				
$3d^4\,{}^3F^4$	2	$3d^4\,{}^3F^4$	2	17.140	17.113				
$3d^4\,{}^3F^4$	2	$3d^4\,{}^3D^4$	3	16.668	16.649				
$3d^4\,{}^3F^4$	2	$3d^4\,{}^3F^2$	3	13.846	13.834				
$3d^4\,{}^3F^4$	2	$3d^4\,{}^3P^2$	1	13.772	13.759				
$3d^4\,{}^3F^4$	2	$3d^4\,{}^1D^4$	2	13.143	13.133				
$3d^4\,{}^3F^4$	2	$3d^4\,{}^1D^2$	2	8.314	8.302				
$3d^4\,{}^3F^2$	2	$3d^4\,{}^1D^2$	2	68.306	68.536				
$3d^4\,{}^3F^2$	2	$3d^4\,{}^3F^4$	3	27.325	27.278				
$3d^4\,{}^3F^2$	2	$3d^4\,{}^3D^4$	1	24.544	24.501				
$3d^4\,{}^3F^2$	2	$3d^4\,{}^3F^4$	2	20.751	20.724				
$3d^4\,{}^3F^2$	2	$3d^4\,{}^3D^4$	3	20.063	20.048				
$3d^4\,{}^3F^2$	2	$3d^4\,{}^3F^2$	3	16.111	16.102				
$3d^4\,{}^3F^2$	2	$3d^4\,{}^3P^2$	1	16.011	16.000				
$3d^4\,{}^3F^2$	2	$3d^4\,{}^1D^4$	2	15.166	15.161				
$3d^4\,{}^3F^2$	2	$3d^4\,{}^1D^2$	2	9.081	9.069				
$3d^4\,{}^3F^2$	4	$3d^4\,{}^5D^4$	4	32.196	32.074				
$3d^4\,{}^3F^2$	4	$3d^4\,{}^3F^4$	3	31.096	30.977				
$3d^4\,{}^3F^2$	4	$3d^4\,{}^3G^4$	5	27.020	26.950				
$3d^4\,{}^3F^2$	4	$3d^4\,{}^3D^4$	3	22.024	21.977				
$3d^4\,{}^3F^2$	4	$3d^4\,{}^3F^2$	3	17.352	17.323				
$3d^4\,{}^3F^2$	4	$3d^4\,{}^1G^2$	4	17.201	17.172				
$3d^4\,{}^3F^2$	4	$3d^4\,{}^3F^2$	4	10.282	10.260				
$3d^4\,{}^1D^2$	2	$3d^4\,{}^3F^4$	3	45.544	45.312				
$3d^4\,{}^1D^2$	2	$3d^4\,{}^3D^4$	1	38.310	38.133				
$3d^4\,{}^1D^2$	2	$3d^4\,{}^3F^4$	2	29.806	29.706				
$3d^4\,{}^1D^2$	2	$3d^4\,{}^3D^4$	3	28.407	28.337				
$3d^4\,{}^1D^2$	2	$3d^4\,{}^3F^2$	3	21.085	21.046				
$3d^4\,{}^1D^2$	2	$3d^4\,{}^3P^2$	1	20.912	20.873				
$3d^4\,{}^1D^2$	2	$3d^4\,{}^1D^4$	2	19.495	19.467				
$3d^4\,{}^1D^2$	2	$3d^4\,{}^1D^2$	2	10.473	10.452				
$3d^4\,{}^3P^4$	0	$3d^4\,{}^3D^4$	1	74.865	74.385				
$3d^4\,{}^3P^4$	0	$3d^4\,{}^3P^2$	1	28.511	28.466				
$3d^4\,{}^5D^4$	4	$3d^4\,{}^3F^4$	3	910.088	905.530				
$3d^4\,{}^5D^4$	4	$3d^4\,{}^3G^4$	5	168.067	168.682				
$3d^4\,{}^5D^4$	4	$3d^4\,{}^3D^4$	3	69.712	69.810				
$3d^4\,{}^5D^4$	4	$3d^4\,{}^3F^2$	3	37.636	37.665				
$3d^4\,{}^5D^4$	4	$3d^4\,{}^1G^2$	4	36.935	36.959				

Table 5. *Cont.*

Label and J for Lower		Label and J for Upper		Present Work $n = 5$	$n = 6$	Expt (Ref. [2])	GRASP	$RMBPT_g$	RCI_g
$3d^4\ ^5D^4$	4	$3d^4\ ^3F^2$	4	15.106	15.086				
$3d^4\ ^3F^4$	3	$3d^4\ ^3F^4$	2	86.253	86.253				
$3d^4\ ^3F^4$	3	$3d^4\ ^3D^4$	3	75.495	75.642				
$3d^4\ ^3F^4$	3	$3d^4\ ^3F^2$	3	39.260	39.300				
$3d^4\ ^3F^4$	3	$3d^4\ ^1G^2$	4	38.498	38.532				
$3d^4\ ^3F^4$	3	$3d^4\ ^1D^4$	2	34.085	34.130				
$3d^4\ ^3F^4$	3	$3d^4\ ^3F^2$	4	15.361	15.342				
$3d^4\ ^3F^4$	3	$3d^4\ ^1D^2$	2	13.600	13.586				
$3d^4\ ^3D^4$	1	$3d^4\ ^3F^4$	2	134.268	134.420				
$3d^4\ ^3D^4$	1	$3d^4\ ^1S^4$	0	64.031	64.317				
$3d^4\ ^3D^4$	1	$3d^4\ ^3P^2$	1	46.049	46.113				
$3d^4\ ^3D^4$	1	$3d^4\ ^1D^4$	2	39.694	39.769				
$3d^4\ ^3D^4$	1	$3d^4\ ^1D^2$	2	14.413	14.399				
$3d^4\ ^3D^4$	1	$3d^4\ ^3P^2$	0	11.160	11.149				
$3d^4\ ^3G^4$	5	$3d^4\ ^3H^4$	6	1997.212	2027.160				
$3d^4\ ^3G^4$	5	$3d^4\ ^1G^2$	4	47.338	47.329				
$3d^4\ ^3G^4$	5	$3d^4\ ^3F^2$	4	16.598	16.568				
$3d^4\ ^3F^4$	2	$3d^4\ ^3D^4$	3	605.307	614.828				
$3d^4\ ^3F^4$	2	$3d^4\ ^3F^2$	3	72.059	72.193				
$3d^4\ ^3F^4$	2	$3d^4\ ^3P^2$	1	70.085	70.193				
$3d^4\ ^3F^4$	2	$3d^4\ ^1D^4$	2	56.355	56.479				
$3d^4\ ^3F^4$	2	$3d^4\ ^1D^2$	2	16.146	16.126				
$3d^4\ ^3D^4$	3	$3d^4\ ^3F^2$	3	81.796	81.797				
$3d^4\ ^3D^4$	3	$3d^4\ ^1G^2$	4	78.556	78.539				
$3d^4\ ^3D^4$	3	$3d^4\ ^1D^4$	2	62.140	62.192				
$3d^4\ ^3D^4$	3	$3d^4\ ^3F^2$	4	19.285	19.245				
$3d^4\ ^3D^4$	3	$3d^4\ ^1D^2$	2	16.589	16.561				
$3d^4\ ^1S^4$	0	$3d^4\ ^3P^2$	1	163.968	162.926				
$3d^4\ ^3F^2$	3	$3d^4\ ^1G^2$	4	1982.892	1971.659				
$3d^4\ ^3F^2$	3	$3d^4\ ^1D^4$	2	258.586	259.474				
$3d^4\ ^3F^2$	3	$3d^4\ ^3F^2$	4	25.235	25.166				
$3d^4\ ^3F^2$	3	$3d^4\ ^1D^2$	2	20.809	20.765				
$3d^4\ ^3P^2$	1	$3d^4\ ^1D^4$	2	287.662	289.068				
$3d^4\ ^3P^2$	1	$3d^4\ ^1D^2$	2	20.980	20.936				
$3d^4\ ^3P^2$	1	$3d^4\ ^3P^2$	0	14.730	14.704				
$3d^4\ ^1G^2$	4	$3d^4\ ^3F^2$	4	25.560	25.491				
$3d^4\ ^1D^4$	2	$3d^4\ ^1D^2$	2	22.630	22.571				
W^{51+} (V-like)									
$3d^5\ ^4P^3$	5/2	$3d^5\ ^6S^5$	5/2	21.290	21.243	21.203(3)	21.317	21.185	21.492
$3d^5\ ^4P^3$	5/2	$3d^5\ ^4G^5$	7/2	17.700	17.674	17.660(3)	17.660	17.655	17.826
$3d^5\ ^4P^3$	5/2	$3d^5\ ^4D^5$	3/2	17.253	17.227	17.215(3)	17.247	17.228	17.249
$3d^5\ ^4P^3$	5/2	$3d^5\ ^4D^5$	5/2	15.368	15.350				
$3d^5\ ^4P^3$	5/2	$3d^5\ ^2F^5$	7/2	14.541	14.529	14.531(3)	14.475	14.537	14.513
$3d^5\ ^4P^3$	5/2	$3d^5\ ^2D^1$	3/2	12.136	12.122				
$3d^5\ ^4P^3$	5/2	$3d^5\ ^6S^5$	5/2	9.747	9.728				
$3d^5\ ^4P^3$	5/2	$3d^5\ ^4D^5$	7/2	9.117	9.102				
$3d^5\ ^4P^3$	5/2	$3d^5\ ^4P^3$	3/2	8.586	8.571				
$3d^5\ ^4P^3$	5/2	$3d^5\ ^2F^5$	5/2	8.511	8.499				

Table 5. *Cont.*

Label and J for Lower		Label and J for Upper		Present Work $n=5$	$n=6$	Expt (Ref. [2])	GRASP	$RMBPT_g$	RCI_g
$3d^5\,^4P^3$	5/2	$3d^5\,^2G^5$	7/2	8.078	8.068				
$3d^5\,^4P^3$	5/2	$3d^5\,^4F^3$	5/2	7.971	7.959				
$3d^5\,^4P^3$	5/2	$3d^5\,^2D^5$	3/2	7.940	7.931				
$3d^5\,^4P^3$	5/2	$3d^5\,^2G^3$	7/2	7.646	7.636				
$3d^5\,^4P^3$	5/2	$3d^5\,^2P^3$	3/2	6.645	6.639				
$3d^5\,^4P^3$	5/2	$3d^5\,^2D^1$	5/2	6.522	6.516				
$3d^5\,^4P^3$	5/2	$3d^5\,^4P^3$	5/2	6.021	6.010				
$3d^5\,^4P^3$	5/2	$3d^5\,^4F^3$	7/2	5.768	5.758				
$3d^5\,^4P^3$	5/2	$3d^5\,^4D^5$	3/2	5.684	5.674				
$3d^5\,^4P^3$	5/2	$3d^5\,^2D^3$	5/2	5.424	5.416				
$3d^5\,^4P^3$	5/2	$3d^5\,^2G^3$	7/2	5.343	5.335				
$3d^5\,^4P^3$	5/2	$3d^5\,^2D^1$	3/2	4.847	4.841				
$3d^5\,^4P^3$	5/2	$3d^5\,^2D^1$	5/2	4.233	4.225				
$3d^5\,^6S^5$	5/2	$3d^5\,^4G^5$	7/2	104.969	105.214				
$3d^5\,^6S^5$	5/2	$3d^5\,^4D^5$	3/2	90.995	91.119				
$3d^5\,^6S^5$	5/2	$3d^5\,^4D^5$	5/2	55.250	55.341				
$3d^5\,^6S^5$	5/2	$3d^5\,^2F^5$	7/2	45.868	45.971				
$3d^5\,^6S^5$	5/2	$3d^5\,^2D^1$	3/2	28.226	28.233				
$3d^5\,^6S^5$	5/2	$3d^5\,^6S^5$	5/2	17.977	17.946				
$3d^5\,^6S^5$	5/2	$3d^5\,^4D^5$	7/2	15.946	15.927				
$3d^5\,^6S^5$	5/2	$3d^5\,^4P^3$	3/2	14.388	14.367				
$3d^5\,^6S^5$	5/2	$3d^5\,^2F^5$	5/2	14.181	14.167				
$3d^5\,^6S^5$	5/2	$3d^5\,^2G^5$	7/2	13.018	13.009				
$3d^5\,^6S^5$	5/2	$3d^5\,^4F^3$	5/2	12.741	12.727				
$3d^5\,^6S^5$	5/2	$3d^5\,^2D^5$	3/2	12.662	12.655				
$3d^5\,^6S^5$	5/2	$3d^5\,^2G^3$	7/2	11.932	11.921				
$3d^5\,^6S^5$	5/2	$3d^5\,^2P^3$	3/2	9.660	9.657				
$3d^5\,^6S^5$	5/2	$3d^5\,^2D^1$	5/2	9.403	9.399				
$3d^5\,^6S^5$	5/2	$3d^5\,^4P^3$	5/2	8.395	8.381				
$3d^5\,^6S^5$	5/2	$3d^5\,^4F^3$	7/2	7.912	7.900				
$3d^5\,^6S^5$	5/2	$3d^5\,^4D^5$	3/2	7.755	7.743				
$3d^5\,^6S^5$	5/2	$3d^5\,^2D^3$	5/2	7.277	7.269				
$3d^5\,^6S^5$	5/2	$3d^5\,^2G^3$	7/2	7.133	7.124				
$3d^5\,^6S^5$	5/2	$3d^5\,^2D^1$	3/2	6.276	6.269				
$3d^5\,^6S^5$	5/2	$3d^5\,^2D^1$	5/2	5.283	5.274				
$3d^5\,^4G^5$	7/2	$3d^5\,^2G^5$	9/2	178.750	179.156				
$3d^5\,^4G^5$	7/2	$3d^5\,^4D^5$	5/2	116.648	116.752				
$3d^5\,^4G^5$	7/2	$3d^5\,^2F^5$	7/2	81.465	81.645				
$3d^5\,^4G^5$	7/2	$3d^5\,^6S^5$	5/2	21.692	21.637				
$3d^5\,^4G^5$	7/2	$3d^5\,^4D^5$	7/2	18.802	18.768				
$3d^5\,^4G^5$	7/2	$3d^5\,^4G^5$	9/2	18.116	18.086				
$3d^5\,^4G^5$	7/2	$3d^5\,^2F^5$	5/2	16.396	16.372				
$3d^5\,^4G^5$	7/2	$3d^5\,^2H^3$	9/2	15.329	15.304				
$3d^5\,^4G^5$	7/2	$3d^5\,^2G^5$	7/2	14.861	14.845				
$3d^5\,^4G^5$	7/2	$3d^5\,^4F^3$	5/2	14.501	14.479				
$3d^5\,^4G^5$	7/2	$3d^5\,^2G^3$	7/2	13.462	13.444				
$3d^5\,^4G^5$	7/2	$3d^5\,^2G^3$	9/2	12.275	12.264				
$3d^5\,^4G^5$	7/2	$3d^5\,^2D^1$	5/2	10.329	10.321				
$3d^5\,^4G^5$	7/2	$3d^5\,^4P^3$	5/2	9.125	9.106				

Table 5. *Cont.*

Label and J for Lower		Label and J for Upper		Present Work $n = 5$	$n = 6$	Expt (Ref. [2])	GRASP	$RMBPT_g$	RCI_g
$3d^5\,^4G^5$	7/2	$3d^5\,^4F^3$	7/2	8.557	8.541				
$3d^5\,^4G^5$	7/2	$3d^5\,^2G^5$	9/2	8.057	8.044				
$3d^5\,^4G^5$	7/2	$3d^5\,^2D^3$	5/2	7.820	7.808				
$3d^5\,^4G^5$	7/2	$3d^5\,^2G^3$	7/2	7.653	7.642				
$3d^5\,^4G^5$	7/2	$3d^5\,^2D^1$	5/2	5.563	5.553				
$3d^5\,^2H^3$	11/2	$3d^5\,^2G^5$	9/2	222.726	223.030				
$3d^5\,^2H^3$	11/2	$3d^5\,^4G^5$	11/2	19.056	19.021	18.996(3)	19.064	19.002	19.185
$3d^5\,^2H^3$	11/2	$3d^5\,^4G^5$	9/2	18.486	18.453				
$3d^5\,^2H^3$	11/2	$3d^5\,^2I^5$	13/2	17.664	17.637				
$3d^5\,^2H^3$	11/2	$3d^5\,^2H^3$	9/2	15.593	15.566				
$3d^5\,^2H^3$	11/2	$3d^5\,^2G^3$	9/2	12.443	12.432				
$3d^5\,^2H^3$	11/2	$3d^5\,^2H^3$	11/2	8.544	8.528				
$3d^5\,^2H^3$	11/2	$3d^5\,^2G^5$	9/2	8.129	8.116				
$3d^5\,^4D^5$	3/2	$3d^5\,^4D^5$	5/2	140.653	140.944				
$3d^5\,^4D^5$	3/2	$3d^5\,^4P^3$	1/2	100.009	100.120				
$3d^5\,^4D^5$	3/2	$3d^5\,^2D^1$	3/2	40.920	40.907				
$3d^5\,^4D^5$	3/2	$3d^5\,^6S^5$	5/2	22.403	22.348				
$3d^5\,^4D^5$	3/2	$3d^5\,^4D^5$	1/2	17.360	17.334				
$3d^5\,^4D^5$	3/2	$3d^5\,^4P^3$	3/2	17.091	17.056				
$3d^5\,^4D^5$	3/2	$3d^5\,^2F^5$	5/2	16.799	16.775				
$3d^5\,^4D^5$	3/2	$3d^5\,^4F^3$	5/2	14.815	14.794				
$3d^5\,^4D^5$	3/2	$3d^5\,^2D^5$	3/2	14.708	14.696				
$3d^5\,^4D^5$	3/2	$3d^5\,^4P^3$	1/2	13.725	13.709				
$3d^5\,^4D^5$	3/2	$3d^5\,^2P^3$	3/2	10.807	10.802				
$3d^5\,^4D^5$	3/2	$3d^5\,^2D^1$	5/2	10.487	10.480				
$3d^5\,^4D^5$	3/2	$3d^5\,^4P^3$	5/2	9.248	9.230				
$3d^5\,^4D^5$	3/2	$3d^5\,^4D^5$	3/2	8.477	8.462				
$3d^5\,^4D^5$	3/2	$3d^5\,^2D^3$	5/2	7.910	7.899				
$3d^5\,^4D^5$	3/2	$3d^5\,^2P^3$	1/2	7.384	7.375				
$3d^5\,^4D^5$	3/2	$3d^5\,^2D^1$	3/2	6.741	6.733				
$3d^5\,^4D^5$	3/2	$3d^5\,^2D^1$	5/2	5.609	5.598				
$3d^5\,^2G^5$	9/2	$3d^5\,^2F^5$	7/2	149.683	150.005				
$3d^5\,^2G^5$	9/2	$3d^5\,^4D^5$	7/2	21.012	20.964				
$3d^5\,^2G^5$	9/2	$3d^5\,^4G^5$	11/2	20.839	20.795				
$3d^5\,^2G^5$	9/2	$3d^5\,^4G^5$	9/2	20.159	20.117				
$3d^5\,^2G^5$	9/2	$3d^5\,^2H^3$	9/2	16.767	16.734				
$3d^5\,^2G^5$	9/2	$3d^5\,^2G^5$	7/2	16.209	16.186				
$3d^5\,^2G^5$	9/2	$3d^5\,^2G^3$	7/2	14.558	14.535				
$3d^5\,^2G^5$	9/2	$3d^5\,^2G^3$	9/2	13.180	13.165				
$3d^5\,^2G^5$	9/2	$3d^5\,^4F^3$	7/2	8.987	8.969				
$3d^5\,^2G^5$	9/2	$3d^5\,^2H^3$	11/2	8.885	8.867				
$3d^5\,^2G^5$	9/2	$3d^5\,^2G^5$	9/2	8.437	8.422				
$3d^5\,^2G^5$	9/2	$3d^5\,^2G^3$	7/2	7.995	7.982				
$3d^5\,^4D^5$	5/2	$3d^5\,^2F^5$	7/2	270.094	271.516				
$3d^5\,^4D^5$	5/2	$3d^5\,^2D^1$	3/2	57.709	57.636				
$3d^5\,^4D^5$	5/2	$3d^5\,^6S^5$	5/2	26.647	26.559				
$3d^5\,^4D^5$	5/2	$3d^5\,^4D^5$	7/2	22.415	22.363				
$3d^5\,^4D^5$	5/2	$3d^5\,^4P^3$	3/2	19.454	19.404				
$3d^5\,^4D^5$	5/2	$3d^5\,^2F^5$	5/2	19.077	19.042				

Table 5. *Cont.*

Label and J for Lower		Label and J for Upper		Present Work		Expt (Ref. [2])	GRASP	$RMBPT_g$	RCI_g
				$n = 5$	$n = 6$				
$3d^5\,{}^4D^5$	5/2	$3d^5\,{}^2G^5$	7/2	17.031	17.007				
$3d^5\,{}^4D^5$	5/2	$3d^5\,{}^4F^3$	5/2	16.560	16.529				
$3d^5\,{}^4D^5$	5/2	$3d^5\,{}^2D^5$	3/2	16.426	16.407				
$3d^5\,{}^4D^5$	5/2	$3d^5\,{}^2G^3$	7/2	15.218	15.194				
$3d^5\,{}^4D^5$	5/2	$3d^5\,{}^2P^3$	3/2	11.706	11.699				
$3d^5\,{}^4D^5$	5/2	$3d^5\,{}^2D^1$	5/2	11.332	11.322				
$3d^5\,{}^4D^5$	5/2	$3d^5\,{}^4P^3$	5/2	9.899	9.876				
$3d^5\,{}^4D^5$	5/2	$3d^5\,{}^4F^3$	7/2	9.234	9.215				
$3d^5\,{}^4D^5$	5/2	$3d^5\,{}^4D^5$	3/2	9.021	9.002				
$3d^5\,{}^4D^5$	5/2	$3d^5\,{}^2D^3$	5/2	8.381	8.368				
$3d^5\,{}^4D^5$	5/2	$3d^5\,{}^2G^3$	7/2	8.190	8.177				
$3d^5\,{}^4D^5$	5/2	$3d^5\,{}^2D^1$	3/2	7.080	7.070				
$3d^5\,{}^4D^5$	5/2	$3d^5\,{}^2D^1$	5/2	5.842	5.830				
$3d^5\,{}^4P^3$	1/2	$3d^5\,{}^2D^1$	3/2	69.257	69.169				
$3d^5\,{}^4P^3$	1/2	$3d^5\,{}^4D^5$	1/2	21.006	20.963				
$3d^5\,{}^4P^3$	1/2	$3d^5\,{}^4P^3$	3/2	20.613	20.559				
$3d^5\,{}^4P^3$	1/2	$3d^5\,{}^2D^5$	3/2	17.245	17.225				
$3d^5\,{}^4P^3$	1/2	$3d^5\,{}^4P^3$	1/2	15.909	15.884				
$3d^5\,{}^4P^3$	1/2	$3d^5\,{}^2P^3$	3/2	12.116	12.109				
$3d^5\,{}^4P^3$	1/2	$3d^5\,{}^4D^5$	3/2	9.262	9.243				
$3d^5\,{}^4P^3$	1/2	$3d^5\,{}^2P^3$	1/2	7.973	7.962				
$3d^5\,{}^4P^3$	1/2	$3d^5\,{}^2D^1$	3/2	7.228	7.218				
$3d^5\,{}^2F^5$	7/2	$3d^5\,{}^6S^5$	5/2	29.564	29.438				
$3d^5\,{}^2F^5$	7/2	$3d^5\,{}^4D^5$	7/2	24.443	24.370				
$3d^5\,{}^2F^5$	7/2	$3d^5\,{}^4G^5$	9/2	23.296	23.233				
$3d^5\,{}^2F^5$	7/2	$3d^5\,{}^2F^5$	5/2	20.527	20.478				
$3d^5\,{}^2F^5$	7/2	$3d^5\,{}^2H^3$	9/2	18.882	18.835				
$3d^5\,{}^2F^5$	7/2	$3d^5\,{}^2G^5$	7/2	18.177	18.143				
$3d^5\,{}^2F^5$	7/2	$3d^5\,{}^4F^3$	5/2	17.641	17.600				
$3d^5\,{}^2F^5$	7/2	$3d^5\,{}^2G^3$	7/2	16.127	16.094				
$3d^5\,{}^2F^5$	7/2	$3d^5\,{}^2G^3$	9/2	14.452	14.432				
$3d^5\,{}^2F^5$	7/2	$3d^5\,{}^2D^1$	5/2	11.828	11.814				
$3d^5\,{}^2F^5$	7/2	$3d^5\,{}^4P^3$	5/2	10.275	10.249				
$3d^5\,{}^2F^5$	7/2	$3d^5\,{}^4F^3$	7/2	9.561	9.539				
$3d^5\,{}^2F^5$	7/2	$3d^5\,{}^2G^5$	9/2	8.941	8.923				
$3d^5\,{}^2F^5$	7/2	$3d^5\,{}^2D^3$	5/2	8.650	8.634				
$3d^5\,{}^2F^5$	7/2	$3d^5\,{}^2G^3$	7/2	8.446	8.431				
$3d^5\,{}^2F^5$	7/2	$3d^5\,{}^2D^1$	5/2	5.971	5.958				
$3d^5\,{}^2D^1$	3/2	$3d^5\,{}^6S^5$	5/2	49.507	49.256				
$3d^5\,{}^2D^1$	3/2	$3d^5\,{}^4D^5$	1/2	30.151	30.080				
$3d^5\,{}^2D^1$	3/2	$3d^5\,{}^4P^3$	3/2	29.348	29.253				
$3d^5\,{}^2D^1$	3/2	$3d^5\,{}^2F^5$	5/2	28.498	28.437				
$3d^5\,{}^2D^1$	3/2	$3d^5\,{}^4F^3$	5/2	23.224	23.175				
$3d^5\,{}^2D^1$	3/2	$3d^5\,{}^2D^5$	3/2	22.962	22.936				
$3d^5\,{}^2D^1$	3/2	$3d^5\,{}^4P^3$	1/2	20.653	20.619				
$3d^5\,{}^2D^1$	3/2	$3d^5\,{}^2P^3$	3/2	14.685	14.678				
$3d^5\,{}^2D^1$	3/2	$3d^5\,{}^2D^1$	5/2	14.101	14.089				
$3d^5\,{}^2D^1$	3/2	$3d^5\,{}^4P^3$	5/2	11.948	11.919				
$3d^5\,{}^2D^1$	3/2	$3d^5\,{}^4D^5$	3/2	10.692	10.668				

Table 5. *Cont.*

Label and J for Lower		Label and J for Upper		Present Work $n=5$	$n=6$	Expt (Ref. [2])	GRASP	$RMBPT_g$	RCI_g
$3d^5\,{}^2D^1$	3/2	$3d^5\,{}^2D^3$	5/2	9.806	9.789				
$3d^5\,{}^2D^1$	3/2	$3d^5\,{}^2P^3$	1/2	9.010	8.997				
$3d^5\,{}^2D^1$	3/2	$3d^5\,{}^2D^1$	3/2	8.071	8.059				
$3d^5\,{}^2D^1$	3/2	$3d^5\,{}^2D^1$	5/2	6.500	6.486				
$3d^5\,{}^6S^5$	5/2	$3d^5\,{}^4D^5$	7/2	141.121	141.559				
$3d^5\,{}^6S^5$	5/2	$3d^5\,{}^4P^3$	3/2	72.074	72.035				
$3d^5\,{}^6S^5$	5/2	$3d^5\,{}^2F^5$	5/2	67.156	67.278				
$3d^5\,{}^6S^5$	5/2	$3d^5\,{}^2G^5$	7/2	47.192	47.288				
$3d^5\,{}^6S^5$	5/2	$3d^5\,{}^4F^3$	5/2	43.743	43.766				
$3d^5\,{}^6S^5$	5/2	$3d^5\,{}^2D^5$	3/2	42.824	42.924				
$3d^5\,{}^6S^5$	5/2	$3d^5\,{}^2G^3$	7/2	35.481	35.505				
$3d^5\,{}^6S^5$	5/2	$3d^5\,{}^2P^3$	3/2	20.879	20.910				
$3d^5\,{}^6S^5$	5/2	$3d^5\,{}^2D^1$	5/2	19.717	19.734				
$3d^5\,{}^6S^5$	5/2	$3d^5\,{}^4P^3$	5/2	15.749	15.723				
$3d^5\,{}^6S^5$	5/2	$3d^5\,{}^4F^3$	7/2	14.130	14.111				
$3d^5\,{}^6S^5$	5/2	$3d^5\,{}^4D^5$	3/2	13.638	13.618				
$3d^5\,{}^6S^5$	5/2	$3d^5\,{}^2D^3$	5/2	12.227	12.217				
$3d^5\,{}^6S^5$	5/2	$3d^5\,{}^2G^3$	7/2	11.824	11.815				
$3d^5\,{}^6S^5$	5/2	$3d^5\,{}^2D^1$	3/2	9.643	9.636				
$3d^5\,{}^6S^5$	5/2	$3d^5\,{}^2D^1$	5/2	7.482	7.470				
$3d^5\,{}^4D^5$	7/2	$3d^5\,{}^4G^5$	9/2	496.548	497.861				
$3d^5\,{}^4D^5$	7/2	$3d^5\,{}^2F^5$	5/2	128.131	128.214				
$3d^5\,{}^4D^5$	7/2	$3d^5\,{}^2H^3$	9/2	82.993	82.924				
$3d^5\,{}^4D^5$	7/2	$3d^5\,{}^2G^5$	7/2	70.903	71.009				
$3d^5\,{}^4D^5$	7/2	$3d^5\,{}^4F^3$	5/2	63.393	63.353				
$3d^5\,{}^4D^5$	7/2	$3d^5\,{}^2G^3$	7/2	47.397	47.392				
$3d^5\,{}^4D^5$	7/2	$3d^5\,{}^2G^3$	9/2	35.359	35.390				
$3d^5\,{}^4D^5$	7/2	$3d^5\,{}^2D^1$	5/2	22.919	22.931				
$3d^5\,{}^4D^5$	7/2	$3d^5\,{}^4P^3$	5/2	17.728	17.688				
$3d^5\,{}^4D^5$	7/2	$3d^5\,{}^4F^3$	7/2	15.703	15.674				
$3d^5\,{}^4D^5$	7/2	$3d^5\,{}^2G^5$	9/2	14.097	14.078				
$3d^5\,{}^4D^5$	7/2	$3d^5\,{}^2D^3$	5/2	13.387	13.371				
$3d^5\,{}^4D^5$	7/2	$3d^5\,{}^2G^3$	7/2	12.906	12.891				
$3d^5\,{}^4D^5$	7/2	$3d^5\,{}^2D^1$	5/2	7.901	7.886				
$3d^5\,{}^4G^5$	11/2	$3d^5\,{}^4G^5$	9/2	617.468	617.572				
$3d^5\,{}^4G^5$	11/2	$3d^5\,{}^2I^5$	13/2	241.792	242.297				
$3d^5\,{}^4G^5$	11/2	$3d^5\,{}^2H^3$	9/2	85.801	85.691				
$3d^5\,{}^4G^5$	11/2	$3d^5\,{}^2G^3$	9/2	35.859	35.885				
$3d^5\,{}^4G^5$	11/2	$3d^5\,{}^2H^3$	11/2	15.488	15.460				
$3d^5\,{}^4G^5$	11/2	$3d^5\,{}^2G^5$	9/2	14.176	14.155				
$3d^5\,{}^4G^5$	9/2	$3d^5\,{}^2H^3$	9/2	99.648	99.496				
$3d^5\,{}^4G^5$	9/2	$3d^5\,{}^2G^5$	7/2	82.713	82.822				
$3d^5\,{}^4G^5$	9/2	$3d^5\,{}^2G^3$	7/2	52.399	52.378				
$3d^5\,{}^4G^5$	9/2	$3d^5\,{}^2G^3$	9/2	38.069	38.098				
$3d^5\,{}^4G^5$	9/2	$3d^5\,{}^4F^3$	7/2	16.215	16.183				
$3d^5\,{}^4G^5$	9/2	$3d^5\,{}^2H^3$	11/2	15.887	15.857				
$3d^5\,{}^4G^5$	9/2	$3d^5\,{}^2G^5$	9/2	14.509	14.487				
$3d^5\,{}^4G^5$	9/2	$3d^5\,{}^2G^3$	7/2	13.250	13.233				
$3d^5\,{}^2I^5$	13/2	$3d^5\,{}^2H^3$	11/2	16.548	16.513				

Table 5. *Cont.*

Label and J for Lower		Label and J for Upper		Present Work $n=5$	$n=6$	Expt (Ref. [2])	GRASP	$RMBPT_g$	RCI_g
$3d^5\ ^4D^5$	1/2	$3d^5\ ^4P^3$	3/2	1102.758	1064.996				
$3d^5\ ^4D^5$	1/2	$3d^5\ ^2D^5$	3/2	96.307	96.583				
$3d^5\ ^4D^5$	1/2	$3d^5\ ^4P^3$	1/2	65.560	65.562				
$3d^5\ ^4D^5$	1/2	$3d^5\ ^2P^3$	3/2	28.630	28.668				
$3d^5\ ^4D^5$	1/2	$3d^5\ ^4D^5$	3/2	16.568	16.532				
$3d^5\ ^4D^5$	1/2	$3d^5\ ^2P^3$	1/2	12.849	12.837				
$3d^5\ ^4D^5$	1/2	$3d^5\ ^2D^1$	3/2	11.021	11.009				
$3d^5\ ^4P^3$	3/2	$3d^5\ ^2F^5$	5/2	984.197	1018.757				
$3d^5\ ^4P^3$	3/2	$3d^5\ ^4F^3$	5/2	111.284	111.525				
$3d^5\ ^4P^3$	3/2	$3d^5\ ^2D^5$	3/2	105.523	106.216				
$3d^5\ ^4P^3$	3/2	$3d^5\ ^4P^3$	1/2	69.704	69.863				
$3d^5\ ^4P^3$	3/2	$3d^5\ ^2P^3$	3/2	29.393	29.461				
$3d^5\ ^4P^3$	3/2	$3d^5\ ^2D^1$	5/2	27.141	27.180				
$3d^5\ ^4P^3$	3/2	$3d^5\ ^4P^3$	5/2	20.153	20.113				
$3d^5\ ^4P^3$	3/2	$3d^5\ ^4D^5$	3/2	16.820	16.793				
$3d^5\ ^4P^3$	3/2	$3d^5\ ^2D^3$	5/2	14.726	14.712				
$3d^5\ ^4P^3$	3/2	$3d^5\ ^2P^3$	1/2	13.001	12.994				
$3d^5\ ^4P^3$	3/2	$3d^5\ ^2D^1$	3/2	11.132	11.124				
$3d^5\ ^4P^3$	3/2	$3d^5\ ^2D^1$	5/2	8.349	8.334				
$3d^5\ ^2F^5$	5/2	$3d^5\ ^2G^5$	7/2	158.747	159.155				
$3d^5\ ^2F^5$	5/2	$3d^5\ ^4F^3$	5/2	125.471	125.234				
$3d^5\ ^2F^5$	5/2	$3d^5\ ^2D^5$	3/2	118.195	118.578				
$3d^5\ ^2F^5$	5/2	$3d^5\ ^2G^3$	7/2	75.224	75.182				
$3d^5\ ^2F^5$	5/2	$3d^5\ ^2P^3$	3/2	30.298	30.339				
$3d^5\ ^2F^5$	5/2	$3d^5\ ^2D^1$	5/2	27.911	27.925				
$3d^5\ ^2F^5$	5/2	$3d^5\ ^4P^3$	5/2	20.575	20.518				
$3d^5\ ^2F^5$	5/2	$3d^5\ ^4F^3$	7/2	17.896	17.857				
$3d^5\ ^2F^5$	5/2	$3d^5\ ^4D^5$	3/2	17.113	17.074				
$3d^5\ ^2F^5$	5/2	$3d^5\ ^2D^3$	5/2	14.949	14.928				
$3d^5\ ^2F^5$	5/2	$3d^5\ ^2G^3$	7/2	14.351	14.331				
$3d^5\ ^2F^5$	5/2	$3d^5\ ^2D^1$	3/2	11.259	11.246				
$3d^5\ ^2F^5$	5/2	$3d^5\ ^2D^1$	5/2	8.420	8.403				
$3d^5\ ^2H^3$	9/2	$3d^5\ ^2G^5$	7/2	486.710	494.218				
$3d^5\ ^2H^3$	9/2	$3d^5\ ^2G^3$	7/2	110.511	110.603				
$3d^5\ ^2H^3$	9/2	$3d^5\ ^2G^3$	9/2	61.605	61.739				
$3d^5\ ^2H^3$	9/2	$3d^5\ ^4F^3$	7/2	19.367	19.327				
$3d^5\ ^2H^3$	9/2	$3d^5\ ^2H^3$	11/2	18.900	18.863				
$3d^5\ ^2H^3$	9/2	$3d^5\ ^2G^5$	9/2	16.982	16.957				
$3d^5\ ^2H^3$	9/2	$3d^5\ ^2G^3$	7/2	15.282	15.263				
$3d^5\ ^2G^5$	7/2	$3d^5\ ^4F^3$	5/2	598.575	587.600				
$3d^5\ ^2G^5$	7/2	$3d^5\ ^2G^3$	7/2	142.974	142.492				
$3d^5\ ^2G^5$	7/2	$3d^5\ ^2G^3$	9/2	70.533	70.553				
$3d^5\ ^2G^5$	7/2	$3d^5\ ^2D^1$	5/2	33.865	33.867				
$3d^5\ ^2G^5$	7/2	$3d^5\ ^4P^3$	5/2	23.638	23.555				
$3d^5\ ^2G^5$	7/2	$3d^5\ ^4F^3$	7/2	20.170	20.113				
$3d^5\ ^2G^5$	7/2	$3d^5\ ^2G^5$	9/2	17.595	17.559				
$3d^5\ ^2G^5$	7/2	$3d^5\ ^2D^3$	5/2	16.503	16.473				
$3d^5\ ^2G^5$	7/2	$3d^5\ ^2G^3$	7/2	15.777	15.750				
$3d^5\ ^2G^5$	7/2	$3d^5\ ^2D^1$	5/2	8.892	8.871				

Table 5. *Cont.*

Label and J for Lower		Label and J for Upper		Present Work $n=5$	$n=6$	Expt (Ref. [2])	GRASP	$RMBPT_g$	RCI_g
$3d^5\,^4F^3$	5/2	$3d^5\,^2D^5$	3/2	2038.287	2231.127				
$3d^5\,^4F^3$	5/2	$3d^5\,^2G^3$	7/2	187.841	188.108				
$3d^5\,^4F^3$	5/2	$3d^5\,^2P^3$	3/2	39.944	40.038				
$3d^5\,^4F^3$	5/2	$3d^5\,^2D^1$	5/2	35.896	35.939				
$3d^5\,^4F^3$	5/2	$3d^5\,^4P^3$	5/2	24.610	24.539				
$3d^5\,^4F^3$	5/2	$3d^5\,^4F^3$	7/2	20.873	20.826				
$3d^5\,^4F^3$	5/2	$3d^5\,^4D^5$	3/2	19.815	19.769				
$3d^5\,^4F^3$	5/2	$3d^5\,^2D^3$	5/2	16.971	16.948				
$3d^5\,^4F^3$	5/2	$3d^5\,^2G^3$	7/2	16.205	16.183				
$3d^5\,^4F^3$	5/2	$3d^5\,^2D^1$	3/2	12.369	12.356				
$3d^5\,^4F^3$	5/2	$3d^5\,^2D^1$	5/2	9.026	9.007				
$3d^5\,^2D^5$	3/2	$3d^5\,^4P^3$	1/2	205.353	204.123				
$3d^5\,^2D^5$	3/2	$3d^5\,^2P^3$	3/2	40.742	40.770				
$3d^5\,^2D^5$	3/2	$3d^5\,^2D^1$	5/2	36.540	36.527				
$3d^5\,^2D^5$	3/2	$3d^5\,^4P^3$	5/2	24.911	24.812				
$3d^5\,^2D^5$	3/2	$3d^5\,^4D^5$	3/2	20.010	19.946				
$3d^5\,^2D^5$	3/2	$3d^5\,^2D^3$	5/2	17.114	17.078				
$3d^5\,^2D^5$	3/2	$3d^5\,^2P^3$	1/2	14.828	14.805				
$3d^5\,^2D^5$	3/2	$3d^5\,^2D^1$	3/2	12.445	12.425				
$3d^5\,^2D^5$	3/2	$3d^5\,^2D^1$	5/2	9.066	9.044				
$3d^5\,^2G^3$	7/2	$3d^5\,^2G^3$	9/2	139.208	139.747				
$3d^5\,^2G^3$	7/2	$3d^5\,^2D^1$	5/2	44.377	44.427				
$3d^5\,^2G^3$	7/2	$3d^5\,^4P^3$	5/2	28.321	28.220				
$3d^5\,^2G^3$	7/2	$3d^5\,^4F^3$	7/2	23.482	23.419				
$3d^5\,^2G^3$	7/2	$3d^5\,^2G^5$	9/2	20.065	20.027				
$3d^5\,^2G^3$	7/2	$3d^5\,^2D^3$	5/2	18.657	18.626				
$3d^5\,^2G^3$	7/2	$3d^5\,^2G^3$	7/2	17.735	17.707				
$3d^5\,^2G^3$	7/2	$3d^5\,^2D^1$	5/2	9.482	9.460				
$3d^5\,^4P^3$	1/2	$3d^5\,^2P^3$	3/2	50.826	50.945				
$3d^5\,^4P^3$	1/2	$3d^5\,^4D^5$	3/2	22.170	22.106				
$3d^5\,^4P^3$	1/2	$3d^5\,^2P^3$	1/2	15.982	15.963				
$3d^5\,^4P^3$	1/2	$3d^5\,^2D^1$	3/2	13.248	13.230				
$3d^5\,^2G^3$	9/2	$3d^5\,^4F^3$	7/2	28.247	28.134				
$3d^5\,^2G^3$	9/2	$3d^5\,^2H^3$	11/2	27.264	27.162				
$3d^5\,^2G^3$	9/2	$3d^5\,^2G^5$	9/2	23.444	23.377				
$3d^5\,^2G^3$	9/2	$3d^5\,^2G^3$	7/2	20.324	20.276				
$3d^5\,^2P^3$	3/2	$3d^5\,^2D^1$	5/2	354.267	351.029				
$3d^5\,^2P^3$	3/2	$3d^5\,^4P^3$	5/2	64.110	63.389				
$3d^5\,^2P^3$	3/2	$3d^5\,^4D^5$	3/2	39.323	39.051				
$3d^5\,^2P^3$	3/2	$3d^5\,^2D^3$	5/2	29.510	29.388				
$3d^5\,^2P^3$	3/2	$3d^5\,^2P^3$	1/2	23.312	23.247				
$3d^5\,^2P^3$	3/2	$3d^5\,^2D^1$	3/2	17.918	17.871				
$3d^5\,^2P^3$	3/2	$3d^5\,^2D^1$	5/2	11.661	11.621				
$3d^5\,^2D^1$	5/2	$3d^5\,^4P^3$	5/2	78.275	77.359				
$3d^5\,^2D^1$	5/2	$3d^5\,^4F^3$	7/2	49.873	49.527				
$3d^5\,^2D^1$	5/2	$3d^5\,^4D^5$	3/2	44.232	43.939				
$3d^5\,^2D^1$	5/2	$3d^5\,^2D^3$	5/2	32.191	32.073				
$3d^5\,^2D^1$	5/2	$3d^5\,^2G^3$	7/2	29.540	29.440				
$3d^5\,^2D^1$	5/2	$3d^5\,^2D^1$	3/2	18.873	18.830				

Table 5. *Cont.*

Label and J for Lower		Label and J for Upper		Present Work $n=5$	$n=6$	Expt (Ref. [2])	GRASP	RMBPT$_g$	RCI$_g$
$3d^5\,{}^2D^1$	5/2	$3d^5\,{}^2D^1$	5/2	12.058	12.019				
$3d^5\,{}^4P^3$	5/2	$3d^5\,{}^4F^3$	7/2	137.450	137.658				
$3d^5\,{}^4P^3$	5/2	$3d^5\,{}^4D^5$	3/2	101.706	101.709				
$3d^5\,{}^4P^3$	5/2	$3d^5\,{}^2D^3$	5/2	54.677	54.789				
$3d^5\,{}^4P^3$	5/2	$3d^5\,{}^2G^3$	7/2	47.445	47.528				
$3d^5\,{}^4P^3$	5/2	$3d^5\,{}^2D^1$	3/2	24.869	24.888				
$3d^5\,{}^4P^3$	5/2	$3d^5\,{}^2D^1$	5/2	14.254	14.230				
$3d^5\,{}^4F^3$	7/2	$3d^5\,{}^2G^5$	9/2	137.868	138.255				
$3d^5\,{}^4F^3$	7/2	$3d^5\,{}^2D^3$	5/2	90.796	91.013				
$3d^5\,{}^4F^3$	7/2	$3d^5\,{}^2G^3$	7/2	72.454	72.591				
$3d^5\,{}^4F^3$	7/2	$3d^5\,{}^2D^1$	5/2	15.903	15.871				
$3d^5\,{}^2H^3$	11/2	$3d^5\,{}^2G^5$	9/2	167.319	167.768				
$3d^5\,{}^4D^5$	3/2	$3d^5\,{}^2D^3$	5/2	118.248	118.767				
$3d^5\,{}^4D^5$	3/2	$3d^5\,{}^2P^3$	1/2	57.255	57.442				
$3d^5\,{}^4D^5$	3/2	$3d^5\,{}^2D^1$	3/2	32.917	32.951				
$3d^5\,{}^4D^5$	3/2	$3d^5\,{}^2D^1$	5/2	16.577	16.545				
$3d^5\,{}^2G^5$	9/2	$3d^5\,{}^2G^3$	7/2	152.705	152.840				
$3d^5\,{}^2D^3$	5/2	$3d^5\,{}^2G^3$	7/2	358.662	358.634				
$3d^5\,{}^2D^3$	5/2	$3d^5\,{}^2D^1$	3/2	45.616	45.602				
$3d^5\,{}^2D^3$	5/2	$3d^5\,{}^2D^1$	5/2	19.280	19.223				
$3d^5\,{}^2G^3$	7/2	$3d^5\,{}^2D^1$	5/2	20.376	20.312				
$3d^5\,{}^2P^3$	1/2	$3d^5\,{}^2D^1$	3/2	77.440	77.283				
$3d^5\,{}^2D^1$	3/2	$3d^5\,{}^2D^1$	5/2	33.396	33.231				
W^{50+} (Cr-like)									
$3d^6\,{}^5D^4$	4	$3d^6\,{}^5D^4$	3	19.752	19.720	19.684(3)	19.796	19.693	19.835
$3d^6\,{}^5D^4$	4	$3d^6\,{}^5D^4$	4	19.291	19.267	19.239(3)	19.303	19.237	19.425
$3d^6\,{}^5D^4$	4	$3d^6\,{}^3G^4$	5	17.162	17.150	17.133(3)	17.118	17.131	17.259
$3d^6\,{}^5D^4$	4	$3d^6\,{}^3D^4$	3	15.390	15.378	15.363(3)	15.370	15.289	15.316
$3d^6\,{}^5D^4$	4	$3d^6\,{}^1G^2$	4	13.147	13.139	13.137(3)	13.114	13.153	13.050
$3d^6\,{}^5D^4$	4	$3d^6\,{}^3F^2$	3	12.785	12.779	12.779(3)	12.739	12.800	12.642
$3d^6\,{}^5D^4$	4	$3d^6\,{}^3H^4$	4	9.024	9.013				
$3d^6\,{}^5D^4$	4	$3d^6\,{}^5D^4$	3	8.809	8.798				
$3d^6\,{}^5D^4$	4	$3d^6\,{}^3H^4$	5	8.756	8.746				
$3d^6\,{}^5D^4$	4	$3d^6\,{}^3F^4$	3	8.356	8.347				
$3d^6\,{}^5D^4$	4	$3d^6\,{}^1G^4$	4	8.111	8.104				
$3d^6\,{}^5D^4$	4	$3d^6\,{}^3F^2$	4	7.274	7.269				
$3d^6\,{}^5D^4$	4	$3d^6\,{}^1G^2$	4	5.668	5.661				
$3d^6\,{}^5D^4$	4	$3d^6\,{}^3F^2$	3	5.461	5.455				
$3d^6\,{}^3D^4$	2	$3d^6\,{}^5D^4$	3	22.546	22.503				
$3d^6\,{}^3D^4$	2	$3d^6\,{}^5D^4$	1	20.709	20.668				
$3d^6\,{}^3D^4$	2	$3d^6\,{}^3F^4$	2	17.384	17.367				
$3d^6\,{}^3D^4$	2	$3d^6\,{}^3D^4$	3	17.035	17.019				
$3d^6\,{}^3D^4$	2	$3d^6\,{}^3D^4$	2	14.214	14.202	14.193(3)	14.184	14.202	14.170
$3d^6\,{}^3D^4$	2	$3d^6\,{}^3P^2$	1	14.198	14.188				
$3d^6\,{}^3D^4$	2	$3d^6\,{}^3F^2$	3	13.900	13.892	13.886(3)	13.843	13.895	13.848
$3d^6\,{}^3D^4$	2	$3d^6\,{}^5D^4$	2	10.042	10.025				
$3d^6\,{}^3D^4$	2	$3d^6\,{}^5D^4$	3	9.324	9.312				
$3d^6\,{}^3D^4$	2	$3d^6\,{}^3F^4$	3	8.818	8.808				

Table 5. *Cont.*

Label and J for Lower		Label and J for Upper		Present Work $n = 5$	$n = 6$	Expt (Ref. [2])	GRASP	RMBPT$_g$	RCI$_g$
$3d^6\,^3D^4$	2	$3d^6\,^3D^4$	1	8.662	8.652				
$3d^6\,^3D^4$	2	$3d^6\,^3F^4$	2	8.468	8.460				
$3d^6\,^3D^4$	2	$3d^6\,^3F^2$	2	7.848	7.839				
$3d^6\,^3D^4$	2	$3d^6\,^1D^2$	2	6.867	6.863				
$3d^6\,^3D^4$	2	$3d^6\,^3P^2$	1	6.248	6.238				
$3d^6\,^3D^4$	2	$3d^6\,^3P^2$	2	5.711	5.704				
$3d^6\,^3D^4$	2	$3d^6\,^3F^2$	3	5.655	5.648				
$3d^6\,^3P^2$	0	$3d^6\,^5D^4$	1	29.559	29.509				
$3d^6\,^3P^2$	0	$3d^6\,^3P^2$	1	17.865	17.861	17.826(3)	17.750	17.837	17.921
$3d^6\,^3P^2$	0	$3d^6\,^3D^4$	1	9.901	9.893				
$3d^6\,^3P^2$	0	$3d^6\,^3P^2$	1	6.868	6.858				
$3d^6\,^5D^4$	3	$3d^6\,^5D^4$	4	827.797	838.107				
$3d^6\,^5D^4$	3	$3d^6\,^3F^4$	2	75.926	76.090				
$3d^6\,^5D^4$	3	$3d^6\,^3D^4$	3	69.700	69.830				
$3d^6\,^5D^4$	3	$3d^6\,^1G^2$	4	39.314	39.365				
$3d^6\,^5D^4$	3	$3d^6\,^3D^4$	2	38.467	38.499				
$3d^6\,^5D^4$	3	$3d^6\,^3F^2$	3	36.246	36.304				
$3d^6\,^5D^4$	3	$3d^6\,^5D^4$	2	18.107	18.080				
$3d^6\,^5D^4$	3	$3d^6\,^3H^4$	4	16.615	16.599				
$3d^6\,^5D^4$	3	$3d^6\,^5D^4$	3	15.900	15.886				
$3d^6\,^5D^4$	3	$3d^6\,^3F^4$	3	14.482	14.472				
$3d^6\,^5D^4$	3	$3d^6\,^1G^4$	4	13.764	13.758				
$3d^6\,^5D^4$	3	$3d^6\,^3F^4$	2	13.562	13.556				
$3d^6\,^5D^4$	3	$3d^6\,^3F^2$	2	12.038	12.029				
$3d^6\,^5D^4$	3	$3d^6\,^3F^2$	4	11.514	11.512				
$3d^6\,^5D^4$	3	$3d^6\,^1D^2$	2	9.875	9.874				
$3d^6\,^5D^4$	3	$3d^6\,^1G^2$	4	7.949	7.940				
$3d^6\,^5D^4$	3	$3d^6\,^3P^2$	2	7.648	7.641				
$3d^6\,^5D^4$	3	$3d^6\,^3F^2$	3	7.547	7.540				
$3d^6\,^5D^4$	4	$3d^6\,^3G^4$	5	155.438	156.080				
$3d^6\,^5D^4$	4	$3d^6\,^3D^4$	3	76.108	76.177				
$3d^6\,^5D^4$	4	$3d^6\,^1G^2$	4	41.274	41.305				
$3d^6\,^5D^4$	4	$3d^6\,^3F^2$	3	37.905	37.948				
$3d^6\,^5D^4$	4	$3d^6\,^3H^4$	4	16.955	16.934				
$3d^6\,^5D^4$	4	$3d^6\,^5D^4$	3	16.211	16.193				
$3d^6\,^5D^4$	4	$3d^6\,^3H^4$	5	16.032	16.018				
$3d^6\,^5D^4$	4	$3d^6\,^3F^4$	3	14.740	14.726				
$3d^6\,^5D^4$	4	$3d^6\,^1G^4$	4	13.997	13.987				
$3d^6\,^5D^4$	4	$3d^6\,^3F^2$	4	11.676	11.672				
$3d^6\,^5D^4$	4	$3d^6\,^1G^2$	4	8.026	8.016				
$3d^6\,^5D^4$	4	$3d^6\,^3F^2$	3	7.617	7.609				
$3d^6\,^5D^4$	1	$3d^6\,^3F^4$	2	108.263	108.730				
$3d^6\,^5D^4$	1	$3d^6\,^3P^4$	0	55.744	55.924				
$3d^6\,^5D^4$	1	$3d^6\,^3D^4$	2	45.325	45.393				
$3d^6\,^5D^4$	1	$3d^6\,^3P^2$	1	45.157	45.248				
$3d^6\,^5D^4$	1	$3d^6\,^5D^4$	2	19.495	19.468				
$3d^6\,^5D^4$	1	$3d^6\,^5D^4$	0	18.606	18.576				
$3d^6\,^5D^4$	1	$3d^6\,^3D^4$	1	14.889	14.882				
$3d^6\,^5D^4$	1	$3d^6\,^3F^4$	2	14.326	14.322				

Table 5. *Cont.*

Label and J for Lower		Label and J for Upper		Present Work $n=5$	$n=6$	Expt (Ref. [2])	GRASP	$RMBPT_g$	RCI_g
$3d^6\,{}^5D^4$	1	$3d^6\,{}^3F^2$	2	12.637	12.629				
$3d^6\,{}^5D^4$	1	$3d^6\,{}^1D^2$	2	10.274	10.274				
$3d^6\,{}^5D^4$	1	$3d^6\,{}^1S^0$	0	8.969	8.968				
$3d^6\,{}^5D^4$	1	$3d^6\,{}^3P^2$	1	8.947	8.934				
$3d^6\,{}^5D^4$	1	$3d^6\,{}^3P^2$	2	7.886	7.878				
$3d^6\,{}^5D^4$	1	$3d^6\,{}^3P^2$	0	5.630	5.622				
$3d^6\,{}^3H^4$	6	$3d^6\,{}^3G^4$	5	33171.897	25414.252				
$3d^6\,{}^3H^4$	6	$3d^6\,{}^3H^4$	5	17.866	17.837				
$3d^6\,{}^3H^4$	6	$3d^6\,{}^1I^4$	6	17.042	17.020				
$3d^6\,{}^3G^4$	5	$3d^6\,{}^1G^2$	4	56.196	56.170				
$3d^6\,{}^3G^4$	5	$3d^6\,{}^3H^4$	4	19.031	18.995				
$3d^6\,{}^3G^4$	5	$3d^6\,{}^3H^4$	5	17.876	17.850				
$3d^6\,{}^3G^4$	5	$3d^6\,{}^1I^4$	6	17.051	17.032				
$3d^6\,{}^3G^4$	5	$3d^6\,{}^1G^4$	4	15.382	15.364				
$3d^6\,{}^3G^4$	5	$3d^6\,{}^3F^2$	4	12.625	12.615				
$3d^6\,{}^3G^4$	5	$3d^6\,{}^1G^2$	4	8.463	8.450				
$3d^6\,{}^3F^4$	2	$3d^6\,{}^3D^4$	3	850.017	848.731				
$3d^6\,{}^3F^4$	2	$3d^6\,{}^3D^4$	2	77.967	77.927				
$3d^6\,{}^3F^4$	2	$3d^6\,{}^3P^2$	1	77.469	77.499				
$3d^6\,{}^3F^4$	2	$3d^6\,{}^3F^2$	3	69.354	69.431				
$3d^6\,{}^3F^4$	2	$3d^6\,{}^5D^4$	2	23.777	23.715				
$3d^6\,{}^3F^4$	2	$3d^6\,{}^5D^4$	3	20.111	20.078				
$3d^6\,{}^3F^4$	2	$3d^6\,{}^3F^4$	3	17.895	17.871				
$3d^6\,{}^3F^4$	2	$3d^6\,{}^3D^4$	1	17.263	17.242				
$3d^6\,{}^3F^4$	2	$3d^6\,{}^3F^4$	2	16.511	16.494				
$3d^6\,{}^3F^4$	2	$3d^6\,{}^3F^2$	2	14.307	14.288				
$3d^6\,{}^3F^4$	2	$3d^6\,{}^1D^2$	2	11.351	11.346				
$3d^6\,{}^3F^4$	2	$3d^6\,{}^3P^2$	1	9.754	9.734				
$3d^6\,{}^3F^4$	2	$3d^6\,{}^3P^2$	2	8.505	8.494				
$3d^6\,{}^3F^4$	2	$3d^6\,{}^3F^2$	3	8.381	8.370				
$3d^6\,{}^3D^4$	3	$3d^6\,{}^1G^2$	4	90.179	90.232				
$3d^6\,{}^3D^4$	3	$3d^6\,{}^3D^4$	2	85.841	85.805				
$3d^6\,{}^3D^4$	3	$3d^6\,{}^3F^2$	3	75.516	75.617				
$3d^6\,{}^3D^4$	3	$3d^6\,{}^5D^4$	2	24.461	24.396				
$3d^6\,{}^3D^4$	3	$3d^6\,{}^3H^4$	4	21.815	21.774				
$3d^6\,{}^3D^4$	3	$3d^6\,{}^5D^4$	3	20.598	20.565				
$3d^6\,{}^3D^4$	3	$3d^6\,{}^3F^4$	3	18.280	18.255				
$3d^6\,{}^3D^4$	3	$3d^6\,{}^1G^4$	4	17.151	17.133				
$3d^6\,{}^3D^4$	3	$3d^6\,{}^3F^4$	2	16.838	16.821				
$3d^6\,{}^3D^4$	3	$3d^6\,{}^3F^2$	2	14.552	14.533				
$3d^6\,{}^3D^4$	3	$3d^6\,{}^3F^2$	4	13.792	13.784				
$3d^6\,{}^3D^4$	3	$3d^6\,{}^1D^2$	2	11.505	11.500				
$3d^6\,{}^3D^4$	3	$3d^6\,{}^1G^2$	4	8.972	8.959				
$3d^6\,{}^3D^4$	3	$3d^6\,{}^3P^2$	2	8.591	8.579				
$3d^6\,{}^3D^4$	3	$3d^6\,{}^3F^2$	3	8.464	8.453				
$3d^6\,{}^3P^4$	0	$3d^6\,{}^3P^2$	1	237.757	237.023				
$3d^6\,{}^3P^4$	0	$3d^6\,{}^3D^4$	1	20.315	20.278				
$3d^6\,{}^3P^4$	0	$3d^6\,{}^3P^2$	1	10.658	10.633				
$3d^6\,{}^1G^2$	4	$3d^6\,{}^3F^2$	3	464.412	466.858				

Table 5. *Cont.*

Label and J for Lower		Label and J for Upper		Present Work $n = 5$	$n = 6$	Expt (Ref. [2])	GRASP	$RMBPT_g$	RCI_g
$3d^6\,{}^1G^2$	4	$3d^6\,{}^3H^4$	4	28.776	28.700				
$3d^6\,{}^1G^2$	4	$3d^6\,{}^5D^4$	3	26.696	26.635				
$3d^6\,{}^1G^2$	4	$3d^6\,{}^3H^4$	5	26.215	26.164				
$3d^6\,{}^1G^2$	4	$3d^6\,{}^3F^4$	3	22.928	22.886				
$3d^6\,{}^1G^2$	4	$3d^6\,{}^1G^4$	4	21.178	21.149				
$3d^6\,{}^1G^2$	4	$3d^6\,{}^3F^2$	4	16.283	16.269				
$3d^6\,{}^1G^2$	4	$3d^6\,{}^1G^2$	4	9.963	9.946				
$3d^6\,{}^1G^2$	4	$3d^6\,{}^3F^2$	3	9.341	9.327				
$3d^6\,{}^3D^4$	2	$3d^6\,{}^3P^2$	1	12130.770	14130.082				
$3d^6\,{}^3D^4$	2	$3d^6\,{}^3F^2$	3	627.813	636.845				
$3d^6\,{}^3D^4$	2	$3d^6\,{}^5D^4$	2	34.209	34.088				
$3d^6\,{}^3D^4$	2	$3d^6\,{}^5D^4$	3	27.101	27.047				
$3d^6\,{}^3D^4$	2	$3d^6\,{}^3F^4$	3	23.226	23.189				
$3d^6\,{}^3D^4$	2	$3d^6\,{}^3D^4$	1	22.172	22.140				
$3d^6\,{}^3D^4$	2	$3d^6\,{}^3F^4$	2	20.946	20.922				
$3d^6\,{}^3D^4$	2	$3d^6\,{}^3F^2$	2	17.522	17.497				
$3d^6\,{}^3D^4$	2	$3d^6\,{}^1D^2$	2	13.285	13.280				
$3d^6\,{}^3D^4$	2	$3d^6\,{}^3P^2$	1	11.148	11.124				
$3d^6\,{}^3D^4$	2	$3d^6\,{}^3P^2$	2	9.546	9.533				
$3d^6\,{}^3D^4$	2	$3d^6\,{}^3F^2$	3	9.390	9.377				
$3d^6\,{}^3P^2$	1	$3d^6\,{}^5D^4$	2	34.306	34.171				
$3d^6\,{}^3P^2$	1	$3d^6\,{}^5D^4$	0	31.646	31.512				
$3d^6\,{}^3P^2$	1	$3d^6\,{}^3D^4$	1	22.213	22.175				
$3d^6\,{}^3P^2$	1	$3d^6\,{}^3F^4$	2	20.982	20.954				
$3d^6\,{}^3P^2$	1	$3d^6\,{}^3F^2$	2	17.548	17.518				
$3d^6\,{}^3P^2$	1	$3d^6\,{}^1D^2$	2	13.300	13.292				
$3d^6\,{}^3P^2$	1	$3d^6\,{}^1S^0$	0	11.192	11.185				
$3d^6\,{}^3P^2$	1	$3d^6\,{}^3P^2$	1	11.158	11.132				
$3d^6\,{}^3P^2$	1	$3d^6\,{}^3P^2$	2	9.554	9.539				
$3d^6\,{}^3P^2$	1	$3d^6\,{}^3P^2$	0	6.432	6.419				
$3d^6\,{}^3F^2$	3	$3d^6\,{}^5D^4$	2	36.181	36.016				
$3d^6\,{}^3F^2$	3	$3d^6\,{}^3H^4$	4	30.677	30.580				
$3d^6\,{}^3F^2$	3	$3d^6\,{}^5D^4$	3	28.324	28.246				
$3d^6\,{}^3F^2$	3	$3d^6\,{}^3F^4$	3	24.119	24.065				
$3d^6\,{}^3F^2$	3	$3d^6\,{}^1G^4$	4	22.190	22.153				
$3d^6\,{}^3F^2$	3	$3d^6\,{}^3F^4$	2	21.669	21.633				
$3d^6\,{}^3F^2$	3	$3d^6\,{}^3F^2$	2	18.025	17.991				
$3d^6\,{}^3F^2$	3	$3d^6\,{}^3F^2$	4	16.875	16.857				
$3d^6\,{}^3F^2$	3	$3d^6\,{}^1D^2$	2	13.572	13.563				
$3d^6\,{}^3F^2$	3	$3d^6\,{}^1G^2$	4	10.182	10.163				
$3d^6\,{}^3F^2$	3	$3d^6\,{}^3P^2$	2	9.694	9.677				
$3d^6\,{}^3F^2$	3	$3d^6\,{}^3F^2$	3	9.532	9.517				
$3d^6\,{}^5D^4$	2	$3d^6\,{}^5D^4$	3	130.436	130.930				
$3d^6\,{}^5D^4$	2	$3d^6\,{}^3F^4$	3	72.345	72.524				
$3d^6\,{}^5D^4$	2	$3d^6\,{}^3D^4$	1	63.013	63.168				
$3d^6\,{}^5D^4$	2	$3d^6\,{}^3F^4$	2	54.026	54.171				
$3d^6\,{}^5D^4$	2	$3d^6\,{}^3F^2$	2	35.921	35.947				
$3d^6\,{}^5D^4$	2	$3d^6\,{}^1D^2$	2	21.720	21.755				
$3d^6\,{}^5D^4$	2	$3d^6\,{}^3P^2$	1	16.537	16.511				

Table 5. *Cont.*

Label and J for Lower		Label and J for Upper		Present Work $n=5$	$n=6$	Expt (Ref. [2])	GRASP	RMBPT$_g$	RCI$_g$
$3d^6\,^5D^4$	2	$3d^6\,^3P^2$	2	13.242	13.233				
$3d^6\,^5D^4$	2	$3d^6\,^3F^2$	3	12.942	12.935				
$3d^6\,^5D^4$	0	$3d^6\,^3D^4$	1	74.520	74.840				
$3d^6\,^5D^4$	0	$3d^6\,^3P^2$	1	17.236	17.213				
$3d^6\,^3H^4$	4	$3d^6\,^5D^4$	3	369.368	370.105				
$3d^6\,^3H^4$	4	$3d^6\,^3H^4$	5	294.589	296.075				
$3d^6\,^3H^4$	4	$3d^6\,^3F^4$	3	112.825	112.959				
$3d^6\,^3H^4$	4	$3d^6\,^1G^4$	4	80.216	80.382				
$3d^6\,^3H^4$	4	$3d^6\,^3F^2$	4	37.506	37.563				
$3d^6\,^3H^4$	4	$3d^6\,^1G^2$	4	15.240	15.221				
$3d^6\,^3H^4$	4	$3d^6\,^3F^2$	3	13.830	13.817				
$3d^6\,^5D^4$	3	$3d^6\,^3F^4$	3	162.444	162.579				
$3d^6\,^5D^4$	3	$3d^6\,^1G^4$	4	102.469	102.683				
$3d^6\,^5D^4$	3	$3d^6\,^3F^4$	2	92.227	92.402				
$3d^6\,^5D^4$	3	$3d^6\,^3F^2$	2	49.574	49.552				
$3d^6\,^5D^4$	3	$3d^6\,^3F^2$	4	41.744	41.805				
$3d^6\,^5D^4$	3	$3d^6\,^1D^2$	2	26.059	26.090				
$3d^6\,^5D^4$	3	$3d^6\,^1G^2$	4	15.896	15.874				
$3d^6\,^5D^4$	3	$3d^6\,^3P^2$	2	14.738	14.721				
$3d^6\,^5D^4$	3	$3d^6\,^3F^2$	3	14.368	14.353				
$3d^6\,^3H^4$	5	$3d^6\,^1I^4$	6	369.318	371.637				
$3d^6\,^3H^4$	5	$3d^6\,^1G^4$	4	110.232	110.337				
$3d^6\,^3H^4$	5	$3d^6\,^3F^2$	4	42.977	43.021				
$3d^6\,^3H^4$	5	$3d^6\,^1G^2$	4	16.071	16.046				
$3d^6\,^3F^4$	3	$3d^6\,^1G^4$	4	277.541	278.717				
$3d^6\,^3F^4$	3	$3d^6\,^3F^4$	2	213.361	214.066				
$3d^6\,^3F^4$	3	$3d^6\,^3F^2$	2	71.346	71.275				
$3d^6\,^3F^4$	3	$3d^6\,^3F^2$	4	56.181	56.276				
$3d^6\,^3F^4$	3	$3d^6\,^1D^2$	2	31.039	31.076				
$3d^6\,^3F^4$	3	$3d^6\,^1G^2$	4	17.620	17.592				
$3d^6\,^3F^4$	3	$3d^6\,^3P^2$	2	16.208	16.187				
$3d^6\,^3F^4$	3	$3d^6\,^3F^2$	3	15.762	15.743				
$3d^6\,^3D^4$	1	$3d^6\,^3F^4$	2	378.830	380.349				
$3d^6\,^3D^4$	1	$3d^6\,^3F^2$	2	83.550	83.418				
$3d^6\,^3D^4$	1	$3d^6\,^1D^2$	2	33.145	33.182				
$3d^6\,^3D^4$	1	$3d^6\,^1S^0$	0	22.557	22.569				
$3d^6\,^3D^4$	1	$3d^6\,^3P^2$	1	22.422	22.355				
$3d^6\,^3D^4$	1	$3d^6\,^3P^2$	2	16.764	16.740				
$3d^6\,^3D^4$	1	$3d^6\,^3P^2$	0	9.053	9.035				
$3d^6\,^1G^4$	4	$3d^6\,^3F^2$	4	70.440	70.514				
$3d^6\,^1G^4$	4	$3d^6\,^1G^2$	4	18.815	18.777				
$3d^6\,^1G^4$	4	$3d^6\,^3F^2$	3	16.711	16.685				
$3d^6\,^3F^4$	2	$3d^6\,^3F^2$	2	107.190	106.853				
$3d^6\,^3F^4$	2	$3d^6\,^1D^2$	2	36.323	36.354				
$3d^6\,^3F^4$	2	$3d^6\,^3P^2$	1	23.832	23.751				
$3d^6\,^3F^4$	2	$3d^6\,^3P^2$	2	17.541	17.511				
$3d^6\,^3F^4$	2	$3d^6\,^3F^2$	3	17.019	16.992				
$3d^6\,^3F^2$	2	$3d^6\,^1D^2$	2	54.939	55.101				
$3d^6\,^3F^2$	2	$3d^6\,^3P^2$	1	30.646	30.539				

Table 5. *Cont.*

Label and J for Lower		Label and J for Upper		Present Work $n=5$	$n=6$	Expt (Ref. [2])	GRASP	RMBPT$_g$	RCI$_g$
$3d^6\ {}^3F^2$	2	$3d^6\ {}^3P^2$	2	20.973	20.943				
$3d^6\ {}^3F^2$	2	$3d^6\ {}^3F^2$	3	20.232	20.205				
$3d^6\ {}^3F^2$	4	$3d^6\ {}^1G^2$	4	25.671	25.592				
$3d^6\ {}^3F^2$	4	$3d^6\ {}^3F^2$	3	21.909	21.857				
$3d^6\ {}^1D^2$	2	$3d^6\ {}^3P^2$	1	69.307	68.508				
$3d^6\ {}^1D^2$	2	$3d^6\ {}^3P^2$	2	33.922	33.783				
$3d^6\ {}^1D^2$	2	$3d^6\ {}^3F^2$	3	32.025	31.905				
$3d^6\ {}^1S^0$	0	$3d^6\ {}^3P^2$	1	3732.652	2353.561				
$3d^6\ {}^3P^2$	1	$3d^6\ {}^3P^2$	2	66.443	66.648				
$3d^6\ {}^3P^2$	1	$3d^6\ {}^3P^2$	0	15.184	15.162				
$3d^6\ {}^1G^2$	4	$3d^6\ {}^3F^2$	3	149.470	149.753				
$3d^6\ {}^3P^2$	2	$3d^6\ {}^3F^2$	3	572.500	573.976				
W^{49+} (Mn-like)									
$3d^7\ {}^4F^3$	9/2	$3d^7\ {}^4F^3$	7/2	18.940	18.908	18.880(3)	19.006	18.901	18.943
$3d^7\ {}^4F^3$	9/2	$3d^7\ {}^4F^3$	9/2	17.138	17.119	17.106(3)	17.149	17.118	17.132
$3d^7\ {}^4F^3$	9/2	$3d^7\ {}^2H^3$	11/2	15.381	15.371	15.368(3)	15.343	15.372	15.380
$3d^7\ {}^4F^3$	9/2	$3d^7\ {}^2F^3$	7/2	14.180	14.170	14.166(3)	14.156	14.187	14.063
$3d^7\ {}^4F^3$	9/2	$3d^7\ {}^2H^3$	9/2	8.434	8.424				
$3d^7\ {}^4F^3$	9/2	$3d^7\ {}^2F^3$	7/2	8.259	8.250				
$3d^7\ {}^2P^3$	3/2	$3d^7\ {}^2D^1$	5/2	174.708	174.056				
$3d^7\ {}^2P^3$	3/2	$3d^7\ {}^4P^3$	3/2	19.753	19.715				
$3d^7\ {}^2P^3$	3/2	$3d^7\ {}^4P^3$	5/2	19.111	19.081	19.047(3)	19.130	19.037	19.271
$3d^7\ {}^2P^3$	3/2	$3d^7\ {}^4P^3$	1/2	18.720	18.692	18.670(3)	18.764	18.680	18.733
$3d^7\ {}^2P^3$	3/2	$3d^7\ {}^2F^3$	5/2	15.597	15.585				
$3d^7\ {}^2P^3$	3/2	$3d^7\ {}^2D^1$	3/2	12.711	12.700				
$3d^7\ {}^2P^3$	3/2	$3d^7\ {}^4F^3$	5/2	9.864	9.847				
$3d^7\ {}^2P^3$	3/2	$3d^7\ {}^4P^3$	3/2	9.561	9.546				
$3d^7\ {}^2P^3$	3/2	$3d^7\ {}^2P^3$	1/2	9.171	9.157				
$3d^7\ {}^2P^3$	3/2	$3d^7\ {}^2D^1$	5/2	7.643	7.636				
$3d^7\ {}^2P^3$	3/2	$3d^7\ {}^2D^1$	3/2	6.060	6.050				
$3d^7\ {}^2D^1$	5/2	$3d^7\ {}^4F^3$	7/2	27.098	27.043				
$3d^7\ {}^2D^1$	5/2	$3d^7\ {}^4P^3$	3/2	22.271	22.233				
$3d^7\ {}^2D^1$	5/2	$3d^7\ {}^4P^3$	5/2	21.458	21.430				
$3d^7\ {}^2D^1$	5/2	$3d^7\ {}^2F^3$	7/2	18.307	18.294	18.276(3)	18.258	18.274	18.425
$3d^7\ {}^2D^1$	5/2	$3d^7\ {}^2F^3$	5/2	17.126	17.117				
$3d^7\ {}^2D^1$	5/2	$3d^7\ {}^2D^1$	3/2	13.709	13.700				
$3d^7\ {}^2D^1$	5/2	$3d^7\ {}^4F^3$	5/2	10.455	10.437				
$3d^7\ {}^2D^1$	5/2	$3d^7\ {}^4P^3$	3/2	10.115	10.099				
$3d^7\ {}^2D^1$	5/2	$3d^7\ {}^2F^3$	7/2	9.507	9.496				
$3d^7\ {}^2D^1$	5/2	$3d^7\ {}^2D^1$	5/2	7.993	7.986				
$3d^7\ {}^2D^1$	5/2	$3d^7\ {}^2D^1$	3/2	6.278	6.268				
$3d^7\ {}^4F^3$	7/2	$3d^7\ {}^4F^3$	9/2	180.107	180.908				
$3d^7\ {}^4F^3$	7/2	$3d^7\ {}^4P^3$	5/2	103.103	103.259				
$3d^7\ {}^4F^3$	7/2	$3d^7\ {}^2F^3$	7/2	56.428	56.550				
$3d^7\ {}^4F^3$	7/2	$3d^7\ {}^2F^3$	5/2	46.537	46.637				
$3d^7\ {}^4F^3$	7/2	$3d^7\ {}^4F^3$	5/2	17.022	16.998				
$3d^7\ {}^4F^3$	7/2	$3d^7\ {}^2H^3$	9/2	15.204	15.193				
$3d^7\ {}^4F^3$	7/2	$3d^7\ {}^2F^3$	7/2	14.645	14.635				

Table 5. *Cont.*

Label and J for Lower		Label and J for Upper		Present Work $n=5$	$n=6$	Expt (Ref. [2])	GRASP	$RMBPT_g$	RCI_g
$3d^7\,^4F^3$	7/2	$3d^7\,^2D^1$	5/2	11.337	11.332				
$3d^7\,^4F^3$	9/2	$3d^7\,^2H^3$	11/2	150.021	150.559				
$3d^7\,^4F^3$	9/2	$3d^7\,^2F^3$	7/2	82.173	82.265				
$3d^7\,^4F^3$	9/2	$3d^7\,^2H^3$	9/2	16.606	16.586				
$3d^7\,^4F^3$	9/2	$3d^7\,^2F^3$	7/2	15.942	15.923				
$3d^7\,^4P^3$	3/2	$3d^7\,^4P^3$	5/2	587.816	593.270				
$3d^7\,^4P^3$	3/2	$3d^7\,^4P^3$	1/2	357.969	360.283				
$3d^7\,^4P^3$	3/2	$3d^7\,^2F^3$	5/2	74.127	74.387				
$3d^7\,^4P^3$	3/2	$3d^7\,^2D^1$	3/2	35.657	35.693				
$3d^7\,^4P^3$	3/2	$3d^7\,^4F^3$	5/2	19.704	19.673				
$3d^7\,^4P^3$	3/2	$3d^7\,^4P^3$	3/2	18.530	18.506				
$3d^7\,^4P^3$	3/2	$3d^7\,^2P^3$	1/2	17.118	17.100				
$3d^7\,^4P^3$	3/2	$3d^7\,^2D^1$	5/2	12.467	12.462				
$3d^7\,^4P^3$	3/2	$3d^7\,^2D^1$	3/2	8.742	8.729				
$3d^7\,^4P^3$	5/2	$3d^7\,^2F^3$	7/2	124.648	125.014				
$3d^7\,^4P^3$	5/2	$3d^7\,^2F^3$	5/2	84.823	85.051				
$3d^7\,^4P^3$	5/2	$3d^7\,^2D^1$	3/2	37.960	37.978				
$3d^7\,^4P^3$	5/2	$3d^7\,^4F^3$	5/2	20.388	20.347				
$3d^7\,^4P^3$	5/2	$3d^7\,^4P^3$	3/2	19.133	19.101				
$3d^7\,^4P^3$	5/2	$3d^7\,^2F^3$	7/2	17.070	17.052				
$3d^7\,^4P^3$	5/2	$3d^7\,^2D^1$	5/2	12.737	12.729				
$3d^7\,^4P^3$	5/2	$3d^7\,^2D^1$	3/2	8.874	8.860				
$3d^7\,^4P^3$	1/2	$3d^7\,^2D^1$	3/2	39.602	39.618				
$3d^7\,^4P^3$	1/2	$3d^7\,^4P^3$	3/2	19.542	19.508				
$3d^7\,^4P^3$	1/2	$3d^7\,^2P^3$	1/2	17.978	17.952				
$3d^7\,^4P^3$	1/2	$3d^7\,^2D^1$	3/2	8.961	8.946				
$3d^7\,^2H^3$	11/2	$3d^7\,^2H^3$	9/2	18.673	18.639				
$3d^7\,^2F^3$	7/2	$3d^7\,^2F^3$	5/2	265.490	266.061				
$3d^7\,^2F^3$	7/2	$3d^7\,^4F^3$	5/2	24.375	24.303				
$3d^7\,^2F^3$	7/2	$3d^7\,^2H^3$	9/2	20.812	20.774				
$3d^7\,^2F^3$	7/2	$3d^7\,^2F^3$	7/2	19.779	19.745				
$3d^7\,^2F^3$	7/2	$3d^7\,^2D^1$	5/2	14.187	14.173				
$3d^7\,^2F^3$	5/2	$3d^7\,^2D^1$	3/2	68.707	68.619				
$3d^7\,^2F^3$	5/2	$3d^7\,^4F^3$	5/2	26.839	26.746				
$3d^7\,^2F^3$	5/2	$3d^7\,^4P^3$	3/2	24.706	24.634				
$3d^7\,^2F^3$	5/2	$3d^7\,^2F^3$	7/2	21.371	21.328				
$3d^7\,^2F^3$	5/2	$3d^7\,^2D^1$	5/2	14.988	14.970				
$3d^7\,^2F^3$	5/2	$3d^7\,^2D^1$	3/2	9.911	9.890				
$3d^7\,^2D^1$	3/2	$3d^7\,^4F^3$	5/2	44.043	43.830				
$3d^7\,^2D^1$	3/2	$3d^7\,^4P^3$	3/2	38.579	38.430				
$3d^7\,^2D^1$	3/2	$3d^7\,^2P^3$	1/2	32.925	32.828				
$3d^7\,^2D^1$	3/2	$3d^7\,^2D^1$	5/2	19.169	19.147				
$3d^7\,^2D^1$	3/2	$3d^7\,^2D^1$	3/2	11.581	11.555				
$3d^7\,^4F^3$	5/2	$3d^7\,^4P^3$	3/2	310.962	311.919				
$3d^7\,^4F^3$	5/2	$3d^7\,^2F^3$	7/2	104.893	105.290				
$3d^7\,^4F^3$	5/2	$3d^7\,^2D^1$	5/2	33.943	34.000				
$3d^7\,^4F^3$	5/2	$3d^7\,^2D^1$	3/2	15.713	15.692				
$3d^7\,^4P^3$	3/2	$3d^7\,^2P^3$	1/2	224.661	225.205				
$3d^7\,^4P^3$	3/2	$3d^7\,^2D^1$	5/2	38.102	38.159				

Table 5. *Cont.*

Label and J for Lower		Label and J for Upper		Present Work $n = 5$	$n = 6$	Expt (Ref. [2])	GRASP	RMBPT$_g$	RCI$_g$
$3d^7\ ^4P^3$	3/2	$3d^7\ ^2D^1$	3/2	16.549	16.524				
$3d^7\ ^2P^3$	1/2	$3d^7\ ^2D^1$	3/2	17.865	17.832				
$3d^7\ ^2H^3$	9/2	$3d^7\ ^2F^3$	7/2	398.243	398.611				
$3d^7\ ^2F^3$	7/2	$3d^7\ ^2D^1$	5/2	50.182	50.215				
$3d^7\ ^2D^1$	5/2	$3d^7\ ^2D^1$	3/2	29.257	29.143				
W^{48+} (Fe-like)									
$3d^8\ ^3F$	4	$3d^8\ ^3F$	3	19.041	19.009	18.988(3)	19.114	19.007	19.027
$3d^8\ ^3F$	4	$3d^8\ ^1G$	4	15.531	15.518	15.511(3)	15.503	15.463	15.525
$3d^8\ ^1D$	2	$3d^8\ ^3F$	3	22.073	22.029				
$3d^8\ ^1D$	2	$3d^8\ ^3P$	2	18.931	18.902	18.878(3)	18.978	18.888	18.966
$3d^8\ ^1D$	2	$3d^8\ ^3P$	1	17.548	17.525	17.502(3)	17.548	17.517	17.489
$3d^8\ ^1D$	2	$3d^8\ ^3F$	2	9.664	9.648				
$3d^8\ ^3P$	0	$3d^8\ ^3P$	1	24.268	24.236				
$3d^8\ ^3F$	3	$3d^8\ ^3P$	2	132.993	133.189				
$3d^8\ ^3F$	3	$3d^8\ ^1G$	4	84.240	84.489				
$3d^8\ ^3F$	3	$3d^8\ ^3F$	2	17.190	17.167				
$3d^8\ ^3P$	2	$3d^8\ ^3P$	1	240.207	240.600				
$3d^8\ ^3P$	2	$3d^8\ ^3F$	2	19.742	19.707				
$3d^8\ ^3P$	1	$3d^8\ ^3F$	2	21.510	21.465				
$3d^8\ ^3P$	1	$3d^8\ ^1S$	0	15.102	15.082				
W^{47+} (Co-like)									
$3d^9\ ^2D$	5/2	$3d^9\ ^2D$	3/2	18.615	18.586	18.567(3)	18.671	18.586	18.580

4. Conclusions

The present study has shown that the inclusion of core correlation effects improves the accuracy of theoretical transition wavelengths for M1 transitions in $3d^k$ configurations of tungsten ions. Omitted in our work were correlation effects arising from the $1s^2$ core. Further studies are needed to determine whether the discrepancy with observation arises from the limited orbital set for core correlation or from the inactive $1s^2$ shell in our present work.

Acknowledgments: Computations were performed on resources at the High Performance Computing Center "HPC Sauletekis" in Vilnius University Faculty of Physics (Lithuania). Per Jönsson acknowledge support from the Swedish Research Council under contract 2015-04842.

Author Contributions: All authors have participated in the development of the programs that made these calculations possible and in preparation of this manuscript.

Conflicts of Interest: The authors declare no conflict of interest.

References

1. Hawryluk, R.J.; Campbell, D.J.; Janeschitz, G.; Thomas, P.R.; Albanese, R.; Ambrosino, R.; Bachmann, C.; Baylor, L.; Becoulet, M.; Benfatto, I.; et al. Principal physics developments evaluated in the ITER design review. *Nucl. Fusion* **2009**, *49*, 065012.
2. Ralchenko, Y.; Draganić, I.N.; Osin, D.; Gillaspy, J.; Reader, J. Spectroscopy of diagnostically important magnetic-dipole lines in highly charged $3d^n$ ions of tungsten. *Phys. Rev. A* **2011**, *83*, 032517.
3. Gu, M.F. The flexible atomic code. *Can. J. Phys.* **2008**, *86*, 675–689.
4. Kramida, A. Recent progress in spectroscopy of tungsten. *Can. J. Phys.* **2011**, *89*, 551–570.

5. Kramida, A.; Ralchenko, Y.; Reader, J.; NIST ASD Team. *NIST Atomic Spectra Database (Ver. 5.2)*; National Institute of Standards and Technology: Gaithersburg, MD, USA, 2014. Available online: http://physics.nist.gov/asd (accessed on 28 December 2014).
6. Quinet, P. Dirac-Fock calculations of forbidden transitions within the $3p^k$ and $3d^k$ ground configurations of highly charge tungsten ions (W^{47+} – W^{61+}). *J. Phys. B: At. Mol. Opt. Phys.* **2011**, *44*, 195007.
7. Norrington, P.H. GRASP^0 Manual. 2009, unpublished.
8. Guo, X.L.; Huang, M.; Yan, J.; Li, S.; Si, R.; Li, C.Y.; Chen, C.Y.; Wang, Y.S.; Zou, Y.M. Relativistic many-body calculations on wavelengths and transition probabilities for forbidden transitions within the $3d^k$ ground configurations in Co- through K-like ions of hafnium, tantalum, tungsten, and gold. *J. Phys. B Atomic Mol. Opt. Phys.* **2015**, *48*, 144020.
9. Lindgren, I. The Rayleigh-Schrodinger perturbation and the linked-diagram theorem for a multi-configurational model space. *J. Phys. B Atomic Mol. Opt. Phys.* **1974**, *7*, 2441.
10. Jönsson, P.; Gaigalas, G.; Bieroń, J.; Froese Fischer, C.; Grant, I.P. New Version: Grasp2K relativistic atomic structure package. *Comput. Phys. Commun.* **2013**, *184*, 2197–2203.
11. Grant, I.P. *Relativistic Quantum Theory of Atoms and Molecules*; Springer: New York, NY, USA, 2007.
12. Froese Fischer, C.; Godefroid, M.; Brage, T.; Jönsson, P.; Gaigalas, G. Advanced multiconfiguration methods for complex atoms: Part I—Energies and wave functions. *J. Phys. B Atomic Mol. Opt. Phys.* **2016**, *49*, 182004.
13. Verdebout, S.; Rynkun, P.; Jönsson, P.; Gaigalas, G.; Froese Fischer, C.; Godefroid, M. A partitioned correlation function interaction approach for describing electron correlation in atoms. *J. Phys. B Atomic Mol. Opt. Phys.* **2013**, *46*, 085003.
14. McKenzie, B.J.; Grant, I.P.; Norrington, P.H. A program to calculate transverse Breit and QED corrections to energy levels in a multiconfiguration Dirac-Fock environment. *Comput. Phys. Commun.* **1980**, *21*, 233–246.
15. Bentley, M; Fischer, C.F. Hypercube conversion of serial codes for atomic structure calculations. *Parallel Comput.* **1992**, *18*, 1023–1031.
16. Fischer, C.F.; Tong, M.; Bentley, M.; Shen, Z.; Ravimohan, C. The Distributed-memory implementation of the MCHF atomic structure package. *J. Supercomput.* **1994**, *8*, 117–134.
17. Safronova, U.I.; Safronova, A.S. Wavelengths and transition rates for $nl - n'l'$ transitions in Be-, B-, Mg-, Al-, Ca-, Zn-, Ag-, and Yb-like tungsten ions. *J. Phys. B Atomic Mol. Opt. Phys.* **2010**, *43*, 074026.

atoms

Article

Calculation of Rates of 4p–4d Transitions in Ar II

Alan Hibbert

CTAMOP, School of Mathematics & Physics, Queen's University, Belfast BT7 1NN, UK; a.hibbert@qub.ac.uk

Academic Editor: Joseph Reader
Received: 29 December 2016; Accepted: 9 February 2017; Published: 21 February 2017

Abstract: Recent experimental work by Belmonte et al. (2014) has given rates for some 4p–4d transitions that are significantly at variance with the previous experimental work of Rudko and Tang (1967) recommended in the NIST tabulations. To date, there are no theoretical rates with which to compare. In this work, we provide such theoretical data. We have undertaken a substantial and systematic configuration interaction calculation, with an extrapolation process applied to ab initio mixing coefficients, which gives energy differences in agreement with experiment. The length and velocity forms give values that are within 10%–15% of each other. Our results are in sufficiently close agreement with those of Belmonte et al. that we can confidently recommend that their results are much more accurate than the early results of Rudko and Tang, and should be adopted in place of the latter.

Keywords: E1 transitions; configuration interaction calculaton; transition rates

---

1. Introduction

Some years ago, we [1–3] studied transitions among Ar II levels arising from configurations $3s^23p^5$, $3s3p^6$, $3p^43d$, $3p^44s$, and $3p^44p$. That work was prompted by a range of conflicting experimental results and a limited amount of theoretical work. We found that our calculations gave transition rates in close agreement with the experimental values recommended by Vujnović and Wiese [4], and gave much closer agreement between length and velocity forms of transition rates than were obtained by the only other major theoretical work, conducted by Luyken[5]. The values cited in the NIST tabulations [6] are taken from Bennett et al. [7] where possible, in agreement with the recommended values given in [4], but for other 4p–4d transitions, it is the data of Rudko and Tang [8] which are quoted.

Recently, Belmonte et al. [9]—building on the work of Aparicio et al. [10]—extended the experimental study to 4p–4d (and a few other) transitions. They also included results for some transitions between the lower-lying levels previously studied in [2–4], and found that they were in much closer agreement with the experimental values recommended by Vujnović and Wiese [4], and with our previous calculations, than with other experimental work. By contrast, they found that their results differed by up to a factor of five from the experimental values of Rudko and Tang [8]. The purpose of the present work is to provide some theoretical corroboration (or otherwise) of the new experimental results.

2. Method of Calculation

The calculations in this work have been undertaken using the code CIV3 [11,12].

2.1. Basic Theory

We express the wave functions in terms of configuration interaction (CI) expansions:

$$\Psi(J) = \sum_{i=1}^{M} a_i \Phi_i(\alpha_i L_i S_i J) \tag{1}$$

where $\{\Phi_i\}$ are single-configuration functions (configuration state functions—CSFs) and the expansions in general include summations over L_i and S_i. For a specific choice of $\{\Phi_i\}$, the expansion coefficients $\{a_i\}$ are the eigenvector components of the diagonalized Hamiltonian with matrix elements $H_{ij} = < \Phi_i|H|\Phi_j >$. In this work, we take the Hamiltonian H to be the Schrödinger Hamiltonian plus the mass correction and Darwin terms, together with a modified spin-orbit term

$$H_{so} = \frac{1}{2}\alpha^2 \sum_{i=1}^{N} \frac{Z\zeta_l}{r_i^3} \mathbf{l_i}.\mathbf{s_i} \tag{2}$$

In (2), the sum is over the electrons, and the parameters $\{\zeta_l\}$ depend on the l-value of the electrons involved in the interaction (Hibbert and Hansen 1989) [2].

The ordered eigenvalues $\{E_i\}$ of the Hamiltonian matrix are upper bounds to the similarly-ordered energy levels:

$$E_i \geq E_i^{\text{exact}} \tag{3}$$

Hence, any of the eigenvalues may be used as the variational functional for optimisation of the radial parts of the one-electron orbitals from which the $\{\Phi_i\}$ are constructed. We express these radial functions as sums of normalised Slater-type orbitals (STOs):

$$P_{nl}(r) = \sum_{j=1}^{k} C_{jnl}\chi_{jnl}(r) \tag{4}$$

where the STOs are of the form

$$\chi_{jnl}(r) = \left[\frac{(2\xi_{jnl})^{2I_{jnl}+1}}{(2I_{jnl})!}\right]^{1/2} r^{I_{jnl}} \exp(-\xi_{jnl}r) \tag{5}$$

Being integers, the $\{I_{jnl}\}$ are kept fixed, but the exponents $\{\xi_{jnl}\}$ and the coefficients $\{C_{jnl}\}$ may be treated as variational parameters in (3), subject to the orthonormality conditions:

$$\int_0^\infty P_{nl}(r)P_{n'l}(r)dr = \delta_{nn'}; \qquad l < n' \leq n \tag{6}$$

2.2. *Radial Function Parameters*

Since we were adding to earlier work [3], we were able to use many of the radial functions we used previously. However, that work did not include 4d levels. The radial function parameters are determined by optimising the energy associated with different states; the optimisation is undertaken in LS coupling. The radial function parameters used in this work were optimised as displayed in Table 1. We comment here on the reasons underpinning the choice of procedure used for the functions new to this work.

- The 6p function was newly introduced in this calculation. While retaining the 4p and 5p functions from previous work, the parameters for 6p were optimised on the ground state to improve the capture of the electron correlation effect in the n = 3 shell, and thereby improve the calculated separation between the ground and excited states.
- We retained the previous 3d and 4d functions, but reoptimised 5d and 6d. We considered the lowering of the energy of several different states brought about by the introduction of 5d. The effect was largest for the $3p^4 4d\ ^4F$ state. Similarly, the lowering of the energy of several different doublet states through the introduction of 6d was noted. There was a substantial difference in the mixings between doublet states, depending on the final LS symmetry chosen for the optimisation. As a consequence, we selected those obtained during the optimisation of the $3p^4(^3P)4d\ ^2D$ state.

- We reoptimised the 6s function on the 5s ^{4}P state, since the energy of that state lay in the region of those of the 4d states.

The set of parameters for all the radial functions used here is displayed in Table 2.

Table 1. Method of determining the radial functions.

Orbital	Process of Optimisation	
1s, 2s, 2p, 3s	Hartree–Fock orbitals of 3p^4 ^{1}D of Ar III (Clementi and Roetti (1974)) [13]	
3p	Exponents taken from the Hartree–Fock orbital of 3p^4 ^{1}D of Ar III; coefficients reoptimised on 3p^{4}4s ^{4}P of Ar II	
	Eigenvalue minimised	Configurations
3d	3s3p^6 ^{2}S	3s3p^6, 3s^{2}3p^{4}3d
4s	3p^{4}4s ^{4}P	3p^{4}4s
4p	3p^{4}4p ^{4}D^o	3p^{4}4p
4d	3p^{4}3d ^{4}D	3p^{4}3d, 3p^{4}4d
4f	3p^{4}3d ^{4}P	3p^{4}4s, 3p^{4}3d, 3p^{4}4d, 3p^{3}3d4f
5s	3p^{4}4p ^{4}D^o	3p^{4}4p, 3p^{3}4s5s
5p	3p^{4}4p ^{4}P^o	3p^{4}4p, 3p^{4}5p
5d	3p^{4}4d ^{4}F	3p^{4}3d, 3p^{4}4d, 3p^{4}5d
5f	3p^{4}4p ^{4}D^o	3p^{4}4p, 3p^{4}4f, 3p^{4}5f
6s	3p^{4}5s ^{4}P	3p^{4}4s, 3p^{4}5s, 3p^{4}6s
6p	3p^5 ^{2}P^o	3p^5, 3p^{4}4p, 3p^{4}5p, 3p^{4}6p
6d	3p^{4}4d(^{3}P) ^{2}D	3p^{4}3d, 3p^{4}4d,3p^{4}5d, 3p^{4}6d

2.3. *Choice of Configurations*

In our previous work [3], we included a limited range of configurations aimed at capturing the main correlation effects in the 3p^{4}3d/4s/4p states. This led to some difficulties, primarily that the degree of correlation included in the ground state was substantially greater than for the excited states, and the order of some 3d and 4s levels was incorrect.

Consequently, in this work, we have included all possible configurations that can be obtained by one- and two-orbital replacements from the 3l and 4l subshells to the full set of orbitals shown in Table 2, from the configurations of the following reference sets.

Odd	3p^5; 3p^{4}4p
Even	3s3p^6; 3p^{4}4s, 3p^{4}5s, 3p^{4}6s; 3p^{4}3d, 3p^{4}4d, 3p^{4}5d, 3p^{4}6d

The configurations of the reference sets were those with a significant CI coefficient in a relatively small CI calculation. For each possible LSπ symmetry, all CSFs were then constructed and combined to give a set of CSFs for each allowed J π symmetry, resulting in Hamiltonian matrices of the following sizes.

	J = 0.5	J = 1.5	J = 2.5	J = 3.5	J = 4.5
Odd	13,082	18,144	17,603	9148	
Even	44,149	75,383	75,964	61,072	28,854

Table 2. Radial function parameters.

nl	C_{jnl}	I_{jnl}	ξ_{jnl}	nl	C_{jnl}	I_{jnl}	ξ_{jnl}
1s	0.926 94	1	17.332 10	2p	−0.011 17	3	3.102 81
	0.058 91	1	25.455 00		0.004 97	3	2.011 93
	0.007 82	2	7.657 68		0.145 75	3	5.190 03
	0.017 65	2	15.623 20		0.824 78	2	6.928 92
	0.000 90	3	3.237 31		0.087 03	2	13.042 40
	−0.000 47	3	2.296 92				
	−0.003 17	3	6.726 86	3p	0.530 23	3	3.102 81
					0.583 91	3	2.011 93
2s	−0.277 90	1	17.332 10		−0.074 28	3	5.190 03
	−0.008 62	1	25.455 00		−0.272 24	2	6.928 92
	0.816 64	2	7.657 68		−0.025 06	2	13.042 40
	−0.127 59	2	15.623 20				
	0.013 06	3	3.237 31	4p	0.744 05	4	0.995 10
	−0.003 71	3	2.296 92		0.297 40	4	0.775 02
	0.331 25	3	6.726 86		−0.296 30	3	2.508 00
					0.082 76	2	7.360 60
3s	−0.094 80	1	17.332 10				
	−0.001 41	1	25.455 00	5p	4.629 07	4	0.861 89
	0.289 14	2	7.657 68		−4.686 80	4	1.000 00
	−0.043 25	2	15.623 20		0.735 21	3	3.221 66
	−0.640 52	3	3.237 31		−0.403 55	2	3.473 69
	−0.494 62	3	2.296 92				
	0.216 65	3	6.726 86	6p	6.828 07	5	0.937 15
					−8.463 99	4	0.892 84
4s	0.487 62	4	1.299 90		3.198 41	4	1.697 71
	0.569 47	4	1.016 95		−1.193 75	3	2.716 06
	−0.384 57	3	2.931 16		0.308 57	2	8.508 55
	0.157 04	2	6.149 39				
	−0.042 15	1	14.064 49	3d	0.249 70	3	3.465 62
					0.821 57	3	1.684 33
5s	1.133 28	5	1.364 91				
	−2.125 95	4	2.019 60	4d	0.216 25	3	2.817 41
	1.487 29	3	2.884 30		0.300 37	3	1.891 60
	−0.468 68	2	5.897 16		−1.104 59	4	0.964 72
	0.113 37	1	14.178 86		0.067 82	4	0.570 20
6s	1.293 76	5	0.685 92	5d	0.438 75	3	2.189 96
	1.084 37	4	2.019 36		−1.653 12	3	0.717 04
	−1.350 12	4	1.180 83		1.981 83	4	0.612 62
	−0.458 59	3	2.844 45				
	0.120 74	2	5.635 72	6d	0.594 56	3	2.112 65
	−0.026 11	1	14.432 79		−3.249 46	3	0.719 43
					4.546 89	4	0.699 99
4f	1.000 00	4	2.154 77		−2.054 58	4	0.429 47
5f	0.522 16	4	2.573 22				
	−1.044 69	5	1.224 76				

2.4. Relativistic Effects

As in our earlier work [3], relativistic effects are included using the Breit–Pauli approximation, retaining in the Hamiltonian the mass correction and Darwin terms and a modified spin-orit term as given in (2). The parameters ζ_l—which depend only on the l-value of the electrons—were chosen to give the best fit to matrix elements of the full spin-orbit plus spin-other-orbit operators with respect to key CSFs. This led to the values 0.0, 0.856, 1.0, 1.0 for $l = 0, 1, 2, 3$, respectively. The d- and f-orbitals contribute little to the fine structure, most of which comes from configurations containing $3p^4(^3P)$.

3. Results

In our earlier work [3], we found that our choice of configurations resulted in the ground state being around 12,000 cm^{-1} too low when compared with the excited states. In the present work, with our more systematic choice of configurations, we find that our ab initio energy separations are in much better agreement with the experimental work of Minnhagen [14] and Saloman [15], given in the tabulations of NIST [6]. Most of the energy separations agree to within 1000 cm^{-1} with these experimental results, the exceptions being a few of the levels associated with states containing a $3p^4\ ^1D$ core (within 3000 cm^{-1}) and those of $3p^4\ ^1S$ 3d (about 4000 cm^{-1}). Moreover, the difficulty we encountered earlier with a very strong mixing between the $3p^4(^1D)4s$ and $3p^4(^3P)3d\ ^2D_{3/2}$ levels is now sufficiently removed to clearly define the lower of the two as belonging to the 4s state, in agreement with experiment.

Before calculating the electric dipole transition rates between all these levels, we refined the CI mixing coefficients by making small adjustments to some diagonal elements of the Hamiltonian matrices, and then rediagonalising the adjusted matrices. In this way, we were able to bring the calculated eigenvalue differences into agreement with the experimental energy separations. From past experience, we have found that, while the mixing coefficients are improved by this process, there is a tendency for the coefficients to be somewhat over-corrected. However, since most of the matrix corrections are quite small, and many of the levels are spectroscopically fairly pure, the principal effect of this fine-tuning process will be to allow the use of experimental energy separations, with some modifications to the interactions between levels in a limited number of cases.

In Table 3, we present our calculated transition rates in both length and velocity gauges for those 4p–4d transitions for which experimental values are given by [9]. The corresponding results from the experimental determinations of [7,8] are also listed. Belmonte et al. [9] also give estimates of the uncertainties in their results, which they obtain not only from the customary standard deviation of experimental measurements, but also from a detailed and careful analysis of a range of other factors which could lead to uncertainties. As a result of this analysis, they are able to provide uncertainties, most of which lie in the 10%–20% range, with a small proportion having higher uncertainties. Table 3 quotes those uncertainties.

Table 3. A-values ($10^8\ s^{-1}$) for 4p–4d transitions in Ar II.

Transition		This Work					
4p *	4d	Wavelength (nm)	A_l	A_v	[9]	[7]	[8]
$^4P^o_{5/2}$	$^4F_{3/2}$	319.423	0.074	0.066	0.086 (12%) †		0.236
$^4P^o_{1/2}$	$^4F_{3/2}$	326.357	0.105	0.094	0.13 (11%)	0.155	0.348
$^4P^o_{5/2}$	$^4F_{7/2}$	326.899	0.0031	0.0026	0.002 (84%)		
$^4P^o_{5/2}$	$^4P_{5/2}$	313.902	0.625	0.551	0.49 (18%)	0.52	1.00
$^4P^o_{3/2}$	$^4P_{5/2}$	316.967	0.524	0.455	0.43 (18%)	0.49	0.817
$^4P^o_{5/2}$	$^4P_{3/2}$	318.104	0.469	0.421	0.36 (12%)	0.37	0.627
$^4P^o_{3/2}$	$^4P_{1/2}$	324.369	1.18	1.05	1.07 (11%)	1.1	1.99
$^4P^o_{1/2}$	$^4P_{3/2}$	324.980	0.763	0.678	0.60 (14%)	0.63	1.00
$^4P^o_{1/2}$	$^4P_{1/2}$	328.170	0.459	0.405	0.41 (11%)	0.42	0.733
$^4D^o_{3/2}$	$^4D_{1/2}$	384.152	0.258	0.235	0.19 (12%)	0.269	0.267
$^4D^o_{5/2}$	$^4D_{7/2}$	384.473	0.051	0.046	0.049 (17%)	0.048	0.047
$^4D^o_{5/2}$	$^4D_{5/2}$	382.681	0.325	0.297	0.30 (15%)	0.281	0.345
$^4D^o_{5/2}$	$^4D_{3/2}$	379.938	0.221	0.199	0.22 (13%)	0.17	0.23
$^2D^o_{3/2}$	$^2P_{3/2}$	320.432	0.176	0.171	0.24 (12%)		0.402
$^2D^o_{3/2}$	$^2P_{1/2}$	327.332	0.172	0.158	0.20 (16%)		0.371
$^2D^o_{3/2}$	$^4D_{1/2}$	403.138	0.039	0.033	0.07 (60%)	0.075	
$^2D^o_{5/2}$	$^2D_{5/2}$	295.539	0.325	0.297	0.19 (13%)		

Table 3. *Cont.*

Transition			This Work				
$^2D^o_{5/2}$	$^2D_{5/2}$	301.448	0.036	0.034	0.039 (19%)		
$^2P^o_{3/2}$	$^4F_{3/2}$	383.017	0.0008	0.0009	0.042 (27%)		
$^2P^o_{3/2}$	$^2F_{5/2}$	365.528	0.326	0.316	0.37 (13%)		0.232
$^2P^o_{3/2}$	$^2P_{3/2}$	329.364	0.899	0.847	0.59 (17%)		1.73
$^2P^o_{1/2}$	$^2P_{1/2}$	330.723	1.44	1.38	1.43 (11%)		3.35
$^2P^o_{3/2}$	$^2P_{1/2}$	336.658	0.271	0.255	0.24 (15%)		0.409
$^2S^o_{1/2}$	$^2P_{3/2}$	338.853	0.761	0.795	0.81 (12%)		1.91
$^2S^o_{1/2}$	$^2D_{3/2}$	316.137	0.370	0.368	0.35 (45%)		1.837
4p′	4d′	Wavelength (nm)	A_l	A_v	[9]	[7]	[8]
$^2F^o_{5/2}$	$^2F_{5/2}$	335.092	0.929	0.815	0.90 (13%)		1.48
$^2F^o_{7/2}$	$^2F_{5/2}$	336.552	0.073	0.066	0.075 (18%)		0.131
$^2F^o_{7/2}$	$^2F_{7/2}$	337.644	0.860	0.764	0.74 (13%)		1.49
$^2P^o_{3/2}$	$^2P_{3/2}$	366.044	0.741	0.693	0.73 (11%)		2.22
$^2P^o_{3/2}$	$^2P_{1/2}$	367.101	0.199	0.191	0.23 (31%)		0.709
$^2P^o_{1/2}$	$^2D_{3/2}$	368.006	0.031	0.007	0.59 (19%)		1.15
$^2P^o_{3/2}$	$^2S_{1/2}$	302.675	0.600	0.679	1.03 (21%)		
$^2D^o_{3/2}$	$^2D_{5/2}$	379.659	0.141	0.132	0.18 (23%)		0.250
$^2D^o_{5/2}$	$^2D_{5/2}$	380.317	0.978	0.902	0.89 (12%)		1.53
$^2D^o_{3/2}$	$^2P_{3/2}$	381.902	0.244	0.172	0.15 (49%)		0.0036
$^2D^o_{5/2}$	$^2P_{3/2}$	382.567	0.384	0.356	0.33 (55%)		0.756

* nl denotes $3p^4(^3P)nl$; nl' denotes $3p^4(^1D)nl$; † estimated uncertainty.

4. Discussion

The accuracy of theoretical energy differences and transition rates can only be estimated: there is no monotonic convergence of these quantities, even as the wave functions are systematically improved. Instead, it is necessary to refer to a number of *indicators* of accuracy, as explained in [16]. These indicators include a comparison between calculated and experimental energy levels, the convergence of results as the wave functions are improved, the degree of agreement between different forms of the transition rates (typically length and velocity), comparison with other calculations, and of course, comparison with experiment.

In this work, we have adopted our fine-tuning process, which ensures that we are using accurate transition energies and that the CI mixing coefficients are as accurate as we can obtain within the limitations of our finite configuration lists. We have not undertaken a sequence of calculations of different complexity, as would be necessary if we were to establish the degree of convergence of the results, but as many of the levels are fairly pure spectroscopically, we do not believe that this would have a major influence on the level of accuracy achieved. There are no other theoretical transition rates available in the literature for these transitions. That leaves two major factors to be taken into account in assessing the accuracy of our calculations.

It can be observed from Table 3 that the length and velocity forms of our calculated transition rates differ fairly consistently by about 10%–15%, the length form mostly giving the larger of the two. This discrepancy is an indication of either insufficient treatment of electron correlation in the $3p^4$ core, or (given the strong state-dependency of the valence orbitals) insufficient flexibility in the form of the radial functions of the valence orbitals; that is, there may be too few basis functions in the expansions (4).

However, in spite of these limitations, the important thing to note is the comparison between our calculated *A*-values and the experimental values recently determined by Belmonte et al. [9]. For most transitions listed in Table 3, our results lie quite close to the experimental values of [9], bearing in mind the uncertainty of both sets of results. Similar good agreement is found with the experimental results of [7], which are the values recommended in the critical compilation of [4].

By contrast, the experimental results of Rudko and Tang [8] are substantially different from both the recent experimental values and our calculations.

In view of these considerations, we would anticipate that for most of the transitions listed in Table 3, our results are accurate to about 20%–25%, or better.

5. Conclusions

We have undertaken a substantial calculation of 4p–4d transitions in Ar II, using a systematic configuration interaction process. These results provide the only theoretical corroboration with which the recent experimental results given in [9] and in other earlier work may be compared. It is clear that our calculations substantially support the results of Belmonte et al. [9], and of Bennett et al. [7] (where comparison is possible), but are in substantial disagreement with the experimental data of Rudko and Tang [8] for many of the transitions considered here. However, until the recent work of [9], the only available data for the doublet transitions was that of [8], and for those transitions, it is the values of [8] which are quoted in the NIST tabulations [6]. We therefore recommend that—where possible—the transition rates of [9] are adopted instead.

Conflicts of Interest: The author declares no conflict of interest.

References

1. Hibbert, A.; Hansen, J.E. Accurate wavefunctions for ^{2}S and ^{2}P^o states of Ar II. *J. Phys. B At. Mol. Phys.* **1987**, *20*, L245–L251.
2. Hibbert, A.; Hansen, J.E. Lifetimes of some 3p^{4}4p levels in Ar II. *J. Phys. B At. Mol. Opt. Phys.* **1989**, *22*, L347–L351.
3. Hibbert, A.; Hansen, J.E. Transitions in Ar II. *J. Phys. B At. Mol. Opt. Phys.* **1994**, *27*, 3325–3347.
4. Vujnović, V.; Wiese, W.L. A critical compilation of atomic transition probabilities for singly ionized argon. *J. Phys. Chem. Ref. Data* **1992**, *21*, 919–939.
5. Luyken, B.F.J. Transition probabilities and radiative lifetimes for Ar II. *Physica* **1972**, *60*, 432–458.
6. Kramida, A.; Ralchenko, Y.; Reader, J.; NIST ASD Team. *NIST Atomic Spectra Database (Ver. 5.3)*; National Institute of Standards and Technology: Gaithersburg, MD, USA, 2015. Available online: http://physics.nist.gov/asd (accessed on 27 December 2016).
7. Bennett, W.R., Jr.; Kindlmann, P.J.; Mercer, G.N. Measurement of excited state relaxation rates. In *Chemical Lasers: Applied Optics Supplement 2*; Howard, J.N., Ed.; OSA Publishing: Washington, DC, USA, 1965; Volume 34.
8. Rudko, R.I.; Tang, C.L. Spectroscopic studies of the Ar$^+$ laser. *J. Appl. Phys.* **1967**, *38*, 4731–4739.
9. Belmonte, M.T.; Djurović, S.; Peláez, R.J.; Aparicio, J.A.; Mar, S. Improved and expanded measurements of transition probabilities in UV Ar II spectral lines. *Mon. Not. R. Astron. Soc.* **2014**, *445*, 3345–3351.
10. Aparicio, J.A.; Gigosos, M.A.; Mar, S. Transition probability measurement in an Ar II plasma. *J. Phys. B At. Mol. Opt. Phys.* **1997**, *30*, 3141–3157.
11. Hibbert, A. A general program to calculate configuration interaction wavefunctions and oscillator strengths of many-electron atoms. *Comput. Phys. Commun.* **1975**, *9*, 141–172.
12. Hibbert, A.; Glass, R.; Froese Fischer, C. A general program for computing angular integrals of the Breit-Pauli Hamiltonian. *Comput. Phys. Commun.* **1991**, *64*, 455–472.
13. Clementi, E.; Roetti, C. Roothaan Hartree-Fock Wavefunctions. *Atom. Data Nucl. Data Tables* **1974**, *14*, 177–478.
14. Minnhagen, L. The spectrum of singly ionized argon, Ar II. *Ark. Fys.* **1963**, *25*, 203.
15. Saloman, E.B. Energy Levels and Observed Spectral Lines of Ionized Argon, Ar II through Ar XVIII. *J. Phys. Chem. Ref. Data* **2010**, *39*, 033101.
16. Hibbert, A. Estimation of inaccuracies in oscillator strength calculations. *Phys. Scr. T* **1996**, *65*, 104–109.

atoms

MDPI

Review

Multiconfiguration Dirac-Hartree-Fock Calculations with Spectroscopic Accuracy: Applications to Astrophysics

Per Jönsson [1,*], Gediminas Gaigalas [2], Pavel Rynkun [2], Laima Radžiūtė [2], Jörgen Ekman [1], Stefan Gustafsson [1], Henrik Hartman [1], Kai Wang [1], Michel Godefroid [3], Charlotte Froese Fischer [4], Ian Grant [5,6], Tomas Brage [7] and Giulio Del Zanna [6]

1 Materials Science and Applied Mathematics, Malmö University, SE-205 06 Malmö, Sweden; jorgen.ekman@mah.se (J.E.); stefan.gustafsson@mah.se (S.G.); henrik.hartman@mah.se (H.H.); kaiwang1128@aliyun.com (K.W.)

2 Institute of Theoretical Physics and Astronomy, Vilnius University, Saulėtekio av. 3, LT-10222 Vilnius, Lithuania; Gediminas.Gaigalas@tfai.vu.lt (G.G.); pavel.rynkun@gmail.com (P.R.); laima.radziute@gmail.com (L.R.)

3 Chimie Quantique et Photophysique, Université libre de Bruxelles, B-1050 Brussels, Belgium; michel.godefroid@ulb.ac.be

4 Department of Computer Science, University of British Columbia, Vancouver, BC V6T 1Z4, Canada; cff@cs.ubc.ca

5 Mathematical Institute, University of Oxford, Woodstock Road, Oxford OX2 6GG, UK; iangrant15@btinternet.com

6 Department of Applied Mathematics and Theoretical Physics, Centre for Mathematical Sciences, University of Cambridge, Wilberforce Road, Cambridge CB3 0WA, UK; gd232@cam.ac.uk

7 Division of Mathematical Physics, Department of Physics, Lund University, 221-00 Lund, Sweden; tomas.brage@fysik.lu.se

* Correspondence: per.jonsson@mah.se; Tel.: +46-40-66-57251

Academic Editor: Joseph Reader

Received: 31 January 2017; Accepted: 7 April 2017; Published: 14 April 2017

Abstract: Atomic data, such as wavelengths, spectroscopic labels, broadening parameters and transition rates, are necessary for many applications, especially in plasma diagnostics, and for interpreting the spectra of distant astrophysical objects. The experiment with its limited resources is unlikely to ever be able to provide a complete dataset on any atomic system. Instead, the bulk of the data must be calculated. Based on fundamental principles and well-justified approximations, theoretical atomic physics derives and implements algorithms and computational procedures that yield the desired data. We review progress and recent developments in fully-relativistic multiconfiguration Dirac–Hartree–Fock methods and show how large-scale calculations can give transition energies of spectroscopic accuracy, i.e., with an accuracy comparable to the one obtained from observations, as well as transition rates with estimated uncertainties of a few percent for a broad range of ions. Finally, we discuss further developments and challenges.

Keywords: transition energies; lifetimes; transition rates; multiconfiguration Dirac-Hartree-Fock

PACS: 31.15.am; 32.30.Jc; 32.70.Cs

1. Introduction

Atomic data, such as wavelengths, spectroscopic labels, broadening parameters, excitation and transition rates, are necessary for many applications, especially in plasma diagnostics, and for interpreting laboratory and astrophysical spectra [1,2]. Plasma diagnostics are commonly applied to measure the physical state of the plasma, e.g., temperatures, densities, ion and chemical abundances. Atomic databases, such as CHIANTI [3,4], are widely used for such diagnostic purposes. Their accuracy relies on a range of atomic rates, the main ones being electron collision rates and transition rates. For the solar corona, lines from highly charged iron ions, emitted in the extreme ultraviolet (EUV) and soft X-ray region, are commonly used for diagnostics, together with those from all other abundant elements. Atomic data and line identifications involving states of the lowest configurations of an ion are now relatively well known and observed. However, much less data are available for lines from higher configurations; one example is the lack of line identifications and rates for transitions from $n = 4$ iron ions in the soft X-rays [5].

Line identification from observed spectra is a very difficult and challenging task. Different methods such as isoelectronic interpolation and extrapolation, perfected by Edlén [6], can be used, but the work is nowadays mostly done with the aid of calculated transition energies and simulated spectra. For calculated transition energies, or wavelengths, to be of practical use, they need to be very accurate with uncertainties of just a few mÅ, placing high demands on computational methodologies.

Transition rates and line ratios are needed for diagnostic purposes. Due to the almost complete lack of accurate experimental data for atoms a few times ionized or more, the bulk of the transition rates must be calculated. Not only the rates themselves should be provided, but also uncertainty estimates that can be propagated in plasma models for sensitivity analysis. Both accurate rates and uncertainty estimates pose a challenge, calling for methods for which computed properties can be monitored as the wave functions are systematically improved.

This review summarizes the results from recent accurate relativistic multiconfiguration calculations for lowly charged ions or more of astrophysical importance. Focus is on the transition energies and their uncertainties, but transition rates and the associated uncertainty estimates are also discussed. The astrophysical background is provided in the individual papers covered by the review. Neutral atoms and ions in the lowest charge states are not covered in the review.

2. Multiconfiguration Methods

Multiconfiguration methods are versatile and can, in principle, be applied to any atomic or ionic system [7]. Multiconfiguration methods generate approximate energies and wave functions for the each of the targeted states in a system. The wave functions can then be used to compute measurable quantities, such as transition rates, hyperfine structures or Landé g-factors [8]. Looking at strengths and weaknesses, multiconfiguration methods capture near degeneracies and valence-valence electron correlation very efficiently. They are however less good at accounting for core-core correlation, and here, perturbative methods relying on a complete orbital basis have advantages. Work has been done to combine multiconfiguration and perturbative methods in different ways [9–12], a development that will open up accurate results also for more complex systems [13].

The relativistic multiconfiguration method, to be described below, is implemented in the GRASP2K program package [14]. The package is generally available and utilizes a message passing interface (MPI) for the most time-consuming programs, allowing for large-scale computing on parallel computers.

2.1. Multiconfiguration Dirac-Hartree-Fock

Atomic calculations are based on a Hamiltonian. In the relativistic multiconfiguration Dirac-Hartree-Fock (RMCDHF) method [7,15], as implemented in the GRASP2K package, the Hamiltonian is taken as the Dirac-Coulomb Hamiltonian:

$$H_{\mathrm{DC}} = \sum_{i=1}^{N} \left(c\, \boldsymbol{\alpha}_i \cdot \boldsymbol{p}_i + (\beta_i - 1)c^2 + V_{\mathrm{nuc}}(r_i) \right) + \sum_{i>j}^{N} \frac{1}{r_{ij}}, \tag{1}$$

where $V_{\text{nuc}}(r_i)$ is the nuclear potential modelled from an extended nuclear charge distribution, r_{ij} is the distance between electrons i and j and $\boldsymbol{\alpha}$ and β are the Dirac matrices. Wave functions $\Psi(\gamma PJM_J)$ for fine-structure states labelled by parity, P, and angular quantum numbers, JM_J, are expanded in antisymmetrized and coupled configuration state functions (CSFs):

$$\Psi(\gamma PJM_J) = \sum_{j=1}^{N_{\text{CSF}}} c_j \Phi(\gamma_j PJM_J). \tag{2}$$

The labels $\{\gamma_j\}$ denote the information of the CSFs, such as orbital occupancy and subshell quantum numbers in the angular momentum coupling tree. The CSFs are built from products of one-electron orbitals, having the general form:

$$\psi_{n\kappa,m}(\mathbf{r}) = \frac{1}{r}\begin{pmatrix} P_{n\kappa}(r)\chi_{\kappa,m}(\theta,\varphi) \\ \imath Q_{n\kappa}(r)\chi_{-\kappa,m}(\theta,\varphi) \end{pmatrix}, \tag{3}$$

where $\chi_{\pm\kappa,m}(\theta,\varphi)$ are two-component spin-orbit functions and where the radial functions are numerically represented on a logarithmic grid. The selection of the CSFs depends on the atomic system at hand and is described in Section 3.

In applications, one often seeks to determine energies and wave functions for a number, sometimes up to a few hundred, of targeted states. This is most conveniently done in the extended optimal level (EOL) scheme [16]. Given initial estimates of the radial functions, the energies E and expansion coefficients $c = (c_1, \ldots, c_M)^t$ for the targeted states are obtained as solutions to the relativistic configuration interaction (RCI) problem:

$$Hc = E\mathbf{c}, \tag{4}$$

where H is the RCI matrix of dimension $M \times M$ with elements:

$$H_{ij} = \langle \Phi(\gamma_i PJM_J)|H_{\text{DC}}|\Phi(\gamma_j PJM_J)\rangle. \tag{5}$$

Once the expansion coefficients have been determined, the radial functions $\{P_{n\kappa}(r), Q_{n\kappa}(r)\}$ are improved by solving a set of differential equations that results from applying the variational principle on a weighted energy functional of the targeted states together with additional terms needed to preserve the orthonormality of the orbitals. Appropriate boundary conditions for the radial orbitals exclude undesired negative-energy solutions [15]. The RCI problem and the solution of the differential equations are iterated until the radial orbitals and the energy are converged to a specified tolerance.

2.2. Configuration Interaction

The RMCDHF calculations are used to generate an orbital basis. Given this basis, the final wave functions for the targeted states are obtained in RCI calculations based on the frequency dependent Dirac-Coulomb-Breit Hamiltonian:

$$H_{\text{DCB}} = H_{\text{DC}} - \sum_{i<j}^{N}\left[\boldsymbol{\alpha}_i \cdot \boldsymbol{\alpha}_j \frac{\cos(\omega_{ij} r_{ij}/c)}{r_{ij}} + (\boldsymbol{\alpha}_i \cdot \boldsymbol{\nabla})(\boldsymbol{\alpha}_j \cdot \boldsymbol{\nabla}) \frac{\cos(\omega_{ij} r_{ij}/c) - 1}{\omega_{ij}^2 r_{ij}/c^2}\right], \tag{6}$$

where $\boldsymbol{\nabla}$ is the gradient operator involving differentiation with respect to $\mathbf{r}_{ij} = \mathbf{r}_i - \mathbf{r}_j$ and $r_{ij} = |\mathbf{r}_{ij}|$ [17]. In the RCI calculations leading quantum electrodynamic (QED) effects, vacuum polarization and self-energy are also taken into account. RCI calculations require less computational effort than do RMCDHF calculations, and currently, expansions with millions of CSFs can be handled. The relativistic multiconfiguration and configuration interaction calculations go together and are referred to as RMCDHF/RCI calculations.

2.3. *Managing Large Expansions*

To manage large expansions, CSFs can a priori be divided into two groups, referred to as a zero- and first-order partitioning. The first group, P, with m elements ($m \ll M$) contains CSFs that account for the major parts of the wave functions. The second group, Q, with $M - m$ elements contains CSFs that represent minor corrections. Allowing interaction between CSFs in group P, interaction between CSFs in groups P and Q and diagonal interactions between CSFs in Q gives a matrix:

$$\begin{pmatrix} H^{(PP)} & H^{(PQ)} \\ H^{(QP)} & H^{(QQ)} \end{pmatrix}, \tag{7}$$

where $H_{ij}^{(QQ)} = \delta_{ij} E_i^Q$. The restriction of $H^{(QQ)}$ to diagonal elements results in a huge reduction in the total number of matrix elements and the corresponding time for RCI calculations [12]. A similar reduction in computational time is obtained when constructing and solving the differential equations obtained from the weighted energy functional. Different computational strategies apply: RMCDHF calculations with limited interactions followed by RCI calculations with full interactions or RMCDHF calculations with limited interactions followed by RCI calculations with limited interaction, possibly with more CSFs in group P.

2.4. *Labelling*

In fully-relativistic calculations, quantum labels for the targeted states are obtained in jj-coupling. Most often, this wave function representation is not pure, i.e., there is no dominant CSF whose quantum numbers can be used to label a state in a proper way. Using the methods developed by Gaigalas and co-workers [18], the wave function representation in jj-coupling is transformed to an approximate representation in LSJ-coupling. This representation is normally more pure and better suited for labelling. One should be aware of the fact that even in LSJ-coupling, the labelling is not straight forward, and several components in the LSJ-coupling representation must be used in a recursive way to find unique labels [19,20]. Programs for transforming wave functions and assigning unique labels are important parts of the GRASP2K package [21].

2.5. *Transition Properties*

Given wave functions from RMCDHF/RCI calculations, transition properties, such as rates, A, line strengths, S, and weighted oscillator strengths, gf, between two states γPJ and $\gamma' P' J'$ are computed in terms of reduced matrix elements:

$$\langle\, \Psi(\gamma PJ) \,\|T^{(\mathrm{EMK})}\|\, \Psi(\gamma' P' J')\,\rangle, \tag{8}$$

where the operator $T^{(\mathrm{EMK})}$ depends on the multipolarity, E1, M1, E2, M2, etc., of the transition. By including Bessel functions in the definition of the operator, GRASP2K accounts for more high-order effects than the usual transition operator used in non-relativistic calculations with Breit–Pauli corrections [15]. Inserting the CSF expansions for the wave functions, the reduced matrix element reduces to a sum over reduced matrix elements between CSFs. Using Racah algebra techniques, these matrix elements are finally obtained as sums over radial integrals [22,23]. The above procedure assumes that the two states γPJ and $\gamma' P' J'$ are built from the same set of orbitals. When this is not the case, e.g., when separate calculations have been done for the even and odd parity states, the representation of the wave functions are changed in such a way that the orbitals become biorthonormal [24,25], in which case the calculation continues along the lines above. For electric transitions, parameters can be computed in both length and velocity gauge [26], where the results in the length gauge are the preferred.

3. General Computational Methodology: The SD-MR Approach

Systematic calculations using multiconfiguration methods follow a determined scheme as described below. Details of the scheme are determined by the shell structure of the atom, the number of targeted states, the desired accuracy of the final results and the available computational resources. The atomic Hamiltonian is invariant with respect to space inversions, and there are no interactions between odd and even parity states. The odd and even parity states are thus often treated in separate sets of calculations. After validation for selected ions and states, computed transition energies and rates can be used to aid the analysis of unknown spectra.

3.1. *Multireference and Gross Features of the Wave Functions*

For highly ionized systems, a natural starting point is the multireference set (MR). In this review, we define the MR as the set of configurations associated with the targeted states of a given parity together with important closely degenerate configurations. Applying rules for the coupling of angular momenta, the configurations in the MR give rise to a set of CSFs that account for the most important gross features of the wave functions. The expansion coefficients of the CSFs and the orbitals are determined in an initial RMCDHF calculation. The orbitals for the initial calculation are called spectroscopic orbitals. They are required to have the same node structure as hydrogenic orbitals, i.e., the node structure is determined by the principal quantum number. The spectroscopic orbitals are kept frozen in all subsequent calculations.

3.2. *Including Electron Correlation and Determining an Orbital Set*

The initial approximation of the wave functions is improved by adding CSFs that account for electron correlation. Guided by a perturbative analysis, the CSFs are generated by the single (S) and double (D) multireference (SD-MR) active space method in which a number of configurations is obtained by SD substitutions of orbitals in the configurations of the MR with orbitals in an active set [7,8]. Again, applying rules for the coupling of angular momenta, the generated configurations give rise to the CSFs. Not all of these CSFs are important, and the CSFs are further required to be such that they interact (have non-zero Hamiltonian matrix elements) with the CSFs of the MR. The expansion coefficients of the CSFs and the radial parts of the orbitals in the active set are determined in RMCDHF calculations where, for large expansions, limited interactions are used.

The active set, often denoted by the number of orbitals with a specified symmetry, so that $\{4s3p2d1f\}$ is a set with four s orbitals, three p orbitals, two d orbitals and one f orbital, is systematically enlarged one orbital layer at the time until the computed excitation energies and transition rates have converged to within some predetermined tolerance. For small systems, SD substitutions are done from all subshells of the configurations in the MR, and the generated CSFs account for valence-valence, core-valence and core-core electron correlation. For larger systems, it becomes necessary to define a core for which restrictions on the substitutions apply. In many cases, the SD-MR substitutions are restricted in such a way that there are only S substitutions from subshells that define a so-called active core. There may also be subshells deep down in the core for which there are no substitutions at all. CSFs obtained from S-MR substitutions from the active core together with SD-MR substitutions from the valence subshells account for valence-valence and core-valence correlation.

3.3. *Final Configuration Interaction Calculations Including the Breit Interaction and QED Effects*

The frequency dependent Breit (transverse photon) interaction and leading QED effects are included in final RCI calculations. To account for higher order correlation effects, the MR is sometimes enlarged at this final step leading to larger expansions. Full interaction is normally used, although limited interactions have been shown effective for including core-valence and core-core effects in larger systems [12,27].

4. Excitation Energies

In this section, RMCDHF/RCI excitation energies are compared with observations for a range of systems in order to illustrate the predictive power of highly accurate calculations. Generally, there are enough observations to validate computational methodologies and to distinguish between different approaches.

4.1. Energies for $2s^22p^n$, $2s2p^{n+1}$ and $2p^{n+2}$ States in the B-, C-, N-, O- and F-Like Sequences

Excitation energies and E1, M1, E2, M2 transition rates between $2s^22p^n$, $2s2p^{n+1}$ and $2p^{n+2}$ states of ions in the B-, C-, N-, O- and F-like sequences were calculated using the RMCDHF/RCI and SD-MR method [28–32]. The range of ions, as well as the details of the calculations are summarized in Table 1. Calculations of Landé g_J factors, hyperfine structures and isotope shifts were done separately for ions in the Be-, B-, C- and N-like sequences [33,34].

Table 1. Multireference (MR), active set, number of generated configuration state functions (CSFs) (N_{CSFs}) and the range of ions for the relativistic multiconfiguration Dirac-Hartree-Fock (RMCDHF) and relativistic configuration interaction (RCI) calculations of the boron-, carbon-, nitrogen-, oxygen- and fluorine-like sequences.

Configuration	MR for RMCDHF	MR for RCI	Active Set	N_{CSFs}
		boron-like, N III to Zn XXVI		
$1s^22s^22p$	$1s^2\{2s^22p, 2p^3\}$	$1s^2\{2s^22p, 2p^3, 2s2p3d, 2p3d^2\}$	$\{9s8p7d6f5g3h1i\}$	200 100
$1s^22p^3$	$1s^2\{2s^22p, 2p^3\}$	$1s^2\{2s^22p, 2p^3, 2s2p3d, 2p3d^2\}$	$\{9s8p7d6f5g3h1i\}$	360 100
$1s^22s2p^2$	$1s^22s2p^2$	$1s^2\{2s2p^2, 2p^23d, 2s^23d, 2s3d^2\}$	$\{9s8p7d6f5g3h1i\}$	300 100
		carbon-like, F IV to Ni XXIII		
$1s^22s^22p^2$	$1s^2\{2s^22p^2, 2p^4\}$	$1s^2\{2s^22p^2, 2p^4, 2s2p^23d, 2s^23d^2\}$	$\{8s7p6d5f4g2h\}$	340 100
$1s^22p^4$	$1s^2\{2s^22p^2, 2p^4\}$	$1s^2\{2s^22p^2, 2p^4, 2s2p^23d, 2s^23d^2\}$	$\{8s7p6d5f4g2h\}$	340 100
$1s^22s2p^3$	$1s^22s2p^3$	$1s^2\{2s2p^3, 2p^33d, 2s^22p3d, 2s2p3d^2\}$	$\{8s7p6d5f4g2h\}$	1 000 100
		nitrogen-like, F III to Kr XXX		
$1s^22s^22p^3$	$1s^2\{2s^22p^3, 2p^5\}$	$1s^2\{2s^22p^3, 2p^5, 2s2p^33d, 2s^22p3d^2\}$	$\{8s7p6d5f4g1h\}$	698 100
$1s^22p^5$	$1s^2\{2s^22p^3, 2p^5\}$	$1s^2\{2s^22p^3, 2p^5, 2s2p^33d, 2s^22p3d^2\}$	$\{8s7p6d5f4g1h\}$	382 100
$1s^22s2p^4$	$1s^22s2p^4$	$1s^2\{2s2p^4, 2p^43d, 2s^22p^23d, 2s2p^23d^2\}$	$\{8s7p6d5f4g1h\}$	680 100
		oxygen-like, F II to Kr XXIX		
$1s^22s^22p^4$	$1s^2\{2s^22p^4, 2p^6\}$	$1s^2\{2s^22p^4, 2p^6, 2s2p^43d\}$	$\{8s7p6d5f4g3h\}$	709 690
$1s^22p^6$	$1s^2\{2s^22p^4, 2p^6\}$	$1s^2\{2s^22p^4, 2p^6, 2s2p^43d\}$	$\{8s7p6d5f4g3h\}$	67 375
$1s^22s2p^5$	$1s^22s2p^5$	$1s^2\{2s2p^5, 2p^53d, 2s^22p^33d\}$	$\{8s7p6d5f4g3h\}$	702 892
		fluorine-like, Si VI to WLXVI		
$1s^22s^22p^5$	$1s^22s^22p^5$	$1s^22s^22p^5$	$\{8s7p6d5f4g3h2i\}$	73 000
$1s^22s2p^6$	$1s^22s2p^6$	$1s^22s2p^6$	$\{8s7p6d5f4g3h2i\}$	15 000

A trend for all atomic structure calculations, including RMCDHF/RCI, is that the accuracy of the excitation energies is, relatively speaking, lower for lowly charged ions and that the accuracy then increases as the effects of electron correlation diminish. For the highly charged ions, the situation is less clear. Often experimental excitation energies are associated with large uncertainties or missing altogether. The situation is illustrated in Tables 2 and 3 for the O-like sequence [31].

In Table 2, excitation energies in Ne III and Fe XIX from different calculations are compared with energies from observations. The most accurate calculations are the RMCDHF/RCI calculation [31] and the multireference second-order Möller-Plesset calculation (MRMP). For Ne III, the relative differences with observation for these two calculations are in the range of 0.2–0.4% (slightly worse for MRMP). For Fe XIX the relative errors go down by an order of magnitude, and now, the calculated energies are accurate enough to detect misidentifications or errors in observational data, but also to serve as a valuable tool for identifying new lines. The usefulness of computed energies is illustrated in Table 3 for Br XXVIII, where the RMCDHF/RCI and MRMP calculations clearly discriminate between observed energies [35] and energies from semiempirical fits [36], being in better agreement with the latter. This suggests that there may be some calibration problems in relation to the observed energies [35].

Table 2. Excitation energies in cm^{-1} for O-like Ne and Fe from observations and different calculations. Relative errors in % for the calculated energies are shown in parenthesis. E_{obs} observation NIST [37], E_{RCI} energies from RMCDHF/RCI [31], E_{MRMP} energies from Möller-Plesset calculation (MRMP) [38], E_{MBPT} energies from many-body perturbation theory [39], E_{BP} energies from multiconfiguration Hartree-Fock-Breit-Pauli [19], E_{SS} energies from super structure [40], E_{MCDF} energies from RMCDHF [41] and E_{FAC} energies from RCI with the FAC code [42].

				Ne III			
Level	J	E_{obs}	E_{RCI}	E_{MRMP}	E_{MBPT}	E_{BP}	E_{SS}
$2s^22p^4\,^3P$	2	0	0 (0.00)	0 (0.00)	0 (0.00)	0 (0.00)	0 (0.00)
	1	643	645 (0.31)	638 (0.77)	645 (0.31)	628 (2.33)	744 (15.70)
	0	921	923 (0.21)	912 (0.97)	926 (0.54)	899 (2.38)	1 069 (16.06)
$2s^22p^4\,^1D$	2	25 841	25 954 (0.43)	26 097 (0.99)	25 573 (1.03)	25 759 (0.31)	29 219 (13.07)
$2s^22p^4\,^1S$	0	55 753	56 058 (0.54)	55 772 (0.03)	55 459 (0.52)	55 382 (0.66)	72 484 (30.00)
$2s2p^5\,^3P^o$	2	204 290	204 608 (0.15)	204 718 (0.20)	200 686 (1.76)	204 635 (0.16)	215 348 (5.41)
	1	204 873	205 200 (0.15)	205 297 (0.20)	201 276 (1.75)	205 236 (0.17)	216 008 (5.43)
	0	205 194	205 603 (0.19)	205 617 (0.20)	201 598 (1.75)	205 539 (0.16)	216 367 (5.44)
$2s2p^5\,^1P^o$	1	289 479	290 315 (0.28)	290 703 (0.42)	288 219 (0.43)	291 659 (0.75)	315 511 (8.99)
				Fe XIX			
Level	J	E_{obs}	E_{RCI}	E_{MRMP}	E_{MBPT}	E_{MCDF}	E_{FAC}
$2s^22p^4\,^3P$	2	0	0 (0.000)	0 (0.000)	0 (0.00)	0 (0.00)	0 (0.00)
	0	75 250	75 313 (0.083)	75 218 (0.042)	74 742 (0.67)	75 446 (0.26)	75 198 (0.06)
	1	89 441	89 434 (0.007)	89 251 (0.212)	87 559 (2.10)	88 791 (0.72)	88 821 (0.69)
$2s^22p^4\,^1D$	2	168 852	168 985 (0.078)	168 792 (0.035)	167 881 (0.57)	170 847 (1.18)	170 578 (1.02)
$2s^22p^4\,^1S$	0	325 140	325 417 (0.085)	324 949 (0.058)	321 124 (1.23)	326 536 (0.42)	325 421 (0.08)
$2s2p^5\,^3P^o$	2	922 890	923 044 (0.016)	922 855 (0.003)	917 435 (0.59)	933 081 (1.10)	929 231 (0.68)
	1	984 740	984 920 (0.018)	984 791 (0.005)	978 242 (0.65)	995 006 (1.04)	991 246 (0.66)
	0	1030 020	1030 199 (0.017)	1029 992 (0.002)	1022 753 (0.70)	1039 692 (0.93)	1036 058 (0.58)
$2s2p^5\,^1P^o$	1	1267 600	1268 093 (0.038)	1267 771 (0.013)	1258 927 (0.68)	1287 773 (1.59)	1282 914 (1.20)
$2p^6\,^1S$	0	2134 180	2134 958 (0.036)	2132 810 (0.064)	2120 211 (0.65)	2175 645 (1.94)	2160 701 (1.24)

Table 3. Excitation energies in cm^{-1} for O-like Br. Comparison between calculations, observations and semiempirical estimates. E_{obs} observation NIST [37] with original data from Kelly [35], E_{SE} semiempirical fit [36] and E_{RCI} energies from RMCDHF/RCI [31], E_{MRMP} energies from MRMP [38]; ΔE_1, difference between calculated energies and E_{obs}; ΔE_2, difference between calculated energies and E_{SE}. The calculations support energies from the semiempirical fit.

Level	J	E_{obs}	E_{SE}	E_{RCI}	ΔE_1	ΔE_2	E_{MRMP}	ΔE_1	ΔE_2
	2	0	0	0	0	0	0	0	0
$2s^22p^4\,^3P$	0	218 800	153 478	151 954	−66 846	−1524	15 2035	−66 765	−1443
	1	379 800	371 663	371 606	−8 194	−57	37 1858	−7 942	195
$2s^22p^4\,^1D$	2	483 040	470 699	470 643	−12 397	−56	47 0804	−12 236	105
$2s^22p^4\,^1S$	0	944 150	912 501	911 968	−32 182	-533	91 2282	−31 868	−219
	2		1 579 903	1 579 537		−366	1 580 945		1042
$2s2p^5\,^3P^o$	1		1 755 028	1 755 196		168	1 756 684		1656
	0		1 986 274	1 985 784		−490	1 987 396		1122
$2s2p^5\,^1P^o$	1		2 229 358	2 230 149		791	2 231 636		2278
$2p^6\,^1S$	0		3 573 416	3 575 415		1999	3 579 486		6070

Summarizing the mean relative errors in the excitation energies for the $2s^22p^n$, $2s2p^{n+1}$ and $2p^{n+2}$ states of B-, C-, N-, O- and F-like Fe from RCI calculations [28–32], we have 0.022% for B-like, 0.022% for C-like, 0.050% for N-like, 0.042% for O-like and 0.011% for F-like Fe.

4.2. Energies of the $2s^22p^6$ and $2s^22p^53l$ States in the Ne-Like Sequence

The transitions connecting the $2s^22p^53l$, $l = 0, 1, 2$ configurations in Ne-like ions give rise to prominent lines in the spectra of many high temperature light sources. Some of these lines are considered for diagnostics of fusion plasmas. Excitation energies and E1, M1, E2, M2 transition rates between states of the above configurations in Ne-like Mg III and Kr XXVII sequences were calculated using the RMCDHF/RCI and SD-MR method [43]. The calculations were done based on expansions from SD substitutions from the $2s^22p^6$ and $2s^22p^53l$ configurations to active sets $\{7s6p5d4f3g2h1i\}$. The $1s^2$ was kept as a closed core. Some triple substitutions were allowed to capture higher order electron correlation effects. In Table 4, the RMCDHF/RCI excitation energies are displayed for Ca XI and Fe XVII. In the same table, the energies are compared with energies from NIST, as well as from MRMP calculations by Ishikawa et al. [44]. Again, the table illustrates the situation when it comes to experiments. For many ions, the excitation energies of the lower states are known from experiments. For other ions, such as Ca XI, energies are only known for a few states. The correlation model from the RMCDHF/RCI calculations predicts the excitation energies extremely well for all of the calculated ions. For Fe XVII, the relative differences with observations are around 0.005%. Calculated energies with this accuracy aid line identification in spectra and can be used to validate previous observations. As can be seen from the table, the RMCDHF/RCI and MRMP calculations both do very well, but the latter lose some of the accuracy at the neutral end of the sequence.

In Table 4, also the LSJ composition is shown for each state. There are many states that are heavily mixed, with terms of almost the same weight. In these cases, labelling becomes difficult, and for many ions in the sequence, there are states that have the same leading term. Labeling is a general problem that needs considerable attention [21].

4.3. Energies for Higher States in the B-, C-, N-, O-, F- and Ne-Like Sequences

In plasma modelling and diagnostics, it is important to provide atomic data for more than just the states of the lowest configurations. To meet this demand, the RMCDHF/RCI and SD-MR calculations for the B-, C-, N-, O-, F- and Ne-like sequences have been extended to hundreds of states in what we refer to as spectrum calculations [45–52]. The range of ions, the targeted configurations and the number of studied states for each sequence are summarized in Table 5. Calculations were done by parity, i.e., odd and even parity states were treated in separate sets of calculations. The targeted configurations define the MR, and the expansions were obtained by SD-MR substitutions from all subshells to increasing active sets of orbitals. In addition to excitation energies, E1, M1, E2 and M2 transition rates were calculated.

Spectrum calculations are challenging for different reasons. The active sets of orbitals often have to be large, since many states with different charge distributions should be represented. The large active sets lead to large CSF expansions, and typically, the number of CSFs are a few millions for each parity. Another challenge is to handle the labelling. With closely degenerate configurations, the states are often not pure, but need to be described by the leading LSJ composition. However, the LSJ composition depends on the details of the calculation and different calculation may lead to different compositions. Thus, it is not unusual that there are inconsistencies in labelling, making comparisons between different sets of calculations, as well as with observations difficult and time consuming.

Table 4. Excitation energies in cm^{-1} for Ne-like Ca and Fe from observations and different calculations. Relative errors in % for the calculated energies are shown in parenthesis. E_{obs} observation NIST [37], E_{RCI} energies from RMCDHF/RCI [43] and E_{MRMP} energies from MRMP [44].

Level	LSJ	Composition	E_{obs}	E_{RCI}	E_{MRMP}
			Ca XI		
$2p^6$	$^1S_0$	1.00	0	0	0
$2p^53s$	$^3P^o_2$	0.99		2 801 989	2 801 819
$2p^53s$	$^3P^o_1$	$0.62 + 0.38\,^1P^o_1$	2 810 900	2 810 834 (0.0023)	2 810 588 (0.011)
$2p^53s$	$^3P^o_0$	0.99		2 831 800	2 831 670
$2p^53s$	$^1P^o_1$	$0.62 + 0.38\,^3P^o_1$	2 839 900	2 839 662 (0.0084)	2 839 386 (0.018)
$2p^53p$	$^3S_1$	0.92		2 953 791	2 953 594
$2p^53p$	$^3D_2$	$0.68 + 0.24\,^1D_2$		2 978 410	2 977 968
$2p^53p$	$^3D_3$	1.00		2 978 650	2 978 276
$2p^53p$	$^3D_1$	$0.43 + 0.36\,^1P_1 + 0.20\,^3P_1$		2 986 908	2 986 513
$2p^53p$	$^3P_2$	$0.65 + 0.34\,^1D_2$		2 993 760	2 993 336
$2p^53p$	$^3D_1$	$0.54 + 0.40\,^1P_1$		3 007 301	3 006 932
$2p^53p$	$^3P_0$	0.98		3 009 345	3 009 000
$2p^53p$	$^1D_2$	$0.41 + 0.31\,^3D_2 + 0.27\,^3P_2$		3 016 749	3 016 378
$2p^53p$	$^3P_1$	$0.68 + 0.24\,^1P_1$		3 017 175	3 016 845
$2p^53p$	$^1S_0$	0.98		3 101 166	3 098 308
$2p^53d$	$^3P^o_0$	0.99		3 196 075	3 195 830
$2p^53d$	$^3P^o_1$	0.95	3 199 300	3 199 045 (0.0080)	3 198 902 (0.012)
$2p^53d$	$^3P^o_2$	0.85		3 205 278	3 205 169
$2p^53d$	$^3F^o_4$	1.00		3 208 351	3 208 165
$2p^53d$	$^3F^o_3$	$0.72 + 0.23\,^1F^o_2$		3 212 392	3 212 144
$2p^53d$	$^3F^o_2$	$0.53 + 0.29\,^1D^o_2 + 0.18\,^3D^o_2$		3 219 655	3 219 428
$2p^53d$	$^3D^o_3$	$0.55 + 0.41\,^1F^o_3$		3 224 394	3 224 078
$2p^53d$	$^3D^o_1$	0.89	3 239 700	3 239 502 (0.0061)	3 239 308 (0.012)
$2p^53d$	$^3F^o_2$	$0.47 + 0.38\,^1D^o_2 + 0.14\,^3D^o_2$		3 244 348	3 244 161
$2p^53d$	$^3D^o_2$	$0.58 + 0.27\,^1D^o_2 + 0.14\,^3P^o_2$		3 248 017	3 247 805
$2p^53d$	$^3D^o_3$	$0.40 + 0.35\,^1F^o_3 + 0.24\,^3F^o_3$		3 248 345	3 248 099
$2p^53d$	$^1P^o_1$	0.91	3 284 300	3 284 444 (0.0044)	3 283 473 (0.025)
			Fe XVII		
$2p^6$	$^1S_0$	1.00	0	0	0
$2p^53s$	$^3P^o_2$	1.00	5 849 490	5 849 108 (0.0065)	5 848 891 (0.0102)
$2p^53s$	$^1P^o_1$	$0.54 + 0.45\,^3P_1$	5 864 760	5 864 469 (0.0049)	5 864 138 (0.0106)
$2p^53s$	$^3P^o_0$	1.00	5 951 478	5 951 003 (0.0079)	5 950 877 (0.0100)
$2p^53s$	$^3P^o_1$	$0.54 + 0.45\,^1P_1$	5 961 022	5 960 633 (0.0065)	5 960 410 (0.0102)
$2p^53p$	$^3S_1$	$0.80 + 0.17\,^3P_1$	6 093 568	6 093 573 (0.0000)	6 093 209 (0.0058)
$2p^53p$	$^3D_2$	$0.58 + 0.30\,^1D_2 + 0.12\,^3P_2$	6 121 756	6 121 769 (0.0002)	6 121 253 (0.0082)
$2p^53p$	$^3D_3$	1.00	6 134 815	6 134 794 (0.0003)	6 134 360 (0.0074)
$2p^53p$	$^1P_1$	$0.51 + 0.25\,^3D_1 + 0.19\,^3P_1$	6 143 897	6 143 898 (0.0000)	6 143 431 (0.0075)
$2p^53p$	$^3P_2$	$0.67 + 0.32\,^1D_2$	6 158 540	6 158 481 (0.0009)	6 158 010 (0.0086)
$2p^53p$	$^3P_0$	0.94	6 202 620	6 202 542 (0.0012)	6 202 238 (0.0061)
$2p^53p$	$^3D_1$	$0.67 + 0.31\,^1P_1$	6 219 266	6 219 185 (0.0013)	6 218 795 (0.0075)
$2p^53p$	$^3P_1$	$0.63 + 0.17\,^1P_1 + 0.13\,^3S_1$	6 245 490	6 245 346 (0.0023)	6 245 018 (0.0075)
$2p^53p$	$^3D_2$	$0.41 + 0.38\,^1D_2 + 0.21\,^3P_2$	6 248 530	6 248 390 (0.0022)	6 248 024 (0.0080)
$2p^53p$	$^1S_0$	0.93	6 353 356	6 353 605 (0.0039)	6 351 136 (0.0349)
$2p^53d$	$^3P^o_0$	0.99	6 464 095	6 463 913 (0.0028)	6 463 611 (0.0074)
$2p^53d$	$^3P^o_1$	0.91	6 471 233	6 471 519 (0.0044)	6 471 317 (0.0012)
$2p^53d$	$^3P^o_2$	$0.72 + 0.18\,^3D^o_2$	6 486 440	6 486 166 (0.0042)	6 485 977 (0.0071)
$2p^53d$	$^3F^o_4$	1.00	6 487 000	6 486 745 (0.0039)	6 486 514 (0.0074)
$2p^53d$	$^3F^o_3$	$0.65 + 0.29\,^1F^o_3$	6 492 924	6 492 689 (0.0036)	6 492 387 (0.0082)
$2p^53d$	$^1D^o_2$	$0.41 + 0.35\,^3F^o_2 + 0.24\,^3D^o_2$	6 506 808	6 506 561 (0.0037)	6 506 276 (0.0081)
$2p^53d$	$^3D^o_3$	$0.64 + 0.34\,^1F^o_3$	6 515 479	6 515 276 (0.0031)	6 514 936 (0.0083)
$2p^53d$	$^3D^o_1$	$0.74 + 0.20\,^1P^o_1$	6 552 221	6 552 697 (0.0072)	6 552 491 (0.0041)
$2p^53d$	$^3F^o_2$	$0.63 + 0.29\,^1D^o_2$	6 594 617	6 594 260 (0.0054)	6 594 099 (0.0078)
$2p^53d$	$^3D^o_2$	$0.50 + 0.27\,^3P^o_2 + 0.21\,^1D^o_2$	6 601 210	6 600 855 (0.0053)	6 600 688 (0.0079)
$2p^53d$	$^1F^o_3$	$0.37 + 0.33\,^3F^o_3 + 0.30\,^3D^o_3$	6 605 469	6 605 078 (0.0059)	6 604 858 (0.0092)
$2p^53d$	$^1P^o_1$	$0.78 + 0.18\,^3D^o_1$	6 660 894	6 661 101 (0.0031)	6 660 232 (0.0099)

Table 5. Sequence, ions and targeted configurations for the RMCDHF/RCI calculations. N is the number of studied states for each ion. In the table, $l = 0,1,2, l' = 0,1,2,3, l'' = 0,\ldots,n-1$.

Sequence	Ions	Configurations	N	Ref.
B-like	Si, Ti-Cu	$2s^22p, 2s2p^2, 2p^3, 2s^23l, 2s2p3l, 2p^23l, 2s^24l', 2s2p4l', 2p^24l'$	291	[45]
B-like	Na	$2s^22p, 2s2p^2, 2p^3, 2s^23l, 2s2p3l, 2p^23l, 2s^24l', 2s2p4s$	133	[46]
C-like	Ar-Zn	$2s^22p^2, 2s2p^3, 2p^4, 2s^22p3l, 2s2p^23l, 2p^33l, 2s^22p4l$	262	[47]
N-like	Cr, Fe, Ni, Zn	$2s^22p^3, 2s2p^4, 2p^5, 2s^22p^23l, 2s2p^33l, 2p^43l$	272	[48]
N-like	Ar-Zn	$2s^22p^3, 2s2p^4, 2p^5, 2s^22p^23l, 2s2p^33l, 2p^43l, 2s^22p^24l'$	359	[49]
O-like	Cr-Zn	$2s^22p^4, 2s2p^5, 2p^6, 2s^22p^33l, 2s2p^43l$	200	[50]
F-like	Cr-Zn	$2s^22p^5, 2s2p^6, 2s^22p^43l, 2s2p^53l, 2p^63l, 2s^22p^44l'$	200	[51]
Ne-like	Cr-Kr	$2s^22p^6, 2s2p^63l, 2s^22p^54l, 2s2p^64l, 2s^22p^55l'', 2s^22p^56l''$	201	[52]

For many ions, excitation energies for lower lying states are known from observations. Going higher, comparatively less data are available, and these are often associated with large uncertainties. The situation is well illustrated for C-like Fe, and in Table 6, the RMCDHF/RCI excitation energies by Ekman et al. [47] are compared with observations. Due to near degeneracies, many states have the same leading LSJ term. In these cases, labelling can be done either by giving the leading terms in the composition or, more simply, introducing an additional index A and B to separate the states. For the 20 first states belonging to the $n = 2$ configurations, observations are available from the NIST [37] and CHIANTI databases [3,4]. There is an agreement between the RMCDHF/RCI and relativistic many body calculations (RMBPT) by Gu [53] and observations at the 0.028–0.032% level (slightly worse for RMBPT). The RCI calculation using the Flexible Atomic Code (FAC) [42] is less accurate. For the higher lying states, experimental data are sparse. In many cases, there is excellent agreement between observations and calculations also for these states, but in some cases, there are obvious disagreements. For State Number 36, the excitation energy from NIST and CHIANTI disagree, and the calculations by Ekman et al. and Gu support the energy from the CHIANTI database. For State 54, all calculations agree, but differ markedly from the energies given by NIST and CHIANTI.

Table 6. Energies in cm^{-1} for levels in Fe XXI. E_{RCI} energies from RMCDHF/RCI calculations [47], E_{RMBPT} energies from RMBPT [53], E_{FAC} energies from RCI calculations with FAC [42], E_{NIST} NIST recommended values [37] and E_{CHI} observed energies from the CHIANTI database [3,4].

No.	Level	E_{RCI}	E_{RMBPT}	E_{FAC}	E_{NIST}	E_{CHI}
1	$2s^22p^2\ ^3P_0$	0	0	0	0	0
2	$2s^22p^2\ ^3P_1$	73 864	73 867	73 041	73 851	73 851
3	$2s^22p^2\ ^3P_2$	117 417	117 372	117 146	117 354	117 367
4	$2s^22p^2\ ^1D_2$	244 751	244 581	245 710	244 561	244 568
5	$2s^22p^2\ ^1S_0$	372 137	372 261	373 060	371 980	371 744
6	$2s2p^3\ ^5S_2$	486 584	487 683	479 658	486 950	486 991
7	$2s2p^3\ ^3D_1$	776 775	777 005	779 724	776 690	776 685
8	$2s2p^3\ ^3D_2$	777 404	777 655	779 963	777 340	777 367
9	$2s2p^3\ ^3D_3$	803 618	803 869	805 768	803 540	803 553
10	$2s2p^3\ ^3P_0$	916 444	916 773	920 272	916 330	916 333
11	$2s2p^3\ ^3P_1$	925 074	925 408	928 822	924 920	924 920
12	$2s2p^3\ ^3P_2$	942 621	942 986	946 135	942 430	942 364
13	$2s2p^3\ ^3S_1$	1 096 019	1 095 820	1 105 578	1 095 670	1 095 679
14	$2s2p^3\ ^1D_2$	1 127 672	1 127 460	1 137 533	1 127 240	1 127 250
15	$2s2p^3\ ^1P_1$	1 261 577	1 261 240	1 272 627	1 261 140	1 260 902
16	$2p^4\ ^3P_2$	1 646 437	1 646 467	1 657 411	1 646 300	1 646 409
17	$2p^4\ ^3P_0$	1 735 823	1 735 813	1 747 301	1 735 700	1 735 715
18	$2p^4\ ^3P_1$	1 740 623	1 740 707	1 750 848	1 740 500	1 740 453
19	$2p^4\ ^1D_2$	1 817 786	1 817 362	1 832 102	1 817 100	1 817 041
20	$2p^4\ ^1S_0$	2 048 512	2 047 850	2 066 463	2 048 200	2 048 056
21	$2s^22p3s\ ^3P_0$	7 663 283	7 664 054	7 654 119		
22	$2s^22p3s\ ^3P_1$	7 671 971	7 672 703	7 663 398		7 661 883

Table 6. *Cont.*

No.	Level	E_{RCI}	E_{RMBPT}	E_{FAC}	E_{NIST}	E_{CHI}
23	$2s^22p3s\ ^3P_2$	7 780 298	7 781 147	7 770 895		
24	$2s^22p3s\ ^1P_1$	7 803 764	7 804 419	7 796 397		
25	$2s^22p3p\ ^3D_1$	7 841 903	7 842 922	7 834 847		
26	$2s^22p3p\ ^3P_1 : A$	7 898 154	7 898 974	7 891 978		
27	$2s^22p3p\ ^3D_2$	7 901 553	7 902 378	7 895 497		
28	$2s^22p3p\ ^3P_0$	7 914 849	7 915 811	7 909 434		7 915 463
29	$2s^22p3p\ ^3P_1 : B$	7 983 446	7 984 350	7 977 011		
30	$2s^22p3p\ ^3D_3$	7 994 588	7 995 388	7 987 318		
31	$2s^22p3p\ ^3S_1$	8 004 987	8 005 793	7 998 341		
32	$2s^22p3p\ ^3P_2$	8 007 326	8 008 319	8 002 052		
33	$2s^22p3p\ ^1D_2$	8 068 537	8 069 071	8 065 382		
34	$2s^22p3d\ ^3F_2$	8 078 540	8 079 119	8 072 911		8 074 160
35	$2s2p^2(4P)3s\ ^5P_1$	8 080 551	8 082 001	8 070 805		
36	$2s^22p3d\ ^3F_3$	8 116 048	8 116 480	8 111 336	8 101 400	8 118 008
37	$2s^22p3d\ ^3P_2 : A$	8 121 922	8 122 529	8 118 025		8 124 085
38	$2s^22p3p\ ^1S_0$	8 128 645	8 129 396	8 126 192		8 143 710
39	$2s2p^2(4P)3s\ ^5P_2$	8 131 973	8 133 460	8 121 258		
40	$2s^22p3d\ ^3D_1$	8 139 290	8 140 735	8 135 992	8 140 000	8 141 785
41	$2s2p^2(4P)3s\ ^5P_3$	8 181 331	8 182 599	8 170 876		
42	$2s2p^2(4P)3s\ ^3P_0$	8 182 172	8 182 844	8 179 292		8 180 254
43	$2s^22p3d\ ^3F_4$	8 202 073	8 202 670	8 195 771		
44	$2s^22p3d\ ^1D_2$	8 208 705	8 209 597	8 204 329		
45	$2s2p^2(4P)3s\ ^3P_1$	8 222 156	8 222 948	8 217 390		
46	$2s^22p3d\ ^3D_3$	8 230 918	8 231 868	8 227 144	(8 195 000)	8 229 642
47	$2s^22p3d\ ^3P_2 : B$	8 245 453	8 246 428	8 241 436	8 230 900	8 229 642
48	$2s^22p3d\ ^3P_1$	8 245 737	8 247 075	8 241 557		
49	$2s^22p3d\ ^3P_0$	8 247 732	8 249 164	8 243 033		
50	$2s2p^2(4P)3p\ ^5D_0$	8 267 963	8 269 220	8 259 742		
51	$2s2p^2(4P)3p\ ^5D_1$	8 270 558	8 272 088	8 262 373		
52	$2s2p^2(4P)3s\ ^3P_2$	8 274 704	8 275 427	8 270 235		
53	$2s^22p3d\ ^1F_3$	8 300 618	8 301 128	8 301 379	8 313 600	
54	$2s^22p3d\ ^1P_1$	8 303 730	8 307 428	8 305 376	8 293 900	8 293 791
55	$2s2p^2(4P)3p\ ^5D_2$	8 305 917	8 309 162	8 297 457		
56	$2s2p^2(4P)3p\ ^3S_1$	8 312 499	8 309 359	8 300 070		
57	$2s2p^2(4P)3p\ ^5P_1$	8 349 456	8 350 857	8 342 324		8 350 731
58	$2s2p^2(4P)3p\ ^5D_3$	8 351 775	8 353 277	8 342 522		
59	$2s2p^2(4P)3p\ ^5P_2$	8 352 117	8 353 731	8 343 801		
60	$2s2p^2(4P)3p\ ^3D_1$	8 379 967	8 380 890	8 373 689		8 376 741
61	$2s2p^2(4P)3p\ ^5P_3$	8 388 634	8 390 292	8 380 281		
62	$2s2p^2(4P)3p\ ^5D_4$	8 399 557	8 401 039	8 390 478		
63	$2s2p^2(4P)3p\ ^3D_2$	8 410 077	8 411 182	8 404 386		
64	$2s2p^2(2D)3s\ ^3D_1$	8 420 588	8 421 426	8 420 569		
65	$2s2p^2(2D)3s\ ^3D_2$	8 428 405	8 429 248	8 428 172		
66	$2s2p^2(2D)3s\ ^3D_3$	8 440 926	8 441 758	8 437 725		
67	$2s2p^2(4P)3p\ ^3P_0$	8 442 813	8 443 776	8 438 241		
68	$2s2p^2(4P)3p\ ^5S_2$	8 443 646	8 445 037	8 440 027		
69	$2s2p^2(4P)3p\ ^3D_3$	8 462 365	8 463 510	8 456 502		
70	$2s2p^2(4P)3p\ ^3P_1$	8 467 690	8 468 680	8 462 413		
71	$2s2p^2(4P)3p\ ^3P_2$	8 470 871	8 471 913	8 466 158		
72	$2s2p^2(4P)3d\ ^5F_1$	8 480 620	8 481 735	8 471 575		
73	$2s2p^2(4P)3d\ ^5F_2$	8 488 782	8 489 971	8 479 898		8 486 331
74	$2s2p^2(2D)3s\ ^1D_2$	8 496 990	8 497 512	8 498 660		
75	$2s2p^2(4P)3d\ ^5F_3 : A$	8 506 111	8 507 343	8 497 311		8 511 385
76	$2s2p^2(4P)3d\ ^5F_4$	8 544 575	8 545 928	8 534 831		
77	$2s2p^2(2P)3s\ ^3P_1$	8 545 485	8 546 507	8 544 603		
78	$2s2p^2(2P)3s\ ^3P_0$	8 553 885	8 554 716	8 545 420		
79	$2s2p^2(4P)3d\ ^5D_0$	8 554 798	8 555 918	8 546 365		
80	$2s2p^2(4P)3d\ ^5D_1$	8 555 297	8 556 534	8 547 553		
81	$2s2p^2(4P)3d\ ^5D_2$	8 555 491	8 556 850	8 558 611		
82	$2s2p^2(4P)3d\ ^5F_3 : B$	8 561 662	8 562 969	8 552 789		8 564 535
83	$2s2p^2(4P)3d\ ^3P_2$	8 581 274	8 582 755	8 576 965		8 575 780

Table 6. *Cont.*

No.	Level	E_{RCI}	E_{RMBPT}	E_{FAC}	E_{NIST}	E_{CHI}
84	$2s2p^2(4P)3d\,^5F_5$	8 586 636	8 588 014	8 577 151		
85	$2s2p^2(4P)3d\,^5D_4$	8 597 735	8 599 106	8 588 388		
86	$2s2p^2(2D)3p\,^3F_2$	8 606 110	8 606 654	8 607 148		8 605 427
87	$2s2p^2(4P)3d\,^3F_2$	8 611 432	8 611 843	8 608 283		
88	$2s2p^2(4P)3d\,^5P_3$	8 619 312	8 620 847	8 611 148		
			⋮			
228	$2p^3(^2P)3d\,^3F_4$	9 735 480	9 736 111	9 746 771		
229	$2p^3(^2P)3d\,^3P_1$	9 740 645	9 742 719	9 754 847		
230	$2p^3(^2P)3d\,^3P_0$	9 748 184	9 749 774	9 758 859		
231	$2p^3(^2P)3d\,^3P_2 : B$	9 757 890	9 758 107	9 770 963		
232	$2p^3(^2P)3d\,^3D_1$	9 765 663	9 766 056	9 781 499		
233	$2p^3(^2P)3d\,^3D_3$	9 780 738	9 781 044	9 794 389		
234	$2p^3(^2P)3d\,^1F_3$	9 800 368	9 800 742	9 819 206		
235	$2p^3(^2P)3d\,^3D_2$	9 800 852	9 801 738	9 819 939		
236	$2p^3(^2P)3d\,^1P_1$	9 879 471	9 879 655	9 902 175		
237	$2s^22p4s\,^3P_0$	10 368 077		10 362 393		
238	$2s^22p4s\,^3P_1$	10 371 121		10 365 585	10 380 000	
239	$2s^22p4p\,^3D_1$	10 442 616		10 437 633		
240	$2s^22p4p\,^3P_1$	10 466 102		10 460 676		
241	$2s^22p4p\,^3D_2$	10 468 322		10 462 978		
242	$2s^22p4p\,^3P_0$	10 470 990		10 465 993		
243	$2s^22p4s\,^3P_2$	10 485 597		10 479 693		
244	$2s^22p4s\,^1P_1$	10 492 966		10 487 317		
245	$2s^22p4d\,^3F_2$	10 532 099		10 526 459		
246	$2s^22p4d\,^3P_2 : A$	10 548 542		10 542 323	(10 547 000)	10 547 249
247	$2s^22p4d\,^3F_3$	10 549 480		10 543 488	10 548 000	10 548 160
248	$2s^22p4d\,^3D_1$	10 554 447		10 548 345	10 553 000	10 553 955
249	$2s^22p4p\,^1P_1$	10 568 810		10 563 327		
250	$2s^22p4p\,^3D_3$	10 574 912		10 569 412	10 664 000	
251	$2s^22p4p\,^3P_2$	10 575 111		10 569 433		
252	$2s^22p4p\,^3S_1$	10 578 203		10 572 657		
253	$2s^22p4p\,^1D_2$	10 597 862				
254	$2s^22p4p\,^1S_0$	10 619 563				
255	$2s^22p4d\,^3F_4$	10 652 979				
256	$2s^22p4d\,^1D_2$	10 653 631			10 675 000	
257	$2s^22p4d\,^3D_3$	10 660 593				
258	$2s^22p4d\,^3P_2 : B$	10 666 807				
259	$2s^22p4d\,^3P_1$	10 666 946			10 688 000	
260	$2s^22p4d\,^3P_0$	10 667 948				
261	$2s^22p4d\,^1F_3$	10 683 984			10 681 000	
262	$2s^22p4d\,^1P_1$	10 687 400				

One should note the excellent agreement between the energies from RMCDHF/RCI and RMBPT, the mean difference being less than 0.013%. Although the energies are in very good agreement, there seems to be a small systematic shift for the higher states, as depicted in Figure 1. The reason for this shift is not known, and further research is needed to shed light on this. To further access the accuracy of the excitation energies, calculations for the N-, O-, F- and Ne-like sequences [49–52] were done using both the RMCDHF/RCI and RMBPT methods. Cross-validations show that the mean energy differences for N-like, O-like, F-like and Ne-like Fe are 0.023%, 0.011%, 0.01% and 0.029%, respectively. The energy differences increase for the ions closer to the neutral end, where the RMBPT method is less efficient in capturing correlation effects.

Obviously, calculations with high accuracy that also give the leading LSJ compositions are indispensable tools for analysing astrophysical observations.

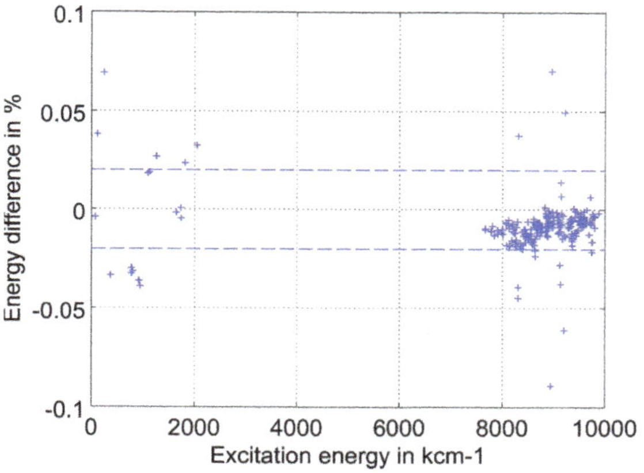

Figure 1. Difference between RMCDHF/RCI and MBPT excitation energies in percent for C-like Fe as a function of the excitation energy in kcm^{-1}. The dashed lines show the 0.02 % levels.

4.4. *Energies for Higher Lying States in the Mg-, Al- and Si-Like Sequences*

For larger atomic systems, one needs to think in terms of a core and a number of valence electrons. In many calculations, only valence-valence (VV) correlation is included. More accurate results are obtained when accounting for the interactions with the core through the inclusion of core-valence correlation (VV + CV). The final step is to include core-core correlation (VV + CV + CC). The situation has been analysed by Gustafsson et al. [12] for $3l3l'$, $3l4l'$, $3s5l$ states in Mg-like Fe where $1s^22s^22p^6$ is taken as the core. The results of the analysis can be inferred from Figure 2 that shows the difference between the computed excitation energies and the observed energies from the NIST database as a function of the excitation energies for the three computational models: VV, VV + CV and VV + CV + CC.

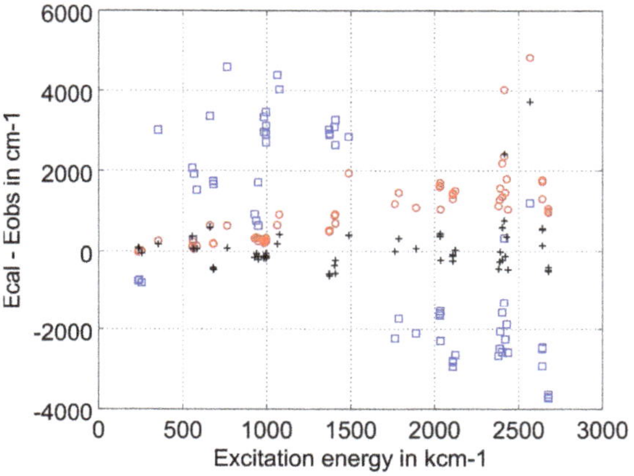

Figure 2. Difference between observed and RMCDHF/RCI excitation energies in cm^{-1} for Mg-like Fe [12] as a function of the excitation energy in kcm^{-1}. valence-valence (VV) (blue squares), VV + core-valence (CV) (red circles) and VV + CV + CC (black+).

From the figure, we see that the differences between the RMCDHF/RCI energies and observed energies are quite large, of the order of several thousand cm^{-1}, for the VV model. For many of the low lying states, calculated energies are too high, whereas for the more highly lying states, calculated energies are too low. Adding core-valence correlation (VV + CV) substantially improves the calculated energies. To explain the difference in behaviour, as shown in the figure, between the low lying states and the more high lying states when core-valence correlation is added, we note that core-valence correlation is a combination of core polarization, an electrostatic long range rearrangement and an electron-electron cusp correcting effect [7,8]. The cusp correcting effect lowers all energies with an amount that depends on the overlap of the valence electron charge distribution and the core. The charge distributions of the low lying states from the $3l3l'$ configurations are to a larger extent overlapping the core region compared to the charge distributions from the higher states of the $3l4l'$ and $3l5l'$ configurations, leading to a more pronounced energy lowering for the former states. The core polarization, in turn, lowers all energies except the $3s^2\ ^1S_0$ ground state for which the valence electron charge density is spherically symmetric. In total, these two effects explain the observed behaviour. Whereas the low lying states are now in very good agreement with observations, the high lying states are still a little high compared to observations. The effect of the core-core correlation (VV + CV + CC) is small for the low lying states, but brings down the more highly states that are now in perfect agreement with observations.

The increased accuracy comes with a price. For an orbital set $\{8s7p6d5f4g3h2i\}$, the valence-valence (VV) expansions sizes are less than 3000 CSFs for each parity. Including the core-valence correlation (VV + CV) increases the expansions sizes to around 650,000 CSFs for each parity. Finally, including also core-core (VV + CV + CC) make the expansion sizes grow to around 6,000,000 CSFs for each parity. For these large expansions, it becomes necessary to use a zero- and first-order partition of the CSFs and include part of the interactions perturbatively as described in Section 2.3.

Based on the valence-valence and core-valence model (VV + CV), RMCDHF/RCI and SD-MR calculations have been done for the Mg-, Al- and Si-like sequences [54–56]. The range of ions, the targeted configurations and the number of studied states for each sequence are summarized in Table 7. Calculations were done by parity, i.e., odd and even parity states were treated in separate sets of calculations. The targeted configurations define the MR, and the expansions were obtained by SD-MR substitutions to increasing active sets of orbitals with the restriction that only one substitution is allowed from the $2s^22p^6$ core. $1s^2$ is treated as an inactive core and is always closed.

Table 7. Sequence, ions and targeted configurations for the calculations. N is the number of studied states for each ion. In the table, $l = 0, \ldots, n-1$, $l' = 0, \ldots, n-1$.

Sequence	Ions	Configurations	N	Ref.
Mg-like	Ca-As, Kr	$3l3l'$, $3l4l'$, $3s5l$	146	[54]
Al-like	Ti-Kr, Xe, W	$3s^2\{3l;4l;5l\}$, $3p^2\{3d;4l\}$, $3s\{3p^2;3d^2\}$, $3s\{3p3d;3p4l;3p5s;3d4l'\}$, $3p3d^2$, $3p^3$, $3d^3$	360	[55]
Si-like	Ti-Ge, Sr, Zr, Mo	$3s^23p^2$, $3s3p^3$, $3s^23p3d$	27	[56]

To illustrate the accuracy of the RMCDHF/RCI calculation accounting for valence-valence and core-valence effects, we look at Si-like Fe [56]. In Table 8, the computed excitation RMCDHF/RCI energies are compared with observed energies from Del Zanna [57], as well as with energies by Vilkas and Ishikawa [58] using the MRMP method. The mean deviation is 0.076% for RMCDHF/RCI and only 0.034% for MRMP. The expansions for the even and odd states contained 1,500,000 and 4,500,000 CSFs, respectively.

The mean energy deviations for Mg-like, Al-like and Si-like iron from RMCDHF/RCI calculation accounting for valence-valence and core-valence effects are 0.051%, 0.039% and 0.076%, respectively. To improve the energies for the RMCDHF/RCI calculations, core-core correlation effects can be included as perturbative corrections, and work is in progress to develop tractable computational methods. For systems with five and more valence electrons, the expansions grow rapidly, and it may be necessary to start with valence-valence correlation and include core-valence effects as perturbative corrections.

Table 8. Comparison of calculated and observed excitation energies in cm^{-1}. E_{RCI} RMCDHF/RCI energies from [56], E_{MRMP} MRMP energies from [58] and E_{DZ} observed energies from [57]. Relative errors in % for the calculated energies are shown in parenthesis.

Fe XIII			
Level	E_{RCI}	E_{MRMP}	E_{DZ}
$3s^2\,3p^2\,{}^3P_0$	0	0	0
$3s^2\,3p^2\,{}^3P_1$	9281 (0.237)	9295 (0.086)	9303.1
$3s^2\,3p^2\,{}^3P_2$	18553 (0.048)	18576 (0.075)	18561.7
$3s^2\,3p^2\,{}^1D_2$	48236 (0.344)	47985 (1.077)	48069.7
$3s^2\,3p^2\,{}^1S_0$	91839 (0.357)	91508 (0.003)	91511.0
$3s\,3p^3\,{}^5S^o_2$	214152 (0.220)	214540 (0.039)	214624.0
$3s\,3p^3\,{}^3D^o_1$	287123 (0.028)	287199 (0.002)	287205.0
$3s\,3p^3\,{}^3D^o_2$	287270 (0.029)	287348 (0.002)	287356.0
$3s\,3p^3\,{}^3D^o_3$	290095 (0.029)	290179 (0.000)	290180.0
$3s\,3p^3\,{}^3P^o_0$	328974 (0.014)	328980 (0.016)	328927.0
$3s\,3p^3\,{}^3P^o_1$	329689 (0.015)	329702 (0.019)	329637.0
$3s\,3p^3\,{}^3P^o_2$	330323 (0.012)	330334 (0.015)	330282.0
$3s\,3p^3\,{}^1D^o_2$	362482 (0.020)	362416 (0.002)	362407.0
$3s\,3p^3\,{}^3S^o_1$	415577 (0.027)	415519 (0.013)	415462.0
$3s^2\,3p\,3d\,{}^3F^o_2$	430277 (0.035)	430129 (0.001)	430124.0
$3s^2\,3p\,3d\,{}^3F^o_3$	437064 (0.033)	436905 (0.003)	436919.0
$3s\,3p^3\,{}^1P^o_1$	438365 (0.063)	438005 (0.018)	438086.0
$3s^2\,3p\,3d\,{}^3F^o_4$	447134 (0.029)	446959 (0.009)	447001.0
$3s^2\,3p\,3d\,{}^3P^o_2$	486542 (0.037)	486403 (0.009)	486358.0
$3s^2\,3p\,3d\,{}^3P^o_1$	495102 (0.032)	495242 (0.060)	494942.0
$3s^2\,3p\,3d\,{}^1D^o_2$	499060 (0.038)	498925 (0.011)	498870.0
$3s^2\,3p\,3d\,{}^3P^o_0$	501676 (0.032)	501667 (0.030)	501514.0
$3s^2\,3p\,3d\,{}^3D^o_1$	506661 (0.030)	506681 (0.034)	506505.0
$3s^2\,3p\,3d\,{}^3D^o_3$	509303 (0.024)	509479 (0.059)	509176.0
$3s^2\,3p\,3d\,{}^3D^o_2$	509394 (0.028)	509441 (0.037)	509250.0
$3s^2\,3p\,3d\,{}^1F^o_3$	557432 (0.093)	557303 (0.070)	556911.0
$3s^2\,3p\,3d\,{}^1P^o_1$	571376 (0.110)	571187 (0.077)	570743.0

5. Transition Probabilities

Whereas there are enough observations to validate calculated excitation energies, the situation is very different for transition rates. For highly charged ions, there are few experimental methods available to determine transition rates. Lifetimes for long-lived states of the ground configuration or the lowest excited configurations have been determined in accurate storage-ring and trapping experiments (see for example, the review by Träbert [59]) and are used for benchmarking. Lifetimes for a large range of short-lived states have been determined using beam-foil spectroscopy [60]. However, even if these beam-foil data are very valuable, they are in general not accurate enough to discriminate between different computational approaches. In addition, lifetimes are dominated by the strong decay channels down to the lower configurations, and the lack of experimental transition rates, including weak transitions, between states of the excited configurations is of a major concern.

5.1. Internal Validation and Uncertainty Estimates

Due to the almost complete lack of experimental transition rates for highly charged ions, internal validation becomes important. For RMCDHF/RCI calculations, the convergence of the transition rates should be monitored as the active set is increased. Then, based on the same logic, the convergence of the transition rates should be monitored as the more involved correlation models are used, e.g., VV, VV + CV and VV + CV + CC. Considering the fact that there often are tens of thousands of transitions for extended spectrum calculations, this validation method is impractical, and only smaller numbers of selected transitions can be monitored. Another internal validation method is based on the accuracy of the transition energy and the agreement between the computed line strength S in the length and velocity gauge. Along these lines, Froese-Fischer [61] has suggested that the uncertainties $\delta A'$ of the calculated transition rates for LS allowed transitions can be estimated according to:

$$\delta A' = (\delta E + \delta S)A', \tag{9}$$

where A' is the energy-scaled transition rate computed from the observed transition energy (E_{obs}), $\delta E = |E_{calc} - E_{obs}|/E_{obs}$ is the relative error in the transition energy and $\delta S = |S_{len} - S_{vel}|/\max(S_{len}, S_{vel})$ is the relative discrepancy between the length and velocity forms of the line strengths. In cases where the transition energies are not known, the expression reduces to:

$$\delta A = (\delta S)A. \tag{10}$$

Based on a statistical analysis of large datasets of accurate E1 transition rates from many independent calculations, Ekman et al. [62] found that the estimated errors from Equation (10) are correlated with and very close to the presumed actual errors. A validation of the method extended to intercombination lines reveals a smaller correlation in the statistical analysis and suggests that the uncertainty estimate in this case should only be used if averaging over a larger sample. The analysis further confirms the well-known fact that the uncertainty is large for weaker transitions, the general explanation being cancellations between the contributions to the matrix elements from different pairs of CSFs [63] or cancellations in the integrands of the transition integrals.

5.2. Transition Rates for the B- to Si-Like Sequences

The RMCDHF/RCI and SD-MR method has been used to compute tens of thousands of E1, M1, E2, M2 transitions rates for the B- to Si-like sequences [28–32,43,45–52,54–56]. The E1 and E2 rates are internally validated by giving $\delta A/A$ along with A. The results for C-like Fe [29], shown in Table 9, illustrate the typical uncertainties. The table displays computed transition energies along with relative uncertainties obtained by comparing with observations from NIST. The uncertainties for the transition energies are all well below 1%, and many of them are around 0.1%, which is highly satisfactory. The transition rates in the length form are given together with the uncertainty estimate $\delta A/A$. The uncertainties for the transition rates are a few percent or less for the strong transitions, but go up to around 20% for some of the weak intercombination transitions. To further shed light on the situation, we compare the RMCDHF/RCI rates for Ne-like S [43] with rates from accurate MCHF-BP calculations [19] and with CI calculations using CIV3 [64] in Table 10. From the table, we see that there is in general a very good agreement between the rates from the different calculations. It is clear that the largest differences are for the weak transitions.

Table 9. Transition energies in cm^{-1} and E1 rates A in s^{-1} in the length gauge for Fe XXI from RMCDHF/RCI calculations [29]. Relative errors in % for the calculated transition energies and rates are shown in parenthesis. For the transition energies, the relative errors were obtained by comparison with observations from NIST. For the transition rates, the relative errors are estimated from Equation (10).

Upper	Lower	ΔE_{calc}	A
		Fe XXI	
$2s2p^3\ ^3D^o_1$	$2s^22p^2\ ^3P_0$	776750 (0.049)	1.156E+10 (0.00)
$2s2p^3\ ^3P^o_1$	$2s^22p^2\ ^3P_0$	925023 (0.120)	4.213E+09 (0.07)
$2s2p^3\ ^3S^o_1$	$2s^22p^2\ ^3P_0$	1096012 (0.089)	9.460E+09 (0.16)
$2s2p^3\ ^1P^o_1$	$2s^22p^2\ ^3P_0$	1261529 (0.205)	2.850E+07 (1.82)
$2s2p^3\ ^5S^o_2$	$2s^22p^2\ ^3P_1$	412701 (0.293)	3.597E+07 (8.72)

Table 9. *Cont.*

Upper	Lower	ΔE_{calc}	A
$2s2p^3\,{}^3D^o_1$	$2s^22p^2\,{}^3P_1$	702930 (0.049)	7.606E+08 (1.64)
$2s2p^3\,{}^3D^o_2$	$2s^22p^2\,{}^3P_1$	703550 (0.048)	9.240E+09 (0.66)
$2s2p^3\,{}^3P^o_0$	$2s^22p^2\,{}^3P_1$	842581 (0.122)	2.200E+10 (0.31)
$2s2p^3\,{}^3P^o_1$	$2s^22p^2\,{}^3P_1$	851203 (0.119)	1.602E+10 (0.12)
$2s2p^3\,{}^3P^o_2$	$2s^22p^2\,{}^3P_1$	868735 (0.120)	3.820E+08 (1.85)
$2s2p^3\,{}^3S^o_1$	$2s^22p^2\,{}^3P_1$	1022191 (0.133)	2.533E+10 (0.11)
$2s2p^3\,{}^1D^o_2$	$2s^22p^2\,{}^3P_1$	1053811 (0.089)	3.907E+08 (1.68)
$2s2p^3\,{}^1P^o_1$	$2s^22p^2\,{}^3P_1$	1187709 (0.205)	4.909E+09 (0.34)
$2s2p^3\,{}^5S^o_2$	$2s^22p^2\,{}^3P_2$	369157 (0.293)	3.272E+07 (11.61)
$2s2p^3\,{}^3D^o_1$	$2s^22p^2\,{}^3P_2$	659387 (0.050)	8.335E+07 (6.39)
$2s2p^3\,{}^3D^o_2$	$2s^22p^2\,{}^3P_2$	660006 (0.050)	4.535E+06 (22.51)
$2s2p^3\,{}^3D^o_3$	$2s^22p^2\,{}^3P_2$	686197 (0.050)	6.105E+09 (0.80)
$2s2p^3\,{}^3P^o_1$	$2s^22p^2\,{}^3P_2$	807659 (0.120)	2.681E+09 (1.30)
$2s2p^3\,{}^3P^o_2$	$2s^22p^2\,{}^3P_2$	825192 (0.121)	2.038E+10 (0.04)
$2s2p^3\,{}^3S^o_1$	$2s^22p^2\,{}^3P_2$	978647 (0.134)	6.104E+10 (0.24)
$2s2p^3\,{}^1D^o_2$	$2s^22p^2\,{}^3P_2$	1010267 (0.090)	8.020E+09 (0.39)
$2s2p^3\,{}^1P^o_1$	$2s^22p^2\,{}^3P_2$	1144165 (0.206)	2.491E+08 (0.16)
$2s2p^3\,{}^5S^o_2$	$2s^22p^2\,{}^1D_2$	241853 (0.714)	1.307E+06 (19.51)
$2s2p^3\,{}^3D^o_1$	$2s^22p^2\,{}^1D_2$	532083 (0.048)	1.808E+08 (6.63)
$2s2p^3\,{}^3D^o_2$	$2s^22p^2\,{}^1D_2$	532703 (0.048)	3.827E+07 (6.27)
$2s2p^3\,{}^3D^o_3$	$2s^22p^2\,{}^1D_2$	558894 (0.049)	9.767E+08 (3.61)
$2s2p^3\,{}^3P^o_1$	$2s^22p^2\,{}^1D_2$	680356 (0.051)	2.577E+08 (2.44)
$2s2p^3\,{}^3P^o_2$	$2s^22p^2\,{}^1D_2$	697888 (0.052)	1.435E+08 (4.80)
$2s2p^3\,{}^3S^o_1$	$2s^22p^2\,{}^1D_2$	851344 (0.084)	3.607E+08 (3.16)
$2s2p^3\,{}^1D^o_2$	$2s^22p^2\,{}^1D_2$	882963 (0.037)	4.485E+10 (0.37)
$2s2p^3\,{}^1P^o_1$	$2s^22p^2\,{}^1D_2$	1016862 (0.170)	6.583E+10 (0.15)
$2s2p^3\,{}^3D^o_1$	$2s^22p^2\,{}^1S_0$	404698 (0.165)	4.070E+07 (2.97)
$2s2p^3\,{}^3P^o_1$	$2s^22p^2\,{}^1S_0$	552971 (0.014)	1.492E+08 (7.10)
$2s2p^3\,{}^3S^o_1$	$2s^22p^2\,{}^1S_0$	723959 (0.008)	6.511E+08 (2.13)
$2s2p^3\,{}^1P^o_1$	$2s^22p^2\,{}^1S_0$	889477 (0.147)	1.727E+10 (0.52)
$2p^4\,{}^3P_2$	$2s2p^3\,{}^5S^o_2$	1159940 (0.180)	1.422E+09 (2.39)
$2p^4\,{}^3P_1$	$2s2p^3\,{}^5S^o_2$	1254085 (0.181)	2.381E+08 (3.65)
$2p^4\,{}^1D_2$	$2s2p^3\,{}^5S^o_2$	1331255 (0.201)	5.282E+07 (2.48)
$2p^4\,{}^3P_2$	$2s2p^3\,{}^3D^o_1$	869710 (0.102)	3.455E+09 (0.75)
$2p^4\,{}^3P_0$	$2s2p^3\,{}^3D^o_1$	959080 (0.137)	3.474E+10 (0.43)
$2p^4\,{}^3P_1$	$2s2p^3\,{}^3D^o_1$	963855 (0.102)	1.434E+10 (0.27)
$2p^4\,{}^1D_2$	$2s2p^3\,{}^3D^o_1$	1041025 (0.103)	4.033E+06 (20.67)
$2p^4\,{}^1S_0$	$2s2p^3\,{}^3D^o_1$	1271738 (0.137)	2.331E+08 (1.32)
$2p^4\,{}^3P_2$	$2s2p^3\,{}^3D^o_2$	869090 (0.103)	1.335E+10 (0.00)
$2p^4\,{}^3P_1$	$2s2p^3\,{}^3D^o_2$	963235 (0.104)	2.124E+10 (0.28)
$2p^4\,{}^1D_2$	$2s2p^3\,{}^3D^o_2$	1040406 (0.106)	7.878E+08 (0.20)
$2p^4\,{}^3P_2$	$2s2p^3\,{}^3D^o_3$	842899 (0.137)	2.822E+10 (0.53)
$2p^4\,{}^1D_2$	$2s2p^3\,{}^3D^o_3$	1014215 (0.260)	5.622E+09 (0.85)
$2p^4\,{}^3P_1$	$2s2p^3\,{}^3P^o_0$	824204 (0.038)	4.813E+09 (0.02)
$2p^4\,{}^3P_2$	$2s2p^3\,{}^3P^o_1$	721437 (0.039)	3.678E+09 (0.32)
$2p^4\,{}^3P_0$	$2s2p^3\,{}^3P^o_1$	810807 (0.040)	1.827E+10 (0.93)
$2p^4\,{}^3P_1$	$2s2p^3\,{}^3P^o_1$	815582 (0.043)	1.273E+08 (3.29)
$2p^4\,{}^1D_2$	$2s2p^3\,{}^3P^o_1$	892752 (0.085)	1.241E+09 (1.85)
$2p^4\,{}^1S_0$	$2s2p^3\,{}^3P^o_1$	1123465 (0.233)	4.222E+09 (1.30)
$2p^4\,{}^3P_2$	$2s2p^3\,{}^3P^o_2$	703905 (0.039)	3.629E+09 (0.88)
$2p^4\,{}^3P_1$	$2s2p^3\,{}^3P^o_2$	798050 (0.040)	1.598E+10 (0.75)
$2p^4\,{}^1D_2$	$2s2p^3\,{}^3P^o_2$	875220 (0.085)	2.512E+09 (1.79)
$2p^4\,{}^3P_2$	$2s2p^3\,{}^3S^o_1$	550450 (0.021)	6.005E+09 (0.76)
$2p^4\,{}^3P_0$	$2s2p^3\,{}^3S^o_1$	639819 (0.020)	1.733E+10 (0.28)
$2p^4\,{}^3P_1$	$2s2p^3\,{}^3S^o_1$	644594 (0.049)	1.290E+10 (0.00)
$2p^4\,{}^1D_2$	$2s2p^3\,{}^3S^o_1$	721764 (0.059)	7.002E+06 (0.68)
$2p^4\,{}^1S_0$	$2s2p^3\,{}^3S^o_1$	952477 (0.060)	5.111E+09 (0.25)
$2p^4\,{}^3P_2$	$2s2p^3\,{}^1D^o_2$	518830 (0.065)	7.538E+08 (1.85)
$2p^4\,{}^3P_1$	$2s2p^3\,{}^1D^o_2$	612975 (0.124)	3.767E+08 (3.31)
$2p^4\,{}^1D_2$	$2s2p^3\,{}^1D^o_2$	690145 (0.311)	3.182E+10 (0.28)
$2p^4\,{}^3P_2$	$2s2p^3\,{}^1P^o_1$	384932 (0.274)	8.409E+07 (3.44)
$2p^4\,{}^3P_0$	$2s2p^3\,{}^1P^o_1$	474302 (0.271)	9.885E+06 (20.99)
$2p^4\,{}^3P_1$	$2s2p^3\,{}^1P^o_1$	479076 (0.265)	6.387E+08 (1.58)
$2p^4\,{}^1D_2$	$2s2p^3\,{}^1P^o_1$	556247 (0.139)	4.499E+09 (1.15)
$2p^4\,{}^1S_0$	$2s2p^3\,{}^1P^o_1$	786959 (0.156)	7.549E+10 (0.09)

Table 10. Transition rates for Ne-like S. A_{RCI} transition rates from RMCDHF/RCI [43], A_{BP} transition rates from multiconfiguration Hartree-Fock-Breit-Pauli [19] and A_{CIV3} transition rates from CI calculations using CIV3 [64].

States		ΔE (cm^{-1})	Type	A_{RCI}	A_{BP}	A_{CIV3}
Upper	Lower					
		S VII				
$2p^53s\ ^3P^o_2$	$2p^6\ ^1S_0$	1371667	M2	7.638E+02	7.617E+02	
$2p^53s\ ^3P^o_1$ $0.81 + 0.18\ ^1P^o_1$	$2p^6\ ^1S_0$	1376084	E1	1.855E+10	1.816E+10	1.989E+10
$2p^53s\ ^1P^o_1$ $0.81 + 0.18\ ^3P^o_1$	$2p^6\ ^1S_0$	1388242	E1	8.421E+10	8.507E+10	8.777E+10
$2p^53p\ ^3D_2$ $0.79 + 0.16\ ^1D_2$	$2p^6\ ^1S_0$	1484530	E2	3.021E+06	2.964E+06	
$2p^53p\ ^3P_2$ $0.56 + 0.42\ ^1D_2$	$2p^6\ ^1S_0$	1492576	E2	8.028E+06	8.008E+06	
$2p^53d\ ^3P^o_1$	$2p^6\ ^1S_0$	1624769	E1	2.160E+09	2.182E+09	2.312E+09
$2p^53d\ ^3P^o_2$	$2p^6\ ^1S_0$	1627240	M2	1.710E+04	1.728E+04	
$2p^53d\ ^3D^o_1$	$2p^6\ ^1S_0$	1644545	E1	6.122E+10	6.206E+10	6.230E+10
$2p^53d\ ^1P^o_1$	$2p^6\ ^1S_0$	1662346	E1	9.452E+11	9.448E+11	9.087E+11
$2p^53s\ ^3P^o_1$ $0.81 + 0.18\ ^1P^o_1$	$2p^53s\ ^3P^o_2$	4417	M1	1.587E+00	1.601E+00	
$2p^53s\ ^1P^o_1$ $0.81 + 0.18\ ^3P^o_1$	$2p^53s\ ^3P^o_2$	16575	M1	1.849E+01	1.867E+01	
$2p^53p\ ^3S_1$	$2p^53s\ ^3P^o_2$	95373	E1	6.393E+08	6.504E+08	6.480E+08
$2p^53p\ ^3D_3$	$2p^53s\ ^3P^o_2$	111609	E1	1.566E+09	1.608E+09	1.594E+09
$2p^53p\ ^3D_2$ $0.79 + 0.16\ ^1D_2$	$2p^53s\ ^3P^o_2$	112863	E1	6.439E+08	6.627E+08	6.418E+08
$2p^53p\ ^3D_1$ $0.72 + 0.17\ ^1P_1$	$2p^53s\ ^3P^o_2$	116432	E1	1.994E+08	2.060E+08	2.023E+08
$2p^53p\ ^3P_2$ $0.56 + 0.42\ ^1D_2$	$2p^53s\ ^3P^o_2$	120909	E1	1.000E+09	1.031E+09	1.006E+09
$2p^53p\ ^1P_1$ $0.55 + 0.27\ ^3D_1$	$2p^53s\ ^3P^o_2$	124214	E1	7.124E+07	7.315E+07	6.807E+07
$2p^53p\ ^1D_2$ $0.42 + 0.38\ ^3P_2$	$2p^53s\ ^3P^o_2$	127448	E1	2.464E+08	2.501E+08	2.832E+08
$2p^53p\ ^3S_1$	$2p^53s\ ^1P^o_1$ $0.81 + 0.18\ ^3P^o_1$	78797	E1	1.188E+07	1.206E+07	1.253E+07
$2p^53p\ ^3D_2$ $0.79 + 0.16\ ^1D_2$	$2p^53s\ ^1P^o_1$ $0.81 + 0.18\ ^3P^o_1$	96287	E1	6.945E+06	7.197E+06	5.508E+06
$2p^53p\ ^3D_1$ $0.72 + 0.17\ ^1P_1$	$2p^53s\ ^1P^o_1$ $0.81 + 0.18\ ^3P^o_1$	99856	E1	3.825E+06	4.290E+06	4.536E+06
$2p^53p\ ^3P_2$ $0.56 + 0.42\ ^1D_2$	$2p^53s\ ^1P^o_1$ $0.81 + 0.18\ ^3P^o_1$	104333	E1	2.932E+08	2.977E+08	3.152E+08
$2p^53p\ ^1P_1$ $0.55 + 0.27\ ^3D_1$	$2p^53s\ ^1P^o_1$ $0.81 + 0.18\ ^3P^o_1$	107638	E1	7.415E+08	7.625E+08	7.557E+08

Table 10. *Cont.*

States		ΔE (cm^{-1})	Type	A_{RCI}	A_{BP}	A_{CIV3}
Upper	Lower					
$2p^53p\ ^3P_0$	$2p^53s\ ^1P^o_1\ 0.81 + 0.18\ ^3P^o_1$	110460	E1	1.908E+08	1.974E+08	2.006E+08
$2p^53p\ ^1D_2\ 0.42 + 0.38\ ^3P_2$	$2p^53s\ ^1P^o_1\ 0.81 + 0.18\ ^3P^o_1$	110872	E1	1.206E+09	1.243E+09	1.199E+09
$2p^53p\ ^3P_1\ 0.70 + 0.27\ ^1P_1$	$2p^53s\ ^1P^o_1\ 0.81 + 0.18\ ^3P^o_1$	112070	E1	7.082E+08	7.307E+08	7.000E+08
$2p^53p\ ^3P_1\ 0.70 + 0.27\ ^1P_1$	$2p^53s\ ^1P^o_1\ 0.81 + 0.18\ ^3P^o_1$	112070	M2	1.139E−01	1.191E−01	
$2p^53p\ ^1S_0$	$2p^53s\ ^1P^o_1\ 0.81 + 0.18\ ^3P^o_1$	165149	E1	5.082E+09	5.118E+09	5.073E+09
$2p^53d\ ^3P^o_1$	$2p^53p\ ^1D_2\ 0.42 + 0.38\ ^3P_2$	125654	E1	1.229E+08	1.235E+08	1.323E+08
$2p^53d\ ^3P^o_2$	$2p^53p\ ^1D_2\ 0.42 + 0.38\ ^3P_2$	128125	E1	3.010E+08	3.017E+08	3.358E+08
$2p^53d\ ^3P^o_2$	$2p^53p\ ^1D_2\ 0.42 + 0.38\ ^3P_2$	128125	M2	1.161E−01	1.175E−01	
$2p^53d\ ^3F^o_3\ 0.77 + 0.20\ ^1F^o_3$	$2p^53p\ ^1D_2\ 0.42 + 0.38\ ^3P_2$	132832	E1	1.471E+06	1.472E+06	1.078E+06
$2p^53d\ ^3F^o_2\ 0.71 + 0.17\ ^1D^o_2$	$2p^53p\ ^1D_2\ 0.42 + 0.38\ ^3P_2$	136157	E1	1.190E+06	1.370E+06	1.643E+06
$2p^53d\ ^1F^o_3\ 0.53 + 0.40\ ^3D^o_3$	$2p^53p\ ^1D_2\ 0.42 + 0.38\ ^3P_2$	138763	E1	1.511E+08	1.519E+08	1.131E+08
$2p^53d\ ^3D^o_1$	$2p^53p\ ^1D_2\ 0.42 + 0.38\ ^3P_2$	145430	E1	3.071E+06	3.159E+06	1.717E+06
$2p^53d\ ^1D^o_2\ 0.52 + 0.28\ ^3F^o_2$	$2p^53p\ ^1D_2\ 0.42 + 0.38\ ^3P_2$	145447	E1	6.172E+08	6.286E+08	6.276E+08
$2p^53d\ ^3D^o_3\ 0.57 + 0.27\ ^1F^o_3$	$2p^53p\ ^1D_2\ 0.42 + 0.38\ ^3P_2$	146716	E1	4.030E+09	4.093E+09	4.183E+09
$2p^53d\ ^3D^o_3\ 0.57 + 0.27\ ^1F^o_3$	$2p^53p\ ^1D_2\ 0.42 + 0.38\ ^3P_2$	146716	M2	1.029E+00	1.054E+00	
$2p^53d\ ^3D^o_2\ 0.65 + 0.27\ ^1D^o_2$	$2p^53p\ ^1D_2\ 0.42 + 0.38\ ^3P_2$	147358	E1	1.521E+08	1.531E+08	1.489E+08
$2p^53d\ ^1P^o_1$	$2p^53p\ ^1D_2\ 0.42 + 0.38\ ^3P_2$	163231	E1	4.854E+07	4.851E+07	5.550E+07

5.3. Systematic Comparisons between Methods

Wang and co-workers have systematically compared large sets of transition rates from accurate RMCDHF/RCI and RMBPT calculations [49–52]. These comparisons show that the rates from the two methods agree within a few percent for the strong transitions and that the agreement gets slightly worse for the weak intercombination and the two-electron, one photon transitions [1]. The comparisons also show that the differences between the methods are large for transitions for which there are large differences between the rates in the length and velocity form, thus confirming the usefulness of $\delta A/A$ as an uncertainty estimate. In Figure 3, we show the results of a comparison between methods for O-like Fe [31]. The figure clearly shows the consistency of the RMCDHF/RCI and RMBPT transitions rates, but also the comparatively large differences with rates from the CHIANTI database. These types of comparisons point to the fact that transition rates can be computed with high accuracy, but that much effort remains in order to make data practically available for astronomers and astrophysicist in updated databases.

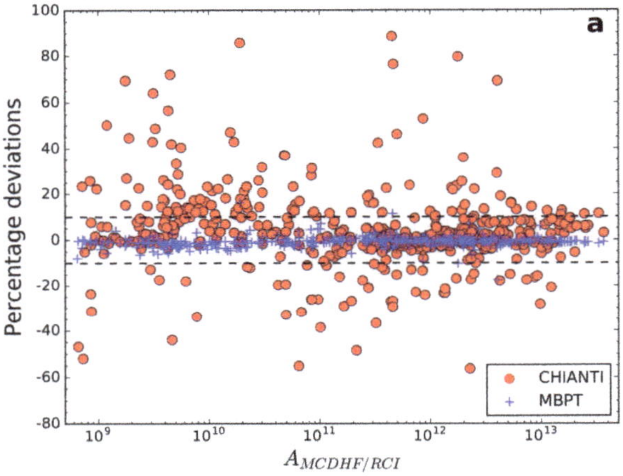

Figure 3. Results of a comparison between methods for O-like Fe [31]. Deviation in percent between RMCDHF/RCI and RMBPT transition rates as a function of the transition rate in s^{-1}. Deviations from the values of the CHIANTI database [3,4] are given in red. The dashed lines give the 10% levels.

6. Conclusions

Current computational methodologies make it possible to compute excitation and transition energies to almost spectroscopic accuracy for many ionized systems. In an astrophysical context, this means that calculated transition energies can be used to unambiguously identify new lines from spectra or correct old identifications. Transition data are lacking for many ions, and calculated values fill this gap. Whereas many of the calculations have been done for systems with relatively few electrons with a full RCI matrix, zero- and first-order methods, allowing for parts of the interactions to be treated perturbatively, have extended the range of applicability, and many calculations with high accuracy are in progress for isoelectronic sequences starting from the third and fourth row of the periodic table.

Accurate and consistent transition rates are essential for collisional and radiative plasma modelling and for diagnostic purposes. Very few experimental data are available for the rates, and thus, the bulk of the data must be computed. The lack of experimental data means that internal validation of

[1] Transitions between two states for which the configurations differ by more than one electron. These transitions are zero in the lowest approximation and are induced by CSFs that enter the calculation to correct for electron correlation effects.

computed data becomes important. For accurate calculations that predict the energy structure at the per millelevel, the differences between E1 rates in the length and velocity forms can be used to estimate the uncertainties. Internal validation based on convergence analysis and agreement between rates in length and velocity, as well as systematic comparisons of rates from RMCDHF/RCI and RMBPT calculations show that the uncertainties of the E1 rates are at the level of a few percent for the strong transitions. For the weakest transitions, the uncertainties are higher and come with a more irregular pattern.

7. Further Developments and Outlook

The time for angular integration is a limiting factor for very large RCI calculations. This time can be cut down by regrouping CSFs from SD-MR expansions in blocks that can be represented symbolically. For example, in non-relativistic notation, $1s^2nsmp\ ^3P$ and $1s^2npmd\ ^3P$ with $n = 2, \ldots, 12$ and $m = 2, \ldots, 12$ represent two blocks where the angular integration between CSFs in the blocks, as well as between CSFs in the different blocks is independent of the principal quantum numbers or can be reduced to only a few cases. For large n and m, the reduction in computing time is substantial. Discussed already decades ago [65], it seems essential that these ideas are now broadly implemented in the generally available computer codes.

With angular integration being a negligible part of the computation comes the possibility to extend the orbital set to higher n. Currently, the orbitals are variationally determined on a grid in RMCDHF calculations. The variational determination is computationally costly, and it would be valuable to augment the variationally-determined orbitals with analytical orbitals or orbitals determined in simplified and fast procedures. Work along these lines is in progress.

Among the targeted systems for improved computer codes are the α-elements, including Mg, Si, Ca and the iron group elements Sc, Ti, Cr, Mn, Fe at lower ionization states. These elements are of key importance for stellar and galactic evolution studies [66].

Acknowledgments: Per Jönsson, Henrik Hartman, Jörgen Ekman and Tomas Brage acknowledge support from the Swedish Research Council under Contract 2015-04842. Laima Radziute is thankful for the high performance computing resources provided by the Information Technology Open Access Center of Vilnius University. Michel R. Godefroid acknowledges support from the Belgian F.R.S.-FNRS Fonds de la Recherche Scientifique under CDR J.0047.16.

Author Contributions: All authors contributed equally to the work presented here.

Conflicts of Interest: The authors declare no conflict of interest.

References

1. Editorial. Nailing fingerprints in the stars. *Nature* **2013**, *437*, 503.
2. Pradhan, A.K.; Nahar, S.N. *Atomic Astrophysics and Spectroscopy*; Cambridge University Press: Cambridge, UK, 2011.
3. Dere, K.P.; Landi, E.; Mason, H.E.; Monsignori Fossi, B.C.; Young, P.R. CHIANTI—An atomic database for emission lines—Paper I: wavelengths greater than 50 Å (Version 1). *Astron. Astrophys. Suppl. Ser.* **1997**, *125*, 149–173.
4. Landi, E.; Young, P.R.; Dere, K.P.; Del Zanna, G.; Mason, H.E. CHIANTI—An atomic database for emission lines. XIII. Soft X-ray improvements and other changes: Version 7.1 of the database. *Astrophys. J.* **2013**, *763*, 86.
5. Del Zanna, G. Benchmarking atomic data for astrophysics: A first look at the soft X-ray lines. *Astron. Astrophys.* **2012**, *546*, A97.
6. Edlén, B. Atomic Spectra. *Encyclopedia of Physics*; Flügge, S., Ed.; Springer-Verlag: Berlin, Germany, 1964.
7. Froese Fischer, C.; Godefroid, M.; Brage, T.; Jönsson, P.; Gaigalas, G. Advanced multiconfiguration methods for complex atoms: Part I—Energies and wave functions. *J. Phys. B At. Mol. Opt. Phys.* **2016**, *49*, 182004.
8. Froese Fischer, C.; Brage, T.; Jönsson, P. *Computational Atomic Structure—An MCHF Approach*; CRC Press: Boca Raton, FL, USA, 1997.

9. Kozlov, M.G.; Porsev, S.G.; Safronova, M.S.; Tupitsyn, I.I. CI-MBPT: A package of programs for relativistic atomic calculations based on a method combining configuration interaction and many-body perturbation theory. *Comput. Phys. Commun.* **2015**, *195*, 199–213.
10. Verdebout, S.; Rynkun, P.; Jönsson, P.; Gaigalas, G.; Froese Fischer, C.; Godefroid, M. A partitioned correlation function interaction approach for describing electron correlation in atoms. *J. Phys. B At. Mol. Opt. Phys.* **2013**, *46*, 085003.
11. Dzuba, V.A.; Berengut, J.; Harabati, C.; Flambaum, V.V. Combining configuration interaction with perturbation theory for atoms with large number of valence electrons. *Phys. Rev. A* **2016**, *95*, 012503.
12. Gustafsson, S.; Jönsson, P.; Froese Fischer, C.; Grant, I.P. Combining Multiconfiguration and Perturbation Methods: Perturbative Estimates of Core–Core Electron Correlation Contributions to Excitation Energies in Mg-Like Iron. *Atoms* **2017**, *5*, 3.
13. Radžiūtė, L.; Gaigalas, G.; Kato, D; Jönsson, P.; Rynkun, P.; Kučas, S.; Jonauskas, V.; Matulianec, R. Energy level structure of Er^{3+}. *J. Quant. Spectrosc. Radiat. Transf.* **2015**, *152*, 94–106.
14. Jönsson, P.; Gaigalas, G.; Bieroń, J.; Froese Fischer, C.; Grant, I.P. New Version: grasp2K relativistic atomic structure package. *Comput. Phys. Commun.* **2013**, *184*, 2197–2203.
15. Grant, I.P. *Relativistic Quantum Theory of Atoms and Molecules*; Springer: New York, NY, USA; 2007.
16. Dyall, K.G.; Grant, I.P.; Johnson, C.T.; Parpia, F.A.; Plummer, E.P. GRASP: A general-purpose relativistic atomic structure program. *Comput. Phys. Commun.* **1989**, *55*, 425–456.
17. Grant, I.P.; Pyper, N.C. Breit-interaction in multi-configuration relativistic atomic calculations. *J. Phys. B* **1976**, *9*, 761–774.
18. Gaigalas, G.; Žalandauskas, T.; Rudzikas, Z. LS-jj transformation matrices for a shell of equivalent electrons. *At. Data Nucl. Data Tables* **2003**, *84*, 99–190.
19. Froese Fischer, C.; Tachiev, G. Breit-Pauli energy levels, lifetimes, and transition probabilities for the beryllium-like to neon-like sequences. *At. Data Nucl. Data Tables* **2004**, *87*, 1–184.
20. Froese Fischer, C.; Gaigalas, G. Multiconfiguration Dirac-Hartree-Fock energy levels and transition probabilities for W XXXVIII. *Phys. Rev. A* **2012**, *85*, 042501.
21. Gaigalas, G.; Froese Fischer, C.; Rynkun, P.; Jönsson, P. JJ2LSJ transformation and unique labeling for energy levels. *Atoms* **2017**, *5*, 6.
22. Gaigalas, G.; Rudzikas, Z.; Froese Fischer, C. An efficient approach for spin-angular integrations in atomic structure calculations. *J. Phys. B At. Mol. Opt. Phys.* **1997**, *30*, 3747–3771.
23. Gaigalas, G.; Fritzsche, S.; Grant, I.P. Program to calculate pure angular momentum coefficients in jj-coupling. *Comput. Phys. Commun.* **2001**, *139*, 263–278.
24. Olsen, J.; Godefroid, M.; Jönsson, P.; Malmqvist, P.Å.; Froese Fischer, C. Transition probability calculations for atoms using nonorthogonal orbitals. *Phys. Rev. E* **1995**, *52*, 4499–4508.
25. Jönsson, P.; Froese Fischer, C. Multiconfiguration Dirac-Fock calculations of the $2s^2\ ^1S_0 - 2s2p\ ^3P_1$ intercombination transition in C III. *Phys. Rev. A* **1998**, *57*, 4967–4970.
26. Grant, I.P. Gauge invariance and relativistic radiative transitions. *J. Phys. B At. Mol. Phys.* **1974**, *7*, 1458–1475.
27. Froese Fischer, C.; Gaigalas, G.; Jönsson, P. Core effects on transition energies for $3d^k$ configurations in Tungsten ions. *Atoms* **2017**, *5*, 7.
28. Rynkun, P.; Jönsson, P.; Gaigalas, G. Energies and E1, M1, E2, M2 transition rates for states of the $2s^22p$, $2s2p^2$, and $2p^3$ configurations in boron-like ions between N III and Zn XXVI. *At. Data Nucl. Data Tables* **2012**, *98*, 481–556.
29. Jönsson, P.; Rynkun, P.; Gaigalas, G. Energies, E1, M1, and E2 transition rates, hyperfine structures, and Landé g_J factors for the states of the $2s^22p^2$, $2s2p^3$, and $2p^4$ configurations in carbon-like ions between F IV and Ni XXIII. *At. Data Nucl. Data Tables* **2011**, *97*, 648–691.
30. Rynkun, P.; Jönsson, P.; Gaigalas, G.; Froese Fischer, C. Energies and E1, M1, E2, and M2 transition rates for states of the $2s^22p^3$, $2s2p^4$, and $2p^5$ configurations in nitrogen-like ions between F III and Kr XXX. *At. Data Nucl. Data Tables* **2014**, *100*, 315–402.
31. Rynkun, P.; Jönsson, P.; Gaigalas, G.; Froese Fischer, C. Energies and E1, M1, E2, and M2 transition rates for states of the $2s^22p^4$, $2s2p^5$, and $2p^6$ configurations in oxygen-like ions between F II and Kr XXIX. *Astron. Astrophys.* **2013**, *557*, A136.

32. Jönsson, P.; Alkauskas, A.; Gaigalas, G. Energies and E1, M1, E2 transition rates for states of the $2s^22p^5$ and $2s2p^6$ configurations in fluorine-like ions between Si VI and W LXVI. *At. Data Nucl. Data Tables* **2013**, *99*, 431–446.
33. Nazé, C.; Verdebout, S.; Rynkun, P.; Gaigalas, G.; Godefroid, M.; Jönsson, P. Isotope shifts in beryllium-, boron-, carbon-, and nitrogen-like ions from relativistic configuration interaction calculations. *At. Data Nucl. Data Tables* **2014**, *100*, 1197–1249.
34. Verdebout, S., Nazé, C.; Jönsson, P.; Rynkun, P.; Godefroid, M.; Gaigalas, G. Hyperfine structures and Landé g_J-factors for $n = 2$ states in beryllium-, boron-, carbon-, and nitrogen-like ions from relativistic configuration interaction calculations. *At. Data Nucl. Data Tables* **2014**, *100*, 1111–1155.
35. Kelly, R.L. Atomic and Ionic Spectrum Lines Below 2000 Angstroms: Hydrogen Through Krypton. *J. Phys. Chem. Ref. Data Suppl.* **1987**, *16*, 1–1698.
36. Edlén, B. Comparison of Theoretical and Experimental Level Values of the n = 2 Complex in Ions Isoelectronic with Li, Be, O and F. *Phys. Scr.* **1983**, *28*, 51–67.
37. Kramida, A.E.; Ralchenko, Y.; Reader, J.; NIST ASD Team (2015). NIST Atomic Spectra Database (ver. 5.3). National Institute of Standards and Technology: Gaithersburg, MD. Available online: http://physics.nist.gov/asd (accessed on 28 December 2016).
38. Vilkas, M.J.; Ishikawa, Y.; Koc, K. Relativistic multireference many-body perturbation theory for quasidegenerate systems: Energy levels of ions of the oxygen isoelectronic sequence. *Phys. Rev. A* **1999**, *60*, 2808–2821.
39. Gaigalas, G.; Kaniauskas, J.; Kisielius, R.; Merkelis, G.; Vilkas, M.J. Second-order MBPT results for the oxygen isoelectronic sequence. *Phys. Scr.* **1994**, *49*, 135–147.
40. Bhatia, A.K.; Thomas, R.J.; Landi, E. Atomic data and spectral line intensities for Ne III. *At. Data Nucl. Data Tables* **2003**, *83*, 113–152.
41. Jonauskas, V.; Keenan, F.P.; Foord, M.E.; Heeter, R.F.; Rose, S.J.; Ferland, G.J.; Kisielius, R.; van Hoof, P.A.M.; Norrington, P.H. Dirac-Fock energy levels and transition probabilities for oxygen-like Fe XIX. *Astron. Astrophys.* **2004**, *424*, 363–369.
42. Landi, E.; Gu, M.F. Atomic Data for High-Energy Configurations in Fe XVII-XXIII. *Astrophys. J.* **2006**, *640*, 1171–1179.
43. Jönsson, P.; Bengtsson, P.; Ekman, J.; Gustafsson, S.; Karlsson, L.B.; Gaigalas, G.; Froese Fischer, C.; Kato, D.; Murakami, I.; Sakaue, H.A.; et al. Relativistic CI calculations of spectroscopic data for the $2p^6$ and $2p^53l$ configurations in Ne-like ions between Mg III and Kr XXVII. *At. Data Nucl. Data Tables* **2014**, *100*, 1–154.
44. Ishikawa, Y.; Lopez Encarnacion, J.M.; Träbert, E. $N = 3 - 3$ transitions of Ne-like ions in the iron group, especially Ca $^{10+}$ and Ti $^{12+}$. *Phys. Scr.* **2009**, *79*, 025301.
45. Jönsson, P.; Ekman, J.; Gustafsson, S.; Hartman, H.; Karlsson, L.B.; du Rietz, R.; Gaigalas, G.; Godefroid, M.R.; Froese Fischer, C. Energy levels and transition rates for the boron isoelectronic sequence: Si X, Ti XVIII – Cu XXV. *Astron. Astrophys.* **2013**, *559*, A100.
46. Jönsson, P.; Ekman, J.; Träbert, E. MCDHF Calculations and Beam-Foil EUV Spectra of Boron-like Sodium Ions (Na VII). *Atoms* **2015**, *3*, 195–259.
47. Ekman, J.; Jönsson, P.; Gustafsson, S.; Hartman, H.; Gaigalas, G.; Godefroid, M.R.; Froese Fischer, C. Calculations with spectroscopic accuracy: Energies, Landé g_J-factors, and transition rates in the carbon isoelectronic sequence from Ar XIII to Zn XXV. *Astron. Astrophys.* **2014**, *564*, A24.
48. Radžiūtė, L.; Ekman, J.; Jönsson P.; Gaigalas, G. Extended calculations of level and transition properties in the nitrogen isoelectronic sequence: Cr XVIII, Fe XX, Ni XXII, and Zn XXIV. *Astron. Astrophys.* **2015**, *582*, A61.
49. Wang, K.; Guo, X.L.; Li, S.; Si, R.; Dang, W.; Chen, Z.B.; Jönsson, P.; Hutton, R.; Chen, C.Y.; Yan, J. Calculations with spectroscopic accuracy: Energies and transition rates in the nitrogen isoelectronic sequence from Ar XII to Zn XXIV. *Astrophys. J. Suppl.* **2016**, *223*, 3.
50. Wang, K.; Jönsson, P.; Ekman, J.; Gaigalas, G.; Godefroid, M.R.; So, R.; Chen, Z.B.; Li, S.; Chen, C.Y.; Yan, J. Extended Calculations of Spectroscopic Data Energy Levels, Lifetimes and Transition Rates for O-like Ions from Cr XVII to Zn XXIII. *Astrophys. J. Suppl.* **2017**, *229*, 37.
51. Si, R.; Li, S.; Guo, X.L.; Chen, Z.B.; Brage, T.; Jönsson, P.; Wang, K.; Yan, J.; Chen, C.Y.; Zou, Y.M. Extended calculations with spectroscopic accuracy: energy levels and transition properties for the fluorine isoelectronic sequence with $Z = 24 - 30$. *Astrophys. J. Suppl.* **2016**, *227*, 16.

52. Wang, K.; Chen, Z.B.; Si, R.; Jönsson, P.; Ekman, J.; Guo, X.L.; Li, S.; Long, F.Y.; Dang, W.; Zhao, X.H.; et al. Extended relativistic configuration interaction and many-body perturbation calculations of spectroscopic data for the $n \leq 6$ configurations in Ne-like ions between Cr XV and Kr XXVII. *Astrophys. J. Suppl.* **2016**, *226*, 14.
53. Gu, M.F. Wavelengths of $2l \rightarrow 3l'$ transitions in L-shell ions of iron and nickel: A combined configuration interaction and many-body perturbation approach. *Astrophys. J. Suppl.* **2005**, *156*, 105–110.
54. Gustafsson, S.; Jönsson, P.; Froese Fischer, C.; Grant, I.P. MCDHF and RCI calculations of energy levels, lifetimes and transition rates for $3l3l'$, $3l4l'$ and $3s5l$ states in Ca IX—As XXII and Kr XXV. *Astron. Astrophy* **2017**, *505*, A76.
55. Ekman, J.; Jönsson, P.; Radžiūtė, L.; Gaigalas, G.; Del Zanna, G.; Grant, I.P. Large-scale calculations of atomic level and transition properties in the aluminium isoelectronic sequence from Ti X through Kr XXIV, Xe XLII, and W LXII. *At. Data Nucl. Data Tables* **2017**, submitted.
56. Jönsson, P.; Radžiūtė, L.; Gaigalas, G.; Godefroid, M.R.; Marques, J.P.; Brage, T.; Froese Fischer, C.; Grant, I.P. Accurate multiconfiguration calculations of energy levels, lifetimes and transition rates for the silicon isoelectronic sequence: Ti IX—Ge XIX, Sr XXV, Zr XXVII, Mo XXIX. *Astron. Astrophys.* **2016**, *585*, A26.
57. Del Zanna, G. Benchmarking atomic data for astrophysics: Fe XIII EUV lines. *Astron. Astrophys.* **2011**, *533*, A12.
58. Vilkas, M.J.; Ishikawa, Y. High-accuracy calculations of term energies and lifetimes of silicon-like ions with nuclear charges $Z = 24 - 30$. *J. Phys. B At. Mol. Opt. Phys.* **2004**, *37*, 1803–1816.
59. Träbert, E. Critical Assessment of Theoretical Calculations of Atomic Structure and Transition Probabilities: An Experimenter's View. *Atoms* **2014**, *2*, 15–85.
60. Träbert, E.; Curtis, L.J. Isoelectronic trends of line strength data in the Li and Be isoelectronic sequences. *Phys. Scr.* **2006**, *74*, C46–C54.
61. Froese Fischer, C. Evaluating the accuracy of theoretical transition data. *Phys. Scr.* **2009**, *134*, 014019.
62. Ekman, J.; Godefroid, M.R.; Hartman, H. Validation and Implementation of Uncertainty Estimates of Calculated Transition Rates. *Atoms* **2014**, *2*, 215–224.
63. Ynnerman, A.; Froese Fischer, C. Multiconfigurational-Dirac-Fock calculation of the $2s\ ^1S_0 - 2s2p\ ^3P_1$ spin-forbidden transition for the Be-like isoelectronic sequence. *Phys. Rev. A* **1995**, *51*, 2020–2030.
64. Hibbert, A.; Ledourneuf, M.; Mohan, M. Energies, Oscillator Strengths, and Lifetimes for Neon-like Ions Up to Kr XXVII. *At. Data Nucl. Data Tables* **1993**, *53*, 23–112.
65. Froese Fischer, C.; Jönsson, P. MCHF calculations for atomic properties. *Comput. Phys. Commun.* **1994**, *84*, 37–58.
66. Pagel, B.E.J. *Nucleosynthesis and Chemical Evolution of Galaxies*; Cambridge University Press: Cambridge, UK, 2009.

atoms

MDPI

Article

Wavelengths of the Self-Photopumped Nickel-Like $4f\ ^1P_1 \rightarrow 4d\ ^1P_1$ X-ray Laser Transitions

Elena Ivanova

Institute of spectroscopy of RAS, Physicheskaya st., 5, Troitsk, 108840 Moscow, Russia; eivanova@isan.troitsk.ru

Academic Editor: Joseph Reader
Received: 31 January 2017; Accepted: 27 June 2017; Published: 13 July 2017

Abstract: The energies for the lower $3d_{3/2}4d_{3/2}$ $[J = 1]$ and upper $3d_{3/2}4f_{5/2}$ $[J = 1]$ working levels in the self-photopumped X-ray laser are analyzed along the Ni-like sequence. We have found some irregularities in these energy levels in the range Z = 42–49. The causes of the irregularities are studied. The list of elements that lase on the self-photopumped transition can be extended much further than originally known. We calculate the wavelengths of this transition in Ni-like sequence to Z = 79 using the relativistic perturbation theory with a zero approximation model potential. We estimate the wavelength accuracy for Z > 50 as $\Delta\lambda/\lambda \leq 0.005$.

Keywords: X-ray lasers; spectroscopy of multicharged ions; self-photo pumped lasers

1. Introduction

Self-photo pumped (SPP) X-ray lasers (XRL) in Ni-like ions were presented in 1996 [1] as an alternative approach to the standard radiative collisional scheme for inversion creation. We use the term SPP following the name given in literature. This is really a collisionally pumped laser assisted by radiation trapping. Both schemes for Ni-like ions are shown in Figure 1. This new class of SPP in Ni-like XRL was first investigated theoretically in [2], where high gain was predicted for the $4f\ ^1P_1$–$4d\ ^1P_1$ transition in Mo^{14+} at 22.0 nm. It was supposed that preplasma was created by a nanosecond pulse followed by a picosecond pulse to control the temperature and density in plasma, and to achieve high gain. This wavelength was calculated using the multiconfiguration Dirac-Fock atomic physics code by Grant and co-workers in the extended average level mode [3]. In the experiment [4], the Ni-like SPP XRL on the $4f\ ^1P_1$–$4d\ ^1P_1$ transition was demonstrated in Zr, Nb, and Mo, and the measured wavelengths for these ions were presented. For Mo^{14+} a gain of 13 cm^{-1} was measured at 22.6 nm for a target up to 1 cm long [4]. The wavelengths of this transition for ions from Z = 36 to 54 were predicted in [4] using the experimental data of this work to provide small corrections to their calculations. In the experiment [5], the progress in the optimization and understanding of the collisional pumping of X-ray lasers using an ultrashort subpicosecond heating pulse was reported. Time-integrated and time-resolved lasing signals at the standard $4d\ ^1S_0$–$4p\ ^1P_1$ XRL line in Ni-like Ag were studied in detail. Under specific irradiation conditions, strong lasing was obtained on the SPP $4f\ ^1P_1$–$4d\ ^1P_1$ transition at 16.1 nm. The strong lasing on the SPP transition in Mo^{14+} was also observed with very modest (less than 1 J) pump energy at a high repetition rate [6]. Recently, lasing on the SPP $3d\ ^1P_1$–$3p\ ^1P_1$ laser line has been observed for Ne-like V, Cr, Fe, and Co, as well as for Ni-like Ru, Pd, and Ag [7]. A strong dependence on the delay between the main and second prepulse was found: the optimum delay shifts towards smaller delays with increasing atomic number Z. Accurate wavelength measurements and calculations were shown to be in excellent agreement. The experiment [7] demonstrated that the list of elements that lase on the SPP transitions can be extended much further than originally known.

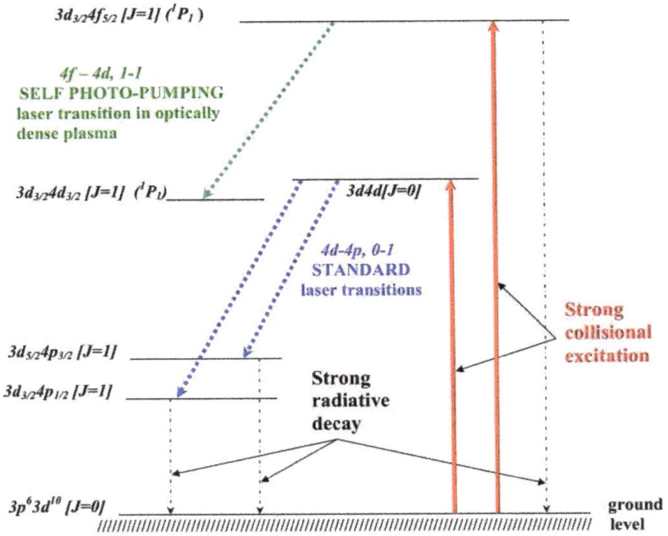

Figure 1. Schematic diagram of three low XRL transitions in Ni-like ions.

Many authors have investigated the spectra of Ni-like ions using vacuum sparks, laser produced plasma and electron beam ion traps as light sources [8–14]. The $3d^94d$ and $3d^94f$ configurations have been analyzed in the Rb X–Mo XV sequence [10,11]. In [10,11], these configurations were investigated using parameter extrapolations within the Generalized-Least-Squares (GLS) method. This method was used in [12,13] to predict for $3d^94d$, $3d^94f$ configuration energy levels in Cd XXI and Ag XX. GLS predictions of $3d^94d$, $3d^94f$ energy levels in the Zr XIII–Pd XIX sequence are tabulated in [14].

Note that lasing wavelength (λ_{las}) in Mo^{14+} was determined theoretically [2] and in the experiment [4] using one and the same atomic physics code [3], but results for λ_{las} were somewhat different (by 4 Å). The $3d_{3/2}4f_{5/2}$ $[J = 1]$ upper working level has the largest oscillator strength and radiative transition probability to the $3d^{10}$ ground level. This fact allows it to achieve high precision in this level energy measurement along the Ni-like sequence up to high Z ~84; in some ions, the energy of the transition to the ground state was accurate up to the fourth significant digit. The wavelengths of resonant radiative transitions in heavy Ni-like ions were calculated by us to Z = 83 in [15]. Moreover, in [15], the wavelengths (for Z within 79–82) were predicted with the same accuracy, although they have not yet been measured experimentally.

In the present paper, we analyze the smoothness of the working energy levels of SPP XRL along the Ni-like sequence. We found some irregularities in Ni-like sequence energies in the region Z = 42 (Mo^{14+}) and in the region Z = 49 (In^{21+}) for the upper $3d_{3/2}4f_{5/2}$ $[J = 1]$ working level. The causes of the irregularities are studied.

The principle purpose of this paper is to predict the wavelengths of SPP XRL lines in Ni-like ions with $Z \leq 79$. The calculations are performed by the Relativistic Perturbation Theory with Model Zero Approximation, (RPTMP). The fundamental principles of the RPTMP approach are given in [16]. Energy levels of the $3p^63d^94l$, $3p^53d^{10}4l$, $(l = 0, 1)$ configurations and radiative transition rates to the $3p^63d^{10}$ ground state in the *Kr IX* ion are calculated by this method in [16]. The stability of calculations on the approximation used is shown in [16].

2. Features of Lower and Upper Working Levels of SPP XRL along the Ni-Like Sequence

The schematic diagram of three strong XRL transitions is shown in Figure 1: two of them are standard $3d4d$ [J = 0]–$3d_{5/2}4p_{3/2}$ [J = 1] and $3d4d$ [J = 0]–$3d_{3/2}4p_{1/2}$ [J = 1] transitions. The classifications of lower working levels in Figure 1 are valid for Z > 42. The $3d_{5/2}4p_{3/2}$ [J = 1] level is the lower working level of an XRL for the entire nickel isoelectronic sequence, the $3d_{3/2}4p_{1/2}$ [J = 1] level is the lower working level for heavy ions starting with Z = 62. The third $3d_{3/2}4p_{3/2}$ [J = 1] level decays to a ground state significantly weaker than the two mentioned above, and does not provide a significant gain. In our recent work [17], the energies of standard XRL transitions in ions of the Ni-like sequence with Z $\leq$ 79 are refined by RPTMP calculations. The calculated energies of the two standard $4d$–$4p$, J = 0–1 XRL transitions are corrected by extrapolation of the experimental differentials of XRL transition energies $dE_Z^{las} = E_Z^{las} - E_{Z-1}^{las}$, i.e., the differences between transition energies of neighboring ions, which weakly depend on Z (especially in the region Z $\leq$ 50). It is proven that the accuracy for the final results for large Z is within the experimental error.

The $3d_{3/2}4f_{5/2}$ [J = 1]–$3d_{3/2}4d_{3/2}$ [J = 1] transition is optically self-photopumped XRL in all Ni-like ions, the positions of working levels vary with respect to other levels along the sequence. Based on our previous studies of XRL [18–20], it can be argued that there are at least four principal differences between standard and self photo-pumped mechanisms:

(1) In the standard scheme, the upper working level is populated by strong monopole electron collisions: in the SPP scheme it is populated by strong dipole electron collisions, which means high oscillator strength and effective photoabsorption.
(2) Effective SPP XRL is possible only in optically thick plasma (large electron density n_e and diameter d), while the standard XRL is possible both in optically thick and in optically thin plasma over a wide range of n_e and d.
(3) In the SPP, the upper working level is quickly emptied due to the large radiative decay rate. Therefore, in this scheme, a laser effect is short-lived; maximum XRL duration may be a few tens of picoseconds. A standard XRL can operate in quasi-continuous mode (under certain conditions).
(4) In the SPP, the lower and upper working levels do not change their classification along the Ni-like sequence; in the standard scheme the upper working level changes its classification: the $3d_{5/2}4d_{5/2}$ [J = 0] state is dominant in the classification of the upper working level at Z $\leq$ 51, and the $3d_{3/2}4d_{3/2}$ [J = 0] state is dominant for Z > 51 [17].

Below, we demonstrate the irregularities in the sequence of both the lower and the upper working levels of SPP XRL. Crossing of each working level with another level causes these irregularities. Level crossing is accompanied by a strong interaction at certain Z points. Figure 2a shows the scaled energies along Z of the $3d_{3/2}4d_{3/2}$ [J = 1] lower working level and the $3d_{3/2}4d_{5/2}$ [J = 1] level close to it. In addition to the energy levels calculated here, Figure 2a also shows the corresponding experimental values [14]. Reference [14] does not indicate classification of $3d4d$ [J = 1] levels, their classification was made earlier in [11]. Note that theoretical and experimental classifications are identical. There are some differences between theoretical and experimental energies, typically a few units in the 4th–5th digits. These differences are conditioned by the shift of the theoretical list of energy levels as a whole, but this shift does not affect the accuracy of λ_{las}. The energy levels in Figure 2a are scaled by dividing by $(Z-23)^2$, so that the behavior of the third and fourth significant digits can be observed. At the beginning of the sequence, the $3d_{3/2}4d_{3/2}$ [J = 1] level is above the $3d_{3/2}4d_{5/2}$ [J = 1] level. The crossing of these levels is in the range 41 < Z < 42 (shown by arrows). The crossing of the corresponding experimental energy levels occurs at exactly the same Z values. At Z = 42, one can observe the "repulsion" of levels caused by their interaction; the "repulsion" is a feature of theoretical and experimental data. Note, that repulsion can be seen due to energy scaling; in fact, the repulsion value is approximately a few thousand cm^{-1}, i.e., a few units in the fourth digit for the $3d_{3/2}4d_{5/2}$ [J = 1] level.

In Figure 2b, we can see hard-to-explain behavior of the $3d_{3/2}4f_{5/2}$ [J = 1] upper working level in the region of Z = 42. The features of this level will be considered below in more detail; however, it is important to note, here, that the energy structure of odd states in the range Z = 40–49 exhibits extremely high instability caused by the interaction of levels with each other, which rapidly changes with Z. In the case at hand, we understand the instability as the ambiguity of the calculation of eigenvectors and eigenenergies. As a result, the calculation in the same approximation leads to different energies at a certain level. The deviation from the smooth curve in Figure 2a is ~10,000 cm^{-1}; however, such a value leads to a sufficiently large deviation from the corresponding experimental values of λ_{las} shown in Figure 3.

At the point Z = 42, λ_{las} calculated here is ~222 Å, which is smaller than the experimental and theoretical values of [4] by 4 Å. In a recent experiment [7], the delay time between preliminary and main pump pulses was optimized to achieve the maximum yield of the X-ray laser. In fact, the electron density was optimized in [7]. X-ray lasing occurs in the Ni-like ion ionization mode, so that the lasing times on both transitions were restricted to the ionization time of Ni-like ions to the Co-like state. Time-resolved measurements in [7] allowed high-accuracy wavelength measurements of the SPP and standard X-ray laser lines. Thus, the calculations of the previous work [4] were confirmed: $\lambda_{las} \approx 22.61$ nm in Ni-like molybdenum (Mo^{14+}, Z = 42). Our calculations are performed for an isolated atom. Based on the studies performed, it can be argued that the interaction of levels at the point Z = 42 is so strong that the energy levels *3d*$_{3/2}$*4d*$_{3/2}$ *[J = 1]*, *3d*$_{3/2}$*4d*$_{5/2}$ *[J = 1]* in dense hot plasma can differ significantly from the corresponding energy levels in an isolated atom.

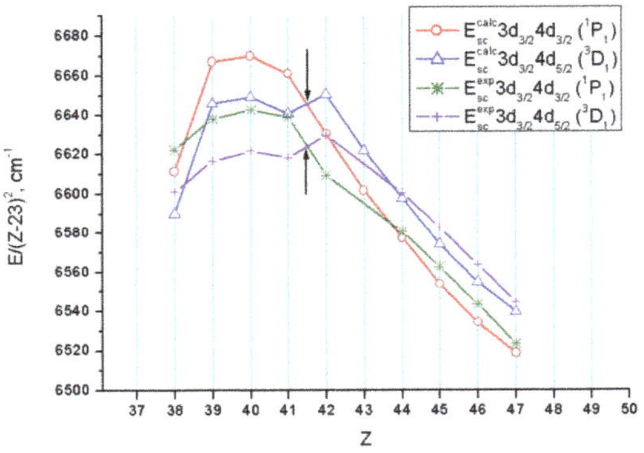

(a)

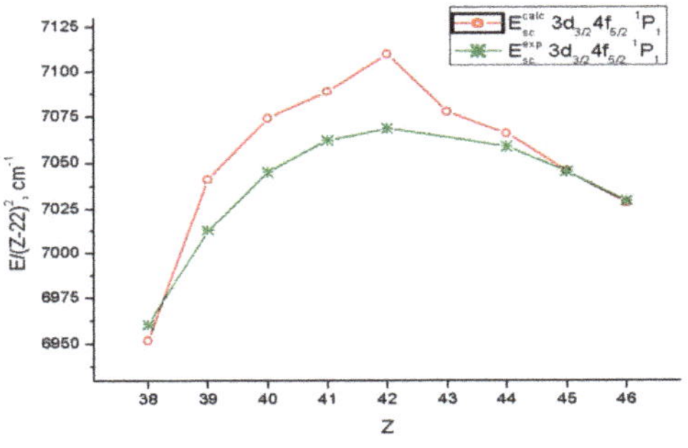

(b)

Figure 2. (**a**) Theoretical and experimental crossing of low working *3d*$_{3/2}$*4d*$_{3/2}$ *[J = 1]* energy level with *3d*$_{3/2}$*4d*$_{5/2}$ *[J = 1]* energy level in Ni-like sequence, shown by scaled energy values along Z.; (**b**) Features of theoretical upper working level *3d*$_{3/2}$*4f*$_{5/2}$ *[J = 1]*, shown by scaled energy along Z in comparison with correspondent experimental data.

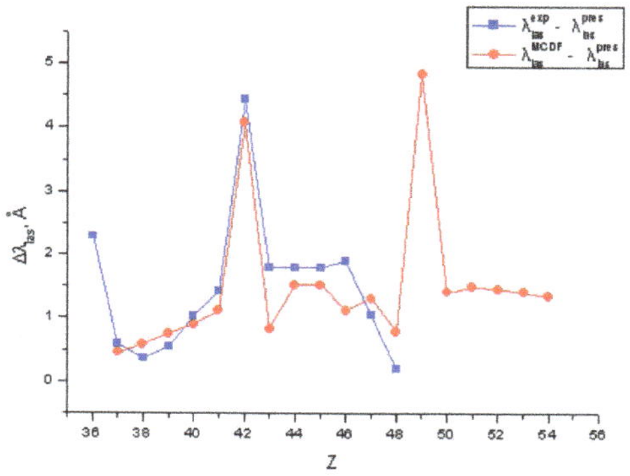

Figure 3. Difference between experimental, predicted from [4] and calculated here, λ_{las} of SPP XRL transitions in Ni-like ions.

The problem is the composition of the $3d_{3/2}4d_{3/2}$ *[J = 1]* working level, which indicates the strength of level interaction. It is shown in Figure 4 for all *3d4d [J = 1]* levels in Ni-like ions with Z = 36–79. Figure 4 shows that contributions of the $3d_{3/2}4d_{3/2}$ *[J = 1]* and $3d_{3/2}4d_{5/2}$ *[J = 1]* levels are almost equal at Z = 42, which could lead to levels' misidentification. Theoretical energies of these levels at Z = 42 are 2,393,554 cm^{-1} and 2,400,846 cm^{-1} (51% and 41%, respectively, are the contributions to the $3d_{3/2}4d_{3/2}$ *[J = 1]* low working level). The contributions of these levels in [11] are 45% and 34%, and the energies are 2,385,902 cm^{-1} and 2,393,229 cm^{-1} respectively. (We note that the theoretical list of energies of Ni- like ions in the range of small Z is shifted as a whole by 5000–8000 cm^{-1}). Figure 4 demonstrates the rapid restructuring of lower working level compositions: so that the $3d_{5/2}4d_{3/2}$ *[J = 1]* level contribution increases by five orders of magnitude in the range Z = 40–42.

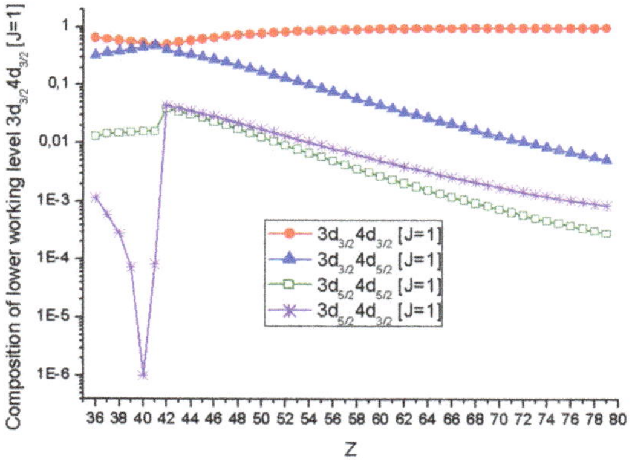

Figure 4. Composition of lower working level $3d_{3/2}4d_{3/2}$ *[J = 1]* along Ni-like sequence on a logarithmic scale.

Figure 5 shows the scaled energies along Z of the $3d_{3/2}4f_{5/2}$ *[J = 1]* upper working level and the close $3p_{3/2}4s_{1/2}$ *[J = 1]* level. Crossing of these levels occurs in the range 48 < Z < 49. At Z = 49 one can see the "repulsion" of levels caused by their interaction; the "repulsion" is a feature of theoretical data. In Figure 5, the corresponding experimental energies for the $3d_{3/2}4f_{5/2}$ *[J = 1]* level are shown [14]. Unfortunately, we have no available data on the experimental $3p_{3/2}4s_{1/2}$ *[J = 1]* levels in the Z region under consideration. The value Z = 49 is the point of an abrupt jump (irregularity) in spectroscopic constants of the $3d_{3/2}4f_{5/2}$ $[J = 1]$ upper working level and the $3p_{3/2}4s_{1/2}$ $[J = 1]$ level crossing it, caused by the strong interaction of these levels at this value of Z. This interaction is shown in Figure 6, where we can see the $3d_{3/2}4f_{5/2}$ $[J = 1]$ level composition. The interaction of levels at the point Z = 49 leads to the so-called effect of oscillator strength transfer we considered in [21] for the Ne-like sequence. At this point, the rate of radiative processes abruptly changes: the probabilities of the transition from the $3d_{3/2}4f_{5/2}$ $[J = 1]$ level to the ground state and to the state of the lower working level slightly decrease. At the same time, these probabilities for the $3p_{3/2}4s_{1/2}$ $[J = 1]$ level increase by an order of magnitude and become almost equal in magnitude to the corresponding values of the $3d_{3/2}4f_{5/2}$ $[J = 1]$ level. It can be assumed that there was an incorrect identification at the point Z = 49 when extrapolating the upper working level in [4], and the $3p_{3/2}4s_{1/2}$ $[J = 1]$ level that is close to the $3d_{3/2}4f_{5/2}$ $[J = 1]$ level in energy was used as the upper working level (see Figure 5). If this assumption is correct, λ_{las} ~144.7 Å for Z = 49, which is identical to [4]. When using our value for $3d_{3/2}4f_{5/2}$ $[J = 1]$, λ_{las} ~140.0 Å (here the energy jump shown in Figure 5 is taken into account). Another argument in favor of the incorrect identification in [4], are large jumps of the differential $d\lambda_{las}(Z) = \lambda_{las}(Z) - \lambda_{las}(Z-1)$ in the range Z = 47–50.

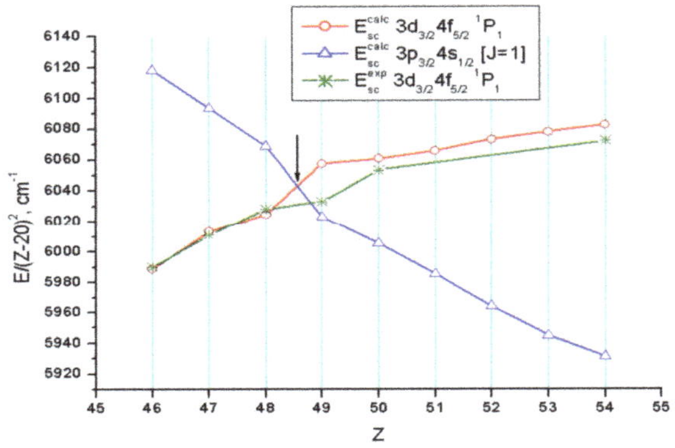

Figure 5. Crossing of upper working $3d_{3/2}4f_{5/2}$ $[J = 1]$ energy level with $3p_{3/2}4s_{1/2}$ $[J = 1]$ energy level in Ni-like sequence, shown by scaled energy values along Z. The corresponding experimental values for $3d_{3/2}4f_{5/2}$ $[J = 1]$ energies are also shown.

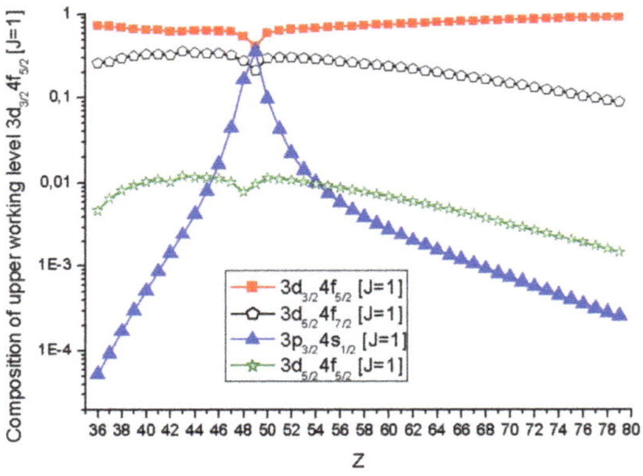

Figure 6. Composition of upper working level $3d_{3/2}4f_{5/2}$ $[J = 1]$ along Ni-like sequence on a logarithmic scale.

3. Wavelengths of the Self-Photopumped Nickel-Like $4f\,^1P_1 \rightarrow 4d\,^1P_1$ X-ray Laser Transitions

A comparison of the wavelengths of the self-photopumped nickel-like $4f\,^1P_1 \rightarrow 4d\,^1P_1$ X-ray laser transitions, calculated using the RPTMP method with corresponding experimental values and shown in Figure 3, exhibits a deviation of $\leq 1\%$ in the range Z = 37–46. For Z $\geq$ 48 Å, our results are identical to experimental data, with an accuracy of several units in the fourth significant digit. Two values of Z are exceptions: (i) the calculation instability point at Z = 42; and (ii) the point Z = 49, where the $3d_{3/2}4f_{5/2}$ $[J = 1]$ and $3p_{3/2}4s_{1/2}$ $[J = 1]$ states are probably incorrectly identified in the calculation by the MCDF method in [4]. We estimated the accuracy of the calculation of the energies of the upper and lower working states for high Z using experimental measurements of various studies. As an example, we compared the experimental energies for Z = 74 (W^{46+}), obtained using the Super EBIT (electron beam ion trap) [22,23], presented in Table 1. There are also listed the theoretical results calculated using the MCDF method called Grasp92 [24]. Here, we do not present earlier calculations of other authors. We also note the impossible comparison to the other calculations [25] in view of the level identification entanglement in this paper.

Table 1. Energy levels (10^3 cm^{-1}) of W XLVII. Comparison of present calculations with experimental data [22,23] and with calculations by GRASP92 [24].

Configuration	Term	J	Experiment	Present Work	GRASP92
$3p^63d^{10}$	$^1S_0$	0	0.0	0.0	0.0
$3p^63d^94s$	(5/2,1/2)	3	12,601.5	12,600.1	
		2	12,616.44	12,615.2	12,591.1
$3p^63d^94s$	(3/2,1/2)	1	13,138.66	13,137.8	13,110.8
		2	13,148.2	13,147.4	13,120.7
$3p^63d^94p$	(5/2,1/2)	2	13,379.05	13,357.5	
		3	13,388.20	13,366.3	
$3p^63d^94p$	(3/2,1/2)	2	13,916.27	13,894.8	
		1	13,940.6	13,922.4	13,930.6
$3p^63d^94p$	(5/2,3/2)	1	14,229.0	14,234.9	14,221.0
$3p^63d^94p$	(3/2,3/2)	1	14,751.0	14,756.2	14,741.1
$3p^63d^94d$	(3/2,3/2)	1		15,935.9	15,924.2
$3p^63d^94d$	(5/2,5/2)	1	15,556.1	15,561.3	15,550.2
		2	15,610.2	15,614.9	15,605.0
$3p^53d^{10}4s$	(3/2,1/2)	1	16,247.0	16,258.9	
$3p^63d^94d$	(3/2,3/2)	0	16,256.2	16,284.7	16,282.9
$3p^63d^94f$	(5/2,7/2)	1	17,045.9	17,042.2	17,030.6
$3p^63d^94f$	(3/2,5/2)	1	17,574.7 17,580.3 *	17,586.5	17,585.6
$3p^53d^{10}4s$	(1/2,1/2)	1	18,727	18,726.4	18,724.4
$3p^53d^{10}4d$	(3/2,3/2)	1	19,044.4	19,041.8	19,057.5
$3p^53d^{10}4d$	(3/2,5/2)	1	19,244.5	19,234.8	19,244.1
$3p^53d^{10}4f$	(3/2,7/2)	2	20,589.0	20,600.1	20,613.8
$3p^53d^{10}4d$	(1/2,3/2)	1	21,561.0	21,547.0	21,614.6

* Data from [23].

Good agreement between experimental and theoretical results for the energy levels in Table 1 may be noted: the maximum deviation is two units in the fourth significant digit. For the problem under study, it is important to ascertain the high accuracy of the calculation of the upper and lower working levels. For the experimental energy of the $3d_{3/2}4f_{5/2}$ $[J = 1]$ level, Table 1 gives two values: one obtained in the experiments [22], and the other later [23]. The difference with our calculation is 6 units in the fifth significant digit. We did not find the experimental energy of the $3d_{3/2}4d_{3/2}$ $[J = 1]$ lower working level for high Z in the literature. The energies of two other states of the $3d4d$ configuration with $J = 1, 2$, given in Table 1, also agree with high accuracy, which indirectly confirms the calculation reliability. Wavelengths of the $3d_{3/2}4f_{5/2}$ $(^1P_1)$–$3d_{3/2}4d_{3/2}$ $(^1P_1)$ SPP laser transitions in Ni-like sequence calculated by RPTMP are listed in Table 2.

Table 2. Wavelengths (λ_{las}, Å) of the $3d_{3/2}4f_{5/2}$ $(^1P_1)$–$3d_{3/2}4d_{3/2}$ $(^1P_1)$ SPP laser transitions in Ni-like sequence calculated by RPTMP.

Z	λ_{las}
50	134.08
51	128.12
52	122.54
53	117.39
54	112.66
55	108.36
56	104.295
57	100.51
58	96.98
59	93.68
60	90.57
61	87.65
62	84.89
63	82.28
64	79.81
65	77.47
66	75.23
67	73.08
68	71.06
69	69.11
70	67.25
71	65.47
72	63.75
73	62.10
74	60.51
75	58.97
76	57.48
77	56.04
78	54.64
79	53.23

4. Conclusions

The data on λ_{las} (see Table 2) were obtained a priori, no fittings were used. The error could be several units in the fourth significant digit. The precision wavelengths of laser transitions are necessary, in particular, to determine ions in which intense laser emission is possible at wavelengths for which multilayer mirrors (MM) with high reflectance are developed. At least three values of λ_{las} are of interest from the viewpoint of the development of XRL-based sources for nanolithography.

(i) For Z = 50, λ_{las} ~134.1 Å. For this wavelength region, MMs for nanolithography were developed as early as in 1993 [26]. The maximum normal incidence reflectivity achieved that time was 66% for a Mo/Si MM at λ = 13.4 nm, the reflectivity can be increased to 70%.
(ii) For Z = 54 λ_{las} ~11.3 nm. A series of normal-incidence reflectance measurements at just longer than the beryllium K-edge (11.1 nm) from Mo/Be MM was reported in [27]. The highest peak reflectance was 68.7 ± 0.2% at =11.3 nm obtained from a MM with 70 bilayers ending in beryllium. Our model of the high efficient monochromatic radiation sources near λ = 13.5 at 11.3 nm obtained in Xe^{26+}, intended for commercial nanolithography, was presented in our recent work [28].
(iii) In Ni-like Ytterbium (Z = 70), $\lambda_{las} \approx$ 67.25 Å. MM for wavelengths of 6.71–6.89 nm were developed in [29]. Summary of measured and calculated reflectivity of La/B MM for these wavelengths is listed in Table 1 of [29]. The largest reflectivity was observed and calculated for λ = 6.71–6.74 nm.

The crossing region of each working level with another level is characterized by their strong effect on each other, which can cause strong instability of the energy structure in the crossing region. In such regions, jumps in functions of energy levels and probabilities of radiative transition on Z are possible (see Figure 2a). The authors of [30], where the level crossing in the Ni-like sequence and associated irregularities in the functions of energies and probabilities of radiative transitions in the range Z = 74–84 were studied, arrived at the same conclusion. From this, the conclusion regarding the possible incorrect identification of levels in their crossing regions follows.

The SPP XRL can be very sensitive to external fields. It is implied that even an insignificant change in the plasma density can affect the emission spectrum. The remarkable phenomenon (see Figure 4) where a rapid increase in the contribution of the $3d_{5/2}4d_{3/2}$ $[J = 1]$ level to the composition of the lower working level is demonstrated could be an indirect confirmation of this. In the interval Z = 40–42, the contribution of this level increases by five orders of magnitude. A similar pattern is observed in Figure 6, where the contribution of $3p_{3/2}4s_{1/2}$ $[J = 1]$ also rapidly increases to Z = 49, where this level strongly interacts with the upper working level. In this case, the oscillator strength is transferred from the upper working level to the $3p_{3/2}4s_{1/2}$ $[J = 1]$ level.

Conflicts of Interest: The authors declare no conflict of interest.

References

1. Nilsen, J. Self photo-pumped neon-like and nickel-like X-ray lasers. In Proceedings of the Fifth International Conference on X-Ray Lasers, Lund, Sweden, 10–14 June 1996; Svanberg, S., Walström, C.-G., Eds.; CRC Press: Boca Raton, FL, USA, 1996.
2. Nilsen, J. Design of a picosecond laser-driven Ni-like Mo X-ray laser near 20 nm. *J. Opt. Soc. Am. B* **1997**, *14*, 1511–1514. [CrossRef]
3. Grant, I.P.; McKenzie, B.J.; Norrington, P.H.; Mayers, D.F.; Pyper, M.C. An atomic multiconfigurational Dirack-Fock package. *Comput. Phys. Commun.* **1980**, *21*, 207–231. [CrossRef]
4. Nilsen, J.; Dunn, J.; Osterheld, A.L.; Li, Y. Lasing on the self-photopumped nickel-like $4f\,^1P_1$–$4d\,^1P_1$ X-ray transition. *Phys. Rev. A* **1999**, *60*, R2677–R2680. [CrossRef]
5. Kuba, J.; Klisnick, A.; Ros, D.; Fourcade, P.; Jamelot, G.; Miquel, J.-L.; Blanchot, N.; Wyart, J.-F. Two-color transient pumping in Ni-like silver at 13.9 and 16.1 nm. *Phys. Rev. A* **2000**, *62*, 043808. [CrossRef]
6. Luther, B.M.; Wang, Y.; Larotonda, M.A.; Alessi, D.; Berrill, M.; Marconi, M.C.; Rocca, J.J.; Shlyaptsev, V.N. Saturated high-repetition rate 18.9 nm table top laser in Ni-like molybdenum. *Opt. Lett.* **2005**, *30*, 165–167. [CrossRef] [PubMed]
7. Siegrist, M.; Staub, F.; Jia, F.; Feuer, T.; Balmer, J.; Nilsen, J. Self-photopumped X-ray lasers from elements in the Ne-like and Ni-like ionization state. *Opt. Commun.* **2017**, *382*, 288–293. [CrossRef]
8. Reader, J.; Aquista, N.; Kaufman, V. Spectrum and energy levels of seven-times-ionized krypton (Kr VIII) and resonance lines of eight-times-ionized krypton. *J. Opt. Soc. Am. B* **1991**, *8*, 538–547. [CrossRef]
9. Chen, H.; Beiersdorfer, P.; Fournier, K.B.; Träbert, E. Soft X-ray spectra of highly charged Kr ions in an electron-beam ion trap. *Phys. Rev. E* **2002**, *65*, 056401. [CrossRef] [PubMed]
10. Churilov, S.S.; Ryabtsev, A.N.; Wyart, J.-F. Identification of n = 4, Δn = 0 transitions in the spectra of Nickel-like and Zn-like ions through tin. *Phys. Scr.* **1988**, *38*, 326–335. [CrossRef]
11. Ryabtsev, A.N.; Churilov, S.S.; Nilsen, J.; Li, Y.; Dunn, J.; Osterheld, A.L. Additional analysis of Ni-like ions spectra. *Opt. Spectrosc.* **1999**, *87*, 197–202. (In Russian)
12. Rahman, A.; Hammarsten, E.C.; Sakadzik, S.; Rocca, J.J.; Wyart, J.-F. Identification of n = 4, Δn = 0 transitions in the spectra of Nickel-like cadmium ions from a capillary discharge plasma column. *Phys. Scr.* **2003**, *67*, 414–419. [CrossRef]
13. Rahman, A.; Rocca, J.J.; Wyart, J.-F. Classification of the Nickel-like silver spectrum (Ag XX) from a fast capillary discharge plasma. *Phy. Scr.* **2004**, *70*, 21–25. [CrossRef]
14. Churilov, S.S.; Ryabtsev, A.N.; Wyart, J.-F. Analysis of the 4–4 transition in the Ni-like Kr IX. *Phys. Scr.* **2005**, *71*, 457–463. [CrossRef]
15. Ivanova, E.P.; Gogava, A.L. Energies of X-ray transitions in heavy Ni-like ions. *Opt. Spectrosc.* **1985**, *59*, 1310–1314. (In Russian)

16. Ivanova, E.P. Energy levels and probability of radiative transitions in the Kr IX ion. *Opt. Spectrosc.* **2014**, *117*, 179–187. [CrossRef]
17. Ivanova, E.P. Wavelengths of the *4d–4p*, 0–1 X-ray laser transitions in Ni-Like ions. *Int. J. Adv. Res. Phys. Sci.* **2016**, *3*, 34–40. [CrossRef]
18. Ivanova, E.P. Proposal for precision wavelength measurement of the Ni-like gadolinium X-ray laser formed during the interaction of nanostructured target with an ultrashort laser beam. *Laser Phys. Lett.* **2015**, *12*, 105801. [CrossRef]
19. Ivanova, E.P.; Zinoviev, N.A.; Knight, L.V. Theoretical investigation of X-ray laser on the transitions of Ni-like xenon in the range 13–14 nm. *Quantum Electron.* **2001**, *31*, 683–688. [CrossRef]
20. Ivanova, E.P.; Ivanov, A.L. A superpowerful source of far-ultraviolet monochromatic radiation. *J. Exp. Theor. Phys.* **2005**, *100*, 844–856. [CrossRef]
21. Ivanova, E.P.; Grant, I. Oscillator strength anomalies in the neon isoelectronic sequence with applications to X-ray laser modeling. *J. Phys. B At. Mol. Opt. Phys.* **1998**, *31*, 2871–2883. [CrossRef]
22. Kramida, A.E.; Shirai, T. Energy levels and spectral lines of tungsten. W III through W LXXIV. *At. Data Nucl. Data Tables* **2009**, *95*, 305–474. [CrossRef]
23. Clementson, J.; Beiersdorfer, P.; Brown, G.V.; Gu, M.F. Spectroscopy of M-shell X-ray transitions in Zn-like through Co-like W. *Phys. Scr.* **2010**, *81*, 015301. [CrossRef]
24. Dong, C.-Z.; Fritzsche, S.; Xie, L.-Y. Energy levels and transition probabilities for possible X-ray laser lines of highly charged Ni-like ions. *J. Quant. Spectrosc. Rad. Transf.* **2003**, *76*, 447–465. [CrossRef]
25. Safronova, U.I.; Safronova, A.S.; Hamasha, S.M.; Beiersdorfer, P. Relativistic many-body calculations of multipole (E1, M1, E2, M2, E3, and M3) transitions wavelengths and rates between $3l^{-1}4l'$excited and ground states in nickel-like ions. *At. Data Nucl. Data Tables* **2006**, *92*, 47–104. [CrossRef]
26. Stearns, D.G.; Rosen, R.S.; Vernon, S.P. Multilayer mirror technology for soft-X-ray projection lithography. *Appl. Opt.* **1993**, *32*, 6952–6960. [CrossRef] [PubMed]
27. Skulina, K.M.; Alford, C.S.; Bionta, R.M.; Makowiecki, D.M.; Gullikson, E.M.; Soufli, R.; Kortright, J.B.; Underwood, J.H. Molybdenum/beryllium multilayer mirrors for normal incidence in the extreme ultraviolet. *Appl. Opt.* **1995**, *34*, 3727–3730. [CrossRef] [PubMed]
28. Ivanova, E.P. X-ray laser near 13.5 and 11.3 nm in Xe^{26+} driven by intense pump laser interacting with xenon cluster jet as a promising radiation source for nanolithography. *Laser Phys.* **2017**, *27*, 055802–055811. [CrossRef]
29. Makhotkin, I.A.; Zoethout, E.; Van de Kruijs, R.; Yakunin, S.N.; Louis, E.; Yakunin, A.M.; Banine, V.; Müllender, S.; Bijkerk, F. Short period La/B and LaN/B multilayer mirrors for 6.8 nm wavelengths. *Opt. Express* **2013**, *21*, 29894–29904. [CrossRef] [PubMed]
30. Dong, C.Z.; Fritzsche, S.; Gaigalas, G.; Jacob, T.; Sienkievicz, J.E. Theoretical level structure and decay dynamics of Ni-like ions: search for laser lines in the soft X-ray domain. *Phys. Scr.* **2001**, *92*, 314–316.

atoms

MDPI

Article

Configuration Interaction Effects in Unresolved $5p^65d^{N+1}-5p^55d^{N+2}+5p^65d^N5f^1$ Transition Arrays in Ions Z = 79–92

Luning Liu [1,2,*], Deirdre Kilbane [2], Padraig Dunne [2], Xinbing Wang [1] and Gerry O'Sullivan [2]

1 Wuhan National Laboratory for Optoelectronics, Huazhong University of Science and Technology, Wuhan 430074, China; xbwang@hust.edu.cn
2 School of Physics, University College Dublin, Belfield, Dublin 4, Ireland; deirdre.kilbane@ucd.ie (D.K.); padraig.dunne@ucd.ie (P.D.); gerry.osullivan@ucd.ie (G.O.)
* Correspondence: luningliu@outlook.com

Academic Editor: Joseph Reader
Received: 12 April 2017; Accepted: 12 May 2017; Published: 21 May 2017

Abstract: Configuration interaction (CI) effects can greatly influence the way in which extreme ultraviolet (EUV) and soft X-ray (SXR) spectra of heavier ions are dominated by emission from unresolved transition arrays (UTAs), the most intense of which originate from $\Delta n = 0$, $4p^64d^{N+1}-4p^54d^{N+2}+4p^64d^N4f^1$ transitions. Changing the principle quantum number n, from 4 to 5, changes the origin of the UTA from $\Delta n = 0$, $4p^64d^{N+1}-4p^54d^{N+2}+4p^64d^N4f^1$ to $\Delta n = 0$, $5p^65d^{N+1}-5p^55d^{N+2}+5p^65d^N5f^1$ transitions. This causes unexpected and significant changes in the impact of configuration interaction from that observed in the heavily studied $n = 4-n = 4$ arrays. In this study, the properties of $n = 5-n = 5$ arrays have been investigated theoretically with the aid of Hartree-Fock with configuration interaction (HFCI) calculations. In addition to predicting the wavelengths and spectral details of the anticipated features, the calculations show that the effects of configuration interaction are quite different for the two different families of $\Delta n = 0$ transitions, a conclusion which is reinforced by comparison with experimental results.

Keywords: configuration interaction (CI); unresolved transition array (UTA); Cowan code

1. Introduction

Laser produced plasmas (LPPs) from tin droplet targets have been adopted as the optimum extreme ultraviolet (EUV) light sources for next generation lithography for high-volume manufacturing (HVM) of semiconductor circuits with feature sizes of 10 nm or less [1,2]. Transitions of the type $4p^64d^{N+1}-4p^54d^{N+2}+4p^64d^N4f^1$ in Sn^{8+}–Sn^{13+} merge to form an unresolved transition array (UTA) [3] which contains thousands of individual lines and emits strongly in such a plasma at an electron temperature of ~30 eV in a narrow wavelength range around 13.5 nm [4,5]. This value coincides with the wavelength of peak reflectance of ~70% of the Mo/Si multilayer mirrors (MLMs) that are used in the scanning tools [6] and tin plasmas are the brightest sources at this wavelength. Other recent research has concentrated on investigating future-generation lithographic sources at shorter wavelengths, in particular at 6.75 nm where an intense UTA is emitted by gadolinium and terbium plasmas with an electron temperature of close to 100 eV [7–9], and where LaB_4C and $LaNB_4C$ MLMs have a peak theoretical reflectivity of close to ~80% [10]. Once more the transitions responsible are predominantly of the type $4p^64d^{N+1}-4p^54d^{N+2}+4p^64d^N4f^1$.

Moving to shorter wavelengths, we encounter the "water window" (2.3–4.4 nm) spectral region lying between the K-edges of carbon and oxygen, where carbon K-edge absorption is strong, but oxygen L edge absorption is weak, and where sources are being developed for in vivo single shot imaging and tomography of biological samples in aqueous environments with nm resolution [11,12]. Initially, sources in this region used strong quasi-monochromatic emission at wavelengths of λ = 2.879 nm and λ = 2.478 nm arising from the $1s^2$–$1s\ 2p$ line in N^{5+} and the $1s$–$2p$ doublet in N^{6+} respectively [13]. However more recently $4d$–$4f$ transitions of the $4p^6 4d^{N+1} - 4p^5 4d^{N+2} + 4p^6 4d^N 4f^1$ UTA in Bi^{37+}–Bi^{46+} have been proposed, and a Bi source based on a plasma heated to a sufficient temperature (T_e > 500 eV) to generate these ion stages is under development for water window imaging [14].

The dominant emission in all of these sources arises from $4p^6 4d^{N+1} - 4p^5 4d^{N+2} + 4p^6 4d^N 4f^1$ transitions and due to the near degeneracy of the $4p^5 4d^{N+2}$ and $4d^N 4f$ configurations, it is well known that it is necessary to allow for configuration interaction (CI) in the upper state [4,15,16]. The effects of CI in any particular ion stage have been shown to cause a strong spectral narrowing and concentrate the available emission intensity at the high energy end of the array. Moreover, although the $4p^5 4d^{N+2}$ configuration must be included in order to obtain the correct energy eigenvalues and eigenvectors, the latter remain sufficiently pure while the emission is dominated by the valence $4p^6 4d^{N+1} - 4p^6 4d^N 4f^1$ transitions and there is little evidence in any spectrum of a sizable contribution in emission from $4p^6 4d^{N+1} - 4p^5 4d^{N+2}$ lines until $N = 1$. This is presumably due to the electron impact excitation rates for valence and sub-valence excitation responsible for populating the upper states being very different. Based on a simple line strengths comparison, the ratio of total lines strengths for p–d transitions to that for d–f should scale as $9-N/N+1$ times the ratio of their respective dipole matrix elements [17]. So one would expect the p–d contribution to overtake that for d–f with increasing ionization around $n = 3$ [17]. Moreover, if one allows for spin orbit splitting of the $4p$ and $4d$ subshells, for $N > 4$ the lowest configuration will be $4p^2{}_{1/2}4p^4{}_{3/2}4d^4{}_{3/2}4d^{N-4}{}_{5/2}$, Thus if it is easier to collisionally excite outer electrons, for $N > 4$ the dominant excitation will involve $4d_{5/2}$ electrons and the transitions expected are: $4p^2{}_{1/2}4p^4{}_{3/2}4d^4{}_{3/2}4d^{N-4}{}_{5/2} - 4p^2{}_{1/2}4p^3{}_{3/2}4d^4{}_{3/2}4d^{N-3}{}_{5/2} + 4p^2{}_{1/2}4p^4{}_{3/2}4d^4{}_{3/2}4d^{N-5}{}_{5/2}4f$.

For $N < 4$, the lowest configuration will be $4p^2{}_{1/2}4p^4{}_{3/2}4d^N{}_{3/2}$ and transitions can now take place to $4p^2{}_{1/2}4p^4{}_{3/2}4d^N{}_{3/2} - 4p^2{}_{1/2}4p^3{}_{3/2}4d^{N+1}{}_{3/2} + 4p^1{}_{1/2}4p^4{}_{3/2}4d^{N+1}{}_{3/2} + 4p^2{}_{1/2}4p^4{}_{3/2}4d^{N-1}{}_{3/2}4f$ with the $4p_{1/2}$–$4d_{3/2}$ contribution appearing on the short wavelength side of the UTA, or, if the 4p spin orbit splitting is sufficiently large, forming a second UTA at a shorter wavelength.

However, both sets of transitions are responsible for absorption by ions in the plasma periphery which is the major problem that must be overcome to attain the maximum conversion efficiency of laser to spectral emission energy in EUV source development. Theoretical studies of the effects of CI in ions from $Z = 50 - Z = 89$ have been reported which showed that CI effects in general diminish as Z increases as the upper state arrays separate in energy [18,19] and recently the corresponding UTA emission in a number of elements at the higher Z end of this sequence has been observed [20,21].

In the absence of CI, according to the UTA formalism, for $4p^6 4d^{N+1} - 4p^6 4d^N 4f^1$ transitions the position of the line strength weighted mean of an array is shifted from the position of the differences in average energies by an amount [22]

$$\delta E = \frac{35}{9}(\mathrm{N})\left[\sum_{k\neq 0} f_k F^k(4d,4f) + g_k G^k(4d,4f)\right] \tag{1}$$

where $F^k(4d,4f)$ and $G^k(4d,4f)$ are Slater Condon direct and exchange integrals respectively and the coefficients f_k and g_k result from integrals over polar and azimuthal angles that, in general, decrease with increasing k [23]. Here $g_1 = 137/2450$ has the largest numerical value and the above formula can be roughly approximated as $\delta E = \frac{2}{9} N G^1(4d,4f)$ [24] so that the position of the emission peak is determined by the degree of $4d$ and $4f$ overlap. In higher ion stages (beyond ~4+) of the rare earths, where $G^1(4d, 4f)$ is almost constant for different ion stages of a given element, the effect of CI is to essentially remove this N dependence and the array is narrowly peaked at around $2G^1(4d,4f)$ above the difference between the average energies of the ground and upper state configurations. Thus the UTAs in successive ion stages overlap with each other to yield a very intense, relatively narrow (ΔE ~ 10 eV), emission band in a low opacity plasma, whose shape is completely modified by increasing opacity [25].

In performing calculations for low ion stages of the lighter lanthanides and the elements preceding them in the periodic table, it is necessary to expand the excited state basis to include higher *nf* orbitals or reduce the effective exchange interaction. This is achieved by scaling the $G^1(4d, 4f)$ parameter, as is done in calculations with the Cowan code, in order to obtain good agreement between calculated and observed results. Mixing of *4f* and *nf* orbitals essentially increases the mean radius of the *4f* wave function and so leads to a reduction in the size of both direct and exchange integrals [26]. The photoionization spectra of low ion stages of these elements are well known to be dominated by *4f* contraction effects and the correct estimation of the *4f* radial wave function is essential if good agreement between theory and experimental spectra is to be obtained [27].

For the elements from Ag to La, *4f* contraction increases with ion stage due to the interplay between the attractive Coulomb and centrifugal repulsion $\left(\frac{l(l+1)}{2\mu r^2}\right)$ terms in the effective radial potential, where l is the orbital angular momentum quantum number and μ is the reduced mass of the electron. In the neutral atom, the effective potential is bimodal with an inner well close to the nucleus, whose depth rapidly increases from $Z = 47$ (silver), where it first appears, to $Z = 58$ (cerium), where it first supports a bound state leading to the formation of the lanthanides [27,28]. This inner well is separated by a centrifugal barrier from a broad outer well with a minimum near the hydrogenic value of $16a_0$. The EUV absorption spectrum of these elements is dominated by a large $4d$–εf shape resonance [29] since depending on Z, $4d$–εf excitation can only occur when the εf photoelectron has sufficient energy to surmount the centrifugal barrier, or the lowest state of the inner well is autotomizing. Due to the lack of any appreciable overlap between the *4d* wave function, which lies in the core, and the bound *nf* wave functions which are eigenstates of the outer well, *4d*–*4f* transitions have vanishing oscillator strength. With increasing ionization, the inner well deepens, the potential barrier decreases, and the outer-well *nf* functions gradually contract into the inner well region. As they do, the *4d*, *4f* overlap increases and the intensity of *4d*–*4f* transitions increases and the oscillator strength, associated with $4d$–εf in the neutral is effectively transferred to *4d*–*4f* excitation [30].

In contrast to the situation for $\Delta n = 0$, $n = 4 - n = 4$ transitions, no systematic study of the equivalent $\Delta n = 0$, $n = 5 - n = 5$ transitions has been reported. From studies of the photoionization cross-sections of neutral elements past $Z = 79$ (gold) it is known that the spectra display strong $5d-\varepsilon f$ resonances and that any difference from their $4d$–εf counterparts can be attributed to the increased influence of spin-orbit effects [30–33]. UTAs due to $5p^65d^{N+1}-5p^55d^{N+2}+5p^65d^N5f^1$ transitions in LPPs of Th and U have been observed and some of the simpler transitions identified [34,35]. Compared to the $4p^64d^{N+1}-4p^54d^{N+2}+4p^64d^N4f^1$ UTAs observed under identical experimental conditions in the homologous elements Ba and Ce, the $n = 5 - n = 5$ UTAs were broader [36]. Spectra from ionized uranium that were recorded following impurity injection into the TEXT Tokamak were found to contain two distinct UTAs which were assigned primarily to $5p_{1/2}$–$5d$ and $5d$–$5f$ component groups of $5p^65d^{N+1}-5p^55d^{N+2}+5p^65d^N5f^1$ transitions in U XV –U XXXI [37]. However, apart from this work, no calculations were performed to elucidate and explore CI effects.

In this paper, we report on the results of calculations for $5p^65d^{N+1}-5p^55d^{N+2}+5p^65d^N5f^1$ transitions in elements from $Z = 79$ to $Z = 92$ to predict the positions and spectral properties of the corresponding UTAs and in particular to compare the effects of CI between $\Delta n = 0$, $n = 5 - n = 5$ transitions in these elements and $n = 4 - n = 4$ transitions in their homologous, lower Z counterparts.

2. Results

2.1. 5p−5d and 5d−5f Unresolved Transition Arrays of Ions with Z = 79–92

Calculations were performed using the Hartree-Fock with Configuration Interaction (HFCI) suite of codes written by Cowan [17]. Because of the high Z of the atoms and ions of interest, relativistic effects which are the mass-velocity and Darwin contributions to the energy were included. The Slater Condon F^k, G^k, and R^k parameters were scaled to 90% of their *ab initio* values while the spin orbit parameters were unchanged. Energies and wavelengths were determined for $5p^65d^{N+1}-5p^55d^{N+2}+5p^65d^N5f^1$

transitions both with and without CI for all ions with $N = 0\text{–}8$ of the elements considered. For the CI calculations, the eigenvectors percentage compositions were used to assign *5d–5f* and *5p–5d* lines within the overall arrays.

The results of these calculations are presented in Figures 1 and 2. Figure 1 shows the calculated spectra for ions of the elements from Au ($Z = 79$) to Po ($Z = 84$), while Figure 2 contains the corresponding data for ions from At ($Z = 85$) to U ($Z = 92$). For each element, the green and red line distributions denote *5p–5d* transitions with and without CI included, respectively, while blue and black denote *5d–5f* transitions with and without CI included. In the case of Au, the most obvious feature of the spectra is that with increasing ion stage, the *5p–5d* transition arrays move slowly towards shorter wavelength while the *5d–5f* transition arrays move more rapidly towards higher energy with increasing ion stage. The arrays never overlap and so CI effects are almost non-existent up to Au^{5+} and from Au^{6+} onwards, CI mainly affects the *5d–5f* transitions where they dramatically alter the line distributions. It should be noted that most or all of the *5p–5d* transitions are autoionizing until we reach Au^{7+} and even if the upper states are populated, they will never appear in emission. The near absence of CI for *5d–5f* transitions in lower stages and the closeness in energy of the *5d–5f* sub-arrays in the higher stages would suggest that the intensity weighted mean positions of these arrays should be given by Equation (1). The fact that the arrays move to shorter wavelength so dramatically is due to the *5f* wave function contraction which leads to both an increase in the separation of average energies of the upper and lower configurations and also a rapid increase in $G^1(5d, 5f)$. Similar behavior, in the case of *4d–4f* transitions has been found in Sn spectra [38].

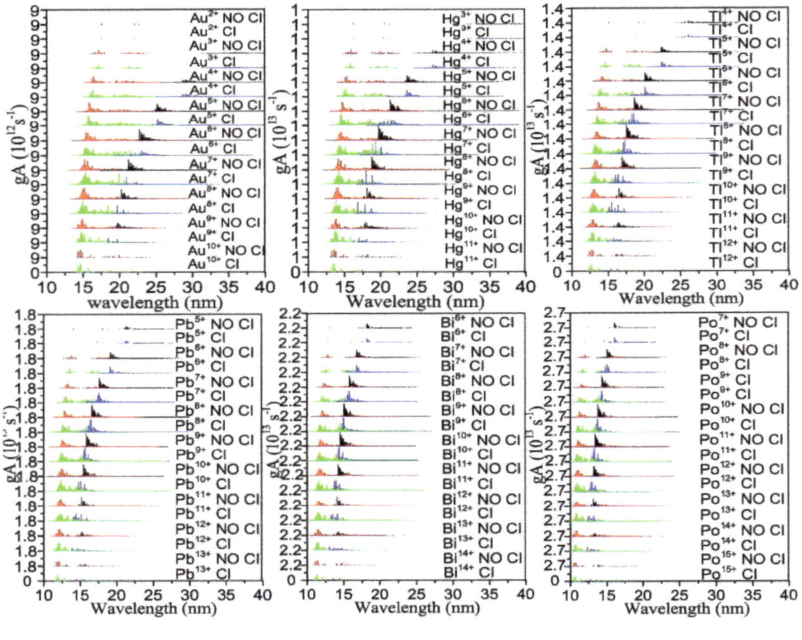

Figure 1. (Color online) Ir-like through Tm-like spectra of Au-Po calculated with the Cowan Code both including Configuration interaction (CI) (green denotes *5p–5d* and blue denotes *5d–5f*) and excluding CI (red denotes *5p–5d* and black denotes *5d–5f*).

In the case of Hg and Tl, CI effects again become important for *5d–5f* spectra at Hg^{6+} and Tl^{6+}. For Pb and Bi the effects of CI on *5d–5f* transitions are predicted to become noticeable at Pb^{7+} and Bi^{7+}, while in all cases the changes in the *5p–5d* sub-arrays only become noticeable when they begin to overlap with the *5d–5f* sub-arrays and where a redistribution of intensity towards the higher energy end of the overall arrays become visible. With increasing *Z*, *5f* contraction effects diminish as the transitions now involve significantly higher charge state ions. As can be seen from Figure 2, the *5p–5d* and *5d–5f* sub-arrays become closer and CI effects cause subtle changes to the spectral profiles of both sub-arrays for situations where the $5p^6 5d^{N+1}$ ground configuration has $N > 3$ and more dramatic effects when $N \leq 3$.

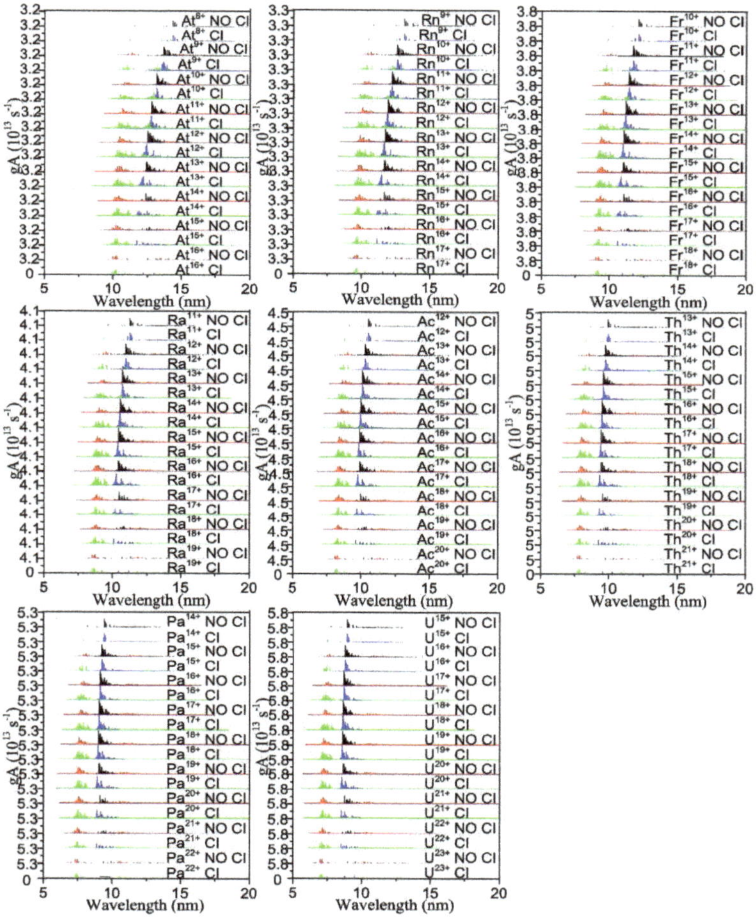

Figure 2. (Color online) Ir-like through Tm-like spectra of At-U calculated with the Cowan Code both including CI (green denotes *5p–5d* and blue denotes *5d–5f*) and excluding CI (red denotes *5p–5d* and black denotes *5d–5f*).

To explore the effects of wave function contraction with increasing ion stage, the radial wave functions $P_{n,l}(r)$ were extracted for *5p*, *5d*, and *5f* electron orbitals for each ion considered. From these the mean radius $\langle r \rangle$ was computed using $\langle r \rangle = \int_0^\infty P_{n,l}^2 r dr$ and the results are presented in Figure 3. It is clear from this figure that the mean radii of the *5p* and *5d* functions decrease slowly with *Z* and charge state. The situation for the *5f* wave function is very different. In Au, for example, $\langle r \rangle$ contracts from $5.4a_0$ in Au^{2+} to $1.5a_0$ in Au^{10+}. With increasing *Z*, the effect is less dramatic and past Ra, the *5f* contacts with increasing ionization much like the *5p* and *5d*. This is mirrored in the spectra by the fact that separation of the *5p–5d* and *5d–5f* arrays becomes essentially constant as the *5d–5f* array does not dramatically move to higher energy with increasing charge.

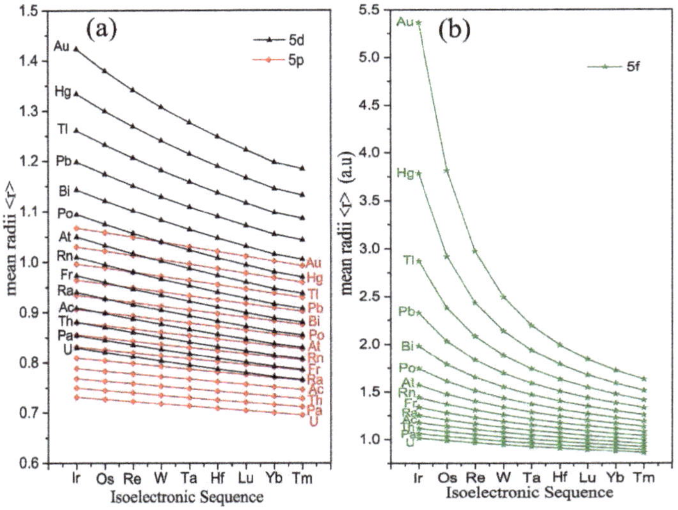

Figure 3. (Color online) mean radii of *5p*, *5d*, and *5f* eigenfunctions for ions of the Ir (ground state $5d^9$) through Tm (ground configuration $5d^1$) for all elements from Au-U.

2.2. 5p−5d and 5d−5f UTA Statistics of Ions with Z = 79–92

In general, the complexity of arrays with $1 < N < 8$, the UTA formalism is suitable for the parameterization of the calculated wavelength data [3,21]. The general *n*th-order moment for a set of N values λ_i with line strengths ω_i reads

$$\mu_n = \sum_{i=1}^{N} \omega_i \lambda_i^n / W \tag{2}$$

where $W = \sum_{i=1}^{N} \omega_i$ is the total line strength. The first-order moment μ_1 gives the intensity weighted average wavelength. The centered second-order moment $\mu_2^c = \mu_2 - \mu_1^2$ gives the variance, ν, which is obtained by the above expression after replacing λ_i by $\lambda_i - \mu_1$. For a Gaussian-shaped distribution, its full width at half maximum (FWHM) is given by $2(2\ln 2)^{1/2}\sigma = 2.355\sigma$, where $\sigma = (\mu_2^c)^{1/2}$. Thus the variance is related to the width of the array. Using the UTA formalism described above, the gA weighted UTA positions and widths for the *5d–5f* and *5p–5d* component sub-arrays of the $5p^6 5d^{N+1} - 5p^5 5d^{N+2} + 5p^6 5d^N 5f^1$ array were calculated and the results are presented in Figure 4 and Tables 1 and 2. Separate UTAs for *5p–5d* and *5d–5f* transitions were identified from their eigenvector compositions and UTA statistics were computed for both sets of transitions with and without CI effects included for comparison. From this figure it is clear that in the case of *5d–5f* transitions, which will be observed in emission from a plasma, the effect of CI is to shift the corresponding sub-array towards higher energy especially for the higher *Z* elements. This trend is also clear from Tables 1 and 2. Interestingly, unlike the corresponding *4−4* arrays, where spectral narrowing is the dominant effect observed, the effect of CI is actually to increase the width of the UTAs. Again, during the rapid contraction phase of the *5f* wave function in lower ion stages of the lighter elements, CI effects are noticeably absent as can be seen from the coincidence in energies in both cases. For *5p–5d* transitions, CI effects are somewhat different also for lighter and heavier elements. For the elements past francium, the mean energies are shifted by CI towards higher values in lower ion stages and gradually converge towards their non-CI value at the highest ion stage.

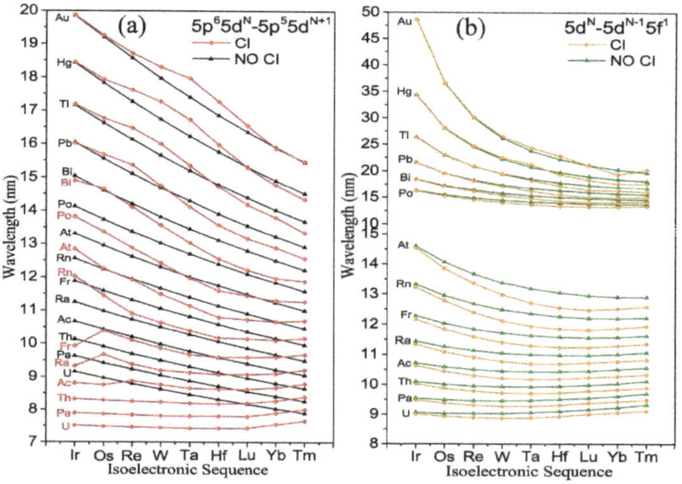

Figure 4. (Color online) Mean wavelength of transition arrays $5p^65d^{N+1}-5p^55d^{N+2}$ and of $5p^65d^{N+1}-5p^65d^N5f^1$ Ir-like through Tm-like ions of gold through uranium (**a**) $5p^65d^{N+1}-5p^55d^{N+2}$ including CI (red) and excluding CI (black); (**b**) $5p^65d^{N+1}-5p^65d^N5f^1$ including CI (orange) and excluding CI (green).

Table 1. Calculated mean wavelength $\overline{\lambda}_{gA}$ (nm) and spectral width $\Delta\lambda_{gA}$ (nm) for the UTA of gold through astatine ions: Ir-like to Tm-like ions for the *5d–5f* arrays without and with the effect of configuration interaction.

***5d–5f* (No CI)**	**Au**		**Hg**		**Tl**		**Pb**		**Bi**		**Po**		**At**	
Ion	$\overline{\lambda}_{gA}$	$\Delta\lambda_{gA}$	$\overline{\lambda}_{gA}$	$\Delta\lambda_{gA}$	$\overline{\lambda}_{gA}$	$\Delta\lambda_{gA}$	$\overline{\lambda}_{gA}$	$\Delta\lambda_{gA}$	$\overline{\lambda}_{gA}$	$\Delta\lambda_{gA}$	$\overline{\lambda}_{gA}$	$\Delta\lambda_{gA}$	$\overline{\lambda}_{gA}$	$\Delta\lambda_{gA}$
Ir-like	48.62	5.99	34.33	3.11	26.40	1.97	21.59	1.47	18.45	1.23	16.23	1.08	14.59	0.98
Os-like	36.54	3.83	28.09	2.39	22.96	1.80	19.58	1.51	17.21	1.34	15.44	1.22	14.06	1.13
Re-like	29.99	2.77	24.48	2.09	20.84	1.77	18.27	1.57	16.36	1.43	14.87	1.32	13.67	1.24
W-like	26.17	2.35	22.23	1.98	19.45	1.76	17.37	1.61	15.75	1.48	14.46	1.38	13.38	1.30
Ta-like	23.77	2.16	20.74	1.91	18.48	1.74	16.72	1.60	15.31	1.49	14.15	1.40	13.18	1.32
Hf-like	22.17	2.00	19.70	1.82	17.79	1.67	16.25	1.55	14.99	1.45	13.94	1.37	13.04	1.29
Lu-like	21.05	1.80	18.96	1.65	17.29	1.53	15.91	1.43	14.77	1.35	13.79	1.28	12.94	1.21
Yb-like	20.25	1.48	18.42	1.37	16.92	1.29	15.67	1.21	14.61	1.15	13.69	1.10	12.90	1.05
Tm-like	19.67	0.87	18.03	0.83	16.67	0.80	15.51	0.78	14.52	0.76	13.65	0.75	12.89	0.74
***5d–5f* (CI)**	**Au**		**Hg**		**Tl**		**Pb**		**Bi**		**Po**		**At**	
Ion	$\overline{\lambda}_{gA}$	$\Delta\lambda_{gA}$	$\overline{\lambda}_{gA}$	$\Delta\lambda_{gA}$	$\overline{\lambda}_{gA}$	$\Delta\lambda_{gA}$	$\overline{\lambda}_{gA}$	$\Delta\lambda_{gA}$	$\overline{\lambda}_{gA}$	$\Delta\lambda_{gA}$	$\overline{\lambda}_{gA}$	$\Delta\lambda_{gA}$	$\overline{\lambda}_{gA}$	$\Delta\lambda_{gA}$
Ir-like	48.64	6.00	34.36	3.16	26.43	2.10	21.60	1.69	18.43	1.45	16.19	1.25	14.53	1.10
Os-like	36.62	3.90	28.17	2.60	23.02	2.17	19.58	2.00	17.09	1.90	15.24	1.56	13.85	1.30
Re-like	30.15	2.99	24.63	2.69	20.86	2.66	18.16	2.51	16.13	2.07	14.55	1.61	13.36	1.36
W-like	26.48	2.89	22.47	3.18	19.43	3.05	17.12	2.58	15.33	2.03	14.00	1.59	12.98	1.37
Ta-like	24.34	3.15	21.21	3.46	18.24	3.00	16.18	2.35	14.75	1.85	13.62	1.55	12.70	1.39
Hf-like	22.79	3.29	19.36	3.60	17.19	2.64	15.46	2.03	14.26	1.68	13.33	1.50	12.54	1.43
Lu-like	21.01	3.24	18.25	2.88	16.35	2.08	15.02	1.73	14.06	1.61	13.20	1.51	12.47	1.43
Yb-like	19.37	2.34	17.60	1.79	16.25	1.42	14.92	1.70	14.06	1.53	13.18	1.37	12.51	1.37
Tm-like	20.19	0.16	17.56	0.12	15.99	1.05	14.98	1.00	14.09	0.97	13.29	0.94	12.58	0.92

Table 2. Calculated mean wavelength $\overline{\lambda}_{gA}$ (nm) and spectral width $\Delta\lambda_{gA}$ (nm) for the UTA of radon through uranium ions: Ir-like to Tm-like ions for the *5d–5f* arrays without and with the effect of configuration interaction.

5d–5f (No CI)	Rn		Fr		Ra		Ac		Th		Pa		U	
Ion	$\overline{\lambda}_{gA}$	$\Delta\lambda_{gA}$	$\overline{\lambda}_{gA}$	$\Delta\lambda_{gA}$	$\overline{\lambda}_{gA}$	$\Delta\lambda_{gA}$	$\overline{\lambda}_{gA}$	$\Delta\lambda_{gA}$	$\overline{\lambda}_{gA}$	$\Delta\lambda_{gA}$	$\overline{\lambda}_{gA}$	$\Delta\lambda_{gA}$	$\overline{\lambda}_{gA}$	$\Delta\lambda_{gA}$
Ir-like	13.32	0.91	12.28	0.85	11.43	0.80	10.71	0.77	10.09	0.73	9.54	0.71	9.06	0.68
Os-like	12.95	1.06	12.03	1.00	11.25	0.95	10.59	0.90	10.01	0.86	9.49	0.83	9.03	0.80
Re-like	12.68	1.16	11.84	1.10	11.13	1.04	10.50	1.00	9.95	0.95	9.46	0.92	9.03	0.88
W-like	12.48	1.22	11.71	1.16	11.04	1.10	10.45	1.05	9.93	1.01	9.46	0.97	9.03	0.94
Ta-like	12.35	1.25	11.62	1.18	10.99	1.13	10.43	1.08	9.92	1.04	9.47	1.00	9.06	0.97
Hf-like	12.25	1.23	11.57	1.17	10.97	1.12	10.43	1.08	9.94	1.04	9.50	1.00	9.10	0.97
Lu-like	12.20	1.16	11.55	1.11	10.97	1.07	10.45	1.03	9.98	1.00	9.55	0.97	9.16	0.94
Yb-like	12.19	1.01	11.56	0.98	11.00	0.95	10.50	0.92	10.03	0.90	9.61	0.88	9.23	0.86
Tm-like	12.21	0.73	11.61	0.72	11.06	0.72	10.56	0.71	10.11	0.71	9.70	0.71	9.32	0.71

5d–5f (CI)	Rn		Fr		Ra		Ac		Th		Pa		U	
Ion	$\overline{\lambda}_{gA}$	$\Delta\lambda_{gA}$	$\overline{\lambda}_{gA}$	$\Delta\lambda_{gA}$	$\overline{\lambda}_{gA}$	$\Delta\lambda_{gA}$	$\overline{\lambda}_{gA}$	$\Delta\lambda_{gA}$	$\overline{\lambda}_{gA}$	$\Delta\lambda_{gA}$	$\overline{\lambda}_{gA}$	$\Delta\lambda_{gA}$	$\overline{\lambda}_{gA}$	$\Delta\lambda_{gA}$
Ir-like	13.22	0.99	12.16	0.84	11.33	0.77	10.63	0.71	10.02	0.68	9.48	0.65	9.01	0.63
Os-like	12.78	1.11	11.84	0.98	11.08	0.90	10.44	0.83	9.87	0.79	9.38	0.77	8.93	0.74
Re-like	12.39	1.19	11.58	1.06	10.90	0.99	10.30	0.94	9.78	0.90	9.31	0.87	8.89	0.85
W-like	12.11	1.22	11.39	1.15	10.77	1.09	10.21	1.05	9.72	1.01	9.27	0.98	8.87	0.96
Ta-like	11.93	1.29	11.28	1.24	10.69	1.19	10.17	1.14	9.70	1.11	9.27	1.01	8.88	1.05
Hf-like	11.84	1.36	11.23	1.30	10.68	1.27	10.18	1.23	9.72	1.18	9.31	1.15	8.93	1.12
Lu-like	11.81	1.38	11.23	1.32	10.72	1.29	10.24	1.26	9.78	1.22	9.38	1.18	9.00	1.14
Yb-like	11.86	1.27	11.28	1.22	10.77	1.17	10.28	1.12	9.83	1.08	9.43	1.06	9.06	1.04
Tm-like	11.93	0.90	11.34	0.89	10.81	0.87	10.33	0.86	9.88	0.85	9.48	0.84	9.11	0.83

3. Comparison of $5p^65d^{N+}-5p^55d^{N+2}+5p^65d^N5f^1$ with $4p^64d^{N+1}-4p^54d^{N+2}+4p^64d^N4f^1$ Arrays

In the case of $4p^64d^{N+1}-4p^54d^{N+2}+4p^64d^N4f^1$ transitions, as already discussed, configuration interaction leads to a strong spectral narrowing and redistribution of oscillator strength towards the high energy end of the resulting UTA. Here, the opposite is true and the widths of the predicted *5d–5f* UTAs is in general slightly greater when CI effects are accounted for. In order to directly compare the results of CI on the spectral distribution rearrangement of $n = 4-n = 4$ UTA and $n = 5-n = 5$ UTA, calculations were performed for $4p^64d^2-4p^54d^3+4p^64d4f^1$ transitions in Sr-like Ag^{9+}, Sn^{12+}, Ba^{18+}, and Nd^{22+} and $5p^65d^2-5p^55d^3+5p^65d5f^1$ transitions in the homologous ions Au^{9+}, Pb^{12+}, Ra^{18+}, and U^{22+} of the Yb-isoelectronic sequence. The results are shown in Figure 5. From this figure it is clear that for $n = 4-n = 4$ transitions, CI completely reallocates the intensity of the *4d–4f* component transitions as well as the lower energy *4p–4d* lines to the higher energy end of the array and that with increasing ionization the resulting spectrum narrows until its FWHM becomes less than 0.5 nm. For $n = 5-n = 5$ transitions, in the absence of CI the *5p–5d* array splits with increasing Z due to spin orbit interaction into $5p_{1/2}$–$5d$ and $5p_{3/2}$–$5d$ sub-arrays. The *5d–5f* sub array overlays the longer wavelength $5p_{3/2}-5d_{5/2}$ sub-array in Au^{9+} and Pb^{12+}, and lies between the $5p_{1/2}$–$5d$ and $5p_{3/2}$–$5d$ sub-arrays in Ra^{18+} and U^{22+}. The effect of CI is to narrow the spectral width of the $5p_{1/2}$–$5d$ sub-array while leaving its mean position essentially unchanged, while mixing the *5d–5f* and $5p_{3/2}$–$5d$ sub-arrays to produce a broader spectral profile that in some instances contains fewer strong individual lines, that is shifted to shorter wavelength by the interaction. Thus, the effect of CI is less dramatic for *5–5* transitions though it still leads to major redistribution of intensity both between and within the resulting two sub arrays.

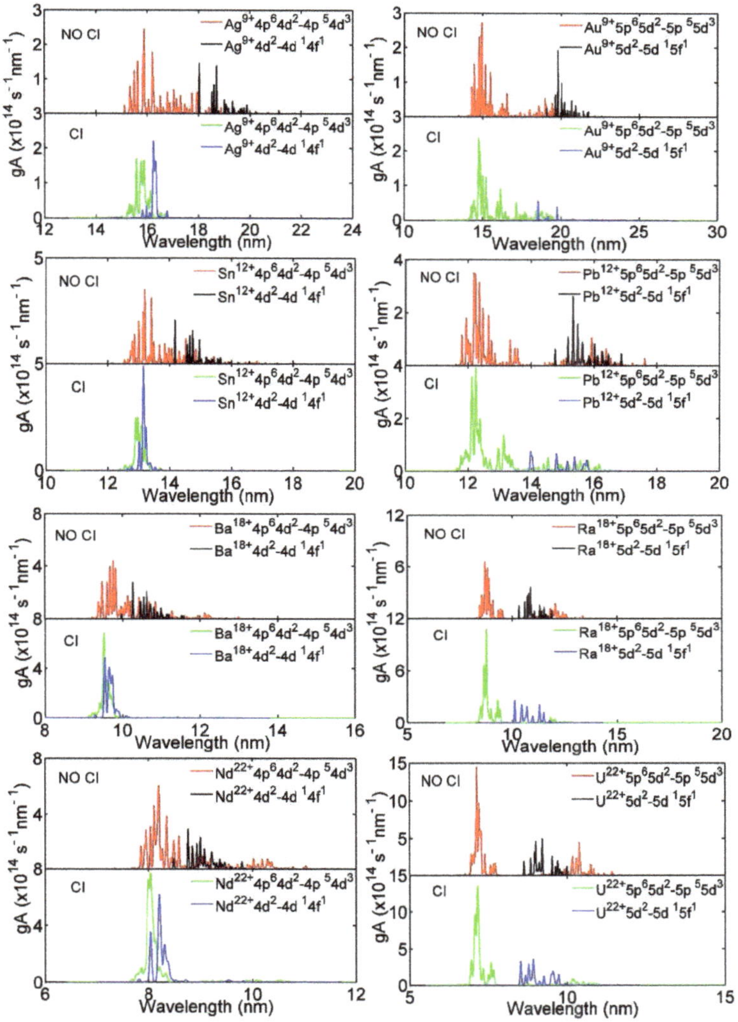

Figure 5. (Color online) Gaussian convolved spectra of $4p^6 4d^2 - 4p^5 4d^3 + 4p^6 4d^1 4f^1$ transitions in Sr-like Ag^{9+}, Sn^{12+}, Ba^{18+}, and Nd^{22+} and $5p^6 5d^2 - 5p^5 5d^3 + 5p^6 5d^1 5f^1$ transitions in the homologous ions Au^{9+}, Pb^{12+}, Ra^{18+}, and U^{22+} of the Yb-isoelectronic sequence.

From the CI calculations, the normalized gA ($gA/\Sigma gA$) distributions for *5d–5f* and *5p–5d* transitions were extracted for each ion stage, i.e., for $0 \leq N \leq 8$ of each of the elements considered here and summed to give an overall profile for both sets of transitions. The results are shown in Figure 6. As in the rare earths, the *d–f* lines are expected to contribute to the emission spectra from hot plasmas of these elements whilst both sets of transitions may be observed in absorption. It is interesting to compare the positions of the strong UTAs observed in LPPs of Th and U [34,35] with the predictions of the present calculations. In the Th spectrum, recorded under essentially optically thin conditions, a UTA extending from approximately 9.5–11.5 nm and peaking near 10.3 nm was observed while in the U spectrum the same feature lay between approximately 9.0 and 10.5 nm and peaked near 9.5 nm. From Table 2, the peak positions are predicted to lie near 9.8 and 9 nm respectively indicating a wavelength shift of approximately 0.5 nm between observed and calculated data for $5p^65d^{N+1}-5p^65d^N5f^1$ transitions. No shorter wavelength UTA corresponding to $5p^65d^{N+1}-5p^55d^{N+2}$ was observed. However, the maximum ionization stages produced in these experiments were around 16 or 17 times ionized and some contribution from $5d^{10}5f^N-5d^95f^{N+1}$ transitions in lower ion stages is also present. When first reported it was assumed that the increased widths of these $5p^65d^{N+1}-5p^65d^N5f^1$ UTAs relative to their $4p^64d^{N+1}-4p^64d^N4f^1$ counterparts in the spectra of the homologous species Ce and Nd was due to increased spin orbit interaction effects [34]. From this work it is clear that the $5p$ spin orbit splitting essentially limits the interaction to the $5p_{3/2}$–$5d$ sub-array and this interaction results in a broadening of the $5d$–$5f$ array. In the more highly ionized spectra of U recorded from the TEXT Tokamak, two distinct UTAs were observed with peaks near 7 and 9 nm which are in excellent agreement with the results obtained in this work. However, the shorter wavelength observed peak also contains a contribution from $5p^n-5p^{n-1}5d$ transitions, which may dominate over $5p^65d^{N+1}-5p^55d^{N+2}$ emission.

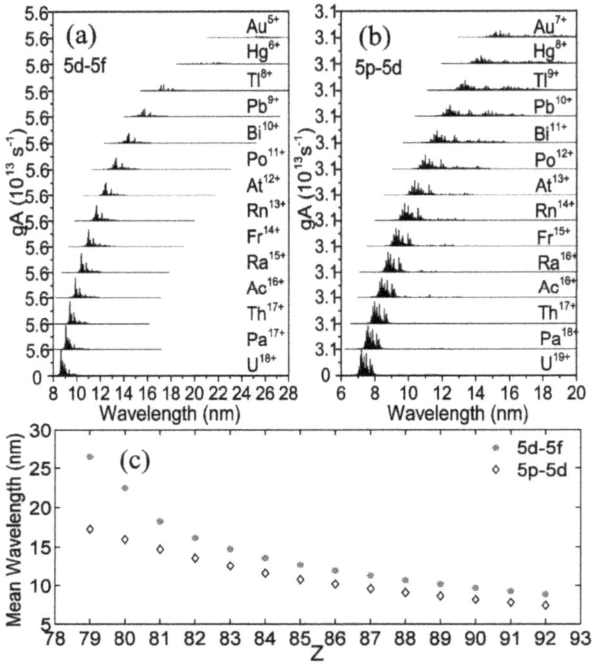

Figure 6. (Color online) Summed peak emission from (**a**) $5d$–$5f$ and (**b**) $5p$–$5d$ UTAs including CI in elements with $Z = 79$–92. (**c**) Dependence of UTA transition energies on atomic number Z, $5d$–$5f$ (red stars) and $5p$–$5d$ (black diamonds).

4. Conclusions

Unresolved transition arrays (UTAs) of the type $\Delta n = 0$, $4p^64d^{N+1}-4p^54d^{N+2}+4p^64d^N4f^1$ have been extensively studied because their intensity and emission bandwidth makes them ideal candidates for applications as radiation sources for a variety of technological applications in the EUV and SXR region. In contrast, the corresponding $\Delta n = 0$, $5p^65d^{N+1}-5p^55d^{N+2}+5p^65d^N5f^1$ UTAs have not been studied in detail. In this paper, the properties of these arrays have been studied theoretically with the aid of Hartree-Fock with configuration interaction (CI) calculations. We report on calculations for $5p$–$5d$ and

5d–5f transitions in elements from Z = *79* to Z = *92* and predict the positions and spectral properties of the corresponding UTAs. We compared the effects of CI between $\Delta n = 0$, $n = 5$–$n = 5$ transitions in these elements and $n = 4-n = 4$ transitions in their homologous, lower Z counterparts and found that the strong spectral narrowing, which is a feature of $\Delta n = 0$, $n = 4-n = 4$ transitions is not expected to be important in these spectra but shifts the position of *5d–5f* arrays to slightly shorter wavelengths and results in a broadening of their spectral profiles. This broadening points to their potential usefulness in the development of broadband sources for future EUV and soft X-ray metrology applications.

Acknowledgments: Luning Liu acknowledges support from UCD and from a Chinese Scholarship Council (CSC) scholarship and from the Fundamental Research Funds for the Central Universities under grant No. HUST: 2016YXMS028. DK acknowledges funding from the Irish Research Council and the Marie Curie Actions ELEVATE fellowship.

Author Contributions: Luning Liu and Gerry O' Sullivan performed the calculations; Gerry O' Sullivan, Deirdre Kilbane, Padraig Dunne, Luning Liu and Xinbing Wang analyzed the data; Gerry O'Sullivan, Deirdre Kilbane and Luning Liu wrote the paper.

Conflicts of Interest: The authors declare no conflict of interest. The founding sponsors had no role in the design of the study; in the collection, analyses, or interpretation of data; in the writing of the manuscript, and in the decision to publish the results.

References

1. Van den Zande, W. EUVL exposure tools for HVM: It's under (and about) control. In Proceedings of the EUV and Soft X-ray Source Workshop, Amsterdam, The Netherlands, 7–9 November 2016.
2. O'Sullivan, G.; Li, B.W.; D'Arcy, R.; Dunne, P.; Hayden, P.; Kilbane, D.; McCormack, T.; Ohashi, H.; O'Reilly, F.; Sheridan, P.; et al. Spectroscopy of highly charged ions and its relevance to EUV and soft X-ray source development. *J. Phys. B At. Mol. Opt. Phys.* **2015**, *48*, 144025. [CrossRef]
3. Bauche-Arnoult, C.; Bauche, J.; Klapisch, M. Variance of the distributions of energy levels and of the transition arrays in atomic spectra. *Phys. Rev. A* **1979**, *20*, 2424–2439. [CrossRef]
4. O'Sullivan, G.; Faulkner, R. Tunable narrowband soft X-ray source for projection lithography. *Opt. Eng.* **1994**, *33*, 3978–3983.
5. Churilov, S.S.; Ryabtsev, A.N. Analysis of the Sn IX-Sn XII spectra in the EUV region. *Phys. Scr.* **2006**, *73*, 614. [CrossRef]
6. Attwood, D.T. *Soft X-rays and Extreme Ultraviolet Radiation: Principles and Applications*; Cambridge University Press: Cambridge, UK, 2000.
7. Churilov, S.S.; Kildiyarova, R.R.; Ryabtsev, A.N.; Sadovsky, S.V. EUV spectra of Gd and Tb ions excited in laser-produced and vacuum spark plasmas. *Phys. Scr.* **2009**, *80*, 045303. [CrossRef]
8. Cummins, T.; Otsuka, T.; Yugami, N.; Jiang, W.; Endo, A.; Li, B.; O'Gorman, C.; Dunne, P.; Sokell, E.; O'Sullivan, G.; et al. Optimizing conversion efficiency and reducing ion energy in a laser-produced Gd plasma. *Appl. Phys. Lett.* **2012**, *100*, 061118. [CrossRef]
9. Yoshida, K.; Fujioka, S.; Higashiguchi, T.; Ugomori, T.; Tanaka, N.; Ohashi, H.; Kawasaki, M.; Suzuki, Y.; Suzuki, C.; Tomita, K.; et al. Efficient extreme ultraviolet emission from one-dimensional spherical plasmas produced by multiple lasers. *Appl. Phys. Express* **2014**, *7*, 086202. [CrossRef]
10. Louis, E.; Mullender, S.; Bijkerk, F. Multilayer development for extreme ultraviolet and shorter wavelength lithography. In Proceedings of the International Workshop on EUV and Soft X-ray Sources, Dublin, Ireland, 7–9 November 2011.
11. Skoglund, P.; Lundström, U.; Vogt, U.; Hertz, H.M. High-brightness water window electron-impact liquid-jet microfocus source. *Appl. Phys. Lett.* **2010**, *96*, 084103. [CrossRef]
12. McDermott, G.; le Gros, M.A.; Larabell, C.A. Visualizing cell architecture and molecular location using soft X-ray tomography and correlated cryo-light microscopy. *Annu. Rev. Phys. Chem.* **2012**, *63*, 225–239. [CrossRef] [PubMed]
13. Wachulak, P.W.; Bartnik, A.; Fiedorowicz, H.; Rudawski, P.; Jarocki, R.; Kostecki, K.; Szczurek, M. "Water window" compact, table-top laser plasma soft X-ray sources based on gas puff target. *Nucl. Instrum. Methods B* **2010**, *268*, 1692–1700. [CrossRef]

14. Higashiguchi, T.; Otsuka, T.; Yugami, N.; Jiang, W.; Endo, A.; Li, B.; Dunne, P.; O'Sullivan, G. Feasibility study of broadband efficient "water window" source. *Appl. Phys. Lett.* **2012**, *100*, 014103. [CrossRef]
15. Mandelbaum, P.; Finkenthal, M.; Schwob, J.L.; Klapisch, M. Interpretation of the quasicontinuum band emitted by highly ionized rare-earth elements in the 70–100-Å range. *Phys. Rev. A* **1987**. [CrossRef]
16. Koike, F.; Fritzsche, S.; Nishihara, K.; Sasaki, A.; Kagawa, T.; Nishikawa, T.; Fujima, K.; Kawamura, T.; Furukawa, H. Precise and Accurate Calculations of Electronic Transitions in Heavy Atomic Ions Relevant to Extreme Ultra-Violet Light Sources. *J. Plasma Fusion Res.* **2006**, *7*, 253.
17. Cowan, R.D. *The Theory of Atomic Structure and Spectra*; University of California Press: Berkeley, CA, USA, 1981.
18. Kilbane, D.; O'Sullivan, G. Ground-state configurations and unresolved transition arrays in extreme ultraviolet spectra of lanthanide ions. *Phys. Rev. A* **2010**, *82*, 062504. [CrossRef]
19. Kilbane, D. Transition wavelengths and unresolved transition array statistics of ions with Z=72–89. *J. Phys. B At. Mol. Opt. Phys.* **2011**, *44*, 165006. [CrossRef]
20. Ohashi, H.; Higashiguchi, T.; Suzuki, Y.; Arai, G.; Li, B.; Dunne, P.; O'Sullivan, G.; Sakaue, H.A.; Kato, D.; Murakami, I.; et al. Characteristics of X-ray emission from optically thin high-Z plasmas in the soft X-ray region. *J. Phys. B At. Mol. Opt. Phys.* **2015**, *48*, 144011. [CrossRef]
21. Wu, T.; Higashiguchi, T.; Li, B.W.; Arai, G.; Harac, H.; Kondo, Y.; Miyazaki, T.; Dinh, T.-H.; O'Reilly, F.; Sokell, E.; et al. Analysis of unresolved transition arrays in XUV spectral region from highly charged lead ions produced by subnanosecond laser pulse. *Opt. Commun.* **2017**, *385*, 143–152. [CrossRef]
22. Bauche-Arnoult, C.; Bauche, J. Statistical approach to the spectra of plasmas. *Phys Scr.* **1992**. [CrossRef]
23. Condon, E.U.; Odabasi, H. *Atomic Structure*; Cambridge University Press: Cambridge, UK, 1980.
24. Bauche-Arnoult, C.; Bauche, J. Variance of the distributions of energy levels and of the transition arrays in atomic spectra. *Phys Rev. A* **1979**, *20*, 2424–2439. [CrossRef]
25. Carroll, P.K.; O'Sullivan, G. Ground-state configuration of ionic species I through XVI for Z=57–74 and the interpretation of 4d-4f emission resonances in laser-produced plasma. *Phys Rev. A* **1982**, *25*, 275. [CrossRef]
26. Connerade, J.P.; Mansfield, M.W.D. Term-dependent Hybridization of the 5f wave functions of Ba and Ba^{++}. *Phys. Rev. Lett.* **1982**. [CrossRef]
27. O'Sullivan, G. The origin of line-free XUV continuum emission from laser-produced plasmas of the elements $62 \leq Z \leq 74$. *J. Phys. B* **1983**. [CrossRef]
28. Connerade, J.P.; Esteva, J.M.; Karnatak, R.C. *Giant Resonances in Atoms, Molecules and Solids*; Springer: New York, NY, USA, 1987.
29. Fano, U.; Cooper, J.W. Spectral distribution of atomic oscillator strengths. *Rev. Mod. Phys.* **1968**. [CrossRef]
30. Cheng, K.T.; Fischer, C.F. Collapse of the 4f orbital for Xe-like ions. *Phys. Rev. A* **1983**. [CrossRef]
31. Band, I.M.; Trzhaskovskaya, M.B. On the 5d photoabsorption spectra in the gaseous and metallic states of uranium and thorium. *J. Phys. B At. Mol. Opt. Phys.* **1992**. [CrossRef]
32. Carroll, P.K.; Costello, J.T. Giant-dipole-resonance absorption in atomic thorium by a novel two-laser technique. *Phys Rev. Lett.* **1986**. [CrossRef] [PubMed]
33. Carroll, P.K.; Costello, J.T. The XUV photoabsorption spectrum of uranium vapor. *J. Phys. B At. Mol. Opt. Phys.* **1987**. [CrossRef]
34. Carroll, P.K.; Costello, J.T.; Kennedy, E.T.; O'Sullivan, G. XUV emission from uranium plasmas: The identification of U XIII and U XV. *J. Phys. B At. Mol. Opt. Phys.* **1984**. [CrossRef]
35. Carroll, P.K.; Costello, J.T.; Kennedy, E.T.; O'Sullivan, G. XUV emission from thorium plasmas; the identification of Th XI and Th XIII. *J. Phys. B At. Mol. Opt. Phys.* **1986**, *19*, L651. [CrossRef]
36. Carroll, P.K.; O'Sullivan, G. The observation of *5d–5f* resonant emission in thorium in high ion stages (≈VIII to XVI). *Phys. Lett. A.* **1981**. [CrossRef]
37. Finkenthal, M.; Lippmann, S.; Moos, H.W.; Mandelbaum, P.; The TEXT Group. Highly ionized uranium emission in the soft-X-ray region 50–100 Å. *Phys. Rev. A* **1989**. [CrossRef]
38. Hayden, P.; Cummings, A.; Murphy, N.; O'Sullivan, G.; Sheridan, P.; White, J.; Dunne, P. 13.5 nm extreme ultraviolet emission from tin based laser produced plasma sources. *J. Appl. Phys.* **2006**. [CrossRef]

MDPI
St. Alban-Anlage 66
4052 Basel
Switzerland
Tel. +41 61 683 77 34
Fax +41 61 302 89 18
www.mdpi.com

Atoms Editorial Office
E-mail: atoms@mdpi.com
www.mdpi.com/journal/atoms

www.ingramcontent.com/pod-product-compliance
Lightning Source LLC
Chambersburg PA
CBHW051706210326
41597CB00032B/5390